Imaging

A LABORATORY MANUAL

ALSO FROM COLD SPRING HARBOR LABORATORY PRESS

IMAGING SERIES

Imaging: A Laboratory Manual
Imaging in Neuroscience: A Laboratory Manual
Imaging in Developmental Biology: A Laboratory Manual

RELATED LABORATORY MANUALS

Basic Methods in Microscopy: Protocols and Concepts from Cells: A Laboratory Manual
Drosophila *Neurobiology: A Laboratory Manual*
Gene Transfer: Delivery and Expression of DNA and RNA, A Laboratory Manual
Live Cell Imaging: A Laboratory Manual, 2nd edition
Single-Molecule Techniques: A Laboratory Manual

OTHER RELATED TITLES

At the Bench: A Laboratory Navigator, Updated Edition
At the Helm: Leading Your Laboratory, Second Edition
Experimental Design for Biologists
Lab Dynamics: Management Skills for Scientists
Lab Math: A Handbook of Measurements, Calculations, and Other Quantitative Skills for Use at the Bench
Lab Ref, Volume 1: *A Handbook of Recipes, Reagents, and Other Reference Tools for Use at the Bench*
Lab Ref, Volume 2: *A Handbook of Recipes, Reagents, and Other Reference Tools for Use at the Bench*
Statistics at the Bench: A Step-by-Step Handbook for Biologists

WEBSITE

Cold Spring Harbor Protocols
http://www.cshprotocols.org/imaging

Imaging

A LABORATORY MANUAL

EDITOR AND SERIES EDITOR

Rafael Yuste

Howard Hughes Medical Institute
Columbia University

www.cshprotocols.org/imaging

COLD SPRING HARBOR LABORATORY PRESS
Cold Spring Harbor, New York • www.cshlpress.com

Imaging: A Laboratory Manual

Printed in China

Publisher	John Inglis
Acquisition Editor	David Crotty
Director of Development, Marketing, & Sales	Jan Argentine
Managing Editor	Michael Zierler
Developmental Editors	Michael Zierler and Kaaren Janssen
Project Manager	Inez Sialiano
Permissions Coordinator	Carol Brown
Production Editor	Kathleen Bubbeo
Desktop Editors	Lauren Heller and Susan Schaefer
Production Manager	Denise Weiss
Book Marketing Manager	Ingrid Benirschke
Sales Account Managers	Jane Carter and Elizabeth Powers
Cover Designer	Ed Atkeson

Front cover artwork: Image of microtubules (green) and clathrin-coated pits (red) in a cultured mammalian cell, acquired with STORM (stochastic optical reconstruction microscopy; see Chapter 35), with a spatial resolution of ~30 nm. STORM is an example of several novel microscopy techniques that have enabled optical imaging of structures with a resolution substantially smaller than the wavelength of light (see Chapters 31–35). These ultra-resolution techniques can be used to follow the dynamics of individual molecules in living cells. (Reprinted, with permission, from Bates et al. 2007. *Science* **317:** 1749–1753. ©AAAS.)

Library of Congress Cataloging-in-Publication Data

Imaging : a laboratory manual / edited by Rafael Yuste.
p. ; cm.
Includes bibliographical references and index.
ISBN 978-0-87969-935-2 (hardcover : alk. paper) -- ISBN 978-0-87969-936-9 (pbk. : alk. paper)
1. Diagnostic imaging--Laboratory manuals. I. Yuste, Rafael.
[DNLM: 1. Diagnostic Imaging--Laboratory Manuals. WN 25 I31 2011]

RC78.7.D53I423 2011
616.07'54--dc22

2010022944

10 9 8 7 6 5 4 3 2 1

Students and researchers using the procedures in this manual do so at their own risk. Cold Spring Harbor Laboratory makes no representations or warranties with respect to the material set forth in this manual and has no liability in connection with the use of these materials. All registered trademarks, trade names, and brand names mentioned in this book are the property of the respective owners. Readers should please consult individual manufacturers and other resources for current and specific product information.

With the exception of those suppliers listed in the text with their addresses, all suppliers mentioned in this manual can be found on the BioSupplyNet website at http://www.biosupplynet.com.

All World Wide Web addresses are accurate to the best of our knowledge at the time of printing.

Procedures for the humane treatment of animals must be observed at all times. Check with the local animal facility for guidelines.

Certain experimental procedures in this manual may be the subject of national or local legislation or agency restrictions. Users of this manual are responsible for obtaining the relevant permissions, certificates, or licenses in these cases. Neither the authors of this manual nor Cold Spring Harbor Laboratory assume any responsibility for failure of a user to do so.

The materials and methods in this manual may infringe the patent and proprietary rights of other individuals, companies, or organizations. Users of this manual are responsible for obtaining any licenses necessary to use such materials and to practice such methods. COLD SPRING HARBOR LABORATORY MAKES NO WARRANTY OR REPRESENTATION THAT USE OF THE INFORMATION IN THIS MANUAL WILL NOT INFRINGE ANY PATENT OR OTHER PROPRIETARY RIGHT.

All Cold Spring Harbor Laboratory Press publications may be ordered directly from Cold Spring Harbor Laboratory Press, 500 Sunnyside Blvd., Woodbury, New York 11797-2924. Phone: 1-800-843-4388 in the Continental U.S. and Canada. All other locations: (516) 422-4100. FAX: (516) 422-4097. E-mail: cshpress@cshl.edu. For a complete catalog of all Cold Spring Harbor Laboratory Press publications, visit our website at http://www.cshlpress.com/.

Contents

ACCOMPANYING MOVIES

Movies are freely available online at www.cshprotocols.org/imaging.

CHAPTER 12
Spectral Methods for Functional Brain Imaging

MOVIE 12.1. Spinning-disk confocal imaging data on Ca^{2+} waves in organotypic culture before denoising. Video clip from 1200-frame sequence of 100 x 100-pixel data. Square field is 115 µm on edge.

MOVIE 12.2. Spinning-disk confocal imaging data on Ca^{2+} waves in organotypic culture after denoising. The same data set as in Movie 12.1 after reconstruction with 25 of the 1200 modes (Equation 29).

CHAPTER 16
Immunoimaging: Studying Immune System Dynamics Using Two-Photon Microscopy

MOVIE 16.1. The T-cell–B-cell border within a lymph node, imaged with two-photon microscopy. B cells are labeled red, and T cells are labeled green. Video dimensions are 295, 270, 52 μm (*x*, *y*, *z*).

MOVIE 16.2. T cells (red) and dendritic cells expressing yellow fluorescent protein (YFP) under control of the CD11c promoter. Movie was taken using two-photon microscopy. Video dimensions are 300, 235, 25 μm (*x*, *y*, *z*).

MOVIE 16.3. Two-photon imaging of an inguinal lymph node from a cyan fluorescent protein (CFP) β-actin mouse in which all of the β-actin-expressing cells also express CFP. Video dimensions are 240, 240, 50 μm (*x*, *y*, *z*).

MOVIE 16.4. Two-photon imaging of cyan fluorescent protein (CFP) β-actin $CD4^+$ T cells in a peripheral lymph node 24 h after injection of 4 x 10^6 total cells. Video dimensions are 240, 240, 50 μm (*x*, *y*, *z*).

CHAPTER 19
Preparation and Use of Retroviral Vectors for Labeling, Imaging, and Genetically Manipulating Cells

MOVIE 19.1. Representative time series for spine motility studies. (*a*) Adult running, 56 dpi. (*b*) Adult, 126 dpi. Each movie consists of four time frames (0, 30, 60, and 90 min). The size of the field of view: (*a*) 9.9 x 6.1 μm; (*b*) 10.8 x 6.4 μm.

CHAPTER 41
Time-Domain Fluorescence Lifetime Imaging Microscopy: A Quantitative Method to Follow Transient Protein–Protein Interactions in Living Cells

MOVIE 41.1. Time lapse of Rac activation followed by fast fluorescence lifetime imaging microscopy (FLIM). Rac activation is visualized by the relative number of interacting particles (NP = $I_0 \times mf_D$) of eGFP-Rac that undergoes Förster resonance energy transfer (FRET) with PBD-mCherry as described in Figure 8. I_0 and mf_D were obtained with the TriM-FLIM (five time-gated images of 2 nsec each; 5-sec acquisition time per NP image; time-lapse duration: 100 sec). The dimensions of the micrograph are 16 x 15 μm.

MOVIE 41.2. Time lapse of eGFP-Rac fluorescence intensity image (I_0 images) corresponding to the time lapse of relative NP of Movie 41.1.

CHAPTER 43
Fluorescent Speckle Microscopy

MOVIE 43.1. Fluorescent speckle microscopy (FSM) of poleward tubulin movement (flux) in a control *Xenopus laevis* egg extract spindle. (*Left panel*) Individual tubulin speckles undergoing antiparallel poleward movement. (*Right panel*) Trajectories of individual speckles are recovered by software for quantitative analysis. Red and green squares mark speckles moving to the left and the right poles, respectively. In each frame, the center of each square denotes the current position of the corresponding speckle being followed. Only those speckles whose trajectories last at least four frames are displayed using squares. Frame rate: 5 sec/frame. Display rate: 15 frames/sec. Field of view, width by height: 83 x 66 μm.

Preface to the Book Series

To train young people to grind lenses... . I cannot see there would be much use...because most students go there to make money out of science or to get a reputation in the learned world. But in lens-grinding and discovering things hidden from our sight, these count for nought.

—Antonie van Leeuwenhoek

Letter to Gottfried Leibniz on 28 September 1715 in response to Leibniz' request that he should open a school to train young people in microscopy

You can observe a lot just by watching. —Yogi Berra

ONE OF THE CENTRAL THEMES OF BIOLOGY IS the constant change and transformation of most biological systems. In fact, this dynamic aspect of biology is one of its most fascinating characteristics, and it draws generation after generation of students absorbed in understanding how an organism develops, how a cell functions, or how the brain works. This series of manuals covers imaging techniques in the life sciences—techniques that try to capture these dynamics. The application of optical and other visualization techniques to study living organisms constitutes a direct methodology to follow the form and the function of cells and tissues by generating two- or three-dimensional images of them and to document their dynamic nature over time. Although it seems natural to use light to study cells or tissues, and microscopists have been doing this with fixed preparations since van Leeuwenhoek's time, the imaging of living preparations has only recently become standard practice. It is not an overstatement to say that imaging technologies have revolutionized research in many areas of biology and medicine. In addition to advances in microscopy, such as differential interference contrast or the early introduction of video technology and digital cameras, the development of methods to culture cells, to keep tissue slices alive, and to maintain living preparations, even awake and behaving, on microscopes has opened new territories to biologists. The synthesis of novel fluorescent tracers, indicator dyes, and nanocrystals and the explosive development of fluorescent protein engineering, optogenetical constructs, and other optical actuators like caged compounds have made possible studies characterizing and manipulating the form and function of cells, tissues, and circuits with unprecedented detail, from the single-molecule level to that of an entire organism. A similar revolution has occurred on the optical design of microscopes. Originally, confocal microscopy became the state-of-the-art imaging approach because of its superb spatial resolution and three-dimensional sectioning capabilities; later, the development of two-photon excitation enabled fluorescence imaging of small structures in the midst of highly scattered living media, such as whole-animal preparations, with increased optical penetration and reduced photodamage. Other nonlinear optical techniques, such as second-harmonic generation and coherent anti-Stokes Raman scattering (CARS), now follow and appear well suited for measurements of voltage and biochemical events at interfaces such as plasma membranes. Finally, an entire generation of novel "superresolution" techniques, such as stimulated emission depletion (STED), photoactivated localization microscopy (PALM), and stochastic optical reconstruction microscopy (STORM), has arisen. These techniques have broken the diffraction limit barrier and have enabled the direct visualization of the dynamics of submicroscopic particles and individual molecules. On the other side of the scale, light-sheet illumination techniques allow the investigator to capture the development of an entire organism, one cell at a time. Finally, in the field of medical imaging, magnetic resonance scanning techniques have provided detailed images of the structure of the living human body and the activity of the brain.

This series of manuals originated in the Cold Spring Harbor Laboratory course on Imaging Structure and Function of the Nervous System, taught continuously since 1991. Since its inception, the course quickly became a "watering hole" for the imaging community and especially for neuroscientists and cellular and developmental neurobiologists, who are traditionally always open to microscopy approaches. The original manual, published in 2000, sprang from the course and focused solely on neuroscience, and its good reception, together with rapid advances in imaging techniques, led to a second edition of the manual in 2005. At the same time, the increased blurring between neuroscience and developmental biology made it necessary to encompass both disciplines, so the original structure of the manual was revised, and many new chapters were added. But even this second edition felt quickly dated in this exploding field. More and more techniques have been developed, requiring another update of the manual, too unwieldy now for a single volume. This is the reasoning behind this new series of manuals, which feature new editors and a significant number of new methods. The material has been split into several volumes, thus allowing a greater depth of coverage. The first book, *Imaging: A Laboratory Manual*, is a background text focused on general microscopy techniques and with some basic theoretical principles, covering techniques that are widely applicable in many fields of biology and also some specialized techniques that have the potential to greatly expand the future horizon of this field. A second manual, *Imaging in Neuroscience: A Laboratory Manual*, keeps the original focus on nervous system imaging from the Cold Spring Harbor Imaging course. A third volume, *Imaging in Developmental Biology: A Laboratory Manual*, now solely deals with developmental biology, covering imaging modalities particularly suited to follow developmental events. There are plans to expand the series into ultrastructural techniques and medical-style imaging, such as functional magnetic resonance imaging (fMRI) or positron emission tomography (PET), so more volumes will hopefully follow these initial three, which cover mostly optical-based approaches.

Like its predecessors, these manuals are not microscopy textbooks. Although the basics are covered, I refer readers interested in a comprehensive treatment of light microscopy to many of the excellent texts published in the last decades. The targeted audience of this series includes students and researchers interested in imaging in neuroscience or developmental or cell biology. Like other CSHL manuals, the aim has been to publish manuals that investigators can have and consult at their setup or bench. Thus, the general philosophy has been to keep the theory to the fundamentals and concentrate instead on passing along the little tidbits of technical knowledge that make a particular technique or an experiment work and that are normally left out of the methods sections of scientific articles.

This series of manuals has only been possible because of the work and effort of many people. First, I thank Sue Hockfield, Terri Grodzicker, Bruce Stillman, and Jim Watson, who conceived and supported the Imaging course over the years and planted the seed blossoming now in these manuals and, more importantly, in the science that has spun out of this field. In addition, the staff at CSHL Press has been exceptional in all respects, with special gratitude to John Inglis, responsible for an excellent team with broad vision, and David Crotty, who generated the ideas and enthusiasm behind this new series. Also, Inez Sialiano, Mary Cozza, Michael Zierler, Kaaren Janssen, Catriona

Simpson, Virginia Peschke, Judy Cuddihy, Martin Winer, Kevin Griffin, Kathleen Bubbeo, Lauren Heller, Susan Schaeffer, Jan Argentine, and Denise Weiss worked very hard, providing fuel to the fire to keep these books moving, and edited them with speed, precision, and intelligence. More than anyone, they are the people responsible for their timely publication. Finally, I honor the authors of the chapters in these books, many of them themselves past instructors of the CSH Imaging course and of similar imaging courses at institutions throughout the world. Teaching these courses is a selfless effort that benefits the field as a whole, and these manuals, reflecting the volunteer efforts of hundreds of researchers, who not only have taken the time to write down their technical knowledge but have agreed to generously share it with the rest of the world, are a beautiful example of such community cooperation. As Leibniz foresaw, "lens grinding" is a profession that is indeed meaningful and needs the training of young people.

— RAFAEL YUSTE

Preface to Book 1

THE PURPOSE OF THIS BOOK IS TO SERVE AS THE introduction to, and common base for, a series of laboratory manuals that cover different aspects of biological imaging. At launch, this series includes a general manual on imaging techniques, a second one on neuroscience applications, and a third one on developmental biology. This first book covers basic microscopy techniques and also some more advanced ones that have not yet become commonplace in the laboratory but that are included because of their great potential.

In organizing the material for this first manual I was aware of the difficulty inherent in splitting this dynamic field into manageable sections. The techniques discussed here span many scales and applications and are based on many different optical principles and on combinations of them. Science is fluid and the reader should be aware that the sections of the book are merely artificial placeholders to help the reader find the relevant material faster.

The book is divided into three main sections. The first (Instrumentation) focuses on the hardware and covers the basics of light microscopy, light sources, cameras, and image processing. This section also covers some novel technologies, such as liquid crystal, acousto-optical tunable filters, ultrafast lasers, and grating systems; discusses different forms of imaging, from DIC to confocal to two-photon—techniques that are becoming relatively standard in biological research institutes; and ends with a chapter that discusses the challenges of making the microscope environment compatible with the survival of common biological preparations.

The second section (Labeling and Indicators) focuses on labeling methods to stain cells, organelles, and proteins or to measure ions or molecular interactions. It includes some well-established methods, such as immunological and nonimmunological staining, and newer genetic engineering techniques where one tags a protein directly or indirectly with a fluorophore. This section also covers fluorescence and luminescent indicators of several intracellular biochemical pathways, with particular emphasis on measurements of calcium dynamics.

The third section of the manual (Advanced Microscopy) covers less established techniques, many of them at the forefront of imaging research. This section is organized by scale, covering first imaging of molecules, then imaging of cells, and then imaging of tissues or entire organisms. In addition, this section has a separate set of chapters dealing with strategies to perform fast laser imaging, an area of rapid development that aims to enhance the slow time resolution arising from the serial scanning by laser microscopes. Finally, there are three chapters on the use of caged compounds, photochemical actuators that enable the optical manipulations of cells and tissues in situ. The ability to optically alter the concentration of a substance in a small region of a cell or a tissue is turning imaging from a descriptive technique into an experimental one.

The manual ends with a series of appendices, including a glossary of imaging terms, useful information on spectra, lenses, and filters, and instructions for handling imaging hardware safely.

Besides the people and institutions already acknowledged in the series preface, a separate thanks goes to the funding agencies that have made my work as "imagist" possible over the years. The research of my group has been supported by the generosity of the National Eye Institute, the Howard Hughes Medical Institute, and the Kavli Foundation. Columbia University, and its Department of Biological Sciences and its Neuroscience Program, has been a wonderful environment in which to work and pursue my dreams as a researcher and scholar. I would also like to thank the members of my laboratory and, in particular, Darcy Peterka, Kira Poskanzer, Roberto Araya, and Alan Woodruff, who helped me in the final copy editing of all the chapters of this and the other books in the series. In addition, I especially thank Fred Lanni and Arthur Konnerth for co-editing the first two editions of the manual and for all the wonderful late-night discussions when we ran the CSH course. Finally, as they say in Basque, *hau etxekoentzat da* (this here is for the people of the house). I dedicate this book to my *etxekoak*, my extended group of family and friends, because it is from them that I gather my strength.

— RAFAEL YUSTE

1 Microscope Principles and Optical Systems

Frederick Lanni[1] and H. Ernst Keller[2]

[1]*Department of Biological Sciences, Molecular Biosensor & Imaging Center, Carnegie Mellon University, Pittsburgh, Pennsylvania 15213;* [2]*Carl Zeiss, Inc., Thornwood, New York 10594 (retired)*

ABSTRACT

Given the 300-year history of the light microscope since the time of van Leeuwenhoek and Hooke, one may expect this invention to have matured. Rather, the past decades have seen a remarkable set of advances in microscope optical systems and the uses to which the microscope has been put. In the decade since this introduction to microscope optical systems first appeared, innovation has continued. Advances in optical materials and fabrication, electronic detectors and cameras, lasers, fluorescent probes, and computers have led to enormous advances in microscopy and the fields of science and technology in which the light microscope is a major tool. Nevertheless, the optics that constitute the core of a microscope embodies a set of principles, understood since the time of Abbe, that continues to spark the imagination and to give rise to new imaging technologies and discoveries.

USE AND CARE OF MICROSCOPES

It is our experience that most users of microscopes learn by trial and error long before reading a book (or a chapter!) on the subject. Therefore, a discussion of use and care of the instrument has been put up front, rather than set as an appendix.

The three essential components of any microscope are the objective, the condenser, and a precision stage system to position the specimen. In a transmitted light microscope, the objective and the condenser are separate components, whereas in an incident-light microscope, the objective serves as its own condenser. Use and care of the instrument is defined by the characteristics and the limitations of these components; for example, objectives differ in an important way from lenses in other types of image-forming instruments such as cameras. To minimize aberration and to maximize resolution, objectives are designed with restricted use conditions. A high-quality camera lens can be adjusted to focus over a large subject range without obvious change in performance over the corresponding range of demagnification. In contrast, microscope objectives are designed to image, at a fixed magnification,

TABLE 1. Standard objective types for biological microscopy

Immersion	n	NA range	Design condition
Dry	1.00	0.13–0.95	Cover glass; standard thickness = 0.17 mm, #1½
			Cell culture dishes or flasks (0–2.0 mm plastic or glass);
			Adjustable corrector or interchangeable window optics
		0.8–0.95	Adjustable corrector for cover glass thickness
Water	1.33	0.3–1.0	Direct; no cover glass
		0.5–1.2	Indirect; adjustable correction for cover-glass thickness
Glycerol	1.47	0.6–1.35	Fused silica cover glass—0.2 mm
Oil	1.52	0.5–1.45	(Cover glass)
Multi-immersion	Variable	0.5–0.9	Adjustable correction for n and cover-glass thickness

a particular plane in an object space onto a detector located at a particular position in an image space. The gain is that the image formed by a microscope is much closer to the diffraction limit of resolution than a camera photograph. Alteration of a microscope optical system in a way that causes the objective to be focused at a distance from the specimen that differs from the design working distance will result in degraded image quality. Likewise, when the refractive index of the specimen differs from the value on which the objective design was based, large losses in contrast and resolution can result when focusing deep within the object. Sensitivity to deviation from design conditions is generally much greater for lenses of higher numerical aperture (NA). In an ideal case, the objective would be designed for use with an immersion fluid in which the refractive index was the same as the index of the specimen. However, the refractive index of most cellular biological specimens is higher than n_{water}, is considerably lower than n_{glass}, and is spatially variant. A practical solution is to design a series of objectives for use under specific optical conditions, using standard immersion fluids (Table 1). Condensers, likewise, are designed for specific uses. A large working distance (2–5 cm) is a common requirement for transilluminating cells in flasks or dishes on an inverted microscope, but this limits the NA to moderate values. Dry high-resolution condensers are designed to operate with 3–4-mm clearance between the lens and a standard microscope slide (1-mm thickness). For the highest-resolution work in transmitted light, oil-immersion condensers are used with 1-mm clearance to the slide.

The most important use rule in microscopy is to avoid mechanical damage to the objective and the condenser. Clearly, optical surfaces are most susceptible to damage by abrasion. However, permanent misalignment is an equally serious problem commonly caused by impact or deformation of the metal or ceramic barrel of the lens. For two reasons, microscope lenses and condensers are generally positioned very close to the specimen. As with any image-forming instrument, high magnification is obtained by placing the object close to the front focal plane of a short focal-length lens or a system of lenses. Second, high resolution depends on capture by the objective of light rays that exit the specimen at a high angle; that is, the imaging system must have high aperture. Although condensers are not usually corrected for image formation, the same rules apply; high demagnification (concentration of light) and high aperture are required. For low-power optics, the working distance can range from millimeters to >1 cm, but for widely used high-NA immersion objectives, the range is 0.06–0.25 mm. Therefore, microscope users must take great care to avoid collision between the objective and the specimen, the condenser and the specimen, the objective and the condenser, the objective and the stage, etc., caused by movement of the focus drive, the turret, and the stage or the condenser focus drive.

Of equal importance is the cleanliness of optical surfaces. Settled dust particles or fingerprints anywhere in the lens system, particularly between the specimen and the final image, will cause degraded image quality and can even create artifactual image features. The most common problem is oil residue or a fingerprint on the front lens surface of a dry objective or dried residue on the top surface of a cover glass. In a fluorescence microscope, a single particle of fluorescent dust that settles onto the final lens surface of the objective can cause a sizable increase in the background. Oil-immersion optics presents a different set of problems. After use, the oiled lens surfaces should be gently wiped with lens tissue, and the objective or the condenser should be stored in a dust-free container. It is generally not necessary to remove all traces of oil between uses. For the same reason, it is very important not to carelessly change oil types because differences in refractive index or actual immiscibility can cause serious image degradation. Microscope manufacturers supply oil of the appropriate index for their lenses. If blending of immersion fluids is required, compatible standard oils can be obtained from Cargille Laboratories (Cedar Grove, NJ). Changeover to an oil of a different index requires complete removal of the residue of the previous oil. If the oils are miscible, this is most easily accomplished simply by the repeated application of the new oil along with the gentle removal using a lens tissue. The other case in which complete removal of oil is necessary is when changing from oil immersion to glycerin or water immersion (WI) with a multi-immersion objective. In this case, the use of a droplet of xylene on the lens tissue will remove the oil.

Spillover of specimen materials poses both cleanliness and damage hazards. In microscopy of living specimens, a common hazard is spillage of culture medium into the condenser on an upright microscope or into everything but the condenser on an inverted microscope. Dried culture medium leaves a heavy residue of salt crystals and protein that can permanently damage lens coatings and surfaces and can quickly corrode metal components, iris leaves, bearings, and more. Especially in perfusion setups, it is well worth using a sheet rubber dam or a shield over at-risk components. Optical and precision mechanical components that get wet with culture medium should be carefully cleaned with distilled water and carefully dried. This may entail disassembly and reassembly of filter sets, condensers, or other components. After use in saline, seawater, pond water, or culture medium, the fluid contact surfaces of a direct-immersion objective should always be gently rinsed with distilled water and dried with a lens tissue. In some cases, direct-immersion objectives will contact medium that contains bioactive compounds, fluorescent vital stains, or microorganisms. In this case, special cleaning procedures may be required to prevent contamination of subsequent specimens.

PRINCIPLES OF LIGHT MICROSCOPY

Paraxial Image Formation

Many of the basic properties of image-forming instruments (telescopes, cameras, and microscopes) can be understood in terms of ray optics. In particular, an optic consisting of nothing more than a hemispherical boundary between two materials such as glass and air can be shown to have the properties of a lens. This means that, for a set of rays that all propagate close to and nearly parallel to a specific axis, the optic will form a focused image of a small object located close to the axis at some distance from the spherical surface. This result also holds for thin lenses and for more complicated optics composed of sets of lenses. The restriction to low-angle near-axial rays is known as the paraxial limit in which the law of refraction can be linearized and lens action can be summarized in two rules, which form the basis for geometric ray tracing.

1. The lens has both a front and a back focal length defined on a common axis through its center.
2. Rays entering the lens parallel to the axis are refracted so as to exit on a path that intersects the opposite focal point.

Using only these two rules, familiar equations can be derived for magnification and conjugate distances (object and image location) (Fig. 1). Multielement lens systems can be represented by a serial application of the rules in which the image from the *n*th lens surface is the real or the virtual object for the following surface. In the paraxial limit, multielement lens systems act like single lenses, possessing front and back focal lengths measured out from mathematically defined front and back principal planes, respectively. An important result, known as the Helmholtz relation, is that the front and back focal lengths are in the ratio of the refractive indices of the object and the image spaces; that is, $f/f' = n/n'$, regardless of the specific sequence of lens elements in between the entrance and the exit surfaces.

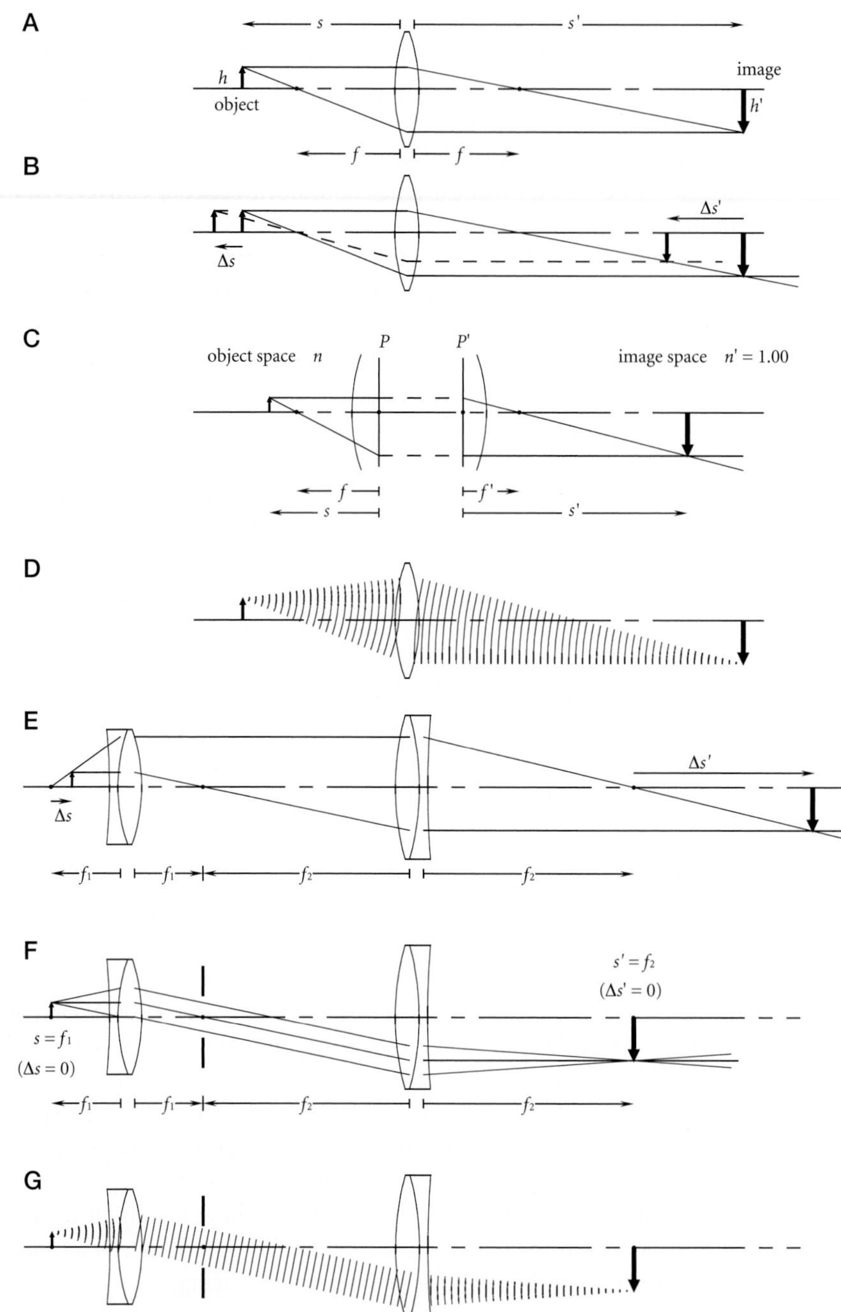

FIGURE 1. Ray and wave diagrams for lens systems. (*A*) Ray trace for a thin lens; the principal ray from the object arrowhead enters the lens parallel to the axis and emerges on a path intersecting the back focal point. Likewise, the ray from the arrowhead passing through the near-focal point emerges parallel to the axis. The intersection of the two rays defines the location and the magnification of the image (Eqs. 4a and 5). (*B*) Focusing by axial displacement of the object changes both the image location and the magnification. (*C*) In the paraxial limit, a lens system composed of an arbitrary number of spherical surfaces on a common axis can be represented by front and rear focal points measured out from a pair of principal planes (*P* and *P'*). Principal rays arriving at the input plane are transferred directly to the output plane as shown. The location of the principal planes and the focal points are functions of the curvature, the spacing, and the refractive indices of all of the lens elements in the assembly. The ratio of the system focal lengths is determined only by the object space and the image space indices; $f/f' = n/n'$ (the Helmholtz relation). (*D*) Wave-optical diagram corresponding to the ray trace in *A*. The action of a lens is to change the radius of curvature of incident spherical waves; in *D*, a diverging spherical wave is converted to a converging wave. (*E*) Two lenses separated by the sum of the focal lengths form an afocal combination in which a principal ray (parallel to axis) entering from one side exits as a principal ray on the other side. Therefore, the magnification is constant and equal to the ratio of the focal lengths (Eq. 7). (*F*) When the object is located in the front focal plane of lens 1, the image is located in the back focal plane of lens 2. (*G*) Wave-optical diagram corresponding to *F*. The diverging wave is first collimated (infinite radius of curvature) and then converged to form an image at a finite conjugate distance. In both *D* and *G*, the wavelength is greatly exaggerated.

TABLE 2. Paraxial optic definitions and formulas

f,f'	Object-side and image-side focal lengths of lens system
s,s'	Object and image conjugate distances
h,h'	Object and image offsets from axis
$f/f' = n/n'$	Helmholtz's relation for any coaxial system of spherical lenses n = refractive index of object immersion medium n' = refractive index of the image space (usually air; $n' = 1.00$) For nonimmersed optics, $n' = n$, and $f' = f$
$M = h'/h$	Definition of transverse magnification For a microscope, $M >> 1$
$M_{ax} = \lvert ds'/ds \rvert$	Definition of axial magnification as the derivative between object and image conjugate distances

A very concise derivation of useful results can be made by reference to the ray diagram in Figure 1. In general, object-space quantities are represented by unmarked symbols: n, f, and s. Image-space variables are primed: n', f', and s' (Table 2). The relation between object and image locations can be derived by considering the similar triangles in the ray diagram:

$$h/(s-f) = h'/f, \tag{1a}$$

$$h'/(s'-f') = h/f'. \tag{1b}$$

Solving for s and s', and using the definition of magnification ($h'/h = M$),

$$s = (1 + 1/M)f, \tag{2a}$$

which shows that the object must be located a short distance outside the front focal length, and

$$s' = (M + 1)f', \tag{2b}$$

which shows that the image will be located a long distance beyond the back focal length. Approximately, $(M + 1)f'$ is equal to the body tube (BT) length in finite-tube systems or the focal length of the tube lens (f_{TL}) in infinite-conjugate (IC) systems. In ratio,

$$\begin{aligned} s'/s &= Mf'/f \\ &= n'M/n \quad \text{(using the Helmholtz relation).} \end{aligned} \tag{3}$$

Solving for the relation between conjugate distances s and s′,

$$1 = f/s + f'/s' \quad \text{(general form).} \tag{4a}$$

For nonimmersed optical systems ($n' = n$ and $f' = f$), Equation 4a can be put in the familiar form

$$1/f = 1/s + 1/s'. \tag{4b}$$

For computation of image distance, the Gaussian optics form is used:

$$s' = f's/(s-f). \tag{4c}$$

Transverse magnification clearly varies with object distance, growing without bound as the object is brought up to the front focal point. Using Equations 3 and 4c,

$$M = f/(s-f). \tag{5}$$

The axial magnification is found by differentiation of Equation 4c:

$$\begin{aligned} ds'/ds &= -(f/f')(s'/s)^2 \\ &= -(n/n')(n'M/n)^2. \end{aligned} \tag{6a}$$

Using the definition of axial magnification (Table 2),

$$M_{ax} = (n'/n)M^2, \tag{6b}$$

showing that it is proportional to the square of the transverse magnification.

Afocal combinations, in which an objective and a converging lens (or tube lens) are separated by the sum of their respective focal lengths (Fig. 1E), are a special case because the system focal length is infinite by the usual definition. However, a geometric ray trace shows that the conjugate relations are finite. In particular, the magnification is constant (independent of s and s') and equal to the ratio of the objective and the tube lens focal lengths:

$$M = h'/h = f_2/f_1 = f_{TL}/f'_{obj}\,. \tag{7}$$

Likewise, the conjugate distances are related by a linear equation (compare with Eq. 4c)

$$s' = f_2 - (n'/n)M^2(s - f_1). \tag{8}$$

When the object is located in the front focal plane of the objective, the image is formed in the rear focal plane of the tube lens. This differs significantly from the Gaussian optics result (Eqs. 2a and 2b), in which the object is located at a slightly greater distance. In the afocal case also, axial magnification is proportional to the square of transverse magnification. If s and s' are measured from the front and rear focal planes, respectively, rather than from the principal planes (i.e., $s - f_1 = \Delta s$ and $s' - f_2 = \Delta s'$), the conjugate-distance relation can be put into the particularly simple form

$$\Delta s' = -(n'/n)M^2\Delta s. \tag{9}$$

When an afocal system is set up so that the object is in-focus when it is in the front focal plane of the objective (i.e., when the camera is located in the back focal plane of the tube lens), rays from each point in the specimen are collimated on the path between the objective and the tube lens. This is an idealized model for IC (infinity-corrected) microscopes (see below). Real IC systems are not necessarily exactly afocal. Generally, the objective and the tube lens are separated by a design distance other than the sum of the focal lengths to optimize compensation of aberration. Nevertheless, the system is designed so that an object will be in-focus when it is located in the front focal plane of the objective, thus, preserving the condition of collimated rays in the IC space between the two lenses.

Measurement of Depth When There Is an Index Mismatch

Because axial position (depth) within a specimen is encoded through sharpness of focus, calibrated movement of the microscope focus drive provides the primary means for quantitative analysis of three-dimensional (3D) structure. The paraxial optics formulas provide the basic relation between stage increment and focus shift (Galbraith 1955). If, for example, a low-power nonimmersed objective is focused on a particular point within a mounted specimen, light diverging from that point is more strongly diverged by refraction on exiting the cover glass (Fig. 2). In the paraxial limit, this has the effect of creating a virtual object at a lesser depth. If the actual source point is at depth z_P, the virtual source appears at z_V, with the two coordinates in proportion to the ratio of the immersion refractive index to specimen index, such that

$$z_V = (n_{imm}/n)z_P\,. \tag{10}$$

When the focus drive is adjusted to move the stage closer to the objective by an increment Δz_s, the virtual source remains fixed. The paraxial convergence point of the actual rays shifts away from the objective so that the increment in focus depth exceeds the stage increment,

$$\Delta z_F = -(n/n_{imm})\Delta z_s\,. \tag{11}$$

An intervening layer of different index, such as a cover glass, does not affect this result. The correction can be significant; for an embedded specimen, $n/n_{imm} = 1.52$. When the immersion index

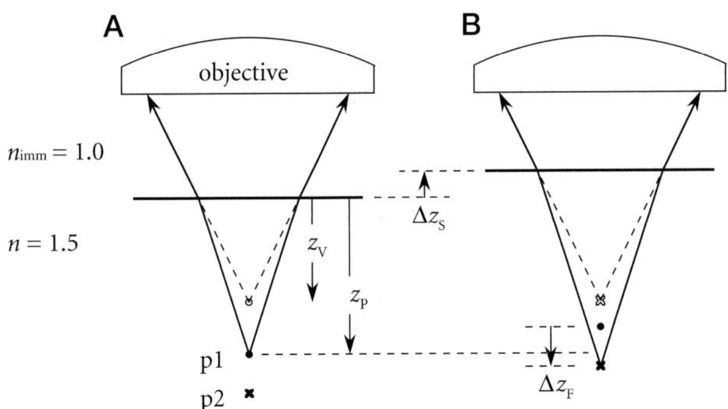

FIGURE 2. Correction for refraction in depth measurement by focus. In this example, a dry objective is used to view a specimen mounted in a high-index medium below a cover glass ($n > n_{imm}$). (*A*) When the microscope is focused on particle p1, the rays exiting the specimen appear to diverge from a virtual object o in the paraxial limit. (*B*) Adjustment of the stage to bring p2 into focus requires a smaller displacement than the actual axial separation of p1 and p2. In the opposite circumstance ($n < n_{imm}$), the stage increment exceeds the focus increment. In either case, the optical focus shift is related to the stage displacement by the ratio of the immersion and specimen indices: $\Delta z_F = -(n/n_{imm})\Delta z_S$.

exceeds the specimen index (such as in the use of an oil-immersion lens to look into living specimens), the virtual source point is deeper into the specimen than the actual source point, and the stage increment exceeds the increment in the focus depth. Equations 10 and 11 apply without modification. Only in the case in which $n = n_{imm}$ are the two increments identical. The difference between stage and focus increment must be corrected to eliminate axial distortion when 3D reconstructions are computed from serial-focus image sets. With high-NA lenses, refractive index mismatch not only causes focus shift, but also introduces focus-dependent spherical aberration. Therefore, an immersion system closest to matching the average index of the object should be optimal, although the situation is rarely ideal for living specimens larger than a small cluster of cells.

Light as a Wave

For discussion of most of the principles and practices of biological light microscopy, it is sufficient to consider light as a scalar-wave field. Clearly, the vector nature of the electromagnetic field cannot be ignored in the discussion of focusing of light by a high-aperture lens or the measurement of polarized components of an image field. Likewise, lasers, absorption, fluorescence, and detection of light are most easily discussed in a quantum picture. There are many matching concepts between wave optics and quantum mechanics, so there is intrinsic value in understanding image-forming instruments in a wave picture.

A light wave has several characteristics that can be represented as a graph of a periodic function such as

$$A(x, t) = A_0 \sin[k(x - vt)]. \tag{12}$$

A_0 scales the sine function, which oscillates between the values ±1. Therefore, A_0 represents the peak amplitude of the wave. The argument of the sine function is linear in both position (x) and time (t) to represent both the spatial pattern of the wave field and the fact that it propagates in the x direction steadily at speed v. The direction of propagation would be in the $+x$ direction for positive values of v and in the $-x$ direction for negative values of v. $A(x, t)$ can be graphed on either a spatial axis (at a specified time) or a time axis (at a specified position), and, in either case, it is a graph of amplitude versus the independent coordinate. It is important to realize that $A(x, t)$ can be graphed in 3D space in which A is then a scalar amplitude value and in which Equation 12 is seen to represent plane waves because $A(x, t)$ is independent of y and z and is, therefore, of constant value on any surface in which the argument of the sine function is constant. In this case, these surfaces are planes normal to the x-axis. In a more general representation,

$$A(\mathbf{r}, t) = A_0 \sin[\mathbf{k} \cdot \mathbf{r} - \omega t + \phi] \tag{13}$$

represents a scalar plane-wave field oriented normal to the wave vector **k** and propagating in the direction defined by **k**.

In addition to amplitude and velocity, the periodic wave field $A(x, t)$ has an exactly defined wavelength and frequency and, therefore, represents an idealized monochromatic wave. Because $\sin(x)$ has a natural period of 2π, the constant k is the scale factor that sets the wavelength λ. Effectively, k is the spatial frequency of the field in radians/unit length,

$$k = ||\mathbf{k}|| = 2\pi/\lambda. \tag{14}$$

The frequency of the wave (ν) is simply the number of cycles that occur per unit time, so that the product of frequency and wavelength must equal the wave speed,

$$\nu\lambda = v, \tag{15a}$$

or, with frequency expressed in radians/sec (ω),

$$\omega/k = v. \tag{15b}$$

For light in vacuum, $v = c = 3.0 \times 10^8$ m/sec. When a wave enters a material medium in which its speed is altered, the frequency must remain constant—otherwise, there would be no conservation of "cycles." Therefore, both λ and v must change proportionally. The phase of $A(x, t)$ at a particular location and time is the shift of the wave relative to some standard graph such as $\sin(kx)$. Phase shift is usually expressed in angular units of degrees or radians or in fractions of a wavelength. For example, $\cos[k(x - vt)]$ lags $\sin[k(x - vt)]$ in time by a phase shift of 90°. In Equation 13, ϕ is a phase angle that represents an offset in the argument of the sine function. It causes a shift in the graph by a distance $(\phi/2\pi)\lambda$. To represent linear and elliptical states of polarization, the amplitude of a wave field can be written as a complex vector quantity, $\mathbf{A}(\mathbf{r}, t)$. Because light is a transverse wave in isotropic materials, the vector representing the amplitude must be oriented normal to the direction of propagation. For example,

$$\mathbf{A}(x, t) = \mathbf{e}_y A_0 \sin[k(x - vt)] \tag{16}$$

represents plane waves propagating in the $+x$ direction, linearly polarized in the y direction. Finally, intensity is a measure of the energy carried by a wave field. In analogy to the alternating current delivery of power to a resistor, which is proportional to the mean square voltage or current, the intensity of a wave field is proportional to the time average of the square of the amplitude. In reduced units,

$$I = \langle A^2 \rangle = A_0^2 \langle \sin^2[\omega t] \rangle = (1/2)A_0^2 . \tag{17}$$

In a quantum description, the energy carried by the field is proportional to the number density of photons and to the energy carried per photon (E_{photon}):

$$E_{\text{photon}} = h\nu = hc/\lambda, \tag{18}$$

where h is Planck's constant, 6.6256×10^{-34} J/sec. A light quantum carries energy far greater than the average thermal energy per mode in a molecule. Therefore, absorption of light by a molecule can initiate many processes that do not occur otherwise.

Huygens' principle of wave propagation is a fundamental and powerful concept that is most dramatically illustrated by attempting to isolate a single ray of light by the use of a collimator and a pinhole in a distant opaque screen. In such an experiment, the observation is that collimated light (plane waves) incident on the pinhole emerges as expanding spherical waves rather than as a thread-like collimated beam—showing that the concept of a ray has meaning only as the local director of an extended wave front. Huygens formulated this experimental result (easily observable in water waves) into a picture in which every differential patch on a wave front acts as a source of an expanding spherical wavelet. At a time increment Δt, each of these wavelets will have expanded to a radius

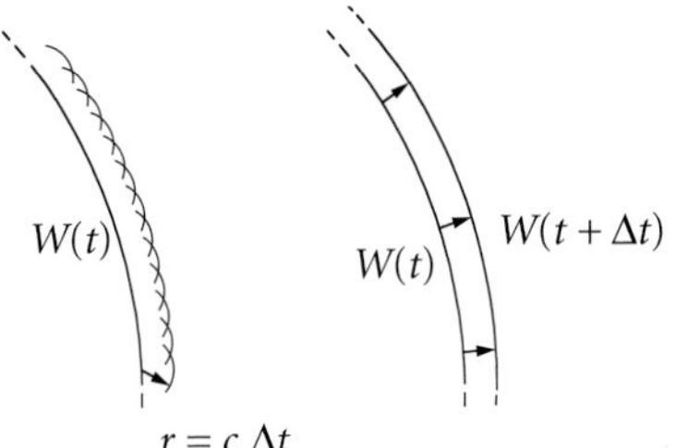

FIGURE 3. Huygens' construction. Each infinitesimal area element on wave front $W(t)$ acts as a source of an expanding hemispherical wavelet proportional in amplitude to the local value of W. After a time interval Δt, each wavelet will have expanded to a radius $v\Delta t$, in which v is the wave speed (here shown as c, the speed of light). The constructive interference of the wavelets in the forward direction defines a continuous wave front $W(t + \Delta t)$ that is advanced and is generally altered in shape and amplitude distribution. An unbounded plane wave maintains its planarity and uniform amplitude. An expanding or contracting spherical wave remains spherical but with decreasing or growing amplitude, respectively. In general, an arbitrary wave front, including partial plane or spherical waves, will undergo diffractive modification as it is propagated by this process.

$c\Delta t$, and the propagated wave front is the net result of the superposition and the interference of all of the wavelets (Fig. 3). Using Huygens' principle, it is straightforward to show that an unbounded plane wave front continues to propagate as such and that a complete spherical wave grows as an expanding sphere (of diminishing amplitude). The principle is rigorously expressed in the Kirchhoff diffraction integral, which is of great use in the computation of scalar fields for real optical systems. For the purposes of the present discussion, the Huygens picture can be used to describe the propagation of plane waves incident on a planar boundary between two materials in which the wave speed is altered from v_1 to v_2 (Fig. 4). The construction shows geometrically that the transmitted wave is refracted according to the rule $v_1/\sin\theta_1 = v_2/\sin\theta_2$. This result corresponds directly to the law of refraction,

$$n_1 \sin\theta_1 = n_2 \sin\theta_2, \tag{19}$$

and shows that the refractive index (n) is the ratio of the speed of light to the speed of the wave in each material (Table 3),

$$n_i = c/v_i. \tag{20}$$

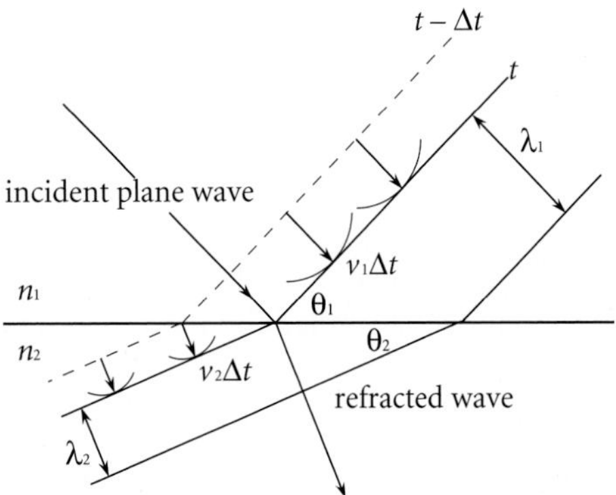

FIGURE 4. The relation between refraction and wave speed. Huygens' construction shows that the direction of propagation of a wave front is altered at a boundary in which wavelets differ in speed. In a time interval Δt, wavelets expand to a radius of $v_1\Delta t$ in medium 1 and to $v_2\Delta t$ in medium 2. A comparison of triangles shows that $(1/v_1)\sin\theta_1 = (1/v_2)\sin\theta_2$. Because the frequency of the wave must be identical in the two media, the wavelength (λ) must change in proportion to v. With the refractive index (n) defined as c/v, Snell's law is obtained directly: $n_1 \sin\theta_1 = n_2 \sin\theta_2$. Also, $\lambda_i = \lambda_0/n_i$ in each medium. The reflected wave in medium 1 is not shown.

TABLE 3. Optical constants relevant to microscopy

Visible light wavelength band (λ)	390–750 nm	
Speed of light in free space (c)	3.0×10^8 m/sec	(30 cm/nsec)
Frequency band of visible light	$7.7–4.0 \times 10^{14}$ Hz	
Important wavelengths		
Peak human scotopic vision	556 nm	(Yellow–green)
Mercury green lamp	546.1 nm	Standard optical design wavelength
Low-pressure sodium lamp	589.0 nm	Sodium D line, a standard yellow
(a spectral doublet)	589.6 nm	wavelength for refractometry
Red helium–neon laser	632.8 nm	
Important refractive indices (n) at ambient temperature:		
Air	1.00028	
Water	1.333	(20°C)
Culture medium or saline	1.335	
Animal cells	1.36	(Average, by refractometry)
Glycerol	1.47	
Silica	1.46	
Crown glass	1.52	
Average refractive index increment for proteins and nucleic acids: 0.0018 per gm/dL		

Equations 15 and 20 lead to the corresponding result for the wavelength, which must also change proportionally with v,

$$\lambda = \lambda_0/n, \tag{21}$$

where λ_0 is the wavelength in vacuum. For light propagating through a sequence of materials that differ in refractive index, the total time of travel for a wave front on a particular ray path is simply the sum of the path length divided by the wave speed in each material,

$$\Delta t = L_1/v_1 + L_2/v_2 + L_3/v_3 + \cdots . \tag{22}$$

The optical path (OP) is defined as $c\Delta t$, the total path length that would be traveled by the wave in vacuum. Through Equation 20,

$$\mathrm{OP}_{1,N} = n_1L_1 + n_2L_2 + n_3L_3 + \cdots + n_NL_N. \tag{23}$$

Clearly, OP can be computed easily on any known ray path. However, in a fundamental sense, OP is a constant of integration resulting from Fermat's principle.

Resolution in Transmitted Light Microscopes: The Abbe Model

The most fundamental difference between microscopes and other types of imaging instruments is the high aperture of the optics. By this, it is meant that the specimen can be illuminated by light rays entering over a wide range of angles and that the objective can form an image from light rays that exit the specimen over a similar angular range. The relationship between aperture and resolution in a microscope was described by Ernst Abbe (1840–1905) in a model in which the specimen is considered to be a planar diffraction grating mask on a slide (Fig. 5). The mask is presumed to be immersed in a medium of refractive index n so that the wavelength is λ_0/n. This picture is the appropriate model for transmitted light microscopy in which there is spatial coherence across the light waves impinging on the specimen from a particular direction. If the grating is illuminated by a beam of monochromatic plane waves from the condenser (also called a light pencil), the transmitted light exits as a set of plane-wave diffraction orders that travel at well-defined angles relative to the incident-light direction. For a grating composed of parallel slits of period p and light of wavelength λ_0/n,

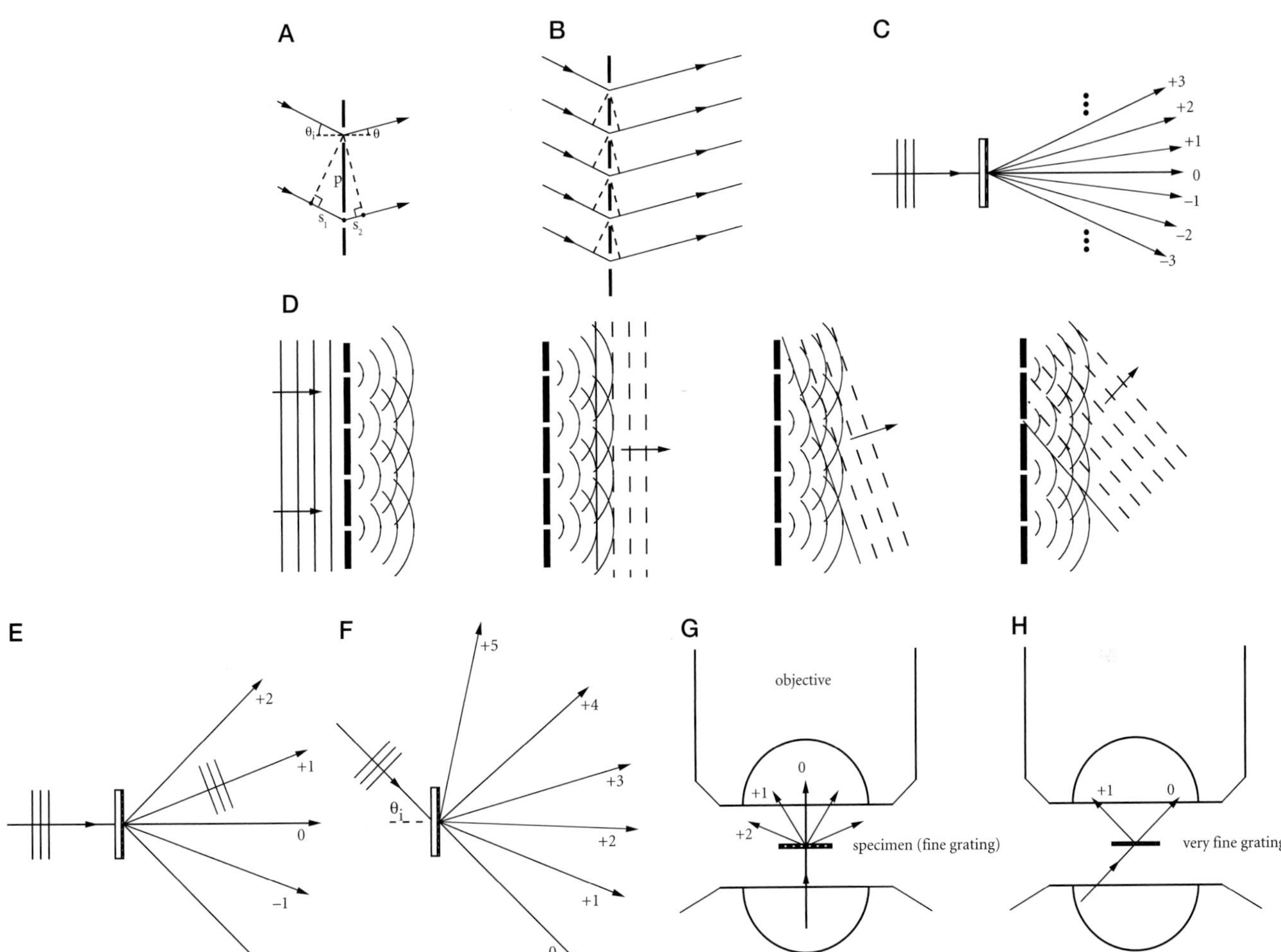

FIGURE 5. The Abbe model—coherent image formation. (*A*) The condition for constructive interference in a diffraction grating is that the OP difference between neighboring slits or sources is an integer number of wavelengths; $s_1 + s_2 = N\lambda$. This is clearly dependent on both the direction from which light is incident on the grating and the direction from which the grating is viewed and leads directly to Bragg-type relations (Eqs. 24 and 26). (*B*) A grating is composed of a large number of equispaced sources in which case there is nearly complete destructive interference for light exiting the grating in directions for which the sources are not all in phase and completely constructive interference in the specific directions from which the sources all appear in phase. (*C*) Plane waves normally incident on a grating are scattered as a set of diffraction orders, in which the neighboring slits differ in phase by 0, 1, 2, ... or more cycles of the wave. Each diffraction order is a set of plane waves exiting the grating at a specific angle (Eq. 25). A finer grating produces more widely divergent orders (compare C and *E*). (*D*) Wave-optical origin of diffraction orders. Each slit produces cylindrical waves that act as Huygens' wavelets. These interfere to produce plane waves at specific angles to the grating. For the wavelength and the grating period shown, three such angles occur. These are shown in *E* as five diffraction orders. (*F*) When light is incident obliquely, the orders are shifted in angle, with the zeroth order propagating in the forward direction. (*G*) A fine diffraction grating as an elementary object in a transmitted light microscope. With normally incident illumination, the zeroth-order and both first-order waves are captured by the objective and interfere to form the image. In *G*, the second-order waves exit the grating at an angle exceeding the aperture angle of the objective. A very fine grating would produce orders so widely divergent that only the zeroth order would enter the microscope. No interference pattern would be formed in the image plane, and the grating would be unresolved. (*H*) With oblique illumination, the widely divergent first order of a very fine grating is captured by the objective along with the zeroth order. Interference between these waves produces a fundamental image of the grating. The Abbe resolution limit is defined by the maximally oblique condition in which the incidence angle is the aperture angle of the objective, and the grating period is such that the first order exits at the opposite of this angle (Eq. 27). In the usual circumstances, the condenser produces waves incident on the grating from a range of directions.

diffraction orders occur at angles (θ_N) for which the OP increment per slit is an integer multiple of the wavelength. This is the condition that guarantees purely constructive interference between wavelets expanding from each slit. In the case in which the plane waves arrive at the grating at normal incidence,

$$p \sin \theta_N = N\lambda_0/n, \quad N = 0, \pm1, \pm2, \pm3, \ldots, \pm N_{\max}, \tag{24}$$

where $N_{\max}$ is defined by the requirement that $|\sin\theta_N| \leq 1$. Each diffraction order is a set of featureless plane waves, showing clearly that the information specifying the structure of the grating must be carried by the angle, the amplitude, and the phase of each order relative to the others. In the Abbe picture, the objective functions to capture and to re-direct the diffraction orders so that they interfere to produce a magnified image of the grating. Equation 24 shows that the orders become more widely divergent (and fewer in number) for finer gratings,

$$\theta_N = \text{Arcsin}[N\lambda_0/np]. \tag{25}$$

In this simple case, the finest grating that would produce an image is the one for which only the three central orders ($N = -1, 0, 1$) could enter the objective (Fig. 5G). In other words, the maximum aperture angle for the objective ($\pm\theta_{\max}$) sets the limit on θ_1. Abbe's insight was that the fundamental periodicity in the grating image was produced by interference between the zeroth order ($N = 0$) and either one of the first orders ($N = \pm1$). When the incident plane waves are oblique (incidence angle = θ_i), the condition for constructive interference is a slight modification of Equation 24,

$$p(\sin \theta_N - \sin \theta_i) = N\lambda_0/n, \tag{26}$$

which is a form of Bragg's law that is fundamental in X-ray diffraction. In particular, when $\theta_i = -\theta_{\max}$, the zeroth order will exit the specimen and will enter the objective at $\theta = -\theta_{\max}$, and the first order can then enter at an angle as large as $\theta = +\theta_{\max}$ (Fig. 5H). In this case, Equation 26 defines the transverse resolution limit of the microscope $p_{\min}$ as the finest grating period that can be resolved,

$$p_{\min}(\sin \theta_{\max} - \sin(-\theta_{\max})) = p_{\min}(2 \sin \theta_{\max}) = \lambda_0/n,$$

or

$$p_{\min} = \lambda_0/(2n \sin \theta_{\max}). \tag{27}$$

The quantity ($n \sin \theta_{\max}$) is of fundamental importance in microscopy and defines the numerical aperture (NA) of the optical system when n is the index of the immersion medium. The relation is, therefore, most often put in the form

$$p_{\min} = \lambda_0/(2\text{NA}), \tag{28}$$

widely known as the Abbe resolution limit. For mid-visible wavelengths, the Abbe resolution can be as high as 0.2 μm (Table 4). In cases in which the condenser NA is less than the objective NA, Equation 26 reduces instead to

$$p_{\min} = \lambda_0/(\text{NA}_{\text{cond}} + \text{NA}_{\text{obj}}). \tag{29}$$

TABLE 4. Transverse resolution and depth of field versus NA (λ_0 = 0.55 μm)

NA	Immersion	$p_{\min}$ (μm)	δ (μm)
0.3	Dry	0.92	6.1
0.5	Dry	0.55	2.2
0.75	Water	0.37	1.3
1.2	Water	0.23	0.51
1.25	Oil	0.22	0.53
1.4	Oil	0.20	0.43

For an idealized objective in which $\theta_{max} = 90°$, the NA would then equal the immersion fluid refractive index n_{imm}. Practical constraints on the lens design limit θ_{max} to <90° and the NA to values close to but less than n_{imm} (Table 1). For example, high-resolution water-immersion objectives can reach NA 1.2, for which $\theta_{max} = 64.45°$, whereas oil-immersion objectives now range from NA 1.3 ($\theta_{max} = 58.9°$) to NA 1.45 ($\theta_{max} = 72.8°$).

The Abbe picture makes evident the essential function of both the condenser and the objective in transmitted light microscopy. The condenser serves not only to concentrate light into the field of view (FOV), but also provides incident light over a wide range of angles. This ensures that the objective aperture is fully utilized in the capture of diffraction orders from all possible elementary gratings constituting the specimen—up to the transverse resolution limit and regardless of grating orientation in the plane of the slide. Biological objects rarely, if ever, look like a grating but can be pictured through Fourier analysis as a superposition of many gratings differing in period, amplitude, phase, and orientation. Less obvious, but just as important, the NA of the condenser and the objective also determine the axial resolution, or depth of field, of the microscope. Again, considering the specimen to be a diffraction grating illuminated by a single set of plane waves, it is the superposition of diffraction orders that produces an interference pattern in the image plane of the microscope. This pattern exists not only in the image plane, but also is extended axially in image space; that is, it has a large depth of focus, which is equivalent to poor axial resolution (Fig. 6). This situation is

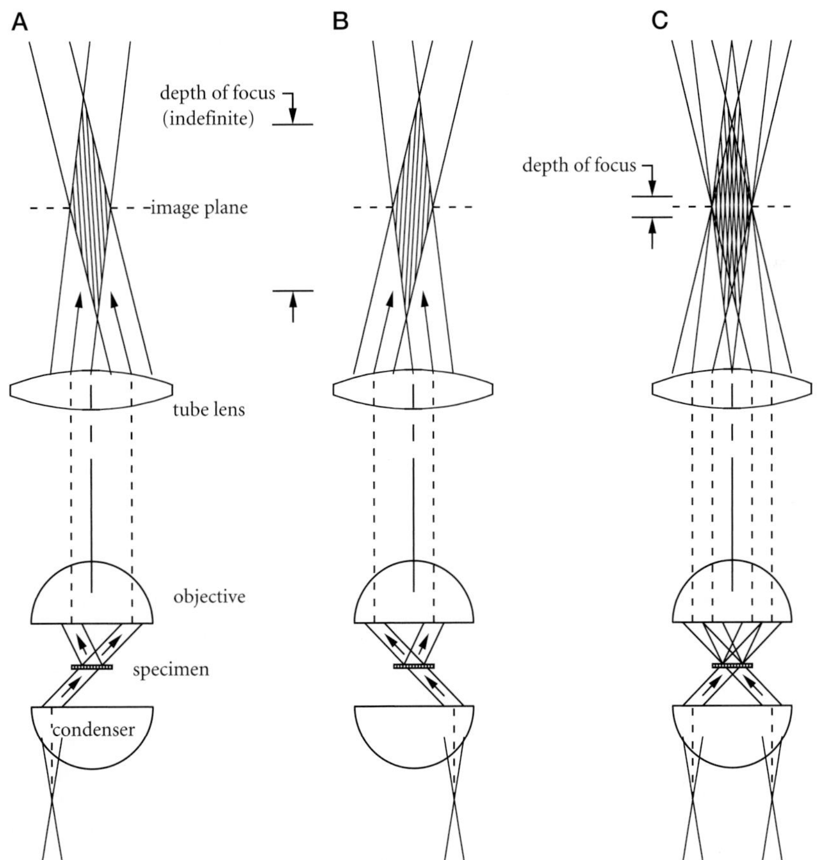

FIGURE 6. Relation between condenser aperture and axial resolution. Superposed interference patterns define the depth of field. (*A*) A fine grating object illuminated obliquely from one direction produces two diffraction orders that interfere in the image plane to produce a fundamental image. The interference pattern has indefinite depth of focus; therefore, the microscope has no axial resolution. (*B*) Light incident from a different part of the condenser pupil produces a similar image. (*C*) The superposition of these two fields reinforces the image only in the neighborhood of the image plane. Use of the full condenser aperture (not shown) superposes many such interference patterns and minimizes the range over which the grating is sharply focused.

changed when a condenser is used to illuminate the grating from many directions simultaneously. Because the light incident from different directions is mutually incoherent, the interference patterns are formed independently in the image space. In the image plane, all of the patterns sum in-register to give a sharp image of the grating. Out of the image plane, the individual patterns are progressively shifted out of register by an increment depending on the actual angles of the incoming diffraction orders. The net result is that the grating image loses sharpness over a finite axial range in image space, which defines the depth of focus (Fig. 6C). In a microscope, where the camera is in a fixed position with respect to the optical system, this must be translated into an equivalent focus range. Normalizing the depth of focus by the axial magnification (Eq. 6b) defines the depth of field (δ),

$$\delta = n\lambda_0/\mathrm{NA}^2. \qquad (30)$$

This equation will vary by a numerical factor depending on the measure of image sharpness used in its derivation, but the important point is that depth of field in a transmitted light microscope depends inversely on the square of the system NA (Table 4). A similar result holds in fluorescence microscopy (see Eq. 42). Reduced depth of field is equivalent to enhanced axial resolution. Increased NA improves both the transverse and the axial resolution and, therefore, strongly improves the 3D resolution of the microscope.

The strong variation of the depth of field with NA is of very practical and routine use. It is often difficult to initially bring a specimen into focus in a high-NA system because of the absence of visible features in the defocused field. Reduction of the condenser aperture by closing the condenser iris greatly increases the depth of field and, therefore, the contrast of out-of-focus features. They can then be discriminated against the background and brought into focus. Once the focus is set, the condenser iris is opened to increase resolution. In general, visibility (a combination of contrast and resolution) is optimized at an intermediate setting of the iris. However, maximum resolution is obtained with maximum aperture. In video camera–based imaging systems, the black level and the gain of the video signal amplifier can be adjusted to greatly increase contrast at full aperture (see Chapter 2).

Immersion Optics

High NA as a prerequisite for obtaining high resolution (both transverse and axial) was the driving force more than a century ago behind the development of highly corrected immersion optics. Light exiting the specimen into the objective is refracted most strongly at the cover glass–air interface in which there is a large difference in refractive index (Fig. 7). There is not only increased reflective loss for rays exiting the cover glass at a high angle, but there is also total internal reflection (TIR) of rays incident above the critical angle, which, for glass–air refraction, is 41.8°. This limitation is overcome by filling the air space between the cover glass and the objective with an immersion fluid having a higher index. In oil-immersion systems, the cover glass, the oil, and the objective front lens are all nearly index matched so that rays propagate without deviation into the objective aperture. In terms of the Abbe model, the objective can then capture a greater number of diffraction orders, thus, increasing resolution. Reduction of the severity of refraction between the specimen and the objective is important more generally because it makes possible the design of lenses having very low aberration at high NA, the key to high-resolution imaging. In general, the ideal circumstance is a lens designed for use with an immersion fluid having the same refractive index as the specimen, although this is generally not possible in practice.

The optical structure of the sample (specimen + mounting materials) affects ray paths into the microscope. Even if the actual biological object is only weakly refractive or absorptive, the index of the mounting medium and the thickness of the cover glass both have optical effects that can lead to focus-dependent spherical aberration in the image (Frisken Gibson 1990; Frisken Gibson and Lanni 1991; McNally et al. 1994; Scalettar et al. 1996). The situation is most severe in the most common case: dry- or air-immersion objectives for biological use. These lenses are designed to compensate for the refraction that occurs as rays exit the specimen through the cover glass–air interface in which there is a relatively large index difference of 0.5. The correction depends on the use of a cover glass of standard thickness, by convention, 0.17 mm or #1½. Furthermore, the specimen must be mounted in high-

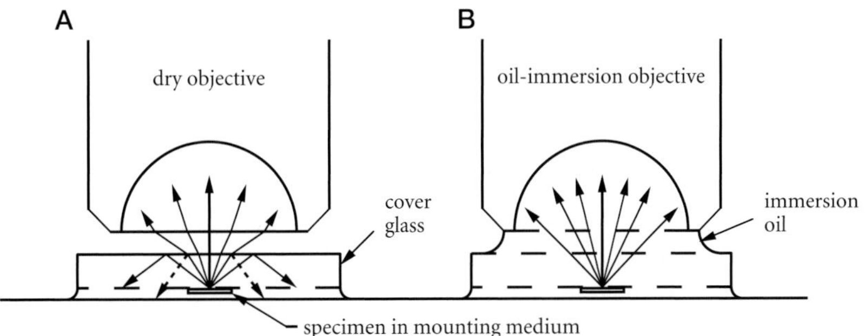

FIGURE 7. Ray diagram for immersion optics. (*A*) Light waves exiting a specimen are refracted at the cover-glass–air interface and again at the air–lens interface. This introduces spherical aberration and causes reflective loss (dotted arrows), deviation of high-angle rays out of the aperture range of the objective, and TIR of rays exiting above the air–glass critical angle. Dry objectives are designed to compensate for the spherical aberration introduced by a cover glass of standard thickness (#1½). (*B*) In an oil-immersion system, the specimen mounting medium, the cover glass, the oil, and the objective are all nearly index matched. In this case, waves propagate without refraction, minimizing aberration and reflective loss. Additionally, diffraction orders that would be lost because of refraction in a dry system enter the immersion objective and contribute to the sharpness of the image.

index medium immediately below the cover glass—if the specimen is not in contact with the cover glass, the intervening layer of mounting medium will essentially act as additional cover-glass thickness, throwing off the correction. Likewise, focusing into a thick specimen located below a cover glass also adds to the ray paths in the high-index medium and causes aberration that grows with depth. In the ideal case, the lens would be designed for use with an immersion fluid that is isorefractive with the specimen—with or without a cover glass. In living cells, the average index is 1.36, only slightly greater than for water, 1.33. Therefore, the ideal case can be approached with WI objectives. Direct-immersion objectives, of most use on upright microscope stands, are of fixed correction. They are designed for contact with culture medium and require no cover glass between cells and lenses. Indirect water-immersion (IWI) objectives are designed for viewing living specimens through a cover glass, with distilled water as the coupling fluid. These lenses incorporate a variable correction that must be set depending on cover-glass thickness over the range 0.15–0.20 mm (#1 to #2). For the best results, cover-glass thickness should be measured with a micrometer caliper before cell culture or specimen assembly. When IWI objectives are used for time-lapse imaging of cells held at 37°C in a culture chamber, evaporation of the immersion water is a practical issue. Fluorocarbon-based immersion fluid that is isorefractive with water, but does not readily evaporate, can be used with certain IWI lenses. Oil-immersion objectives (n = 1.51–1.52) are designed for use with specimens mounted in a high-index medium. Cover-glass thickness is less important in this case because of the near-index match between immersion oil and glass. Other types of correction exist, such as for viewing cells on inverted microscopes through plasticware of various thickness. Glycerol-immersion objectives and variable-correction multi-immersion objectives (water/direct water/cover glass–glycerol/direct glycerol/cover glass–oil/direct or with cover glass) are less common but are useful optical systems (see Table 1).

The Sine Condition, Body Tube Length, and Infinite-Conjugate Systems

Given the importance of NA in the performance of a microscope, it is clear that paraxial optics cannot provide a basis for deriving resolution limits or for minimizing aberration. The diffraction-limited performance provided by a microscope over a flat field is a result of design to satisfy a wave-optical restriction known as the Abbe sine condition. In essence, the relation requires not only that the OP (or propagation delay) between a source point (A) and its corresponding geometric image point (A′) be constant for all possible rays connecting A and A′, but also that the path be stationary for any choice of A in the design object plane and within the design field of view. The mathematical expression of the sine condition maps the angle (U) of a ray entering the objective to the angle (U') of that ray as it arrives at the image plane,

$$n \sin U = M \sin U'. \tag{31}$$

It is straightforward to show that a microscope that satisfies the sine condition for a specimen located exactly at the design working distance will not satisfy the condition for a specimen shifted toward or away from the lens—even if the camera, the film, or the eyepiece is correspondingly shifted to the position of best focus. To accommodate this restriction in a practical way, microscopes are designed to a standard tube length, traditionally defined as the distance behind the objective at which the primary image is formed + 10 mm (Fig. 8). For biological microscopes, the standard body tube (BT) length was 160 mm. In the past 20 years, there has been a major design change to infinite-conjugate (IC) systems (Keller 1995) in which the objective is followed by a tube lens, usually a high-quality singlet, doublet, or triplet (Table 5). In IC systems, the objective is designed so that light from a point source located exactly in the front focal plane emerges from the back pupil of the objective as collimated light. This is equivalent to the formation of the image of the point source at infinity. The function of the tube lens is to converge each collimated set of rays to form the primary image at a distance of one focal length (Fig. 8). The primary advantage of an IC system is that a space is provided between the objective and the tube lens in which flat optics (filters, wave plates, dichroic reflectors, crystal prisms, polarizers, electro-optic devices) can be inserted without causing significant image shift, focus shift, or aberration. A

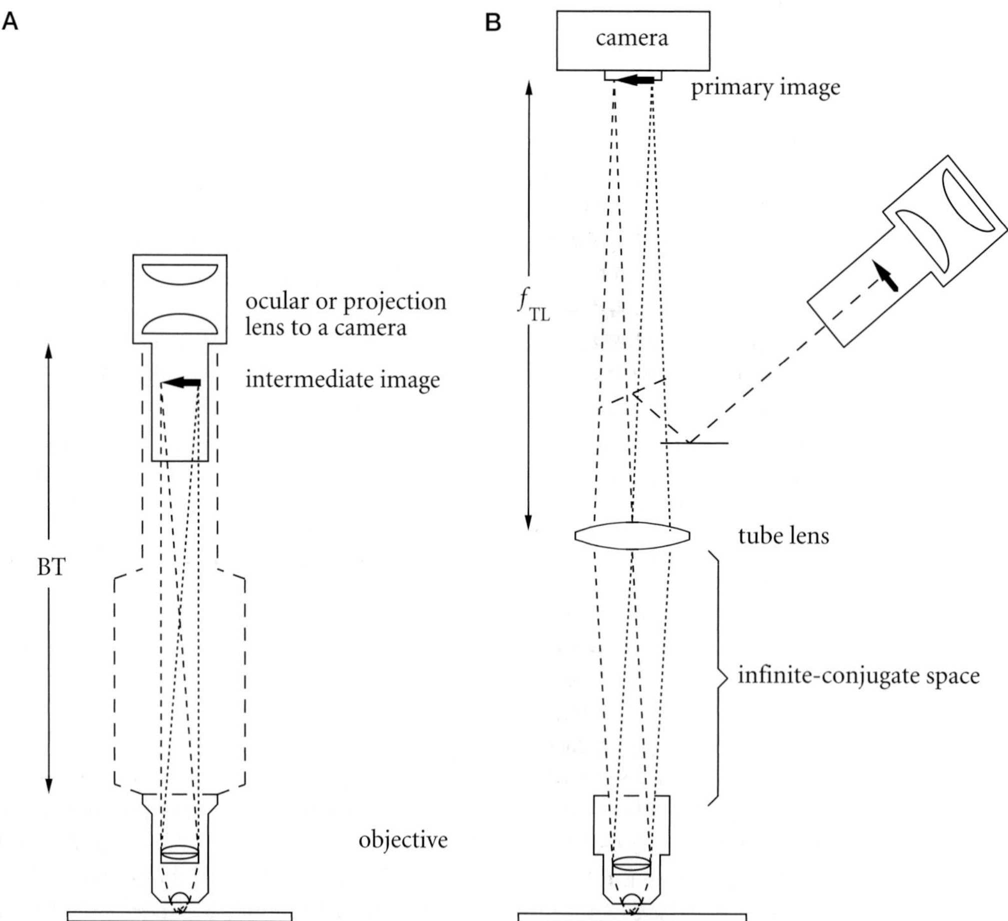

FIGURE 8. Standard layout for 160-mm and infinite-conjugate (IC) microscopes. (*A*) In a finite tube-length system, the objective converges light from an in-focus object, forming an intermediate image, by design, 10 mm below the rim of the body tube (BT). A projection lens is used in place of the ocular to form a secondary image in a camera. For biological microscopes, the standard BT length has traditionally been 160 mm. (*B*) In an IC system, the objective collimates light from point sources in an in-focus object. The primary image is then formed in the back focal plane of a separate tube lens. Flat optics can be inserted into the IC space with a minimal aberrating effect.

TABLE 5. Tube lens focal length in IC systems

Leica	200 mm
Nikon	200 mm
Olympus	180 mm
Zeiss	160 mm

Use of an infinite-conjugate (IC) objective on a microscope stand that incorporates a tube lens of focal length (f_2) different from the design value (f_2*) will have the effect of altering the objective magnification by the factor f_2/f_2*.

secondary advantage is that the focus drive can move the objective over a limited range relative to the microscope (with a fixed stage) without causing magnification error or aberration.

The sine condition applies to both finite-conjugate and IC objectives. When U' is a marginal ray (i.e., when $U = \theta_{max}$ and $U' = U'_{max}$), then Equation 31 can be put in a particularly useful form,

$$n \sin \theta_{max} = \text{NA} = M \sin U'_{max}. \tag{32}$$

Because $M >> 1$ in microscope systems, U'_{max} is generally a small angle. Therefore, to a good approximation, $\sin U'_{max}$ is simply the ratio of half the diameter of the objective pupil ($d_P/2$) to the tube length or the focal length of the tube lens (f_{TL}). Therefore,

$$d_P = 2(\text{NA}/M)f_{TL}. \tag{33}$$

This simple relation shows that the diameter of the clear pupil of different objectives, as seen from the back, increases with NA and decreases with M. It is important in fluorescence microscopy in which the specimen is illuminated through the objective and in which there should be a good match between the projected size of the light source and the diameter of the objective back pupil. Alternatively, in laser-scanning microscopes, this relation sets the optimum laser beam expansion ratio to match the Gaussian beam diameter to each objective back pupil.

The most important practical implication of the sine condition is that aberration will be minimized when the use conditions of the microscope match the design conditions. A problem such as loss of parfocality between eyepieces and camera often results from misadjustment of the camera tube length. If the microscope is simply refocused to compensate for this problem, the specimen will be offset from its design position, and the image will show increased aberration. With the growth in use of microscopes for optical sectioning or imaging at depth, a common problem is spherical aberration caused by refractive index mismatch between immersion fluid and specimen. The index range in living cells (1.34–1.45, average 1.36) is closer to that of water (1.33) than to immersion oil (1.52). Therefore, whereas high-NA oil-immersion objectives generally give outstanding performance imaging adherent living cells on a cover glass, focusing into the specimen causes spherical aberration to accumulate at a rate of one-third wave per micrometer (Frisken Gibson and Lanni 1991). At a depth of 10 µm, the loss in image sharpness is easily noticeable. This problem can be minimized by use of WI objectives (direct or indirect) in which there is a better match between specimen and immersion index. Design parameters and use conditions are permanently marked on objective barrels (Table 6).

It has long been known that the Abbe sine condition has broad significance. For example, it can be put in a form that shows that the back focal plane of the objective is a Fourier map of the object in transmitted light imaging. Specifically, for a lens designed to the condition, parallel rays entering an objective (from the object) at an angle θ converge to a spot in the back focal plane of that lens that is offset from the lens axis in proportion to the sine of the entrance angle:

$$y = f \sin \theta. \tag{34}$$

As depicted in Figure 9, if the specimen is again taken to be a flat diffraction grating illuminated straight-on by parallel rays, its diffraction orders satisfy Bragg's condition (Eqs. 24–26) and make an angle $\theta_N = \text{Arcsin}[N\lambda/p]$ relative to the microscope axis. It can be seen immediately that the sine con-

TABLE 6. Objective identifier markings

Identifier	Symbol	Typical value/explanation
Degree of correction		
Achromat	–	Basic correction for chromatic aberration (CA)
Fluar	–	Moderate correction for CA
Apochromat	–	High correction for CA
Plan-	–	Flat field correction
Magnification	(*M*,x)	0.5x–40x (dry), 10x–100x (water), 40x–100x (oil)
Numerical aperture	(NA)	0.02–0.75 (dry), 0.3–1.2 (water), 0.5–1.4 (oil)
Immersion type		
Dry	(No marking)	
Water (direct)	W, WI	
Water (cover glass)	W Korr	Adjustable correction for cover glass thickness
Glycerol	G, Glyc	Fused silica cover glass—0.2 mm
Oil	Oil, Oel	
Multi	Imm	Adjustable correction for water, glycerol, oil
Tube length/cover-glass thickness		
160-mm body tube (BT)	160/0.17	Standard cover glass
	160/–	Unspecified or none
Infinite-conjugate (IC)	∞/0.17	Standard cover glass
	∞/–	Unspecified or none
Specialized use		
Phase contrast	Ph1,Ph2,Ph3	Standard phase annuli/phase plates
Polarized light	Pol,DIC	Strain-free lens elements
UV fluorescence	U-,U340/380	UV transmissive lens elements
Dark field	Iris	Internal iris for variable NA

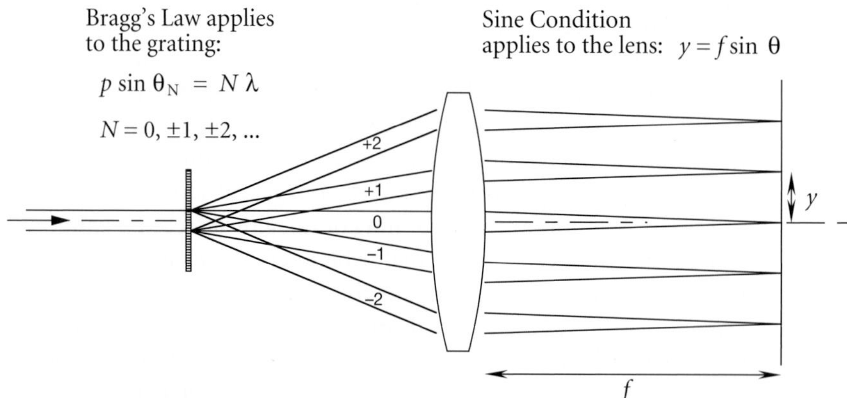

FIGURE 9. The sine condition must hold for a lens to produce an aberration-free image of a small field of view that is normal to the optic axis, with no restriction on the angle that rays make with respect to the axis. It is the essential nonparaxial design restriction for a microscope. In the figure, the lens represents the objective of a microscope. The specimen is a diffraction grating of period p normally illuminated by monochromatic plane waves. Bragg's law (Eqs. 24 and 26) gives the angles of the diffraction orders produced by the grating: $p \sin \theta_N = N\lambda$, $N = 0, \pm 1, \pm 2, \ldots$, whereas the sine condition applies to the lens: $y = f \sin \theta$. Because the spatial frequency of the grating is $k_G = 2\pi/p$, Bragg's law can be restated as $\sin \theta_N = N\lambda k_G/2\pi$, and, by substitution, $y = f(N\lambda/2\pi)k_G$ for the Nth diffraction order. Thus, the sine condition exactly compensates Bragg's law. Because y is linear in both N and k, (1) the diffraction orders of the grating appear as equispaced dots across the back focal plane of the lens, and (2) the 2D coordinates in the back focal plane of the lens are linear in k (i.e., the back focal plane is a Fourier map). In the example shown, the grating has five orders, but a true sine-wave grating would have only two orders +1 and –1 (a phase grating could be made with this property). In this case, $y = (f\lambda/2\pi)k_G$. A square-wave grating such as a Ronchi ruling has only odd orders plus a zeroth order, 0, ±1, ±3, ±5, ..., which correspond to the harmonics in its Fourier series.

dition exactly compensates Bragg's law,

$$y_N = f \sin \theta_N = f \sin(\text{Arcsin}[N\lambda/p]) = fN\lambda/p, \tag{35}$$

which means that monochromatic diffraction orders appear as a row of equispaced dots ($-N_{max} \leq N \leq +N_{max}$) across the diameter of the back pupil of the objective. For a finer grating, the dots are more widely spaced, and fewer appear within the circular limit of the pupil, likewise, for longer wavelength. Furthermore, because $2\pi/p = k_G$, the spatial wave vector for the fundamental sine wave in the grating, there is a linear relation between radial position in the back pupil and grating transverse spatial frequency:

$$y_1 = f(\lambda/2\pi)k_G. \tag{36}$$

This shows that, for 2D objects, spatial frequency maps linearly into offset in the objective back pupil; therefore, the diffraction pattern seen in the back pupil is an intensity map of the 2D spatial Fourier transform (FT) of the flat object.

Principal Optical Components and Köhler Illumination

Best utilization of the full aperture and field of a microscope is usually achieved in Köhler illumination, the condition in which (a) the condenser and the specimen are set in their correct axial positions relative to the objective and (b) the light source is imaged into the back focal plane of the condenser. This involves basic setup steps such as centering of the light source and focusing the collector lens and routine adjustments such as approximately focusing the specimen, then centering and focusing the condenser. The importance of Köhler illumination can be seen in a ray diagram showing light propagation through the principal components of the microscope. In the simplest transmitted light system, these would be, in order: (1) light source, (2) collector lens, (3) field iris, (4) condenser aperture iris, (5) condenser lens, (6) specimen, (7) objective, (8) BT lens (IC system), and (9) oculars, or (10) camera primary image plane (Fig. 10). The collector lens is usually of moderately high aperture to obtain high brightness and forms a magnified image of the light source in the back focal plane of the condenser. This guarantees that every point in the image of the light source is completely defocused over the entire field of view by the condenser so that the illumination of the specimen is very even. As can be seen in Figure 10, the image of the light source at 4 fills the entire condenser pupil so that the full illuminating numerical aperture (INA) can be used. The variable iris at 4 masks the light source image that is allowed to enter the condenser and sets the maximum angle at which light enters the specimen. The condenser iris, therefore, is an aperture stop, which sets the INA. In contrast, the diameter of the variable iris at 3 sets the cone angle of each converging ray bundle that forms the source image and, therefore, sets the diameter of the illuminated field in the specimen. Iris 3, therefore, is a field stop. In practice, the condenser is moved axially to put the image of iris 3 exactly on the specimen. Setup of the microscope can be quickly accomplished by the following sequence of steps.

1. Center, and focus the image of the light source projected into the condenser.
2. Focus the microscope on a feature within the specimen that is easily visible. This properly sets the position of the specimen relative to the objective.
3. Close the field stop enough to put its image, at least partially, within the FOV. If the condenser is far from focus or far off center, the image of the field stop may be highly blurred or may be completely out of view, initially.
4. Center, and focus the condenser to get a sharp centered image of the field stop on the specimen.
5. Open the field and aperture irises. For visual use of the microscope, the aperture is often not opened fully because this improves contrast at the expense of some resolution. The best setting is specimen dependent. For maximum resolution, the aperture iris must be fully opened.
6. Focus on the object of interest. When this is performed, check to see whether the image of the field iris is still sharp, and adjust the condenser focus, if necessary. In most microscopes, the

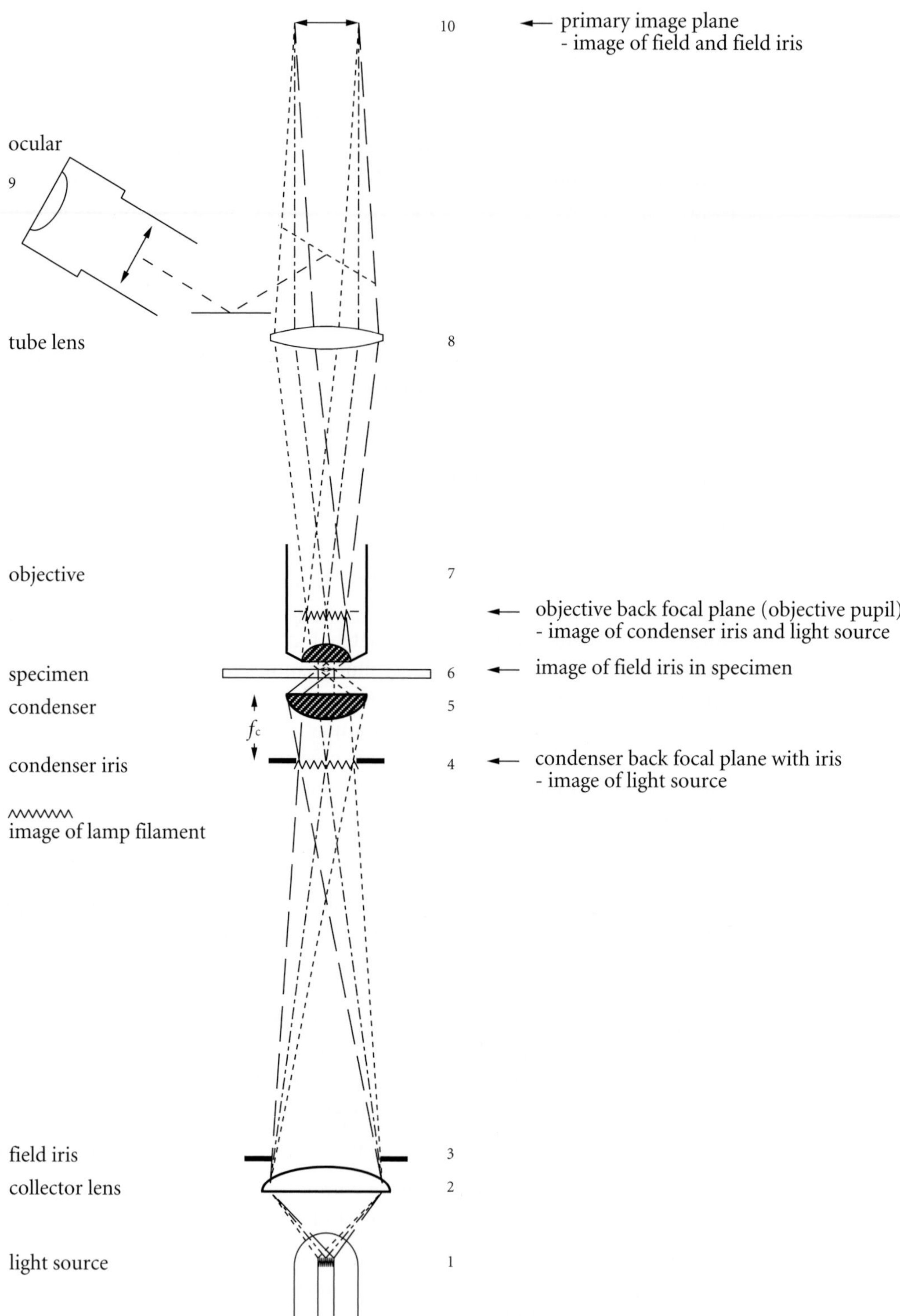

FIGURE 10. Principal optical components of the transmitted light microscope. Köhler illumination is shown in an IC system. The collector optics (2) form an image of the light source in the back focal plane of the condenser (4). This guarantees that the lamp will be completely defocused and the specimen will be evenly illuminated. The source image is masked by the condenser iris, which controls the illumination aperture. Additionally, the field iris (3) is located so that its demagnified image is projected into the specimen plane. The collimated components of the transmitted illumination are converged by the objective (7) to form an image of the light source in its back focal plane. Spherical waves originating in the specimen are collimated by the objective and converged by the tube lens to form an image in its focal plane. The primary magnification is the ratio of tube lens to objective focal lengths.

main focus drive moves the stage and the condenser as a unit relative to the objective or moves the objective relative to the stage and the condenser. In either case, the set position of the condenser relative to the objective is not maintained as focus is changed. Ideally, the focus drive would move the stage relative to the fixed condenser and the objective.

It can now be seen that, in Köhler illumination, light is sent into the specimen over a wide angular range. Each collimated pencil of light passing through the specimen represents an essentially independent set of plane waves because each pencil originates in a different part of the image of the lamp filament. In other words, the pencils are mutually incoherent. In the Abbe model, the image formed by a single light pencil was considered. In Köhler illumination, $\sim 10^6$–10^7 independent light pencils pass through the specimen when a high-NA condenser is used at full aperture. For most biological specimens such as living cells, the zeroth-order diffraction is much stronger than higher orders, resulting in a bright even background. The net result is that the highest-resolution images formed by a microscope in transmitted light also have low contrast and, therefore, low visibility. The development of video-enhanced contrast (VEC) has revolutionized transmitted light microscopy by enabling instruments to be used at full aperture (Allen et al 1981; Inoue 1981; Inoue and Spring 1997).

In the Köhler setup (Fig. 10), two sets of optical conjugate planes can be easily identified. An image of the lamp filament is formed in the back focal plane of the condenser and, along with an image of the condenser iris, in the back focal plane (pupil) of the objective. The filament is completely defocused in the planes of the field stop, the specimen, and the primary image. An image of the field stop is formed in the specimen and in the primary image plane. Microscope stands are designed with this setup in mind, so it is important to note that large alterations in the system, such as would be caused by moving the light source away from its design position, could result in reduced INA and/or uneven illumination of the field of view. It can also be seen that irregularities in the image of the light source, such as the bright filament wires in an incandescent lamp, will not cause unevenness in the field of illumination but instead will cause unevenness in the angular distribution of the incoming light. Because this represents underutilization of the available NA, both transverse and axial resolutions will be affected. Use of a ground-glass light diffuser will help but is optically inefficient. Ideally, the image of the light source in the condenser pupil should be uniformly bright and should be matched to the pupil diameter. In a light scrambler, this is accomplished by focusing the source into a 1-mm fiber optic. A lens is then used to magnify the uniformly bright output end of the fiber so that its image fills the condenser pupil. Use of a scrambler produces high field and angular uniformity, and maximum resolution (Inoue and Spring 1997; Ellis fiber-optic light scrambler; http://www.technicalvideo.com). For certain applications in which a laser is used in place of a conventional light source, active scrambling is required to eliminate speckle due to coherence. This is most easily accomplished by the use of a rotating diffuser (Hard et al. 1977) or by vibrating the fiber in an Ellis scrambler.

In fluorescence or other incident-light microscopes, the objective also acts as the condenser. In this case, it is not practical to place the aperture iris in the objective pupil because it would needlessly reduce the NA and, therefore, the resolution and the brightness of the image. In this case, the aperture iris is placed between the light source and the field iris along with a lens that images the aperture iris into the objective pupil (see Fig. 18).

OPTICAL SYSTEMS FOR IMAGING OF LIVING CELLS

The natural or experimentally induced optical properties of living cells provide the basis for contrast generation in a wide variety of special-purpose optical systems for the microscope. These optical systems can be placed in two main groups, those dependent on the refractive properties of cells and those dependent on the absorptive and luminescence properties (Table 7). This does not exhaust all possibilities, an example being second-harmonic generation microscopy, which depends on a nonlinear optical property of the specimen.

TABLE 7. Microscope optical systems and modes used in biological imaging

Contrast caused by refractive structure within the object:	
	Phase contrast (Zernike phase contrast)
	DIC (Nomarski phase contrast)
	Polarization
	Reflection interference
	Confocal scanning reflectance and interference
	Hoffman modulation contrast, Varel contrast
	Dark field
	Single sideband (Ellis SSB edge-enhancement)
Contrast caused by absorption of light within the object:	
	Bright field/transmittance contrast/amplitude contrast
	Spectral reflectance
	Fluorescence
	Multiband fluorescence
	Confocal scanning fluorescence
	Fluorescence polarization
	Total internal reflection fluorescence (TIRF)/evanescent wave excitation
	Time-resolved fluorescence (fluorescence lifetime, anisotropy)
	Delayed fluorescence
	Fluorescence resonance energy transfer (FRET)
	Multiphoton scanning fluorescence
	Near-field scanning fluorescence
	Structured illumination fluorescence
	Fluorescence image interferometry
	Fluorescence stimulated emission/depletion (STED)
	Fluorescence speckle
	Fluorescence photoactivation single-molecule imaging (PALM, iPALM)
	Fluorescence stochastic single-molecule imaging (STORM)
	Phosphorescence

Phase Contrast

The study of living cells by light microscopy was revolutionized by Zernike's development of phase-contrast optics, which makes the refractive structure of nearly transparent objects visible. With a few notable exceptions such as the erythrocyte and the chloroplast, most cells or organelles present a very low optical absorbance to visible light. Additionally, the average refractive index (1.36–1.38) of most cells in culture is not much greater than that of the surrounding medium (typically 1.335), although refractive heterogeneity within the cells and in the extracellular matrix adds up to a formidable light-scattering cross section in dense tissue. Therefore, the passage of light waves through a single cell results mainly in the distortion of the wave fronts because of all of the structures within the cell that differ in refractive index relative to its immediate surroundings. No significant change in amplitude or intensity (brightness) occurs. More-refractive organelles slightly retard the passage of each wave front. This phase delay, usually expressed as a difference in OP ($\Delta n \times L$; see Eq. 23) is a result of the lower speed of the wave in the region of greater index. For single cells, this difference is usually small. The average phase delay for a cell of 5-µm thickness and average index 1.36 in culture medium of index 1.335 is $(1.36 - 1.335) \times 5 = 0.125$ µm or approximately one-quarter wavelength. A subcellular structure will cause a much smaller retardation. The lamellipod of fibroblasts, which contains mainly actin at a concentration of 40 mg/mL (4 gm/dL) (Abraham et al. 1999) causes a wave-front delay of only 1.3 nm or $\lambda/430$. Because, essentially, no amplitude reduction of the light waves occurs to create contrast and because the phase delays are usually very small, the generation of sufficient contrast for visual discrimination requires conversion of phase shifts into brightness variation.

Zernike phase contrast is based on the general principle that refracted or locally phase-shifted wave fronts can be represented as the summed or superposed amplitudes of scattered waves and the unperturbed waves (or direct waves) that illuminate the specimen. In other words, the phase-distorted wave fronts exiting the specimen into the microscope consist of (1) the plane-wave field that would be present with no specimen in place (or with a perfectly index-matched specimen in place) and (2) the light scattered as spherical waves by the point-like refractive features in the specimen. In general, the scattered waves will be weak relative to the direct light because the net phase shift is small. Additionally, Zernike recognized that the coherent superposition of all of the scattered spherical waves from an elementary plane in the specimen must lag in phase by 90° the direct light to account for the absence of absorption (180° lag) or stimulated emission (no lag). Generation of a phase-contrast image occurs by optically processing the direct and scattered light differently, then allowing the two fields to interfere. In a phase-contrast microscope, two optical elements effect this transformation: a mask or annulus inserted into the back focal plane of the condenser and a phase plate located in the back focal plane of the objective (Fig. 11). Because this plane is usually located within the multilens structure, phase plates are integral parts of phase-contrast objectives. In the Köhler setup, the annulus masks the image of the light source so that illumination is directed into the specimen from a restricted set of directions forming a hollow cone, rather than from all of the directions subtended by the full condenser aperture. An image of the annulus is formed inside the objective coincident with the phase plate so that 100% of the direct light passes through a matching annular zone on the plate. On the other hand, the scattered light, being in the form of spherical waves expanding out of the specimen, passes mainly through the complementary zones of the plate, which constitute most of its area. The annular zone in the phase plate affects the direct light in two ways. It is semitransparent to attenuate the direct light to an amplitude comparable to the strongest

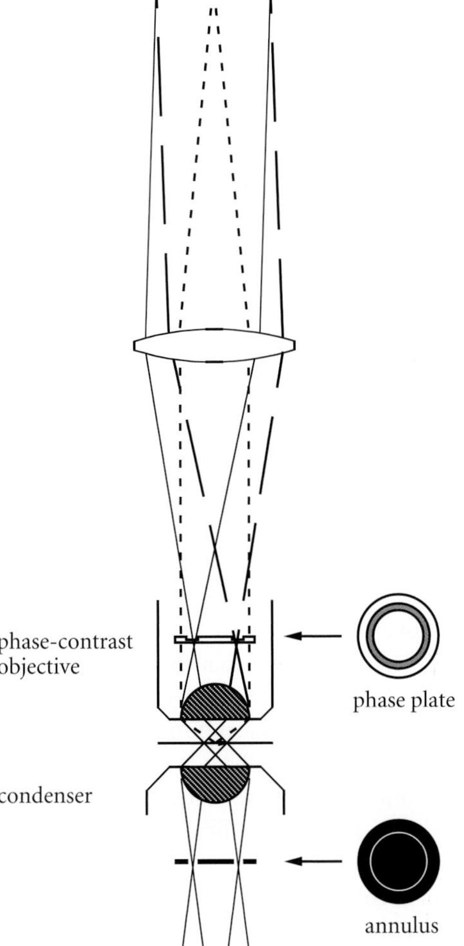

FIGURE 11. Optical components of a phase-contrast microscope. The principal components of the transmitted light microscope (Fig. 10) are augmented by a mask in the condenser back focal plane and a phase plate in the objective. The mask is usually in the form of an annulus, and the phase plate is in the form of a semitransparent ring. In use, the annulus restricts the illumination to be incident on the specimen from a set of directions that form a hollow cone, as shown. Because the annulus is completely defocused by the condenser, illumination of the specimen is uniform. In the objective, the phase plate is coincident with the image of the annulus. This allows for attenuation and phase shifting of the unscattered light. Most of the scattered light passes through the transparent zones of the phase plate (dotted lines). The phase-shifted direct light and the scattered light interfere in the image plane to produce a high-contrast image.

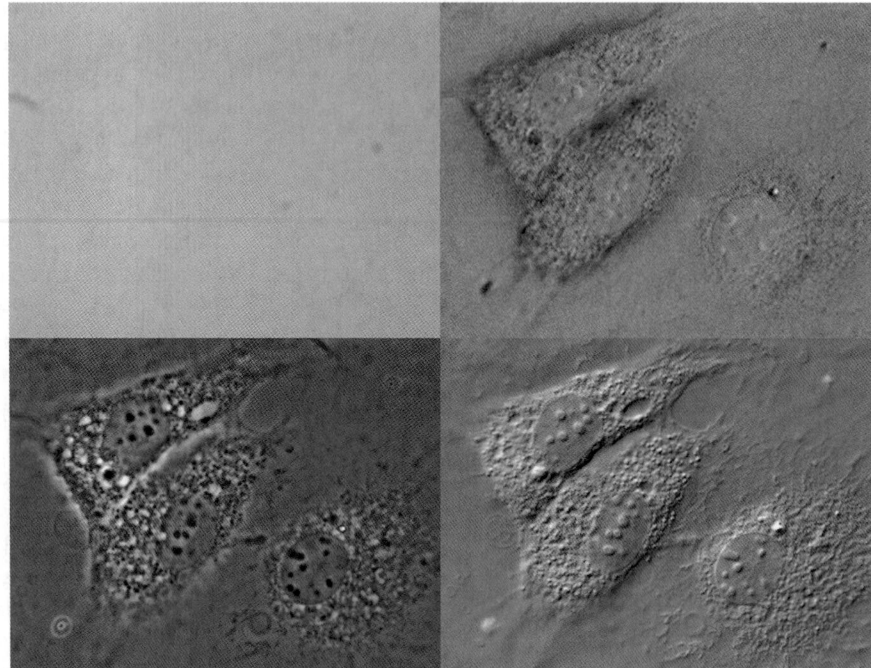

FIGURE 12. Fibroblasts in bright field, phase contrast, and VEC-DIC (differential interference contrast). (*Upper left*) Bright-field image of adherent fibroblasts in physiological medium, 100x 1.3 NA. No features produce sufficient contrast for visibility. (*Upper right*) Digitally contrast-stretched bright-field image shows weak amplitude effects of scatter in the object. (*Lower left*) Zernike phase-contrast image, no digital enhancement. (*Lower right*) Nomarski phase-contrast image with digital enhancement (VEC-DIC). FOV width, 120 µm.

expected scattered light, and it phase shifts the direct light to change the phase difference from 90° to 180°. This creates the condition under which phase delay can be seen as brightness variation. In practice, the direct light is phase advanced by 90° by making the phase plate thinner in the attenuation zone. In the image plane of the microscope, the phase-advanced direct light provides uniform reference waves over the entire FOV. The scattered light amplitude field, which would produce a dark-field image if the direct light were blocked in the objective, interferes with the direct light to produce the phase-contrast image. Because of the 180° phase difference between the two fields, the interference process results in attenuation or gain in brightness and, therefore, an image with significant contrast. In most phase-contrast microscopes, the phase shifting is designed so that more-refractive structures show up as darker features in the image (Fig. 12). This is known as positive phase contrast. In high-NA phase-contrast systems, cellular structures as thin as lamellipods can be discriminated under typical conditions.

The beauty of phase contrast is its simplicity, ease of alignment, insensitivity to polarization and birefringence effects (a major advantage for viewing cells through plasticware), and isotropy of response to different in-plane orientations of a particular feature in the specimen. Its difficulties include the need for a phase-contrast objective, the fact that the full INA is not used because of the annular mask in the condenser, the optical effect of the phase plate on other modes of microscopy, and a well-known halo artifact in the image around sharp boundaries of the object. The limitation on INA causes an extended out-of-focus response. The halo is caused by the passage of some of the scattered light through the direct-light zone of the phase plate. This is an unavoidable consequence of the need for an annulus of finite width to allow for sufficient illumination. The halo often significantly increases the dynamic range of the image, making it difficult to apply VEC methods to this mode of microscopy. Experimental phase-contrast systems exist that suppress the halo and remove the restriction on VEC (G.W. Ellis, pers. comm.). Given the limitations of phase contrast, it remains a remarkably reliable and useful optical mode.

In practice, alignment of phase-contrast optics is a simple addendum to the Köhler setup procedure. After focusing and centration (Steps 1–6), do the following.

7. Switch condenser turret or slider to the phase annulus corresponding to the objective in use.
8. Remove one ocular, and insert a phase telescope or a Bertrand lens, or switch to an internal telescope setting. Focus the telescope (not the microscope!) to obtain a sharp image of the phase annulus and the phase plate in the back focal plane of the objective.
9. Adjust the centration of the annulus (not the centration of the condenser!) relative to the phase plate so that the bright image of the annulus falls exactly within the annular zone of the phase plate.

Mismatched diameters of the two annular components signify that the incorrect phase annulus is in place in the condenser. Vignetting of the image of the light source within the annular mask will occur if the condenser is out of focus.

Differential Interference Contrast (DIC)

For many live cell imaging applications, Nomarski DIC has supplanted the Zernike system. The main advantage of DIC over phase contrast is that the optical elements required for DIC do not mask or obstruct the condenser or objective pupils so that the instrument can be used at full NA and INA. This improves resolution—particularly axial resolution, obviates halo artifacts, and produces an image well suited for electronic enhancement of contrast. However, DIC is essentially a phase-contrast imaging mode. In this case, phase shifts caused by the refractive structure of the specimen are encoded in a field of polarized light, the two superposed components of which are then mutually offset and analyzed to show refractive index gradients.

The essential DIC optical components consist of (1) a linear polarizer inserted between the light source and the condenser, (2) a modified Wollaston prism mounted close to the iris in the condenser back focal plane, (3) a Nomarski prism inserted immediately behind the objective, and (4) a linear polarizer (analyzer) mounted before the tube lens and the image plane (Fig. 13). In some DIC systems, a compensator is used in place of the polarizer or the analyzer. The polarizer and the analyzer are crossed; that is, if the polarizer is oriented at 0° (east–west), the analyzer is oriented at 90° (north–south) as shown in Figure 13B. With the Wollaston and Nomarski prisms removed, this optical configuration is equivalent to a polarizing microscope set for maximum extinction of transmitted light. In this state, birefringent features in the specimen would show up bright against a dark background. The prisms are inserted at 45° (northwest–southeast) and can greatly change the optical response, depending on their relative adjustment. The prisms are each composed of two thin crystal-optic wedges in which the optical axis in one wedge is normal to the optical axis in the other wedge. The complete prism is formed by cementing the wedges to form a thin plate, which is anisotropic. One particular direction in the plane of the plate is the prism shear axis; a beam of light entering normal to one face of the prism is split into two emergent beams, which are orthogonally polarized and are deviated by an angle (Fig. 14). The prism shear axis and the normal axis define the shear plane in which the emergent beams separate, and they define the polarization axes of the two beams. One beam will be polarized in the shear plane (p), and the other beam will be polarized across the shear plane (s). The angle of beam deviation is set by the design of the prism. For microscopes, this angle is so small that there is no observable separation of the emergent light. In terms of wave fronts, Figure 14 shows that the angular deviation of the two beams is equivalent to a constant phase shift per unit length introduced by the prism across its face, in a direction parallel to the shear axis. The phase shift per unit length is equal but opposite for the p- and s-waves. The relative intensities of the emergent p- and s-beams depend on the polarization axis of the incoming light relative to the prism shear axis. If the incoming beam is p polarized (polarized parallel to the shear axis), only a p-polarized beam will emerge. In the opposite case, only an s-polarized beam will emerge. A balanced case occurs when light that is linearly polarized at 45° to the shear axis enters the prism. A 45° linear polarization is equivalent to the in-phase vector summation of two fields, one polarized parallel to the shear axis (p), and

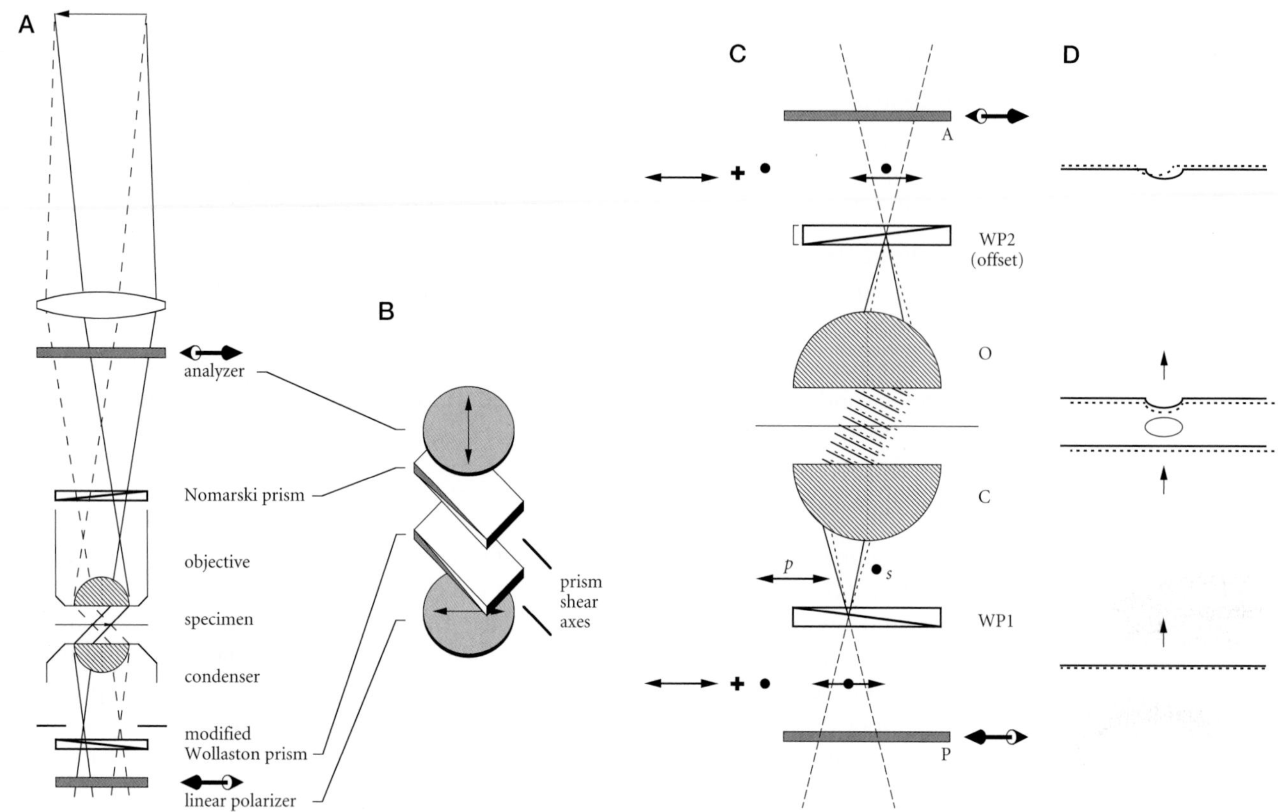

FIGURE 13. Optical components of a differential interference contrast (DIC) microscope. (*A*) The principal layout of Figure 10 is augmented with a polarizer, a modified Wollaston prism, a Nomarski prism, and an analyzer. (*B*) The polarizer and the analyzer are set nominally to extinction (crossed), and the birefringent prisms are inserted at 45° to the polarizer axes. (*C*) Effect of the principal DIC components; for simplicity, Wollaston prisms (WP1, WP2) located in the condenser and the objective back focal planes are shown. In practice, Nomarski-type prisms are used. The prism shear axes are parallel to the page. Considering only light diverging from a single point in the condenser back focal plane, WP1 slightly deviates the *p*- and *s*-polarized waves. This angular deviation (shear) is converted into a small spatial offset when the condenser collimates the light. The offset *p*- and *s*-polarized plane waves traversing the specimen are modified in phase by its refractive structure. The objective reverts the differential offset of the plane waves to angular convergence, which is then compensated by an inverted Wollaston prism (WP2). Transverse offset of the inverted prism has the effect of adding a uniform phase bias across the objective pupil. In the figure, this is shown as phase lag of *p* relative to *s*. The analyzer (A) extinguishes the original component of linear polarization. (*D*) DIC-wave optics; the uniform linearly polarized plane wave below WP1 is sheared by the prism and the condenser and is phase distorted by the specimen. When the shear is removed by the objective and WP2, a phase shift is produced between the *p* and *s* components where the waves interacted with refractive index gradients in the specimen. The phase shift produces elliptically polarized light, which is not completely extinguished by the analyzer. Therefore, the DIC image highlights refractive index boundaries in the specimen oriented across the shear axis. With zero phase bias (WP2 compensates WP1 exactly), the background appears dark, and the index boundaries appear bright. With sufficient phase bias, the background is midrange gray with index boundaries appearing bright or dark depending on the sign of the index gradient.

the other polarized across the shear axis (*s*), each with 70.7% amplitude. The emergent *p*- and *s*-beams are then of equal brightness. Circularly polarized incident light would also produce equally bright emergent linearly polarized beams, as would 45°-polarized light of any intermediate degree of ellipticity. The same is true for incident nonpolarized light. In addition to being anisotropic, Wollaston and Nomarski prisms are directional; that is, collimated light sent in from the opposite side of the prism will also split into two beams but with exchanged *p*- and *s*-polarizations.

Basic DIC image formation can be visualized apart from the microscope. Consider first a collimated beam of light into which is placed a polarizer, followed by a thin specimen, a lens, a Wollaston prism with its shear axis at 45° to the polarization, an analyzer crossed to the polarizer, and a screen on which to project the image (Fig. 15). The prism is placed in the back focal plane of the lens. The

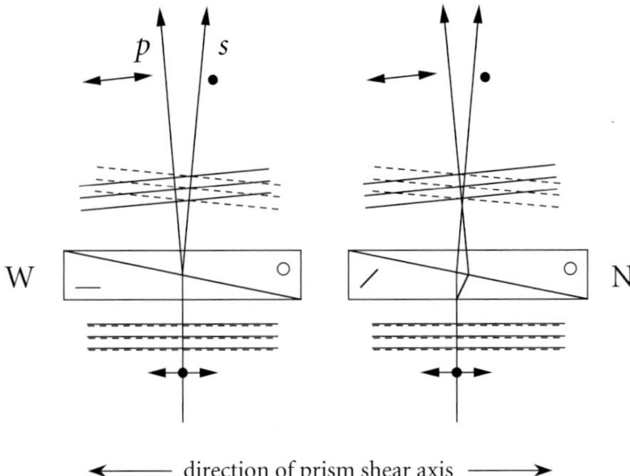

FIGURE 14. Ray and wave diagrams for Wollaston and Nomarski prisms. Prisms of this type are usually made from wedges of quartz, a positively birefringent crystal. The angular deviation between *p*- and *s*-polarized light produced by double refraction in the prism is a function of its wedge angle. For DIC microscopy, this angle is so small that separation of the *p* and *s* waves is not seen. In both prism types, the optical axis of the quartz in one wedge is normal to the axis in the other wedge. The main difference between the prisms is that the convergence point for deviated rays is within the Wollaston prism but is outside the Nomarski prism. This allows the Nomarski prism to be used with conventional condensers and objectives in which the back focal plane is not easily accessible.

polarized plane-wave field entering the specimen gets phase delayed in regions of higher refractive index and emerges with wave-front distortion but with essentially unchanged polarization. The lens brings the wave fronts to an approximate focus at the location of the Wollaston prism, after which the wave fronts then expand toward the analyzer and the screen. In the absence of the Wollaston prism, the entire field would be extinguished by the analyzer, and the image would be dark. With the prism in place, the *p* and *s* components of the field are deviated slightly in angle and arrive at the analyzer

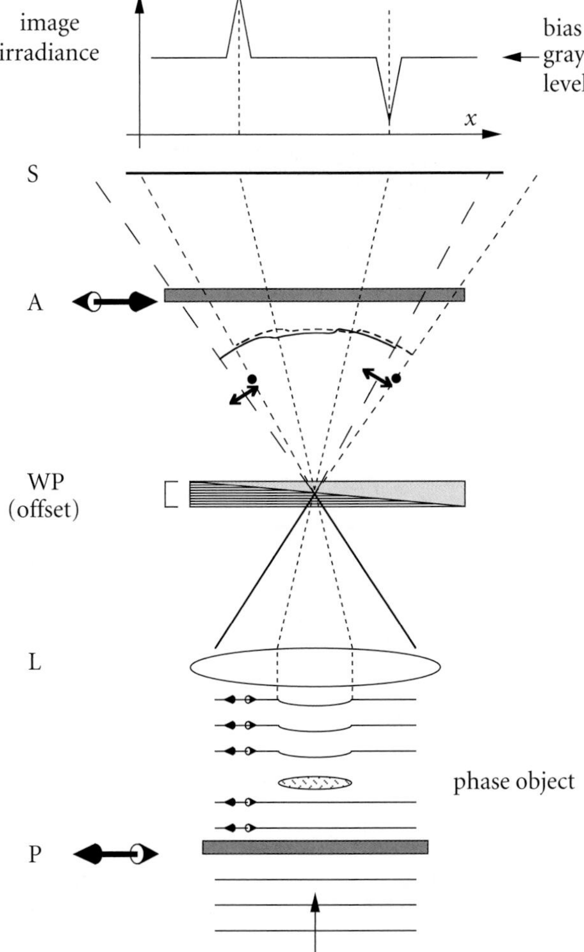

FIGURE 15. Elementary DIC optics. DIC can be obtained by simply shearing polarized collimated light that traverses the specimen. Components: (P) linear polarizer; (L) lens; (WP) Wollaston prism; (A) analyzer; (S) image receiving screen or film. As shown, the refractive structure of the object causes a phase delay in the wave fronts. The lens and the Wollaston prism convert the delay into an angular shift between *p*- and *s*-polarized components, giving the linearly polarized light a small elliptical component. If the prism is offset, a uniform phase bias is also introduced, here shown as *p* lagging *s*. At every point in the field, the analyzer passes the linear component normal to the original polarization. In the schematic, this creates an image in which one edge of the phase object appears brighter than the background and the other edge appears darker than the background. The INA in this elementary system is zero (only collimated light is sent into the specimen from the condenser); therefore, no axial resolution and reduced transverse resolution would result. Finite INA would produce a different bias for every set of plane waves traversing the specimen; therefore, contrast would be lost.

with a small spatial offset along the direction of the prism shear axis. Across all parts of the field in which the wave fronts are undistorted, the offset causes no change in the net state of polarization. The *p* and *s* components are added in-phase to produce the original linear polarization, which is then extinguished by the analyzer. However, across parts of the field in which the wave fronts show a phase gradient, the offset can produce a phase difference between the *p* and *s* components. This changes the local polarization of the field from linear to elliptical so that the analyzer passes some light. The net result is that the image will show features in the specimen in which there is a change in OP caused by a difference in refractive index or thickness. The image is differential because the offset is very small; that is, phase differences from neighboring points are compared by interference. More specifically, the idealized DIC image will show OP gradients in the specimen with maximum contrast occurring when the gradient is along the direction of the shear axis of the prisms. Mathematically, the gradient of the OP in the specimen is a vector field, and the DIC image shows, at each point in the image, the (scalar) component of the gradient parallel to the direction of the prism shear axis and within the depth of field. For a thorough technical discussion of DIC optical systems, see Pluta (1989).

In a real DIC microscope, the full set of optics—polarizer, Wollaston prism, condenser, specimen, objective, Nomarski prism, and analyzer—enables the DIC image to be formed in a high-NA system. First, the polarizer and the Wollaston prism are nominally set to balance the *p* and *s* fields; that is, the polarizer defines 0°, and the prism is set at 45°. The location of the modified Wollaston prism close to the back focal plane of the condenser is fundamentally important because the lens then converts the angular deviation of the *p*- and *s*-polarized light into a fixed transverse spatial offset of the fields that exit the condenser and propagate through the specimen (Fig. 16). The offset distance is differential in

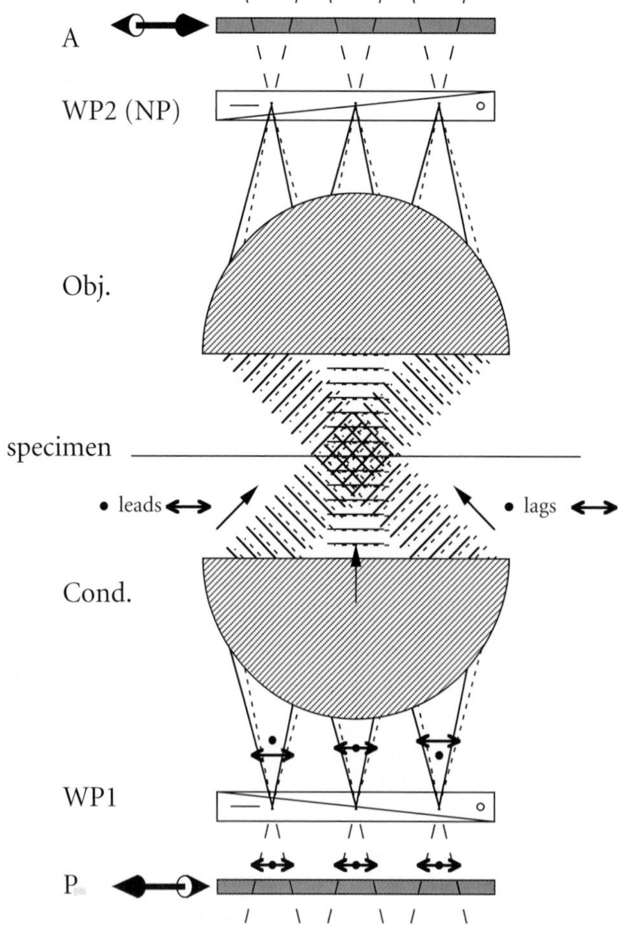

FIGURE 16. High-NA DIC microscope—ray and wave optics. The essential feature of a high-NA DIC system is the use of two polarization-shearing prisms to compensate for linear variation in bias across the condenser pupil. To simplify the diagram, Wollaston prisms are shown in the back focal planes of the condenser and the objective. Three source points are shown in the condenser pupil (coincident with WP1), radiating into the condenser. Each gives rise to a pencil of plane waves that traverses the specimen at a specific angle. In addition to an angular deviation between *p*- and *s*-polarized light, the prism introduces a bias phase shift that is different for each location along the prism shear axis. In the left-most pencil, the *s*-polarized field leads the *p*-polarized field. In the on-axis pencil, *s* and *p* are in-phase. In the right-most pencil, *s* lags *p*. The spatially variant bias is exactly compensated by the inverted prism (WP2) that follows the objective. Essentially, WP1 is imaged onto WP2 so that the bias is matched at each point. A transverse shift in either prism along its shear axis causes a bias mismatch that is uniform across the pupil. The design of the prisms is specific to the focal lengths, the pupil diameters, and the pupil locations of the condenser and the objective.

that it is, by design, approximately equal to one-half the Abbe resolution limit for the objective. After passage of the light through the specimen in which phase delays are caused by its refractive structure, the objective reconverts the spatial offset of the *p* and *s* fields to angular convergence. Ideally, an inverted Wollaston prism located in the back focal plane (pupil) of the objective, and also oriented at 45°, would recombine the converging *p* and *s* fields, as shown in Figure 16. In most objectives, however, the pupil is located inaccessibly within the lens groups. Nomarski's solution to this problem was to modify the prism so that it could be located immediately behind the objective (see Fig. 14). In this location, the Nomarski prism recombines the *p* and *s* fields, effectively removing the offset. The phase gradients impressed on the offset *p* and *s* fields in passage through the specimen are, therefore, converted into phase differences, creating elliptical polarization in the recombined field. When the Nomarski prism exactly compensates for the effect of the Wollaston prism, the action of the analyzer would be to extinguish the light at all locations where there was no phase gradient. The resulting image would be dark in the background with bright features showing refractive edges in the specimen, similar to dark-field optics. Like a dark-field image, both increasing and decreasing index boundaries would be brighter than the background. To resolve this ambiguity in the sign of the gradient, the Nomarski prism serves a second important function. Offsetting the prism along its shear axis (by turning a fine screw to slide the prism) uniformly shifts the relative phase of the recombining beams so that the polarization of the emergent light arriving at the analyzer can be varied from linear through elliptical to circular, etc. The phase shift of the *p* wave relative to the *s* wave is referred to as the bias. With zero bias, the analyzer causes the image background to appear dark and phase gradients to appear bright. As the bias is increased, the background becomes uniformly brighter, and phase gradients appear brighter or darker than the background, depending on the direction of the phase gradient with respect to the direction of the prism offset. Changing the sign of the bias reverses the directional response. For weakly refractive objects such as cells, the signal-to-noise ratio (SNR) of the DIC image peaks at a bias setting slightly greater than the largest phase shift caused by the specimen. However, the SNR is generally not much less for bias settings as high as 90° in which the linearity of DIC is highest, and the brighter image can be detected and can be processed as a video-rate signal. This is the basis of the widely used method of VEC developed for DIC systems by R.D. Allen and N.S. Allen and for polarization microscopy by S. Inoue. Additionally, the total light dose to the specimen is lower than would be needed to get a usable image at zero bias. In microscopes in which the Nomarski prism is fixed, the bias can be set by use of a variable compensator (polarizer and wave plate) to produce elliptically polarized light before the Wollaston prism in the condenser, or inverted, in place of the analyzer. Work on the use of a compensator in place of the analyzer in a DIC system along with algebraic image processing (geometric phase-shift DIC; Cogswell et al. 1997) has shown precise linearization of the imaging response.

The use of two prisms in the Nomarski DIC system is the key feature that enables sharp images to be formed at high NA. As shown in the Köhler setup in Figure 10, the condenser and the objective together transfer an image of the light source and the first shearing prism onto the second prism (which is inverted). Across the image of the light source, the second prism introduces a phase shift per unit length that exactly compensates the linear phase shift between *p* and *s* polarizations introduced by the first prism. Offsetting the second prism along its shear axis does not change the phase shift per unit length but simply adds or subtracts a constant phase difference across the pupil, hence, changing the bias. Therefore, the use of paired prisms allows DIC image formation to occur with the same bias for every light pencil in the condenser aperture, regardless of the direction from which it traverses the specimen. In this condition, maximum contrast and resolution are attained. The sharp depth of field that results from the use of the full condenser and objective NA gives DIC systems a strong optical sectioning characteristic.

In addition to operation at full aperture and formation of a relatively low-noise image, DIC systems provide a number of other advantages. Although the lenses must be free of strain birefringence, objectives used in DIC systems do not have to contain a special element such as a phase plate. This is an advantage not only in manufacture and cost, but also because the phase plate in phase-contrast objectives reduces light transmission and complicates the point-spread function (PSF) in fluorescence microscopy. The phase plate also increases back reflection of light in the incident-light illuminators used in both fluorescence and reflection-interference systems. In general, switchover from DIC to fluorescence requires only the removal of the analyzer to avoid attenuation of the fluorescence image. For high-resolution fluorescence microscopy, the Nomarski prism should also be removed because it shears the fluorescence image into two very slightly offset polarized components. This can significantly lower the peak brightness of sharply defined features (Kontoyannis et al. 1996). One disadvantage of DIC is that image quality is degraded by the birefringence of common plastic dishes, plastic multiwell plates, and plastic culture flasks (and their covers). Recent developments have overcome this problem in low- to moderate-NA DIC microscopes (Plas-DIC; Danz et al. 2004). The more fundamental disadvantage of DIC is that its contrast transfer function is directional; that is, phase gradients oriented along the shear axis show up with much greater relief than gradients that are nearly normal to the shear axis. In some cases, this can be used as an advantage, for example, to reduce the contrast of a highly anisotropic cell process, such as an axon, to make visible internal organelles of much lower contrast (P. Forscher, pers. comm.). Using a high-precision centerable rotation stage, DIC images can be obtained for any orientation of the shear axis relative to the object, and the information in this image set can be used to compute an orientation-independent OP (phase) image (Preza 2000; Shribak and Inoue 2006). However, specimen rotation is often not practical under experimental conditions. Very recent advances (Shribak 2009) in electro-optic selection between orthogonal shear axes in a composite Wollaston prism may enable orthogonal-axis DIC imaging at high speed with no moving parts or specimen rotation.

Fluorescence and Fluorescence Optical Systems

At the present time, fluorescence microscopy has developed into the most widely used method for the study of both fixed and living specimens. This is due to the high specificity possible with a vast array of fluorescent molecules ranging from stains and labeled antibodies to molecular analogs, physiological indicators, and expressible markers such as fluorescent proteins (green fluorescent protein [GFP] and many variants), phytofluors, and fluorogen-activating proteins (Szent-Gyorgyi et al. 2008). It is also due to the high sensitivity of fluorescence detection. Under practical conditions, very low background is possible in fluorescence microscopy. So, it is, therefore, not surprising that four of the five demonstrated single-molecule tracking/imaging techniques are fluorescence-based (Betzig and Chichester 1993; Nie et al. 1994; Funatsu et al. 1995; Sase et al. 1995; Vale et al. 1996; Smith et al. 1999; Moerner and Orrit 1999; Weiss 1999).

Unlike phase-contrast methods, fluorescence microscopy requires the absorption of light by molecules in the specimen. For brevity, all fluorescent molecules will be referred to here as dyes. The full range includes synthetic dyes, modified and native biochemical fluors, and semiconductor quantum dots. As shown in a Jablonski diagram (Fig. 17), this excitation process usually brings the dye from its singlet electronic ground state to its lowest excited singlet state. Each of these electronic states is composed of a manifold of vibrational substates, which broadens the range of photon energies (wavelengths), which can cause the transition. At normal temperatures, dye molecules are virtually always in the lowest vibrational level of the ground state, so excitation occurs from this level into the vibrational manifold of the lowest excited state. This excitation band is usually graphed as an absorption spectrum (extinction coefficient vs. wavelength). For organic dyes, extinction coefficients range from 50,000 to 250,000 per M·cm. Vibrational relaxation of electronic states (internal conversion; IC in Fig. 17) occurs with great rapidity, so that an excited state dye quickly ends up in the lowest vibrational level of the excited electronic state. Excited state lifetimes are usually in the nanosecond range but can be affected by the relative rates of nonradiative relaxation, spontaneous emission, stimulated emission, intersystem crossing, quenching, or direct photochemistry. Fluorescence is observed when a photon is emitted by the dye as it returns directly to the ground state. Because of the energy loss that accompanies the initial internal conversion, fluorescence occurs at a lower photon energy (longer wavelength) than excitation. The return to the electronic ground state may leave the dye transiently in a vibrationally excited state before rapid internal conversion to the lowest ground-state level. This allows for a range of emitted photon energies, observed as a fluorescence emission spectrum (Fig. 17B). In most cases, the absorption spectrum is closely related to the fluorescence excitation spectrum, which is determined by measuring emission flux

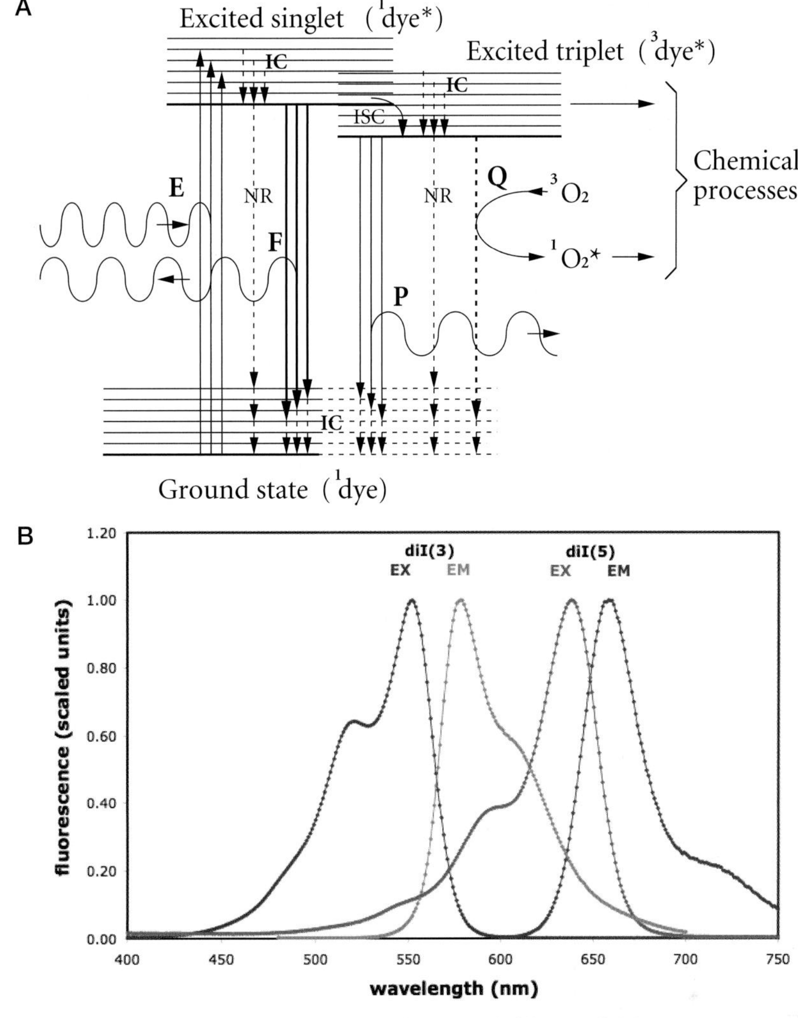

FIGURE 17. (*A*) Jablonski energy-level diagram of photophysical processes in organic dyes. Each set of horizontal lines represents vibrational levels in a single electronic stationary state of the molecule. The electronic ground state (lowest energy) is shown along with two excited electronic states. Vertical solid lines represent radiative transitions between states, whereas vertical dotted lines represent nonradiative transitions: (E) excitation; (IC) internal conversion; (F) fluorescence; (NR) nonradiative transition; (ISC) intersystem crossing; (P) phosphorescence; (Q) quenching by energy transfer. The ground state is usually an electronic singlet (all electrons spin paired). At ambient temperature, very few molecules would be in any state other than the lowest vibrational level of the ground state, therefore, excitation processes originate in this level. Absorption of a photon results in excitation of the dye to a stationary state of greater energy. In general, the energy of the photon must match the energy difference between the initial and final states (multiphoton absorption is not shown here). Closely spaced vibrational levels plus thermal motion in the molecule allow for a range of photon energies to match a transition and to cause the excitation to be seen spectroscopically as an absorption band rather than as a sharp absorption line. Excitation by absorption normally occurs with no change in spin pairing, therefore, the excited state is also a singlet. Fast relaxation processes (IC) convert vibrational energy into thermal motion and bring the molecule to the lowest vibrational level of the excited state. Under normal conditions, the lifetime of this state is in the nanosecond range. Fluorescence originates from this level and can be the main relaxation process in a bright dye. The closely spaced vibrational levels of the ground state, along with thermal motion, allow for a range of emitted photon energies. Fluorescence (and phosphorescence) is, therefore, normally seen as an emission band. A number of processes or conditions can cause spin unpairing and conversion (intersystem crossing) of the excited singlet to a triplet state. The radiative transition from the triplet state to the ground state (phosphorescence) occurs with low probability. Therefore, triplet states generally persist until quenched or until the dye is involved directly in a chemical reaction. In biological specimens, dissolved oxygen (O_2) is a highly effective quencher of dye triplet states. In this photophysical reaction, the ground-state oxygen molecule (which is a triplet) can be excited to a reactive singlet state. Singlet oxygen can initiate reactions leading to bleaching of the dye and to phototoxicity. (*B*) Excitation and fluorescence spectra of two cyanine dyes, diI(3) and diI(5). The excitation spectrum is similar to the absorption spectrum but rolls off more sharply in the overlap region. For diI(3), peak excitation occurs at 552 nm (green), peak emission at 578 nm (orange). For diI(5), peak extinction occurs at 638 nm (red), with peak emission at 658 nm (far red). Both dyes show very high extinction (150,000 and 250,000 per M•cm, respectively) and good quantum yield (QY) when conjugated to protein (>15% and >28%, respectively). (Spectra provided courtesy of Nathaniel Shank, Carnegie Mellon University, Molecular Biosensor & Imaging Center.)

as a function of excitation wavelength. Internal conversion, although seemingly a minor feature of the Jablonski diagram, profoundly affects fluorescence because it decouples excitation from subsequent processes. As a result, fluorescence has the following characteristics: (1) The emission spectrum is independent of the precise excitation wavelength, (2) the emitted photon is of lower energy (longer wavelength) than the absorbed photon, (3) the excitation spectrum is independent of the precise emission wavelength, and (4) emission from a population of excited dye molecules is incoherent (i.e., the molecules emit independently). Under special conditions, such as two-photon excitation or stimulated emission, important differences are observed.

When expressed on a quantum basis of photons fluoresced versus photons absorbed, the quantum yield (QY) of the dye is the fractional probability with which fluorescence will occur relative to other relaxation modes. For bright dyes, QY usually exceeds 0.1 and may range as high as 0.9 (Tsien and Waggoner 1995). A number of nonradiative processes act to reduce QY. For example, molecular collisions can cause the excited state to lose its energy to the surrounding solvent as heat (NR in Fig. 17). Spontaneous spin unpairing of two electrons in the excited state also can occur. This process, known as intersystem crossing (ISC in Fig. 17), is enhanced by spin-orbit coupling when the dye contains one or more heavy atoms (Cl, Br, I). The resulting triplet state is reactive and is highly susceptible to quenching because of its long intrinsic lifetime. If not quenched, a triplet can relax by phosphorescence, emission of a photon at a longer wavelength than the fluorescence. Triplets are very effectively quenched by dissolved molecular oxygen (O_2), which is prevalent and highly diffusible, and which has a triplet ground state. Oxygen quenching of dye triplets can result in the production of singlet oxygen, a long-lived excited state of O_2, or several forms of oxygen radical. Dye triplets can also react directly with other organic molecules, especially intracellular redox intermediates. Two results of this photochemistry relevant to microscopy of living cells are photobleaching (or fading) and phototoxicity. Evolution has put the intrinsic chromophore of green fluorescent protein (GFP) in the core of a β-barrel structure (Ormo et al. 1996). In that environment, the dye is protected from collisions with water molecules and is isolated from frequent encounters with dioxygen. GFP is one of the most stable fluorescent labels known. Synthetic dyes held as cryptands in soluble macromolecules such as cyclodextrins have also been shown to be significantly stabilized against photochemical degradation (Guether and Reddington 1997). In fixed specimens, antioxidants can be used to slow fading. Some, such as Trolox, can be used with living cells (Scheenen et al. 1996). In some living preparations, oxygen scavengers can be used to remove dissolved O_2 for limited periods of time. Glucose oxidase along with glucose and catalase is a highly effective oxygen scavenger that can be added to culture medium at the time of infusion.

Because fluorescence is the result of absorption of a photon by a molecule, followed by emission of a photon of longer wavelength, the essential feature of any fluorescence microscope is a means to excite the specimen with color-filtered light and to separate the excitation from the emission wavelength band by use of a second filter. This allows the fluorescence image to be formed in a dark background to give maximum sensitivity. Normally, the degree of labeling in a biological specimen is so low that only a small fraction of the illumination traversing the specimen will be absorbed by the fluorescent marker molecules and a fraction of this re-emitted. In short, the fluorescence image will be weak relative to the brightness of the illumination. Therefore, the central problem in fluorescence microscopy is obtaining high-efficiency illumination of the specimen but very high rejection of the illumination band from the image along with efficient capture of the emission. The most common optical configuration giving the required high performance is incident-light illumination combined with a filter set that includes (1) an excitation filter, (2) a dichroic or wavelength-selective reflector, and (3) an emission filter (Fig. 18). Broadband (white) light from an intense lamp is directed into the microscope through the excitation filter, which is chosen to match the excitation band of the dye. The resulting filtered light is deflected by a matched dichroic reflector into the back pupil of the objective. Therefore, in this "epi-fluorescence" setup, the objective functions as a condenser in addition to its usual role. High-NA objectives, therefore, function as more powerful condensers as well as collect a greater fraction of the fluoresced light. High performance also results from the fact that the illumination exits the objective into the specimen and is, therefore, directed away from the microscope. Backscattered and back-reflected light that re-enters the objective is of unchanged wavelength and, therefore, is mostly deflected out of the microscope by the

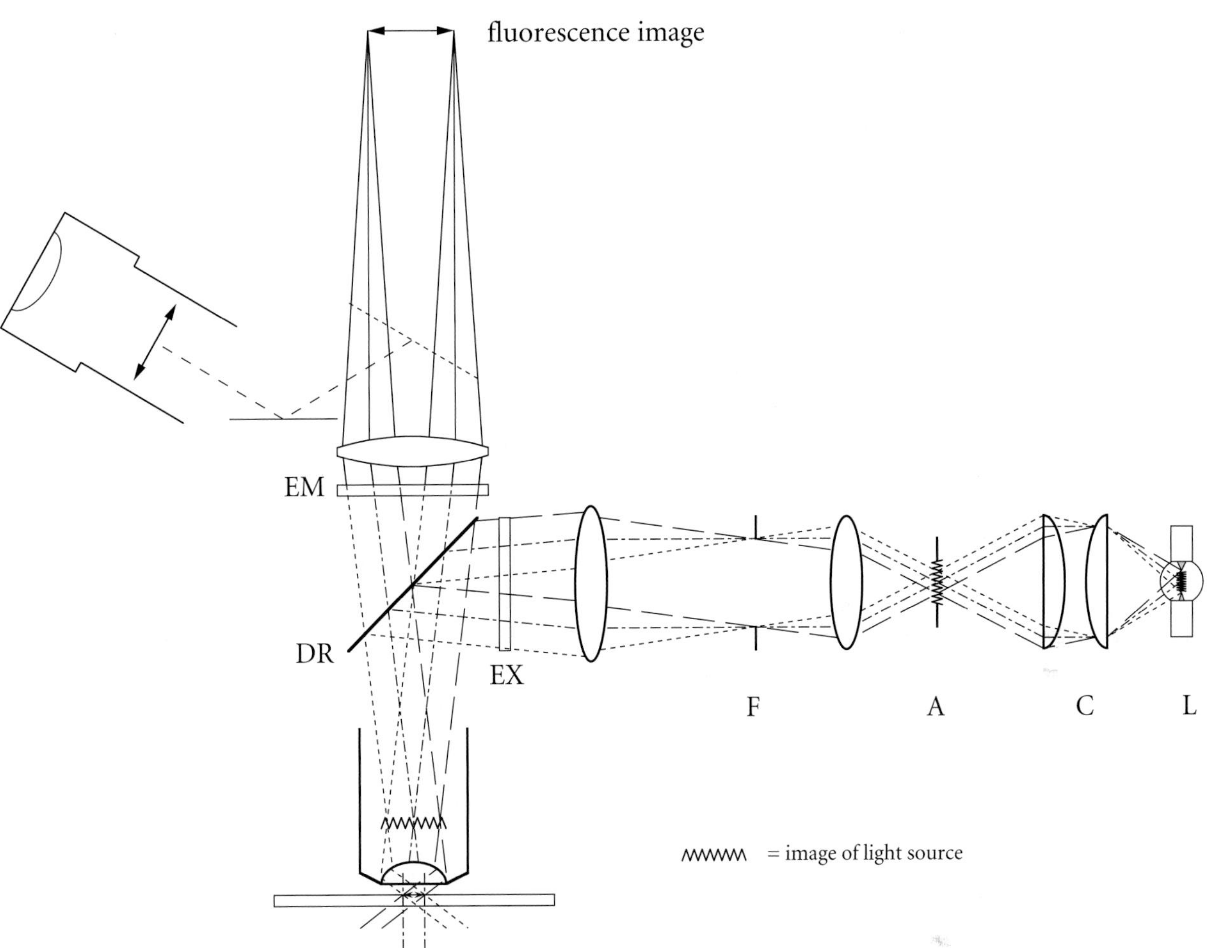

FIGURE 18. Principal optical components of a fluorescence microscope. Incident-light illuminator components: (L) arc lamp; (C) collector lens assembly; (A) illumination aperture iris (coincident with image of light source); (F) field iris. Fluorescence filter set: (EX) exciter filter; (DR) dichroic reflector; (EM) emission filter. In this optical system, the objective functions also as the condenser. Unlike a transmitted light microscope, the aperture iris is imaged into the condenser (objective) back pupil, rather than being physically located at this position. In this way, there is no obstruction of the image-forming light path. The fluorescence microscope differs from a reflectance microscope only in the wavelength selectivity of the filters and the dichroic reflector. Use of a green band-pass excitation filter (such as for rhodamine) along with a blue-band dichroic reflector (such as for fluorescein) suffices to generate a reflection-interference image of moderate to good quality. In this case, the band-pass filter functions to increase longitudinal coherence.

dichroic reflector. Fluorescence emission captured by the objective is efficiently transmitted by both the dichroic reflector and the matched emission filter and is focused to form an image. The small fraction of the backscattered illumination that passes the dichroic reflector is yet many times brighter than the fluorescence, so the emission filter must effectively block this light, which would otherwise appear as a background in the image. The second function of the emission filter is to block out the fluorescence of other dyes that may interfere with the tracer being imaged.

Overall, stringent conditions are placed on the optical quality of the excitation, dichroic, and emission filters. The dichroic reflector (DR in Fig. 18) is usually a multilayer thin-film interference filter formed on the surface of a thin glass substrate and is designed for a 45° angle of incidence. In general, these are not overcoated and, hence, are very easily damaged by careless handling. A small scratch will cause significant transmission of unwanted light. An ideal DR will be highly reflective at the excitation wavelength and highly transmissive at the emission wavelength. In general, a DR will also have some polarization selectivity caused by the nonzero angle of incidence. Excitation and emission filters are designed for normal incidence, and many are composite, consisting of a colored glass substrate on which one or more multilayer interference filters are deposited. The interference films can provide sharp-cut band-pass characteristics, back-reflecting out-of-band light. However, simple interference filters transmit weak sidebands. Colored glass, on the other hand, usually shows long-pass transmission or relatively broad band-pass characteristics but can have extremely high absorbance outside the transmission range. The composite filter can, therefore, be designed as a band pass with extremely high blocking power (rejection) at other wavelengths. In modern band-pass filters, high in-band transmission and high out-of-band rejection is achieved without the use of colored glass by combining two or more band-pass films that block each other's sidebands on a glass substrate that absorbs mainly infrared (IR) light. Dichroics and band-pass filters usually are designed to be set in a particular orientation relative to the light source (and are usually marked as such by the manufacturer). The correct orientation maximizes wavelength selection by back-reflection and minimizes absorption within the filter, thus minimizing filter heating and filter luminescence. In addition to high-performance band-pass and band-reject properties, emission filters (and DRs) must not degrade wave-front quality because image-forming rays pass through these optics. All of the strata within these filters must be optically flat, especially the entrance and exit surfaces. However, it is very difficult to make these two planes parallel to better than one wavelength across the diameter of the optic. Therefore, even filters with excellent band-pass and wave-front quality may have residual wedge; that is, the optic will act like a very weak prism and will cause the image to shift laterally by one or more pixels. This leads to image-registration errors because the wedge in each filter set is usually different. Fortunately, the effect is highly reproducible (as long as the emission filter is not rotated in its mount), so the lateral shift can be measured for each installed filter set and can be corrected by image processing before overlay.

Fluorescence filter sets of three basic types are widely used in microscopy (Fig. 19). For specimens labeled with a single fluorochrome, the highest optical throughput can often be obtained with a simple long-pass emission filter. For example, the universal fluorescein ("FITC") set consists of a blue band-pass exciter, a blue-reflecting dichroic, and a yellow glass long-pass emitter. Likewise, a rhodamine filter set consists of a green band-pass exciter, a green-reflecting dichroic, and a red–orange glass long-pass emitter. However, when fluorescein is used along with rhodamine in doubly labeled specimens, the yellow long-pass filter is replaced by a green band-pass filter. This is necessary to block the orange emission from rhodamine that is weakly excited by blue light. Without this change, both dyes would contribute to the FITC image. No change is required in the rhodamine filter set because fluorescein is not excited by green light. Five to seven sequential band-pass sets can be overlapped to span the spectrum from ultraviolet (UV) excitation/violet emission to far-red excitation/near-IR emission (Waggoner et al 1989). In some applications, a panel of two or more spectrally distinct dyes is used along with a multiband filter set. For direct fluorescence color imaging with a color camera or a spectrographic camera, a multiband exciter is used along with a complementary multiband dichroic and emitter pair. In this type of filter set, the dichroic/emitter pass bands interleave on the long-wavelength side of the exciter bands. For high-spatial precision multicolor imaging, a single multiband dichroic and emitter pair can be used along with a set of single-band exciters in an external filter changer. In this way, the optics in the imaging light path remain fixed as the excitation filter is changed, eliminating a source of alignment error in the superposition of images.

The most common light sources for fluorescence microscopy are high-pressure mercury and xenon arc lamps. These sources both appear to output intense white light but actually differ greatly in spectral distribution. Mercury arcs produce light concentrated in bands centered around the atomic emission lines of the gas-phase mercury atom, most notably at the following wavelengths: 254 nm (UV), 265 nm (UV), 365 nm (UV), 405 nm (violet), 436 nm (deep blue), 546 nm (green), and 578 nm (yellow). Therefore, this source is ideal if there is good coincidence between one of these bands and the excitation band of a specific dye. Outside of these bands, the output power of a high-pressure mercury arc is still very significant, as evidenced by the fact that mercury arcs are used routinely with fluorescein-like dyes (480–490-nm peak excitation), as well as for in-band rhodamine-like dyes. In contrast, xenon arcs emit a less-intense continuum over the range 250–800 nm, with superimposed emission lines in the 450–500-nm and 800–1050-nm ranges. Therefore, good output can be obtained at any wavelength by the use of the appropriate band-pass filter. For

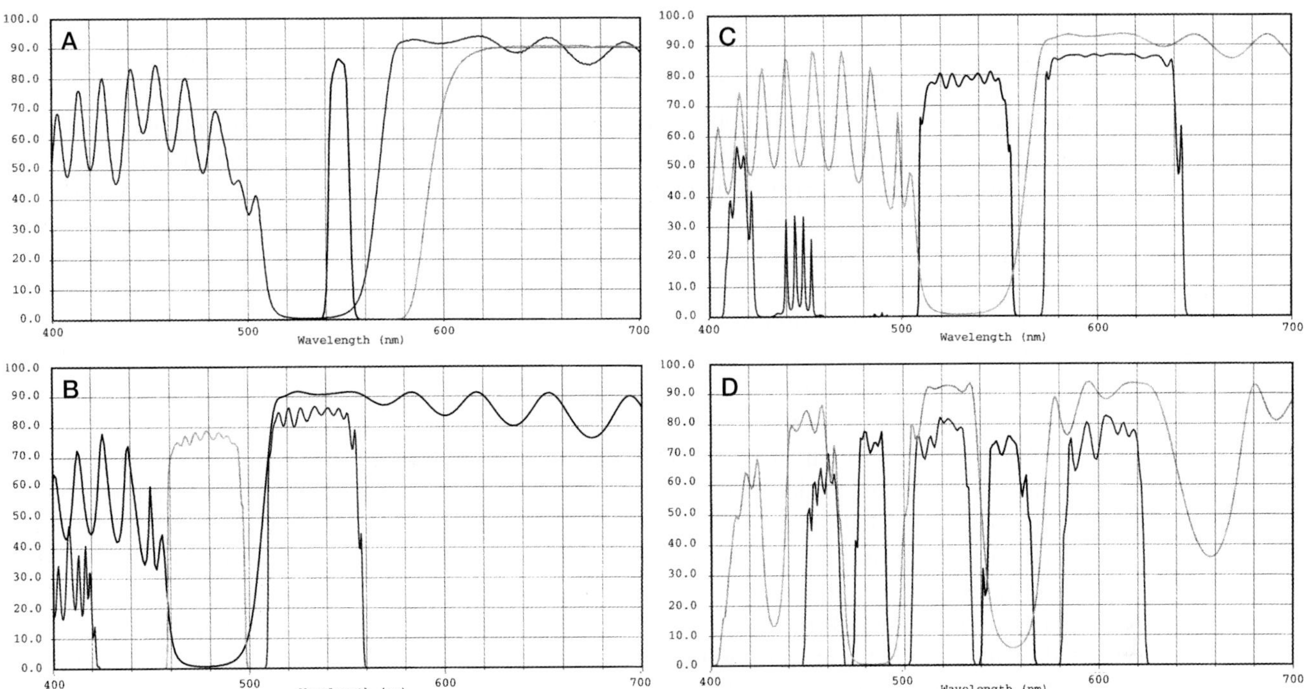

FIGURE 19. Spectra of representative fluorescence filter sets. Spectra are plotted as percent transmittance for the filters (normal incidence) and for the dichroic reflector (45° incidence). Each set consists of three elements: an excitation filter, a dichroic reflector, and an emission filter. The essential feature of each set is the complete separation of the excitation and the emission pass bands and the intervening transition from high reflectivity to high transmissivity in the dichroic reflector. Because of its multilayer structure, the dichroic reflector will generally have complex transmission bands at short wavelengths that are of no consequence in filter set performance. In contrast, the exciter and the emitter are generally strongly blocked out of band. In the simplest type of set (shown in *A*), a band-pass excitation filter is used along with a long-pass emission filter. This maximizes the detection of fluorescence in the image but would give mixed color images for a doubly labeled specimen in which both dyes were excited to some degree within the single illumination band. The set shown, green excitation/orange–red emission, would be suitable for rhodamine- and cyanine3-type dyes. In the sets shown in *B* and *C*, sharp-cutoff band-pass filters are used for both excitation and emission in the fluorescein and rhodamine bands, respectively. Although blue light, which excites fluorescein efficiently, also excites rhodamine weakly, rhodamine emission is blocked from the fluorescein-band image by the band-pass emission filter in *B*. In *C*, the green exciter filter cuts off above the fluorescein excitation band. Filter sets of this type, therefore, give outstanding performance with doubly labeled specimens. More complex multiband filters are used with color cameras or in cases in which movement of optical elements is disadvantageous. In *D*, a dual-band exciter is used along with a dual-band emitter; the pass bands interdigitate to eliminate cross talk. The intervening reflector is designed to be highly reflective in the two excitation bands and is highly transmissive in the emission bands. The filter set can also be used with separate excitation filters in a motorized wheel or with another type of external wavelength-selecting device. Filter curves provided courtesy of P. Millman and M. Stanley, Chroma Technology Corporation, Rockingham, VT. Key to Chroma filter sets: (*A*) #11002 Basic set for green-excitation dyes; (*B*) #41001 Hi-Q set for fluorescein (FITC), redshifted GFP, Bodipy, Fluo-3, and diO; (*C*) #41002 Hi-Q set for rhodamine, tetramethylrhodamine (TRITC), and diI; (*D*) #51004v2 multiband set for FITC and TRITC.

certain important applications, the continuum has advantages. For example, in ratio imaging of the calcium indicator dye Fura-2 (Grynkiewicz et al. 1985), the largest response is obtained by UV excitation at 340 and 380 nm. High power is available at these wavelengths in a mercury arc, but both excitation filters must be able to block the much more intense emission at the 365-nm peak of the band. This is much less a problem with the xenon spectrum. In quantitative imaging applications, chronic spatial instability of the arc (particularly in mercury lamps) can cause image-to-image brightness variation. For both types of arcs, good ventilation is required for removal of heat, removal of ozone, and for safety in the event of mercury lamp breakage.

Recently, light-emitting diodes (LEDs) have reached power and brightness levels useful not only for traffic signals and automobile tail lights, but also for routine fluorescence microscopy. Very inexpensive single LEDs operating at a few volts and 350-mA DC are available in specific color bands ranging from deep blue to red. Because these are not laser diodes, the spectral bandwidth of the LED output is generally in the range 25–50 nm, measured at the half-power points to either side of the peak wavelength. This is very well suited for fluorescence microscopy but still requires the use of a conventional excitation filter to eliminate both the short- and the long-wavelength tails of the LED emission. The emissive area of a single LED junction typically ranges from 0.3–1.0-mm square, mounted on a small heat sink under a hemispheric plastic lens. Both the size and the aperture (emission cone angle) of such LEDs are well matched to the collector optics of a fluorescence microscope arc lamp housing, so efficient use of the LED can be obtained simply by mounting it exactly in place of the arc. As a caution: "ultra-bright" versions of an LED may be side-by-side arrays of diodes under one hemispheric lens and may have lower intrinsic surface brightness compared with single diode junctions. A unique feature of LED sources is that they can be turned on or off with submicrosecond speed by the use of a transistor-controlled power supply.

The physical size of an arc (or an LED junction) affects not only its brightness, but also the efficiency with which it can deliver light to the illuminated field of a microscope. The brightest sources are the widely used DC short-arc lamps—usually 100-W mercury or 75-W xenon—whereas larger lamps produce more light overall but lower brightness within the arc projected area. In a Köhler-illuminated system, the image of the light source should be matched exactly to the back pupil of the objective (see Eq. 33). If the image diameter is less, the full INA will be underutilized; if it is greater, light will be wasted, and stray light will be needlessly introduced into the microscope. Because the NA of the lamp collector lens (NA_{coll}) effectively sets the maximum diameter of the field of view (FOV) in the specimen, and the diameter of the light source (d_s) sets the INA, a simple relation exists between the four quantities when the full aperture is used ($INA = NA_{obj}$),

$$d_s = (NA_{obj}/NA_{coll})(FOV), \tag{37}$$

which is an example of the optical invariant for image-forming systems. With $NA_{coll} = 0.5$ and with FOV = 150 μm for a 100x 1.3-NA objective, the optimal source diameter is 0.4 mm. In such a case, short-arc lamps are very efficient. For low-magnification objectives (particularly those of high NA), the large back pupil can be most effectively filled by a larger arc. Optical scramblers with an output zoom lens can be used to produce a circular homogeneous source matched to the diameter of any pupil.

The combination of a 75-W or 100-W short-arc lamp and high-NA optics can produce intense irradiation of the specimen, a situation requiring attention particularly for live cell imaging. After collimation and filtering, 1–10 mW can be delivered to the back pupil of a 100x objective. Focused into a 150-μm FOV, a 5-mW flux produces an irradiance of 30 W/cm^2, 300-fold greater than bright sunlight integrated over all wavelengths. Therefore, it is a general rule in fluorescence microscopy to minimize light exposure in any wavelength band in which the specimen or a dye within the specimen has a significant extinction coefficient. As a comparison, this level of irradiance is still far less than in a laser-scanning confocal microscope, in which 10 μW focused into the objective diffraction limit (0.5 μm) produces 50,000 times the solar flux.

Many variations exist on the basic epifluorescence microscope. Use of laser excitation, such as is common in scanning confocal microscopes, obviates the need for an excitation filter. Use of acousto-optic tunable filters (AOTFs), liquid-crystal tunable filters (LCTFs), interferometers, diffraction gratings, and other wavelength-selective devices can replace one or more of the basic elements of the fluorescence filter set.

Principles of Fluorescence Microscopy

Image formation in a fluorescence optical system differs in a fundamental way from transmitted light and reflected light microscopes. The molecular excited states that lead to fluorescence have a finite average lifetime in the nanosecond range so normally do not maintain any degree of coherence with the excitation field. Essentially, dye molecules in the specimen emit fluorescence independently and, therefore, are mutually incoherent. In place of the Abbe picture, in which the

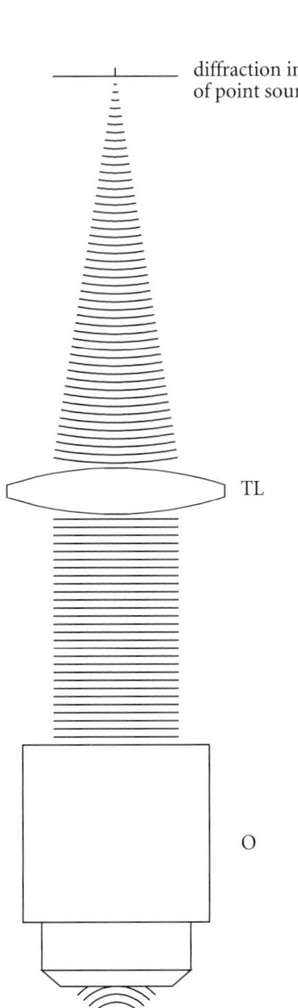

FIGURE 20. Wave-front capture by an objective. The Airy pattern—diffraction image of a point source. Resolution in a fluorescence microscope is set by the effect of diffraction on the image of a point source. Because of the finite aperture of an objective lens, a microscope captures only part of the spherical wave front emitted by a dye molecule in the specimen. The finite transverse extent of the wave front and its finite wavelength cause phase and amplitude variations to occur as it converges to a focus in the image plane. The result of this fundamental process, known as diffraction, is that the light is not concentrated to a point but is distributed in a characteristic pattern consisting of a central bright disk surrounded by faint rings. Ideally, this is the Airy pattern, in which 84% of the incoming power is encircled by the first dark ring. Aberration (including defocus) causes the pattern to become more diffuse and less bright. In the image plane, the diameter of the central disk is (magnification) x $1.22\lambda/\text{NA}$, usually several tens of micrometers. In unmagnified coordinates, the radius of the Airy disk defines the Rayleigh resolution limit $0.61\lambda/\text{NA}$. For a high-NA lens, this can be as small as 0.22 µm. (*Inset*) A highly magnified view of the Airy pattern, shown in a logarithmic visibility gray scale, marked to show the diameter of the disk.

specimen is represented as a collection of diffraction gratings that interact with plane-wave illumination, it is now more appropriate to consider point sources of spherical waves representing the spontaneous emission of light from individual molecules. This model was first developed by Rayleigh (J.W. Strutt) in the analysis of telescope performance in the imaging of stars (equivalent to point sources because of their great distance). The second important result of the mutual incoherence of emission from dye molecules is that the fluorescence image is simply a superposition or a sum of the individual brightness patterns contributed by every fluorescent molecule in the specimen. This situation can be changed by stimulated emission, which has been used in fluorescence microscopy (Dong et al 1995; Klar et al 2000), but spontaneous emission predominates under usual conditions. However, the independence of emission from dye molecules does not obliterate all traces of coherence; the finite bandwidth ($\Delta\lambda$) of fluorescence emission guarantees temporal or longitudinal correlation, usually expressed as a propagation distance over which a wave train holds a single frequency; $L_{||} = (\lambda_{avg})^2/(\Delta\lambda)$, typically 5–10 µm for dyes with a 50-nm emission bandwidth. Additionally, the waves emitted by each dye molecule are transversely coherent. This self-coherence has been used to gain axial resolution in interferometric 4Pi microscopes (Hell and Stelzer 1992) and by direct fluorescence image interferometry in I^2M, I^5M, and interferometric photoactivated localization microscopy (iPALM) microscopes (Gustafsson et al. 1995; Shtengel et al. 2009).

In the Rayleigh picture, a dye molecule emitting light can be pictured as a point source of scalar spherical waves. It is more accurate to treat the molecule as a dipole antenna radiating polarized light (Axelrod 1989), but it is not necessary for the present discussion. Only a circular section of the diverging wave fronts is captured by the objective (Fig. 20), and the cone angle of this set of waves is defined by the NA of the lens. Essentially, the objective (or objective plus tube lens in modern IC microscopes) captures a circular section of each expanding wave front and transforms it into a converging wave front headed for a focus in the image plane or in a neighboring plane, if the object is not in focus. Because the wave front has been limited by the circular pupil of the objective and is no longer a complete spherical wave, diffraction modifies the converging wave in both amplitude and phase. The result is that convergence to a point does not occur; rather, the image of the point source consists of a bright central spot surrounded by a set of progressively attenuated rings (Fig. 20, inset). This diffraction image, derived (for telescopes) by GB Airy in 1834, sets the transverse resolution limit in fluorescence microscopy. Rayleigh defined the limiting resolution condition by the offset superposition of the diffraction images of two stars, in which the shift (s_{min}) was equal to the radius of the first dark ring of the Airy pattern. In a fluorescence microscope, this would correspond to viewing a specimen in which two point-like fluorescent particles were close enough so that their in-focus diffraction images overlapped to that same degree. For microscopes, the mathematical form of the normalized Airy pattern intensity,

$$i_A = [2J_1(u)/u]^2, \tag{38}$$

contains the argument $u = 2\pi \mathrm{NA}r/\lambda_0$, where r is a demagnified radial coordinate. The first dark ring in the Airy pattern occurs where $u = 3.831706$, the first nonzero root of the first-order Bessel function $J_1(x)$. This defines the "Airy disk." The corresponding value of r is, therefore, $r_A = (3.83/2\pi)\lambda_0/\mathrm{NA}$, which directly gives the Rayleigh transverse resolution formula:

$$s_{min} = r_A = 0.61\lambda_0/\mathrm{NA}. \tag{39}$$

For high-NA objectives, this limit can approach 0.2 μm (200 nm). Because it is based on a subjective criterion, the Rayleigh formula differs from the Abbe diffraction result for transmitted light imaging (Eq. 28). Fourier transformation of the Airy pattern shows that it is, in fact, a band-limited function with the same spectral range as the Abbe limit. In that sense, there is no difference in the resolution limits for transmitted light and fluorescence microscopes, although the optical models differ greatly. The actual scale of an Airy pattern in the image plane of the microscope will be enlarged by the magnification ($r' = Mr$). In terms of the diameter of the central spot, or Airy disk,

$$D_{min} = 2Ms_{min} = 1.22M\lambda_0/\mathrm{NA}. \tag{40}$$

For a 100X 1.3-NA objective, the Airy disk diameter would be 50 μm, and for a 10X 0.3-NA lens, the Airy disk diameter would be 20 μm. It is this dimension relative to the detector element size in the imaging device (e.g., a charge-coupled device [CCD] element or film grain size) that determines how accurately the recorded image matches the true image field (see Nyquist sampling, below).

Clearly, the Airy pattern represents only the in-focus image of a fluorescent point object. If the point source is defocused (i.e., is axially displaced from the design focus plane), the geometric image point will also be axially displaced. The converging wave front in the image space will either (a) impinge on the detector before contracting to an Airy pattern, or (b) will fully contract and will partially diverge before reaching the detector. In either case, a less-bright and more-diffuse image will be seen. The Airy pattern represents the idealized in-focus 2D point-spread function (PSF) for a fluorescence microscope. The entire through-focus series of stacked images, in which the Airy pattern is simply the central plane, constitutes the 3D PSF for the microscope. The depth of field in fluorescence (δ_F) is determined by the loss in brightness of the Airy disk with defocus. This is conservatively defined as the stage increment that causes the Airy disk to become a dark spot surrounded by concentric rings. Use of the Kirchhoff diffraction integral to derive the on-axis variation of the 3D PSF shows that its normalized form is

$$i_{ax} = [\sin(\mathrm{v})/\mathrm{v}]^2 , \tag{41}$$

in which the argument, $\mathrm{v} = \pi \mathrm{NA}^2 \Delta z/2n\lambda_0$, is linear in the focus shift Δz. The first axial zero of the 3D PSF occurs where $\mathrm{v} = \pi$, resulting in an axial Rayleigh resolution formula,

$$\delta_F = 2n\lambda_0/\mathrm{NA}^2. \tag{42}$$

As in the transmitted light case (Eq. 30), the depth of field depends strongly on NA. For the highest-NA objectives, δ_F is in the range 0.7–0.9 µm. Equation 42 represents a conservative criterion, distinguishing two collinear axial point sources by serial adjustment of focus. When observing distinct point sources, much smaller differences in axial location can be discriminated. In that case, the Rayleigh quarter-wave criterion (RQWC) defines the depth of field by the effect of focus adjustment on the radius of curvature of the spherical wave front that converges to form the point-source image. When the curvature change is such that there is one-quarter wavelength of phase difference at the margin of the objective pupil (relative to zero difference at the center), there will be significant destructive interference at the center of the image. Formulated in terms of object–space ray angles and coordinates (Taylor and Salmon 1989),

$$\delta_{RQWC} = \lambda_0/(8n \sin^2[(1/2) \sin^{-1}(NA/n)]). \tag{43}$$

For low- or moderate-NA systems, the RQWC formula gives a depth of field equal to one-quarter of the axial Rayleigh criterion (Eq. 42). For high-NA systems, the RQWC focus range is as small as 0.13 µm. This limit is generally not attained in fluorescence microscopy, most likely because of signal-to-noise limitations, but has been approached in transmitted light systems (Inoue 1989). Because transverse resolution sharpens as 1/NA in both dimensions and axial resolution sharpens as $1/NA^2$ (or better), the volume resolution of the microscope sharpens as $1/NA^4$.

This particular example in fluorescence microscopy illustrates an important point: As the Airy pattern is defocused, there is little change in the total flux of light arriving at the image plane from the point source. The detected photons are simply distributed over a larger region in the camera. In phase-contrast or DIC microscopy, defocused features blend into the relatively bright incoherent background. In fluorescence, the background is dark so that out-of-focus features are noticeable. The out-of-focus features in fluorescence have a more serious effect because of the much lower total photon flux and lower SNR. The need for quantitative accuracy in fluorescence imaging of cells has driven the development of systems for fluorescence optical sectioning microscopy, most notably through confocal or multiphoton scanning, but also through computational deblurring and encoding methods.

The Importance of Optical Efficiency and NA

Unlike transmitted light imaging in which the light flux is high, fluorescence microscopy is usually photon limited. This is particularly true for imaging of living cells in which photobleaching of dyes and photochemical toxicity almost always place severe limits on the allowable light dose to the specimen. Therefore, in fluorescence, it is always best to maximize optical efficiency through careful selection of light source, filter sets, and objectives. Because both light-condensing and light-collecting efficiencies of an objective increase with NA, it is always best to use the highest-NA objective compatible with the application.

The relation between NA and light collection efficiency can be derived by considering a small source located at the focus of a lens. The fraction of radiated power that enters the lens within the cone angle defined by θ_{max} is proportional to the solid angle taken up by the lens pupil as seen from the source,

$$P = P_0(1 - \cos \theta_{max})/2 \tag{44a}$$

$$= (P_0/2)(1 - [1 - \sin^2 \theta_{max}]^{1/2}) \tag{44b}$$

$$= (P_0/2)(1 - [1 - (NA/n)^2]^{1/2}). \tag{44c}$$

For lenses of low to moderate NA, the square root term can be accurately simplified:

$$P = 0.32P_0(NA/n)^2, \tag{44d}$$

showing that the efficiency grows as the square of the NA or better.

The overall effect of NA on optical efficiency and image brightness is a result of three factors: (1) efficiency of concentration of illumination, (2) efficiency of collection of emission, and (3) sharpness of image formation. In the usual Köhler setup, the incident-light illuminator forms an image of the light source in the back pupil of the objective, which also functions as the condenser. Therefore, the total power focused into the field of view will be equal to the brightness of the source image multiplied by the area of the pupil. Using the result (Eq. 33) that the pupil radius is proportional to NA/M, the total power will be proportional to the square of this ratio. The excitation intensity is proportional to the total power normalized by the FOV area, which is proportional to the square of 1/M. Therefore,

$$I_{ex} = B(\mathrm{NA}/M)^2/(1/M)^2 \tag{45a}$$

$$= B\mathrm{NA}^2, \tag{45b}$$

where B is an effective source brightness. A more detailed analysis shows that, ideally

$$I_{ex} = (1/4)B_0\mathrm{NA}^2\,, \tag{45c}$$

where B_0 is the brightness of the filtered source. (The factor 1/4 is derived from a thermodynamic restriction on image brightness.) The fluorescence emitted by a point source in the specimen will be proportional to I_{ex}. The total collected emission power (P_{em}) reaching the camera will be proportional to both I_{ex} and the light collection efficiency of the objective. Using the previous results, Equations 44a and 45c,

$$P_{em} = k_F I_{ex}(\,1 - \cos\theta_{max})/2$$

$$= (k_F/4)B_0\mathrm{NA}^2(1 - [1 - (\mathrm{NA}/n)^2]^{1/2})/2. \tag{46a}$$

With a quadratic approximation to the square root for low to moderate NA,

$$P_{em} = 0.32(k_F B_0/4n^2)\mathrm{NA}^4, \tag{46b}$$

showing that the total detected fluorescence is independent of magnification and increases as the fourth power of NA, or better. Of greater significance is the peak brightness of the Airy pattern image of a point source. This can be estimated by normalizing P_{em} to the area of the Airy disk in the image plane,

$$B_F = P_{em}/\pi r_A^{\,2}$$

$$= 0.32(k_F B_0/4n^2)\mathrm{NA}^4/\pi(0.61M\lambda_0/\mathrm{NA})^2$$

$$= B_{em}\mathrm{NA}^6/M^2\lambda_0^{\,2}, \tag{47}$$

which shows the remarkable importance of NA in fluorescence microscopy. Sixth-order dependence would be seen only in the limit in which the camera pixel density was better than or equal to the optical bandwidth, and the object was much smaller than the resolution limit (See Nyquist sampling, below) (Fig. 21). As pointed out above, alteration of M or NA changes the size of the Airy disk relative to the detector pixel elements and may lead to low signal per pixel because of oversampling or to loss of recorded resolution because of undersampling. If an auxiliary magnifier is adjusted to maintain the detector pixel density equal to the optical bandwidth as the NA is changed, the peak brightness increases simply as NA^4 (Eq. 46b). For a diffraction-limited line object, such as a labeled microtubule or an actin filament, the image brightness should increase as NA$^5/M^2$. When the source of fluorescence is spatially extended (such as when imaging an indicator dye in a cell) and the image features are not diffraction limited, the total detected light flux will depend on NA^4, and the brightness will depend on NA$^4/M^2$. Finally, if the fluorescence is collected over the entire FOV (such as from a labeled cell monolayer), both the total signal and the brightness will depend roughly on NA$^4/M^2$ because the number of illuminated cells in the FOV will decrease as $1/M^2$. This strong dependence in all situations is the main reason behind the use of high-NA lenses for fluorescence microscopy.

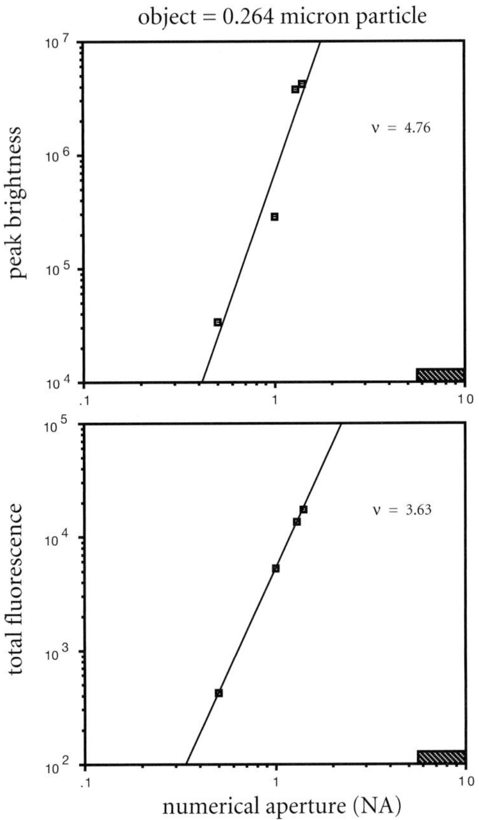

FIGURE 21. Total flux and peak brightness versus NA—point source. In the fluorescence microscope, the objective captures emitted light but also serves as the condenser. The efficiency of the lens in each function increases approximately as the square of the NA (Eqs. 44–47). Therefore, the total light flux in the image of a point source should increase with the fourth power of NA. Additionally, the size of the Airy disk formed by the microscope when imaging a point source decreases as 1/NA. Therefore, at constant magnification, the brightness of a point-source image should increase as the total flux normalized by the area of the Airy disk (i.e., as the sixth power of NA). In practice, this will be reduced by the finite spatial extent of the test particle. Measurements made using fluorescent 0.26-µm particles embedded in optical cement show 3.63-power dependence for the total flux and 4.76-power dependence for the brightness, clearly demonstrating the importance of NA in high-resolution fluorescence imaging.

OPTICAL SECTIONING MICROSCOPY

The Optical Transfer Function in Fluorescence Microscopy

In fluorescence microscopy, the brightness of the image represents a photon count in each pixel and is usually directly related to a measurement of interest, such as local density of labeled molecules, calcium ion concentration, rate of energy transfer, etc. Generally, quantitative interpretation of pixel values is confounded by the simple fact that a specimen is three dimensional (and irregular) and its image is two dimensional. In other words, the image is only a projection of the specimen, and that projection is usually composed of superposed in-focus and out-of-focus contributions. Several options exist for partially or fully correcting this situation: (1) Reduction of the system NA to increase depth of field so that the entire specimen is in focus. This alternative is equivalent to reducing the specimen to a 2D model, and is usually unattractive because of loss of image brightness and resolution. (2) Ratio imaging of the fluorescence signal of interest against a reference dye distribution. The ratio method is widely used to partially correct the images of soluble physiological indicator dyes for the irregular shape of a cell but is much less useful in structural studies. (3) Removal of out-of-focus contributions from one or more images by optical spatial filtering or computational methods. This approach is known as optical sectioning microscopy (OSM). By far, the most widely used methods providing direct optical sectioning are confocal scanning fluorescence microscopy (CSFM) (White et al. 1987) and multiphoton scanning fluorescence microscopy (Denk et al. 1990). Optical sections can also be obtained by computational deconvolution, also known as computational optical sectioning microscopy (COSM) applied to a serial-focus set of conventional fluorescence images. More recently, very simple optical sectioning systems for conventional fluorescence microscopes have been developed based on incoherent illumination of a moveable Ronchi grating mask to project a sharply focused periodic pattern into the specimen to encode in-focus specimen features differently from out-of-focus features (Neil et al. 1997; Lanni and Wilson 2000; Lanni 2005). Much more powerful 3D-superresolving systems based on illumination with a set of crossed beam periodic interference patterns (and fluorescence image interferometry) also have been developed in an approach now known as structured illumination microscopy (I^5S) (Gustafsson 2000; Gustafsson et al. 2008; Shao et al. 2008). When optical sections can be obtained throughout a serial-focus image set, the result is effectively a 3D reconstruction of the object. In many cases, however, it is the elimination of the out-of-focus component from a single image more than getting a 3D reconstruction that is important.

Not surprisingly, the degree to which a microscope can produce optical sections (directly or indirectly) is related to the sharpness of its depth of field, which is a strong function of the NA and of the absence of aberration. This is clearly related to both the transverse and axial sharpnesses of the PSF of the microscope. The characteristics of the PSF and its effect on image formation is most apparent in the optical transfer function (OTF), a mathematical expression of the spectrum and weighting of spatial frequency information that can be captured by the microscope in the form of images of the object. In general, there is a Fourier transform relationship between images and spectra, although the exact form of the relation will depend on the optical system in use: transmitted light, fluorescence, reflected light, or interference. The OTF is related to the Ewald sphere in X-ray crystallography, which sets a limit on the number of Fourier coefficients that can be determined from the X-ray diffraction pattern and can be used to reconstruct the unit cell of the crystal. Both the OTF and the Ewald sphere are defined in reciprocal space, in which the k_x, k_y, and k_z coordinate axes are in units of spatial frequency. However, in crystallography, there is no imaging step. The diffraction order angles and intensities are measured individually and directly as the crystal is rotated, phases determined indirectly, and the resulting set of Fourier coefficients inverse-transformed to produce a model of the unit cell. In light microscopy, the situation is slightly different. First of all, light exits the specimen into the microscope simultaneously over the full range of angles defined by the NA. Second, the fixed cone of ray directions defined by the NA and the optic axis selectively transmits, or filters, the diffracted or fluoresced light on its way to the image plane. The filtering or weighting function constitutes the OTF, and its effect is to limit resolution and to alter contrast. In a formal sense, the chief aim in optical sectioning microscopy is to reverse the filtering effect of the OTF as much as possible and to recover an accurate model of the true object.

In fluorescence, the relationship between the true object and the image is particularly straightforward. As described above, individual dye molecules in the specimen fluoresce independently and, therefore, lack mutual coherence. As a result, the fluorescence image is a linear superposition of the intensity patterns caused by every point source constituting the object. Sources in the plane of focus will contribute Airy patterns to the image, each centered on the corresponding geometric image point. A source displaced from the plane of focus by a distance z will contribute a blurred pattern that is a slice cut from the 3D PSF at a distance z from the Airy pattern plane. Mathematically, the contribution to the image at location (x, y) can be expressed differentially:

$$\delta i(x, y; \Delta z) = \mathrm{obj}(x', y', z')\mathrm{PSF}(x - x', y - y'; \Delta z - z')\delta V', \tag{48}$$

where $\mathrm{obj}(\mathbf{r}')$ is the fluorescent label density in the object at (x', y', z'), and the PSF is centered on that source point and is evaluated at coordinates that are both transversely and axially offset to the image point (x, y). The PSF is weighted by the object source density because fluorescence is linearly proportional to the concentration of the dye label. The excitation field strength is treated as a constant over the region of interest (this would be different in the case of confocal, multiphoton, or other spatial encoding systems). The image is then the integrated contribution of all source points in the object and has the form of a convolution of the true object with the PSF,

$$i(x, y; \Delta z) = \iiint \mathrm{obj}(x', y', z')\mathrm{PSF}(x - x', y - y'; \Delta z - z')dV'. \tag{49a}$$

In this expression, the variables x and y represent demagnified image-plane coordinates, whereas Δz is the focus drive increment. With that definition, the convolution can be written compactly as

$$i(\mathbf{r}) = \iiint \mathrm{obj}(\mathbf{r}')\ \mathrm{PSF}(\mathbf{r} - \mathbf{r}')dV', \tag{49b}$$

where $i(\mathbf{r})$ can represent a 3D data set consisting of a stack of serial-focus images. The Fourier transform is generally a deconvolver; in a formal sense, 3D transformation into the spatial frequency coordinate system (k_x, k_y, k_z) deconvolves the object from the PSF,

$$I(\mathbf{k}) = \mathrm{FT}[i(\mathbf{r})] \tag{50a}$$

$$= \mathrm{FT}[\iiint \mathrm{obj}(\mathbf{r}')\ \mathrm{PSF}(\mathbf{r}-\mathbf{r}')dV'] \tag{50b}$$

$$= \mathrm{FT}[\mathrm{obj}(\mathbf{r})] \cdot \mathrm{FT}[\mathrm{PSF}(\mathbf{r})], \tag{50c}$$

and defines the fluorescence microscopy OTF as the Fourier transform of the fluorescence PSF:

$$\mathrm{OTF}(\mathbf{k}) = \mathrm{FT}[\mathrm{PSF}(\mathbf{r})]. \tag{51}$$

Therefore, the transformed image set is formally equivalent to the transform of the true object weighted (multiplied) by the OTF,

$$I(\mathbf{k}) = \mathrm{OBJ}(\mathbf{k}) \cdot \mathrm{OTF}(\mathbf{k}). \tag{52}$$

In this relation, the filtering effect of the OTF can be clearly seen. Spatial frequency components (Fourier coefficients) of the object for which the OTF is small valued will be attenuated in the data. Where the OTF is zero, information on object Fourier coefficients is not at all present in the data.

The transform of the true object, OBJ($\mathbf{k}$), is a Fourier spectrum, each value of which is a Fourier coefficient that represents a fixed plane-wave grating that contributes to the structure of the object. For example, the coefficient OBJ($\mathbf{k}''$) represents the differential contribution:

$$\delta\ \mathrm{obj}(\mathbf{r}) = \mathrm{OBJ}(\mathbf{k}'') \cdot \{\cos[\mathbf{k}'' \cdot \mathbf{r}] - i \sin[\mathbf{k}'' \cdot \mathbf{r}]\} \cdot \delta \mathrm{k}_x \delta \mathrm{k}_y \delta \mathrm{k}_z/(2\pi)^3, \tag{53}$$

which is a fixed sinusoidal plane-wave field oriented with wave fronts normal to the wave vector $\mathbf{k}''$, with spatial frequency $k''/2\pi$ cycles/mm and period $2\pi/k''$ mm (where $k'' = ||\mathbf{k}''||$). In the Abbe model of transmitted light microscopy, the object is pictured as a superposition of 2D gratings composed of periodic transmittance or OP variations. This model can be directly extended to 3D (Streibl 1985), in which case it becomes equivalent to Bragg diffraction from crystal planes. In the case of fluorescence, actual gratings do not exist as such but only represent virtual periodic distributions of label density in the specimen. The main point is that, when formulated in terms of Fourier spectra, the physical meaning of the transfer function is readily apparent.

The PSF (or its Fourier transform, the OTF) clearly determines the resolution and the optical sectioning capability of the fluorescence microscope. Fourier transformation of the PSF shows that it is a band-limited function; that is, the OTF turns out to be nonzero only within a sharply defined volume in reciprocal space (Fig. 22). For direct image formation, that volume has the shape of a chord torus, rotationally symmetric about the k_z-axis. The most notable feature of the OTF band limit is the axial missing-cone region, a result of the limited range of angles encompassed by the objective aperture. The geometry of the torus in this reciprocal space is significant. The simple fact that it is oblate signifies that axial resolution is inferior to transverse resolution. Object spatial frequency components, which are purely transverse (plane-wave analogs of the flat diffraction gratings considered in the Abbe theory) map into the (k_x, k_y) plane, which cuts through the OTF in its largest diameter. The radius of the OTF, therefore, corresponds to the highest transverse spatial frequency (k_{tr}) admitted by the objective, a grating pattern for which the period equals the Abbe resolution limit (Eq. 28):

$$k_{tr} = (4\pi n/\lambda_0)\sin\theta_{max} \tag{54a}$$

$$= 2\pi/(\lambda_0/2\mathrm{NA}). \tag{54b}$$

Quite a different situation exists for the Fourier coefficients that fall on the k_z-axis. These correspond to purely axial spatial frequency components in the object, fixed plane-wave gratings that are

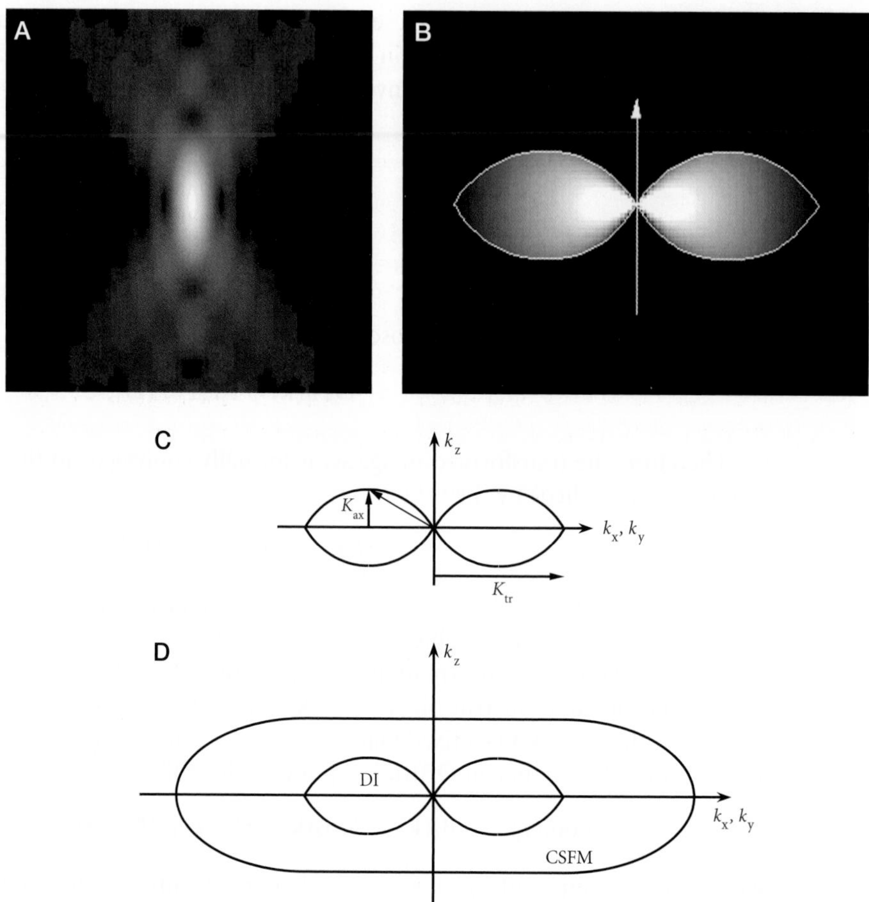

FIGURE 22. The point-spread function (PSF) and the 3D optical transfer function (OTF)—direct image (DI) formation. (*A*) Central axial section through a computed PSF for a fluorescence microscope. The PSF is rotationally symmetric about the vertical (z-) axis. A horizontal slice through the center of the PSF would show the Airy pattern as the focused image. A horizontal slice through any plane offset from the center would show a defocused image consisting of diffuse rings but representing the same total light flux. (*B*) Fourier transformation of the PSF shows that it is a band-limited function; the OTF is nonzero only within a toroidal region of reciprocal space symmetric around the k_z-axis. In this figure, the OTF was drawn for an objective of NA 1.25. Within the passband, the OTF is a strong low-pass filter. The central missing-cone region causes loss of spatial frequency information that defines the axial structure of the object. The OTF inflates with increasing NA, reducing the relative volume of the missing cone. (*C,D*) Band limits of the OTF for DI formation and for confocal scanning (CSFM). Both figures are rotationally symmetric around the k_z-axis. The radius of the toroid (K_{tr}) defines the Abbe resolution limit. The oblique wave vector in C has the highest spatial frequency axial component within the passband (K_{ax}). (*D*) The confocal OTF is also a low-pass band-limited filter but without a missing-cone region. Its volume is approximately eight times that of the DI OTF.

periodic along the axis of the microscope. Because of the missing-cone region in the OTF, every one of these axial components is zeroed out in Equation 52 except for the "DC" Fourier coefficient that maps to the origin of reciprocal space. The same is true for the off-axis coefficients that fall into the missing-cone region, and which correspond to near-axial tilted gratings. As a result of this severe loss of axial information, true optical sectioning does not occur directly. Point objects blur when defocused, but the fluoresced photons are still detected in the image field. Within the OTF passband, the Fourier coefficients with the highest frequency axial projection (k_{ax}) map to the apex of the torus,

$$k_{ax} = (2\pi n/\lambda_0)(1 - \cos\theta_{max}) \tag{55a}$$

$$= (2\pi n/\lambda_0)(1 - [1 - (\mathrm{NA}/n)^2]^{1/2}). \tag{55b}$$

For lenses of low to moderate NA, the square-root term can be accurately simplified to show that the corresponding axial period is similar to the classical depth of field,

$$k_{ax} = 2\pi/(1.56n\lambda_0/\mathrm{NA}^2) \tag{55c}$$

(compare the denominator of Eq. 55c to Eqs. 30 and 42). Within the toroidal volume, the OTF is a low-pass filter, much more strongly weighting the coefficients close to the origin ($k = 0$). Therefore, regardless of the 3D range of reciprocal space over which object Fourier coefficients are distributed, Equation 52 restricts transfer to only those coefficients that fall within the OTF passband, progressively attenuating those of higher spatial frequency within the band. Not surprisingly, the volume enclosed by the OTF passband increases as NA^4 or better, whereas the volume lost in the missing-cone region decreases. In confocal scanning microscopes, optical spatial filtering greatly modifies the PSF and the OTF. The confocal OTF is also band limited but has no missing-cone region and encloses approximately an eightfold greater volume in reciprocal space (Fig. 22). In multiphoton scanning microscopes, nonlinear excitation of fluorescence leads to a similar result, eightfold expansion of the OTF and elimination of the missing-cone region. In both types of scanning microscopes, optical sectioning is obtained directly.

Deconvolution and Estimation

The transfer relation for conventional fluorescence imaging (Eq. 52) suggests that a refined estimate of the true object can be obtained by compensating for the filtering effect of the OTF. The simplest computational scheme is some form of inverse filter applied to the data, much in the same way that an audio equalizer is used to boost frequencies attenuated in the playback of recorded music. In an idealized noiseless case, the inverse filter would have the form $1/\mathrm{OTF}(\mathbf{k})$ in reciprocal space and would be applied to $I(\mathbf{k})$. However, because of the band limitation, the best possible result of inverse filtering would be perfect recovery of all object Fourier coefficients within the OTF volume. This still leaves the severe effect of the missing cone. In practice, the result is far worse. In peripheral zones of the OTF in which its transmittance is low, 1/OTF will be very large. Because real image data will always have spatially distributed noise in it, inverse filters cause high-frequency noise to dominate the object estimate. Therefore, any practical computational method of deconvolution must be stable when operating on noisy data (Preza et al. 1992). An alternative to the simple inverse is the Wiener filter or the Wiener inverse (Agard et al. 1989), which is the linear filter that minimizes the integrated squared error between object and estimate when averaged over all possible square-integrable objects in an unbiased way. The mathematical form of the Wiener inverse in reciprocal space depends on both the OTF and the noise power spectrum, $\langle|N(\mathbf{k})|^2\rangle$,

$$W(\mathbf{k}) = \begin{cases} \mathrm{OTF}^*(\mathbf{k})/(|\mathrm{OTF}(\mathbf{k})|^2 + \langle|N(\mathbf{k})|^2\rangle) & \text{within the band limit,} \\ 0 & \text{outside the band limit.} \end{cases} \tag{56}$$

Where the noise level is very low, $W(\mathbf{k})$ is similar to 1/OTF. In zones in which the OTF has very low amplitude relative to the noise power spectral density, $W(\mathbf{k})$ rolls off to balance accuracy against stability. In practice, the object estimate ($E(\mathbf{r})$) is computed as an inverse Fourier transform of the inverse-filtered data:

$$E(\mathbf{r}) = \mathrm{FT}^{-1}[W(\mathbf{k})I(\mathbf{k})]. \tag{57}$$

However, even in the best of circumstances, the Wiener inverse can only give good estimates of in-band object Fourier coefficients. Therefore, it is not particularly useful for dealing with the OTF in conventional DI systems, although it should perform well with confocal or multiphoton scanning data.

As pointed out by pioneering groups (Harris 1964; Agard et al. 1989; Fay et al. 1989; Carrington 1990), computational deconvolution of DI data requires both regularization for stability and extrapolation in reciprocal space for estimation of out-of-band Fourier coefficients. Extrapolation provides mainly the axial spatial frequency information that is cut out by the missing-cone region of the OTF.

Mathematically, extrapolation is possible when the image set contains a sufficiently large number of independent measurements (pixel values) of sufficient precision to overdetermine the in-band coefficients. A number of elegant computational deconvolution methods have been devised that iteratively converge on an optimum estimate of obj(**r**), given a serial-focus image set as data, along with a measured or computed system PSF or OTF. All of the methods make essential use of constraints or a priori knowledge for reliable extrapolation. The most important of these is nonnegativity of the object label density function that is being estimated. Another is the finite field in which the object is located. For typical image sets that may consist of 8–128 focus planes, each 1024 x 1024 pixels x 10–16 bits, iteration to convergence requires significant processing time on a powerful computer. A number of deconvolution algorithms have undergone commercial development with or without associated computerized digital microscope systems. However, one set of algorithms, originally known as X-Windows computational optical sectioning microscopy (XCOSM), was developed and was placed in the public domain by the National Institutes of Health Resource in Biomedical Computing formerly located at Washington University, St. Louis. XCOSM files are accessible at www.essrl.wustl.edu/~preza/xcosm/. The COSMOS software package, http://cirl.memphis.edu/cosmos/index.html, developed in the Computational Imaging Research Laboratory (CIRL) at The University of Memphis, is an extended and advanced version of the XCOSM project. COSMOS includes two regularized linear least-squares (RLLS) algorithms based on an advanced Wiener filter (Preza et al. 1992), two maximum-likelihood estimators based on the iterative expectation-maximization (E-M) algorithm and a Jansson–van Cittert algorithm. Although E-M is computationally very demanding, it readily incorporates the important constraints of nonnegativity and finite field (Fig. 23). Additionally, it can be adapted to situations in which the data sets are incomplete or in which the PSF is shift variant (i.e., varies from location to location within the FOV). The form of the E-M algorithm used in fluorescence microscopy was specifically derived for Poisson processes, which is a realistic model for the independent fluorescence of dye molecules in an illuminated specimen (Vardi et al. 1985; Holmes 1988; Krishnamurthi et al. 1995).

In practice, E-M takes the form of an array of correction factors used to modify an existing estimate of the object,

$$E_{(n+1)}(\mathbf{r}) = E_{(n)}(\mathbf{r})\left\{\iiint\left\{i(\mathbf{r}')\left[\iiint E_{(n)}(\mathbf{r}'')\mathrm{PSF}(\mathbf{r}'-\mathbf{r}'')dV''\right]^{-1}\right\}\mathrm{PSF}(\mathbf{r}'-\mathbf{r})dV'\right\}. \tag{58}$$

In this iteration formula, $E_{(n)}(\mathbf{r})$ is the current estimate of the 3D object, $i(\mathbf{r})$ is the 3D image data, PSF(**r**) is the 3D PSF, and $E_{(n+1)}(\mathbf{r})$ is the new estimate. In each iteration cycle, the data are weighted by the inverse of the convolution of the current estimate with the PSF, and then back projected by convolution with the coordinate-inverted PSF. This generates a 3D array of weighting coefficients, which is used to update the current estimate by multiplication. Therefore, each iteration requires four 3D operations: (1) convolution of the current estimate with the PSF, (2) division of the data by the blurred estimate, (3) convolution of the weighted data with the coordinate-inverted PSF, and (4) multiplication of the current estimate by the resulting coefficient array. The initial estimate can be a smoothed version of the data set or can be as simple as a nonzero constant field. Because the correction is multiplicative, any pixels in the initial estimate initially set to zero will retain that value. This provides a means for directly incorporating the object field constraint. Additionally, because the fluorescence PSF and data are nonnegative everywhere, negative values cannot be generated by the composed multiplication, division, and convolution operations, provided that the initial estimate is also nonnegative. Thus, E-M algorithms incorporate the universal nonnegativity constraint directly. The E-M iteration, as well as the other rigorous deconvolution algorithms, requires computation of discrete convolutions in every cycle. Practically, these are implemented by utilizing the equivalence between convolution and multiplication of Fourier transforms. The discrete Fourier transform (DFT) is computed for each of the two functions to be convolved, the transforms are multiplied, and the inverse DFT is computed.

Reliable deconvolution in fluorescence microscopy requires attention to detail in every stage of data acquisition and processing. The microscope must be free from stage drift, and the focus drive must be computer controlled and free from backlash. Feedback stabilization of focus (more correctly, of axial setting) can null drift to nanometer levels (Lanni 1993). The high-NA objective, the

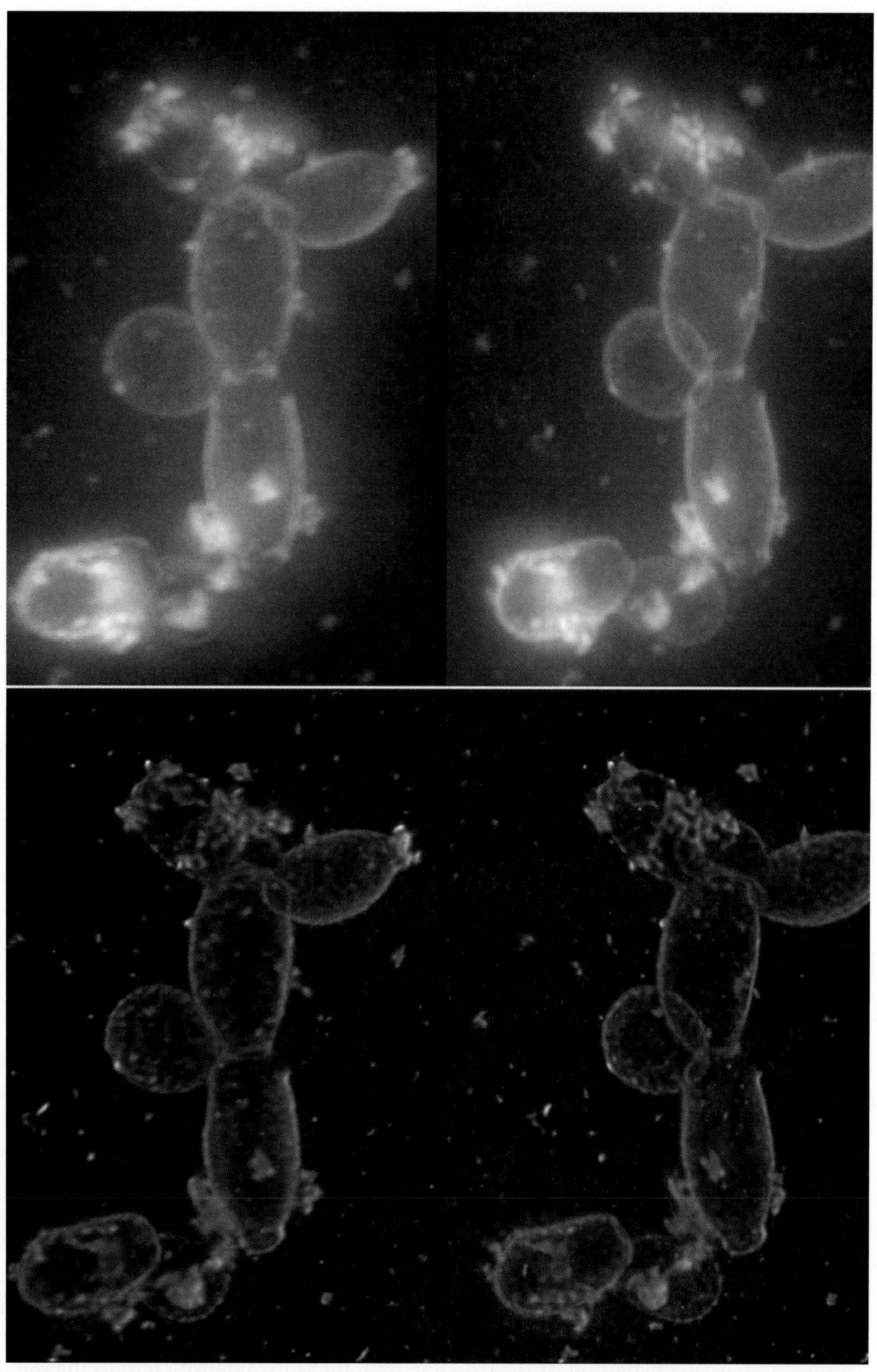

FIGURE 23. Computational deconvolution of serial-focus fluorescence image data. Image stacks showing yeast cells labeled with a fluorescent marker of the plasma membrane. The stacks are displayed as stereo pairs. An automated microscope system equipped with a 100x 1.35-NA objective was used, with a focus increment of 0.1 µm and with the magnification at the camera set for Nyquist sampling (10 pixels/µm). (*Upper panel*) Image data before deconvolution. (*Lower panel*) Object estimate after 1000 rounds of E-M (Eq. 58). The algorithm used is archived at www.essrl.wustl.edu/~preza/xcosm/. An advanced version of the algorithm is available in the COSMOS software package developed by C. Preza, http://cirl.memphis.edu/cosmos/index.html. FOV width, 33 µm. (Images provided courtesy of J.-A. Conchello and W. Goldman, Washington University, Institute for Biomedical Computing and Department of Molecular Biology.)

dichroic reflector, and the emission filter must be of the highest quality, and the camera must be parfocal with the design ocular settings to ensure that no spherical aberration is introduced by body tube length error. High geometric precision, linearity, and dynamic range are all essential camera characteristics. Cooled CCD cameras providing 10–16 bits of dynamic range and very low dark count rates have become the standard detector in this application (Hiraoka et al. 1988). Additional magnification must be incorporated based on the objective and the size of the CCD elements in the camera array (see Nyquist sampling). Depending on the chip used in the camera, individual CCD elements range in size from 6.45 µm to 23.0 µm. Division of this number by the total system magnification gives the effective pixel size in object–space coordinates. The fluorescence illuminator should be IR blocked and should be adjusted to match the objective INA and to maximize uniformity of the excitation field. This provides best utilization of the available camera dynamic range. Before deconvolution, images should be preprocessed to compensate for (1) gain and baseline differences between CCD elements in the camera array, (2) light source instability, and (3) systematic effects such as photobleaching and focus-dependent background caused by luminescence from immersion oil. For a given objective, background light in a fluorescence microscope increases with illuminated field diameter. This wastes dynamic range and is a source of spatial noise. Reduction of the field iris diameter to encircle the object of interest or, at least, to encircle the camera FOV (which is usually smaller than the ocular FOV) minimizes this contribution. For small objects, a reduced illumination field also produces partial confocality in the excitation, which is advantageous in deconvolution (Hiraoka et al. 1990). If the PSF of the microscope is measured for use in deconvolution of reduced field data, it is important that it be measured under the same field iris setting.

Determination of PSFs and OTFs for Deconvolution

The PSF can be derived from basic principles for model systems with or without aberrations (Frisken Gibson and Lanni 1991; Kontoyannis and Lanni 1996) or can be measured by the imaging of subresolution fluorescent particles. Neither route is a trivial exercise because precision and accuracy are needed. Computation has the advantage of absence of noise and unlimited dynamic range but the disadvantage of being only as accurate as the mathematical model of the optical system. Measurement has the advantage of accurately including the aberrations of a real system but the disadvantages of measurement error and spatial noise. Because the optical properties of the specimen itself can introduce aberration, the highest-precision work in deconvolution uses PSFs that are computationally refined from measurements in situ (Carter et al. 1993; Carrington et al. 1995; Scalettar et al. 1996). In this procedure, 0.1–0.2-µm fluorescent polystyrene latex microspheres are added sparsely to the specimen to serve as pseudopoint sources. If test slides are made for PSF measurement, it is extremely important that the slide simulate the refractive structure of the actual specimen. Microspheres embedded in optical cement mimic the refractive structure of fixed specimens embedded in high-index mounting medium. For microscopy of living cells, microspheres in 2% agarose gel sealed under a cover glass provide a more realistic optical test slide. (On a practical note, it is important to use agarose that has a low gelling temperature and to prepare the aqueous gel-based slides on the day they are to be used.) PSF image stacks are registered on-axis and averaged to reduce noise. Processing may include rotational averaging around the centroid, which improves precision but obliterates any actual transverse asymmetry that may be intrinsic to the microscope or may be caused by the specimen. Because the test particles have a small but finite diameter, the measured PSF is a convolution of the true PSF with the source density within the microsphere. The true PSF can be estimated by deconvolving a uniform sphere from the measured and averaged PSF data.

Alternatively, the OTF can be obtained through measurement or computation. Whereas the PSF is computed via numerical evaluation of a modified Kirchhoff diffraction integral, OTF formulas derived from first principles are generally expressible in terms of analytic functions within the OTF band limit (Frieden 1967; Wang and Frieden 1990). In principle, a computed OTF can be converted to its corresponding PSF via the inverse FT. In the more usual case, a computed or measured PSF is Fourier transformed to obtain a discretized OTF, which is then used directly in an iterative deconvolution algorithm.

Nyquist Sampling and Restoration of the Image Field

When viewed by eye, the image field of a microscope is perceived as a continuous distribution of irradiance. This is true, although the retina is essentially a discrete array of photodetector elements—in which light is detected as discrete photocounts. When the image field is detected with an electronic camera, the output is sampled and digitized (also true for scanned film). Under these conditions, pixellation and quantization become apparent. Overly coarse pixellation degrades resolution; overly fine pixellation (empty magnification) needlessly reduces the FOV and increases variance by reducing the number of photocounts per pixel. Clearly, optimum pixellation should be determined by the resolution limit of the optical system that forms the image.

Pixellation and resolution are related by the sampling theorem (Goldman 1968), which sets an important restriction on the information content of images. The theorem relates the frequency content of a signal to the density of measured values that is required for exact reconstruction of the signal in the absence of noise. The signal can be a time series, as in communications, or can be a spatial pattern such as a spectrum or an image. Specifically, if a signal or an image $f(x)$ is limited to spatial frequencies lower than a band limit value F_0 (cycles/mm, for example), then $f(x)$ can be reconstructed exactly at all values of x if the signal is known at sample points spaced at an increment $\Delta x = 1/\text{bandwidth}$. In this case, the bandwidth is $2F_0$ because the Fourier transform of $f(x)$ would show nonzero coefficients in the range $-F_0$ to F_0. Therefore, $\Delta x = 1/(2F_0)$ mm. In principle, the sample points must extend throughout the spatial domain, but a finite set is usually sufficient to determine $f(x)$ accurately for an object well within a finite FOV. The basis of the sampling theorem is in representation of a finite frequency spectrum as a Fourier series truncated to a single period by a window function of width $2F_0$. Inverse Fourier transformation of this representation shows that the nth coefficient in the series is actually a sample value of the image $f(n\Delta x)$ at pixel location $x = n\Delta x$. Furthermore, each sample value is the weighting coefficient of a continuous sampling function centered at $x = n\Delta x$:

$$f(x) = \Sigma_n\, f(n\Delta x)\, \sin[2\pi F_0(x - n\Delta x)]/2\pi F_0(x - n\Delta x) \quad (59a)$$

$$= \Sigma_n\, f(n\Delta x)\, \mathrm{sinc}[2\pi F_0(x - n\Delta x)]. \quad (59b)$$

Together, the set of terms constitutes an exact interpolation series for $f(x)$. The quantity 1/bandwidth represents a maximum interval between samples known as the Nyquist limit. Undersampling ($\Delta x > 1/\text{bandwidth}$) results in loss of resolution and can result in aliasing—the generation of artifactual spatial frequency components.

For microscopes, it is the Abbe resolution limit (p_{min}; see Eq. 28) that sets the information density. Specifically, the Abbe limit is the period of the highest spatial frequency transferred to the image field by the optical system ($k_{tr}/2\pi$; see Eqs. 54). The bandwidth of a microscope is, therefore, $2(k_{tr}/2\pi)$, which is also the diameter of the OTF. A noiseless 2D image field is, therefore, completely defined by point samples in a square array spaced at 1/bandwidth,

$$\Delta s_{Nyq} = 1/[2(k_{tr}/2\pi)] \quad (60a)$$

$$= \lambda_0/4\text{NA}, \quad (60b)$$

that is, the Nyquist sample interval equals one-half the Abbe resolution limit (Lanni and Baxter 1992). For high-NA objectives, Δs_{Nyq} is, therefore, close to 0.1 μm (Nyquist density = 10 pixels/μm). In real terms, the sampling interval is the camera pixel size divided by the total system magnification. For Nyquist sampling, this number must not exceed $\lambda_0/4$NA. A CCD camera with 19-μm square elements used with a 100x 1.3-NA objective, for example, requires a total system magnification of 190x for Nyquist sampling in the mid-visible range. Therefore, additional magnification ≥1.9x must be used. Needless to say, most biological imaging with electronic cameras is performed sub-Nyquist, in which the recorded image is of significantly lower resolution than the actual optical field of the microscope. This is usually dictated by the need for a sufficient FOV (an advantage still held by photographic film!). However, there is no getting around meeting or exceeding the Nyquist density to properly match camera to microscope. The fact that the OTF is circular rather than square means that there is some redundancy in image data sampled at the Nyquist density over a square array as defined above. However, there has yet been no practical use made of this latter fact in microscopy.

With Nyquist sampling, it is possible to computationally convert pixellated data into a continuous image field by the use of a 2D interpolation series analogous to Equation 59,

$$f(x, y) = \Sigma_m \Sigma_n f(m\Delta s, n\Delta s) \operatorname{sinc}[2\pi F_0(x - m\Delta s)] \operatorname{sinc}[2\pi F_0(y - n\Delta s)]. \qquad (61)$$

Although rarely performed, this has the advantage of producing a field that is independent of arbitrary offsets between the sampling array (or Nyquist lattice) and the image features of interest. As the simplest possible example, the Airy pattern image of a point source could fall exactly centered on a pixel, centered at the adjoining corners of four pixels, or centered at an arbitrary offset in between. In each case, the pixellated image will appear differently, with 16–21 pixels located within the Airy disk (4–5 pixels across). The band-limited reconstruction, on the other hand, should produce the same Airy pattern in every case.

Real cameras strictly do not produce image data in the form of point samples as required for direct application of the sampling theorem. At the present time, use of cooled CCD cameras provides the greatest optical efficiency, dynamic range, and geometric precision in image capture. Pixel values in the digitized image represent summed photocounts in the contiguous-square bins that constitute the CCD array. In contrast, the interpolation series (Eqs. 59 and 61) are based on point-sampled values of the image field. The difference is illustrated by the example of the Airy pattern image. Although the Airy disk is defined by a prominent circular node at which the brightness drops to zero, no pixel in the CCD image will ever read zero because of the finite area over which each CCD element captures photons. Direct use of CCD camera output in the interpolation formula results in a blurred reconstruction of the Airy pattern. The effect of square-pixel summing can be reversed, in the absence of noise, by the use of an inverse filter to estimate point samples from real square-pixel data. The directly usable form of this filter (Lanni and Baxter 1992) is a discrete 2D convolution kernel, centered at K_{00},

$$K_{00} = (16/\pi^2)(\lambda_c)^2 ,$$

$$K_{m'n'} = (16/\pi^2)(\lambda_c + \Sigma_m(-1)^m/(2m-1)^2)(\lambda_c + \Sigma_n(-1)^n/(2n-1)^2), \qquad (62)$$

where λ_c is Catalan's constant (0.91596559), and where the summations run 1 to $|m'|$ or 1 to $|n'|$ or are zero when $m'=0$ or $n'=0$, respectively. This is applied directly to square-pixel image data to estimate point samples for use in an interpolation series (Eq. 61). The interpolation series can be computed to any degree of fineness, but the main advantage is that the result is, in principle, independent of the original pixellation.

Not surprisingly, in optical sectioning microscopy, the Nyquist criterion applies to axial sampling or focus increment as well as to pixellation. Using Equation 55b for the axial spatial frequency limit of the OTF, the Nyquist focus increment equals 1/(axial bandwidth):

$$\Delta z_{\mathrm{Nyq}} = 1/2(k_{ax}/2\pi) \qquad (63a)$$

$$= \lambda_0/2n(1 - [1 - (\mathrm{NA}/n)^2]^{1/2}) \qquad (63b)$$

$$= 2\,\delta_{\mathrm{RQWC}} \qquad \text{(see Eq. 43).} \qquad (63c)$$

For a high-NA objective, Δz_{Nyq} is ~0.3 µm. However, for direct image formation, this situation is not strictly analogous to transverse sampling requirements because the axial range in a serial-focus image set is usually much less than needed to completely blur out-of-focus features. Therefore, the sample points do not run out to infinity or to where the signal drops to arbitrarily small values as required by the sampling theorem. In a confocal or multiphoton scanning microscope, the object is strongly attenuated with defocus so that a finite number of sample points (focus planes) can accurately represent the axial structure of the object in an interpolation series representation.

An additional point regarding confocal and multiphoton scanning microscopes is that the transverse and axial cutoffs of the OTF are doubled relative to direct image formation. Nyquist densities are, therefore, doubled in each dimension. In other words, resolution is increased but so is the sampling requirement that must be satisfied to capture the additional information.

PRACTICAL LIMITS ON IMAGE QUALITY

Aberration

Both optical aberration and background light cause loss of contrast and resolution in all modes of microscopy and can cause a microscope to perform far below the fundamental limits to which it is designed. Aberration can originate in the optical system of the microscope or in the optical properties of the mounted specimen. For example, mixing of dissimilar immersion oil residues can cause irregular refraction of light rays exiting the specimen. Likewise, because glycerol is highly hygroscopic, its uniformity as an immersion fluid is affected by ambient humidity. As discussed above, a difference between the immersion and specimen refractive indices leads to focus-dependent spherical aberration (SA), most noticeable as axial asymmetry in the PSF. When SA is severe, the fluorescence image of a point particle will show prominent concentric rings on one side of the focus and featureless blur on the other side of the focus. The central maximum of the in-focus image will be attenuated, and the brightness of the surrounding rings will be increased in comparison to the Airy pattern. This reduces contrast in any image. SA also causes elongation of the central maximum of the PSF, which dramatically reduces axial resolution. In fluorescence systems, dichroic reflectors and emission filters must be selected to minimize wave-front distortion. To begin with, a high-quality filter with perfectly flat surfaces can cause image offset because of wedge—the condition in which the flat surfaces are not exactly parallel. This condition makes the flat optics into a weak prism (flat optics made with precisely parallel surfaces are known as etalons) and is not considered an aberration because a wedge does not degrade the image. Much more serious is surface curvature, a common condition in general-purpose filters composed of two or more cemented glass substrata. In this case, the filter will act as a weak irregular lens, causing astigmatism, spherical aberration, or other distortions in the image. Manufacturers of fluorescence filter sets know that emission filters for microscopy must be of imaging grade, preferably selected for flatness by examination of the filter in a transmission interferometer. Clearly, the same high quality is needed in any optics inserted between the objective and the image detector: prisms, analyzers, wave plates, or electro-optic devices. In fluorescence optical systems used simultaneously with DIC, the Nomarski prism will shear the fluorescence image into two slightly offset orthogonally polarized components (Kontoyannis et al. 1996). This is not an aberration in the usual sense and is of little consequence when imaging extended fluorescent features in an object. However, the shear can significantly reduce the brightness over background of punctate and line objects at the diffraction limit.

Background Light

Sources of background in fluorescence microscopy can be found in both the microscope and the specimen. Emission filters must block, with great efficiency, light at the excitation wavelength that is back-reflected into the detector path of the instrument. Even with incident-light illumination in which most of the excitation propagates away from the microscope, the back-reflected light flux can be orders of magnitude greater than the fluorescence. Paired excitation and emission filters for fluorescence microscopy characteristically attenuate by factors $>>10^4$ in complementary passbands. A heat filter should be used in the illuminator to block the intense IR light produced by incandescent and arc lamps. Stray IR light in the 750–1000-nm band is not directly visible but may be well within the spectral range of an electronic camera. Luminescence can also be excited within objectives and other glass optical components, particularly by UV light. Over the past two decades, objectives have become greatly improved by the use of low-absorbance low-luminescence glasses and optical cements. Some of the best colored glasses used in emission filters are also luminescent. Used alone, these glasses make excellent long-pass filters but will cause a background glow when there is a significant flux of back-reflected excitation to be blocked. This can be minimized by the use of immersion objectives to reduce reflection. Modern band-pass interference filters generally do not include a colored glass element to minimize filter luminescence. However, some excellent band-pass filters may contain colored glass. In this case, luminescence is minimized when the light is first incident on the reflective multilayer side of the filter.

The most common sources of background light in fluorescence microscopy are in the assembled specimen, including immersion fluid, cover glass, and slide. Only immersion oil of fluorescence grade should be used. Contamination of the oil by marker inks, stains, and dust should be avoided (lint is often fluorescent because of the brighteners used in laundry detergents!). Background due to the immersion fluid will depend on the amount of fluid in the illuminated volume between cover glass and objective. This is clearly less for high-NA lenses having short working distances and is also dependent on the depth to which the objective is focused. Cover glasses are rarely a significant source of luminescence but should be carefully cleaned of dried culture medium before immersion fluid is applied. Dried medium causes severe light scatter and is intensely fluorescent. Slides are six times thicker than cover glasses and, if not of good quality, can be a significant source of luminescence. Fused silica slides can be used when extremely low background is required. For most live cell microscopy, culture dishes or chambers with cover glass windows are used, eliminating the slide as a factor. By far, the living specimen and the culture medium are the major contributors of background light. In culture medium, Phenol Red used as a visual pH indicator is moderately fluorescent and should be omitted for use in microscopy. Base medium and serum also contain many weakly fluorescent biochemicals and proteins. Luminescence from this source can be reduced by the use of a thin culture chamber that minimizes the volume of medium that is illuminated. In general, background light increases with the diameter of the illuminated field; therefore, the field iris should be reduced when the object of interest is much smaller than the design FOV. Because background sources are less strongly excited outside of the blue end of the spectrum, the use of long-wavelength fluorescent dyes has grown rapidly. A fundamental limitation on background reduction is set by Raman emission, an inelastic scattering process that converts a small fraction of the incident light to isotropic emission at longer wavelengths. Raman emission generally originates in water within the specimen or in the immersion fluid. Although it is well beyond the scope of this discussion, Raman scatter, backscatter, and phosphorescence can all be gated out through time-resolved fluorescence imaging (Herman 1998). Because of its noise content, background light is one of the factors that ultimately limits sensitivity in fluorescence microscopy (photobleaching and excited state lifetime are the other factors). Nevertheless, the fluorescence microscope has emerged as an unmatched instrument for molecular level imaging, including single-molecule tracking and single-molecule spectroscopy (Gross and Webb 1986; Kron and Spudich 1986; Betzig and Chichester 1993; Nie et al. 1994; Funatsu et al. 1995; Sase et al. 1995; Vale et al. 1996; Dickson et al. 1997; Femino et al. 1998; Waterman-Storer et al. 1998; Waterman-Storer and Salmon 1998; Moerner and Orrit 1999; Smith et al. 1999; Weiss 1999).

CONCLUSION

In this chapter, we laid out principles and practical considerations on basic light microscopy as applied to biological specimens. Much of the discussion centers on the effect of high aperture in the image formation process and a description of the main phase-contrast and fluorescence contrast systems. Not covered are the many interesting and useful special purpose instruments and methods, such as confocal scanning fluorescence, multiphoton scanning fluorescence, reflection-interference, total internal reflection fluorescence (TIRF), near-field scanning, polarization, fluorescence polarization, resonance energy transfer, structured illumination, and interferometric fluorescence microscopies. We have not covered microscope automation, which has revolutionized the data-gathering power of the core instrument. In the bibliography accompanying this chapter are references that more extensively cover special topics in optics and applications. We have every reason to expect that these remarkable developments will continue.

ACKNOWLEDGMENT

F. Lanni thanks the National Science Foundation in particular for past support through the Science and Technology Centers program.

GENERAL REFERENCES ON MICROSCOPY AND OPTICS

Born M, Wolf E. 1980. *Principles of optics*, 6th ed. Pergamon, New York.

Haugland RD. 1996. *Handbook of fluorescent probes and research chemicals*, 6th ed. Molecular Probes, Inc., Eugene, OR.

Herman B. 1998. *Fluorescence microscopy*, 2nd ed. Springer/BIOS Scientific, Oxford.

Herman B, Jacobson K, eds. 1990. *Optical microscopy for biology.* Wiley, New York.

Inoue S, Spring KR. 1997. *Video microscopy: The fundamentals*, 2nd ed. Plenum, New York.

Longhurst RS. 1973. *Geometrical and physical optics*, 3rd ed. Longman Group, London.

Matsumoto B, ed. 1993. *Cell biological applications of confocal microscopy. Methods in Cell Biology*, Vol. 38. Academic, New York.

Pawley JB, ed. 1995. *Handbook of confocal fluorescence microscopy*, 2nd ed. Plenum, New York.

Piller H. 1977. *Microscope photometry*. Springer-Verlag, Heidelberg.

Pluta M. 1989. *Advanced light microscopy: Principles and basic properties*, Vol. 1. *Specialized methods*, Vol. 2. *Measuring techniques*, Vol. 3. Elsevier, New York.

Rebhun LI, Taylor DL, Condeelis JS, eds. 1988. Optical approaches to the dynamics of cellular motility. A symposium in honor of Robert Day Allen. October 5–8, 1987, Woods Hole, Massachusetts. Proceedings. *Cell Motil Cytoskel* **10:** 1–348.

Slayter EM. 1976. *Optical methods in biology*. Krieger, Huntington, NY.

Spector D, Goldman RD, Leinwand LA. 1998. *Cells: A laboratory manual. Light microscopy and cell structure*, Vol. 2. Cold Spring Harbor Laboratory Press, Cold Spring Harbor, NY.

Taylor DL, Wang Y-L, eds. 1989. *Fluorescence microscopy of living cells in culture.* Part B: *Quantitative fluorescence microscopy: Imaging and spectroscopy. Methods in Cell Biololgy*, Vol. 30. Academic, New York.

Wang Y-L, Taylor DL, eds. 1989. *Fluorescence microscopy of living cells in culture.* Part A: *Fluorescent analogs, labeling cells, and basic microscopy. Methods in Cell Biology*, Vol. 29. Academic, New York.

Wilson T, Sheppard C. 1984. *Theory and practice of scanning optical microscopy.* Academic, London.

RESEARCH AND REVIEW ARTICLES

Abraham VA, Krishnamurthi V, Taylor DL, Lanni F. 1999. The actin-based nanomachine at the leading edge of migrating cells. *Biophys J* **77:** 1721–1732.

Agard DA, Hiraoka Y, Shaw P, Sedat JW. 1989. Fluorescence microscopy in three dimensions. *Methods Cell Biol* **30:** 353–377.

Allen RD, Allen NS, Travis JL. 1981. Video-enhanced contrast, differential interference contrast (AVEC-DIC) microscopy: A new method capable of analyzing microtubule-related motility in the reticulopodial network of *Allogromia laticollaris. Cell Motil Cytoskel* **1:** 291–302.

Axelrod D. 1989. Fluorescence polarization microscopy. *Methods Cell Biol* **30:** 333–352.

Betzig E, Chichester RJ. 1993. Single molecules observed by near-field scanning optical microscopy. *Science* **262:** 1422–1425.

Carrington WA. 1990. Image restoration in 3D microscopy with limited data. *Proc Soc Photo-optical Instr Eng* **1205:** 72–83.

Carrington WA, Lynch RM, Moore EDW, Isenberg G, Fogarty KE, Fay FS. 1995. Superresolution three-dimensional images of fluorescence in cells with minimal light exposure. *Science* **268:** 1483–1487.

Carter KC, Bowman D, Carrington W, Fogarty K, McNeil JA, Fay FS, Lawrence JB. 1993. A three-dimensional view of precursor messenger RNA metabolism within the mammalian nucleus. *Science* **259:** 1330–1335.

Cogswell C, Smith NI, Larkin KG, Hariharan P. 1997. Quantitative DIC microscopy using a geometric phase shifter. *Proc Soc Photo-optical Instr Eng* **2984:** 72–81.

Danz R, Vogelgsang A, Kathner R. 2004. PlasDIC: A useful modification of the differential interference contrast according to Smith/Nomarski in transmitted light arrangement. Photonik www.zeiss.com/C1256F8500454979/0/366354E1E8BA8703C1256F8E003BBCB9/$file/plasdic_photonik_2004march_e.pdf.

Denk W, Strickler JH, Webb WW. 1990. Two-photon laser scanning fluorescence microscopy. *Science* **248:** 73–76.

Dickson RM, Cubitt AB, Tsien RY, Moerner WE. 1997. On/off blinking and switching behaviour of single molecules of green fluorescent protein. *Nature* **388:** 355–358.

Dong CY, So PT, French T, Gratton E. 1995. Fluorescence lifetime imaging by asynchronous pump-probe microscopy. *Biophys J* **69:** 2234–2242.

Fay FS, Carrington W, Fogarty KE. 1989. Three-dimensional molecular distributions in single cells analysed using the digital imaging microscope. *J Microsc* **153:** 133–149.

Femino AM, Fay FS, Fogarty K, Singer RH. 1998. Visualization of single RNA transcripts in situ. *Science* **280:** 585–590.

Frieden BR. 1967. Optical transfer of the three-dimensional object. *J Opt Soc Am* **57:** 56–66.

Frisken Gibson SF. 1990. "Modeling the 3d imaging properties of the fluorescence light microscope." PhD thesis, Carnegie Mellon University, Pittsburgh.

Frisken Gibson S, Lanni F. 1991. Experimental test of an analytical model of aberration in an oil-immersion objective lens used in three-dimensional light microscopy. *J Opt Soc Am A* **8:** 1601–1613 (reprinted in *J Opt Soc Am A* **9:** 154–166 [1992]).

Funatsu T, Harada Y, Tokunaga M, Saito K, Yanagida T. 1995. Imaging of single fluorescent molecules and individual ATP turnovers by single myosin molecules in aqueous solution. *Nature* **374:** 555–559.

Galbraith W. 1955. The optical measurement of depth. *Q J Microsc Sci* **96:** 285–288.

Goldman S. 1968. *Information theory*. Dover, New York.

Gross D, Webb WW. 1986. Molecular counting of low-density lipoprotein particles as individuals and small clusters on cell surfaces. *Biophys J* **49:** 901–911.

Grynkiewicz G, Poenie M, Tsien RY. 1985. A new generation of Ca^{2+} indicators with greatly improved fluorescence properties. *J Biol Chem* **260:** 3440–3450.

Guether R, Reddington MV. 1997. Photostable cyanine dye β-cyclodextrin conjugates. *Tetrahedron Lett* **38:** 6167–6170.

Gustafsson MGL. 2000. Surpassing the lateral resolution limit by a factor of two using structured illumination microscopy. *J Microsc* **198:** 82–87.

Gustafsson MGL, Agard DA, Sedat JW. 1995. Sevenfold improvement of axial resolution in 3D widefield microscopy using two objective lenses. *Proc Soc Photo-optical Instr Eng* **2412:** 147–156.

Gustafsson MG, Shao L, Carlton PM, Wang CJ, Golubovskaya IN, Cande WZ, Agard DA, Sedat JW. 2008. Three-dimensional resolution doubling in wide-field fluorescence microscopy by structured illumination. *Biophys J* **94:** 4957–4970.

Hard R, Zeh R, Allen RD. 1977. Phase-randomized laser illumination for microscopy. *J Cell Sci* **23:** 335–343.

Harris JL. 1964. Diffraction and resolving power. *J Opt Soc Am* **54:** 931–936.

Hell S, Stelzer EHK. 1992. Properties of a 4Pi confocal fluorescence microscope. *J Opt Soc Am A* **9:** 2159–2166.

Hiraoka Y, Sedat JW, Agard DA. 1988. The use of a charge-coupled device for quantitative optical microscopy of biological structures. *Science* **238:** 36–41.

Hiraoka Y, Sedat JW, Agard DA. 1990. Determination of three-dimensional imaging properties of a light microscope system. Partial confocal behavior in epifluorescence microscopy. *Biophys J* **57**: 325–333.

Holmes TJ. 1988. Maximum-likelihood image restoration adapted for noncoherent optical imaging. *J Opt Soc Am A* **5:** 666–673.

Inoue S. 1981. Video image processing greatly enhances contrast, quality, and speed in polarization-based microscopy. *J Cell Biol* **89:** 346–356.

Inoue S. 1989. Imaging of unresolved objects, superresolution, and precision of distance measurement with video microscopy. *Methods Cell Biol* **30:** 85–112.

Keller HE. 1995. Objective lenses for confocal microscopy. In *Handbook of biological confocal microscopy*, 2nd ed. (ed. Pawley JB), pp. 111–126. Plenum, New York.

Klar TA, Jakobs S, Dyba M, Egner A, Hell SW. 2000. Fluorescence microscopy with diffraction resolution barrier broken by stimulated emission. *Proc Natl Acad Sci* **97:** 8206–8210.

Kontoyannis NS, Lanni F. 1996. Measured and computed point spread functions for an indirect water-immersion objective used in three-dimensional fluorescence microscopy. *Proc Soc Photo-optical Instr Eng* **2655:** 34–42.

Kontoyannis NS, Krishnamurthi V, Bailey B, Lanni F. 1996. Three-dimensional fluorescence microscopy of cells. *Proc Soc Photo-optical Instr Eng* **2678:** 6–14.

Krishnamurthi V, Liu Y-H, Bhattacharyya S, Turner JN, Holmes TJ. 1995. Blind deconvolution of fluorescence micrographs by maximum-likelihood estimation. *Appl Opt* **34:** 6633–6647.

Kron SJ, Spudich JA. 1986. Fluorescent actin filaments move on myosin fixed to a glass surface. *Proc Natl Acad Sci* **83:** 6272–6276.

Lanni F. 1993. Feedback-stabilized focal plane control for light microscopes. *Rev Sci Instr* **64:** 1474–1477.

Lanni F. 2005. Fluorescence grating imager systems for optical-sectioning microscopy. In *Imaging in neuroscience and development* (ed. Yuste RM, Konnerth A), pp. 805–813. Cold Spring Harbor Laboratory Press, Cold Spring Harbor, NY.

Lanni F, Baxter GJ. 1992. Sampling theorem for square-pixel image data. *Proc Soc Photo-optical Instr Eng* **1660:** 140–147.

Lanni F, Wilson T. 2000. Grating image systems for optical sectioning fluorescence microscopy of cells, tissues and small organisms. In *Imaging neurons: A laboratory manual* (ed. Yuste R, et al.), pp. 8.1–8.9. Cold Spring Harbor Laboratory Press, Cold Spring Harbor, NY.

McNally JG, Preza C, Conchello J-A, Thomas LJ. 1994. Artifacts in computational optical sectioning microscopy. *J Opt Soc Am A* **11:** 1056–1067.

Moerner WE, Orrit M. 1999. Illuminating single molecules in condensed matter. *Science* **283:** 1670–1676.

Neil MAA, Juskaitis R, Wilson T. 1997. Method of obtaining optical sectioning by using structured light in a conventional microscope. *Opt Lett* **22:** 1905–1907.

Nie S, Chiu DT, Zare RN. 1994. Probing individual molecules with confocal fluorescence microscopy. *Science* **266:** 1018–1021.

Ormo M, Cubitt AB, Kallio K, Gross LA, Tsien RY, Remington SJ. 1996. Crystal structure of the *Aequorea victoria* green fluorescent protein. *Science* **273:** 1392–1395.

Piston DW. 1998. Choosing objective lenses: The importance of numerical aperture and magnification in digital optical microscopy. *Biol Bull* **195:** 1–4.

Preza C. 2000. Rotational-diversity phase estimation from differential-interference-contrast microscopy images. *J Opt Soc Am A* **17:** 415–424.

Preza C, Miller MI, Thomas LJ, McNally JG. 1992. Regularized linear method for reconstruction of three-dimensional microscopic objects from optical sections. *J Opt Soc Am A* **9:** 219–228.

Reichman J. 2000. *Handbook of optical filters for fluorescence microscopy.* HB1.2 (rev. 2007). Chroma Technology Corporation, Bellows Falls, VT.

Sase I, Miyata H, Corrie JET, Craik JS, Kinosita K. 1995. Real time imaging of single fluorophores on moving actin with an epifluorescence microscope. *Biophys J* **69:** 323–328.

Scalettar BA, Swedlow JR, Sedat JW, Agard DA. 1996. Dispersion, aberration, and deconvolution in multi-wavelength fluorescence images. *J Microsc* **182:** 50–60.

Scheenen WJ, Makings LR, Gross LR, Pozzan T, Tsien RY. 1996. Photodegradation of indo-1 and its effect on apparent Ca2+ concentrations. *Chem Biol* **3:** 765–774.

Shao L, Isaac B, Uzawa S, Agard DA, Sedat JW, Gustafsson MG. 2008. I5S: Wide-field light microscopy with 100-nm-scale resolution in three dimensions. *Biophys J* **94:** 4971–4983.

Shribak M. 2009. Orientation independent differential interference contrast microscopy technique and device. U.S. Patent # 7,564,618 B2.

Shribak M, Inoue S. 2006. Orientation-independent differential interference contrast microscopy. *Appl Opt* **45:** 460–469.

Shtengel G, Galbraith JA, Galbraith CG, Lippincott-Schwartz J, Gillette JM, Manley S, Sougrat R, Waterman CM, Kanchanawong P, Davidson MW, et al. 2009. Interferometric fluorescent super-resolution microscopy resolves 3D cellular ultrastructure. *Proc Natl Acad Sci* **106:** 3125–3130.

Smith DE, Babcock HP, Chu S. 1999. Single-polymer dynamics in steady shear flow. *Science* **283:** 1724–1727.

Streibl N. 1985. Three-dimensional imaging by a microscope. *J Opt Soc Am A* **2:** 121–127.

Szent-Gyorgyi C, Schmidt BF, Creeger Y, Fisher GW, Zakel KL, Adler S, Fitzpatrick JA, Woolford CA, Yan Q, Vasilev KV, et al. 2008. Fluorogen-activating single-chain antibodies for imaging cell surface proteins. *Nat Biotechnol* **26:** 235–240.

Taylor DL, Salmon ED. 1989. Basic fluorescence microscopy. *Methods Cell Biol* **29:** 207–237.

Tsien RY, Waggoner AS. 1995. Fluorophores for confocal microscopy. In *Handbook of biological confocal microscopy*, 2nd ed. (ed. Pawley JB), pp. 267–279. Plenum, New York.

Vale RD, Funatsu T, Pierce DW, Romberg L, Harada Y, Yanagida T. 1996. Direct observation of single kinesin molecules moving along microtubules. *Nature* **380:** 451–453.

Vardi Y, Shepp LA, Kaufman L. 1985. A statistical model for positron emission tomography. *J Am Stat Assoc* **80:** 8–20.

Waggoner AS, DeBiasio R, Conrad P, Bright GR, Ernst L, Ryan K, Nederlof M, Taylor D. 1989. Multiple spectral parameter imaging. *Methods Cell Biol* **30:** 449–478.

Wang S-I, Frieden BR. 1990. Effects of third-order spherical aberration on the 3-D incoherent optical transfer function. *Appl Opt* **29:** 2424–2432.

Waterman-Storer CM, Salmon ED. 1998. How microtubules get fluorescent speckles. *Biophys J* **75:** 2059–2069.

Waterman-Storer CM, Desai A, Bulinski JC, Salmon ED. 1998. Fluorescent speckle microscopy, a method to visualize the dynamics of protein assemblies in living cells. *Curr Biol* **8:** 1227–1230.

Weiss S. 1999. Fluorescence spectroscopy of single biomolecules. *Science* **283:** 1676–1683.

White JG, Amos WB, Fordham M. 1987. An evaluation of confocal versus conventional imaging of biological structures by fluorescence light microscopy. *J Cell Biol* **105:** 41–48.

2 Video Microscopy, Video Cameras, and Image Enhancement

Masafumi Oshiro,[1] Lowell A. Moomaw,[1] and H. Ernst Keller[2]

[1]*System Division, Hamamatsu Photonics K.K., Hamamatsu City, Shizuoka 430-8587, Japan;* [2]*Carl Zeiss, Inc., Thornwood, New York 10594 (retired)*

ABSTRACT

Video microscopy is the application of video technology to microscopy, resulting in two fields of microscopy called video-enhanced contrast microscopy (VEC) and video-intensified microscopy (VIM). VEC involves the production of an image from a specimen that is invisible to the eye, either because of a lack of contrast or because of its spectral characteristics (UV or infrared). VIM involves imaging a specimen when the light levels are too low for standard cameras or, in some cases, even for the eye. Images are produced by VIM using image analysis computers.

INTRODUCTION

The major components for video-enhanced contrast microscopy (VEC) are a differential interference contrast (DIC) or phase-contrast microscope, a standard video rate high-resolution camera, a real-time image processor, and a high-resolution video monitor. The use of a video rate camera in conjunction with a real-time image processor greatly improves the contrast of the image and enables the use of the maximum aperture of the optical system. As a result, maximum resolution of the microscope is achieved with sufficient contrast to render the information perceivable.

The video-intensified microscopy (VIM) method usually uses a fluorescence microscope, a low-light-level video camera, a real-time image processor, and a high-resolution video monitor. The low-light-level video camera and real-time image processor act together to integrate the image. As a result, detailed structure within the specimen is revealed, and photo damage to the sample is minimized by reducing illumination intensity. Even in ultralow-light situations, such as bioluminescence and chemiluminescence, a photon-counting camera detects single photon events, and an image processor integrates the image to generate gray-level information based on photon counting per pixel.

Modified slightly, with permission, from Spector et al. (1998c). (Material provided by M. Oshiro, L.A. Moomaw, and E. Keller.)

This rapidly growing field is well on its way to replacing conventional photomicroscopy. Technical advances in cameras, camera electronics, and image processors, as well as printers, have reached a point where the resolution of the video image becomes indistinguishable from that of photographic fine-grain film. Although it is substantially more expensive, the added technical benefits of video microscopy are considerable. These include:

- direct display and optional recording of dynamic events, including time-lapse studies of live cells
- availability of a wide range of analog and digital cameras for every budget and to suit many specific applications
- electronic contrast enhancement via gain and black-level control, or digital image processing, reveals image information inaccessible to eye or film
- recording of exceptionally low light levels with time integrating or intensified (e.g., ICCD) cameras (see below)

  This is important in studies involving weak fluorescence emissions in vivo (e.g., following microinjection of fluorochrome-labeled proteins or transfection of cells with green fluorescent protein [GFP] constructs) (see Spector et al. 1998a) or for minimizing the phototoxic effects of illuminating cells with high light intensities.

- electronic color balancing with single, three-tube, or chip color cameras
- electronic image transfer to other locations
- optional image storage for further processing or image analysis

To take full advantage of video-enhanced microscopy, it is important to know a few basic facts about video cameras and image processors. There are many versions of each, and each version has certain characteristics that must be considered for different applications.

DEFINITIONS

Several terms used throughout this chapter that may be unfamiliar to the reader are defined below.

Analog and digital cameras: An analog video signal is generated from the detector itself. This video signal is processed and converted to certain formats to interface to display monitors, image storage devices, and image processors. If this interface signal is analog, the camera is called an analog camera. If the interface signal is digital, the camera is called a digital camera. Analog cameras normally generate a standard video signal such as RS-170.

Integration camera: The standard video camera's frame rate is 30 frames/sec. This means that incoming light is integrated in a detector for 1/30 sec, and the signal is read out after the integration. To realize low-light sensitivity, integration cameras integrate incoming light for more than one frame, and the integrated signal is then read out. The sensitivity is increased in proportion with the number of integrating frames. However, maximum integration frames are limited by the dark noise of the camera itself.

Video rate camera: A camera that generates a standard analog video signal, such as RS-170, is called a video rate camera.

Image processor: An image processor modifies, analyzes, displays, and stores images.

Readout speed: The speed at which each pixel is read out of a charge-coupled device (CCD) camera. The unit is megahertz per pixel or kilohertz per pixel.

Linearity versus lag: Linearity is the relationship between incoming light to a video camera and the output signal from the video camera. If some signal generated in a detector remains in the detector after the reading, the leftover signal appears as the lag.

Frame buffer: A frame buffer (or memory plane) stores one or more complete images for further image processing.

Horizontal versus vertical resolution: The horizontal and vertical spatial resolutions are not always the same. In the case of video rate cameras, the horizontal resolution is higher than the vertical. Normally, slow-scan cooled CCDs have square pixels, and the resolution for both directions is the same.

Geometric distortion: Tube-type cameras and intensified cameras have geometric distortion. The distortion of tube cameras is due to the mechanism of the beam scanning. The distortion of intensified cameras is due to the intensifiers and relay optics, such as relay lens and tapered fiber plates.

Spectral range: The spectral sensitivity range of detectors.

Temporal resolution: The speed at which one complete image is acquired determines the temporal resolution. The temporal resolution of video rate is 33 msec. The temporal resolution of slow-scan cooled CCD cameras is low because of the slow readout speed and the time of signal integration.

Camera nonuniformity: Detectors do not have uniform sensitivity over an entire area. The camera nonuniformity of tube cameras and intensified cameras is higher than that of CCD cameras.

VIDEO CAMERAS

The following are two basic classifications of cameras: high-light-level cameras and low-light-level cameras. A common component to the two classifications is a CCD camera. This is an electronic device that incorporates a two-dimensional (2D) photodetector, an amplifier, and a timing device that produces a signal that conforms to one of the industry standards. These standards are indicated by the terms RS-170, RS-330, RS-343, NTSC, RGB, and others. This signal transfers an image from the detector to an image processor, a storage device, or display device. The details of this operation can be found in various sources.

High-Light-Level Cameras

Although most low-light-level cameras work in high light levels, these cameras are expensive. If high-light-level images are to be recorded, the use of a high-light-level camera offers benefits in both cost and convenience. The following are characteristics of high-light/CCD cameras.

- Detectors are available in 1/3-in, 1/2-in, 2/3-in, and larger formats.
- The size of the detector does not determine the number of horizontal pixels.
- Horizontal resolution is a function of the number of horizontal pixels.
- Spectral sensitivity ranges from <200 nm to 1000 nm.
- The linearity of the photometric response is very good.
- There is virtually no lag.
- They will not be damaged by high light levels.
- There is no geometric distortion.
- Horizontal resolution is greater than vertical resolution.

The following special features of the high-light camera are essential for optimal VEC:

- analog contrast enhancement to "stretch the gray levels" for low-contrast images
- analog shading correction used to compensate for illumination and camera nonuniformity
- video level indicator to adjust the proper light level for the camera

Low-Light-Level Cameras

Low-light-level cameras use one of two methods to produce an image in situations in which the number of photons is too low to create a meaningful image for the human eye. The camera must have an intensifier of some sort or the ability to integrate the photoelectrons on the surface of the detector. There are two approaches to intensification: One is the combination of image intensifiers and CCD cameras (referred to as "intensified CCD camera" [ICCD]), and the other is the inclusion of special detectors—having electron-multiplication mechanisms inside the detector (referred to as an "electron-multiplying CCD camera" [EMCCD]). The integrated approach is almost always used in CCD cameras that have cooling capability and a mechanism to minimize the readout noise (referred to as "integrating cooled CCD camera").

Intensified CCD Cameras

Advantages of ICCD Cameras

- Standard interface (RS-170, CCIR): The standard video format makes it easy to use video peripherals such as video recorders, image processors, and image processing software.
- The photocathode or the microchannel plate may be turned on and off at very high frequencies for recording very-high-speed events or for very fast time resolution.
- Fast frame rate (30 Hz for RS-170, 25 Hz for CCIR): The frame rate is fast enough for most real-time applications and focusing.

Disadvantages of ICCD Cameras

ICCD cameras are usable only in low light levels. The camera can be used for high-light applications such as DIC and phase contrast only if the light source level is adjusted low enough for the detector. Under these conditions the image quality is not as good as that of high-light detectors. The photocathode and the microchannel plate are at risk for burn-in when excessive light is applied. The damage can be minimized with a protection circuit; however, special caution is required.

ICCD cameras are a combination of a Gen. II intensifier and a CCD camera. The two components can be coupled either by a fiber optic bundle or by a set of lenses. In either case, the intensifier uses a device called a microchannel plate (MCP) as the low-noise electron amplifier. The MCP provides more than one order higher gain compared to the SIT camera.

Characteristics of ICCD Cameras

- The detector size is 2/3 in or 1 in.
- The spectral range can be 200–950 nm.
- Horizontal resolution depends on the combination of coupling and the pixel dimensions and number of pixels in the attached CCD.
- The best sensitivity is achieved in low-light applications.
- There is a possibility of damage from too much light.
- There are minimal problems with geometric distortion of images.
- There is very good linearity.
- There are very low lag characteristics.
- Horizontal resolution is greater than vertical resolution when a video camera is used, but it may be the same if a digital camera is used.
- Intensification and integration may be combined in one camera.

Photon-counting cameras are special versions of the ICCD cameras. The primary differences are the incorporation of a very-high-gain MCP device to detect single photons and a special photocathode to reduce the camera background noise, which limits detectability. This design produces tremendous gain but relatively low resolution when compared to a regular ICCD.

Characteristics of Photon-Counting Cameras

- The detector size is 2/3 in or 1 in.

- The spectral range is 200–950 nm.
- The horizontal resolution is up to 400 TV lines.
- There is very low camera background.
- The best sensitivity is achieved in low-light applications.
- There is a possibility of damage from more intense light.
- There is very low geometric distortion of images.
- There is very good linearity.
- There are very low lag characteristics.
- Real-time image processing is essential to eliminate camera noise.

The following special features for the low-light-intensified camera are essential for VIM:

- intensifier protection circuit to avoid damage from excessive light
- manual and auto sensitivity adjustment for the intensifier
- signal level indicator to adjust the proper light level for the camera or the sensitivity of the camera

EMCCD Cameras

An EMCCD camera is a CCD that has an electron-multiplying mechanism inside the detector. The mechanism of the electron multiplication is either electron bombardment (EBCCD) or electron-multiplying (EMCCD). Both provide adjustable electron gain. Because of the electron multiplication, the signal becomes relatively larger than the readout noise of the CCD, and it enables faster readout with "relatively" lower readout noise. However, the noise related to the dark current and signal itself cannot be reduced by this approach, and the camera has other noise sources such as fluctuation and variation of each pixel gain. If the electron-multiplication gain is adjusted to minimum, the detector behaves like a regular CCD camera.

Advantages of EBCCD and EMCCD Cameras

- All advantages of cooled CCDs apply to EBCCD and EMCCD cameras.
- High-speed frame rate: Electron multiplication increases the signal compared to the noise caused by fast readout. With subarray and/or binning, the camera can achieve much faster frame rates than standard video rate cameras. This is useful for applications that require a higher frame rate.

Disadvantages of EMCCD Cameras

EMCCD cameras have no major disadvantages as compared to other cameras. A minor disadvantage, however, is that the variety of cameras available, in terms of number of pixels, chip size, spectrum responses, and the like, is somewhat limited, as compared to cooled CCD cameras.

EBCCD cameras are a recent technological advance, providing a CCD back-illuminated behind a photocathode within a vacuum tube. This arrangement provides high-gain signal without the limitations of a microchannel plate. Other benefits include high spatial resolution and little chance of damage from high light exposure.

Characteristics of Intensified EBCCD Cameras

- The detector size is 2/3 in or 1 in.
- The spectral range can be 200–900 nm, depending on the photocathode material.
- Resolution depends on the pixel size and number in the CCD.

- It is best used in low-light applications, but moderate-light applications are also acceptable.
- There is a slight possibility of damage to photocathode from too much light.
- There is minimal geometric distortion of images.
- There is very good linearity.
- There are very low lag characteristics.
- There is good signal-to-noise ratio (SNR) characteristics in images.
- Intensification and integration may be combined in one camera.

EMCCD cameras are a new technology for signal amplification directly on the CCD without the requirement of a photocathode. Direct signal amplification provides high versatility, wide dynamic range, and no possibility of damage in high-light situations.

Characteristics of EMCCD Cameras

- The detector size is 2/3 in.
- Spectral range depends only on the CCD itself (usually from 350 nm to 1000 nm).
- Resolution depends on the pixel size and number in the CCD.
- They may be used in any light level.
- There is no possibility of damage to detector from too much light.
- There are no problems with geometric distortion of images.
- There is very good linearity.
- There are no lag characteristics.
- There are good SNR characteristics in images in low light, but high camera noise in high light.
- Intensification and integration may be combined in one camera.
- Camera noise and gain characteristics are extremely sensitive to camera temperature.

Integrating Cooled CCD Cameras

Integrating cameras are almost always CCD cameras. In most instances, these cameras are cooled to keep the dark current low during integration and require a slow readout that can help to keep the readout noise low to prevent the loss of low-light-level signals. This camera is also excellent for intense light conditions. Traditionally, the major limitation was the low temporal resolution inherent in their operation. The temporal (time) resolution is limited by the necessity of waiting until enough photoelectrons accumulate on the detector to make an image and the time involved in reading out all the pixel rows at the degree of precision required. Recent improvements in readout speed with low noise realize faster frame rates, and temporal resolution is not a problem for most applications. A frame buffer or computer is required to hold an image with this type of camera, because the image can be produced only after integration (not continuously).

Advantages of Cooled CCD Cameras

- Low light to high light: Cooled CCDs can be used to obtain low-light and high-light images. At both levels, cooled CCDs generate a similar quality of image, and there is no risk of detector burn in high light.
- Pixel manipulation (binning, subarray): Binning combines the pixels and handles multiple pixels as one. This increases sensitivity and reduces data size at the expense of spatial resolution, which is useful for very low-light imaging and for applications that require a higher frame rate. Subarray scan allows partial images to be read from the CCD. This maintains the original resolution and reduces data size, which is useful for applications that require a higher frame rate.

Disadvantages of Cooled CCD Cameras

- Nonstandard interface: A cooled CCD generates a digital video signal, and each cooled CCD has its own interface to a computer. This makes it difficult to use standard video peripherals such as video recorders, image processors, and image processing software. A movement to standardize interfaces, such as IEEE1394 and Camera Link, will make compatibility less problematical.
- Cooling required: CCDs are normally cooled by a thermoelectric cooler; however, secondary cooling may be required, such as water circulation for longer exposures.
- Slow frame rate: Because of slow readout and the existence of a mechanical shutter, the frame rate of a cooled CCD is slow. This limits the temporal resolution and makes focusing difficult under low light conditions. Frame transfer-type CCDs and interline CCDs are available that generate faster frame rates.

Characteristics of Cooled CCD Cameras

Because cooled CCD cameras have no standards for detector sizes or specifications, each device may have a variable number of pixels in both the horizontal and vertical directions. Resolution can be estimated by multiplying the number of pixels in each direction by a value between 0.7 and 0.9 (Kell Factor), with 0.7 being used most for microscopy applications. The number of pixels multiplied by their individual dimensions will provide the total detector area.

- Detector sizes are not standardized.
- The size of the detector does not determine the number of horizontal or vertical pixels.
- The resolution is related to the number of horizontal and vertical pixels.
- The size of the pixels is a factor in the sensitivity of the detector.
- The number of electrons that can be stored in each pixel (well depth) is a factor of the dynamic range and SNR at saturation.
- The temperature of the detector is a factor of the dynamic range and SNR.
- The readout speed is a factor of the dynamic range and SNR.
- The spectral range is 200–1000 nm.
- There is very linear photometric response.
- Slow readout may limit applications because of low frame rate.
- There is no damage from high light levels.
- There is no geometric distortion.
- Integration time may limit applications.
- Groups of pixels may be summed (binned) to provide greater sensitivity and speed at the expense of spatial resolution.
- Discrete portions of the array may be read out (subarrayed) to increase speed at the expense of field of view without changing resolution or sensitivity.
- Horizontal and vertical resolution may be equal.

IMAGE PROCESSORS

Image processors are used to improve SNR and to change contrast on the images. Several types of image processors are available for video microscopy. They are classified into three groups.

- Hardware-based real-time image processor to enable real-time integration, background subtraction, frame averaging, and LUT (lookup table). LUT converts input digital value to other digital values based on the content of the LUT. LUT is used to stretch the gray level and invert intensity (negative image). This is mainly used to improve image quality at up to 30 frames/sec and make the image perceivable. Video rate cameras generate images every 1/30 sec. If those images are processed synchronizing to the video rate, this is called real-time image processing.

- Software-based image analyzer to enable quantitative intensity measurement and/or morphological measurement.
- Combination of real-time image processor and image analyzer. If the original image quality from the camera is poor owing to low light or low contrast, the result of the image analysis will also be poor. In this case, the combination of the image processor and the image analyzer will give the best result for the analysis.

VEC and VIM require an image processor with the following features:

- real-time image averaging with background subtraction
- digital contrast enhancement using LUT

VIDEO ADAPTERS

Several considerations are important in determining the mechanical and optical interface between TV or CCD camera and microscope, and sensible transfer factors to the video detector and monitor. First, the optical alignment of the microscope must be as close to perfect as one would require for critical photomicrography (see Spector et al. 1998b). Although some deficiencies in the microscope, such as shading across the field or poor contrast, can be electronically compensated, and although even simple image-processing techniques permit removal of dust and dirt on the optics, the best video images are still obtained by presenting the best possible optical image to the camera.

Most black-and-white or single-chip color cameras are equipped with so-called C-mounts, a female thread of 1-in diameter, 32 threads per inch, and a shoulder at a distance of 0.690 in from the detector surface. A corresponding C-mount adapter is provided by the microscope manufacturers, fits onto the camera port, and will, if all tolerances (also on the camera) are maintained, place the video detector in an image plane that is parfocal to other cameras or visual observation. Similarly, a so-called ENG-mount with bayonet will parfocalize three-tube or three-chip color cameras. Some special cameras use the Nikon F-F-mount, also a bayonet.

Most modern microscopes generate a fully color-corrected intermediate image (see Spector et al. 1998b) and make it directly available for video pickup. Only older microscopes require correcting optics in the video adapter, and most of the new C-mounts with 1x transfer factor contain no optics. Only objectives and tube lenses (in the infinity system) generate the video image. Potential internal reflections are kept to a minimum.

Detector Field of View Compared with Visual Field of View

Tube and the ever-more-popular chip cameras come in different sizes, with a clear trend toward smaller and smaller chips with more densely packed picture elements to retain similar resolutions. The following table lists some typical chip sizes and the actually used diagonals.

Chip size (in)	2/3	1/2	1/3	1
Diagonal (mm)	10.7	8	5.3	15.9

Selecting such small areas from a total field of 20–25-mm diameter relieves the microscope optical system from attaining a high degree of correction for off-axis aberrations, but also severely limits the area recorded. For this reason, video adapters with C- or ENG-mounts have been developed with transfer factors of 0.5x, 0.63x, and 0.8x. Such reducing adapters can also be very useful to increase the image brightness in low-light-level conditions. Bear in mind, however, that the resolved detail in the intermediate image may no longer be recorded because the fixed pixel size of the detector may be larger than the point-to-point resolution in the image.

To ensure that the optical resolution is fully transferred to the video image, 2.5x or 4x video adapters are often used. Magnification changers or zoom systems on the microscope, or all in combination, can be very useful. The following are just two examples.

- A 10x/0.3 objective has a point-to-point resolution of ~1 mm. At a transfer factor of 1x, the pixel size must be <10 μm, and for good sampling, 5 μm. Because the individual pixels for most arrays are 10–12 μm in size, information would be lost or a higher transfer factor would be required.
- A 100x/1.3 objective resolves 0.25 μm; in the intermediate image, the point-to-point resolution required for the detector is better than 25 μm, relatively easily accomplished by most modern chip cameras or CCDs.

The electronic magnification from video detector or target to final image on the monitor naturally depends on the size of the monitor. A 14-in monitor and a 0.5-in CCD camera would result in a 43x factor. Because the video monitor is always viewed from a distance of 1–2 m, the magnification to the eye becomes again 5x–10x, and close to what one would see through the eyepieces.

VIDEO MICROSCOPY AND IMAGING PROCEDURES

Described here are the applications, system requirements, and methods for video-enhanced microscopy, fluorescence imaging, and luminescence and photon-counting imaging.

Video-Enhanced Contrast Microscopy

VEC microscopy greatly improves the contrast of the image and allows maximal resolution of the microscope. This system is used in numerous applications of cell biology, including the following:

- axonal transport studies
- monitoring neuronal growth cone activity
- cytoskeletal motion
- chromosomal motion
- near-infrared brain-section imaging
- checking electron microscope thin sections

The system requirements for VEC include the following:

- high mechanical stability
- very high optical magnifications (2x–40,000x)
- maximum-numerical-aperture (NA) objectives
- maximum-NA condensers
- minimum-strain or strain-free optics
- high-quality polarizers and analyzers (e.g., for DIC and polarized light microscopy)
- high-quality infrared blocking filters (except in the case of infrared brain-section imaging)
- microscope stages with very fine gear ratios
- rotating stage mount
- high-intensity, stabilized, uniform illumination
- high-resolution video camera
- analog gain and offset video controls
- real-time image processor with background subtraction and digital contrast enhancement

Procedure A: VEC Employing DIC Microscopy

1. Establish Köhler illumination (see Chapter 1).

   Use color LUTs in the image processor to help evaluate the evenness of the illuminator because the eye is more sensitive to color changes than to gray-level changes.

   The use of fiber optic scramblers (see Spector et al. 1998b) is highly recommended.

2. Establish DIC (see Chapters 1 and 3); the condenser aperture diaphragm may be closed if needed at this point.

   Closing the condenser aperture diaphragm will help increase not only the contrast, but also the depth of field to help find the sample.

3. Switch the beam splitter to video position.

   A beam splitter with 100% transmission to the video port is recommended for two reasons. It will increase the signal to the camera and eliminate possible external images from the eyepieces (due to room lights, etc.) from being superimposed on the image of the sample.

4. Open the aperture diaphragm to its maximum to obtain the highest NA (see Spector et al. 1998b).

5. Adjust analog gain and offset knobs to increase video contrast as needed.

6. Recheck illuminator evenness and adjust as needed (see notes to Step 1).

7. Increase the magnification with optical or mechanical means to final required magnification.

   It is sometimes advisable to initially set up at a lower magnification to aid in finding the sample because of the larger field of view.

   Increasing optical magnification can be done by changing the relay eyepiece or an optical intermediate lens, called an optovar.

   It is also possible to increase the magnification by moving the camera farther away from the microscope. This "draw tube" method has the advantage of continuous variable magnification, but at the expense of optical corrections, parfocality, and free working distance. Because monochromatic illumination is generally used, the loss of optical correction is not important.

   The loss of parfocality is only of concern when changing objectives. The operator must take precautions to prevent the possible striking of objectives into the sample or coverslip.

   The change in working distance at the front of the objective can be important when using thick samples or coverslips.

8. Carefully move the sample to an adjacent, but specimen-free, area and defocus slightly.

   If the working distance of the objective permits, focus into the glass of the slide below the sample. This area is sure to be free of dirt, scratches, or other defects that will adversely affect the background image.

9. Acquire the background image with an image processor.

10. Return to the live image, refocus the specimen, and return to the area of interest.

11. Start background subtraction function from the live image. Any shading, dirt, or illumination defects inherent in the optical system should disappear from the final "background subtracted image."

12. Use the digital contrast enhancement mode of the image processor to "STRETCH" the gray levels of the image to maximize contrast.

    Use the arrows next to "HI" and "LOW" to change the values. Decreasing the value of "HI" will increase the brightness of light areas in the image. Increasing the value of "LOW" will decrease the brightness of dark areas. This combination results in the remaining gray levels within the image being separated by more shades, thus increasing contrast.

    Stretching is accomplished by adjusting the gray levels in the intensity histogram of the image. Reassign the brightest available gray level from the image—gray level 100, for example—to the brightest available gray level in the output histogram, gray level 255. Then reassign the darkest gray level in the image—gray level 50, for example—to the darkest gray level in the output histogram, gray level 0. The gray levels between 100 and 50 will now be automatically assigned new gray levels in a linear manner between 0 and 255 in the output. This change or stretching separates the previously similar gray levels, making them easier to perceive.

FLUORESCENCE IMAGING

The equipment used for VIM intensifies and integrates the image, even in low-light situations. The applications include the following:

- immunofluorescence
- autofluorescence of plant tissues
- forensic medicine
- contaminant inspection
- bone-growth studies using tetracycline
- microcrack evaluation of materials

The following are the system requirements:

- high mechanical stability
- maximum-NA objectives
- high-quality excitation filters, dichroic mirrors, and barrier filters

  Quality is a function of spectral selectivity, plane parallel surfaces, and precise mounting angles.
- good infrared blocking filters
- high-intensity, stabilized, and uniform illumination
- high-sensitivity video camera with fast overload protection circuit *or* a cooled CCD camera
- analog gain and offset video controls
- real-time image processor with image averaging and digital contrast enhancement

Procedure B: Epifluorescence Microscopy

1. Establish Köhler illumination in reflected light using a sample with a large area of fluorescing tissue or cells.

   A routine pathology slide with liver tissue stained with hematoxylin and eosin is a good choice because this stain will fluoresce with almost any filter set, and liver tissue is very homogeneous.

   The adjustment of the illuminator for evenness is critical and can be best imaged using color LUTs to help the eye perceive subtle intensity differences (i.e., use color contrast rather than gray scales).
2. Change to the sample to be studied, select appropriate objective, and focus.

   The ideal objective for fluorescence is the one that has the lowest ratio of magnification divided by NA, because brightness increases with the square of decrease in magnification and the square of increase in NA.
3. Make sure to adjust the field diaphragm of the epi-illumination system (see Chapter 1) to illuminate only the area visible with the video camera to prevent fading in the surrounding areas.
4. Adjust the sensitivity control of intensified type cameras until just before video saturation *or* adjust the exposure time of cooled CCD until just before full well capacity is reached.
5. Adjust the analog gain and offset of the camera (if possible) to maximize contrast in the image before digitization in digital cameras.
6. Recheck the illuminator evenness and adjust it as needed (see note to Step 2).

7. Using an image processor, pick the frame-averaging function and select the number of frames that are required to increase the SNR to an acceptable level or apply binning, subarray, and/or gain as needed.

   The SNR in the image can be seen as rapidly flashing bright points in each video frame when using intensified cameras. Although commonly referred to as "noise," these points actually reflect the photon variability within the image under very low light conditions.

8. Use the digital contrast enhancement of the image processor to "STRETCH" the gray levels of the image to maximize the contrast or adjust the LUT of software to increase the contrast as needed.

   Stretching is accomplished by adjusting the gray levels in the intensity histogram of the image. Reassign the brightest available gray level from the image—gray level 100, for example—to the brightest available gray level in the output histogram, gray level 255. Then reassign the darkest gray level in the image—gray level 50, for example—to the darkest gray level in the output histogram, gray level 0. The gray levels between 100 and 50 will now be automatically assigned new gray levels in a linear manner between 0 and 255 in the output. This change, or stretching, separates the previously similar gray levels, making them easier to perceive.

LUMINESCENCE AND PHOTON-COUNTING IMAGING

Photon-counting imaging is particularly useful for detecting single-photon events that occur, for example, in biological luminescence systems. Applications include the following:

- monitoring gene expression with luciferase reporter genes
- calcium ion imaging using aequorin (see Chapter 27)
- ATP or glucose imaging in tissue
- real-time visualization of oxyradical burst activities

The following are the requirements of the system:

- a microscope that has 100% transmission to video camera port
- maximum-NA objectives with high transmission at the luminescent wavelength
- photon-counting video camera with single-photon detection capability and low camera background
- image processor with photon-counting and image-overlay capability
- RGB color monitor
- light-tight room for the microscope to minimize stray light-noise problems

The photon-counting camera generates images that have spots corresponding to the position of photon hits. An image processor uses the threshold method to extract the spots, and the spots are converted to the digital value of "1," whereas the other area is converted to the value of "0" (binary image). The "slice" image is the accumulation of the extracted spots. The spots spread to more than 1 pixel, and the total counts of the slice image are more than the number of actual hits of the photons. If the center of gravity is calculated for each spot and the results are accumulated, the image is called a "gravity" image. The total counts of the gravity image and the number of actual hits are the same. The photon-counting camera generates some spots even in dark conditions. The accumulated image in the dark condition is called the "dark" image. Pixel depth is the number of bits of the frame buffer or memory plane; 8 bits and 16 bits are standard. The number of bits determines the number of gray levels (8 bits, 256 gray levels; 16 bits, 65,536 gray levels).

Procedure C: Photon-Counting Imaging

1. Turn off the high voltage of the photon-counting video camera and focus on the specimen through the eyepieces using whichever illumination technique is appropriate for the sample.

The photon-counting camera is very sensitive and fragile. Keep the high voltage OFF except when actually imaging.

2. Reduce to minimum the illuminating source in the microscope, and switch 100% of the light to the camera using the beam splitter.
3. Make sure that there are no unnecessary optics (polarizer, fluorescence filter, beam splitter, etc.) between the objective lens and the video camera port.
4. Set the sensitivity of the photon-counting video camera to 0 and turn on the high voltage. If the image is too bright, the automatic protection circuit shuts off the high voltage. In this case, reduce the illumination light level using neutral density or color filters (see Fig. 94.1 in Spector et al. 1998b) and turn on the high voltage again. If the image is too dark, increase the sensitivity of the photon-counting camera.
5. Choose the resolution of the image.

   The resolution selected for the image depends on the resolution required to see the details and the amount of intensity within the object. It is possible to use a smaller array of pixels if the object does not require higher resolution. This means that available memory can be subdivided into more individual memory planes, and less disk space will be needed for storage of each image. If the number of photons per pixel is <256, it is also possible to change the depth of each pixel to 8 bits from the default value of 16 bits. This has a similar effect on the memory by allowing more memory planes and requiring less disk space for storage.

6. Integrate 64 frames of the image and make contrast changes using histogram stretching. Then save the image as the reference or background image.
7. Turn off the illumination of the microscope, and increase the sensitivity of the photon-counting camera to 10.
8. Adjust the light path of the microscope so that no light will reach the camera.
9. Set the discrimination level of photon counting at 100, and select an integration time of 1 min to create an image of the "dark noise" of the camera.
10. Select a memory number in which to store this "dark" image.
11. Integrate this dark image for 1 min, and make a note of the total count of photons or events.

    The camera background is usually <10 counts/sec. More background is usually caused by a light leak or illuminator afterglow. Check the box by draping it with a dark cloth and turning off all the room lights. If the count goes down, there is a light leak. If not, try putting an opaque object over the field diaphragm and a drape over the lamp housing vents, as a test to try to eliminate background light. *Do not operate the illuminator in this!* The drape will burn.

12. Set the light path back to 100% to the camera.
13. Integrate the sample without the reporter (e.g., luciferin or aequorin) for 1 min to check stray light conditions. Counts should be similar to the dark count reading.

    The same checks should be made for light leaks and illuminator problems if the count is now much different from the dark count. If the illuminator is the problem, it will be necessary to provide an additional light baffle between the microscope and the illuminator.

14. Select the integration time for the sample and set memory output locations.

    The integration time for a typical sample must usually be determined by trial and error. The simplest way is to integrate for periods of a minute or two at a time, recording the counts and continuing for another period until the area of the photon concentration can be distinguished from the background. If no object is distinguished after 30 min of integration, there is very little chance that luminescent objects will be detected with additional integration.

15. Select memory locations in which to store images of the samples.
16. After integration, use histogram stretch to enhance the image.
17. Superimpose the photon-counting image onto the reference image if necessary.

REFERENCES

Spector DL, Goldman RD, Leinwand LA, eds. 1998a. Heterologous expression of the green fluorescent protein. In *Cells: A laboratory manual*, Vol. 2. *Light microscopy and cell structure*, pp. 78.1–78.21. Cold Spring Harbor Laboratory Press, Cold Spring Harbor, NY.

Spector DL, Goldman RD, Leinwand LA, eds. 1998b. Light microscopy. In *Cells: A laboratory manual*, Vol. 2. *Light microscopy and cell structure*, pp. 94.1–94.53. Cold Spring Harbor Laboratory Press, Cold Spring Harbor, NY.

Spector DL, Goldman RD, Leinwand LA, eds. 1998c. Video microscopy and image enhancement. In *Cells: A laboratory manual*, Vol. 2. *Light microscopy and cell structure*, pp. 95.1–95.15. Cold Spring Harbor Laboratory Press, Cold Spring Harbor, NY.

3 Differential Interference Contrast Imaging of Living Cells

Noam E. Ziv and Jackie Schiller

Rappaport Institute and the Technion Faculty of Medicine, Haifa 31096, Israel

ABSTRACT

Differential interference contrast (DIC) microscopy is a contrast method for resolving fine details in biological samples that are nearly transparent and are, therefore, ill suited for viewing by conventional (bright-field) light microscopy. In DIC microscopy, local differences in the index of refraction and specimen thickness are converted into changes in brightness, so that borders of cells and organelles, as well as their fine structure, stand out with high contrast. DIC microscopy can be applied to cell cultures, brain slices, and even intact organisms (such as embryos). In combination with infrared (IR) light and video microscopy, DIC microscopy is very useful for resolving individual neurons in brain slices. Although fluorescence microscopy and genetically encoded fluorescent moieties have become the tools of choice for live imaging, DIC microscopy is still very useful for complementing fluorescence microscopy data and for performing electrophysiological recordings and micromanipulations in living cells and tissues. This chapter describes the use of electronically enhanced DIC. Protocols for DIC imaging of cells using an inverted microscope and for IR imaging of brain slices using an upright microscope are presented.

INTRODUCTION

Ever since Robert Hooke and Antonie van Leeuwenhoek peered into their handcrafted microscopes, biology and microscopy have been practically inseparable. Ironically, however, important types of biological samples, in particular, those of mammalian origin, are not ideally suited for conventional (bright-field) microscopy, simply because they are nearly transparent, and thus the contrast provided by the differential absorbance/scattering of light passing through such samples is grossly unsatisfactory. This situation has driven the development of microscopy contrast methods that are based on other changes that occur to light as it passes through biological material. One of the most successful methods is differential interference contrast (DIC) microscopy, also known as Nomarski microscopy. In DIC microscopy, local differences in the index of refraction and specimen thickness are converted into changes in brightness in such a manner that the borders of cells and organelles (i.e., steep gradients of index of refractions and/or tissue thickness), as well as their fine structure, stand out with striking contrast. DIC microscopy is based on a pair of special prisms (Wollaston or Nomarski prisms) that split each light wave from the illumination source into two waves that are slightly displaced relative to one another (less than one Airy disk diameter) and have orthogonal polarizations. After passing through the specimen, the two waves are recombined in such a manner that differences in the effective paths they experienced (that alter their phase relative to each other) are translated into changes in amplitude (intensity). As a result, changes in the quantity ([index of refraction × tissue thickness] per unit length) (hence the term differential) are translated into positive or negative deviations from background intensity levels, resulting in images with a distinctive shadow-cast appearance, giving the impression that the specimen is illuminated obliquely from above.

Traditionally, DIC microscopy is combined with electronic imaging devices (mainly video cameras) that provide a second stage of contrast enhancement. These enhancements include electronic signal gain (which allows the amplification of weak optical contrast), electronic subtraction of background signals (which allows the use of optimal DIC settings without the penalty of overly bright backgrounds), and digital image processing.

For further details concerning the manner by which DIC microscopy is realized and for in-depth explanations of DIC principles, please see Chapter 1 or visit the Molecular Expressions website (http://www.microscopy.fsu.edu/primer/techniques/dic/dichome.html).

DIC microscopy can be applied to cell cultures, brain slices, and even intact organisms (such as embryos). It is extremely useful for resolving individual cells and cellular organelles in live unstained tissue. Furthermore, high-quality images of living cells in brain slices can also be obtained using a combination of DIC optics and IR video microscopy (Dodt and Zieglgänsberger 1994). The data obtained are valuable in their own right mainly for studying development, cellular dynamics, and intracellular trafficking. Recently, however, with the widespread use of fluorescence microscopy, DIC microscopy has also become very useful for complementing fluorescence microscopy data or for visually directing microelectrodes and micropipettes used for electrophysiological recordings and micromanipulation.

This chapter describes the use of electronically enhanced DIC imaging techniques for the study of living cells. Protocols for DIC imaging of cells using an inverted microscope and for IR imaging of brain slices using an upright microscope are presented.

IMAGING SETUP

Glass Substrates

Contrast in DIC microscopy is derived from differences in the refractive index and in the optical path length experienced by pairs of light waves whose polarization planes are perpendicular to each other. Thus, in general, materials that affect light polarity cannot be used. In practice, this means that specimens cannot be grown or mounted on standard plastic substrates (such as polystyrene Petri dishes or culture flasks) and must be grown and mounted on nonpolarizing materials, typically glass.

Cultured cells are usually grown on the following:

- no. 1 glass coverslips, usually maintained in six-well plates or 35-mm Petri dishes until used for experiments
- glass-bottomed Petri dishes from commercial sources (such as MatTek, World Precision Instruments, or Ted Pella, Inc.) or made in-house by attaching coverslips with Sylgard 184 (Dow Corning Corp.) or a hot-melt glue gun to the bottom of polystyrene Petri dishes in which a hole has been bored
- glass-bottomed chambers (such as Lab-Tek chamber slides and chambered cover glass, Nunc/Thermo Scientific)

Brain slices are usually mounted on glass-bottomed chambers, which can be custom made or procured from commercial sources (e.g., Luigs & Neumann, Scientific Systems Design, Inc., or Harvard Apparatus).

Microscopes

Microscopes, both inverted and upright, with suitable DIC optics can be purchased from all major manufacturers. The major components are as follows:

- a polarizer in front of the light source (typically a halogen lamp)
- an analyzer (another polarizer) in front of the imaging device
- objectives suitable for DIC microscopy
- modified Wollaston (Nomarski) prisms for each objective
- matched Wollaston prisms within the condenser

Electronic Imaging Devices

The imaging devices used for DIC are usually high-quality monochrome video cameras (typically charge-coupled devices [CCDs]), with high refresh rates (video rates or similar) and manual gain and offset adjustments. For IR DIC, the imaging device must have adequate sensor sensitivity in the IR range. Recent developments in CCD-based imaging devices (namely, electron-multiplying CCDs), have led to the appearance of extremely sensitive CCD cameras that can be used for both fluorescence and DIC microscopy, reducing the alignment problems associated with the use of separate cameras for each imaging mode. Note, however, that on laser-scanning confocal microscopes, alternative sensors can be used (see the section Simultaneous DIC and Fluorescence Imaging).

Protocol A

DIC Imaging of Cells in Culture Using an Inverted Microscope

DIC microscopy is an excellent imaging method for rendering contrast in transparent specimens.

MATERIALS

Reagents

Cells grown on glass substrate

Equipment

Camera
Coverslips, glass
Inverted microscope with DIC optics

EXPERIMENTAL METHOD

1. Mount the chosen cells (grown on a glass substrate) onto the microscope stand.
2. Place a coverslip over the cells, avoiding air bubbles and liquid spillover. If necessary, use spacers to keep the coverslip away from the cells.
3. Verify that all DIC optical elements (polarizer, analyzer, and Wollaston prisms) are introduced into the optical path.
4. Observe the cells through the oculars, focus, and adjust the condenser for Köhler illumination (Chapter 1), as follows: Close the field diaphragm, and adjust the condenser height until the field diaphragm is in focus. Center the diaphragm, and reopen it.
5. Adjust the bias retardation using the objective Wollaston/Nomarski prism (slider), or the de Sénarmont compensator, if the microscope is equipped with one.
6. Adjust the camera offset and gain settings until most of the camera's dynamic range is used.
7. Collect images.
8. Optionally, at the end of the experiment, move the specimen to a relatively featureless region, defocus the image, and collect an image (a mottle image) at the same settings used throughout the experiment.
9. Subtract the image collected in Step 8 from each of the other images collected during the experiment to digitally correct for shadowing and blemishes within the field of view.

Protocol B

IR DIC Imaging of Brain Slices Using an Upright Microscope

DIC microscopy can be combined with IR video to generate high-quality images of living cells in brain slices.

MATERIALS

Reagents

Brain tissue slices

Equipment

Imaging chamber
Upright microscope with IR DIC optics
U-shaped grid, made from platinum wire
Video camera

EXPERIMENTAL METHOD

1. Mount the tissue slice in the chamber, and fix it in place using a metal U-shaped grid made from platinum wire.
2. Adjust the condenser for Köhler illumination, and open the field aperture to illuminate the whole field of view. This step is usually performed while observing the specimen through the oculars.
3. Keep the aperture iris diaphragm open—contrast will be enhanced using the video camera contrast enhancement functions.
4. Place the IR filter in position, and switch the light path to the video camera.
5. Increase the offset and gain of the camera to maximize the use of its dynamic range. Avoid saturation.
6. Adjust the camera's shading corrections to obtain an evenly illuminated field of view.
7. Collect images.

   To obtain further magnification of the image, an extra magnification lens can be inserted into the light path (depending on the microscope configuration) or in front of the camera.

SIMULTANEOUS DIC AND FLUORESCENCE IMAGING

Although DIC microscopy can provide images at an excellent level of detail, the technique is more informative for cellular structures than for specific molecules. Fluorescence microscopy, on the other hand, provides detailed information on the distribution patterns of specific molecules and ions, but this information is often difficult to interpret in the absence of additional structural information. Combining fluorescence microscopy with DIC imaging is an excellent way to establish the structural context of fluorescence data. Unfortunately, combining DIC imaging with fluorescence microscopy using video cameras can be awkward. Light sources must be changed, various components must be switched in and out of the light path, and a separate camera is often used for each imaging mode.

On a laser-scanning confocal microscope, however, DIC and fluorescence imaging can be performed simultaneously at practically no cost in terms of exposure (Ryan et al. 1990, 1993). A photodiode or a photomultiplier is placed at the rear end of the condenser (at a conjugate location on the illumination lamp), and the light used for epifluorescence excitation is collected after it travels backward through the condenser (Fig. 1A). Most lasers used for confocal microscopy emit polarized light, so the DIC analyzer is not required (which is fortunate as this component would absorb much of the fluorescence emitted by the specimen). Because the objective Wollaston/Nomarski prism has only a marginal impact on the fluorescence signal, it can be left in the light path. The signal generated by the transmitted light detector is fed into the digitizing circuitry of the confocal microscope and is displayed. As both the fluorescence and the DIC images are generated by raster scanning the focused laser beam over the specimen, the DIC and fluorescence images generated are in perfect registration. Images obtained in this fashion are shown in Figure 1, B–G. After contrast adjustment (Fig. 1C), the fluorescence and DIC images can be merged (Fig. 1G), revealing the structural context of the fluorescence data.

Utilization of the raster-scanned excitation light for generating contrast images has recently been extended to two-photon microscopy, where it is particularly useful for visualizing neuronal cell bodies and dendrites in brains slices and for targeting micropipettes for electrophysiological recordings. One contrast method used in such a fashion is called Dodt gradient contrast (Dodt et al. 1998; Wimmer et al. 2004; Nevian et al. 2007; components can be purchased from Luigs & Neumann). The IR excitation light of the pulsed laser, primarily used to excite fluorescent molecules, is also collected via a high-numerical-aperture condenser, passed to a photodetector, and used to create a contrast image. Here, too, the fluorescence and contrast images are in perfect registration, enabling high-accuracy electrode positioning and identification of small structures. It should be noted that the DIC-like appearance of images obtained using Dodt gradient contrast stems from preferential collection of oblique light, not from interference of orthogonally polarized beam pairs as in true DIC. Because no contrast-generating elements are placed in the epifluorescence light path, fluorescence light collection remains optimal while contrast images are of sufficient quality for detecting neuronal structures and guiding microelectrodes.

ADVANTAGES AND LIMITATIONS OF DIC MICROSCOPY

The major disadvantage of DIC microscopy is its requirement for glass substrates. More convenient substrates, such as polystyrene, are not compatible with DIC because of their anisotropic polarization optical properties. A relatively recent development is PlasDIC (developed and sold by Zeiss), which is a modification of DIC microscopy that can be used to image cells growing on plastic dishes (Danz et al. 2004). However, PlasDIC (in common with phase contrast and Hoffman modulation contrast imaging techniques) involves the placement of special inserts in the illumination path. This differs from classical DIC, in which the objective and condenser apertures are not obstructed, allowing for maximal image resolution. Nevertheless, new contrast methods compatible with conventional cell culture dishes are clearly useful under many circumstances and are nearly guaranteed to increase in popularity.

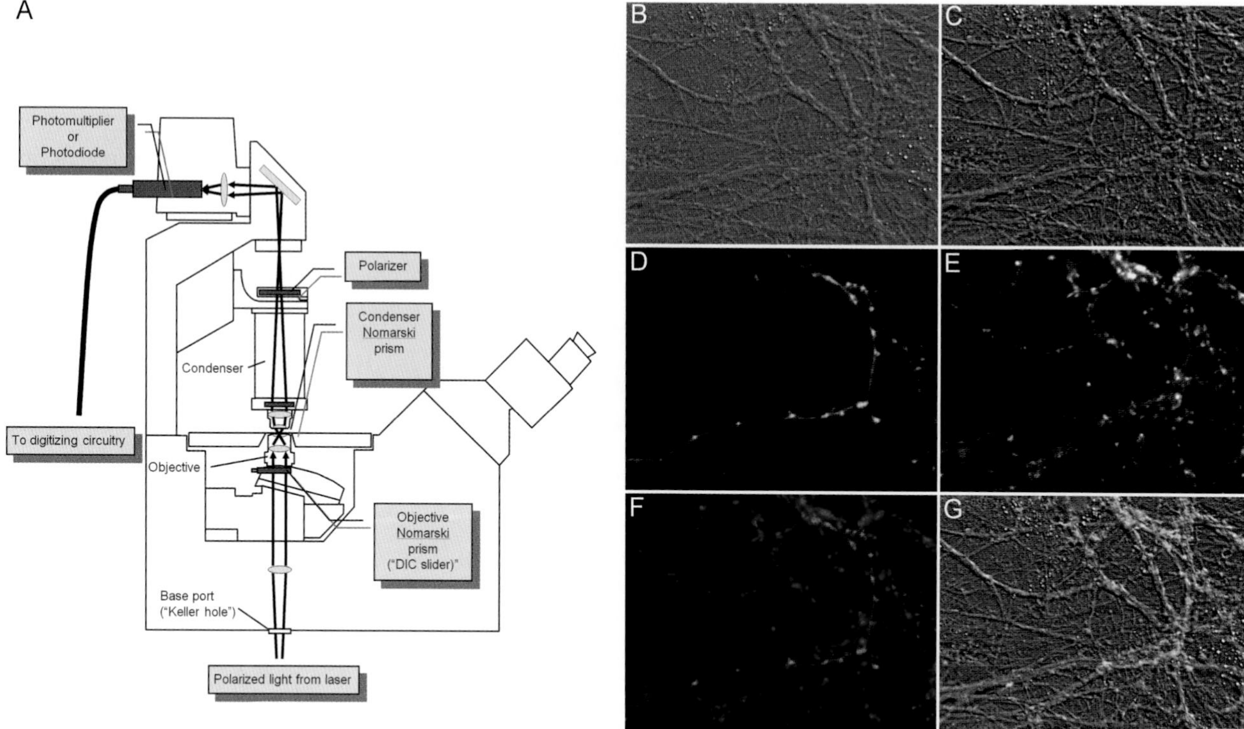

FIGURE 1. Simultaneous laser-scanning epifluorescence and DIC imaging. (*A*) A laser-scanning confocal microscope equipped for DIC microscopy. As in many confocal systems, the specimen is raster scanned by a focused beam of laser light that serves to excite fluorescent substances in the specimen. The emitted light is collected by the objective and is routed to photomultipliers that convert changes in light intensity to changes in an electrical signal (not shown). Here, the light passing through the specimen is also collected by the microscope condenser and is focused onto a separate photodetector (a photodiode or, preferably, a photomultiplier). The signal from this detector is then fed into the digitizing circuitry concomitantly with signals arising from the photomultipliers used for measuring the fluorescence. This results in the DIC and epifluorescence images being in perfect registration. (*B*) A DIC image of cultured rat hippocampal neurons (3 wk in culture). Note the intercrossing network of dendrites and axons growing on top of a glial monolayer. (*C*) The same image after contrast enhancement. (*D*) A fluorescence image of the same field of view showing axons originating from a neuron expressing a green fluorescent protein–tagged (GFP-tagged) variant of the presynaptic protein Rim1 (generously provided by Magdalena Zurner and Susanne Schoch, University of Bonn, Germany). (*E*) A fluorescence image of the same field of view showing presynaptic boutons labeled with the fluorescent dye FM4-64. (*F*) A merged image of GFP-tagged Rim1 (green) and FM4-64 (red). (*G*) The enhanced DIC image overlaid with the fluorescence data. Note that all images in *B–G* were collected simultaneously during one raster scan, ensuring their perfect registration, minimizing exposure to the excitation light source, and demonstrating the ability to collect DIC images at no extra cost in terms of specimen illumination. Scale bar, 10 μm. (Images collected by Magdalena Zurner and Noam Ziv.)

Images produced in DIC microscopy have a distinctive shadow-cast appearance, as if they were illuminated from an angle and from above. It is important to be aware that the highlights and shadows do not necessarily represent actual topographical structures.

Although DIC microscopy provides little information on the molecular level, it can provide extraordinarily detailed information on subcellular structures and their dynamics (see Forscher et al. 1987), and thus DIC microscopy is particularly well suited for studying cell motility and organelle dynamics.

REFERENCES

Danz R, Vogelgsang A, Käthner R. 2004. PlasDIC—A useful modification of the differential interference contrast according to Smith/Nomarski in transmitted light arrangement. *Photonik* 1/2004.

Dodt HU, Zieglgänsberger W. 1994. Infrared videomicroscopy: A new look at neuronal structure and function. *Trends Neurosci* **17:** 453–458.

Dodt HU, Frick A, Kampe K, Zieglgansberger W. 1998. NMDA and AMPA receptors on neocortical neurons are differentially distributed. *Eur J Neurosci* **10:** 3351–3357.

Forscher P, Kaczmarek LK, Buchanan JA, Smith SJ. 1987. Cyclic AMP induces changes in distribution and transport of organelles within growth cones of *Aplysia* bag cell neurons. *J Neurosci* **7:** 3600–3611.

Nevian T, Larkum ME, Polsky A, Schiller J. 2007. Properties of basal dendrites of layer 5 pyramidal neurons: A direct patch-clamp recording study. *Nat Neurosci* **10:** 206–214.

Ryan TA, Sandison DR, Webb WW. 1990. Simultaneous DIC and fluorescence in laser scanning confocal microscopy. *Biophys J* **57:** A374–A374.

Ryan TA, Reuter H, Wendland B, Schweizer FE, Tsien RW, Smith SJ. 1993. The kinetics of synaptic vesicle recycling measured at single presynaptic boutons. *Neuron* **11:** 713–724.

Wimmer VC, Nevian T, Kuner T. 2004. Targeted in vivo expression of proteins in the calyx of Held. *Pflugers Arch* **499:** 319–333.

WWW RESOURCE

http://www.microscopy.fsu.edu/primer/techniques/dic/dichome.html Molecular Expressions Optical Microscopy Primer, Specialized Techniques.

4 Infrared Video Microscopy

Hans-Ulrich Dodt,[1,2] Klaus Becker,[1,2] and Walter Zieglgänsberger[3]

[1]*Department of Bioelectronics, FKE, Vienna University of Technology, 1040 Vienna, Austria;* [2]*Bioelectronics, Center of Brain Research, Medical University of Vienna, 1090 Vienna, Austria;* [3]*Neural Network Dynamics, Max-Planck-Institute of Psychiatry, 80804 Munich, Germany*

ABSTRACT

This chapter describes how neurons and neuronal excitation can be visualized in brain slices. Infrared video microscopy can be used with all kinds of brain slices up to a thickness of 500 µm. Infrared imaging can also be used for visualizing the intrinsic optical signal. To image the intrinsic optical signal, it is necessary to preserve long-range axonal projections, for example, in slices of the neocortex.

INTRODUCTION

Infrared Video Microscopy Setup

Single neurons in thick brain slices cannot be seen with standard microscopy because the neuronal network consists of a large number of neurons packed closely together. These cells act as birefringent-phase objects that scatter light very effectively. Any method for visualization of neurons in brain slices must, therefore, reduce this scattering of light. Light scattering can be reduced by increasing the wavelength of illumination, by the optics used for contrast generation, and, indirectly, by electronic contrast enhancement (Dodt 1992).

Infrared Illumination

The first reduction of light scattering is achieved by the use of near-infrared (near-IR) radiation instead of visible light (Dodt and Zieglgänsberger 1990). IR radiation is scattered to a lesser extent than visible light, because of its longer wavelength. Standard halogen lamps serve adequately as light sources for IR radiation because their peak emission is in the near-IR range. The wavelength of illumination can be selected by placing a broadband interference filter in the filter holder of the microscope.

Microscope Setup

The microscope setup is built around an Axioskop FS microscope (Zeiss). For patch-clamping, water-immersion objectives with long working distances and high numerical apertures (NAs) are required. The procedure discussed here uses an Olympus 60x (NA 0.9) water-immersion objective with a 2-mm working distance. This objective can be used without any noticeable reduction in image sharpness on Zeiss microscopes.

CONTRAST SYSTEM

Unstained neurons in brain slices are phase objects. To render them visible, their phase gradients have to be converted into amplitude gradients by the optics, which also have to provide optical sectioning.

The investigators use the gradient-contrast system (Dodt et al. 1998, 2002). The aperture plane of the condenser is reimaged with a lens system between the rear of the microscope body and the lamphouse, making it accessible for spatial filtering (Luigs & Neumann). A light stop in the form of a quarter annulus is positioned in the illumination beam path. At a small distance downstream from the annulus, a diffuser is placed, generating a "gradient" of illumination across the condenser aperture plane. No spatial frequencies in the illuminating light are completely filtered out so that the image remains similar to the object. In addition, the curved form of the slit gives a gradient of illumination in two perpendicular directions in the image, left to right and up to down. This can be very helpful for visualization of dendritic branches running in different directions in the image. Because the light stop blocks much of the illuminating light, only a part of the normal, illuminating light cone is used and, therefore, less stray light is generated in the slice. This is the same principle used in the slit lamps for ophthalmology. The contrast generated this way is so high that gradient contrast alone allows visualization of neurons in thick slices even with visible light by the naked eye.

ELECTRONIC CONTRAST ENHANCEMENT

Contrast enhancement in real time is most easily achieved by the use of modern low-noise, cooled charge-coupled device (CCD) cameras (e.g., C9300-201 for patch-clamp alone or ORCA R2 if fluorescence is also needed). Both cameras are manufactured by Hamamatsu. Similar IR cameras are also available from DAGE.

MICROMANIPULATORS

The experimental setup must allow simultaneous visualization and patch-clamping of neurons. After patch-clamping a neuron, it must be possible to move the microscope relative to the brain-slice chamber. This can be achieved by setting the microscope on a two-axis translation stage (Luigs & Neumann). The micromanipulator holding the patch pipette must also be motorized. Motorizing the translation stage of the brain-slice chamber, and the translation stage and focus drive of the microscope, is helpful.

INFRARED–DARK-FIELD MICROSCOPY SETUP

A further development of IR video microscopy allows the spread of neuronal excitation to be visualized directly at a macroscopic level. The brain slice is illuminated with a dark-field condenser, and the image of the slice in IR scattered light is projected by a low-power objective onto the target of the video camera (Dodt and Zieglgänsberger 1994). Because light scattering of brain slices changes during neuronal excitation, the spread of neuronal excitation can be visualized by purely optical means, without the use of any dyes (MacVicar and Hochman 1991). To visualize the small (a few percent) light-scattering changes, a differential subtraction technique is used. The image of the brain slice at rest is stored in the computer memory and subtracted, online, from the image of the slice during electrical stimulation. By strong digital contrast enhancement, areas of neuronal excitation in the brain slice become visible. These areas are then overlaid onto the black and white image of the brain slice by custom-made software.

Protocol A

Patch-Clamping in the Infrared

Patch-clamping in the infrared has become extremely popular. Use of the gradient contrast system allows for electrical recording from the finest neuronal processes such as basal dendrites.

MATERIALS

Reagents

Brain slice

Equipment

Axioskop FS microscope (Zeiss)
IR broadband interference filter ($\lambda = 780 \pm 50$ nm, type KMZ 50-2, Schott)
IR light source
IR video camera
Patch-clamp rig

EXPERIMENTAL METHOD

1. Adjust the microscope for correct Köhler illumination, as follows.
   i. Bring a test object (e.g., a thread of the grid made for holding the brain slice) into focus with the 60x objective.
   ii. Close the field diaphragm at the foot of the microscope, and adjust the position of the condenser with centering screws and focusing mechanisms. In the correct position, the diaphragm appears as a sharp round hole in the center of the field of view. Do this without the IR filter in visible light, while looking through the eyepieces of the microscope.
2. Rotate the IR broadband interference filter ($\lambda = 780 \pm 50$ nm) in the filter holder of the microscope into place.
3. Place the brain slice in the recording chamber and bring the neurons into focus.
4. For patch-clamping, search first at a lower magnification for "good" neurons in the slice, then use higher magnification for approaching the selected neuron with the patch pipette. This can be achieved by placing an additional magnification changer in front of the video camera, allowing additional magnifications of 1x and 4x the magnification of the objective (Luigs & Neumann).

Protocol B

IR–Dark-Field Microscopy to Visualize the Intrinsic Optical Signal

The Axioskop is also used for IR–dark-field microscopy with a low-power objective.

MATERIALS

Reagants

Brain slice
Krebs-Ringer solution

Equipment

Axioskop FS microscope (Zeiss)
Concentric stimulation electrode (SNX-100, Rhodes Medical Instruments, Canada)
Dipping cone
IR broadband interference filter ($\lambda = 780 \pm 50$ nm, type KMZ 50-2, Schott)
IR light source
IR video camera
Patch-clamp rig

EXPERIMENTAL METHOD

1. Place a dipping cone on a 2.5x objective (NA 0.12) to avoid fluctuations of light intensity by changes of the fluid level in the slice chamber.
2. Use the condenser with a NA of 0.32.
3. Place a circular light stop (diameter 5 mm) in the condenser before the 0.32-NA condenser lens.
4. Submerge the slice in the recording chamber in Krebs-Ringer solution, and lower a concentric stimulation electrode onto the slice.
5. Use tetanic stimulation (50 Hz for 2 sec) of a few volts to elicit the intrinsic optical signal (IOS).

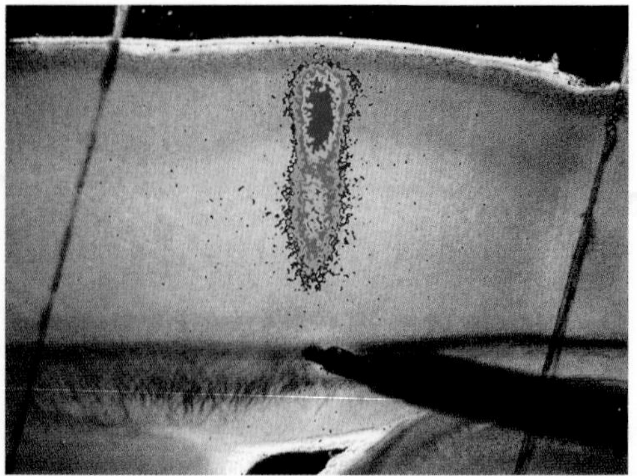

FIGURE 1. (*A*) Neuronal network in lamina 2/3 of the adult rat somatosensory neocortex, as seen with IR video microscopy using gradient contrast. Two pyramidal neurons, covered with small beaded structures, are visible on the right-hand side of the picture. A dendritic bundle on the left-hand side is overlaid with a dense fiber network, presumably axons. By focusing the microscope through different focal planes, a three-dimensional impression of the neuronal network can be obtained. Image width, 100 μm. (*B*) Neuronal excitation in a neocortical slice visualized by IR dark-field video microscopy. Dark-field image of a sagittal neocortical slice overlaid with the IOS in the slice after tetanic stimulation (50 Hz for 2 sec) in layer VI. Image width, 5 mm.

EXAMPLE OF APPLICATION

Figure 1A gives some idea of the complexity of structure that can be imaged with IR-gradient contrast. Because no light-consuming optical elements have to be placed in the beam path after the objective, gradient contrast can be combined with techniques such as fluorescence and photostimulation.

An example of an IOS, which, in the neocortex, shows a column-like shape, is given in Figure 1B. This technique allows investigation of the modulatory influences of many kinds of neuroactive substances on the spatial spread of neuronal excitation (Dodt et al. 1996; D'Arcangelo et al. 1997). Even the enhancement of inhibitory neurotransmission by a neuroactive steroid can be visualized (Dodt et al. 1996).

ADVANTAGES AND LIMITATIONS

The clear advantage of IR video microscopy and IOS imaging is that there is no risk of phototoxicity. Because no staining is necessary, all the neuronal elements that can be visualized by IR video microscopy can be seen at the same time. In contrast to imaging with voltage-sensitive dyes, IOS imaging can be performed over extended periods (hours) with no deleterious side effects of the dyes.

A disadvantage of IR video microscopy is the limitation set by the low inherent contrast of very fine neuronal structures like spines. To date, staining is the only way to reliably visualize spines. This may change as new techniques, such as transmission confocal microscopy, are developed.

REFERENCES

D'Arcangelo G, Dodt H-U, Zieglgänsberger W. 1997. Reduction of excitation by interleukin-1 β in rat neocortical slices visualized using infrared-darkfield videomicroscopy. *Neuroreport* **8:** 2079–2083.

Dodt H-U. 1992. Infrared videomicroscopy of living brain slices. In *Practical electrophysiological methods* (ed. Kettenmann H, Grantyn R), pp. 6–10. Wiley-Liss, New York.

Dodt H-U, Zieglgänsberger W. 1990. Visualizing unstained neurons in living brain slices by infrared DIC-videomicroscopy. *Brain Res* **537:** 333–336.

Dodt H-U, Zieglgänsberger W. 1994. Infrared videomicroscopy: A new look at neuronal structure and function. *Trends Neurosci* **17:** 453–458.

Dodt H-U, D'Arcangelo G, Pestel E, Zieglgänsberger W. 1996. The spread of excitation in neocortical columns visualized with infrared-darkfield videomicroscopy. *Neuroreport* **7:** 1553–1558.

Dodt H-U, Frick A, Kampe K, Zieglgänsberger W. 1998. NMDA and AMPA receptors on neocortical neurons are differentially distributed. *Eur J Neurosci* **10:** 3351–3357.

Dodt HU, Eder M, Schierloh A, Zieglgänsberger W. 2002. Infrared-guided laser stimulation of neurons in brain slices. *Sci STKE* **120:** PL2.

MacVicar BA, Hochman D. 1991. Imaging of synaptically evoked intrinsic signals in hippocampal slices. *J Neurosci* **11:** 1458–1469.

5 Confocal Microscopy
Principles and Practice

Alan Fine

Neuroscience Institute, Dalhousie University, Halifax, Nova Scotia B3H 1X5, Canada

ABSTRACT

Since its introduction more than 20 years ago, the confocal microscope has become recognized as an invaluable tool for high-resolution fluorescence microscopy. This chapter outlines the basic principles of confocal microscopy, relevant practical considerations, and various approaches to its implementation, principally from the perspective of visualizing rapid small-scale phenomena in living tissue.

Much of the interest in confocal microscopy stems from the increasing reliance on fluorescent probes in contemporary biology. Fluorescently labeled antibodies and ligands are essential tools for localizing specific molecules. Intracellular or membrane-bound fluorescent dyes are widely used to follow morphological changes in cells, and retrograde transport of fluorescent markers has been used to identify living neurons with particular projections for subsequent electrophysiological or structural study. Voltage- and ion-sensitive indicator dyes have been used to observe patterns of electrical activity in large networks of neurons and in structures too small to be monitored with classical electrode techniques. Green fluorescent protein and related fluorescent proteins make it possible to visualize specific proteins, including engineered functional probes, in essentially unperturbed living tissue.

Unfortunately, fluorescence images are often severely degraded by light that is scattered or emitted by structures outside of the plane of focus. This problem is particularly severe for thick specimens, such as brain slices or whole embryos, and is exacerbated by the poor depth discrimination of conventional (wide-field) light microscopy. These limitations have been only partly overcome by video image processing and deconvolution techniques (Inoué 1986) but have been greatly reduced by confocal optics.

THEORY OF CONFOCAL OPTICS

Principle

Confocal optics provides an alternative solution to the problem of image degradation by out-of-focus light, improving resolution by physical rather than by electronic or computational means. The principle of confocal microscopy was described by Minsky half a century ago (Minsky 1961). The

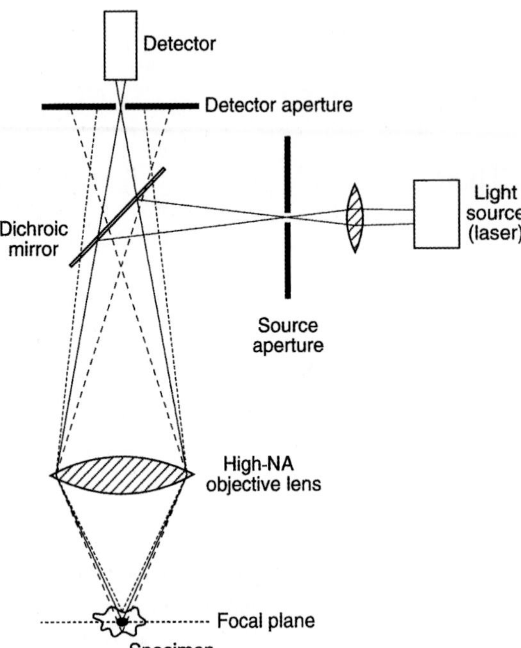

FIGURE 1. The confocal principle. Light from a point source is focused on a point in the specimen. Light (here, fluorescence) from the illuminated point is focused on, and passed through, a detector aperture; fluorescence from all other points is excluded.

crucial element of confocal optics is the projection of an image of a focally illuminated point in the specimen onto a small aperture in a conjugate focal plane (see Fig. 1). Light from the illuminated point can pass through the aperture and can be detected; light from out-of-focus structures will be spread out and will be blocked by the aperture and, therefore, will be largely eliminated. By scanning the specimen across the point (or the point across the specimen; see below; see also Fig. 1), a 2D or 3D confocal image can be generated (Fine et al. 1988; Pawley 2006).

Factors Influencing Resolution

To appreciate the optical improvements provided by confocal microscopy, it is useful to consider some of the factors influencing optical performance. A discussion of these factors is given below.

Point-Spread Function

The resolution of a refractive system such as a microscope is set by its point-spread function (PSF), the 3D intensity distribution of an imaged point. For ideal optics (i.e., unaffected by spherical or other aberrations) and incoherent illumination, the PSF extends (to its first minimum) in the object focal plane to a radius of $r_A = 0.61\lambda/\text{NA}$, where λ is the wavelength of light and NA is the numerical aperture of the objective. This central zone of the PSF is referred to as the Airy disk. The distance to the first minimum of the PSF along the optical axis is $2\lambda n/(\text{NA})^2$ (where n is the index of refraction for the object; see also Chapter 1). The PSF establishes the classical theoretical resolution limit of the system because the images of two points closer than these distances begin to overlap and, thus, cannot be distinguished. (Note that it is important to distinguish resolution from detection and localization. Subresolution particles can be detected if they are sufficiently bright. Also, given sufficient signal levels, and sparseness, they can be localized to much greater precision than the PSF size. Similarly, for high-contrast structures larger than the resolution limit, it is possible to detect changes in size less than the resolution limit, if the same "edge" criterion [e.g., position at which intensity is half-maximum] is applied to the "before and after" images.)

Specimen Thickness and Depth of Field

The axial dimension of the PSF ordinarily establishes the depth of field of the system. Although a large depth of field is useful in some circumstances, high-resolution applications generally require thin optical sectioning. For an infinitely thin specimen containing minute elements spaced more widely than r_A, the theoretical depth of field is defined as one-half of the axial extent of the PSF. Real specimens, on the other hand, may be relatively thick and may contain numerous fluorescent structures; for such specimens, the depth of field is greater because the PSFs generated by these structures or by light scattering can overlap, both within and beyond the focal plane. Under such circumstances, the wide-field fluorescence image is obscured by flare, decreasing the contrast between image features (and thus resolution) and increasing the apparent depth of field. The confocal aperture can eliminate this light from out-of-focus elements, reducing the depth of field toward its theoretical limit. *It is this elimination of out-of-focus light that is the most significant aspect of confocal optics* (see Fig. 2). The extent of this improvement depends on the size of the confocal aperture. Decreasing the aperture size improves resolution and rejection of out-of-focus light up to a limit, but it also decreases signal intensity; optimal signal-to-noise ratio (SNR) is generally achieved when the diameter of the aperture is ~75% that of the Airy disk (Wilson 1989).

By eliminating out-of-focus light and decreasing depth of field, confocal optics permits optical sections of <1 µm to be imaged even in relatively thick scattering tissue. Serial optical sections can be generated simply by changing focal depth; because the specimen remains intact, these sections are necessarily in proper register, greatly simplifying subsequent computational reconstruction of a 3D structure.

Object Illumination and Detector Aperture

The classical resolution limit (the Rayleigh criterion) assumes wide-field illumination such that the PSFs from nearby image points are present simultaneously and interfering. It is possible to exceed this limit by various superresolution methods that bypass this assumption. In near-field fluorescence microscopy, for example, this is performed by illuminating and detecting light, point by point, through the submicrometer tip of a fiber optic placed close to the object. In confocal microscopy, a similar result is obtained by use of point illumination and conjugate-point detection. Actual improvements in resolution depend on the size and shape of the illumination spot and the detector aperture; a 1.4-fold improvement over the

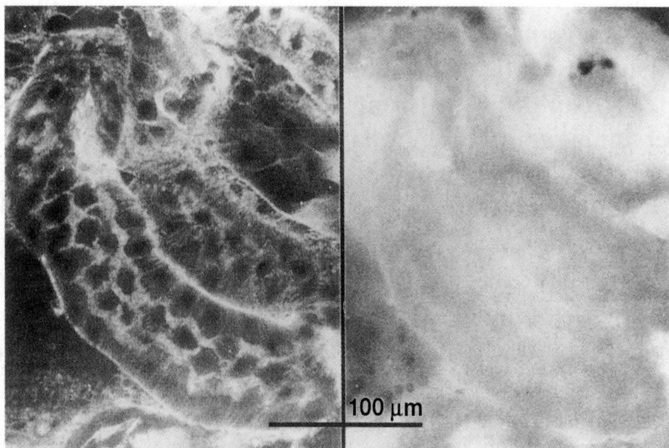

FIGURE 2. The confocal advantage. Cells close to the cut surface of a kidney glomerulus, stained with a fluorescent lipophilic styryl dye, are clearly visible when imaged by confocal laser-scanning microscopy (*top*) but are obscured by out-of-focus light when viewed by ordinary wide-field fluorescence optics (*bottom*). Both images were obtained with the same 0.7-NA objective, focused at the same depth.

Rayleigh criterion has been obtained with circular apertures, and even greater improvements can be obtained at the expense of signal intensity by using annular apertures (Slater and Slayter 1992). Greater improvements have recently been achieved by reducing the size of the illumination spot through stimulated-emission depletion. In this method, a redshifted torroidal spot is superimposed on the illumination spot, quenching fluorescence from the superimposed regions; this effectively reduces the illumination spot size to the central zero of the torroid, enhancing axial resolution more than 10-fold with larger improvements in the specimen plane (Donnert et al. 2006).

Deconvolution

The resolution of confocal images (as with ordinary wide-field images) can be further improved by deconvolution methods (Agard et al. 1989; Hosokawa et al. 1994). The image of a point is the PSF; thus, the image of a complex structure represents the convolution of the points in that structure with the PSF. Therefore, if the PSF of the imaging system can be estimated from known properties (e.g., NA) or empirically determined by imaging subresolution particles, the true structure of the object may be approximated by dividing the image point by point by the PSF. Various algorithms have been developed for efficient implementation of this computationally intensive procedure; through their use, resolution in raw images can be improved by a factor of 2 or more (Shaw 2006).

Other Factors

Other benefits of using confocal microscopes result from the nature of the electronic light detectors used, which facilitates digital image enhancement and can permit sensitive quantitative measurement of small rapid changes in even low levels of fluorescence. In addition, computer control of many commercial instruments allows imaging to be easily integrated with other experimental manipulations (e.g., electrical stimulation).

SIGNAL OPTIMIZATION: PRACTICAL CONSIDERATIONS

To exploit the advantages offered by confocal optics, a number of important practical considerations must be addressed.

Signal Intensity and SNR

The foremost consideration is the problem of signal intensity, along with the associated issue of SNR. Confocal imaging depends on an aperture to exclude out-of-focus light, but, as a result of scattering and optical aberration, the aperture invariably also eliminates some light emanating from the focal point. Particularly for dim fluorescent objects, only a very small number of photons may reach the detector during the sampling interval; indeed, in many cases, the number of photons detected per sample point (pixel) will be none or a few. Under these conditions, the inescapable random statistical fluctuations in photon flux can dominate the signal, eliminating contrast and resolution.

Because this intrinsic photon noise increases only with the square root of the mean light intensity, to increase the SNR, it is essential to increase the detected light intensity. This can be done by decreasing the imaging rate and/or by averaging multiple images; however, for some applications, such as detection of fast transient events, these are not possible. The detector aperture cannot be opened too far without loss of confocality, so other means must be found to achieve this. Higher dye concentrations can be used, but beyond a certain concentration, self-quenching may reduce fluorescence, and high dye concentrations may perturb the behavior of the system under study (e.g., calcium indicator dyes can buffer the free-calcium concentration). Dyes with higher quantum yield or with excitation maxima more closely matched to the available laser lines can be substituted where available. Increasing illumination intensity is an obvious strategy, and most confocal microscopes incorporate lasers for this purpose. However, this approach is limited by several factors. At a certain illumination intensity, essentially all the fluorophore molecules in the illuminated spot will be excited even during brief exposure to the beam. At this point, the dye is saturated, and further increases in excitation-light intensity yield no additional fluorescence, although higher-power excitation lines may permit the use of fluorophores whose excitation spectra are not well matched to the laser emission. With most commercial laser-scanning confocal microscopes, dye saturation occurs with single-line laser power of a few milliwatts; more powerful lasers confer little advantage. Furthermore, increasing illumination increases the rate at which fluorophores bleach, limiting the useful duration of investigation and generating toxic free radicals. (Note that bleaching, which is an oxidative process, can be significantly reduced by the inclusion of antioxidants in the medium surrounding the preparation. For living cells, a particularly useful antioxidant is the water-soluble vitamin E derivative Trolox [6-hydroxy-2,5,7,8-tetramethyl-chroman-2-carbonate].) These considerations are usually dominant, so it is generally preferable to use the minimum possible illumination intensity.

A further important consideration concerning light sources is the instability of output power, which can introduce additional noise and can impair reproducibility. The best available lasers have output noise on the order of 0.1%. Some noise reduction can be obtained, at the expense of intensity, by feedback-regulated devices such as electro-optic modulators, but their frequency response may limit their usefulness to slow-speed scanning.

Optical-Transfer Efficiency

Other routes to increasing detected light intensity involve increasing optical-transfer efficiency at all points along the optical path and increasing detector sensitivity. Significant amounts of light are lost to scattering or absorption when light passes through lenses or is reflected from mirrors. Laser-scanning confocal microscopes may have eight or more mirrors for scanning and beam folding; thus, even 95% reflectance mirrors can lead to >30% reduction in image intensity. Mirrors with ≥99% reflectance are available and should be used where possible. Dirt and misalignment can dramatically reduce optical-transfer efficiency and must be avoided. High-NA objectives not only improve resolution, but also increase light gathering; epifluorescence intensity for spatially extended structures varies with the fourth power of NA. Thus, as a general rule, for any magnification, the highest-NA objective should be used. Conversely, for any NA, the lowest-magnification objective should be used to provide maximal field of view and working distance. Objectives of similar nominal NA can vary substantially in optical-transfer efficiency at different wavelengths, however, and should be directly tested and compared.

Detector Quantum Efficiency

Most commercial confocal microscopes use photomultiplier tubes (PMTs) as light detectors because of their high gain and sensitivity. Simultaneous imaging at several wavelengths is easily achieved with dichroic mirrors and additional detectors. Quantum efficiency (QE; the percentage of incident photons that generate a signal) of all detectors is a function of wavelength. QE for PMTs with standard alkali photocathodes is seldom >25% and can be much lower at longer (red) wavelengths. QE can be improved by using individually selected tubes and matching photocathode material to the intended wavelength with prismatic windows or similar arrangements. Recently, GaAs and GaAsP photocathode PMTs have become available with substantially higher QE, particularly at red wavelengths. Silicon detectors such as charge-coupled devices (CCDs) or avalanche photodiodes can have much higher QE (up to 90%), but generally have lower gain, higher background-noise levels (dark noise), and inferior high-frequency response (important when detecting several photons at each of a million pixels per second). Thermal electron emission contributes to dark noise, which can be reduced by cooling the detector. In addition, this noise is stochastic and uncorrelated from pixel to pixel, so it can be reduced digitally or by low-pass filtration of the detector output prior to digitization.

Image Digitization

In most instruments, it is necessary (and useful) to digitize the detector output, generating discrete image-intensity samples corresponding to pixels. (Once digitized, images can be summed, averaged, and subtracted from or added to other images, and a wide range of image-enhancement and volume-reconstruction methods can be applied.) An important issue is the optimal spatial frequency of these samples. For nonperiodic structures, a sampling frequency at least 2.3 times the maximum spatial frequency in the image (the Nyquist criterion) is necessary to retain all image details. Thus, if r_A for the objective in use is 300 nm, pixel size should be <130 nm to obtain maximal detail in the x–y plane. For laser-scanning confocal microscopes (see below), the temporal sampling rate (pixels/sec) may be constant, but the spatial sampling rate (pixels/μm) can be varied by changing the zoom factor. Sampling at spatial frequencies below the Nyquist criterion will lose spatial information, but can reduce bleaching and phototoxicity, considerations that may dominate for live cell imaging; pixel sizes smaller than the Nyquist criterion may aid visualization, but provide no additional information and needlessly increase bleaching and phototoxicity.

Scanning Mode

Temporal resolution becomes a serious consideration when the aim is to monitor fast phenomena. The Nyquist criterion applies equally in the time domain. Some scanning methods are too slow for these purposes. Even with fast-scanning methods (see below), it may not be possible to collect sufficient photons from small pixels over short intervals.

IMPLEMENTATION

Confocal imaging can be implemented in a variety of ways (see Fig. 3). The main designs used are described below.

Specimen-Scanning Design

The light path can be fixed, and the specimen can be scanned under the beam. This has the virtue of simplicity, permits scanning of large fields, and avoids lateral aberrations in the optics; however, it is slow and impractical when the specimen is massive or attached to other (e.g., electrical recording) apparatuses.

Nipkow Disk Design

Imaging can also be achieved by coordinated scanning of apertures, arranged in a spiral pattern as a Nipkow disk across conjugate focal planes in the incident and image light paths (Fig. 3A). Disks with suitably spaced apertures permit confocal monitoring of many points at once, in turn allowing imaging at up to several hundred frames per second. In simple disk configurations, adequate simultaneous illumination of the multiple points is difficult to achieve, thus limiting sensitivity. This design has been substantially improved by incorporation in the incident light path of an array of microlenses, spinning together with the Nipkow disk, and aligned so as to focus virtually all of the illumination onto the array of pinholes (Ichihara et al. 1996). This design is particularly useful for imaging rapidly changing specimens that are relatively bright or easily bleached. Disks with linear apertures have also been used. They increase light throughput but at the expense of confocality and, thus, of image quality.

Laser-Scanning Design

Mirror Based

More intense illumination is possible by using a laser to scan a single focused spot across the specimen. A common strategy is illustrated in Figure 3B. The laser intensity is adjusted with a neutral-density filter, and the appropriate laser line is selected by a band-pass filter. A microscope, often equipped for ordinary epifluorescence, is used to focus the illumination spot on the specimen with a dichroic mirror serving to introduce the illuminating beam into the microscope optical axis. A pair

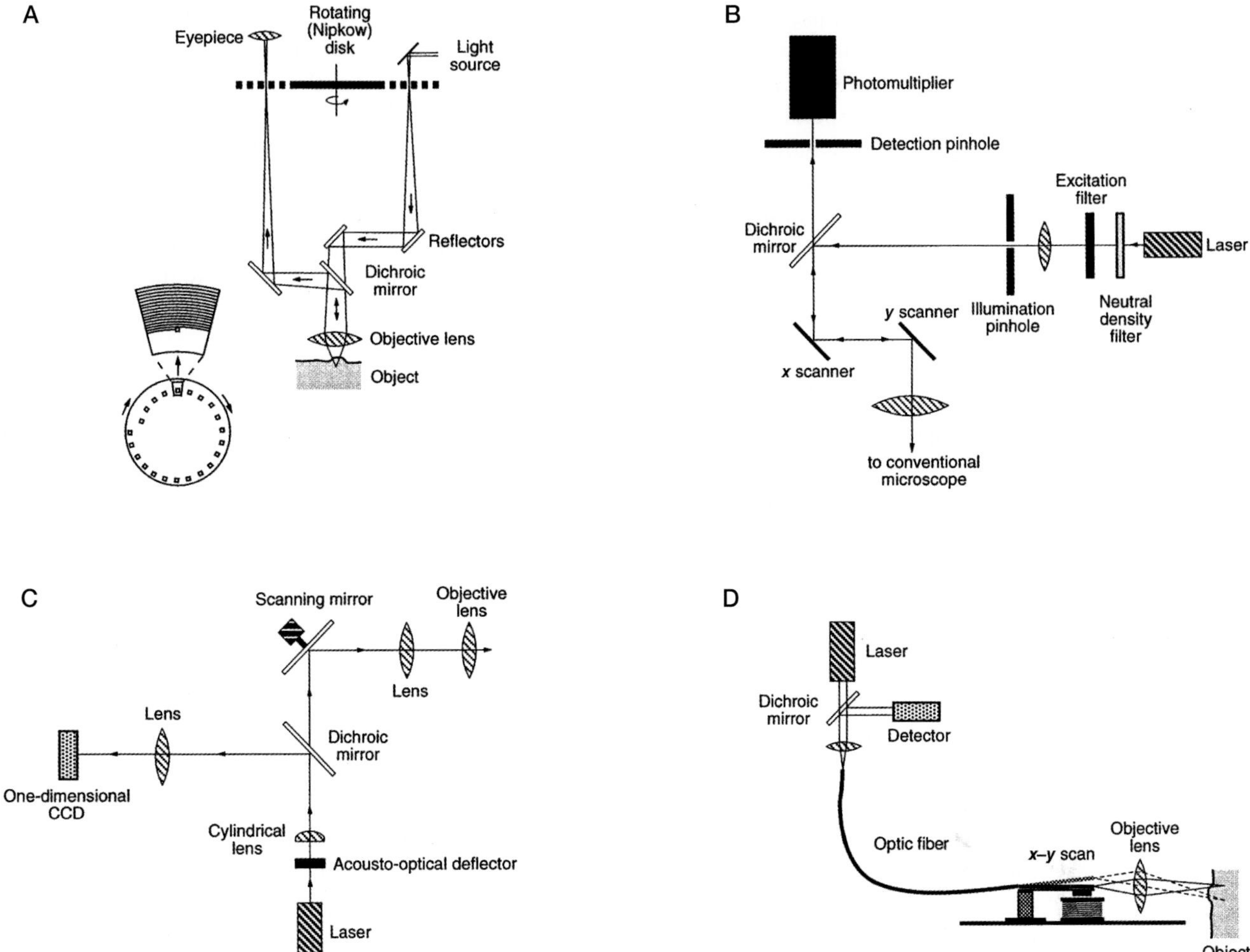

FIGURE 3. Confocal implementation. (*A*) A 2D image can be generated by spinning a spiral array of illumination apertures (a Nipkow disk) over a window on the specimen, thereby illuminating points on sequential lines (*inset*). A confocal image of the illuminated points is viewed through a similarly moving array in a conjugate image plane. (*B*) A spot of light from a fixed-point source can be scanned across the specimen by mirrors. Longer-wavelength fluorescence from the point is reflected back along the same light path through a dichroic reflector to a conjugate fixed-point aperture. The time-varying signal from the detector behind this aperture is converted by a computer into a 2D image. (*C*) Scanning can also be performed by an acousto-optic deflector, but the wavelength dependency of these devices prevents their use for descanning the longer-wavelength fluorescence. Instead, a linear detector array can be used, sampling each element in sequence corresponding to the moving point. (*D*) A simpler configuration uses an optical fiber, vibrating in an image plane, as both point source and detector aperture.

of computer-controlled galvanometer mirrors beyond the dichroic mirror steer the spot in a raster pattern over the specimen. Fluorescent light from the illuminated spot is returned and descanned by the same galvanometer mirrors, passed by the dichroic mirror, and focused onto an aperture in front of a detector. The digitized output of the detector is correlated in time with the position of the spot, permitting digital construction of an x–y image. Usefully, the true magnification of the image can be increased by compressing the raster into a smaller region about the optical axis (zoom).

Acousto-Optic Modulator Based

The mass of the galvanometer mirror (discussed above) makes it difficult to scan fast enough to generate images at the rates needed for observing very rapid phenomena (e.g., ion redistribution during action potentials). Higher scan speeds can be achieved with acousto-optic deflectors (see

Chapter 54), which use sound to induce refraction waves that behave like a diffraction grating. The degree of refraction is wavelength dependent, however, so that the (longer-wavelength) fluorescence cannot be descanned to a fixed point by the same device. A slit can therefore be used as the confocal aperture, with acceptable rejection of out-of-focus light, or a linear detector array such as a CCD can be used to select the moving point on the acousto-optically scanned line that corresponds to the stationary focal point (see Fig. 3C). Acousto-optic deflectors or the related acousto-optic modulators can also serve as ultrafast shutters or intensity regulators, useful for switching and balancing laser lines, minimizing unnecessary illumination and bleaching (e.g., during changes in beam scan direction and flyback during raster scanning), or reducing noise due to laser intensity fluctuations. These applications are compatible with all scanning designs.

Alternative Laser-Scanning Designs

Simpler designs are possible. A single optical fiber can be used as both the point source and the detector aperture. The point can be scanned by mirrors in the light path as described above or, more simply, by vibrating the fiber in an image plane (see Fig. 3D). High-speed confocal imaging with no moving parts is also possible, using spatial light modulators to generate moving or stationary patterns of transmittance or reflectance (Liang et al. 1997; Verveer et al. 1998). All laser-scanning designs can be miniaturized, so that head-mounted or endoscopic confocal imaging devices could be used for in vivo applications.

ADVANCES IN SPECTROSCOPY

Until recently, commercial instruments offered only a small number of detector channels for simultaneous imaging at different wavelengths; the wavelengths usually being determined by preselected interference filters with fixed spectral properties. Spectral dispersion by prism or diffraction grating offers greater flexibility. In one configuration (Leica TCS SP), an arbitrary spectral band can be selected by an adjustable slit formed between mirrored plates in front of the PMT; rejected portions of the spectrum are reflected to additional adjustable slits (Calloway 1999). In an alternative configuration (Zeiss META), the dispersed spectrum is projected onto a 32-channel PMT array, allowing simultaneous imaging within ~10-nm-wide emission bands across the visible spectrum. Signals derived from multiple fluorophores with known overlapping emission spectra can be resolved from such data sets with suitable linear unmixing algorithms (Dickinson et al. 2001).

CONCLUSION

Confocal microscopy is an established technique for obtaining high-resolution images with good optical searching. Although other methods for high-resolution imaging, such as atomic force microscopy, near-field optical microscopy, and two-photon fluorescence microscopy, are becoming progressively more powerful and accessible, not all can be applied to living tissue, and others can be used in conjunction with confocal microscopy. With the ongoing improvements in instrumentation and fluorescent probes, confocal microscopy will continue to be a powerful method for biological imaging.

REFERENCES

Agard DA, Hiraoka Y, Shaw PJ, Sedat JW. 1989. Fluorescence microscopy in three dimensions. *Methods Cell Biol* **30:** 353–378.

Calloway CB. 1999. A confocal microscope with spectrophotometric detection. *Microsc Microanal* (suppl 2) **5:** 460–461.

Dickinson ME, Bearman G, Tille S, Lansford R, Fraser SE. 2001. Multi-spectral imaging and linear unmixing add a whole new dimension to laser scanning fluorescence microscopy. *BioTechniques* **31:** 1272–1278.

Donnert G, Keller J, Medda R, Andrei MA, Rizzoli SO, Lührmann R, Jahn R, Eggeling C, Hell SW. 2006. Macromolecular-scale resolution in biological fluorescence microscopy. *Proc Natl Acad Sci* **103:** 11440–11445.

Fine A, Amos WB, Durbin RM, McNaughton PA. 1988. Confocal microscopy: Applications in neurobiology. *Trends Neurosci* **11:** 346–351.

Hosokawa T, Bliss TVP, Fine A. 1994. Quantitative three-dimensional confocal microscopy of synaptic structures in living brain tissue. *Microsc Res Tech* **29:** 290–296.

Ichihara A, Tanaami T, Isozaki K, Sugiyama Y, Kosugi Y, Mikuriya K, Abe M, Uemura I. 1996. High-speed confocal fluorescence microscopy using a Nipkow scanner with microlenses for 3-D imaging of single fluorescent molecule in real-time. *Bioimages* **4:** 57–62.

Inoué S. 1986. *Video microscopy.* Plenum, New York.

Klar TA, Jakobs S, Dyba M, Egner A, Hell SW. 2000. Fluorescence microscopy with diffraction resolution barrier broken by stimulated emission. *Proc Natl Acad Sci* **97:** 8206–8210.

Liang M, Stehr RL, Krause AW. 1997. Confocal pattern period in multiple-aperture confocal imaging systems with coherent illumination. *Opt Lett* **22:** 751–753.

Minsky M. 1961. Microscopy apparatus. U.S. Patent No. 3,013,467.

Pawley JB, ed. 2006. *Handbook of biological confocal microscopy*, 3rd ed. Springer SBM, New York.

Shaw PJ. 2006. Comparison of wide-field/deconvolution and confocal microscopy for three-dimensional imaging. In *Handbook of biological confocal microscopy*, 3rd ed (ed. JB Pawley), pp. 453–467. Springer SBM, New York.

Slater EM, Slayter HS. 1992. *Light and electron microscopy*. Cambridge University Press, Cambridge.

Verveer PJ, Hanley QS, Verbeek PW, Van Vliet LJ, Jovin TM. 1998. Theory of confocal fluorescence imaging in the programmable array microscope (PAM). *J Microsc* **189:** 192–198.

Willig KI, Keller J, Bossi M, Hell SW. 2006. STED microscopy resolves nanoparticle assemblies. *New J Phys* **8:** 106.

Wilson T. 1989. Optical sectioning in confocal fluorescent microscopes. *J Microsc* **154:** 143–156.

6 Spinning-Disk Systems

Tony Wilson
Department of Engineering Science, University of Oxford, Oxford OX1 3PJ, United Kingdom

ABSTRACT

In this chapter, we discuss the origin of optical sectioning in optical microscopy in terms of the structure of the illumination and the structure of the detection. This parallel approach to image formation allows the introduction of high-speed light efficient approaches to obtaining optically sectioned images in real time, using conventional microscope illumination systems.

INTRODUCTION

The popularity of the confocal microscope in life science laboratories around the world is undoubtedly due to its ability to permit volume objects to be imaged and to be rendered in three dimensions. It is important to realize that the confocal microscope itself does not produce three-dimensional images. Indeed, it does the opposite. The critical property that the confocal microscope possesses, which the conventional microscope does not, is its ability to image efficiently (and in-focus) only those regions of a volume specimen that lie within a thin section in the focal region of the microscope. In other words, it is able to reject (i.e., vastly attenuate) light originating from out-of-focus regions of the specimen. To image a three-dimensional volume of a thick specimen, it is necessary to take a whole series of such thin optical sections as the specimen is moved axially through the focal region. Once this through-focus series of optically sectioned images has been recorded, it is a matter of computer processing to decide how the three-dimensional information is to be presented.

Any optical microscope that is to be used to produce three-dimensional images must have the ability to record a thin optical section. There are many methods for producing optical sections, of which the confocal optical system is just one. We shall review these methods and shall describe a number of convenient methods of implementation that can lead to, among other things, real-time image formation.

OPTICAL SECTIONING

In the following discussion, we shall restrict our attention to bright-field or (single-photon) fluorescence imaging in which the optical sectioning results from the optical system of the microscope rather

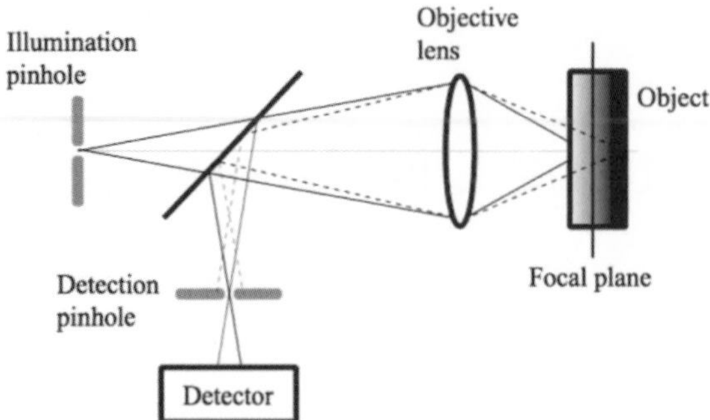

FIGURE 1. The physical origin of the depth discrimination or optical sectioning property of the confocal optical system. A barrier in the form of a pinhole aperture is used to physically prevent (block) light originating from regions in the specimen from passing through the pinhole and, hence, being detected by the photodetector and thus contributing to the final image contrast.

than by any nonlinear interaction between the probe light and the specimen. To be able to make general remarks about various optical systems, we will describe the design in terms of the structure of the illumination and in terms of the structure of the detection. To put these terms in context, we note that, in the conventional fluorescence microscope, we essentially illuminate the specimen uniformly and image the fluorescence emitted by the specimen to an image plane in which we view the image intensity either directly by eye or via a charge-coupled device (CCD) camera. In this case, the structure of the illumination is uniform as is the structure of the detection, and the microscope does not show optical sectioning. In the confocal microscope, on the other hand, we use point illumination and point detection to introduce optical sectioning. The optical principle can be seen in Figure 1, where we see that the action of the point detector is to block light that originates in out-of-focus regions from passing through the pinhole (Wilson and Sheppard 1984; Wilson 1990). Its efficacy in achieving this, which also determines the axial width of the optical section, clearly depends on the size and the shape of the pinhole used. An infinitely large pinhole, for example, would block no light and, hence, provide no optical sectioning. This effect is discussed in detail elsewhere (Wilson 1989, 1995; Wilson and Carlini 1987). The system illustrated in Figure 1 might be regarded as the ultimate in structured illumination and detection—point illumination and point detection—and has resulted in the desired optical sectioning but has only produced an image of a single point of the specimen. To produce an image of a finite region of the specimen, it is necessary to introduce scanning so as to probe the entire specimen. In general terms, we have introduced a particular structure to both the illumination and the detection, which we might also refer to as modulation, such that the optical system shows optical sectioning. We must then remove any undesirable side effects of this modulation to obtain the desired image. In this particular case, the modulation results in a restriction of the field of view to a single point; hence, a demodulation stage consisting of scanning is required to restore the field of view. We shall return to practical implementations of the demodulation below, but we note that there are two basic approaches. In the first, a single point source–point detector confocal system is used together with a scanning mechanism designed to scan a single focused spot of light with respect to the specimen. In the second approach, a number of confocal systems is constructed in parallel. These serve to produce many focused spots of light, which are used to image different parts of the specimen simultaneously. This is achieved by using an aperture disk consisting of many pinholes.

Another way to think about optical sectioning is in terms of the way in which the spatial frequencies present in the specimen are imaged. In essence, we describe the fluorescence distribution within the specimen in terms of its spatial frequency spectrum (Fourier content) and ask how each of these spatial frequency components is imaged by the optical system. The optical transfer function of the optical system provides the answer to this question because it describes how efficiently each spatial frequency is imaged. A requirement that the system show optical sectioning might be that the contrast of all spatial frequencies must attenuate as the microscope is defocused. Figure 2A shows

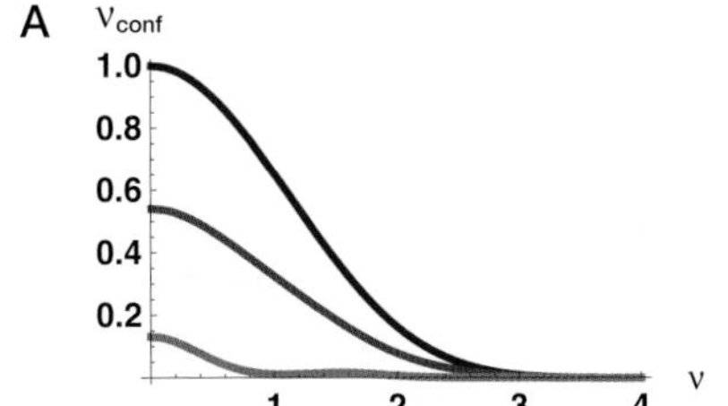

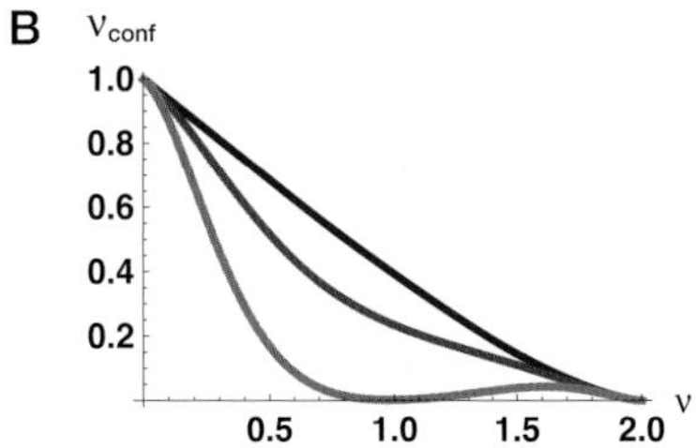

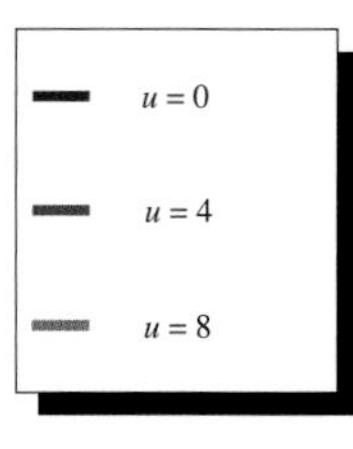

FIGURE 2. (*A*) The confocal optical transfer function as a function of the normalized spatial frequency for a number of values of defocus u. The normalized defocus u is related to the actual defocus z by $u = 4knz \sin^2(\alpha/2)$ in which the numerical aperture (NA) is given by $NA = n \sin(\alpha)$. The normalized spatial frequency v is related to the actual spatial frequency f measured in the focal plane via $v = f\lambda/NA$. We note that all spatial frequencies attenuate with increasing defocus. (*B*) The optical transfer function of the conventional microscope as a function of the normalized spatial frequency for a number of values of defocus u. Note that all spatial frequencies attenuate with increasing defocus apart from the zero spatial frequency case.

the optical transfer function of a confocal fluorescence microscope, in which we see that the contrast of all spatial frequencies attenuates with increasing defocus. Figure 2B, on the other hand, shows the equivalent function for the conventional fluorescence microscope. In this case, we see that it is only the zero spatial frequency whose contrast does not attenuate with increasing defocus. The contrast of all other spatial frequencies is seen to reduce as the degree of defocus increases.

Although the aperture disk consisting of many pinholes was described above as a natural way to parallelize many confocal microscopes, it may also be thought of as acting like a mask that causes the whole specimen to be illuminated by a particular structure. It is natural, therefore, to ask whether there are other simple forms of structure to the illumination that may be used to introduce optical sectioning. If we modify the illumination system of the microscope so as to project a single spatial frequency grid pattern onto the object, the microscope will then image efficiently only that portion of the object where the grid pattern is in focus (Fig. 2B). We will thus obtain an optically sectioned image of the object but with the (unwanted) grid pattern superimposed. The rate of attenuation with defocus or optical sectioning strength will, of course, depend on the particular spatial frequency that is projected onto the object (Fig. 2B). For example, a 40-μm pitch grid imaged using a 63x, 1.4-NA objective lens with light of wavelength 0.5 μm yields $v = 0.56$, whereas an 80-μm pitch grid yields $v = 0.28$ and a 20-μm pitch grid gives $v = 1.12$. Here we have used the structure of the illumination (harmonic modulation) to introduce optical sectioning. The price is that the optical section is now delineated or labeled by that portion of the image where the superimposed grid pattern is visible. It is now necessary to introduce a demodulation stage whereby the out-of-focus regions as well as the grid pattern are removed from the "raw" image to reveal the desired optically sectioned image. This may be done in two ways, computationally or optically. The computational approach typically requires that three raw images be taken, corresponding to three different spatial positions of the illumination grid. This is the approach taken in several commercial structured illumination systems such as the OptiGrid system from Qioptiq (Neil et al. 1997). The alternative optical demodulation technique, which will be discussed in this chapter, is to combine harmonic structured illumination with harmonic structured detection. In this case an identical mask is used for both the illumination and detection. Demodulation is carried out by scanning the masks in synchronism.

We conclude this section by noting that a system with uniform structure of illumination and detection—the conventional microscope—does not exhibit optical sectioning, whereas one with point illumination and point detection—the confocal system—does exhibit optical sectioning. An equivalent way of saying this is to say that the conventional system employs zero spatial frequency illumination and detection whereas the point source/detector confocal system employs full spatial frequency illumination and detection. The harmonic approach we have just discussed, on the other hand, lies somewhere between these approaches, because only one spatial frequency is used for both the illumination and detection. The nature of the illumination/detection used in the last two cases requires that a further demodulation step—often achieved by scanning—be performed to provide a full field optically sectioned image.

PRACTICAL REALIZATION OF OPTICAL SECTIONING MICROSCOPES

We shall now discuss the practical implementation of these two approaches to achieve optical sectioning. We begin with the traditional confocal system.

It is clear from the previous discussion that an optical system consisting of a single point source and single point detector serves to discriminate against light originating from out-of-focus planes. Figure 1 shows the generic optical system. The light source is typically a laser, because traditional microscope illumination systems are insufficiently bright. A photomultiplier tube has usually been used as the photodetector. Because this system probes only one point of the specimen, scanning must be used to obtain an image of a whole optical section. This may be achieved in a variety of ways. The specimen may be physically scanned with respect to the fixed focal spot. Alternatively the objective lens may also be scanned. These approaches have advantages from both the optical performance and optical design points of view but are generally considered to be impractical. In most commercial designs, therefore, the specimen is fixed and the scanning is achieved by scanning the focused spot of light across the fixed specimen by the use of galvonometer mirror scanners. This allows an optical section to be easily recorded. To record the next optical section, however, it is necessary to physically move the specimen axially to bring the next region into the focal volume of the confocal microscope. Commercial systems do not allow this important *z*-scanning step to be performed quickly, so this represents a bottleneck in the speed with which a through-focus set of images may be obtained. Recent work has shown that high-speed optical refocusing can be achieved, and hence this bottleneck may be removed (Botcherby et al. 2007). Although the layout of Figure 1 is typical, there are problems to be overcome relating to system alignment in the sense that the detector pinhole must be located in a position optically equivalent to the source pinhole. These problems may be resolved if a reciprocal geometry is employed in which the same pinhole is used both as source and detector pinhole. In practice these systems are often more easily implemented when a single-mode optical fiber replaces the pinhole (Wilson and Kimura 1991). However, all of these approaches involve the use of one confocal optical system, and the image is obtained serially by the appropriate scanning of the spot in three dimensions with respect to the specimen.

An advantage may be gained by building an optical layout consisting of many confocal systems lying side by side. In this way many parts of the specimen will be imaged confocally at the same time. This has the advantage of increasing image acquisition speed as well as dispensing with the need to use laser illumination. Each pinhole acts as both the illumination and detection pinhole and so the system acts rather like a large number of parallel, reciprocal geometry, confocal microscopes, each imaging a specific point on the object. However, we need to remember that the confocal system achieves depth discrimination by blocking out-of-focus light reaching the image by the use of a limiting pinhole detector. This observation leads us to conclude that the neighboring confocal systems must be placed sufficiently far apart that any out-of-focus light from one confocal system is not collected by an adjacent system. In other words, we must prevent cross talk between neighboring confocal systems. In practice this means that the pinholes must be placed on the order of 10 times their diameters apart, which has two immediate consequences. First, only a small amount—typically 1%—of the available light is used for imaging, and, second, the wide spacing of the pinholes means that the object is only sparsely probed. To probe—and hence image—the whole object, it is usual to arrange the pinhole apertures in a series of Archimedean spirals and to rotate the (Nipkow) disk. The generic layout of any system that is designed to contain many confocal systems operating in parallel is shown in Figure 3A. The original idea for such an approach goes back to Mojmir Petrán in the late 1960s (Petrán et al. 1968). A single-sided variant was subsequently introduced by Kino and his colleagues (Xiao and Kino 1987). The key element to these systems is a spinning Nipkow disk containing many pinholes. These systems are capable of producing high-quality images without the need to use laser illumination in real time at both television rate and higher imaging speeds. A further development, which does, however, require the use of laser illumination and, hence, restricts the use to fluorescence imaging, is to introduce an array of microlenses to concentrate the illumination laser light into the source pinholes (Ichihara et al. 1996).

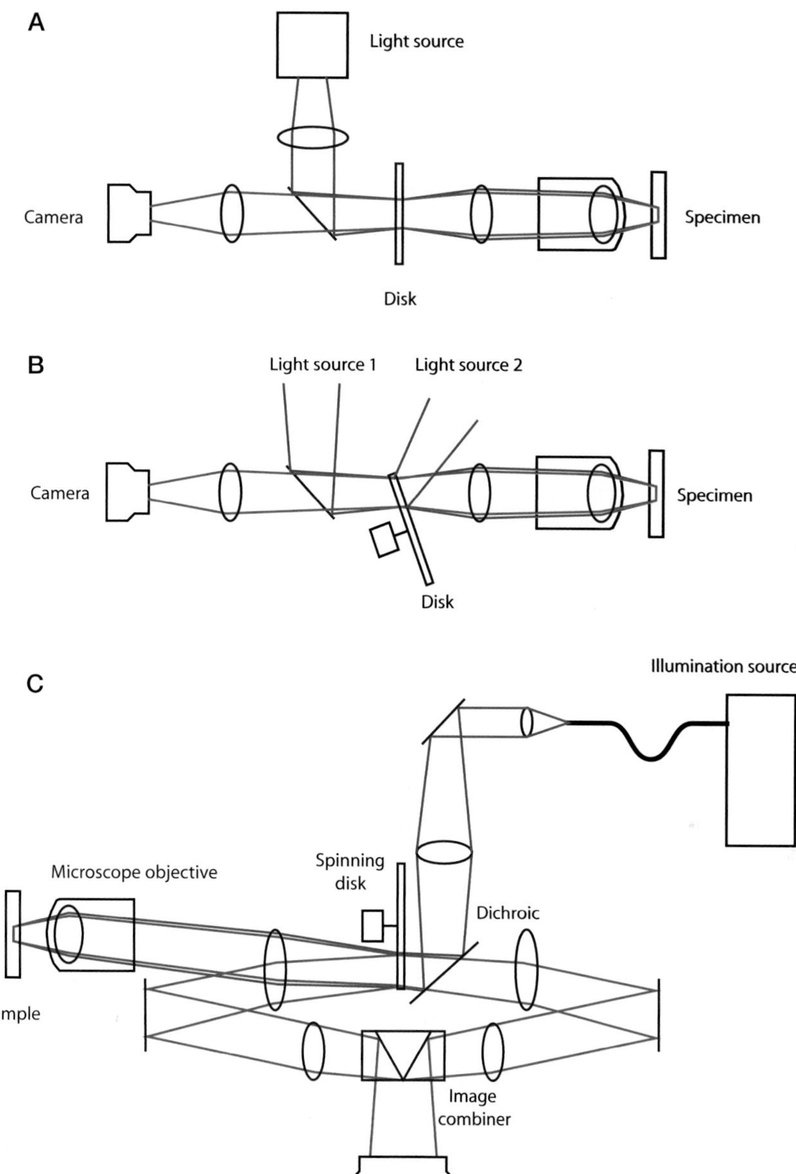

FIGURE 3. (*A*) The generic form of a spinning-disk-based microscope system. In the case of the tandem-scanning microscope, the disk consists of a large number of pinholes, each of which acts as a reciprocal geometry confocal system. The pinholes, which are placed far apart to prevent cross talk between adjacent confocal systems, are usually arranged in an Archimedean spiral on a Nipkow disk. Rotation of the disk causes the entire specimen to be imaged. In the case of the correlation code microscope, the appropriate codes are impressed photolithographically onto the disk, which is then rotated to effect the desired time-dependent correlation codes. In the harmonic illumination and detection case, a stripe pattern is imposed on the disk. In this case, the disk is spun to achieve the desired demodulation. (*B*) An example of a double-sided operation in which two light sources are used to sequentially generate the two composite images: conventional plus confocal and conventional minus confocal. (*C*) An optical configuration that permits both composite images to be recorded simultaneously, thereby removing any motion artifacts.

One approach to make greater use of the available light is to place the pinholes closer together. However, this means that cross talk between the neighboring confocal systems inevitably occurs; hence, a method must be devised to prevent this. To achieve this goal, the Nipkow disk of the tandem-scanning microscope is replaced with an aperture mask consisting of many pinholes placed as close together as possible. This aperture mask has the property that any of its pinholes can be opened and closed independently of the others in any desired time sequence. This might be achieved, for example, by using a liquid-crystal spatial light modulator. Because we require there to be no cross talk between the many parallel confocal systems, it is necessary to use a sequence of openings and closings of each pinhole that is completely uncorrelated with the openings and the closings of all the other pinholes. There are many such orthonormal sequences available. However, they all require the use of both positive and negative numbers, and, unfortunately, we cannot have a negative intensity of light! The pinhole is either open, which corresponds to 1, or closed, which corresponds to 0. There is no position that can correspond to –1. The way to avoid the dilemma is to obtain the confocal signal indirectly. To use a particular orthonormal sequence $b_i(t)$ of plus and minus 1s, for the ith pinhole, we must add a constant offset to the desired sequence to make a sequence of positive numbers, which can be encoded in terms of pinhole opening and closing. Thus, we encode each of the pinhole openings and closings as $(1 + b_i(t))/2$, which will correspond to open (1) when $b_i(t) = 1$ and to close (0) when $b_i(t) = -1$. The effect of adding the constant offset to the desired sequence is to produce a composite image that will be partly confocal because of the $b_i(t)$ terms and partly conventional because of the constant term. The method of operation is now clear. We first take an image with the pinholes encoded as we have just discussed and so obtain a composite conventional plus confocal image. We then switch all the pinholes to the open state to obtain a conventional image. It is then a simple matter to subtract the two images in real time using a computer to produce the confocal image.

Although this approach may be implemented using a liquid-crystal spatial light modulator, it is cheaper and simpler merely to impress the correlation codes photolithographically on a disk and to rotate the disk so that the transmissivity at any picture point varies according to the desired orthonormal sequence. A blank sector may be used to provide the conventional image. If this approach is adopted, then all that is required is to replace the single-sided Nipkow disk of the tandem-scanning microscope with a suitably encoded aperture disk (Juskaitis et al. 1996a,b). We note that the coded sector on the disk may be coded so as to provide the appropriate correlation codes, or, alternatively, it may consist of a pattern of grid lines to simulate the harmonic illumination/detection case (Neil et al. 1998).

Double-Sided Operation

We have seen that the image obtained from the coded sector of the disk is a composite image from which the conventional image needs to be removed. If, rather than use a blank sector, we were to encode the whole disk such that the image we obtained may be written as $I_1 = I_{conv} + I_{conf}$, we would need to find another way to remove the conventional image. One approach is suggested in Figure 3B in which a second light source is used. In this case, a composite image of the form $I_2 = I_{conv} - I_{conf}$ is obtained. It is clear that a confocal image $I_1 - I_2 = 2I_{conf}$ may be readily extracted with a more efficient use of light than in the single-sided disk case in which I_1 and I_{conv} are obtained sequentially.

Although such an approach is entirely feasible, it does require extremely careful design so as to provide equivalent uniform illumination at each flip of the mirror. A preferred approach might be to use a single light source and a single camera. The optical system (Fig. 3C) would be such that the camera recorded the two required images simultaneously, thus eliminating any possibility of motion and other artifacts between the capture of the two required raw images. We note that these systems operate well with standard microscope illuminators (e.g., Exfo Inc.) and standard CCD cameras, producing images that are directly comparable to those taken with traditional laser-based confocal systems. Figure 4 shows a typical through-focus series of images together with a standard three-dimensional rendering.

CONCLUSION

The confocal microscope is now firmly established as a workhorse instrument in laboratories throughout the world because of its ability to enable volume specimens to be imaged in three dimensions. However, there is still much work to be performed to make these instruments suitable for high-speed

FIGURE 4. (*A*) A through-focus series of images taken of the eye of a fly. (*B*) A three-dimensional rendering of the eye of the fly constructed from the through-focus series of images presented in *A*.

imaging of living specimens. Further advances will require a combination of new contrast mechanisms together with advances in instrument design. This chapter has described a parallelization of the traditional confocal principle so as to permit real-time confocal imaging without the need for laser illumination. The concept of structured illumination and structured detection has been introduced to stimulate the search for alternative methods of image encoding to reveal optical sectioning.

REFERENCES

Botcherby EJ, Juskaitis R, Wilson T. 2007. Real-time scanning microscopy in the meridional plane. *Opt Express* **34:** 1504–1506.

Botcherby EJ, Juskaitis R, Wilson T. 2009. Real-time extended depth of field microscopy. *Opt Express* **16:** 21843–21848.

Ichihara A, Tanaami T, Isozaki K, Sugiyama Y, Kosugi Y, Mikuriya K, Abe M, Uemura I. 1996. High-speed confocal fluorescence microscopy using a Nipkow scanner with microlenses for 3-D imaging of single fluorescent molecule in real-time. *Bioimages* **4:** 57–62.

Juskaitis R, Neil MAA, Wilson T, Kozubek M. 1996a. Efficient real-time confocal microscopy with white light sources. *Nature* **383:** 804–806.

Juskaitis R, Neil MAA, Wilson T, Kozubek M. 1996b. Confocal microscopy by aperture correlation. *Opt Lett* **21:** 1879–1881.

Neil MAA, Juskaitis R, Wilson T. 1997. Method of obtaining optical sectioning using structured light in a conventional microscope. *Opt Lett* **22:** 1905–1907.

Neil MAA, Wilson T, Juskaitis R. 1998. A light efficient optically sectioning microscope. *J Microsc* **189:** 114–117.

Petrán M, Hadravsky M, Egger M, Galambos R. 1968. Tandem scanning reflected light microscope. *J Opt Soc Am* **58:** 661–664.

Wilson T. 1989. Optical sectioning in confocal fluorescent microscopes. *J Microsc* **154:** 143–156.

Wilson T. 1990. *Confocal microscopy*. Academic, London.

Wilson T. 1995. The role of the pinhole in confocal imaging system. In *Handbook of biological confocal microscopy*, 2nd ed. (ed. Pawley JB), pp. 167–182. Plenum, New York.

Wilson T, Carlini AR. 1987. Size of the detector in confocal imaging systems. *Opt Lett* **12:** 227–229.

Wilson T, Kimura S. 1991. Confocal scanning optical microscopy using a single mode fiber for signal detection. *Appl Opt* **30:** 2143– 2150.

Wilson T, Sheppard CJR. 1984. *Theory and practice of scanning optical microscopy*. Academic, London.

Xiao GQ, Kino GS. 1987. A real-time scanning optical microscope. *Proc SPIE* **809:** 107–113.

7 Introduction to Multiphoton-Excitation Fluorescence Microscopy

Winfried Denk

Department of Biomedical Optics, Max-Planck Institute for Medical Research, D-69120 Heidelberg, Germany

ABSTRACT

Fluorescence microscopy has been steadily gaining importance in quantitative biological research as dramatic improvements have been seen in fluorophores, optical systems, light sources, and detectors. In particular, the invention (Minsky 1961) and implementation of the confocal fluorescence microscope (for a useful collection of reprints, see Masters 1996), usually by laser scanning, has, for the first time, allowed the observation of biological processes with high spatial resolution inside intact living tissue. This ability, often called optical sectioning, allows spatial reconstruction of three-dimensional specimens without the use of a microtome. This chapter reviews the physical mechanisms on which the properties of multiphoton microscopy are based and discusses some practical aspects of its implementation. These issues are discussed further in Denk et al. (1994, 1995a); Williams et al. (1994); Denk (1996); Potter (1996); Denk and Svoboda (1997); and original papers referenced therein.

THE THEORY OF MULTIPHOTON MICROSCOPY: A COMPARISON WITH CONFOCAL MICROSCOPY

Out-of-Focus Light Rejection

For all of its virtues, confocal fluorescence microscopy has a major flaw: It excites fluorophores with abandon but detects only a small fraction of the generated fluorescence. This is unavoidable with one-photon absorption because the average intensity of the excitation light does not really vary with the distance from the focal plane (see the next subsection). In one-photon microscopy, optical sectioning is instead based on the rejection, by the confocal pinhole, of all fluorescence except that coming from the focus. For example, in a specimen several hundred micrometers thick, >99% of the generated fluorescence is wasted (even before taking into account the losses, by another factor of ~20, due to the limited solid angle of detection and quantum efficiency [QE]). This loss of fluorescence is particularly unfortunate in the case of living specimens in which certain chemical techniques for the avoidance of photobleaching and photodamage cannot be used because they involve the removal of oxygen.

This chapter first appeared in *Imaging in neuroscience and development: A laboratory manual* (ed. Yuste R, Konnerth A). Cold Spring Harbor Laboratory Press, Cold Spring Harbor, NY (2005). This is the original version.

Localized Excitation

Multiphoton microscopy (Denk et al. 1990) combines most of the advantages of confocal microscopy, including its optical sectioning properties and background rejection, with an almost complete utilization of the excited fluorescence (to the extent possible, given the limited acceptance angle and transmission of the collection optics, as well as the limited detector QE). The central idea—exciting only those fluorophores from which one wants to collect fluorescence—is simple enough but impossible to achieve with one-photon (also called linear) absorption. The reason excitation confinement is impossible with one-photon absorption can be understood by considering the distribution of the fluorescence excitation in a focused laser beam, traversing a 3D object that is uniformly stained (for tissue, uniform staining is meant in a statistical sense). Now imagine the specimen divided into slices (perpendicular to the optical axis) of uniform thickness. For one-photon excitation, each slice produces the same amount of fluorescence, irrespective of the slice location relative to the focal point. In a thick specimen, extending along the optical axis for a large multiple (as high as several hundred times) of the focal depth (~1 μm for a high-numerical-aperture [high-NA] lens), only a small fraction of slices, corresponding to the focal depth or imaging depth, will intersect the focal volume. The small fraction of the total fluorescence that comes from the focal volume (roughly equivalent to the observation volume of the confocal microscope) is all that is used for image generation. For one-photon excitation, this behavior is entirely unavoidable: The excitation light has to reach the focus and, on the way there, it intersects all the slices in front of the focal point and beyond because most excitation light continues to propagate. Of course, the incident light has to get to the focus, even for multiphoton excitation (Goeppert-Mayer 1931; Kaiser and Garrett 1961), which, here, specifically means the simultaneous absorption of a number (≥2) of photons, combining their quantum energies to bridge the gap between ground and excited molecular state. For one-photon excitation, the decay in excitation rate per fluorophore ($\propto z^{-2}$), as one moves away from the focus (by z), is balanced by the increase in beam cross-sectional area ($\propto z^2$); for two-photon (and greater than two-photon) absorption, the excitation rate decreases much more quickly ($\propto z^{-2N}$), because it now depends on the light intensity to the Nth power, in which N is the number of simultaneously absorbed photons. This decay of the excitation rate is so rapid that, for $N = 2$ and for all $N > 2$, even for an infinitely thick specimen, most of the fluorescence excitation occurs near the focus. It is important to appreciate that there is a qualitative difference in the excitation distribution as one moves from single to multiphoton excitation.

Detection

Once excitation has been successfully localized (to a volume roughly the size of the wavelength cubed, 1 fL or less), there is no longer any great need for spatially selective detection. It follows that high-resolution objectives need not be used for collection of the fluorescence; a high-NA condenser, for example, will do just as well, if not better. There is no reason, however, to discard the fluorescence entering the objective, which, for a high-NA lens, can be a significant portion. This fluorescence is often sufficient to generate an image and is sometimes used exclusively, even in multiphoton laser-scanning microscopy. Threading the fluorescence light back through the scanning pathway (descanning), on the other hand, which is essential in a confocal microscope, can certainly be dispensed with (Denk et al. 1995a). Indeed, confocal detection (i.e., use of a confocal aperture) is often quite counterproductive because scattered fluorescence light (see below), which, in the case of localized excitation, carries perfectly good information (Denk et al. 1994), is lost.

Scattered Light

In scattering samples, confocal schemes are particularly wasteful (Denk et al. 1994; Denk 1996; Denk and Svoboda 1997). In a clear specimen, the photodynamic damage in the focal slice is comparable for confocal and multiphoton microscopies (assuming detection through the objective only) because a large fraction of the focal plane fluorescence is used, even in the confocal case. However, this is not the case for scattering tissue, in which even focal-slice fluorescence is increasingly lost in the confocal microscope. This is unavoidable, simply because a scattered focal photon's provenience is tainted, as it has become indistinguishable from out-of-focus light; such a photon will and should be rejected by the confocal pinhole. For multiphoton excitation, incident light is also scattered, but such light can no longer excite because it is too dilute in time and space (Denk and Svoboda 1997). To make up for scattering losses, the excitation power can be increased (in the absence of significant one-photon absorption) without exacerbating damage. For whole-field detection, the detection efficiency does not decrease significantly with depth (Oheim et al. 2001); fluorescence photons eventually emerge from the tissue (quite possibly after being scattered many times) because most biological tissues, even if strongly scattering, usually do not absorb much light. Incidentally, but quite fortuitously, better depth penetration is seen in multiphoton excitation because light scattering is typically reduced (Svaasand and Ellingsen 1983) as the wavelength increases.

Resolution

The theoretical spatial resolution of the multiphoton microscope is reduced, as compared with an ideal confocal microscope using the same fluorophore. This is because the excitation wavelength in the multiphoton instrument is roughly double that in the confocal microscope (Sheppard and Gu 1990; Gu 1996). In practice, however, the difference is much less, if present at all. This is because the finite pinhole size (which is needed for efficient detection; Gu and Sheppard 1991), chromatic aberration, and imperfect alignment of laser focus and detector pinhole all degrade resolution in the confocal microscope. The resolution of a multiphoton microscope can be improved by using confocal detection (Stelzer et al. 1994), but this comes at the cost of reduced collection efficiency, particularly for scattering samples, and with more severe chromatic aberration. Without confocal detection, chromatic aberration in the objective is only a minor concern when using multiphoton excitation. However, for highly chromatic systems (e.g., acousto-optical modulators), even the small spread of excitation wavelength that results from the shortness of the laser pulses needed to achieve efficient multiphoton excitation (Denk et al. 1990, 1995a) cannot be ignored.

Laser Pulse Width

Other potential problems in multiphoton microscopy are the increase in laser pulse width and the resultant decrease in multiphoton excitation efficiency that is caused by group velocity dispersion (GVD) in conjunction with ultrashort pulses. GVD is the result of the difference in propagation speed for wave packets of different wavelengths in different optical materials (Denk et al. 1995a). The effect can be partially compensated for by using a prism sequence (Gordon and Fork 1984). In practice, GVD has not been a real problem because pulses with widths of ~100 fsec (as produced by commercial titanium:sapphire lasers) broaden only by a fraction of the original pulse width after passage through the microscope optics. However, if shorter excitation pulses (fractional broadening depends on the square of the inverse pulse width) and higher-order (≥3 photons) nonlinear processes are to be used, GVD compensation may become useful and even necessary.

INSTRUMENTATION

Practical Considerations

Excitation Light Source

The main difference between multiphoton excitation and confocal microscopy is the excitation light source. The average multiphoton excitation rate depends, like any nonlinear optical process, on the temporal structure of the excitation light and not just—as in the case of linear (one-photon) absorption, which is a special case in this respect—on the average intensity. In particular, the use of pulsed light with a very small duty cycle enhances N-photon processes by the inverse duty cycle to the $(N-1)$th power. The use of mode-locked lasers (e.g., with pulse lengths of around 100 fsec and repetition rates of ~100 MHz) increases the multiphoton excitation rate enormously, as compared with a continuous-wave laser at the same average power: Rates are increased by factors of 10^5 and 10^{10} for two-photon and three-photon excitation, respectively. It is not only the average available laser power that dictates the use of mode-locked lasers. These lasers are often required to minimize residual one-photon absorption at the fundamental wavelength. The development of ultrashort-pulse mode-locked lasers was the breakthrough that made two-photon microscopy truly practical (Denk et al. 1990). (For a collection of useful reprints on these types of lasers, see Gosnell and Taylor 1991; see also Chapter 13.)

Wavelength Selection

Because the excitation wavelengths for two-photon microscopy and one-photon confocal laser-scanning microscopy are different (roughly on the order of 2:1, respectively), different dichroic beam splitters must be used. Sometimes the emission filter must also be changed to suppress the new excitation wavelength efficiently. Usually, however, elimination of the excitation light actually becomes easier because the two-photon excitation wavelength is longer and is separated from the fluorescence wavelengths by several hundred nanometers. Discrimination against excitation light can often be accomplished by using colored glass filters and is helped by the decrease in the sensitivity of photomultiplier tubes toward longer wavelengths.

SETTING UP A MULTIPHOTON MICROSCOPY SYSTEM

There are several ways of setting up a system for multiphoton excitation microscopy. Some of the options are discussed below.

- *Buy a complete commercial instrument* (e.g., Bio-Rad Laboratories, Zeiss, Olympus, LaVision [femtosecond instruments], Leica [picosecond instrument]). The commercial solution will work right away, but it will be expensive and may not be optimal for the intended application.
- *Convert an existing laser-scanning microscope.* This will be the most sensible approach if a (under-used) confocal microscope is already available. The main changes necessary are the replacement of dichroic beam splitters, filters, and those excitation path steering mirrors that are of the multilayer dielectric type. Detection efficiency can usually be gained by adding a whole-field (also called external) detector (Denk et al. 1995a). This is essential if scattering samples are to be imaged (see Chapter 8).
- *Build an instrument from scratch.* This can be the most economical approach and, more importantly, allows the instrument to be tailored to achieve the best performance for a particular application. Technical expertise is required, however, and the resulting instrument may not be very user friendly.

EXAMPLES OF APPLICATIONS

The following list contains selected publications from my own and other laboratories that have used multiphoton microscopy in neuroscience and related applications:

- two-photon imaging and microspectrofluorometry of synaptic spines (Denk et al. 1995b, 1996; Yuste and Denk 1995; Koester and Sakmann 1998; Sabatini and Svoboda 2000; Sabatini et al. 2001; Oertner et al. 2002), calcium in hair-cell stereocilia (Denk et al. 1995c)
- two-photon measurement of dendritic activity in vivo (Svoboda et al. 1997, 1999; Helmchen et al. 1999)

- imaging in behaving rats using a miniaturized fiber-coupled two-photon microscope (Helmchen et al. 2001)
- photochemical release of caged compounds (Denk 1994; Svoboda et al. 1996; Furuta et al. 1999; Matsuzaki et al. 2001)
- imaging of green fluorescent protein (GFP)-labeled cells (Niswender et al. 1995; Potter et al. 1996; Kohler et al. 1997), nicotinamide adenine dinucleotide plus hydrogen (NADH) in tissue (Piston et al. 1995; Bennett et al. 1996), and chromosomes in embryos (Summers et al. 1996)
- three-photon imaging of intrinsic (Maiti et al. 1997) and extrinsic (Wokosin et al. 1996) fluorophores
- observation of long-term changes in neurite morphology (Grutzendler et al. 2002; Trachtenberg et al. 2002)
- measurement of visual stimulus-induced calcium signals in the retina (Denk and Detwiler 1999; Euler et al. 2002)

REFERENCES

Bennett BD, Jetton TL, Ying G, Magnuson MA, Piston DW. 1996. Quantitative subcellular imaging of glucose metabolism within intact pancreatic islets. *J Biol Chem* **271:** 3647–3651.

Denk W. 1994. Two-photon scanning photochemical microscopy: Mapping ligand-gated ion channel distributions. *Proc Natl Acad Sci* **91:** 6629–6633.

Denk W. 1996. Two-photon excitation in functional biological imaging. *J Biomed Opt* **1:** 296–304.

Denk W, Detwiler PB. 1999. Optical recording of light-evoked calcium signals in the functionally intact retina. *Proc Natl Acad Sci* **96:** 7035–7040.

Denk W, Svoboda K. 1997. Photon upmanship: Why multiphoton imaging is more than a gimmick. *Neuron* **18:** 351–357.

Denk W, Strickler JH, Webb WW. 1990. Two-photon laser scanning fluorescence microscopy. *Science* **248:** 73–76.

Denk W, Delaney KR, Gelperin A, Kleinfeld D, Strowbridge BW, Tank DW, Yuste R. 1994. Anatomical and functional imaging of neurons using 2-photon laser scanning microscopy. *J Neurosci Methods* **54:** 151–162.

Denk W, Piston DW, Webb WW. 1995a. Two-photon molecular excitation in laser scanning microscopy. In *The handbook of confocal microscopy* (ed. J Pawley), pp 445–458. Plenum, New York.

Denk W, Sugimori M, Llinas R. 1995b. Two types of calcium response limited to single spines in cerebellar Purkinje cells. *Proc Natl Acad Sci* **92:** 8279–8282.

Denk W, Holt JR, Shepherd GMG, Corey DP. 1995c. Calcium imaging of single stereocilia in hair cells: Localization of transduction channels at both ends of tip links. *Neuron* **15:** 1311–1321.

Denk W, Yuste R, Svoboda K, Tank DW. 1996. Imaging calcium dynamics in dendritic spines. *Curr Opin Neurobiol* **6:** 372–378.

Euler T, Detwiler PB, Denk W. 2002. Directionally selective calcium signals in dendrites of starburst amacrine cells. *Nature* **418:** 845–852.

Furuta T, Wang SSH, Dantzker JL, Dore TM, Bybee WJ, Callaway EM, Denk W, Tsien RY. 1999. Brominated 7-hydroxycoumarin-4-ylmethyls: Photolabile protecting groups with biologically useful cross-sections for two photon photolysis. *Proc Natl Acad Sci* **96:** 1193–1200.

Göppert-Mayer M. 1931. Ueber Elementarakte mit zwei Quantenspruengen. *Ann Phys* **9:** 273.

Gordon JP, Fork RL. 1984. Optical resonator with negative dispersion. *Opt Lett* **9:** 153–155.

Gosnell TR, Taylor AJ, eds. 1991. *Selected papers on ultrafast laser technology*. SPIE, Bellingham, WA.

Grutzendler J, Kasthuri N, Gan WB. 2002. Long-term dendritic spine stability in the adult cortex. *Nature* **420:** 812–816.

Gu M. 1996. Resolution in three-photon fluorescence scanning microscopy. *Opt Lett* **21:** 988–990.

Gu M, Sheppard CJR. 1991. Effects of finite-sized detector on the OTF of confocal fluorescent microscopy. *Optik* **89:** 65–69.

Helmchen F, Svoboda K, Denk W, Tank DW. 1999. In vivo dendritic calcium dynamics in deep-layer cortical pyramidal neurons. *Nat Neurosci* **2:** 989–996.

Helmchen F, Fee MS, Tank DW, Denk W. 2001. A miniature head-mounted two-photon microscope. High-resolution brain imaging in freely moving animals. *Neuron* **31:** 903–912.

Kaiser W, Garrett CBG. 1961. Two-photon excitation in CaF_2:Eu^{2+}. *Phys Rev Lett* **7:** 229–231.

Koester HJ, Sakmann B. 1998. Calcium dynamics in single spines during coincident pre- and postsynaptic activity depend on relative timing of back-propagating action potentials and subthreshold excitatory postsynaptic potentials. *Proc Natl Acad Sci* **95:** 9596–9601.

Kohler RH, Cao J, Zipfel WR, Webb WW, Hanson MR. 1997. Exchange of protein molecules through connections between higher plant plastids. *Science* **276:** 2039–2042.

Maiti S, Shear JB, Williams RM, Zipfel WR, Webb WW. 1997. Measuring serotonin distribution in live cells with three-photon excitation. *Science* **275:** 530–532.

Masters BR, ed. 1996. *Selected papers on confocal microscopy*. SPIE Milestone Series, SPIE, Bellingham, WA.

Matsuzaki M, Ellis-Davies GCR, Nemoto T, Miyashita Y, Iino M, Kasai H. 2001. Dendritic spine geometry is critical for AMPA receptor expression in hippocampal CA1 pyramidal neurons. *Nat Neurosci* **4:** 1086–1092.

Minsky M. 1961. Microscopy apparatus. U.S. Patent No. 3013467.

Niswender KD, Blackman SM, Rohde L, Magnuson MA, Piston DW. 1995. Quantitative imaging of green fluorescent protein in cultured cells: Comparison of microscopic techniques, use in fusion proteins and detection limits. *J Microsc* **180:** 109–116.

Oertner TG, Sabatini BL, Nimchinsky EA, Svoboda K. 2002. Facilitation at single synapses probed with optical quantal analysis. *Nat Neurosci* **10:** 10.

Oheim M, Beaurepaire E, Chaigneau E, Mertz J, Charpak S. 2001. Two-photon microscopy in brain tissue: Parameters influencing the imaging depth. *J Neurosci Methods* **111:** 29–37.

Piston DW, Masters BR, Webb WW. 1995. Three-dimensionally resolved NAD(P)H cellular metabolic redox imaging of the in situ cornea with two-photon excitation laser scanning microscopy. *J Microsc* **178:** 20–27.

Potter SM. 1996. Vital imaging: Two photons are better than one. *Curr Biol* **6:** 1595–1598.

Potter SM, Wang CM, Garrity PA, Fraser SE. 1996. Intravital imaging of green fluorescent protein using two-photon laser-scanning microscopy. *Gene* **173:** 25–31.

Sabatini BL, Svoboda K. 2000. Analysis of calcium channels in single spines using optical fluctuation analysis. *Nature* **408:** 589–593.

Sabatini BL, Maravall M, Svoboda K. 2001. Ca^{2+} signaling in dendritic spines. *Curr Opin Neurobiol* **11:** 349–356.

Sheppard CJR, Gu M. 1990. Image-formation in two-photon fluorescence microscopy. *Optik* **86:** 104–106.

Stelzer EHK, Hell S, Lindek S, Stricker R, Pick R, Storz C, Ritter G, Salmon N. 1994. Nonlinear absorption extends confocal fluorescence microscopy into the ultra-violet regime and confines the illumination volume. *Opt Commun* **104:** 223–228.

Summers RG, Piston DW, Harris KM, Morrill JB. 1996. The orientation of first cleavage in the sea urchin embryo, *Lytechinus variegatus*, does not specify the axes of bilateral symmetry. *Dev Biol* **175:** 177–183.

Svaasand LO, Ellingsen R. 1983. Optical properties of human brain. *Photochem Photobiol* **38:** 293–299.

Svoboda K, Tank DW, Denk W. 1996. Direct measurement of coupling between dendritic spines and shafts. *Science* **272:** 716–719.

Svoboda K, Denk W, Kleinfeld D, Tank DW. 1997. In vivo dendritic calcium dynamics in neocortical pyramidal neurons. *Nature* **385:** 161–165.

Svoboda K, Helmchen F, Denk W, Tank DW. 1999. Spread of excitation in layer 2/3 pyramidal neurons in rat barrel cortex in vivo. *Nat Neurosci* **2:** 65–73.

Trachtenberg JT, Chen BE, Knott GW, Feng G, Sanes JR, Welker E, Svoboda K. 2002. Long-term in vivo imaging of experience-dependent synaptic plasticity in adult cortex. *Nature* **420:** 788–794.

Williams RM, Piston DW, Webb WW. 1994. Two-photon molecular excitation provides intrinsic 3-dimensional resolution for laser-based microscopy and microphotochemistry. *FASEB J* **8:** 804–813.

Wokosin DL, Centonze VE, Crittenden S, White J. 1996. Three-photon excitation fluorescence imaging of biological specimens using an all-solid-state laser. *Bioimaging* **4:** 1–7.

Yuste R, Denk W. 1995. Dendritic spines as basic functional units of neuronal integration. *Nature* **375:** 682–684.

8 How to Build a Two-Photon Microscope with a Confocal Scan Head

Volodymyr Nikolenko and Rafael Yuste

Department of Biological Sciences, Howard Hughes Medical Institute, Columbia University, New York, New York 10027

ABSTRACT

This chapter provides practical guidelines for the conversion of a standard confocal microscope into a two-photon microscope. This conversion enables the investigator to have access to two-photon microscopy without the large budget necessary to purchase a commercial instrument. Two-photon fluorescence microscopy allows deep-tissue imaging in highly scattering preparations and long-term imaging of live tissue without the photodamage that is caused by out-of-focus light (see Chapter 7). It is, therefore, an essential tool for imaging cells under physiologically relevant conditions such as acute or cultured brain slices or in vivo.

IMAGING SETUP

Laser

The key component of the system is a femtosecond-pulsed laser that can generate reasonable power in the near-infrared (NIR) spectral region needed for convenient practical imaging (average power >50 mW). Our current setup uses the Chameleon laser from Coherent, Inc. This is a fully automated turnkey laser, which can be tuned to any wavelength between 690 and 1080 nm.

Optical Table

The laser beam is delivered to a modified Olympus FluoView confocal laser-scanning system through the set of optical elements on the optical table (Fig. 1). The optical elements are intermediate mirrors, a spatial filter, a retardation wave plate, and a Pockels cell. BB1-E02 dielectric mirrors from Thorlabs, Inc. are used as intermediate mirrors. These reflect >99% of light between 700 and 1150 nm at a 45° angle of incidence for all polarizations and do not introduce additional group velocity dispersion of ultrafast pulses. An optical spatial filter, which has two plano-convex lenses and a pinhole in the focus of the first lens, acts as a simple telescope. It is used to restore a smooth Gaussian profile to the intensity of the laser beam cross section and also to modify the beam size to ensure proper overfilling of the back aperture of the microscope objective (Tsai et al. 2002). A retardation wave plate ($\lambda/2$ or $\lambda/4$) is used for complex experiments in which the polarization of the scanning laser beam must be controlled.

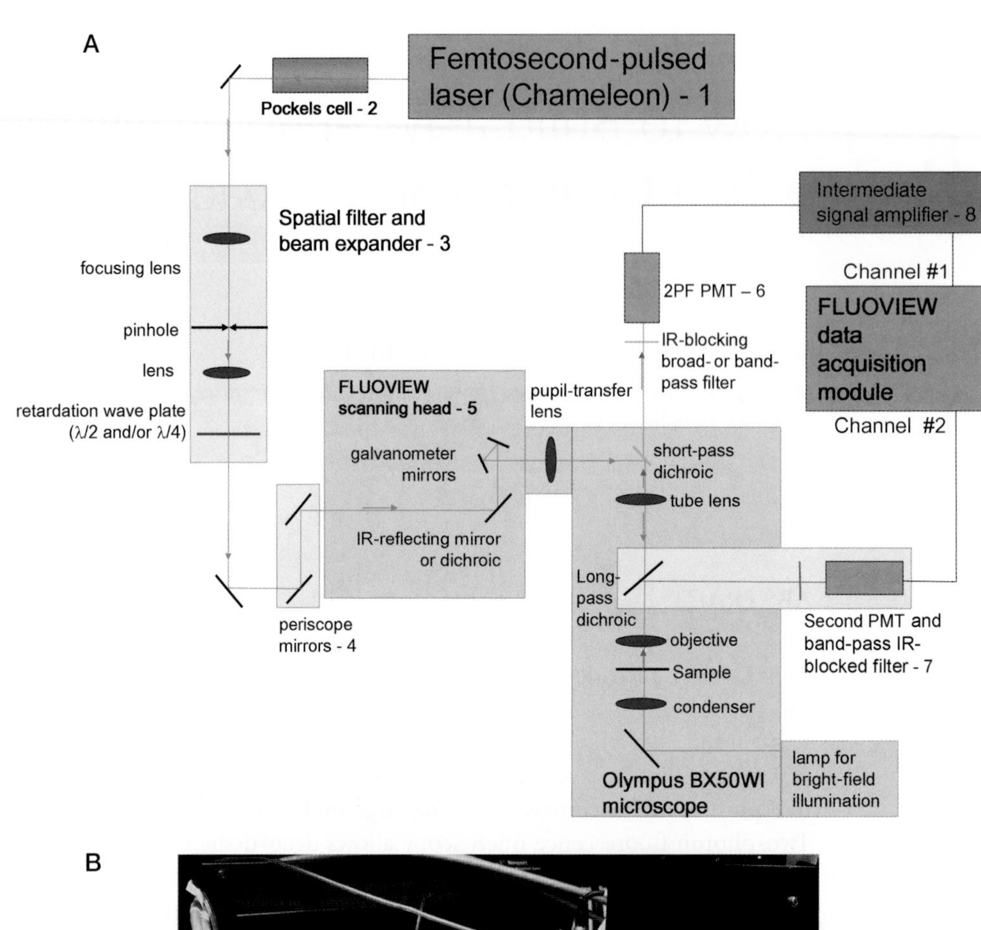

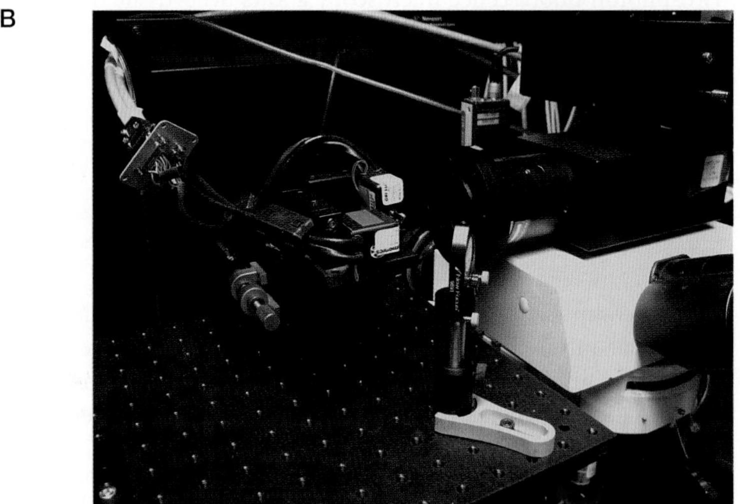

FIGURE 1. (*A*) Optical design of the instrument. Major components: (1) NIR femtosecond-pulsed laser; (2) Pockels cell modulator; (3) system of lenses, which works as a spatial filter and a beam expander (~1.2x in our case); (4) periscope mirrors (which deliver the laser beam from the optical table level to the scanning box of the FluoView, which is raised for the upright microscope); (5) modified Olympus FluoView scanning unit; (6) external photomultiplier tube (PMT) detector for the 2P-fluorescence signal, attached to the camera port of the microscope; (7) second external PMT attached to the microscope through a custom-made adapter; (8) intermediate signal amplifier for matching the dynamic range of the signal source (PMT) and the FluoView data acquisition module. (*B*) A photograph of the modification of the FluoView confocal system (element 5 in *A*): The original galvanometer mirrors are completely removed from the Olympus scanning unit and are placed on a generic breadboard in front of the pupil transfer (scan) lens.

Pockels Cell

A 350-160 Pockels cell, a nonlinear optical modulator from Conoptics, is included to allow the dynamic modulation of laser light intensity with excellent contrast (>200:1) and submicrosecond temporal resolution. The temporal resolution of the Pockels cell is limited, in practical terms, only by the electronics of the high-voltage driver. The model in our current setup (275 linear amplifier from Conoptics) can work in a DC-8-MHz modulation range (thus providing 125-nsec temporal resolution).

Scanning Microscope

A BX50WI Olympus upright scanning microscope is coupled to a modified FluoView scanning unit. The Olympus FluoView platform relieves the need for a separate, custom-built scanning system and software package (Majewska et al. 2000).

The scanning box contains only one essential component: a set of galvanometer mirrors, which steer the laser beam and scan the image (see Fig. 1B). The scanning box is optically linked to the infinity-corrected BX50WI microscope through a pupil transfer lens (or scan lens), which is part of the original FluoView system. This lens, together with the tube lens of the microscope (original part of the microscope), forms a telescope, which provides collimated light for the infinity-corrected objective lens (see optical scheme in Fig. 1). This telescope also approximately images the scanning mirrors onto the back aperture of the objective. This minimizes the movement of the laser beam at the back aperture, thus reducing variation of laser power at the sample. The laser beam is reflected downward by a short-pass dichroic mirror (650DCSP or similar from Chroma Technology) placed inside the standard trinocular tube of the Olympus microscope.

Fluorescence Detection

In the case of two-photon absorption, excitation of fluorescence is essentially limited to the diffraction-limited spot in the focal plane. This provides the 3D sectioning characteristic of two-photon microscopy. Because fluorescence from the excited region irradiates in all directions, it is important to use a high-numerical-aperture objective to collect as many fluorescence photons as possible. Our system uses an external photomultiplier tube (PMT) as a detector. The tube is mounted on the camera port of the microscope's trinocular tube—along with the dichroic mirror mentioned above, which transmits visible fluorescent light collected by the objective. An additional infrared (IR)-blocking filter (et700sp-2p8 from Chroma Technology) is placed in front of the PMT to filter out residual IR fluorescent light reflected from the excitation path. The external PMT could also be positioned right next to the objective (see schematic in Fig. 1). In this case, the PMT would have to be mounted via a custom-made adapter with a long-pass dichroic mirror, which transmits excitation IR light and reflects visible fluorescence to the detector. Positioning the PMT on top of the objective improves light collection efficiency but compromises the convenient positioning of the micromanipulators.

By choosing additional appropriate dichroic mirrors, it is possible to use multiple channels of fluorescence imaging. Placing additional band-pass filters in front of the detectors can also efficiently separate the emissions of different fluorescent dyes. The Olympus FluoView data acquisition board comes already equipped with two independent input channels (see Chapter 1 and Appendix 2 for example applications and for advice on choosing filters and dichroic mirrors).

The PMT used in our instrument is the H7422P-40 assembly from Hamamatsu, which comprises a GaAsP, a head-on PMT, a high-voltage power supply, and a thermoelectric cooler. This PMT provides a current signal that is proportional to the light intensity. The signal is fed into a current preamplifier (SRS570, Stanford Research Systems [SRS]) and then into the Olympus FluoView data acquisition board. The SR570 has variable gain and adjustable high/low-pass filters and can be controlled through a user-friendly computer interface. The preamplifier is very convenient for proper signal conditioning because it allows proper filling of the dynamic range of the FluoView signal analog-to-digital converter, something crucially important for imaging weak signals and quantitative measurements. The standard Olympus FluoView software is used in the scanning mode for signal acquisition and reconstruction of a digital image. For calibration curves of the available dynamic range of the Olympus FluoView signal inputs, see Nikolenko et al. (2003).

EXPERIMENTAL METHOD

The full description of the practical changes in the standard Olympus FluoView confocal system can be found in Majewska et al. (2000) and Nikolenko et al. (2003). A brief summary of the necessary modification is presented here.

1. Install the pulsed femtosecond NIR laser.
2. Build the external optical pathway from the laser source to the laser-scanning microscope with the optional spatial filter and the Pockels cell.
3. Modify the scanning unit of a confocal laser-scanning microscope for scanning by the IR beam from the external laser source. This essentially requires drilling a hole in the back of the scanning unit and replacing the internal dichroic mirror with an IR-reflecting mirror (see the full list of Olympus FluoView modifications in Majewska et al. 2000 and Nikolenko et al. 2003). An alternative and more radical approach to drilling the hole in the original FluoView scan unit is to remove the scanning mirrors with their control cables altogether and to mount them on a separate optical breadboard as shown in Figure 1B.
4. Place a short-pass dichroic mirror into the trinocular tube of the optical microscope, and install the external detector (PMT) on the camera-imaging port of the trinocular tube.
5. Connect the detector signal output to the data acquisition input of the confocal system through the intermediate signal amplifier.

DISCUSSION

Advantages and Limitations

The two-photon system described here, based on the Olympus FluoView confocal system, successfully combines a customized homemade system with a reliable commercial instrument. Although tailor-made for the chosen application, the individual elements of a homemade system can be difficult to maintain in optimal working condition. However, recent advances are easing this problem. For example, the appearance on the market of turnkey femtosecond-pulsed tunable lasers, such as the Chameleon, eliminates the need for realignment of the optical path. Potential users, however, should be aware of the correct procedures for cleaning optical surfaces that are exposed to dust, changes in humidity, etc., and should follow proper laser safety guidelines (see Appendix 6).

Application of the Two-Photon Microscope with a Confocal Scan Head

The custom-made two-photon microscope described here has been used successfully for long-term imaging of the action potential activity in large (>1000) populations of neocortical neurons in acute brain slices loaded with Ca^{2+} fluorescent indicators such as Fura-2-acetoxymethyl ester (Fura-2AM) (Cossart et al. 2003) or Indo-1AM (Fig. 2).

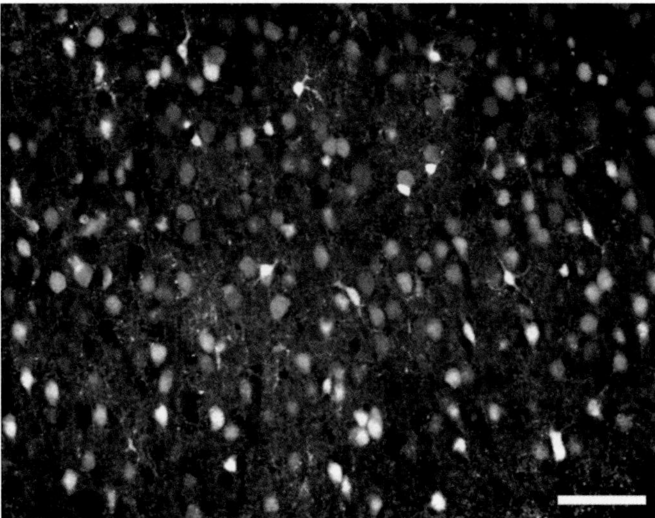

FIGURE 2. Example of imaging. A neocortical brain slice, taken from a postnatal day-13 mouse, loaded with the calcium indicator Indo-1AM. Two-photon fluorescence image acquired with ~730-nm excitation wavelength. Scale bar, 50 μm.

REFERENCES

Cossart R, Aronov D, Yuste R. 2003. Attractor dynamics of network UP states in the neocortex. *Nature* **423:** 283–288.

Majewska A, Yiu G, Yuste R. 2000. A custom-made two-photon microscope and deconvolution system. *Pflügers Arch* **441:** 398–408.

Nikolenko V, Nemet B, Yuste R. 2003. A two-photon and second-harmonic microscope. *Methods* **30:** 3–15.

Tsai PS, Nishimure N, Yoder EJ, Dolnik EM, White GA, Kleinfield D. 2002. Principles, design and construction of a two photon scanning microscope for in vitro and in vivo studies. In *Methods for in vivo optical imaging* (ed. Frostig R), pp. 113–171. CRC, New York.

9 Arc Lamps and Monochromators for Fluorescence Microscopy

Rainer Uhl

BioImaging Zentrum (BIZ), Ludwig-Maximilians-Universität München, 82152 Martinsried, Germany

ABSTRACT

Fluorescence microscopy requires high photon-flux densities in the specimen plane. These intensities are only achieved by lasers, arc lamps, and, most recently, light-emitting diodes (LEDs). Lasers and LEDs, however, are restricted to a limited number of wavelength regions, whereas with arc lamps it is possible to select arbitrary wavelengths and wavelength regions. (Recently so-called white light lasers have become commercially available, but their photon fluxes—although sufficient for laser scanning applications—are still not high enough for applications where extended areas need to be illuminated.) Moreover, the lower cost of arc lamps, compared with lasers, makes them the light source of choice for the majority of fluorescence microscopy applications. This chapter discusses arc lamps and the design and performance of an arc lamp–based illumination system for fluorescence microscopy that allows the user to choose any wavelength from ultraviolet to infrared, and permits rapid switching speed between colors, while maintaining quite stable and homogenous emissions.

ADVANTAGES OF ARC LAMPS AND MONOCHROMATORS FOR FLUORESCENCE MICROSCOPY

Arc lamps offer a sunlike spectrum reaching from the UV to the infrared, are relatively inexpensive, and, if used in the right configuration, can obtain short- and long-term stability values better than any other bright light source. With fiber coupling, they can be positioned at a distance from the microscope, thus minimizing detrimental effects due to heat and vibrations. By using critical illumination, a degree of homogeneity can be achieved in the specimen plane that is not attained when using any other light source.

To exploit the wide spectral coverage of arc lamps, the desired wavelength range is selected from the white spectrum by placing interference filters into the beam path. Filter wheels, which are often used for this task, are relatively slow, requiring >50 msec to switch from one filter position to the next. A faster alternative is galvanometer technology, which directs the beam path through different filters (e.g., the Lambda DG-4 [Sutter Instrument] or the Oligochrome [TILL Photonics]). Although it requires only 1–2 msec to move between filters, galvanometer-driven mirror systems limit the number of available colors to four or five and do not permit bandwidth adjustment. The most versatile filter system is a rapidly scanning monochromator, which selects the desired wavelength by tilting a grating in the illumination beam and can jump from one wavelength to the next in 1–2 msec.

CHARACTERISTICS OF FIBER-COUPLED ARC LAMPS

Spectrum

Three classes of arc lamps are used in fluorescence microscopy: mercury, xenon, and metal halide lamps. All three cover a wide spectral range (Fig. 1), but only xenon lamps show a modest line structure, which is essential for use in a monochromator. The spectral composition delivered by an illumination device depends not only on the spectrum of the map but also on the band-pass characteristics of a given filter or the grating and the optical design used in the monochromator. Although filters show very steep spectral transitions from blocking to transmitting wavelengths, monochromators roll off less steeply outside their chosen wavelength band; thus, if a narrow spectral band in the spectrum of a lamp falls onto the slope of the monochromator transmission, the spectral maximum may not be where it is thought to be. The monochromator design described below thus uses a xenon arc lamp as its light source, because its spectrum does not exhibit narrow spectral bands.

The output power of Xe-arc lamps does not vary by >50% when operating in the range from 320 to 700 nm. Photon fluxes of $>10^{23}$ $m^{-2}sec^{-1}$ can be achieved in the specimen plane, implying a photon transfer efficiency close to the theoretical limits for the lamp. A typical fluorophore, like fluorescein isothiocyanate (FITC) or green fluorescent protein (GFP), emits $>10^3$ photons/sec under these conditions, which is enough for single-molecule detection.

Spectral Purity

Filter-based systems show better than 6 orders of magnitude blocking outside their transmission band. This level of spectral purity cannot be achieved with a monochromator, which typically shows stray-light levels in the range of 3.5–4 OD. The cure is simple: Equip the filter cube used for a given

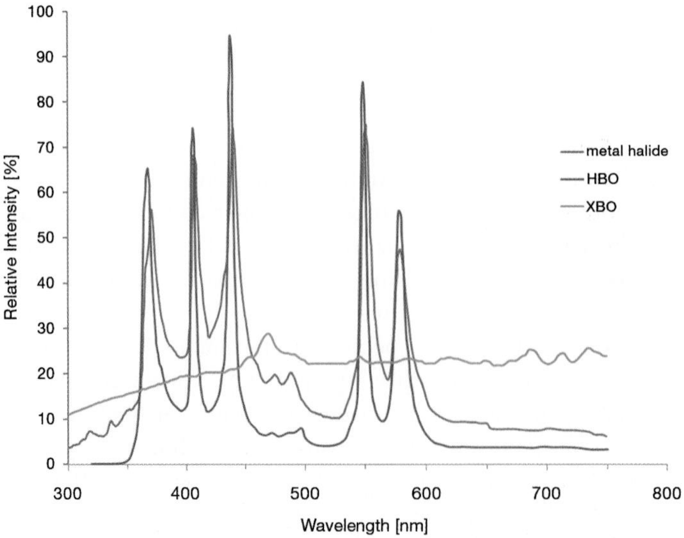

FIGURE 1. Spectral characteristics of three types of arc lamps.

experiment with a suitable short-pass filter in the excitation beam, which transmits all light below the cut-on wavelength of the emission filter and blocks all light in the spectral region where the emission filter transmits. This preserves the full spectral flexibility of the monochromator design while at the same time allowing for the same >6 OD background rejection of a filter-based system.

Usable Output Power

In arc lamps, radiation originates from spatially confined plasma, whose dimensions increase with increasing lamp power. This has consequences for the maximum photon flux that can be concentrated onto a given sample area. According to the Helmholtz–Lagrange invariability principle, every attempt to concentrate light originating from a spot with diameter d_1 onto a smaller spot of diameter d_2 (i.e., $d_2 < d_1$), invariably increases the NA at the target spot over the NA of the source spot. (The invariability principle states that $n \times \mathrm{NA} \times d$ [called the éntendu] remains constant throughout an optical system, where n is the index of refraction of the medium, NA is the numerical aperture, and d is the diameter of a source of radiation or a receiving element.) In a microscope, the NA of the objective that is in use limits the NA of the "target side." Hence, increasing the wattage of a lamp (and hence its diameter) beyond a certain value only increases the total photon flux but not the flux density at the sample. This is illustrated by the following example: Imagine using a 40× oil-immersion objective with NA 1.3. The goal is to concentrate as much light as possible onto the field that is covered by a 0.65-in camera chip positioned under the microscope (225 µm). If we assume that the lamp condenser collects light with high NA (e.g., 0.65), then the usable spot of the arc lamp has a diameter of $1.3/0.65 \times 225\ \mu\text{m} = 450\ \mu\text{m}$. Thus, every arc lamp with a hot spot greater than 450 µm will create light that cannot be used. Figure 2 shows why the hot spot size of a 150-W xenon arc lamp is a good match for the experiment described.

Stability

Until very recently, mercury arc lamps have been the most commonly used light source for fluorescence microscopy. They show the smallest arc dimensions and hence can provide the highest photon-flux densities of all arc sources, particularly within the lines of the typical mercury spectrum. However, they show poor long-term stability; output power drops 30% within 200 h. This not only means a tedious lamp change but also a high cost of ownership. Thus metal halide lamps (which were first used in consumer projection devices) have recently replaced mercury arc lamps as the most popular light source for fluorescence microscopy. The bulbs are easily exchanged and they show lifetimes of >1000 h. Whereas long-term stability is good, short-term stability is poor owing to arc flicker, which occurs irregularly every few seconds and causes short-term fluctuations of several percent. Although this does not matter for longer time-lapse studies that allow for longer inte-

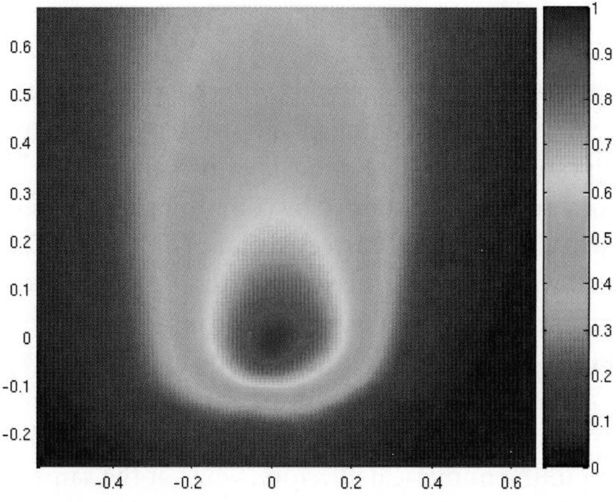

FIGURE 2. Intensity distribution (false colors) of the arc of a 150-W xenon arc lamp.

gration times, or for nonquantitative experiments in which temporal stability is not essential, more demanding quantitative experiments are best served by xenon arc lamps, which provide the best short-term and long-term stability of all arc lamps. They operate for 2000–3000 h and, provided their constant current power supply does not degrade performance, RMS (root mean square) intensity fluctuations remain in the range of 3×10^{-3}. To achieve this degree of stability, xenon lamps need to be convection cooled. Forced-air cooling aggravates arc wander, which is caused by convection inside the quartz envelope, and is the major source of lamp instability in xenon lamps.

Fiber Delivery

Most commercially available microscopes have been designed with an epifluorescence condenser optimized for direct coupling of an attached mercury arc lamp. These lamps usually have a power consumption of 50–100 W; hence they can still be convection cooled. The higher wattage and optical packaging of metal halide lamps renders forced-air cooling a necessity, and to avoid unwanted heat and vibration transfer from the lamp to the microscope, metal halide lamps are usually coupled to the microscope by means of a liquid light guide. However, a metal halide lamp with its built-in reflector has not been designed for concentrating light onto the smallest possible spot. Therefore, to collect a large fraction of its radiation, liquid light guides of 5 mm diameter must be used. Their high NA of >0.5 brings the Helmholtz–Lagrange invariability principle into play again, and thus only a small fraction of the light carried by the light guide is usable. As an example, consider a 60× oil-immersion objective, whose maximum NA for illumination is 1.33 (every greater angle suffers total internal reflection and thus does not reach the sample). The maximum demagnification achievable without loss is 1.33/0.5 = 2.66; thus, the image size of the light guide cannot be reduced below 5/2.66 = 1.88 mm. With a 60× objective this would correspond to a field size of 112 mm, in contrast to the standard microscope field size of 22 mm. As a consequence only 3.8% of the radiation transported by the light guide is usable. The advantage of the large diameter of the liquid light guide is that no sophisticated optics are required to couple its light into a microscope. By overilluminating both the field and pupil of the objective, a homogeneous illumination is almost always guaranteed.

A very different situation arises if quartz fibers are used. They have much smaller diameters and a smaller NA (usually 0.22), and maximum homogeneity is only achieved by means of critical illumination, that is, by imaging the exit window of the fiber into the specimen plane. This requires sophisticated condenser designs, one for each type of microscope, and careful alignment. However, if set up correctly, such a system provides the best transfer efficiency and maintains it over an indefinite period of time, because quartz fibers (unlike liquid light guides, which show significant aging) do not degrade with age.

EXPERIMENTAL SETUP

The Illumination System

As noted above, transferring light from a source to a target without sacrificing brightness (photon-flux density) requires an optical system with good imaging quality. Moreover, for a monochromator system covering an extended wavelength range, this level of optical performance has to be maintained over the spectral range in question. The best results are obtained by combining an inherently achromatic reflective design (toroidal mirror) with a chromatically corrected pair of aplanatic lenses (Fig. 3). Light from a xenon arc lamp (XA) is collected with a high numerical aperture of 0.56. A positive quartz aplanatic lens (PAL) reduces this numerical aperture by 1/1.46 (where 1.46 is the refractive index of quartz) while at the same time creating a virtual 1.46× magnified image of the arc, which is reimaged by a toroidal mirror (TM). The toroidal surface is the approximation of an elliptical surface, which provides 1:1 imaging for source and target spots separated from each other. A second (negative) aplanatic lens (NAL) causes a further reduction in the numerical aperture, while at the same time increasing image size accordingly. The material of the NAL (LF-5, refractive index 1.6) is chosen so that it forms an

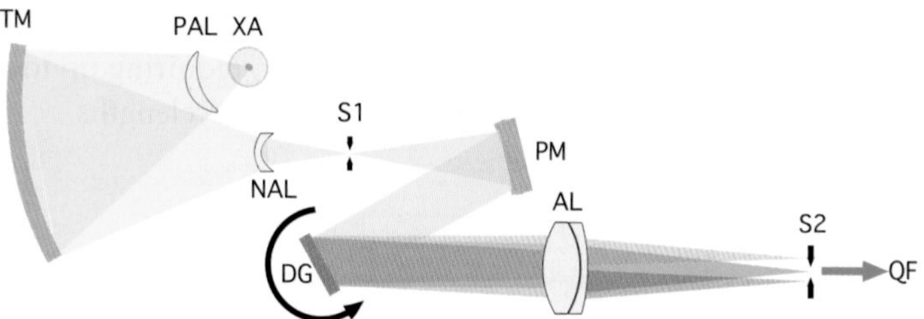

FIGURE 3. Schematic diagram of the optical setup of a monochromator-based illumination system. See text for details.

achromatic pair with the PAL. The combination of the three optical elements PAL, TM, and NAL, reduces the numerical aperture to 0.56/(1.46 × 1.6) = 0.24 while at the same time increasing image size 2.33 times. This magnified image fills the entrance slit (S1) of the monochromator.

The Monochromator

White light passing entrance slit S1 is collimated by an off-axis parabolic mirror (PM) and directed onto a diffraction grating (DG), having 1302 lines/mm. The diffracted light is collected by an UV-transmitting achromatic lens (AL), which projects the resulting spectrum onto an exit slit (S2). By adjusting the tilt of DG, the spectrum moves over the exit slit and different spectral regions are allowed to pass. The light that passes through S2 is coupled into a quartz–quartz fiber (QF) with a 1.25-mm diameter and NA of 0.22.

Standard systems are equipped with an entrance slit width of 1.5 mm and an exit slit width of 2.4 mm. Given a focal length of the parabolic mirror of 50 mm and a focal length of the achromatic lens of 76 mm, the bandwidth (half-width) is 14 nm. By changing both the entrance and exit slit widths, the spectral bandwidth can be changed. By changing only one of them, the intensity can be controlled while maintaining half-width.

Dynamic Performance

When mounted on a galvanometric scanner, the angle of the monochromator grating can be adjusted, which alters the wavelength that passes through the exit slit. An adjustment occurs in 1–1.5 msec, depending on the galvo-excursion required for a given wavelength change. Scanner accuracy is maintained to better than 0.01 nm over the lifetime of the system.

By choosing a resting wavelength not transmitted by the microscope optics (e.g., 260 nm), the monochromator can be used as a shutter, too. An alternative is to choose an experiment-dependent resting wavelength, in which the blocked wavelength is determined by the blocking region of the shortpass filter within the filter cube, to the desired wavelength and back.

With a fast scanner, the exposure time can be set to millisecond precision. Unlike with rotating filter wheels, the exposure time can be set independently for each wavelength, and complex wavelength protocols can be executed under computer control. As with most modern single-lens reflex (SLR) cameras, image brightness can be determined by the exposure time. Because photodamage depends much more on the integral photon dosage and not on the intensity, neutral density filters are unnecessary.

One drawback is the nonzero transition time of the scanner during which illumination with undesired wavelengths occurs. The relative error increases as exposure time decreases. This can become an issue with calcium ratio experiments, in which for an exposure time of 3 msec the error is 7% and can be exacerbated if the dye has more absorption in the transition wavelengths than at the final wavelength (Grynkiewicz et al. 1985). Correct calcium concentrations may still be derived from ratio experiments, provided the calibration values R_{min} and R_{max} are determined using the same exposure times that are used in the experiment (Messler et al. 1996). Another cure is to "blindfold" the detector during wavelength transitions. This works with photomultipliers, gatable image intensifiers, or charge-coupled device cameras with fast electronic shutter capabilities (interline transfer sensors). Frame transfer cameras are slower, requiring up to a millisecond for this shift, so that pixels are likely to be illuminated by transition wavelengths.

CONCLUSION

Xenon arc lamps remain the light source of choice for wide-field fluorescence microscopy when wavelength flexibility, broad spectral coverage, and high-intensity stability are required. Monochromators based on galvanometers that carry a diffraction grating provide the most flexible means of achieving rapid shuttering and wavelength selection. The arc lamp and monochromator should be used together when performing quantitative time-lapse fluorescence imaging.

REFERENCES

Grynkiewicz G, Poenie M, Tsien RY. 1985. A new generation of Ca^{2+} indicators with greatly improved fluorescence properties. *J Biol Chem* **260:** 3440–3450.

Messler P, Harz H, Uhl R. 1996. Instrumentation for multiwavelengths excitation imaging. *J Neurosci Methods* **69:** 137–147.

10 Light-Emitting Diodes for Biological Microscopy

Tomokazu Sato[1] and Venkatesh N. Murthy[2]

[1]Graduate Program in Biophysics, Harvard Medical School, Boston, Massachusetts 02115; [2]Department of Molecular and Cellular Biology and Center for Brain Science, Harvard University, Cambridge, Massachusetts 02138-3800

ABSTRACT

Various technological advances have made imaging an increasingly useful tool in the life sciences. Imaging techniques that move away from the limitations of wide-field microscopy have allowed deeper, higher-resolution imaging of thick biological tissue. Even within wide-field microscopy, advancements such as structured and sheet illumination, as well as improvements of biological probes, have led to better visualization of cells and their subcellular structures. The illumination source for wide-field microscopy, however, has remained relatively unchanged for decades, relying mainly on xenon, mercury, and halogen lamps. Light-emitting diodes (LEDs) have existed for more than 80 years, but their spectral range and output light flux have only recently become large enough for use in biological applications. This chapter presents the basic information necessary to build and to use an LED-based illumination source for microscopy. It also provides some useful resources about LED advancements. Although commercial LED-based illuminators for microscopy are available, custom-built illumination systems can incorporate the latest LED chips into microscopes more quickly and inexpensively than is possible through the retail market.

INTRODUCTION

An LED is made of a combination of semiconducting materials (the so-called LED chip or die) that are doped with impurities to allow conversion of electrical energy into light through a reversible process called injection electroluminescence (Schubert 2006). The wavelength of generated light is determined by the energy bandgap at the p–n junction in the semiconductor. Various advances have been made in the growth and the choice of the semiconductor materials, allowing for brighter LEDs over a broader emission spectrum. Although there is a gap in high-efficiency LEDs within the yellow–green range, some LEDs emit in this wavelength by converting the primary emission through phosphors or semiconductor materials (Schubert 2006).

An LED package is not simply a small piece of the right mix of semiconductor materials. A major issue is the extraction efficiency, or how much of the produced light can be brought out of the LED package. Much of the light emitted from the junction is either reabsorbed on its path out to the surface or, even if it reaches the surface, reflected back inward and eventually reabsorbed. The geometry of the LED die, as well as the built-in optics, can affect how much of the light escapes. Although some LED die shapes can minimize surface reflection losses, they are not always very practical to produce. Instead, epoxy domes have traditionally been used to lessen the refractive index mismatch and reduce internal reflection. More recently, lens arrays, photonic lattice waveguides, antireflection coatings, transparent substrates, and reflectors have been built into LED chips to further enhance the efficiency of light output. LEDs typically come in one of two emission types: surface emitting or edge emitting. In surface-emitting LEDs, the light exits over the entire area of the p–n junction. The radiation pattern is typically close to Lambertian (in which intensity is proportional to the cosine of the angle measured from the axis perpendicular to the surface of the junction). In edge-emitting LEDs, the light is guided along in the plane of the junction and comes out from all sides of the LED in a highly directional flux. The characteristics of the optical output from LEDs are important for at least two reasons. First, the amount of useful light that can be brought to the specimen depends not only on the total amount of light emitted, but also on the spatial pattern of emission, the wavelength dependence on junction temperature and current, and individual variability (even among the same model of LEDs). A second important characteristic of an LED is its efficiency: If more energy is converted to electromagnetic radiation, the less the LED will heat up. Heat management is an important issue that will be discussed later. Currently, LEDs may convert around 5%–10% of the electrical energy to optical output; the rest will mostly end up as heat. Future advancements may help improve the percent conversion, but as of now, careful thermal management is required for high-power LEDs.

Most LED producers provide data sheets containing important spectral, mechanical, and electrical characteristics, as well as more detailed application briefs that help users integrate these LEDs into various illumination systems. Many of these application briefs delve into far more detail on the implementation of LED circuits, thermal management, mounting options, and other important issues for specific LEDs than space in this chapter allows; we direct the interested reader to these sources, some of which are listed at the end of this chapter.

COMPARING LEDs WITH OTHER ILLUMINATION SOURCES

One of the advantages of an LED over more traditional filament-based or arc-based illumination systems is that the timescale of its operation is much faster: on the order of a microsecond. There is no need to wait tens of minutes for a lamp to heat up and to reach thermal equilibrium. Because of their much faster response, LEDs do not need a high-speed shutter system (which is at least a few orders of magnitude slower) as long as the on/off control is integrated into the LED control circuitry. Even without that circuitry, LED light can be controlled from a power supply located at a distance from the experimental apparatus, thereby eliminating vibrations caused by the use of shutters or other light-blocking methods on the microscope itself. An electric circuit or power supply can control the intensity of the LED light; there is no need to swap neutral density filters or to change the gain and the contrast on a camera. Furthermore, LEDs can be synchronized with the acquisition camera to rapidly shutter the excitation light and thus to reduce phototoxicity (Nishigaki et al. 2006).

Another advantage of an LED is that its emission lies within a narrow bandwidth of the spectrum, thereby minimizing unwanted light. A visible-light LED will emit very little ultraviolet or infrared (IR) radiation, thus minimizing toxicity to or heating of samples. In addition, even within the visible spectrum, because most of the light will be within a particular band, there is less concern about undesirable light bleeding through filter sets and being detected by a camera. Thus, although there may be less total emitted power compared with a more traditional illumination source, with an LED, the signal to background is often greater. With careful electronic design, LEDs emissions are stable and reproducible, making it much simpler to correct for background and to compare image brightness among different specimens.

LEDs are susceptible to fast noise in the controlling circuitry because there is no thermal inertia to rely on for smoothing. The noise in illumination is generally related to noise in the power source, and oscillations or ripples in light intensity can sometimes be seen. However, other illumination sources, such as arc lamps, can have fast noise as well. In general, the problem can be resolved by using well-designed circuitry and well-regulated current supplies. Because an LED's output power, peak wavelength, emission spectrum, and voltage characteristics depend heavily on the applied current and temperature of the LED, careful design of the LED control system is needed to avoid drifting of the output light. Unless the LED undergoes catastrophic failure, this change in light output is a slow drift. However, an LED with a stable driver can have significantly lower low-frequency noise compared to arc lamps (although high-frequency noise may be comparable) and therefore offers an overall improvement in signal-to-noise ratio (Rumyantsev et al. 2004).

Another light source that is increasingly being used is the laser. LEDs produce light that has far less spatial and temporal coherences than does a laser. In some applications, the laser simply cannot be replaced by an LED. However, there are instances, such as spinning-disk confocal microscopy, where LEDs may potentially replace lasers. LED use could reduce speckle noise and wash out some reflection-induced interference patterns. Cost and maintenance will also be improved significantly. However, LED brightness is still an issue, and the lack of collimation of the emitted light places restrictions on optical coupling to a fiber or the optical train that must be used to bring the light to the specimen.

LEDs are made completely out of solid-state components, so they are more robust than light sources that use filaments or highly pressurized gas. In addition, they are safer with no risk of explosions. Typically, LEDs will start to break down at ~125°C, which places some constraints on illuminator design and in some ways makes them safer than tungsten lamps, which operate at much higher temperatures. Although LEDs can undergo catastrophic breakdown when the temperature is not controlled or when there are large current spikes from the power source, a well-designed system provides continuous light for thousands of hours with very little chance of spontaneous failure. LEDs, even when mounted on heat sinks, are usually much smaller than other light sources. Some small LEDs are powerful enough to be mounted inside a microscope (as opposed to in an external lamphouse), directly under the specimen, or even on a live animal (Moser et al. 2006; Huber et al. 2007; Lang et al. 2008). LEDs can be used in cold environments in which other illumination systems may have difficulties. Thus, the long-term costs of an LED-based system, when considering that shutters are not needed and bulb replacement is rare, tend to be much lower than other illumination sources.

CHOOSING AN LED FOR IMAGING

The following factors should be considered when choosing an LED as an illumination source for an imaging system.

- The most important is the wavelength(s) of light needed.
- Other important features are the LED's brightness, emission directivity, thermal manageability, and the size of the LED package.
- Consider the dependence of the optical power and spectra on junction temperature and drive current.
- Other important factors are the maximum allowable current, voltage–current relationship, allowable duty cycle, and other electrical specifications needed for control circuitry design.
- If possible, obtain diagrams showing the physical dimensions as well as the mounting options.

Most LED manufacturers provide detailed data sheets containing information for a typical LED package. Note, however, that because of the complicated production process, there is a lot of variability between individual LEDs of the same model (Benavides and Webb 2005). For most experiments, when using LEDs for simple epifluorescence or other less demanding applications, this variability is unimportant. Some manufacturers bin their LEDs based on brightness and spectral characteristics, but be aware that there can still be significant variability within bins. Currently available high-power LEDs can provide radiance nearing that of mercury lamps in many parts of the spectrum. If desired, the user can perform calibrations of various operating parameters such as the forward voltage, spectral output, and thermal resistance from the junction. A list of major LED manufacturers and their websites is provided at the end of this chapter.

LED Microscope Illuminators

LED-based illuminators are available that have been designed to mount to specific microscope models from Olympus, Zeiss, Leica, Nikon, and other manufacturers. Thorlabs, Inc., Newport Corporation, FRAEN, CoolLED, and Zeiss are some of the sources for these LED-based illuminators. These illuminators come fully aligned, usually mount easily onto a microscope, and are packaged with easy-to-use software and hardware controls.

BUILDING AN LED ILLUMINATION SYSTEM FOR IMAGING

Next, we introduce the concept of building your own LED illumination system. Why might this be desirable? Although some companies use proprietary technologies to produce LED units having unusual spectra, many rely on existing LED chips. Even if a new LED package comes out with highly desirable properties, these may not be integrated into commercially available LED illuminators. Specific power and spectral combinations desired by a researcher may not be available from a single company. It can be easier to repair a homemade LED illumination system if it breaks. The skills involved in building an LED illuminator for a microscope can be adapted to other uses, such as LEDs mounted on animals for in vivo applications, something that is not currently available commercially. Finally, there may not be any commercial LED microscope illuminators that are compatible with the specific microscope in the laboratory.

LED Light

LED packages emit either single-color or white light. LEDs that emit white light do so either through a suitable combination of colored LEDs or through wavelength conversion. White LEDs can be used for trans-illumination and for oblique or reflectance imaging, and the white color may be more pleasant to a viewer who is looking directly through the eyepiece of a microscope. However, white LEDs may not be optimal for these forms of imaging when used with a camera. For example, differential interference contrast (DIC) imaging of thick acute tissue slices often uses IR as opposed to visible light because IR is scattered less by tissue. One report indicates that blue LEDs provide better image quality than red LEDs for DIC imaging of microtubules (Bormuth et al. 2007), whereas another compares images of various color illumination for DIC and oblique illumination (Safronov et al. 2007). Intrinsic optical signal imaging also benefits from the use of certain colors of light. For example, blood vessels stand out more from the surface of the brain when green light is used. In addition, white LEDs are unlikely to be suitable replacements for traditional broad-spectrum light sources in fluorescence microscopy. With LEDs that emit white light by combining the emissions of several colored LEDs, the white light is not a continuous spectrum of wavelengths but has peaks of red, green, and blue and large gaps in spectral power within the other regions. Density constraints limit the number of LEDs per unit area, so for a given specific color, the amount of usable light available from a composite white LED package will be lower than that available from a single-color LED package of the appropriate wavelength. With LEDs that emit white light through wavelength conversion, a broad spectrum of light can be obtained, but the inefficiency of wavelength conversion reduces the usefulness of these LEDs. More spectral power can be obtained by combining the beam paths of single-color LEDs. Wavelength-conversion-based LEDs should be chosen only when there is no LED package available that directly produces light in the desired spectral band.

The light from a single-color LED is pseudomonochromatic, with a spectral full width at half-maximum (FWHM) that is dependent on the wavelength, the die temperature, the drive current, and whether the light was converted via a phosphor or a secondary semiconducting device. If the FWHM is >20 nm and fluorophores do not show a significant Stokes shift, then excitation filters become necessary. This somewhat broad spectrum does provide an opportunity to excite a broader range of fluorophores.

When only a single color is needed, optical coupling between the illumination source and the microscope is relatively simple. This is described below in the sections on LED optics and illumination. Multicolor imaging requires a more sophisticated optical coupling process because the most useful LEDs are pseudomonochromatic. One option is to have a housing that combines multiple LED lights through a combination of collection lenses and dichroic mirrors that merge the light paths, as in Figure 1A. This light can be coupled to a microscope lamphouse via direct mechanical attachment at the back of the lamphouse or through a light guide. Excitation filters may be fitted into the microscope filter cubes, or a suitable filter can be placed in front of each LED. Alternatively, light from all of the sources can be combined, resulting in a single multiband excitation. Although image quality may suffer when using this latter method due to cross excitation of fluorophores, it is an economical option because each excitation filter costs ~$250.

When vibration, speed, and pixel shift are not of concern, one can choose a color channel by simply rotating the microscope filter turret, which contains excitation, dichroic, and emission filters mounted in cubes. However, it is often advantageous to move different optical elements independently, and there are three alternatives: (1) a single stationary multiband dichroic and multiple excitation and emission filters that are on fast wheels; (2) a configuration in which the emission and dichroics are both multiband but with multiple excitation filter sets on a wheel; and (3) a full multiband configuration in which all of the components are fixed and multiband (Fig. 1B).

For LED-based light sources, a modified configuration with an excitation filter in front of each LED color can be used without a wheel as switching can be performed through the driving circuit that controls the LEDs. Traditionally, the full multiband configuration requires a color camera if the sample is stained with multiple fluorophores because each one would be excited and would emit at the same time. However, LED-based illumination sources permit switching between narrow bandwidth excitation light sources that correspond to only one fluorophore excitation color, provided there is enough spectral separation between the fluorophores being used. In those cases, the full multiband configuration can be used with a black-and-white camera. These LED-specific configurations are effective at avoiding pixel shift, a phenomenon in which different filters cause images as seen by a charge-coupled device (CCD) camera to shift. Correcting pixel shift requires that captured images be processed to realign the images.

LED light intensity can be controlled either via current control or by pulse-width modulation (PWM), a method in which the LED is rapidly turned on and off so that the effective brightness is reduced. Current control of LED light intensity can change the spectral properties of the emitted light even without any temperature change in the LED, although, for most imaging applications, this is not of concern. With PWM, unintentionally altering spectral properties is largely avoided, but the LED temperature can shift, which might lead to changes in the characteristic of the output light. PWM allows for greater linear control over the effective intensity. In situations in which turning the light on and off is not desirable, one should not use PWM. Note that because of the normally sublinear relationship between voltage, current, and efficiency of converting electrical power to light, control of light intensity by PWM may not be as thermally efficient for the same brightness as using the current limitation method.

LED Optics

Many LEDs come with built-in optics, which enhance light extraction efficiency. In addition, many companies sell optics made for collecting light from LEDs. When choosing additional optics for LEDs, consider the following.

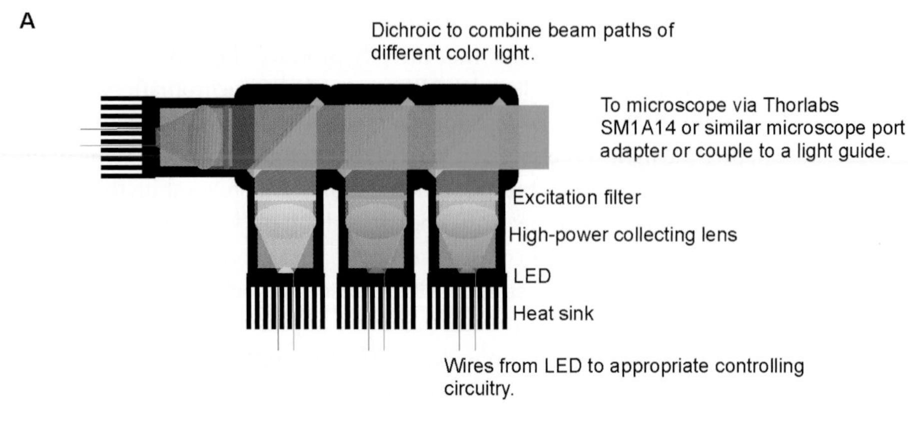

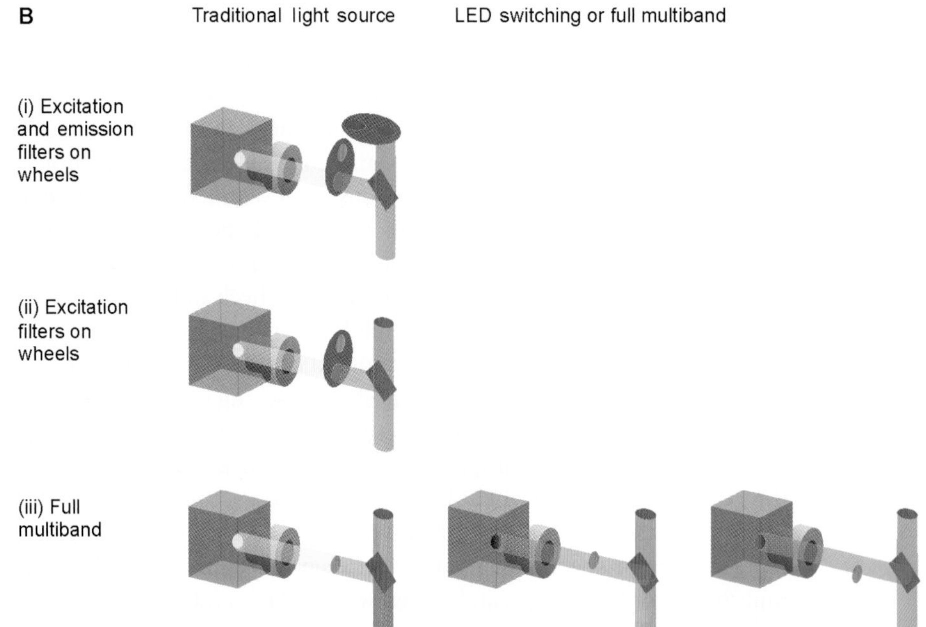

FIGURE 1. LED as a light source. (*A*) The light from LEDs of different colors is collected by a strong lens, and the beam paths are combined by appropriate dichroic filters. To minimize light loss, the distance from each LED to the output is kept as short as possible. The output light can be directly coupled to the back of a microscope lamphouse (in the investigator's experience, there is good coupling if the LED is focused to infinity by the collection lenses), or it can be focused into a light guide. (*B*) Comparisons of multicolor imaging. (i) Configuration requiring filter wheels for both excitation and emission filters, (ii) configuration with one excitation filter wheel, and (iii) the full multiband configuration for a traditional light source requires a color camera because multiple fluorophores will be activated at the same time. LEDs, however, can be used in LED-specific configurations in which either multiple individual excitation filters in front of each LED or a single multiband excitation filter after all of the LED lights are combined are used. Because the different colors of the LEDs can be turned on and off individually and the narrow bandwidth will typically lead to low bleed-through into the neighboring excitation band, configurations with no filter changing can be used.

First, consider the concept of geometric and angular losses, which involves the optical or Lagrange invariant and its related quantity, étendue. Étendue is proportional to the product of source size and divergence angle of emission. This value limits the amount of light that is available for coupling into any optical system.

Second, consider the concept of Fresnel reflection. Light passing between media having different refractive indices will experience partial reflection, even when the angle is less than the critical angle for total internal reflection. For coupling a single-color LED directly into a fiber, index matching gels or glues should be used to minimize Fresnel losses at the fiber input. The end of a polished fiber is placed flat against the smoothed flat LED surface. Some machining may be required to shape the epoxy dome so that the fiber and the LED surface are properly coupled. For optical trains of lenses, antireflection coatings can help. A good discussion on coupling LEDs to optical fibers is provided by Doric and Tubic (2004).

Homogeneous illumination of a specimen is usually achieved by Köhler illumination. Not all LEDs, however, may be able to provide perfect Köhler illumination because of their radiation pattern (although this is rarely a problem). Depending on the type of LED and the illumination collection optics used in the microscope, it may be worthwhile to use a diffuser or critical/Nelsonian illumination. In the latter method, a homogenous light source, or the end of a light guide, is imaged directly onto the sample. Many LED chips, however, do not provide homogenous light because the pattern of electrical contacts block or separate the light-emitting areas. Light guides and bent fiber optics can scramble coupled light, although the amount of light lost in either the coupling or during the transmission through the light guide might be significant enough to cause problems. Diffusers can also be used to provide light homogenization. With holographic diffusers, there is less loss of light than that from ground glass or opal diffusers.

Mounting and Heat-Sinking LEDs

Thermal management is an important consideration when building an LED illuminator. The electrical power required to run a high-power LED far exceeds the emitted optical power. The rest of the energy is converted to heat, and depending on the LED used, this amount could easily exceed 50 W. Many LED data sheets show the maximum allowable temperature for the LED junction as well as the maximum current at a given LED temperature or ambient room temperature. The heat sink should be designed to ensure that the LED does not burn out. Although there are theoretical methods for estimating the necessary characteristics of the heat sink, these do not account for convection or the efficiency of heat spreading within the heat sink. There are ways of measuring or estimating the LED junction temperature by measuring the temperature of the LED package or the temperature dependence of the forward voltage, and some LED packages have built-in temperature readout lines. The websites of many LED producers (e.g., OSRAM Opto Semiconductors, Luminus Devices, Inc., and Philips Lumileds Lighting Company) have application notes and data sheets that discuss the details of heat-sink selection, temperature measurement, and mechanical attachment methods of LEDs to other surfaces.

The heat generated by an LED must be effectively dissipated using heat sinks. It is not enough to simply place the LEDs onto a heat sink; the two must be glued together with a thermal adhesive designed for the task. Effective heat transfer also requires that there are no air gaps between the LED and heat-sink surfaces. One option is to use screws to attach the LED to the heat sink after having placed thermal grease (and possibly an electrically isolating, thermally conductive mica sheet) between the LED and the heat sink. For some LEDs, another option is soldering the LED and heat sink together. A thermally conductive adhesive pad or glue can also be used, although without some method for applying pressure to the area of contact, thermal transfer is compromised. Refer to the LED manufacturer's website, or contact them about acceptable attachment methods.

In general, heat-sink efficiency is a function of the material used; its total surface area, shape and surface finish; flow of fluid over the heat sink; and the ambient room and fluid temperatures. Most heat sinks are made of aluminum, although more expensive and efficient ones are made of copper or an alloy. Typically, a larger surface area means greater efficiency. Although finned heat sinks are less efficient per unit of surface area than flat heat sinks (as a result of less radiative heat loss and less convection), they are more efficient when considering the overall footprint. The temperature around the heat sink affects how much (or the rate at which) heat can be dissipated. A fan greatly improves heat-sink efficiency because heat is transferred to the ambient environment by convection. Many heat-sink fans generate sufficiently small vibrations so that experiments are not affected. If a heat-sink fan cannot be used, orient the heat sink vertically so that rising warm air creates convection through the fins. Low-temperature circulating fluids may also help.

LED Electronics

The simplest solution for powering an LED is to connect it to an alternating-current to direct-current converter, although this method is fraught with problems and is not recommended. There is no control over light intensity. Nonisolating transformers are susceptible to spikes in power. Importantly, because temperature is unregulated, thermal runaway can occur when as an LED heats up, the forward voltage drops, leading to an increase in drive current, which further heats the LED. This could lead to catastrophic failure of the LED. There is also no electronic control for turning the light on and off, which could be a minor inconvenience or a critical flaw depending on experimental requirements. A voltage source with a resistor in series is a simple way to limit current but can waste power through the resistor, and in the end, the current, not the voltage, is the primary determinant of optical output.

The simplest functional solution for powering and controlling an LED is a high-quality, regulated, adjustable current supply. This will at least permit some control for intensity, put a cap on current, and allow for some on/off switching (either manual or perhaps through digital inputs depending on the power supply), all for as little as $200. Alternatively, many companies now offer drivers for LEDs, a simple and cost-effective choice. Homemade control circuits can be built for <$50, but much time can be lost designing and building the circuitry. Potentiometers could be used in a circuit for light intensity control, and transistors could be used for switching on and off the light either as part of an experiment or to shut down an LED before it reaches critical temperature. The use of op-amps or feedback from photodiodes can further stabilize light output. Note that the circuitry may have a time constant slower than that of the LED itself, but again, this can be limited to the order of microseconds. Figure 2A illustrates a simple circuit for driving a Philips Luxeon V blue LED. The light intensity is adjusted by a potentiometer, and the light can be turned on and off by a low-current 5-V transistor–transistor-logic (TTL) input (Albeanu et al. 2008). Figure 2B shows the square wave voltage input and the resulting intensity of the LED light. The time constant is on the order of 10 msec in this simple but suboptimal circuit.

With LEDs that require very high currents, it may be necessary to use high-power dissipation potentiometers (it may not be easy to vary their resistance), and even the use of a Darlington pair transistor may not be enough if the current drive of the TTL input is too low. There is not necessarily one simple circuit design that could be easily adapted to all LEDs currently on the market because of the highly variable electrical ratings and the voltage–current relationships of the devices. Also, in the investigators' experience, switching very high current devices on and off can produce substantial

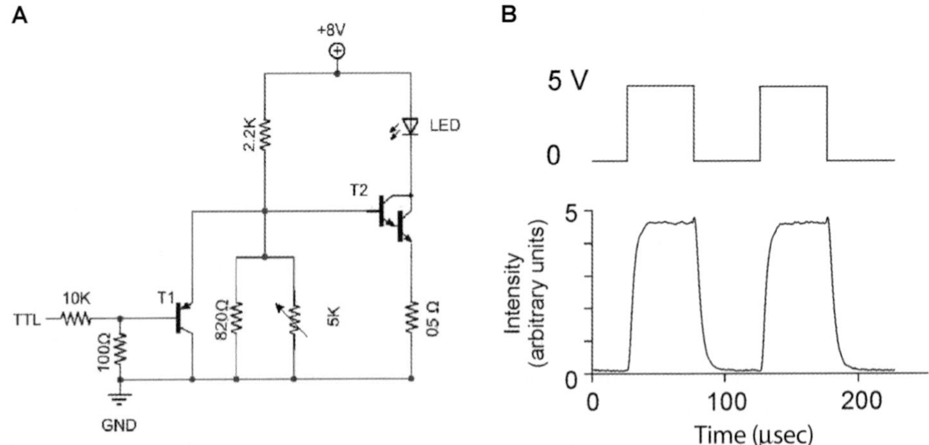

FIGURE 2. A high-intensity Luxeon V LED driving circuitry and timing. (*A*) A simple electronic circuit that can be used to drive a single blue LED is shown. Its output intensity is adjusted by a potentiometer; the light can be turned on and off by a low current TTL input. Note that the specific resistors should be optimized for the specific drive current required by the LED. (*B*) An LED driven by a high square wave input follows the signal closely. The speed of a typical LED array is clearly illustrated by its response to a voltage step. The driver circuit was that shown in *A*. (Reproduced, with permission, from Albeanu et al. 2008.)

artifacts on electrophysiological recordings. These artifacts can be reduced by electrically isolating the electrophysiological recording area or by using long light guides or fiber optics to transmit the light to the microscope from a source placed some distance away.

LED-Based Illuminators

Uncoupling the LED illuminator and the microscope can be beneficial, especially if light loss through beam splitters is undesirable, dichroics cannot be used, one wants to bypass the other coupling optics in a microscope, or one desires oblique illumination. The LED light will move from the illuminator(s) through fiber-optic cables or liquid light guides to some collection optics that focus the light onto the target. The end of the light guide will generally be outfitted with a lens to either spread or focus the light to the appropriate distance and spot. (Such an illuminator can also be used for nonimaging microscopes such as surgery scopes.) For cell culture and slice imaging, it might be possible to use an LED module that fits under the sample stage of an upright microscope, eliminating the need for light guides. A microscope-mounted LED might not work for experiments in which a bulky condenser is aligned for some form of contrast imaging. As another alternative, the LED illuminator can be coupled to the microscope through one of the lamphousing ports or via a removable filter, field stop, or aperture stop port (Albeanu et al. 2008). This type of coupling can be performed either with a fiber into which the different LED light is coupled (Fig. 3A) or from a homemade lamphouse that collects LED light of different wavelengths into one path and then emits it (Fig. 3B). For multicolor LED sources, the size of the illuminator will most likely constrain the coupling options to either light guides or through the rear lamphouse port. Finally, it is worth noting that the low-noise characteristics of LEDs (Herman et al. 2001; Rumyantsev et al. 2004; Salzberg et al. 2005) compared to arc lamps make them useful for imaging voltage-sensitive dyes, which typically exhibit low modulation of signals for physiological voltage fluctuations (Grinvald and Hildesheim 2004).

ADVANCED USES FOR LEDs

LEDs can be used for more than simply replacing the light source of a commercial microscope. Technical advances continually provide new opportunities for using LEDs. One example is patterned illumination either for advanced imaging techniques that rely on patterned illumination or for patterned stimulation of neurons (Fukano et al. 2004; Wang et al. 2007; Poher et al. 2008). One very simple way to create a patterned LED illuminator with the desired wavelength is to purchase a digital light processing (DLP) projector and to swap its light source with a suitable LED. Most higher-end projectors use halogen or metal halide bulbs and may include complicated systems to ensure that the unit is functioning properly. In these projectors, after replacing the bulb with an LED, it may be necessary to trick the projector into thinking that the bulb is still present, which may be performed through a combination of electronic and optical controls. Most projectors that use broad-spectrum bulbs have color filters and wheels that will have to be removed to optimize the projector for use with an LED. However, many of the cheapest projectors, costing $500 or less, are easier to modify. They are often LED based, although the pre-installed LEDs may not be optimal for imaging or neuronal stimulation. A simple solution is to move the original LEDs away from the system by extending the connecting wires, thus maintaining the electrical characteristics. Another point to note is that almost all DLP projectors will have timing issues because a single light chip is used to sequentially create multiple repeated red, green, blue, and white (brightness) frames within one frame of a video signal. The technical aspects of adapting a DLP projector for use as a pattern illuminator are beyond the scope of this chapter. We simply wish to point out that LEDs can provide an additional source of temporal and intensity control, with much lower noise than arc lamps.

LEDs can also be used to optically stimulate awake and moving animals. Although this may not be an imaging development per se, recent advances in optogenetic techniques for controlling neurons with light (Boyden et al. 2005) have opened the doors for research using animals with head-mounted illumination sources (Huber et al. 2007). Without LEDs, these techniques would be

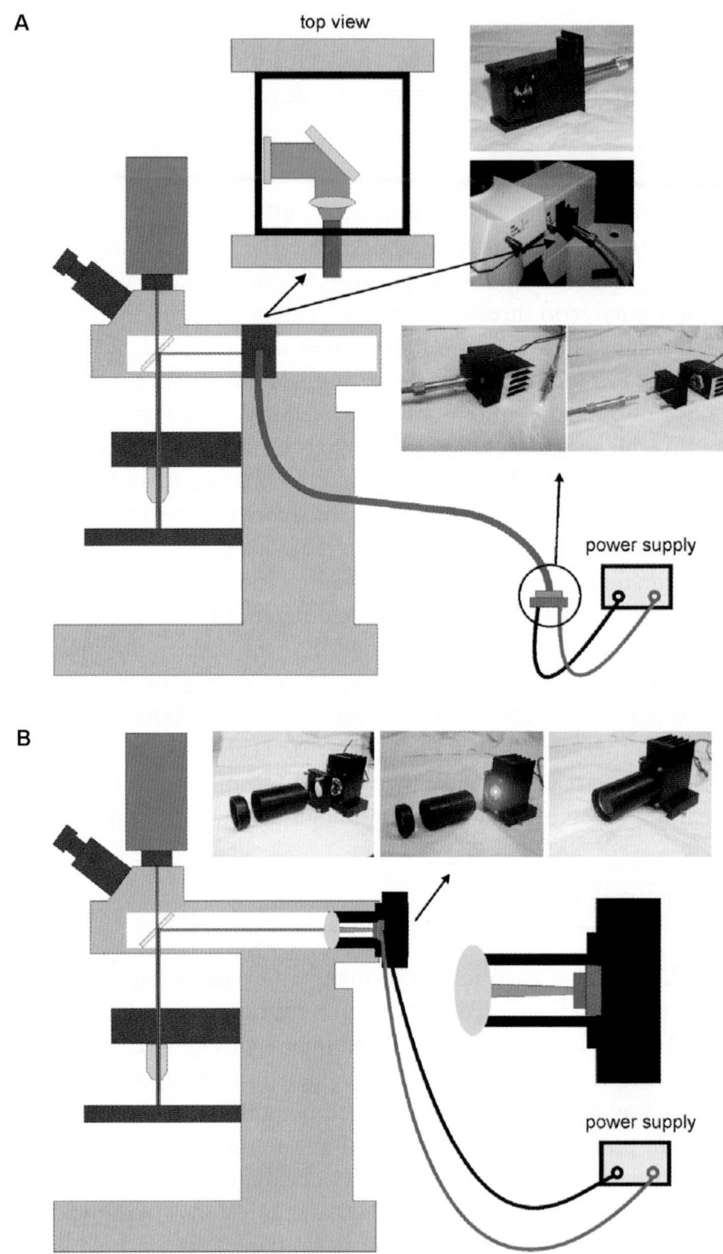

FIGURE 3. Two methods for coupling an LED-based illuminator to a microscope. (*A*) An optical fiber is placed on the surface of an LED to direct light into a small box consisting of a focusing lens and mirror that is inserted into a side port on the microscope. (*B*) A homemade lamp consisting of an LED, a heat sink, a tube, and a lens is coupled to the lamphousing port in the back of the microscope. (Reproduced, with permission, from Albeanu et al. 2008.)

unthinkable because the size, thermal management, control of timing, and spectral broadness of more traditional light sources would make rodent-mounted light sources highly impractical. Small LEDs mounted onto an animal may not need a heat sink if the duration of illumination pulses is short enough. The circuitry to control an animal-mounted LED can be quite small and simple and, thus, can be integrated into the mounting apparatus.

CONCLUSION

Recent advances in LED technology have made LEDs both practical and valuable in biological microscopy. As a wider array of spectra from high-power LEDs becomes available, LEDs will be able to replace even mercury lamps and will be able to expand the use of all of the available fluorophores. Further enhancements in LED efficiency will simplify heat management, making it possible to use smaller heat sinks. Further reductions in the size of LED illumination modules could be useful for building smaller imaging systems or for mounting onto small animals. One exciting recent advance is in the field of organic and polymer LEDs, which are now used to make display systems and can be particularly bright. By having LED chips that provide pixel-by-pixel control, structured illumination may be possible without the need for external light patterning technologies.

REFERENCES

Albeanu DF, Soucy E, Sato TF, Meister M, Murthy VN. 2008. LED arrays as cost effective and efficient light sources for widefield microscopy. *PLoS ONE* **3:** e2146. doi: 10.1371/journal.pone.0002146.

Benavides JM, Webb RH. 2005. Optical characterization of ultrabright LEDs. *Appl Opt* **44:** 4000–4003.

Bormuth V, Howard J, Schaffer E. 2007. LED illumination for video-enhanced DIC imaging of single microtubules. *J Microsc* **226:** 1–5.

Boyden ES, Zhang F, Bamberg E, Nagel G, Deisseroth K. 2005. Millisecond-timescale, genetically targeted optical control of neural activity. *Nat Neurosci* **8:** 1263–1268.

Doric S, Tubic M. 2004. Fiber-coupling of LEDs depends on emitter type. *Laser Focus World* **October:** 2004.

Fukano T, Hama H, Miyawaki A. 2004. Similar diffusibility of membrane proteins across the axon-soma and dendrite-soma boundaries revealed by a novel FRAP technique. *J Struct Biol* **147:** 12–18.

Grinvald A, Hildesheim R. 2004. VSDI: New era in functional imaging of cortical dynamics. *Nat Rev Neurosci* **5:** 874–885.

Herman P, Maliwal BP, Lin HJ, Lakowicz JR. 2001. Frequency-domain fluorescene microscopy with the LED as a light source. *J Microsc* **203:** 176–181.

Huber D, Petreanu L, Ghitani N, Ranade S, Hromadka T, Mainen Z, Svoboda K. 2007. Sparse optical microstimulation in barrel cortex drives learned behaviour in freely moving mice. *Nature* **451:** 61–66.

Lang DS, Zeiser T, Schultz H, Stellmacher F, Vollmer E, Zabel P, Goldmann T. 2008. LED-FISH: Fluorescence microscopy based on light emitting diodes for the molecular analysis of Her-2/neu oncogene amplification. *Diagn Pathol* **3:** 49–52.

Moser C, Mayr T, Klimant I. 2006. Filter cubes with built-in ultrabright light-emitting diodes as exchangeable excitation light sources in fluorescence microscopy. *J Microsc* **222:** 135–140.

Nishigaki T, Wood CD, Shiba K, Baba SA, Darszon A. 2006. Stroboscopic illumination using light-emitting diodes reduces phototoxicity in fluorescence cell imaging. *BioTechniques* **41:** 191–197.

Poher V, Grossman N, Kennedy GT, Nikolic K, Zhang HX, Gong Z, Drakakis EM, Gu E, Dawson MD, French PMW, et al. 2008. Micro-LED arrays: A tool for two-dimensional neuron stimulation. *J Phys D* **41:** 1–9.

Rumyantsev SL, Shur MS, Bilenko Y, Kosterin PV, Salzberg BM. 2004. Low frequency noise and long-term stability of noncoherent light sources. *J Appl Phys* **96:** 966–969.

Safronov BV, Pinto V, Derkach VA. 2007. High-resolution single-cell imaging for functional studies in the whole brain and spinal cord and thick tissue blocks using light-emitting diode illumination. *J Neurosci Methods* **164:** 292–298.

Salzberg BM, Kosterin PV, Muschol M, Obaid AL, Rumyantsev SL, Bilenko Y, Shur MS. 2005. An ultra-stable noncoherent light source for optical measurements in neuroscience and cell physiology. *J Neurosci Methods* **141:** 165–169.

Schubert EF. 2006. *Light-emitting diodes*. Cambridge University Press, Cambridge.

Wang S, Szobota S, Wang Y, Volgraf M, Liu Z, Sun C, Trauner D, Isacoff EY, Zhang X. 2007. All optical interface for parallel, remote, and spatiotemporal control of neuronal activity. *Nano Lett* **7:** 3859–3863.

WWW AND OTHER USEFUL RESOURCES

General Background

http://www.ecse.rpi.edu/~schubert/Light-Emitting-Diodes-dot-org/ The light-emitting diodes website, maintained by the research group of E. Fred Schubert at Rensselaer Polytechnic Institute.
http://www.olympusmicro.com/primer/lightandcolor/ledsintro.html LED information from Olympus.
http://zeiss-campus.magnet.fsu.edu/articles/lightsources/leds.html LED information from Zeiss.

Application Notes and Design Considerations

http://catalog.osram-os.com/catalogue/catalogue.do?favOid=000000000001fab4000100b7&act=showBookmark OSRAM Opto Semiconductors.
http://www.doriclenses.com/lire/39.html Doric Lenses, Inc.
http://www.lunaraccents.com/ Lunar Accents Design.
http://www.philipslumileds.com/docs/docs.cfm?docType=4 Philips Lumileds Lighting Company.

Latest Information

http://www.ledsmagazine.com/ *LEDs Magazine.*
http://www.photonics.com/ Photonics Media, Laurin Publishing Company.

Major LED Producers and Distributors

http://www.acriche.com/en/ Seoul Semiconductor, Inc.
http://www.cree.com Cree.
http://www.enfis.com/ Enfis.
http://www.epitex.com/ Epitex, Inc.
http://lsgc.com/ Lamina Ceramics, Inc.
http://www.lumex.com/ Lumex, Inc.
http://www.luminus.com/ Luminus Devices, Inc.
http://www.tech-led.com/index.shtml/ Marubeni America Corporation.
http://www.nichia.com/ Nichia Corporation.
http://www.osram-os.com/ OSRAM Opto Semiconductors.
http://www.philipslumileds.com/ Philips Lumileds Lighting Company.
http://www.roithner-laser.com/ Roithner Lasertechnik.
http://www.toyoda-gosei.com/ TOYODA GOSEI Co., Ltd.

The investigators have experience using LEDs from Philips Lumileds Lighting Company, Luminus Devices, Enfis, and Roithner Lasertechnik for imaging various biological probes such as green fluorescent protein and its cousins, Oregon Green BAPTA, rhodamine derivatives, various Alexa dyes, as well as for channelrhodopsin-2 stimulation, and in all applications, optical power has been sufficient.

Optics and Optomechanics

Lens tubing systems and microscope adapters were obtained from Thorlabs, Inc. (Newton, NJ). Other parts were obtained from Thorlabs, Inc., Edmund Optics (Barrington, NJ), Newport Corporation (Irvine, CA), and LINOS Photonics GmbH & Co. (Munich, Germany).

Filters

Filters were obtained from Semrock (Rochester, NY), Chroma Technology Corp. (Rockingham, VT), and Omega Optical, Inc. (https://www.omegafilters.com/), as well as the previously mentioned optical components suppliers.

11 Lasers for Nonlinear Microscopy

Frank Wise

School of Applied and Engineering Physics, Cornell University, Ithaca, New York 14853

ABSTRACT

Various versions of nonlinear microscopy are revolutionizing the life sciences, almost all of which are made possible because of the development of ultrafast lasers. In this chapter, the main properties and technical features of short-pulse lasers used in nonlinear microscopy are summarized. Recent research results on fiber lasers that will impact future instruments are also discussed.

INTRODUCTION

Multiphoton microscopy (Denk et al. 1990) has become a powerful technique for the visualization and the quantitative investigation of dynamic cellular processes. Microanalytical chemistry and micropharmacology are other techniques that require nonlinear optical excitation. Despite these advanced capabilities, further instrumentation development is needed before these techniques will find widespread acceptance. Extremely stable, user-friendly sources of femtosecond-duration optical pulses are required, with the output wavelength tunable in the visible or the near-infrared (IR) regions of the spectrum.

A revolution in short-pulse sources occurred in 1990, with the discovery of Kerr-lens mode-locking in titanium-doped sapphire (Ti:sapphire; Spence et al. 1991). Mode-locked Ti:sapphire lasers are extremely stable solid-state sources of high-power 100-fsec pulses at wavelengths in the near IR, and commercial versions have been available since 1991. With a Ti:sapphire source, it is possible to construct microscope workstations that can be operated by reasonably knowledgeable users. Without the concomitant development of mode-locked solid-state lasers (mainly Ti:sapphire), multiphoton microscopy would likely have been limited to the handful of laboratories with expertise in the development of short-pulse lasers.

Coherent anti-Stokes Raman scattering (CARS) microscopy (Evans and Xie 2008; see Chapter 48) is a nonlinear microscopy based on the cubic nonlinear optical susceptibility of molecules. The demands on the short-pulse source for CARS are even more demanding than the requirements for multiphoton microscopy: CARS is typically performed with synchronized pairs of ultrashort pulses of distinct wavelengths. The nonlinear dependence of the signal that produces the image is a feature that CARS and multiphoton microscopy have in common. Third-harmonic generation microscopy (Barad et al. 1997; Squier et al. 1998) is another example of a nonlinear microscopy that has attracted significant attention.

In this chapter, I discuss the features of existing laser sources used in nonlinear microscopy and summarize recent developments in short-pulse lasers that may have significant impact on the practice of nonlinear microscopy in the future.

INTRODUCTION TO LASERS

There are two key elements of a laser: the gain medium and the optical resonator (Fig. 1). The gain medium is an amplifier for light, and the optical resonator (two mirrors in the simplest case) feeds the light back into the gain medium for continued amplification of the laser beam.

Ordinarily, light is not amplified when it passes through matter; it is either transmitted with no gain or loss, or it is absorbed. This is a property of matter in thermal equilibrium and is the reason that optical amplifiers do not occur in nature. If the energy of an incident light photon (a packet of electromagnetic energy) matches the difference between allowed energy levels of an atom, the light is absorbed because more atoms occupy the lower of the two energy levels than the higher level. For amplification to occur, we must arrange for the population of the upper level to be greater than that of the lower level. This never occurs in equilibrium and is referred to as a population inversion. An inversion can be generated by exciting some of the atoms from their ground state up to an excited state. This is referred to as the pumping process and obviously requires energy.

If the excited state has a reasonably long lifetime, the atoms accumulate in that state while it is being pumped, thus creating an inversion. Once an inversion is created in an optical resonator, laser action starts. A single photon of light propagating along the axis of the laser is amplified by the process of stimulated emission—the presence of that photon in the gain medium stimulates an atom in the excited state to make a transition to the lower state. At that point, there are two photons (the original one and the stimulated one), so the light has been amplified. The initially weak light is reflected from the mirrors of the resonator so that it passes repeatedly through the gain medium, becoming amplified each time. Eventually the amplification must decrease (referred to as saturation) because a finite amount of energy is available to create the population inversion. Thus, the beam builds up to a steady high intensity. One of the mirrors of the resonator (the output coupler) allows a fraction of the light to leak out, and that produces the output beam.

Because the emission wavelength of an amplifying medium depends on its energy levels, each material tends to produce a unique wavelength. Thus, lasers are specialized tools, with each one limited to a relatively narrow range of emission wavelengths. Lasers have been made of thousands of dif-

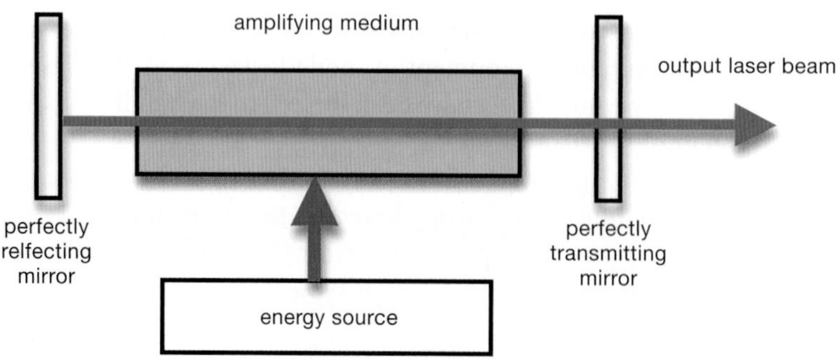

FIGURE 1. Schematic of a laser.

ferent materials, in solid, liquid, and gas phases. However, only a small fraction of these have found common use.

Interested readers should consult one of the many books on lasers for further information; a well-written textbook on lasers was written by Silvfast (1996). For facts about many kinds of lasers, see Hecht (1992).

FEMTOSECOND-PULSE LASERS

The operation of a continuous-wave laser, which produces a steady output beam of constant power, is described above. It is often more useful to produce the output power in the form of pulses because a pulse can produce very high instantaneous power. In particular, ultrashort pulses can be used to expose atoms, molecules, and materials to high instantaneous power (thousands of watts [kilowatts, kW] to millions of watts [megawatts, MW]), whereas the average power incident on the sample is low—only milliwatts—because of the low duty cycle. Another benefit of short pulses is that they can be used to initiate and record dynamic processes as fast as the timescale of the pulse.

Pulse generation in the most common femtosecond lasers is based on soliton formation. Solitons are pulses that are stable despite the presence of processes that, if operating individually, would make them decay. The relevant processes are ordinary linear group-velocity dispersion and nonlinear self-phase modulation. Dispersion arises from the frequency dependence of the index of refraction of any material, and the nonlinearity reflects the intensity dependence of the refractive index. When the pulse has the correct energy, these two processes impress phase shifts of equal magnitude but opposite sign on the pulse, so the pulse propagates without change. A device or structure that transmits more of the light incident on it when the intensity is higher is referred to as a saturable absorber, and such a device is needed in every short-pulse laser to start the process of pulse formation and to help stabilize solitons after they form. As a pulse builds in intensity, the peak of the pulse is transmitted more than the low-intensity wings. The saturable absorber, thus, sharpens and narrows the pulse with each passage. When the pulse duration reaches the subpicosecond range, the processes of dispersion and nonlinearity dominate the pulse shaping.

Because the nonlinearity is positive or self-focusing in transparent materials (the nonlinear phase shift is directly proportional to the pulse intensity), anomalous group-velocity dispersion is required for soliton formation. Prism pairs or chirped mirrors provide adjustable anomalous dispersion and are incorporated into femtosecond lasers to compensate for the normal dispersion of the gain medium and other components. The net dispersion is anomalous, as the prism pairs slightly overcompensate the dispersion of other components.

If picosecond pulses are desired (e.g., as is the case for CARS microscopy; see Chapter 48), the saturable absorber alone may be able to provide adequate pulse shaping, so that soliton pulse formation is not needed. A common saturable absorber is the semiconductor saturable absorber mirror (SESAM) (Keller et al. 1992).

Group-velocity dispersion is a major concern in work with femtosecond-duration pulses. As an example, a transform-limited 100-fsec pulse will broaden appreciably after passage through a few centimeters of ordinary glass. If it is desirable to maintain the short-pulse duration or to deliver a pulse through a significant length of dispersive material, dispersion control/compensation is needed. Devices that provide the required anomalous dispersion with little loss include the prism pairs and chirped mirrors discussed above in the context of intracavity dispersion compensation. In ultrafast optics research, it has long been common practice to place a prism pair or a quartet between a laser and any experiments for this purpose. The high-numerical-aperture objectives used in nonlinear microscopy commonly contain elements fabricated of highly dispersive glasses (e.g., dense flint glasses) to control chromatic aberration. A 100-fsec pulse incident on such a device can be broadened to several hundred femtoseconds if dispersion compensation is not used. Such broadening reduces the peak power by the same factor, which, in turn, reduces the two-photon excitation rate by the square of that factor.

PERFORMANCE CHARACTERISTICS OF THE LASER SOURCE

It is straightforward to list the critical performance characteristics of a source for multiphoton-excitation laser-scanning microscopy. The following capabilities and characteristics are desired.

- The pulse duration should be ~100 fsec. Longer pulses (up to 10 psec) are useful for CARS microscopy and may be useful for two-photon microscopy, although with a reduction in performance. Excitation with three or more photons places a premium on peak power and, thus, minimum pulse duration.
- The pulse repetition rate should be between 10 and 500 MHz for microscopy, to achieve reasonable scan rates. The upper limit of 500 MHz is determined by typical chromophore lifetimes of a few nanoseconds. The repetition rate can be reduced to ~1 MHz in nonimaging applications (e.g., uncaging of bioeffector molecules).
- The peak power of each pulse must be ~10 kW to achieve good signal-to-noise ratios at fast scan rates. This implies an emitted pulse energy of 1 nJ for a 100-fsec pulse and an average power of 100 mW for a 100-MHz repetition rate.
- Ideally, the wavelength should be tunable from 500 to 1000 nm. It is highly unlikely that a single source will be able to cover this entire range.
- Fluctuations in the power must be low (<1%), and the source must supply a high-quality Gaussian beam with good pointing stability.
- The source must be compact and robust and should require little maintenance. Fiber coupling of the source to the microscope would eliminate the need for some alignment of the source. Turnkey operation is important for most groups.
- It should ultimately be possible to construct the source for <$50,000. A price much higher than this will put the source beyond the reach of many research groups.

EXISTING LASER SOURCES

Ti:Sapphire Lasers

The great majority of work in multiphoton microscopy to date has used Ti:sapphire lasers operating at a pulse repetition rate near 80 MHz. These lasers are based on the soliton-type pulse shaping described above. Initially, the term Kerr-lens mode-locking was used to describe Ti:sapphire lasers. It refers to one common implementation of the saturable absorber based on self-focusing. Use of this phrase seems to be diminishing. Although the availability of Ti:sapphire lasers was a crucial factor in the development of multiphoton microscopy, it does not completely solve the illumination problem. Ti:sapphire lasers are tunable from 700 to 1000 nm and so provide two-photon illumination for dyes with single-photon absorption between ~350 and 500 nm. In fact, it is known that the two-photon cross sections of many dyes peak at wavelengths somewhat shorter than twice that of the linear absorption peak (Xu and Webb 1996). New sources will be needed to excite the numerous useful chromophores with linear absorptions <350 nm and >500 nm. Ti:sapphire lasers can pump optical parametric oscillators to generate broadly tunable light at wavelengths in the 1100–1600-nm range. A Ti:sapphire laser is pumped by a solid-state laser or by an argon-ion laser in older installations. Both of these are large and expensive to purchase, operate, and maintain. Widespread use of multiphoton excitation is unlikely if the cost of the illumination source is a significant fraction of the cost of the microscope, as is the case with current Ti:sapphire laser systems.

The Ti:sapphire laser still offers the greatest performance in terms of power and wavelength tunability over the range of 700–1000 nm. Femtosecond Ti:sapphire lasers are available from Spectra-Physics, Coherent, Inc., and others. Current commercial versions such as the Chameleon from Coherent, Inc. offer outstanding performance with push-button wavelength tunability. These are likely to be the choice of research laboratories desiring the greatest flexibility for at least the next few years. Second-harmonic generation extends the range of a high-power Ti:sapphire laser to include wavelengths between 350 and 500 nm. The current state-of-the-art commercial systems offer outstanding performance with broad wavelength tunability and excellent stability and reliability. Their relatively high prices (>$100,000) are justified by the flexibility they offer to the users (mainly researchers) that purchase them. Although commercial Ti:sapphire lasers are stable enough and acceptably user friendly for research laboratories, cheap turnkey instruments will be needed if multiphoton microscopy is to become a routine tool or to have an impact in clinical settings.

As discussed above, dispersion control is generally needed to avoid broadening of ~100-fsec pulses during propagation. Commercial instruments that provide this control are available. The Deep See from Spectra-Physics, for example, is a completely automated device for controlling dispersion at wavelengths generated by a Ti:sapphire laser. Similar capabilities are offered by other vendors.

Lasers Based on Neodymium and Ytterbium Ions

Lasers that operate on essentially the same principles as Ti:sapphire lasers can be constructed at other wavelengths. Neodymium (Nd) (Aus der Au et al. 1997) and ytterbium (Yb) (Druon et al. 2003) ions emit near 1000-nm wavelengths and are commonly incorporated in glass and crystal hosts. These lasers are commercially available (from, e.g., High-Q, Inc., Time-Bandwidth Products, Inc., and Amplitude Systemes). They typically offer high average power (up to ~5 W, compared with the 1–2 W of Ti:sapphire lasers) but also somewhat longer pulses, ~200 fsec. Their outputs are tunable over narrow ranges, typically 1020–1070 nm at best. In some cases, a saturable absorber, usually a SESAM, provides the main pulse shaping. These instruments are beginning to find application in nonlinear microscopy. Picosecond versions have become a favored source for CARS microscopy.

Other Solid-State Gain Media

Since 1990, researchers have devoted significant effort to the development of femtosecond lasers that could complement the performance of Ti:sapphire. Chromium-doped lithium strontium aluminum fluoride (Cr:LISAF) received substantial attention (and commercial development) because it can support femtosecond pulses over a tuning range similar to that of Ti:sapphire with the advantage that it can be directly diode pumped (Kopf et al. 1994). (Ti:sapphire lasers are pumped by diode-pumped solid-state lasers, the output of which is frequency doubled. When high-power diode lasers that emit in the green part of the spectrum become available, they may replace the solid-state pump laser.) Cr:forsterite emits in the range of 1160–1360 nm and so complements the wavelengths available with Ti:sapphire (Pang et al. 1993; Liu et al. 1998). In recent years, work on these materials has diminished, as Ti:sapphire, Nd, and Yb lasers showed major practical advantages and received commercial development.

Picosecond and Femtosecond Fiber Lasers

Fiber lasers offer a number of major practical advantages over the solid-state lasers discussed above.

- With the light contained in fiber, a laser will not be susceptible to misalignment. The wave-guiding properties of fiber ensure good spatial mode quality.
- Thermal effects are reduced because fibers have large surface-to-volume ratios.
- Fiber laser systems are scalable to high average powers; kilowatt devices exist.
- Fiber components are used in the telecommunications industry, so manufacturing costs will benefit from economies of scale.

Well-developed technological infrastructures exist for fiber lasers based on the Yb and erbium ions, which emit near 1000 nm and 1550 nm, respectively. Fiber-based femtosecond sources that compete with solid-state lasers will be compact and will require minimal utilities, will never require realignment, and ultimately should be much cheaper than solid-state versions. These features have motivated substantial research in the area of short-pulse fiber lasers and amplifiers, and several reviews of the field have been published (Duling and Dennis 1995; Fermann et al. 1997).

Despite much progress, short-pulse fiber lasers currently are having little impact compared with solid-state (mostly Ti:sapphire) lasers. This is largely due to the fact that fiber lasers have lagged behind their solid-state counterparts in key performance parameters: pulse energy and duration. Historically, the pulse energy available from femtosecond fiber lasers was an order of magnitude below the 10-nJ level of standard Ti:sapphire lasers. Amplification of lower-energy 100-fsec pulses from a fiber laser is not trivial. The gain bandwidth of Yb-doped fiber, which is most relevant to nonlinear microscopy, limits the pulse duration to ~120 fsec. Amplification to even the 10-nJ level in an ordinary single-mode fiber (SMF) requires chirped pulse techniques to avoid excessive nonlinear distortion of the pulse (Ilday et al. 2003). The use of large-mode-area fiber reduces, but does not solve, the problem, and introduces further complications. As a result, fiber sources still have limited impact in nonlinear microscopy. Another relative disadvantage of fiber lasers is their lack of wavelength tunability. For example, applications that require the broad emission range of Ti:sapphire will need to generate the new wavelengths by nonlinear frequency conversion.

Soliton fiber lasers are simple and reliable sources, but the pulse energy is limited to ~0.1 nJ (about 100 times below the energy of a routine Ti:sapphire laser) by the intrinsic property of the soliton itself. The soliton forms by the precise balance of nonlinear and linear phase accumulations, which only happens at a specific energy. Above that energy, the soliton splits into multiple unstable pulses. (The same effect limits the pulse energy in solid-state lasers.) Commercial fiber lasers based on solitons and dispersion-managed solitons (Tamura et al. 1993) are available from several vendors now. Fiber amplifiers can supply femtosecond pulses at high energies, but they typically sacrifice pulse duration and quality and operate at repetition rates below the standard 80 MHz. Nevertheless, if the energy is high enough, it will still be possible to attain high peak power even with a long or a distorted pulse, and the use of such sources is being explored by several groups.

Advances in pulse-propagation physics in the past few years have enabled order-of-magnitude increases in the pulse energy and the peak power from femtosecond fiber lasers. As a result, it is now realistic to design oscillators based on an ordinary SMF that can compete with the performance of solid-state lasers. Future instruments based on large-mode-area fibers will significantly out-perform solid-state lasers. Thus, there is reason to expect that short-pulse fiber devices may begin to supplant solid-state lasers in nonlinear microscopy in the reasonably near future, at least in settings in which broad wavelength tunability is not crucial.

The first major advance was the demonstration of the so-called self-similar evolution of a pulse in a fiber laser (Ilday et al. 2004). In a self-similar oscillator, a highly chirped and, thus temporally stretched, pulse propagates around the cavity. The pulse duration increases and decreases as the pulse propagates, but the pulse is always very long, at least 10 times the transform limit. The pulse is compressed back to the transform limit outside the laser. Intuitively, the long-pulse duration reduces the nonlinear phase accumulation that severely limits soliton pulses. Self-similar lasers were the first fiber lasers to reach the 10-nJ and 100-fsec performance levels. Like all prior femtosecond lasers, a self-similar laser has a segment with anomalous dispersion. The net cavity dispersion is near 0, but in contrast to prior lasers, the net dispersion is actually normal. As the dispersion is made even larger and normal, the pulse energy increases.

The second major development was the generation of ultrashort pulses from a laser without any anomalous dispersion in the cavity (Chong et al. 2006). The so-called all-normal-dispersion (ANDi) lasers go against the design of femtosecond lasers that was established over the previous 20 years. With large normal dispersion, soliton-like pulse shaping is not possible. A highly chirped pulse propagates in the ANDi laser, and such a pulse can be thought of as mapping frequency to time: The edges of the pulse temporal profile correspond to the edges of the frequency spectrum. Thus, a filter cuts off the edges of the pulse in time, and this is the dominant pulse-shaping process in an ANDi laser. Because this loss process is crucial to pulse formation and gain is required to compensate loss, the pulse in an ANDi laser is considered a dissipative soliton (Renninger et al. 2008). Dissipative refers to the system that supports the pulse, not to a property of the pulse. Dissipative solitons are more stable at higher nonlinear phase shifts than any known pulse evolution in a laser. A fiber laser based on dissipative solitons recently achieved the several-watt output level of a Ti:sapphire laser (Kieu et al. 2009a).

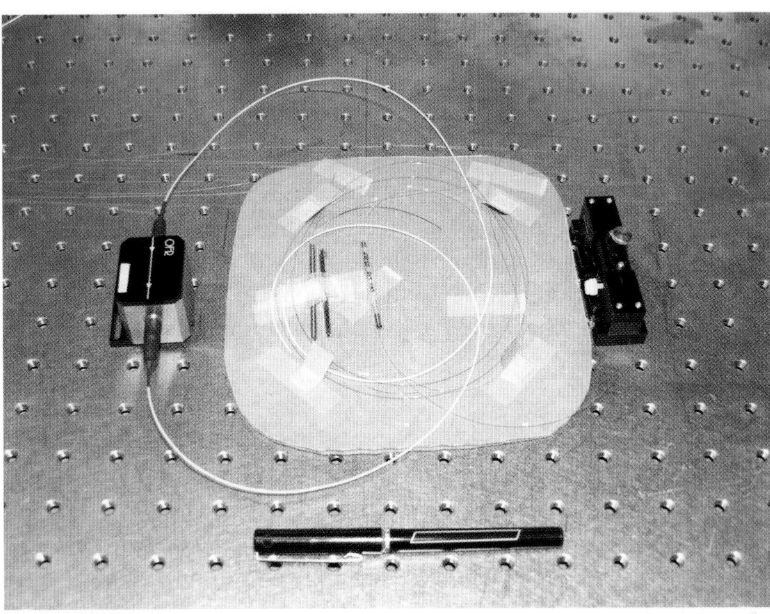

FIGURE 2. Photograph of an all-fiber version of a dissipative-soliton laser. Holes in the table are 1 in apart.

Femtosecond fiber lasers based on self-similar evolution or dissipative solitons can now compete with the performance of solid-state lasers (in a narrow wavelength range), and it will be possible to construct them for a small fraction of the cost. As a result, commercial versions are likely to appear in the near future. A tutorial introduction to normal-dispersion fiber lasers is presented in Wise et al. (2008).

The elimination of an anomalous-dispersion segment simplifies the construction of dissipative-soliton lasers. The potential applications of these lasers can be appreciated from Figure 2, which shows an all-fiber version of a dissipative-soliton laser. This laser generates 250-fsec pulses with 150 mW average power (Kieu and Wise 2008). Current work is aimed at achieving the 100-fsec and 1-W performances shown by dissipative-soliton lasers in integrated devices as in Figure 2.

The generation of high-energy picosecond pulses from a fiber source is also quite challenging. Standard mechanisms for managing nonlinearity such as chirped pulse amplification cannot be practically implemented with the narrow spectrum of a picosecond pulse. CARS microscopy is most commonly performed with synchronized picosecond pulses generated parametrically from a single high-energy pulse. Recent work shows that a carefully designed fiber laser and an amplifier can replace the typical solid-state laser for the parametric generation (Kieu et al. 2009b), which is an encouraging first step toward all-fiber sources for Raman microscopy.

CONCLUSION

The development of stable and reliable femtosecond-pulse lasers was crucial to the development of nonlinear microscopy as a field. Ti:sapphire lasers remain the workhorse of the field, and with 20 years of engineering, they have become outstanding scientific instruments. Solid-state lasers based on Nd and Yb ions emit near 1000 nm and, thus, complement the capabilities of Ti:sapphire lasers. Picosecond lasers in this category are commonly used to pump parametric oscillators for CARS imaging. Although the solid-state lasers currently dominate nonlinear microscopy, there is a growing need to develop short-pulse sources that are cheaper and more user friendly. These are needed for multiuser facilities and clinical applications of nonlinear microscopy, among other things. Recent demonstrations of new ways to shape pulses in fiber lasers have greatly enhanced their performance, to the point where laboratory versions now offer similar power and pulse duration to standard solid-state lasers, but with the potential benefits of fiber construction. Commercial versions of self-similar and dissipative-soliton fiber lasers can be expected in the near future, and these should be quite attractive for nonlinear microscopy.

ACKNOWLEDGMENTS

Portions of this work were supported by the National Science Foundation (ECS-0500956) and the National Institutes of Health (EB002019). I thank Professor W. Webb, Professor C. Schaffer, Professor Z. Chen, and Professor S. Xie for encouragement and numerous stimulating discussions.

REFERENCES

Aus der Au J, Kopf D, Morier-Genoud F, Keller U. 1997. 60-fs pulses from a diode-pumped Nd:glass laser. *Opt Lett* **22:** 307–309.

Barad Y, Eisenberg H, Horowitz M, Silberberg Y. 1997. Nonlinear scanning laser microscopy by third harmonic generation. *Appl Phys Lett* **70:** 922–924.

Chong A, Buckley J, Renninger W, Wise F. 2006. All-normal-dispersion femtosecond fiber laser. *Opt Express* **14:** 10095–10100.

Denk W, Strickler JH, Webb WW. 1990. Two-photon laser scanning fluorescence microscopy. *Science* **248:** 73–76.

Duling IN, Dennis ML. 1995. Modelocking of all-fiber lasers. In *Compact sources of ultrashort pulses* (ed. Duling IN), pp. 140–178. Cambridge University Press, Cambridge.

Druon F, Balembois F, Georges P. 2003 Laser crystals for the production of ultra-short laser pulses. *Ann Chim Sci Mat* **28:** 47–72.

Evans C, Xie XS. 2008. Coherent anti-Stokes Raman scattering microscopy: Chemical imaging for biology and medicine. *Annu Rev Anal Chem* **1:** 883–909.

Fermann ME, Galvanauskas A, Sucha G, Harter D. 1997. Fiber-lasers for ultrafast optics. *Appl Phys B: Lasers Opt* **B65:** 259–271.

Hecht J. 1992. *The laser guidebook*, 2nd ed. McGraw-Hill, Blue Ridge Summit, PA.

Ilday F, Lim H, Buckley J, Wise FW. 2003. Practical, all-fiber source of high-power, 120-fs pulses at 1 micron. *Opt Lett* **28:** 1362–1364.

Ilday FO, Buckley J, Wise FW. 2004. Self-similar evolution of parabolic pulses in a laser. 2004. *Phys Rev Lett* **92:** 213902.

Keller U, Miller DAB, Boyd GD, Chiu TH, Ferguson JF, Asom MT. 1992. Solid-state low-loss intracavity saturable absorber for Nd:YLF lasers: An antiresonant semiconductor Fabry-Perot saturable absorber. *Opt Lett* **17:** 505–507.

Kieu K, Wise FW. 2008. All-fiber normal-dispersion femtosecond laser. *Opt Express* **16:** 11453–11456.

Kieu K, Renninger W, Chong A, Wise FW. 2009a. Sub-100-fs pulses at watt-level powers from a dissipative-soliton fiber laser. *Opt Lett* **34:** 593–595.

Kieu K, Saar B, Holtom GR, Xie XS, Wise FW. 2009b. High-power picosecond fiber source for coherent Raman microscopy. *Opt Lett* **34:** 2051–2053.

Kopf D, Weingarten K, Brovelli L, Kamp M, Keller U. 1994. Diode-pumped 100-fs passively modelocked Cr:LiSAF laser using an anti-resonant Fabry-Perot saturable absorber. *Opt Lett* **19:** 2143–2145.

Liu X, Qian LJ, Wise FW, Zhang Z, Itatani T, Sugaya T, Nakagawa T, Torizuka K. 1998. Femtosecond Cr:forsterite laser diode pumped by a double-clad fiber. *Opt Lett* **23:** 129–131.

Pang Y, Yanovsky V, Wise F, Minkov B. 1993. Self-mode locked Cr:forsterite laser. *Opt Lett* **18:** 1168–1170.

Renninger W, Chong A, Wise F. 2008. Dissipative solitons in normal-dispersion fiber lasers. *Phys Rev A* **77:** 023814-1–023814-4.

Silvfast WT. 1996. *Laser fundamentals.* Cambridge University Press, Cambridge.

Spence D, Kean P, Sibbett W. 1991. 60-femtosecond pulse generation from a self mode-locked Ti:sapphire laser. *Opt Lett* **16:** 42–44.

Squier J, Muller M, Brakenhoff G, Wilson KR. 1998. Third harmonic generation microscopy. *Opt Express* **3:** 315–324.

Tamura T, Jacobson J, Haus HA, Ippen EP, Fujimoto JG. 1993. 77-fs pulse generation from a stretched-pulse mode-locked all-fiber ring laser. *Opt Lett* **18:** 1080–1082.

Wise FW, Chong A, Renninger WH. 2008. High-energy femtosecond fiber lasers based on pulse propagation at normal dispersion. *Laser Photon Rev* **2:** 58–73.

Xu C, Webb WW. 1996. Measurement of two-photon excitation cross-sections of molecular fluorophores with data from 690 to 1050 nm. *J Opt Soc Am B* **13:** 481–489.

12 Spectral Methods for Functional Brain Imaging

David Kleinfeld[1] and Partha P. Mitra[2]

[1]*Department of Physics and Graduate Program in Neurosciences, University of California, La Jolla, California 92093-0374;* [2]*Cold Spring Harbor Laboratories, Cold Spring Harbor, New York 11724*

ABSTRACT

Dynamic functional imaging experiments typically generate large, multivariate data sets that contain considerable spatial and temporal complexity. The goal of this chapter is to present signal processing techniques that allow the underlying spatiotemporal structure to be readily distilled and that also enable signal versus noise contributions to be separated.

INTRODUCTION

This chapter presents multivariate signal processing techniques that help reveal the spatiotemporal structure of optical imaging data and also allow signal versus noise contributions to be separated. These techniques typically assume that the underlying activity may be modeled as stationary stochastic processes over short analysis windows; that is, the statistics of the activity do not change during the analysis period. This requires selection of an appropriate temporal window for the analysis, which can be checked in a self-consistent manner.

The following worked examples are provided that serve to show the utility and implementation of these spectral methods:

1. deduction of rhythmic components of the dilation and constriction of a cortical penetrating arteriole in rat to illustrate basic frequency-domain concepts
2. deduction of synaptic connectivity between neurons in the leech swim network to emphasize the notions of spectral coherence and the associated confidence limits
3. the denoising of imaging data in the study of calcium waves in brain slice to introduce the concept of singular value decomposition (SVD) in the time domain and to illustrate the notion of space–time correlation in multisite measurements
4. the delineation of wave phenomena in turtle visual cortex to illustrate spectrograms, along with the concept of SVD in the frequency domain to determine the dominant patterns of spatial coherence in a frequency localized manner

Much of our exposition involves spectral analysis. Why work in the frequency domain?

- Many physiological phenomena have rhythmic components, ranging from electrical rhythms in the brain to visceral functions like breathing and heartbeat. The time series of these phenomena may appear very complicated, yet the representation in the frequency domain may be relatively simple and readily connected with underlying physiological processes.
- The calculation of confidence intervals requires that the number of degrees of freedom are known. Determining this number is complicated in the time domain, where all but white noise processes lead to correlation between neighboring data points. In contrast, counting the number of degrees of freedom is readily established in the frequency domain, as neighboring frequency bins are uncorrelated under stationarity assumptions.

Our emphasis is on the explanation and applications of signal processing methods and not on scientific questions per se.

Some relevant signal processing texts include Papoulis (1962), Ahmed and Rao (1975), and Percival and Walden (1993). The latter book includes sections on multitaper spectral analysis methods, developed originally by Thomson (1982) and used extensively in our analysis. Applications of modern signal processing methods to problems from neuroscience can be found in the book by Mitra and Bokil (2008) and in numerous reviews (Mitra and Pesaran 1998; Mitra et al. 1999; Pesaran et al. 2005; Kleinfeld 2008). Our notation follows that in Mitra and Bokil (2008).

BACKGROUND

The process of data collection involves sampling a voltage or current so that the signals are represented as an ordered set of points, called a time series, that are collected at a regular time interval. Let us denote the sampled time series as $V(t)$, the sampling time interval as Δt, and the length of sampling as T. In spectral analysis, one re-expresses this time series in the frequency domain by decomposing $V(t)$ into a weighted sum of sinusoids. We must first understand the range of frequencies that may be represented in the data.

The lowest resolvable frequency interval is given by the inverse of the length of the analysis window and is denoted the Rayleigh frequency, $\Delta f_{\text{Rayleigh}} = 1/T$. In multitaper spectral analysis the resolution bandwidth is typically denoted as $2\Delta f$, where Δf is an adjustable parameter. The resolution bandwidth $2\Delta f$ is also parameterized by the dimensionless product, p, of the half-bandwidth Δf and the length of the window T, such that

$$\Delta f T = p, \tag{1}$$

with $p \geq 1$.

Sampling a continuous signal at discrete time intervals will in general lead to a loss of signal. However, there is an important class of signals, the so-called band-limited signals whose spectral transforms vanish outside of a frequency range whose highest frequency is denoted B. For this case the signal can be perfectly reconstructed from discrete samples at uniform intervals Δt, as long as the sampling interval satisfies $\Delta t < 1/(2B)$. An alternative way of representing this criterion is to define the so-called Nyquist frequency, $f_{\text{Nyquist}} = 1/(2\Delta t)$. Then the criterion for perfect reconstruction of the original band-limited signal from sampled data becomes $f_{\text{Nyquist}} > B$.

Neural signals are not naturally band-limited, although physiological mechanisms such as the membrane time constant of neurons provide natural cutoff frequencies. It is customary to low-pass filter the original analog signal so that it becomes effectively band-limited and the Nyquist criterion may be applied. Typically, f_{Nyquist} is chosen to be significantly greater than B, which is called "oversampling." However, if $\Delta t > 1/(2B)$, the sampling is not sufficiently rapid and signals at frequencies greater than B are reflected back, or "aliased," into the sampled interval that ranges from 0 to f_{Nyquist}. For example, a signal at $1.4 \times f_{\text{Nyquist}}$ is aliased to appear at $0.6 \times f_{\text{Nyquist}}$. The experimentally imposed low-pass filters to prevent aliasing are often called "antialiasing filters." Such filtering is not always possible and aliasing cannot always be avoided.

The discrete Fourier transform of a data segment is defined by

$$\tilde{V}(f) = \frac{1}{\sqrt{T}} \sum_{t=\Delta t}^{T} \Delta t \; e^{-i2\pi f t} V(t). \tag{2}$$

Qualitatively, the time series $V(t)$ is projected against all possible sinusoids, indexed by frequency f, to form a set of weights $\tilde{V}(f)$. An immediate complication is that the finite extent of our data is equivalent to multiplying an infinite data series with a square pulse of width T. The effect of such a window on the Fourier transform is to produce oscillations for each estimate of $\tilde{V}(f)$ that extend into neighboring frequency bands (Fig. 1A,B). This is known as leakage, and is minimized by multiplying the time series with a function of time, denoted a taper, that smoothes the sharp edges of the pulse. A half-sine taper offers improvement, but a special function devised specificially to minimize leakage and known as a Slepian taper is optimal. The cost of reduced leakage is decreased spectral resolution through an increase in the resolution bandwidth (Fig. 1B). There are considerable advantages to computing a family of independent estimates, $\tilde{V}^{(k)}(f)$, rather than a single estimate, and in weighting the data with a set of orthonormal Slepian tapers to form this family (Fig. 1C). We denote each taper in the set by $w^{(k)}(t)$ and compute the estimates

$$\tilde{V}^{(k)}(f) = \frac{1}{\sqrt{T}} \sum_{t=\Delta t}^{T} \Delta t \; e^{-i2\pi f t} w^{(k)}(t) V(t). \tag{3}$$

The maximum number of tapers, denoted K, that supports this minimization, and which is used throughout our presentation, is

$$K = 2p - 1. \tag{4}$$

The lower spectral resolution, or equivalently a larger resolution bandwidth, is offset by a greater number of spectral estimates $\tilde{V}^{(k)}(f)$. The increase in number of independent estimates minimizes the distortion of the value in one frequency band by the value in a neighboring band and thus improves the statistical reliability of quantities that depend on the Fourier transform of the original signal.

Numerical processing of sampled data requires that we work in dimensionless units. We normalize time by the sample time, Δt, so that the number of data points in the time series is given by $N \equiv T/\Delta t$. We further normalize the resolution bandwidth by the sample time, Δt, and define the unitless half-bandwidth $W \equiv \Delta t \Delta f$. Then the time-frequency product $T\Delta f = p$ is transformed to

$$NW = p. \tag{5}$$

Given sampled data,

$$\tilde{V}^{(k)}(f) = \frac{1}{\sqrt{N}} \sum_{t=1}^{N} e^{-i2f\pi t} w_t^{(k)} V_t, \tag{6}$$

where time is now an index that runs from 1 to N in steps of 1 rather than a discrete variable that runs from 0 to T in steps of Δt, whereas frequency runs from –1/2 to +1/2 in steps of $1/N$ (Table 1).

TABLE 1. Relation of laboratory and computational units

Quantity		Units	
Name	Description	Sampled data	Computational
Record length	Longest time	T	N
Sample time	Shortest time	$\Delta t = T/N$	1
Resolution half-bandwidth	Lowest frequency	$\Delta f = p/T$	$W = p/N$
Nyquist frequency (f_{Nyquist})	Highest frequency	$1/2\Delta t = N/2T$	1/2
Temporal range		$[\Delta t, T]$	$[1, N]$
Spectral range		$[-N/2T, N/2T]$	$[-1/2, 1/2]$
Time-bandwidth product (p)	$p \geq 1$	$T \cdot \Delta f$	$N \cdot W$

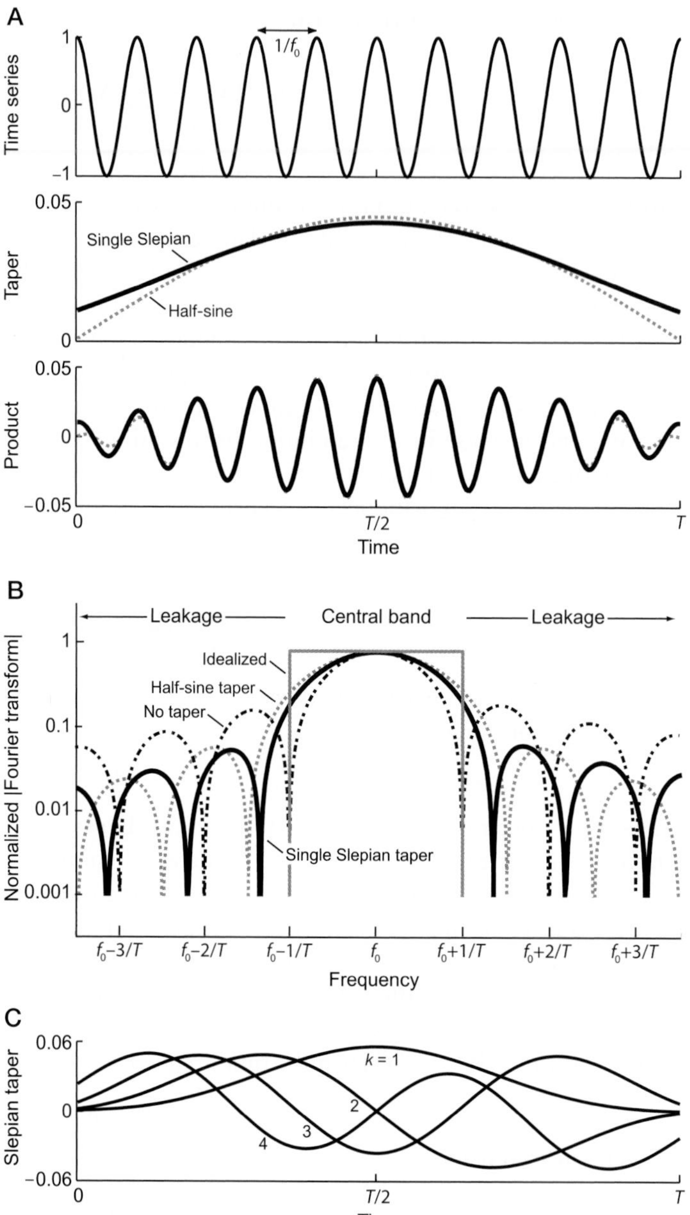

FIGURE 1. Basics of Fourier transforms and tapers. (*A*) Example of the process of tapering data. *Top* panel shows the time series of a sine wave with center frequency $f_0 = T/10$. *Middle* panel shows a half-sine taper, defined as $\sin\{\pi t/T\}$, and a single Slepian taper with $p = 1$ and $K = 1$; the norm of both functions are set to 1. *Bottom* panel shows the products of the tapers and the time series. (*B*) Magnitude of the Fourier transforms of the untapered data, the data tapered with a single taper, and the data tapered with a single Slepian taper. Also shown is the "ideal" representation with power only in the interval $[-1/T, 1/T]$ surrounding the center frequency. For the case of no taper, the transform is

$$\tilde{V}(f) \sim \sin\{\pi(f - f_0)/T\}/[\pi(f - f_0)/T],$$

and the first zero is at $f = \pm 1/T$ relative to f_0, whereas for the case of the half-sine taper,

$$\tilde{V}(f) \sim \cos\{\pi(f - f_0)/T\}/\{1 - [2(f - f_0)/T]^2\},$$

and the first zero is at $f = \pm 3/(2T)$ relative to f_0. There is no analytical expression for the transform of the Slepian taper. (*C*) The family of four Slepian tapers, $w^{(k)}(t)$, for the choice $p = 2.5$.

Implementation of the algorithms can be in any programming environment, but the use of the MatLab-based programming environment along with packaged routines in Chronux (www.chronux.org) is particularly convenient.

CASE ONE: SPECTRAL POWER

As a means of introducing spectral estimation, we analyze the rhythms that give rise to motion of the wall of a penetrating arteriole that sources blood to cortex (Fig. 2A). These arterioles are gateways that transfer blood from the surface of cortex to the underlying microvasculature (Nishimura et al. 2007). Past work has established that isolated arterioles can generate myographic activity in the 0.1-Hz range (Osol and Halpern 1988), similar to that seen in vivo by noninvasive imaging techniques of brain tissue (Mayhew et al. 1996) and two-photon imaging of capillaries (Kleinfeld et al. 1998). Here we look at the relative contribution of vasomotion, as well as breathing and heart rate, to penetrating vessels.

The raw signal is the diameter of the vessel, denoted $D(t)$ (Fig. 2B). The mean value is removed to give

$$\delta D_t = D_t - \frac{1}{N}\sum_{t=1}^{N} D_t. \tag{7}$$

Our goal is to understand the spectral content of this signal—with confidence limits! The Fourier transform of this signal, with respect to the kth taper, is (Equation 6)

$$\delta\tilde{D}^{(k)}(f) = \frac{1}{\sqrt{N}}\sum_{t=1}^{N} e^{-i2\pi ft}\, w_t^{(k)}\, \delta D_t, \tag{8}$$

where, as noted previously (Equation 3), $w^{(k)}(t)$ is the kth Slepian taper, whose length is also T. We compute the spectral power density, denoted $\tilde{S}(f)$, with units of distance^2/frequency, in terms of an average over the index of tapers, that is,

$$\tilde{S}(f) \equiv \frac{1}{K}\sum_{k=1}^{K} \left|\delta\tilde{D}^{(k)}(f)\right|^2, \tag{9}$$

where $|\delta\tilde{D}^{(k)}(f)|^2 = \delta\tilde{D}^{(k)}(f)[\delta\tilde{D}^{(k)}(f)]^*$; we further average over all trials if appropriate. Note that the "$1/\sqrt{N}$" normalization satisfies Parseval's theorem, that is,

$$\sum_{f=0}^{f_{\text{Nyquist}}} \tilde{S}(f) = \frac{1}{N}\sum_{t=1}^{N} D_t^2. \tag{10}$$

The spectrum in this example has strong features, yet has a trend to decrease as roughly $1/f^2$ that tends to obscure the peaks (insert in Fig. 2C). We remove the trend by computing the spectrum of the temporal derivative of δD_t,

$$\delta D'_t \equiv \frac{\delta D_{t+1} - \delta D_t}{\Delta t}, \tag{11}$$

as a means to flatten or "prewhiten" the spectrum. We now observe a multitude of peaks on a relatively flat background (Fig. 2C). A broad peak is centered at 0.2 Hz and corresponds to vasomotion. A sharper peak near 1 Hz corresponds to breathing; the nonsinusoidal shape of variations in diameter caused by breathing leads to the presence of second and third harmonics. Finally, a sharp peak at 7 Hz corresponds to heart rate and also includes a harmonic. No additional peaks are observed beyond the second harmonic of breathing. Note that the resolution half-bandwidth was chosen to be $\Delta f = 0.03$ Hz ($p = 16$), which is narrower than the low-frequency band.

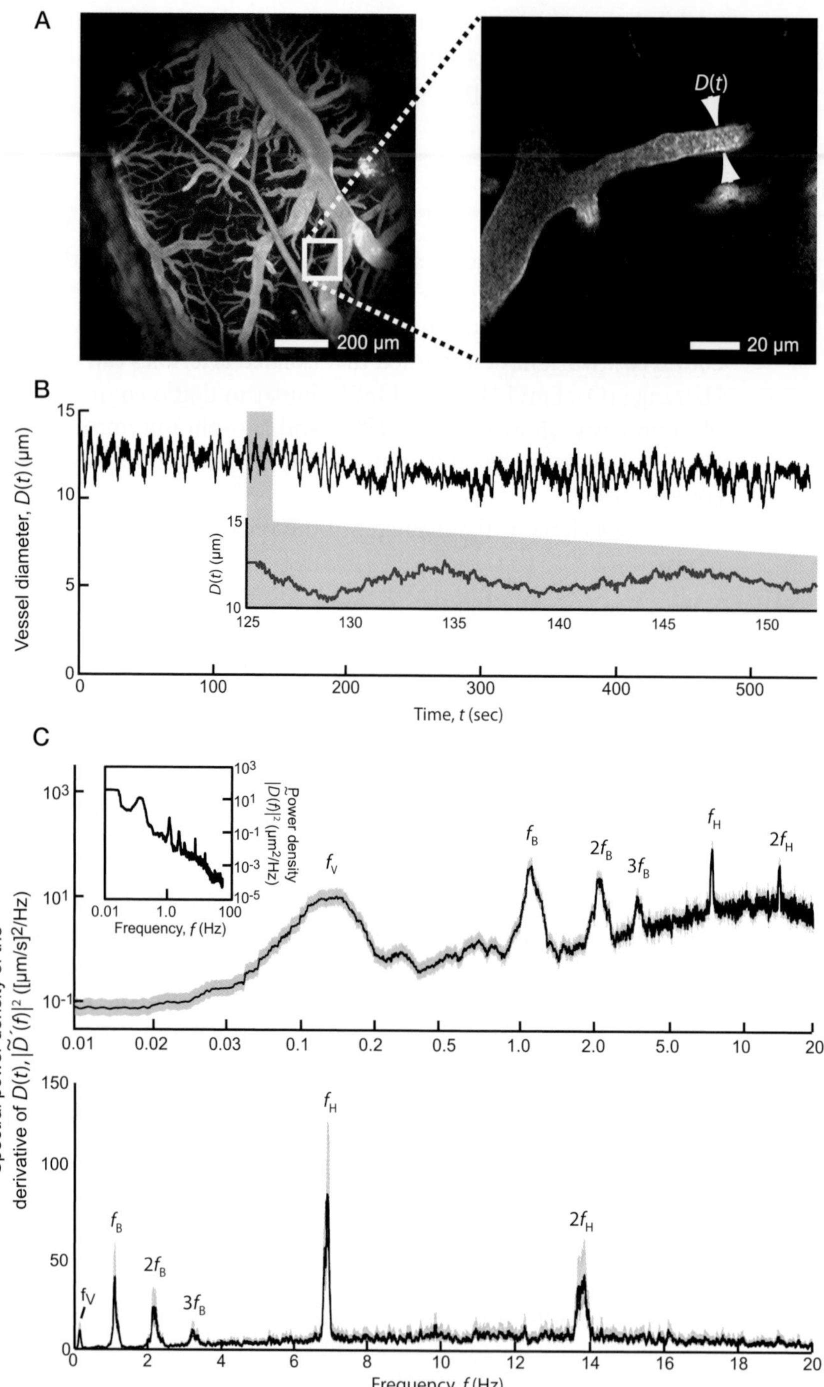

FIGURE 2. Analysis of the intrinsic motion of the diameter of a penetrating arteriole from rat. (*A*) Two-photon line-scan data through the center of a penetrating vessel over parietal cortex was obtained as described (Shih et al. 2009); $f_{Nyquist} = 500$ Hz. (*B*) Time series of the diameter as a function of time, as derived from the line-scan data as described (Devor et al. 2007) ($T = 540$ sec). The *insert* shows an expanded region to highlight the multiplicity of rhythmic events present in the signal. (*C*) Spectrum (Equations 8 and 9 with $p = 24$) of the time derivative of the diameter, $D'(t)$ (Equation 11) plotted on log–log (*top*) and linear–linear (*bottom*) axes. The gray bands encompass the 95% confidence bands (Equations 13–16) and appear symmetric on a log scale (*top*) but asymmetric on the linear scale (*bottom*). The frequencies are labeled f_V for vasomotion, f_B for breathing, and f_H for heartbeat. The *insert* in the *top* figure shows the spectrum of the diameter $D(t)$; note the steep, $\sim 1/f^2$ trend that is removed by taking the spectrum of $D'(t)$ (A.Y. Shih, unpubl. data).

The next issue is the calculation of confidence intervals so that the uncertainty in the power at each peak may be established and the statistical significance of each peak may be assessed. Confidence limits may be estimated analytically for various asymptotic limits. However, the confidence intervals may also be estimated directly by a jackknife, where we compute the standard error in terms of "delete-one" means (Thomson and Chave 1991). In this procedure, we exploit the multiple estimates of the spectral power density and calculate K different mean spectra in which one term is left out, that is,

$$\tilde{S}^{(n)}(f) \equiv \frac{1}{K-1} \sum_{\substack{k=1 \\ k \neq n}}^{K} \left| \delta \tilde{D}^{(k)}(f) \right|^2 . \tag{12}$$

Estimating the standard error of the spectral power density requires an extra step becuase spectral amplitudes are defined on the interval $[0, \infty)$, whereas Gaussian variables exist on the full interval $(-\infty, \infty)$. Taking the logarithm leads to variables defined over the full interval; thus we transform the delete-one estimates, $S^{(n)}(f)$, according to

$$g\left\{\tilde{S}^{(n)}(f)\right\} \equiv \ln\left\{\tilde{S}^{(n)}(f)\right\} \tag{13}$$

or

$$\tilde{S}^{(n)}(f) = e^{g\left\{\tilde{S}^{(n)}(f)\right\}} . \tag{14}$$

The mean of the transformed variable is

$$\tilde{\mu}(f) \equiv \frac{1}{K} \sum_{k=1}^{K} g\left\{\tilde{S}^{(n)}(f)\right\} \tag{15}$$

and the standard error is

$$\tilde{\sigma}(f) = \sqrt{\frac{K-1}{K} \sum_{n=1}^{K} \left[g\left\{\tilde{S}^{(n)}(f)\right\} - \mu(f) \right]^2 } . \tag{16}$$

The 95% confidence limit for the transformed spectral density is given by $2\tilde{\sigma}(f)$, so that one visualizes $\tilde{S}(f)$ by plotting the mean value of $\tilde{S}(f)$ (i.e., $e^{\tilde{\mu}(f)}$) along with the lower and upper bounds (i.e., $e^{\tilde{\mu}(f) - 2\tilde{\sigma}(f)}$ and $e^{\tilde{\mu}(f) + 2\tilde{\sigma}(f)}$, respectively). The confidence bands are symmetric about the mean when spectral power is plotted on a logarithmic scale (upper trace in Fig. 2C) rather than on a linear scale (lower trace in Fig. 2C).

CASE TWO: COHERENCE BETWEEN TWO SIGNALS

To introduce coherence, a measure of the tracking of one rhythmic signal by another, we consider the use of optical imaging to determine potential pair-wise connections between neurons (Cacciatore et al. 1999). We focus on imaging data taken from the ventral surface of a leech ganglion and seek to identify cells in the ganglion that receive monosynaptic input from neuron Tr2 in the head (Fig. 3A). This cell functions as a toggle for the swim rhythm in these animals. Rather than serially impale each of the roughly 400 cells in the ganglion and look for postsynaptic currents induced by driving Tr2, a parallel strategy was adopted (Taylor et al. 2003). The cells in the ganglion were stained with a voltage-sensitive dye (Fig. 3B), which transforms changes in membrane potential into changes in the intensity of fluorescent light. The emitted light from all cells is detected with a charge-coupled device (Fig. 3B), from which time series for the change in fluorescence are calculated for each neuron in the field (Fig. 3C). Presynaptic cell Tr2 was stimulated with a periodic signal, at frequency f_{Drive}, with the assumption that candidate postsynaptic followers of Tr2 would fire with the same periodicity (Fig. 3D). The phase of the coherence relative to the drive depends on the sign of the synapse, propagation delays, and filtering by postsynaptic processes.

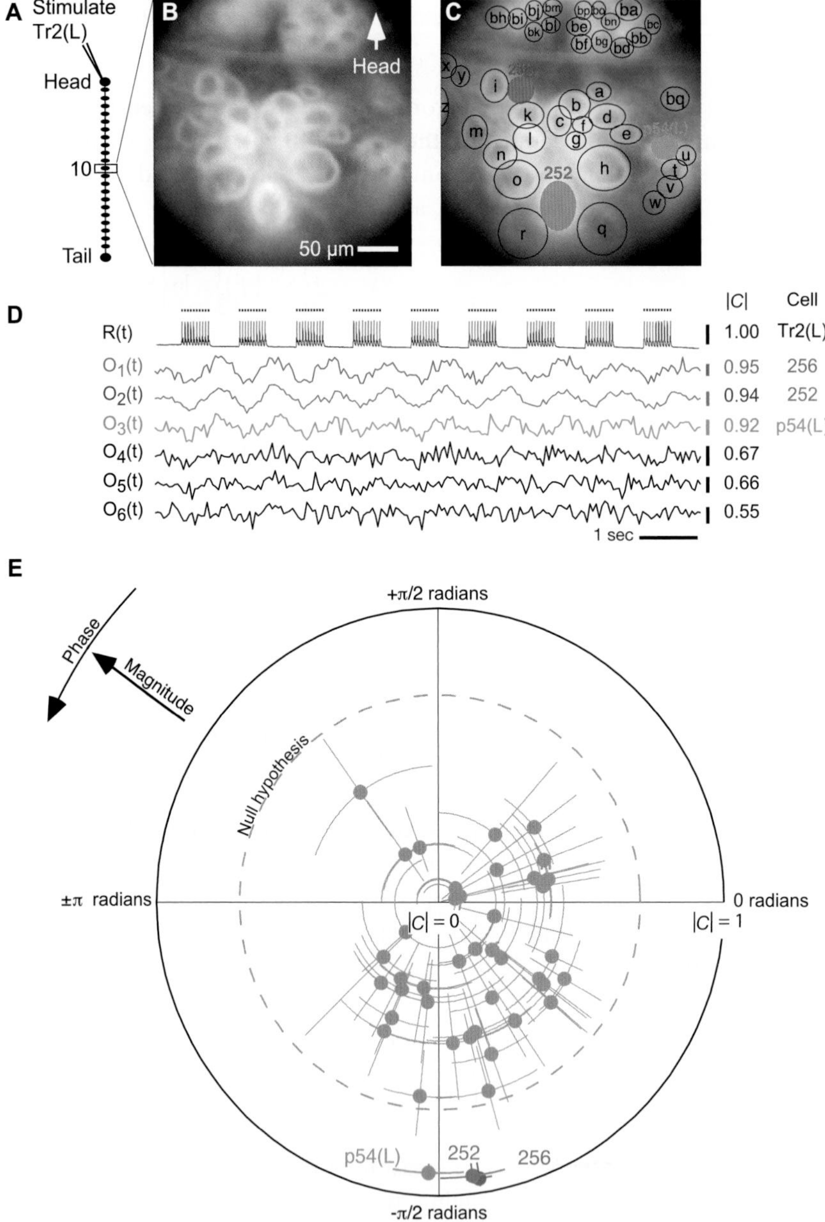

FIGURE 3. Analysis of voltage-sensitive dye imaging experiments to find followers of Tr2. (*A*) Cartoon of the leech nerve cord; input to Tr2 forms the drive signal $U(t)$. (*B*) Fluorescence image of ganglion 10 stained with dye. (*C*) Ellipses drawn to encompass individual cells and define regions whose pixel outputs were averaged to form the optical signals $V_i(t)$. (*D*) Simultaneous electrical recording of Tr2 (i.e., $U(t)$), and optical recordings from six of the cells shown in panel C ($T = 9$ sec) (i.e., $V_1(t)$ through $V_6(t)$), along with $|\tilde{C}_i(f_{Drive})|$ (Equation 17 with $p = 6$). (*E*) Polar plot of $\tilde{C}_i(f_{Drive})$ between each optical recording and the cell Tr2 electrical recording for all 43 cells in panel C. The dashed line indicates the $\alpha = 0.001$ threshold for significance (Equations 24 and 25); error bars one standard error (Equations 18–25). (Modified from Taylor et al. 2003.)

The coherence between the response of each cell and the drive, a complex function denoted $\tilde{C}_i(f)$, was calculated over the time period of the stimulus. We denote the measured time series of the optical signals as $O_i(t)$ and the reference drive signal as $R(t)$. The spectral coherence is defined as

$$\tilde{C}_i(f) = \frac{\frac{1}{K}\sum_{k=1}^{K}\tilde{O}_i^{(k)}(f)\left[\tilde{R}_i^{(k)}(f)\right]^*}{\sqrt{\left(\frac{1}{K}\sum_{k=1}^{K}\left|\tilde{O}_i^{(k)}(f)\right|^2\right)\left(\frac{1}{K}\sum_{k=1}^{K}\left|\tilde{R}_i^{(k)}(f)\right|^2\right)}}. \tag{17}$$

To calculate the standard errors for the coherence estimates, we again use the jackknife (Thomson and Chave 1991) and compute delete-one averages of coherence, denoted $\tilde{C}_i^{(n)}(f)$, where n is the index of the deleted taper, that is,

$$\tilde{C}_i^{(n)}(f) = \frac{\frac{1}{K-1}\sum_{\substack{k=1\\k\neq n}}^{K}\tilde{O}_i^{(k)}(f)\left[\tilde{R}_i^{(k)}(f)\right]^*}{\sqrt{\left(\frac{1}{K-1}\sum_{\substack{k=1\\k\neq n}}^{K}\left|\tilde{O}_i^{(k)}(f)\right|^2\right)\left(\frac{1}{K-1}\sum_{\substack{k=1\\k\neq n}}^{K}\left|\tilde{R}_i^{(k)}(f)\right|^2\right)}}. \tag{18}$$

Estimating the standard error of the magnitude of $\tilde{C}_i(f)$, as with the case for the spectral power, requires an extra step because $|\tilde{C}_i(f)|$ is defined on the interval [0, 1], whereas Gaussian variables exist on $(-\infty, \infty)$. The mean value of the magnitude of the coherence for each postsynaptic cell (i.e., $|\tilde{C}_i(f)|$) and the delete-one estimates, $|\tilde{C}_i^{(n)}(f)|$, are replaced with the transformed values

$$g\left\{\left|\tilde{C}_i^{(n)}(f)\right|\right\} = \ln\left(\frac{\left|\tilde{C}_i^{(n)}(f)\right|^2}{1-\left|\tilde{C}_i^{(n)}(f)\right|^2}\right) \tag{19}$$

or

$$\left|\tilde{C}_i^{(n)}(f)\right| = \frac{1}{\sqrt{1+\mathrm{e}^{-g\left\{\left|\tilde{C}_i^{(n)}(f)\right|\right\}}}}. \tag{20}$$

The means of the transformed variables are

$$\tilde{\mu}_{i;\mathrm{Mag}}(f) = \frac{1}{K}\sum_{n=1}^{K} g\left\{\tilde{C}_i^{(n)}(f)\right\} \tag{21}$$

and their standard errors are

$$\tilde{\sigma}_{i;\mathrm{Mag}}(f) = \sqrt{\frac{K-1}{K}\sum_{n=1}^{K}\left|g\left\{\tilde{C}_i^{(n)}(f)\right\} - \tilde{\mu}_{i;\mathrm{Mag}}(f)\right|^2}. \tag{22}$$

The 95% confidence interval for $|\tilde{C}_i(f)|$ corresponds to values within the interval

$$\left[\sqrt{1+\mathrm{e}^{-(\tilde{\mu}_{i;\mathrm{Mag}} - 2\tilde{\sigma}_{i;\mathrm{Mag}})}}^{\,-1}, \sqrt{1+\mathrm{e}^{-(\tilde{\mu}_{i;\mathrm{Mag}} + 2\tilde{\sigma}_{i;\mathrm{Mag}})}}^{\,-1}\right].$$

For completeness, an alternate transformation for computing the variance is

$$g\left\{\tilde{C}_i(f)\right\} = \tanh^{-1}\left\{\tilde{C}_i(f)\right\}.$$

We now consider an estimate of the standard deviation of the phase of $|\tilde{C}_i(f)|$. Conceptually, the idea is to compute the variation in the relative directions of the delete-one unit vectors (i.e., $\tilde{C}_i(f)/|\tilde{C}_i(f)|$). The standard error is computed as

$$\tilde{\sigma}_{i;\mathrm{Phase}}(f) = \sqrt{2\,\frac{K-1}{K}\left(K - \left|\sum_{n=1}^{K} \frac{\tilde{C}_i^{(n)}(f)}{\left|\tilde{C}_i^{(n)}(f)\right|}\right|\right)}. \tag{23}$$

Our interest is in the values of $\tilde{C}_i(f)$ for $f = f_{\mathrm{Drive}}$ and the confidence limits for this value. We choose the resolution bandwidth so that the estimate of $|\tilde{C}_i(f_{\mathrm{Drive}})|$ is kept separate from that of the harmonic $|\tilde{C}_i(2f_{\mathrm{Drive}})|$; the choice $\Delta f = 0.4f_{\mathrm{Drive}}$ works well. We graph the magnitude and phase of $\tilde{C}_i(f_{\mathrm{Drive}})$ for all neurons, along with the confidence intervals, on a polar plot (Fig. 3E).

Finally, we consider whether the coherence of a given cell at f_{Drive} is significantly >0 (i.e., larger than one would expect by chance from a signal with no coherence) as a means to select candidate postsynaptic targets of Tr2. We compared the estimate for each value of $\tilde{C}_i(f_{\mathrm{Drive}})$ with the null distribution for the magnitude of the coherence, which exceeds

$$\left|\tilde{C}_i\left(f_{\mathrm{Drive}}\right)\right| = \sqrt{1 - \alpha^{1/(K-1)}} \tag{24}$$

only in a fraction α of the trials (Hannan 1970; Jarvis and Mitra 2001). We used $\alpha = 0.001$ in our experiments to avoid false positives. We also calculated the multiple comparisons α level for each trial, given by

$$\alpha_{\mathrm{multi}} = 1 - (1 - \alpha)^N, \tag{25}$$

where N is the number of cells in the functional image, and verified that it did not exceed $\alpha_{\mathrm{multi}} = 0.05$ on any trial (Fig. 3E).

The result of the above procedure was the discovery of three postsynaptic targets of cell Tr2, two of which were functionally unidentified neurons (Taylor et al. 2003).

CASE THREE: SPACE–TIME SINGULAR-VALUE DECOMPOSITION AND DENOISING

A common issue in the analysis of optical imaging data is the need to remove "fast" noise, that is, fluctuations in intensity that occur on a pixel-by-pixel and frame-by-frame basis. The idea is that the imaging data contains features that are highly correlated in space, such as underlying cell bodies, processes, etc., and highly correlated in time, such as long-lasting responses. The imaging data may thus be viewed as a space–time matrix of random numbers (i.e., the fast noise) with added correlated structure. The goal is to separate the fast, uncorrelated noise from the raw data so that a compressed image file with only the correlated signals remains (Figs. 4A,B show single frames; for the complete movies, see Movies 12.1 and 12.2). With this model in mind, we focus on the case of intracellular Ca^{2+} oscillations in an organotypic culture of rat cortex, which contains both neurons and glia. All cells were loaded with a fluorescence-based calcium indicator, and spontaneous activity in the preparation was imaged with a fast-framing (Δt = 2 msec), low-resolution (100 x 100 pixels) confocal microscope (Fig. 4A).

Imaging data is in the form of a three-dimensional array of intensities, denoted $V(x, y, t)$. We consider expressing the spatial location in terms of a pixel index, so that each $(x, y) \rightarrow s$ and the data is now in the form of a space–time matrix (i.e., $V(s, t)$). This matrix may be decomposed into the outer product of functions of pixel index with functions of time. Specifically,

$$V(s,t) = \sum_{n=1}^{\mathrm{rank}\{V\}} \lambda_n F_n(s)\, G_n(t), \tag{26}$$

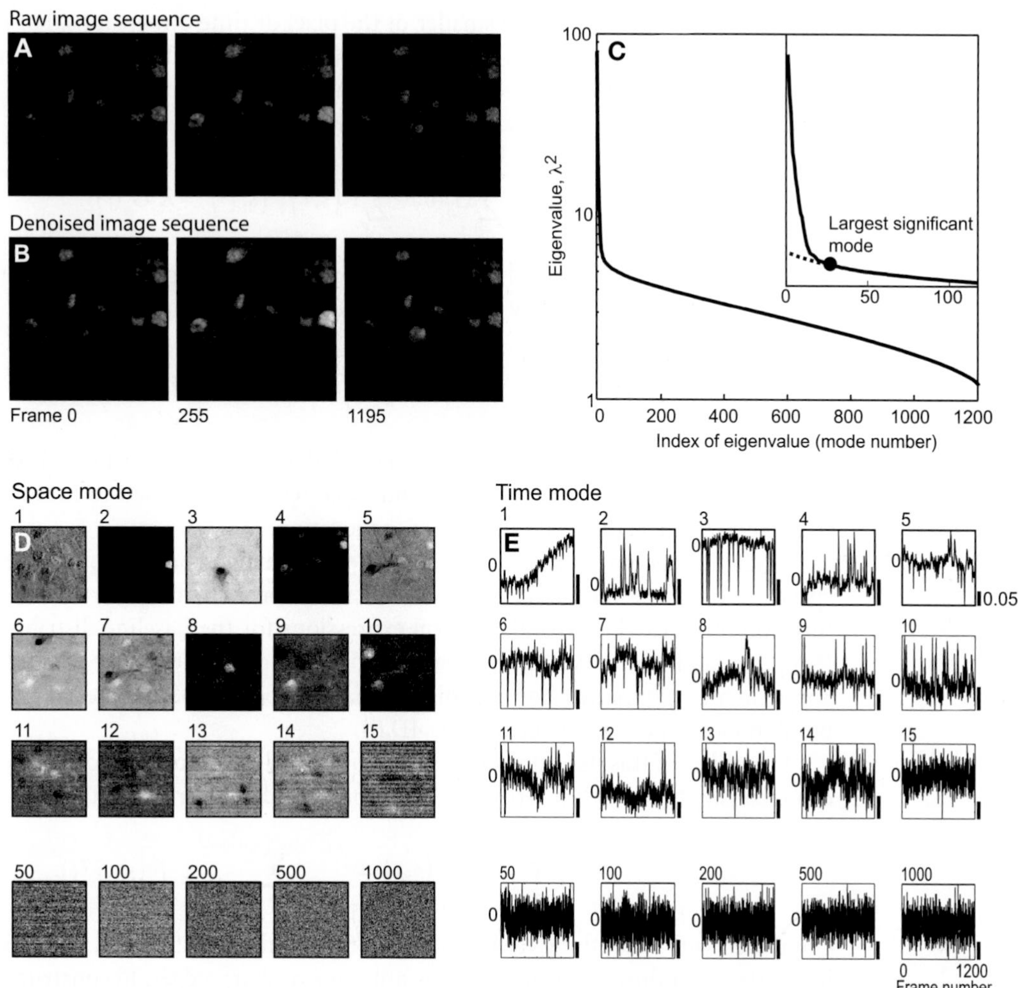

FIGURE 4. Denoising of spinning-disk confocal imaging data on Ca^{2+} waves in organotypic culture. (*A*) Selected frames from a 1200-frame sequence of 100 x 100-pixel data. (*B*) The same data set after reconstruction with 25 of the 1200 modes (Equation 29). Denoising is particularly clear when the data is viewed as video clips. (*C*) Singular value decomposition of the imaging sequence in *A*. The spectrum for the square of the eigenvalues for the space and time modes. Note the excess variance in the roughly 25 dominant modes (Equations 27 and 28). (*D*) The top 15 spatial modes, $F_n(s)$, plus high-order modes. Light shades correspond to positive values and dark shades negative values. The amplitude of the modes is set by the orthonormal condition

$$\sum_{t=1}^{N_t} F_m(t)\, F_n(t) = \delta_{nm}.$$

(*E*) The top 15 temporal modes of $G_n(t)$. The amplitude of the modes is set by the orthonormal condition

$$\sum_{t=1}^{N_t} G_m(t)\, G_n(t) = \delta_{nm}$$

(J.T. Vogelstein, unpubl. data). Fields in *A*, *B*, and *D* are 115 μm on edge.

where the rank of $V(s, t)$ is the smaller of the pixel or time dimensions. For example data of Figure 4A, there are $N_t = 1200$ frames or time points and $N_s = 10{,}000$ pixels, so that rank$\{V(s, t)\} = N_t$. The above decomposition is referred to as a singular-value decomposition (Golub and Kahan 1965). The temporal functions satisfy an eigenvalue equation, that is,

$$\sum_{t'=1}^{N_t} G_n(t') \sum_{s=1}^{N_s} V(s,t)V(s,t') = \lambda_n^2 G_n(t), \tag{27}$$

where the functions $F_n(s)$ and $G_n(t)$ are orthonormal. The spatial function that accompanies each temporal function is found by inverting the defining equation, so that

$$F_n(s) = \frac{1}{\lambda_n} \sum_{t=1}^{N_t} V(s,t)\, G_n(t). \tag{28}$$

For completeness, note that this is equivalent to determining the principal components of responses recorded from a single location across multiple trials, as opposed to multiple locations in a single trial, where s labels the trial rather than the location.

When this decomposition is applied to the Ca^{2+} imaging data (Fig. 4A), we see that the eigenvalue spectrum has large values for the low-order modes and then rapidly falls to a smoothly decreasing function of index (Fig. 4B); theoretical expressions for the baseline distribution have been derived (Sengupta and Mitra 1999). The spatial and temporal modes show defined structure for the first approximately 20 modes; beyond this the spatial modes appear increasingly "grainy" and the temporal modes appear as fast noise (Fig. 4D,E).

The utility of this decomposition is that only the lower-order modes carry information. Thus we can reconstruct the data matrix from only these modes and remove the "fast" noise, that is,

$$V^{\text{reconstructed}}(s,t) = \sum_{n=1}^{\text{largest significant mode}} \lambda_n F_n(s)\, G_n(t). \tag{29}$$

Compared with smoothing techniques, the truncated reconstruction respects all correlated features in the data and thus, for example, does not remove sharp edges. Reconstruction of the intracellular Ca^{2+} oscillation data highlights the correlated activity by removing fast, grainy variability (Fig. 4B).

CASE FOUR: SPECTROGRAMS AND SPACE-FREQUENCY SINGULAR-VALUE DECOMPOSITION

The final example concerns the characterization of coherent spatiotemporal dynamics, such as waves of activity. We return to the use of voltage-sensitive dyes, this time to image the electrical dynamics of turtle visual cortex in response to a looming stimulus. Early work had shown that a looming stimulus led to the onset of ~20-Hz oscillations, the g-band for turtle, in visual cortex (Prechtl and Bullock 1994, 1995). The limited range of cortical connections led to the hypothesis that this oscillation might be part of wave motion. We investigated this issue by direct electrical measurements throughout the depth of cortex at selected sites (Prechtl et al. 2000) and, of relevance for the present discussion, by imaging the spatial patterns from cortex using voltage-sensitive dyes as the contrast agent (Prechtl et al. 1997).

The electrical activity is expected to evolve between prestimulus versus poststimulus epochs and possibly over an extended period of stimulation. Thus the spectral power is not stationary over long periods of time and we must consider a running measure of the spectral power density, denoted the spectrogram, that is a function of both frequency and time. We choose a restricted interval of time, denoted T_{window}, with N_{window} data points, compute the Fourier transforms $\tilde{V}^{(k)}(f;\, t_0)$, and spectrum $\tilde{S}(f;\, t_0)$ over that interval, where t_0 indexes the time at the middle of the epoch, and then step forward in time and recalculate the transforms and spectrum. Thus

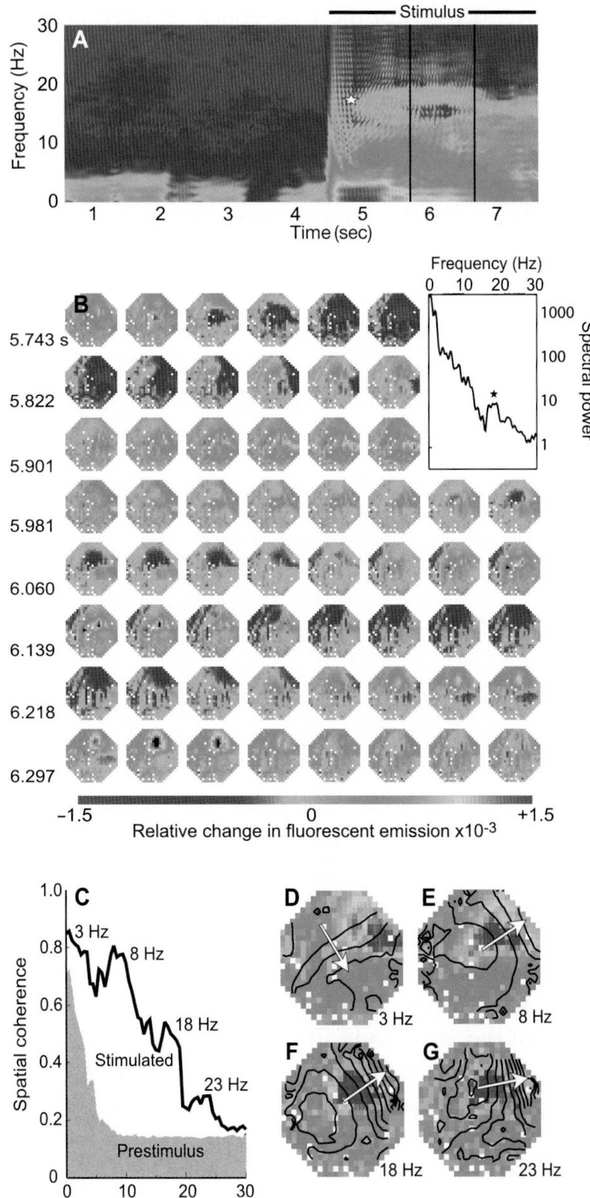

FIGURE 5. Analysis of single-trial voltage-sensitive dye imaging data to delineate collective excitation in visual cortex of turtle. (*A*) Spectrogram of the response averaged over all active pixels in the image (Equations 30 and 31). (*B*) Space–time response during the period when the animal was presented with a looming stimulus. The data were denoised (Equation 29), low-pass filtered at 60 Hz, and median filtered (400 msec width) to remove a stimulus-induced depolarization. We show every eighth frame (126 Hz); note the flow of depolarization from left to right. The *insert* is the spectrum for the interval 4.7–5.7 sec and is the power spectrum over the T = 1 sec interval that encompasses this epoch (black band in *A*). (*C*) Coherence, $\tilde{C}(f_0)$, over intervals both before and after the onset (T = 3 sec; K = 7) estimated at successive frequency bins; $\tilde{C}(f_0) > 0.14$ indicates significance (Equations 33–35). (*D–G*) Spatial distribution of amplitude (red for maximum and blue for zero) and phase (π/12 radians per contour; arrow indicates dominant gradient) of the coherence at f_0 = 3, 8, 18, and 22 Hz, respectively, during stimulation. Fields in *B* and *D–G* are 3.5 mm in diameter. (Modified from Prechtl et al. 1997.)

$$\tilde{V}^{(k)}\left(f;t_0\right) = \frac{1}{\sqrt{N}} \sum_{t=t_0-(1/2)N_{\text{window}}}^{t_0+(1/2)N_{\text{window}}-1} e^{-i2\pi f t}\, w_t^{(k)}\, V_t \tag{30}$$

and

$$\tilde{S}(f;t_0) \equiv \frac{1}{K}\sum_{k=1}^{K}\left|\tilde{V}^{(k)}\left(f;t_0\right)\right|^2 . \tag{31}$$

The resolution half-bandwidth is now p/N_{window} and, as a practical matter, the index is shifted in increments no larger than $N_{\text{window}}/2$. For the case of the summed optical signal from turtle cortex, we observe low-frequency activity before stimulation and multiple bands of high-frequency oscillations upon stimulation (Fig. 5A). A particularly pronounced band occurs near 18 Hz; this is clearly seen in a line plot of the spectral power density for the 1-sec epoch centered in the middle of the stimulation period (insert in Fig. 5B).

The image data, even after denoising (Equation 29) and broad-band filtering, appears complex (Fig. 5B), with regions of net depolarization sweeping across cortex, but no simple pattern emerges. One possibility is that cortex supports multiple dynamic processes, each with a unique center frequency, that may be decomposed by a singular value decomposition in the frequency domain. In this method, proposed by Mann and Park (1994), the space–time data $V(s, t)$ is first projected into a local temporal frequency domain by transforming with respect to a set of tapers, that is,

$$\tilde{V}^{(k)}(s,f) = \frac{1}{\sqrt{N}} \sum_{t=1}^{N} e^{-i2\pi ft} w_t^{(k)} V_t(s). \tag{32}$$

The index k defines a local frequency index in the band $[f_0 - \Delta f/2, f_0 + \Delta f/2]$. For a fixed frequency, f_0, a SVD is performed on the complex matrix

$$\tilde{V}(s,k;f_0) \equiv \tilde{V}^{(1)}(s,f_0), \dots, \tilde{V}^{(K)}(s,f_0) \tag{33}$$

to yield

$$\tilde{V}(s,k;f_0) = \sum_{n=1}^{\mathrm{rank}\{\tilde{V}\}} \lambda_n \tilde{F}_n(s) \tilde{G}_n(k), \tag{34}$$

where the rank is invariably set by K. A measure of coherence is given by the ratio of the power of the leading mode to the total power (Fig. 5C), that is,

$$\tilde{C}(f_0) = \frac{\lambda_1^2(f_0)}{\sum_{k=1}^{K} \lambda_k^2(f_0)}. \tag{35}$$

A completely coherent response leads to $\tilde{C}(f_0) = 1$, whereas for a uniform random process $\tilde{C}(f_0) = 1/K$. Where $\tilde{C}(f_0)$ has a peak, it is useful to examine the largest spatial mode, $\tilde{F}_1(s)$. The magnitude of this complex image gives the spatial distribution of coherence that is centered at frequency f_0, whereas gradients in the phase of the image indicate the local direction of propagation.

For the example data (Fig. 5B), this analysis revealed linear waves as the dominant mode of electrical activity. Those with a temporal frequency centered at $f_0 = 3$ Hz are present with or without stimulation (Prechtl et al. 1997) (Fig. 5D), whereas those centered at $f_0 = 8$, 18, and 23 Hz are seen only with stimulation and propagate orthogonal to the wave at 3 Hz (Fig. 5E–G). It is of biological interest that the waves at $f_0 = 3$ Hz track the direction of thalamocortical input, whereas those at higher frequencies track a slight bias in axonal orientation (Cosans and Ulinski 1990) that was unappreciated in the original work (Prechtl et al. 1997).

CONCLUSION

This chapter covered a number of key applications of spectral methods to optical imaging data. The choice of topics is representative but by no means exhaustive. An additional application that is likely to be of utility is the fitting of line spectra to signals with relatively pure periodic contributions, such as may occur from physiological rhythms, from the response to a periodic stimulus, or from environmental contaminants like line power (Mitra et al. 1999; Pesaran et al. 2005). A second application of note is demodulation of a spatial image in response to periodic stimulation either at the fundamental drive frequency (Borst 1995; Kalatsky and Stryker 2003; Sornborger et al. 2005) or the second harmonic of the drive (Benucci et al. 2007). Demodulation also is valuable for delineating wave dynamics in systems with rhythmic activity (Kleinfeld et al. 1994; Prechtl et al. 1997). In general, spectral techniques are an essential tool for the statistical analysis of imaging data.

ACKNOWLEDGMENTS

We thank Bijan Pesaran and David J. Thomson for many useful discussions, Andy Y. Shih for acquiring the unpublished data in Figure 2, Joshua T. Vogelstein for acquiring the unpublished data in Figure 4, and Pablo Blinder, Adrienne L. Fairhall, and Karel Svoboda for comments on a preliminary version of the chapter. The material is derived from a presentation at the Society for Neuroscience short course on "Neural Signal Processing: Quantitative Analysis of Neural Activity" as well as presentations at the "Neuroinformatics" and "Methods in Computational Neuroscience" schools at the Marine Biology Laboratories and the "Imaging Structure and Function in the Nervous System" school at Cold Spring Harbor Laboratory. The development of spectral tools in the Chronux library was funded by the National Institutes of Health (Grant MH071744 to PPM). The application of spectral methods to imaging data sets was also funded by the National Institutes of Health (Grants EB003832, MH085499, and NS059832 to DK and MH062528 to PPM).

REFERENCES

Ahmed N, Rao KR. 1975. *Orthogonal transforms for digital signal processing*. Springer Verlag, New York.

Benucci A, Frazor RA, Carandini M. 2007. Standing waves and traveling waves distinguish two circuits in visual cortex. *Neuron* **55:** 103–117.

Borst A. 1995. Periodic current injection (PCI): A new method to image steady-state membrane potential of single neurons in situ using extracellular voltage-sensitive dyes. *Z Naturforsch* **50:** 435–438.

Cacciatore TW, Brodfueher PD, Gonzalez JE, Jiang T, Adams SR, Tsien RY, Kristan Jr WB, Kleinfeld D. 1999. Identification of neural circuits by imaging coherent electrical activity with FRET-based dyes. *Neuron* **23:** 449–459.

Cosans CE, Ulinski PS. 1990. Spatial organization of axons in turtle visual cortex: Intralamellar and interlamellar projections. *J Comp Neurol* **296:** 548–558.

Devor A, Tian P, Nishimura N, Teng IC, Hillman EM, Narayanan SN, Ulbert I, Boas DA, Kleinfeld D, Dale AM. 2007. Suppressed neuronal activity and concurrent arteriolar vasoconstriction may explain negative blood oxygenation level–dependent signaling. *J Neurosci* **27:** 4452–4459.

Golub GH, Kahan W. 1965. *Calculating singular values and pseudo-inverse of a matrix.* Society for Industrial and Applied Mathematics, Philadelphia.

Hannan EJ. 1970. *Multiple time series.* Wiley, New York.

Jarvis MR, Mitra PP. 2001. Sampling properties of the spectrum and coherency of sequences of action potentials. *Neural Comput* **13:** 717–749.

Kalatsky VA, Stryker MP. 2003. New paradigm for optical imaging: Temporally encoded maps of intrinsic signal. *Neuron* **38:** 529–545.

Kleinfeld D. 2008. Application of spectral methods to representative data sets in electrophysiology and functional neuroimaging. In *Syllabus for Society for Neuroscience Short Course III on "Neural Signal Processing: Quantitative Analysis of Neural Activity"* (ed. Mitra P), pp. 21–34. Society for Neuroscience, Washington, DC.

Kleinfeld D, Delaney KR, Fee MS, Flores JA, Tank DW, Gelperin A. 1994. Dynamics of propagating waves in the olfactory network of a terrestrial mollusk: An electrical and optical study. *J Neurophysiol* **72:** 1402–1419.

Kleinfeld D, Mitra PP, Helmchen F, Denk W. 1998. Fluctuations and stimulus-induced changes in blood flow observed in individual capillaries in layers 2 through 4 of rat neocortex. *Proc Natl Acad Sci* **95:** 15741–15746.

Mann ME, Park J. 1994. Global-scale modes of surface temperature variability on interannual to centuries timescales. *J Geophys Res* **99:** 25819–25833.

Mayhew JEW, Askew S, Zeng Y, Porrill J, Westby GWM, Redgrave P, Rector DM, Harper RM. 1996. Cerebral vasomotion: 0.1 Hz oscillation in reflectance imaging of neural activity. *Neuroimage* **4:** 183–193.

Mitra PP, Bokil HS. 2008. *Observed brain dynamics.* Oxford University Press, New York.

Mitra PP, Pesaran B. 1998. Analysis of dynamic brain imaging data. *Biophys J* **76:** 691–708.

Mitra PP, Pesaran B, Kleinfeld D. 1999. Analysis of dynamic optical imaging data. in *Imaging neurons: A laboratory manual* (ed. Yuste R, et al.), pp. 9.1–9.9. Cold Spring Harbor Laboratory Press, Cold Spring Harbor, NY.

Nishimura B, Schaffer CB, Friedman B, Lyden PD, Kleinfeld D. 2007. Penetrating arterioles are a bottleneck in the perfusion of neocortex. *Proc Natl Acad Sci* **104:** 365–370.

Osol G, Halpern W. 1988. Spontaneous vasomotion in pressurized cerebral arteries from genetically hypertensive rats. *Am J Physiol* **254:** H28–H33.

Papoulis A. 1962. *The Fourier integral and its applications.* McGraw-Hill, New York.

Percival DB, Walden AT. 1993. *Spectral analysis for physical applications: Multitaper and conventional univariate techniques.* Cambridge University Press, Cambridge.

Pesaran B, Sornborger A, Nishimura N, Kleinfeld D, Mitra PP. 2005. Spectral analysis for dynamical imaging data. in *Imaging in neuroscience and development: A laboratory manual* (ed. Yuste R, Konnerth A), pp. 439–444. Cold Spring Harbor Laboratory Press, Cold Spring Harbor, NY.

Prechtl JC, Bullock TH. 1994. Event-related potentials to omitted visual stimuli in a reptile. *Electroencephal Clin Neurophys* **91:** 54–66.

Prechtl JC, Bullock TH. 1995. Structure and propagation of cortical oscillations linked to visual behaviors in the turtle. In *Proceedings of the 2nd Joint Symposium on Neural Computation* (ed. Sejnowski TJ), pp. 105–114. Institute for Neural Computation, University of California, San Diego and California Institute of Technology.

Prechtl JC, Cohen LB, Mitra PP, Pesaran B, Kleinfeld D. 1997. Visual stimuli induce waves of electrical activity in turtle cortex. *Proc Natl Acad Sci* **94:** 7621–7626.

Prechtl JC, Bullock TH, Kleinfeld D. 2000. Direct evidence for local oscillatory current sources and intracortical phase gradients in turtle visual cortex. *Proc Natl Acad Sci* **97:** 877–882.

Sengupta AM, Mitra PP. 1999. Distributions of singular values for some random matricies. *Phys Rev E* **60:** 3389–3392.

Shih AY, Friedman B, Drew PJ, Tsai PS, Lyden PD, Kleinfeld D. 2009. Active dilation of penetrating arterioles restores red blood cell flux to penumbral neocortex after focal stroke. *J Cerebr Blood Flow Metab* **29:** 738–751.

Sornborger A, Yokoo T, Delorme A, Sailstad C, Sirovich L. 2005. Extraction of the average and differential dynamical response in stimulus-locked experimental data. *J Neurosci Meth* **141:** 223–229.

Taylor AL, Cottrell GW, Kleinfeld D, Kristan WB. 2003. Imaging reveals synaptic targets of a swim-terminating neuron in the leech CNS. *J Neurosci* **23:** 11402–11410.

Thomson DJ. 1982. Spectral estimation and harmonic analysis. *Proc IEEE* **70:** 1055–1096.

Thomson DJ, Chave AD. 1991. Jackknifed error estimates for spectra, coherences, and transfer functions. in *Advances in spectrum analysis and array processing* (ed. Shykin S), pp. 58–113. Prentice Hall, Englewood Cliffs, NJ.

MOVIE LEGENDS

Movies are freely available online at www.cshprotocols.org/imaging.

MOVIE 12.1. Spinning-disk confocal imaging data on Ca^{2+} waves in organotypic culture before denoising. Video clip from 1200-frame sequence of 100 x 100-pixel data. Square field is 115 µm on edge. Also see Figure 4A.

MOVIE 12.2. Spinning-disk confocal imaging data on Ca^{2+} waves in organotypic culture after denoising. The same data set as in Movie 12.1 after reconstruction with 25 of the 1200 modes (Equation 29). Also see Figure 4B.

13 Preparation of Cells and Tissues for Fluorescence Microscopy

Andrew H. Fischer,[1] Kenneth A. Jacobson,[2] Jack Rose,[3] Pascal Lorentz,[4] and Rolf Zeller[4]

[1]Emory University Hospital, Atlanta, Georgia 30322; [2]University of North Carolina, Chapel Hill, North Carolina 27599; [3]Yale University School of Medicine, New Haven, Connecticut 06510; [4]Department of Biomedicine, University of Basel, 4058 Basel, Switzerland

ABSTRACT

Fluorescence microscopy is one of the most widely used approaches for localizing proteins and subcellular compartments at the light microscopic level. The strategies described in some of the following chapters take advantage of the sensitivity and the specificity of nonimmunological as well as immunological-based fluorescent probes for revealing structure–function relationships. A selected group of protocols is provided here to serve as representative examples for preparing cells and tissues for fluorescent studies; many others may be found in literature.

INTRODUCTION

Nonimmunological fluorescent probes may be used to directly label specific subcellular components and macromolecules such as the Golgi apparatus or the components of the nuclear pore complex. Typical fluorescent groups associated with probes used in these studies are rhodamine, fluorescein, boron-dipyrromethane (BODIPY), nitrobenzoxazole (NBD), carbocyanines, and the quantum dot (Qdot) bioconjugates (available from Molecular Probes). Many of these probes are directly taken up into living cells and are incorporated and are concentrated in specific organelles that can then be examined using the fluorescence microscope.

Immunofluorescence, another widely used application of fluorescence microscopy, involves the use of antibodies, obtained by conventional methods of antibody generation and purification (see Harlow and Lane 1988) or derived from the serum of individuals with a variety of autoimmune disorders to localize a particular protein or other antigens. Direct or one-step immunofluorescence involves the conjugation of a fluorochrome (e.g., fluorescein isothiocyanate [FITC]) directly to the primary antibody. Indirect immunofluorescence, which is mostly used, involves the initial binding of the primary antibody to the antigen in a fixed cell. Subsequently, a fluorescently labeled secondary antibody is incubated with the specimen to form a fluorescent sandwich at the site of the target. The secondary antibody was chosen to react with the host species in which the primary antibody was raised. For example, if the primary antibody was raised in a mouse, the secondary antibody may be X-species antimouse (e.g., goat antimouse IgGs). Secondary antibodies are available as affinity-purified reagents from a variety of companies and can be purchased coupled to many different fluorochromes (e.g., Alexa Fluor Dyes or Qdot secondary antibody conjugates, Molecular Probes). Typically, the overall scheme for localization of a cellular protein by indirect immunofluorescence involves:

- fixation (unless live cells are to be studied)
- permeabilization of fixed cells to allow penetration of antibodies
- blocking sites prone to nonspecific interactions
- labeling the fixed cells/tissues with specific antibodies
- mounting the sample for microscopic examination

In general, the choice of conditions for pretreatment and labeling of the sample is dictated both by the nature of the sample and by the type of labeling procedure to be used. This chapter presents an overview of the methods for preparing slides and coverslips for fixed and live specimens, a discussion of various approaches for fixation of cells and tissues (with subsequent sectioning of tissue specimens), and a collection of procedures used for mounting live and fixed cells. For extensive protocols on approaches for live cell imaging, see Goldman and Spector (2005).

FIXATION AND PERMEABILIZATION OF CELLS AND TISSUES

Fluorescence microscopy is used to visualize specific cellular components in as native a state and organization as possible. To preserve cellular structure, the specimen is fixed chemically to retain the cells or tissue in a state as near to life as possible by rapidly terminating all enzymatic and other metabolic activities to minimize postfixation changes. Sample fixation is one of the most crucial steps in assuring the accuracy of detection protocols and is, therefore, decisive in determining the subsequent success or failure. Underfixation of the sample leads to poor morphological preservation and/or loss of signal, whereas overfixation may lead to fixation artifacts, loss of signal, and/or increased nonspecific background (noise). An ideal fixative should preserve a given antigen such as to reflect the in vivo distribution (no diffusion or rearrangement). Ideally, cell morphology should be preserved, the antigen of interest should remain accessible to the probe, and the fixation should cause minimal denaturation of the antigen. However, several of these criteria are mutually incompatible; therefore, a workable compromise must often be reached.

Glutaraldehyde, formaldehyde, and methanol/acetone are the most commonly used fixatives. Glutaraldehyde, a five-carbon dialdehyde, provides for best preservation of fine structure and is the fixative of choice for electron microscopy and some immunofluorescence studies (such as, e.g., localization of microtubules). However, it is also the harshest of the fixatives, and, frequently, epitopes are altered such as to interfere with binding of specific antibodies. Glutaraldehyde forms a Schiff's base with amino groups of proteins and polymerizes via Schiff's base-catalyzed reactions. The ability to polymerize allows glutaraldehyde to form extended crosslinks. For immunocytochemistry, 0.01%–0.5% glutaraldehyde fixatives have been useful in some studies. Two percent glutaraldehyde can penetrate ~700 μm into tissue in 1 h at room temperature. Glutaraldehyde fixation also contributes to nonspecific fluorescence at some excitation wavelengths. This autofluorescence may be partially reduced by treating cells with 1.0–1.5 mg/mL sodium borohydride in phosphate-buffered saline (PBS), twice for 5 min after fixation (Tagliaferro et al. 1997).

Formaldehyde, a one-carbon monoaldehyde made freshly from paraformaldehyde (PFA), is widely used and is an excellent general fixative suitable for the localization of most proteins.

Formaldehyde interferes with epitope recognition to a lesser extent than glutaraldehyde. In most cases, it is the best choice of fixative for fluorescence microscopy. Formaldehyde crosslinks proteins by forming methylene bridges between reactive groups. Formaldehyde is a less effective crosslinker than glutaraldehyde because of its smaller size, which results in many reactive groups that escape crosslinking. In addition, formaldehyde crosslinking is partially reversible; thus, formaldehyde-fixed tissue can be become unfixed again when exposed to certain buffers during postfixation processing. Because it is a milder fixative, it is extensively used in histochemical and immunocytochemical studies, and it is the most effective fixative for nucleic acids. Four percent formaldehyde penetrates ~2 mm in 1 h at room temperature. In addition, formaldehyde does not contribute significantly to autofluorescence. Cultured cells are usually fixed in 2%–4% formaldehyde in PBS, pH 7.4, at 20°C for 15 min. Because most commercially available bulk liquid forms of formaldehyde contain methanol, they should not be used; rather, formaldehyde solutions should be prepared freshly from PFA. A combination of glutaraldehyde and formaldehyde fixative can be used for optimized penetration and fixation of the tissue, which quickly stabilizes cellular constituents and fixes them thoroughly.

Cold methanol or acetone solutions provide more rapid fixation than aldehydes and have been used in a variety of studies to examine, for example, components of the cytoskeleton. However, because these fixatives precipitate proteins and carbohydrates, they are more likely to alter the localization of certain antigens. In addition, they fix and permeabilize cells at the same time, which may result in the loss of some of the more soluble antigens. Furthermore, dehydration and fixation occur simultaneously, causing possible shrinkage.

If intracellular components are to be visualized after aldehyde fixation, it is necessary to permeabilize the cell, either with detergents or with organic solvents that remove lipids from the plasma membrane and nuclear envelope and thereby allow probes such as antibodies to penetrate and to bind to the subcellular structures of interest (Fig. 1). Some of the more commonly used permeabilization agents include Triton X-100, NP-40, Tween 20, and Brij-58, which solubilize phospholipid membranes. These detergents are most commonly used at concentrations ranging between 0.1% and 0.5% at room temperature or on ice for 5 min and are used after fixation. Permeabilization is not required to localize cell surface or extracellular matrix antigens. However, some protocols suggest fixing and permeabilizing cells at the same time or permeabilizing cells before fixation. Saponin is a detergent that solubilizes cholesterol in membranes and is less damaging to membranes than the other detergents. However, it must also be included during the probe (e.g., antibody) incubation steps and subsequent washes because the membrane reseals in the absence of the saponin. Saponin is mostly used in studies in which preservation of membrane structures is important, and it is commonly used at 0.5% at room temperature. Digitonin is a detergent that selectively permeabilizes the plasma membrane by binding cholesterol and other 3-β-hydroxysterols (Stearns and Ochs 1982). Digitonin has

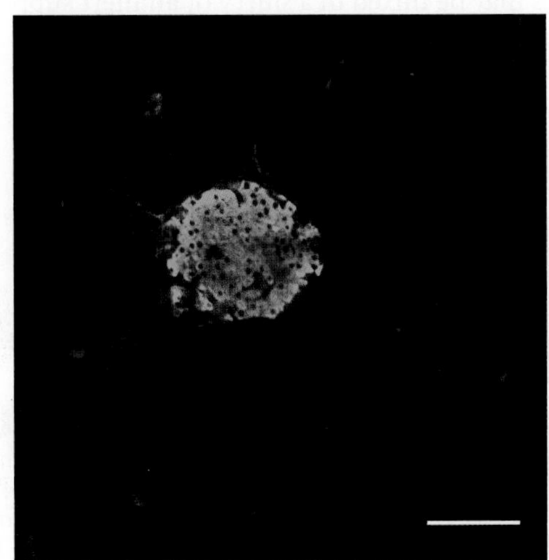

FIGURE 1. β cells of an islet of Langerhans of a wild-type C57BL/6 mouse stained for insulin. The organ was fixed in 4% PFA at 4°C overnight and was embedded in paraffin. Then, 5-mm sections were cut, deparaffinized, and rehydrated. After that, the tissue was permeabilized using 0.2% Triton X-100 for 10 min and was stained with insulin antiswine (Dako A0564) primary antibody followed by detection with Alexa 488 secondary antibodies (Molecular Probes). Scale bar, 100 μm.

been used for fluorescence-based nuclear transport assays at a concentration of 50 μg/mL for 5 min at room temperature (concentration may need to be optimized for various cell types).

Methanol and acetone are also used to permeabilize cells, usually at low temperatures ranging from –20°C to +14°C. These solvents have the advantage of providing a rapid one-step fixation and permeabilization procedure. They are most useful for analysis of, for example, the cytoskeleton of cells such as actin-containing stress fibers and intermediate filament networks. However, it is important to remember that these solvents are harsh and are not the optimal choice for studies of more labile cellular antigens.

For a particular study, the best fixation and permeabilization scheme must be determined empirically. The following chapters in this manual provide recommendations and guidelines for fixation and permeabilization of specific organelles or components of interest.

SECTIONING OF TISSUES

Three sectioning techniques are commonly used for analyzing the morphology of tissues at light microscopical levels: cryosections (frozen sections), paraffin sections, and plastic (or methacrylate) sections (protocols for preparing cryosections and paraffin sections are presented at the end of the chapter). The major advantage of cryosections is the use of unfixed tissue, which saves time and allows different portions of a frozen tissue to be used for immunohistochemistry, genetic, or biochemical analysis. However, the quality of cryosections is usually inferior to paraffin sections, and certain tissues (e.g., adipose tissue, dense or calcified tissues) are difficult or impossible to section. Plastic is firmer than paraffin, allowing preparation of exceptionally thin sections (e.g., 1 μm for light microscopy and much thinner for electron microscopy) and sectioning of hard tissues such as calcified bone. Both paraffin and plastic require extensive fixation and processing, which results in superior morphology in comparison to cryosections. Paraffin sections of bone usually require a decalcification step following fixation. The advantages of cryo- over paraffin sections have decreased as techniques for paraffin-embedded tissues have improved. Sectioning of paraffin blocks requires experience that should be acquired from an experienced histology technician or a colleague. Some institutions have service facilities for sectioning. Typically, specimens in paraffin blocks are cut into thin (5–8-mm) tissue sections that are then mounted on coated slides for immunodetection (Fig. 2).

Cryosections are rapidly prepared before fixation and provide a good tool for visualizing fine details of cells (Fig. 3). Although cryosections are physically less stable than sections of paraffin- or resin-embedded material, the preservation of antigenicity is usually superior. The preparation of cryosections does not involve the dehydration steps typical for embedded material, and the whole procedure of sectioning and detection of a particular antigen in a specimen can usually be completed in 1 d. In general, the sample is frozen quickly in either isopentane or liquid nitrogen (small samples such as cells and small tissues may be mixed in a slurry of an inert support medium such as

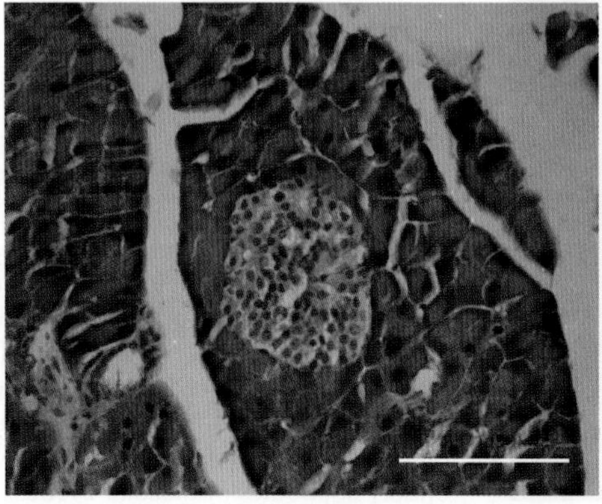

FIGURE 2. Hematoxylin and eosin staining of an islet of Langerhans of a Rip1Tag2 transgenic mouse (Hanahan 1985). The organ was fixed in 4% PFA at 4°C overnight and was embedded in paraffin. Then, 5-mm sections were cut, deparaffinized, and rehydrated. Scale bar, 100 μm.

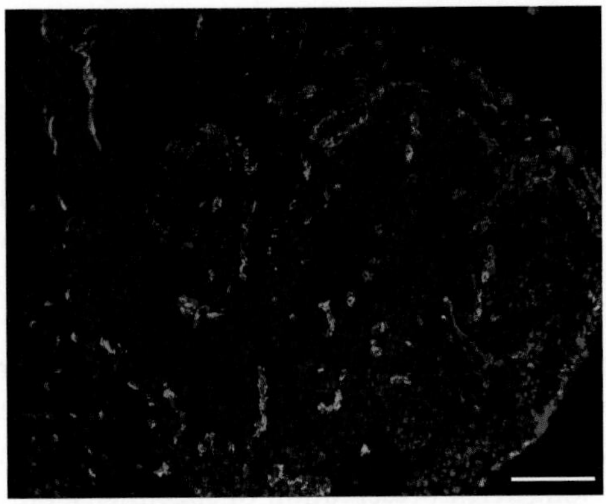

FIGURE 3. A pancreatic tumor of a Rip1Tag2 transgenic mouse (Hanahan 1985) was stained with DAPI (blue), CD31 for endothelial cells (red), and NG2 for pericytes (green).The organ was freshly frozen in 3-octanol using liquid nitrogen. Then, 5-mm sections were cut and were directly fixed in 70% ETOH (ethyl alcohol) for 1 min. They were dried at room temperature, and after that, the tissue was permeabilized by 0.2% Triton X-100 for 10 min. The following antibodies were used to detect endothelial cells and pericytes: rabbit anti-NG2 (Chemicon AB5320) and rat anti-CD31 (BD Pharmigen 550274) as primary antibodies; Alexa 488 and Alexa 568 as secondary antibodies (Molecular Probes). Scale bar, 100 µm.

3-octanol compound before freezing). This rapid freezing reduces ice crystal formation and minimizes morphological damage. Frozen sections may be used for a variety of procedures, including immunochemistry, enzymatic detection, and in situ hybridization.

Water-soluble methyl methacrylate enables the production of semithin sections that improve resolution. The ability to infiltrate and to polymerize both soft and hard tissues at room temperature without prior removal of water results in better preservation and improved tissue morphology, which reduces artifacts. Plastic embedding enables the production of sections with considerably less distortion and shrinkage than paraffin. Because processing does not include harsh organic solvents or heat, the preservation of delicate biological structures and histochemical detection result in superior results in comparison to paraffin sections. The main disadvantage of methyl-methacrylate processing is the initial investment for equipment. However, the costs can be somewhat reduced by purchasing a microtome that can be used for preparing both paraffin and plastic sections. Thereafter, the only expense will be the costs for the embedding materials. The majority of reusable materials can be used for both plastic- and paraffin-embedding procedures. An additional difficulty is the toxicity and the skin-irritating properties of the methyl-methacrylate chemicals. Good laboratory practice is essential, and the use of a fume hood and gloves is recommended.

CELLULAR AUTOFLUORESCENCE

Autofluorescence often interferes with the detection of fluorescent probes in cells and tissues. Appropriate unlabeled control slides must be used to assess the extent of autofluorescence. Autofluorescence can be minimized by spectral discrimination, which encompasses selecting probes and optical filters that maximize the probe fluorescence signal relative to the nonspecific autofluorescence. Autofluorescence of mammalian cells is largely caused by flavin coenzymes (FAD [flavin adenine dinucleotide] and FMN [flavin mononucleotide]: absorption, 450 nm; emission, 515 nm) and reduced pyridine nucleotides (NADH [nicotinamide adenine dinucleotide]: absorption, 340 nm; emission, 460 nm; Aubin 1979; Benson et al. 1979). When using fixed cells (especially with glutaraldehyde), the autofluorescence can be significantly diminished by washing the slides with 0.1% sodium borohydride in PBS for 30 min (Beisker et al. 1987; Bacallao et al. 1995) before first antibody incubation.

MOUNTING MEDIA

After the tissue sections have been labeled (see the following chapters for specific labeling and detection protocols), the glass slides are overlaid with a coverslip. A number of recipes for the preparation of several commonly used mounting media are presented at the end of the chapter, but mounting media are also available from a number of commercial sources (FluorSave, Calbiochem; Slowfade or Prolong, Molecular Probes; Vectashield, Vector Laboratories).

Fading and/or bleaching of a labeled specimen can be a major problem during analysis by fluorescence microscopy. The rate of fading of fluorescent signals can be retarded by the addition of free-radical scavengers, such as para-phenylenediamine (*p*-phenylenediamine) (Johnson and Nogueira Araujo 1981) or *n*-propyl gallate (Giloh and Sedat 1982). Mounting media with a pH ≥ 8.0–8.5 increase the initial intensity of FITC fluorescence and reduce fading. Antifading reagents included in the mounting media will slow the rate of fading, which, in turn, allows for longer observation times. Many factors influence the fluorescence intensity and the bleaching of fluorochromes, including the intensity of the excitatory light, the pH, the embedding medium, and the presence of other substances that may quench fluorescence. In general, mounting media are based on glycerol in a buffer and include an antifade reagent. For long-term preservation of specimens, polyvinyl-based alcohol mounting media are better than glycerol-based media. Generally, fluorescently labeled slides are best stored in the dark at 4°C; slides mounted in *p*-phenylenediamine are stored at –70°C in the dark and remain in excellent condition for up to 6 mo.

Photobleaching of live samples (e.g., in cells expressing GFP-fusion constructs) can be minimized by including antioxidants in the mounting medium. Oxyrase (Oxyrase, Inc., Mansfield, OH) is an enzyme additive used to deplete oxygen and can be used at 0.3 units/mL to reduce photodynamic damage during observation of cells (Waterman-Storer et al. 1993). Alternatively, ascorbic acid can be added as a reducing agent to the mounting medium at concentrations of 0.1–3.0 mg/mL.

Protocol A

Preparation of Slides and Coverslips

It is imperative that the slides and the coverslips used in fluorescence procedures be extremely clean. Although coverslips look clean, especially when a new box is first opened, they may have a thin film of grease on them that will not allow tissue culture cells to adhere well and that may interfere with some processing steps in certain protocols. Therefore, coverslips should routinely be washed with acid or base solutions to rid them of this film. Commercial precleaned slides are also likely to be dirty and must be washed before use.

Primary cells do not attach well to glass slides or coverslips. Therefore, coverslips can be coated with different growth substrates that enhance the adhesion of cells to glass (such as MatriGel or rat tail collagen) or they can be grown on ACLAR plastic (Ted Pella) that is not autofluorescent. However, it should be noted that plastic coverslips are not optimal for most light microscopic preparations; therefore, glass should be used whenever possible.

The following procedures describe various approaches for cleaning slides and coverslips and sterilizing them for cell culture, followed by methods for coating slides and coverslips with a solution that will promote the adhesion of cells or tissues to the glass surface.

MATERIALS

CAUTION: See Appendix 6 for proper handling of materials marked with <!>.

Reagents

Ethanol <!> (70%)
HCl <!>
Liquid detergent
NaOH <!> (2 N)
Nitric acid <!>

Equipment

Coverslips (#1.5)
#1.5 coverslips are of a thickness that is compatible with achieving focus and high resolution with most objective lenses, and they are less likely to break than are #1 coverslips during handling.
Glass slides, 25 x 75-mm (precleaned)

METHOD 1: CLEANING COVERSLIPS WITH AN ACID

1. In a glass beaker in the hood, make up 300 mL of two parts nitric acid to one part HCl (solution will turn orange–red).
2. Place 10 oz of #1.5 coverslips into the acid solution, a few at a time, so that they are separated and do not break, then allow them to sit for ~2 h with occasional swirling.
3. Decant the acid carefully into a waste receptacle.

4. Wash the coverslips thoroughly in running tap water until the pH of the wash water is reduced to ~5.5–6.0.
5. Store the coverslips in a covered container submerged in 70% ethanol, and carefully flame each coverslip in the tissue culture hood before use, or autoclave, and store them in a sterile Petri dish.

METHOD 2: CLEANING COVERSLIPS WITH A BASE

1. Incubate coverslips for 2 h in 2 N NaOH.
2. Rinse extensively in dH_2O.
3. Follow Step 5 of Method 1.

METHOD 3: CLEANING SLIDES WITH A DETERGENT

1. Wash slides (25 x 75-mm precleaned slides) with a liquid detergent for a few minutes.
2. Rinse slides in H_2O for 30 min, and air dry.

Protocol B

Coating Slides and Coverslips

Slides or coverslips can be coated with a solution that promotes adhesion of cells or tissue to the surface, such as gelatin, aminoalkyl silane, and poly-L-lysine. Gelatin or aminoalkyl silane is usually used for tissue sections or small organisms, whereas poly-L-lysine is routinely used for cultured cells. After extensive cleaning as outlined in Protocol A1, slides or coverslips are coated. If coverslips are to be used for cell culture, they should be sterilized after air drying by exposure to ultraviolet light for 45 min in, for example, a laminar flow hood. In general, the coated coverslips and slides can be stored at room temperature for several weeks.

MATERIALS

CAUTION: See Appendix 6 for proper handling of materials marked with <!>.

Reagents

3-aminopropyltriethoxysilane <!> (aminoalkyl silane)
Acetone <!> (Sigma-Aldrich)
Chromium potassium sulfate
Gelatin (type Bloom 225; Sigma-Aldrich G 9382)
HCl <!> (2 N)
Laminin (Invitrogen), optional (see Method 3)

Equipment

Glass staining dish

METHOD 1: GELATIN COATING

1. Prepare the coating solution.
   i. Dissolve gelatin in H_2O at 60°C to make a 0.2% solution.
   ii. Cool the solution to 40°C, and add chromium potassium sulfate to 0.02%.
   iii. Cool the solution to 4°C, and use immediately, or store at 4°C for up to several weeks.
2. Place cleaned slides or coverslips in appropriate racks, and immerse into a glass staining dish filled with the gelatin coating solution at 4°C for 2 min. Perform the coating carefully to avoid formation of bubbles on the slide surface.
3. Remove the slide rack from the coating solution, and set it on its side to allow the excess solution to drain off.
4. Dry the slides overnight before use.
   Slides and coverslips coated with gelatin are stable at room temperature for several weeks.

METHOD 2: SILANIZATION

1. Acid clean the slides in 2 N HCl for 5 min.
2. Rinse thoroughly in dH_2O.
3. Rinse in acetone, and air dry.
4. Prepare a fresh 2% solution of 3-aminopropyltriethoxysilane (aminoalkyl silane) in acetone (Sigma-Aldrich). Immerse slides in this solution with agitation for 2 min.
5. Rinse slides in dH_2O, and air dry.

   Slides may be stored for >5 yr (Nuovo 1997).

METHOD 3: POLY-L-LYSINE COATING

1. Prepare a suitable amount of 100–500 mg/mL poly-L-lysine (molecular weight >150,000; Sigma-Aldrich P 1399) in dH_2O. For some cell types (e.g., for the culture of neurons), a higher concentration of poly-L-lysine (1 mg/mL) may be more appropriate.
2. Coat slides or coverslips by dipping into the poly-L-lysine solution or by applying enough solution to cover the glass surface.
3. Incubate slides/coverslips for 10 min at room temperature, for 1 h at 37°C, or overnight at 4°C; then rinse them three times in sterile dH_2O.
4. Air dry the slides in the hood.
5. (Optional) To enhance cell attachment to poly-L-lysine-coated coverslips, include an additional treatment to coat the coverslips with laminin (2–5 mg/cm^2).
   i. Incubate several hours at room temperature.
   ii. Aspirate to remove laminin and rinse coverslips with media or PBS.

      Coated coverslips should be air dried at least 45 min before plating of cells. It is not advisable to store laminin-coated coverslips for longer times.

Protocol C

Cryosectioning

This protocol describes the preparation of cryosections (frozen sections) of tissue samples.

MATERIALS

Reagents

Fixative

See the section Fixation and Permeabilization of Cells and Tissues preceding the protocols section.

Staining solution (toluidine blue [1% to 2% in H_2O], hematoxylin, and eosin or any aqueous stain)

Tissue sample, fresh unfixed

Tissue Tek 3-octanol Compound (Sakura Finetek U.S.A., Inc., Torrance, CA)

Equipment

Cryostat

A cryostat essentially consists of a –20° freezer enclosing a microtome. Cryostats are expensive, but in many medical centers, hospital pathology laboratories have cryostats that can be used or rented.

Glass microscope slides, coated as described in Protocol B

Poly-L-lysine-coated or silanized slides improve the adherence of the section. Alternatively, commercially available coated slides such as Superfrost Plus (Thermo Scientific) can be used.

Tissue mold: plastic or metal molds of various sizes are sold by many suppliers (such as Fisher Scientific, VWR International, Shandon-Lipshaw, Inc.)

METHOD

1. Freeze a fresh unfixed tissue sample, up to 2.0 cm in diameter, in 3-octanol compound on special metal grids that fit onto the cryostat.

   The 3-octanol compound is viscous at room temperature and can be mixed with water but freezes into a solid support at –20°C.

2. Place the grid with frozen tissue sample into the cryostat at –20°C, and cut sections that are 5–15-µm thick. The temperature of the cutting chamber may have to be adjusted ±5°C for some tissues. A camel hair brush is useful to help guide the emerging section over the knife blade.

   If the tissue frozen in 3-octanol does not cut in a smooth thin sheet, the knife is probably dull. Watery tissues, fatty tissues, or tissues with variable textures are difficult to section.

3. Transfer the sections to a microscope slide. Using a microscope slide (at room temperature), touch the tissue section; the tissue section will melt onto the slide. To avoid freeze drying of the tissue section, it must be transferred to the slide within a minute.

4. Stain the sections on the first slide quickly with a staining solution to assess tissue preservation and orientation.

5. Immediately immerse the slide into the appropriate fixative.

    See the section Fixation and Permeabilization of Cells and Tissues for choosing the appropriate fixative.

    To maximize the adherence of sections to slides, some researchers allow the section to air dry onto the slide at room temperature for 30 min before fixing the sample. The disadvantage of air drying the sample is that surface tension forces distort the cells, causing a loss in high-resolution detail. Air drying may also cause some changes in immunostaining results. If you do not air dry the sections, however, the tissue might not attach properly to the slide and may come off during subsequent treatments. Therefore, the exact procedure must be established by pilot experiments.

6. Cover any unused tissue with a layer of 3-octanol compound to prevent freeze drying, and store the reminder of the sample at –70°C. For long-term storage, a moistened tissue should be added to the container with the 3-octanol blocks to prevent desiccation (particularly in a frost-free freezer).

Protocol D

Paraffin Sections

The following protocol describes fixation of tissues, decalcification (of fixed bone tissue), embedding of tissues in paraffin, and the sectioning of paraffin-embedded tissues.

MATERIALS

CAUTION: See Appendix 6 for proper handling of materials marked with <!>.

Reagents

Alcohol series for dehydration steps (70%, 95%, and 100%)

Ethanol <!>, or mixtures of methanol <!> and isopropyl alcohol (e.g., Flex alcohols, Richard-Allen Scientific, Kalamazoo, MI) are most often used. The alcohol is mixed with dH_2O to reach the required concentration.

Collodion solution (Electron Microscopy Sciences), for preparing tiny tissue fragments of cell suspensions (see Step 1)

Fixative (typically 4%–10% freshly prepared formaldehyde)

Formaldehyde <!>

Formic acid <!>

Gelatin (Fisher Scientific)

HCl <!>

Histoclear (from National Diagnostics) or xylene <!>

Important: Xylene is very toxic; work in a fume hood. Long-term exposure to xylene can lead to serious health problems, therefore, the use of histoclear is preferred.

Paraffin (e.g., Surgipath Medical Industries, Inc.)

Equipment

Embedding molds: commercial stainless-steel or vinyl molds (HistoPrep Base molds, Fisher Scientific)

Glass centrifuge tube, 15-mL

Glass slides, uncoated or coated, as described in Protocol B, and labeled

Conveniently coated slides with a built-in spacer that permits capillary action to hold reagents during the reaction are commercially available (Biotech Probe-On Plus, Fisher Scientific). Slides with a frosted end for labeling are available from many companies.

Microtome (Leica, Fisher Scientific, VWR International, Shandon-Lipshaw, Inc.).

Alternatively, a hospital pathology laboratory may be willing to cut sections for a fee.

Tissue cassettes (e.g., Simport biopsy cassettes, Fisher Scientific)

Large cassettes are available to embed whole organs of small animals without the need for fine dissection (e.g., entire mouse brains).

Water bath for tissue sectioning (at least 6 in in diameter): prepare a clean dH_2O bath, prewarmed to 42°C–48°C.

Water baths designed for paraffin sectioning are commercially available (Fisher TissuePrep flotation bath, Fisher Scientific).

METHOD

Fixing Tissues for Paraffin Embedding

1. Prepare the tissue.

   For normal tissue samples:

   Cut the tissue into blocks of ~2 mm in thickness and up to 2 cm in length and width. Place the tissue into the tissue cassettes.

   For large pieces of calcified (e.g., bone) tissue:

   Cut the fresh tissue into 2–3-mm-thick slices with a saw.

   This step improves the speed of fixation and eventual decalcification. It is essential to fix the tissue well before decalcification because the acids used for decalcification could otherwise cause tissue damage.

   For tiny tissue fragments or cell suspensions to be paraffin embedded:

   i. Prepare a 15-mL glass centrifuge tube by adding a small amount of collodion solution into the tube, swirling the solution to completely coat the inside, then inverting and drying the collodion for 10–15 min.

   ii. Centrifuge the tissue sample or cell suspension in the coated centrifuge tube for 15 min at 500*g*–1000*g*.

   iii. Withdraw the collodion-coated cell pellet from the tube as a thin sac, and process as for tissue or paraffin embedding and sectioning.

   Alternatively, small samples can be concentrated and embedded in 2% agar. Automatic processors for paraffin embedding are available but are only useful when routinely processing large numbers of samples.

2. Fix the tissue by immersion in at least 10 volumes of fixative. Fixation times can vary between 2 h to overnight at 4°C.

   Fixation should be standardized for a given procedure because increasing fixation times will alter immunoreactivity. For small pieces of tissue (≤1-mm thickness), all incubation times given can be cut in half.

Decalcifying Tissues for Paraffin Embedding

If the fixed tissue does not require decalcification, proceed to Step 7.

3. Rinse the fixed calcified tissue for 10 min in tap water.

4. Prepare a decalcifying solution as follows:

88% formic acid stock	100 mL
Concentrated HCl	80 mL
dH_2O	820 mL

   Important: Do not mix the solution with formaldehyde because this would result in the production of bis-chloromethyl ether, which is a carcinogen.

5. Immerse the tissue in ~100-fold excess of decalcifying solution with gentle stirring. Typically, an overnight incubation is required; monitor the decalcification by trying to flex the tissue.

   Excessive decalcification leads to extensive depurination of the DNA, which will interfere with nuclear stains. As decalcification can interfere with immunodetection, any adverse effects must be determined empirically for each antigen.

6. Rinse the fixed sample with cold running tap water for 1 h, and proceed to Step 7 (dehydration).

Paraffin Embedding of Tissue Samples

7. Dehydrate the fixed sample by immersion (with gentle stirring) at room temperature through a series of increasing ethanol concentrations, using at least 10 volumes of ethanol each time.
   i. Incubate the tissue in 70% alcohol for 20 min; repeat twice.
   ii. Rinse the tissue in 95% alcohol, then incubate in 95% ethanol for 1 h; repeat once.
   iii. Incubate the tissue in 100% alcohol for 1 h; repeat twice.

      If the H_2O is not completely removed from the tissue, then the subsequent processing steps will fail. Therefore, it is important to carry out the final dehydration in 100% alcohol.

8. Remove lipids from the tissue by immersing it in at least 10 volumes of histoclear or xylene. Incubate with stirring at room temperature for 1 h, then change the solution, and incubate for another hour. Repeat once.

   This step is essential to remove all of the alcohol from the previous step and the fat that would otherwise render the paraffin block soft and difficult to cut.

9. Infiltrate the samples with paraffin.
   i. Melt the paraffin by heating it to 58°C–60°C.
   ii. Immerse the cassette containing the dehydrated defatted tissue in 10 volumes of melted paraffin and change the paraffin every hour. Repeat this procedure three times. Monitor the temperature, and make sure it remains at 58°C–60°C.
10. Embed the tissue in paraffin.
   i. Place a small volume of molten paraffin into the embedding mold, and transfer the tissue into the mold using heated forceps or another appropriate tool (e.g., metal spoon or cut-off pasteur pipette).
   ii. Carefully push the tissue flat against the bottom of the mold, and completely fill the mold with paraffin wax. Invert the original tissue cassette over the mold, and pour in more paraffin wax to cover the base of the cassette.

      Make certain that the base of the cassette is parallel to the tissue because it will later be used to fix the block in the holder of the microtome.

   iii. After the paraffin hardens, remove the mold. Paraffin tissue blocks can be stored for long periods at room temperature with little effect on immunoreactivity or nucleic acids.

Cutting Paraffin Sections

11. Place the water bath for sectioning next to the microtome. If using uncoated slides, add 50 mg of gelatin per liter of H_2O in the bath.

    Gelatin increases the adhesion of the sections to the glass. Gelatin should not be used in combination with slides that are already coated.

12. Place the tissue block on ice, or cool it to 0°C–4°C.

    This makes the paraffin harder and, therefore, easier to cut into thin ribbons.

13. Lock the microtome handwheel, and move the microtome knife out of the way before loading the paraffin block.

    Careful: There is potential risk of serious cutting injuries in performing Steps 13–15. Please get trained by an experienced user or a histology technician.

14. Mount the tissue cassette bearing the embedded tissue sample into the tissue cassette holder of the microtome. Trim the edges of the paraffin block if necessary, and assure that the edge of the knife is parallel to the upper and lower edges of the tissue block, otherwise ribbons cannot be cut.

15. Advance the microtome knife using coarse adjustments to within ~1 mm of the block. Fine-adjust the knife and/or block position so that the block face is parallel to the sweep of the knife blade. Set the angle of the microtome blade to about 3°–8° from the face of the tissue block. This is the minimal angle, measured from the block to the edge of the backside of the knife bevel.
16. Set the thickness of the sections to be cut. The lower limit for most paraffin-embedded tissues is ~3 μm, but, in general, sections of 6–8 μm are used to achieve good resolution and morphology.
17. Face the block (i.e., trim it to form a smooth surface), and then prepare a ribbon of tissue sections of the desired thickness.
18. Pick up the ribbon carefully by hand, and float it onto the water bath. Use wooden sticks to manipulate the floating sections and to stretch the ribbon to remove any wrinkles.

    See *Troubleshooting*.
19. Transfer the tissue sections onto the glass slides.
    i. Dip a clean microscope slide under the meniscus of the water bath.
    ii. Position one or more sections of the tissue toward the labeled end of the slide. Slowly pull the slide out of the water at about a 45° angle to place the section(s) onto the slide.
20. Air dry the sections overnight at room temperature or at 39°C to promote the sections attaching firmly to the slide.

    Note that immunoreactivity may decrease if the sections are not used within 1–2 wk after sectioning.
21. Deparaffinize and rehydrate the sections: Before using the slides, remove the paraffin by immersion in xylene or histoclear, followed by immersion (with gentle stirring) through a series of decreasing alcohol concentrations.
    i. Incubate the sections in xylene or histoclear for 5 min; repeat twice.
    ii. Incubate the sections in 100% alcohol for 5 min; repeat twice.
    iii. Incubate the sections in 70% alcohol for 5 min; repeat twice.
    iv. Incubate the sections in the appropriate buffer for 2 min; repeat twice.

TROUBLESHOOTING

Problem (Step 18): If the tissue sections fail to form nice ribbons, or if wrinkles cannot be removed without tearing the tissue, one or several problems are likely.

- The knife is dull.
  Solution: Replace or sharpen the knife on the microtome.
- The knife angle is improper.
  Solution: Adjust the angle of the knife on the microtome closer to a 45° angle.
- The tissue is not properly infiltrated with paraffin.
  Solution: Poorly infiltrated tissue shrinks within the wax block, dipping below the surrounding surface. It is best to begin again with a fresh tissue sample if there are more samples of the same kind; otherwise reprocessing may be a last attempt to save the experiment. Reprocessing the tissues is not recommended because deterioration of the morphology is almost inevitable and stochastic changes in immunoreactivity are possible. If reprocessing is required, the steps for paraffin embedding must be systematically reversed to bring the tissue back to 100% alcohol, and the tissue must then be defatted and re-embedded in paraffin again.
- The water bath temperature is above or below 42°C–48°C.
  Solution: Readjust the temperature.

- Tissues that are very dense or of variable textures may be difficult to section.
  Solution: Increasing the thickness of the sections may help in these cases.
- The tissue face has become dry.
  Solution: The tissue may need to be rehydrated for proper sectioning. Soak a Kimwipe in H_2O, and apply to the block face for a few minutes.

Protocol E

Mounting of Live Cells Attached to Coverslips

Live cells, grown on coverslips, may be mounted.

MATERIALS

See the end of the chapter for recipes for reagents marked with <R>.

Reagents

Live cell culture, grown on coverslips in Petri dishes (Method 1)
Live cells on coverslips in PBS (containing 1.0 mM Ca^{2+} and 0.5 mM Mg^{++})
Mounting medium <R>
Nail polish
PBS (containing 1.0 mM Ca^{2+} and 0.5 mM Mg^{++})
Valap <R>

Equipment

Coverslips
Parafilm strips

Parafilm spacers can be useful to avoid the squashing of life cells during mounting.

Whatman #1 filter paper

METHOD 1

1. Carefully remove coverslips with live cells from the Petri dishes with forceps.
2. Wick excess buffer from the coverslip by carefully touching the edges of the coverslip with a piece of Whatman #1 filter paper, and dry the top of the coverslip (the side with no cells).

   If you are likely to lose track of which side of the coverslip the cells are attached to, then mark a corner with a felt-tip marker. Examine the wet coverslip in a Petri dish containing buffer using a microscope to determine whether the mark is on the same side (or not) as the cells.
3. Place a drop (~20 μL) of the mounting medium on a clean microscope slide.
4. Gently lower the coverslip onto the mounting medium, cell-side down, so that no air bubbles are trapped.
5. Blot away the excess mounting medium with filter paper.
6. Seal the edges of the coverslip with Valap (live cells) or nail polish (fixed cells).

METHOD 2

1. Place narrow strips of Parafilm slightly apart on a microscope slide so that the coverslip will fit between them. This results in two edges of the coverslip being supported by Parafilm, which acts as a physical spacer.

Alternatively, small pieces of broken #1 coverslips can be used.

2. Mount coverslips containing the live cells in PBS (containing 1.0 mM Ca^{2+} and 0.5 mM Mg^{++}) by following Steps 1–3 of Method 1.

   It is important to use PBS with Ca^{2+} and Mg^{++} to avoid live cells from rounding up and detaching from the coverslip.

3. Seal the coverslip around its edges with melted Valap.

   For more long-term observation, live cell chambers are available that control the temperature on the microscope stage (see Goldman and Spector 2005).

RECIPES

CAUTION: See Appendix 6 for proper handling of materials marked with <!>.
Recipes for reagents marked with <R> are included in this list.

Carbonate/Bicarbonate Buffer

Reagent	Concentration
Sodium carbonate (Solution A)	4.24 g/100 mL dH_2O (0.4-M anhydrous)
Sodium bicarbonate (Solution B)	3.36 g/100 mL dH_2O (0.4 M)
dH_2O	

Mix 4 mL of Solution A with 46 mL of Solution B, bring the volume to 200 mL dH_2O, and adjust the final pH to 9.2.

Gelvatol Mounting Medium

Reagent	Amount
Gelvatol (Monsanto, St. Louis, MO)	0.35 g
Glycerol	1.5 mL
dH_2O (or PBS)	3 mL

Add gelvatol to the water and the glycerol, and stir in a boiling water bath until completely dissolved. Add antifade agents as desired (see Antifade reagent recipes).

Gelvatol is a polyvinyl alcohol-based mounting medium used for semipermanent mounting. It is viscous, not autofluorescent, and hardens slowly. Preparations mounted in this way can be stored in the dark at either 4°C or –20°C. If stored at 4°C for long periods (months), however, contamination with molds may be a problem.

Glycerol Antifade Mounting Medium (adapted from Shuman et al. 1989)

Reagent	Amount
n-propyl gallate	5 g
1,4-diazabicyclo-[2,2,2]-octane (DABCO) <!>	0.25 g
p-phenylenediamine <!>	2.5 mg
Glycerol	100 mL

Dissolve the first three reagents in glycerol, add several pellets of NaOH <!> to increase the pH to above neutral. Stir thoroughly for ~1 d, and store in aliquots at –20°C (wrapped in alufoil).

- The effectiveness of this glycerol-based antifade mounting medium is greatly diminished by small amounts of H_2O. To minimize residual H_2O in the sample, drain all washing solution before mounting, cover the specimen with mounting medium, let it sit for ~15 min, drain again, and mount in fresh glycerol antifade mounting medium.
- Small amounts of residual Triton X-100 <!> (used for permeabilization) and perhaps other detergents convert this antifade solution into a very powerful quenching agent. If the fluorescence disappears when you apply this solution, remove the coverslip from the slide by flooding it with buffer and placing it cell-side up in a Petri dish. Rinse the sample well with an excess of detergent-free buffers, and remount in fresh mounting medium.

Mowiol Mounting Medium for Fluorescent Samples

Reagent	Amount
Mowiol 4-88 (Calbiochem No. 475904)	2.4 g
Glycerol (100%, water-free)	6 g
Tris-HCl <!> (0.2 M, pH 8.5)	12 mL
dH_2O	6 mL

1. Combine glycerol and 2.4 g of Mowiol 4-88 in a 50-mL Falcon tube, mix well, and add dH_2O. Continue mixing for 2 h at room temperature.
2. Add 0.2-M Tris-HCl (pH 8.5), and incubate the solution with occasional stirring at 53°C until completely dissolved.
3. Centrifuge the solution at 4000 to 5000 rpm at room temperature for 20 min, aliquot the clear supernatant into glass vials, and store in the dark at –20°C or at 4°C. Defrost an aliquot, and use within a few days; do not refreeze.

n-*Propyl Gallate Antifade Medium (Giloh and Sedat 1982)*

Dissolve 2% (w/v) ratio *n*-propyl gallate in glycerol, and adjust the pH to 8.0.

p-*Phenylenediamine Mounting Medium (modified from Johnson and Nogueira Araujo 1981)*

Reagent	Amount
Glycerol (Polysciences, Inc., 00084)	9 mL
PBS (1x, pH 7.4)	1 mL
p-phenylenediamine <!> (Fisher Scientific, AC13057-5000)	10 mg
Carbonate/bicarbonate buffer <R>	

1. Use a graduated serological pipette to deliver the glycerol and the PBS into a 15-mL Falcon tube. Mix by vortexing, place Parafilm around the cap of the Falcon tube, and incubate for 20 min at 37°C to remove air bubbles.
2. Wrap a 20-mL glass scintillation vial containing a stir bar with foil, and add *p*-phenylenediamine and 9.75 mL of the glycerol/PBS solution. Stir the solution at room temperature for ~4 h, until the *p*-phenylenediamine is dissolved. The final color of the medium should be pink/yellow.
3. Add carbonate/bicarbonate buffer stepwise to the medium to reach a final pH of 8.0. Check the pH carefully with pH paper as a final pH of 7.0 would cause fading. Store the mounting media in aliquots at –20°C, wrapped in aluminum foil.

- *p*-phenylenediamine oxidizes readily in air when exposed to light, yielding a fluorescent product that binds to nuclei. If the antifade mounting medium increases the general autofluorescence or stains the nuclei nonspecifically (particularly noticeable with fluorescein filter sets), it is important to prepare new mounting medium or to use *p*-phenylenediamine of a higher purity grade. The oxidation products appear dark brown, whereas pure *p*-phenylenediamine appears nearly white. *p*-Phenylenediamine is largely insoluble in cold water, whereas the oxidation products are readily soluble. Washing the *p*-phenylenediamine powder about five times its weight in cold water will remove most of the autofluorescent contaminants.
- The *p*-phenylenediamine containing mounting medium is not compatible with cyanine-conjugated antibodies (i.e., Cy2, Cy3, and Cy5).

- Suppliers are suggested for the most critical reagents as the purity of the compounds is crucial for the preparation of a good mounting medium. Triton X-100 <!> quenches the antifade agents, so do not include Triton X-100 in the final washes before mounting specimens for light microscopy.

Valap

Weigh out a mixture of Vaseline, lanolin, and paraffin (1:1:1w/w/w), and place the component in a glass or a ceramic vessel. Melt the mixture on a hot plate over medium to low heat until completely liquid. The wax mixture should spread smoothly and should dry quickly on a glass slide. If it hardens too quickly, then add more Vaseline and lanolin. If it does not harden fast enough, then add more paraffin. Valap is solid at room temperature; just before use, warm it at low setting on a hot plate.

REFERENCES

Aubin JE. 1979. Autofluorescence of viable cultured mammalian cells. *J Histochem Cytochem* **27:** 36–43.

Bacallao R, Kiai K, Jesaitis L. 1995. Guiding principles of specimen preservation for confocal fluorescence microscopy. In *Handbook of biological confocal microscopy* (ed. Pauley JB). Plenum, New York.

Beisker W, Dolbeare F, Gray JW. 1987. An improved immunocytochemical procedure for high-sensitivity detection of incorporated bromodeoxyuridine. *Cytometry* **8:** 235–239.

Benson RC, Meyer RA, Zaruba ME, McKhann GM. 1979. Cellular autofluorescence—Is it due to flavins? *J Histochem Cytochem* **27:** 44–48.

Giloh H, Sedat JW. 1982. Fluorescence microscopy: Reduced photobleaching of rhodamine and fluorescein protein conjugates by *n*-propyl gallate. *Science* **217:** 1252–1255.

Goldman RD, Spector DL, eds. 2005. *Live cell imaging: A laboratory manual.* Cold Spring Harbor Laboratory Press, Cold Spring Harbor, NY.

Hanahan D. 1985. Heritable formation of pancreatic β-cell tumours in transgenic mice expressing recombinant insulin/simian virus 40 oncogenes. *Nature* **315:** 115–122.

Harlow E, Lane DL. 1988. *Antibodies: A laboratory manual.* Cold Spring Harbor Laboratory Press, Cold Spring Harbor, NY.

Johnson TJ. 1987. Glutaraldehyde fixation chemistry. *Eur J Cell Biol* **45:** 160–169.

Johnson GD, Nogueira Araujo GM. 1981. A simple method of reducing the fading of immunofluorescence during microscopy. *J Immunol Methods* **43:** 349–350.

Nuovo GJ. 1997. *PCR in situ hybridization: Protocols and applications*, 3rd ed. Lippincott-Raven, Philadelphia.

Shuman H, Murray JM, DiLullo C. 1989. Confocal microscopy: An overview. *BioTechniques* **7:** 154–163.

Stearns ME, Ochs RL. 1982. A functional in vitro model for studies of intracellular motility in digitonin-permeabilized erythrophores. *J Cell Biol* **94:** 727–739.

Tagliaferro P, Tandler CJ, Ramos AJ, Pecci Saavedra J, Brusco A. 1997. Immunofluorescence and glutaraldehyde fixation. A new procedure based on the Schiff-quenching method. *J Neurosci Methods* **77:** 191–197. Erratum: 1998. *J Neurosci Methods* **82:** 235–236.

Waterman-Storer CM, Sanger JW, Sanger JM. 1993. Dynamics of organelles in the mitotic spindles of living cells: Membrane and microtubule interactions. *Cell Motil Cytoskel* **26:** 19–39.

14 Labeling Cell Structures with Nonimmunological Fluorescent Dyes

Brad Chazotte

Department of Pharmaceutical Sciences, College of Pharmacy and Health Sciences, Campbell University, Buies Creek, North Carolina 27506

ABSTRACT

This chapter provides a brief overview of methods for studying static or dynamic cell organization and function using various nonantibody-based fluorescent-labeling approaches. The focus is primarily on preparing cells for imaging, although the labeled cells can also be used with fluorescence polarization, fluorescence recovery after photobleaching, fluorescence lifetime imaging, and resonance energy transfer. Fluorescent tags on cells are also used in flow cytometry. Protocols are provided for labeling the nucleus, the endoplasmic reticulum, the Golgi, the mitochondria, the lysosomes, the pinocytotic vesicles, the cytoskeleton, and the plasma membrane. Various fluorescent moieties such as rhodamines, fluorescein, NBD (*N*-(7-nitrobenz-2-oxa-1,3-diazole-4-yl)), carbocyanines, BODIPY (boron-dipyrromethene), and Alexa Fluor are used.

INTRODUCTION

The increasing commercial availability of fluorescent moieties and probe molecules conjugated to these fluorophores has greatly enhanced the experimental flexibility and the tools available to researchers. Fluorescent probes used to monitor cell metabolism such as intracellular ion concentrations (e.g., FURA-2 for calcium or SNARFs [seminaphthorhodafluors] for pH), the burgeoning use of green fluorescent protein and related fluorescent constructs (e.g., Conn 1999a), or immunofluorescence are not dealt with here. However, the interested researcher should be aware that there are other sources (Taylor and Wang 1989; Wang and Taylor 1989; Mason 1999; Haugland 2002; see Chapters 15–23) with which they may wish to consult on the use of such probes or im - munospecific fluorescent probes not presented here. Likewise, the development of fluorescent semiconductor nanocrystals or quantum dots is relatively recent, offering advantageous increases in brightness and photostability (for reviews, see Jasiwal and Simon 2004; Alivisatos et al. 2005; Drbohlavova et al. 2009), but unless properly coated, these can be cytotoxic. The conjugation of quantum dots to certain peptides or antibodies (Jaiswal and Simon 2004; see Chapter 36) has been the primary way to achieve specific labeling of cells. The reader should be aware that many companies (e.g., Olympus, Nikon, Bio-Rad Laboratories, and Molecular Probes) also have helpful websites for fluorescence methods and probes. The author has necessarily been selective in the choice of techniques and probes. Readers, if needed, may find the text by Alberts et al. (2008) helpful as a general reference on cellular structures.

Cell Organization and Choosing a Fluorescent Probe

The eukaryotic cell sustains life by organizing and compartmentalizing its functions and by transporting various metabolites among compartments. Fluorescence techniques have helped revolutionize the imaging and the manipulation of cells, particularly for studying structure/function relationships and dynamic processes within live cells.

When studying cell organization, it may be necessary to distinguish between static and dynamic organization. For example, the plasma membrane is a static entity that is always present in a living cell, but the lipid and protein components of the membrane are frequently undergoing dynamic movement—lipid rafts may be present, the shape and the composition of the membrane itself may also change over time, and phospholipid-based vesicles may be added to or removed from the plasma membrane during pinocytosis, receptor-mediated endocytosis, secretion, and other cellular processes. Both the choice of fluorescent probes and the length of time spent imaging a live cell are frequently dictated by the static or dynamic nature of the molecules, structures, cells, or tissue one wishes to study.

The specific fluorophore used will also depend on a number of factors. First, if the intent is to view multiple probes within the same cell or to view them simultaneously, then by careful selection of the fluorescent moieties and filter sets, it is possible to use multiple probes targeted to several parts of the cell and to image them independently. For example, fluorescein and rhodamine can be incorporated together into a single cell and can be imaged separately. The Alexa Fluor probes (Molecular Probes) are a class of fluorescent moieties that span a range of excitation and emission wavelengths that facilitate the simultaneous use of multiple probes. Second, the mechanism for incorporating the fluorophore into the cell can impact the choice of probe. There are cell-permeant probes and probes that can be actively taken up by cells. In other cases, getting the probe into the cells requires microinjecting them or using chemicals to permeabilize the cell. If cells are permeabilized or fixed, then structure but (usually) not function can be studied.

Instrumentation

Because of the number of instruments available to carry out one or more of the existing fluorescence techniques, the reader should follow the specific instrument instruction manual. For general principles of fluorescence, microscopy, imaging, fluorescence recovery after photobleaching (FRAP), fluorescence polarization, fluorescence lifetime, etc., the reader is referred to Taylor and Wang (1989); Wang and Taylor (1989); Chazotte and Hackenbrock (1991); Chazotte (1994); Pawley (1995); Conn (1999b); Mason (1999); and, in this book, see Section 3.

Protocols for Labeling Cells with Fluorescent Probes

This chapter contains a large number of straightforward protocols for labeling subcellular components with fluorescent molecules. One extensive source for all types of fluorescent probes and related technical information is Molecular Probes (a division of Invitrogen); their catalog (Haugland 2002) and website have been used as a general reference for this chapter. Wide-field and confocal microscope manufacturers such as Olympus, Nikon, Zeiss, and Bio-Rad Laboratories have websites with detailed instructions, tutorials, and interactive applications for microscopy, plus suggestions for selecting fluorophores and filter sets. Commonly used fluorescent moieties include rhodamines, fluoresceins, Alexa Fluors, BODIPY, NBD, and the carbocyanines, although this is by no means a complete list. For details on cell culture and preparation of cells on coverslips for microscopic imaging, please see Chapter 13.

The reader should note that for single-cell analysis techniques, such as microscopic imaging and FRAP, the following protocols assume that cells were grown on glass microscope coverslips and were immersed in small Petri dishes containing culture medium. *At no time in the following protocols should the cells be allowed to dry out.* Bulk techniques in which cells are suspended in solution require centrifugation of cells and aspiration of the solutions used in the protocols. Generally, labeling conditions vary by cell type, and the readers may find it necessary to alter the protocols for their own particular use. In addition, in some of the following sections, separate protocols are given for labeling live or fixed cells.

Autofluorescence, from endogenous cellular molecules, such as NADH (nicotinamide adenine dinucleotide, reduced form) or $FADH_2$ (flavin adenine dinucleotide, reduced form), can interfere with imaging by reducing the signal-to-noise ratio. This occurs when the excitation and/or emission wavelengths of the probe and the autofluorescing molecules are similar. This occurs frequently with excitation wavelengths shorter than 500 nm, particularly at ultraviolet (UV) wavelengths. Autofluorescence can be reduced by careful selection of the excitation and the emission wavelengths used, by treatment of fixed cells with reducing agents such as $NaBH_4$ (1% solution for 20 min), and by comparison between the experimental images and the unlabeled control slides. Avoid fixation with glutaraldehyde because it can increase interference from cellular autofluorescence, most frequently at wavelengths <500 nm. Note that autofluorescence has, on occasion, been used to study mitochondrial function based on NADH or FAD (flavin adenine dinucleotide) intrinsic fluorescence, sometimes called redox fluorometry (e.g., Pawley 1995 and chapters therein; Huang et al. 2002).

Protocol A

Mounting Live Cells onto Microscope Slides

Cells are grown on cleaned coverslips in cell culture dishes containing the appropriate medium and supplements for the cell type being used. Labeling is performed with the coverslips placed in culture dishes. After cells are labeled, the coverslips can be mounted onto microscope slides to be viewed for short-term observation of live cells or longer-term observation of fixed cells. For viewing live cells for longer periods of time, specialized chambers should be used in which suitable growth conditions can be maintained (see Chapter 13).

MATERIALS

See the end of the chapter for recipes for reagents marked with <R>.

Reagents

Cells of interest grown on coverslips and labeled with a fluorescent probe(s)

The remaining protocols in this chapter are labeling protocols.

HBSS$^+$ (Hanks' buffered salt solution with Ca^{2+} and Mg^{++}) (Invitrogen/Gibco 14025-092)
Phosphate-buffered saline (PBS$^+$) <R>
Valap <R>

Equipment

Cell culture dishes, sterile
Cotton swabs for applying Valap
Fluorescent microscope
Hot plate to keep Valap at the proper temperature
Incubator for cell culture, water-jacketed and CO_2-regulated
Microscope slides, cleaned, or specialized commercial chamber (to mount coverslips for imaging)
Parafilm
Tweezers, fine for handling the coverslips

EXPERIMENTAL METHOD

1. Transfer live cells that have been grown on coverslips and labeled to a dish of PBS$^+$ or desired cell culture medium.

   Some cell culture media supplements, such as phenol red, can interfere with fluorescence observation. PBS, PBS$^+$, HBSS, or HBSS$^+$ is frequently used when observing fluorescence to reduce or to eliminate this problem. If cells are fixed before mounting, they can be mounted in commercially available aqueous mounting medium.

2. Place narrow strips of Parafilm onto a microscope slide, spaced so that the coverslip will fit between them with two edges of the coverslip extending onto the Parafilm strips. The Parafilm serves as a spacer, which prevents live cells from being crushed.

   Cells that are fixed before mounting do not require spacers.

3. Seal the coverslip around its edges with melted Valap.

   Avoid using commercial slide sealants that contain organic solvents (e.g., acetone), because they can be detrimental to live cells.

4. Image the cells using fluorescent microscopy.

   See *Troubleshooting*.

TROUBLESHOOTING

Problem (Step 4): Images are blurred, distorted, or lacking in contrast.

Solution: Poor-quality images can result from using plastic. This is particularly a problem when phase-contrast or differential interference contrast microscopy is used. It can also be a problem for transmitted light images on confocal microscopes. Some plastics interfere with fluorescence only at certain wavelengths. To avoid this problem, grow cells on optical-quality glass. Microscope coverslips # 1.5 are optimal for most uses. There are optical-quality cell culture dishes available for situations when coverslips and slides are not feasible.

Protocol B

Labeling Membranes with Carbocyanine Dyes (DiIs) as Phospholipid Analogs

Cell membranes can be labeled with fluorescent carbocyanine dyes that function as phospholipid analogs. These dyes, such as DiI-C_{16}(3) (1,1′-dihexadecyl-3,3,3′,3′-tetramethylindocarbocyanine perchlorate), label the plasma membrane and (eventually) label all of the membranes within a living cell. The dyes can be also used with model membranes. Carbocyanine dyes are commercially available in different carbon chain lengths (e.g., C_{12}, C_{14}, C_{18}, and C_{22}). Depending on the fatty acid composition of the phospholipids in the membrane studied with respect to chain length and degree of saturation, different chain length DiIs may prove optimal for incorporation and for accurate reporting of membrane motions. DiI-labeled cells can be studied using fluorescence microscopy or FRAP as well as other techniques.

IMAGING SETUP

For DiI-C_{16}(3): $\lambda_{excitation}$ maximum ~ 550 nm; $\lambda_{emission}$ maximum ~ 565 nm (using methanol as the solvent). Maxima may vary slightly for different carbon chain lengths and especially in different solvents.

Filter set: rhodamine/Texas Red

FRAP and confocal microscopy laser excitation: 514-nm line argon-ion laser; 543-nm line of a green He:Ne laser

MATERIALS

See the end of the chapter for recipes for reagents marked with <R>.

Reagents

Cells of interest grown on coverslips

DiI-C_{16}(3) (molecular weight [mwt] = 877.77; Invitrogen/Molecular Probes D-384)

Prepare a 1-mg/mL stock solution of DiI-C_{16}(3) in absolute ethanol. Store sealed and protected from light at –20°C. Some other analogs are DiI-C_{12}(3), mwt = 765.56; DiI-C_{18}(3), mwt = 933.88.

PBS^+ <R>

Equipment

Cell culture dishes, sterile

Fluorescent microscope

EXPERIMENTAL METHOD

1. Dilute the stock DiI-C_{16}(3) solution 1:500 in PBS^+ for use in labeling.
2. Aspirate cell medium from the cells grown on coverslips, and rinse the cells three times with PBS^+.

   Do not allow the cells to dry out.

3. Incubate the cells at room temperature for 30 sec with DiI-C_{16}(3) solution.

   Incorporation of the label should be performed above the main-phase transition of the membrane, which, for most cell membranes, is a broad transition below 10°C. Labeling at room temperature or higher should ensure that cell membranes are in the liquid-crystalline state without temperature-induced lateral phase separations.

4. Rinse the cells three times with PBS^+.
5. Mount the coverslips as described in Protocol A.
6. Image the cells.

   See *Troubleshooting*.

TROUBLESHOOTING

Problem (Step 6): The fluorescence signal is too weak.
Solution: Enough dye should be present for a strong signal but not too much to perturb the plasma membrane itself. For experiments such as FRAP, a probe/phospholipid ratio between 1:1000 and 1:10,000 is desirable. The plasma membrane composition, the incubation temperature and time, and the dye concentration in the incubating medium all affect the amount of dye incorporated. The use of ethanolic solution and the above concentration is to ensure that the dye is intercalated in the bilayer and not adsorbed on the surface or micellized in the medium.

Problem (Step 6): The fluorescent intensity of the plasma membrane changes over time.
Solution: The length of usable observation time will be affected, in part, by the rate at which the plasma membrane is altered in the normal course of cellular function. DiI tends to locate in the outer leaflet of the plasma membrane and may take several days to appear in cytoplasmic vesicles in unfixed cells.

Problem (Step 6): Other cellular membranes become labeled.
Solution: DiI probes are not specific for the plasma membrane and will label most phospholipid membranes provided they can come into contact with the membrane. They are cationic (positively charged) lipophilic probes. Fixed cells tend to show more uniform staining because they are permeabilized. In healthy cells labeled and observed at 37°C, significant endocytosis occurs.

DISCUSSION

Many plasma membrane proteins are studied using immunofluorescent probes, although some quasispecific chemical labeling is possible (e.g., eosin maleimide conjugation to Band 3 protein of erythrocytes; Nigg and Cherry 1979). The bulk of membrane lipids are phospholipids, along with sterols (e.g., cholesterol), glycolipids, and proteolipids. Lipid organization and dynamics have been studied by introducing fluorescent lipid analogs such as DiI, perylene, and pyrene (e.g., Kok and Hoekstra 1999; Maier et al. 2002). DiI carbocyanine dyes are good choices because they resist fading. They are available in different carbon chain lengths (e.g., C_{12}, C_{14}, C_{18}, and C_{22}), which affect the probe's miscibility in the phospholipids of the plasma membrane. For example, an 18-carbon chain may not mix well in a phase containing primarily 12-carbon chains (Spink et al. 1990). For dynamic studies of a membrane, the probe selected should have carbon chain lengths similar to those of the membrane studied, and the probe concentration should be kept low enough that the probe does not perturb the membrane properties. DiI probes have also been used to label the mitochondrial inner membrane and to report on phospholipid diffusion (Fig. 1).

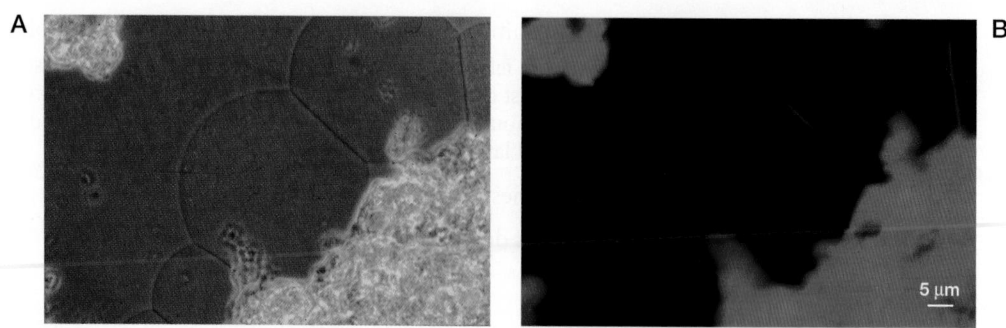

FIGURE 1. Large, osmotically active, fused mitochondrial inner membranes from rat liver (Chazotte et al. 1985) labeled with DiI-C_{16}(3). (*A*) Phase-contrast image. (*B*) Fluorescent image using a rhodamine filter set. (Courtesy of B. Chazotte, Department of Pharmaceutical Sciences, College of Pharmacy and Health Sciences, Campbell University.) Scale bar, 5 µm.

Protocol C

Labeling Membranes with Fluorescent Phosphatidylethanolamine

The phospholipid, phosphatidylethanolamine (PE), can be conjugated via its head group to a number of fluorophores, including rhodamine, BODIPY, and NBD. These probes can be used to label biological membranes and to study phospholipids within membranes. Rhodamine-PE, which does not readily exchange between lipid bilayers, has also been used in membrane fusion assays. Rhodamine-DHPE (lissamine rhodamine B 1,2-dihexyldecanoyl-sn-glycero-3-phosphoethanolamine, triethylammonium salt) and NBD-PE have been used together in resonance energy transfer studies in membranes.

IMAGING SETUP

	$\lambda_{excitation}$ maximum (nm)	$\lambda_{emission}$ maximum (nm)	Solvent	Filter set	FRAP and confocal microscopy laser excitation
NBD-PE	~460	~534	Ethanol	Fluorescein	488-nm line argon-ion laser
Rhodamine-DHPE	560	581	Methanol	Rhodamine	514-nm line argon-ion laser 543-nm line of a green He:Ne laser
BODIPY-DHPE	505	511	Methanol	Fluorescein	488-nm line argon-ion laser

MATERIALS

See the end of the chapter for recipes for reagents marked with <R>.

Reagents

Cells of interest grown on coverslips

Fluorescent PE:

- NBD-PE, mwt = 956.25 (Invitrogen/Molecular Probes N-360)
- Rhodamine-DHPE, mwt = 1333.81 (Invitrogen/Molecular Probes L-1392)
- BODIPY-FL DHPE, mwt = 1067.23 (Invitrogen/Molecular Probes D-3800)

  Prepare a stock solution of a fluorescent PE probe at a concentration of 1 mg/mL in absolute ethanol. Store sealed and protected from light at –20°C. Some other NBD-PE sources are 16:0 NBD-PE (mwt = 872.08; Avanti Polar Lipids, Inc. 8101440) and 14:0 NBD-PE (mwt = 815.9; Avanti Polar Lipids, Inc. 810143).

PBS^+<R>

Equipment

Cell culture dishes, sterile

Fluorescent microscope

EXPERIMENTAL METHOD

1. Prepare a 1:500 dilution of the stock fluorescent probe.
2. Aspirate cell medium from the cells grown on coverslips, and rinse them three times with PBS^+.
   Do not allow the cells to dry out.
3. Cover the cells with the dilute fluorescent probe, and incubate for 30 sec at room temperature.
   Incorporation should be performed above the main-phase transition of the membrane, which for most cell membranes is a broad transition below 10°C. Labeling at room temperature or higher should ensure that cell membranes are in the liquid-crystalline state without temperature-induced lateral phase separations.
4. Rinse the cells three times with PBS^+.
5. Mount the coverslips as described in Protocol A.
6. Image the cells.
   See *Troubleshooting.*

TROUBLESHOOTING

Problem (Step 6): The sample is nonfluorescent or very dim on initial observation.
Solution: Protect the stock probe solution from light, and perform the cell labeling under minimal light conditions.

Problem (Step 6): The sample fades rapidly on observation when using NBD-PE.
Solution: NBD photobleaches rapidly, so minimize the sample's exposure to light, and use the lowest laser excitation intensity possible to obtain a clear image. Rhodamine and BODIPY are more resistant to photobleaching than NBD and may be preferred for longer-term observation.

DISCUSSION

Lipid organization and dynamics in membranes, such as the plasma membrane, are frequently studied by the reintroduction of fluorescently labeled phospholipid molecules (see also a comparative study by Mazeres et al. 1996). For recent reviews of fluorescent lipids and their uses in fluorescence quenching, resonance energy transfer, membrane fusion studies, membrane phase behavior, and lipid transport studies, see Kok and Hoekstra (1999) and Maier et al. (2002). The use of fluorescently labeled lipids is preferred over lipid analogs for some studies because it is assumed that, with the former, there is less perturbation of a membrane's physical properties. Similarly, labeling the lipid head group is often preferred to a fluorescent label located within the hydrocarbon region of the membrane, as this reduces perturbation of the membrane structure (and function).

Protocol D

Labeling the Plasma Membrane with TMA-DPH

TMA-DPH (trimethylamine-diphenylhexatriene) is a fluorescent membrane probe that has classically been used to label the outer leaflet of a membrane bilayer, to label the outer leaflet of the plasma membrane in cells, and to report on membrane dynamics using the techniques of fluorescence polarization and/or fluorescence lifetime. This probe has also been used to follow exocytosis and endocytosis of labeled plasma membranes. The interaction of the aqueous environment with mitochondrial inner membrane dynamics has also been studied following the fluorescence polarization and the lifetime of TMA-DPH.

IMAGING SETUP

For bulk fluorescence polarization and lifetime measurements, the excitation wavelength is 360 nm with a 5-nm (monochromator) slit width, and the emission wavelength is monitored at 430 nm with a 10-nm (monochromator) slit width. For microscopy and imaging, a UV filter set should be used. For fluorescence polarization or fluorescent lifetime measurements, the specific microscopic imaging settings to use depend on the particular microscope that is used.

MATERIALS

CAUTION: See Appendix 6 for proper handling of materials marked with <!>.
See the end of the chapter for recipes for reagents marked with <R>.

Reagents

Cells of interest grown on coverslips
TMA-DPH (mwt = 461.62; Invitrogen/Molecular Probes T-204)
N,N-dimethylformamide <!>
PBS$^+$ <R>

Equipment

Cell culture dishes, sterile
Cuvette suitable for fluorescence measurements
Fluorescent microscope
Fluorometer

EXPERIMENTAL METHOD

1. Prepare a 1-mM stock solution of TMA-DPH in DMF (dimethylformamide). Store it sealed and protected from light at –20°C.

For Conventional Fluorescence Polarization Measurements

For bulk fluorescence polarization and lifetime measurements, corrections for light-scattering artifacts may be necessary. Control specimens omitting the fluorophore should be prepared simultaneously (see Chazotte 1994).

2. To a bulk suspension of cells at 10^6 cells/mL, add enough TMA-DPH for a final concentration of 1 μM, and incubate for 5 min at 37°C.

   Centrifuging and rinsing the cells should not be necessary.

3. Place the labeled cells in a fluorescence cuvette.
4. Stir the sample while in the cuvette to prevent settling during bulk measurements using a fluorometer.

For Imaging Using a Fluorescent Microscope

5. Prepare a labeling solution as a 1000-fold dilution (1 μM) in PBS^+.
6. Aspirate off the cell medium from cells grown on coverslips, and wash the cells three times in PBS^+.

   Do not allow the cells to dry out at any time.

7. Immerse the cells in the labeling solution for 5 min at 37°C.
8. Aspirate off the labeling solution, and rinse the cells twice with PBS^+.
9. Mount the coverslips as described in Protocol A.
10. Image the cells.

DISCUSSION

TMA-DPH, a fluorophore closely related to diphenylhexatriene (DPH), has been used to study the motional dynamics of membranes (Chazotte 1994) and has also been used to follow endocytosis from the plasma membrane of live cells (Illinger 1990; Illinger and Kuhry 1994, and references therein). TMA-DPH, which locates in the outer leaflet of the membrane bilayer because of its charged head group, is preferable to DPH. DPH, in contrast, is freely permeable and labels all cellular membranes in a living or permeabilized cell and reports on the phospholipid hydrocarbon (acyl chain) region. With respect to cells, TMA-DPH is better suited to study the motional dynamics of the plasma membrane.

Typically, a bulk suspension of cells in a cuvette is used for fluorescence polarization measurements to obtain information about the anisotropy of the phospholipid head group region in the plasma membrane or for fluorescence lifetime measurements of TMA-DPH to report on the local membrane motional dynamics. Chazotte (1994) has also studied isolated mitochondrial inner membranes using these techniques. Fluorescence polarization and fluorescent lifetime measurements have also been adapted to microscopic imaging (Herman and Jacobson 1990; Periasamy and Herman 1994; Mason 1999, and references therein; Pawley 1995, and references therein). These measurements would be performed on cells attached to glass coverslips.

Protocol E

Labeling Membrane Glycoproteins or Glycolipids with Fluorescent Wheat Germ Agglutinin

Glycoprotein and glycolipids are found as part of the outer leaflet of cellular plasma membranes. Those glycoproteins or glycolipids that contain sialic acid and *N*-acetylglucosamine residues can be labeled with wheat germ agglutinin (WGA), a plant lectin that exists as a dimer (mwt = ~36,000) and is normally cationic. Fluorescently labeled WGA is commercially available with fluorescein, Alexa Fluor, or rhodamine moieties. Fluorescent WGA can also be used to stain the Golgi (the *trans*-Golgi) in fixed cells. Fluorescently labeled concanavalin A (Con A), which selectively binds to α-mannopyranosyl and α-glucopyranosyl residues, can also be used to label plasma membranes.

IMAGING SETUP

	$\lambda_{excitation}$ maximum (nm)	$\lambda_{emission}$ maximum (nm)	Solvent	Filter set	FRAP and confocal microscopy laser excitation
NBD-PE	~460	~534	Ethanol	Fluorescein	488-nm line argon-ion laser
FITC-WGA	~495[a]	~519[a]	Water	Fluorescein	488-nm line argon-ion laser
Alexa Fluor 488-WGA[b]	~495	~519	—	Fluorescein	488-nm line argon-ion laser
Rhodamine-WGA or rhodamine-Con A	~555	~580	Water	Rhodamine	514-nm line argon-ion laser 543-nm line of a green He:Ne laser

FITC, fluorescein-5-isothiocyanate.
[a]Maxima may vary on binding and especially in different solvents or pH values.
[b]Other wavelength Alexa Fluor WGAs are available.

MATERIALS

CAUTION: See Appendix 6 for proper handling of materials marked with <!>.
See the end of the chapter for recipes for reagents marked with <R>.

Reagents

Cells of interest grown on coverslips
Fluorescent WGA or Con A
- FITC-WGA (Invitrogen/Molecular Probes W-834)
- Alexa Fluor 488-WGA (Invitrogen/Molecular Probes W-11262)
- Tetramethylrhodamine-WGA (Invitrogen/Molecular Probes W-849)
- Tetramethylrhodamine-Con A (Invitrogen/Molecular Probes C-860)
  - Prepare a stock solution of the desired fluorophore at a concentration of 2 mg/mL. Store protected from light at 4°C.

Glutaraldehyde <!>, 0.5% (or 3.7% formaldehyde <!>) in PBS <R>
- Prepare fresh just before use.

PBS[+] <R>

Equipment

Cell culture dishes, sterile
Fluorescent microscope

EXPERIMENTAL METHOD

Staining Plasma Membrane of Live Cells

1. Prepare a 1:200 dilution of the stock fluorescent WGA (or Con A) in PBS^+.
2. Rinse the cells grown on coverslips three times in PBS^+.
3. Immerse the cell-containing coverslip in a small Petri dish containing the labeling solution. Incubate for 5–10 min at 37°C.
4. Rinse the cells three times in PBS^+.

   Do not allow the cells to dry out.
5. Mount the coverslips as described in Protocol A.
6. Image the cells.

Fixed Cells (for Staining Golgi)

7. Prepare a 1:200 dilution of the stock fluorescent WGA in PBS^+.
8. Fix the cells for 10 min with 0.5% glutaraldehyde (or 3.7% formaldehyde) in PBS.
9. Aspirate off the fixative, and wash the cells three times with PBS^+ for 5 min each.
10. Stain the cells from 5 min to 10 min at room temperature.
11. Aspirate off the labeling medium. Rinse the cells three times with PBS^+.
12. Mount the coverslips as described in Protocol A.
13. Image the cells.

DISCUSSION

WGA is a useful probe for the plasma membrane, although not a highly specific one. The Golgi contains a high concentration of glycosylated lipids and proteins, particularly the *trans*-Golgi; however, the Golgi is inaccessible for labeling unless cells have been permeabilized. Thus, in live cells, WGA will stain the plasma membrane; and, in fixed cells, the *trans*-Golgi will also be stained. Con A can also be used to probe plasma membrane function. Both Con A and WGA have been used to follow the lateral diffusion of glycolipids and glycoproteins in plasma membranes (Swaisgood and Schindler 1989).

Protocol F

Labeling Membranes with Fluorescent Cholesterols

Cholesterol, an essential component of most cellular membranes, is present in the plasma membrane and is enriched within lipid rafts. It is, however, virtually absent from the mitochondrial inner membrane, and there is little within the endoplasmic reticulum. Dehydroergosterol (DHE), a fluorescent cholesterol analog, has been found to effectively label membranes and has been used to monitor cholesterol distribution and trafficking within live cells. Other cholesterol-based probes, such as NBD-cholesterol, are commercially available (from Molecular Probes). Reports indicate, however, that cell labeling using the NBD-cholesterol analogs frequently does not show the same labeling pattern as for DHE and native cholesterol (Mukherjee et al. 1998). This protocol is based on the method of Mukherjee et al. (1998).

IMAGING SETUP

	$\lambda_{excitation}$ maximum (nm)	$\lambda_{emission}$ maximum (nm)	Solvent	Filter set	FRAP and confocal microscopy laser excitation
DHE	~324–328	~375	Methanol	For UV excitation with a 355-nm, 20-nm band-pass excitation filter, a 365-nm long-pass dichromatic filter, and a 405-nm, 40-nm band-pass emission filter (Chroma Technology)	UV line of argon-ion laser, if of sufficient intensity, or UV laser
22-NBD-cholesterol	~467	~538	Methanol	Fluorescein	488-nm line argon-ion laser

MATERIALS

CAUTION: See Appendix 6 for proper handling of materials marked with <!>.
See the end of the chapter for recipes for reagents marked with <R>.

Reagents

Cells of interest grown on coverslips
Cell culture medium
Ethanol <!>
Fluorescent cholesterol or its analog, DHE
  DHE (Ergosta-5,7,9(11),22-tetraen-3β-ol; mwt = 394.63; Sigma-Aldrich E2634)
    Prepare a 5-mM stock solution in ethanol, and store protected from light under argon at –80°C.
  22-NBD-cholesterol (22-(*N*-(7-nitrobenz-2-oxa-1,3-diazol-4-yl)amino)-23,24-bisnor-5-cholen-3β-ol); mwt = 575.75; Invitrogen/Molecular Probes N-1148)
    Prepare a 5-mM stock solution in ethanol, and store at –80°C and protect from light.
Lipoprotein-depleted serum (LPDS)
PBS$^+$ <R>
Trypsin/EDTA <!>

Equipment

Cell culture dishes, sterile
Centrifuge, tabletop
Coverslips, # 1.5 (170-μm thick; square 18 x 18-mm or 22 x 22-mm, or round coverslips)
Fluorescent microscope
Incubator for cell culture

EXPERIMENTAL METHOD

1. Prepare a 2.5-μM working solution of DHE in LPDS. Alternatively, prepare a 1.5-μM working solution of 22-NBD-cholesterol in LPDS.
2. Supplement an appropriate cell culture medium with either the DHE probe solution or the NBD-cholesterol solution prepared in Step 1 to a final LPDS concentration of 5%.

   LPDS is used in lieu of complete serum to prevent back exchange of DHE out of the plasma membrane caused by the presence of lipoproteins.
3. Place semiconfluent cells grown on coverslips into Petri dishes containing the DHE-supplemented (or 22-NBD-cholesterol) and LPDS-supplemented cell culture medium. Incubate the cells for 16–20 h in a cell culture incubator.
4. Rinse the cells three times with PBS^+.
5. Treat the cells with trypsin/EDTA to detach the cells from the coverslip. Centrifuge the cells at 135*g*, and wash them in PBS^+. Repeat twice.

   This step is necessary to separate particles of unincorporated DHE (background) from the DHE incorporated into the plasma membrane.
6. Replate the cells onto coverslips, and incubate them in a Petri dish in the desired culture medium with 5% LPDS (but no fluorescent probe) for 16–20 h in a cell incubator, until the cells are semiconfluent.
7. Rinse the cells three times with PBS^+.
8. Mount the coverslips as described in Protocol A.
9. Image the cells.

DISCUSSION

Fluorescent cholesterols have been used to report on cholesterol distribution within cell membranes, its trafficking among cell membranes, and its metabolism (e.g., Mukherjee et al. 1998). The method presented here is that of Mukherjee et al. (1998). These investigators, as well as others, report that, although DHE mirrors the distribution and the movement of native cholesterol within the cell, other commercially available fluorescent cholesterols such as 25-NBD-cholesterol and 22-NBD-cholesterol distribute differently than native cholesterol, which limits their utility for trafficking studies. An alternative, and less complicated, procedure has been reported by Frolov et al. (1996) in which mouse L-cells were used. The cells were subcultured, and stock DHE (5 mg/mL in 95% nondenaturated grain ethanol) was added to the medium to obtain a final concentration of 20-μg/mL medium and cultured for 3 d before observation.

Protocol G

Fluorescent Labeling of Membrane Lipid Rafts

Lipid rafts are structures within the plasma membrane that have a different lipid composition than the bulk membrane, specifically being enriched in certain lipids. Lipid rafts are believed to have physiologically important functions. Commercially available kits have been developed (Invitrogen/Molecular Probes) for labeling lipid rafts using a cholera toxin subunit that has been fluorescently tagged with one of several Alexa Fluor dyes. Cells are labeled while in suspension, after which they can be mounted for imaging.

IMAGING SETUP

	$\lambda_{excitation}$ maximum (nm)	$\lambda_{emission}$ maximum (nm)	Solvent	Filter set	FRAP and confocal microscopy laser excitation
Alexa Fluor 488[a]	~495	~519	—	Fluorescein	488-nm line argon-ion laser 543-nm line of a green He:Ne laser
Alexa Fluor 555[a]	~555	~565	—	Rhodamine	514-nm line argon-ion laser 543-nm line of a green He:Ne laser

[a]Other wavelength Alexa Fluor probes are available.

MATERIALS

CAUTION: See Appendix 6 for proper handling of materials marked with <!>.

Reagents

Cell culture medium, 4°C
Cells of interest
Formaldehyde <!>, 4% in PBS, 4°C (optional, for fixing cells)
Mounting medium (for fixed cells; see Step 8)
PBS, 4°C (10x PBS stock included with Vybrant kit)
Vybrant Alexa Fluor lipid raft labeling kits (Invitrogen/Molecular Probes, V-34403, V-34405)
Follow the manufacturer's instructions for preparation and storage of the reagents.

Equipment

Centrifuge
Centrifuge tubes
Fluorescent microscope

EXPERIMENTAL METHOD

1. Prepare a fresh 1-µg/mL working solution of the fluorescent cholera toxin subunit B (CT-B) conjugate by adding 2 µL of 1-mg/mL CT-B conjugate stock solution to a final volume of 2 mL of the desired complete cell culture medium (medium is at 4°C).

   This working solution is sufficient for one labeling using a 2-mL incubation volume.

2. Centrifuge the cells, and gently resuspend the cell pellet in the desired chilled (4°C) complete cell culture medium.
3. Centrifuge the cells to pellet them. Gently resuspend the cells in 2 mL of the fluorescent CT-B conjugate working solution, and incubate them for 10 min at 4°C.
4. Wash the cells gently three times with chilled PBS.
5. Centrifuge the cells, and gently resuspend the cell pellet in 2 mL of chilled (4°C) anti–CT-B antibody working solution. Incubate the cells for 15 min at 4°C.
6. Wash the cells gently three times with chilled PBS.
7. (Optional) Fix the cells in chilled PBS containing 4% formaldehyde for 15 min at 4°C. Wash the cells three times with PBS.
8. Mount live cells in 4°C PBS. Mount fixed cells using an appropriate mounting medium.
9. Image the cells. The appropriate filter set to use will depend on the Alexa Fluor dye that was selected.

DISCUSSION

There are areas of the plasma membrane that have been found to be enriched in sphingolipids, cholesterol, and some membrane proteins such as glycosylphosphatidylinositol (GPI)-linked proteins. For a recent review, see Mukherjee and Maxfield (2004). Rather than being totally random, there can be some lateral organization of plasma membranes. These lipid rafts are microdomains that tend to exist transiently but are of physiological importance (e.g., see Brown and England 1998). These microdomains, because of their composition, tend to be thicker with their longer hydrocarbon chains interacting with the other monolayer of the bilayer and also have a more ordered arrangement of their more saturated hydrocarbon chains. Functionally, it is thought that certain proteins can be concentrated in the lipid raft to facilitate protein–protein interactions such as for cell signaling and for the transport of proteins in small vesicles. Many GPI-linked proteins are present in these rafts. Several viruses such as influenza, measles, and HIV are found to localize to lipid rafts during infection and budding. This labeling protocol permits the visualization of lipid rafts by microscopy, but does use antibody cross linking to give patches of fluorescence due to the small size of individual "rafts."

Protocol H

Labeling F-Actin of the Cytoskeleton with Rhodamine Phalloidin or Fluorescein Phalloidin

Cytoskeletal F-actin can be fluorescently labeled in vivo using the mushroom-derived toxin phalloidin that has been modified by the addition of a fluorescent moiety such as rhodamine or fluorescein.

IMAGING SETUP

	$\lambda_{excitation}$ maximum (nm)	$\lambda_{emission}$ maximum (nm)	Solvent	Filter set	FRAP and confocal microscopy laser excitation
Fluorescein phalloidin	~496	~516[a]	Water	Fluorescein	488-nm line argon-ion laser
Rhodamine phalloidin	~542	~565	Water	Rhodamine	514-nm line argon-ion laser 543-nm line of a green He:Ne laser

[a]Other wavelength probes are available such as the Alexa Fluor probes available from Molecular Probes.

MATERIALS

CAUTION: See Appendix 6 for proper handling of materials marked with <!>.
See the end of the chapter for recipes for reagents marked with <R>.

Reagents

Cells of interest grown on coverslips
Formaldehyde <!>, 3.7% in PBS
PBS (pH 7.4) <R>
Rhodamine phalloidin <!> (300 units/mL) (Invitrogen/Molecular Probes R-415) or fluorescein phalloidin <!> (300 units/mL) (Invitrogen/Molecular Probes F-432)
Do not store either phalloidin solution in water, particularly in dilute solution.
Triton X-100 <!>, 0.2% in PBS

Equipment

Cell culture dishes, sterile
Fluorescent microscope

EXPERIMENTAL METHOD

1. Prepare a 1:200 dilution of the phalloidin stock solution in PBS.
2. Aspirate the cell medium from cells grown on coverslips, and rinse the cells three times with PBS.

3. Fix the cells for 10 min in 3.7% formaldehyde.
4. Rinse the fixed cells three times for 5 min each in PBS.
5. Permeabilize the cells for 5 min in 0.2% Triton X-100.
6. Rinse the cells three times with PBS.
7. Label the cells with fluorescent phalloidin at room temperature for 5–10 min.
8. Rinse the cells three times in PBS for 5 min each.
9. Mount the coverslips as described in Protocol A.
10. Image the cells.

DISCUSSION

Immunofluorescent staining has been most frequently used to study cytoskeletal components. However, it is also possible to carefully label isolated cytoskeletal proteins and either to microinject them back into the cell or to add them to fixed permeabilized cells. Alternatively, it is possible to use fluorescently labeled versions of the mushroom-derived toxins, phalloidin or phallacidin, to label F-actin of the cytoskeleton. Rhodamine is more resistant to photobleaching than fluorescein. In Figure 2, the F-actin within a fibroblast has been labeled with fluorescein phalloidin.

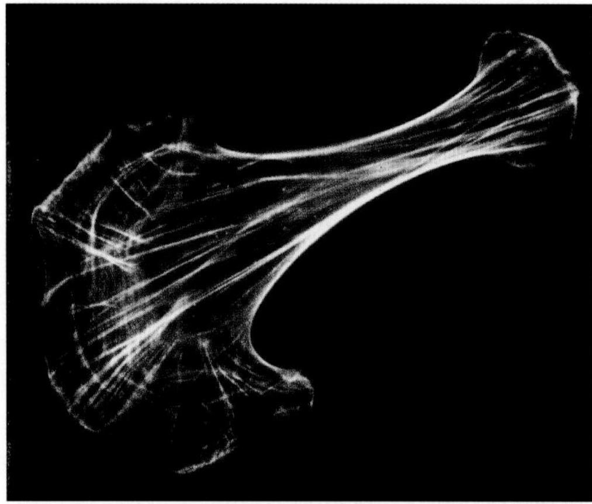

FIGURE 2. Fluorescent image of C3H 10T1/2 fibroblast cell labeled with fluorescein phalloidin. (Courtesy of K.A. Jacobson, Department of Cell and Developmental Biology, School of Medicine, University of North Carolina at Chapel Hill.)

Protocol 1

Labeling Nuclear DNA Using DAPI

A number of fluorescent stains are available that label DNA and allow easy visualization of the nucleus in interphase cells and chromosomes in mitotic cells. These stains include Hoechst, DAPI (4′,6-diamidino-2-phenylindole), ethidium bromide, propidium iodide, and acridine orange. It is believed that DAPI associates with the minor groove of double-stranded DNA with a preference for the adenine–thymine clusters. Cells must be permeabilized and/or fixed for DAPI to enter the cell and to bind DNA. Fluorescence increases approximately 20-fold when DAPI is bound to double-stranded DNA.

IMAGING SETUP

DAPI, λ_{ex} ~ 359 nm; λ_{em} ~ 461 (when bound to DNA)
Filter set: UV
Confocal microscopy laser excitation: UV line of an argon-ion laser, if sufficient intensity or UV laser

MATERIALS

CAUTION: See Appendix 6 for proper handling of materials marked with <!>.
See the end of the chapter for recipes for reagents marked with <R>.

Reagents

Cells of interest grown on coverslips
DAPI (Invitrogen/Molecular Probes D-1306)

Prepare a stock solution at 10 mg/mL in distilled H_2O, protect from light, and store at 4°C. Prepare a 5000-fold dilution in PBS to be used for labeling. The lactate salt of DAPI is more water soluble than the chloride salt, but DAPI is not very soluble in PBS; therefore, use distilled H_2O to prepare the stock solution.

Formaldehyde <!>, 3.7%, prepared fresh
PBS^+ <R>
Triton X-100 <!>, 0.2%

Equipment

Cell culture dishes, sterile
Fluorescent microscope

EXPERIMENTAL METHOD

1. Aspirate the cell medium from cells grown on coverslips.
2. Rinse the cells three times with PBS^+.

3. Fix the cells for 10 min in 3.7% formaldehyde.
4. Aspirate and rinse the cells three times for 5 min each in PBS$^+$.
5. Permeabilize the cells by immersion in 0.2% Triton X-100 for 5 min.
6. Aspirate and rinse the cells three times for 5 min each in PBS$^+$.
7. Incubate the cells at room temperature for 1–5 min in DAPI labeling solution.

   Although not as bright as the vital Hoechst stains for DNA, DAPI has greater photostability. Cells that have been immunolabeled can be stained with DAPI by starting at this step.
8. Aspirate the labeling medium.
9. Rinse the cells three times in PBS$^+$.
10. Mount the coverslips as described in Protocol A.
11. Image the cells.

Protocol J

Labeling Nuclear DNA with Hoechst 33342

A number of fluorescent stains are available that label DNA and allow easy visualization of the nucleus in interphase cells and chromosomes in mitotic cells. One advantage of Hoechst 33342 is that it is membrane permeant and, thus, can stain live cells. Hoechst 33342 binds to adenine–thymine-rich regions of DNA in the minor groove. On binding to DNA, the fluorescence greatly increases.

IMAGING SETUP

Hoechst 33342, λ_{ex} ~ 353 nm; λ_{em} ~ 483
Filter set: UV
Confocal microscopy laser excitation: UV line of an argon-ion laser, if sufficient intensity or UV laser

MATERIALS

See the end of the chapter for recipes for reagents marked with <R>.

Reagents

Cells of interest grown on coverslips
Hoechst 33342 (Invitrogen/Molecular Probes H-1399)

Prepare a 10-mg/mL stock solution in distilled H_2O, protect from light, and store at 4°C.

PBS^+ <R>

Equipment

Cell culture dishes, sterile
Fluorescent microscope

EXPERIMENTAL METHOD

1. Prepare a 100-fold dilution of the stock Hoechst 33342 solution in distilled H_2O to be used for labeling.

   Hoechst 33342 can also be used to stain fixed cells by following Protocol I and by substituting Hoechst 33342 for DAPI.

2. Aspirate the cell medium from cells grown on coverslips.
3. Rinse the cells three times with PBS^+.
4. Incubate the cells in the Hoechst-labeling solution for 10–30 min at room temperature.
5. Aspirate the labeling medium, and rinse the cells three times in PBS^+.
6. Mount the coverslips as described in Protocol A.
7. Image the cells.

Protocol K

Labeling Mitochondria with Rhodamine 123

Rhodamine 123 is a cationic fluorescent dye that is used to specifically label respiring mitochondria. The dye distributes according to the negative membrane potential across the mitochondrial inner membrane. Loss of potential will result in loss of the dye and, therefore, the fluorescence intensity. The dye has been used to monitor mitochondrial function in living cells.

IMAGING SETUP

Rhodamine-123, λ_{ex} ~ 504 nm; λ_{em} ~ 534 (methanol as solvent)

Maxima may vary somewhat depending on the concentration of the dye in the mitochondrion and especially in different solvents. Isolated, energized mitochondria show a redshift in the Rhodamine-123 emission and also at high concentrations of fluorescence quenching.

Filter set: rhodamine

FRAP and confocal microscopy laser excitation: 514-nm line argon-ion laser; 543-nm line of a green He:Ne laser

MATERIALS

See the end of the chapter for recipes for reagents marked with <R>.

Reagents

Cells of interest grown on coverslips
PBS$^+$ <R>
Rhodamine 123 (Invitrogen/Molecular Probes R-302)

Prepare a 1-mg/mL stock solution in dH_2O, protect from light, and store at 4°C in a sealed vial.

Equipment

Cell culture dishes, sterile
Fluorescent microscope
Incubator set at 37°C

EXPERIMENTAL METHOD

1. Prepare a 10-µg/mL Rhodamine-123-labeling solution (a 100-fold dilution) in PBS$^+$.
2. Aspirate the culture medium from cells grown on coverslips, and wash the cells three times with PBS$^+$.
3. Incubate the cells in the rhodamine-labeling solution for 15 min at 37°C.
4. Aspirate the labeling solution, and rinse the cells three times in PBS$^+$.

5. Mount the coverslips as described in Protocol A.
6. Image the cells.

   See *Troubleshooting*.

TROUBLESHOOTING

Problem (Step 6): Fluorescence fades shortly after labeling.
Solution: If the dye concentration is too high internally, it will inhibit mitochondrial function (i.e., oxidative phosphorylation) and reduce the membrane potential. Reduce the initial dye-labeling concentration.

DISCUSSION

Rhodamine 123, like TMRE (tetramethylrhodamine ethyl ester) and TMRM (tetramethylrhodamine methyl ester), is a membrane potential–sensitive, cationic fluorophore that follows a Nernstian distribution across the mitochondrial inner membrane in response to the negative membrane potential (e.g., Emaus et al. 1986). Labeling requires that the mitochondrion must be functioning and generating a membrane potential to attract and to retain the dye. The dye has been used as a sensitive indicator of mitochondrial membrane potential. The effects of mitochondrial inhibitors (such as KCN, rotenone, and antimycin a) and uncouplers of oxidative phosphorylation (such as carbonylcyanide-*m*-chlorophenyl hydrazone and the related FCCP [carbonyl cyanide 4-(trifluoromethoxy)phenylhydrazone]) can be examined in isolated mitochondria or mitochondria within live cells using submicromolar concentrations.

Protocol L

Labeling Mitochondria with TMRM or TMRE

TMRM (tetramethylrhodamine ethyl ester) or the related TMRE (tetramethylrhodamine methyl ester) is extensively used for labeling and measuring the membrane potential (and the function) of mitochondria in living cells. Because of its optical and chemical properties, it has been one of the main dyes used for mitochondria studied with confocal microscopy to carry out time-resolved and spatially resolved studies of mitochondria. Quantitative observation of mitochondrial membrane potential should be performed with confocal microscopy or spot photometry–based microscopy.

IMAGING SETUP

TMRM or TMRE, λ_{ex} ~ 548 nm; λ_{em} ~ 573 (methanol as solvent)

Maxima may vary somewhat depending on the concentration of the dye in the mitochondrion and especially in different solvents.

Filter set: rhodamine

FRAP and confocal microscopy laser excitation: 514-nm line argon-ion laser; 543-nm line of a green He:Ne laser

MATERIALS

CAUTION: See Appendix 6 for proper handling of materials marked with <!>.

Reagents

Cells of interest grown on coverslips

TMRM (Invitrogen/Molecular Probes T-668) or TMRE (Invitrogen/Molecular Probes T-669)

Prepare a stock solution of TMRM or TMRE at a concentration of 600 µM in DMSO (dimethylsulfoxide) <!>, protect from light, and store sealed at –20°C. Prepare a 1000-fold dilution (600 nM) of the stock TMRM or TMRE solution in cell culture medium. Also prepare a 4000-fold dilution (150 nM) of the stock TMRM or TMRE in cell culture medium.

Equipment

Cell culture dishes, sterile

Fluorescent microscope

Incubator set at 37°C

EXPERIMENTAL METHOD

1. Aspirate the culture medium from cells grown on coverslips.
2. Immerse the cells in 600 nM TMRM or TMRE for 20 min at 37°C to load the cells with dye.
3. Aspirate the labeling medium, and immerse the cells in 150-nM TMRM or TMRE to maintain the equilibrium distribution of the fluorophore.

4. Mount the coverslips as described in Protocol A or in an appropriate chamber for observation.
5. Image the cells.
   See *Troubleshooting*.

TROUBLESHOOTING

Problem (Step 5): Mitochondria lose fluorescence after labeling.
Solution: If the concentration of these membrane-potential-driven dyes is too high, mitochondrial function will become impaired over time. As a rule, the dye concentration should be kept below 1 mM.

DISCUSSION

The cationic fluorophores TMRM and TMRE are membrane-potential-dependent probes of a functioning mitochondrion. A number of their properties has made them useful for confocal microscopy and the quantification of mitochondrial membrane potential. TMRM and TMRE do not inhibit mitochondrial function, like other potential-sensitive dyes. They do not readily show concentration-dependent fluorescence quenching, like other dyes. Membrane-potential measurements can be made using spot photometry with a pinhole to limit depth of field but not by simple wide-field microscopy. The effect of submicromolar concentrations of mitochondrial inhibitors (such as KCN, rotenone, and antimycin a) and uncouplers of oxidative phosphorylation (such as carbonylcyanide-*m*-chlorophenyl hydrazone and the related FCCP) can be examined on isolated mitochondria or mitochondria in live cells. A confocal membrane-potential image of a human fibroblast cell shows the brightly labeled mitochondrion (Fig. 3A) and a decrease in fluorescence after exposure to α-interferon (Fig. 3C).

Chazotte (2001) presented a quantitative technique for membrane-potential analysis of confocal images based on the Nernstian distribution of TMRM in labeled mitochondria and cells. The membrane potential is determined based on the ratio of the mitochondrial/extracellular fluorescent probe intensity according to the Nernst equation $\Delta\Psi = -60 \log F_{in}/F_{out}$, where F refers to the fluorescent intensity (see, e.g., Lemasters et al. 1999). Pseudocolored images of the cell in Figure 3, A and C, depicting the membrane potentials using the Nernst equation are shown in Figure 3, B and D, respectively. A key to the membrane-potential mapping is shown in Figure 3E. A histogram analysis (Chazotte 2001) is applied to quantitatively analyze individual cells or any part of the image by selecting a specific area and by counting the number of pixels (the small dots that make up a digital image) at each membrane potential using image analysis software. The number of these pixels at a given potential are counted to determine the area potentials, integrated potential density, and mean (density) potential functions (see Chazotte 2001 for details). These functions together provide powerful analytical tools to quantify the membrane potentials of the cell and its mitochondria. Typically, experience has shown 56 of approximately 190 parameters calculated for each cell to provide sufficient information for analysis. The parameters of many individual cells are examined further in standard statistical analyses to calculate means, standard deviations, medians, and minima and maxima for populations of cells, individual cells with respect to a population of cells, and/or populations with populations.

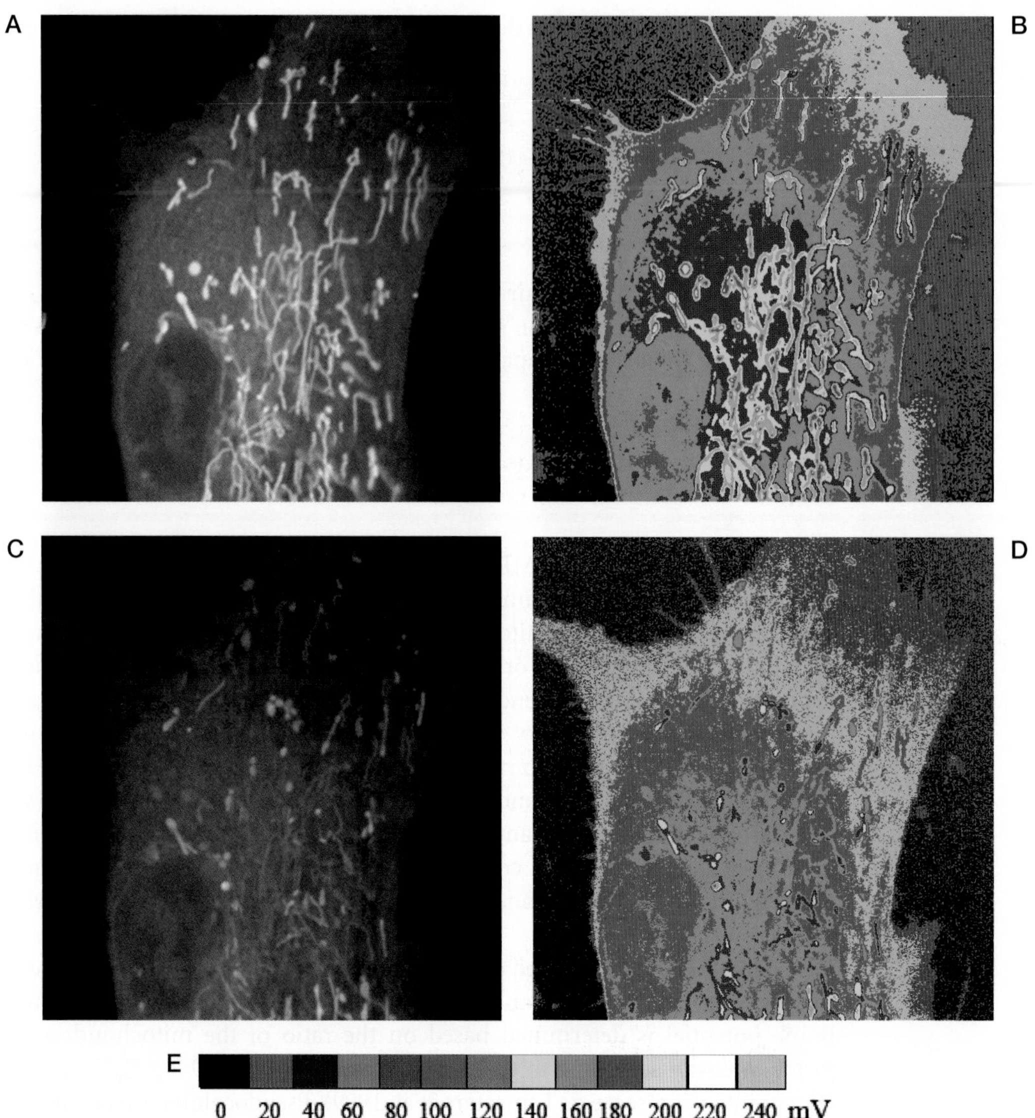

FIGURE 3. Confocal image of human BG-9 fibroblast cell in HBSS$^+$ medium labeled with TMRM as per protocol. (*A*) Membrane-potential confocal image. (*B*) Pseudocolored membrane-potential mapped image of the same cell in *A*. (*C*) Membrane-potential image of the same cell 3 min after exposure to 180-ng/mL α-interferon. (*D*) Pseudocolored membrane-potential mapped image of the same cell in *C*. (*E*) Key for colors mapped to membrane potentials in millivolts for panels *B* and *D*. (Courtesy of B. Chazotte, Department of Pharmaceutical Sciences, College of Pharmacy and Health Sciences, Campbell University.)

Protocol M

Labeling Mitochondria with MitoTracker Dyes

Membrane-potential-dependent dyes such as Rhodamine 123, TMRM, and TMRE are useful as long as the mitochondrion maintains its negative membrane potential. MitoTracker is a commercially available fluorescent dye (Invitrogen/Molecular Probes) that, like the aforementioned dyes, labels mitochondria within live cells utilizing the mitochondrial membrane potential. However, MitoTracker is chemically reactive, linking to thiol groups in the mitochondria. The dye becomes permanently bound to the mitochondria and, thus, remains after the cell dies or is fixed. In addition, it can be used in experiments in which multiple labeling diminishes mitochondrial function.

IMAGING SETUP

	$\lambda_{excitation}$ maximum (nm)	$\lambda_{emission}$ maximum (nm)	Solvent	Filter set	FRAP and confocal microscopy laser excitation
MitoTracker Green FM	~490[a]	~516[a]	Methanol	Fluorescein	488-nm line argon-ion laser
MitoTracker Orange[a]	~551	~576	Methanol	Rhodamine	514-nm line argon-ion laser 543-nm line of a green He:Ne laser

[a]Maxima may vary somewhat depending on how much the dye is concentrated in the mitochondrion and especially in different solvents.

MATERIALS

CAUTION: See Appendix 6 for proper handling of materials marked with <!>.
See the end of the chapter for recipes for reagents marked with <R>.

Reagents

Cell culture medium

Cells of interest grown on coverslips

MitoTracker Green FM, mwt = 671.88 (Invitrogen/Molecular Probes M-7514)

MitoTracker Green FM is nonfluorescent in aqueous medium. Caution: Subsequent cold acetone permeabilization will disturb the staining pattern.

MitoTracker Orange CMTMRos, oxidized form, mwt = 427.37 (Invitrogen/Molecular Probes M-7510)

Prepare a 1-mM stock solution of the dye in DMSO <!>. Protect from light, and store at –20°C.

MitoTracker Orange CMTM-H_2TMRos, reduced form, mwt = 392.93 (Invitrogen/Molecular Probes M-7511)

Prepare a 1-mM stock solution of the dye in DMSO. Protect from light, and store at –20°C. The reduced form of the dye must additionally be stored in nitrogen or argon atmosphere. The reduced form will not fluoresce until it enters an active respiring cell in which it is oxidized and subsequently accumulates in the mitochondrion.

PBS$^+$ <R>
PBS <R>
Formaldehyde <!>, 3.7% in PBS

Equipment

Cell culture dishes, sterile
Fluorescent microscope
Incubator set at 37°C

EXPERIMENTAL METHOD

Live Cells

1. Prepare a labeling solution at 25–500 nM in the desired cell culture medium, and warm it to the cell culture temperature.
2. Aspirate the culture medium from cells grown on coverslips.
3. Immerse the cells in the labeling medium for 15–45 min at 37°C.
4. Aspirate the labeling medium, and rinse the cells three times with culture medium.
5. Mount the cells as described in Protocol A.
6. Check the cells using a fluorescence microscope. If the fluorescence staining is too low, either incubate for an additional 30 min in the normal culture medium to allow the thiol conjugation to proceed, or try a higher initial concentration of the labeling solution.

   See *Troubleshooting*.
7. When the fluorescence is sufficient, image the cells.

Fixed Cells

Live cells are labeled, and then the cells are fixed.

1. Carry out Steps 1–4, but use a labeling solution at a working concentration between 100 and 1000 nM.
2. Rinse the cells three times with PBS$^+$.
3. Fix the cells for 10 min in 3.7% formaldehyde in PBS (pH 7.4).
4. Aspirate and rinse the cells three times for 5 min each in PBS.
5. Mount the coverslips as described in Protocol A.
6. Image the cells.

TROUBLESHOOTING

Problem (Live Cells, Step 6): Mitochondria do not take up the dye.
Solution: As a rule, the dye concentration should be kept below 1 mM. If the membrane-potential-driven dye concentration is too high, mitochondrial function will be increasingly impaired.

Protocol N

Labeling Mitochondria with JC-1

JC-1 (5,5′,6,6′-tetrachloro1,1′,3,3′-tetramethylbenzimidazolylcarbocyanine iodide) dye has been used by Lan Bo Chen and colleagues to monitor mitochondrial potential (e.g., Smiley et al. 1991). The monomeric form has an emission maximum of ~526 nm. The dye at higher concentrations or potentials forms red fluorescent J-aggregates with an emission maximum at 590 nm. The ratio of this green/red fluorescence is independent of mitochondria shape, density, or size but depends only on the membrane potential. It has been used to study whether all mitochondria in the same cell are at the same potential and whether membrane potential in a single long mitochondrion is uniform. JC-1 has also found useful in flow cytometry studies because the membrane potential can be followed without the need for confocal microscopy.

IMAGING SETUP

	$\lambda_{excitation}$ maximum (nm)	$\lambda_{emission}$ maximum (nm)	Solvent	Filter set[a]	FRAP and confocal microscopy laser excitation
JC-1 monomeric dye	514	529	Methanol	Fluorescein	488-nm or 514-nm line argon-ion laser depending on the filter set used
J-aggregate form of JC-1	~485–585	~590	Methanol	Rhodamine	488-nm or 514-nm line argon-ion laser depending on the filter set used

[a]Alternative filter set: Both monomer and aggregate can be viewed simultaneously with a standard long-pass filter set for fluorescein.

MATERIALS

CAUTION: See Appendix 6 for proper handling of materials marked with <!>.

Reagents

Cell culture medium containing 10% calf serum
Cells of interest grown on coverslips
JC-1, mwt = 652.23 (Invitrogen/Molecular Probes T-3168)
Prepare a 1-mg/mL stock solution of JC-1 in DMSO <!>. Protect from light, and store at –20°C.

Equipment

Cell culture dishes, sterile
Fluorescent microscope
Incubator set at 37°C

EXPERIMENTAL METHOD

1. Prepare a small aliquot of fresh 10-µg/mL JC-1 in the appropriate culture medium with 10% calf serum at 37°C.
   Staining solution older than 3 min should be discarded.

2. Aspirate the cell culture medium from cells grown on a coverslip.

   Do not allow the cells to dry out.

3. Apply 50 μL of the JC-1-labeling solution to the cells.
4. Incubate the cells with the JC-1-labeling solution in a covered Petri dish for 10 min at 37°C in a cell culture incubator.
5. Aspirate the labeling solution, and rinse the cells three times in cell culture medium without JC-1.
6. Mount the coverslips as described in Protocol A.
7. Image the cells.

DISCUSSION

JC-1 is a delocalized lipophilic cation that distributes in a membrane-potential-dependent manner (Smiley et al. 1991). Because potential measurements can be performed on a ratiometric basis of the J-aggregate to the monomer, the image measurement is independent of the mitochondrion's size, shape, or density in contrast to fluorescent measurements based on a single component. This property can be particularly helpful when confocal microscopy is not available. JC-1 has also been used in flow cytometry applications (e.g., Barbier et al. 2004).

Protocol O

Labeling the Golgi with BODIPY-FL-Ceramide (C_5-DMB-Ceramide)

The Golgi may be considered a principal organizer of macromolecular traffic in the cell because many molecules, such as secreted proteins, glycoproteins, glycolipids, and plasma membrane glycoproteins, pass through the Golgi during their maturation. The fluorescent probes used to tag the Golgi make use of this function for labeling by using a fluorescent ceramide. In this protocol, two probes are presented, a classic NBD-ceramide and a BODIPY-ceramide probe that is more resistant to photobleaching.

IMAGING SETUP

	$\lambda_{excitation}$ maximum (nm)	$\lambda_{emission}$ maximum (nm)	Solvent	Filter set	FRAP and confocal microscopy laser excitation
BODIPY-FL-ceramide	~464	~532	Methanol	Fluorescein	488-nm line argon-ion laser
NBD-C_6-ceramide	~466	~536	Methanol	Fluorescein	488-nm line argon-ion laser

MATERIALS

CAUTION: See Appendix 6 for proper handling of materials marked with <!>.
See the end of the chapter for recipes for reagents marked with <R>.

Reagents

BODIPY-FL C_5-ceramide (*N*-(4,4-difluoro-5,7-dimethyl-4-bora-3a,4a-diaza-*s*-indacene-3-pentanoyl)sphingosine; mwt = 601.6, Invitrogen/Molecular Probes D3521) or NBD-C_6-ceramide (6-((*N*-(7-nitrobenz-2-oxa-1,3-diazol-4-yl)amino)hexanoyl)sphingosine; mwt = 575.75, Invitrogen/Molecular Probes N-1154)

For either probe, prepare a 1-mM stock solution in cell culture medium. Protect from light in a sealed vial, and store at –20°C.

Cells of interest grown on coverslips

Dulbecco's modified Eagle's medium (DMEM) containing fetal bovine serum (FBS), penicillin <!>, and streptomycin <!>

PBS$^+$ <R>

Equipment

Cell culture dishes, sterile

Fluorescent microscope

Incubator set at 37°C

EXPERIMENTAL METHOD

1. Aspirate the cell culture medium from cells grown on coverslips.
2. Wash the cells three times with PBS^+.
3. With the cells on coverslips in a small Petri dish, immerse the cells in fluorescent-ceramide solution (without dilution). Incubate the cells for 10 min at 37°C.
4. Aspirate the labeling solution, and incubate the cells for 30 min at 37°C in DMEM with 10% FBS and penicillin/streptomycin.
5. Mount the coverslips as described in Protocol A.
6. Image the cells.

DISCUSSION

When labeling cells with BODIPY-FL-ceramide or NBD-C_6-ceramide, if the cells are first incubated at low temperature (2°C) with the probe and then subsequently washed and warmed to 37°C, the result is a strongly fluorescent Golgi apparatus, followed over time by a fluorescent plasma membrane (for details, see, e.g., Pagano et al. 1989). The latter result probably reflects the intracellular synthesis of fluorescent sphingomyelin and glucosylceramide analogs from the added fluorescent ceramide and their subsequent transport through the Golgi to the plasma membrane. Using BODIPY-FL-ceramide provides some advantages. At the high membrane concentrations that BODIPY-FL-ceramide achieves in the *trans*-Golgi, the probe forms excimers, which allow a cleaner visualization of the Golgi when a long-pass red filter is used. It is also brighter and more fade resistant than NBD-ceramide. For more detailed information on labeling with C_6-NBD-ceramide, see the review by Pagano (1989).

Protocol P

Labeling Lysosomes in Live

Lysosomes, although anous sacs containing more than 40 different acid hydrola e Alberts et al. 2008; van Meel and
Klumperman 2008). acidic pH (~5) found in the lysosome. A
number of the fluo osomes make use of their acidic pH.
Commonly used pro 3-[2,4-dinitrophenyl amino] propyl)-*N*-
(3-aminopropyl)met A stain).

IMAGING SETUP

Neutral Red, λ_{ex} ~ 54
Filter set: rhodamine
FRAP and confocal r he argon-ion laser or 543-nm line of a
green He:Ne laser.

MATERIALS

See the end of the chapter for recipes for reagents marked with <R>.

Reagents

Cells of interest grown on coverslips
PBS⁺ <R>
Neutral Red (mwt = 288.78, Invitrogen/Molecular Probes N2-364)
Prepare a 2-mM labeling solution in PBS⁺.

Equipment

Cell culture dishes, sterile
Fluorescence microscope

EXPERIMENTAL METHOD

1. Aspirate the culture medium from cells grown on coverslips, and wash the cells three times with PBS⁺.
2. Immerse the cells in the labeling solution for 10 min at room temperature.
   Incubation time may need to be adjusted, depending on the cell type used.
3. Aspirate the labeling solution, and wash the cells three times in PBS⁺.
4. Mount the coverslips as described in Protocol A.
5. Image the cells.
   See *Troubleshooting*.

TROUBLESHOOTING

Problem (Step 5): Cells are stained nonspecifically.
Solution: Reduce the staining time: The longer the staining time, the greater the chance for nonspecific staining.

Protocol Q

Labeling Lysosomes in Live Cells with LysoTracker

The LysoTracker probes (Invitrogen/Molecular Probes) are more selective for acidic organelles than the classically used neutral red or acridine orange dyes (Haugland 2002). They are freely membrane permeant at neutral pH and effectively label lysosomes at nanomolar concentrations, with ~50 nM being optimal for selectivity. It has been reported that in larger acidic compartments, the staining pattern is preserved even after fixation with aldehydes (Haugland 2002). LysoTracker dyes can be visualized at different wavelengths, so they are useful when multiple fluorescent dyes are used to label cells simultaneously.

IMAGING SETUP

	$\lambda_{excitation}$ maximum (nm)	$\lambda_{emission}$ maximum (nm)	Solvent	Filter set	FRAP and confocal microscopy laser excitation
LysoTracker Green FM	~504	~511	Methanol	Fluorescein	488-nm line argon-ion laser
LysoTracker Red	~577	~590	Methanol or aqueous buffer	Rhodamine	514-nm line argon-ion laser 543-nm line of a green He:Ne laser

MATERIALS

Reagents

Cells of interest grown on coverslips

HBSS$^+$ (HBSS with Ca^{2+} and Mg^{++}; Invitrogen/Gibco 14025-092)

LysoTracker Red, mwt = 399.25 (Invitrogen/Molecular Probes L-7528) or LysoTracker Green, mwt = 368.69 (Invitrogen/Molecular Probes L-7526)

Prepare a 1-mM stock solution of the desired probe. Store at –20°C.

Equipment

Cell culture dishes, sterile

Fluorescence microscope

Incubator set at 37°C

EXPERIMENTAL METHOD

1. Make a dilution of the 1-mM probe stock solution to a final working concentration of 50–75 nM in HBSS$^+$. Warm to 37°C.

   The dye concentration should be kept as low as possible (<50 nM) to minimize potential artifacts from overloading.

2. Aspirate the culture medium from cells grown on coverslips, and wash the cells in the culture dish three times with HBSS$^+$.
3. Cover the cells with prewarmed (37°C) probe-containing medium. Incubate for 30 min to 2 h under the desired growth conditions for the cells under study.
4. Replace the labeling solution with fresh medium, and observe the cells using a fluorescent microscope fitted with the correct filter set, to determine if cells are sufficiently fluorescent.

   See *Troubleshooting*.

TROUBLESHOOTING

Problem (Step 4): Cells are not sufficiently labeled.
Solution: Either increase the labeling concentration, or increase the time allowed for the dye to accumulate in the lysosomes.

Problem (Step 4): A decrease in fluorescent signal and/or cell blebbing occurs after incubating the cells in dye-free culture medium.
Solution: Maintain dye in the culture medium after labeling.

Problem (Step 4): When LysoTracker Green is also used as a pH reporter of the lysosomes, it gives alkaline readings.
Solution: LysoTracker Green is rapidly internalized (within seconds) but can raise the pH of the lysosomes. When it is used for pH determination, incubate the cells for no more than 1–5 min at 37°C.

Protocol R

Labeling Endoplasmic Reticulum with DiO-C_6(3)

DiO-C_6(3) is an effective label for the endoplasmic reticulum (ER) because of the distinctive intracellular morphology of the ER that distinguishes it from other intracellular organelles. The DiO-C_6(3) probe has also been used to visualize the ER. For more detailed information, see articles by Terasaki (1989, 1993).

IMAGING SETUP

DiO-C_6(3), λ_{ex} ~ 484 nm; λ_{em} ~ 501 (methanol as solvent)
Filter set: fluorescein
FRAP and confocal microscopy laser excitation: 488-nm line argon-ion laser

MATERIALS

CAUTION: See Appendix 6 for proper handling of materials marked with <!>.
See the end of the chapter for recipes for reagents marked with <R>.

Reagents

Cell culture medium
Cells of interest grown on coverslips
DiO-C_6(3), mwt = 573 (Invitrogen/Molecular Probes D-273)

Prepare a 0.5-mg/mL stock solution of the cationic dye in absolute ethanol. Protect the solution from light in a sealed vial at room temperature.

Glutaraldehyde <!>, 0.5% in PBS (for fixing cells)

Alternatively, 3.7% formaldehyde <!> in PBS can be used (but see the note to Step 8).

PBS$^+$ <R>

Equipment

Cell culture dishes, sterile
Fluorescence microscope

EXPERIMENTAL METHOD

Live Cells

1. Dilute the dye in cell culture medium.

   A general starting point would be a dye concentration of 0.5 µg/mL in cell culture medium. Depending on the particular cell type to be used, the specific conditions may vary.

2. Aspirate the cell culture medium from cells grown on coverslips, and rinse the cells three times with PBS$^+$.

3. Stain the cells for 10 min at room temperature.

   To improve staining of a particular cell type, try shorter or longer staining times.

4. Aspirate the dye-containing cell culture medium. Rinse the cells three times with PBS^+.
5. Mount the coverslips as described in Protocol A.
6. Image the cells.

Fixed Cells

7. Prepare the labeling solution at a concentration of 2.5 µg/mL (a 2000-fold dilution of the stock).
8. Fix the cells for 10 min with 0.5% glutaraldehyde (or 3.7% formaldehyde) in PBS.

   Do not use methanol for fixation because it may extract membranes of the ER so that no pattern can be seen. Formaldehyde fixation can cause vesiculation of the ER.

9. Aspirate the fixative, and wash the cells three times with PBS, allowing 5 min per wash.
10. Stain the cells for 10 sec at room temperature.
11. Aspirate the labeling medium. Rinse the cells three times with PBS.
12. Mount the coverslips as described in Protocol A.
13. Image the cells.

Protocol S

Labeling Endoplasmic Reticulum with ER-Tracker

The ER of cells can be labeled with ER-Tracker probes (Invitrogen/Molecular Probes). Two different probes are available, depending on the emission wavelength desired. The fluorescent moiety for these probes is BODIPY. The probe uses the drug molecule, glibenclamide, which binds to the sulphonylurea receptors of ATP-sensitive K^+ channels that are found throughout the ER membranes. It has been reported that $HBSS^+$ provides the best labeling results (Molecular Probes, 2005, ER-Tracker Product Information). For live cells, however, it is possible that the probe's pharmacological properties can alter ER function over time. Some specialized cells can express the sulphonylurea receptors, giving rise to some non-ER labeling.

IMAGING SETUP

	$\lambda_{excitation}$ maximum (nm)	$\lambda_{emission}$ maximum (nm)	Solvent	Filter set	FRAP and confocal microscopy laser excitation
ER-Tracker Green	~504	~514	Methanol	Fluorescein	488-nm line argon-ion laser
ER-Tracker Red	~587	~615	Methanol	Rhodamine	543-nm line of a green He:Ne laser

MATERIALS

CAUTION: See Appendix 6 for proper handling of materials marked with <!>.

Reagents

Cell culture medium
Cells of interest grown on coverslips or in dishes
$HBSS^+$ (HBSS with Ca^{2+} and Mg^{++}; Invitrogen/Gibco 14025-092)
ER-Tracker Green, mwt = 783.1 (Invitrogen/Molecular Probes E-34251) or ER-Tracker Red, mwt = 915.23 (Invitrogen/Molecular Probes E-34250)

Prepare a 1-mM stock solution of the appropriate dye by dissolving the lyophilized reagent in DMSO <!>. Separate it into aliquots, and store frozen with desiccant.

Equipment

Cell culture dishes, sterile
Fluorescence microscope
Incubator set at 37°C

EXPERIMENTAL METHOD

1. Warm the staining solution to 37°C before using.

2. Remove the cell culture medium from the cells grown in either a culture dish or on coverslips.
3. Rinse the cells three times with HBSS$^+$.
4. Add the prewarmed staining solution, and incubate the cells for ~15–30 min at 37°C in a 5% CO_2 incubator.
5. Remove the staining solution, and replace it with fresh probe-free cell culture medium.
6. Mount the coverslips as described in Protocol A.
7. Image the cells.

   See *Troubleshooting*.

TROUBLESHOOTING

Problem (Step 7): There is nonspecific staining of other cellular components.
Solution: Use the lowest dye concentration possible.

Protocol T

Labeling Acetylcholine Receptors Using Rhodamine α-Bungarotoxin

The rhodamine fluorophore is resistant to fading. Integral protein receptors within cell membranes can be studied by using toxins or hormones that bind to specific receptors. To image these receptors, the ligand can be modified by the covalent attachment of a fluorescent moiety. The α subunit of the nicotinic acetylcholine receptor of the neuromuscular junction strongly binds α-bungarotoxin. A rhodamine conjugate of α-bungarotoxin is used to selectively label the receptor. This procedure for labeling the acetylcholine receptor using α-bungarotoxin is based on Axelrod et al. (1976). Other receptors for which there are fluorescent probes include the benzodiazepine receptor, the Na^+-K^+-ATPase using anthroyl ouabain to bind to the protein's α subunit, and the epidermal growth factor (EGF) receptor using fluorescent EGF (Haugland 2002). For more detailed information, see Angelides (1989), Maxfield (1989), and Haugland (2002).

IMAGING SETUP

Rhodamine α-bungarotoxin, $\lambda_{ex} \sim 550$ nm; $\lambda_{em} \sim 575$ (methanol as solvent)
Filter set: rhodamine
Confocal microscopy laser excitation: 514-nm line argon-ion laser or 543-nm line of a green He:Ne laser

MATERIALS

CAUTION: See Appendix 6 for proper handling of materials marked with <!>.
See the end of the chapter for recipes for reagents marked with <R>.

Reagents

α-bungarotoxin <!> (see Step 2)
Cells of interest grown on coverslips
PBS^+ <R>
PBS^+ containing 2-mg/mL bovine serum albumin (BSA)
Rhodamine α-bungarotoxin <!>, mwt = ~8400 (Invitrogen/Molecular Probes T-1175)
Prepare a 0.1-mM labeling solution in PBS^+.

Equipment

Cell culture dishes, sterile
Fluorescence microscope
Incubator set at 37°C

EXPERIMENTAL METHOD

Live Cells

1. Aspirate the culture medium from cells grown on coverslips, and wash the cells three times with PBS^+.
2. Add the labeling solution, and incubate the cells for 30 min at 37°C.

   As a control, pretreat cells with 5×10^{-8} M α-bungarotoxin for 1 h to prevent fluorescent labeling.
3. Aspirate the labeling solution, and rinse the cells three times with PBS^+.
4. Incubate the cells for 30 min in PBS^+ containing 2-mg/mL BSA.
5. Mount the coverslips as described in Protocol A.
6. Image the cells.

   See *Troubleshooting*.

TROUBLESHOOTING

Problem (Step 6): Receptors disappear after extended observation.
Solution: In living cells, receptors will remain on the membrane surface for only several hours after labeling.

Protocol U

Labeling Pinocytotic Vesicles and Cytoplasm with Fluorescently Labeled Ficoll or Dextran

Dextran, a high molecular weight poly-D-glucose, and ficoll, a synthetic polymer of epichlorhydrin and sucrose, are electroneutral hydrophilic polysaccharides whose size can be varied. This variability in size coupled with their membrane impermeability can make them useful for studying fluid phase pinocytosis, the size of membrane pores such as nuclear membrane pores, or the environmental conditions and the size of a cell compartment. In this protocol, two fluorophores based on fluorescein and rhodamine are used to label either dextran or ficoll. The resulting probes can then be used to label cells. For more detailed information, see Luby-Phelps (1989) and Berlin and Oliver (1980) on probe use and preparation.

IMAGING SETUP

	$\lambda_{excitation}$ maximum (nm)	$\lambda_{emission}$ maximum (nm)	Solvent	Filter set[a]	FRAP and confocal microscopy laser excitation
FITC-labeled polysaccharides[a]	~460	~534	Methanol	Fluorescein	488-nm line argon-ion laser
TRITC-labeled dextran	~555	~580	—	Rhodamine	514-nm line argon-ion laser 543-nm line of a green He:Ne laser

TRITC, tetramethylrhodamine-5-(and-6)-isothiocyanate.
[a]The emission spectrum of fluorescein depends on its local pH and ionic strength.

MATERIALS

CAUTION: See Appendix 6 for proper handling of materials marked with <!>.
See the end of the chapter for recipes for reagents marked with <R>.

Reagents

Aminoethylcarboxymethyl (AECM)-ficoll or AECM-dextran
Carbonate–bicarbonate buffer (pH 9.2) <R>
Cells of interest grown on coverslips
Dimethylformamide <!>
DMEM containing 10% FBS
FITC, mwt = 389.38 (Invitrogen/Molecular Probes F-143) or TRITC, mwt = 443.52 (Invitrogen/Molecular Probes T-490)

Alternatively, commercial versions of fluorescently labeled dextrans and ficolls are available.

NaOH <!>, 0.1 N
Paraformaldehyde <!>, 4%
PBS+ <R>
Sephadex G-25

Equipment

Beaker, 1-L
Cell culture dishes, sterile
Chromatography column, 1 x 30-cm
Dialysis tubing
Fluorescence microscope
Incubator set at 37°C
Lyophilizer
Magnetic stirrer
Stir bar
Water bath set at 40°C

EXPERIMENTAL METHOD

FITC or TRITC Labeling of AECM-Ficoll or AECM-Dextran

This portion of the protocol is based on the Inman method as per Luby-Phelps (1989) and can be used if commercially labeled material is unavailable.

1. Prepare 100 mg of AECM-dextran or AECM-ficoll in 2 mL of 10 mM carbonate–bicarbonate buffer (pH 9.2).
2. Prepare 15 mg of dye in 1 mL of 10 mM carbonate–bicarbonate buffer, as follows.
   i. TRITC: Add 15 mg of TRITC to 400 mL of dimethylformamide, and then titrate the solution into 1 mL of 10 mM carbonate–bicarbonate buffer with constant stirring. Adjust the pH to 9.0 with 0.1 N NaOH.
   ii. FITC: Add 15 mg of FITC to 1 mL of 10 mM carbonate–bicarbonate buffer with stirring. Continue stirring as the pH is adjusted to 9.0 with 0.1 N NaOH.
   It is likely that the FITC will not completely dissolve until the proper pH has been reached.
3. Add the desired dye solution in a dropwise manner with constant stirring to either the ficoll or the dextran solution.
4. Incubate the resultant solution for 30 min at 40°C.
5. Use a 1 x 30-cm column of Sephadex G-25 to remove unreacted dye by desalting.
6. Dialyze the fluorescently labeled material twice versus 1 L of dH_2O.
7. Lyophilize the material for storage, and protect it from light at –20°C until used.

Pinocytotic Studies

8. Prepare 5 mg/mL of labeled dextran or ficoll in a complete cell medium (e.g., DMEM with 10% FBS).
9. Immerse the cells grown on glass coverslips with labeling medium in a small Petri dish. Incubate for 30 min at 37°C.
   Pinocytotic uptake is concentration, time, and cell cycle dependent. There is greater uptake during mitosis than during interphase.

For fixing cells:

10. Aspirate off the labeling medium, and wash the cells three times with PBS^+.
11. Immediately incubate the cells with 4% paraformaldehyde for 10 min.

For live cells:

12. Wash the cells three times with DMEM containing 10% FBS.
13. Mount fixed or live cells as described in Protocol A.
14. Image the cells.

RECIPES

CAUTION: See Appendix 6 for proper handling of materials marked with <!>.
Recipes for reagents marked with <R> are included in this list.

Carbonate–Bicarbonate Buffer

1. Prepare a 0.2-M solution of anhydrous sodium carbonate (2.2 g/100 mL).
2. Prepare a 0.2-M solution of sodium bicarbonate (1.68 g/100 mL).
3. Combine 4 mL of carbonate solution and 46 mL of bicarbonate solution.
4. Bring to 200 mL with dH_2O. Final pH will be 9.2.

PBS and PBS$^+$

PBS may be obtained commercially or prepared as follows. Dissolve the following in 800 mL of dH_2O: 8 g of NaCl, 0.24 g of KH_2PO_4, 0.2 g of KCl <!>, and 1.44 g of Na_2PO_4. Adjust the pH to 7.4 with HCl <!>, and then dilute to 1000 mL with dH_2O. The solution should be autoclaved in an appropriate sized container for 20 min at 15 psi. Store the solution at room temperature.

PBS$^+$ is prepared as for PBS above, but it also contains 1.0 mM $CaCl_2$ <!> and 0.5 mM $MgCl_2$. This solution allows cells to adhere to each other and to the substrate. If cells are in medium containing no Ca^{2+} or Mg^{++}, they will round up and detach from the substrate.

Valap

Prepare a stock supply of Valap (mixture of Vaseline, lanolin, and paraffin, 1:1:1) to seal coverslips. Weigh the mixture components, combine, and melt over low to medium heat on a hot plate. The wax mixture should spread smoothly and dry quickly on a glass slide. If it hardens too quickly, add more Vaseline and lanolin; if it does not harden quickly enough, add more paraffin.

ACKNOWLEDGMENTS

I thank my wife, Nancy, and my daughter, Bryanna, for their patience while I was writing this chapter. I dedicate this chapter in memory of my mother, Cozette Chazotte, 1919–2009.

REFERENCES

Alberts B, Johnson A, Lewis J, Raif M, Raff M, Roberts K, Walter P. 2008. *Molecular biology of the cell*, 5th ed. Garland, New York.

Alivisatos AP, Gu W, Larabell C. 2005. Quantum dots as cellular probes. *Annu Rev Biomed Eng* **7:** 55–76.

Angelides KJ. 1989. Fluorescent analogues of toxins. *Methods Cell Biol* **29:** 29–58.

Axelrod D, Ravadin P, Koppel DE, Schlessinger J, Webb WW, Elson EL, Podleski TR. 1976. Lateral motion of fluorescently labeled acetylcholine receptors in membranes of developing muscle fibers. *Proc Natl Acad Sci* **73:** 4594–4598.

Barbier M, Gray BD, Muirhead KA, Ronot X, Boutonnat J. 2004. A flow cytometric assay for simultaneous assessment of drug efflux, proliferation, and apoptosis. *Cytometry B Clin Cytom* **59B:** 46–53.

Berlin RD, Oliver JM. 1980. Surface functions during mitosis. II. Quantitation of pinocytosis and kinetic characterization of the mitotic cycle with a new fluorescence technique. *J Cell Biol* **85:** 660–670.

Brown DA, London E. 1998. Functions of lipid rafts in biological membranes. *Annu Rev Cell Biol* **14:** 111–136.

Chazotte B. 1994. Comparisons of the relative effects of polyhydroxyl compounds on local versus long-range motions in the mitochondrial inner membrane. Fluorescence recovery after photobleaching, fluorescence lifetime, and fluorescence anisotropy studies. *Biochim Biophys Acta* **1194:** 315–328.

Chazotte B. 2001. *Mitochondrial dysfunction in chronic fatigue syndrome in mitochondria in pathogenesis* (ed. Lemasters JJ, Nieminen A-L), pp. 393–410. Plenum, New York.

Chazotte B, Hackenbrock CR. 1991. Lateral diffusion in mitochondrial inner membranes is unaffected by membrane folding or matrix density. *J Biol Chem* **266:** 5973–5979.

Chazotte B, Wu E-S, Höchli M, Hackenbrock CR. 1985. Calcium-mediated fusion to produce ultra large osmotically active mitochondrial inner membranes of controlled protein density. *Biochim Biophys Acta* **818:** 87–95.

Conn PM, ed. 1999a. *Green fluorescent protein. Methods enzymol*, Vol. 302. Academic, New York.

Conn PM, ed. 1999b. *Confocal microscopy. Methods enzymol*, Vol. 307. Academic, New York.

Drbohlavova J, Adam V, Kizek R, Hubalek J. 2009. Quantum dots— Characterization, preparation and usage in biological systems. *Int J Mol Sci* **10:** 656–673.

Emaus RK, Grunwald R, Lemasters JJ. 1986. Rhodamine 123 as a probe of transmembrane potential in isolated rat-liver mitochondria: Spectral and metabolic properties. *Biochim Biophys Acta* **850:** 436–442.

Frolov A, Woodford JK, Murphy EJ, Billheimer JT, Schroeder F. 1996. Spontaneous and protein-mediated sterol transfer between intracellular membranes. *J Biol Chem* **271:** 16075–16083.

Haugland R. 2002. *Handbook of fluorescent probes and research products*, 9th ed. Molecular Probes, Invitrogen, Carlsbad, CA. http://www.invitrogen.com/site/us/en/home/References/Molecular-Probes-The-Handbook.html.

Herman B, Jacobson K. 1990. *Optical microscopy for biology*. Wiley, New York.

Huang S, Heikal AA, Webb WW. 2002. Two-photon fluorescence spectroscopy and microscopy of NAD(P)H and flavoprotein. *Biophys J* **82:** 2811–2825.

Illinger D, Kuhry JG. 1994. The kinetic aspects of intracellular fluorescence labeling with TMA-DPH support the maturation model for endocytosis in L929 cells. *J Cell Biol* **125:** 783–794.

Illinger D, Poindron P, Fonteneau P, Modollel M, Kuhry JG. 1990. Internalization of the lipophilic fluorescent probe trimethylamino-diphenylbexatriene follows the endocytosis and recycling of the plasma membrane in cells. *Biochim Biophys Acta* **1030:** 73–81.

Jaiswal JK, Simon SM. 2004. Potentials and pitfalls of fluorescent quantum dots for biological imaging. *Trends Cell Biol* **14:** 497–504.

Kok JW, Hoekstra D. 1999. Fluorescent lipid analogs: Applications in cell and membrane biology. In *Fluorescence and luminescent probes for biological activity*, 2nd ed. (ed. Mason WT), pp. 136–155. Academic, New York.

Lemasters JJ, Trollinger DR, Qian T, Cascio WE, Ohata H. 1999. Confocal imaging of Ca^{2+}, pH, electrical potential, and membrane permeability in single cells. *Methods Enzymol* **302:** 341–358.

Luby-Phelps K. 1989. Preparation of fluorescently labeled dextrans and ficolls. *Methods Cell Biol* **29:** 59–74.

Mason WT. 1999. *Fluorescence and luminescent probes for biological activity*, 2nd ed. Academic, New York.

Maier O, Oberle V, Hoekstra D. 2002. Fluorescent lipid probes: Some properties and applications (a review). *Chem Phys Lipids* **116:** 3–18.

Maxfield FR. 1989. Fluorescent analogs of peptides and hormones. *Methods Cell Biol* **29:** 103–124.

Mazeres S, Schram V, Tocanne J-F, Lopez A. 1996. 7-Nitrobenz-2-oxa-1,3-diazole-4-yl-labeled phospholipids in lipid membranes: Differences in fluorescence behavior. *Biophys J* **71:** 327–335.

Mukherjee S, Maxfield FR. 2004. Membrane domains. *Annu Rev Cell Dev Biol* **20:** 839–866.

Mukherjee S, Zha X, Tabas I, Maxfield FR. 1998. Cholesterol distribution in living cells: Fluorescence imaging using dehydroergosterol as fluorescent cholesterol analog. *Biophys J* **75:** 1915–1925.

Nigg EA, Cherry RJ. 1979. Influence of temperature and cholesterol on the rotational diffusion of band 3 in the human erythrocyte membrane. *Biochemistry* **18:** 3457–3538.

Pagano RE. 1989. A fluorescent derivative ceramide: Physical properties and use in studying the Golgi apparatus of animal cells. *Methods Cell Biol* **29:** 75–87.

Pagano RE, Sepanski MA, Martin OC. 1989. Molecular trapping of a fluorescent ceramide analogue at the Golgi apparatus of fixed cells: Interaction with endogenous lipids provides a trans-Golgi marker for both light and electron microscopy. *J Cell Biol* **109:** 2067–2079.

Pawley JB. 1995. *Handbook of confocal microscopy,* 2nd ed. Plenum, New York.

Periasamy A, Herman B. 1994. Computerized fluorescence microscopic vision in the biomedical sciences. *J Comput Assist Microsc* **6:** 1–26.

Smiley ST, Reers M, Mottola-Hartshorn C, Lin M, Chen A, Smith TW, Steele GD, Chen LB. 1991. Intracellular heterogeneity in mitochondrial membrane potentials revealed by a J-aggregate-forming lipophilic cation JC-1. *Proc Natl Acad Sci* **88:** 3671–3675.

Spink CH, Yeager MD, Figenson GW. 1990. Partitioning behavior of indocarbocyanine probes between coexisting gel and fluid phases in model membranes. *Biochim Biophys Acta* **1023:** 25–33.

Swaisgood M, Schindler M. 1989. Lateral diffusion of lectin receptors in fibroblast membranes as a function of cell shape. *Exp Cell Res* **180:** 515–528.

Taylor DL, Wang Y-L, eds. 1989. *Fluorescence microscopy of living cells in culture. Part B. Quantitative fluorescence microscopy—Imaging and spectroscopy. Methods in Cell Biology*, Vol. 30. Academic, New York.

Terasaki M. 1989. Fluorescence labeling of endoplasmic reticulum. *Methods Cell Biol* **29:** 125–136.

Terasaki M. 1993. Probes for the endoplasmic reticulum. In *Fluorescence and luminescent probes for biological activity* (ed. Mason WT), pp. 120–123. Academic, New York.

van Meel E, Klumperman JK. 2008. Imaging and imagination: Understanding the endo-lysomal system. *Histochem Cell Biol* **129:** 253–266.

Wang Y-L, Taylor DL, eds. 1989. *Fluorescence microscopy of living cells in culture. Part A. Fluorescent analogs, labeling cells and basic microscopy. Methods in cell biology*, Vol. 29. Academic, New York.

15 Introduction to Immunofluorescence Microscopy

George McNamara,[1] Leonid (Alex) Belayev,[2] Carl Boswell,[3] and Jose Santos Da Silva Figueira[4]

[1]Analytical Imaging Core Facility, Miller School of Medicine, University of Miami, Miami, Florida 33136; [2]Miami, Florida 33136; [3]Department of Molecular and Cellular Biology and The Arizona Cancer Center, University of Arizona, Tucson, Arizona 85721; [4]W.H. Coulter Center for Translational Research, Miller School of Medicine, University of Miami, Miami, Florida 33136

ABSTRACT

Immunofluorescence microscopy of cells and tissues is a mainstay of modern biology because of its relative simplicity and power. Coupling the sensitivity and specificity of antibodies with a fluorescence microscopy system enables the user to both observe and quantify antibody–antigen interactions, making it possible to potentially localize any molecule within a cell. The fluorescence microscope is also capable of imaging antibody surrogates, such as major histocompatability complex (MHC)-peptide tetramer binding to an antigen, fluorescent streptavidin binding to a biotin-linked target, carbohydrate-binding lectins, fluorescent toxins binding to specific molecules, and in situ hybridization of DNA and RNA. The same microscope with minor modifications also excels for live cell imaging of fluorescent proteins and organic biosensors. Research-grade fluorescence microscopes are now evolving into nanoscopes, able to count, localize, and colocalize molecules with near-nanometer precision. For drug discovery and biomarker validation, the fluorescence microscope has morphed into a high-content screening and analysis instrument. Within clinical pathology, immunofluorescence is being used as a special stain to complement immunohistochemistry. Clinical fluorescence microscopes are evolving into automated high-throughput digital slide imaging systems, as the need grows to simultaneously identify tissue compartments and to localize and to quantify biomarkers, and then relate this information to patient-specific molecular medicine. The focus in this chapter is on immunofluorescence microscopy in the research laboratory and how its evolution into a quantitative tool impacts basic research, drug discovery, and molecular medicine. Two protocols on immunofluorescence specimen preparation are included as examples. In addition, a procedure is provided for configuring a confocal microscope for quantitative immunofluorescence imaging.

INTRODUCTION

Immunofluorescence microscopy was first reported by Coons et al. (1942) and has seen significant improvements every decade since. The latest development, fluorescence nanoscopes with single voxel sizes ranging from ~100 x 100 x 250 nm to ~40 x 40 x 40 nm, enable researchers to count and localize single fluorescent molecules with unprecedented precision. For comparison, a high-resolution confocal volume is ~200 x 200 x 500 nm. By using multiple fluorescent antibodies, molecular complexes can be enumerated at every location within cells, tissues, and whole organisms. In addition, many types of antibodies are now used to detect and resolve single antigens at spatial scales ranging from nanometers to entire organisms, including humans. With humans, the antibody–antigen interactions are usually detected by positron emission tomography or technetium-99 body scanning, but with the success of Herceptin and Rituxan therapeutic antibodies, the addition of fluorophores for clinical research in human patients and mouse scanning is becoming useful.

Immunofluorescence microscopy is a complementary technique to hematoxylin and eosin tissue staining, enzyme-linked immunohistochemistry, RNA and DNA fluorescence in situ hybridization, quantitative polymerase chain reaction and quantitative real-time polymerase chain reaction, immunofluorescence flow cytometry and flow sorting, western blotting, enzyme-linked immunoassays, in vivo bioluminescence and fluorescence imaging, and various array methods such as reverse phase protein array, RNA microarray, and tissue microarray. Important reference works include Shapiro (2003) on flow cytometry and fluorophores; Pawley (2006) on microscopy; Russ and Russ (2007) on image processing and analysis, and *The Handbook—A Guide to Fluorescent Probes and Labeling Technologies* (http://www.invitrogen.com/site/us/en/home/References/Molecular-Probes-The-Handbook.html).

The focus of this chapter is on immunofluorescence of cells and tissues imaged using wide-field, confocal, and multiphoton excitation fluorescence microscopy in the research setting. Reagents include both conventional IgG antibodies and other specific detection reagents such as MHC-peptide tetramers, avidin, streptavidin, lectins, and DNA and RNA in situ hybridization. A wealth of information about antibodies, immunology, and microscopes is available online and from antibody vendors (for a jump page on microscopy, see Molecular Expressions at http://micro.magnet.fsu.edu/).

ANTIBODIES FOR IMMUNOFLUORESCENCE

Most research antibodies are of one of the IgG subclasses, which are ~150-kDa glycoproteins made of two light chains (κ or λ) of about 25 kDa each and two heavy chains each of about 50 kDa organized as light-heavy–heavy-light (LH–HL) heterotetramers, with the unique antigen-binding site formed from one or both variable domains within the amino-terminal half of the light chains (V_L) and the amino-terminal quarter of the heavy chains (V_H). The V_H and V_L domains are each followed by constant regions, with dimerization by the carboxy-terminal half of the light chain (C_L) to the C_H1 domain of the heavy chain and H–H dimerization in the hinge region and paired C_H3 and C_H4 domains. The light chain and $V_H + C_H1$ regions are referred to as the fragment antigen-binding (Fab) domain, and the hinge + C_H3 + C_H4 regions are referred to as the fragment complement-binding (Fc) domain (Fig. 1). Fab and Fc domains are named for proteolyzed fragments. An alternative fragment is the Fab_2, which is bivalent in both LH–HL arms after cleaving the $C_H3 + C_H4$ domain below the hinge region. In a mouse, the IgG subclasses are IgG_1, IgG_{2a}, IgG_{2b}, and IgG_3, and in a human,

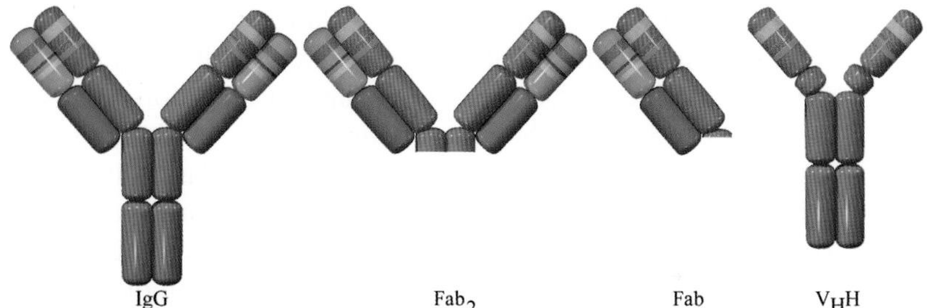

FIGURE 1. IgG antibodies and related fragments. Light (green) and heavy (orange) chains, shown with variable (stripes) and constant (solid) regions. In the Fab_2 fragment most of the C_H2–C_H3 domain has been removed by proteolysis, except for the interheavy chain cysteine–cysteine bridge segment. In the Fab fragment, the proteolytic site was above (amino terminal) to the S–S bridge (only one of the two Fabs is shown). In camel, llama, and mouse V_HH variant IgGs, a deletion of most of the C_H1 domain eliminates light-chain binding and enables the dimerized heavy chain to be secreted (and prevents the B-cell from committing suicide). V_HH antibodies tend to have longer hypervariable regions, enabling some to insert into clefts of target molecules. Not shown are scFv, diabodies, and many other variants. (Artwork modified from redistributable BioGeek graphic by the authors [www.biogeek.com].)

they are IgG_1, IgG_2, IgG_3, and IgG_4 (for humans, see http://www.researchd.com/rdikits/subbk23.htm). Recently, van der Neut Kolfschoten et al. (2007) discovered that human IgG_4 molecules exchange LH arms continuously in solution (i.e., serum), resulting in bispecific antibodies with each arm potentially having a different specificity (that can change over time in solution).

In most immunofluorescence experiments, an antibody will bind with an antigen at only one of the two Fab sites either on or in the live or fixed cells or tissue. Unfortunately, immunoglobulins often stick to other immunoglobulins and to other molecules and, thus, cannot be assumed to be free in solution (see Bennett et al. 2009). This means that the antibody being used as a probe may stick to immunoglobulins of other specificities, which can stick to the cells or tissues under study for reasons other than the desired specific antibody–antigen-binding event. The Fc domain of some classes (IgG, IgE) of immunoglobulins binds to specific Fc receptors on cells. Further confounding interpretation of immunofluorescence results is the potential binding by reduced sulfhydryl groups on antibodies/immunoglobulins to other reduced sulfhydryls in the antiserum or in the cells or tissues (Rogers et al. 2006).

The intended function of the Fc domain in immunofluorescence experiments is to provide labeling sites for fluorophores. With improvements to immunofluorescence sample preparation and the quality of fluorophore labeling, with a decrease in the autofluorescence background of cells and tissues, and with the development of spectral unmixing and other methods to separate fluorophore signals from unwanted noise, the Fc domain may become dispensable. The monoclonal Fab fluorophore is beginning to contribute to fluorescence nanoscopy because the Fab arm of ~5-nm length is smaller than an intact IgG of ~10-nm length, and recent fluorescence nanoscopes such as isoSTED and nanoSIM (20 x 20 x 20-nm voxels) are approaching these dimensions in their ability to localize single fluorophores.

In the meantime, the best choice for a specific reproducible antibody reagent is a well-characterized monoclonal antibody obtained from a reliable vendor. A high-affinity antibody-binding site may bind its specific antigen with a K_d of 10^{-8} M. In the case of a protein antigen, an epitope of about seven to nine amino acids makes contact with a similar number of amino acids in the complementarity determining regions of the light and/or heavy chain (often mostly or entirely the heavy chain). (Note that some epitopes are 3D juxtapositions of noncontiguous amino acids.) A single amino acid mismatch may reduce the affinity by an order of magnitude or more. Nonetheless, given enough antibody or antigen, moderate affinity binding of related molecules can occur. Furthermore, a user should always be aware of the many sources of both specific but unwanted interactions and nonspecific interactions. Failure to include an Fc-blocking reagent results in a spectacularly bright specific detection of cells expressing Fc receptors but will not help localize or quantify the antigen of interest.

In the context of a microscopy or flow cytometry experiment, what do these affinity values mean? In immunofluorescence microscopy, typically, a solution of antibody is applied to a cell suspension, attached cells, or a tissue. A ballpark calculation is to use a 10-µL volume with 15 ng of IgG antibody of 150 kDa. The concentration of this solution is 10 nM. A typical immunofluorescence experiment is conducted on a 20 x 20-mm cover glass. A typical adherent mammalian cell is ~4000 μm^2 surface area (e.g., HeLa cells, see http://bionumbers.hms.harvard.edu/), so 10^6 cells fit on a cover glass. Cell-surface proteins range in density from hundreds to millions per cell. For example, from 400 ErbB2 (Her-2/neu) on the low end (the limit of detection in a flow cytometry experiment using fluorescein or Alexa Fluor 488 direct-labeled antibody) to 400,000 on the high end (ErbB2 on an SKBR3 breast carcinoma cell), this range is 0.1–100 molecules/μm^2, or 4×10^8 to 4×10^{11} molecules per cover glass. A 10-nM solution is (10×10^{-9} M) x (6×10^{23} molecules/mol) or 6×10^{10} IgG molecules in 10 µL. Therefore, an excess of antibody can be applied to the cells on the cover glass for a cell-surface protein. Even for a cell with exceptionally high protein expression, such as the N87 breast cancer cell line with 2×10^6 to 4×10^6 ErbB2/cell, an antibody can be in excess.

A popular protein for normalizing western blot loading is actin because it is from 1% to 10% of a cell's protein. There is ~1 µM actin in the cytoplasm of a eukaryotic cell (Wu and Pollard 2005), which is ~8×10^7 per HeLa cell (~5000 μm^3 volume, ignoring organelle exclusion) or ~8×10^{13} actin monomers per 10^6 HeLa cells on a confluent cover glass). The upshot of this exercise is that, in a typical experiment, a specific antibody is in excess and has a high enough affinity to bind a target antigen, that is, a saturating antibody concentration. The downside is that cells are made of many molecules that may be sticky and that may bind to that same excess of antibody to produce nonspecific binding. With two to five fluorophores on a direct-labeled primary antibody, or four to 20 fluorophores in a complex of antigen, primary antibody, and a few secondary antibodies, nonspecific binding can easily spoil an experiment. The ability to optimize antigen–antibody–fluorophore-specific binding while minimizing nonspecific interactions becomes even more important as fluorescent nanoscopy becomes widespread (see Ji et al. 2008). Actin immunofluorescence has also been used for normalizing immunofluorescence of tissues (Kai et al. 2007).

In the next decade, improvements in microscope and detector sensitivities may enable routine use of direct-labeled Fabs (50-kDa monomeric antigen-binding fragments) bearing one or a few fluorophores and showing less nonspecific binding, which are either photostable (wide-field and confocal microscopy) or show experimenter-controlled photoconversion/photoswitching for fluorescence nanoscopy. If the nonspecific binding of antibodies to cell and/or extracellular matrix molecules is linearly proportional to antibody concentration, then using a large excess of antibody requires the use of blocking and/or enhancing reagents (e.g., MaxBlock, Active Motif; PeroxAbolish, Biocare Medical; and Image-iT FX signal enhancer, Invitrogen/Molecular Probes). Alternatively, use a subsaturating concentration of antibody. However, the problems with a subsaturating concentration of antibody are that (1) by definition, some antigens will not be labeled, (2) the total amount of antigen on the specimen will dictate how much is missed, and (3) control peptide spots will be more variable between specimens (Bogen et al. 2009). We suggest that when using photostable fluorophores, cost-effective production of a defined number of directly labeled Fabs and sensitive instrumentation (i.e., moderate- to high-numerical-aperture [high-NA] objective lens and high-sensitivity detectors) that experiments should be performed with saturating amounts of a fluorophore–Fab conjugate to enable routine quantitation on a per μm^2 or per cell basis.

Simultaneous Imaging of Multiple Antigens

Increasingly, there is interest in imaging multiple antigens simultaneously using an array of antibody probes. Early examples include Tsurui et al. (2000), who used spectral imaging microscopy to image seven antigens, and Hermiston et al. (1992), who imaged six antigens. Perfetto et al. (2004) achieved 17-color flow cytometry. Schubert et al. (2009) have developed the multiepitope ligand cartography/toponome imaging system (MELC/TIS) in which eight cycles of fluorescein and phycoerythrin immunofluorescence plus one cycle of propidium iodide DNA detection were used on a tissue section, generating images having nine colors. Schubert et al. (2009) indicated that they used subsaturating concentration of antibodies, so MELC/TIS may be regarded as a plus-or-minus test rather than as a method to quantify the amount of antigen per cell. The MELC/TIS method could be continued to >>16 immunocycles. However, 2^{16} binary combinations of antigens either present or absent on a cell is a lot of possible combinations (but maybe not so far out of the realm of all possible cell phenotypes in a mouse or a human).

The key question, however, is how many probes are needed to answer your biological question. Although simultaneous imaging of multiple antigens has many benefits, bear in mind that the more probes in use, the greater the number of controls that have to be included (Perfetto et al. 2004 and see the section, Controls, below). As described by Bennett et al. (2009), the higher the aggregate concentration of antibodies, or when one or more of the antigens is at a much greater concentration than the others, the more opportunities arise for artifact(s) to confound an experiment.

Choosing Antibodies and Fluorophores for Immunofluorescence Imaging

Fluorescein has long been a popular fluorophore. However, we no longer recommend fluorescein as a label because it is pH sensitive (e.g., the fluorophore is dim if the mounting medium pH is <7.5) and photobleaches rapidly. Instead, use a pH-independent photostable fluorophore conjugated to a secondary antibody. As an interesting alternative, there are fluorophore-conjugated antifluorescein antibodies available, so you could use fluorescein as a hapten and detect it with, for example, Alexa Fluor 488-antifluorescein or DyLight 488-antifluorescein.

Table 1 presents one example of the reagents required for a multiple immunofluorescence experiment, in this case, looking at seven protein antigens and DNA, and utilizing eight different fluorophores. This experiment could easily be performed on a current-generation confocal microscope, or with enough filter sets or wheels, on a research-grade fluorescence microscope equipped with a good detector such as a CRi Nuance, a LightForm, Inc. PARISS, or an Applied Spectral Imaging SKY spectral imager (see Tsurui et al. 2000).

However, there are a number of risks associated with an experiment as proposed in Table 1 (see Bennett et al. 2009). First, although the detection of a set of antigens, all of which are in high abundance (in at least some cells), is fine, replacement of anti-CD31, for example, with anti-Rad51, might lead to a misleading result such as an apparent colocalization artifact (see, e.g., Bennett et al. 2009). Second, there is a possibility of protein dimerization, especially when pools of IgGs are present (e.g., if a mixture of secondary antibodies was derived from a blend of goat antisera) or if the pool of secondary antibodies listed in Table 1 was premixed (Yoo et al. 2003; Bennett et al. 2009). Goat IgGs may be especially prone to forming Fc–Fc dimers in solution or on slides. As a substitute, consider using donkey antibodies or direct-labeled antibodies.

Immunofluorescence could be simplified by moving to direct-labeled antibodies, as is commonly performed in flow cytometry (e.g., Perfetto et al. 2004). BD Biosciences has taken the lead in validating many of their flow cytometry direct-labeled monoclonal mouse IgG antibodies as bioimaging certified reagents. Many other antibody companies are also producing either niche-targeted or wide-

TABLE 1. An example of multiple fluorescence immunocytochemistry

Antigen	Primary antibody	Secondary antibody	Fluorophore (conjugated to secondary antibody)	Emission color
DNA	–		DAPI	Blue
Tubulin	Mouse IgG1	Goat–anti-mouse IgG1	Alexa Fluor 488	Green
Actin	Guinea pig IgG	Goat–anti-guinea pig-IgG	Lucifer Yellow	Yellow
Vimentin	Rat mAb IgM	Goat–anti-rat-IgM	Cy3B	Orange
Cytokeratin	Mouse IgG2a	Goat–anti-mouse IgG2a	DyLight 594 (or Texas Red)	Red
Neurofilament	Chicken IgY	Goat–anti-chicken IgY	HiLyte Fluor 647 (or Cy5)	Far red
Nestin	Rabbit IgG	Goat–anti-rabbit IgG	Atto 680	Infrared
CD31	Donkey IgG	Goat–anti-donkey IgG	IRDye800	Infrared

DAPI, 4′,6-diamidino-2-phosphate.

spectrum direct-labeled monoclonal antibody lines. Alternatively, labeling kits can be purchased to make customized direct-labeled antibodies in the laboratory. As fluorescence nanoscopes become more widely available, it might be possible to use direct-labeled Fabs exclusively, each conjugated to either one or a few fluorophores per Fab or to one quantum dot (QD) per Fab (see Chapter 36).

It might seem that fluorophores that share a brand name would have similar chemical structures or characteristics. However, the only current trademark family of fluorophores that are all chemically related is the boronpyrromethene (BODIPY) family (Invitrogen/Molecular Probes) (Loudet and Burgess 2007). The BODIPY fluorophores have common physical characteristics that make them excellent probes for lipids and certain other molecules. In contrast, the Alexa Fluor dyes are a family in name only. Different family members may have very different chemical structures. Realize that, if family members do have similar chemical structures, this could be a disadvantage if they tend to dimerize.

Tables 2 and 3 list several current-generation fluorophores, including classic small organic dyes, fluorescent proteins, and QD nanocrystals. The tables are not all-inclusive, and the same fluorophore

TABLE 2. Fluorophore families for IF and live cell imaging

		Organic fluorophores						
Emission band	Emission (nm)	Alexa Fluor[a]	Atto[b]	BODIPY[c]	CF[d]	Cy[e]	DyLight[f]	HiLyte Fluor[g]
Blue	440	405			405			
Cyan	470	350	390					
Cyan–green	500		465	493/503				
Green	520	488	488	FL	488A		488	488
Yellow	540	514	495	R6G				
Orange	560	532	550	530/550				
Red	580	555	Rho3B	564/570	555	3B	549	555
Redder	600	594	Rho11	581/591				594
Far red	620	610	Rho13	TR				
NIR	640			630/650				
NIR	660	647	647N	650/655		5	649	647
NIR	690	660	680		660			680
NIR	720	700	700		680	5.5	680	
NIR	760	750	740			7		750
NIR	780				750			
NIR	800	790			770		794	
Vendor(s)		Invitrogen/ Molecular Probes	Atto-Tec	Invitrogen/ Molecular Probes	Biotium, Inc.	GE Healthcare	Rockland Immuno-chemicals, Inc.	AnaSpec

Spectral data for most fluorophores are available online at http://www.sylvester.org/AICF/PubSpectra.zip (PubSpectra xlsx data file download with more than 2000 fluorophore and filter spectra), http://www.photochemcad.com/ (spectra calculator; all PhotoChemCAD 2.1 spectra are in PubSpectra.xlsx). Online spectral graphing sites include http://www.mcb.arizona.edu/ipc/fret/index.html (University of Arizona interactive spectral viewer), http://www.invitrogen.com/site/us/en/home/support/Research-Tools/Fluorescence-SpectraViewer.html, http://www.bdbiosciences.com/research/multicolor/tools/index.jsp?CMP=KNC-PQ3922888249, http://www.coulterflow.com/bciflow/tools.php (Spectrios viewer), https://www.micro-shop.zeiss.com/us/us_en/spektral.php?f=db (commercial interactive graphing sites/downloads). Three major filter companies, Chroma, Omega Optical, Inc. (www.omegafilters.com/front/curvomatic/spectra.php), and Semrock also have online spectra comparison tools.

NIR, near infrared.

[a]Not all Alexa Fluor products are listed. Alexa Fluor 555 and Alexa Fluor 568 are comparable in brightness and are preferred to Alexa Fluor 546 (see confocal list server). Alexa Fluor 430 (=Lucifer Yellow) has a large Stokes shift (433-nm excitation maximum, 540-nm emission maximum) that has a similar emission maximum to the Alexa Fluor 514 selected for inclusion in the table.

[b]Not all Atto dyes are listed (see https://www.atto-tec.com).

[c]See also Umezawa et al. (2009) for Keio Fluor BODIPY fluorophores spanning 517-nm to 701-nm emission with high brightness (highest extinction coefficient * quantum yield/1000 = 284 for KFL-7 with 662-nm excitation, 671-nm emission maxima).

[d]Biotium CF dye information (http://www.biotium.com/product/product_info/Newproduct/CF_dyes.asp). CF405 is reported to be brighter than Alexa Fluor 405, making the former more useful for OMX structured illumination microscopy–fluorescence nanoscopy (P. Goodwin, Applied Precision, pers. comm.). Biotium online literature reports a higher degree of labeling with CF555 compared with Alexa Fluor 555 or Cy3. A better comparison would have been to Cy3B.

[e]Cy2 is not competitive with the other green reagents. Cy3B, a rigidized Cy3 variant, is reported by the manufacturer to be ~7X brighter than Cy3. Patents suggest the existence of Cy5B, but GE Healthcare has not introduced such a product.

[f]DyLight fluorophores are a subset of a much larger family of dyes available from Dyomics (www.dyomics.com).

[g]AnaSpec also offers HiLyte Fluor Plus products, which are brighter and more expensive.

TABLE 3. Fluorescent proteins and QD nanocrystals families for IF and live cell imaging

		Fluorescent proteins				QDs		
Emission band	Emission (nm)	Evrogen[a]	Invitrogen[b]	Other[c]	Phycobiliproteins, tandem dyes[d]	eFluor NC, EviTag[e]	QD	TriLite[f]
Blue	440	TagBFP	BFP	EBFP2				420
Cyan	470	TagCFP	CFP	Cerulean				
Cyan–green	500	CopGFP				490 (L.P.B.)		Blue 490
Green	520	TagGFP2	EmGFP	mWasabi		520 (A.G.)	525	Green 525
Yellow	540	TagYFP	YFP	mVenus		540 (C.G.)	545	
Orange	560	TurboRFP	OFP	mOrange2	R-phycoerythrin	560 (H.Y.)	565	
Red	580	TagRFP	RFP	DsRed		580 (B.Y.)	585	Yellow 575
Redder	600	TurboFP602			R-PE-Texas Red	605 NC	605	
Far red	620	mKate2		mKeima	Allophycocyanin	625 NC	625	Red 630
NIR	640							
NIR	660				APC-Cy5	650 NC	655	Deep Red 665
NIR	690			PR1		680 (J.R.)		685
NIR	720			IFP1.4	APC-Cy5.5		705	
NIR	760				APC-Cy7			
NIR	780							
NIR	800							
Vendor(s)		Evrogen	Invitrogen/ Molecular Probes	Clontech, Allele, Biotech, other[b]	Becton, Dickinson and Company, Beckman Coulter, Inc., Columbia Biosciences, Corp., Invitrogen	eBioscience, Evident Technologies, Inc.	Invitrogen/ Molecular Probes	Crystalplex

See the footnotes in Table 2 concerning websites for spectral data downloads, calculator, and interactive graphing.

BFP, blue fluorescent protein; CFP, cyan fluorescent protein; YFP, yellow fluorescent protein; OFP, orange fluorescent protein; RFP, red fluorescent protein.

[a]Evrogen is currently the major innovator of commercial fluorescent proteins. Evrogen column does not list all products (http://www.evrogen.com/products/basicFPs.shtml). Table does not attempt to show photoactivatable/photoconvertible fluorescent proteins (http://www.evrogen.com/products/photoactivatableFPs.shtml) or the photosensitizer Killer Red (http://www.evrogen.com/products/KillerRed/KillerRed.shtml).

[b]Invitrogen does not disclose the identity of all its fluorescent protein products.

[c]We will not attempt to be all inclusive with respect to fluorescent proteins. See Chapters 22 and 26 for details, or the PubSpectra.xlsx and Fluorescent Proteins.xls in http://www.sylvester.org/AICF/PubSpectra.zip. Roger Tsien's laboratory (www.ucsd.edu), Addgene.com, Allele Biotech, Clontech, Evrogen, MBL International, Promega, and others market fluorescent proteins. Addgene.com is a nonprofit repository in which research laboratories can deposit plasmids for a nominal fee (less expensive than shipping hundreds of orders as Professor Tsien's laboratory does, but not everyone has Howard Hughes Medical Institute financial backing to provide a full time technician), and users can purchase plasmid(s) at cost ($65/plasmid). PR1 and IFP1.4 are phytofluors (fluorescent phytochromes) developed by Lagarias and Tsien laboratories, respectively.

[d]Phycobiliproteins and their FRET conjugates list is not all inclusive. Phycoerythrin (R-PE, B-PE) and allophycocyanin (APC), or their FRET tandem conjugates (e.g., R-PE→Texas Red) are more popular in flow cytometry but can be used in imaging by minimizing exposure time and illumination power. See Columbia Biosciences (http://www.columbiabiosciences.com) or your favorite antibody vendor or flow cytometry instrument/reagent vendors (BD Biosciences, Beckman Coulter) for specifications. mKeima is of special note because of its large Stokes shift (440-nm excitation peak, 620-nm emission peak). Unfortunately, mKeima has relatively low extinction coefficient and quantum yield. A better choice for large Stokes shift fluorescent proteins may be to modify Professor Miyawaki's CY11.5 (Cyan→Yellow) with similar linker (or lack of linker) and different donor and/or acceptor. Multimering mKeima (re: 5xGFP, Bulinski and Waterman's laboratories) may be another way to increase intensity of an mKeima_n-fusion protein.

[e]Evident Technologies EviTags have been discontinued with respect to the life science market and/or are being transferred to eBioscience as eFluor NCs. The abbreviations in parentheses are for the EviTag color names Lake Placid Blue, L.P.B.; Adirondack Green, A.G.; Catskill Green, C.G.; Hops Yellow, H.Y.; Birch Yellow, B.Y.; Fort Orange, Maple Red-Orange, Macoun Red, Jonamac Red, J.R. In July 2009, Evident Technologies filed in bankruptcy court for Chapter 11 reorganization. As of early 2010, eBioscience still offers eFluor NC products.

[f]Crystalplex (www.crystalplex.com) TriLite nanocrystals are alloyed gradient construction and all are 10 nm in diameter, even with bioprotective coating (synthesis developed by Crystalplex based on Bailey and Nie 2003, Bailey et al. 2004). Crystalplex offers custom emission peak Trilite nanocrystals from 420 to 685 nm, all reported be the same high intensity.

may be sold under several different names (e.g., Dyomics GmbH DY-610, Rockland Immuno - chemicals, Inc. DyLight 610, and PromoKine PromoFluor 610 are equivalent as are Alexa Fluor 405 and Cascade Blue). We have chosen to list American vendors with whom we are familiar. We have ignored certain fluorophores that are subperformers (e.g., fluorescein and Cy2). The world of available fluorophores and fluorescently labeled compounds is dizzying, and space prevents us from listing many additional products, such as Chromeo fluorophores (Active Motif), specialty reagents such as CypHer5 (GE Healthcare's pH-sensitive version of Cy5), the hundreds of products at Sigma-Aldrich, or the thousands of fluorophore products at Invitrogen/Molecular Probes. Tables 2 and 3 are organized by emission color and wavelength range because (1) many fluorophores can be excited at shorter wavelengths, especially in the ultraviolet (UV) range, and (2) some immunofluorescence experiments are now conducted with multiphoton excitation fluorescence microscopes, which are clearly advantageous for imaging thick specimens, especially with the recent development of multiple nondescanned detectors (e.g., NDD4 developed by Dr. Owen Schwartz of the National Institutes of Health and commercialized by Leica). A periodically updated list of stimulated emission depletion (STED) fluorophores for fluorescence nanoscopy can be found at the Hell laboratory website (http://www.mpibpc.mpg.de/groups/hell/STED_Dyes.html).

Fluorophore–Fab Conjugates

Because an ~50-kDa Fab is about 5 nm long, most of the common fluorophores conjugated to it will be within fluorescence resonance energy transfer (FRET) distance of each other. FRET is not guaranteed, however, because the orientation parameter κ^2 can range from 0 to 4 (the standard κ^2 value of 0.667 is based on the assumption of free rotation of dipole moments). Thus, a Fab with two identical fluorophores may undergo homo-FRET, which typically leads to a reduction in fluorescence intensity but has the advantage that when one fluorophore does photobleach, the other is still available. If Fabs could be routinely conjugated with two different fluorophores (e.g., one on each of the heavy- and light-chain carboxy termini), a large palette of donor$_x$–acceptor$_y$ fluorophore pairs could greatly expand the available color palette. In all practicality, if the yield of purified Fab fragments obtained from a valuable IgG antibody is low, then the manufacturer would have to charge a premium to go through the effort of making fluorophore-Fab conjugates. On the other hand, if a fluorescence nanoscope is used in a microscopy system such as STED, PALM (photoactivated light microscopy; see Chapter 34), or STORM (stochastic optical reconstruction microscopy; see Chapter 35), then the significant increase in resolution that is possible with direct fluorophore-Fabs versus traditional IgG antibodies might justify the increased cost of the fluorophore–Fab conjugates. For example, a single microscope slide might use $5.00 worth of IgG antibodies and might result in 60-nm resolution, whereas it would require $25.00 worth of direct fluorophore–Fab conjugate but might yield 40-nm resolution. This increase in reagent price may be trivial compared with the 33% improvement in resolution. The benefit is even more dramatic when comparing a conventional confocal microscope with a nanoscope because a 500% improvement in resolution is possible (200 nm vs. 40 nm). In summary, the hope is that by the year 2020, the reagents and microscopes or nanoscopes to perform immunofluorescence will be readily available so that the complexity of having to use indirect antibody methods can be replaced by the simplicity of direct methods using fluorophore-conjugated Fab fragments.

Increasing Fluorescence Intensity

The best fluorescence results come from maximizing signal and minimizing background. Ideally, the background should be black. This requires a clean microscope objective lens, a clean and nonfluorescent slide and cover glass, minimized cell/tissue autofluorescence, optimized antibody incubations, optimized washing, optimal DNA counterstaining, and nonfluorescent mounting medium. Focusing here on the antibody incubations and wash steps, we can do no better than to emphasize three points that impressed us in reading Burry (2010): (1) Keep the cells or tissue submerged throughout the incubation and wash steps. Allowing your antibodies to dry on the specimen will greatly increase fluorescent background by causing antibodies to aggregate and stick as well as being detrimental to good morphology. (2) Rinses after the primary antibody are critical. Burry (2010, its Fig 10.3) uses seven rinses and mentions elsewhere that some researchers use 10 rinses. Burry shows insufficient washes after the primary antibody result in weak signal. He suggests that unbound primary antibodies present in the cells or tissue intercept incoming secondary antibody and prevent the latter from reaching the antigen–primary antibody complexes. (3) Rinses after the secondary antibody (Burry 2010, its Fig 10.4) are needed to decrease background of nonbound fluorescent secondary antibodies in order to reveal the location of antigen–primary antibodies–secondary antibodies complexes. The more rinses the better. For rinses, Burry (2010) advocates leaving 10% of the fluid over the specimen, to enable point #1 on submergence, and doing enough rinses to achieve thorough dilution. PBS or other buffers are inexpensive—use them. Be gentle exchanging solutions, to avoid knocking off your cell or tissue, but wash all the excess antibody away. Antibodies are large molecules and take time to diffuse out of cells or tissues. If you use a 12-h incubation to allow antibody to diffuse in, do not expect the antibody to diffuse out in a few minutes.

A powerful way to increase fluorescence intensity is to use enzymatic amplification. The two most commonly used enzymes are alkaline phosphatase and horseradish peroxidase (HRP). HRP is the focus here because of its historical use in chromogenic immunohistochemistry and the availability of commercial fluorescent substrates that make covalent, hence permanent, bonds in target cells and tissues. When performing immunofluorescence with two or more primary antibodies, each HRP–primary antibody can be applied, can be detected, and then can be inactivated before the next one is applied to the specimen. Alternatively, tyramide signal amplification (TSA; see Protocol B) can be used to sequentially detect the antigens by incorporating HRP–secondary antibodies and the appropriate steps to inactivate the HRP between rounds of detection. There are many fluorophore–tyramides with a range of emission spectra available from commercial suppliers (e.g., Invitrogen and PerkinElmer), and it is also possible to make your own (see Vize et al. 2009).

Storing and Testing Antibodies

Like many proteins, antibodies are labile and can inactivate within days or weeks when stored at 4°C. Even newly purchased antibodies should be tested to determine their efficacy at binding antigens and with what specificity (Couchman 2009; see also Holsmeth et al. 2006; Burry 2010). Proper handling and storage of antibodies are essential for maintaining antibody function. When antibodies are ordered, be sure to receive them promptly, determine that they have been shipped under conditions in which they remain stable, aliquot them immediately, and store them at the recommended temperature. Avoid unnecessary freeze–thaw cycles when using antibodies. Every antiserum and every monoclonal antibody is a unique reagent. Attention to details may not guarantee excellent results, but inattention will likely lead to failure.

CONTROLS FOR IMMUNOFLUORESCENCE MICROSCOPY

Ideally, antibody-labeled cells or tissues are bright, and negative controls are dim. However, often, samples and controls are either all bright or all dim. Literature is full of good advice on setting up and evaluating immunohistochemistry and immunofluorescence experiments, and we recommend Mighell et al. (1998); Burry (2000); Saper and Sawchenko (2003); Ward (2004); Saper (2005, 2009); Holmseth et al. (2006); Fritschy (2008); Couchman (2009), and references therein. Suffice it to say that (1) a lot can go wrong, and (2) there is no consensus on what the best controls are (one of our mottos: "if you don't do the controls, you should not publish; if you do *all* the controls, you'll never publish"). For example, when studying a cell-surface antigen with antibodies that work in a western blot, flow cytometry, and microscopy, you should be able to see one band (or a few known bands) on a western blot, to tell positive specimens from no-antibody or isotype controls by flow cytometry, and to get pretty pictures using a fluorescence, deconvolution, confocal, or multiphoton microscope. However, to do so and to trust the results, then the proper controls need to be identified and included. Even when results between laboratories differ, or results within a laboratory using different antibodies differ, it still makes sense to optimize the protocol(s) and to identify the best reagent(s), and then to publish the results. It is worth reading Wilson et al. (1999) for an example of a consensus report on which anti-BRCA1 antibody was most useful at the time; also Press et al. (1994), who compared 28 antibodies against ErbB2 (Her-2/neu) and identified three that were adequate with respect to specificity and sensitivity on immunohistochemistry tissue sections. Finally, look at Bogen et al. (2009), who discuss the next step in ErbB2 immunoquantitation.

Sensitivity and Specificity

Specificity and sensitivity are closely tied together in immunofluorescence. If your system is capable of single-molecule sensitivity (Ji et al. 2008, Bennett et al. 2009)—that is, every single fluorophore-labeled antibody molecule in a cell can be detected—then excellent specificity is critical. Any non-specific interaction of the primary antibody or fluorophore-conjugated secondary antibody could result in a false positive signal in the wrong location within or on the specimen. Likewise, signal amplification (Protocol B) is only useful if the signal is specific.

If your immunofluorescence experiments suffer from low sensitivity, it is necessary to distinguish between problems with the microscopy hardware, the reagents, and the specimen. To possibly resolve hardware problems: Align the lamp, run a pinhole calibration routine (and test to make sure it made an improvement), and if necessary, invest in a new arc lamp bulb (if it is old), new filter set(s), and new objective lenses. If possible, run the experiment on a different fluorescence microscope to determine whether it is your microscope or your specimen that is causing sensitivity problems.

Sometimes, investing in new hardware is unnecessary; you just need to spend the time to understand your equipment and how to best use it. For example, on a scientific-grade monochrome digital camera system, sensitivity can be improved by increasing pixel binning (e.g., 2 × 2 binning is four times more sensitive than unbinned). On a confocal microscope, you can (1) increase laser power, (2) increase gain, (3) open the pinhole beyond the 1-Airy-unit theoretical optimal resolution setting, (4) adjust the image window display contrast/brightness control (not the computer monitor!), and/or (5) increase frame averaging. On a confocal microscope (and on some digital cameras) you can—and should—also make sure that your detector offset is always >0. This is easy: Set the illumination power to 0, and acquire an image. If some or all the pixels are intensity level 0, adjust the offset to just above 0 so that you can quantify the lowest specimen intensity while maximizing dynamic range. For example, intensity level should be 12–25 for an 8-bit detector (maximum intensity 255) or 100–200 for a 12-bit detector (maximum intensity 4095). If you change the detector gain, you may also have to change the offset. In our laboratory, we have found that with a 1-Airy-unit pinhole size on a confocal microscope, we can usually visualize cellular autofluorescence by increasing laser power and/or detector gain, followed by adjusting the image window contrast/brightness.

If no antibody is available for the protein under study, add a DNA sequence for a fluorescent reporter or an epitope tag (GFP, DsRed, or a fruity derivative, TagBFP, mKate2, FLAG, myc, 6His, etc.) to the cDNA, and express a fusion protein. This is a reliable method for introducing fluorescent probes into cells, tissues, and whole organisms.

BRIEF REVIEW OF A RESEARCH-GRADE FLUORESCENCE MICROSCOPE

Microscope theory and use are reviewed in Chapter 1. Here, we discuss what is needed on a basic immunofluorescence microscope. An inverted microscope is preferred over an upright one. For live cell imaging, it will be necessary to have a temperature controller, thermal stabilization for constant focus, provisions for controlling humidity and gas composition, and if high resolution is needed, a water-immersion objective lens.

Purchasing a Fluorescence Microscopy System

A typical research-grade fluorescence microscope, including a set of Plan-Apochromat differential interference contrast (DIC)/fluorescence objective lenses, may cost $50,000. Each fluorescence filter set will cost about $1000. A good-quality charge-coupled device (CCD) camera, such as the Hamamatsu ORCA-ER, costs $10,000–$15,000 (although a $5000 camera may be fine for tyramide signal amplification). An excellent CCD camera (such as the ORCA-R2) will be over $15,000, and a back-illuminated electron-multiplying charge-coupled device (EMCCD) camera costs $28,000–$36,000. A computer is necessary, and an old personal computer (PC) will generally do the job, but a new PC with plenty of processing speed and memory will be able to deconvolve data fast (e.g., ImageJ has a plug-in for deconvolution). Software ranges from free (Microsoft Biology Foundation or Fiji ImageJ download + μManager) to expensive (MetaMorph, Volocity, or LAS AF, ZEN 2008). There are alternatives, however, to spending $70,000 or more. Protasenko et al. (2005) assembled a single-molecule fluorescence microscope, including laser and camera, for $3000.

Maintaining a Fluorescence Microscope

Proper maintenance of your fluorescence microscope, especially its objectives, epi-illuminator, and filter sets, is essential for proper function and reliable results. If the microscope has oil- or glycerol-immersion objective lenses, be sure to train all users to avoid transferring the oil or glycerol to a dry lens. If immersion oil or glycerol does get on a dry lens, we recommend careful cleaning with a sheet of lens paper (folded over twice for mechanical strength) wrapped around a Q-tip cotton swab and moistened with 70% ethanol. As soon as the moistened lens paper is moved away from the lens, immediately blow the lens dry (with a clean handheld puffer, not a can of compressed air) so that the ethanol does not evaporate, leaving spots. Several cycles of cleaning may be needed. Cleaning dirty lenses should be performed by the microscope manager or a microscope service person, not by an untrained beginner. Users also need to be careful not to get mounting media or sealant (e.g., wet nail polish) on any of the lenses. Consult with the manufacturer about the best way to remove such reagents from the lens without damaging the lens.

We typically clean the lens on an inverted microscope equipped with a Bertrand lens so we can inspect the lens surfaces optically during cleaning. When a user is finished using any immersion objective lens, that lens should be immediately wiped clean of immersion media by holding one clean sheet of lens paper along parallel edges, folding the sheet twice for mechanical strength (not touching the surfaces that will be used for cleaning), and lightly sliding the lens paper over the lens in one direction, one time, then flipping the paper over to a clean surface and lightly sliding the clean surface over the lens. We recommend against having users apply purple, green, or even transparent lens cleaning fluids because they are likely to leave fluid on the lens, which will evaporate and will leave a residue.

Lamps and Filters

The heart of the research fluorescence microscope is shown in Figure 2. An epi-illuminator, typically a mercury or xenon arc lamp, is used as the light source. Mercury bulbs usually last 200–400 h at full power. Xenon bulbs should last 300–400 h.

The purpose of the epi-illuminator is to provide an intense illumination that can be launched into the microscope and can be filtered by an excitation filter positioned parallel to the light beam. Modern excitation filters use high-tech interference coatings to select one or more wavelengths to be transmitted to the dichroic beam splitter. Over the last five years, improvements have been made to microscope interference filter set technology such that a new filter set should deliver a black background as long as the immersion and mounting media are nonfluorescent and the slides and cover glass are clean. The dichroic beam splitter reflects short wavelengths to the specimen and transmits long wavelengths to the emission filter and detector(s). The emission filter then cleans up stray excitation light and provides additional wavelength selection for the emitted light that is allowed to reach the detectors. The emission filter is usually needed because excitation light is usually many orders of magnitude more intense than the emitted fluorescence. Some dichroic beam splitters are simple short reflection/long emission (as in the inset to Fig. 2). The filter set acts together to provide specific excitation and emission ranges. In addition to the simple single filter (e.g., the DAPI filter set shown in the Fig. 2 inset), dual, triple, and quadruple filter sets are available that selectively reflect specific wavelength bands and transmit other bands. For example, a blue/green/orange triple-pass filter set acts to excite the specimen with appropriate UV, blue, and yellow wavelengths for reagents such as DAPI, Alexa Fluor 488, and Alexa Fluor 568 (or Hoechst, mWasabi, and mCherry) and then filter the light that comes back through the objective lens so that only blue, green, and orange–red light reaches the detector.

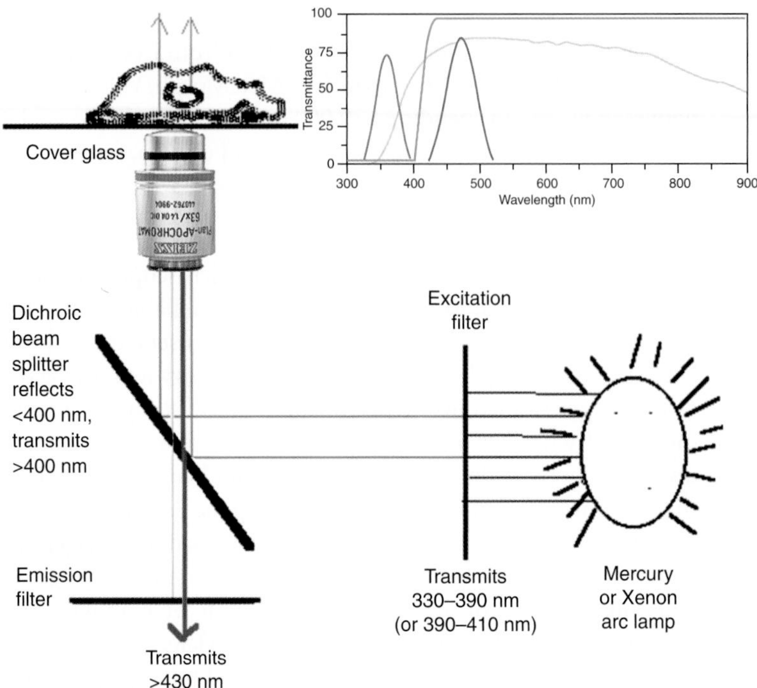

FIGURE 2. Key components of a research fluorescence microscope (not to scale), showing example of a 4′,6-diamino-2-phenylindole (DAPI) filter set and a Zeiss Plan-Apochromat 63x/1.4-NA oil-immersion DIC objective lens (Zeiss part 440762-9904, https:// www.micro-shop.zeiss.com). This objective lens is excellent for wide-field microscopy but may not have sufficient transmission for exciting DAPI or other fluorophores sufficiently when using the 351- and 364-nm UV confocal laser lines (investigator's Zeiss LSM510/UV). This lens transmission is fine for newer confocal microscopes using new 405-nm lasers. (*Inset*) Transmission curves of DAPI excitation filter (purple), dichroic filter (green), and emission filter (blue), and the Zeiss 63x/1.4-NA objective lens (cyan).

If you mostly look at tissue culture cells, having an epi-illumination arc lamp on your tissue culture microscope may be sufficient. Most tissue culture microscopes have phase-contrast objective lenses, which will result in a small amount of light loss and resolution loss, but this can be compensated in several ways: by turning off the room lights and by letting your eyes dark adapt; by using a longer camera exposure time during image acquisition; and by utilizing tyramide signal amplification. However, turning off the lights in a busy tissue culture room may be impractical.

Lenses

The objective lens is the most important optical component within the light path of a fluorescence microscope. Take all necessary steps to protect the lenses and to keep them clean. This includes making sure that the specimen has equilibrated to room temperature (or microscope temperature) before using. A slide moved from a refrigerator to the bench will soon be spotted with condensation. If the water touches the immersion oil, the result will be an optical mess. For experiments that are performed at 37°C with an oil-immersion lens, consider testing Cargille 37DF immersion oil (www.cargille.com, see Oomen et al. 2008). For fixed cell imaging, 37DF oil is not necessary, unless your microscope is kept at 37°C. With live cell imaging, when imaging cells (average refractive index ~1.38) in an aqueous culture mounting medium (refractive index ~1.33), avoid using an oil-immersion objective lens, which will be a bad refractive index (RI) match. Instead, find an appropriate glycerol-objective lens, and if possible, use an appropriate additive (e.g., bovine serum albumin [BSA]) so that the RIs of the glycerol-immersion medium (i.e., ~1.42, although it could be diluted with water to decrease this value somewhat) and the live cell mounting medium have a value appropriate for the objective lens. Some multi-immersion lenses have an RI correction collar enabling you to dial in the lens for the immersion and mounting media RI that works for your experiment.

If you are planning to perform quantitative deconvolution, the precise RIs and specimen temperatures are important variables for the deconvolution equations. Even if you do not intend to deconvolve image data, the better matched the RIs of the immersion and mounting media, the better the image quality will be throughout your specimen (as long as both media are nonfluorescent).

When working with fixed specimens, you have nearly complete control over what mounting medium to use. Take the time to evaluate different media to see which one best fits your needs and provides the best results. With a thick specimen (>20 µm), the RI may make it possible to image with a high-NA lens throughout the entire working distance of the objective lens (e.g., 190 µm for the Zeiss 63x/1.4-NA lens). When using a total internal reflection fluorescence (TIRF) lens for wide-field or confocal immunofluorescence, be aware that the working distance may be narrow. For example, the Zeiss α Plan-Fluar 100x/1.45 oil-immersion lens has a working distance of 110 µm. If your specimen is attached to the slide (instead of the cover glass), and if there is an appreciable layer of mounting medium (e.g., when using a Lab-Tek II chamber slide) and standard #1.5 cover glass, the microscope may be unable to focus on the specimen. The same considerations apply when looking at a retina or a similar whole mount, and you want to image throughout the entire specimen. If the mounting and immersion media both have an RI of 1.515 or 1.52, then a #00 (~70-µm), a #0 (~100-µm), or a #1 (~150-µm) could be used instead of a conventional #1.5 (~170-µm) cover glass to be able to image deeper into the sample. Cover-glass thickness can be measured by looking at the edge with a stereomicroscope or a 10x objective lens, by marking the upper and lower surfaces with a Sharpie marking pen, and by focusing through or in the reflection confocal mode. For the latter, it is necessary to account for imaging through a medium with RI = 1.515 rather than air, RI = 1.0. If a large number of cover glasses need to be measured, purchase precision calipers. The important point to remember is: Do not assume a #1.5 cover glass is 170.0-µm thick.

Table 4 lists the relative brightness, the *x–y* and *z* resolutions, and the prices of different objective lenses. To obtain the full performance of a high-NA oil-immersion objective lens, the RIs of the immersion and mounting media should be as well matched as possible. Oil-immersion objective lenses are designed for imaging at, and a few micrometers from, the cover glass. Imaging with an oil-immersion lens into 10 µm of saline mounting medium results in a large drop in intensity and resolution (see Staudt et al. 2007). The live cell imaging niche of TIRF (~100-nm optical section thickness at the cover-glass-mounting medium interface) has led to wider availability of >1.40-NA lenses. TIRF lenses have little advantage in theoretical relative brightness and resolution for RI-matched fixed cell immunofluorescence experiments. If a TIRF lens outperforms a standard confocal-rated 40x/1.3 or a 63x/1.4-NA oil-immersion lens, it may simply be that the latter needs a good cleaning.

TABLE 4. Objective lenses: numerical aperture (NA), brightness, resolution, and price

NA	Immersion medium (RI)	Relative brightness (NA^4)	x–y resolution (µm)	z resolution (µm)	Price (approximate)
0.5	Air (1.0)	0.06	0.60	1.80	$1500
0.75	Air (1.0)	0.32	0.40	1.20	$3000
1.0	Water (1.33)	1.00	0.30	0.90	$4000 to $18,000[a]
1.25	Water, glycerol, or oil	2.44	0.24	0.72	$6000
1.4	Oil (1.515)	3.84	0.21	0.64	$8000
1.45	Oil (1.515)	4.42	0.21	0.62	$10,000
1.49	Oil (1.515)	4.93	0.20	0.60	$8300 (Olympus)
1.57[b]	Oil (1.515)	6.08	0.19	0.57	$$$
1.65[c]	Special (1.78)	7.41	0.18	0.55	$$$

Resolution calculated for 500-nm wavelength (in vacuum) using $d = 0.6\ \lambda/\text{NA}$.

RI, refractive index.

[a]The high end is a price for specialty lenses such as the Leica 20x/1.0 saline dipping immersion lens for confocal and multiphoton excitation fluorescence live tissue and live animal imaging.

[b]Zeiss exclusive TIRF lens (standard immersion medium and cover glass, only available with PAL-m or HR-SIM nanoscope).

[c]Olympus exclusive TIRF lens (special immersion medium and RI 1.78 sapphire cover glass required).

TABLE 5. Immunofluorescence mounting media

Mounting medium	RI	Comments
PBS, pH 7.4	1.34	Fluorescein photobleaches rapidly. Ideal for live cell imaging.
VectaShield (Vector Laboratories)	1.458	Glycerol:buffer plus proprietary antifade reagent(s). Becomes fluorescent over time. Quenches CyDyes.
Glycerol/PBS, 90%/10%	1.458	Similar to VectaShield, but should remain nonfluorescent
Glycerol, 100%	1.474	Too viscous for practical use.
Prolong Gold (Invitrogen)	~1.50	Hardens in a day or two (at room temperature), increasing RI slightly. Use for most organic fluorophores.
Qmount (Invitrogen)	1.50	Hardens in 12 h. Use for Invitrogen QDs. Do not use with organic fluorophores.
TDE	1.52	See Staudt et al. (2007). Somewhat miscible with water (aqueous bubbles may form around cells if applied directly). Best used with gradual infiltration.

This is a partial list of mounting media. To learn more, go to http://www.uhnres.utoronto.ca/facilities/wcif/fdownload2.html for WCIF Mountants and Antifades guide.

RI, refractive index; PBS, phosphate-buffered saline; TDE, 2,2′-thiodiethanol.

For simplicity, in Table 4, z resolution has been calculated as three times x–y resolution. This rule of thumb is close enough. The Nyquist sampling theorem states that a sine wave should be sampled at 2.3 times its frequency. Images are 2D and are usually nonperiodic, so they need to be sampled at ~3X resolution. This means that the z-step size should be one-third of the z resolution, conveniently, identical to the x–y resolution in the table. The x–y pixel size should be one-third of the x–y resolution, ~0.070 µm (70 nm) for a 1.4+-NA objective lens. Optical aberrations, immersion versus mounting medium mismatch, or photobleaching issues may compel the use of larger z-step size and/or pixel size.

Deconvolution has the potential to improve image quality by 1.2 times in x–y and z ($1.2^3 = 1.728$X per voxel) for both wide-field and confocal microscope images. Different algorithms may be needed for wide-field versus confocal deconvolution, and small details, such as using an exactly 170-µm thick cover glass, may have a big impact on deconvolution performance.

Mounting Media

Table 5 contains a short list of mounting media used in immunofluorescence microscopy experiments. The mounting medium should possess the following properties: (1) It does not fluoresce, (2) it has a known RI, and (3) if using any pH-sensitive fluorophores, the mounting medium has a known pH. Many antifade additives have the unattractive property of being or becoming fluorescent. If you have not used a bottle of mounting medium in a long time, test it by placing a drop of the medium on a slide, add a cover glass, and look at the slide. If the drop fluoresces at any of the wavelengths being used for an experiment, then discard the mounting medium, and replace it with fresh medium that does not fluoresce. Some manufacturers now offer hardening mounting media. Be certain that the medium has hardened completely before placing the slide on the microscope because transfer onto an objective lens could lead to an expensive lens replacement.

QUANTITATIVE IMMUNOFLUORESCENCE MICROSCOPY

Historically, research immunofluorescence microscopy has been limited to qualitative images (i.e., pretty pictures), and clinical immunohistochemistry has focused on using antibody-based detection as specialty stains (Taylor and Levenson 2006). We believe that immunohistochemistry and immunofluorescence methods could be quantitative. For this to be successful, it will be necessary to standardize specimen collection and preparation, detection and image acquisition, and image processing and analysis, to minimize variability in each step. It will also be necessary to provide internal reference standards to relate the measurements of cells and tissues to standards on the same slide(s). Finally, the internal reference standards must be related to external reference standards using the reagents known to operate in different formats.

This also applies to quantifying all of the fluorescent proteins in which it can be performed (Niswender et al. 1995; Furtado and Henry 2002; Wu and Pollard 2005; Gordon et al. 2007; Digman et al. 2008; Joglekar et al. 2008; Wu et al. 2008). Unfortunately, quantitation on a per cell basis is currently the exception, not the rule. For a useful example, see Kai et al. (2007), who took a pragmatic double fluorescence immunohistochemistry approach, in which actin per pixel (per μm^2) was used to normalize antigen concentration. This is, basically, a single-cell version of a biochemical western blot normalization.

The new fluorescence nanoscopes (i.e., precision localization microscopy, stimulated emission depletion microscopy, structured illumination microscopy, and the newest TIRF microscopy) have the potential to provide much more quantitative data than has been possible in the past. The ability to turn single-molecule fluorescence speckle microscopy (Danuser and Waterman-Storer 2003) into quantitative data hints at the potential of the new nanoscopes.

Quantitative Immunofluorescence Microscopy: Basic Steps

Quantitative immunofluorescence microscopy should be possible on any current fluorescence microscope. Ideally, aim for absolute quantitation in photons/sec/pixel, or number of fluorophores/pixel, or number of antigens/cell. Several publications have shown that calibration is possible, including Cho and Lockett (2006), Young et al. (2006), Bernas et al. (2007), Digman et al. (2008), and Grunwald et al. (2008). The pragmatic quantitation of arbitrary intensity units per pixel can be made more useful if the instrument (1) is shown to be linear, (2) is stable over time, and (3) does not have undersaturated or oversaturated pixels. Test linearity by adjusting exposure times (CCD) or by changing intensity power (neutral-density filters or laser power control). Validate stability over multiple timescales. An instrument that is not stable over seconds is not going to be stable over minutes or hours. On the other hand, a system that is stable over a few seconds may drift over minutes and/or hours. Stability, especially for a confocal microscope system, should be tested for illumination stability, *x–y* stage drift, and *z*-focus drift. Arc lamps or lasers may fluctuate in power over several timescales, the *x–y* stage may drift, and/or the *z*-focus may drift such that measuring only one variable may result in data confounded by others. If the *x–y* stage does not drift, the simplest test for the confocal microscope is to perform *x–z* scans over several hours with a reflection slide. Using multiple scan tracks, each with one laser line, measure how much the intensity of each laser changes over time (for a particular gain setting), and observe *z*-focus drifts and parfocality. Using reflection from a metal-coated or even a bare glass or slide, avoids having fluorescence photobleaching as a variable. Be alert to the potential for warping of the slide as an alternative explanation for *z*-focus drift. For example, we have observed that the popular Lab-Tek chamber slides, after removing the chamber and mounting a cover glass, tend to warp during and after focus shifts.

Confocal microscope laser acousto-optical tunable filter (AOTF) intensity controls are nonlinear, although most current confocal control software has a calibration for this feature to enable linearity between 2% and 98% (e.g., Zeiss LSM510). The photomultiplier tubes (PMTs) have a relative sensitivity that scales to the seventh power: $[(A - V_0)/(B - V_0)]^7$, where V_0 is the null voltage (which unfortunately is not provided by the manufacturers). On a Zeiss LSM510, possible voltage values range from 78 to 1250, with most fluorescence signals (at 1-Airy-unit pinhole size) typically being in the range of 600–1000. If $V_0 = 0$ V (which is probably slightly low), and we use *B* as 600 V (typical for a bright fluorescence signal), then plugging in *A* of 700, 800, 900, and 1000 V results in relative sensitivities of 2.9, 7.5, 17.1, and 35.7, which are roughly compatible with the Hamamatsu R6357 PMT gain chart values of 600 V = 10^5 gain, 700 V ~3 x 10^5 gain, 800 V ~8 x 10^5 gain, 900 V ~1.7 x 10^6 gain, and 1000 V ~4 x 10^6 gain (a complete table is available from gmcnamara@med.miami.edu. Additional information on the Hamamatsu R6357 PMT datasheet is available at http://sales.hamamatsu.com/en/products/electron-tube-division/detectors/photomultiplier-tubes/part-r6357.php). Hamamatsu does not report quantum efficiency (QE) (photons detected/photons reaching the detector), but standard PMTs have a QE of ~10%, and new GaAsP PMTs have peak QE of ~40%. The upshot is that a user can usefully adjust the gain of a typical confocal microscope PMT around 40-fold (600–1000 V), but unless special care is taken throughout the image acquisition steps, a user is unlikely to be able to equate intensity values taken on the same confocal microscope at different gain settings or to be able to equate intensity values taken on different confocal microscopes at any gain values. These claims are made because additional variables, including detector offset, scan speed, laser power, and total instrument calibrations come into play (not even counting objective lens performance and RI-induced aberrations).

Likewise, first- and second-generation EMCCDs typically are nonlinear in their dial settings versus gain values, and many users appear to fail to adjust the electron-multiplication (EM) gain correctly. Briefly, an EMCCD operated at zero EM gain is just an unoptimized big pixel front- or back-illuminated CCD. Conversely, an EMCCD operated at maximum gain (e.g., 1000x) is mostly amplifying thermal electron noise. The EM gain should be adjusted so that photoelectrons are amplified to be above the readout noise floor. For a given specimen, the optimum EM gain will depend on photon flux (photons/pixel/time), so effectively it will depend on exposure time. A simple way to acclimate to an EMCCD is to first observe a fluorescence specimen at low EM gain and relatively long exposure (e.g., 500 msec), then shorten the exposure time (e.g., to 100, 50, and 10 msec) and adjust EM gain to get comparable image quality from each exposure time. Many EMCCD cameras also have a conventional gain, which we recommend you leave at 1x and use the EM gain. Most scientific-grade CCD cameras have conventional gain controls. For example, the Hamamatsu ORCA-ER CCD has a maximum 10x gain (in many software programs this is confusingly a setting of 255). The ORCA-ER gain's main use is to enable a shorter exposure time; for example, you will get comparable intensities with a 5000-msec exposure at 1x gain and a 500-msec exposure at 10x gain, although the latter will have more noise.

In the future, we hope users and manufacturers will agree on standards so that every image can be displayed and can be quantified with real world units, such as those suggested at the beginning of this section (e.g., antigens/cell). In the meantime, the following procedure explains how to adjust confocal microscope settings, using a Zeiss LSM510/UV confocal microscope (four R6357 PMTs) with AIM.exe software as an example.

Procedure: Configuring a Confocal Microscope for Quantitative Immunofluorescence Microscopy

1. Use mounting medium of known RI and (if possible) verify that your #1.5 cover glass is really 170-µm thick and is scrupulously clean. Some cover-glass manufacturers have very loose tolerances for cover-glass thickness, and many provide dirty cover glasses. You can measure a cover glass by standing it on end and by imaging with a 10x objective lens. You can easily measure the thickness of each cover glass by transmitted light using a Sharpie marker, focusing up and down both sides, and reading *z* values from the microscope fine-focus vernier or from the *z* motor in the image acquisition software (then cleaning the ink off with 70% ethanol). Or you can measure thickness by using a confocal microscope in the reflected light mode. A reflection confocal mode with a high-NA dry-objective lens and a pinhole set to 1.0 Airy unit will be more precise than white transmitted light and marks. With either method, you will need to account for the RI of the cover glass versus the mechanical travel of the lens through air. For example, a #1.5 cover glass with an apparent thickness of 118.57 µm multiplied by an $RI_{\text{cover glass}}$ of 1.518 = 170.0-µm thick. (Note that different cover-glass materials may have different RIs, and the RI may be wavelength dependent.) The cover-glass thickness is an important value to input into deconvolution software. Correctly deconvolving confocal microscope data can improve *x–y* and *z* resolutions by about 20% in each axis. Check the literature or a confocal list server (http://lists.umn.edu/cgi-bin/wa?A0=confocalmicroscopy) for cleaning recipes.
2. Adjust the specimen focus by eye. If using an immersion lens, match the lens type and the immersion medium as closely as possible to the RI of your mounting medium. For example, if all of your confocal microscope lenses are oil-immersion lenses (RI 1.518 for Zeiss Immersol 518F), an ideal mounting medium is 97% thiodiethanol (TDE):3% saline (Staudt et al. 2007. Note that TDE is miscible with water but may require gradual concentration steps to best infiltrate tissue, see http://lists.umn.edu/cgi-bin/wa?A2=ind0901&L=CONFOCALMICROSCOPY&D=0&P=2450). Consider using neat (100%) TDE because any additive may impact fluorescence and the immersion oil may have an RI closer to 1.52.

3. Use enough immersion medium to provide good optical contact between the specimen and the cover glass but not so much that it oozes down the outside of the objective lens.

4. If imaging multiple fluorophores, set up multiple tracks with enough wavelength separation to minimize cross talk. Although you could have four separate tracks for Alexa Fluor 647, Alexa Fluor 555, Alexa Fluor 488, and DAPI (see Tables 2 and 3 for alternatives), we find that two tracks are sufficient: Alexa Fluor 647 plus Alexa Fluor 488 excited with 488 and 633 nm, and Alexa Fluor 555 and DAPI excited with 543 and 351 + 364 nm, respectively. Using two tracks instead of four reduces the scan time, requires fewer hardware changes, and may limit sensitivity if one label is weak. However, if a signal is weak, why multiplex? Consider optimizing detection, for example, with tyramide signal amplification, before trying to differentiate the real signal from autofluorescence.

5. Adjust each channel gain, laser power, beam splitter, and filter path to obtain good quality images with no obvious cross talk between fluorophores. For example, in the DAPI and Alexa Fluor 555 track, if the 543-nm laser line is turned off, practically no DAPI nuclear fluorescence should be visible in the Alexa Fluor 555 detection channel. If DAPI does appear in the wrong channel, lower the DAPI laser (UV or 405-nm) power and/or decrease the Alexa Fluor 555 channel detector gain. The offset should be adjusted, so, if the laser lines are off, no pixels are undersaturated (intensity level 0). This can be performed by inspecting pixels and/or by selecting a saturation warning marker lookup table, and verifying that there are no saturated pixels. (Note that live fast *x–y* scans and all image acquisition scan speed settings may not be calibrated to the same intensity. Avoid saturation in the production image.) To help determine the settings, use any of several spectra viewing/selection tools (reviewed in McNamara et al. 2006), or download the free open data access PubSpectra data set (http://www.sylvester.org/AICF/PubSpectra) and evaluate in Microsoft Excel (data are in Excel 2007 xlsx format [if you do not wish to own Excel 2007, you can download the free Excel 2007 file viewer and copy/paste data columns of interest into another spreadsheet program or to MATLAB, etc.]). The Zeiss ZEN software includes a Smart Setup command that uses PubSpectra data, and researchers and vendors are welcome to use the PubSpectra data (data are not subject to copyright).

6. Set the pinhole size of the longest wavelength channel (e.g., Alexa Fluor 647-nm emission) to be 1.0 Airy unit. On the Zeiss LSM510 with the 63x/1.4-NA oil-immersion objective lens, this corresponds to an optical slice thickness of 1.0 µm. Adjust the other fluorescence channels to the same 1.0-µm optical slice thickness. This way any colocalization measurements will be made on the same optical slice thickness because it makes no sense to try to colocalize a green object in a 0.7-µm optical slice with a near-infrared object in a 1.0-µm optical slice.

7. Most confocal microscope software calculate the optimal *x–y* pixel size for each objective lens and wavelength(s). When in doubt, click the optimal *x–y* button. On the Zeiss LSM510, the maximum image dimension is 2048 x 2048 pixels. The 40x/1.3-NA oil-immersion lens requires a zoom of 1.3x or higher to match optimal *x–y* pixel size. Using higher zooms with optimal *x–y* settings enables faster image acquisition but with a smaller field of view.

8. If any channel is noisy, increase frame averaging, and if necessary, open the channel pinhole to greater than optimum. Two-frame averaging is typical, but use 4, 8, or 16 (maximum on LSM510) if needed. On the LSM510, the DAPI channel is dim with the 63x/1.4-NA lens because the lens does not transmit the 351-nm or 364-nm laser lines well. We, therefore, use a high gain (1000 V), increase pinhole size, and use a higher frame averaging setting.

9. Turn off all laser lines, and verify that the offset for each channel is adjusted so that every pixel is above zero intensity level (not undersaturating) using the production scan settings with no frame averaging (e.g., 1024 x 1024 pixels, maximum scan speed). Acquire multiple frames (e.g., short time series). We usually adjust the offsets so each channel is between intensity levels 100 and 200 (when in 12-bit mode, maximum 4095) with no laser light. Then acquire with maximum frame averaging (or average the time series), and use for background subtraction in post-

processing later (e.g., save as 16-bit tagged image file formats [TIFFs], and process in ImageJ or MetaMorph). Note that if you change the gain on any channel, you will need to reacquire, verify no 0s, and save new dark reference images. The no-laser-light image also provides an excellent image to check whether the instrument produces any fixed pattern noise. Adjust the image digital contrast and brightness to check this.

10. Turn off each laser line, one at a time, and verify that there is no (or at least tolerable) cross talk between channels. Adjust the laser power and/or detector gain if needed (readjust the offset if necessary so all values are >0 with no laser power). Iterate as needed. If necessary, use separate tracks, one per laser line.

11. Optionally, add a reflected light/interference reflection microscopy (IRM) channel. IRM is the reflection from the cover-glass-culture medium interface (which should be adjusted to be mid-intensity) plus the interference patterns from cover-glass substratum-cell interfaces, especially focal adhesion sites. The main benefit in using IRM in immunofluorescence is that it enables consistent focusing. Focusing at the cover-glass-cell surface results in consistent focus throughout the experiment. This is of special value for negative immunofluorescence control cells—because they should be dark (but still >0) in the immunofluorescence channel(s). Live focusing can be performed with IRM, then uncheck the IRM detection channel for fluorescence acquisition if this channel is unimportant for the experiment. IRM is free, meaning it does not add any time to your acquisition and only a little additional time when saving images, so we encourage you to keep it on when imaging cells. If the experiment involves the cell–substratum interface, then the locations can be determined for focal adhesion sites and close contacts, filopodia/tunneling nanotubes touching the cover glass and the edge of the cell (as well as, often, reflections of the nucleus). If the planes of interest are above the cover glass (inverted microscope configuration), consistently acquire *z*-series starting at 0 µm (cover-glass IRM image plane) and moving in appropriate step size up (always acquire *z*-series focusing against gravity). IRM is less useful for imaging tissues, but the reflected confocal images enable visualization of collagen fibrils, cells versus voids, and other features in the specimen. For an IRM channel, we typically adjust the gain so that empty areas between cells of the cover glass are medium brightness, the focal adhesion sites of cells are dark (destructive interference) but above intensity level 0, and the bright features (constructive interference) are not saturated (<4095 in the 12-bit mode, typically aim for intensity level 3500). In the 12-bit detector mode with gain adjusted for IRM, the reflections from the plasma membrane, nuclear membrane, and other membrane organelles will typically be dim on the monitor, but by not under- or oversaturating the original image, these features can be made visible by digital contrast adjustment. We typically use the IRM channel to obtain the most consistent possible focus of the cover-glass–cell interface. We, therefore, set its optical slice thickness to 1.0 Airy unit (e.g., use a 488-nm laser line with either no transmitting emission filter or a 488-nm transmitting emission filter). Usually the 488-nm laser line is much brighter than the 633-nm laser line so no emission filter is needed in the 488 + 633-nm laser scan track (see Ch2-T1 in Fig. 3) to get the best possible focus of the bright cover glass. IRM makes redundant any of the perfect focus system accessories for any confocal microscope that have reflection light capability. For additional details on IRM and the closely related reflection interference contrast microscopy, see Barr and Bunell (2009) and Limozin and Sengupta (2009).

12. Optionally, add a transmitted light channel track (ChD on an LSM510). If you want to use transmitted light DIC, make sure the objective lens has a Wollaston prism, the condenser focus is set adjusted to be parfocal with the specimen plane, the condenser turret has a matching Wollaston prism (e.g., DIC III for a Zeiss 63X/1.4-NA lens), and the condenser side polarizer is in the light path and is correctly crossed with respect to the laser polarization. We usually adjust gain so that no features oversaturate (intensity level 4095) and no cellular features undersaturate (no intensity level 0 during scan; okay in this case if the image is 0 when the laser line is off because the DIC image will not be used for intensity quantitation). Note that the objective lens DIC Wollaston prism may affect confocal resolution. This is most easily seen in IRM images with

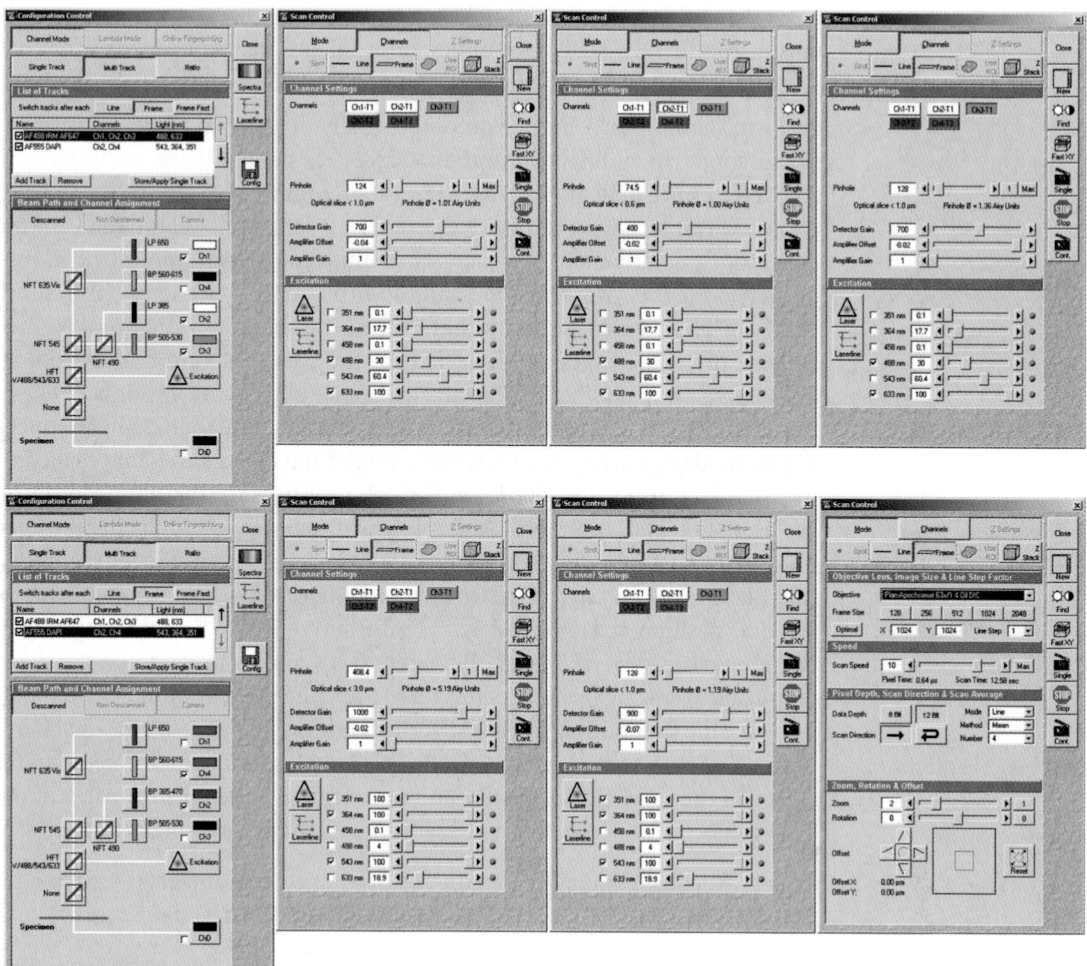

FIGURE 3. The Configuration Control dialog (*left* column) is used to select the dichroic filters, emission filters, and channels. In the top row, Ch1 is Alexa Fluor 647, Ch2 is IRM, and Ch3 is Alexa Fluor 488. In the bottom row, Ch2 is DAPI and Ch4 is Alexa Fluor 555. We get better fluorescence separation by having two tracks than by trying to save a factor of 2 in scan time by combining all four fluorescence channels in a single track (which would also result in not having a channel available for IRM). The Scan Control dialogs are shown in channel mode for all five detection channels. The pinhole sizes for Alexa Fluor 488, 555, and 647 are adjusted so that they all have a common optical slice thickness of 1.0 µm (1.0 Airy unit for the Alexa Fluor 647 detection channel). The IRM channel (Ch2-T1 in top row) is set to its theoretical optimum slice thickness (0.6 µm) to enable the best possible focus of the cover glass (which acts like a mirror and is brightest at best focus). The DAPI channel pinhole is open to 3-µm optical slice thickness because this objective lens transmits UV poorly and, therefore, produces a dim DAPI signal. Opening the pinhole increases light collection at the expense of making the DAPI counterstained nuclei look thicker than they are (it is better to collect an imperfect signal than none). In the bottom right panel, Scan Control is set to the Mode dialog, frame size is 1024 x 1024 pixels, Scan Speed is 10 (maximum for these settings and also the brightest scan speed on this LSM510 because of the way the system is calibrated), scan depth is 12 bit, scan direction is unidirectional, frame averaging (mean) is 4, and zoom is 2.0x. Note, we could equivalently use 2048 x 2048 pixels and zoom of 1.0x to obtain the same pixel resolution at the expense of longer scan time and larger file size. The Zeiss LSM510 can zoom down to as low as 0.7x (the inverse of the guitar amplifier scene in *This Is Spiñal Tap*). Zeiss and other manufacturers include an optimal frame size button (see last panel) to calculate the theoretical optimal Nyquist limited number of pixels based on NA and zoom. For the 63x/1.4-NA lens with 2x zoom, optimal is 904 x 904 pixels, so the default of 1024 x 1024 pixels is oversampled by 10%. Most confocal microscopes do not require square image dimensions; if you only need 1000 x 300 pixels, go for it.

filopodia. If fluorescence (and IRM) resolution are top priority, the DIC Wollaston prism(s), and, optionally, the DIC analyzer, can be removed; and the transmitted light channel track gain and offset, and, optionally, the condenser NA, can be adjusted for bright-field contrast.

13. Perform shading correction if needed (usually performed post acquisition). Most fluorescence microscopes, including confocal microscopes, do not have perfectly uniform illumination across the entire field of view (e.g., 1× or 0.7× zoom). Some objective lenses perform better than others. Clean lenses work better than dirty lenses. Well-maintained microscopes work better than poorly maintained microscopes. Using the correct amount of immersion medium (for immersion lenses) works better than too little. At higher zoom (>2×), most Plan-Apochromat and Plan-Fluorite lenses should have very little shading in the middle of the field of view (all bets are off at the edges and especially at the corners). Chroma Technology and Applied Precision give away plastic fluorescence reference slides (the Chroma slides can be purchased at http://www.microscopyeducation.com/fluorrefslides.html), although the thickness and the RI may not be an ideal match to real specimens. Model and colleagues found that concentrated fluorescent dye solutions (e.g., 10% fluorescein) work well because the high concentration of fluorophore replaces photobleached molecules rapidly while absorption limits fluorescence to a thin layer very close to the surface, producing an optical sectioning-like effect (Model and Burkhardt 2001; Model and Blank 2008). The fluorescence reference image should be acquired with the same zoom, image window position, scan speed, detector gain, and objective lens as the specimen images. We recommend acquiring several images of different fields of view to minimize the risk of debris marring the result and averaging the images together (using the median would be even better). It should be fine to adjust the laser power, provided that the setting does not result in wild laser fluctuations (i.e., very low AOTF power setting). We typically do shading correction off-line using ImageJ or MetaMorph. Test that the correction is making an improvement by acquiring images at the same, one-half, and one-quarter laser power compared with your fluorescence reference images. The equation is

$$\text{corrected image} = \text{scaling factor} \times (\text{specimen image(s)} - \text{dark reference})/(\text{fluorescence reference} - \text{dark reference}).$$

This equation applies to both confocal and standard fluorescence microscopy images. If you determine your fluorescence reference images are less than a few percent across the field of view, you may not need to perform correction (unless you are trying to quantify smaller than a few percent differences). If you are using image-processing software that includes a floating-point image type, you can use 1.0 for the scaling factor (e.g., ImageJ). Most imaging software (e.g., MetaMorph) have only integer image types. Typical scaling factors are 100 or 200 for 8-bit images, 1000 for 12-bit images, or 10,000 for 16-bit images. We recommend against using the maximum pixel value of either the specimen image (because it would change for every image) or the fluorescence reference (because it would change every session). In general, avoid breaking the Borisy law: Never rely on a single number. Some digital deconvolution software automatically calculate shading correction for you. For example, with a scaling factor of 1000 (12-bit images), a single image of the fluorescence reference target acquired with the same intensity as the fluorescence reference (average of many images) should be ~1000, one-half laser intensity should be ~500, and one-quarter laser power should be ~250 (reporting these with every quantitative immunofluorescence paper would enable reviewers and readers to assess instrument performance). Perform the test acquisition with the same slide used for the fluorescence reference, at the beginning, the middle, and the end of the imaging session. If the result is not as expected, re-evaluate whether the imaging system is stable (e.g., overnight *x–z* stability test) and whether the image arithmetic was correct.

Quantitative Immunofluorescence Microscopy: New Hardware Options

Most confocal microscopes and many research microscopes are now automated. Many microscope hardware configurations and peripherals are discussed in Chapters 1–12, as well as in more specialized or customized configurations elsewhere in the book (see also Pawley 2006). Two complementary technologies have diverged from the basic microscope stand in the past decade: digital slide imaging (DSI) and high-content screening (HCS) systems. Table 6 summarizes DSI, HCS, and related specialized microscope platforms that may facilitate quantitative immunofluorescence microscopy. Among the DSI and HCS systems, currently the HistoRx PM-2000 scanner coupled with the AQUA score data output comes closest to true quantitation in molecules/cell or molecules/mm^2 of tissue, but even AQUA stops short with an arbitrary 0–255 unitless scale (Camp et al. 2002; their two-plane unsharp mask image processing also implies a lack of true quantitation; AQUA is being ported to 16-bit in 2010, D.L. Rimm, pers. comm.). The few tissue microarray slides needed per publication suggests to us that taking the time to scan in 3D and to perform quantitative deconvolution on each tissue spot is a logical step if true quantitation is desired. Among research microscope systems, we consider the Applied Precision DeltaVision deconvolution workstations and the PerkinElmer/Improvision Volocity deconvolution software to be closest to being able to provide molecular units (as should the new fluorescence nanoscopes). Within the small animal in vivo molecular imaging field, Caliper Life Sciences (formerly Xenogen Corporation) has pioneered the use of units of photons/sec/cm^2/str (steradians, the metric unit for solid angle) in reporting firefly luciferase output when imaging from live mice, dissected organs, or cells in multiwell plates. This has made it possible to compare the results from all of the publications that used the Xenogen IVIS system and reported in these units. The fluorescence molecular tomography platform (FMT; VisEn Medical Systems) comes the closest to reporting real fluorescence quantitation in vivo.

At this time, microscope vendors rarely provide calibration standards and software features, so we encourage users to take up the effort to calibrate DSI and HCS systems. The most pragmatic units might be photons emitted/sec/cell (for a particular excitation power and assuming no saturation, photobleaching, or blinking) because the detector can be calibrated for photons to intensity level, the exposure time is easily found, and the number of cells can be counted, or at least the number of nuclei can be estimated. Even better would be to report in fluorophores/cell (assuming no saturation, etc.), and better still, units could be the number of antigens in each cell (mean and standard deviations) because there is no reason to think that all cells in a population have the same number of antigens. In preparations involving whole cells, cell-surface antigen(s) could be labeled in one color, then permeabilized, then labeled with additional colors. The use of fluorescent protein–target protein fusions (especially if knocked-in to the endogenous gene and shown to be as functional) provides an alternative handle for detection.

Among the current generation of DSI and HCS systems, most use conventional detectors (i.e., standard PMTs in point-scanning confocal systems, and line or area CCD cameras for camera-based systems). A number of interesting new detectors are or will soon be available. The Zeiss LSM710 confocal microscope is available with the new Quasar spectral detector, which Zeiss claims is superior to the META detector. EMCCDs are becoming more common, and each generation appears to be superior to previous generations. An interesting new technology is the hybrid CCD-CMOS (complementary metal-oxide semiconductor), scientific (sCMOS; see www.sCMOS.com and http://hamamatsu cameras.com/orca-flash/index.php) that combines the quality features of a current generation interline CCD (microlenses to achieve good fill factor, ~65% QE, small pixel size, and binning) with the good qualities of a CMOS readout (individual pixels are addressable and have fast low-noise readout, high dynamic range, and large number of pixels), to produce a prototype 5.5 megapixel, 30–100-fps frame rate with sensitivity between that of a conventional scientific-grade CCD (e.g., ORCA-ER or ORCA-R2) and a large pixel back-illuminated EMCCD. If the sCMOS is successful (performance and price between scientific-grade CCD and back-illuminated EMCCD), then the next challenge will be how to deal with a data stream of 5.5 megapixels × 2 byte/pixel × 100 fps = 1.1 gigabyte/sec of raw data. The sCMOS may be an especially good fit to the needs of fluorescence nanoscopy by taking advantage of parallel processing on graphic processor units, to process the data stream into a single final image reporting number of fluorophores per pixel at, for example, 10-nm precision localization over an area of 25 × 20 µm.

TABLE 6. Specialized microscope platforms facilitating quantitative immunofluorescence

Manufacturer	Platform	Hardware	Software
3DHISTECH Ltd./Zeiss	DSI	Mirax Scan	
Amnis		ImageStreamX	IDEAS
Aperio	DSI	ScanScope's (bright field)	
		ScanScope FL (fluorescence)	Spectrum
Applied Precision	–	ArrayWoRx Biochip Reader	
	AM	DeltaVision	
	Nanoscope	OMX, OMX-SR	
Applied Spectral Imaging Ltd	DSI	SD-300 spectral imager	PathEx, ScanView,
		AutoMate Tray Loader	SKYview
BD Biosciences	HCS	Pathway 855 Bioimager	
	HCS	Pathway 435 Bioimager	
	DSI	BD FocalPoint GS Imaging System	
BioView	AM	Duet-3	
Bioimagene	DSI	iScan	
Biomedical Photometrics Inc.	DSI	TissueScope	
CellProfiler.org		–	CellProfiler
CompuCyte Corporation	LSC	iCys, iCyt laser-scanning cytometers	
CRi	DSI	Vectra (Nuance + Inform + 200 slide loader)	Inform
Cyntellect	AM	LEAP	
		Celigo microplate cytometer	
Dako	DSI	ACIS III	
Definiens		–	Cellenger, TissueMap
GE Healthcare	HCS	In Cell 2000	
Genetix Ltd./Applied Imaging	AM	Ariol	SlidePath
Corporation	–	SL-50, GSL-120 slide loader	
		ClonePix	
HistoRX	DSI	PM-2000 Image Analyzer	AQUA
Human Protein Atlas http://www.proteinatlas.org			Tissue microarray resource
Leica	Confocal	TCS LSI Large Scale Imaging	
	DSI	SCN400	
	Nanoscope	STED	
Ludl Electronics Products	–	LEP Slide Handler	
MDS Analytical Technologies	HCS	ImageXpress	
	HCS	IsoCyte	
MelTec GmbH	AM	Toponome Imaging Cycler	Multiepitope ligand cartography
Meyer Instruments, Inc.	–	Pathscan IV	QuickScan, SilverFast Ai
Micro-Manager.org	AMs	Automated microscopes and peripherals	ImageJ add-on
Nikon	AM	CoolScope II	
	Confocal	A1	
	Confocal	C1si	
Olympus/Bacus Laboratories	DSI	NanoZoomer II, RT (Hamamatsu)	NDPExplorer
		VS110 virtual microscopy	BLISS, WebSlide
		FluoView 1000, FV10i	
		WI-CDEVA Condenser TIRF	
Omnyx LLC (GE/UPMC)	DSI	(No product as of early 2010)	
PerkinElmer	HCS	Opera	
PerkinElmer/Improvision	AM	UltraView VOX	
	–	Volocity deconvolution	
Prior Scientific	–	PL-200 slide loader	
Thermo Fisher Scientific, Inc. Cellomics	HCS	ArrayScan, KineticScan	
Ventana Medical Systems, Inc. (Roche)	DSI	VIAS	
Visiopharm	–	–	VIS/Visiomorph
VisiTech International	Confocal	VT-Infinity3, VT-Hawk	
Zeiss	Confocal	LSM710, LSM780	ZEN
	Nanoscope	HR-SIM	
	Nanoscope	PAL-m	

Note: This table makes no claims of completeness.

See Rojo et al. (2006) for a comparison of 31 commercial DSI systems.

DSI, digital slide imaging system; AM, automated microscope; HCS, high-content screening; LSC, laser-scanning cytometer; UPMC, University of Pittsburgh Medical Center.

Guidelines for Quantitative Immunofluorescence Imaging

Quantitation of a fluorescence specimen is not trivial. There are many variables that are not under user control, such as laser power drift over time. The user or confocal microscope manager can test laser power stability (see Zucker and Price 1999 for additional tests). A simple way to gain confidence in the confocal system is to compare intensities of a specimen acquired with 96%, 48%, 25%, 12%, and 6% laser power (keep gain and offset constant). After accounting for the detector offset (dark reference), these should be close to twofold intensity changes.

If you retest by acquiring images of fluorescent beads (e.g., Invitrogen InSpeck beads, Souchier et al. 2004) in different locations in the field of view, you may not get the same answer. This may be for several reasons, including size of the beads, optical sectioning settings, RI of the mounting media, immersion media and the beads, photobleaching if using the same beads many times, and the bead brightness and density (number of beads in field of view). At high bead density, strange results may appear from reflections from one bead bouncing off another bead (and vice versa) to the detector (in the wide-field mode) and increasing apparent brightness compared with lower density (we have also observed reflections with very bright fluorescent tyramide-stained cells as sources). See Barlow and Guerin (2006) for a comparison of quantifying the wide-field versus the OptiGrid structured illumination optical sectioning devices (similar to Zeiss Apotome).

We typically use the 12-bit data mode (Zeiss LSM510 or LSM710, Leica SP5), recognizing the Pawley rule that a bright fluorescent confocal microscope voxel may represent a maximum of 10–20 photons/msec voxel dwell time for standard confocal PMTs (e.g., 63x/1.4-NA objective lens, 1 Airy unit, 0.64 µsec/pixel) (Pawley 2000; Pawley 2006; see also Protasenko et al. 2005). Note that this is the number of photons in a microsecond of scan time, not necessarily the number of fluorophores in the confocal volume.

For large data sets, such as the 32-channel Zeiss LSM710 QUASAR spectral detector, the large stage tiling, the *z*-series, and the time series, using the 8-bit mode reduces memory and disk space usage by twofold because 12-bit data are stored in computer RAM (random-access memory) and on the disk as 16-bit values. For fluorescent tyramide signal amplification, there may be a large number of fluorescent tyramides in the brightest voxels, and none away from the cells, which may call for low (but stable) laser power, low gain, and 12-bit or 16-bit data modes to best take advantage of the dynamic range. Ideally, use the confocal microscope (or EMCCD, scientific-grade CCD, or avalanche photodiode) on specimens in which the imaging system is photon-shot-noise limited, rather than dark current, readout noise, or EM gain noise limited. In a photon shot noise–limited image, the signal-to-noise ratio (SNR) is $\sqrt{n}/n$, where n is the number of photons (not the intensity value). Table 7, column 2 shows the real SNR for reasonable photon signals for a confocal microscope experiment operating in the Pawley rule mode.

TABLE 7. Photon signal-to-noise ratio (SNR) is the real SNR

# photons	Photon SNR	Photon SNR %	8-bit intensity	8-bit apparent SNR	8-bit apparent SNR %	12-bit intensity	12-bit apparent SNR	12-bit apparent SNR %	16-bit intensity	16-bit apparent SNR	16-bit apparent SNR %
0	0.00		0	0.00		0	0.00		0	0.00	
1	1.00	100.0	8	2.83	35	128	11.31	9	2048	45.25	2.21
2	1.41	70.7	16	4.00	25	256	16.00	6	4096	64.00	1.56
3	1.73	57.7	24	4.90	20	384	19.60	5	6144	78.38	1.28
4	2.00	50.0	32	5.66	18	512	22.63	4	8192	90.51	1.10
8	2.83	35.4	64	8.00	13	1024	32.00	3	16384	128.00	0.78
9	3.00	33.3	72	8.49	12	1152	33.94	3	18432	135.76	0.74
15	3.87	25.8	120	10.95	9	1920	43.82	2	30720	175.27	0.57
16	4.00	25.0	128	11.31	9	2048	45.25	2	32768	181.02	0.55
24	4.90	20.4	192	13.86	7	3072	55.43	2	49152	221.70	0.45
25	5.00	20.0	200	14.14	7	3200	56.57	2	51200	226.27	0.44

Table 7 shows that the best case for SNR for a confocal voxel on a new PC with a better light budget than the Pawley rule is 20%. Most commercial confocal microscopes have internal calibration settings adjusted so that approximately the same intensity value is reached for every scan speed. That is, a pixel (voxel) dwell time of 0.64 µsec/pixel and 160 µsec/pixel may produce the same intensity on the computer. The shorter dwell time pixel is collecting fewer photons, but the digitization process multiplies this value to look good on the computer. In evaluating brief versus long dwell time images, the former are much noisier. Brief dwell times may be advantageous if the photophysics of the fluorophore(s) has favorable triplet state recovery rates. Ideally, confocal microscope vendors should provide an option to turn off the dwell time intensity equalization feature, or even better, to implement a photon-counting mode, as on some older confocal microscopes (which unfortunately had grossly inefficient light paths) or are available on time-correlated single-photon-counting (TCSPC) fluorescence lifetime imaging microscopes (e.g., the Becker & Hickl add-on to Zeiss or Leica; see Chapters 28 and 41).

IN VIVO IMMUNOFLUORESCENCE MICROSCOPY OF THE EYE

In vivo imaging of cells expressing fluorescent proteins or fluorescently labeled blood vessels is now routine (e.g., Montet et al. 2005; Fukumura and Jain 2007; Levenson et al. 2008; Gligorijevic et al. 2009; Ogawa et al. 2009). In addition, immunohistochemistry and immunofluorescence of a target antigen using monoclonal antibody therapeutics (e.g., Rituximab [anti-CD20], Bevacizumab [blocks human VEGF-A], Trastuzumab [anti-Her-2/neu (ErbB2)] and the HercepTest) are now routine, as is determining how much antibody is binding both target and nontarget tissues in clinical specimens and research models (see Camp et al. 2002).

What has not been routine is in vivo immunofluorescence at the single-cell scale over long time periods. The eye is an excellent specimen for in vivo imaging because the cornea acts as a cover glass. Scheiffarth et al. (1990) published immunofluorescence images of the retinal fundus after ear-vein injection. Sharma and Shimeld (1997) topically applied fluorescent anti-HSV-1 virus antibody onto the cornea and imaged infected cells in the cornea. Our colleagues at the University of Miami have shown that an excellent site for in vivo imaging is the mouse eye anterior chamber (Speier et al. 2008a,b). They also found that transplanted mouse or human islets stimulate new blood vessel formation and are innervated from nerves of the iris within a few days. A mouse iris has a large enough surface area to support several transplanted islets (ranging in diameter from 50 to 300 µm, with most ~150 µm). The iris and the transplanted tissue can be documented with a 5x dry lens and either a confocal scanner (i.e., pinhole open wide, reflection, and fluorescence channels) or with a wide-field camera on a camera port or through the eyepiece port. For high-resolution in vivo imaging, use of a saline electrophysiology-type dipping water-immersion lens is recommended (e.g., Leica HCX 20x/1.0-NA W [saline], or Olympus XLUMPlanFl 20x/0.95-NA water [saline] lens). Saline electrophysiology dipping lenses can be recognized by their Teflon nosepiece—these lenses are designed for immersion in saline culture medium (be sure to clean the lens according to manufacturer's instructions after each session to avoid salt and protein buildup). Do not use conventional metal nosepiece water-immersion lenses—these are designed for distilled water-immersion imaging through a cover glass.

The eye anterior chamber is an immune privileged site as long as wounds or irritations do not persist. If the cornea turns cloudy, use multiphoton excitation and nondescanned detection (Z. Chen and J. Suzuki, pers. comm.). Depending on the mouse host strain and the transplant source, autografts, allografts, and xenografts can be transplanted and can be accepted (human pancreatic islets in NOD [nonobese diabetic]-SCID [severe combined immunodeficiency] or nude mice), and if desired, rejection can be imaged in real time (e.g., allograft rejection or autograft rejection in the NOD mouse type 1 diabetes model). Using bone marrow transplants of vital dyes (i.e., carboxyfluorescein diacetate, succimidyl ester [CFSE]) or fluorescent protein expressing hematopoietic cells, graft rejection can be imaged by cell type–specific technicolor. The islets produce enough insulin to cure diabetic mice. These findings set the stage for the same group to introduce immune system cells to study autoimmunity in vivo (see Cahalan and Parker 2008, for a review of in vivo cellular immunology). With respect to this chapter, the same group have been using a few microliters of direct-labeled anti-CD antibodies in 10 µL of an Fc-blocking reagent injected directly into the aqueous humor to perform in vivo immunofluorescent immunophenotyping of GFP-labeled cells (M.H. Abdulreda et al., in prep.).

If additional in vivo landmark(s) are needed, we recommend adapting the DiI method (Li et al. 2008) as an inexpensive method for making blood and endothelial cells fluorescent ($0.50/mouse for the DiI fluorophore, standard grade from Sigma-Aldrich). The published protocol is for labeling mouse blood vessels post sacrifice. The protocol has been modified for in vivo imaging by using tail-vein injection perfusion of a smaller volume of DiI (R. Wen, pers. comm.). DiI is a bright orange–red fluorophore with broad excitation (561-nm, 543-nm, or 532-nm excitation works well, but 488-nm light will also excite) and emission (some green, bright orange and red, dim infrared), and can be substituted with DiD or DiR (far red, near infrared), DiO (green), or DiA (large Stokes shift: blue 457-nm excitation, red 590-nm peak emission). DiI and similar dyes can be retained in tissues post fixation by careful selection of permeabilization reagent and concentration, before immunostaining. See Matsubayashi et al. (2008) for the use of digitonin, instead of the more common Triton X-100 or saponin, for DiI-compatible permeabilization.

Protocol A

Immunofluorescence of Tissue Sections for Subcellular Localization of Two or More Antigens

This protocol is adapted from Kim et al. (2006) and has been provided by Nirupa Chaudhari, Michael Sinclair, and Guennadi Dvoriantchikov.

IMAGING SETUP

Any confocal microscope is suitable for immunofluorescence microscopy. To configure the microscope, follow the steps in the section, Procedure: Configuring a Confocal Microscope for Quantitative Immunofluorescence Microscopy. For subcellular localization, use Plan-Apochromat 63x/1.4-NA or 100x/1.4-NA oil-immersion objective lens, optionally with DIC prisms and an analyzer polarizer. Adjust the image offset so that all pixel values are >0 when the laser(s) are unchecked. Green (Alexa Fluor 488 or equivalent) and orange/red (Alexa Fluor 555 or equivalent) are preferred emission colors so that tissue sections can also be evaluated by eye. These require 488-nm and 561-nm (or 532-nm or 543-nm) laser excitation. A UV (351/364-nm) or violet (405-nm) laser is only needed if localization of nuclei is desired. An alternative fluorophore is DRAQ5 or To-Pro-3 excited by a 633-nm or a 647-nm laser.

MATERIALS

CAUTION: See Appendix 6 for proper handling of materials marked with <!>.
See the end of the chapter for recipes for reagents marked with <R>.

Reagents

Animal, source of tissue
IF-blocking solution <R>
Mounting medium suitable for fluorescence microscopy (e.g. 90% glycerol/10% PBS pH 7.6, 100% thiodiethanol or Prolong Gold without DAPI [Invitrogen]) <!>
 Verify that the mounting medium is nonfluorescent at all wavelengths (VectaShield with DAPI and Prolong Gold with DAPI each become fluorescent with prolonged storage).
Optimal cutting temperature (OCT) compound <!>
Paraformaldehyde <!>, 4%
Phosphate-buffered saline (PBS), pH 7.6
Primary antibodies:
 Mouse anti-GFP, diluted 1:1000 (11814460001, Roche)
 Rabbit anti-PLCβ2, diluted 1:1000 (SC-206, Santa Cruz Biotechnology, Inc.)
Secondary antibodies:
 Donkey antirabbit IgG, conjugated to Alexa Fluor 594, diluted 1:1000 (Invitrogen)
 Donkey antimouse IgG, conjugated to Alexa Fluor 488, diluted 1:1000 (Santa Cruz Biotechnology, Inc.)
Sucrose, 30% (w/v)
Triton X-100 <!>, 1% (v/v)

Equipment

Cover glasses, either #1.5, #1, or #0

The #1.5 specification is 160–190-µm thickness, but high-resolution microscopy or nanoscopy (e.g., PALM, STORM) requires 170-µm cover glasses. For critical experiments, use premium-specification cover glass, such as Zeiss 170 ± 5 µm (474030-9000) and measure thickness with calipers or reflection confocal microscopy. The fluorescence nanoscopes may have limited specimen thickness capability (≤20 µm).

Cryostat
Dissection tools
Fluorescence microscopy system
Glass slides or multiwell plates (see Step 8)
Platform oscillator

EXPERIMENTAL METHOD

1. (Optional) If antigen requires it, perfuse the animal intracardially with PBS followed by 4% paraformaldehyde to internally fix the tissues.
2. Dissect the desired tissue.
3. Immerse the tissues in 4% paraformaldehyde for at least 1 h at 4°C.
4. Rinse the tissues of fixative by washing in PBS three times for 20 min each.

   Cryoprotect the tissues by immersing them overnight in 30% (w/v) sucrose at 4°C.
5. Embed the tissues in OCT or another tissue-embedding matrix. Make sure that the tissues are in the proper orientation.
6. Section the tissues on a cryostat to the desired thickness.
7. Either mount the sections on slides, or transfer them to wells filled with PBS to proceed with free-floating sections.
8. After rinsing with PBS, permeabilize the sections for 15 min in 1% (v/v) Triton X-100. For this and subsequent steps, very gentle agitation of free-floating sections on a platform oscillator may be helpful.
9. Block for 2 h in IF-blocking solution.
10. Rinse for 10 min in PBS.
11. Incubate overnight in primary antibody diluted appropriately in IF-blocking solution.
12. Rinse primary antibody from the tissue by washing in PBS three times for 20 min each.
13. Incubate the sections for 2 h in secondary antibody diluted appropriately in IF-blocking solution.
14. Rinse secondary antibody by washing in PBS three times for 20 min each.
15. If sections are free floating, transfer them to slides. Allow sections to dry completely, and then apply mounting medium designed for fluorescent microscopy onto the slide, and mount the coverslip.
16. Image the tissue sections.

DISCUSSION

Figure 4 provides an example of results in which approximately 25 vallate and foliate taste buds (5–15 images) were imaged to quantify expression patterns (e.g., are GFP and PLCβ2 expressed in the same cells?). The antibodies used in Figure 4 are different than the ones used in this protocol.

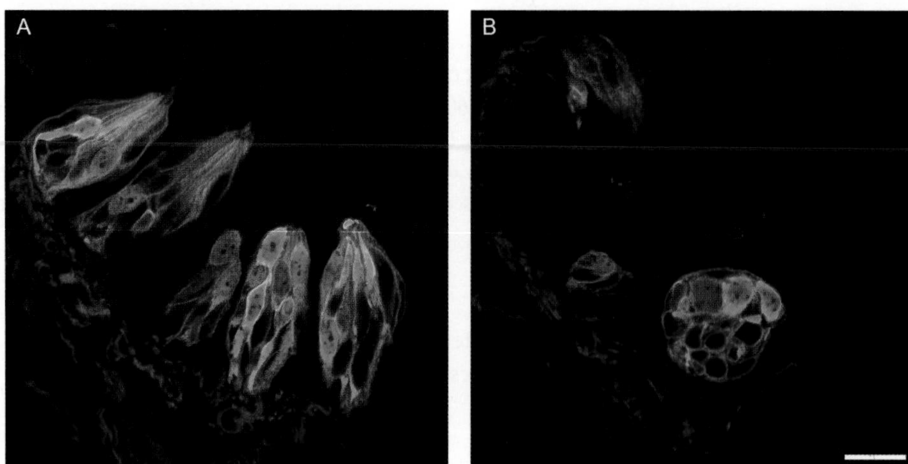

FIGURE 4. Immunofluorescence imaged with confocal microscopy reveals intricate spatial relationships between different types of cells in taste buds. (*A*) Mouse taste buds immunostained for NTPDase2 (red, Alexa Fluor 594), a marker of glial-like taste cells. GFP-labeled receptor cells (green), a distinct taste cell subpopulation; the GFP signal is enhanced by immunostaining for GFP using an Alexa Fluor 488-conjugated secondary antibody. (*B*) Mouse taste buds immunostained as in *A* but viewed in cross section. Here, it can be clearly seen that glial-like cells are distinct from receptor cells but wrap their membranes around the receptor cells, ensheathing them. Scale bar, 20 µm; 1-µm thick optical confocal sections. (Images courtesy of Michael Sinclair and Nirupa Chaudhari, University of Miami.)

For live cell imaging followed by immunofluorescence, consider using 35-mm imaging dishes (e.g., P35G-1.5-10-C or P35G-1.5-20-C from www.glassbottomdishes.com; FluoroDish http://www.wpi-inc.com/index.php/vmchk/FD35-100.html, and CELLview http://www.gbo.com/en/index_3455.php). Smaller diameter dishes hold fewer cells but offer some advantages. For example, smaller reagent volumes are required, which for expensive antibody reagents can provide a considerable savings. A smaller imaging area also means less time searching for cells worth imaging. All small-volume incubation steps should be performed in a 100% humidity environment, or at least, cover the dish with something that will reduce evaporation (e.g., Parafilm). The wash steps can be performed using the entire dish volume. If loss of cells during washing is a concern, acquire bright-field or phase-contrast images, and count the cells. Cells can be retained by trapping them under a thin sheet of nonfluorescing agarose. In that case, account for additional antibody dilution and longer diffusion times by extending the incubation and washing times (Yumura and Fukui 1985; http://dictybase.org/techniques/geneex/agaroverlay.htm). The agarose can be left on throughout the experiment or can be removed gently (verify that most cells remain attached to the cover glass), and standard mounting medium can be applied. The mounting medium can then optionally be overlaid by a standard cover glass, being careful to avoid trapping air bubbles.

Protocol B

Tyramide Signal Amplification of Two-Antigen Immunofluorescence of Tissue Sections

This protocol is adapted from H. Takahashi et al. (in prep.). Tissue is fixed, embedded, and sectioned, and then is incubated simultaneously with two primary antibodies, raised in different species. One species-specific HRP-polymer secondary antibody is then applied. The signal from the HRP-conjugated antibody is amplified using fluorophore-conjugated tyramide. HRP is then inactivated, and the process is repeated beginning with the second species-specific HRP secondary antibody. A variation of this protocol will also work with two or more primary antibodies of different isotypes (e.g., mouse IgG1 and mouse IgG2a). In this case, one primary antibody is applied, followed by the detection and inactivation steps. Then, the next primary antibody is applied, is detected, and so on.

MATERIALS

CAUTION: See Appendix 6 for proper handling of materials marked with <!>.

Reagents

Alexa 488-conjugated tyramide and Alexa 647-conjugated tyramide (Invitrogen)
Antibody diluent (e.g., Van Gogh Yellow, Biocare Medical, Inc.)
Antigen retrieval reagent, pH 9.5 (e.g., Borg Decloaker, Biocare Medical, Inc.)

See antibody manufacturer's recommendation for optimal buffer reagent and pH—the majority of antigens are retrieved with citrate buffer pH 6.0. The antigens detected in this protocol were retrieved with Tris buffer pH 9.5 (http://www.biocare.net/decloaking-chamber.html has Tris buffer, pH 9.5).

Ethanol <!> Tris buffer (e.g., 10%, 25%, 50%, 70%, 95% ethanol; see Vitha et al. 2010)
Formalin <!>, 10% neutral buffered
Hydrogen peroxide<!>, 3%
Mounting medium suitable for fluorescence microscopy (e.g., ProLong, Invitrogen) <!>
Normal goat serum, 1%
PBS, pH 7.6
Peroxidase-blocking reagent (PeroxAbolish, Biocare Medical, Inc.)
Primary antibodies:
- Mouse monoclonal antihuman CD4 antibody (mouse clone BC/1F6, IgG1, Biocare Medical. Inc.)
- Rabbit polyclonal antihuman CD8 (Abcam plc)

Propidium iodide diluted 1:50 with antibody diluent (Invitrogen)

Alternative DNA counterstains can be used. See Step 18.

Secondary antibodies:
- IgG (anti-CD4), polymer-HRP-conjugated antimouse secondary antibody (EnVision, Dako)
- Polymer-HRP-conjugated antirabbit secondary antibody (EnVision, Dako)

Sodium azide <!> (optional; see Step 5)
Tissue to be imaged
Xylene <!>

Equipment

Coated glass slides (SuperFrost Plus, positively electrostatically charged slides; available from many suppliers)
Coverslips
Fluorescence microscopy system
Humidified 4°C chamber
Tissue processor (e.g., Tissue-Tek Express, Sakura Color Products Corporation) or materials to process tissue manually

EXPERIMENTAL METHOD

1. Fix tissue in 10% neutral buffered formalin for 24 h.
2. Process the tissue, ideally using a rapid tissue processor. This includes embedding the tissue in a paraffin block and slicing it into serial 4-µm sections.
3. Place the sections on a coated glass slide, and bake them in an oven for 30 min.
4. Deparaffinize the sections with xylene, and rehydrate them using an ethanol series into PBS.
5. Block endogenous peroxidase activity (e.g., hemoglobin) using a peroxidase-blocking reagent.

   Alternative methods to inactivate peroxidase activity include 3% H_2O_2 or sodium azide.
6. Perform antigen retrieval by placing the slides into a pressure cooker with Borg Decloaker or another antigen retrieval reagent at pH 9.5. Incubate for 10 min at 120 psi.
7. Block protein by incubating the slides in 1% normal goat serum for 20 min.
8. Dilute mouse monoclonal antihuman CD4 antibody 1:25 with antibody diluent. Also, dilute rabbit polyclonal antihuman CD8 1:50 with antibody diluent.
9. Mix the diluted antibodies together, immerse the tissue sections in the antibody solution, and incubate them overnight at 4°C in a humidified chamber. Wash extensively—at least five times (Burry 2010 recommends seven times)—and keep tissue fully submerged during changes.
10. Remove the primary antibodies, add mouse IgG (anti-CD4), polymer-HRP-conjugated anti-mouse secondary antibody, and incubate for 45 min at room temperature. Wash extensively—at least five times (Burry 2010 recommends seven times)—and keep tissue fully submerged during changes.
11. Add Alexa 647-conjugated tyramide plus 0.03% H_2O_2, and incubate for 10 min at room temperature.
12. Remove secondary and fluorophore-conjugated antibodies. Inactivate HRP activity with a peroxidase blocker for 30 min at room temperature.
13. Add polymer-HRP-conjugated anti-rabbit secondary antibody, and incubate for 45 min at room temperature. Wash extensively—at least five times (Burry 2010 recommends seven times)—and keep tissue fully submerged during changes.
14. Add Alexa 488-conjugated tyramide plus 0.03% H_2O_2, and incubate for 10 min at room temperature.
15. (Optional) If additional rounds of tyramide immunofluorescence are to be performed, repeat Step 12 followed by appropriate HRP-secondary antibodies and desired fluorophore-conjugated tyramide.
16. (Optional) Repeat Steps 8–12 using additional primary and HRP-polymer-secondary antibodies.
17. (Optional) Apply conventional primary antibodies and fluorophore-labeled secondary antibodies, being careful to use reagents that will not bind those used in previous cycles.

18. Counterstain the nucleus by incubating the tissue sections with propidium iodide diluted 1:50 with antibody diluent for 10 min.

    Propidium iodide is preferred for the laser-scanning cytometry. DAPI, Hoechst 333258, or Hoechst 33342 (at 1 μg/mL) would be used more often as a DNA stain when imaging with a fluorescence microscope or a DSI system.

19. Add mounting medium, and apply a coverslip to the stained slide. Store the slides in the dark at 4°C until they are used for imaging.

    Prolong and Prolong Gold solidify in one to a few days, although sometimes it takes longer. When using hardening mounting medium such as Prolong, Prolong Gold, or VectaMount (Vector Laboratories), or when sealing with nail polish, be sure that the mountant or sealant has completely solidified before imaging on a microscope. This avoids ruining lenses by transferring mountant onto the objective.

20. Image the immunofluorescence slides as soon as possible after slide preparation.

DISCUSSION

Takahashi et al. (2009) have published a brief summary of this method without technical details. For additional protocols using HRP tyramide signal amplification immunofluorescence (TSA-IF), along with detailed discussions, see van Gijlswijk et al. (1997); van Tine et al. (2005); Toth et al. (2007); Toth and Mezey (2007); van der Loos (2008). Note that the same HRP-secondary antibody or HRP-polymer-secondary antibody can be used for both tyramide signsal amplification immunofluorescence (TSA-IF) and conventional immunohistochemistry (IHC) using diaminobenzidine (DAB) or another HRP chromogenic substrate, with hematoxylin counterstaining. This means that serial section TSA-IF and DAB-IHC could be performed (e.g., Ince et al. 2008). If these two methods are combined, it may be necessary, however, to adjust the dilution of the primary and/or HRP-secondary antibodies for the tyramide detection because TSA-IF is more sensitive than DAB-IHC. The tyramide reagent is also more expensive than DAB, so avoid using more than is necessary per slide. If a relatively large volume is needed to cover a slide, try diluting the tyramide reagent and extending the incubation time, or repeat the application of dilute solutions.

Takahashi et al. (2009, and in prep.) used formalin-fixed paraffin-embedded tissue sections of human tonsils as a positive control (Fig. 5). Their primary imaging system was an iCys laser-scanning cytometer (CompuCyte Corporation). Similar (or better) quality data can be obtained from a conventional research microscope or a confocal microscope, optionally equipped with a motorized stage. State-of-the-art tissue imaging is exemplified by the newest generation DSI systems, such as

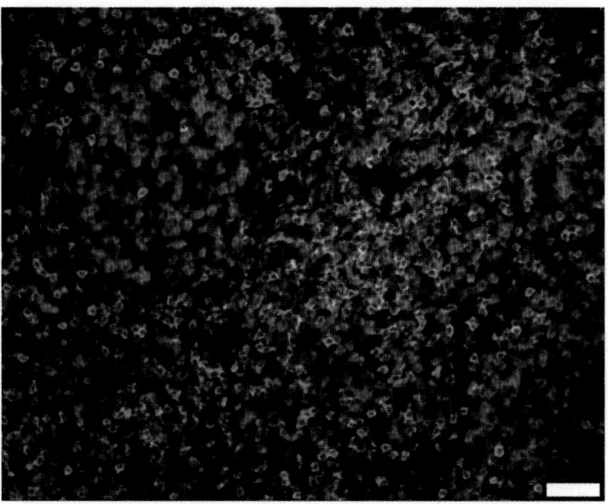

FIGURE 5. Double immunofluorescent staining on formalin-fixed paraffin-embedded human tonsil section stained for CD4 (Alexa 647-tyramide, purple) and CD8 (Alexa 488-tyramide, green). Specimen was counterstained with propidium iodide (not shown). Courtesy of Professor H. Ichii (see Takahashi et al. 2009, and in prep.). Scale bar, 50 μm.

the 3DHistech Mirax Scan, Aperio ScanScope, Bioimagene iScan, and Hamamatsu NanoZoomer (distributed in the United States by Olympus/Bacus Laboratories, Inc.), all of which are available with fluorescence scanning capability. Each system can scan more than 50 slides overnight. A typical DSI system uses an Olympus Plan-Apochromat 20x/0.75-NA objective lens, which enables ~0.5-μm *x–y* resolution and should be used with an ~4-μm or thinner tissue section. The DSI systems also benefit from image-processing and analysis software to automatically analyze all the data. A confocal microscope with a motorized stage can be used for stage tiling. Depending on spatial resolution and time constraints, either image at full *x–y* resolution at a confocal pinhole setting of 1 Airy unit and optimal *z*-step size, or open the pinhole to increase light collection and to enable the use of a larger *z*-step size. Additional tips for confocal tile scanning include: Make sure the microscope stage and slide are as flat as possible, use a low *x–y* resolution (e.g., 128 x 128-pixel tiles), do not average, use a single-channel scout scan, and optimize the detection channels and laser power to scan simultaneously. A motorized stage confocal microscope can also be used to image hematoxylin and eosin or IHC slides using sequential tracks of two or more lasers to the transmitted light detector (for hematoxylin and diaminobenzidine, try 633-nm and 457-nm laser lines, respectively, coloring red and cyan for a red–green–blue [RGB] display). This is particularly useful for comparing serial sections of conventional HRP/H_2O_2/DAB immunohistochemistry with HRP/H_2O_2/fluorescent tyramide detection.

PUBLISHING IMMUNOFLUORESCENCE MICROSCOPY IMAGES

After all of the effort preparing samples, tittering antibodies, optimizing protocols, and obtaining images, the resulting images are likely to become postage-stamp-sized figures in a journal article. We recommend that you include images of a usable size in the supplemental file and urge you to publish the single-channel images in monochrome gray. With three-color immunofluorescence images that also contain DAPI counterstain, publish the immunochannels in RGB (or cyan, magenta, and yellow), and use Adobe Photoshop or image-processing software to make the DAPI-stained nuclei either with cyan or with white outlines. We are wary of the argument that colocalization of two antigens occurs wherever red and green pixels overlap to form yellow ones (see Fig. 5a–d of Bennett et al. 2009 for one of the many ways yellow pixels can mislead).

Many scientific publishers now accept RGB image files as artwork. We recommend 300-dpi (dots per inch) or 600-dpi uncompressed TIFF images. Note that a confocal micrograph with 1-µm pixel size is 25,400 pixels/in and would print as such. Input whatever dpi value is appropriate for your printer. Many journals still ask that hard copies of color artwork be provided to guide the printers. Perfectly good prints can be made on an inexpensive inkjet printer if the right paper and settings are used. We have obtained nice results with "HP premium photo paper, glossy," specifying paper type as such, selecting Print Quality = Best, and keeping the paper in the plastic sleeve until ready to print to avoid edge of ink problems. High-resolution image display may eventually require switching from paper to the Internet to be able to explore the image data at full resolution. For example, see the DSIs at http://imagearchive.compmed.ucdavis.edu/publications/Shachaf/figure1.py (Shachaf et al. 2004).

CONCLUSIONS

Immunofluorescence microscopy will continue to be a critical research tool. The reagents, protocols, and instrumentation have come a long way since Coons et al. (1942). The ambition of Coons et al. was to detect bacteria by specific fluorescent-antibody–antigen binding on the surface of bacteria in 10-µm liver tissue sections. Although the technology has evolved significantly, the ambition of most current users of immunofluorescence is similar: Detect the presence or absence and cellular or subcellular localization of one or more antigens. With the appropriate microscope(s), a user can go from imaging single fluorescent antibody-labeled cells almost 1-mL deep in a live animal to precisely localizing and counting single molecules on and in a live cell. With the aid of a macroscope or an in vivo imaging system, fluorescently labeled cells can be quantified throughout a live mouse. Bioluminescence in vivo imaging systems has been routinely calibrated in absolute units (photons/sec/cm^2/str), but recent improvements in in vivo scattering, absorption, and tomographic modeling provide more precise and accurate estimates of source size and intensity. Now that fluorescence microscopes—really nanoscopes—can count and can localize single molecules with near-nanometer precision, there is no further downscale to reach. Therefore, for the future direction in immunofluorescence, we look forward to the improvement of the quantitation of immunofluorescence data to routinely report number of molecules either per pixel, μm^2, voxel, μm^3, macromolecular complex, organelle, cell, or tissue compartment.

Fluorescence microscopy works best with a bright signal and a black background. Whether the background is from highly autofluorescent cover glasses or slides, nonspecific primary antibody–secondary antibody complexes, nonspecific binding of secondary antibody, or fluorescent components of the mounting medium, bright background is bad. Black background is beautiful; work to achieve this consistently. Modern confocal microscopes and digital camera widefield microscopes are sensitive enough that endogenous autofluorescence of cells or tissue should be detectable above the electronic noise level. You should be able to show autofluorescent cells as negative controls, in your lab meetings and manuscripts, with proper acquisition settings and contrast adjustments.

RECIPE

Recipes for reagents marked with <R> are included in this list.

IF-Blocking Solution

2% (w/v) BSA
2% (v/v) reconstituted normal serum from the species in which the secondary antibody was obtained

ACKNOWLEDGMENTS

We apologize to our many mentors, to colleagues who have helped us with immunofluorescence and microscopy, and to the authors of the many papers we did not have space to discuss or cite. We thank Zbigniev Iwinski of Zeiss for information about the PMT voltage equation on page 245 and Justin Price (HIHG, University of Miami) for proofreading. When citing the protocols used in this chapter, please cite, by name, the contributors of the protocol.

REFERENCES

Auerbach R, Arensman R, Kubai L, Folkman J. 1975. Tumor-induced angiogenesis: Lack of inhibition by irradiation. *Int J Cancer* **15:** 241–245.

Bailey RE, Nie S. 2003. Alloyed semiconductor quantum dots: Tuning the optical properties without changing the particle size. *J Am Chem Soc* **125:** 7100–7106.

Bailey RE, Strausburg JB, Nie S. 2004. A new class of far-red and near-infrared biological labels based on alloyed semiconductor quantum dots. *J Nanosci Nanotech* **2004:** 569–574.

Barlow AL, Guerin CJ. 2006. Quantization of widefield fluorescence images using structured illumination and image analysis software. *Microsc Res Tech* **70:** 76–84.

Barr VA, Bunnell SC. 2009. Interference reflection microscopy. *Curr Prot Cell Biol* **4.23:** 1–19.

Bennett BT, Bewersdorf J, Knight KL. 2009. Immunofluorescence imaging of DNA damage response proteins: Optimizing protocols for super-resolution microscopy. *Methods* **48:** 63–71.

Bernas T, Barnes D, Asem EK, Robinson JP, Rajwa B. 2007. Precision of light intensity measurement in biological optical microscopy. *J Microsc* **226:** 163–174.

Bogen SA, Vani K, McGraw B, Federico V, Habib I, Zeheb R, Luther E, Tristram C, Sompuram SR. 2009. Experimental validation of peptide immunohistochemistry controls. *Appl Immunohistochem Mol Morphol* **17:** 239–246.

Burry RW. 2000. Specificity controls for immunocytochemical methods. *J Histochem Cytochem* **48:** 163–166.

Burry RW. 2010. *Immunocytochemistry: A practical guide for biomedical research*. Springer Science+Business Media, New York.

Cahalan MD, Parker I. 2008. Choreography of cell motility and interaction dynamics imaged by two-photon microscopy in lymphoid organs. *Annu Rev Immunol* **26:** 585–625.

Camp RL, Chung GG, Rimm DL. 2002. Automated subcellular localization and quantification of protein expression in tissue microarrays. *Nat Med* **8:** 1323–1327.

Cho EH, Lockett SJ 2006. Calibration and standardization of the emission light path of confocal microscopes. *J Microsc* **223:** 15–25.

Coons AH, Creech HJ, Jones RN, Berliner E. 1942. The demonstration of pneumococcal antigen in tissue by the use of fluorescent antibody. *J Immunol* **45:** 159–170.

Couchman JR 2009. Commercial antibodies: The good, bad, and really ugly. *J Histochem Cytochem* **57:** 7–8.

Danuser G, Waterman-Storer CM. 2003. Quantitative fluorescent speckle microscopy: Where it came from and where it is going. *J Microsc* **211:** 191–207.

Digman MA, Dalal R, Horwitz AF, Gratton E. 2008. Mapping the number of molecules and brightness in the laser scanning microscope. *Biophys J* **94:** 2320–2332.

Fritschy JM. 2008. Is my antibody-staining specific? How to deal with pitfalls of immunohistochemistry. *Eur J Neurosci* **28:** 2365–2370.

Fukumura D, Jain RK. 2007. Tumor microvasculature and microenvironment: Targets for anti-angiogenesis and normalization. *Microvasc Res* **74:** 72–84.

Furtado A, Henry R. 2002. Measurement of green fluorescent protein concentration in single cells by image analysis. *Anal Biochem* **310:** 84–92.

Gligorijevic B, Kedrin D, Segall JE, Condeelis J, van Rheenen J. 2009. Dendra2 photoswitching through the Mammary Imaging Window. *JoVE* 28. http://www.jove.com/index/details.stp?id=1278, doi: 10.3791/1278.

Gordon A, Colman-Lerner A, Chin TE, Benjamin KR, Yu RC, Brent R. 2007. Single-cell quantification of molecules and rates using open-source microscope-based cytometry. *Nat Methods* **4:** 175–181.

Grunwald D, Shenoy SM, Burke S, Singer RH. 2008. Calibrating excitation light fluxes for quantitative light microscopy in cell biology. *Nat Protoc* **3:** 1809–1814.

Hermiston ML, Latham CB, Gordon JI, Roth KA. 1992. Simultaneous localization of six antigens in single sections of transgenic mouse intestine using a combination of light and fluorescence microscopy. *J Histochem Cytochem* **40:** 1283–1290.

Holmseth S, Lehre KP, Danbolt NC. 2006. Specificity controls for immunocytochemistry. *Anat Embryol* **211:** 257–666.

Ince TA, Ward JM, Valli VE, Sgroi D, Nikitin AY, Loda M, Griffey SM, Crum CP, Crawford JM, Bronson RT, Cardiff RD. 2008. Do-it-yourself (DIY) pathology. *Nat Biotechnol* **26:** 978–979.

Ji N, Shroff H, Zhong H, Betzig E. 2008. Advances in the speed and resolution of light microscopy. *Curr Opin Neurobiol* **18:** 605–616.

Joglekar AP, Salmon ED, Bloom KS. 2008. Counting kinetochore protein numbers in budding yeast using genetically encoded fluorescent proteins. *Methods Cell Biol* **85:** 127–151.

Kai K, Kitajima Y, Hiraki M, Satoh S, Tanaka M, Nakafusa Y, Tokunaga O, Miyazaki K. 2007. Quantitative double-fluorescence immunohistochemistry (qDFIHC), a novel technology to assess protein expression: A pilot study analyzing 5-FU sensitive markers thymidylate synthase, dihydropyrimidine dehydrogenase and orotate phosphoribosyl transferases in gastric cancer tissue specimens. *Cancer Lett* **258:** 45–54.

Kim JW, Roberts C, Maruyama Y, Berg S, Roper S, Chaudhari N. 2006. Faithful expression of GFP from the PLCβ2 promoter in a functional class of taste receptor cells. *Chem Senses* **31:** 213–219.

Langer R, Brem H, Falterman K, Klein M, Folkman J. 1976. Isolations of a cartilage factor that inhibits tumor neovascularization. *Science* **193:** 70–22.

Levenson RM, Lynch DT, Kobayashi H, Backer JM, Backer MV. 2008. Multiplexing with multispectral imaging: From mice to microscopy. *ILAR J* **49:** 78–88.

Li Y, Song Y, Zhao L, Gaidosh G, Laties AM, Wen R. 2008. Direct labeling and visualization of blood vessels with lipophilic carbocyanine dye DiI. *Nat Protoc* **3:** 1703–1708.

Limozin L, Sengupta K. 2009. Quantitative reflection interference contrast microscopy (RICM) in soft matter and cell adhesion. *Chemphyschem* **10:** 2752–2768.

Loudet A, Burgess K. 2007. BODIPY dyes and their derivatives: Syntheses and spectroscopic properties. *Chem Rev* **107:** 4891–4932.

Matsubayashi Y, Iwai L, Kawasaki H. 2008. Fluorescent double-labeling with carbocyanine neuronal tracing and immunohistochemistry using a cholesterol-specific detergent digitonin. *J Neurosci Methods* **174:** 71–81.

McNamara G, Gupta A, Reynaert J, Coates TD, Boswell C. 2006. Spectral imaging microscopy web sites and data. *Cytometry A* **69:** 863–871.

Mighell AJ, Hume WJ, Robinson PA. 1998. An overview of the complexities and subtleties of immunohistochemistry. *Oral Dis* **4:** 217–223.

Model MA, Blank JL. 2008. Concentrated dyes as a source of two-dimensional fluorescent field for characterization of a confocal microscope. *J Microsc* **229:** 12–16.

Model MA, Burkhardt JK. 2001. A standard for calibration and shading correction of a fluorescence microscope. *Cytometry* **44:** 309–316.

Montet X, Ntziachristos V, Grimm J, Weissleder R. 2005. Tomographic fluorescence mapping of tumor targets. *Cancer Res* **65:** 6330–6336.

Namimatsu S, Ghazizadeh M, Sugisaki Y. 2005. Reversing the effects of formalin fixation with citraconic anhydride and heat: A universal antigen retrieval method. *J Histochem Cytochem* **53:** 3–11.

Niswender KD, Blackman SM, Rohde L, Magnuson MA, Piston DW. 1995. Quantitative imaging of green fluorescent protein in cultured cells: Comparison of microscopic techniques, use in fusion proteins and detection limits. *J Microsc* **180:** 109–116.

Ogawa M, Kosaka N, Choyke PL, Kobayashi H. 2009. In vivo molecular imaging of cancer with a quenching near-infrared fluorescent probe using conjugates of monoclonal antibodies and indocyanine green. *Cancer Res* **69:** 1268–1272.

Oomen LC, Sacher R, Brocks HH, Zwier JM, Brakenhoff GJ, Jalink K. 2008. Immersion oil for high-resolution live-cell imaging at 37°C: Optical and physical characteristics. *J Microsc* **232:** 353–361.

Pawley J. 2000. The 39 steps: A cautionary tale of quantitative 3-D fluorescence microscopy. *Biotechniques* **28:** 884–886, 888.

Pawley JB, ed. 2006. *Handbook of biological confocal microscopy*, 3rd ed. Springer, New York.

Perfetto SP, Chattopadhyay PK, Roederer M. 2004. Seventeen-colour flow cytometry: Unravelling the immune system. *Nat Rev Immunol* **4:** 648–655.

Press MF, Hung G, Godolphin W, Slamon DJ. 1994. Sensitivity of HER-2/neu antibodies in archival tissue samples: Potential source of error in immunohistochemical studies of oncogene expression. *Cancer Res* **54:** 2771–2777.

Protasenko V, Hull KL, Kuno M. 2005. Demonstration of a low-cost, single-molecule capable, multimode optical microscope. *Chem Educ* **10:** 1–19.

Rogers AB, Cormier KS, Fox JG. 2006. Thiol-reactive compounds prevent nonspecific antibody binding in immunohistochemistry. *Lab Invest* **86:** 526–533.

Rojo MG, García GB, Mateos CP, García JG, Vicente MC. 2006. Critical comparison of 31 commercially available digital slide systems in pathology. *Int J Surg Pathol* **14:** 285–305.

Russ C, Russ JC. 2007. *The image processing handbook*, 5th ed. CRC/Taylor & Francis, Boca Raton, FL.

Saper CB. 2005. An open letter to our readers on the use of antibodies. *J Comp Neurol* **493:** 477–478.

Saper CB. 2009. A guide to the perplexed on the specificity of antibodies. *J Histochem* Cytochem **57:** 1–5.

Saper CB, Sawchenko PE. 2003. Magic peptides, magic antibodies: Guidelines for appropriate controls for immunohistochemistry. *J Comp Neurol* **465:** 161–163.

Scheiffarth OF, Zrenner E, Disko R, Stefani FH, Brabander B. 1990. Intraocular in vivo immunofluorescence. *Invest Ophthalmol Vis Sci* **31:** 272–276.

Schubert W, Gieseler A, Krusche A, Hillert R. 2009. Toponome mapping in prostate cancer: Detection of 2000 cell surface protein clusters in a single tissue section and cell type specific annotation by using a three symbol code. *J Proteome Res* **8:** 2696–2707.

Shachaf CM, Kopelman AM, Arvanitis C, Karlsson A, Beer S, Mandl S, Bachmann MH, Borowsky AD, Ruebner B, Cardiff RD, et al. 2004. MYC inactivation uncovers pluripotent differentiation and tumour dormancy in hepatocellular cancer. *Nature* **431:** 1112–1117.

Shapiro HM. 2003. *Practical flow cytometry*, 4th ed. Wiley-Liss, New York.

Sharma A, Shimeld C. 1997. In vivo immunofluorescence to diagnose herpes simplex virus keratitis in mice. *Brit J Ophthalmol* **81:** 785–788.

Souchier C, Brisson C, Batteux B, Robert-Nicoud M, Bryon PA. 2004. Data reproducibility in fluorescence image analysis. *Meth Cell Sci* **25:** 195–200.

Speier S, Nyqvist D, Cabrera O, Yu J, Molano RD, Pileggi A, Moede T, Köhler M, Wilbertz J, Leibiger B, et al. 2008a. Noninvasive in vivo imaging of pancreatic islet cell biology. *Nat Med* **14:** 574–578.

Speier S, Nyqvist D, Köhler M, Caicedo A, Leibiger IB, Berggren PO. 2008b. Noninvasive high-resolution in vivo imaging of cell biology in the anterior chamber of the mouse eye. *Nat Protoc* **3:** 1278–1286.

Staudt T, Lang MC, Medda R, Engelhardt J, Hell SW. 2007. 2,2′-thiodiethanol: A new water soluble mounting medium for high resolution optical microscopy. *Microsc Res Tech* **70:** 1–9.

Takahashi H, Ruiz P, Ricordi C, Miki A, Mita A, Barker S, Tzakis A, Ichii H. 2009. In situ quantitative immunoprofiling of regulatory T cells using laser scanning cytometry. *Transplant Proc* **41:** 238–239.

Taylor CR, Levenson RM. 2006. Quantification of immunohistochemistry—Issues concerning methods, utility and semiquantitative assessment II. *Histopathology* **49:** 411–424.

Toth ZE, Mezey E. 2007. Simultaneous visualization of multiple antigens with tyramide signal amplification using antibodies from the same species. *J Histochem Cytochem* **55:** 545–554.

Toth ZE, Shahar T, Leker R, Szalayova I, Bratincsák A, Key S, Lonyai A, Németh K, Mezey E. 2007. Sensitive detection of GFP utilizing tyramide signal amplification to overcome gene silencing. *Exp Cell Res* **313:** 1943–1950.

Tsurui H, Nishimura H, Hattori S, Hirose S, Okumura K, Shirai T. 2000. Seven-color fluorescence imaging of tissue samples based on Fourier spectroscopy and singular value decomposition. *J Histochem Cytochem* **48:** 653–662.

Umezawa H, Matsui A, Nakamura Y, Citterio D, Suzuki K. 2009. Bright, color-tunable fluorescent dyes in the VIS/NIR region: Establishment of new "tailor-made" multicolor fluorophores based on borondipyrromethene. *Chemistry* **15:** 1096–1106.

van der Loos CM. 2008. Multiple immunoenzyme staining: Methods and visualizations for the observation with spectral imaging. *J Histochem Cytochem* **56:** 313–328.

van der Neut Kolfschoten M, Schuurman J, Losen M, Bleeker WK, Martínez-Martínez P, Vermeulen E, den Bleker TH, Wiegman L, Vink T, Aarden LA, et al. 2007. Anti-inflammatory activity of human IgG4 antibodies by dynamic Fab arm exchange. *Science* **317:** 1554–1557.

van Gijlswijk RP, Zijlmans HJ, Wiegant J, Bobrow MN, Erickson TJ, Adler KE, Tanke HJ, Raap AK. 1997. Fluorochrome-labeled tyramides: Use in immunocytochemistry and fluorescence in situ hybridization. *J Histochem Cytochem* **45:** 375–382.

Van Tine BA, Broker TR, Chow LT. 2005. Simultaneous in situ detection of RNA, DNA, and protein using tyramide-coupled immunofluorescence. *Methods Mol Biol* **292:** 215–230.

Vitha S, Bryant VB, Zwa A, Holzenburg A. 2010. 3D confocal imaging of pollen. *Microscopy Today* **18:** 26–28.

Vize PD, McCoy KE, Zhou X. 2009. Multichannel wholemount fluorescent and fluorescent/chromogenic in situ hybridization in *Xenopus* embryos. *Nat Prot* **4:** 975–983.

Ward JM. 2004. Controls for immunohistochemistry: Is "brown" good enough? *Toxicol Pathol* **32:** 273–274.

Wilson CA, Ramos L, Villaseñor MR, Anders KH, Press MF, Clarke K, Karlan B, Chen JJ, Scully R, Livingston D, et al. 1999. Localization of human BRCA1 and its loss in high-grade, non-inherited breast carcinomas. *Nat Genet* **21:** 236–240.

Wu JQ, Pollard TD. 2005. Counting cytokinesis proteins globally and locally in fission yeast. *Science* **310:** 310–314.

Wu JQ, McCormick CD, Pollard TD. 2008. Counting proteins in living cells by quantitative fluorescence microscopy with internal standards. *Methods Cell Biol* **89:** 253–273.

Young IT, Garini Y, Vermolen B, Liqui Lung G, Brouwer G, Hendrichs S, el Morabit M, Spoelstra J, Wilhelm E, Zaal M. 2006. Absolute fluorescence calibration. *Proc SPIE* **6088:** 1–9.

Yoo EM, Wins LA, Chan LA, Morrison SL. 2003. Human IgG2 can form covalent dimers. *J Immunol* **170:** 3134–3138.

Yumura S, Fukui Y. 1985. Reversible cyclic AMP-dependent change in distribution of myosin thick filaments in *Dictyostelium*. *Nature* **314:** 194–196.

Zucker RM, Price OT. 1999. Practical confocal microscopy and the evaluation of system performance. *Methods* **18:** 447–458.

WWW RESOURCES

http://dictybase.org/techniques/geneex/agaroverlay.htm Agar-overlay technique for *Dictyostelium* contributed by Yoshio Fukui, July 2000, Northwestern University Medical School, Chicago, IL.

http://www.mcb.arizona.edu/ipc/fret/default.htm Fluorescent Spectra: An Interactive Exploratory Database.

http://www.invitrogen.com/site/us/en/home/References/Molecular-Probes-The-Handbook.html *The handbook—A guide to fluorescent probes and labeling technologies* is a comprehensive resource for fluorescence technology and its applications. Invitrogen Corporation.

http://micro.magnet.fsu.edu/ Molecular Expressions website, maintained by the Graphics & Web Programming Team in collaboration with Optical Microscopy at the National High Magnetic Field Laboratory, ©1995–2010 Michael W. Davidson and The Florida State University.

http://www.mpibpc.mpg.de/groups/hell/STED_Dyes.html Fluorescent dyes used in STED (stimulated emission depletion) microscopy, Stefan Hell laboratory, Max Planck Institute for Biophysical Chemistry.

http://www.researchd.com/rdikits/subbk23.htm Properties of Human IgG subclasses, RDI Division of Fitzgerald Industries Intl.

http://www.sylvester.org/AICF/PubSpectra PubSpectra Data zip file archive contains data for more than 2000 fluorophores, filters, filter sets, and illuminators for biomedical fluorescence microscopy and flow cytometry.

16 Immunoimaging
Studying Immune System Dynamics Using Two-Photon Microscopy

Melanie P. Matheu,[1] Michael D. Cahalan,[1,2] and Ian Parker[1,3]

[1]Department of Physiology and Biophysics, University of California, Irvine, California 92697; [2]Center for Immunology, University of California, Irvine, California 92697; [3]Department of Neurobiology and Behavior, University of California, Irvine, California 92697

ABSTRACT

Cells of the immune system explore a wider territory than any other cells in our bodies. Responses to a pathogen typically require long-range migration of cells, short-range communication by local chemical signaling, and direct cell–cell contact. The adoption of two-photon microscopy by immunologists has, for the first time, allowed these processes to be visualized within native tissue environments. Immunoimaging is rapidly developing from a merely descriptive technique into a set of methods and analytical tools that can be used to quantify and to characterize an immune response at the cellular level. Here we outline the hardware requirements for immunoimaging, describe methods of cell labeling, describe immune response induction, outline preparations for imaging in tissue explants and in vivo murine models, and discuss methods for quantitative analysis of multidimensional image stacks.

INTRODUCTION

The immune system evolved to detect foreign antigens and deliver a set of coordinated effector responses including mobilization of effector T cells, cytokine delivery, antibody production, and development of immunological memory while minimizing "friendly fire" that could result in autoimmune disease. Until recently, studies were limited to static measurements on cell populations in living animals or to in vitro experiments utilizing individual cells in artificial environments. Whereas the latter approach allowed real-time imaging as an experimental approach, the results were often difficult to relate to the in vivo setting. The adoption of two-photon microscopy by immunol-

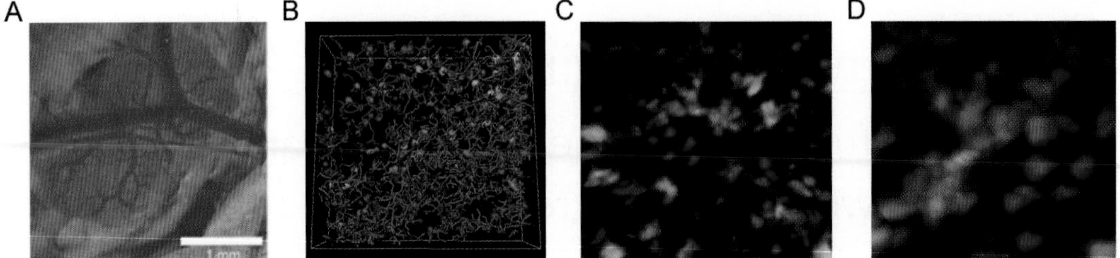

FIGURE 1. Lymph node images. (*A*) Murine inguinal LN showing the intricate vasculature of the LN (scale bar, 1 mm). (*B*) Two-photon image of the T-cell–B-cell border in the LN where representing B cells and corresponding tracks are labeled red and T cells and corresponding tracks are labeled green (tracks represent 20 min of imaging time) (see Movie 16.1). (*C*) Two-photon image of T cells (red) and DCs expressing yellow fluorescent protein (YFP) under the CD11c promoter (see Movie 16.2). (*D*) Zoomed-in image of a YFP-CD11c DC (yellow–green) and CD4$^+$ T cells (red).

ogists, following earlier applications in neurobiology and developmental biology (Denk et al. 1990; Yuste and Denk 1995; Periasamy et al. 1999; Blancaflor and Gilroy 2000) enables the visualization of cell motility, cell–cell interactions, and cell-signaling responses deep within intact organs and tissues. When compared with other four-dimensional (*x, y, z,* time) imaging techniques, such as confocal microscopy, the primary advantages of two-photon microscopy are greater than fivefold deeper tissue penetration, resulting from the reduced light scattering of infrared excitation wavelengths, along with the reduction of photodamage and photobleaching that come with the strong local confinement of the multiphoton excitation volume. Together, these advantages allow for real-time four-dimensional imaging of lymphocyte behavior in the native physiologic environment of the lymph node (LN) (Fig. 1; Movies 16.1 and 16.2). Since the first applications of two-photon immunoimaging in 2002, the technique has rapidly evolved beyond merely phenomenological description to address ever more nuanced immunological questions at a quantitative level and is now routinely performed in a variety of tissues including LNs, skin, spinal cord, brain, gut, bone marrow, and liver (Miller et al. 2002; Cavanagh et al. 2005; Kawakami et al. 2005; Chieppa et al. 2006; Egen et al. 2008; Matheu et al. 2008b; Kim et al. 2009).

The overall goal in immunoimaging studies is to study events such as lymphocyte motility and distribution, cellular interactions, chemotaxis, and antigen recognition in the physiological context of the native tissue environment, all the while maximizing resolution and minimizing disruption. Several recent reviews discuss techniques and applications of two-photon immunoimaging (Bousso 2008; Cahalan and Parker 2008). In this chapter, we focus on imaging in secondary lymphoid organs, but the techniques and technical considerations we discuss are broadly applicable. Among these are the selection of appropriate fluorescent probes and endogenous labels to provide specific and bright labeling while minimizing photodamage, the choice between simple explanted tissue preparations with excellent imaging characteristics and more complex in vivo preparations that maintain intact blood and lymphatic circulation, and the methods for analysis and quantification of cell movement and interactions from multidimensional data sets. Figure 2 summarizes important considerations in designing and performing immunoimaging experiments, which we describe in the following sections.

INSTRUMENTATION

Imaging the single-cell dynamics of the immune system within an intact environment requires the ability to look deep inside intact tissues and organisms with spatial and temporal resolutions adequate to track cell morphology, motility, and signaling processes, all the while minimizing perturbation of the system under study. Fluorescence techniques are highly suited for this purpose, permitting both labeling of specific cells, organelles, or proteins and functional readout of physiological events. Moreover, two-photon microscopy involving near-simultaneous absorption of two photons by a fluorophore molecule has considerable advantages over linear techniques such as con-

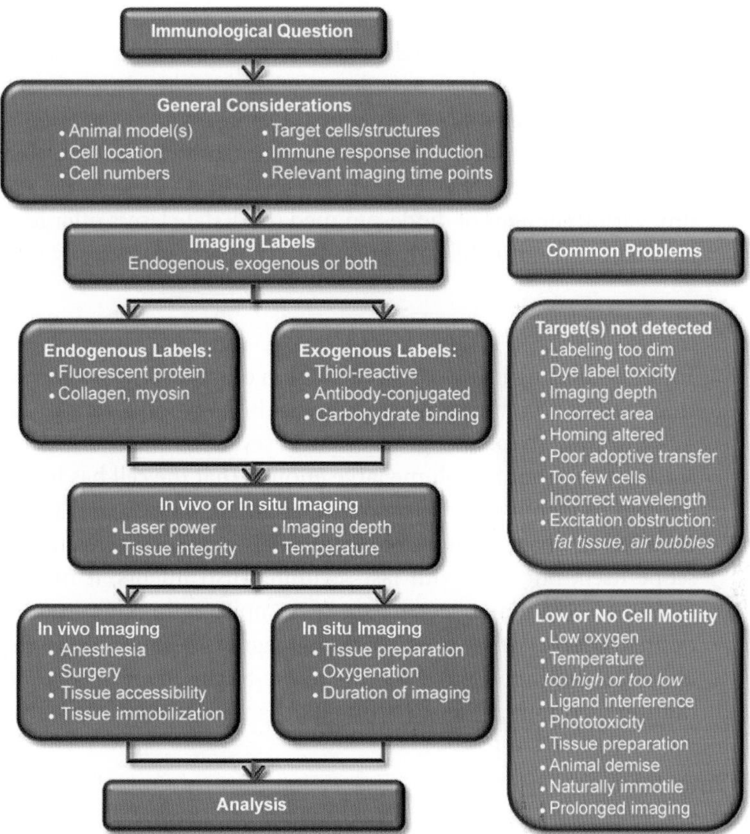

FIGURE 2. Flow chart for designing and performing immunoimaging experiments.

focal microscopy. Because excitation varies as the square of incident light intensity, fluorescence is essentially confined to the small focal volume formed under the microscope objective, thereby providing an inherent optical sectioning effect. In addition, two-photon excitation has other major advantages that arise because the excitation wavelengths are roughly twice as long as would be used for conventional linear excitation (Nguyen et al. 2001). A longer excitation wavelength enables deeper imaging into highly scattering biological tissues because scattering decreases with increasing wavelength, and absorption by hemoglobin and other proteins is minimized. Second-harmonic generation (SHG), produced by nonlinear interactions at spatially ordered molecules, has also proved useful for imaging ordered structural proteins such as collagen fibers and microtubules and simply involves detecting emitted light of half the wavelength of the excitation light, thus obviating the need for extrinsic fluorescent probes (Rigacci et al. 2000; Matheu et al. 2008b).

The requirements for two-photon immunoimaging are largely similar to those for other applications such as developmental biology or neuroscience, and most systems based on an upright microscope should be suitable. We thus limit our description below to those aspects of particular relevance for immunoimaging.

Lasers

Two-photon imaging requires a laser source emitting subpicosecond ultrashort pulses at a repetition rate of around 100 mHz. This temporal compression achieves the high instantaneous power needed for multiphoton excitation with a relatively low average power. It is desirable that the laser tune over a wide range (700–1100 nm) for optimal excitation of diverse probes and fluorescent proteins (e.g., 700 nm for the Ca^{2+} indicators Fura-2 and Indo-1, 780 nm for carboxyfluorescein succinimidyl ester [CFSE], and 900 nm for green fluorescent protein [GFP]). Although the incident laser power at the specimen is typically only a few tens of milliwatts or less, lasers with maximum outputs >1 W are preferable so as to compensate for inevitable losses in the scanning and microscope optics, to have power in reserve for scattering losses when imaging deeply into tissues, and to have adequate reserve at wavelengths near the extremes of the tuning curve. At present, only Ti:sapphire femtosecond lasers fully meet these criteria. These are a mature technology, requiring no specialized user expertise. Their main disadvantages are the high cost (typically ~$150,000) and the inability to tune beyond ~1100 nm. Various designs of diode-pumped solid-state lasers offer less expensive alternatives at long wavelengths, but their inability to tune over a broad range greatly limits their versatility.

Image Scanning

Immunoimaging involves the acquisition of multidimensional (*x, y, z*, time) information from within thick tissues or intact organs. The most highly motile cells (T cells) migrate with mean velocities of ~14 µm/min. To follow a given cell over several minutes, it is necessary to acquire image sequences (*z*-stacks) throughout a volume large enough that the cell has little chance of escaping and at a rate fast enough to identify a cell from one frame to the next with little ambiguity. A limit is set by the time required to image a single *x–y* plane. This may be determined by the scanning speed of the microscope itself; but even with fast-scanning instruments, it may be necessary to average over several frames to obtain an adequate signal-to-noise ratio. The overall acquisition speed is then set by the desired axial (*z*) depth of the desired imaging volume and the *z*-spacing between planes. A spacing of about 2.5 µm (appreciably less than the ~8-µm diameter of a T cell) is sufficient, yielding a time resolution of around 18 sec per *z*-stack for an axial depth of 50 µm at 0.5 sec per plane.

Fluorescence Detection

Photodamage (readily evident as cessation or reduction of cell motility) increases nonlinearly with the laser power incident on the specimen, setting a hard limit on the maximal excitation that can be used. It is thus essential that the microscope is optimized to collect and to detect as many emitted photons as possible so as to obtain an adequately bright signal while minimizing laser power. The microscope objective is critical regarding this and should have both good infrared transmission and good correction together with a high numerical aperture (NA) so as to form a smaller diffraction-limited excitation spot and to maximize light collection (which improves as the square of the NA). Water-dipping objectives are a good choice because the preparation can be superfused with warmed oxygenated medium, and spherical aberration associated with oil-immersion objectives is avoided. Optimal detection is achieved by the use of wide area nondescanned detectors placed close to the back aperture of the objective to collect even highly scattered photons. This effectively limits the choice of detectors to photomultiplier tubes (PMTs) which, although having lower quantum efficiency (QE) than semiconductor devices (typically, 0.25 vs >0.9), are available with wide active areas and low dark counts. PMTs with gallium arsenide photocathodes offer higher QE (~0.5) than conventional PMTs but are less robust. Operation in the photon-counting mode is ideal, but many systems achieve satisfactory results with analog detection. Confocal microscopes that have been modified for two-photon use with the only change being the addition of a femtosecond laser are likely to have poor detection efficiency, even with the confocal pinhole wide open. Multiple PMTs are typically used together with dichroic mirrors and (optionally) band-pass filters to monitor simultaneously the fluorescence from SHG signals and from multiple fluorophores having differing emission spectra.

VISUALIZING CELLS AND STRUCTURES OF INTEREST

Monitoring lymphocyte behavior at the single-cell level via immunoimaging has a broad range of applications from infectious disease to transplant rejection. Cell–cell communication is necessary for the orchestration of an effective immune response; thus, immunoimaging has the capability to

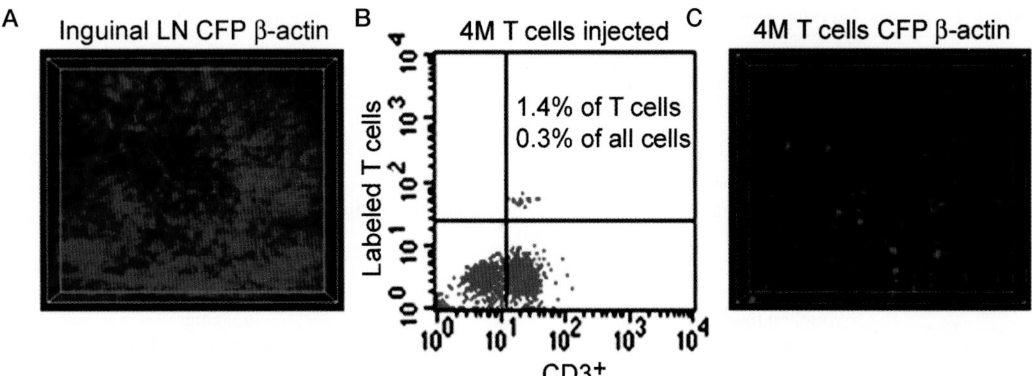

FIGURE 3. Labeled cell density greatly affects imaging quality and data analysis. (*A*) Image of an inguinal LN from a cyan fluorescent protein (CFP) β-actin mouse, where all β-actin-expressing cells also express CFP, shows the high cellular and structural densities in an LN in which individual cells cannot be clearly resolved (see Movie 16.3). (*B*) Injection of 4 × 10^6-labeled T cells into an adult mouse results in labeled T cells comprising 1.4% of the total T-cell population ($CD3^+$) in a peripheral LN. (*C*) CFP-actin $CD4^+$ T cells in a peripheral LN 24 h after injection of 4 × 10^6 total cells, a density of cells that can be clearly resolved (see Movie 16.4).

reveal the necessary components and underlying mechanisms that may require cell–cell communication in a host's immunological response. Secondary lymphoid organs are dense organs filled with a variety of lymphocytes and are traversed by a scaffold of structural tissues. Intrinsic signals can be used to image structures (e.g., autofluorescence, SHG), but most applications require that specific cell populations be fluorescently labeled, most often by staining with vital cell tracker dyes or by expressing genetically encoded fluorescent proteins in a cell-lineage-specific manner (Table 1). Generally, only a small proportion of cells are labeled, so that these can be clearly identified against an apparently empty background. At high densities, the behavior of individual cells becomes indistinguishable from neighboring cells; but, on the other hand, the collection of statistically meaningful data becomes slow and laborious if too few cells are within the imaging volume (Fig. 3; Movies 16.3 and 16.4).

TABLE 1. Commonly used fluorophores and labeling techniques for two-photon imaging

Probe class	Labeling mechanism	Labeling conditions	Duration and example use	Representative probes
Exogenous				
Thiol reactive	Irreversible thiol coupling following esterase cleavage of dye-label ester	37°C 5–15-μM RPMI or CO_2-independent medium (serum free) 30–45 min	5–7-d labeling for two-photon imaging	CMAC CMFDA CMTMR
Amine reactive	Irreversible amine coupling following esterase cleavage of dye-label acetate	37°C 2–8 μM RPMI or CO_2-independent medium (serum free) 30–40 min	5–7-d labeling for two-photon imaging Imaging dividing cell populations In vivo labeling of DCs	CMDA (CFSE) SNARF-1
Loading by membrane-permeant ester	Passive diffusion of small molecule dye label into live cells; esterase cleavage renders labels membrane impermeant	Follow product recommendations	1–4 h	Fura-2 Indo-1
Active uptake and internalization	Peptide-conjugated probe taken up and internalized by local antigen-presenting cells	Injection of concentrated peptide-probe conjugate under the skin at the site of imaging or near the draining LN; up to 100 μg/imaging site	1–3-d labeling of local antigen-presenting cells or antigen-presenting/collecting cells in a draining LN, antigen distribution tracking	Antigen-conjugated probe of choice
Dye-conjugated dextran	Diffusion, nonspecific blood, or lymphatic flow	Intravenous or subcutaneous injection	Tracking blood flow (70–100 kDa, 0.5 mg/100 μL) or lymphatic flow (20 kDa, 0.25 mg /30–50 μL)	Texas Red, FITC, or rhodamine conjugates
QDs	Free or antigen conjugated	Intravenous or subcutaneous injection	Blood flow and perfusion monitoring or induction of an immune response/antigen distribution tracking	20–100-μL i.v. or 20-pM QD in 50 μL + antigen of choice
Carbohydrate binding (whole tissue labeling)	Sialic acid and lectin-binding dye-conjugated proteins	Soak tissue at 4°C in 100-μg/mL lectin resuspended in RPMI or CO_2-independent medium; for 2–4 h	Same-day labeling and imaging to define tissue structures in vitro	WGA (Alexa 633)
Probe conjugated to antibody	Labeling of specific cells in live tissue presumably by antibody binding to cell surface epitopes	Preinjection of label-conjugated antibody 100–200 μg near the draining LN of interest	Technique has been applied in labeling some cells and tissues; not tested extensively; likely varies with antibody and probe set used	Antibody-probe conjugate
Endogenous				
Promoter-driven FP[a] expression	Cell-specific label or all cell types when FP is expressed under a ubiquitously expressed protein promoter	Produce transgenic animal or order readily available animal model Some promoters may not induce high enough expression of the FP to be visible with two-photon imaging	Permanent nontoxic label that allows for imaging the lifetime of the cell Adoptive transfer of purified cells or imaging endogenous cell behavior Production of chimeric animals (Witt et al. 2005) Photoconvertible FPs for tracking the behavior of single cells (Tomura et al. 2008)	CFP, GFP, YFP, mCherry, HcRed, Kaede, various FPs
SHG	Endogenous highly ordered tissues emit wavelength approximately half of two-photon excitation	Emission best when excitation is perpendicular to structure orientation. 880–920-nm excitation are preferable blue/violet channel detection	Long-term endogenous structure imaging in various tissues including LN, skin, lung, and muscle	Myosin, collagen, and tendon

CMAC, cerebellar model articulation controller; CMFDA, 5-chloromethylfluorescein diacetate; FITC, fluorescein isothiocyanate; QDs, quantum dots; i.v., intravenous; WGA, wheat germ agglutinin; FP, fluorescent protein; GFP, green fluorescent protein; SHG, second-harmonic generation.

Protocol A

General Approach to Adoptive Transfer and Cell Labeling

Adoptive transfer is the generally preferred method for introducing labeled cells of interest into a host animal for immunoimaging. Cells are derived from a donor animal with identical genetic background to the host and either may be endogenously fluorescent (isolated from a transgenic mouse expressing fluorescent protein) or may be labeled before transfer. Typically, transfer of 2×10^6 to 6×10^6 labeled cells of a given type results in an appropriate cell density for two-photon imaging.

MATERIALS

CAUTION: See Appendix 6 for proper handling of materials marked with <!>.

All procedures must be approved by the local animal use and care committee. Use sterile needles and syringes, and change needles for each injection.

Reagents

Antibodies to check for cell population purity by fluorescence-activated cell sorting (FACS)
Cell separation kit and associated equipment
Dye label <!>, appropriate for animal model
LNs, bone marrow, or other appropriate tissue for lymphocyte isolation
Mice, hosts for adoptively transferred cells
RPMI (Roswell Park Memorial Institute)-1640 or CO_2-independent medium, serum free
RPMI or CO_2-independent medium with 10% fetal bovine serum (FBS)

Equipment

Dissection tools
FACS cell sorter
Incubator, 37°C
Needles and syringes, 18–20 gauge, sterile

EXPERIMENTAL METHOD

Isolate Lymphocyte(s) of Interest (2 h to 11 d)

Cell isolation can be rapidly accomplished with a variety of commercially available kits, which produce a population of cells that is >90% pure (e.g., Matheu and Cahalan 2007). Cell lines can also be used (Matheu et al. 2008b). Additionally, an in vitro culture of certain lymphocytes is useful, for example, for the study of antigen presentation by monocyte-derived dendritic cells (DCs) that require several days to mature (Wei et al. 2007; Matheu et al. 2008a). All cells should be checked for population purity by FACS analysis before adoptive transfer.

Antibodies used for positive selection can alter the behavior of the isolated lymphocyte population after injection (Garrod et al. 2007).

Cell Labeling (30 min to 1 h)

1. Label cells according to recommended time and label concentration (Table 1). Endogenously fluorescent cells may be counterstained. For amine-reactive or thiol-reactive labels, treat cells in pre-warmed (37°C) serum-free, CO_2-independent medium or RPMI 1640.

   If imaging several days after cell injection, increase the dye-label incubation time by 10–20 min. This may increase cell loss during the labeling process.

2. Wash the cells in 10 times the original volume with cell culture medium containing 10% serum.
3. Centrifuge cells, and inject them into a host animal as soon as possible.
4. Confirm the intensity of the dye label by FACS before injection; it should be at least 50-fold above background cell fluorescence.

   When using multiple fluorescent markers, adjust the loading conditions so that all labels are similarly bright.

Cell Injection

Adoptively transfer cells using the appropriate route of injection (Table 2). Premixing and co-injection of two or more labeled cell types is useful to minimize the number of injections required and to synchronize the introduction of cells.

5. Resuspend cells in the desired volume for injection (Table 2).
6. Inject, and allow appropriate time for homing or immune response (Table 2).

Tissue Preparation for Imaging

7. When cells have equilibrated, harvest tissue for explant imaging, or prepare animal for intravital imaging (Protocol C or D).

TABLE 2. Adoptive transfer by injection route and cell type (mouse)

Cell type	Injection route(s)	Injection volume and needle gauge	Homing time	Procedure recommendations
T cell: naïve or central memory	Intravenous (i.v.): retro-orbital or tail vein	0.1–0.5 mL (adult animal) 26–28 gauge (i.v.)	LN/spleen: Early: 15 min Equilibrated: 6–8 h	Prewarm tail, and gently restrain animal for tail vein injections. Anesthetize animals for retro-orbital injections.
T cell: effector/ memory	Intraperitoneal (i.p.) or i.v.	0.25–1.5 mL or 0.1–0.5 mL Up to 20 gauge (i.p.) 26–28 gauge (i.v.)	Early (i.v.): 1–3 h Equilibrated: 24 h	Practice firm gentle grip on animal for i.p. injections; train with skilled professional before attempting.
B cell	i.v. tail vein or retro-orbital	0.1–0.5 mL 26–28 gauge (i.v.)	LN/spleen: Early: 2–4 h Equilibrated: 12 h	i.v. injection procedure same as T cell injection
DCs	Subcutaneous (s.c.) or i.v.	0.05–0.1 mL per site 24–28 gauge (s.c.) 0.1–0.5 mL 26–28 gauge (i.v.)	LN: Early (s.c.): 18–24 Peak (s.c.): 36 h i.v. injections: 12–23 h lower cell numbers home	Anesthetize or restrain for subcutaneous injection. The number of DCs found in the LN is greatly improved by subcutaneous injections at sites where lymphatics drain to the LN of interest.

LN, lymph node; DC, dendritic cell.

INDUCTION OF AN IMMUNE RESPONSE

The nature of lymphocyte behavior can change drastically in the presence of a cognate antigen or during inflammation. The study of lymphocyte behavior during an immune response has enhanced our understanding of immune response kinetics (Cahalan and Parker 2008). When inducing an immune response, it is important to consider the immunological implications of the method. An important caveat is the large numbers of antigen-specific cells (relative to those involved in an endogenous immune response) necessary to acquire statistically significant data. It is currently unknown how or if an increased number of responder cells affects cell behavior. Selection of an appropriate method for immune response induction is a crucial step in setting up an experimental protocol, and several methods have been developed for characterizing the immune response by two-photon imaging (Cahalan and Parker 2008). Here, we provide three well-tested methods for the introduction of antigen-pulsed antigen-presenting cells (APCs) for T-cell activation.

Protocol B

Induction of an Immune Response for Imaging APC–T-Cell Interactions

Activated DCs are the most efficient APCs in the immune system. In this protocol, we describe experiments using (1) bone-marrow-derived DCs, (2) endogenous dermal DCs that drain to the local LN, and (3) tissue-resident APCs in a delayed type hypersensitivity (DTH) response to activate antigen-specific $CD4^+$ T cells.

Practical considerations: Methods for purification, labeling, and cell injection will depend on the particular experiment. In most experiments, antigen-specific cells are used. The amount of antigen needed to elicit an immune response should be determined empirically for each of the methods. Also, ensure that cells naturally home to the expected tissue during the immune response by using FACS analysis of whole tissue before imaging is performed. Note that the amount and avidity of the antigen can alter the time course of an immune response (Henrickson et al. 2008).

Method 1: Imaging T-Cell Priming Using Bone-Marrow-Derived DCs

This is a simple and robust method for producing antigen-loaded DCs (up to 50×10^6 per donor animal) and eliciting a DC-initiated immune response.

MATERIALS

CAUTION: See Appendix 6 for proper handling of materials marked with <!>.
See the end of the chapter for recipes for reagents marked with <R>.

Reagents

Antigen of choice
CFSE <!> or dye label of choice
Ethanol, 70%
FLT-3 ligand (optional, see Step 17)
H_2O, sterile
Lipopolysaccharide (LPS)
Mice to be used as hosts for injected labeled DCs
Mice to be used as source of femurs
Phosphate-buffered saline (PBS), sterile 10×
Primary DC medium <R>
RPMI medium
Secondary DC medium <R>

Equipment

Centrifuge
Dissection tools, sterile
Incubator, tissue culture set at 37°C
Needles, sterile 26–28-gauge and insulin syringes
Tissue culture equipment and pipettes
Tubes, sterile plastic

EXPERIMENTAL METHOD

Removal of Femur Bones and Preparation of Bone-Marrow Single-Cell Suspension

1. Expose the femur by cutting away muscle above and below the knee and hip joints.
2. Hold the femur with dissection tweezers, and cut above and below the joints leaving the epiphyses intact.
3. Clean off as much muscle as possible using small dissection scissors, and transfer the femur into a dish of RPMI. Do not let the bone dry out at this point.

   The remaining procedures should be performed in a tissue culture hood using sterile conditions.
4. Sterilize the femur bones by transferring from the RPMI to a small culture dish filled with 70% ethanol for 5 min, keep on ice.
5. Transfer bones to a small culture dish containing sterile filtered primary DC medium.
6. Cut off the epiphysis, exposing the marrow. Using a 26-gauge or 28-gauge needle attached to an insulin syringe filled with sterile medium, flush the marrow out of the bone and into a new dish of sterile medium. The marrow washes out either in small pieces or as a single piece. Repeat as necessary.
7. Gently pipette intact marrow up and down in the culture dish to create a single-cell suspension. This may take several minutes and should be performed slowly.
8. Transfer the cell suspension to a sterile plastic tube, and centrifuge the cells.

Lysis and Removal of Red Blood Cells

There are several methods for red blood cell lysis. Here, we outline the water lysis method, which is relatively quick and highly effective.

9. Decant the medium, and resuspend the cell pellet in remaining volume (<0.5 ml).
10. Take up 900 μL of sterile filtered water in one pipette and 100 μL of 10x PBS in a second pipette.
11. Add 900 μL of water to the resuspended cells, wait 2–5 sec, and add 100 μL of 10x PBS. Mix gently.
12. Immediately add 5–10 mL of sterile primary DC medium.

Plating and Maintenance of Cultures

13. (Day 1) Plate cells at a density of 1×10^6/mL.
14. (Day 4) Remove 75% of the medium and nonadherent cells. Replace with fresh primary DC medium.
15. (Day 7) Passage the cells, collecting both adherent and nonadherent cells. Count, and plate the cells at a density of 1×10^6/mL using 10-mL secondary DC medium per 10-cm tissue culture plate.
16. (Day 10/11) DCs should be ready for stimulation and pulsing with antigen.

Stimulation and Pulsing with Antigen

17. Pulse DCs by adding an additional 10 mL of secondary DC medium containing double the final desired antigen concentration, LPS, 100-ng/mL final concentration is sufficient, and/or FLT-3 ligand (optional) to the 10 mL of DCs in the culture from Steps 15 and 16.
18. Incubate the cells for 18–24 h to allow for antigen uptake and processing.

Labeling and DC Injection

19. Wash DCs twice in 50-mL volume (RPMI) to remove LPS and free antigen.
20. Label DCs with the appropriate dye label (Table 1).
21. Inject subcutaneously up to 20 x 10^6 cells per site (Table 2). Pretransfer $CD4^+$ T cells, or transfer them at an appropriate time to induce an immune response of choice.

    This protocol creates DCs that resemble mature monocyte-derived DCs (Inaba et al. 1992).

Method 2: Stimulating and Antigen-Pulsing Endogenous DCs

This approach more closely replicates the immune response evoked by native DCs draining from the skin (Miller et al. 2004) but typically results in fewer labeled antigen-presenting DCs draining to the LN. A subcutaneous injection mixture is prepared for DC activation of ova-specific $CD4^+$ T cells. The mixture is injected into the ears and neck so as to target cervical LNs.

Practical considerations: Review the practical considerations outlined in the introduction to this protocol. Identify the target LN or organ, and practice injection at sites that drain to this LN or organ using just the dye label and DC stimulant to ensure that the injections are being performed correctly and can elicit DC maturation and migration.

MATERIALS

CAUTION: See Appendix 6 for proper handling of materials marked with <!>.

Reagents

Alum, 1.2%
CMDA <!> (CFSE <!>) or dye label of choice
DC maturation stimulant, such as FLT-3L or tumor necrosis factor (TNF)
Mice to be used as hosts for DC activation of ova-specific $CD4^+$ T cells
Ovalbumin (OVA) peptide (1 mg/mL) or antigen of choice

Equipment

Incubator set at 37°C
Needles, sterile 26–28-gauge and insulin syringes

EXPERIMENTAL METHOD

1. Mix 50 μg of 1-mg/mL OVA peptide with 50 μL of 1.2% alum, and warm at 37°C for 1 h.
2. Add 10 μL of 10-μM CFSE, and mix thoroughly.
3. (Optional) Add 2 μg of TNF and/or 5 μg of Flt-3L just before injection.
4. Inject a total of 20 μL of the mixture subcutaneously into each ear and up to 50 μL into the scruff of the neck to target cervical LNs.

    A large influx of DCs to the draining LN is seen 24 h after injection with fewer cells present at 72 h (Miller et al. 2004).

Method 3: Imaging Antigen-Presenting Cells in the Periphery

Eliciting a cellular response by injecting antigen under the skin can be skewed in a variety of ways by including particular adjuvants and immune response modulators with the injection of the labeled peptide. In previous studies of a DTH-primed animal and effector memory T-cell responses, no adjuvant was necessary to elicit a robust cell response (Matheu et al. 2008b).

Practical considerations: Review the practical considerations outlined in the introduction and Method 2 (if applicable). Labeled cells are typically a mixture of monocytes, macrophages, and DCs. FACS analysis of a single-cell suspension of tissue (collagenase treatment is recommended) can confirm the phenotype of the APCs that have taken up the antigen. If the dye-label antigen conjugate is too bright in the imaging field, a 1:1 or a 1:2 mixture of dye-labeled and unlabeled antigens can be used.

MATERIALS

Reagents

Antigen conjugated to the dye label of choice
Mice to serve as sources of labeled cells

Equipment

Needles, sterile 26–28-gauge and insulin syringes
Dissection tools, sterile
Tissue culture equipment and pipettes

EXPERIMENTAL METHOD

1. Subcutaneously inject the antigen of choice (up to 100 µg) conjugated to the desired dye label for imaging near the chosen site of imaging or tissue harvest.
2. Harvest the tissues, or set up an in vivo experiment at the desired time point as described in Protocol C or D.

In Situ and In Vivo LN Imaging Preparations

Immunoimaging preparations fall into three categories: (i) explanted tissues and organs maintained by superfusion with warmed oxygenated medium (in situ); (ii) intravital imaging, requiring invasive surgery to access the tissue/organ; and (iii) noninvasive intravital imaging of accessible surfaces such as skin, footpad, or cornea. Explant preparations are undoubtedly more physiological than in vitro cell systems and are easy to work with—they can be cleaned of surrounding tissue, oriented as desired with respect to the microscope objective, are readily accessible for administration of pharmaceutical agents, provide a robust and relatively drift-free preparation for long-term imaging, and minimize animal welfare issues. Explant imaging can yield results indistinguishable from intravital imaging (Miller et al. 2002; Matheu et al. 2008b), but intravital preparations are essential if intact blood and lymphatic flow are essential for the experimental question.

Practical Considerations for In Situ and In Vivo Tissue Imaging

Tissue harvesting: All dissections should be performed with care to prevent unnecessarily disrupting or damaging tissues. After tissue harvest, keep samples on ice in CO_2-independent medium or RPMI. Tissue harvest or intravital preparations should be performed immediately before imaging. Intravital tissue preparations should not disrupt blood flow to the tissue of interest.

Dissection tools: Stainless steel, 5-mm forceps and dissection scissors are recommended (Dumont). Forceps may be straight, curved, or bent as preferred. Keep forceps from becoming blunted or damaged.

Anesthesia: Animal sacrifice and anesthesia protocols should be approved by the local campus animal care and use committee. Keep anesthetized animals oxygenated (a modified mouse-sized oxygen mask can be made from the body of a syringe).

Protocol C

In Situ LN Imaging

A video demonstration of a portion of this protocol can be found online (Matheu et al. 2007b).

MATERIALS

CAUTION: See Appendix 6 for proper handling of materials marked with <!>.

Reagents

Ethanol <!>, 70%
Mice to serve as sources of LNs
RPMI or CO_2-independent medium
Tissue-safe cyanoacrylate adhesive (Vetbond, 3M)

Equipment

Coverslips, plastic
Dissection board
Dissection tools, sterile
Microscope, dissecting
Microscope, two photon

The microscope should include an imaging chamber perfused with medical grade carbogen <!> and maintained at 37°C by using an in-line heater and a thermistor sensor.

EXPERIMENTAL METHOD

Dissection of Peripheral LNs

1. Sacrifice the animal using an animal-protocol-approved method.
2. Pin the animal to a dissection board in the supine position, and wet the fur with 70% ethanol.
3. Make a midline incision from the groin to the jaw, and separate the skin from the peritoneum in a way that exposes the peripheral LNs.
4. Locate the peripheral LNs highlighted in Figure 4, grip the underlying connective tissue with forceps, and gently remove it by pulling away from the body. Do not pull the LN itself as this can damage the node. For experiments involving subcutaneous injections, only those nodes draining the injection sites will have high numbers of cells.
5. Place the LNs in CO_2-independent medium or RPMI, and keep on ice.
6. If necessary, remove additional tissue using a dissecting microscope before imaging.

Preparing and Mounting an Excised LN for Imaging

7. Cut a plastic coverslip into a strip wide enough to accommodate the LN or tissue piece of choice.

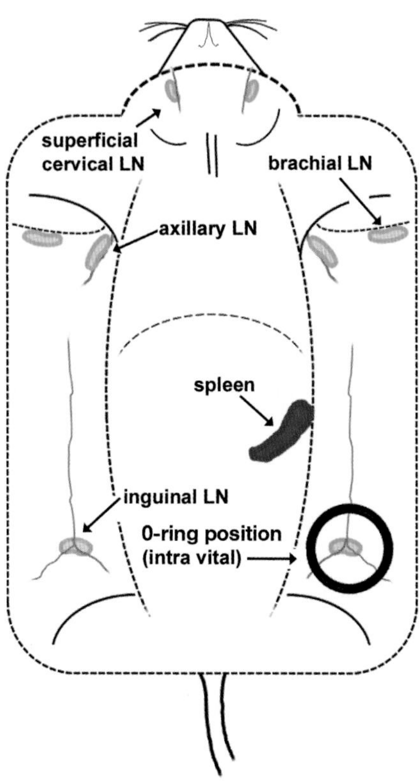

FIGURE 4. Diagram of mouse dissection for in situ and intravital imagings. There are eight easily accessible, large peripheral LNs that are easily removed and imaged in situ, these include the superficial cervical nodes, axillary LNs, brachial LNs, and the inguinal LNs. LNs are often found along major blood vessels (blue). The inguinal LN is located in a skin flap in the lower abdomen and is also accessible for intravital imaging where, after exposure of the node, an O-ring can be gently secured over the node (*lower right*).

8. Dab a small amount of tissue-safe cyanoacrylate adhesive on one end of the plastic coverslip. Wick away excess glue so that only a thin film is left.
9. Hold the connective tissue (not the node itself), and first touch one side of the LN to the glue-covered coverslip. Next, touch the side being held to gently stretch the node across the coverslip.

   For T, B, natural killer (NK), and DC imaging in their respective regions, glue the LN medullary-side down. For medullary imaging, be sure to remove all excess fat tissue, and glue the LN cortex-side (top) down.

10. Flip the coverslip upside down, and dip it into the medium at a 45° angle.

    Tissue-safe super glue (cyanoacrylate based) will cure immediately on making contact with something wet or moist. Excess cured glue covering any part of the LN will obscure imaging.

11. Trim the coverslip to fit within the imaging chamber, dab the underside with a small amount of silicone grease, and place the coverslip onto the glass bottom of the imaging/superfusion chamber.
12. Visually locate the LN under the microscope, and focus on the top using a light microscope, then extinguish all lights and switch to two-photon imaging as appropriate for your microscope.
13. Oxygenate the perfusion medium by bubbling medical grade carbogen through the reservoir using an aquarium air stone. Keep the temperature at the LN at a constant 37°C using an in-line heater and a thermistor sensor.

    Too high a temperature will result in irreversible damage and slow cell motility. Too low a temperature will slow cell motility, but proper motility can be restored if the sample is brought back to physiologic temperatures.

Protocol D

In Vivo LN Imaging

For in vivo imaging of the inguinal LN, a rubber O-ring is glued onto a skin flap so as to form a fluid chamber for the dipping microscope objective. This arrangement also helps mechanically stabilize the tissue without interrupting circulation. Keep the tissue moist while preparing it for imaging. A large variety of anesthesia will work for intravital imaging; however, an injectable one, such as a ketamine/xylazine mixture or avertin, is more easily set up and contained than is a vapor-based system (e.g., isofluorane).

MATERIALS

CAUTION: See Appendix 6 for proper handling of materials marked with <!>.

Reagents

Anesthesia <!> (varies with animal protocol and system requirements)
Mice that have had their immune cells activated and labeled (Protocols A and B)
PBS, 1×

Equipment

Carbogen <!>, medical grade
Dissection tools (scissors, forceps, fine forceps ~0.3-mm tip), sterile
Gauze, sterile
In-line heater for medium temperature control
Microscope, dissecting
Microscope, two-photon
O-ring holder (custom built; metal or Plexiglas)
O-rings, 1–2 cm diameter
Peristaltic pump for medium flow
Syringe to make oxygen mask
Tissue-safe cyanoacrylate adhesive (Vetbond, 3M)
Tubing for oxygen mask or medium perfusion
Warming plate or heating pad

EXPERIMENTAL METHOD

Exposure of the Inguinal LN

1. Make a midline incision in the skin just below the ribs, and continue toward the tail.
2. Separate a skin flap from the body to expose the inguinal LN.
3. Place a small amount of tissue-safe glue around the edge of an O-ring (1–2 cm in diameter), and glue in place over the inguinal LN (Fig. 4).

Removing the Fat Pad (if Necessary)

This technique requires practice to achieve good results. Younger mice typically have a smaller fat pad covering the inguinal LN, and sometimes this is absent. It is critical to remove any fat before imaging because it strongly obscures the laser excitation.

4. Place the animal under a dissecting microscope keeping the tissue moist with 1x PBS.
5. Fill the secured O-ring with 1x PBS (this makes it possible to check for leaks and to fix them before imaging).
6. Carefully remove the transparent connective tissue over the LN, and then gently begin to clean away fat cells, removing them from the PBS that is within the O-ring reservoir.

   Avoid rupture or constriction of any blood vessels, if this does occur, the LN is no longer useful for in vivo imaging but can be salvaged for an in situ experiment.

Stabilizing the Inguinal LN

7. Secure the O-ring into a holder that fits under the microscope objective.
8. Position the skin flap and the O-ring away from the body of the animal to minimize respiratory movements. Keep the peritoneum moist by covering it with PBS-soaked gauze.
9. Place the animal and, in particular, the LN on a warmed stage. Place the animal under oxygen perfusion using a modified syringe base that comfortably fits over the nose and mouth of the animal.

Imaging

10. Fill the secured O-ring with pre-warmed 1x PBS, and ensure that there are no leaks.

    Most leaks can be repaired with silicone grease on the inside edge of the O-ring or by addition of more tissue-safe glue on the outside edge of the O-ring.

11. Locate the LN as described in Protocol C, and proceed with imaging.
12. Monitor the animal for signs of distress, and adjust the level of anesthesia as appropriate.

    Tissue drift and breathing artifact most often disrupt intravital imaging, and care should be taken to minimize both by keeping temperatures steady, making sure the tissue is well secured and the imaged tissue is not disturbed by movement of the diaphragm.

MULTIDIMENSIONAL DATA ANALYSIS

Data sets produced by the microscope are a time series of 3D image stacks containing one or more fluorescence intensity channels. These are then analyzed to yield quantitative data on cell movements and interactions. Cell tracking requires identification and location of the centroid of each cell within the 3D image at successive time points. This can be performed manually or, less laboriously, by automated software packages such as Imaris (Bitplane, Zurich, Switzerland) and Volocity (PerkinElmer, Coventry, UK). In the latter case especially, it is important that labeling is bright enough that cells can unambiguously be discriminated from background signals by simple thresholding algorithms and that bleed-through between channels does not result in cells labeled with one fluorophore (e.g., 5-(and -6-)-{[(4-chloromethyl)benzoyl]amino}tetramethylrhodamine [CMTMR], red fluorescence) being detected in another channel (e.g., the green channel used for CFSE). Nevertheless, automated tracking is not totally reliable, and software-generated tracks should be edited or deleted as required. A computer system with at least 4 GB of RAM and a high-capacity video card is required for multidimensional analysis and visualization. The following terms are illustrated in Figure 5.

Instantaneous velocity: Instantaneous velocity is measured as the displacement of the cell centroid between successive time points in the 3D image stack. Mean velocity of a given cell is an average of instantaneous velocity measurements throughout the track of that cell. Velocities are

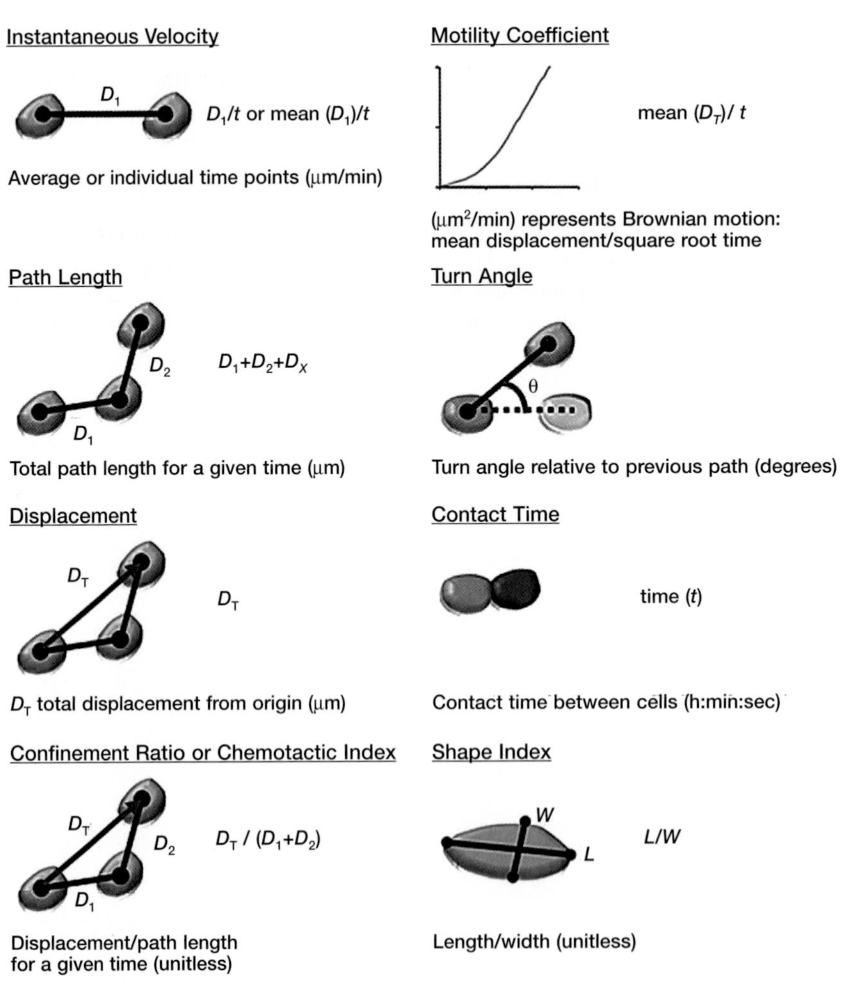

FIGURE 5. Two-photon multidimensional data analysis parameters.

usually expressed as microns per minute (μm/min) and provide a convenient index of cell type and behavior.

Path length and displacement: The path length of a cell is represented by the total distance a cell travels in a given period of time. The overall displacement of the cell is the measurement of the distance between the origin of the cell track and the location of the cell at a given time point.

Confinement ratio or chemotactic index: This measurement is defined as the distance traveled divided by the overall displacement of the cell from the origin of the track. A ratio of 1 indicates movement in a straight line, for example, as a result of directed cell movement or confinement.

Motility coefficient: The displacement from the origin of a cell undergoing random (Brownian) motion with constant velocity increases as a square root function of time. A plot of displacement^2 versus time for a cell (or population of cells) thus follows a straight line, whereas directed motion yields an upwardly curved plot. The slope of the fitted line provides an estimate of the cell motility coefficient analogous to the diffusion coefficient for molecules (Miller et al. 2002).

Turn angle: The turn angle of a cell is measured by comparing the previous direction of the cell and the angle of the new direction adopted by that cell.

Contact time: Contact time (h:min:sec) indicates the duration of time a cell spends in contact with another cell or structure and can indicate cell–cell communication.

Shape index: The ratio of the lengths of the long/short axis of a cell. As a cell moves, cell length and width change with a correlative relationship to instantaneous cell velocity. Suppression of motility due to internal signaling defects may also lead to cell polarization (Matheu et al. 2007a).

Cell volume: The size of a cell can change dramatically during an immune response and can indicate the activation state of a cell when compared with control cells (Matheu et al. 2008b). Cell volume can be automatically calculated by volume-rendering features in 3D analysis programs. Artifacts of dye-label intensity and imaging depth can make a brighter shallow cell appear larger than a deep dim cell and should be controlled for in analysis.

SUMMARY

The application of two-photon microscopy techniques for immunoimaging has opened a window to visualize cell motility and interaction dynamics of immune cells in lymphoid organs, as reviewed by Cahalan and Parker (2008). The protocols for cell labeling, injections, and introduction of antigen presented here are applicable to a wide variety of immunoimaging models and can be modified accordingly. In addition to lymphocytes and DCs, immunoimaging has been applied to investigate the role of subcapsular macrophages and follicular DCs in antigen capture (Junt et al. 2007; Phan et al. 2007; Phan et al. 2009; Roozendaal et al. 2009; Suzuki et al. 2009), NK cells in patrolling lymphoid organs and rejecting foreign cells (Garrod et al. 2007), and fibroblastic reticular cells that provide a scaffold for T-cell motility in the node (Bajenoff et al. 2006). In addition to LN, two-photon imaging has revealed behavior of macrophages, neutrophils, and effector T cells in peripheral tissue sites during inflammation (Chtanova et al. 2008; Egen et al. 2008; Matheu et al. 2008b). Immunoimaging has already provided important insights on mechanisms of immunosuppressive agents in vivo (Wei et al. 2005; Matheu et al. 2008b) and will continue to remain a valuable tool for the exploration of infectious disease models. Increasingly, as gene expression reporters, cell-signaling reporters, and photoconvertible proteins (Tomura et al. 2008) come on line as transgenic mice with expression in the immune system, these will be incorporated as powerful genetically encoded probes of immune cell function.

RECIPES

CAUTION: See Appendix 6 for proper handling of materials marked with <!>.
Recipes for reagents marked with <R> are included in this list.

Primary DC Medium

To prepare 500 mL, combine the following.

IMDM (Iscove's modification of DMEM)	~450 mL
Heat-inactivated FBS	50 mL
200-mM L-Gln (2 mM final concentration)	5 mL
Penicillin <!>	100 IU/mL
Streptomycin <!>	100 µg/mL
β-mercaptoethanol <!>	50 µM
GM-CSF	20–30 ng/mL
IL-4 (~10–40 ng/mL)	100–400 IU/mL

Sterilize the mixture. Medium can be stored for ~3 mo.

Secondary DC Medium

Add 100 ng/mL of TNF to primary DC medium.

ACKNOWLEDGMENTS

We thank Luette Forrest for helpful comments as well as past members of the Cahalan Laboratory, especially Mark J. Miller and Sindy H. Wei, for their valuable contributions to these protocols.

REFERENCES

Bajenoff M, Egen JG, Koo LY, Laugier JP, Brau F, Glaichenhaus N, Germain RN. 2006. Stromal cell networks regulate lymphocyte entry, migration, and territoriality in lymph nodes. *Immunity* **25:** 989–1001.

Blancaflor EB, Gilroy S. 2000. Plant cell biology in the new millennium: New tools and new insights. *Am J Bot* **87:** 1547–1560.

Bousso P. 2008. T-cell activation by dendritic cells in the lymph node: Lessons from the movies. *Nat Rev Immunol* **8:** 675–684.

Cahalan MD, Parker I. 2008. Choreography of cell motility and interaction dynamics imaged by two-photon microscopy in lymphoid organs. *Annu Rev Immunol* **26:** 585–626.

Cavanagh LL, Bonasio R, Mazo IB, Halin C, Cheng G, van der Velden AW, Cariappa A, Chase C, Russell P, Starnbach MN, et al. 2005. Activation of bone marrow-resident memory T cells by circulating, antigen-bearing dendritic cells. *Nat Immunol* **6:** 1029–1037.

Chieppa M, Rescigno M, Huang AY, Germain RN. 2006. Dynamic imaging of dendritic cell extension into the small bowel lumen in response to epithelial cell TLR engagement. *J Exp Med* **203:** 2841–2852.

Chtanova T, Schaeffer M, Han SJ, van Dooren GG, Nollmann M, Herzmark P, Chan SW, Satija H, Camfield K, Aaron H, et al. 2008. Dynamics of neutrophil migration in lymph nodes during infection. *Immunity* **29:** 487–496.

Denk W, Strickler JH, Webb WW. 1990. Two-photon laser scanning fluorescence microscopy. *Science* **248:** 73–76.

Egen JG, Rothfuchs AG, Feng CG, Winter N, Sher A, Germain RN. 2008. Macrophage and T cell dynamics during the development and disintegration of mycobacterial granulomas. *Immunity* **28:** 271–284.

Garrod KR, Wei SH, Parker I, Cahalan MD. 2007. Natural killer cells actively patrol peripheral lymph nodes forming stable conjugates to eliminate MHC-mismatched targets. *Proc Natl Acad Sci* **104:** 12081–12086.

Henrickson SE, Mempel TR, Mazo IB, Liu B, Artyomov MN, Zheng H, Peixoto A, Flynn MP, Senman B, Junt T, et al. 2008. T cell sensing of antigen dose governs interactive behavior with dendritic cells and sets a threshold for T cell activation. *Nat Immunol* **9:** 282–291.

Inaba K, Inaba M, Romani N, Aya H, Deguchi M, Ikehara S, Muramatsu S, Steinman RM. 1992. Generation of large numbers of dendritic cells from mouse bone marrow cultures supplemented with granulocyte/macrophage colony-stimulating factor. *J Exp Med* **176:** 1693–1702.

Junt T, Moseman EA, Iannacone M, Massberg S, Lang PA, Boes M, Fink K, Henrickson SE, Shayakhmetov DM, Di

Paolo NC, et al. 2007. Subcapsular sinus macrophages in lymph nodes clear lymph-borne viruses and present them to antiviral B cells. *Nature* **450:** 110–114.

Kawakami N, Nagerl UV, Odoardi F, Bonhoeffer T, Wekerle H, Flugel A. 2005. Live imaging of effector cell trafficking and autoantigen recognition within the unfolding autoimmune encephalomyelitis lesion. *J Exp Med* **201:** 1805–1814.

Kim JV, Kang SS, Dustin ML, McGavern DB. 2009. Myelomonocytic cell recruitment causes fatal CNS vascular injury during acute viral meningitis. *Nature* **457:** 191–195.

Matheu M, Sen D, Cahalan MD, Parker I. 2008a. Generation of bone marrow derived murine dendritic cells for use in 2-photon imaging. *J Vis Exp* **2008:** pii: 773. doi: 10.3791/773.

Matheu MP, Beeton C, Garcia A, Chi V, Rangaraju S, Safrina O, Monaghan K, Uemura MI, Li D, Pal S, et al. 2008b. Imaging of effector memory T cells during a delayed-type hypersensitivity reaction and suppression by Kv1.3 channel block. *Immunity* **29:** 602–614.

Matheu MP, Cahalan MD. 2007. Isolation of CD4+ T cells from mouse lymph nodes using Miltenyi MACS purification. *J Vis Exp* **2007:** 409. doi: 10.3791/409.

Matheu MP, Deane JA, Parker I, Fruman DA, Cahalan MD. 2007a. Class IA phosphoinositide 3-kinase modulates basal lymphocyte motility in the lymph node. *J Immunol* **179:** 2261–2269.

Matheu MP, Parker I, Cahalan MD. 2007b. Dissection and 2-photon imaging of peripheral lymph nodes in mice. *J Vis Exp* **2007:** 265.

Miller MJ, Hejazi AS, Wei SH, Cahalan MD, Parker I. 2004. T cell repertoire scanning is promoted by dynamic dendritic cell behavior and random T cell motility in the lymph node. *Proc Natl Acad Sci* **101:** 998–1003.

Miller MJ, Wei SH, Parker I, Cahalan MD. 2002. Two-photon imaging of lymphocyte motility and antigen response in intact lymph node. *Science* **296:** 1869–1873.

Nguyen QT, Callamaras N, Hsieh C, Parker I. 2001. Construction of a two-photon microscope for video-rate Ca^{2+} imaging. *Cell Calcium* **30:** 383–393.

Periasamy A, Skoglund P, Noakes C, Keller R. 1999. An evaluation of two-photon excitation versus confocal and digital deconvolution fluorescence microscopy imaging in *Xenopus* morphogenesis. *Microsc Res Tech* **47:** 172–181.

Phan TG, Green JA, Gray EE, Xu Y, Cyster JG. 2009. Immune complex relay by subcapsular sinus macrophages and noncognate B cells drives antibody affinity maturation. *Nat Immunol* **10:** 786–793.

Phan TG, Grigorova I, Okada T, Cyster JG. 2007. Subcapsular encounter and complement-dependent transport of immune complexes by lymph node B cells. *Nat Immunol* **8:** 992–1000.

Rigacci L, Alterini R, Bernabei PA, Ferrini PR, Agati G, Fusi F, Monici M. 2000. Multispectral imaging autofluorescence microscopy for the analysis of lymph-node tissues. *Photochem Photobiol* **71:** 737–742.

Roozendaal R, Mempel TR, Pitcher LA, Gonzalez SF, Verschoor A, Mebius RE, von Andrian UH, Carroll MC. 2009. Conduits mediate transport of low-molecular-weight antigen to lymph node follicles. *Immunity* **30:** 264–276.

Suzuki K, Grigorova I, Phan TG, Kelly LM, Cyster JG. 2009. Visualizing B cell capture of cognate antigen from follicular dendritic cells. *J Exp Med* **206:** 1485–1493.

Tomura M, Yoshida N, Tanaka J, Karasawa S, Miwa Y, Miyawaki A, Kanagawa O. 2008. Monitoring cellular movement in vivo with photoconvertible fluorescence protein "Kaede" transgenic mice. *Proc Natl Acad Sci* **105:** 10871–10876.

Wei SH, Rosen H, Matheu MP, Sanna MG, Wang SK, Jo E, Wong CH, Parker I, Cahalan MD. 2005. Sphingosine 1-phosphate type 1 receptor agonism inhibits transendothelial migration of medullary T cells to lymphatic sinuses. *Nat Immunol* **6:** 1228–1235.

Wei SH, Safrina O, Yu Y, Garrod KR, Cahalan MD, Parker I. 2007. Ca^{2+} signals in $CD4^{+}$ T cells during early contacts with antigen-bearing dendritic cells in lymph node. *J Immunol* **179:** 1586–1594.

Witt CM, Raychaudhuri S, Schaefer B, Chakraborty AK, Robey EA. 2005. Directed migration of positively selected thymocytes visualized in real time. *PLoS Biol* **3:** e160. doi: 10.1371/journal.pbio.0030160.

Yuste R, Denk W. 1995. Dendritic spines as basic functional units of neuronal integration. *Nature* **375:** 682–684.

MOVIE LEGENDS

Movies are freely available online at www.cshprotocols.org/imaging.

MOVIE 16.1. The T-cell–B-cell border within a lymph node, imaged with two-photon microscopy. B cells are labeled red, and T cells are labeled green. Video dimensions are 295, 270, 52 µm (*x*, *y*, *z*).

MOVIE 16.2. T cells (red) and dendritic cells expressing yellow fluorescent protein (YFP) under control of the CD11c promoter. Movie was taken using two-photon microscopy. Video dimensions are 300, 235, 25 µm (*x*, *y*, *z*).

MOVIE 16.3. Two-photon imaging of an inguinal lymph node from a cyan fluorescent protein (CFP) β-actin mouse in which all of the β-actin-expressing cells also express CFP. Video dimensions are 240, 240, 50 µm (*x*, *y*, *z*).

MOVIE 16.4. Two-photon imaging of cyan fluorescent protein (CFP) β-actin $CD4^{+}$ T cells in a peripheral lymph node 24 h after injection of 4×10^{6} total cells. Video dimensions are 240, 240, 50 µm (*x*, *y*, *z*).

MOVIE LEGENDS

17 Biarsenical Labeling of Tetracysteine-Tagged Proteins

Asit K. Pattnaik and Debasis Panda

Department of Veterinary and Biomedical Sciences and Nebraska Center for Virology, University of Nebraska-Lincoln, Lincoln, Nebraska 68583-0900

ABSTRACT

The use of fluorescent proteins fused to proteins of interest has greatly advanced our knowledge of protein function, intracellular localization, and protein–protein interactions. Coupled with microscopy, fluorescent-protein tags have been invaluable in understanding protein dynamics within living cells. Fluorescent proteins, such as green fluorescent protein, are large (approximately 240 amino acids in length) and are generally placed at the amino or carboxyl terminus of a protein of interest, although they can also function when inserted in-frame within a protein of interest. Sometimes, these fluorescent protein tags perturb the biological function of the protein of interest. To avoid this problem, proteins can be tagged genetically with a short amino acid sequence containing a tetracysteine motif, and the recombinant protein can then be labeled with membrane-permeable biarsenical dyes. The tetracysteine tagging and biarsenical dye labeling method is suitable for investigating protein dynamics that are otherwise difficult to study with fluorescent tagging methods. For example, it can be used for visualizing the trafficking of newly synthesized and existing proteins, because these pools of proteins can be distinguished by labeling them with different biarsenical dyes, one that fluoresces red and the other green.

INTRODUCTION

Genetic tagging of a protein of interest with fluorescent proteins like green fluorescent protein (GFP) has been a very valuable method for studying protein localization and interactions in live cells. However, the relatively large size of fluorescent fusion proteins is a major limitation with this approach and sometimes the fluorescent protein tag makes the protein of interest nonfunctional. Tetracysteine (TC) tagging coupled with biarsenical labeling can be used to overcome this limitation. Although binding of the biarsenical dye FlAsH to TC tag may sometimes affect the function of the protein, the method is powerful for visualizing both newly synthesized proteins in real time and protein trafficking because the existing and newly synthesized pools of proteins can be sequentially labeled with different biarsenical dyes. The nonfluorescent biarsenical derivative of fluorescein,

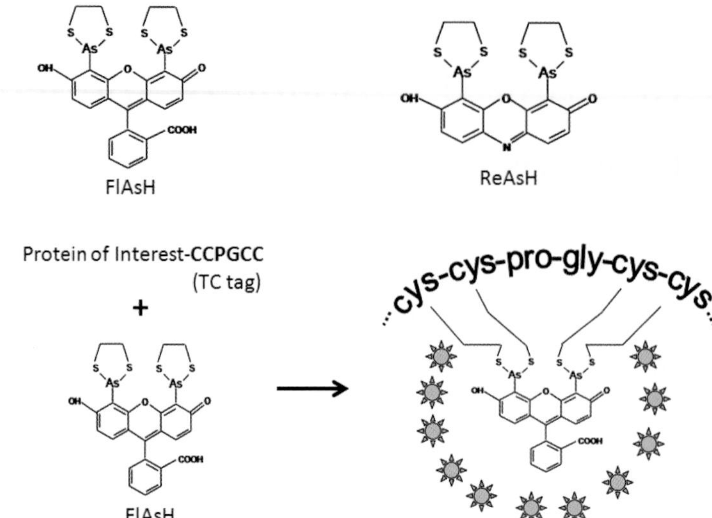

FIGURE 1. Chemical structures of FlAsH and ReAsH (*top*). Binding of FlAsH (nonfluorescent) to TC-tagged protein of interest through the cysteine residues in the tag results in the detection of green fluorescence (*bottom*).

FlAsH-EDT_2, and the nonfluorescent biarsenical derivative of the red fluorophore resorufin, ReAsH-EDT_2, bind to TC tags in a protein of interest with high specificity and high affinity (Adams et al. 2002). Once bound, they become strongly fluorescent, emitting either green or red light, respectively, allowing the TC-tagged proteins to be easily imaged in either fixed or living cells (Fig. 1). The TC tagging and biarsenical labeling technique described here was developed by Roger Tsien and colleagues (Adams et al. 2002; Griffin et al. 1998). The method relies on the introduction of a small motif, 6–20 amino acids in length and containing four cysteines, to a protein at either its amino or carboxyl terminus. Cells expressing a protein with this tag are then treated with the membrane-permeable dyes FlAsH-EDT_2 or ReAsH-EDT_2. The two arsenic groups of FlAsH or ReAsH each bind to two thiols in the tetracysteine motif. Once bound, FlAsH and ReAsH are converted to a fluorescent state and can be detected by fluorescent microscopy. Additionally, TC-tagged proteins can also be selectively inactivated by chromophore-assisted light inactivation (Tour et al. 2003). It should be noted that the labeling of TC-tagged proteins with biarsenical dyes requires a reducing environment (Adams et al. 2002) and therefore may limit detection of certain proteins by this approach.

Protocol

Labeling Proteins with Tetracysteine and Biarsenicals for Live Cell Imaging

A method is presented for labeling proteins for live cell imaging using fluorescence microscopy. The protein(s) of interest are tagged genetically with a tetracysteine peptide and subsequently labeled with a membrane-permeable biarsenical dye. This protocol also includes a method for labeling virally produced proteins. Only labeling conditions for cultured cells (virus-infected or uninfected) have been included in this protocol; for work done in tissue or in vitro, see Marek and Davis (2002), Poskanzer et al. (2003), and Tour et al. (2003).

IMAGING SETUP

FlAsH has an excitation maximum at 508 nm, has an emission maximum at 528 nm, and produces green fluorescence. ReAsH has an excitation maximum at 593 nm, an emission maximum at 608 nm, and fluoresces red. The recommended filter sets to visualize FlAsH-labeled protein and ReAsH-labeled protein are fluorescein isothiocyanate and Texas Red, respectively. Use an appropriate (60×, 100×, or other) objective lens to visualize the protein. To reduce photobleaching or inactivation of proteins (Tour et al. 2003), minimize exposure of the cells to light.

MATERIALS

CAUTION: See Appendix 6 for proper handling of materials marked with <!>.
See the end of the chapter for recipes for reagents marked with <R>.

Reagents

BHK-21 cell culture medium <R>
Cells, BHK-21 (ATCC)
Hank's balanced salt solution (optional, see Step 12)
Lipofectamine 2000 (Invitrogen)
Mammalian expression vector containing a T7 promoter (e.g., pCDNA 3.1; Invitrogen)
Mounting medium (aqueous; for fixed cells only)
Opti-MEM (Invitrogen)
Paraformaldehyde <!> (4%, prepared in 1× phosphate-buffered saline [PBS]; for fixed cells only)
Polymerase chain reaction primers specific for the TC insert
  Prepare primers encoding the 12-amino-acid sequence FLNCCPGCCMEP. See Step 1.
Plasmid Midi Kit (QIAGEN)
Reagents for plasmid cloning and sequencing
Reagent for sucrose density gradient centrifugation
TC-FlAsH II In-Cell Tetracysteine Tag Detection Kit (T34561, Invitrogen)
TC-ReAsH II In-Cell Tetracysteine Tag Detection Kit (T34562, Invitrogen)
  Alternatively, FlAsH-EDT2 and ReAsH-EDT2 can be prepared as described in Adams and Tsien (2008).
Vaccinia virus (recombinant)-expressing T7 polymerase
Vesicular stomatitis virus expressing TC-tagged M-protein (for viral particle labeling only)

Equipment

Aluminum foil
Equipment for plasmid cloning and sequencing
Equipment for sucrose density gradient centrifugation
Incubator preset to 37°C
Micropipettor with tips
Microscope (fluorescence, preferably confocal)
Microscope slides (for fixed cells only)
Tissue culture plates (12-well) with a round coverslip (No. 1 thickness) in each well

Alternatively, glass-bottomed tissue culture dishes can be used (see Step 6).

EXPERIMENTAL METHOD

Construction of Plasmid Encoding Tetracysteine-Tagged Protein

1. Using polymerase chain reaction (PCR), generate a copy DNA (cDNA) encoding the protein of interest fused with the peptide FLNCCPGCCMEP.

   This sequence has been optimized to improve fluorescence (Martin et al. 2005).

2. Clone the cDNA in a vector with a T7 promoter.
3. Verify the sequence of the plasmid by sequencing the entire insert with the TC tag.
4. Purify endotoxin-free DNA plasmid using a plasmid purification kit according to manufacturer's instruction.
5. Drive protein expression using a recombinant vaccinia virus-expressing T7 polymerase.

   Alternatively, use a cellular polymerase II–dependent expression vector to drive protein expression.

Biarsenical Labeling

This method requires several washings and labeling at room temperature over a long period of time, so a relatively adherent cell type, such as BHK-21, should be used. Labeling should be performed when 80%–95% of the cells are expressing the protein of interest. It is important, therefore, to coordinate the timing of transfection/infection with dye labeling for optimal efficiency.

6. Grow BHK-21 cells on coverslip in a 12-well tissue culture plate in BHK-21 culture medium for 14–16 h at 37°C (i.e., to 50%–60% confluency).

   Alternatively, use glass-bottomed tissue culture dishes.

7. Infect cells with recombinant vaccinia virus-expressing T7 polymerase at a multiplicity of infection (MOI) of 5. Using Lipofectamine 2000, transfect the cells with 1–1.5 µg of the plasmid encoding the TC-tagged protein (from Step 5) for the desired length of time (e.g., 4–6 h). Transfect cells with an empty vector as a negative control.

   Alternatively infect the cells with 3–5 MOI of a virus encoding the TC-tagged protein.

8. Dilute the biarsenical dye (FlAsH or ReAsH) 1:1600 to 1:800 in Opti-MEM (i.e., dye concentration from 1.25–2.5 µM).
9. Remove the growth media from the wells. Wash the cells with Opti-MEM.
10. Add the diluted dye to the cells (400 µL/well). Cover the plates with aluminum foil to protect them from light. Incubate the cells for 30–60 min at room temperature.
11. Remove the dye. Wash the cells once with Opti-MEM.
12. Dilute the BAL washing buffer stock solution (included in the kits) to 1x with Opti-MEM or Hank's balanced salt solution.

13. Process as desired:

    For single labeling:

    i. Wash the cells three times with BAL wash buffer (400 μL/well), 10 min for each wash, at room temperature to remove residual dye.

    ii. Proceed to Step 14.

    For double labeling:

    i. Wash each well once with 400 μL of BAL wash buffer for 10 min at room temperature.

    ii. Incubate the cells for 30–60 min in complete medium at 37°C.

    iii. Remove the growth media from the wells. Wash the cells with Opti-MEM.

    iv. Add 400 μL of the second dye (diluted to 1.25–2.5 μM) to each well. Cover the plate with aluminum foil. Incubate the cells for 30 min at room temperature.

    v. Remove the dye. Wash the cells once with Opti-MEM.

    vi. Wash the cells three to four times with BAL wash buffer (400 μL/well), 10 min for each wash, at room temperature to remove residual dye.

14. Visualize the cells as follows:

    For live cells:

    i. Visualize the cells directly by fluorescence microscopy. Collect the images using the parameters described in the imaging setup section.

    See *Troubleshooting.*

    For fixed cells:

    i. Fix the cells in 4% paraforlmaldehyde in PBS at room temperature for 15 min.

    ii. Mount the coverslip containing the fixed cells with aqueous mounting medium.

    iii. Visualize the cells by fluorescence microscopy.

    Paraformaldehyde-fixed cells can also be processed for immunofluorescence staining for other proteins.

    See *Troubleshooting.*

Labeling Viral Particles

This technique is described in greater detail in Das et al (2009).

15. Infect BHK-21 cells with virus particles, encoding TC-tagged M-protein, at an MOI of 3–5. Incubate for 4 h.
16. Add the dye of choice at a concentration of 1.25 μM.
17. Collect the culture supernatant 16–20 h postinfection.
18. Purify the viral particles by sucrose density gradient centrifugation.
19. Resuspend the viral pellet in PBS.

TROUBLESHOOTING

Problem (Step 14): Fluorescence signal is low.
Solution: Consider the following.

- Optimize transfection efficiency to achieve greater protein expression.
- Increase the amount of plasmid and the time cells are in culture before labeling.

- Increase the concentration of the labeling reagent.
- Increase the labeling time.
- Avoid multiple freezing and thawing of the dye.
- Prepare the dye labeling solution fresh before use.
- Protect the dye labeling solution from light.
- Use the appropriate filter set to visualize the fluorescent signal.

Problem (Step 14): There is high background compared with negative control.
Solution: Consider the following.

- Increase the washing time. Instead of 10-min washes, use a 15-min wash three to five times.
- Use BAL wash buffer supplied in the kit.
- Reduce the dye concentration and labeling time.
- Use serum-free medium to label and wash the cells.
- Use a shorter exposure time while imaging.

Problem (Step 14): Fewer numbers of cells are present at imaging.
Solution: Consider the following.

- Handle the cells more gently while adding or removing reagents.
- For live cell imaging use poly-L-lysine-coated plates.

Problem (Step 14): Clumps of fluorescence are seen while imaging viral particles.
Solution: Consider the following.

- Remove any residual solution from the wall of the centrifuge tube.
- Resuspend the viral pellet thoroughly in PBS.

SUMMARY AND FUTURE DIRECTIONS

Recent advances in imaging technique coupled with the availability of fluorescent protein tags have greatly facilitated studies in cell biology. After being synthesized, a fluorescent protein folds and autooxidizes to form its functional chromophore. A variety of fluorescent proteins are now available whose emission wavelengths cover most of the visible spectrum. Moreover, photoactivable fluorescent proteins and photoconvertible fluorescent proteins have provided new ways to better understand the activities, interactions, localization, and trafficking of proteins within cells. Because fluorescent proteins are relatively large, tagging with them may sometimes disturb the biological function of the protein of interest. The TC-tagging method described in this chapter minimizes this problem because it uses a relatively small peptide tag of approximately 6–20 residues. One advantage of TC tagging is that the biarsenical dyes bind quickly, specifically, and irreversibly to TC hairpin motifs, resulting in fluorescence that can be imaged by fluorescence microscopy. The biarsenical labeling reagents can also be applied sequentially to separately label existing protein pools and nascently synthesized pools of proteins, thus providing a way to visualize protein dynamics within living cells. Moreover, this is the only technique by which existing and newly synthesized proteins can be tracked and visualized by fluorescent microscopy in live cells. Another significant advantage of TC-tagging proteins is that when ReAsH is bound to the TC tag, it can be used to photoconvert diaminobenzidine, which can be used for direct correlation of live-cell fluorescent images with high-resolution electron microscopic images (Gaietta et al. 2002, 2006). Additionally, calcium dynamics and calcium nanodomains can be probed in cultured cells with the use of a biarsenical Ca^{2+} indicator called Calcium Green FlAsH (CaGF) and TC-tagged proteins (Tour et al. 2007). Such studies with combined optical and electron microscopic imaging techniques are likely to generate invaluable information concerning fundamental aspects of protein trafficking and localization within tissues and cells. In addition, TC tagging can be especially useful in detecting unfolded and folded protein conformations, as it has been shown recently that the P and G residues within the CCPGCC motif can be replaced with one or more protein domains without any significant reduction in affinity to biarsenical dyes. This technique is termed bipartite TC display (Luedtke et al. 2007) and is conceptually similar to the split-GFP technology (Ghosh et al. 2000) used for detection of protein–protein interactions. The bipartite TC display method has some advantages over split-GFP again because the TC molecule is smaller than GFP.

RECIPE

CAUTION: See Appendix 6 for proper handling of materials marked with <!>.
Recipes for reagents marked with <R> are included in this list.

BHK-21 Cell Culture Medium

Fetal bovine serum, 5%
Kanamycin <!>, 20 units/mL
Minimal essential medium
Penicillin <!>, 100 units/mL
Streptomycin <!>, 20 units/mL

ACKNOWLEDGMENTS

This work was supported by Public Health Service grant AI-34956 from the National Institute of Allergy and Infectious Diseases, National Institute of Health.

REFERENCES

Adams SR, Tsien RY. 2008. Preparation of the membrane-permeant biarsenicals FlAsH-EDT2 and ReAsH-EDT2 for fluorescent labeling of tetracysteine-tagged proteins. *Nat Protoc* **3:** 1527–1534.

Adams SR, Campbell RE, Gross LA, Martin BR, Walkup GK, Yao Y, Llopis J, Tsien RY. 2002. New biarsenical ligands and tetracysteine motifs for protein labeling in vitro and in vivo: Synthesis and biological applications. *J Am Chem Soc* **124:** 6063–6076.

Das SC, Panda D, Nayak D, Pattnaik AK. 2009. Biarsenical labeling of vesicular stomatitis virus encoding tetracysteine-tagged M protein allows dynamic imaging of M protein and virus uncoating in infected cells. *J Virol* **83:** 2611–2622.

Gaietta G, Deerinck TJ, Adams SR, Bouwer J, Tour O, Laird DW, Sosinsky GE, Tsien RY, Ellisman MH. 2002. Multicolor and electron microscopic imaging of connexin trafficking. *Science* **296:** 503–507.

Gaietta GM, Giepmans BN, Deerinck TJ, Smith WB, Ngan L, Llopis J, Adams SR, Tsien RY, Ellisman MH. 2006. Golgi twins in late mitosis revealed by genetically encoded tags for live cell imaging and correlated electron microscopy. *Proc Natl Acad Sci* **103:** 17777–17782.

Ghosh I, Hamilton AD, Regan L. 2000. Antiparallel leucine-zipper directed protein assembly: Application to the green fluorescent protein. *J Am Chem Soc* **122:** 5658–5659.

Griffin BA, Adams SR, Tsien RY. 1998. Specific covalent labeling of recombinant protein molecules inside live cells. *Science* **281:** 269–272.

Luedtke NW, Dexter RJ, Fried DB, Schepartz A. 2007. Surveying polypeptide and protein domain conformation and association with FlAsH and ReAsH. *Nat Chem Biol* **3:** 779–784.

Marek KW, Davis GW. 2002. Transgenically encoded protein photoinactivation (FlAsH-FALI): Acute inactivation of synaptotagmin I. *Neuron* **36:** 805–813.

Martin BR, Giepmans BN, Adams SR, Tsien RY. 2005. Mammalian cell-based optimization of the biarsenical-binding tetracysteine motif for improved fluorescence and affinity. *Nat Biotechnol* **23:** 1308–1314.

Poskanzer KE, Marek KW, Sweeney ST, Davis GW. 2003. Synaptotagmin I is necessary for compensatory synaptic vesicle endocytosis in vivo. *Nature* **426:** 559–563.

Tour O, Meijer RM, Zacharias DA, Adams SR, Tsien RY. 2003. Genetically targeted chromophore-assisted light inactivation. *Nat Biotechnol* **21:** 1505–1508.

Tour O, Adams SR, Kerr RA, Meijer RM, Sejnowski TJ, Tsien RW, Tsien RY. 2007. Calcium Green FlAsH as a genetically targeted small-molecule calcium indicator. *Nat Chem Biol* **3:** 423–431.

18 Bimolecular Fluorescence Complementation (BiFC) Analysis of Protein Interactions and Modifications in Living Cells

Tom K. Kerppola

Howard Hughes Medical Institute, and Department of Biological Chemistry, University of Michigan Medical School, Ann Arbor, Michigan 48109-0650

ABSTRACT

Bimolecular fluorescence complementation (BiFC) analysis enables direct visualization of protein interactions and modifications in living cells. The BiFC assay is based on the facilitated association of two nonfluorescent fragments of a fluorescent protein fused to proteins of interest. The association of the fragments can be facilitated by an interaction between proteins fused to the fragments, by the covalent conjugation of an ubiquitin-family peptide to a substrate protein, or by fusion protein binding to cellular scaffolds. The BiFC assay provides an exquisitely sensitive method for the detection of protein interactions and modifications in living cells at high spatial resolution.

INTRODUCTION

Protein interactions are fundamental for the control of cellular functions. Complexes formed by different combinations of proteins produce the cell type–specific differences in functions that are required for multicellular life. Numerous methods for the investigation of protein interactions have been devised. Many of these methods require removal of the interaction partners from their normal cellular environment. These methods have the advantage that the interactions can be studied in great detail, even at atomic resolution using X-ray crystallography or nuclear magnetic resonance (NMR). However, these methods cannot be used to determine whether proteins interact in a particular cell type or to identify the subcellular localization of the complexes. Alternatively, interactions between the protein products of particular genes can be inferred from the identification of suppressor mutations or by using two-hybrid assays. These approaches enable identification of genes that have nonadditive functional effects in cells. However, these methods do not define whether the functional effects are caused by complex formation or whether they are caused by independent effects of the gene products.

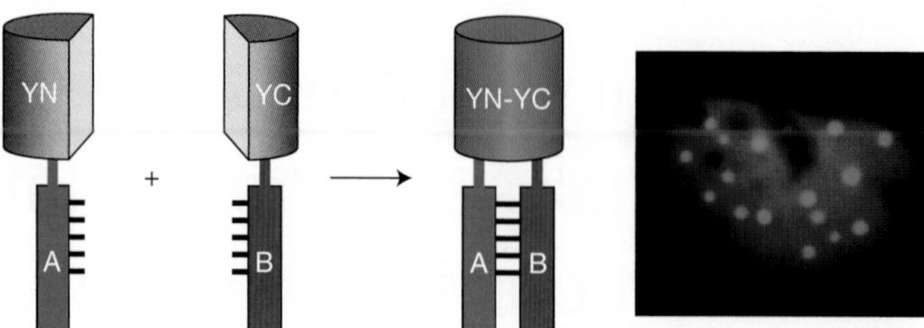

FIGURE 1. Principle of bimolecular fluorescence complementation analysis (BiFC). The red and blue rectangles represent putative interaction partners (A and B). They are fused to fragments of a fluorescent protein, represented by gray half-cylinders (YN and YC). If the proteins interact with each other, they can facilitate association of the fluorescent protein fragments to produce a fluorescent complex (green cylinder YN-YC). The image shows an example of a complex formed by nuclear proteins. (Reprinted from Rajaram and Kerppola 2004, ©American Society for Microbiology.)

Methods that enable direct detection of protein interactions in living cells are generally based on detection of the molecular proximity of tags attached to putative interaction partners. These methods share the limitations that the proteins whose interactions are investigated are not the native proteins and that the tags that are added to the proteins, as well as the methods that are used to produce the tagged proteins, can influence the results. Many such methods can be used to detect interactions in single cells, enabling investigation of variations in protein interactions among individual cells within the population. Some of these methods also enable determination of the subcellular localization of the complex, which can provide insight into its functional roles and regulation. When deciding among alternative methods for imaging protein interactions, it is important to consider the sensitivity, the signal-to-background ratio, and the spatial and temporal resolutions of different assays.

The classical proximity-based approach is Förster/fluorescence resonance energy transfer (FRET) analysis (Förster 1959). This approach has been widely used to study the architecture of macromolecular complexes and to visualize protein interactions in live cells (Hink et al. 2002; Majoul et al. 2002; Giepmans et al. 2006; see also Chapter 22). The major advantage of FRET is its ability to visualize rapid changes in complex formation. Significant limitations of this method include its relatively low sensitivity for interactions in which a majority of the partners are associated with other cellular proteins and the requirement for a short distance (<10 nm) between the fluorescent labels.

Other, less commonly used methods for visualizing protein interactions include bioluminescence resonance energy transfer, which has characteristics similar to FRET, fluorescence lifetime imaging, and fluorescence correlation spectroscopy and related methods that are based on measurement of the diffusion of individual fluorescent molecules (Hink et al. 2002; Pfleger and Eidne 2006; see also Chapters 39–41). Fluorescence photobleaching also measures molecular interactions but generally the number and the identities of the interaction partners cannot be deduced from the data (Ren et al. 2008; see also Chapter 42).

Protein interactions can also be detected by taking advantage of the facilitated association of protein fragments that are fused to putative interaction partners. Fragments of many proteins, including ubiquitin, β-galactosidase, dihydrofolate reductase, β-lactamase, luciferases, and green fluorescent protein (GFP) family members can associate with each other to form active complexes. The association can be facilitated when the fragments are brought in proximity to each other by tethering them to interaction partners (Kerppola 2009). All assays that are based on the facilitated association of protein fragments share some characteristics, but unique properties of the different fragments and complexes produce important differences. The protocols described in this chapter make use of the facilitated association of fragments of the GFP family of fluorescent proteins. This approach, known as bimolecular fluorescence complementation (BiFC) analysis, can be used to detect interactions between proteins fused to nonfluorescent fragments of fluorescent proteins (Fig. 1) (Hu et al. 2002; Kerppola 2008).

Advantages and Limitations of BiFC Analysis

BiFC analysis has several unique advantages compared with other complementation assays and assays of protein interactions in general (Kerppola 2008, 2009). The intrinsic fluorescence of the complex enables detection of protein interactions with high sensitivity, fine spatial resolution, and minimal perturbation of the cells (see Protocol A). Studies in intact cells avoid the changes in protein interactions that can occur on lysis of the cell and mixing of the contents of different cellular compartments. Visualization of protein interactions in individual cells also enables investigation of differences in cellular processes among different cells in the population. Because interactions with different partners can occur in different subcellular locations, determination of the location of a protein complex can provide insight into its functional roles and regulation.

The excitation and emission spectra of BiFC complex fluorescence can be used to encode information about the protein interaction that produced the BiFC complex. This enables, among other things, visualization of several different complexes simultaneously in the same cell (see Protocol B). BiFC analysis can be used to study interactions without the need for structural information about the interaction partners. This approach can be used in virtually any aerobically grown cell type or organism that can be genetically modified to express fusion proteins. The fluorescence of BiFC complexes can be detected using widely available fluorescence microscopes, flow cytometers, and other instruments that are capable of detecting fluorescence. BiFC analysis, therefore, does not require elaborate instrumentation, data processing, or exogenous reagents apart from the fusion proteins.

BiFC analysis has several limitations. Chief among these are the potential for the fluorescent protein fragments to associate with each other independent of an interaction between proteins fused to the fragments and the stabilization of protein complexes by association of the fluorescent protein fragments. The latter, together with the time required for the chemical reactions that produce the fluorophore, make studies of the timing of protein interactions using BiFC analysis difficult. Most of these limitations can be overcome by the use of appropriate controls and by exercising due care in interpretation of the results.

Many proteins can interact with several alternative partners. These proteins can form many different complexes, often in the same cell. These interactions are frequently mutually exclusive such that an interaction with one partner excludes simultaneous interactions with other partners. The competition among alternative interaction partners is important for the specificity of protein interactions as favorable partners can displace less favorable partners. The fact that most proteins form complexes with a large number of alternative interaction partners poses a problem for many methods that are used to detect protein interactions because complexes formed with other partners reduce the sensitivity and increase the background of these assays. In the case of FRET analysis, fusion proteins that interact with other partners produce donor and acceptor fluorescence that may be affected by their interaction partners but no FRET signal. The interactions with other partners are less disruptive for BiFC analysis because these complexes are invisible to this assay, although they can reduce the efficiency of BiFC complex formation.

The multicolor BiFC assay enables simultaneous visualization of multiple protein interactions in the same cell. This assay is based on fusion of fragments of fluorescent proteins that form spectrally distinct BiFC complexes to different interaction partners. These complexes can thereby be spectrally resolved and the relative amounts of complexes formed with different interaction partners can be determined. The multicolor BiFC assay enables comparison of the subcellular distributions of complexes formed with different interaction partners and allows analysis of the competition between mutually exclusive interaction partners for binding a shared partner present at limiting concentration.

VISUALIZATION OF PROTEIN INTERACTIONS IN LIVE CELLS

The direct visualization of protein interactions in live cells is an important objective in many areas of research. The BiFC assay can be used to image protein complexes in their normal cellular environment. The assay is based on the association of two nonfluorescent fragments of a fluorescent protein, each of which is fused to a putative interaction partner. The fusion proteins used for BiFC analysis must be designed to reproduce the characteristics of their endogenous counterparts. The levels of fusion protein expression should be ideally equal to or lower than the levels of expression of their endogenous counterparts. The specificity of BiFC complex formation must be established by testing the effects of mutations that disrupt the interaction between unmodified proteins on BiFC complex formation. Many fluorescent protein fragments can be used for BiFC analysis. The choice of which fragments to use involves a trade-off between the fluorescence intensity and the potential for spontaneous association independent of a specific interaction between the proteins fused to the fragments. BiFC analysis can be used to detect an interaction within minutes after complex formation, but it does not provide a real-time measurement of the amount of complexes formed, and it does not allow for the detection of complex dissociation. BiFC analysis provides a technically straightforward strategy for visualization of protein complexes at high spatial resolution in live cells.

Principle of BiFC Analysis

BiFC analysis is based on the facilitated association of two fragments of a fluorescent protein (Fig. 1; Hu et al. 2002; Kerppola 2008). The association of these fragments can be facilitated by fusing them to proteins whose interaction tethers the fragments in proximity to each other. Thus, protein interactions can be visualized by fusing the putative interaction partners to two fragments of a fluorescent protein and by monitoring the fluorescence of the complex formed by the fusion proteins.

Design of Constructs for Fusion Protein Expression

BiFC analysis requires careful consideration of the design and expression of the fusion proteins for the results to be interpretable. The fusion proteins should reproduce the characteristics of their endogenous counterparts and should be expressed at levels comparable to the endogenous proteins.

Positions and Testing of Fusions

The fluorescent protein fragments can be fused to either the amino- or the carboxy-terminal end of each interaction partner. Thus, eight different permutations of fusions containing both fluorescent protein fragments can be tested (Fig. 2). The combinations of fusions to be tested and to be used in BiFC experiments should meet the following criteria.

1. The fusion proteins should reproduce the characteristics of the endogenous proteins. Ideally, this should be tested by replacing the endogenous protein with the fusion and testing the effects

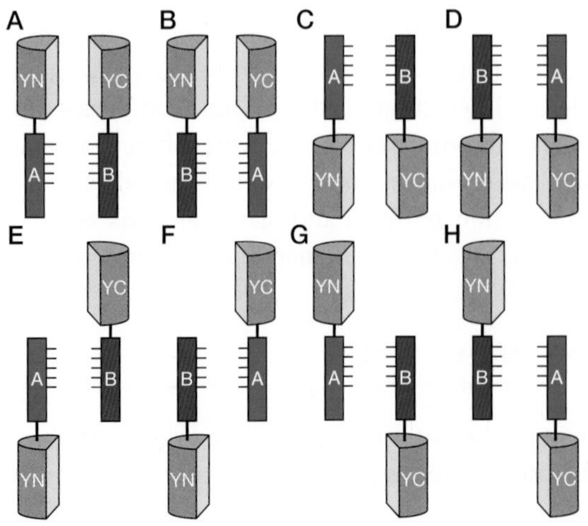

FIGURE 2. Multiple combinations of fusion proteins should be tested for BiFC. Amino- and carboxy-terminal fusions can be used to test eight distinct combinations (*A–H*). Although it may appear that combinations *E–H* would not be favorable for bimolecular complex formation, this will depend on the precise structures and the flexibilities of the fusion proteins, which are difficult to predict. The labels are the same as in Figure 1. (Modified from Hu and Kerppola 2005.)

of this replacement on relevant phenotypes. When this is not possible or practicable, the functions of the fusion proteins should be compared with those of the endogenous proteins using all available assays. In particular, the localizations and turnover rates of the fusion proteins and their endogenous counterparts should be compared.

2. The fusions should produce BiFC complexes when, and only when, the putative interaction partners associate with each other. Mutations and experimental conditions that block the interaction between the unmodified proteins should reduce or eliminate BiFC fluorescence.

Steric constraints can influence BiFC complex formation. However, the multiple permutations of fusion proteins that can be used generally include several combinations that produce detectable BiFC complexes (Fig. 2). Steric constraints can be further alleviated by adding flexible linker sequences between the fluorescent protein fragments and the putative interaction partners. Many different linker sequences have been used, but no systematic comparison of the effects of differences in linker length or sequence on BiFC complex formation has been reported.

Expression System

The fusion proteins should be expressed at levels comparable to their endogenous counterparts. Ideally, the fusion proteins should be expressed at the same levels in all cells. When this is not possible or practicable, an internal standard that reflects differences in fusion protein expression in individual cells should be included. Transgenic cells and organisms with appropriate levels of fusion protein expression can be produced using regulated expression vectors. The concentrations of the inducers used to activate fusion protein expression should be adjusted to produce levels of fusion protein expression comparable to those of the endogenous proteins. The fusion proteins can also be transiently expressed in cells. However, the levels of transiently expressed proteins generally vary among individual cells.

Controls

Interpretation of results from BiFC analysis requires several controls. Most importantly, the effects of mutations that are predicted to eliminate the interaction should be tested. The relative BiFC fluorescence intensities produced by fusions to wild-type and mutant interaction partners should be compared. Ideally, point mutations or small deletions that are not predicted to alter other properties of the fusion proteins, or the positions of the fluorescent protein fragments, should be used. If the interaction interface between the partners has not been established, it is possible to identify it using BiFC analysis. In this case, it is desirable to examine interactions with several different partners to avoid the possibility that mutations that alter protein folding are misinterpreted to constitute an interaction interface. If the same mutation affects interactions with several unrelated proteins, it is possible that it affects global protein folding rather than a specific interaction interface.

The fluorescent protein fragments alone or the unrelated fusion proteins are not valid controls because fusion of putative interaction partners to the fluorescent protein fragments can alter their potential for spontaneous association. The levels of expression and localization of the wild-type and mutant fusion proteins should be compared with each other to establish that any differences in BiFC complex formation are caused by altered protein interactions rather than by changes in fusion protein expression or localization.

Choice of Fluorescent Protein Fragments

A large number of different fluorescent protein fragments can be used for BiFC analysis. Different fragments have distinct characteristics, and it is important to consider these when deciding on the fragments to use in a particular study. Two different positions in GFP family members have been identified that enable facilitated BiFC complex formation when fused to interaction partners (Hu and Kerppola 2003). Because the fragments that contain partially overlapping sequences can also form BiFC complexes, the two positions increase the number of permutations of fusion proteins that can be used for BiFC analysis threefold. No marked differences in BiFC complex formation or in the characteristics of BiFC complexes formed by either of these combinations of fluorescent protein fragments have been reported.

Fluorescence Intensity and Spontaneous Fragment Association

Many, but not all, fragments of GFP family members that are homologous to yellow fluorescent protein (YFP) fragments can form BiFC complexes (Hu and Kerppola 2003). This provides an abundance of fragments that can be used for BiFC analysis. The importance of maintaining low levels of fusion protein expression and the ability to rapidly visualize the BiFC complexes place a high premium on the brightness of BiFC complexes.

A great deal of effort has been devoted to identifying bright BiFC complexes. The brightest complexes identified to date are based on fragments of Venus (Rackham and Brown 2004; Shyu et al. 2006). Venus is the brightest fluorescent protein identified by mutagenesis of intact fluorescent proteins (Nagai et al. 2002). However, analysis of the specificity of BiFC complex formation by Venus fragments has produced different results for different fusion proteins analyzed under different experimental conditions (Shyu et al. 2006; Saka et al. 2007; Robida and Kerppola 2009). Fusion of the Venus fragments to several different interaction partners has produced BiFC complexes whose fluorescence is significantly reduced by mutation of the regions of the proteins that mediate the interaction (Rackham and Brown 2004; Shyu et al. 2006). However, the Venus fragments form BiFC complexes efficiently in *Xenopus* embryos even when they are not fused to other proteins (Saka et al. 2007). Likewise, cells that express the amino-terminal fragment of Venus and the complementary carboxy-terminal fragments fused to the FK506 binding protein (FKBP) and the FKBP–rapamycin binding domain (FRB) show an increase in constitutive BiFC complex formation and correspondingly lower inducibility than cells that express fusions to the corresponding YFP fragments (Robida and Kerppola 2009). Based on these observations, special precautions should be taken to ensure that BiFC analysis using the amino-terminal fragment of Venus reflects a specific protein interaction.

The spontaneous BiFC complex formation by Venus fragments is reduced by mutation of any one of the four amino acid residues that differ between Venus and YFP (Saka et al. 2007). The Venus fragments containing the M153T substitution fused to Smad2 and Smad4 form inducible BiFC complexes in *Xenopus* embryos. The amino-terminal fragment of Venus containing the M153T substitution also forms inducible complexes with complementary carboxy-terminal fragments when fused to the FKBP and FRB interaction partners (Robida and Kerppola 2009). Cells that express these fusions have an order of magnitude higher fluorescence than cells that express fusions containing the corresponding YFP fragments both in the absence and in the presence of an inducer. The amino-terminal fragment of Venus containing the M153T substitution, therefore, provides a compromise with higher fluorescence intensity but also a higher background. Table 1 compares the advantages and disadvantages of different fluorescent protein fragments for BiFC analysis.

Quantifying the Efficiency of BiFC Complex Formation

The fluorescence intensities of cells that express fusions to fluorescent protein fragments vary for numerous reasons, complicating comparison of the efficiencies of protein interactions using BiFC analysis. It is possible to compare the efficiencies of BiFC complex formation by closely related proteins or by mutants containing single amino acid substitutions or small deletions that do not alter the positions of the fluorescent protein fragments (Hu et al. 2002; Hu and Kerppola 2003; Grinberg et al. 2004). Such comparisons require normalization for the level of protein expression in individual cells. The levels of fusion protein expression in individual cells can be estimated by coexpressing an intact fluorescent protein that has a spectrum that does not overlap with the BiFC complex together with the fusions to fluorescent protein fragments. The mean levels of fusion protein expression can be estimated by immunoblotting using antibodies to the same epitope on wild-type and mutant proteins. One such epitope is the fluorescent protein fragment itself. If the mean levels of

TABLE 1. Advantages and disadvantages of different combinations of fluorescent protein fragments in BiFC analysis

#	Amino-terminal fragment[a]	Carboxy-terminal fragment[a]	Advantages	Disadvantages	Reference(s)[b]
1	YN155	YC155	Low spontaneous association	Weak fluorescence intensity	Hu et al. 2002; Robida and Kerppola 2009
2	YN173	YC173			Hu and Kerppola 2003
3	CN155	CC155	Low spontaneous association, multicolor analysis	Weak fluorescence intensity	Hu et al. 2002; Robida and Kerppola 2009
4	YN155	CC155			
5	VN155 or VN173	YC155 or VC155 or VC173	High fluorescence intensity versus #1, #4, #7, and #8	High spontaneous association intensity versus #1, #4, #7, and #8	Shyu et al. 2006; Saka et al. 2007; Robida and Kerppola 2009
6	VyN155	VC155	Reduced spontaneous association versus #5	None noted	Saka et al. 2007
7	VyN155	YC155	Reduced spontaneous association versus #5; increased fluorescence intensity versus #1 and #4	Reduced fluorescence intensity increased spontaneous association versus #1 and #4	Robida and Kerppola 2009
8	VyN155	CC155			

[a]The letters indicate the fluorescent proteins from which each of the fragments were derived (YN and YC, YFP; CN and CC, CFP; VN and VC, Venus; CrN, Cerulean; VyN, Venus with M153T substitution). The numbers indicate the positions in which the fragments were truncated.

[b]The list of references is not comprehensive.

fusion protein expression are equivalent, the efficiencies of BiFC complex formation can then be determined by dividing the fluorescence intensity produced by the BiFC complex by the fluorescence intensity produced by the intact fluorescent protein.

Protocol A

BiFC Analysis of Protein Interactions in Live Cells

This protocol focuses on imaging protein interactions in cultured mammalian cells but can be readily adapted to any cell type or aerobically grown organism that can be genetically modified to express the fusion proteins.

Imaging Setup

The fluorescence of cells containing BiFC complexes can be detected using any instrument capable of detecting fluorescence emitted by GFPs. The subcellular localization of BiFC complexes can be visualized using a fluorescence microscope with confocal or epi-illumination. Given the low fluorescence intensity characteristic of BiFC experiments, a high-numerical-aperture (high-NA) objective and a high-performance charge-coupled device (CCD) camera are recommended. The filters used to select the excitation and emission wavelengths should be optimized for detection of the fluorescence of the BiFC complexes. Generally, filters optimized for detection of the intact fluorescent proteins can be used because the spectra of the BiFC complexes that have been characterized closely match those of the intact proteins from which the fragments have been derived.

The fluorescence intensities of BiFC complexes can be measured in individual cells using flow cytometry. Because the fluorescence intensity under conditions of low-level fusion protein expression is frequently only moderately higher than the signal due to light scatter, it is important to design the experiments so as to minimize and to correct for irrelevant signals.

MATERIALS

Reagents

Cell culture medium or other medium appropriate for maintaining normal functions of cells/organisms

Cells or organism of interest

Note that protein interactions in normal cells and tissues may not be faithfully reproduced in immortalized cell lines.

Expression vectors containing the DNA sequences encoding each fusion protein

The expression vectors encode the proteins of interest fused to the complementary fluorescent protein fragments. For many purposes, the amino- and carboxy-terminal fragments of YFP are recommended.

Alternatively, retroviral or lentiviral expression vectors can be used to introduce DNA into primary cells or other cell types that cannot be readily transfected.

Transfection reagents or other strategies to introduce DNA into the cells or organism of interest

If the goal is stable transfectants, then reagents that enable selection for and identification of cells or organisms that express the fusion proteins are needed.

Equipment

Culture chambers that maintain conditions for normal function of cells/organisms

The culture vessels should be designed for imaging if the cells are to be visualized by microscopy.

Flow cytometer (see Imaging Setup above for details)

Fluorescence microscope with confocal or epi-illumination, high-NA objectives, and a CCD camera (see Imaging Setup above for details)
Incubator for cell culture

EXPERIMENTAL METHOD

1. Culture cells to ~50% confluence in appropriate medium to allow vigorous growth subsequent to transfection or transduction. Culture the cells in dishes that are suitable for the intended fluorescence detection method.

   Adherent cells are easiest to visualize by microscopy, whereas nonadherent cells can be analyzed by flow cytometry. Parallel analysis using both methods enables both visualization of the subcellular localization of the complexes and measurement of their fluorescence intensities. Coverslip chambers are convenient for imaging cells by microscopy using short working-distance objectives.

2. Transfect the cells with appropriate amounts of plasmids, or transduce with viruses encoding the putative interaction partners fused to complementary fluorescent protein fragments (e.g., A-YN155 and B-YC155). Use the minimal amount of plasmid or virus necessary to produce a detectable signal.

   As a control, use plasmids encoding fusions to proteins that contain mutations that eliminate the interaction (i.e., A(mutant)-YN155 and B-YC155). To compare BiFC complex formation by wild-type and mutant proteins, cotransfect or transduce a plasmid or virus encoding an intact fluorescent protein with the fluorescent protein fragment fusions.

3. Culture the cells to allow the proteins to be expressed. Visualize the complexes as soon as possible after transfection to minimize the possibility that expression of the fusion proteins affects the properties of the cells.

   Generally 12–36 h is sufficient to visualize specific complexes. Longer expression times generally result in overexpression of the fusion proteins and increase the potential for nonspecific BiFC complex formation.

   Incubation of cells at 30°C before imaging often increases the fluorescence intensities of BiFC complexes. However, this incubation can also alter protein interactions or the properties of protein complexes. In some cases, this incubation can also increase BiFC complex formation by noncognate interaction partners and reduce the selectivity of BiFC complex formation.

4. If needed, wash the cells to remove dead cells and cell debris, which can produce spurious signal.

5. Visualize BiFC complexes using a fluorescence microscope with an appropriate objective. Confirm that the cells show normal and/or expected morphology.

   Higher-NA objectives increase the intensity of illumination and generally allow more sensitive detection of fluorescence.

   To ensure that subjective factors do not influence interpretation of the data, perform the imaging using double blind experimental design.

   See *Troubleshooting*.

6. Compare the fraction of cells that has detectable BiFC complexes with the proportion of cells that expresses the fusion proteins determined by coexpression of an intact fluorescent protein.

   It is common for the fraction of cells that produces detectable BiFC fluorescence to be smaller, but a difference of more than 10-fold may indicate that additional factors or signals are required for the interaction.

   See *Troubleshooting*.

7. Determine the distribution of BiFC complex fluorescence in a representative subpopulation (tens or hundreds depending on the heterogeneity of the distribution) of cells.

8. Using either fluorescence microscopy or flow cytometry, quantify BiFC complex formation by measuring the fluorescence intensities of BiFC complexes and of the intact fluorescent protein in the same cells. The ratio of the fluorescence intensities can be used to compare BiFC complex formation by wild-type and mutant proteins.

   It is important to ensure that there is no cross talk between the signals by analyzing cells that express the BiFC complexes and intact fluorescent proteins in separate cells. Although fluorescence intensities can be measured by microscopy, it is often preferable to measure them by flow cytometry to sample a representative subpopulation of cells.

9. Measure the levels of fusion protein expression by preparing cell extracts and analyzing them by immunoblotting.

   The extracts can generally be stored for a few days when denatured and frozen.

   It is important to compare the levels of fusion protein expression with that of their endogenous counterparts. The levels of expression should be corrected for the fraction of cells that expresses the fusion proteins.

   The levels of expression of different fusion proteins can be conveniently compared by the use of antibodies that recognize the fluorescent protein fragments. Note that the amino-terminal and carboxy-terminal fragments of fluorescent proteins are generally not recognized with the same efficiency by antibodies.

   See *Troubleshooting.*

10. Compare the distributions of the wild-type and mutant fusion proteins as well as their endogenous counterparts by indirect immunofluorescence analysis.

    It is important to determine if ectopic expression of the fusion proteins alters their distributions compared with their endogenous counterparts.

    See *Troubleshooting.*

TROUBLESHOOTING

Problem (Step 5): The expression of some fusion proteins causes changes in cell growth or morphology.
Solution: Generally, it is best to visualize the protein complexes before any changes in cell growth or morphology occur. If this is not possible, it is advisable to use conditional expression vectors for production of the fusion proteins to control the level and the timing of fusion protein expression. If this does not solve the problem, another cell type or a different permutation of fusion proteins should be tested.

Problem (Step 6): No fluorescence is detected, or the fluorescence intensity of the BiFC complex is similar to that produced by the negative control in which the interaction interface is mutated.
Solution: It is important to test several different permutations of fusion proteins to determine the combination of fusions that produces the highest signal relative to the negative control. If none of the combinations produces a BiFC signal higher than the negative control, then BiFC analysis does not support the interpretation that the proteins under investigation interact in the cells used.

Problem (Step 9): High levels of expression of the intact fluorescent proteins used for normalization influence BiFC complex formation.
Solution: The intact fluorescent proteins should be expressed at the lowest levels that are sufficient to quantify their fluorescence intensities. Generally, 10-fold lower concentrations of the plasmids used for normalization can be used.

Problem (Steps 9 and 10): The levels and distributions of the fusion proteins differ from those of the endogenous proteins, or the wild-type and mutant protein differ from each other.

Solution: It is generally best to analyze BiFC complex formation by the lowest levels of fusion proteins that can be detected. The levels of fusion protein expression can be controlled by using regulatable expression vectors and by visualizing complex formation as soon after protein expression as possible. The effects of mutations that alter the levels or distributions of fusion protein expression are difficult to interpret. It is generally best to make additional mutations that only affect BiFC complex formation.

DISCUSSION

Evaluation of Results

If fluorescence is detected when wild-type proteins fused to the fluorescent protein fragments are expressed, and this signal is eliminated or significantly reduced by mutations that do not affect the expression or localization of the protein, it is likely that the signal reflects a specific interaction. If mutations that are known to eliminate the interaction of the wild-type proteins do not eliminate the fluorescence, then the bimolecular complementation is due to nonspecific interactions between the chimeric fusion proteins. The levels of fusion protein expression may be too high. A different combination of fusion proteins or linkers could also be tested.

If no fluorescence is detected, no conclusions about the possibility of an interaction can be drawn. If fluorescence can be induced by a change in conditions, it is possible that the interaction is regulated. However, it is important to test if BiFC complex formation by other fusion proteins is affected by the same change in conditions to eliminate the possibility that the altered conditions directly affect BiFC complex formation.

Comparison with Alternative Assays of Protein Interaction

Because of the minimal perturbation caused by fluorescence imaging, cells can be monitored for extended periods as long as culture conditions necessary to sustain normal cellular functions can be maintained. Because of the relatively low fluorescence intensity of BiFC complexes, care should be taken to avoid unnecessary bleaching of the fluorescence. The cells should be illuminated only for the time needed for imaging, and the time intervals between images should be adjusted based on the total period of imaging needed.

BiFC analysis is often the most direct approach for visualizing an interaction between proteins of interest. BiFC analysis is generally most valuable in cases in which determination of the subcellular localization of the complex is of particular importance. However, if the goal is to determine the timing of complex formation or to detect complex dissociation, other assays of protein interactions such as FRET analysis are more useful. Similarly, if it is essential to detect the interaction between endogenous proteins such as in the case of clinical samples in which genetic modification is not possible, assays such as proximity ligation can be used.

A particular advantage of BiFC analysis is that the approach does not require any processing of the image data. This avoids the possibility that assumptions required for such image processing corrupt the interpretation. It is often necessary to compare the fluorescence intensities of cells that express different combinations of fusion proteins. In this case, measurement of mean fluorescence intensities relative to an internal normalization standard is often sufficient. To classify the subcellular localizations of different complexes, it is often convenient to compare the distribution of the BiFC fluorescence with the distributions of markers of specific subcellular structures. In this case, the comparison is often subjective, especially when the distributions do not overlap perfectly. It is possible to avoid such indirect comparisons by direct comparison of the distributions of two protein complexes within the same cell (see Protocol B).

MULTICOLOR BiFC ANALYSIS OF INTERACTIONS WITH ALTERNATIVE PARTNERS

Many cellular proteins have a large number of alternative interaction partners. Often these interactions are mutually exclusive, resulting in competition for shared interaction partners in cells that express several alternative partners. This competition is likely to be a critical determinant of the specificity of protein interactions in the normal cellular environment. The multicolor BiFC assay can be used to simultaneously visualize interactions between multiple combinations of proteins within the same cell. This assay is based on the formation of BiFC complexes with different spectra. Each complex forms through the association of a different combination of fluorescent protein fragments. These fragments are fused to alternative interaction partners such that complexes formed with different partners can be visualized independently in the same cell using different excitation and emission wavelengths.

Multicolor BiFC analysis can be used to compare the subcellular distributions of multiple complexes in the same cell. Comparison of complex distributions within the same cell eliminates problems with variations in the localization of fiduciary markers and potential changes in cellular organization caused by protein expression. Multicolor BIFC analysis can also be used to compare the efficiencies of BiFC complex formation by several proteins with a common interaction partner. Two alternative methods for quantification of the relative efficiencies of protein interactions using BiFC analysis are described. Parallel analysis using both approaches is recommended. The multicolor BiFC assay provides a unique approach for visualization of the coupled networks of protein interactions in living cells.

Principle of Multicolor BiFC Analysis

Multicolor BiFC analysis is based on complementation between different combinations of fluorescent protein fragments that produce BiFC complexes with distinct spectra (Fig. 3) (Hu and Kerppola 2003; Grinberg et al. 2004). To investigate the competition between two alternative interaction partners (e.g., A and B) for a shared partner (e.g., Z), the three proteins are fused to fragments of different fluorescent proteins that can form spectrally distinct bimolecular fluorescent complexes (i.e. A-YN155, B-CN155, and Z-CC155 can interact to form A-YN155–Z-CC155 and B-CN155–Z-CC155, see Fig. 3). The distributions and the fluorescence intensities of the two bimolecular fluorescent complexes formed by the alternative interaction partners can be compared in the same cells. For the purposes of this chapter, the phrase fluorescence intensities of BiFC complexes and variations thereof will be used to refer to the fluorescence intensities of individual cells containing specific BiFC complexes. In the following sections, we discuss the design and optimization of a multicolor BiFC experiment in live cells. Protocol B outlines the steps for carrying out such an experiment.

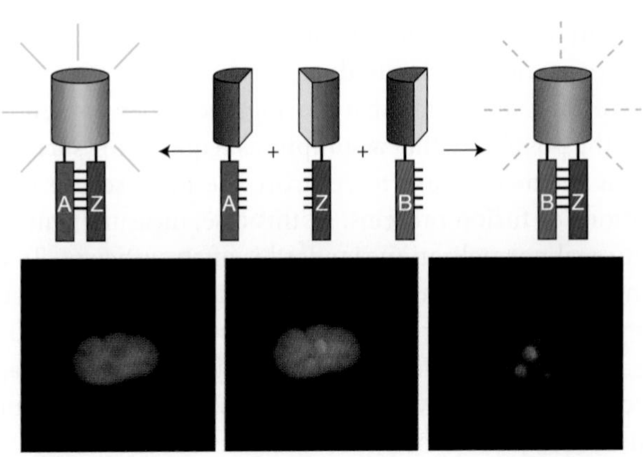

FIGURE 3. Principle of multicolor BiFC analysis. Two alternative interaction partners A and B are fused to fragments of different fluorescent protein fragments (tinted half-cylinders). These fusions are coexpressed in cells with a shared interaction partner Z, fused to a complementary fragment (gray half-cylinder). Complexes formed by A and Z can be distinguished from complexes formed by B and Z based on the differences in their fluorescence spectra. The images under the diagrams show an example of the visualization of two protein complexes in the same cell, one nucleolar (cyan) and the other nucleoplasmic (yellow). (Reproduced from Hu and Kerppola 2003, ©2003, Nature Publishing Group.)

Design of Constructs for Multicolor BiFC Analysis

The basic principles for the design of fusion proteins for BiFC analysis also apply to multicolor BiFC analysis. In particular, the distributions and functions of the fusion proteins must correspond to those of their endogenous counterparts for the data from multicolor BiFC analysis to reflect the properties of native proteins.

Choice of Compatible Fluorescent Protein Fragments

To visualize multiple BiFC complexes within the same cell, these complexes must have distinct spectra. It is advantageous if the complexes are detected with similar sensitivity so that the signal from one complex does not overwhelm the signal from another complex. Many combinations of fluorescent protein fragments produce spectra that can be separated using spectral deconvolution. Moreover, some combinations of fragments produce BiFC complexes with spectra that have so little overlap that cross talk between the signals can be eliminated simply by choosing excitation and emission filters that selectively detect different BiFC complexes. In particular, complexes formed by the carboxy-terminal fragment of cyan fluorescent protein (CFP) (CC) with the amino-terminal fragment of yellow fluorescent protein (YFP) (YN) versus the amino-terminal fragment of cyan fluorescent protein (CN) have well-separated excitation and emission spectra.

Fusion Proteins for Analysis of the Efficiency of Complex Formation

The multicolor BiFC approach can be used to compare the relative efficiencies of complex formation by alternative interaction partners that are related to each other and that share the same interaction interface. Fusion proteins that are to be used for quantitative analysis of the relative efficiencies of complex formation must meet several additional requirements. The fluorescent protein fragments must be fused to the alternative interaction partners in the same manner. The fluorescent protein fragments must be fused to the same positions relative to the interaction interfaces of the alternative interaction partners using the same linker sequences. It is generally advantageous to use long and flexible linker sequences so as to minimize the effects of steric constraints on the efficiencies of BiFC complex formation.

Expression of Fusion Proteins

To compare the relative efficiencies of BiFC complex formation, the fusion to the common interaction partner must be expressed at a lower concentration than the fusions to the alternative partners. To simplify interpretation of the data, it is generally advantageous to express the alternative interaction partner fusions at similar concentrations. It is also beneficial if the levels of fusion protein expression are similar to or lower than the levels of expression of their endogenous counterparts.

Controls

To calibrate the multicolor BiFC assay, the different fluorescent protein fragments should be fused to the same interaction partners, and their efficiencies of BiFC complex formation should be measured within the same cell. This allows correction for differences in quantum yields and detection efficiencies of BiFC complexes with different spectral characteristics as well as any intrinsic differences in the efficiencies of BiFC complex formation by different fluorescent protein fragments. To determine if the identities of the fusion proteins affect the relative efficiencies of BiFC complex formation, it is essential to test the effect of exchanging the fluorescent protein fragment fusions between the alternative interaction partners. It is also important to test the effects of mutations in the interaction partners whose effects on their binding specificities can be predicted. Such mutations should be selected so as to not alter the levels of expression or the localization of the fusion proteins.

Visualizing the Localization of Multiple Complexes within the Same Cell

The distributions of BiFC complexes formed by different combinations of fusion proteins can be compared by coexpression of the fusions within the same cells. It is important to compare the distributions of the BiFC complexes in cells that express alternative interaction partners with the distributions of BiFC complexes in cells that express only one pair of interaction partners. Differences between these distributions suggest that the complexes affect the distributions of each other. If the cells already contain the endogenous counterparts of the fusion proteins, such effects of BiFC complexes on the distributions of each other suggest that either overexpression or other effects of the fusion proteins affect BiFC complex distributions.

Quantifying the Relative Efficiencies of BiFC Complex Formation by Alternative Interaction Partners

We have developed two strategies for quantifying the relative efficiencies of complex formation by alternative interaction partners using multicolor BiFC analysis. Both methods can provide information about the relative efficiencies of complex formation by mutually exclusive interaction partners in living cells. The first approach, designated as absolute competition, is based on comparison of the absolute fluorescence intensity of BiFC complexes formed in the presence and absence of the competitor fusion. This approach can be applied to essentially any competitor, but the method can reliably detect only large differences in the efficiencies of complex formation between alternative partners because the fluorescence intensities of the BiFC complexes in the presence and absence of the competitor must be quantified in separate populations of cells.

The second approach, designated as relative competition, is based on comparison of the fluorescence intensities of BiFC complexes formed by different combinations of fluorescent protein fragments fused to alternative interaction partners within the same cells. The relative competition approach is more precise, but it can only be used to analyze competition among closely related interaction partners. It is affected by any difference in steric constraints associated with the fusion of fluorescent protein fragments to alternative interaction partners.

It should be noted that neither the absolute nor the relative competition approach provides information about the absolute amount of complexes formed or the binding affinities of the complexes but only about the relative efficiencies of BiFC complex formation. The absolute and relative competition approaches make use of the same fusion proteins and can be used to analyze results from the same experiment if the controls and calibration standards necessary for each approach are performed. The absolute and relative competition approaches should ideally be used in parallel to eliminate some of the caveats inherent in each approach.

Limitations of Multicolor BiFC Analysis

Effects of Differences between the Fluorescent Protein Fragments on Multicolor BiFC Analysis

The fluorescent protein fragments fused to the interaction partners can influence complex formation. Because different fluorescent protein fragments are fused to alternative partners, differences between the effects of these fragments on BiFC complex formation might affect the results. It is, therefore, essential to exchange fluorescent protein fragments between the alternative partners to determine if the identities of the fusions affect the relative efficiencies of complex formation. If the exchange of fluorescent protein fragments between alternative interaction partners does not produce the reciprocal result, then the identities of the fluorescent protein fragments could affect the specificities of BiFC complex formation.

The effects of different fluorescent protein fragments on BiFC complex formation can also be compared in vitro using purified fusion proteins (Hu and Kerppola 2003). Because the experimental conditions in vitro can be more carefully controlled, it is possible to produce a known ratio of alternative interaction partners and to measure whether the ratio of BiFC complexes formed agrees with the prediction. If the amounts of different BiFC complexes formed are proportional to the amounts of the fusion proteins added to the reaction, then differences between the fluorescent protein fragments do not influence the relative efficiencies of BiFC complex formation.

Effects of Complex Stabilization by Fluorescent Protein Fragment Association

Because BiFC complex formation is generally irreversible, multicolor BiFC analysis can be used to compare the efficiencies of complex formation by alternative partners but not to determine the relative binding affinities of alternative partners at equilibrium (Hu and Kerppola 2003). Because association of the fluorescent protein fragments is slow relative to the rates of exchange of most regulatory protein complexes, the interactions between the fusion proteins can equilibrate before BiFC complex formation. However, the proportions of the complexes already formed do not change in response to shifts in this equilibrium. Differences in the efficiency of fragment association or the degree of stabilization conferred by formation of BiFC complexes by different fragments could distort the competition between alternative partners. To establish if the identities of the fluorescent protein fragments affect the relative efficiencies of complex formation, it is necessary to exchange the fluorescent protein fragments between the alternative partners.

In addition to the rates of complex formation, the relative amounts of BiFC complexes in cells are affected by the rates of fusion protein synthesis and degradation. It is essential to establish the steady-state levels of the alternative interaction partners in cells and to take these into account when interpreting results from multicolor BiFC analysis. It is also necessary to consider the possibility that differences in the levels of BiFC complexes are caused by differences in turnover of BiFC complexes formed by different interaction partners.

Advantages and Limitations of Absolute and Relative Competition Approaches to Quantify Relative Efficiencies of Complex Formation

The difference between the absolute and relative methods for quantification of multicolor BiFC data is that in the absolute approach, the effects of competitors on the efficiencies of bimolecular fluorescent complex formation are compared in separate cells, whereas in the relative approach, the relative efficiencies of formation of two different bimolecular fluorescent complexes are compared in the same cell. The former is subject to uncertainties caused by variations between different cells, whereas the latter is subject to variations between different bimolecular fluorescent complexes. Whenever possible, it is best to use both the absolute and the relative competition methods in parallel to eliminate potential caveats inherent in each approach. If the two methods produce consistent results, then the relative efficiencies of competition between the interaction partners can be inferred from the data. However, if the two approaches give inconsistent results, then the proteins under investigation may not meet the criteria for analysis by the multicolor BiFC approach.

The absolute fluorescence intensities of bimolecular fluorescent complexes formed by different structurally unrelated interaction partners cannot be used to compare their efficiencies of complex formation because of steric effects on BiFC complex formation. The absolute competition approach compares the efficiencies of BiFC complex formation by the same fusions in the absence and presence of different competitors. This circumvents the effects of differences in the steric arrangements of the fluorescent protein fragments between different interaction partners. The relative competition approach compares the relative efficiencies of BiFC complex formation by structurally related alternative interaction partners in the same cell. The structural similarity of interactions between different partners enables comparison of their efficiencies of BiFC complex formation in the same cell.

The absolute competition approach is well suited for analysis of large differences in efficiency of complex formation between mutually exclusive interaction partners. The advantage of the absolute competition approach is that it is not affected by differences in the efficiencies of fluorescent protein fragment association because the level of only one BiFC complex is measured. The disadvantage of the absolute competition approach is that it is subject to error caused by variations between different cell populations because it involves comparison of BiFC complex formation between cell populations that express different competitors. Because the levels of protein expression in transient transfection experiments often vary over a large range, it is necessary to compare the average fluorescence intensities of large and representative cell populations. It is, therefore, only possible to detect large differences in complex formation using the absolute competition approach.

The relative competition approach is well suited for comparison of the efficiencies of complex formation by structurally related alternative interaction partners. The advantage of the relative competition approach is that relatively small differences in the efficiencies of complex formation can be detected. The disadvantage of the relative competition approach is that the efficiencies of complex formation by structurally unrelated alternative interaction partners cannot be compared because the differences in steric constraints can dominate the efficiencies of BiFC complex formation.

Protocol B

Simultaneous Visualization of Multiple Protein Interactions Using Multicolor BiFC Analysis

This multicolor BiFC protocol focuses on analysis of alternative protein interactions in cultured mammalian cells but can be readily adapted to any cell type or aerobically grown organism that can be genetically modified to express the fusion proteins.

IMAGING SETUP

The imaging setup is similar to that used in Protocol A. The imaging equipment or flow cytometer needs to be able to separate the fluorescent signals produced by the BiFC complexes formed by the alternative interaction partners.

MATERIALS

Reagents

Cell culture medium or other medium appropriate for maintaining normal functions of cells/organisms

Cells or organism of interest

Expression vectors containing the DNA sequences encoding each fusion protein

This protocol requires the construction of plasmid or viral expression vectors encoding the alternative interaction partners fused to fragments of different fluorescent proteins and the common interaction partner to a complementary fragment such that BiFC complexes formed by different combinations of the fusions have distinct excitation and emission spectra.

Transfection reagents or other strategies for introduction of DNA into the cells or organism of interest

If the goal is stable transfectants, then reagents that enable selection for and identification of cells or organisms that express the fusion proteins are needed.

Washing medium or buffer

Equipment

Culture chambers that maintain conditions for normal function of cells/organisms

The culture vessels should be designed for imaging if the cells are to be visualized by microscopy.

Flow cytometer (see Imaging Setup above for details)

Fluorescence microscope with confocal or epi-illumination, high-NA objectives, and a CCD camera (see Imaging Setup above for details)

Incubator for cell culture

EXPERIMENTAL METHOD

1. Seed the cells at an appropriate density to allow for growth after the cells are transfected or transduced.

Adherent cells are easiest to visualize by microscopy, whereas nonadherent cells can be analyzed by flow cytometry.

2. Transfect the cells with appropriate amounts of plasmids, or transduce with viruses encoding alternative interaction partners fused to fragments of different fluorescent proteins (i.e., A-YN155 and B-CN155) and a common interaction partner fused to a complementary fragment (i.e., Z-CC155).

   As a reference for absolute competition, transfect or infect each pair of expression vectors (A-YN155 and Z-CC155, as well as B-CN155 and Z-CC155) into separate cells. As a calibration standard for relative competition, transfect or infect each fluorescent protein fragment fused to the same interaction partners (i.e., A-YN155, A-CN155, and Z-CC155) into the same cells.

3. Culture the cells to allow the proteins to be expressed. Visualize the complexes as soon as possible after transfection to minimize the possibility that expression of the fusion proteins affects the properties of the cells.

4. If needed, wash the cells to remove dead cells and cell debris, which can produce a spurious signal.

5. Visualize the BiFC complexes within the cells by fluorescence microscopy. Compare the distributions of BiFC complexes formed by alternative interaction partners. Also compare the distributions of BiFC complexes in cells that do not express alternative interaction partners with those that do express alternative partners.

   The fluorescence emissions of A-YN155–Z-CC155 and B-CN155–Z-CC155 complexes can generally be separated by using 436/10-nm excitation and 470/30-nm emission windows for CN155–CC155 complexes and 500/20-nm excitation and 535/30-nm emission windows for YN155–CC155 complexes.

   The number of cells that must be imaged depends on the variability in the distributions of the BiFC complexes within different cells.

   See *Troubleshooting*.

6. Determine the amounts of BiFC complexes formed between alternative partners by measuring the fluorescence intensities of BiFC complexes formed in the absence and presence of alternative interaction partners as well as the ratio between the fluorescence intensities of BiFC complexes formed by alternative interaction partners in the same cell.

   It is preferable to measure the fluorescence intensities of BiFC complexes using excitation and emission filters that minimize the cross talk between the signals from BiFC complexes formed by alternative interaction partners. If such cross talk cannot be avoided, it can be corrected for by measuring the ratio of the fluorescence intensities when the BiFC complexes are produced in separate cells.

   The fluorescence intensities can be measured by microscopy or flow cytometry. Flow cytometry is often preferable because it allows sampling of a representative subpopulation of cells.

   See *Troubleshooting*.

7. Calculate the change in the fluorescence intensity of BiFC complexes in the presence versus the absence of competitor for absolute competition analysis.

   See *Troubleshooting*.

8. Calculate the ratio of the fluorescence intensities of BiFC complexes formed by the alternative interaction partners, and compare this ratio to the ratio of the fluorescence intensities produced by BiFC complex formation by different fluorescent protein fragments fused to the same interaction partners.

   The ratio of fluorescence intensities is most effectively compared by plotting the data as a scatterplot and by comparing the slope and spread of the resulting pattern.

9. Determine the levels of fusion protein expression in the cells by preparing cell extracts and analyzing them by immunoblotting using antibodies directed against epitope tags common to the alternative interaction partners.

The frozen lysates can generally be stored for a few days when denatured and frozen.

If necessary, the amounts of fusion proteins that are expressed can be adjusted by changing the amount of plasmid or virus introduced into the cells. The levels of protein expression can be more accurately controlled by the use of promoters whose levels of transcription can be regulated.

TROUBLESHOOTING

Problem (Step 5): The fluorescence intensities of complexes formed with alternative interaction partners differ so much that they cannot be accurately measured under the same conditions.

Solution: The fluorescence intensities of the two complexes can be measured using different camera settings. Although it is ideal to measure the fluorescence intensities of both BiFC complexes accurately, it is not essential. If the fluorescence intensity of one complex is reduced to the point of being low or even being undetectable in the presence of the competitor, it indicates that the complex does not form efficiently.

Problem (Steps 6 and 7): Exchange of the fluorescent protein fragments between the interaction partners does not produce reciprocal results.

Solution: Nonreciprocal results from the reciprocal exchange of fluorescent protein fragments indicates that the fluorescent protein fragments influence the interactions between the partners. To uncouple the interactions between the partners from those of the fluorescent protein fragments, increase the length of the linkers connecting the two. It may also be necessary to use different fluorescent protein fragments.

Problem (Steps 6 and 7): Absolute and relative competition approaches produce different results.

Solution 1: Differences between the structures of the alternative interaction partners may cause a difference in the steric constraints with association of the fluorescent protein fragments. Long and flexible linkers may help reduce the steric constraints.

Solution 2: The plasmid or viral expression vectors may affect the expression of each other. Reduce the amounts of plasmids or viruses introduced into the cells, or change the promoters used to express the fusion proteins.

DISCUSSION

Evaluation of Results

In the absolute and relative competition approaches, the relative efficiencies of complex formation should be consistent with each other. Moreover, exchange of the fluorescent protein fragments between the interaction partners should produce reciprocal results. Thus, in the absolute competition approach, exchange of the fragments should not affect the change in the fluorescence intensity. In other words, if A-YN155–Z-CC155 fluorescence is reduced 90% by coexpression of B-CN155, then A-CN155–Z-CC155 fluorescence should also be reduced 90% by coexpression of B-YN155. Likewise, when using the relative competition approach, the relative fluorescence intensities of the YN–CC and CN–CC complexes fused in one combination should be the converse of the relative intensities when the fragments are exchanged. Thus, the normalized, relative fluorescence intensities of cells coexpressing A-YN155–Z-CC155 and B-CN155–Z-CC155 should be reciprocal to those of cells coexpressing B-YN155–Z-CC155 and A-CN155–Z-CC155.

If exchange of the fluorescent protein fragments does not result in the predicted changes in relative fluorescence intensities, then it is possible that the fluorescent protein fragments influence the relative efficiencies of competition by the alternative interaction partners, and the relative efficiencies of complex formation cannot be determined using multicolor BiFC analysis.

Comparison with Alternative Approaches to Compare Different Protein Complexes

The subcellular locations of complexes formed by different interaction partners can also be compared between different cells by using fiduciary markers whose locations in different cells are assumed to be invariant. It is often difficult or impossible to identify fiduciary markers that have the same distribution as the complex under investigation. Comparison of complex localization using fiduciary markers with imperfectly overlapping distributions is less reliable. The ability to compare the distributions of multiple complexes in the same cell is a significant advantage of multicolor BiFC assays over alternative methods for comparison of the distributions of protein complexes.

VISUALIZATION OF UBIQUITIN CONJUGATES USING UBIQUITIN-MEDIATED FLUORESCENCE COMPLEMENTATION

Ubiquitin-family peptide conjugation regulates the functions and stabilities of many proteins. Numerous cellular proteins are modified by covalent conjugation of ubiquitin-family peptides to specific lysine residues. These modifications provide a flexible means for regulating the properties of the substrate proteins. Because ubiquitin can be conjugated to substrate proteins at many different sites and in many topological configurations, these modifications have the potential to confer a wide range of functional states to the modified proteins.

Ubiquitin conjugation is typically detected by immunoprecipitation of a putative substrate protein followed by immunoblotting to detect ubiquitin conjugated to the substrate. This assay cannot be used to detect ubiquitin conjugates in live cells. It is also difficult to determine the subcellular distribution of a specific ubiquitin conjugate using this approach. To visualize ubiquitin conjugates in live cells, we developed the UbFC assay. This assay is based on the association of fragments of fluorescent proteins when ubiquitin fused to one fragment is conjugated to a substrate protein fused to a complementary fragment (Fig. 4; Protocol C; Fang and Kerppola 2004).

Ubiquitin conjugation to different substrates often has distinct effects on their functions. Likewise, conjugation of different ubiquitin-family peptides to the same substrate can have distinct effects on its properties. Multicolor UbFC analysis enables simultaneous visualization of conjugates formed by different ubiquitin-family peptides as well as conjugates on different substrates within the same cell (Fig. 4) (Fang and Kerppola 2004). When ubiquitin conjugation to different substrates or the conjugation of different ubiquitin-family peptides to the same substrate is studied in different cells, it is often difficult to determine if distinct effects are caused by indirect effects of the expression of the substrates or the ubiquitin-family peptides in different cells. UbFC analysis enables direct comparison of the effects of different modifications on the properties of the conjugates.

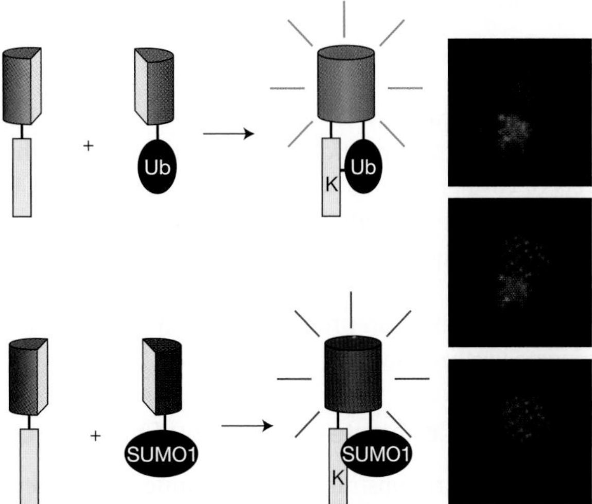

FIGURE 4. Principle of ubiquitin-mediated UbFC analysis. The UbFC assay is based on the association between fluorescent protein fragments that are brought together by covalent conjugation of ubiquitin, which is fused to one fragment, to a substrate that is fused to the complementary fragment. This assay can also be used to visualize conjugation of other ubiquitin-family peptides (e.g., SUMO1) to substrate proteins. The fluorescent protein fragment must be fused to the amino-terminal ends of ubiquitin-family peptides because their carboxy-terminals ends are essential for conjugation to substrate proteins. Fusions to the amino termini of ubiquitin and the small ubiquitin-like modifier-1 (SUMO-1) do not interfere with the activities of the enzyme complexes that conjugate these peptides to protein substrates. It is essential to determine if the conjugates that contain the fluorescent protein fragments retain the biological functions of the unmodified conjugates. Because ubiquitin can be conjugated in different monomeric and polymeric configurations to substrates, it is important to establish that the stoichiometry and the configuration of the ubiquitin conjugate are not altered by the fusions. The images on the right show multicolor UbFC analysis of ubiquitin (green) and SUMO-1 (blue) conjugation in the same cell. (Reproduced from Fang and Kerppola 2004, ©2004, the National Academy of Sciences.)

Ubiquitin-family peptide conjugation serves many different functions in cells. The diverse effects of ubiquitin conjugation are determined, in part, by the nature of the isopeptide bonds that connect the ubiquitin protomers in polyubiquitin chains. Ubiquitin contains seven lysine residues, each of which can serve as an acceptor for ubiquitin chain extension. Ubiquitin chains are removed and are edited by ubiquitin hydrolase/protease enzymes. It is, therefore, difficult to determine if the ubiquitin chains detected in cell extracts represent the native configuration of the chains present in living cells. The UbFC assay can be used to visualize the effects of mutations in lysine residues in ubiquitin as well as in the substrate protein on the formation and localization of ubiquitin conjugates in live cells. This approach can, therefore, be used to identify the roles of specific isopeptide bonds and the sites of ubiquitin conjugation on the properties of ubiquitin-family peptide conjugates.

Design of Constructs for UbFC Analysis

The fluorescent protein fragments must be fused to the amino termini of ubiquitin-family peptides to allow isopeptide bind formation by the carboxyl termini of the peptides. The principles for the design of fusion proteins for BiFC and multicolor BiFC analysis (Protocols A and B) also apply to the design of fusions to the substrate protein investigated by UbFC analysis. The levels of fusion protein expression should also be controlled as described in these protocols.

Effects of Fluorescent Protein Fragment Fusions on Ubiquitin-Family Peptide Conjugation to Substrate Proteins

The effects of mutations that affect ubiquitin-family peptide conjugation to substrate proteins lacking the fluorescent protein fragment fusions should be tested on UbFC complex formation. Fusion of fluorescent protein fragments to the amino termini of ubiquitin and the SUMO-1 appear not to alter their conjugation to the substrates that have been tested (Fang and Kerppola 2004; Ikeda and Kerppola 2008). Mutations in ubiquitin and the c-Jun transcription regulatory protein have the same effects on the formation of conjugates by proteins fused to the fluorescent protein fragments as they have on conjugate formation by proteins lacking these fragments (Ikeda and Kerppola 2008).

Controls

Deletion of the carboxy-terminal glycine residue(s) in ubiquitin-family peptides eliminates covalent conjugation of these peptides to substrate proteins. These mutations can be used to distinguish the effects of covalent versus noncovalent interactions on the formation and localization of fluorescent complexes in UbFC analysis. Noncovalent interactions formed by ubiquitin-family peptide fusions frequently produce detectable signal in UbFC analysis (Fang and Kerppola 2004; Ikeda and Kerppola 2008). However, differences in the distributions of the fluorescence produced by covalent and noncovalent complexes can provide insight into the specific roles of covalent ubiquitin conjugation on conjugate localization.

Simultaneous Visualization of Conjugates Formed by Different Ubiquitin-Family Peptides

Many proteins are substrates for the conjugation of several different ubiquitin-family peptides. The multicolor UbFC approach can be used to visualize conjugates formed by different ubiquitin-family peptides in the same cell using multicolor UbFC analysis analogous to the multicolor BiFC analysis described above (Protocol B). Comparison of the distribution patterns of UbFC conjugates formed by different ubiquitin-family peptides can serve as a control to establish the specificity of conjugate localization.

Protocol C

Visualization of Ubiquitin Conjugates Using UbFC Analysis

This protocol focuses on visualization of ubiquitin conjugated in cultured mammalian cells, but it can be adapted to any cell type or aerobically grown organism that can be genetically modified to express the fusion proteins.

IMAGING SETUP

The cells can be analyzed by microscopy or flow cytometry. If multiple ubiquitin-family conjugates are to be visualized simultaneously, use excitation and emission filters that are able to separate the fluorescence produced by BiFC complexes formed by different fluorescent protein fragments. In addition, if multiple ubiquitin-family conjugates are to be visualized simultaneously within the same cell, then the fusions to fluorescent proteins that produce UbFC complexes must have distinct spectra.

MATERIALS

Reagents

Cell culture medium or other medium appropriate for maintaining normal functions of cells/organisms

Cells or organism of interest

Note that protein interactions in normal cells and tissues may not be faithfully reproduced in immortalized cell lines.

Expression vectors containing the DNA sequences encoding each fusion protein

The expression vectors encode the proteins of interest fused to the complementary fluorescent protein fragments. Recommended choices for many purposes are the amino- and carboxy-terminal fragments of YFP. One set of expression vectors must encode fluorescent protein fragments fused to the amino-terminal ends of the ubiquitin-family peptides that are to be investigated.

Transfection reagents

If the goal is stable transfectants, then reagents that enable selection for and identification of cells or organisms that express the fusion proteins are needed.

Washing medium or buffer

Equipment

Culture chambers that maintain conditions for normal function of cells/organisms

The culture vessels should be designed for imaging if the cells are to be visualized by microscopy.

Flow cytometer

Fluorescence microscope with confocal or epi-illumination, high-NA objectives, and a CCD camera

Incubator for cell culture

EXPERIMENTAL METHOD

1. Seed the cells at appropriate density to allow for growth after the cells are transfected or transduced.

   Adherent cells are easiest to visualize by microscopy, whereas nonadherent cells are better analyzed by flow cytometry.

2. Transfect the cells with appropriate amounts of plasmids, or transduce with viruses encoding ubiquitin-family peptides and putative substrates fused to complementary fluorescent protein fragments (e.g., YN155-Ub and substrate-YC155).

   As a control, analyze fusions to ubiquitin-family peptides lacking the carboxy-terminal glycine residues as well as substrates lacking the region required for conjugation (e.g., YN155-Ub(ΔG76) and substrate-YC155 as well as YN155-Ub and substrate(mut)-YC155.

3. Culture the cells to allow the proteins to be expressed. Image the complexes as soon as possible after fusion protein expression to minimize the possibility that expression of the fusion proteins affects the properties of the cells.

   See *Troubleshooting.*

4. If necessary, wash the cells before measuring fluorescence to remove dead cells and cell debris, which can produce spurious signal.

   See *Troubleshooting.*

5. Image the cells by fluorescence microscopy. Determine the fraction of cells that display detectable fluorescence.

   Higher-NA objectives increase the intensity of illumination and generally allow more sensitive detection of fluorescence. Compare the fraction of cells that display detectable fluorescence with the proportion of cells that express the fusion proteins determined by coexpression of an intact fluorescent protein. It is common for the fraction of cells that produce detectable UbFC fluorescence to be smaller than the fraction of transfected or transduced cells, but a difference of more than 10-fold may indicate that additional factors or signals are required for ubiquitin conjugation.

   See *Troubleshooting.*

6. Determine the distribution of UbFC complex fluorescence in a representative subpopulation (tens or hundreds depending on the heterogeneity of the distribution) of cells.

   To ensure that subjective factors do not influence interpretation of the data, it is important to perform the imaging using double blind experimental design.

7. Using either fluorescence microscopy or flow cytometry, quantify UbFC conjugate formation by measuring the fluorescence intensities of UbFC complexes and of the intact fluorescent protein normalization control in the same cells. The ratio of the fluorescence intensities can be used to compare UbFC conjugate formation between wild-type and mutant proteins.

   It is important to ensure that there is no cross talk between the signals by analyzing cells that express the UbFC conjugates and intact fluorescent proteins in separate cells. Although the fluorescence intensities can be measured by microscopy, it is often preferable to measure them by flow cytometry to sample a representative subpopulation of cells.

   See *Troubleshooting.*

8. Measure the levels of fusion protein expression by preparing cell extracts and analyzing them by immunoblotting.

   The frozen lysates can generally be stored for a few days when denatured and frozen. It is important to compare the level of substrate fusion expression with that of its endogenous counterpart. The level of expression should be corrected for the fraction of cells that express the fusion protein.

   See *Troubleshooting.*

9. Compare the distributions of the wild-type and mutant substrate fusion proteins as well as their endogenous counterparts by indirect immunofluorescence analysis.

It is important to determine if ectopic expression of the fusion proteins alters their distributions compared with their endogenous counterparts.

See *Troubleshooting*.

TROUBLESHOOTING

Problem (Steps 3 and 4): Many cells die as a result of fusion protein expression.

Solution: Lower levels of fusion protein expression are likely to be less toxic. Fusion of the fluorescent protein fragment to a different position of the substrate protein could also help. Other cell types may be less sensitive to the effects of the fusions.

Problem (Step 5): No fluorescence is detected.

Solution: The position in the substrate protein at which the fluorescent protein fragment is fused can affect UbFC complex formation. Fusion of the fragment to a different position or the use of a different linker may alter UbFC-complex fluorescence intensity.

Problem (Step 7): Mutations in the ubiquitin-family peptide or the substrate that eliminate conjugation do not reduce the fluorescence intensity.

Solution: Noncovalent interactions between ubiquitin and putative substrate proteins can facilitate fluorescent protein fragment association. Thus, comparison of the distribution of UbFC complexes formed by the wild-type proteins and the complexes formed by the mutant proteins can provide information about the effect of covalent conjugation on conjugate localization. In this case, it can be difficult to quantify the efficiency of UbFC conjugate formation.

Problem (Steps 8 and 9): The levels or distribution of the fusion proteins differ from those of the endogenous proteins, or the wild-type and mutant proteins differ from each other.

Solution: It is generally best to analyze UbFC complex formation using the lowest levels of fusion proteins that can be detected. The levels of fusion protein expression can be controlled by using expression vectors that can be regulated and by visualizing complex formation as soon after protein expression as possible. If necessary, the amounts of fusion proteins that are expressed can be adjusted by changing the amounts of plasmids or viruses introduced into the cells. The effects of mutations that alter the levels or the distributions of fusion protein expression are difficult to interpret. It is generally best to make additional mutations that only affect UbFC complex formation.

REFERENCES

Fang DY, Kerppola TK. 2004. Ubiquitin-mediated fluorescence complementation reveals that Jun ubiquitinated by Itch/AIP4 is localized to lysosomes. *Proc Natl Acad Sci* **101:** 14782–14787.

Förster T. 1959. 10th Spiers Memorial Lecture. Transfer mechanisms of electronic excitation. *Faraday Discuss* **27:** 7–17.

Giepmans BN, Adams SR, Ellisman MH, Tsien RY. 2006. The fluorescent toolbox for assessing protein location and function. *Science* **312:** 217–224.

Grinberg AV, Hu CD, Kerppola TK. 2004. Visualization of Myc/Max/Mad family dimers and the competition for dimerization in living cells. *Mol Cell Biol* **24:** 4294–4308.

Hink MA, Bisselin T, Visser AJ. 2002. Imaging protein–protein interactions in living cells. *Plant Mol Biol* **50:** 871–883.

Hu CD, Kerppola TK. 2003. Simultaneous visualization of multiple protein interactions in living cells using multicolor fluorescence complementation analysis. *Nat Biotechnol* **21:** 539–545.

Hu CD, Kerppola TK. 2005. Direct visualization of protein interactions in living cells using bimolecular fluorescence complementation analysis. In *Protein–protein interactions*, 2nd ed. (ed. Golemis EA, Adams PD), pp 673–693. Cold Spring Harbor Laboratory Press, Cold Spring Harbor, NY.

Hu CD, Chinenov Y, Kerppola TK. 2002. Visualization of interactions among bZIP and Rel family proteins in living cells using bimolecular fluorescence complementation. *Mol Cell* **9:** 789–798.

Ikeda H, Kerppola TK. 2008. Lysosomal localization of ubiquitinated Jun requires multiple determinants in a lysine-27-linked polyubiquitin conjugate. *Mol Biol Cell* **19:** 4588–4601.

Kerppola TK. 2008. Bimolecular fluorescence complementation (BiFC) analysis as a probe of protein interactions in living cells. *Annu Rev Biophys* **37:** 465–487.

Kerppola TK. 2009. Visualization of molecular interactions using bimolecular fluorescence complementation analysis: Characteristics of protein fragment complementation. *Chem Soc Rev* **38:** 2876–2886.

Majoul I, Straub M, Duden R, Hell SW, Soling HD. 2002. Fluorescence resonance energy transfer analysis of protein–protein interactions in single living cells by multifocal multiphoton microscopy. *J Biotechnol* **82:** 267–277.

Nagai T, Ibata K, Park ES, Kubota M, Mikoshiba K, Miyawaki A. 2002. A variant of yellow fluorescent protein with fast and efficient maturation for cell-biological applications. *Nat Biotechnol* **20:** 87–90.

Pfleger KD, Eidne KA. 2006. Illuminating insights into protein–protein interactions using bioluminescence resonance energy transfer (BRET). *Nat Methods* **3:** 165–174.

Rackham O, Brown CM. 2004. Visualization of RNA–protein interactions in living cells: FMRP and IMP1 interact on mRNAs. *EMBO J* **23:** 3346–3355.

Rajaram N, Kerppola TK. 2004. Synergistic transcription activation by Maf and Sox and their subnuclear localization are disrupted by a mutation in *maf* that causes cataract. *Mol Cell Biol* **24:** 5694–5709.

Ren X, Vincenz C, Kerppola TK. 2008. Changes in the distributions and dynamics of polycomb repressive complexes during embryonic stem cell differentiation. *Mol Cell Biol* **28:** 2884–2895.

Robida AM, Kerppola TK. 2009. Bimolecular fluorescence complementation analysis of inducible protein interactions: Effects of factors affecting protein folding on fluorescent protein fragment association. *J Mol Biol* **394:** 391–409.

Saka Y, Hagemann AI, Piepenburg O, Smith JC. 2007. Nuclear accumulation of Smad complexes occurs only after the midblastula transition in *Xenopus*. *Development* **134:** 4209–4218.

Shyu YJ, Liu H, Deng XH, Hu CD. 2006. Identification of new fluorescent protein fragments for bimolecular fluorescence complementation analysis under physiological conditions. *Biotechniques* **40:** 61–66.

19 Preparation and Use of Retroviral Vectors for Labeling, Imaging, and Genetically Manipulating Cells

Ayumu Tashiro,[1] Chunmei Zhao,[2] Hoonkyo Suh,[2] and Fred H. Gage[2]

[1]*Kavli Institute for Systems Neuroscience and Centre for the Biology of Memory, Norwegian University of Science and Technology, Trondheim NO-7491 Norway;* [2]*Laboratory of Genetics, the Salk Institute for Biological Studies, La Jolla, California 92037*

ABSTRACT

Retroviral vectors are a powerful technology for achieving long-term genetic manipulation. This chapter provides some background on replication-deficient retroviral vectors based on Moloney murine leukemia virus and lentivirus. Details, examples, and protocols are provided for using these vectors to fluorescently label, genetically alter, and image both live and fixed murine brain tissue.

INTRODUCTION

Replication-Incompetent Retrovirus Vectors

Retroviral vectors are engineered gene vectors based on retroviruses whose RNA genomes are converted by their reverse transcriptase into double-stranded DNA on entering cells. The double-stranded DNA then integrates into the host chromosome and becomes a provirus, marking the host cell permanently. This property makes retroviral vectors suitable for long-term genetic manipulations, including visualization by reporter gene expression. Two different kinds of retroviruses have been engineered for cell labeling: oncoretroviruses and lentiviruses.

Oncoretroviruses can enter a cell nucleus only during the prophase–prometaphase transition at the onset of mitosis; therefore, they only transduce dividing cells. We have been using retroviral vectors based on a type of oncoretrovirus called the Moloney murine leukemia virus (Mo-MLV) to genetically label and to manipulate new neurons in the adult dentate gyrus. The retroviral vector CAG-GFP used in our study is based on Mo-MLV (Zhao et al. 2006). CAG-GFP uses the backbone of pCLNCXv.2, and the expression of green fluorescent protein (GFP) is driven by the compound promoter CAG (Naviaux et al. 1996). Typically, a recombinant replication-incompetent retroviral vector contains the following components: 5′-LTR (long terminal repeat), virus packaging signal, ectopic promoter (CAG promoter in our vector) and the reporter gene (GFP) under its control, woodchuck

FIGURE 1. Schematic of retroviral vector CAG-GFP. The viral backbone is based on the Mo-MLV. P_{CMV}, CMV promoter. The CMV promoter is placed 5′ to the virus LTR for efficient transcription of the recombinant virus in virus packaging cells such as HEK293T cells. LTR, long terminal repeat; Ψ, viral packaging signal; CAG, compound promoter containing CMV immediate early enhancer, chicken β-actin promoter, and a synthetic intron; eGFP, enhanced green fluorescent protein; WPRE, posttranscriptional regulatory element.

hepatitis virus posttranscriptional element (WPRE) sequence, and 3′-LTR (Fig. 1). WPRE is frequently used in viral vectors to increase protein production by stabilizing the transcript (Zufferey et al. 1999). In contrast, a lentiviral vector efficiently transduces the transgene in both proliferating and fully differentiated cells (Naldini et al. 1996a,b). The interaction between viron proteins and the nuclear import machinery is attributed to lentivirus-mediated gene transduction in nondividing cells.

To generate replication-incompetent retroviral vectors, a transient transfection of three plasmids (for Mo-MLV vectors) or four plasmids (for lentiviral vectors) is commonly used (Verma and Weitzman 2005; Tiscornia et al. 2006). These plasmids include a transfer vector containing transgenes and all necessary *cis*-elements required for RNA viral genome production and packaging. Production of Mo-MLV vectors requires two additional vectors expressing the *trans*-acting factors, including *gag-pol* and vesicular stomatitis virus glycoprotein (VSV-G). *gag* encodes viral core structural proteins such as capsid, matrix, and nucleocapsid proteins, and *pol* produces viral enzymes including protease, integrase, and reverse transcriptase. The former is absolutely required for the production of viron particles, and the latter has an important role in viral infections. To enhance tropism of viral infection, VSV-G is used to make pseudotyped retroviral vectors. In addition to these two vectors, production of lentiviral vectors requires a vector containing *Rev*, which mediates nuclear export of an unspliced full-length viral genome.

Use of Retroviral Vectors for Cell Labeling and Morphological Visualization

Retrovirus-mediated expression of marker proteins such as GFP in host cells allows the visualization of labeled cells in live and fixed samples without further histological treatments. For example, we used Mo-MLV-based CAG-GFP to label proliferating cells and their progeny in the adult dentate gyrus (Zhao et al. 2006; Protocol A). By imaging newborn granule cells at different times after cell birth (Protocols B and C), we found that these cells go through distinct stages of development, including (1) initial dendritic and axonal growth during the first two weeks after cell birth (Fig. 2), (2) spine formation and growth starting from the third week after cell birth, and (3) spine maturation over the next several weeks, which is shown by an increase in mushroom spine formation and a decrease in spine motility (Fig. 3). In addition, we found that this process can be regulated by the animals' experience. When the mice were housed with unlimited access to a running wheel, mushroom spine formation by newborn granule cells was accelerated, and the density of mushroom spines reached plateau much faster than those in mice housed in standard sedentary conditions (Fig. 3C). Moreover, spine motility of newborn granule cells reached peak level at 21 d postinjection (dpi) and quickly decreased to base level at 28 dpi (Fig. 3E). Both measures indicate faster stabilization of dendritic spines of newborn granule cells when mice were housed in running wheel cages.

Mo-MLV vectors also have been used to analyze specific neuronal populations using different techniques. For slice electrophysiology, reporter GFP expression achieved by Mo-MLV vectors is used to perform patch-clamp recordings specifically targeting new neurons in the adult dentate gyrus (van Praag et al. 2002; Laplagne et al. 2006, 2007). The same approach has been used to help find new neurons for electron microscopic analyses of synaptic structures (Toni et al. 2007, 2008)

Use of Retroviral Vectors for Imaging and Genetic Manipulations

An advantage of viral vectors is the ability to combine visualization by reporter gene expression with other genetic manipulations, making it possible to detect and to morphologically analyze manipulated

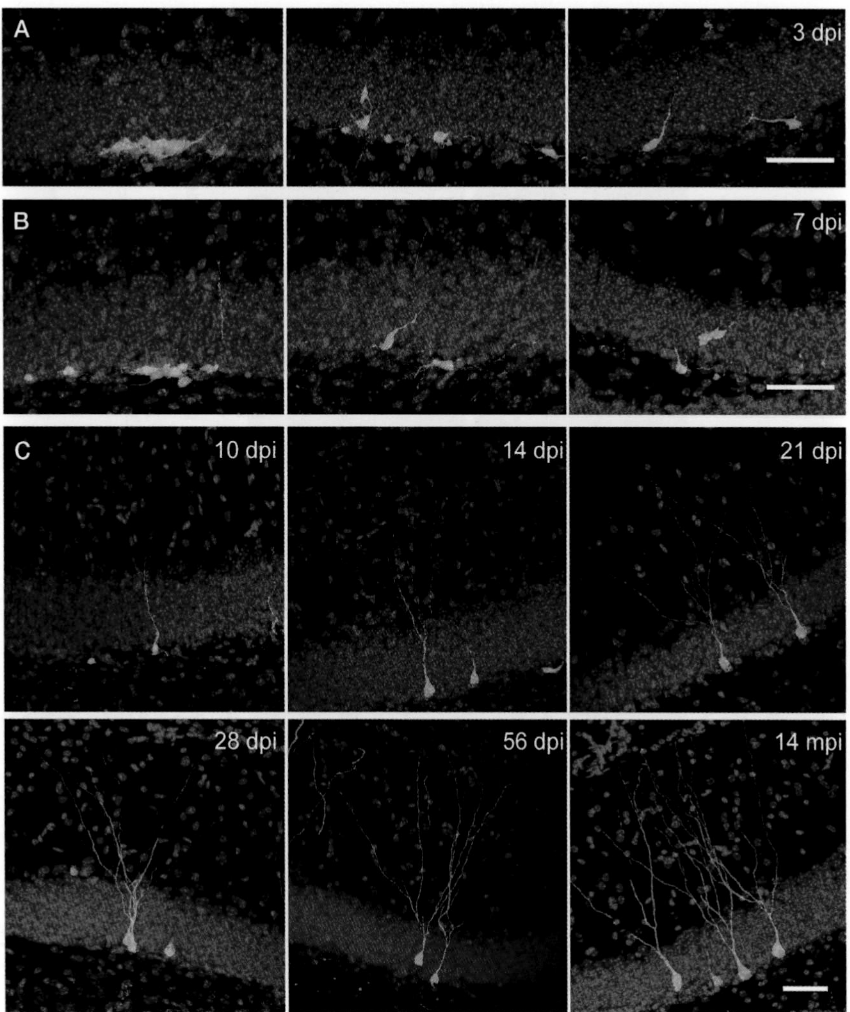

FIGURE 2. The morphological development of neurons born in the adult mouse brain. Dividing cells in the dentate gyrus of 7-wk-old to 10-wk-old C57Bl/6 mice were labeled with GFP through retrovirus-mediated gene transduction. Mice were killed at different time points as indicated in the figure (mpi, months postinjection). Images are oriented with the molecular layer on the upside. (*A*) Cells at 3 dpi. (*B*) Cells at 7 dpi. (*C*) Cells at 10, 14, 21, 28, and 56 dpi, and also 14 mpi. Scale bars, 50 µm. Red, 4′,6-diamino-2-phenylindole (DAPI); green, GFP. (Reprinted, with permission, from Zhao et al. 2006, ©Society for Neuroscience.)

cells among surrounding unaffected cell populations. For example, we developed a single-cell gene knockout technique for new neurons in the adult dentate gyrus using a Mo-MLV vector (Fig. 4). A Mo-MLV vector expressing GFP-fused Cre recombinase (CAG-GFP/Cre) was constructed and was injected into the dentate gyrus of adult floxed NR1 mice. In these mice, transduced neurons become deficient for the NR1 gene, which encodes an essential subunit for the functional *N*-methyl-D-aspartate (NMDA) receptor because the gene is removed by the activity of the Cre recombinase. Using nuclear GFP fluorescence from GFP/cre as an indication of NR1 gene knockout neurons, the number of survived knockout neurons can be detected and counted. Using this approach, we found that the survival of NR1-knockout (NR1KO) new neurons is dramatically reduced 2–3 wk after neuronal birth, indicating that NMDA receptor–mediated input activity determines the survival of new neurons specifically during this critical period. Using overexpression or ectopic expression of genes, other studies have examined the effects of other genes on properties of neural precursor cells, new neurons, and astrocytes in the adult dentate gyrus (Lie et al. 2005; Jessberger et al. 2008a,b; Marumoto et al. 2009).

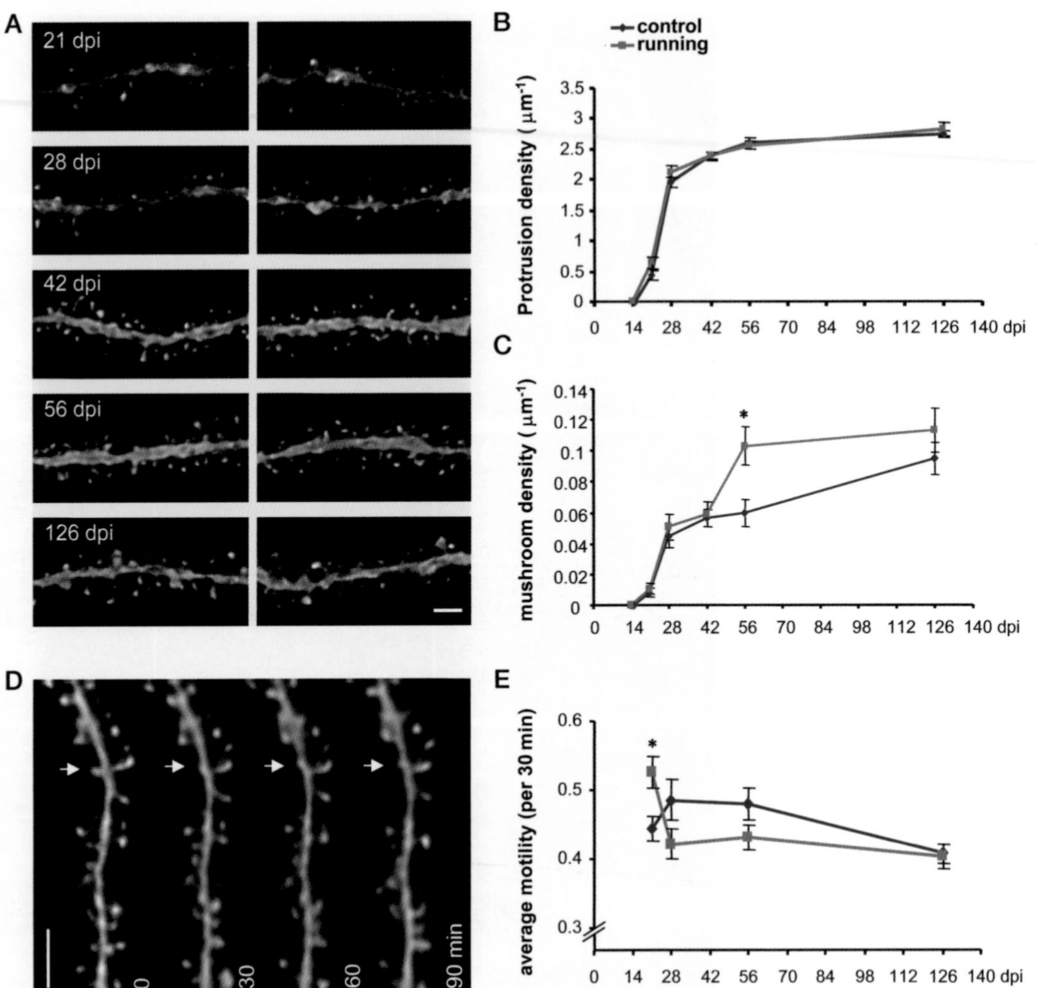

FIGURE 3. The maturation of granule cells born in the adult mouse brain and dendritic spine analyses. (*A*) Representative images of dendritic segments from newborn neurons at 21, 28, 42, 56, and 126 d after viral infection. (*B*) Quantification of total protrusion density. Blue, control; purple, running. The density of protrusions is expressed as the number of protrusions per micrometer dendritic length. (*C*) Quantification of mushroom spine density. Mushroom spines were identified if the estimated surface area ($= \pi \times D_{major} \times D_{minor}/4$) was 0.4 μm^2. The density of mushroom spines is expressed as the total number of mushroom spines per micrometer dendritic length. (*D*) An example of a time-lapse series of a dendritic segment for spine motility analysis. The arrowheads indicate partial retractions of a spine over the time series. (*E*) Quantification of spine motility. Data for *B*, *C*, and *E* are presented as mean ± S.E.M. Asterisks indicate where significant differences were seen between control and running mice ($p < 0.01$). Scale bars, 2 μm. (Reprinted, with permission, from Zhao et al. 2006, ©Society for Neuroscience.)

A lentivirus was also used to examine the fate analysis of neural stem cells (NSCs) in the subgranular zone (SGZ), asking whether Sox2-positive cells can represent NSCs in the SGZ of the hippocampus (Suh et al. 2007). There is an emerging view that both slowly dividing and actively proliferating NSCs contribute to adult neurogenesis, and it is expected that the property of a lentiviral vector that can infect both mitotic and nonmitotic cells can be used to trace the fate of both slowly and actively dividing NSCs. A genetic binary system was used to permanently mark Sox2-positive cells and their progeny. First, a lentiviral vector harboring Cre recombinase was generated. The expression of Cre recombinase is controlled by a well-characterized Sox2 promoter. In this experiment, the targeted lentiviral vector genetically labeled Sox2-positive cells and their daughter cells by activating reporter gene expression in ROSA26 mice. Moreover, the fate of progeny of Sox2-positive cells was determined by cell type–specific markers that can be identified by confocal microscopy. This experiment strongly suggested that Sox2-positive cells in the SGZ can self-renew and differen-

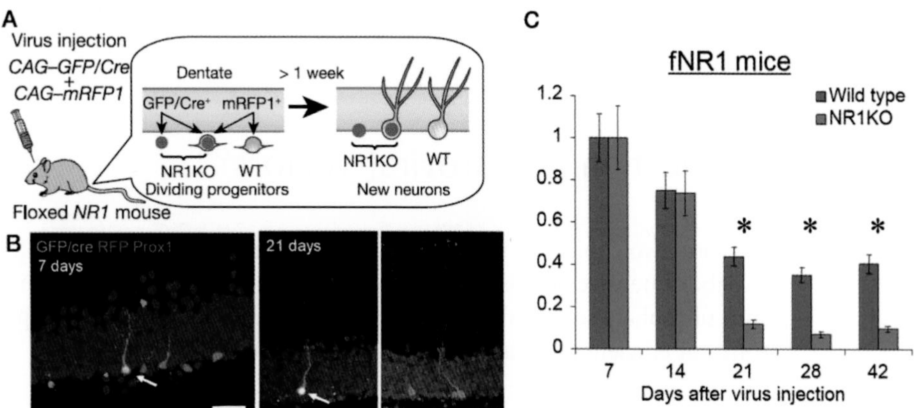

FIGURE 4. Retrovirus-mediated single-cell knockout technique for new neurons in adult dentate gyrus. (*A*) A schematic of the experimental principle. The viral DNA is integrated into dividing neuronal progenitor cells, which become either GFP/Cre-positive (GFP/Cre$^+$; green) only, mRFP1-positive (mRFP1$^+$; red) only, or double positive. In GFP/Cre-positive cells, Cre-mediated recombination occurs, and they become NR1KO cells. These cells eventually become neurons. (*B*) GFP/Cre-positive (green), mRFP1-positive (red), and double-positive (arrows) new neurons at 7 and 21 d after virus injection. Blue, Prox1 immunostaining. (*C*) The survival rate of wild-type (mRFP1-positive only) and NR1KO (GFP/Cre-positive only and double positive) new neurons in fNR1 mice (*$P < 0.0002$, two-sided *t* test, $n = 8$ mice for each time point). (Reproduced from Tashiro et al. 2006, ©Macmillan.)

tiate into both neurons and astrocytes. One caveat of this experiment is that GFP expression controlled by the Sox2 promoter has been found in some differentiated neurons in independent experiments. It is not clear whether this expression represents a residual GFP expression in the newly generated neurons that Sox2-positive cells gave rise to, or whether it is due to the positional effect of the transgene in the infected cells.

Although the methods described above attempted to achieve maximal targeting specificity using a virus-mediated gene transfer, there is room to argue that the progeny may not arise clonally. One way to improve the fidelity of the clonal relationship is to use a transgenic mouse line. Livet et al. (2007) used Brainbow mice, in which different fluorescent proteins (XFPs: green, red, and/or cyan) were floxed and were integrated into the genome in a random manner. On the transient induction of Cre expressions, a stochastic combination of XFP will be expressed, generating different colors. About 90 distinct color combinations can be distinguished with a confocal microscopic technique. If Brainbow mice are infected with a Cre-expressing retrovirus in a clonal dilution, confocal imaging technology can dissect out the origin of the progeny of NSCs and reveal their clonal relationship.

Protocol A

Purification and Injection of Retroviral Vectors

There are two choices of transfection method. Transfection with Lipofectamine 2000 is reliable and less labor intensive because it requires a smaller number of cell culture plates. Transfection with calcium phosphate solution requires selecting good batches of calcium phosphate solution to achieve sufficient efficiency of transfection but with lower costs.

MATERIALS

CAUTION: See Appendix 6 for proper handling of materials marked with <!>.
See the end of the chapter for recipes for reagents marked with <R>.

Reagents

Anesthesia <!>
$CaCl_2$ <!> (2 M) (filter-sterilized)
Culture medium <R>
Dulbecco's phosphate-buffered saline (D-PBS) (Invitrogen 14287-072)
Ethanol <!> (70%)
293FT cells (Invitrogen R70007)
HEBS (2x) <R>
Lipofectamine 2000 (Invitrogen 11668-019)
Mice
Opti-MEM I (optimodified Eagle's medium I), reduced serum medium (Invitrogen 31950-062)
Puralube vet ointment (Pharmaderm NDC 0462-0211-38)
Sucrose (20%) (optional; see Step 31)Tissue adhesive (3M Vetbond, 1469SB)
Viral DNA mix for lentiviral vectors <R>
Viral DNA mix for MML retroviral vectors <R>

Equipment

Biosafety level-2 facility
Centrifuge
Conical tubes, 50 mL
Culture plates (10-cm diameter for Lipofectamine or 15-cm size for $CaCl_2$ transfections)
Drill bur, size no. 1 (Henry Schein Dental 100-7176)
Electric drill (Dremel 395 T6)
Electric hair trimmer or small scissors
Filtration units, 0.45-µm
Fluorescence microscope
Marker pen
Microsyringes (Hamilton Company 87925)
Needles for microsyringes, 33-gauge (Hamilton Company 7762-06)
Polyallomer tubes, 13 x 51-mm (Beckman Coulter, Inc. 326819)
Polyallomer tubes, 25 x 89-mm (Beckman Coulter, Inc. 326823)

Rotors for ultracentrifuge SW32 Ti or SW28, SW55 Ti (Beckman Coulter, Inc.)
Stereotaxic frame for small animals (David Kopf Instruments)
Surgical instruments, sterile (e.g., scissors, forceps, suture)
Tissue culture incubator, set at 37°C and 5% CO_2
Tissue culture plates, 24-well
Tubes, 0.5-mL, sterile
Tubes, 14-mL, polypropylene (Falcon 352059)
Ultracentrifuge (e.g., Optima L-90K; Beckman Coulter, Inc.)
Warming blanket and mouse cage

EXPERIMENTAL METHOD

Transfection of Viral Vector Plasmids

Complete either Procedure A (1–10) or Procedure B (11–20), and then continue with Step 21.

(Procedure A) Transfection with Lipofectamine 2000

Day 1 (30 min)

1. Passage 293FT cells into twelve 10-cm cell culture plates at 5 x 10^6 cells per plate. Use 10 mL of complete medium without antibiotics per plate.

Day 2 (6 h)

2. Add lentiviral or retroviral DNA mix and 2.4 mL of Opti-MEM into four 14-mL polypropylene tubes (A tubes). Mix by tapping each tube.
3. Add 2.4 mL of Opti-MEM and 150 µL of Lipofectamine 2000 into four additional 14-mL polypropylene tubes (B tubes). Mix by tapping each tube. Incubate at room temperature (20°C–25°C) for no longer than 5 min.
4. Pipette the contents of one B tube into one A tube. Mix by inverting the tubes. This results in four tubes of 4.8-mL DNA/Lipofectamine mix.
5. Incubate the tubes at room temperature for 25–30 min.
6. Add 1.6 mL of the DNA/Lipofectamine mix dropwise onto each plate of 293FT cells. The drops should evenly cover the surface of the culture medium. Mix by gently rocking the plate.
7. Incubate the cells for 5 h at 37°C and 5% CO_2.
8. Replace the supernatant with 10 mL of fresh complete medium containing antibiotics.
9. Incubate the plates for 48 h until day 4.

Day 3 (10 min)

10. Check GFP fluorescence to evaluate the transfection efficiency. Proceed to Step 21.

(Procedure B) Transfection with Calcium Phosphate Precipitation

Day 1 (1 h)

11. Passage 293FT cells into sixteen 15-cm cell culture plates at 10 x 10^6 cells per plate. Use 15 mL of complete medium with antibiotics per plate.

Day 2 late afternoon (30 min)

12. Add lentiviral or retroviral DNA mix (*X* mL) and sterile ddH_2O (5.1 – *X* mL) into two 50-mL polypropylene tubes. Mix by tapping the tubes.
13. Add 2.9 mL of $CaCl_2$ solution to each tube. Mix by pipetting.
14. Add 8 mL of 2x HEBS solution dropwise into the DNA/$CaCl_2$ tubes.
15. Mix by pipetting and then by bubbling air from the pipette for 30 sec.
16. Incubate for 10 min at room temperature.
17. Add 2 mL of the mix dropwise onto each plate of 293FT cells. The drops should evenly cover the surface of the culture medium. Mix by gently rocking the plate.
18. Incubate the plates overnight at 37°C and 5% CO_2.

Day 3 morning (10 min)

19. Check GFP fluorescence to evaluate transfection efficiency.

    See *Troubleshooting*.
20. Replace the medium with 15 mL of fresh complete medium with antibiotics.

Concentration of Viral Vectors

Day 4 (20 min)

21. Collect the medium containing viral vectors into sterile containers. Provide each plate of cells with 10 (Procedure A) or 15 (Procedure B) mL of fresh complete medium with antibiotics.
22. Store the collected medium at 4°C.

Day 5 (8–10 h)

23. Collect the medium containing viral particles into sterile containers.
24. Divide the collected medium from days 4 and 5 into 50-mL conical tubes. To remove cell debris, centrifuge the tubes at 1000*g* for 3 min.
25. Filter the supernatant through a 0.45-µm filter unit.
26. Transfer the filtered supernatant into 25 x 89-µm ultracentrifuge tubes (A tubes).

    Prepare 6 A tubes if using the Lipofectamine 2000 method, or prepare 12 A tubes if using the calcium phosphate method.
27. Centrifuge the filtered supernatant in an SW32 Ti rotor at 65,000g for 2 h at 4°C.

    If using the calcium phosphate transfection method, two 2-h centrifugations are required to accommodate all the A tubes.
28. Aspirate off the supernatant.
29. Add PBS (0.5 mL for the lipofectamine method or 0.25 mL for the calcium phosphate method) to the precipitated viral vectors into each A tube. Resuspend the precipitate by pipetting it up and down 20 times. Pool the resuspended viral vector in PBS from all the A tubes into a single 4-mL 13 x 51-mm ultracentrifuge tube (B tube).
30. Sequentially wash the A tubes with 0.5 mL of PBS, and transfer the wash into the 4-mL B tube.

    There should be 3.5 mL in the B tube.
31. (Optional) Add 0.5 mL of a 20% sucrose cushion (made in PBS and filter sterilized) to the B tube.
32. Centrifuge the B tube in an SW55 Ti rotor at 65,000g for 2 h at 4°C.
33. Aspirate off the supernatant.

34. Resuspend the final pellet in 80 µL of PBS by vortexing for 30 sec and then by pipetting 20 times. This procedure is enough to resuspend the virus. Leave the remaining sticky pellet, and transfer the suspended viral vector solution to a sterile 0.5-mL tube (C tube).
35. Wash the B tube once with 20 µL of PBS, and transfer the wash to the C tube.
36. Briefly centrifuge the C tube, and transfer the supernatant to a new 0.5-µL tube.
37. Aliquot the viral vectors into 5–10 µL portions, and store them at –80°C.

Determine Virus Titer

38. Seed 1 x 10^5 293FT cells per well in a 24-well plate, and incubate the cells overnight at 37°C and 5% CO_2.
39. Perform a serial dilution (e.g., 10^4, 10^5, 10^6, and 10^7 for CAG promoter retroviral vectors; 10^5, 10^6, 10^7, and 10^8 for CMV promoter lentiviral vectors).
40. Transfer 10 µL of each dilution to individual wells.
41. Three to five days later, view the infected cells with a fluorescent microscope, and count the number of fluorescent clusters. Calculate the number of colony-forming units per milliliter (cfu/mL) of original virus solution.

    An alternative or additional method for determining virus titer is the p24 measurement by real-time polymerase chain reaction or enzyme-linked immunosorbent assay.

Injection of Viral Vectors into the Brain

42. Deeply anesthetize mice.
43. Shave a small area on the head with an electric trimmer. Apply eye ointment to prevent the eyes from drying out too much.
44. Make a 1-cm incision in the skin over the skull within the shaved patch. Clean the incision of blood.
45. Mount the mouse onto a stereotaxic frame.

    Proper orientation of the head, insuring the alignment of the anteroposterior axis, is important for consistent injection into the target area.
46. Move the tip of the injection needle to the bregma.
47. Move the needle tip according to the lateral and posterior coordinates from the bregma.

    The coordinates for the brain region of interest have to be predetermined for specific ages and strains.
48. Mark the position with a marker pen.
49. Move the needle out, and using an electric drill, make a small hole in the skull at the marked site.
50. Dip the needle tip into the virus solution (from Step 37), and load 1.5 µL of solution into the syringe.
51. Move the needle tip to the hole in the skull, and set it level with the surface of the skull.
52. Move the needle tip down to a predetermined dorsoventral coordinate from the level of the skull surface.
53. Inject a total of 1.5 µL of virus solution.
54. After finishing the injection, wait for 1 min to prevent the injected solution from flowing back through the needle track.
55. Move the needle tip up 1 mm, and then wait for another 1 min. Slowly retract needle from mouse brain.
56. Rinse the needle first with 70% ethanol and then with PBS.

57. Close the skin using tissue glue, or suture it with thread.
58. Place the mouse in a cage on top of a warm blanket until it is fully alert. Transfer the mouse back to its home cage.
59. After a certain survival period (minimum of 3 d for GFP expression), analyze the properties of interest using suitable methodologies.

    See *Troubleshooting.*

TROUBLESHOOTING

Problem (Step 19): Transfection efficiency is low with calcium phosphate precipitation.
Solution: pH of 2x HEBS may not be optimal. Test several different preparations of 2x HEBS for their transfection efficiency.

Problem (Step 59): No reporter gene expression is detected.
Solution: The problem may be caused by a low level of reporter GFP expression. Test immunostaining with an anti-GFP antibody. If this does not help, the problem may be caused by a low titer of viral preparation. Prepare a new batch of vector with a higher titer. Unfortunately, viral vectors encoding genes that interfere with viral production processes in 293FT cells or that are toxic to cells may never be produced with high titers.

Problem (Step 59): GFP-positive cells are only found outside of the targeted area.
Solution: The stereotaxic coordinates may not be optimal. Inject a dye (e.g., electrophoresis gel-loading dye) to aid in finding reliable coordinates. The coordinates may vary among strains and ages of animals. In addition, be sure that the mouse is mounted correctly on the stereotaxic frame.

DISCUSSION

For CAG promoter retroviral vectors (Tashiro et al. 2006; Zhao et al. 2006), we typically get titers of 10^7–10^8 cfu/mL. With this range of titers, a single injection of 1.5 μL into the dorsal dentate gyrus of adult mice transduces genes into more than 400 newly generated neurons covering the dorsal half of the hippocampus. For CMV promoter lentiviral vectors, we typically get titers of 10^8–10^9 cfu/mL.

Gene-modified mice can be an alternative to viral vectors for most applications. The advantages of viral vectors are that (1) different transgenes can be easily incorporated by cloning them into viral plasmids, (2) viral vectors can be used with any mammalian species, and (3) region specificity is readily achieved by infusing viral vector solutions specifically into areas of interest within the brain. Mo-MLV vectors have an additional advantage. Because they require cell division for transduction, it is possible to introduce specific manipulations of newly generated cells (e.g., newly generated granule cells in the adult dentate gyrus).

On the other hand, the limitations of viral vectors are (1) limited capacity of viral genome size, making it difficult to accommodate long promoter elements, which are required to achieve high cell type specificity, and (2) covering large areas of the brain (e.g., the whole neocortex or hippocampus) is difficult because of limited diffusion of viral vectors in the brain, requiring multiple injections.

Protocol B

Imaging Newborn Granule Cells in Fixed Sections

Retroviral vector CAG-GFP is injected into the dentate gyrus of young adult mice (C57Bl/6, female, 6–10-wk old) according to the procedures in Protocol A. Brain tissue from newborn mice is dissected, the tissue is fixed, and the sections are prepared. The fixed sections are imaged using fluorescent confocal microscopy, and newborn granule cells containing GFP are visualized and are characterized.

MATERIALS

CAUTION: See Appendix 6 for proper handling of materials marked with <!>.
See the end of the chapter for recipes for reagents marked with <R>.

Reagents

Anesthesia <!>
DAPI <!>, concentrated
Mice infected with recombinant retroviral vector (from Protocol A)
NaCl (0.9% w/v)
Paraformaldehyde <!> (4%, w/v) in phosphate buffer (pH 7.4)
Sucrose (30%, w/v) in phosphate buffer (pH 7.4)
Tissue collecting solution (TCS) <R>

Equipment

Confocal microscope with a 40x objective
Dissection/surgical instruments
Microscope slides
Microtome <!>
Plates, 96-well
Syringes, pump, and other equipment for transcardial perfusion
Tracing software (e.g., IGL Trace program; http://synapses.clm.utexas.edu/tools/trace/trace.stm)

EXPERIMENTAL METHOD

Preparation of Fixed Brain Tissue

1. At the desired time point after virus injection, give a mouse a lethal dose of anesthesia.
2. Fix the mouse tissue by transcardial perfusion with 0.9% (w/v) NaCl, followed by 4% (w/v) paraformaldehyde in phosphate buffer (pH 7.4).
3. Postfix the mouse brain in 4% paraformaldehyde overnight, and then transfer it into 30% (w/v) sucrose in phosphate buffer.
4. After the mouse brain sinks to the bottom of the 30% sucrose solution, use a microtome to section the brain into 40-μm sections. Collect serial sections in TCS-filled 96-well plates. Store the sections at –20°C until use.

It takes 1–2 d for the sample to be saturated with 30% sucrose solution. Mouse brains can be stored in 30% sucrose solution for a short period of time (several weeks), but it is better to store them at −20°C after sectioning to minimize contamination and protein degradation.

5. Stain the sections with concentrated DAPI for 30 min, and mount the sections on slides for imaging.

Imaging and Measuring Newborn Granule Cells

The development of newborn granule cells can be measured according to several parameters, including dendritic length, branching points, total spine density, and mushroom spine density. In addition, the general growth of mossy fiber axons can be roughly determined by tracking the axon terminals in the CA3 area.

6. Measure dendritic length and branching points. Image the entire cell with a 40x objective using a confocal microscope. Use tracing software to determine total dendritic length by tracing all of the dendrites originating from the cell of interest. Manually count the branching points.
7. Measure total spine density. Image a segment in the outer third of the molecular layer with a 40x objective and a digital zoom of 6. Use the tracing program to quantify spine density.
8. Measure mushroom spine density. Set a threshold of 0.4 μm^2 for classification of mushroom spines. Calculate the spine head area based on the length of the short axis (d_x) and the long axis (d_y) of the spine head (spine head area = $0.25\pi d_x d_y$). To speed up the process, measure only relatively larger spines. Spines with an estimated surface area >0.4 μm^2 are considered to be mushroom spines.

Protocol C

Analysis of Spine Motility of Newborn Granule Cells Using Acute Brain Slices

Newborn granule cells in mice are labeled using retroviral vector technology. Acute brain slices are prepared from the mice. Using a live-cell imaging stage and confocal microscopy coupled to imaging software, dendritic spines are analyzed.

MATERIALS

CAUTION: See Appendix 6 for proper handling of materials marked with <!>.
See the end of the chapter for recipes for reagents marked with <R>.

Reagents

Anesthesia <!>
Artificial cerebrospinal fluid (ACSF) at 30°C <R>
Carbogen <!>
Dissection buffer, ice cold (see note to Step 3) <R>
Mice infected with recombinant retroviral vector (from Protocol A)

Equipment

Confocal microscope with temperature-controlled stage (10x and 60x) (numerical aperture [NA] 0.9) water-immersion objectives

We use an open-dish temperature-controlled chamber (Delta T4 culture dish system; Bioptechs) on an upright Olympus BX51 connected to a Bio-Rad Laboratories R2100 confocal system.

Dissection instruments
Imaging analysis software (e.g., AutoDeblur from AutoQuant, Watervliet, NY; Image-Pro from Media Cybermetics, Silver Spring, MD; and MetaMorph software from Molecular Devices, Sunnyvale, CA)
Vibratome

EXPERIMENTAL METHOD

Sectioning and Imaging Brain Tissue

1. Prepare dissection buffer and ACSF for acute brain slices. Oxygenate buffers by bubbling with carbogen (95% O_2/5% CO_2).
2. Give a mouse a lethal dose of anesthesia, remove the mouse brain, and transfer it to an ice-cold dissection buffer.
3. Section the mouse brains with a vibratome in ice-cold dissection buffer. Prepare 300–400-μm sections. Transfer sections immediately to 30°C ACSF for recovery.

   The dissection buffer needs to be ice cold. We typically freeze the buffer in a –80°C freezer, take it out shortly before sectioning, and thaw it briefly in a water bath so that the buffer is a mixture of liquid and ice when it is used.

4. Set up the imaging system on the stage of the confocal microscope. Continuously perfuse the open-dish chamber with ACSF. Set the perfusion speed so that there is no obvious drifting in the *x*- and *y*-axes during imaging acquisition for each time point.
5. Screen acute brain sections for labeled cells. Transfer the slice with labeled cells onto the open dish, and cover the slice with a wired weight.
6. Use a 10x objective to locate the cell(s) of interest, and switch to a 60x water-immersion lens (NA 0.9; Olympus).
7. Locate the dendritic process at the outer molecular layer, and image the process every 15 min for 2 h.

   When imaging with 512 x 512 resolution, we scan 166 lines per second, so that each image requires 3.08 seconds, and an entire *z*-series is acquired in 1–2 min with adequate resolution. In addition, because the intensity of the GFP signal is quite strong with our labeling, we are able to acquire all of the images with a 488-nm laser using minimum laser power. However, a two-photon laser would be more appropriate.

Analysis of Imaging Data

Analyze the images using image-processing software. We use four time points at 30-min intervals over the initial 90 min of imaging. If a cell appears degenerated at the 120-min imaging time point, the time-lapse series is discarded and not included for data analyses. See Movie 19.1.

8. Deconvolute the image files for five iterations using the AutoDeblur software.
9. Align the image stacks using the Image-Pro alignment function.
10. Create maximum intensity projections for each time point using Image-Pro.
11. Create an image stack for the four time points. Align the stacks first with Image-Pro, and then align manually by means of Photoshop with the dendritic shaft as a reference.
12. Quantify spine motility using the measure colocalization function of the MetaMorph software. First, crop out each spine from the image stack. The motility of each spine is the average of the nonoverlapping areas between time points divided by the average areas of all time points. Because of the overlapping between spines in 2D projections, only those that are reasonably set apart from neighboring spines should be analyzed to minimize noise in the measurement. Quantify a total of at least 100 spines from four to eight time series for each data point.

RECIPES

CAUTION: See Appendix 6 for proper handling of materials marked with <!>.
Recipes for reagents marked with <R> are included in this list.

ACSF

NaCl	125 mM
KCl	2.5 mM
$NaH_2PO_4 \cdot H_2O$	1.3 mM
$NaHCO_3$	25 mM
Dextrose	10 mM
Ascorbate	1.28 mM
Pyruvate	0.6 mM
$CaCl_2$ <!>	2 mM
$MgCl_2$	1.3 mM

Culture Medium (Store at 4°C)

Dulbecco's modified Eagle's medium (Invitrogen 41965-039)	500 mL
Fetal bovine serum (Invitrogen 16000-044)	50 mL
L-glutamine, 200 mM (Invitrogen 25030-024)	5 mL
Sodium pyruvate, 100 mM (Invitrogen 11360-039)	5 mL
Nonessential amino acids, 10 mM (Invitrogen 11140-035)	5 mL

Add 5 mL penicillin <!>–streptomycin <!> as needed (Invitrogen 15070-063).

Dissection Solution

KCl	2.5 mM
$NaH_2PO_4 \cdot H_2O$	1.3 mM
$NaHCO_3$	25 mM
Dextrose	20 mM
Ascorbate	1.3 mM
Pyruvate	0.6 mM
Choline	110 mM
Kynurenic acid (stock solution of kynurenic acid is made with 1 M NaOH <!>)	5.5 mM
$CaCl_2$ <!>	0.5 mM
$MgCl_2$	7.0 mM

HEBS (2x) (Store at 4°C)

NaCl	280 mM
KCl	10 mM
Na_2HPO_4	1.5 mM
D-glucose	12 mM
HEPES buffer (acid-free)	50 mM

Adjust to pH 7.05 with either NaOH <!> or HCl <!>. Filter sterilize the solution.

TCS

Glycerin	25% (v/v)
Ethylene glycol	30%
NaH_2PO_4	1.38 g/L
Na_2HPO_4	5.48 g/L

Viral DNA Mix for Lentiviral Vectors

Purify all of the plasmids with Maxiprep kits (QIAGEN).

For Lipofectamine 2000 transfection (three 10-cm plates):

pCSC-SP-GFP (or other lentiviral plasmid)	22.5 µg
pMDL	15 µg
pRSV-REV	6 µg
pCMV-vsv-g	7.5 µg

For calcium phosphate transfection (eight 15-cm plates):

pCSC-SP-GFP (or other lentiviral plasmid)	240 µg
pMDL	160 µg
pRSV-REV	64 µg
pCMV-vsv-g	80 µg

Viral DNA Mix for MML Retroviral Vectors

Purify all of the plasmids with Maxiprep kits (QIAGEN).

For Lipofectamine 2000 transfection (three 10-cm plates):

CAG-GFP (or other MML retroviral plasmid)	22.5 µg
pCMV-gp	15 µg
pCMV-vsv-g	7.5 µg

For calcium phosphate transfection (eight 15-cm plates):

CAG-GFP (or other MML retroviral plasmid)	240 µg
pCMV-gp	160 µg
pCMV-vsv-g	80 µg

REFERENCES

Jessberger S, Aigner S, Clemenson GD Jr, Toni N, Lie DC, Karalay O, Overall R, Kempermann G, Gage FH. 2008a. Cdk5 regulates accurate maturation of newborn granule cells in the adult hippocampus. *PLoS Biol* **6:** e272. doi: 10.1371/journal.pbio.0060272.

Jessberger S, Toni N, Clemenson GD Jr, Ray J, Gage FH. 2008b. Directed differentiation of hippocampal stem/progenitor cells in the adult brain. *Nature Neurosci* **11:** 888–893.

Laplagne DA, Esposito MS, Piatti VC, Morgenstern NA, Zhao C, van Praag H, Gage FH, Schinder AF. 2006. Functional convergence of neurons generated in the developing and adult hippocampus. *PLoS Biol* **4:** e409. doi: 10.1371/journal.pbio.0040409.

Laplagne DA, Kamienkowski JE, Esposito MS, Piatti VC, Zhao C, Gage FH, Schinder AF. 2007. Similar GABAergic inputs in dentate granule cells born during embryonic and adult neurogenesis. *Eur J Neurosci* **25:** 2973–2981.

Lie DC, Colamarino SA, Song HJ, Desire L, Mira H, Consiglio A, Lein ES, Jessberger S, Lansford H, Dearie AR, et al. 2005. Wnt signalling regulates adult hippocampal neurogenesis. *Nature* **437:** 1370–1375.

Livet J, Weissman TA, Kang H, Draft RW, Lu J, Bennis RA, Sanes JR, Lichtman JW. 2007. Transgenic strategies for combinatorial expression of fluorescent proteins in the nervous system. *Nature* **450:** 56–62.

Marumoto T, Tashiro A, Friedmann-Morvinski D, Scadeng M, Soda Y, Gage FH, Verma IM. 2009. Development of a novel mouse glioma model using lentiviral vectors. *Nat Med* **15:** 110–116.

Naldini L, Blomer U, Gage FH, Trono D, Verma IM. 1996a. Efficient transfer, integration, and sustained long-term expression of the transgene in adult rat brains injected with a lentiviral vector. *Proc Natl Acad Sci* **93:** 11382–11388.

Naldini L, Blomer U, Gallay P, Ory D, Mulligan R, Gage FH, Verma IM, Trono D. 1996b. In vivo gene delivery and stable transduction of nondividing cells by a lentiviral vector. *Science* **272:** 263–267.

Naviaux RK, Costanzi E, Haas M, Verma IM. 1996. The pCL vector system: Rapid production of helper-free, high-titer, recombinant retroviruses. *J Virol* **70:** 5701–5705.

Suh H, Consiglio A, Ray J, Sawai T, D'Amour KA, Gage FH. 2007. In vivo fate analysis reveals the multipotent and self-renewal capacities of Sox2+ neural stem cells in the adult hippocampus. *Cell Stem Cell* **1:** 515–528.

Tashiro A, Sandler VM, Toni N, Zhao C, Gage FH. 2006. NMDA-receptor-mediated, cell-specific integration of new neurons in adult dentate gyrus. *Nature* **442:** 929–933.

Tiscornia G, Singer O, Verma IM. 2006. Production and purification of lentiviral vectors. *Nat Protoc* **1:** 241–245.

Toni N, Teng EM, Bushong EA, Aimone JB, Zhao C, Consiglio A, van Praag H, Martone ME, Ellisman MH, Gage FH. 2007. Synapse formation on neurons born in the adult hippocampus. *Nat Neurosci* **10:** 727–734.

Toni N, Laplagne DA, Zhao C, Lombardi G, Ribak CE, Gage FH, Schinder AF. 2008. Neurons born in the adult dentate gyrus form functional synapses with target cells. *Nat Neurosci* **11:** 901–907.

van Praag H, Schinder AF, Christie BR, Toni N, Palmer TD, Gage FH. 2002. Functional neurogenesis in the adult hippocampus. *Nature* **415:** 1030–1034.

Verma IM, Weitzman MD. 2005. Gene therapy: Twenty-first century medicine. *Annu Rev Biochem* **74:** 711–738.

Zhao C, Teng EM, Summers RG Jr, Ming GL, Gage FH. 2006. Distinct morphological stages of dentate granule neuron maturation in the adult mouse hippocampus. *J Neurosci* **26:** 3–11.

Zufferey R, Donello JE, Trono D, Hope TJ. 1999. Woodchuck hepatitis virus posttranscriptional regulatory element enhances expression of transgenes delivered by retroviral vectors. *J Virol* **73:** 2886–2892.

MOVIE LEGEND

Movies are freely available online at www.cshprotocols.org/imaging.

MOVIE 19.1. Representative time series for spine motility studies. (*a*) Adult running, 56 dpi. (*b*) Adult, 126 dpi. Each movie consists of four time frames (0, 30, 60, and 90 min). The size of the field of view: (*a*) 9.9 x 6.1 µm; (*b*) 10.8 x 6.4 µm.

20 Nonviral Gene Delivery

David A. Dean[1] and Joshua Z. Gasiorowski[2]

[1]Departments of Pediatrics and Biomedical Engineering, School of Medicine and Dentistry, University of Rochester, Rochester, New York 14642; [2]Department of Surgery, Pritzker School of Medicine, University of Chicago, Chicago, Illinois 60637

ABSTRACT

Modern cell biology increasingly relies on molecular tools to facilitate the study of cellular processes. With the advent of recombinant DNA technology, RT-PCR (reverse transcription–polymerase chain reaction) to amplify almost any gene, and fluorescent proteins that can be fused to any desired target protein, functional studies designed to determine the roles of proteins within cells have exploded. Creating the appropriate fusion protein plasmid (or any plasmid for that matter) is a key step, but the DNA must also be delivered to the cell for expression and function studies. There are several common approaches for delivering DNA, including liposome-mediated and polymer-mediated transfection, electroporation, and direct DNA delivery by microinjection. This chapter provides an overview of several of these nonviral gene delivery methods, with an emphasis on directly injecting plasmids into cells.

NONVIRAL TRANSFECTION

A number of chemical and physical methods for introducing DNA and RNA into cells have been developed over the years. The common thread to all of the chemical techniques is that they rely on cationic carriers to complex with negatively charged nucleic acids for their uptake by cells. Because plasmids and RNAs are intrinsically negatively charged, and the plasma membrane carries a net positive charge, cells are largely impermeable to uncomplexed nucleic acids. The steps of essentially all chemical methods for transfection include dilution/preparation of the carrier, formation of DNA-carrier complexes, addition to cells, and subsequent removal of the nonendocytosed excess complexes (Elouahabi and Ruysschaert 2005). The first generation transfection reagents included calcium phosphate and diethylamino ethanol (DEAE)-dextran, which are inexpensive, are extremely simple to use, and continue to show good efficacy in certain cell types. However, as a number of liposomal reagents were introduced in the 1980s (Felgner et al. 1987), the use of calcium phosphate and DEAE-dextran has decreased. Cationic liposomes are the most widely used class of carrier for transfection. Among the available products, most of which differ by performance among cell types, there are Lipofectin (a 1:1 mixture of DOTMA:DOPE [*N*-[1(2,3-dioleoyloxy)propyl]-*N*,*N*,*N*-trimethylammonium chloride:dioleoyl phosphatidylethanolamine]; Invitrogen), Transfectam (Promega Corporation), and DOTAP [1,2-dioleoyloxy-3-trimethylammonium propane]:DOPE. All of these reagents tend to be rather costly. Consequently, a cheaper alternative is to purchase purified lipids (DOTAP, DOPE, and DOTMA) and formulate homemade transfection reagents. Avanti Polar Lipids, Inc. produces all of these lipids in very high purity, and the methods for formulation (add equal molar amounts of each lipid suspended in chloroform, mix, dry under inert gas, store, and reconstitute in appropriate buffer) are simple.

There are also polymer-based systems that can be used to transfect cells, some of which rely on one type of molecule, whereas others contain a mixture of polymers and lipids. Reagents such as SuperFect or PolyFect (QIAGEN) use activated dendrimers, which are positively charged and resemble snowflakes in structure. Multicomponent reagents, including Lipofectamine 2000 (Invitrogen), FuGENE 6 (Roche), and TransIT (Mirus), use mixtures of polymers and lipids that work together to condense the DNA and form micelles for a more uniform size distribution of particles. The choice of transfection reagent many times comes down to personal preference or empirical data on its utility with a desired cell type. Determining which reagent is best for a particular set of experiments and cells often requires trial and error to optimize the conditions. It is also important to note that for many experiments, especially those relying on imaging and expression of fluorescently tagged proteins, it is not necessary to achieve 100% gene transfer efficiency because the transfected cells can be selected before imaging. In some cases, as little as 10% of the transfection efficiency of the cells may be sufficient to make detailed studies possible.

Choosing Where in the Cell Cycle to Transfect

The success of these methods requires that cells be actively dividing to obtain high levels of transfection (Brunner et al. 2000). Plasmid-carrier complexes must be endocytosed, escape the endosome, traffic through the cytoplasm, and cross the nuclear envelope, all before any transcription can occur. The nuclear envelope represents a major barrier to DNA delivery, and the primary way that plasmids cross this barrier is during mitosis and the concomitant breakdown of the nuclear envelope. Thus, for most protocols, it is advisable to transfect cells when they are between 50% and 70% confluent and are very likely to undergo one round of division before the experimental analysis of gene expression.

Preparation of Plasmids

Although, in the early days of transfections, it was considered necessary to use highly purified plasmids obtained from cesium chloride density ultracentrifugation, the introduction of resin-based purification kits and columns over the past 15 years or so has greatly simplified DNA preparation for transfections. Any number of commercially available plasmid purification kits will generate sufficiently pure DNA to use in transfections. Further, the amount of endotoxin present in any of these kits (even those that are not considered endotoxin-free kits) is so low as to not cause any problems with the cells. The only exception may be the use of plasmid from minipreparation spin columns or other minipreparation kits. Although the isolated DNA from minipreparation-scale kits may be adequate for some transfections, transfection results are inconsistent, and often cell death or poor gene transfers results.

Liposome-Mediated and Dendrimer-Mediated Transfection

Generalized protocols are provided for liposome-mediated transfection (Protocol A) and dendrimer-mediated transfection (Protocol B).

Multicomponent-System-Mediated Transfection

As with dendrimers, a major advantage of multicomponent systems is that they can transfect cells in the presence of serum. As with the other gene delivery approaches, transfection conditions will need to be optimized and will vary slightly for each reagent and manufacturer. Common reagent preparation steps include diluting the DNA in serum-free medium, diluting the transfection reagent in medium (with or without serum), combining diluted DNA and reagent, incubating them for 10–30 min to allow complex formation, and finally adding growth medium for plating on washed cells. Because of the proprietary nature of these reagents, follow the specific protocol steps provided by the manufacturers.

ELECTROPORATION

Electroporation is the most widely used physical method to transfect populations of cells. When cells are exposed to brief electrical fields, transient membrane destabilization results, allowing uncomplexed plasmids or other nucleic acids to cross the plasma membrane (Escoffre et al. 2009). Once the electric field is removed, the membrane "seals," trapping the nucleic acids inside the cell. If the applied electric field is too low, no membrane destabilization occurs; and, hence, no DNA enters the cells. Conversely, if the applied field is too high, pores fail to reseal, and cell death occurs. Thus, there is a fine balance between DNA delivery and cell death that must be optimized for each cell type that is being used. Electroporation is simple, is rapid (<5 min start to finish), and can yield very high transfection efficiencies (up to 95% of cells in some cases). The disadvantages, however, are that each cell type requires its own set of pulse parameters, and optimization for new cell types may require much trial and error unless suboptimal levels of transfection are sufficient.

A number of commercially available buffers are available for electroporation, many of which boast 100% transfection efficiency, but again, no buffer is best for all cell types; optimization and testing must be performed on each cell type and electroporation system. Phosphate-buffered saline (PBS) or cell medium (modified Eagle's medium [MEM], Dulbecco's modified Eagle's medium [DMEM], etc.) with or without serum, is also often a good choice for an electroporation buffer, although the presence of serum can result in bubbling or popping at high field strengths.

Electroporators

Two types of electroporators are available: square wave and exponential decay. Both types are suitable for electroporation of mammalian cells in culture, and again, as with other gene delivery methods, optimization is necessary to determine which works best for desired cell types. Square wave electroporators deliver pulses to cells of a set voltage for a defined amount of time. These pulses typically last 0.1–20 msec, and fields are usually 250–1000 V/cm (i.e., 100–400 V in a 0.4-cm cuvette). Exponential decay electroporators deliver a peak of energy that dissipates exponentially, giving a time constant (τ) that is a function of the resistance of the sample and the capacitance set on the instrument. This corresponds to the time necessary for the charge to decrease to ~37% of the initial voltage. Neither type of waveform is better than the other, so access to individual instruments or published protocols usually drives the choice of system.

There are two approaches for delivering nucleic acids into cells using electroporation. The first, and more commonly used, method involves mixing purified plasmid with cells in suspension (usually trypsinized cells), placing the suspension in a 0.4-cm gap electrode, and delivering the pulse(s).

Immediately following electroporation, the cells are removed from the cuvette, complete growth medium is added, and everything is dispensed into the appropriate plates. This method provides relatively uniform delivery of DNA, and transfected cells are randomly and uniformly distributed in the plates for later analysis. The drawback to this approach is that genes are delivered to trypsinized cells that must then settle and attach to tissue culture dish substrates and re-establish appropriate cytoskeletal and cellular structures.

The second electroporation method delivers genes directly to adherent cells using a novel type of electrode called a Petri Pulser (BTX Instrument Division, Harvard Apparatus). This electrode is a series of parallel plates, each separated by several millimeters, that fits into a single well of a six-well plate or an individual 35-mm dish. Cells are washed with PBS, plasmid is added to the well in 1 mL of buffer/medium, and the electrode is placed into the well so that the plates rest ~0.5 mm above the monolayer of cells. One square wave pulse of 10-msec duration at 100–160 V is delivered. The electrode is removed, and complete medium is added to the cells. The correct distance of the electrode above the cells that maximizes expression may have to be determined empirically. If the electrodes touch the cells, they will be killed (evident by equally spaced rows of missing cells across the dish).

DNA MICROINJECTION

Nuclear versus Cytoplasmic Microinjection

Direct microinjection of genetic material into cells began in the 1970s with the demonstration that mRNA isolated from one cell type could be translated in other cells after it was microinjected into the recipients (Graessmann and Graessmann 1971; Gurdon et al. 1971; Lane et al. 1971). In 1980, it was shown that when a plasmid expressing thymidine kinase was microinjected into the nuclei of thymidine kinase–deficient mouse fibroblasts, between 50% and 100% of cells showed enzyme activity at 24 h postinjection (Capecchi 1980). In contrast, no gene expression was detected in more than 1000 cytoplasmically injected cells during the same time frame. Similar results have been obtained in numerous systems (Graessman et al. 1989; Mirzayans et al. 1992; Thornburn and Alberts 1993; Zabner et al. 1995; Dean et al. 1999). These results show that the site of injection can be critically important. Implicit in these experiments is the fact that the microinjected cells did not divide during the course of the experiments. During mitosis, the nuclear envelope breaks down, eliminating a major barrier to gene transfer. If plasmids are present in the cytoplasm, they have full access to the nuclear compartment during this stage of the cell cycle. Although these experiments would suggest that plasmids are incapable of entering the nuclei of nondividing cells, this is not the case. Data from a number of laboratories suggest that certain DNA sequences can promote nuclear entry and gene expression even in nondividing cells in a sequence-specific manner (Graessman et al. 1989; Dean 1997; Dean et al. 1999; Vacik et al. 1999; Mesika et al. 2001). However, to obtain maximal expression of desired transgenes in cells that will not undergo mitosis during the course of the experiment, microinjection of the DNA into the nucleus is strongly recommended.

DNA Concentration

Gene expression is dependent on the copy number of the gene. It has been shown that as few as one to three cytomegalovirus (CMV) promoter-driven, green fluorescent protein (GFP)-expressing plasmids injected into the nucleus of a cell can result in detectable, albeit low, GFP expression in a reasonable percentage of injected cells (Dean et al. 1999). Increasing the number of plasmids delivered to the cell increases both the percentage of cells expressing the gene product and the amount of gene product (Fig. 1) (Graessman et al. 1989; Dean et al. 1999; Ludtke et al. 2002). In typical experiments, DNA should be injected at a starting concentration of ~300 ng/µL but can be used at concentrations between 50 ng/µL and 500 ng/µL, if sufficient expression is not obtained at first. A typical microinjection system is designed to deliver <10% of the total cell volume, which corresponds to an injection

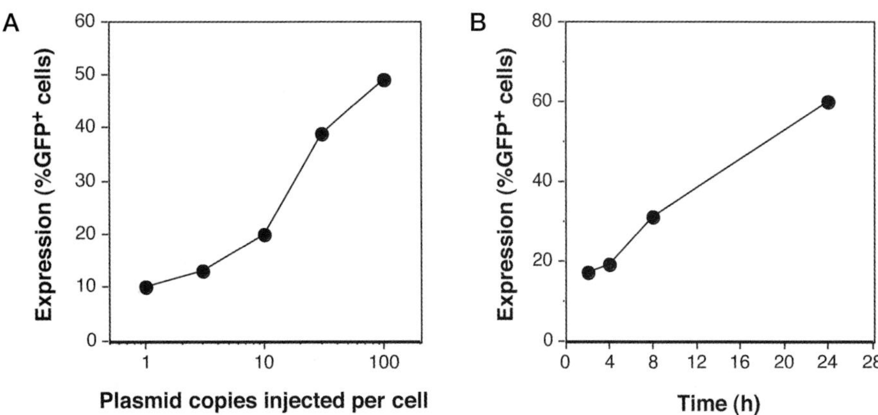

FIGURE 1. Dose dependency and time dependency of transgene expression in microinjected cells. (*A*) Human umbilical vein endothelial cells (HUVECs) were grown on etched coverslips and were microinjected with various copy numbers of pEGFP-N1 (Clontech)-expressing enhanced green fluorescent protein (eGFP) from the CMV immediate early promoter. Eight hours later, eGFP-expressing cells were counted and expressed as a percentage of cells injected. (*B*) HUVECs grown on etched coverslips were microinjected with 10 copies of pEGFP-N1, and GFP expression was assessed at the indicated times following injection. (Reprinted, with permission, from Dean 2005.)

volume of ~10^{-14} L to 10^{-12} L (0.01–1 pL). At a concentration of 300 ng/μL, roughly 5000 copies of a 6-kb plasmid would be delivered to the cell, assuming that 0.1 pL is delivered. Greater concentrations of DNA can be delivered to the cell, but two problems arise. First, gene expression saturates above several hundred to a thousand copies of DNA per cell, so the benefit of delivering more plasmids will be lost. Second, at concentrations above 1 mg/mL, DNA becomes technically difficult to inject because of the viscosity and the aggregation within injection needles. Further, additional problems of injecting too much DNA include potential toxicity of the gene product and other problems associated with overexpression. Thus, lower concentrations are a better starting point.

Timing of Gene Expression

Most studies assess gene expression at 4–24 h post nuclear injection. Unless very low copy numbers of plasmids are injected into the cell, significant expression can be detected by the earlier time point. One parameter that can affect the timing of gene expression is promoter strength: Weak promoters will usually take longer to produce sufficient protein to visualize. However, when using strong promoters, expression at early times after injection is readily detectable. Indeed, using a DNA concentration of 300 ng/μL, GFP expression from the CMV promoter (e.g., pEGFP-N1 from Clontech) can be visually detected in cells easily within 40 min of microinjection (Figs. 1 and 2). Similar results have been obtained using plasmids expressing products from the simian virus 40 (SV40) early promoter and the CMV immediate early promoter/enhancer.

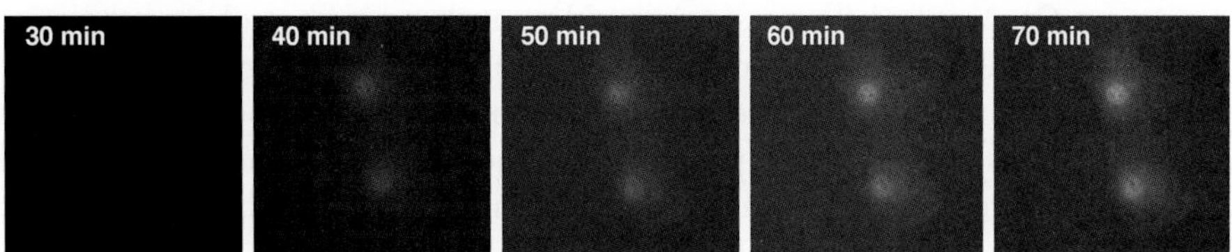

FIGURE 2. Early time course of gene expression in microinjected cells. TC7 cells (African green monkey kidney epithelial cells) were microinjected with pEGFP-N1 at 300 ng/μL and assessed for eGFP expression at the indicated times. All photographs were taken with the same exposure time. As can be seen, eGFP is first detected in these cells at 40 min postinjection, and the expression increases with time. (Reprinted, with permission, from Dean 2005.)

Microinjection Needles

Microinjection needles can either be pulled from glass capillaries on a pipette puller in the laboratory or be purchased premade and sterile from a number of companies. The advantage to pulling needles in the laboratory is that a variety of different needle types can be pulled, depending on the samples and the cells being injected. Protocols for using two pipette pullers are below, one using a high-end Flaming/Brown pipette puller from Sutter Instruments (Protocol D) and one using a less expensive alternative that produces fine needles but requires a little more user input (Protocol E). An added advantage of pulling needles in the laboratory is cost; once a pipette puller has been purchased, boxes of glass capillaries are inexpensive (a box of 500 usually costs <$50). The advantages to buying preformed and sterilized needles (~$5 per needle) include increased uniformity of needles from one to another, ease of use (open the packet, fill the needle, and inject), high quality, and not having to invest in a pipette puller.

When pulling needles, there are several variables that need to be addressed. These include filament design, heat, pull strength (tension), and delay time between heating and pulling. There are several types of filaments that will heat the capillaries, including box (surrounds the capillary in a box) and trough (a U shape) filaments. On the less expensive pullers, the trough is sufficient for most needle types, but, on the high-end instruments, refer to the manufacturer for recommendations according to the desired pipettes. The heat setting will affect the length and the tip size of the needle; high heat will typically produce longer needles and finer tips. The pull strength will also affect length and tip size, with greater pull strength producing longer tapered needles with finer tips. Finally, shorter delay times between heating and pulling can result in longer tapers and finer needles; but, if the time is made too short, the glass will form fibers resembling glass wool, and no needle will be formed.

Micromanipulation/Microinjection Systems

There are two common types of microinjection systems, one that provides a constant flow of sample and the other that provides a pulsed flow. The former is very simple and can be assembled on a relatively low budget. In this system, a constant flow of sample is delivered from the tip of the pipette, and the amount of sample injected into the cell is determined by how long the pipette remains in the cell (Graessman and Graessman 1986). Although this means that each cell will receive a slightly different amount of sample, with practice, microinjections can become highly reproducible. A typical system is composed of a pressure regulator that can be adjusted for two pressures: back pressure and injection pressure (e.g., World Precision Instruments [WPI] pneumatic PicoPump PV830), a capillary holder, and a coarse and fine micromanipulator (e.g., Narishige Group, WPI or Stoelting Co.) (see Protocol H). In this system, using a manual micromanipulator, the needle is positioned above the cell to be injected and lowered into the cell (Fig. 3). As the needle is lowered, the cell is slightly deformed because the tip is entering at an angle. Because sample is constantly flowing out of the needle, this may not be suitable for precious samples, although experience shows that even 5 µL of sample is more than enough to inject 1000 cells using this method (D.A. Dean, unpubl.).

The second type of microinjection system uses a pulsed flow. The most commonly used pulsed-flow system is the Eppendorf FemtoJet injector coupled with the Eppendorf InjectMan (see Protocol G). A pulsed-flow system provides much more control over the injection parameters, and hence, variability in injections is reduced. Another nice feature of this system is the dynamics of the injection itself. The needle is positioned over the site to be injected, and when the injection button is pressed, the needle is pulled back in the *x–y* direction to allow a diagonal insertion of the needle into the cell, causing a direct piercing of the cell (Fig. 3). This method is fast and may do less damage to the cell than the constant-flow method.

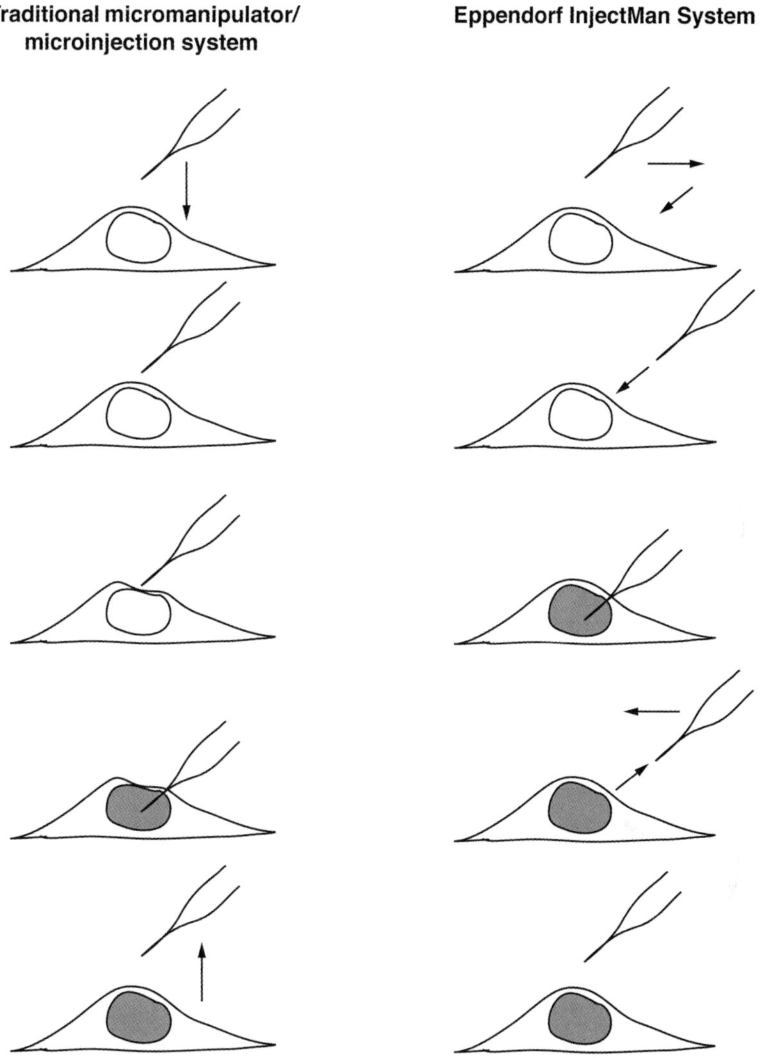

Traditional micromanipulator/ microinjection system

Eppendorf InjectMan System

FIGURE 3. Microinjection systems. A traditional constant pressure microinjection system using a manual micromanipulator is shown on the *left*. With this system, the needle is positioned over the site of injection (the nucleus), and the needle is lowered directly down into the cell. When the needle touches the cell at an angle, the membranes are distorted slightly until the needle enters the cell to deliver its contents (volume delivered depends on time inside the cell). The needle is then lifted out of the cell to complete the process. On the *right* is a cartoon of the Eppendorf InjectMan system using a motorized micromanipulator. After the needle has been positioned, the controller pulls the needle back in the *x*–*y* direction and then lowers the needle on a diagonal so that the tip directly pierces the cell. After the contents have been delivered (based on the time set on the FemtoJet injector), the needle exits the cell on the same diagonal and returns to its original position. (Reprinted, with permission, from Dean 2005.)

Protocol A

Liposome-Mediated Transfection

Although each manufacturer will provide detailed instructions for the use of its reagent, the following general steps are used for essentially all formulations, although times, concentrations, and volumes may differ slightly. The following volumes would be for use in a 35-mm dish or a well of a six-well dish.

MATERIALS

CAUTION: See Appendix 6 for proper handling of materials marked with <!>.

Reagents

Buffer (PBS or 10 mM Tris <!>, 1 mM EDTA [TE]) or medium without serum or antibiotics (e.g., DMEM or MEM)
Cells to be transfected
Liposome transfection kit
Medium, complete
PBS or medium without serum or antibiotics, warmed to 37°C
Plasmid containing desired DNA/RNA for transfection

Equipment

Tissue culture incubator
Tubes, polystyrene
Avoid using polypropylene tubes; see Step 4.

EXPERIMENTAL METHOD

1. Mix lipids in their tube using either a brief vortexing or shake.
2. Dilute the liposomes into either buffer or serum-free/antibiotic-free medium. Typically, 2–20 μL of liposomes into 100 μL final volume is used. Incubate for 30–45 min at room temperature.

   A good starting place is to use 5-μL lipid. If transfection efficiency is sufficient for experiments, no optimization is needed; if not, test several volumes of lipid and several concentrations of DNA.

   It is vital to allow the liposomes to incubate for the prescribed time. If liposomes are diluted and used immediately, transfection efficiency usually decreases 10-fold to 20-fold. However, longer incubations do not further increase efficiency.
3. Dilute plasmid (0.5–2 μg) into 100 μL of the same buffer/medium used for liposome dilution. No incubation time is needed.

   For initial transfections, try 1 μg of plasmid.
4. Mix diluted plasmid and diluted liposomes in a polystyrene tube. Gently mix by flicking the tube several times with a finger. Incubate 10–15 min at room temperature.

   Avoid using polypropylene tubes (including microfuge tubes) because the liposomes will adhere much more strongly to the surface of these tubes, thus decreasing the amounts available for addition to cells.

5. Remove medium from cells to be transfected, and wash twice with prewarmed PBS or growth medium without serum or antibiotics. Remove all liquid from the cells.

6. Add 0.8 mL of serum-free/antibiotic-free medium to the DNA–liposome mixture. Briefly mix by flicking the tube, and add to adherent cells.

   Steps 6 and 7 must be performed within a brief time period so that the cells do not dry out. Many times, a ring of dead cells toward the outside edge of the plate may be seen, indicating that too much time has elapsed between media removal/washing and addition of transfection complexes. Alternatively, the 0.8 mL can be added immediately before washing the cells, if this is easier.

7. Incubate the cells at 37°C in a CO_2 incubator for 4–5 h. Remove the medium, and replace it with serum-containing and antibiotic-containing growth medium.

   Many transfection reagents show some level of toxicity to cells, so removal of excess uninternalized reagent helps to keep cells viable and unperturbed. However, many cells have no problem with the presence of lipid, so it may be possible to add additional serum-containing medium directly to the transfection medium without removal.

8. Incubate for an additional 16–48 h, and evaluate gene expression.

Protocol B

Dendrimer-Mediated Transfection

One advantage of using dendrimer-based reagents is that they appear to be less sensitive to the presence of serum during transfection than liposomal reagents. Thus, this may be a better choice for cells that show greater dependence on the continual presence of serum (to avoid initiation of stress responses or induction of differentiation). As for liposome-mediated transfections, concentrations and volumes may need to be optimized for each cell type used.

MATERIALS

Reagents

Cells to be transfected
Dendrimer transfection kit
Medium, complete
PBS or medium without serum or antibiotics, warmed to 37°C
Plasmid containing desired DNA/RNA for transfection

Equipment

Tissue culture incubator

EXPERIMENTAL METHOD

1. Dilute 0.5–5 µg of plasmid into 100 µL of buffer or serum-free/antibiotic-free medium. No incubation time is needed.

   For initial transfections, try 1 µg of plasmid. Although dendrimer-mediated transfections are not inhibited by the presence of serum, the DNA should not be diluted in serum-containing medium because there are proteins in the serum that may degrade the DNA.

2. Add 10–20 µL (30–60 µg) of dendrimer solution directly to the DNA solution. Mix by pipetting or vortexing. Incubate for 5–10 min at room temperature.
3. Remove medium from cells to be transfected, and wash twice with prewarmed PBS or growth medium without serum or antibiotics. Remove all liquid from the cells.
4. Add 0.8-mL growth medium (containing serum and antibiotics) to DNA-dendrimer mixture, mix by inversion or pipetting, and add to cells.
5. Incubate the cells at 37°C in a CO_2 incubator for 2–4 h. Remove medium, and replace with fresh growth medium.
6. Incubate the cells for an additional 1–48 h, and evaluate gene expression.

Protocol C

Plating Cells for Microinjection

Coverslips for microinjection need to be marked so that microinjected cells can be identified at desired time points after injection. Coverslips can be etched by the user, as described here, or pre-etched coverslips can be purchased (Bellco Glass, Inc.; Fig. 4). Once the coverslips have been etched and sterilized, cells can be plated onto them and allowed to grow.

MATERIALS

CAUTION: See Appendix 6 for proper handling of materials marked with <!>.

Reagents

Cell culture medium (see note to Step 8)
Cells to be injected
Matrigel or rat tail collagen (optional, see note to Step 6)
NaOH <!>, 2 N (optional, see note to Step 6)

Equipment

Aluminum foil
Coverslips, 25 x 25-mm, # 1

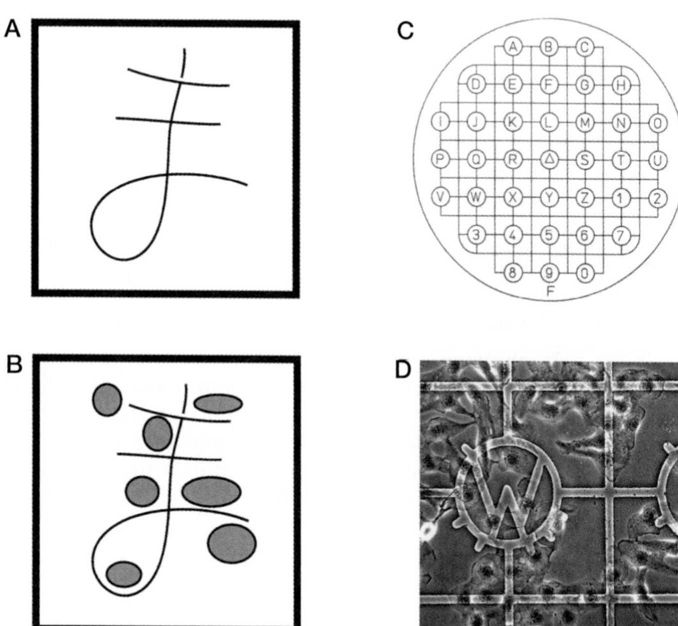

FIGURE 4. Etched coverslips for microinjection. (*A*) A sample of an asymmetric figure etched onto a coverslip. (*B*) Areas for microinjection. Because of the asymmetric nature of the etching, different plasmids can be injected at these sites and can be identified later by their position relative to the figure. (*C*) A photoetched coverslip. (*D*) Cells growing in one area of the photoetched coverslip. (Reprinted, with permission, from Dean 2005.)

Diamond pen (VWR International 52865-005)
Microscope outfitted for microinjection
Petri dish, glass, 100-mm
Tissue culture dishes, 60-mm
Tissue culture plates, six-well

EXPERIMENTAL METHOD

Etching Coverslips

1. Place 25 x 25-mm # 1 coverslips on a solid clean surface (benchtop or in a tissue culture hood).
2. Use a diamond pen to lightly etch an asymmetric figure onto the coverslip (Fig. 4).

   If too much pressure is applied while etching, the coverslip will snap in half or will break later in the experiment (typically after much time and energy has been spent completing all microinjections). However, if too little pressure is applied, the etched figure will be extremely difficult to detect under the microscope, both during microinjection and later when cells are being visualized.
3. Layer etched coverslips onto aluminum foil that is cut to fit into a 100-mm glass Petri dish. Approximately 10 square coverslips can be placed onto one layer (do not overlap, or they will stick together, rendering them useless). Layer aluminum foil and coverslips to fill the dish, and cover with the glass top (approximately 15 full layers in a dish).
4. Autoclave the coverslips on dry cycle (20 min sterilize, 30 min dry).

Plating Cells onto Coverslips

5. Place the sterile etched coverslips (or sterile photoetched coverslips) into the wells of a six-well plate (one coverslip per well).
6. Plate cells into the wells as would normally be performed if the coverslip was absent.

   Although most cells will adhere to the coverslips, some cells will not. Two approaches can be taken to promote cell adherence. First, the coverslips may need to be cleaned. Coverslips and glass slides may have a thin film of grease on them that can reduce cell adherence. To clean etched coverslips, incubate them in 2 N NaOH for 2 h, rinse them extensively with dH_2O, and layer them into the Petri dish before autoclaving in Step 4. A second way to get cells to attach is to coat the coverslips with extracellular matrix proteins, such as Matrigel or rat tail collagen, after autoclaving.
7. When cells are at the desired confluency, remove the coverslips from the six-well dishes, and transfer them to a 60-mm tissue culture dish containing 5 mL of the appropriate medium.
8. Transfer the culture dish to a microscope outfitted for microinjection.

   If injections and time out of the incubator are limited to <20–30 min, cells can be maintained without damage in standard medium containing or lacking serum (e.g., DMEM containing 10% fetal bovine serum). However, because the lack of CO_2 will raise the pH of the medium, many investigators perform microinjections in a buffered medium. This is especially important if longer injection times are required or if the cells are especially sensitive to pH. In these cases, buffer the medium by the addition of 20-mM HEPES buffer (pH 7.4).

Protocol D

Preparing Injection Pipettes on a Flaming/Brown Pipette Puller

For a laboratory performing even a moderate number of microinjection experiments, a Flaming/Brown-style pipette puller, such as the Sutter Instruments P-97, is a cost-effective choice for manufacturing quality micropipettes quickly and reproducibly.

MATERIALS

Equipment

Capillaries for making micropipettes

To produce Femtotip-like pipettes, use 10-cm-long, 1.0-mm outer diameter (OD), 0.78-mm inner diameter (ID) thin-wall borosilicate glass capillaries with filament (Sutter Instruments BF100-78-10). To make long-taper pipettes, use 10-cm-long, 1.0-mm OD, 0.75-mm ID thin-wall borosilicate glass capillaries with filament (WPI, TW100F-4).

Filament for the P-97 pipette puller

Use either a 2.5 x 4.5-mm box filament (Sutter Instruments FB245B) or a 2.5 mm x 2.5-mm box filament (FB255B); see Step 3.

Pipette puller (Model P-97, Sutter Instruments)

Pipette storage container (e.g., WPI E210, with a 1.0-mm needle insert)

EXPERIMENTAL METHOD

1. Push the pipette holders together into their locked positions.
2. Place borosilicate glass capillaries into the capillary holders, in locked position.
3. Select the program to be run:

   For Femtotip-like microinjection pipettes: Use 10-cm-long, 1.0-mm OD, 0.78-mm ID thin-wall borosilicate glass capillaries with filament. The filament on the P-97 pipette puller should be a 2.5 x 4.5-mm box filament.

Step	Heat	Pull	Velocity	Time	Pressure
1	Ramp	100	10	250	500

   This program will loop three times to produce a needle with a 0.5-µm ID tip that is ~100–150-µm long.

   For common, long-taper microinjection pipettes: Use 10-cm-long, 1.0-mm OD, 0.75-mm ID thin-wall borosilicate glass capillaries with filament. Use a 2.5 x 2.5-mm box filament to create these needles.

Step	Heat	Pull	Velocity	Time	Pressure
1	580	none	50	145	500
2	570	115	30	145	500

   This program will produce a tip with a 0.5-µm ID that is ~150–200-µm-long.

These parameters should be used as starting values and will most likely need to be adjusted to obtain the desired tips.

4. Carefully remove the needles (two will be produced), and place them in a pipette storage container with tips down.

   For best results, pull the needles within several hours of use. In climates with high humidity, they should be used within 1 h. In addition, allow the pipettes to rest for at least 5 min before filling them with DNA or protein solution.

Protocol E

Preparing Injection Pipettes on a PUL-1 Micropipette Puller

The PUL-1 micropipette puller is a robust and inexpensive machine that can be found in many laboratories around the world. New machines are no longer being manufactured, although used ones can still be purchased.

MATERIALS

Equipment

Capillaries for making micropipettes
Pipette puller (PUL-1, WPI)
Pipette storage container (e.g., WPI E210, with a 1.0-mm needle insert)

EXPERIMENTAL METHOD

1. Push the pipette holders together into their locked positions.
2. Place borosilicate glass capillaries into the capillary holders, in locked position.

   It is very important to make sure that the capillaries are held tightly in the holder. If there is any play, the needles can slip a little, resulting in poor or no needles.
3. Set the heat to setting 6 and the delay to setting 3 for the initial needles, using a 1.5-mm U-shaped filament. The tension knob on the back should be moved to the middle of the range.
4. Pull needles by pressing the Auto button.
5. Remove needles, and observe them by phase-contrast microscopy.
6. Adjust the settings to obtain a needle that has a uniform taper as in Fig. 4.

   Defining the appropriate pulling parameters is an empirical operation that will require pulling multiple test needles. However, once the settings have been established, this machine will produce highly reproducible needles for months at a time.
7. Carefully remove the needles (two will be produced), and place them in a pipette storage container with tips down.

Protocol F

DNA Sample Preparation and Loading Sample into Pipettes

Plasmid DNA is purified using standard procedures. The resulting preparation can then be delivered into microinjection needles either by a backfilling or by a forward-filling approach.

MATERIALS

Reagents

PBS, 0.5x
Plasmid purification kits, resin-based (e.g., QIAGEN, Promega Corporation, Invitrogen, or Bio-Rad Laboratories)
Plasmids
TE (pH 8.0)

Equipment

Hamilton syringe (10 μL) outfitted with a 1.5–2-in-long needle (Hamilton Company microliter #901) or Eppendorf microloader pipette tip (Eppendorf 5242-956.003) on a 10-μL Eppendorf pipette
Microfuge tubes
Microinjection needles (prepared in either Protocol D or E, or purchased)
Parafilm
Pipette storage container (e.g., WPI E210, with a 1.0-mm needle insert)
Spin-X centrifuge tube filter, 0.22-μm cellulose acetate (Corning 8160)

EXPERIMENTAL METHOD

1. Purify plasmids using commercially available resin-based plasmid purification kits. Resuspend the DNA in TE (pH 8.0).

   Alternatively, high-quality plasmids can be purified by alkaline lysis and cesium chloride gradient centrifugation.

2. Dilute plasmid to a final concentration of 50–500 ng/μL in 0.5x PBS.

   Although 5 μL of plasmid solution is sufficient to fill more than 10 needles and inject several thousand cells, it is best to prepare 50–100 μL of DNA solution to facilitate subsequent steps. DNA can be suspended in sterile ddH_2O, 1x PBS, or other buffered solutions, instead of 0.5x PBS, with similar results.

3. Place DNA solution into the cup of a 0.22-μm cellulose acetate Spin-X centrifuge tube filter, and centrifuge at 16,000*g* for 1 min in a microcentrifuge.

   Alternatively, if enough DNA solution has been made, any 0.22-μm syringe filter will work. For precious samples, the low retention volume of the Spin-X tubes will prevent solution loss. For volumes of <50 μL, prewet the Spin-X filter with 0.5x PBS to reduce sample loss.

4. Remove the filter from the Spin-X tube, and centrifuge the filtrate at 16,000*g* for 15 min at 4°C to remove any particulate or aggregated matter.

5. Transfer the DNA to a clean tube.

   If extra DNA solution is stored for use at later times (i.e., days), it should always be recentrifuged as in Step 4 before use.

6. Proceed to either backfill the microinjection needles (Steps 7–10) or front fill them (Steps 11–14).

Backfilling Microinjection Needles

7. Withdraw 5 µL of particulate-free DNA in 0.5x PBS into either a 10-µL Hamilton syringe outfitted with a 1.5–2-in-long needle or an Eppendorf microloader pipette tip on a 10-µL Eppendorf pipette.

8. Carefully insert the syringe (or tip) into the back end of the microinjection needle, and dispense ~0.2–0.5 µL of solution while slowly rotating or twisting the microinjection needle between index finger and thumb.

   The twisting of the needle aids in the transfer of the DNA solution and helps pull it down into the tip.

9. Remove the syringe (or tip) from the microinjection needle, and place the needle into a second pipette storage container filled with 1 cm of H_2O.

   The humidified chamber will help to prevent evaporation of the small volumes of sample within the needles.

10. Use filled needles for microinjection within 1 h, if possible.

Front-Filling Microinjection Needles

11. Withdraw 5 µL of particulate-free DNA in 0.5x PBS using a pipettor, and place onto a piece of Parafilm.

12. Gently touch the tip of the microinjection needle to the surface of the drop of fluid, and allow the fluid to wick into the needle by capillary action.

    Many investigators feel that front filling draws any particulates or dust in the needle up away from the tip and to the top surface of the sample, away from the opening. In addition, fewer air bubbles are usually present when front filling. Thus, many feel that with front filling, needle clogging is reduced; however, if samples are adequately filtered and centrifuged, this is not an issue.

13. Place the filled needle into a second pipette storage container filled with 1 cm of H_2O.

14. Use filled needles for microinjection within 1 h, if possible.

Protocol G

Microinjecting Using a Pulsed-Flow Microinjection System

A pulsed-flow microinjection system, such as the Eppendorf InjectMan and FemtoJet systems, permits a high level of control over the microinjection parameters, ensuring reproducibility. The system is also easy to use, is rapid, and is less likely to damage cells than the constant-flow method.

MATERIALS

Reagents

Cells to be microinjected
Growth medium appropriate for microinjected cells

Equipment

FemtoJet system
Microinjection needles filled with solution to be injected (prepared in Protocol F)
Microscope
Tissue culture incubator

EXPERIMENTAL METHOD

1. Set starting parameters on the FemtoJet system:

Injection pressure (P_i)	145 hPa
Compensation (holding) pressure (P_c)	35 hPa
Time (t_i)	0.3 sec

   These values will change with the solutions being injected (more viscous solutions need a higher injection pressure) and the individual needles. Although each needle pulled with the same program should have the same internal diameter at the tip, even slight differences can necessitate greatly different injection pressures. The pressures and times indicated are good starting points for the two types of needles pulled above (or for purchased Femtotips).

2. Press the Menu button on the FemtoJet to allow a new pipette to be loaded onto the device.
3. Place a needle into the pipette holder on the micromanipulator by inserting the large end into the grip head so that the butt of the needle just goes past the compression O-ring in the grip head. Screw the grip head–needle assembly onto the universal capillary holder. For Femtotips, screw the Femtotip into the Femtotip adapter on the capillary holder.
4. Rotate the capillary holder back into a position parallel with the micromanipulator.
5. Press the Menu button again to pressurize the system for microinjection.

   Caution: When the system is pressurized, it can act like a rocket launcher; if the needle is not fully secured into the grip head, it can shoot off of the capillary holder when pressure is reapplied. Although small, the needle can cause severe damage to eyes and other tissues, and if biohazardous solutions or samples are in the needle, contamination can be a serious concern.

6. Move the capillary forward until the needle tip appears above the objective, and then lower the needle into the medium. Use Fast speed for these motions, until the needle touches the medium.
7. Focus the microscope on the cell monolayer. Start with a 10X objective for microinjections. If there is not enough resolution to distinguish nuclei from cytoplasm, change to a 20X objective for the injections themselves.
8. Switch the speed to Slow, and continue to lower the needle until a blurry image appears above the cells. If necessary, move the needle in the *x–y* plane to position the tip slightly above the area of cells to be injected. The needle should be ~100 μm above the cells.
9. Press the Clean button to ensure that the needle is not clogged and that the sample will flow easily. Differences in the refractive index of the injected solution and the medium will allow a fan-shaped plume of solution to be seen exiting the tip of the needle.

   If the solution exiting the needle flows in an irregular pattern, the needle is more likely to be bad and may cause cellular damage.
10. Set the injection plane by slowly lowering the needle until it touches and then depresses and pierces a test cell. When the needle is inside the cell, press the Z-Limit button on the micromanipulator controller, and then raise the needle to just above the cells (again, ~100 μm).
11. Position the tip of the needle directly above the nucleus to be injected, and press the Inject button on the top of the micromanipulator joystick to initiate the microinjection (Fig. 3).
12. Continue until all injections are completed or the needle clogs and has to be cleared or replaced.

    See *Troubleshooting.*
13. Incubate microinjected cells under appropriate growth conditions (i.e., medium, serum, temperature, and pH)

    See *Troubleshooting.*

TROUBLESHOOTING

Problem (Step 12): The needle becomes blocked so that the solution will not flow into the cell.
Solution: Even with painstaking attention to filtering samples and centrifuging them to remove particulates, clogging will occur. Consider the following.

1. Make sure that the needle-pulling process produced quality needles from which solution can flow. If not, discard the needle, and replace with another one.
2. Increase the injection pressure to get the needle flowing again (i.e., blow out the blockage). Press the Clean button on the FemtoJet to flush out the needle. If this does not work, use the Clean function on the micromanipulator to rapidly pull the needle out of the medium and return it to its original position. This uses the surface tension of the medium to remove attached debris.
3. Tap the tip to try to dislodge any particulates (which can include cell debris or extracellular matrix that becomes attached to needles during the injection process).
4. If these methods do not work, replace the blocked needle with a new one.

Problem (Step 13): Microinjected cells become damaged or die.
Solution: Although great care is taken during the microinjection process to treat the cells gently, a fraction of the injected cells will die or will fail to express the gene product(s). Three parameters can be adjusted to try to decrease cellular damage and death.

1. Reduce the amount of time a microinjection needle is left in a cell. This will reduce the injected sample volume. Excess sample volume within a cell may result in damage.

2. Reduce the injection pressure. This too will reduce the injected sample volume. In addition, excessive injection pressures can exert a damaging physical force on the cell causing it to be blown up. Fluid exiting a 0.5-μm diameter tip is expelled at a pressure great enough to damage cellular architecture and function.
3. Adjust the depth of needle penetration in injected cells. Coverslips are uneven; and, consequently, cells grow at slightly different heights across the coverslip. An injection depth that may be fine for one group of cells could push too far into another set of cells, damaging them. To circumvent this, pay close attention to the injection *z*-limit.

Protocol H

Microinjecting Using a Constant-Flow Microinjection System

A constant-flow microinjection system can be assembled at relatively low cost using components such as WPI's pneumatic PicoPump and manual micromanipulator. A constant-flow system, by design, injects less reproducible cell volumes than does a pulsed-flow system. However, with practice, this difference can be made negligible.

MATERIALS

Reagents

Cells to be injected
Growth medium appropriate for microinjected cells

Equipment

Foot switch for controlling injection pressure (WPI, 3260)
Microinjection needles filled with solution to be injected (prepared in Protocol F)
Micromanipulator, manual (e.g., WPI M3301R)
Micropipette holder, 1.0 mm (WPI, MPH6S)
Microscope
Nitrogen gas
PicoNozzle (WPI, 5430-10) nitrogen gas
PicoPump (WPI, PV820 or PV830)
Tissue culture incubator

EXPERIMENTAL METHOD

1. Attach a tank of nitrogen gas to the pressure inlet on the PicoPump.
2. Ensure that the injection pressure is off. Attach a foot switch to the PicoPump via the Remote plug.

   The foot switch will be used to control injection pressure and thus solution flow. Depressing the switch activates the injection pressure.
3. Set the holding pressure to 4 psi and the injection pressure to 40 psi as a starting point.
4. Insert a needle into the 1.0-mm micropipette holder so that the end of the needle just goes through the gasket, and screw the head to the base, locking the needle into the holder.

   CAUTION: When the system is pressurized, the needle can become a hazardous projectile. Make sure that the needle is secured into the holder and the injection pressure is not activated before attaching the needle in Step 5.
5. Attach the needle–micropipette holder to a PicoNozzle, and place the PicoNozzle onto the micromanipulator.
6. Advance and lower the needle so that it is just above the medium and over the objective.

7. Lower the needle into the medium, and make sure that the hold pressure is high enough to prevent the medium from entering the needle but low enough so that the sample is not being wasted.
8. Focus the microscope on the cell monolayer.

   Start with a 10x objective for microinjections. If there is not enough resolution to distinguish nuclei from cytoplasm, change to a 20x objective for the injections themselves.
9. Lower the needle to just above the cells to be injected, and depress the foot switch to ensure that the needle is not clogged and that the sample will flow easily. The injection pressure may need to be increased or decreased, depending on the sample flow that is seen in the medium.
10. When injections are to begin, press the foot switch to start the sample flowing, and then lower the needle rapidly into the cell to be injected and raise it as soon as fluid has been delivered.

    Because the time of injection is not set or uniform, leave the needle in the cell until a very slight change in refractive index of the cell or fluid delivery can be detected. It is important not to leave the needle in the cell too long or the cell will die (ideally, it should be in the cell for <0.5 sec; with practice, this is easy to master). The easiest way is to turn the micromanipulator joystick one-quarter turn so that the needle goes down into the cell and almost immediately turn it one-quarter turn in the opposite direction to raise the needle. All micromanipulators can be adjusted to vary the distance traveled by one turn.
11. Continue until all injections are completed or the needle clogs and has to be cleared or replaced.

    See *Troubleshooting.*
12. Incubate microinjected cells under appropriate growth conditions (i.e., medium, serum, temperature, and pH)

    See *Troubleshooting.*

TROUBLESHOOTING

Problem (Step 11): The needle becomes blocked so that the solution will not flow into the cell.
Solution: Even with painstaking attention to filtering samples and centrifuging them to remove particulates, clogging will occur. Consider the following.

1. Make sure that the needle-pulling process produced quality needles from which solution can flow. If not, discard the needle, and replace with another one.
2. Increase the injection pressure to get the needle flowing again (i.e., blow out the blockage), and tap the tip to try to dislodge any particulates (which can include cell debris or extracellular matrix that becomes attached to needles during the injection process). Try increasing the injection pressure while gently tapping the needle against a cell-free area of the coverslip.
3. Drag the needle across one of the etched spots on the coverslip. Dragging the needle, however, can cause it to break (but a broken needle and a clogged one are equally useless).
4. If these methods do not work, replace the blocked needle with a new one.

Problem (Step 12): Microinjected cells become damaged or die.
Solution: Although great care is taken during the microinjection process to treat the cells gently, a fraction of the injected cells will die or will fail to express the gene product(s). Three parameters can be adjusted to try to decrease cellular damage and death.

1. Reduce the amount of time a microinjection needle is left in a cell. This will reduce the injected sample volume. Excess sample volume within a cell may result in damage.
2. Reduce the injection pressure. This too will reduce the injected sample volume. In addition, excessive injection pressures can exert a damaging physical force on the cell causing it to be

blown up. Fluid exiting a 0.5-μm diameter tip is expelled at a pressure great enough to damage cellular architecture and function.

3. Adjust the depth of needle penetration in injected cells. Coverslips are uneven, and consequently, cells grow at slightly different heights across the coverslip. An injection depth that may be fine for one group of cells could push too far into another set of cells, damaging them. To circumvent this, pay close attention to the injection *z*-limit.

REFERENCES

Brunner S, Sauer T, Carotta S, Cotten M, Saltik M, Wagner E. 2000. Cell cycle dependence of gene transfer by lipoplex, polyplex and recombinant adenovirus. *Gene Therapy* **7:** 401–407.

Capecchi MR. 1980. High efficiency transformation by direct microinjection of DNA into cultured mammalian cells. *Cell* **22:** 479–488.

Dean DA. 1997. Import of plasmid DNA into the nucleus is sequence specific. *Exp Cell Res* **230:** 293–302.

Dean DA. 2005. Gene delivery by direct injection and facilitation of expression by mechanical stretch. In *Live cell imaging: A laboratory manual* (ed. Goldman RD, Spector DL), pp. 51–66. Cold Spring Harbor Laboratory Press, Cold Spring Harbor, NY.

Dean DA, Dean BS, Muller S, Smith LC. 1999. Sequence requirements for plasmid nuclear entry. *Exp Cell Res* **253:** 713–722.

Elouahabi A, Ruysschaert JM. 2005. Formation and intracellular trafficking of lipoplexes and polyplexes. *Mol Ther* **11:** 336–347.

Escoffre JM, Portet T, Wasungu L, Teissie J, Dean D, Rols MP. 2009. What is (still not) known of the mechanism by which electroporation mediates gene transfer and expression in cells and tissues. *Mol Biotechnol* **41:** 286–295.

Felgner PL, Gadek TR, Holm M, Roman R, Chan HW, Wenz M, Northrop JP, Ringold GM, Danielsen M. 1987. Lipofection: A highly efficient, lipid-mediated DNA-transfection procedure. *Proc Natl Acad Sci* **84:** 7413–7417.

Graessmann A, Graessmann M. 1971. The formation of melanin in muscle cells after the direct transfer of RNA from Harding-Passey melanoma cells. *Hoppe Seylers Z Physiol Chem* **352:** 527–532.

Graessmann M, Graessmann A. 1986. Microinjection of tissue culture cells using glass microcapillaries: Methods. In *Microinjection and organelle transplantation techniques: Methods and applications* (ed. Celis JE, et al.), pp. 3–37. Academic, London.

Graessmann M, Menne J, Liebler M, Graeber I, Graessmann A. 1989. Helper activity for gene expression, a novel function of the SV40 enhancer. *Nucleic Acids Res* **17:** 6603–6612.

Gurdon JB, Lane CD, Woodland HR, Marbaix G. 1971. Use of frog eggs and oocytes for the study of messenger RNA and its translation in living cells. *Nature* **233:** 177–182.

Lane CD, Marbaix G, Gurdon JB. 1971. Rabbit haemoglobin synthesis in frog cells: The translation of reticulocyte 9 s RNA in frog oocytes. *J Mol Biol* **61:** 73–91.

Ludtke JJ, Sebestyen MG, Wolff JA. 2002. The effect of cell division on the cellular dynamics of microinjected DNA and dextran. *Mol Ther* **5:** 579–588.

Mesika A, Grigoreva I, Zohar M, Reich Z. 2001. A regulated, NFκB-assisted import of plasmid DNA into mammalian cell nuclei. *Mol Ther* **3:** 653–657.

Mirzayans R, Remy AA, Malcom PC. 1992. Differential expression and stability of foreign genes introduced into human fibroblasts by nuclear versus cytoplasmic microinjection. *Mutat Res* **281:** 115–122.

Thornburn AM, Alberts AS. 1993. Efficient expression of miniprep plasmid DNA after needle micro-injection into somatic cells. *Biotechniques* **14:** 356–358.

Vacik J, Dean BS, Zimmer WE, Dean DA. 1999. Cell-specific nuclear import of plasmid DNA. *Gene Ther* **6:** 1006–1014.

Zabner J, Fasbender A J, Moninger T, Poellinger KA, Welsh M J. 1995. Cellular and molecular barriers to gene transfer by a cationic lipid. *J Biol Chem* **270:** 18997–19007.

COMPANIES

Avanti Polar Lipids, Inc.
700 Industrial Park Drive
Alabaster, AL 35007-9105
(800) 227-0651
http://www.avantilipids.com

Bellco Glass, Inc.
340 Edrudo Road
Vineland, NJ 08360
(800) 257-7043
http://www.bellcoglass.com/

Bio-Rad Laboratories
1000 Alfred Nobel Drive
Hercules, CA 94547
(800) 424 6723
http://www.biorad.com

BTX Instrument Division
Harvard Apparatus, Inc.
84 October Hill Road
Holliston, MA 01746-1388
(800) 597-0580
http://www.btxonline.com/

Corning
45 Nagog Park
Acton, MA 01720
(800) 492-1110
http://www.corning.com/lifesciences/

Brinkmann Eppendorf
One Cantiague Road
P.O. Box 1019
Westbury, NY 11590-0207

(800) 645-3050
http://www.eppendorf.com/en/

Hamilton Company
P.O. Box 10030
Reno, NV 89520-0012
(800) 648-5950
http://www.hamiltoncomp.com

Invitrogen Corporation
1600 Faraday Avenue
P.O. Box 6482
Carlsbad, CA 92008
(800) 955- 6288
http://www.invitrogen.com

Mirus
545 Science Drive, Suite A
Madison, WI 53711
(888) 530-0801
http://www.mirusbio.com

Narishige Group International USA, Inc.
1710 Hempstead Turnpike
East Meadow, NY 11554
(800) 445-7914
http://usa.narishige-group.com/

Promega Corporation
2800 Woods Hollow Rd.
Madison, WI 53711
(800) 356-9526
http://www.promega.com

QIAGEN
28159 Avenue Stanford
Valencia, CA 91355
(800) 426-8157
http://www1.qiagen.com/

Research Products International (RPI)
10 N. Business Center Drive
Mount Prospect, IL 60056-2190
(800) 323-9814
http://www.rpicorp.com

Roche
Roche Diagnostics Corporation
P.O. Box 50414
9115 Hague Road
Indianapolis, IN 46250-0414
(800) 262-1640
http://www.roche-applied-science.com/

Stoelting Co.
Physiology Division
620 Wheat Lane
Wood Dale, IL 60191
(630) 860-9700
http://www.stoeltingco.com/physio/

Sutter Instruments
51 Digital Drive
Novato, CA 94949
(415) 883-0128
http://www.sutter.com/

VWR International
1310 Goshen Parkway
West Chester, PA 19380
(800) 932-5000
http://www.vwr.com

World Precision Instruments (WPI)
175 Sarasota Center Boulevard
Sarasota, FL 34240
(941) 371-1003
http://www.wpiinc.com

21 Cellular Bioluminescence Imaging

David K. Welsh[1,2] and Takako Noguchi[1]

[1]*Department of Psychiatry, University of California, San Diego, La Jolla, California 92093;*
[2]*Veterans Affairs San Diego Healthcare System, San Diego, California 92161*

ABSTRACT

Bioluminescence imaging of live cells has recently been recognized as an important alternative to fluorescence imaging. Fluorescent probes are much brighter than bioluminescent probes (luciferase enzymes) and, therefore, provide much better spatial and temporal resolution and much better contrast for delineating cell structure. However, with bioluminescence imaging there is virtually no background or toxicity. As a result, bioluminescence can be superior to fluorescence for detecting and quantifying molecules and their interactions in living cells, particularly in long-term studies. Structurally diverse luciferases from beetle and marine species have been used for a wide variety of applications, including tracking cells in vivo, detecting protein–protein interactions, measuring levels of calcium and other signaling molecules, detecting protease activity, and reporting circadian clock gene expression. Such applications can be optimized by the use of brighter and variously colored luciferases, brighter microscope optics, and ultrasensitive, low-noise cameras. We present here a review of how bioluminescence differs from fluorescence, its applications to cellular imaging, and available probes, optics, and detectors, as well as practical suggestions for optimal bioluminescence imaging of single cells.

INTRODUCTION: BIOLUMINESCENCE VERSUS FLUORESCENCE

Bioluminescence is emission of light as a result of an enzymatic reaction in a living organism (see Fig. 1) (Wilson and Hastings 1998; Shimomura 2006). As in fluorescence, electrons are excited to a higher energy level, and photons are emitted as the electrons return to their resting level. However, in bioluminescence, the energy to excite the electrons comes from a chemical reaction rather than from exogenous illumination. Bioluminescent enzymes are known as luciferases, and their substrates are known as luciferins.

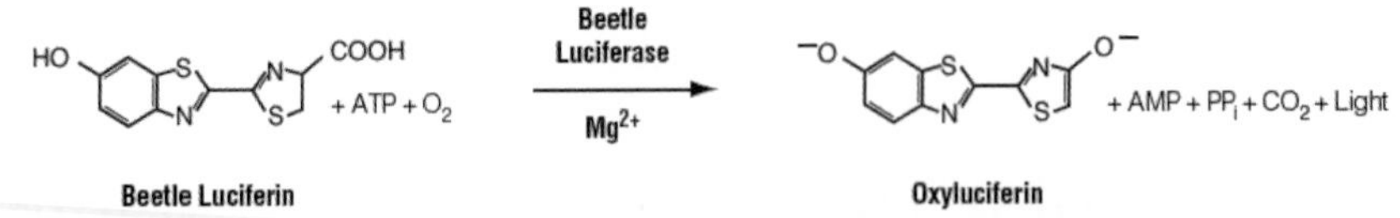

HO ... COOH + ATP + O_2 —Beetle Luciferase, Mg^{2+}→ $^-$O ... O$^-$ + AMP + PP_i + CO_2 + Light

Beetle Luciferin   Oxyluciferin

FIGURE 1. Bioluminescent reaction of beetle luciferin. Beetle luciferin substrate is oxidized in a bioluminescent reaction catalyzed by various beetle luciferases such as firefly luciferase (FLuc). The reaction consumes ATP and oxygen, requires the presence of Mg^{++}, and produces light. In the case of FLuc, the light is yellow–green (peak 560 nm) at 20°C and orange (peak ~612 nm) at 37°C (Zhao et al. 2005). (Reprinted, with the kind permission of Promega Corporation, from the Chroma-Glo™ Luciferase Assay System Technical Manual #TM062 [http://www.promega.com/tbs/tm062/tm062.html].)

Longitudinal studies of single cells are powerful because they capture dynamic processes as well as the inherent variability among cells. Live cell imaging using specific fluorescent probes has been especially useful in delineating cell structure and function (Giepmans et al. 2006). Bioluminescent probes (luciferases) have received relatively little attention until recently because they are exceedingly dim by comparison. There is no doubt that the much brighter fluorescent probes are preferable for cellular-imaging applications requiring fine spatial or temporal resolution, or good contrast, all of which require the collection of many photons.

Cellular imaging, however, is increasingly directed toward detecting and quantifying low abundance molecules, their interactions, and their functional activities in live cells over extended periods of time. For such applications, bioluminescence imaging has some important advantages, largely related to the fact that it does not require exogenous illumination. Unlike in fluorescence imaging, there is no photobleaching of emitting molecules, no phototoxicity, and no artificial perturbation of light-sensitive cells (e.g., in the retina). Furthermore, although it is much dimmer, bioluminescence can be up to 50X more sensitive than fluorescence because the background is so low (Arai et al. 2001; Choy et al. 2003; Troy et al. 2004; Dacres et al. 2009).

The exceedingly low background of bioluminescence imaging derives from two factors. First, relative to endogenous fluorescence (autofluorescence), which can sometimes be as bright as the signal itself (Billinton and Knight 2001), endogenous bioluminescence (autoluminescence) of most cells is extremely low (Troy et al. 2004). For example, from a 500-µm hippocampal slice, autoluminescence (related to oxidative metabolism) is only ~3–4 photons/mm^2/sec (Isojima et al. 1995). Second, with bioluminescence, there are no excitation photons, which contribute greatly to background in fluorescence imaging because of scattering and spectral overlap with emission photons. As a result of these two factors, background levels in bioluminescence imaging are exceedingly low, and the signal-to-noise ratio (SNR) can be very high despite the dim signals.

In summary, fluorescence is usually better suited than bioluminescence for precise localization of cellular components or processes over short time periods or for studying processes with rapid dynamics. Bioluminescence, on the other hand, is more sensitive (because of lower background) and less toxic than fluorescence, making it suitable for many live cell applications in which high spatial and temporal resolution is not critical, particularly long-term studies of biological processes with slower dynamics or light-sensitive components.

APPLICATIONS

Tracking Molecules and Cells

In recent years, bioluminescent probes have been used for a wide variety of imaging applications (Greer and Szalay 2002; Welsh et al. 2005), including tracking molecules and cells. Firefly luciferase (FLuc) is commonly used for ATP assays, exploiting the ATP requirement for its bioluminescent reaction, and this strategy has been adapted to image ATP release from mammalian cells (Zhang et al. 2008). Protein secretory pathways in cells have been imaged using the secreted *Gaussia* luciferase

(Suzuki et al. 2007). Although it has not yet achieved single-cell resolution, a very active application area is tracking cells in whole mice in vivo by imaging (e.g., tumor cells, immune cells, stem cells, bacteria, and viruses) (Dothager et al. 2009). In such applications, the cells are typically engineered to express a luciferase reporter, but the luciferase can also be fused to antibodies directed against desired cell markers (Venisnik et al. 2007).

Protein–Protein Interactions: Bioluminescence Resonance Energy Transfer and Luciferase Complementation Imaging

Protein–protein interactions in single cells can be quantified by two different bioluminescence imaging techniques. In bioluminescence resonance energy transfer (BRET), one protein is tagged with a luminescent photon donor, another protein is tagged with a fluorescent photon acceptor (at a longer wavelength), and close proximity of the two proteins allows the transfer of photons, detected as a change in the emission spectrum (Subramanian et al. 2004). BRET is similar to fluorescence resonance energy transfer (FRET), except that the donor is luminescent rather than fluorescent, so no exogenous illumination is required. Recent developments in BRET include brighter and redshifted probes (Hoshino et al. 2007; De et al. 2009), improved dynamic range (De et al. 2007), and achievement of subcellular resolution (Coulon et al. 2008). In an alternative approach known as luciferase complementation imaging (LCI), two proteins are tagged with complementary luciferase fragments, and close proximity of the proteins reconstitutes luciferase activity, detected as luminescence (Villalobos et al. 2008). For example, LCI has recently been used to detect epidermal growth factor receptor dimerization (Yang et al. 2009).

Calcium Levels: Aequorin

Calcium can be imaged in cells using variants of aequorin, a calcium-sensitive luciferase found in the jellyfish *Aequorea victoria* (Shimomura et al. 1993). Aequorin acts on its substrate (coelenterazine) in a calcium-dependent manner to produce coelenteramide and the emission of blue (470-nm) light. In the jellyfish, aequorin associates with green fluorescent protein (GFP), and BRET transfer to GFP results in the emission of brighter green light (509 nm), brighter because of higher quantum yield. Mimicking nature, the Brulet group engineered a GFP-aequorin fusion probe that is much brighter and redshifted compared with aequorin alone, can be genetically targeted to particular cell types or organelles, and allows easy preliminary focusing and identification of cells using fluorescence (Baubet et al. 2000; also see Chapter 27). This GFP-aequorin has relatively fast kinetics and a wide dynamic range and is relatively insensitive to pH (Rogers et al. 2005; Curie et al. 2007). Further redshifted variants have been generated recently: Venus-aequorin and red fluorescent protein (RFP)-aequorin (Curie et al. 2007; Manjarrés et al. 2008). Such probes have been used to image calcium in *Drosophila* brain (Martin et al. 2007) and whole mice (Curie et al. 2007; Rogers et al. 2007). Single-cell resolution has been achieved in mammalian neurons and cell lines (Baubet et al. 2000; Rogers et al. 2005, 2008) and in mouse retina (Agulhon et al. 2007). The coelenterazine substrate can be chemically altered to change its sensitivity to calcium, allowing measurement over different concentration ranges (Manjarrés et al. 2008).

Circadian Gene Expression

Luciferases are excellent reporters of gene expression, as illustrated by studies of circadian clock function in cells from an impressive variety of species. This approach was pioneered in the 1950s with studies of the naturally occurring circadian bioluminescence rhythm of *Gonyaulax*, a marine dinoflagellate responsible for some red tides and associated bioluminescent waves (Hastings 1989). In the 1990s, engineered bioluminescent reporters of clock gene expression were used to image circadian clock function in plants (Millar et al. 1992), cyanobacteria (Kondo et al. 1993), and flies (Brandes et al. 1996; Plautz et al. 1997). More recently, single-cell resolution has been achieved in cyanobacteria (Mihalcescu et al. 2004), zebrafish cells (Carr and Whitmore 2005), mouse fibroblasts

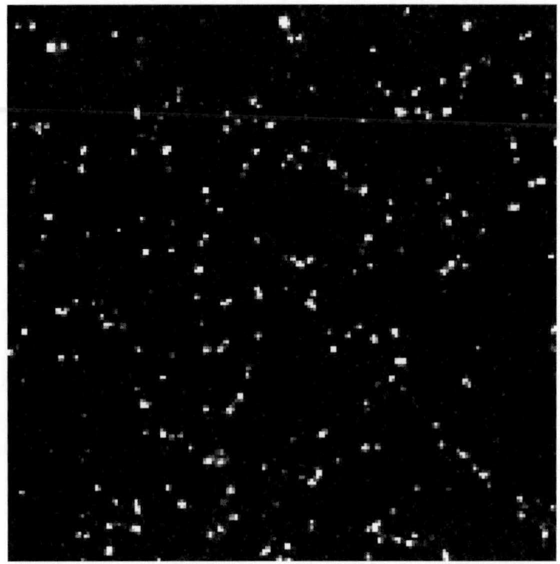

FIGURE 2. Bioluminescent neurons dissociated from the suprachiasmatic nucleus (SCN) of a PER2::LUC knockin mouse (Yoo et al. 2004). Primary SCN neurons were dissociated and cultured for 3 wk under standard conditions (Welsh et al. 1995). The cells were then transferred to HEPES-buffered medium containing 1-mM luciferin and imaged on an inverted Olympus IX70 microscope using a UPlanApo 4x objective and a Spectral Instruments, Inc. SI800 charge-coupled device (CCD) camera cooled to –90°C, with 4 x 4 binning. To eliminate spurious events, this image was constructed by pixelwise minimization of two consecutive 29.9-min exposures. Note the clear luminescence from individual cells, and the visible pixelation (1 pixel = 13 μm), reflecting a deliberate sacrifice of spatial resolution to maximize SNR.

(Welsh et al. 2004), and neurons from the master circadian pacemaker in the mouse suprachiasmatic nucleus (SCN; see Fig. 2) (Yamaguchi et al. 2003; Liu et al. 2007). In all of these studies, circadian (~24-h) rhythms of bioluminescence could be monitored longitudinally, for days or weeks at a time, and with single-cell resolution in the more recent studies.

Retina

Bioluminescence imaging is particularly well suited for studies of the retina because it does not perturb the retina's light-sensitive physiology with exogenous illumination. A few studies have begun to explore this application (Agulhon et al. 2007; Ruan et al. 2008).

Multiple Colors, Multiple Parameters

Luciferases with different spectral properties have been used to monitor multiple parameters simultaneously in cell populations. Most studies have used spectrophotometry or luminometry (Almond et al. 2003; Nakajima et al. 2004b; Branchini et al. 2005, 2007; Nakajima et al. 2005; Michelini et al. 2008), including one recent study measuring calcium selectively in different organelles with targeted GFP-aequorin and RFP-aequorin (Manjarrés et al. 2008). In the study of circadian clocks, differently colored luciferases have been used to monitor circadian expression of two genes at once in cyanobacteria (Kitayama et al. 2004), to measure effects of RORα on both *Per1* and *Bmal1* gene expression (Nakajima et al. 2004a), and to test for interactions between circadian clocks of two separately labeled populations of Rat-1 fibroblasts (see Fig. 3) (Noguchi et al. 2008). A few studies with differently colored luciferases have used CCD cameras to image cell cultures (Gammon et al. 2006; Davis et al. 2007), plants (Ogura et al. 2005), or *Xenopus* embryos (Hida et al. 2009), but discrimination of multiple luciferase signals has not yet been achieved with single-cell resolution.

Engineered Luciferases: Protease Activity, Ligand Binding, Cyclic AMP

As with fluorescent probes (Giepmans et al. 2006), bioluminescent probes can be engineered to expand their range of uses. For example, protease activity can be assayed by attaching to FLuc an inhibitory peptide, which is cleaved by a specific protease (O'Brien et al. 2005). Other modifications allow FLuc to change its conformation and luminescence activity as a result of protease activity, or on binding rapamycin or cyclic AMP (cAMP)(Fan et al. 2008). These probes have not yet been used for imaging but illustrate the potential for future applications.

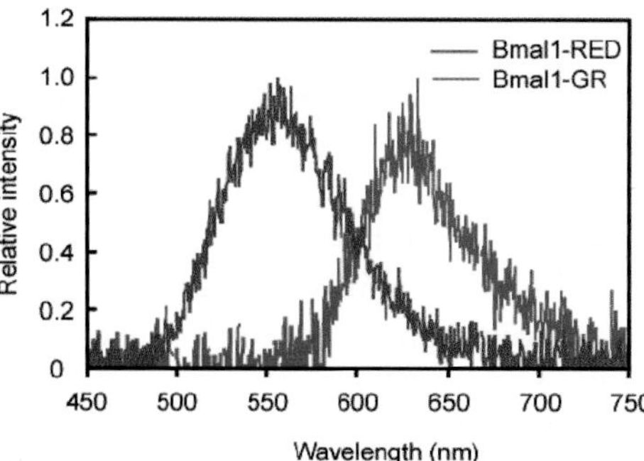

FIGURE 3. Bioluminescence emission spectra of two fibroblast cell lines, one transfected with a green-emitting luciferase from the Japanese luminous beetle (*Rhagophthalmus ohbai*), and the other with a red-emitting luciferase from the railroad worm (*Phrixothrix hirtus*). Luciferases were expressed under control of a promoter from the circadian clock gene *Bmal1* and served as reporters of circadian clock function. Interactions among cellular clocks were tested by simultaneous measurement of circadian oscillations in two separate populations of cells. (Reprinted from Noguchi et al. 2008, with permission, from BioMed Central.)

LUCIFERASES

Firefly Luciferase

The best known bioluminescent probe is FLuc (Fraga 2008), a 61-kDa monomeric enzyme that catalyzes oxidation of its substrate, beetle luciferin, with the emission of yellow–green light (peak ~560 nm at 25°C; maximum quantum yield 41%; see Fig. 1) (Ando et al. 2008). This reaction requires ATP and oxygen. Optimal pH is 7.8 (Baggett et al. 2004), with reduced light emission at low pH because of decreased quantum yield (Ando et al. 2008). The reaction rate is also temperature dependent, yielding brighter bioluminescence at 37°C than at 25°C when FLuc is expressed in mammalian cells (Zhao et al. 2005). Interestingly, the emission spectrum is also substantially redshifted under these conditions (peak ~612 nm [orange] at 37°C) (Zhao et al. 2005). In cells, newly synthesized FLuc is rapidly inactivated by reaction products such as dehydroluciferyl-adenylate (L-AMP) (Fraga et al. 2005), so its functional half-life (as measured by luminescence) is determined partly by this product inhibition (Day et al. 1998) as well as by protein turnover (Leclerc et al. 2000). Estimates of FLuc luminescence half-life in mammalian cells, in the presence of protein synthesis inhibitors, range from ~1 to 4 h (Nguyen et al. 1989; Thompson et al. 1991; Day et al. 1998; Leclerc et al. 2000; Yamaguchi et al. 2003; Baggett et al. 2004). Certain small molecules can also inhibit FLuc, which is a potential source of false positives in high-throughput screens (Auld et al. 2008; Heitman et al. 2008).

Other Luciferases

Many structurally diverse luciferases are found in a wide variety of species, ranging from bacteria to fungi, fish, and insects (Greer and Szalay 2002; Shimomura 2006), but most luciferases used so far in mammalian cells are from beetles (Viviani 2002; Luker and Luker 2008) or marine species. The bacterial *lux* operon conveniently encodes both luciferase and the enzyme for making its substrate and is used widely for studies in prokaryotes, but has not been successfully expressed in mammalian cells. Among beetles (order Coleoptera), luciferases have been cloned from fireflies (family Lampyridae), click beetles (Elateridae), railroad worms (Phengodidae; named for the paired luminous organs on body segments resembling the lighted windows of a train), and Japanese luminous beetles (Rhagophthalmidae). FLuc is the most extensively studied; other beetle luciferases have different structures but use the same luciferin substrate. Among marine species, luciferases have been cloned from the sea pansy (*Renilla reniformis*), the crystal jellyfish (*A. victoria*), a copepod (*Gaussia princeps*), an ostracod (*Vargula hilgendorfii*), and others. The marine luciferases are smaller, ATP independent, and use a different substrate (coelenterazine), which for several reasons is less desirable for cellular-imaging applications. Compared with beetle luciferin, coelenterazine is more expensive, less soluble, less stable, more toxic, and more autoluminescent, although there are improved synthetic versions (e.g., ViviRen and EnduRen, Promega). *Renilla* luciferase is commonly used with a constitutive promoter as a control in gene expression assays. Jellyfish aequorin can be used to monitor cellular calcium as discussed above (also see Chapter 27). *Gaussia* luciferase is secreted, which makes it suitable only for specialized applications.

Optimizing FLuc

Performance of native FLuc in mammalian cells can be improved in several ways (Paguio et al. 2005). Cytoplasmic expression is obtained by removing a peroxisomal targeting site. Expression levels can be increased by mammalian codon optimization, adding enhancer elements (e.g., SV40) (Yoo et al. 2008) or chimeric introns (Hermening et al. 2004), and introducing multiple copies of the gene (e.g., by transfection, electroporation, viral vectors, or random insertion transgenesis), resulting in brighter bioluminescence. Anomalous expression can be reduced by removing sequences in the FLuc gene and vector backbone that might bind mammalian transcription factors. Sensitivity to rapid dynamics of gene expression, which requires rapid luciferase protein turnover (Wood 1995), can be improved by adding degradation signals (e.g., PEST [proline glutamic acid serine threonine]), although this also sacrifices brightness (Leclerc et al. 2000). Many of these improvements have been incorporated in the *luc2* gene and the PGL4 vectors available from Promega (Paguio et al. 2005).

Faithfulness of bioluminescent reporters of gene expression can be optimized by using the largest possible promoter sequence or (even better) fusing the luciferase gene to the gene of interest and introducing it into its native chromosomal site by homologous recombination, as has been performed for the circadian clock gene *Per2* (Yoo et al. 2004). This knockin strategy includes all transcriptional and posttranscriptional regulatory elements, avoids artifactual effects of the insertion site, and achieves expression in 100% of targeted cells, but reduces brightness compared with other methods that introduce multiple copies of the gene.

Brighter Luciferases?

Brighter bioluminescent probes would be very useful for cellular imaging, allowing greater spatial and temporal resolutions for a given sensitivity. In the case of fluorescent probes, brighter versions have been engineered by error-prone polymerase chain reaction and directed evolution (Shaner et al. 2004). Theoretically, this same approach could be taken for FLuc, to improve quantum yield, to increase catalytic rate, or to reduce product inhibition. Such efforts have so far been unproductive, however, so FLuc may be nearly optimized already by natural evolution (K. Wood, Promega, pers. comm.). An alternative approach is to search for naturally brighter luciferases from other species (see Table 1). Caribbean click beetle luciferases (CBG [click beetle green], CBR [click beetle red]) appear somewhat brighter than FLuc in mammalian cells, even when carefully normalizing for expression levels (Miloud et al. 2007). Brazilian click beetle luciferase (Emerald luciferase [ELuc]) is claimed to be ~3x brighter than native FLuc, but this may be because of improved expression. *Renilla* and *Gaussia* luciferases (RLuc, GLuc) may also be brighter than FLuc (Stables et al. 1999; Tannous et al. 2005), but they have limitations as discussed above, including higher background autoluminescence. CBG, CBR, and RLuc (like FLuc) are all brighter at 37°C than at 25°C (Zhao et al. 2005). A third approach for obtaining brighter bioluminescent probes is to fuse a luciferase to a fluorescent protein, creating an autoilluminated fluorescent probe in which higher quantum yield is achieved through BRET. Two such probes have been generated: GFP-aequorin (Baubet et al. 2000) and enhanced yellow fluorescent protein (EYFP)-RLuc (Hoshino et al. 2007), both of which have been used for single-cell imaging but that also have limitations associated with their coelenterazine substrate. In principle, FLuc might be similarly fused with an infrared fluorescent protein (Shu et al. 2009); however, FLuc's quantum yield is already fairly high (41%) (Ando et al. 2008), so improvements in brightness are not likely to be dramatic.

Variously Colored Luciferases

For simultaneous monitoring of multiple parameters, luciferases with different spectral properties are available. The emission spectrum of FLuc can be shifted by mutation (Branchini et al. 2005), and luciferases of various colors are also available from click beetles, railroad worms, Japanese luminous beetles, *Renilla*, and *Gaussia* (see Table 1). Unlike FLuc, the spectral properties of CBG, CBR, and RLuc do not change with temperature (Zhao et al. 2005). Separate signals can be discriminated from

TABLE 1. Luciferases for bioluminescence imaging

Luciferase	Organism	Species name	Emission peak	Supplier	Reference(s)
LuxAB	Bacteria	*Vibrio harveyi*	490 nm	N/A	Kondo et al. 1993
Luc2, optimized FLuc (firefly)	North American firefly	*Photinus pyralis*	560 nm (~612 nm at 37°C)	Promega (Madison, WI)	Paguio et al. 2005
CBG	Caribbean click beetle	*Pyrophorus plagiophthalamus*	537 nm	Promega (Madison, WI)	Wood et al. 1989
CBR	Caribbean click beetle	*P. plagiophthalamus*	613 nm	Promega (Madison, WI)	Wood et al. 1989
RLuc (*Renilla*)	Sea pansy	*Renilla reniformis*	480 nm	Promega (Madison, WI)	Lorenz et al. 1991
GLuc (*Gaussia*)	Gaussia	*Gaussia princeps*	480 nm	N/A	Tannous et al. 2005
GFP-aequorin	Crystal jellyfish	*Aequorea victoria*	509 nm	N/A	Baubet et al. 2000
EYFP-RLuc	Sea pansy	*R. reniformis*	525 nm	N/A	Hoshino et al. 2007
ELuc[a] (emerald)	Brazilian click beetle	*Pyrearinus termitilluminans*	538 nm	Toyobo (Osaka, Japan)	Viviani et al. 1999b
SLG[a]	Japanese luminous beetle	*Rhagophthalmus ohbai*	550 nm	Toyobo (Osaka, Japan)	Nakajima et al. 2005
SLO[a]	Japanese luminous beetle	*R. ohbai*	580 nm	Toyobo (Osaka, Japan)	Viviani et al. 2001
SLR[a]	Railroad worm	*Phrixithrix hirtus*	630 nm	Toyobo (Osaka, Japan)	Viviani et al. 1999a; Nakajima et al. 2004b

N/A, not applicable; CBG, click beetle green; CBR, click beetle red; SLG, stable luciferase green; SLO, stable luciferase orange; SLR, stable luciferase red.
[a]Not commercially available in the United States.

differently colored luciferases by using long-pass filters and a spectral unmixing algorithm (see http://www.promega.com/chromacalc). Signals are usually too dim to use the standard narrow-band filters commonly used with fluorescent probes.

OPTICS

Lens Brightness

Bioluminescent probes are much dimmer than fluorescent probes. For optimal bioluminescence imaging of cells, microscope optics must collect and transmit as much light as possible, with minimal magnification so as to concentrate it on a minimal number of camera pixels (Geusz 2001; Christenson 2002; Welsh et al. 2005; Karplus 2006; also see Chapter 1).

The amount of light collected by an objective lens is specified by its numerical aperture (NA),

$$\mathrm{NA} = n \cdot \sin\theta, \tag{1}$$

where n is the refractive index of the material between the lens and the sample ($n = 1.00$ for air, 1.33 for water, 1.52 for oil) and θ is the half-angle of the cone of light collected. The fraction of emitted light collected by the lens is its collection efficiency (Q):

$$\mathrm{Q} = \mathrm{NA}^2/2n^2. \tag{2}$$

Thus, higher-NA lenses collect more light; they also have better resolving power (distinguish more closely spaced points) but less depth of field. Given the physical constraints of microscope lens design, the practical limit for θ is ~72°, so maximal NA = 0.95 (air), 1.26 (water), 1.44 (oil); and maximal Q ≈ 0.45. Specialized fiber-optic systems can theoretically collect more light ($Q = 0.5$) but require custom fabrication, whereas lenses used for standard 35-mm photography collect much less light ($Q < 0.2$; Karplus 2006).

The proportion of incident light transmitted through a lens is specified by its transmittance, or transmission efficiency. A typical microscope objective lens is composed of approximately eight lens

TABLE 2. Microscope objective lenses for bioluminescence imaging

Lens	NA	Mag	*T* (600 nm)	*B*
Nikon				
Plan Apo 4x	0.20	4	0.92	23.0
Nikon				
Plan Apo 10x	0.45	10	0.88	17.8
Olympus				
XLFLUOR 4x[a]	0.28	4	0.96	47.0
Olympus				
UPLSAPO 10x	0.40	10	0.91	14.6
Zeiss				
FLUAR 5x	0.25	5	0.95	23.8
Zeiss				
FLUAR 10x	0.50	10	0.93	23.3

[a]Requires nonstandard mounting.

elements, and light is lost to reflection at each optical surface as well as by absorption within each lens element. Reflective losses are minimized by antireflection coatings. Transmittance (*T*) should be 90% at visible wavelengths.

The brightness (*B*) of the image is related not only to NA and *T* but also to magnification (Mag), which determines how widely light is spread across the detector. Higher Mag lenses generally collect more light (because of higher NA) but also spread light over more pixels of the detector, reducing the brightness and the SNR. Thus, lower Mag lenses are brighter for a given NA and also provide greater field of view. Still, modest magnification (4x–10x) is helpful for initial focusing on cells in bright field and for discriminating between adjacent cells:

$$B(\text{luminescence or bright field}) = (\text{NA}/\text{Mag})^2 \times T \times 10^4, \quad (3)$$

$$B(\text{epifluorescence}) = (\text{NA}^2/\text{Mag})^2 \times T^2 \times 10^4. \quad (4)$$

Note that high NA and *T* are more important for fluorescence than for luminescence. This is because fluorescence brightness depends on how efficiently the objective lens collects and transmits excitation light to the sample as well as how efficiently it collects and transmits emitted light to the detector. Note also that high NA is more important than low Mag for fluorescence, but they are equally important for luminescence. Hence, whereas the brightest lenses for fluorescence tend to be high Mag, the brightest lenses for luminescence are usually low Mag.

Optics to Maximize SNR

Use the brightest possible objective lens (see Table 2), or use a higher Mag lens with even higher NA and then reduce final Mag with a demagnifying camera adapter. Minimize the number of optical elements in the light path (i.e., no filters or mirrors). Mounting the camera on the bottom port of an inverted microscope instead of on a side port, for example, avoids 2% to 3% attenuation of luminescence by a mirror (R. Nazar, Olympus, pers. comm.). Carefully clean all optical surfaces.

CAMERAS

Detection of Photons

The best detectors for low-light imaging are charge-coupled device (CCD) cameras (Christenson 2002; Karplus 2006; also see Chapter 2), which rely on the photoelectric effect to detect light. When photons impact the silicon chip of a CCD camera, they generate free electrons, which are then channeled in a controlled fashion to a readout amplifier. An analog-to-digital (A/D) converter converts voltage values for perhaps 10^6 individual picture elements (pixels) to numerical brightness values in arbitrary A/D units, which can be converted to electrons using the gain value supplied by the manufacturer. The sensitivity of the camera is expressed as the quantum efficiency (QE), which is the proportion of incident photons actually detected. Sometimes free electrons are generated in the absence of incident photons (thermal electrons, giving rise to dark current). Dark current (D) is usually reported in units of electrons/pixel/sec.

Camera Noise, Signal-to-Noise Ratio, and Standard Deviation/Mean

Noise in a conventional CCD camera (N_{camera}) originates from two principal sources: the readout process (read noise, N_{read}) and fluctuation of dark current (dark noise, N_{dark}, which increases with exposure duration, t):

$$(N_{camera})^2 = \sqrt{(N_{read})^2 + (N_{dark})^2}, \tag{5}$$

$$N_{dark} = \sqrt{Dt}. \tag{6}$$

Useful measures of camera performance for a given sample are the signal-to-noise ratio (SNR; which quantifies the ability to detect faint signals) and the standard deviation (SD)/mean (which quantifies the uncertainty of estimated brightness). The number of incident photons ($P_{incident}$) depends on the sample. The number of detected photons ($P_{detected}$) increases linearly with QE, but the noise associated with the signal (shot noise, N_{shot}) increases only with $\sqrt{P_{detected}}$, so both SNR and SD/mean improve with brighter samples or higher QE. Of course, SNR also improves with lower camera noise (N_{camera}), so the ideal camera has QE = 100% and $N_{camera} = 0$:

$$P_{detected} = P_{incident} \cdot \text{QE}, \tag{7}$$

$$N_{shot} = \sqrt{P_{detected}}, \tag{8}$$

$$N_{total} = \sqrt{(N_{camera})^2 + (N_{shot})^2}, \tag{9}$$

$$\text{SNR} = P_{detected} / N_{total}, \tag{10}$$

$$\text{SD/mean} = N_{shot}/P_{detected} = \sqrt{P_{detected}}/P_{detected} = 1/\sqrt{P_{detected}}. \tag{11}$$

Conventional CCDs

Conventional CCD cameras designed for low-light imaging have high QE and low dark current but significant read noise. In the back-thinned design, CCD chips are thinned to transparency and illuminated from behind so that photons do not have to penetrate the electron channeling structures on the front of the chip. Also, an antireflection coating is applied to the back of the chip to minimize reflective losses. With this design, QE typically exceeds 90% at wavelengths in which luciferase emission peaks. Cooling the CCD reduces dark current by ~50% for every 7°C–8°C, to values as low as 0.0001 electrons/pixel/sec at temperatures of –90°C or –100°C. Thus, N_{dark} can be <1 electron even for a 60-min exposure. Read noise, however, is typically at least two to three electrons per exposure. Read noise can be minimized by slower readout. Another technique to reduce the impact of read noise is on-chip binning, in which arrays of 2 x 2, 4 x 4, or 8 x 8 pixels are combined and read out as single superpixels. Binning can greatly improve SNR (at the expense of spatial resolution) because superpixels detect more photons than single pixels do but with the same read noise.

Intensified CCDs

Intensified charge-coupled device cameras (ICCDs) use an image intensifier to preamplify the signal above the level of read noise but at the expense of much lower QE in the visible range. ICCDs are composed of a photocathode for detecting incident photons, a microchannel plate (MCP; which is like an array of miniature photomultiplier tubes) for amplification, a phosphor-coated plate to convert amplified electrons back to photons, and a CCD chip to image those amplified photons. QE of the photocathode is only ~40%–50% at best, but amplification by the MCP is so great (>10,000x) that individual photon events greatly exceed the level of CCD read noise, so that a relatively high threshold can be set for counting photons (photon-counting mode), and the read noise is effectively excluded. In the Stanford Photonics XR/Mega-10Z ICCD camera, the GaAsP photocathode has low intrinsic dark current, reduced further by cooling to –20°C, to levels even lower than those of conventional CCD cameras. Amplification noise, arising from fluctuations in intensifier gain, can be reduced by using two MCPs in series such that their fluctuations tend to cancel. Moreover, amplification noise becomes unimportant in photon-counting mode, in which pixel brightness is estimated by the number of photon events above a threshold rather than by the brightness of individual events. ICCDs do have drawbacks: They are expensive and complex, the photocathode and phosphor plate slowly degrade over time, image burn-in can occur, and the camera can be destroyed by accidental exposure to high light levels (although the Stanford Photonics camera has built-in safeguards to prevent this).

Electron-Multiplying CCDs

Like ICCDs, electron-multiplying charge-coupled device cameras (EMCCDs) preamplify the signal above the level of read noise, but they also preserve high QE by using a CCD as the primary detector and an extended gain register within the CCD itself for amplification, instead of a separate image intensifier (Ives 2009). Before readout, free electrons are channeled through approximately 600 high-voltage stages to generate additional electrons by impact ionization. This preamplification effectively excludes read noise in photon-counting mode, as in ICCDs, but impact ionization introduces additional noise arising from clock-induced charge (CIC), which is amplified along with the signal. This CIC noise (N_{CIC}), which occurs with every readout, is typically much lower than a conventional CCD camera's read noise, but it increases with binning and can add up quickly with the frequent exposures needed to avoid overlapping events in photon counting. As in ICCDs, there also is amplification noise, which can be overcome by photon counting. However, because amplification levels in EMCCDs are relatively modest (e.g., 500x–1000x), photon counting may not be as clean in EMCCDs as it is in ICCDs. Dark current also tends to be higher than in other types of CCDs:

$$N_{CIC} = \sqrt{CIC}. \qquad (12)$$

Comparisons

Conventional CCDs, ICCDs, and EMCCDs have different strengths and limitations for low-light imaging (see Table 3). Conventional CCDs are simple to operate and perform well when long exposures are feasible, but the high read noise penalty with each exposure limits their use for studying rapidly changing cellular processes. In contrast, ICCDs and EMCCDs are well suited for studying dynamic processes because they preamplify the signal above the read noise, which eliminates or reduces the penalty for frequent exposures. In fact, in the photon-counting mode, exposures must be kept fairly short to avoid overlapping events. This is potentially cumbersome if one needs to do long integrations, constructed from many separate exposures, but allows great flexibility in choosing the integration interval, even after the experiment ends. ICCDs are more complex, more expensive, more difficult to use, and less sensitive than conventional CCDs (QE ≤ 50% vs. > 90%), but they make up for their reduced sensitivity by eliminating read noise, thereby preserving high SNR. The Stanford Photonics ICCD camera also has very low dark current, making its SNR performance competitive with conventional CCDs even at long exposure durations. The low QE might still be a disadvantage, however, because collecting fewer photons increases the variability of brightness estimates (SD/mean). Like ICCDs, EMCCDs can effectively eliminate read noise by preamplification, and yet they still have high QE comparable to conventional CCDs. However, EMCCDs introduce a significant new type of noise due to CIC, cannot count photons as cleanly as ICCDs, and have relatively high dark current.

TABLE 3. Cameras for bioluminescence imaging

Manufacturer	Model	Type	QE (600 nm)	N_{read} or N_{CIC} (electron, rms)	D (electrons/pixel/sec)	Website
Andor Technology	iKon-M 934 (DU934N-BV)	CCD	0.92	2.5	0.000120	andor.com
Spectral Instruments, Inc.	Series 800/850	CCD	0.92	2.9	0.000200	specinst.com
Stanford Photonics	XR/Mega-10Z	ICCD	0.40	0	0.000038	stanfordphotonics.com
Andor Technology	iXon+ 888	EMCCD	0.92	0.071	0.001000	andor.com, emccd.com
Hamamatsu Photonics	ImagEM-1K (C9100-14)	EMCCD	0.92	0.071	0.001000	hamamatsu.com
Photometrics	Evolve	EMCCD	0.92	0.067	0.001000	evolve-emccd.com

rms, root mean square.

Testing

Before purchasing a low-light camera, one should test it in a specific application, because performance cannot always be predicted from published specifications. Artifacts due to subtle defects in the design of camera control hardware or software may be apparent only at very low light levels. For example, some cameras suffer from nonuniform bias (pixel intensities show a distinct pattern on short exposures with shutter closed), latent image (cooled electrons fail to move off the chip during readout), or an excessive number of hot pixels (artifactually bright values).

Measure read noise (N_{read}; N_{CIC} for an EMCCD camera) by taking a 0-sec exposure with the shutter closed and computing the SD of pixel intensity values across the image. Convert from analog-to-digital units (ADUs) to electrons using the gain of the camera supplied by the manufacturer. Measure total camera noise (N_{camera}) by taking a 30–60-min exposure with the shutter closed and computing the noise in a similar fashion. Dark noise (N_{dark}) and dark current (D) can then be calculated from Equations 5 and 6.

For ICCD or EMCCD cameras in photon-counting mode, count dark events (shutter closed) for various exposure durations (e.g., 0, 1, and 10 sec). For EMCCD cameras, noise in very short exposures is primarily due to CIC. In photon-counting mode, exposure duration must be kept short to avoid overlapping events, so simulate long exposures by adding events from a series of short exposures. Calculate noise for each exposure duration as the square root of the number of dark events per pixel.

Camera Settings to Maximize SNR

Even with an ideal detector, one must collect enough photons to overcome shot noise. To maximize the signal from dim samples, use the greatest possible on-chip binning and the longest possible exposure (i.e., sacrifice spatial and temporal resolution). To minimize camera read noise, use on-chip binning and the slowest readout speed. To minimize dark current, cool the CCD to the lowest possible temperature. With ICCDs and EMCCDs, turn amplification up high enough so that single photon events are well above read noise, and use the photon-counting mode.

Integrated Systems

Recently, integrated bioluminescence-imaging systems have become available in Japan and Europe. These consist of a small microscope in a light-tight box, an environmental chamber for temperature control and gassing of cultures on the microscope stage, optimized optics, and an EMCCD camera. The optics consist of a high-NA lens system and a straight optical path to the camera. The Olympus Luminoview LV200 and the ATTO Cellgraph systems are both well suited for cellular bioluminescence imaging but are not yet sold in the United States.

CELL CULTURE AND ENVIRONMENT

Cell Culture

Optimal bioluminescence requires well-oxygenated, healthy cells. For imaging with FLuc reporters, we typically culture tissue explants or dissociated cells in a 35-mm dish containing HEPES-buffered, air-equilibrated Dulbecco's modified Eagle's medium (DMEM; GIBCO 12100-046), supplemented with 1.2 g/L $NaHCO_3$, 10 mM HEPES buffer, 4 mM glutamine, 25 U/mL penicillin, 25 µg/mL streptomycin, 2% B-27 (GIBCO 17504-044), and 1 mM luciferin (BioSynth L-8220). Neurons are cultured in wells of glass-bottomed dishes (MatTek). Brain slices are cultured on Millicell-CM membrane inserts (Fisher Scientific PICMORG50). The dish is covered by a 40-mm circular coverslip (Erie Scientific 40CIR1) sealed with vacuum grease to prevent evaporation. It is placed inside a heated lucite chamber (Solent Scientific, UK) custom engineered to fit around the stage of our inverted microscope (Olympus IX71), which rests on an antivibration table (Technical Manufacturing Corporation). The environmental chamber keeps the stage at a constant 36°C. The chamber also accommodates gassing with 5% CO_2 for pH control of bicarbonate-buffered media, if necessary.

Luciferin

Luciferin remains active for up to 6 wk in culture medium at 36°C, without replenishment (D.K. Welsh and S.A. Kay, unpubl.). We recommend D-luciferin from BioSynth or Promega, stored in 100-mM aliquots at –20°C in the dark, used at 0.1–1 mM. Poorer quality products may contain L-luciferin, which inhibits the bioluminescence reaction (Nakamura et al. 2005). For neurons, the sodium salt of luciferin may be less toxic than the potassium salt, which is depolarizing. The optimal concentration of luciferin is ~1 mM, which is near saturating for FLuc (1–5 mM) but below the toxic range for most cells (≥2 mM; E. Hawkins, Promega, pers. comm.).

Phenol Red

The pH indicator phenol red can attenuate luminescence by absorbing photons; but, in practice, the effect is minimal for imaging applications. Phenol red has an absorbance peak at ~560 nm, near the emission peaks of FLuc (560 nm) and CBG (537 nm), and very little absorbance in the red at any pH (Duggleby and Northrop 1989). Accordingly, in cell lysates assayed at 20°C in a luminometer, phenol red (15 mg/L) attenuates FLuc (yellow–green) and CBG (green) luminescence by 40%–45% but does not attenuate CBR (red) luminescence appreciably (Promega Bright-Glo and Chroma-Glo manuals). In fibroblasts at pH 7.4 and 37°C, however, FLuc luminescence is redshifted (Zhao et al. 2005), and in luminometer experiments under these conditions, we find that phenol red (15 mg/L) attenuates FLuc luminescence more modestly (25%; T. Noguchi and D.K. Welsh, unpubl.). Furthermore, in imaging experiments using an inverted microscope, much less medium is interposed between cells and detector than is the case in a luminometer. In this configuration, attenuation of FLuc luminescence by phenol red is negligible (T. Noguchi and D.K. Welsh, unpubl.). We often include phenol red (15 mg/L) in culture media for both luminometer and imaging experiments.

Dark Room

Stray light can add significantly to noise levels in bioluminescence imaging. Therefore, it is important to place the imaging setup in a completely dark room with black walls, no windows, and a tightly fitting door with a floor-to-ceiling curtain inside. Avoid any phosphorescent materials (e.g., some paints and plastics). After focusing, cover the culture dish with a small black lucite box, drape the microscope with black cloth (Thorlabs BK5), and turn off room lights and computer monitor. Cover instrument LEDs with black electrical tape. Test for light leaks by eye after 20-min dark adaptation.

IMAGE PROCESSING

Artifacts

Low-light images should be corrected for various camera artifacts. Artifacts unrelated to the signal that do not change over time (hot pixels, patterned bias, patterned dark current) can be removed by subtracting a dark image (shutter closed, same exposure duration). A nonuniform pattern of QE or amplification can be removed using a flat field image (uniform illumination): Multiply all pixels by the average intensity of the flat field image, and then divide (pixelwise) by the flat field image. Bright spot artifacts are caused by cosmic rays (Butt 2009) or camera-related spurious events. These bright spots are randomly placed and can, therefore, be removed by pixelwise minimization of successive images, that is, by constructing a new image from each pair of images in the time series, taking each pixel of the new image from the dimmer of the two corresponding pixels in the original images. Alternatively, filter out cosmic rays by excluding events above a certain brightness threshold. The latter method works best with ICCD/EMCCD photon counting.

Cell Brightness

Bioluminescence images can be analyzed to produce a time series of luminescence intensity values for multiple individual cells. This requires defining a region of interest for each cell, in each image. The region can be moved, if necessary, to track cells over time but should not change size. For each cell and each image, measure the average intensity in the cell region (Cell), subtract the average intensity of a background region with no cells (Background), and multiply by the area of the cell region (Area, in pixels), to give cell brightness in A/D units (ADUs). Cell brightness can then be converted to incident photons/min using the camera's gain (Gain, in ADUs/electron) and QE (electrons/photon), and the exposure duration (Exp, in minutes):

$$\text{ADU} = (\text{Cell} - \text{Background}) \cdot \text{Area}, \tag{13}$$

$$\text{photons/min} = \text{ADU}/(\text{Gain} \cdot \text{QE} \cdot \text{Exp}). \tag{14}$$

CONCLUSION

In summary, bioluminescence imaging is a highly sensitive, nontoxic analytical technique that has proven its utility in a wide range of live cell studies. For optimal results, it requires careful selection and use of luciferase probes, microscope optics, and low-light cameras. In the future, brighter, variously colored luciferases engineered for specific applications, and more sensitive detectors, will further expand the range of applications for bioluminescence imaging.

ACKNOWLEDGMENTS

Thanks to Woody Hastings and Hugo Fraga (Harvard), Keith Wood (Promega), Gary Sims (Spectral Instruments), Mike Buchin (Stanford Photonics), Butch Moomaw (Hamamatsu), Chris Campillo (Andor Technology), and Deepak Sharma (Photometrics) for helpful discussions. Supported by National Institutes of Health grant R01 MH082945 (DKW) and a V.A. Career Development Award to DKW.

REFERENCES

Agulhon C, Platel JC, Kolomiets B, Forster V, Picaud S, Brocard J, Faure P, Brulet P. 2007. Bioluminescent imaging of Ca^{2+} activity reveals spatiotemporal dynamics in glial networks of dark-adapted mouse retina. *J Physiol* **583:** 945–958.

Almond B, Hawkins E, Stecha P, Garvin D, Paguio A, Butler B, Beck M, Wood M, Wood K. 2003. A new luminescence: Not your average click beetle. *Promega Notes* **85:** 11–14.

Ando Y, Niwa K, Yamada N, Enomoto T, Irie T, Kubota H, Ohmiya Y, Akiyama H. 2008. Firefly bioluminescence quantum yield and colour change by pH-sensitive green emission. *Nat Photonics* **2:** 44–47.

Arai R, Nakagawa H, Tsumoto K, Mahoney W, Kumagai I, Ueda H, Nagamune T. 2001. Demonstration of a homogeneous noncompetitive immunoassay based on bioluminescence resonance energy transfer. *Anal Biochem* **289:** 77–81.

Auld DS, Thorne N, Nguyen DT, Inglese J. 2008. A specific mechanism for nonspecific activation in reporter-gene assays. *ACS Chem Biol* **3:** 463–470.

Baggett B, Roy R, Momen S, Morgan S, Tisi L, Morse D, Gillies RJ. 2004. Thermostability of firefly luciferases affects efficiency of detection by in vivo bioluminescence. *Mol Imaging* **3:** 324–332.

Baubet V, Le Mouellic H, Campbell AK, Lucas-Meunier E, Fossier P, Brulet P. 2000. Chimeric green fluorescent protein-aequorin as bioluminescent Ca^{2+} reporters at the single-cell level. *Proc Natl Acad Sci* **97:** 7260–7265.

Billinton N, Knight AW. 2001. Seeing the wood through the trees: A review of techniques for distinguishing green fluorescent protein from endogenous autofluorescence. *Anal Biochem* **291:** 175–197.

Branchini BR, Southworth TL, Khattak NF, Michelini E, Roda A. 2005. Red- and green-emitting firefly luciferase mutants for bioluminescent reporter applications. *Anal Biochem* **345:** 140–148.

Branchini BR, Ablamsky DM, Murtiashaw MH, Uzasci L, Fraga H, Southworth TL. 2007. Thermostable red and green light-producing firefly luciferase mutants for bioluminescent reporter applications. *Anal Biochem* **361:** 253–262.

Brandes C, Plautz JD, Stanewsky R, Jamison CF, Straume M, Wood KV, Kay SA, Hall JC. 1996. Novel features of *Drosophila period* transcription revealed by real-time luciferase reporting. *Neuron* **16:** 687–692.

Butt Y. 2009. Beyond the myth of the supernova-remnant origin of cosmic rays. *Nature* **460:** 701–704.

Carr AJ, Whitmore D. 2005. Imaging of single light-responsive clock cells reveals fluctuating free-running periods. *Nat Cell Biol* **7:** 319–321.

Choy G, O'Connor S, Diehn FE, Costouros N, Alexander HR, Choyke P, Libutti SK. 2003. Comparison of noninvasive fluorescent and bioluminescent small animal optical imaging. *BioTechniques* **35:** 1022–1026, 1028–1030.

Christenson MA. 2002. Detection systems optimized for low-light chemiluminescence imaging. In *Luminescence biotechnology: Instruments and applications* (ed. Van Dyke K, et al.), pp. 469–480. CRC, Boca Raton, FL.

Coulon V, Audet M, Homburger V, Bockaert J, Fagni L, Bouvier M, Perroy J. 2008. Subcellular imaging of dynamic protein interactions by bioluminescence resonance energy transfer. *Biophys J* **94:** 1001–1009.

Curie T, Rogers KL, Colasante C, Brulet P. 2007. Red-shifted aequorin-based bioluminescent reporters for in vivo imaging of Ca^{2+} signaling. *Mol Imaging* **6:** 30–42.

Dacres H, Dumancic MM, Horne I, Trowell SC. 2009. Direct comparison of bioluminescence-based resonance energy transfer methods for monitoring of proteolytic cleavage. *Anal Biochem* **385:** 194–202.

Davis RE, Zhang YQ, Southall N, Staudt LM, Austin CP, Inglese J, Auld DS. 2007. A cell-based assay for IκBα stabilization using a two-color dual luciferase-based sensor. *Assay Drug Dev Technol* **5:** 85–103.

Day RN, Kawecki M, Berry D. 1998. Dual-function reporter protein for analysis of gene expression in living cells. *BioTechniques* **25:** 848–856.

De A, Loening AM, Gambhir SS. 2007. An improved bioluminescence resonance energy transfer strategy for imaging intracellular events in single cells and living subjects. *Cancer Res* **67:** 7175–7183.

De A, Ray P, Loening AM, Gambhir SS. 2009. BRET3: A red-shifted bioluminescence resonance energy transfer (BRET)-based integrated platform for imaging protein-protein interactions from single live cells and living animals. *FASEB J* **23:** 2702–2709.

Dothager RS, Flentie K, Moss B, Pan MH, Kesarwala A, Piwnica-Worms D. 2009. Advances in bioluminescence imaging of live animal models. *Curr Opin Biotechnol* **20:** 45–53.

Duggleby RG, Northrop DB. 1989. Quantitative analysis of absorption spectra and application to the characterization of ligand binding curves. *Experientia* **45:** 87–92.

Fan F, Binkowski BF, Butler BL, Stecha PF, Lewis MK, Wood KV. 2008. Novel genetically encoded biosensors using firefly luciferase. *ACS Chem Biol* **3:** 346–351.

Fraga H. 2008. Firefly luminescence: A historical perspective and recent developments. *Photochem Photobiol Sci* **7:** 146–158.

Fraga H, Fernandes D, Fontes R, Esteves da Silva JC. 2005. Coenzyme A affects firefly luciferase luminescence because it acts as a substrate and not as an allosteric effector. *FEBS J* **272:** 5206–5216.

Gammon ST, Leevy WM, Gross S, Gokel GW, Piwnica-Worms D. 2006. Spectral unmixing of multicolored bioluminescence emitted from heterogeneous biological sources. *Anal Chem* **78:** 1520–1527.

Geusz ME. 2001. Bioluminescence imaging of gene expression in living cells and tissues. In *Methods in cellular imaging* (ed. Periasamy A), pp. 395–408. Oxford University Press, Oxford.

Giepmans BN, Adams SR, Ellisman MH, Tsien RY. 2006. The fluorescent toolbox for assessing protein location and function. *Science* **312:** 217–224.

Greer LF 3rd, Szalay AA. 2002. Imaging of light emission from the expression of luciferases in living cells and organisms: A review. *Luminescence* **17:** 43–74.

Hastings JW. 1989. Chemistry, clones, and circadian control of the dinoflagellate bioluminescent system. The Marlene DeLuca memorial lecture. *J Biolumin Chemilumin* **4:** 12–19.

Heitman LH, van Veldhoven JP, Zweemer AM, Ye K, Brussee J, IJzerman AP. 2008. False positives in a reporter gene assay: Identification and synthesis of substituted *N*-pyridin-2-ylbenzamides as competitive inhibitors of firefly luciferase. *J Med Chem* **51:** 4724–4729.

Hermening S, Kugler S, Bahr M, Isenmann S. 2004. Increased protein expression from adenoviral shuttle plasmids and vectors by insertion of a small chimeric intron sequence. *J Virol Methods* **122:** 73–77.

Hida N, Awais M, Takeuchi M, Ueno N, Tashiro M, Takagi C, Singh T, Hayashi M, Ohmiya Y, Ozawa T. 2009. High-sensitivity real-time imaging of dual protein–protein interactions in living subjects using multicolor luciferases. *PLoS ONE* **4:** e5868. doi: 10.1371/journal.pone.0005868.

Hoshino H, Nakajima Y, Ohmiya Y. 2007. Luciferase-YFP fusion tag with enhanced emission for single-cell luminescence imaging. *Nat Methods* **4:** 637–639.

Isojima Y, Isoshima T, Nagai K, Kikuchi K, Nakagawa H. 1995. Ultraweak biochemiluminescence detected from rat hippocampal slices. *NeuroReport* **6:** 658–660.

Ives D. 2009. Electron multiplication CCDs for astronomical applications. *Nucl Instrum Methods Phys Res A* **604:** 38–40.

Karplus E. 2006. Advances in instrumentation for detecting low-level bioluminescence and fluorescence. In *Photoproteins in bioanalysis* (ed. Daunert S, Deo SK), pp. 199–223. Wiley-VCH, Weinheim, Germany.

Kitayama Y, Kondo T, Nakahira Y, Nishimura H, Ohmiya Y, Oyama T. 2004. An in vivo dual-reporter system of cyanobacteria using two railroad-worm luciferases with different color emissions. *Plant Cell Physiol* **45:** 109–113.

Kondo T, Strayer CA, Kulkarni RD, Taylor W, Ishiura M, Golden SS, Johnson CH. 1993. Circadian rhythms in prokaryotes: Luciferase as a reporter of circadian gene expression in cyanobacteria. *Proc Natl Acad Sci* **90:** 5672–5676.

Leclerc GM, Boockfor FR, Faught WJ, Frawley L. 2000. Development of a destabilized firefly luciferase enzyme for measurement of gene expression. *BioTechniques* **29:** 590–598.

Liu AC, Welsh DK, Ko CH, Tran HG, Zhang EE, Priest AA, Buhr ED, Singer O, Meeker K, Verma IM, et al. 2007. Intercellular coupling confers robustness against mutations in the SCN circadian clock network. *Cell* **129:** 605–616.

Lorenz WW, McCann RO, Longiaru M, Cormier MJ. 1991. Isolation and expression of a cDNA encoding Renilla reniformis luciferase. *Proc Natl Acad Sci* **88:** 4438–4442.

Luker KE, Luker GD. 2008. Applications of bioluminescence imaging to antiviral research and therapy: Multiple luciferase enzymes and quantitation. *Antiviral Res* **78:** 179–187.

Manjarrés IM, Chamero P, Domingo B, Molina F, Llopis J, Alonso MT, Garcia-Sancho J. 2008. Red and green aequorins for simultaneous monitoring of Ca^{2+} signals from two different organelles. *Pflugers Arch* **455:** 961–970.

Martin JR, Rogers KL, Chagneau C, Brulet P. 2007. In vivo bioluminescence imaging of Ca^{2+} signalling in the brain of *Drosophila*. *PLoS ONE*. **2:** e275. doi: 10.1371/journal.pone.0000275.

Michelini E, Cevenini L, Mezzanotte L, Ablamsky D, Southworth T, Branchini B, Roda A. 2008. Spectral-resolved gene technology for multiplexed bioluminescence and high-content screening. *Anal Chem* **80:** 260–267.

Mihalcescu I, Hsing W, Leibler S. 2004. Resilient circadian oscillator revealed in individual cyanobacteria. *Nature* **430:** 81–85.

Millar AJ, Short SR, Chua NH, Kay SA. 1992. A novel circadian phenotype based on firefly luciferase expression in transgenic plants. *Plant Cell* **4:** 1075–1087.

Miloud T, Henrich C, Hammerling GJ. 2007. Quantitative comparison of click beetle and firefly luciferases for in vivo bioluminescence imaging. *J Biomed Opt* **12:** 054018.

Nakajima Y, Ikeda M, Kimura T, Honma S, Ohmiya Y, Honma K. 2004a. Bidirectional role of orphan nuclear receptor RORα in clock gene transcriptions demonstrated by a novel reporter assay system. *FEBS Lett* **565:** 122–126.

Nakajima Y, Kimura T, Suzuki C, Ohmiya Y. 2004b. Improved expression of novel red- and green-emitting luciferases of *Phrixothrix* railroad worms in mammalian cells. *Biosci Biotechnol Biochem* **68:** 948–951.

Nakajima Y, Kimura T, Sugata K, Enomoto T, Asakawa A, Kubota H, Ikeda M, Ohmiya Y. 2005. Multicolor luciferase assay system: One-step monitoring of multiple gene expressions with a single substrate. *BioTechniques* **38:** 891–894.

Nakamura M, Maki S, Amano Y, Ohkita Y, Niwa K, Hirano T, Ohmiya Y, Niwa H. 2005. Firefly luciferase exhibits bimodal action depending on the luciferin chirality. *Biochem Biophys Res Commun* **331:** 471–475.

Nguyen VT, Morange M, Bensaude O. 1989. Protein denaturation during heat shock and related stress. *J Biol Chem* **264:** 10487–10492.

Noguchi T, Ikeda M, Ohmiya Y, Nakajima Y. 2008. Simultaneous monitoring of independent gene expression patterns in two types of cocultured fibroblasts with different color-emitting luciferases. *BMC Biotechnol* **8:** 40.

O'Brien MA, Daily WJ, Hesselberth PE, Moravec RA, Scurria MA, Klaubert DH, Bulleit RF, Wood KV. 2005. Homogeneous, bioluminescent protease assays: Caspase-3 as a model. *J Biomol Screen* **10:** 137–148.

Ogura R, Matsuo N, Wako N, Tanaka T, Ono S, Hiratsuka K. 2005. Multi-color luciferases as reporters for monitoring transient gene expression in higher plants. *Plant Biotechnol* **22:** 151–155.

Paguio A, Almond B, Fan F, Stecha P, Garvin D, Wood M, Wood K. 2005. pGL4 vectors: A new generation of luciferase reporter vectors. *Promega Notes* **89:** 7–10.

Plautz JD, Kaneko M, Hall JC, Kay SA. 1997. Independent photoreceptive circadian clocks throughout *Drosophila. Science* **278:** 1632–1635.

Rogers KL, Stinnakre J, Agulhon C, Jublot D, Shorte SL, Kremer EJ, Brulet P. 2005. Visualization of local Ca^{2+} dynamics with genetically encoded bioluminescent reporters. *Eur J Neurosci* **21:** 597–610.

Rogers KL, Picaud S, Roncali E, Boisgard R, Colasante C, Stinnakre J, Tavitian B, Brulet P. 2007. Non-invasive in vivo imaging of calcium signaling in mice. *PLoS ONE* **2:** e974. doi: 10.1371/journal.pone.0000974.

Rogers KL, Martin JR, Renaud O, Karplus E, Nicola MA, Nguyen M, Picaud S, Shorte SL, Brulet P. 2008. Electron-multiplying charge-coupled detector-based bioluminescence recording of single-cell Ca^{2+}. *J Biomed Opt* **13:** 031211.

Ruan GX, Allen GC, Yamazaki S, McMahon DG. 2008. An autonomous circadian clock in the inner mouse retina regulated by dopamine and GABA. *PLoS Biol* **6:** e249. doi: 1071.1371/journal.pbio.0060249.

Shaner NC, Campbell RE, Steinbach PA, Giepmans BN, Palmer AE, Tsien RY. 2004. Improved monomeric red, orange and yellow fluorescent proteins derived from *Discosoma* sp. *Nat Biotechnol* **22:** 1567–1572.

Shimomura O. 2006. *Bioluminescence: Chemical principles and methods.* World Scientific, Hackensack, NJ.

Shimomura O, Musicki B, Kishi Y, Inouye S. 1993. Light-emitting properties of recombinant semi-synthetic aequorins and recombinant fluorescein-conjugated aequorin for measuring cellular calcium. *Cell Calcium* **14:** 373–378.

Shu X, Royant A, Lin MZ, Aguilera TA, Lev-Ram V, Steinbach PA, Tsien RY. 2009. Mammalian expression of infrared fluorescent proteins engineered from a bacterial phytochrome. *Science* **324:** 804–807.

Stables J, Scott S, Brown S, Roelant C, Burns D, Lee MG, Rees S. 1999. Development of a dual glow-signal firefly and *Renilla* luciferase assay reagent for the analysis of G-protein coupled receptor signalling. *J Recept Signal Transduct Res* **19:** 395–410.

Subramanian C, Xu Y, Johnson CH, von Arnim AG. 2004. In vivo detection of protein–protein interaction in plant cells using BRET. *Methods Mol Biol* **284:** 271–286.

Suzuki T, Usuda S, Ichinose H, Inouye S. 2007. Real-time bioluminescence imaging of a protein secretory pathway in living mammalian cells using *Gaussia* luciferase. *FEBS Lett* **581:** 4551–4556.

Tannous BA, Kim DE, Fernandez JL, Weissleder R, Breakefield XO. 2005. Codon-optimized *Gaussia* luciferase cDNA for mammalian gene expression in culture and in vivo. *Mol Ther* **11:** 435–443.

Thompson JF, Hayes LS, Lloyd DB. 1991. Modulation of firefly luciferase stability and impact on studies of gene regulation. *Gene* **103:** 171–177.

Troy T, Jekic-McMullen D, Sambucetti L, Rice B. 2004. Quantitative comparison of the sensitivity of detection of fluorescent and bioluminescent reporters in animal models. *Mol Imaging* **3:** 9–23.

Venisnik KM, Olafsen T, Gambhir SS, Wu AM. 2007. Fusion of *Gaussia* luciferase to an engineered anti-carcinoembryonic antigen (CEA) antibody for in vivo optical imaging. *Mol Imaging Biol* **9:** 267–277.

Villalobos V, Naik S, Piwnica-Worms D. 2008. Detection of protein–protein interactions in live cells and animals with split firefly luciferase protein fragment complementation. *Methods Mol Biol* **439:** 339–352.

Viviani VR. 2002. The origin, diversity, and structure function relationships of insect luciferases. *Cell Mol Life Sci* **59:** 1833–1850.

Viviani VR, Bechara EJ, Ohmiya Y. 1999a. Cloning, sequence analysis, and expression of active *Phrixothrix* railroad-worms luciferases: Relationship between bioluminescence spectra and primary structures. *Biochemistry* **38:** 8271–8279.

Viviani VR, Silva AC, Perez GL, Santelli RV, Bechara EJ, Reinach FC. 1999b. Cloning and molecular characterization of the cDNA for the Brazilian larval click-beetle *Pyrearinus termitilluminans* luciferase. *Photochem Photobiol* **70:** 254–260.

Viviani V, Uchida A, Suenaga N, Ryufuku M, Ohmiya Y. 2001. Thr226 is a key residue for bioluminescence spectra determination in beetle luciferases. *Biochem Biophys Res Commun* **280:** 1286–1291.

Welsh DK, Kay SA. 2005. Bioluminescence imaging in living organisms. *Curr Opin Biotechnol* **16:** 73–78.

Welsh DK, Logothetis DE, Meister M, Reppert SM. 1995. Individual neurons dissociated from rat suprachiasmatic nucleus express independently phased circadian firing rhythms. *Neuron* **14:** 697–706.

Welsh DK, Yoo SH, Liu AC, Takahashi JS, Kay SA. 2004. Bioluminescence imaging of individual fibroblasts reveals persistent, independently phased circadian rhythms of clock gene expression. *Curr Biol* 14: 2289–2295.

Welsh DK, Imaizumi T, Kay SA. 2005. Real-time reporting of circadian-regulated gene expression by luciferase imaging in plants and mammalian cells. *Methods Enzymol* **393:** 269–288.

Wilson T, Hastings JW. 1998. Bioluminescence. *Annu Rev Cell Dev Biol* **14:** 197–230.

Wood KV. 1995. Marker proteins for gene expression. *Curr Opin Biotechnol* **6:** 50–58.

Wood KV, Lam YA, Seliger HH, McElroy WD. 1989. Complementary DNA coding click beetle luciferases can elicit bioluminescence of different colors. *Science* **244:** 700–702.

Yamaguchi S, Isejima H, Matsuo T, Okura R, Yagita K, Kobayashi M, Okamura H. 2003. Synchronization of cellular clocks in the suprachiasmatic nucleus. *Science* **302:** 1408–1412.

Yang KS, Ilagan MX, Piwnica-Worms D, Pike LJ. 2009. Luciferase fragment complementation imaging of conformational changes in the epidermal growth factor receptor. *J Biol Chem* **284:** 7474–7482.

Yoo SH, Yamazaki S, Lowrey PL, Shimomura K, Ko CH, Buhr ED, Siepka SM, Hong HK, Oh WJ, Yoo OJ, et al. 2004. PERIOD2::LUCIFERASE real-time reporting of circadian dynamics reveals persistent circadian oscillations in mouse peripheral tissues. *Proc Natl Acad Sci* **101:** 5339–5346.

Yoo S-H, Yamazaki S, Buhr E, Park J, Kojima S, Green CB, Takahashi JS. 2008. An SV-40 polyadenylation signal sequence in the 3′UTR of Per2::Luciferase(SV) mice lengthens free-running period and increases PER2 (#6). *Soc Res Biol Rhythms Abstr* **20:** 64.

Zhang Y, Phillips GJ, Li Q, Yeung ES. 2008. Imaging localized astrocyte ATP release with firefly luciferase beads attached to the cell surface. *Anal Chem* **80:** 9316–9325.

Zhao H, Doyle TC, Coquoz O, Kalish F, Rice BW, Contag CH. 2005. Emission spectra of bioluminescent reporters and interaction with mammalian tissue determine the sensitivity of detection in vivo. *J Biomed Opt* **10:** 41210.

WWW RESOURCES

Bioluminescence

bioluminescentbeetles.com
explorations.ucsd.edu/biolum
www.lifesci.ucsb.edu/~biolum

Luciferases (Promega)

www.promega.com/chromacalc Chroma-Luc calculator.
www.promega.com/paguide/chap8.htm Protocols & Applications Guide, Chapter 8. Bioluminescence reporters.
www.promega.com/tbs/tm059/tm059.html, www.promega.com/tbs/tm062/tm062.html Chroma-Luc and Chroma-Glo technical manuals.

Camera Tutorials

andor.com/learn Andor.
micro.magnet.fsu.edu
microscopyu.com Nikon.
olympusmicro.com Olympus.
qimaging.com/webinars/webinar_library.php QImaging.

22 Introduction to Indicators Based on Fluorescence Resonance Energy Transfer

Roger Y. Tsien

Howard Hughes Medical Institute and Departments of Pharmacology and Chemistry and Biochemistry, University of California, San Diego, La Jolla, California 92093

ABSTRACT

Fluorescence (Förster) resonance energy transfer (FRET) is a popular and powerful method for monitoring the dynamics of protein conformations and interactions in intact cells, although standard FRET is limited to distances up to ~8 nm, which is not enough to span large proteins or complexes. Engineered modifications to the fluorescing molecules have expanded the ways in which FRET can be used to image cells and their dynamic processes. This chapter provides an introduction to the variety of probes that can be used and the kinds of cellular processes that can be observed using FRET.

INTRODUCTION

One of the major trends in the design of indicators for optically imaging biochemical and physiological functions of living cells has been the exploitation of fluorescence resonance energy transfer (FRET). The underlying process is "Förster resonance energy transfer," of which fluorescence resonance energy transfer is one, and perhaps the most important, manifestation. FRET is a well-known spectroscopic technique for monitoring changes in the proximity and mutual orientation of pairs of chromophores. It has long been used in biochemistry and cell biology to assess distances and orientations between specific labeling sites within a single macromolecule or between two separate molecules (Stryer 1978; Lakowicz 1983; Uster and Pagano 1986; Herman 1989; Jovin and Arndt-Jovin 1989; Tsien et al. 1993). More recently, macromolecules or molecular pairs have been engineered to change their FRET in response to biochemical and physiological signals such as membrane potential, cyclic AMP (cAMP), protease activity, free Ca^{2+} and Ca^{2+}-CaM (calmodulin) concentrations, protein–protein heterodimerization, phosphorylation (Wouters et al. 2001), and reporter-gene expression (Table 1, Fig. 1). Because FRET is general, nondestructive, and easily imaged, it has proven to be one of the most versatile spectroscopic readouts available to the designer of new probes. FRET is particularly amenable to emission ratioing, which is more reliably quantifiable than single-wavelength monitoring and better suited than excitation ratioing to high-speed and laser-excited imaging.

TABLE 1. Physiological indicators using fluorescence resonance energy transfer

Analyte or process	Donor and acceptor[a]	ΔFRET[b]	Wavelength (nm)[c] Donor excitation	Donor emission	Acceptor emission	Maximum emission ratio change[d]	Reference(s)
Depolarization	Coumarin-phosphatidylethanolamine; bis(thiobarbiturate)trimethine-oxonol	↓	414	450	560	1.8/(100 mV)	Gonzalez and Tsien 1997
cAMP	PKA catalytic subunit-FITC; PKA regulatory subunit-TROSu[e]	↓	495	520	580	1.6–2.2	Adams et al. 1991, 1993
Trypsin	BFP-(trypsin-sensitive linker)-GFP	↓	380	445	507	4.6	Heim and Tsien 1996
Factor X_a	BFP-(factor Xa-sensitive linker)-GFP	↓	385	450	505	1.9	Mitra et al. 1996
Caspase-3	BFP-(caspase-3-sensitive linker)-GFP	↓	380	440	511	?	Xu et al. 1998
Ca^{2+}-CaM	GFP-CBSM-BFP	↓	380	448	505	5.7	Romoser et al. 1997
Ca^{2+}	GFP-CBSM-BFP-CaMCN	↓	380	440	505	1.67	Persechini et al. 1997
Ca^{2+}	eCFP-CaM-M13-eYFP	↑	433	476	527	2.1	Miyawaki et al. 1997, 1999
Ca^{2+}	eCFP-CaM; M13-eYFP	↑	433	476	527	4	Miyawaki et al. 1997, 1999
β-lactamase expression	Coumarin-cephalosporin-fluorescein (CCF_2)	↓	409	447	520	70	Zlokarnik et al. 1998

[a]Hyphens indicate covalent conjugation or fusion. Interacting donor and acceptor molecules are separated by semicolons.

[b]FRET indicates whether the efficiency of fluorescence resonance energy transfer is increased (↑) or decreased (↓) by the analyte or process.

[c]Donor excitation wavelength, donor emission wavelength, and acceptor emission wavelength, all in nanometers. Small differences (up to 8 nm) in the wavelengths cited by different laboratories for BFP and GFPs are probably not significant.

[d]Maximum factor by which emission ratio changes from zero to saturating levels of the analyte or process, except for depolarization of 100 mV amplitude.

[e]Tetramethylrhodamine *N*-hydroxysuccinimide.

PHOTOPHYSICAL PRINCIPLES OF FRET

FRET is the quantum-mechanical transfer of energy from the excited state of a donor fluorophore to the ground state of a neighboring acceptor chromophore or fluorophore. The acceptor must absorb light at roughly the same wavelengths as the donor emits. If the donor and acceptor are located within a few nanometers of each other and the mutual orientation of the chromophores is not too unfavorable, FRET becomes probable.

Measurements of FRET

The efficiency of FRET in a population can be measured as a reduction in the fluorescence, excited-state lifetime, and susceptibility to photochemical reactions of the donor relative to the corresponding parameters when acceptors are remote or absent. If the acceptor is fluorescent, it can re-emit the transferred energy as its own fluorescence, which will be at longer wavelengths than those of the donor. The most robust and convenient way to image FRET is usually to ratio the acceptor and donor emission amplitudes. Such emission ratioing cancels out variations in the excitation intensity, the overall collection efficiency, and the number of donor–acceptor pairs in the sample-volume element, because these factors perturb the acceptor and donor signals equally. Measurements of donor excited-state lifetime require picosecond to nanosecond time resolution in excitation and detection, which are expensive and rather rare capabilities in imaging systems. Measurements of donor photobleaching rate (Jovin and Arndt-Jovin 1989) can be accurate but are destructive and have poor time resolution.

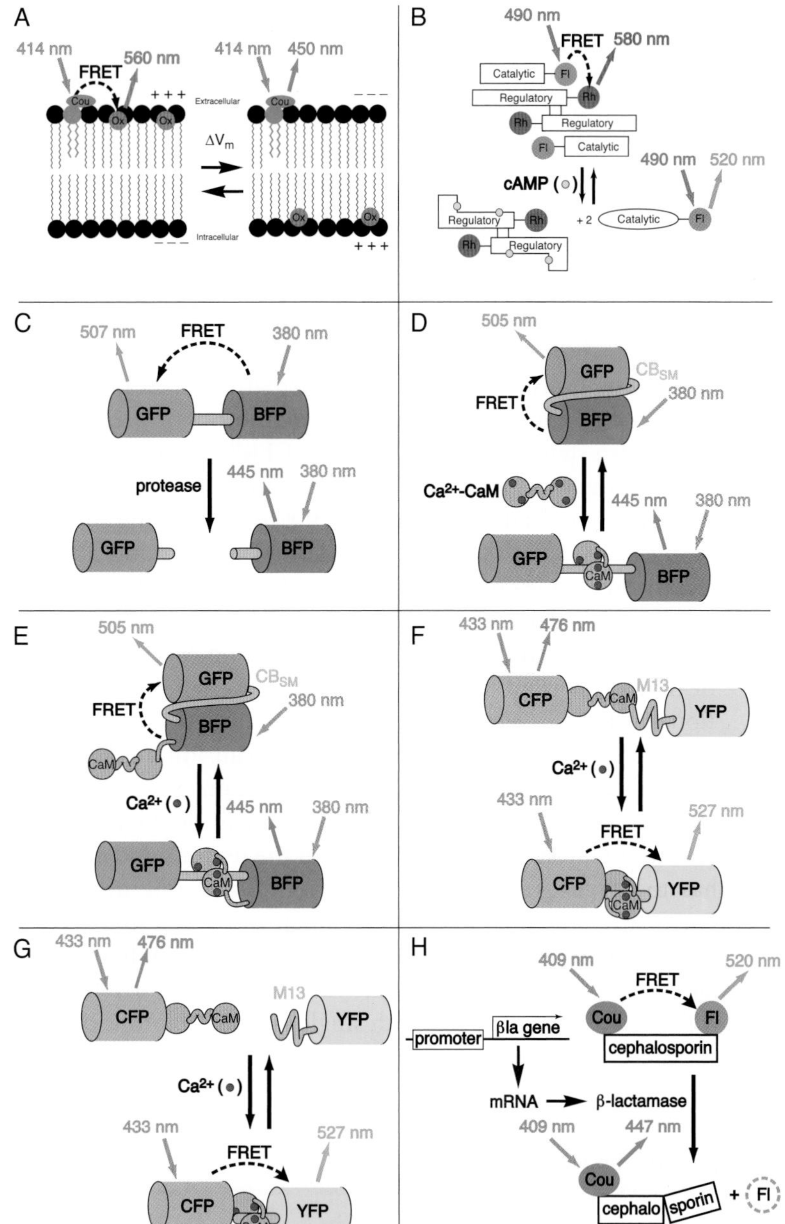

FIGURE 1. Schematic mechanisms by which FRET-based indicators respond to biochemical signals. Drawings are not to scale. Incoming and outgoing arrows represent fluorescence excitation and emission, peaking at (but not confined to) the stated wavelengths. (*A*) Membrane potential indicators. The plasma membrane lipid bilayer (black) is loaded with a coumarin-labeled phospholipid (blue) and a membrane-permeant oxonol dye (orange). The illustration at *left* represents the normal negative resting potential in which FRET from the donor coumarin to the acceptor oxonol is favored. The illustration at *right* shows how depolarization pulls the negatively charged oxonol to the inner leaflet of the membrane and decreases FRET. (*B*) cAMP indicators. The black outlines represent catalytic and regulatory subunits of cAMP-dependent protein kinase, labeled with fluorescein (Fl, green circles) and rhodamine (Rh, red circles), respectively. cAMP (yellow circles) binds to the regulatory subunits, changing their conformation and releasing the catalytic subunits. (*C*) Protease substrates. The green and blue cylinders, respectively, represent GFP and blue mutants (BFP). The protease-cleavable linker is shown as a pink tube. (*D*) Indicator for Ca^{2+}-CaM (brown dumbbell, with red dots representing the four Ca^{2+}) introduced by Romoser et al. (1997). The CaM-binding peptide from smooth muscle (CB_{SM}) connecting the GFP and BFP is shown in pink. Binding of Ca^{2+}-CaM to the CBSM straightens the latter and disrupts intramolecular dimerization of the BFP and GFP. (*E*) Indicator for Ca^{2+} (Persechini et al. 1997) in which the CaM (with amino and carboxyl halves reversed) is fused via a short linker to the carboxyl terminus of the BFP. (Colors as in *D*.) (*F*) Indicator for Ca^{2+} (Miyawaki et al. 1997) in which a cyan mutant (CFP) of GFP, CaM (brown), a CaM-binding peptide ("M13," pink), and a yellowish mutant (YFP) of GFP are fused in order. (*G*) Indicator for Ca^{2+} and model for general intermolecular protein–protein interaction (Miyawaki et al. 1997), similar to *F* except that there is no covalent link between the CaM and the M13. (*H*) Indicator for gene expression using β-lactamase as a reporter. Transcription of the β-lactamase gene and expression of the enzyme cause cleavage of a cephalosporin conjugated to a coumarin (Cou, blue) and fluorescein (Fl, green) and disruption of intramolecular FRET. The hydrolysis of the cephalosporin β-lactam ring indirectly releases the fluorescein. The free thiol group left on the fluorescein (dashed circle) quenches the fluorescence of the latter.

Considerations for the Design of Probes

Space does not permit recapitulation of the physical basis and equations governing FRET, but some practical constraints and common misconceptions are worth mentioning. FRET should not be confused with donor emission of a real photon followed by reabsorption by an acceptor; the latter "trivial" process does not require molecular proximity, but rather, a sufficient optical density of acceptors anywhere within the path of the outgoing donor emission. FRET does not require that the fluorophores actually touch; in fact, such direct contact is usually undesirable because other very short-range processes such as electron transfer or exciplex formation can then compete. Because both distance and mutual orientation affect FRET, it is hard to convert a FRET measurement into an absolute distance between fluorophores. However, if it is known that the probe exists in just two states, whose efficiencies of FRET have been empirically calibrated, measurement of FRET in the test specimen immediately indicates the relative occupancies of the two states. Even if one reference state has zero FRET, the corresponding emission ratio is usually nonzero and must be measured empirically. This undesired background arises because any wavelength that efficiently excites the donor has at least some slight ability to excite the acceptor as well, and any wavelength band that collects acceptor emission will also receive some donor emission as well. Minimizing the overlaps of the two excitation spectra and of the two emission spectra is equally or more important than maximizing the overlap of the donor emission and the acceptor excitation spectra.

Multiphoton excitation of the donor can initiate FRET just as one-photon excitation does. The only practical concern is that multiphoton excitation spectra are often unpredictable and not merely the one-photon spectra scaled to two or three times longer wavelengths. Therefore, the extent of undesirable overlap between the donor's and acceptor's multiphoton excitation spectra has to be checked empirically and is not reliably deducible from their one-photon spectra. Fortunately, the cyan and yellow mutants of green fluorescent protein (GFP), which are among the most promising donor–acceptor pairs for FRET, have proven well suited to two-photon excitation. The effect of FRET on the ratio of 535-nm to 480-nm emissions was essentially the same whether the excitation was with single 440-nm photons or pairs of 770–810-nm photons (Fan et al. 1999).

TYPES OF INDICATORS USING FRET

Membrane Potential Indicators

Membrane voltage can be reported by its effect on the proximity of two fluorophores. One dye is a fluorescent ion, typically negative, that adsorbs strongly to the membrane and translocates from one side of the membrane to the other in response to membrane potential. The other dye is localized to one side of the membrane, typically the extracellular face of the plasma membrane. Hyperpolarization repels the mobile anion to that same face, encouraging FRET, whereas depolarization attracts the anion to the cytoplasmic interface, increasing the average distance between the dyes and weakening FRET (Gonzalez and Tsien 1995, 1997). The major advantages of this two-fluorophore system are that it gives a much better combination of large signals and reasonably fast response kinetics than previous single-dye systems, especially when the readout is measured as the ratio of the donor and acceptor fluorescences. The ratio can change by as much as twofold for ~60 mV depolarization (J.E. Gonzalez, pers. comm.), and the response-time constant can be as little as 380 µsec at room temperature (Gonzalez and Tsien 1997), although not yet simultaneously. Such designed potentiometric indicators are used to image information processing in neural networks such as the swimming rhythm pattern generator in leech (Cacciatore et al. 1998). The main obstacle to wider application is that the two fluorophores are currently quite hydrophobic and can have difficulty reaching the desired target neurons in a complex tissue.

cAMP Indicators

cAMP can be imaged by monitoring the extent of FRET between the fluorescein-labeled catalytic subunit and rhodamine-labeled regulatory subunit of cAMP-dependent protein kinase (PKA). cAMP causes the dissociation of the two subunits and thereby disrupts FRET (Adams et al. 1991, 1993). This indicator is the only method currently available for imaging the dynamics of cAMP and has revealed subcellular diffusion gradients of cAMP in invertebrate neurons during synaptic stimulation and plasticity (Bacskai et al. 1993; Hempel et al. 1996). Now that it is commercially available, the major remaining deficiency is the dexterity required to introduce the 170-kDa PKA holoenzyme complex into cells by pressure microinjection or diffusion from a patch pipette in whole-cell mode.

GFP-based Systems

Proteins can be fluorescently tagged in vivo by fusion to GFP of the jellyfish *Aequorea victoria* (for a review of GFP-based systems, see Tsien 1998). Mutagenesis has made GFPs suitable for FRET by concentrating the excitation amplitude in a single, much amplified peak, and by generating new colors: blue, cyan, and yellowish green (Cubitt et al. 1995; Heim et al. 1995; Heim and Tsien 1996; Ormö et al. 1996; Tsien and Prasher 1998). The first pairs of GFPs suitable for FRET were blue mutants as donors and improved green mutants as acceptors. FRET was shown by linking the two GFPs with a protease-sensitive linker and showing the loss of FRET on addition of the protease, which was either trypsin or Factor Xa (Heim and Tsien 1996; Mitra et al. 1996). Fortunately, the two GFPs themselves are quite resistant to most proteases. Such genetically encodable, tandem fusions with protease-specific cleavage sites offer the means to monitor protease activity inside transfected cells and organisms. For example, caspase-3 activity can be detected in apoptotic cells by flow cytometry (Xu et al. 1998).

Persechini Systems

Instead of using a protease to cleave a flexible linker, Romoser et al. (1997) made the linker a calmodulin (CaM)-binding peptide, CB_{SM}, derived from smooth muscle myosin–light-chain kinase. Binding of Ca^{2+}-CaM to the linker peptide should straighten the latter and increase the distance between the amino-terminal blue mutant and the carboxy-terminal GFP. Indeed, addition of Ca^{2+}-CaM markedly decreased the efficiency of FRET and increased the ratio of blue to green emissions by 5.7-fold. The GFP-CB_{SM}-BFP construct was expressed in bacteria, purified, and microinjected into mammalian cells, where it responded to cytosolic Ca^{2+} transients, albeit with signals much smaller than in vitro. The signals could be enhanced by coinjection of exogenous CaM, suggesting that the latter was limiting (Romoser et al. 1997). To create a self-contained Ca^{2+} sensor, the carboxyl terminus of the GFP-linker-BFP was fused to CaM, in which the amino- and carboxy-terminal halves were swapped to avoid certain steric constraints ("CaMCN"). The complete GFP-CB_{SM}-BFP-CaMCN fusion showed a ratio change of ~1.67-fold from zero to saturating Ca^{2+}, significantly less than that of GFP-CB_{SM}-BFP responding to unfused CaM (Persechini et al. 1997).

Cameleons

An independent approach to genetically encoded Ca^{2+} indicators was to include CaM within the fusion, consisting of cyan GFP-CaM-M13-yellow GFP, in which M13 is another CaM-binding peptide domain from skeletal muscle myosin–light-chain kinase (see Chapter 26). Here the binding of Ca^{2+} to the CaM makes it wrap around the M13, which decreases the distance between the terminal GFP mutants and increases the efficiency of FRET. Cyan GFP (e.g., "eCFP") has become preferred because it is a much brighter and more photostable donor than blue GFP. The acceptor then has to be a yellowish GFP (e.g., "eYFP") to maintain enough separation between the donor and acceptor excitation spectra. The "e"s in these acronyms mean "enhanced" and refer to the inclusion of additional mutations that confer improved expression and folding in mammalian cells. These technical improvements made it possible to image these indicator molecules ("cameleons") biosynthesized in situ by transfected mammalian cells, thus obviating the need for microinjection. The Ca^{2+} affinity was easily tuned by mutation of the CaM, and the constructs were targetable to organelles or other privileged sites by appropriate trafficking signals (Miyawaki et al. 1997). More recently, the eYFP has been further mutated to reduce its sensitivity to acidification (Miyawaki et al. 1999).

Intermolecular Protein–Protein Interaction

Ca^{2+}-triggered intermolecular association of two separate fusions, eCFP-CaM and M13-eYFP, gives even larger FRET changes than the four-part chimera eCFP-CaM-M13-eYFP. The increased spectral signal presumably arises because in the absence of Ca^{2+}, the two GFPs can get much farther apart from one another than they can in the unimolecular cameleons. However, the cameleons are more reliable indicators of Ca^{2+} because their sensitivity to Ca^{2+} is independent of their absolute concentration, and because the CaM and M13 fused within a cameleon are almost completely indifferent to unlabeled bystander molecules of CaM and CaM-binding proteins (Miyawaki et al. 1999). Nevertheless, eCFP-CaM and M13-eYFP show the feasibility of monitoring other dynamic protein–protein interactions in situ, by fusion with donor and acceptor GFPs (Miyawaki et al. 1997; Tsien and Miyawaki 1998).

Further examples of intermolecular protein–protein interactions detected in live single cells by FRET between GFP fusions include homodimerization of the nuclear transcription factor Pit-1 and heterodimer formation between Pit-1 and Ets-1, whereas interaction between Pit-1 and the estrogen receptor was undetectable (Day 1998). Similarly, association between the apoptosis-regulating proteins Bcl-2 and Bax was detectable in mitochondria by FRET, whereas Bcl-2 did not seem to interact with cytochrome c or with human papillomavirus E6 (Mahajan et al. 1998). In both these studies, the donor and acceptor were relatively primitive BFPs and GFPs, and no dynamic modulation of FRET was shown. Nevertheless, they raise hopes for generalizing the use of FRET to image protein–protein interactions with spatial and temporal resolution in live cells.

β-Lactamase

Gene expression can be visualized in single living cells by using β-lactamase as a reporter enzyme to cleave novel membrane-permeant substrates and change their fluorescence from green to blue by disrupting FRET. This enzymatically amplified readout is several orders of magnitude more sensitive than GFP as a transcriptional reporter, although it does not track subcellular protein localization. It permits flow-cytometric selection and training of mammalian cell lines, and high-throughput screening of pharmaceutical candidate drugs (Zlokarnik et al. 1998). A cell line thus selected enabled demonstration that gene expression could be tuned to the frequency of inositol-1,4,5-trisphosphate and Ca^{2+} oscillations (Li et al. 1998). Novel cell clones and genetic elements responsive to acute stimuli can be found by enhancer trapping with β-lactamase as the reporter gene (Whitney et al. 1998).

CONCLUSION

It is safe to predict that applications of FRET will continue to expand. Detection of protein–protein interactions in live cells is increasingly important and popular. There is no more general way of designing detectors for physiological signals. Novel ways to attach the requisite donors and acceptors have been and continue to be developed (Griffin et al. 1998; Tsien and Miyawaki 1998). Nevertheless, much trial and error is currently necessary to get FRET to work. Therefore, a major challenge will be to understand the empirical features and make FRET routinely and reliably transferable to new biological problems, circumstances, and laboratories.

REFERENCES

Adams SR, Harootunian AT, Buechler YJ, Taylor SS, Tsien RY. 1991. Fluorescence ratio imaging of cyclic AMP in single cells. *Nature* **349:** 694–697.

Adams SR, Bacskai BJ, Taylor SS, Tsien RY. 1993. Optical probes for cyclic AMP. In *Fluorescent probes for biological activity of living cells—A practical guide* (ed. Mason WT), pp. 133–149. Academic, New York.

Bacskai BJ, Hochner B, Mahaut-Smith M, Adams SR, Kaang B-K, Kandel ER, Tsien RY. 1993. Spatially resolved dynamics of cAMP and protein kinase A subunits in *Aplysia* sensory neurons. *Science* **260:** 222–226.

Cacciatore TW, Brodfuehrer P, Gonzalez JE, Tsien RY, Kristan WB Jr, Kleinfeld D. 1998. Neurons that are active in phase with swimming in leech, and their connectivity, are revealed by optical techniques. *Soc Neurosci Abstr* **24:** 1890.

Cubitt AB, Heim R, Adams SR, Boyd AE, Gross LA, Tsien RY. 1995. Understanding, using, and improving green fluorescent protein. *Trends Biochem Sci* **20:** 448–455.

Day RN. 1998. Visualization of Pit-1 transcription factor interactions in the living cell nucleus by fluorescence resonance energy transfer microscopy. *Mol Endocrinol* **12:** 1410–1419.

Fan GY, Fujisaki H, Miyawaki A, Tsay R-K, Tsien RY, Ellisman MH. 1999. Video-rate scanning two-photon excitation fluorescence microscopy and ratio imaging with yellow cameleons. *Biophys J* **76:** 2412–2420.

Gonzalez JE, Tsien RY. 1995. Voltage-sensing by fluorescence resonance energy transfer in single cells. *Biophys J* **69:** 1272–1280.

Gonzalez JE, Tsien RY. 1997. Improved indicators of cell membrane potential that use fluorescence resonance energy transfer. *Chem Biol* **4:** 269–277.

Griffin BA, Adams SR, Tsien RY. 1998. Specific covalent labeling of recombinant protein molecules inside live cells. *Science* **281:** 269–272.

Heim R, Tsien RY. 1996. Engineering green fluorescent protein for improved brightness, longer wavelengths, and fluorescence energy transfer. *Curr Biol* **6:** 178–182.

Heim R, Cubitt AB, Tsien RY. 1995. Improved green fluorescence. *Nature* **373:** 663–664.

Hempel CM, Vincent P, Adams SR, Tsien RY, Selverston AI. 1996. Spatio-temporal dynamics of cAMP signals in an intact neural circuit. *Nature* **384:** 166–169.

Herman B. 1989. Resonance energy transfer microscopy. *Methods Cell Biol* **30:** 219–243.

Jovin TM, Arndt-Jovin DJ. 1989. Luminescence digital imaging microscopy. *Annu Rev Biophys Biophys Chem* **18:** 271–308.

Lakowicz JR. 1983. *Principles of fluorescence spectroscopy.* Plenum, New York.

Li W, Llopis J, Whitney M, Zlokarnik G, Tsien RY. 1998. Cell-permeant caged $InsP_3$ ester shows that Ca^{2+} spike frequency can optimize gene expression. *Nature* **392:** 936–941.

Mahajan NP, Linder K, Berry G, Gordon GW, Heim R, Herman B. 1998. Bcl-2 and Bax interactions in mitochondria probed with green fluorescent protein and fluorescence resonance energy transfer. *Nat Biotechnol* **16:** 547–552.

Mitra RD, Silva CM, Youvan DC. 1996. Fluorescence resonance energy transfer between blue-emitting and red-shifted excitation derivatives of the green fluorescent protein. *Gene* **173:** 13–17.

Miyawaki A, Llopis J, Heim R, McCaffery JM, Adams JA, Ikura M, Tsien RY. 1997. Fluorescent indicators for Ca^{2+} based on green fluorescent proteins and calmodulin. *Nature* **388:** 882–887.

Miyawaki A, Griesbeck O, Heim R, Tsien RY. 1999. Dynamic and quantitative Ca^{2+} measurements using improved cameleons. *Proc Natl Acad Sci* **96:** 2135–2140.

Ormö M, Cubitt AB, Kallio K, Gross LA, Tsien RY, Remington SJ. 1996. Crystal structure of the *Aequorea victoria* green fluorescent protein. *Science* **273:** 1392–1395.

Persechini A, Lynch JA, Romoser VA. 1997. Novel fluorescent indicator proteins for monitoring free intracellular Ca^{2+}. *Cell Calcium* **22:** 209–216.

Romoser VA, Hinkle PM, Persechini A. 1997. Detection in living cells of Ca^{2+}-dependent changes in the fluorescence emission of an indicator composed of two green fluorescent protein variants linked by a calmodulin-binding sequence. *J Biol Chem* **272:** 13270–13274.

Stryer L. 1978. Fluorescence energy transfer as a spectroscopic ruler. *Annu Rev Biochem* **47:** 819–846.

Tsien RY. 1998. The green fluorescent protein. *Annu Rev Biochem* **67:** 509–544.

Tsien RY, Miyawaki A. 1998. Seeing the machinery of live cells. *Science* **280:** 1954–1955.

Tsien RY, Prasher DC. 1998. Molecular biology and mutation of GFP. In *GFP: Green fluorescent protein: Properties, applications, and protocols* (ed. Chalfie M, Kain S), pp. 97–118. Wiley-Liss, New York.

Tsien RY, Bacskai BJ, Adams SR. 1993. FRET for studying intracellular signalling. *Trends Cell Biol* **3:** 242–245.

Uster PS, Pagano RE. 1986. Resonance energy transfer microscopy: Observations of membrane-bound fluorescent probes in model membranes and in living cells. *J Cell Biol* **103:** 1221–1234.

Whitney M, Rockenstein E, Cantin G, Knapp T, Zlokarnik G, Sanders P, Durick K, Craig FF, Negulescu PA. 1998. A genome-wide functional assay of signal transduction in living mammalian cells. *Nat Biotechnol* **16:** 1329–1333.

Wouters FS, Verveer PJ, Bastiaens PI. 2001. Imaging biochemistry inside cells. *Trends Cell Biol* **11:** 203–211.

Xu X, Gerard ALV, Huang BCB, Anderson DC, Payan DG, Luo Y. 1998. Detection of programmed cell death using fluorescence energy transfer. *Nucleic Acids Res* **26:** 2034–2035.

Zlokarnik G, Negulescu PA, Knapp TE, Mere L, Burres N, Feng L, Whitney M, Roemer K, Tsien RY. 1998. Quantitation of transcription and clonal selection of single living cells with β-lactamase as reporter. *Science* **279:** 84–88.

23 How Calcium Indicators Work

Stephen R. Adams

Department of Pharmacology, University of San Diego, La Jolla, California 92093

ABSTRACT

In the last three decades, imaging of fluorescent indicators specific for Ca^{2+} has revealed, in unprecedented detail, its spatial dynamics, such as rhythmic oscillations or standing gradients, within single groups or individual cells. This short review describes how the more widely used organic or genetic indicators work, and it discusses some new developments from the laboratory of R.Y. Tsien. More detailed information concerning the biological use of such indicators can be found in other chapters in this manual (Chapters 22–24) and in a number of reviews (Tsien 1993, 1999; Kao 1994). The currently used Ca^{2+} indicators have a modular design consisting of a metal-binding site (or sensor) coupled in some way to a fluorescent dye. Combining different sensors with different dyes results in numerous indicators suited to a wide range of experiments and equipment.

INTRODUCTION

All chemical indicators are based on 1,2-bis-(2-aminophenoxy)ethane-*N*,*N*,*N*′,*N*′-tetra-acetic acid (BAPTA) (Fig. 1), a pH-insensitive homolog of EGTA (ethylene glycol tetra-acetic acid; Tsien 1980). BAPTA retains its high selectivity for Ca^{2+} (K_d ~100 nM at pH 7.0) in the presence of competing millimolar concentrations of Mg^{++} and has fast on–off rates for metal binding as a result of aromatizing the aliphatic nitrogens of EGTA (Tsien 1980). Indicators based on BAPTA can be incorporated into cells by temporarily masking the carboxylates as acetoxymethyl (AM) esters. These hydrophobic uncharged molecules passively diffuse across membranes but release the impermeant polycarboxylate indicator after cleavage of the AM esters by intracellular esterases.

BAPTA-Based Chelators: Approaches for Modifying Affinity for Ca^{2+}

The binding site of BAPTA-based chelators can be further modified to change its affinity, allowing sensitivity to Ca^{2+} in a wider range (nanomolar to millimolar) by three general methods (see Table 1).

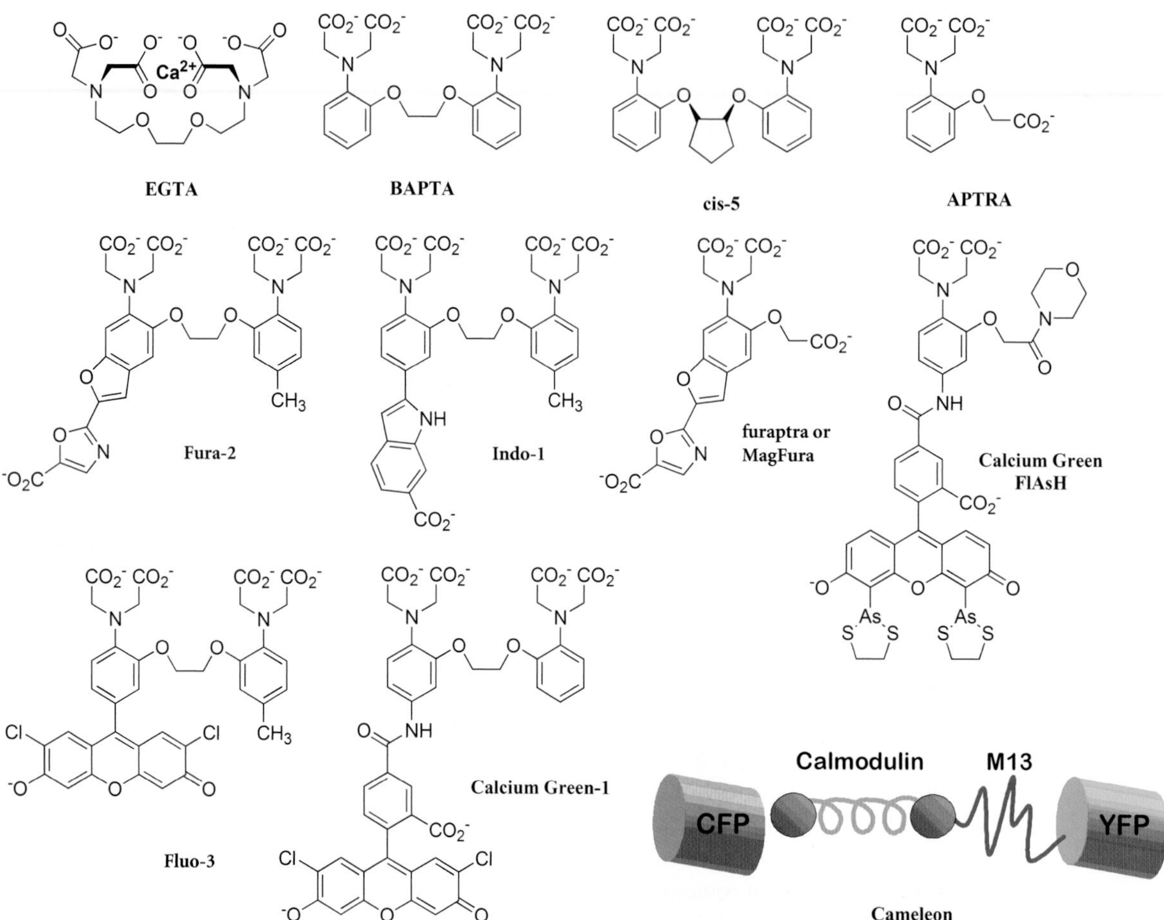

FIGURE 1. Structures of Ca^{2+} chelators and indicators. The Ca^{2+} complex of EGTA is shown to illustrate the complexation of the metal by the four carboxylates, two nitrogens, and two oxygens. The remaining chelators and indicators are shown in their Ca^{2+}-free form, but they bind Ca^{2+} in a similar manner, apart from APTRA and cameleon. Abbreviations: (APTRA) 2-amino phenol-*N,N,O*-triacetic acid; (BAPTA) 1,2-bis-(2-aminophenoxy)ethane-*N,N,N′,N′*-tetra-acetic acid; (CFP) cyan-emitting mutant of green fluorescent protein (GFP); (CaM) calmodulin; (M13) calmodulin-binding peptide of myosin light-chain kinase; (YFP) yellow-emitting mutant of GFP; (cpYFP) circularly permuted YFP.

Addition of Electron-Donating or Electron-Withdrawing Groups to the Aromatic Ring(s) of BAPTA

Electron-withdrawing groups, such as nitro ($-NO_2$), fluoro (–F), or chloro (–Cl) on the aromatic ring(s) in the parapositions or metapositions to the nitrogen, weaken the binding for Ca^{2+} to 42 μM (for paranitro substitution on one ring) and 7.4 mM (paranitro substitution on both rings; Pethig et al. 1989). Electron-donating groups, such as $-CH_3$, result in a modest increase in affinity and are used in many Ca^{2+} indicators, including Fura-2 and Fluo-3.

Addition of Modifying Groups that Sterically Alter Ca^{2+} Binding

Another approach is to sterically alter Ca^{2+} binding by incorporating the ethylene ($-CH_2CH_2-$) group of BAPTA into a carbocycle or heterocycle or by substituting these hydrogen atoms for methyl groups (Adams et al. 1988). For example, Trans-5 binds Ca^{2+} with an affinity of 6 μM, making it useful as a low-affinity Ca^{2+} buffer (Ranganathan et al. 1994). Cis-5 (Fig. 1) has a fourfold increased affinity for Ca^{2+} and is incorporated into Nitr-7, a structurally related photolabile chelator that releases Ca^{2+} on photolysis (Adams et al. 1988). However, these modifications are hard to predict in advance, and the increased hydrophobicity decreases loading of cells by AM-ester derivatives.

TABLE 1. BAPTA-based and APTRA-based chelators: Approaches for modifying affinity for Ca^{2+}

Ca^{2+} chelator[a]	Wavelength (mM)[b]		Ca^{2+} affinity (μM)[c]	Mg^{++} affinity (mM)[c]	Comment/ Reference[d]
	Excitation	Emission			
Electron-withdrawing substituents					
Fura-FF	330/370	510	20	>100	London et al. 1994
Fluo-5N	491	516	90	>100	Orange and red versions available; Haugland 2005
Calcium Green-5N	506	532	14	>100	Oregon Green, orange and red versions available; Haugland 2005
Calcium Ruby	579	598	30	>100	Luccardini et al. 2009
Sterically modifying groups					
Trans-5	NA	NA	6	>100	Ca^{2+} buffer
Cis-5	NA	NA	0.020	4.2	Ca^{2+} buffer
Decreased ligands					
Furaptra (MagFura)	330/369	510	25	1.9	
MagIndo	350	390/480	35	2.7	
Magnesium Green	506	531	6	1.0	Orange and red versions available; Haugland 2005
Calcium Green FlAsH	508	530	100	10	Tour et al. 2007

NA, not applicable.

[a]Ca^{2+} buffers and indicators with altered affinity for Ca^{2+} sorted by the method of modification.

[b]Excitation and emission maxima for indicators except Fura-FF are from Haugland (2005). Two values for excitation or emission refer to the Ca^{2+}-bound and Ca^{2+}-unbound states for ratiometric indicators.

[c]Ca^{2+} and Mg^{++} affinities measured in 0.1–0.15-M ionic strength, pH 7.0–7.4, 20°C–23°C, except those for Fura-FF, are from Haugland (2005).

[d]References are provided only for indicators not described in the text.

Reduction in the Number of Coordinating Ligands for Ca^{2+} Chelation

The final and most successful approach for decreasing Ca^{2+} binding has been to decrease the number of coordinating ligands involved in metal chelation. The APTRA (Fig. 1) series of chelators have about half of the metal-binding sites of BAPTA with the addition of a further carboxy ligand (Levy et al. 1988). Ca^{2+} binding is reduced to about 18 μM, but the affinity for Mg^{++} is increased to 1.45 mM (Adams SR, Tsien RY, unpubl.) within its physiological range. Ca^{2+} prefers eight ligands (APTRA has only five), unlike Mg^{++}, which is hexavalent. APTRA and its fluorescent derivatives (furaptra or MagFura, MagIndo, and Magnesium Green) were originally used as indicators of cytosolic Mg^{++} (Raju et al. 1989; Haugland 2005) and only later used to measure high Ca^{2+} transients or Ca^{2+} in the intracellular stores. Such measurements are always susceptible to misinterpretation if there are concurrent Mg^{++} changes. A further advantage to using APTRA-based indicators is the decrease in molecular weight and the loss of a hydrophobic aromatic ring that should aid loading of AM esters into cells. Recent modifications of APTRA and BAPTA have produced chelators and indicators sensitive to Ca^{2+} concentrations in the low-micromolar to low-millimolar range and are suitable for measuring the elevated Ca^{2+} found in stimulated cells, Ca^{2+} stores, or serum (Tour et al. 2007; S.R. Adams, R.Y. Tsien, unpubl.).

Changes in the Affinity of Cameleons for Ca^{2+}

The newest class of indicators, the cameleons (Miyawaki et al. 1997; Chapter 26), use calmodulin (CaM) fused to color variants of the GFP. The affinity of such indicators can be conveniently decreased by mutation of key chelating glutamate residues to glutamines in the first and third Ca^{2+}-binding loops of CaM. Cameleons responsive to Ca^{2+} concentrations ranging from 0.1 mM to low millimolar now exist. With other related genetically encoded indicators, such as camgaroos (Baird et al. 1999) and pericams (Nagai et al. 2001), a similar approach should be feasible.

COUPLING OF CA^{2+} BINDING TO CHANGES IN FLUORESCENCE

Ca^{2+} binding to an indicator is coupled to changes in fluorescence by mechanisms leading to the following three outcomes.

1. The binding of Ca^{2+} results in a shift in excitation, and sometimes emission peaks, allowing excitation or emission ratioing (e.g., Quin-2, Fura-2, Indo-1, and ratiometric pericam).
2. The binding of Ca^{2+} leads to a change in fluorescence intensity but no wavelength shifts (e.g., Fluo-3, Rhod-2, the Calcium Green family of indicators, and pericams and camgaroos).
3. The binding of Ca^{2+} results in changes in fluorescent resonance energy transfer (FRET; e.g., cameleons).

Quin-2, Fura-2, and Indo-1

The first indicators (e.g., Quin-2, Fura-2, and Indo-1; see Fig. 1) integrated the Ca^{2+}-binding site with small fluorophores (such as benzofurans and indoles) that are excitable in the ultraviolet (UV) region of the spectrum (Tsien 1980; Grynkiewicz et al. 1985). Direct conjugation of the chelating aminodiacetate moiety, $-N(CH_2CO_2-)_2$, with the planar fluorophores results in shifted excitation peaks (and sometimes also emission) on Ca^{2+} binding. Exciting these peaks alternately while collecting the emission allows excitation ratioing, whereas emission ratioing requires excitation at a single wavelength with collection at two separate emission peaks. Ratioing helps correct for variations in indicator concentration, cell shape, lamp and detector fluctuations, and photobleaching and enables calibration of the fluorescent signal in terms of free-Ca^{2+} concentration. Mechanistically, in the absence of Ca^{2+}, the aminodiacetate group lies planar to the aromatic ring system, thereby maximizing electron donation by the nitrogen. Ca^{2+} coordination twists this group out of planarity and causes a decrease in the electron density of the fluorophore, thereby decreasing the excitation wavelength. Rigidization of the aminodiacetate group by bound Ca^{2+} also probably decreases its radiationless deactivation of the excited state, thereby increasing the fluorescent quantum yield.

Fluo-3, Rhod-2, and the Calcium Green Family

Successive indicators such as Fluo-3, Rhod-2 (Minta et al. 1989), and the Calcium Green family (Kuhn 1993; Haugland 2005) incorporated the more fluorescent fluorescein and rhodamine fluors, which operate at visible wavelengths that cause less cellular autofluorescence than UV excitation. All these long-wavelength Ca^{2+} indicators use a different photochemical mechanism for coupling the Ca^{2+} chelation to changes in fluorescence, which results in essentially no excitation or emission shifts but only results in changes in emission intensity on binding Ca^{2+}.

In these molecules, the BAPTA moiety is twisted out of plane with the fluorophore (through steric interaction between the juxtaposed aromatic rings), preventing direct electronic coupling. However, on excitation by light, in the absence of Ca^{2+}, the electron-rich nitrogen of the aminodiacetate group can transfer an electron to the electron-poor fluorophore before a photon can be emitted as fluorescence. The products of electron transfer, a cation and anion radical, cannot fluoresce and rapidly recombine directly or indirectly to reform the ground state. When Ca^{2+} binds to the aminodiacetate group, the nitrogen's electron density is decreased, electron transfer occurs less, and the fluorescence is not quenched. Binding Ca^{2+}, therefore, leads to a change in fluorescence intensity but no wavelength shifts. The degree of quenching and the extent of relief by Ca^{2+} reflect the distance and types of bonds between the Ca^{2+}-binding site and the fluorophore. Long-wavelength indicators can be readily synthesized by reacting simple BAPTA derivatives with fluorescent labeling reagents (Kuhn 1993). Provided the two components are sufficiently close or appropriately linked, the resulting molecule's fluorescence becomes Ca^{2+} sensitive. Using this principle, an infrared Ca^{2+} indicator, cyan-4 (Ozmen and Akkaya 2000; S.R. Adams, R.Y. Tsien, unpubl.), has been synthesized that may be useful in certain cells and tissues containing molecules, which strongly absorb or fluoresce in the visible part of the spectrum such as hemoglobin (e.g., blood-containing tissues), chlorophyll (plants), and rhodopsin (photoreceptors).

Genetically Encoded Indicators: Cameleon, Camgaroo, G-CaMP, and Pericam

Cameleons (see Fig. 1) were the first genetically encoded Ca^{2+} indicators. They are protein based, resulting from the fusion of CaM and M13 (a CaM-binding peptide from myosin light-chain kinase) with cyan and yellow mutants of GFP (CFP and YFP) at the amino and carboxyl termini, respectively (Miyawaki et al. 1997). They use a third mechanism for coupling Ca^{2+} binding to changes in fluorescence. When CFP is excited at 430 nm, FRET occurs from CFP to YFP, resulting in emission as yellow light. (For a detailed discussion on the principles of FRET and the photochemical mechanisms of various indicators utilizing FRET, see Chapters 22, 31, and 32.) FRET is exquisitely sensitive to changes in the distance (in the range of 20–100 nm) and the orientation of the two fluorophores involved, so the well-known conformation change of calmodulin on binding Ca^{2+} results in a significant change in FRET and emission ratio. In low Ca^{2+}, the more extended conformation of CaM inhibits FRET and gives a low ratio of YFP to CFP emission, whereas high Ca^{2+} results in more FRET and a higher ratio. One limitation of cameleons, however, is their requirement for 430-nm excitation, not a convenient laser line. This excitation wavelength is now available on newer imaging systems, although two-photon infrared excitation appears to work well (Fan et al. 1999). Recent developments have improved the expression and folding of cameleons in cells and have decreased their sensitivity to pH and chloride anions (Griesbeck et al. 2001).

New indicators in this class are camgaroo (Baird et al. 1999), G-CaMP (Nakai et al. 2001), and pericams (Nagai et al. 2001), all of which involve insertion of CaM into a single fluorescent protein (usually YFP) at a site close to the chromophore. Ca^{2+} binding alters the pH sensitivity of the YFP, resulting in large changes in fluorescence intensity. Different versions of pericam either increase or decrease fluorescence intensity at a single wavelength or are excitation ratiometric indicators.

Genetically encoded indicators have been targeted using molecular and cell biology techniques (transfection and transgenics) to different tissue and cell types and various subcellular locations such as the endoplasmic reticulum, mitochondria, and the nucleus. This contrasts greatly with conventional indicators that can often be easily loaded into cells as AM esters but are not restricted to specific cytoplasmic compartments or tissues. Genetically encoded indicators have been particularly successful in organisms that do not load AM esters such as plants (Allen et al. 2001), worms (Kerr et al. 2000; Suzuki et al. 2003), flies (e.g., Reiff et al. 2002; Liu et al. 2003), and fish (Higashijima et al. 2003). Recent improvements in the dynamic range and sensitivity of genetically encoded calcium indicators have allowed the measurement of neuronal calcium transients in worms, flies, and mice (Hoogland et al. 2009; Tian et al. 2009).

Recently, small molecule Ca^{2+} indicators have been targeted in mammalian cells to specific cytoplasmic or membrane proteins fused to genetically encoded tags. An AM derivative of a novel low-affinity Calcium Green biarsenical was used to report the localized Ca^{2+} concentrations and dynamics at gap junctions and Ca^{2+} channels that had been tagged with a tetracysteine peptide (Tour et al. 2007). Similarly, nuclear-targeted SnapTag (a 186-aa protein derived from O^6-alkylguanine-DNA alkyltransferase) was labeled in living muscle cells with an indo-1 derivative conjugated to benzyl guanine (Bannwarth et al. 2009).

CONCLUSION

A wide variety of Ca^{2+} indicators with a range of affinities is now available that can be excited by UV or visible light and readily incorporated into many cells as AM esters (chemical indicators) or by molecular biology techniques (cameleons, pericams, and camgaroos).

REFERENCES

Adams SR, Kao JPY, Grynkiewciz G, Minta A, Tsien RY. 1988. Biologically useful chelators that release Ca^{2+} upon illumination. *J Am Chem Soc* **110:** 3212–3220.

Allen GJ, Chu SP, Harrington CL, Schumacher K, Hoffmann T, Tang YY, Grill E, Schroeder JI. 2001. A defined range of guard cell calcium oscillation parameters encodes stomatal movements. *Nature* **411:** 1053–1057.

Baird GS, Zacharias DA, Tsien RY. 1999. Circular permutation and receptor insertion within green fluorescent proteins. *Proc Natl Acad Sci* **96:** 11241–11246.

Bannwarth M, Correa IR, Sztretye M, Pouvreau S, Fellay C, Aebischer A, Royer L, Rois E, Johnsson K. 2009. Indo-1 derivatives for local calcium sensing. *ACS Chem Biol* **4:** 179–190.

Fan GY, Fujisaki H, Miyawaki A, Tsay RK, Tsien RY, Ellisman MH. 1999. Video-rate scanning two-photon excitation fluorescence microscopy and ratio imaging with cameleons. *Biophys J* **76:** 2412–2420.

Griesbeck O, Baird GS, Campbell RE, Zacharias DA, Tsien RY. 2001. Reducing the environmental sensitivity of yellow fluorescent protein. Mechanism and applications. *J Biol Chem* **276:** 29188–29194.

Grynkiewicz G, Poenie M, Tsien RY. 1985. A new generation of Ca^{2+} indicators with greatly improved fluorescence properties. *J Biol Chem* **260:** 3440–3450.

Haugland RP. 2005. *The handbook. A guide to fluorescent probes and labeling technologies*, 10th ed. Invitrogen, Carlsbad, CA.

Higashijima SI, Masino MA, Mandel G, Fetcho JR. 2003. Imaging neuronal activity during zebrafish behavior with a genetically encoded calcium indicator. *J Neurophysiol* **90:** 3986–3997.

Hoogland TM, Kuhn B, Göbel W, Huang W, Nakai J, Helmchen F, Flint J, Wang SS. 2009. Radially expanding transglial calcium waves in the intact cerebellum. *Proc Natl Acad Sci* **106:** 3496–3501.

Kao JPY. 1994. Practical aspects of measuring $[Ca^{2+}]$ with fluorescent indicators. *Methods Cell Biol* **40:** 155–181.

Kerr R, Lev-Ram V, Baird G, Vincent P, Tsien RY, Schafer WR. 2000. Optical imaging of calcium transients in neurons and pharyngeal muscle of *C. elegans*. *Neuron* **26:** 583–594.

Kuhn MA. 1993. 1,2-bis(2-aminophenoxy)ethane-*N,N,N′,N′*-tetraacetic acid conjugates used to measure intracellular Ca^{2+} concentration. In *Fluorescent chemosensors for ions and molecule recognition* (ed. Czarnik AW), pp. 147–161. American Chemical Society, Washington, DC.

Levy LA, Murphy E, Raju B, London RE. 1988. Measurement of cytosolic free magnesium ion concentration by ^{19}F NMR. *Biochemistry* **27:** 4041–4048.

Liu L, Yermolaieva O, Johnson WA, Abboud FM, Welsh MJ. 2003. Identification and function of thermosensory neurons in *Drosophila larvae*. *Nat Neurosci* **6:** 267–273.

Luccardini C, Yakovlev AV, Pasche M, Gaillard S, Li D, Rousseau F, Ly R, Becherer U, Mallet JM, Feltz A, et al. 2009. Measuring mitochondrial and cytoplasmic Ca^{2+} in EGFP expressing cells with a low-affinity Calcium Ruby and its dextran conjugate. *Cell Calcium* **45:** 275–283.

London RE, Rhee CK, Murphy E, Gabel S, Levy LA. 1994. Nmr-sensitive fluorinated and fluorescent intracellular calcium ion indicators with high dissociation constants. *Am J Physiol* **266:** C1313–C1322.

Minta A, Kao JPY, Tsien RY. 1989. Fluorescent indicators for cytosolic calcium based on rhodamine and fluorescein chromophores. *J Biol Chem* **264:** 8171–8178.

Miyawaki A, Llopis J, Heim R, McCaffery JM, Adams JA, Ikura M, Tsien RY. 1997. Fluorescent indicators for Ca^{2+} based on green fluorescent proteins and calmodulin. *Nature* **388:** 882–887.

Nagai T, Sawano A, Park ES, Miyawaki A. 2001. Circularly permuted green fluorescent proteins engineered to sense Ca^{2+}. *Proc Natl Acad Sci* **98:** 3197–3202.

Nakai J, Ohkura M, Imoto K. 2001. A high signal-to-noise Ca^{2+} probe composed of a single green fluorescent protein. *Nat Biotechnol* **19:** 137–141.

Ozmen B, Akkaya EU. 2000. Infrared fluorescence sensing of submicromolar calcium: Pushing the limits of photoinduced electron transfer. *Tetrahedron Lett* **41:** 9185–9188.

Pethig R, Kuhn M, Payne R, Adler E, Chen TH, Jaffe LF. 1989. On the dissociation constants of BAPTA-type calcium buffers. *Cell Calcium* **10:** 491–498.

Raju B, Murphy E, Levy LA, Hall RD, London RE. 1989. A fluorescent indicator for measuring cytosolic free magnesium. *Am J Physiol* **256:** C540–C548.

Ranganathan R, Bacskai BJ, Tsien RY, Zuker CS. 1994. Cytosolic calcium transients: Spatial localization and role in *Drosophila* photoreceptor cell function. *Neuron* **13:** 837–848.

Reiff DF, Thiel PR, Schuster CM. 2002. Differential regulation of active zone density during long-term strengthening of *Drosophila* neuromuscular junctions. *J Neurosci* **22:** 9399–9409.

Suzuki H, Kerr R, Bianchi L, Frokjaer-Jensen C, Slone D, Xue J, Gerstbrein B, Driscoll M, Schafer WR. 2003. In vivo imaging of *C. elegans* mechanosensory neurons demonstrates a specific role for the MEC-4 channel in the process of gentle touch sensation. *Neuron* **39:** 1005–1017.

Tian L, Hires SA, Mao T, Huber D, Chiappe ME, Chalasani SH, Petreanu L, Akerboom J, McKinney SA, Schreiter ER, et al. 2009. Imaging neural activity in worms, flies and mice with improved GCamp calcium indicators. *Nat Methods* **6:** 875–881.

Tour O, Adams SR, Kerr RA, Meijer RM, Sejnowski TJ, Tsien RW, Tsien RY. 2007. Calcium Green FlAsH as a genetically targeted small-molecule calcium indicator. *Nat Chem Biol* **3:** 423-31.

Tsien RY. 1980. New calcium indicators and buffers with high selectivity against magnesium and protons: Design, synthesis, and properties of prototype structures. *Biochemistry* **19:** 2396–2404.

Tsien RY. 1993. Fluorescent and photochemical probes of dynamic biochemical signals inside living cells. In *Fluorescent chemosensors for ions and molecule recognition* (ed. Czarnik AW), pp. 130–146. American Chemical Society, Washington, DC.

Tsien RY. 1999. Monitoring cell Ca^{2+}. In *Calcium as cellular regulator* (ed. Carafoli E, Klee C), pp. 28–54. Oxford University Press, Oxford.

24 Calibration of Fluorescent Calcium Indicators

Fritjof Helmchen

Department of Neurophysiology, Brain Research Institute, University of Zurich, CH-8057 Zurich, Switzerland

ABSTRACT

During the past decades, many different fluorescent indicators have been developed for measuring intracellular ion concentrations. Of particular interest are fluorescent calcium indicators because of the fundamental role of Ca^{2+} in various cellular processes such as contraction, secretion, and gene activation. For a quantitative understanding of the physiological roles of Ca^{2+}, fluorescence signals measured with calcium indicators have to be converted to intracellular free calcium concentration ($[Ca^{2+}]_i$). Similarly, changes in $[Ca^{2+}]_i$ and the underlying calcium fluxes need to be inferred from the corresponding fluorescence changes. This chapter describes the theoretical background, the various principal methods, and the protocols for the calibration of calcium imaging data.

INTRODUCTION

Fluorescent calcium indicators are universal tools for studying intracellular signaling and therefore are used in all fields of cell biology. In neurobiology they have been applied to study many functional aspects, for example, the role of Ca^{2+} in neurotransmitter release and in postsynaptic signal integration. For each specific application the most suitable indicator dye can be chosen from a large palette of indicators, which differ with respect to their affinity, mobility, solubility, and fluorescence properties. Two large groups of indicators exist: (1) synthetic organic molecules (mostly derived from the fast calcium buffer BAPTA; Tsien 1980), which can be loaded into cells and subcellular compartments using various techniques (Tsien 1989; Göbel and Helmchen 2007); and (2) genetically encoded calcium indicators based on fluorescent proteins (for reviews, see Miyawaki 2003; Hires et al. 2008; Mank and Griesbeck 2008) (see Chapter 26; Tian et al. 2011). Common to all indicators is that their fluorescence is sensitive to $[Ca^{2+}]_i$ in at least one respect; Ca^{2+} binding may cause changes in fluorescence intensity, spectral shifts, or changes in fluorescence resonance energy transfer (FRET) efficiency (Fig. 1). In addition, the fluorescence lifetime may change upon Ca^{2+} binding. In the following, the theoretical background and the basic principles of converting fluorescence signals to $[Ca^{2+}]_i$ changes are described. Note that equivalent considerations apply to measurements using fluorescent indicators of other ion species (e.g., H^+, Na^+, Cl^-).

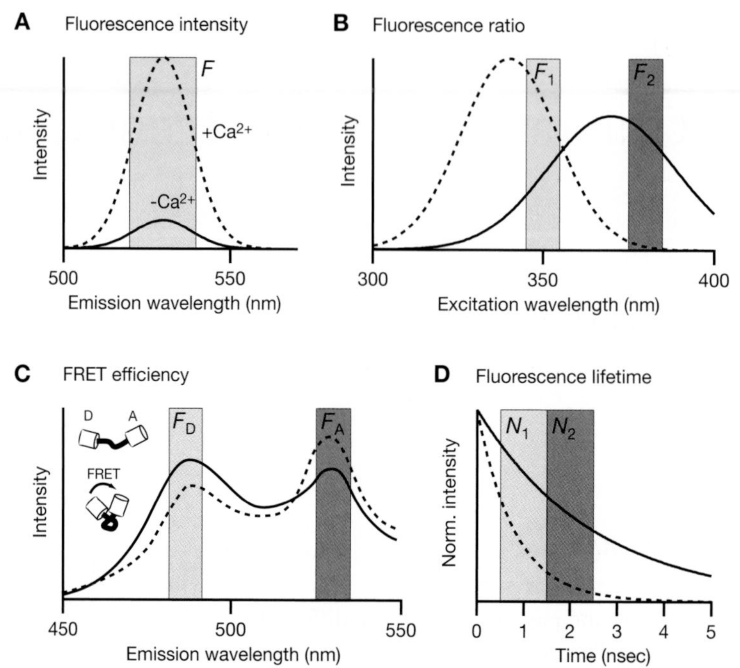

FIGURE 1. Changes in fluorescence properties of calcium indicators used for calibration. (*A*) Changes in the fluorescence intensity result from changes of the quantum yield and the absorption of a dye upon Ca^{2+} binding. An emission spectrum similar to that of Calcium Green-1 is schematically shown with higher intensities of the Ca^{2+} bound form (*dashed line*) compared to the unbound form (solid line). Other indicators (e.g., Fura-Red [not shown]) show decreases in the intensity. In single-wavelength measurements, the dye is excited at a single wavelength, and the emission intensity *F* is collected in a spectral window around the peak of the emission spectrum. (*B*) Spectral shifts allow ratiometric measurements because the ratio of the fluorescence intensities F_1 and F_2 measured at two different wavelengths in this case is sensitive to Ca^{2+} binding. The drawing schematically shows an excitation spectrum similar to that of Fura-2 at zero Ca^{2+} concentration (solid line) and at saturating Ca^{2+} levels (dashed line). The peak of the spectrum shifts to shorter wavelength upon Ca^{2+} binding. Note that excitation at 360 nm results in $[Ca^{2+}]_i$-insensitive fluorescence emission, defining the so-called isosbestic wavelength. (*C*) Changes in fluorescence resonance energy transfer (FRET) efficiency are used in the case of tandem GFP-based indicators, which use a donor (D) fluorescent protein linked via a Ca^{2+}-sensitive spacer to an acceptor (A) fluorescent protein. The emission spectrum of a yellow cameleon (with ECFP and EYFP as donor and acceptor, respectively) is shown schematically at low (solid line) and high (dashed line) $[Ca^{2+}]_i$ level (excitation ~430 nm). The distance between the two fluorescent proteins decreases upon Ca^{2+} binding, which, as a result of enhanced FRET efficiency, causes the donor emission (F_D) to decrease and the acceptor fluorescence (F_A) to increase, thus permitting ratiometric measurements. (*D*) Alternatively, changes in the fluorescence lifetime have been used for calibration. Fluo-3, for example, shows a shorter fluorescence lifetime in the Ca^{2+} bound form (dashed line) compared to the unbound form (solid line) as illustrated schematically. Other indicators show an increase of fluorescence lifetime upon Ca^{2+} binding. As a result, the ratio of the number of photons N_1 and N_2 that are detected in two time windows during the fluorescence decay is sensitive to $[Ca^{2+}]_i$. Not all indicators necessarily are $[Ca^{2+}]_i$ sensitive with respect to all the properties shown in *A*–*D*.

Changes in Fluorescence Intensity

The emission intensity *F* arising from an observation volume *V* (e.g., a single cell or a cell compartment loaded with a fluorescent dye) depends on the number of dye molecules, the illumination intensity I_0, the dye absorption α, the quantum yield of the dye Q_F, the photon-collection efficiency Φ of the optical setup, and the quantum efficiency (QE) of the detector Q_D:

$$F = \Phi Q_D Q_F \alpha I_0 n = S \cdot n, \tag{1}$$

where *n* is the molar amount of dye molecules in *V*. Here, all factors that depend on dye properties or the experimental setup have been "lumped" together in a single proportionality constant *S*. In the case of calcium indicators, we have to consider separately the molar amounts n_f and n_b of the free and the Ca^{2+}-bound indicator forms, respectively. They differ with respect to their quantum yield and their absorption and therefore contribute to F with different factors S_f and S_b, respectively:

$$\begin{aligned} F &= S_f n_f + S_b n_b \\ &= F_{min} + (S_b - S_f) n_b \\ &= F_{max} - (S_b - S_f) n_f. \end{aligned} \tag{2}$$

The equation has been rewritten using the definitions for the fluorescence at zero Ca^{2+} concentration $F_{min} = S_f n_{tot}$ and at saturating $[Ca^{2+}]_i$ levels, $F_{max} = S_b n_{tot}$, with $n_{tot} = n_f + n_b$, and assuming a fluorescence increase upon Ca^{2+} binding. The indicator fluorescence thus depends on the relative amounts of the free and bound forms. Assuming 1:1 complexation of Ca^{2+} with the dye, n_f and n_b vary according to the law of mass action:

$$K_d = \frac{n_f [Ca^{2+}]_i}{n_b}, \tag{3}$$

where K_d is the dissociation constant of the indicator. Equations 1–3 are the basis for all conversion equations that relate fluorescence signals to $[Ca^{2+}]_i$ values. Before introducing these equations, however, we first have to clarify some important issues concerning the dye concentration and the subtraction of background fluorescence.

Dye Concentration and Background Subtraction

A central assumption underlying Equation 1 is that all dye molecules sense the same illumination intensity, meaning that no inner filtering occurs. In general, the intensity of light that is absorbed by a fluorescent layer of thickness l is given by the Beer-Lambert law,

$$I_{abs} = I_0(1 - 10^{-\varepsilon lc}) \approx I_0 \ln(10)\varepsilon lc, \tag{4}$$

where ε is the molar extinction coefficient and c is the dye concentration. The approximation of a linear relationship between I_{abs} and c as used in Equation 1 is only valid if $c << [\ln(10)\varepsilon l]^{-1}$. This sets an upper limit to the useful dye concentration range. Extinction coefficients of calcium indicators typically are in the range of 20,000–100,000 M^{-1} cm^{-1} (*Molecular Probes Handbook*, Invitrogen). Thus, for measurements on small cells with 10-µm diameter, the indicator concentration should be well below 5–20 mM. In thick cuvettes of ~1 cm path length, which are sometimes used for in vitro calibration, much smaller concentrations have to be used (typically 1 µM). At very high concentration, dye fluorescence in addition may be reduced because of self-quenching. Furthermore, for measurements of intracellular calcium dynamics, the choice of indicator concentration is constrained by the Ca^{2+}-buffering effect exerted by the indicator itself. Depending on the endogenous Ca^{2+}-buffering capacity, even a relatively low concentration of a high-affinity indicator (100 µM) may reduce and prolong $[Ca^{2+}]_i$ changes several-fold (Chapter 25; Neher 1995; Helmchen and Tank 2011). Low-affinity dyes can circumvent this problem partly, because they can be applied in relatively high concentration without altering $[Ca^{2+}]_i$ dynamics significantly. The resulting fluorescence signals, however, may be very small (see below).

Optical components, the bathing solution, and endogenous fluorophores all add background to the indicator fluorescence given in Equation 2. Because background fluorescence increases the noise level, it should be minimized (Moore et al. 1990). Importantly, any background fluorescence F_{bkg} must be subtracted from the observed fluorescence ($F = F_{obs} - F_{bkg}$) before applying equations for conversion to $[Ca^{2+}]_i$. This supposedly simple step in practice often is not so easy. In imaging experiments on brain slices, a relatively large background arises from endogenous fluorophores of the surrounding tissue, especially with UV excitation. Also, the background may change during an experiment because of bleaching (Eilers et al. 1995). Therefore, it is necessary to perform background measurements throughout the experiment in a slice region nearby the cell of interest, for example, directly before and after each recording episode (Helmchen et al. 1996). A more severe problem occurs when tissue is bulk-loaded with AM-esters of calcium indicators leading to rather diffuse staining (Stosiek et al. 2003). In this case, the best possible region for background estimation is a relatively dye-free region near the cell (e.g., a blood vessel lumen) because the stained neuropile immediately next to the cell would provide a poor background estimate as it contains numerous stained and $[Ca^{2+}]_i$-sensitive axons and dendrites. A more rigorous and sophisticated approach is to estimate the background fluorescence level within a particular region of interest based on the temporal dynamics of the individual pixels within this region (Chen et al. 2006).

CALIBRATION METHODS

Single-Wavelength Measurements

One class of calcium indicators responds to Ca^{2+}-binding with an up- or downscaling of the fluorescence intensity without showing appreciable spectral shifts (Fig. 1A). Almost all synthetic indicators excited in the visible-wavelength range belong to this group. As a result, all available information on Ca^{2+} concentration is obtained from dye excitation at a single wavelength.

In principle, the fluorescence signal F measured at a single excitation wavelength can be converted to $[Ca^{2+}]_i$ by assuming equilibrium between Ca^{2+} and the indicator and combining Equations 2 and 3:

$$[Ca^{2+}]_i = K_d \frac{n_b(S_b - S_f)}{n_f(S_b - S_f)} = K_d \frac{F - F_{min}}{F_{max} - F}. \tag{5}$$

Although this equation is readily applicable to bulk measurements of cuvette solutions (e.g., of cell suspensions), it is impractical for imaging experiments because F_{min} and F_{max} would need to be determined independently for each observation volume (i.e., each pixel), which is impossible. If, however, ratios of intensities are used instead of absolute intensities, all spatial variations in cell thickness, total dye concentration, and detection efficiency or illumination intensity cancel out and thus are normalized (see Equation 1). For time-dependent measurements, normalization by ratioing most easily is achieved by expressing the signal as relative fluorescence change:

$$\Delta F/F = (F - F_0)/F_0, \tag{6}$$

where F_0 denotes the background-subtracted pre-stimulus fluorescence level. For $\Delta F/F$ ("delta F over F"), the following conversion equation can be derived (Lev-Ram et al. 1992):

$$[Ca^{2+}]_i = \frac{[Ca^{2+}]_{rest} + K_d\,(\Delta F/F)/(\Delta F/F)_{max}}{(1-(\Delta F/F)/(\Delta F/F)_{max})}. \tag{7}$$

Here, $[Ca^{2+}]_{rest}$ is the resting calcium concentration and $(\Delta F/F)_{max}$ is the maximal change upon dye saturation, which, for example, can be estimated using very strong stimulation or cell destruction at the end of an experiment. The major drawback of this and other related single-wavelength equations (Vranesic and Knöpfel 1991; Neher and Augustine 1992; Wang et al. 1995) is that an a priori knowledge of $[Ca^{2+}]_{rest}$ is required, which in principle needs to be obtained independently, for example, using an initial ratiometric measurement (see below). Note also that $(\Delta F/F)_{max}$ depends on $[Ca^{2+}]_{rest}$, which therefore is presumed constant throughout the experiment. Despite these difficulties, a reasonable value of $[Ca^{2+}]_{rest}$ may be assumed (50–100 nM) under certain conditions, in particular, when the health of the cell is monitored in parallel by additional methods like electrophysiology.

An alternative single-wavelength approach that circumvents the necessity for an independent measurement of $[Ca^{2+}]_{rest}$ is based on rearranging Equation 5 differently (Maravall et al. 2000):

$$[\mathrm{Ca}^{2+}]_\mathrm{i} = K_\mathrm{d} \frac{F/F_\mathrm{max} - 1/R_\mathrm{f}}{1 - F/F_\mathrm{max}}, \tag{8}$$

where R_f denotes the dynamic range F_{max}/F_{min} of the indicator. The idea is that the ratio of the actual fluorescence to the saturating fluorescence F/F_{max} reflects the $[Ca^{2+}]_i$ level, given that the K_d and R_f are known. R_f thus needs to be determined initially for an indicator (e.g., using an in vitro calibration procedure), and it turns out that for large R_f values, the conversion to $[Ca^{2+}]_i$ is relatively robust against uncertainties in its exact value. The remaining calibration procedure is then to estimate the saturating fluorescence level F_{max} (or the related $(\Delta F/F)_{max}$, respectively) intermittently during the experiment and/or at its end, which for high-affinity indicators can be achieved by inducing trains of high-frequency action potentials (Maravall et al. 2000).

If only small fluorescence changes are evoked and if the indicator is far from saturation (e.g., in the case of low-affinity calcium indicators), the single-wavelength equations can be linearized to provide an estimate of the change in $[Ca^{2+}]_i$:

$$\Delta[\mathrm{Ca}^{2+}]_\mathrm{i} = \frac{K_\mathrm{d}}{(\Delta F/F)_\mathrm{max}} (\Delta F/F) \qquad (\Delta F/F << (\Delta F/F)_\mathrm{max}), \tag{9}$$

For low-affinity indicators, however, it may be difficult to induce large enough calcium influx to determine the saturating fluorescence changes.

Dual-Wavelength Ratiometric Measurements

A second group of indicators undergoes shifts in the excitation or emission spectrum upon Ca^{2+} binding (Fig. 1B). These spectral shifts can be exploited for $[Ca^{2+}]_i$ calibration because in this case, the ratio $R = F_1/F_2$ of the intensities measured at two wavelengths depends on $[Ca^{2+}]_i$. The ratiometric method is the most often applied calibration method because R is independent of dye concentration, optical path length, and illumination intensity. The intensities F_1 and F_2 are given according to Equation 2:

$$F_1 = S_{f1} n_f + S_{b1} n_b,$$
$$F_2 = S_{f2} n_f + S_{b2} n_b. \tag{10}$$

Note that for actual measurements, both intensities have to be corrected for background fluorescence at the corresponding excitation or emission wavelengths before taking the ratio. From Equations 3 and 10, the standard equation for ratiometric measurements can be derived (Grynkiewicz et al. 1985):

$$[\mathrm{Ca}^{2+}]_\mathrm{i} = K_\mathrm{eff} \frac{R - R_\mathrm{min}}{R_\mathrm{max} - R}, \tag{11}$$

with the ratios at zero Ca^{2+} concentration $R_{min} = (S_{f1}/S_{f2})$, at saturating Ca^{2+} concentrations $R_{max} = (S_{b1}/S_{b2})$, and an effective binding constant $K_{eff} = K_d(S_{f2}/S_{b2})$.

The design and use of ratio imaging systems have been described in several reviews (see, e.g., Neher 1989; Tsien and Harootunian 1990). Dual-wavelength measurements with high time resolution require rapid switching of wavelengths. Alternatively, Equation 11 can be used to obtain an initial value of $[Ca^{2+}]_{rest}$ for subsequent single-wavelength measurement and application of Equation 7. For a more detailed description of how to use Fura-2, the most popular ratiometric dye, see Chapter 25. The ratiometric method also has been extended to mixtures of non-ratiometric dyes that result in Ca^{2+}-sensitive fluorescence ratios (Lipp and Niggli 1993; Oheim et al. 1998).

Changes in FRET Efficiency

Genetically encoded calcium indicators consist either of a single modified green fluorescent protein (GFP) exhibiting Ca^{2+}-sensitive fluorescence (Baird et al. 1999; Nakai et al. 2001), or they are based on changes in fluorescence resonance energy transfer (FRET) between two spectral variants of fluorescent proteins. For example, cameleons are fusion proteins of two fluorescent proteins linked via a spacer consisting of calmodulin and the calmodulin-binding peptide M13 (Miyawaki et al. 1997). Ca^{2+} binding to calmodulin causes a conformational change, which brings the two fluorescent proteins closer together and thereby causes an increase in the FRET efficiency. This change can be read out by measuring the change in the ratio of the donor and acceptor emission intensities in two appropriate spectral windows (Fig. 1C). Because the emission spectra of donor and acceptor typically have some overlap, cross talk may have to be taken into account (Gordon et al. 1998). As genetically encoded calcium indicators often display cooperativity of multiple Ca^{2+}-binding sites, conversion equations incorporating a Hill coefficient >1 may need to be applied (Palmer et al. 2006). For further information on genetically encoded calcium indicators, see Chapter 26.

Changes in Fluorescence Lifetime

As an alternative to the intensity, the fluorescence lifetime can be used as an indicator of Ca^{2+} binding (Lakowicz et al. 1992; Draaijer et al. 1995). After the end of excitation, the fall-off in fluorescence intensity reflects the lifetime of the excited state of the dye molecules (Fig. 1D). In the simplest case, the decay is described by a single exponential curve with a fluorescence lifetime constant τ_L, which typically is in the nanosecond range; in general, multiple decay components are present. Several calcium indicators show useful Ca^{2+}-dependent fluorescence lifetime changes (Draaijer et al. 1995; Wilms and Eilers 2007).

One method to measure the fluorescence lifetime is time-gated photon detection following a brief exciting laser pulse. The ratio of the numbers of photons N_1 and N_2 that are detected in two time windows following the excitation pulse provides an effective fluorescence lifetime $\tau_{eff} = \Delta t/\log(N_1/N_2)$, where Δt is the width of the windows (Fig. 1D). τ_{eff} can also be obtained from the phase shift and the change in amplitude when a modulated light source is used (Lakowicz et al. 1992). Both the free and the Ca^{2+}-bound forms of the indicator contribute to the effective fluorescence lifetime, which relates to $[Ca^{2+}]_i$ via

$$[\mathrm{Ca}^{2+}]_i = K_{app} \frac{\tau_{eff} - \tau_{min}}{\tau_{max} - \tau_{eff}}, \tag{12}$$

where τ_{min} and τ_{max} denote the lifetimes of the bound and free indicator forms, respectively, assuming a lifetime decrease, and K_{app} denotes an apparent dissociation constant. Quantification of $[Ca^{2+}]_i$ from lifetime measurements can be further improved by using time-correlated single-photon counting with a pulsed laser source and multi-exponential fitting routines (Wilms et al. 2006).

Because lifetimes are independent of dye concentration and illumination intensity, a calibration according to Equation 12 is readily applied to imaging experiments. K_{app} depends on the relative intensities of the free and bound indicator forms. Interestingly, the apparent affinity of an indicator showing a spectral shift therefore can be tuned in a wide range simply by selecting the excitation wavelength (Szmacinski and Lakowicz 1995). A fast fluorescence lifetime imaging (FLIM) microscope enabling $[Ca^{2+}]_i$ measurements with frame rates of >50 Hz has been presented in Agronskaia et al. (2003). Lifetime-based Ca^{2+} imaging is also relatively straightforward to add on to microscopic techniques that inherently use pulsed laser sources for excitation, such as two-photon laser scanning microscopy (Wilms et al. 2006; Kuchibhotla et al. 2009). Finally, fluorescence lifetime measurements of the combined donor/acceptor emission can be used as an alternative means to determine changes in FRET efficiency (Harpur et al. 2001).

Total Calcium Flux Measurements

Calcium indicators are mainly used for measuring $[Ca^{2+}]_i$. However, as calcium chelators they inevitably and sometimes significantly alter the magnitude and the time course of $[Ca^{2+}]_i$ changes (see Chapter 25; Helmchen and Tank 2011). This problem of indicator buffering is separate from the problem of an accurate calibration: Even if the concentration values calculated are correct, they may not reflect the physiological $[Ca^{2+}]_i$ levels reached in the absence of the indicator. On the other hand, buffering by the indicator can be exploited to measure Ca^{2+} fluxes (Schneggenburger et al. 1993; Neher 1995). When loaded in excess into a cell compartment, the indicator molecules outcompete the endogenous Ca^{2+} buffers and capture virtually all ions that enter the cytosol. Under such an "overload condition," the change in absolute fluorescence intensity is proportional to the total calcium charge Q_{Ca} injected:

$$\Delta F = (S_b - S_f)\Delta n_b = \frac{(S_b - S_f)}{2F_c} Q_{Ca} = f_{max} Q_{Ca}, \tag{13}$$

where F_c is Faraday's constant. This overload method has been used to determine fractional Ca^{2+} currents through ligand-gated ion channels (Schneggenburger et al. 1993; Neher 1995; Bollmann et al. 1998) and the total calcium influx during an action potential (Helmchen et al. 1997; Bollmann et al. 1998).

Protocol

Calibration Protocols

All the methods we have discussed thus far for converting a fluorescence signal to $[Ca^{2+}]_i$ require the determination of a set of three calibration parameters: (K_{eff}, R_{min}, R_{max}), (K_d, $\Delta F/F_{max}$, $[Ca^{2+}]_{rest}$), or (K_d, R_f, F_{max}) or (K_{app}, τ_{min}, τ_{max}). Here we describe the classical procedure for calibration of ratiometric measurements for both in vivo and in vitro calibrations, which is also useful for determining K_d and R_f. The $[Ca^{2+}]_i$ dependence of the fluorescence ratio is measured using a set of at least three calibration solutions with known $[Ca^{2+}]_i$ levels.

MATERIALS

CAUTION: See Appendix 6 for proper handling of materials marked with <!>.

Reagents

Calcium solutions:

- "Zero calcium" solution: Intracellular solution containing ≥10 mM K_2EGTA and a low concentration (0.1 mM) of calcium indicator.
- "High calcium" solution: Intracellular solution containing ≥20 mM $CaCl_2$ <!> and 0.1 mM calcium indicator.
- "Intermediate calcium" solutions: A mixture of two intracellular solutions containing CaEGTA and K_2EGTA. Mixed at concentrations in the 5–20 mM range at different ratios (1:2, 1:1, or 2:1), these solutions can adjust (buffer) the $[Ca^{2+}]_i$ level to value(s) around the K_d of the indicator. Add a low concentration (0.1 mM) of calcium indicator. These mixture solutions are critical as the $[Ca^{2+}]_i$ level reached directly depends on the assumed K_d of EGTA. Because of its lower pH dependence, it may be preferable to use BAPTA instead of EGTA.

  Calcium calibration buffer kits for preparation of such standard calibration solutions are available from Molecular Probes (Invitrogen). Note, however, that dissociation constants of Ca^{2+} buffers as well as other indicator properties depend on temperature, pH, and the ionic strength of the solution (Groden et al. 1991). Therefore, it may be sensible to prepare custom buffered solutions based on the specific intracellular solution used in order to mimic experimental conditions as closely as possible. For further guidelines on the preparation of calibration solutions, see Tsien and Pozzan (1989).

Cells under study
Silicon grease

Equipment

Glass capillaries or microslides
Fluorescence microscope

METHOD 1: IN VITRO CALIBRATION OF $[Ca^{2+}]_i$

1. Fill glass capillaries or microslides with the calibration solutions. For measurements under a water-immersion objective, seal the ends of the microslides (e.g., with silicone grease).
2. Perform fluorescence measurements under conditions as close as possible to the experimental conditions, that is, using the same microscope objective, filter sets, and detector settings as in

the imaging experiments. Measure fluorescence at both excitation wavelengths for each calibration solution.

3. In addition, measure background fluorescence values at each wavelength using capillaries filled with solution containing no dye.
4. Subtract background fluorescence values and calculate fluorescence ratios R for each $[Ca^{2+}]_i$ level.
5. Calculate the three parameters (K_{eff}, R_{min}, R_{max}) from the set of calibration measurements using Equation 11. If more than one intermediate calcium solution has been used, parameters are obtained from a fit of the data set with Equation 11.

   See *Troubleshooting.*

TROUBLESHOOTING

Problem (Step 5): The parameters obtained from in vitro calibrations may yield unreliable results because the behavior of fluorescent dyes is altered in the viscous cytosolic environment and by intracellular binding or uptake (Moore et al. 1990).

Solution: Apparent negative $[Ca^{2+}]_i$ values are a clear indication of this problem when in vitro calibration parameters are applied to fluorescence data from cells. Poenie (1990) suggested that viscosity can be corrected for by multiplying R_{min} and R_{max} by a factor of 0.7–0.85. Whenever possible, an in vivo calibration is preferable.

METHOD 2: IN VIVO CALIBRATION OF $[Ca^{2+}]_i$

In this case, the cells under investigation are filled directly with the buffered $[Ca^{2+}]_i$ solutions. For example, patch-clamp experiments on cultured cells or cells in brain slices provide direct access to the cytosol (Neher 1989).

1. Start with R_{min}. Fill a patch pipette with the "zero calcium" calibration solution.
2. Obtain a whole-cell recording from the cell type under investigation. Use the same experimental settings as in the experiments.
3. After diffusional equilibration, measure fluorescence at both excitation wavelengths in a region of interest on the filled cell.

   See *Troubleshooting.*
4. Measure appropriate background fluorescence values, for example, from a region next to the filled cell.
5. Subtract background fluorescence values and calculate the fluorescence ratio R_{min}.
6. Repeat the measurement for at least three cells and average data.
7. Proceed in the same way (Steps 1–6) for determining R_{max} and K_{eff}.

TROUBLESHOOTING

Problem (Step 3): Small cells such as chromaffin cells are readily filled with the calibration solutions (within minutes). For large cells with extensive neural processes, however, loading of the cell may take significantly longer. In addition, strong extrusion mechanisms may prevent $[Ca^{2+}]_i$ from reaching and maintaining the concentration level in the pipette solution.

Solution: In these cases, calibration fluorescence measurements should be performed very close to the pipette tip and with low access resistance (Eilers et al. 1995). In general, when using whole-cell recordings reliable values are easily obtained for R_{min}, whereas the determination of R_{max} and

K_{eff} is more susceptible to the problem of an insufficient clamp of $[Ca^{2+}]_i$ to the pipette level. R_{max}, however, can also be estimated at the end of or during an experiment by applying a strong stimulation (e.g., a long high-frequency train of action potentials) to saturate the indicator.

METHOD 3: CALIBRATION OF ABSOLUTE FLUORESCENCE INTENSITY

In the case of calcium flux measurements, the calibration consists in the determination of the proportionality constant f_{max} (Equation 13). f_{max} may be inferred from electrical recordings of pure calcium currents and simultaneous measurements of the evoked fluorescence changes (Helmchen et al. 1997). Once f_{max} is known, changes in absolute fluorescence can be directly converted to calcium charge. To account for long-term changes in the illumination intensity or the detection efficiency in the course of these experiments, absolute fluorescence intensities should be normalized to a fluorescent standard such as 4.5-μm fluorescent beads (Polysciences, Inc., Warrington, PA) and thus expressed in "bead units" (BU) (Schneggenburger et al. 1993). The bead unit should be measured on each experimental day as the mean fluorescence of at least three beads.

EXAMPLE OF APPLICATION

To illustrate several of the calibration techniques described above, Figure 2 shows examples of calcium measurements from large presynaptic terminals ("calyces of Held") in the medial nucleus of the trapezoid body (MNTB) (see also Helmchen et al. 1997). Using an acute brain slice preparation, nerve terminals were loaded with various indicators and using different indicator concentrations via whole-cell patch pipettes. In the left panels the raw fluorescence data are shown, including pre- and poststimulus measurements of the background fluorescence. These data were then evaluated as either relative fluorescence change $\Delta F/F$, fluorescence ratio, or absolute fluorescence change (middle panels). Finally, fluorescence changes were converted to calcium concentration or charge applying the set of predetermined calibration parameters.

Single-wavelength measurements using the low-affinity indicator MagFura-2 were performed to obtain an estimate of the presynaptic $[Ca^{2+}]_i$ dynamics with minimal distortion by indicator buffering (Fig. 2A). At 380-nm wavelength, MagFura-2 fluorescence decreases upon Ca^{2+} binding. For calibration a dissociation constant of $K_d = 45$ µM was assumed and $(-\Delta F/F)_{max}$ was determined to be 78% using long trains of action potentials at 200 Hz. Because a single action potential induced only a very small fluorescence change, the linearized Equation 9 could be applied, revealing a large (several hundred nanomolar) and brief (decay time constant 50 msec) calcium transient.

In a separate set of experiments, Fura-2 was used at a moderate concentration (160 µM) and ratiometric measurements were performed using fast 360/380-nm wavelength switching with a monochromator-based illumination system (see Chapter 9). Calibration parameters were determined using an in vivo calibration procedure by loading terminals with internal solutions clamped to zero, intermediate, and high $[Ca^{2+}]_i$, respectively (resulting in $R_{min} = 0.77$; $R_{max} = 3.15$; $K_{eff} = 1117$ nM). Note that in the case of using the isosbestic wavelength in the nominator, the K_d of the dye is given by $K_d = K_{eff}R_{min}/R_{max}$, yielding $K_d = 273$ nM for Fura-2 in our case. Conversion using these parameters yielded an estimate of ~50 nM for $[Ca^{2+}]_{rest}$. The amplitude (26 nM) and the decay time constant (364 msec) of the single action potential-induced calcium transient were, however, clearly altered compared to the MagFura-2 measurement (Fig. 2B). This is attributed to the relatively large added Ca^{2+}-buffering capacity compared to the endogenous Ca^{2+}-buffering capacity (see Helmchen and Tank 2011).

Finally, Fura-2 was used at high (1 mM) concentration to overload the presynaptic terminal with added buffer. In this case, the change in absolute fluorescence intensity is evaluated (Fig. 2C). To normalize for changes in the imaging system over time (e.g., aging of the arc lamp), absolute fluorescence intensities were expressed in BU. The proportionality constant f_{max} was determined in separate experiments using pure calcium currents as $f_{max} = 0.0144$ BU pC^{-1} (Helmchen et al. 1997). It was found that ~1 pC of Ca^{2+} enter the nerve terminal per action potential. In summary, these examples demonstrate how different methods of calcium calibration can be used to quantify various aspects of the calcium signaling system in small neuronal compartments.

DISCUSSION

In summary, many different fluorescent calcium indicators are meanwhile available, and different methods exist for quantifying various Ca^{2+}-signaling aspects from the observed fluorescence signals. For quantitative measurements a careful understanding of the indicator properties under the experimental conditions and of potential interferences of other binding partners (e.g., H^+, Mg^{++}) is required. As shown in the examples in Figure 2, employment of different types of indicators at appropriate concentration can be highly informative. Single-wavelength measurements are well suited to measure fast signals as they provide the highest possible temporal resolution. Combined with initial ratiometric measurements or following careful determination of the dissociation constant and the dynamic range of the indicator, they can also be used for calibrated $[Ca^{2+}]_i$ measurements.

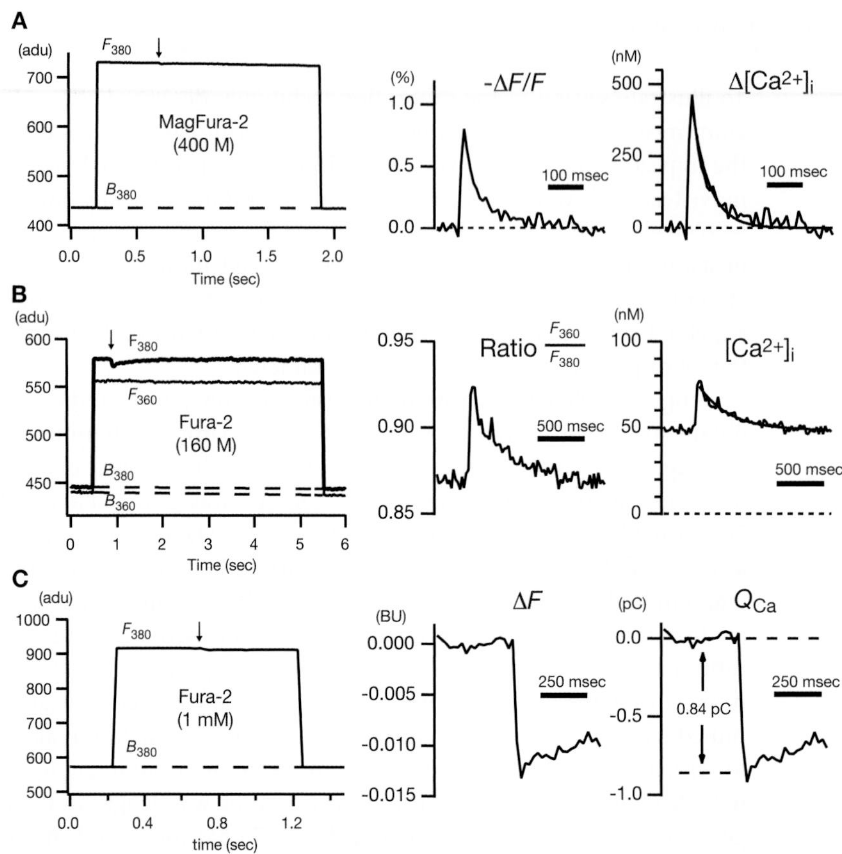

FIGURE 2. Calibration examples from measurements of single action potential-induced fluorescence changes in calyx-type presynaptic terminals in the MNTB. The fluorescence (*F*) averaged over the entire terminal was measured using a fast CCD camera (expressed in analog-to-digital units [adu]). Background fluorescence (*B*) from a nearby region was determined immediately before and after each measurement (interpolated by dashed lines). Timing of single action potentials is indicated by arrows. The temperature was 35°C. (*A*) Single-wavelength measurement using the low-affinity indicator MagFura-2. A single action potential caused a small $\Delta F/F$ change of <1%, which was converted to $\Delta[Ca^{2+}]_i$ using Equation 9. Average of 50 traces. (*B*) Ratiometric measurement using Fura-2. The ratio between the fluorescence intensities at the isosbestic wavelength (F_{360}) and at 380 nm (F_{380}) was evaluated and converted to $[Ca^{2+}]_i$ using Equation 11. Note that an estimate of $[Ca^{2+}]_{rest}$ is obtained, but that the $[Ca^{2+}]_i$ transient is profoundly reduced in amplitude and prolonged owing to the added Fura-2 Ca^{2+}-buffering capacity. No averaging. (*C*) Calcium flux measurement using Fura-2 overload. At high Fura-2 concentration a single action potential induces only a small fluorescence decrement ΔF_{380} of ~5 adu (≈1% $\Delta F/F$), which is expressed in bead units (BU) and then converted to the total calcium charge Q_{Ca} using Equation 13. Average of 20 traces. For further details, see text.

A general problem of calcium measurements using imaging systems is that they represent $[Ca^{2+}]_i$ levels averaged over sizable cytosolic volumes, in the best case, over the diffraction-limited focal volume. They are therefore blind to highly localized $[Ca^{2+}]_i$ gradients and may underestimate the actual $[Ca^{2+}]_i$ level reached at the site of action, for example, the binding to a Ca^{2+} sensor. Two approaches may overcome this problem: (1) Ca^{2+} uncaging by flash photolysis (see Chapter 60), causing spatially homogeneous $[Ca^{2+}]_i$ elevations, which can be quantified using the methods described here; and (2) the application of indicator forms that are targeted to the intracellular sites of interest (e.g., by genetic means).

REFERENCES

Agronskaia AV, Tertoolen L, Gerritsen HC. 2003. High frame rate fluorescence lifetime imaging. *J Phys D Appl Phys* **36:** 1655–1662.

Baird GS, Zacharias DA, Tsien RY. 1999. Circular permutation and receptor insertion within green fluorescent proteins. *Proc Natl Acad Sci* **96:** 11241–11246.

Bollmann JH, Helmchen F, Borst JGG, Sakmann B. 1998. Postsynaptic Ca^{2+} influx mediated by three different pathways during synaptic transmission at a calyx-type synapse. *J Neurosci* **18:** 10409–10419.

Chen TW, Lin BJ, Brunner E, Schild D. 2006. In situ background estimation in quantitative fluorescence imaging. *Biophys J* **90:** 2534–2547.

Draaijer A, Sanders R, Gerritsen HC. 1995. Fluorescence lifetime imaging, a new tool in confocal microscopy. In *Handbook of biological confocal microscopy*, 2nd ed. (ed. Pawley JB), pp. 491–505. Plenum, New York.

Eilers J, Schneggenburger R, Neher E. 1995. Patch clamp and calcium imaging in brain slices. In *Single-channel recording,* 2nd ed. (ed. Sakmann B, Neher E), pp. 213–229. Plenum, New York.

Göbel W, Helmchen F. 2007. In vivo calcium imaging of neural network function. *Physiology* **22:** 358–365.

Gordon GW, Berry G, Liang XH, Levine B, Herman B. 1998. Quantitative fluorescence resonance energy transfer measurements using fluorescence microscopy. *Biophys J* **74:** 2702–2713.

Groden DL, Guan Z, Stokes BT. 1991. Determination of Fura-2 dissociation constants following adjustment of the apparent Ca-EGTA association constant for temperature and ionic strength. *Cell Calcium* **12:** 279–287.

Grynkiewicz G, Poenie M, Tsien RY. 1985. A new generation of Ca^{2+} indicators with greatly improved fluorescence properties. *J Biol Chem* **260:** 3440–3450.

Harpur AG, Wouters F, Bastiaens PI. 2001. Imaging FRET between spectrally similar GFP molecules in single cells. *Nat Biotechnol* **19:** 167–169.

Helmchen F, Tank DW. 2011. A single-compartment model of calcium dynamics in nerve terminals and dendrites. In *Imaging in neuroscience: A laboratory manual* (ed. Helmchen F, Konnerth A). Cold Spring Harbor Laboratory Press, Cold Spring Harbor, NY (in press).

Helmchen F, Imoto K, Sakmann B. 1996. Ca^{2+} buffering and action potential-evoked Ca^{2+} signaling in dendrites of pyramidal neurons. *Biophys J* **70:** 1069–1081.

Helmchen F, Borst JGG, Sakmann B. 1997. Calcium dynamics associated with a single action potential in a CNS presynaptic terminal. *Biophys J* **72:** 1458–1471.

Hires SA, Tian L, Looger LL. 2008. Reporting neural activity with genetically encoded calcium indicators. *Brain Cell Biol* **36:** 69–86.

Kuchibhotla KV, Lattarulo CR, Hyman BT, Bacskai BJ. 2009. Synchronous hyperactivity and intercellular calcium waves in astrocytes in Alzheimer mice. *Science* **323:** 1211–1215.

Lakowicz JR, Szmacinski H, Nowaczyk K, Johnson ML. 1992. Fluorescence lifetime imaging of calcium using Quin-2. *Cell Calcium* **13:** 131–147.

Lev-Ram V, Miyakawa H, Lasser-Ross N, Ross WN. 1992. Calcium transients in cerebellar Purkinje neurons evoked by intracellular stimulation. *J Neurophysiol* **68:** 1167–1177.

Lipp P, Niggli E. 1993. Ratiometric confocal Ca^{2+}-measurements with visible wavelength indicators in isolated cardiac myocytes. *Cell Calcium* **14:** 359–372.

Mank M, Griesbeck O. 2008. Genetically encoded calcium indicators. *Chem Rev* **108:** 1550–1564.

Maravall M, Mainen ZF, Sabatini BL, Svoboda K. 2000. Estimating intracellular calcium concentrations and buffering without wavelength ratioing. *Biophys J* **78:** 2655–2667.

Miyawaki A. 2003. Visualization of the spatial and temporal dynamics of intracellular signaling. *Dev Cell* **4:** 295–305.

Miyawaki A, Llopis J, Heim R, McCaffery JM, Adams JA, Ikura M, Tsien RY. 1997. Fluorescent indicators for Ca^{2+} based on green fluorescent proteins and calmodulin. *Nature* **388:** 882–887.

Moore EDW, Becker PL, Fogarty KE, Williams DA, Fay FS. 1990. Ca^{2+} imaging in single living cells: Theoretical and practical issues. *Cell Calcium* **11:** 157–179.

Nakai J, Ohkura M, Imoto K. 2001. A high signal-to-noise Ca^{2+} probe composed of a single green fluorescent protein. *Nat Biotech* **19:** 137–141.

Neher E. 1989. Combined fura-2 and patch clamp measurements in rat peritoneal mast cells. In *Neuromuscular junction* (ed. Sellin L, et al.), pp. 65–76. Elsevier, Amsterdam.

Neher E. 1995. The use of fura-2 for estimating Ca buffers and Ca fluxes. *Neuropharmacology* **34:** 1423–1442.

Neher E, Augustine G. 1992. Calcium gradients and buffers in bovine chromaffin cells. *J Physiol* **450:** 273–301.

Oheim M, Naraghi M, Müller TH, Neher E. 1998. Two dye two wavelength excitation calcium imaging: Results from bovine adrenal chromaffin cells. *Cell Calcium* **24:** 71–84.

Palmer AE, Giacomello M, Kortemme T, Hires SA, Lev-Ram V, Baker D, Tsien RY. 2006. Ca^{2+} indicators based on computationally redesigned calmodulin-peptide pairs. *Chem Biol* **13:** 521–530.

Poenie M. 1990. Alteration of intracellular Fura-2 fluorescence by viscosity: A simple correction. *Cell Calcium* **11:** 85–91.

Schneggenburger R, Zhou Z, Konnerth A, Neher E. 1993. Fractional contribution of calcium to the cation current through glutamate receptor channels. *Neuron* **11:** 133–143.

Stosiek C, Garaschuk O, Holthoff K, Konnerth A. 2003. In vivo two-photon calcium imaging of neuronal networks. *Proc Natl Acad Sci* **100:** 7319–7324.

Szmacinski H, Lakowicz JR. 1995. Possibility of simultaneously measuring low and high calcium concentrations using Fura-2 and lifetime-based sensing. *Cell Calcium* **18:** 64–75.

Tian L, Hires SA, Looger LL. 2011. Imaging neuronal activity with genetically encoded calcium indicators. In *Imaging in neuroscience: A laboratory manual* (ed. Helmchen F, Konnerth A). Cold Spring Harbor Laboratory Press, Cold Spring Harbor, NY (in press).

Tsien RY. 1980. New calcium indicators and buffers with high selectivity against magnesium and protons: Design, synthesis, and properties of. *Biochemistry* **19:** 2396–2404.

Tsien RY. 1989. Fluorescent probes of cell signaling. *Annu Rev Neurosci* **12:** 227–253.

Tsien RY, Harootunian AT. 1990. Practical design criteria for a dynamic ratio imaging system. *Cell Calcium* **11:** 93–109.

Tsien RY, Pozzan T. 1989. Measurement of cytosolic free Ca^{2+} with quin-2. *Methods Enzymol* **172:** 256–262.

Vranesic I, Knöpfel T. 1991. Calculation of calcium dynamics from single wavelength fura-2 fluorescence recordings. *Pflügers Archiv* **418:** 184–189.

Wang SS-H, Alousi AA, Thompson SH. 1995. The lifetime of inositol 1,4,5-triphosphate in single cells. *J Gen Physiol* **105:** 149–171.

Wilms CD, Eilers J. 2007. Photo-physical properties of Ca^{2+}-indicator dyes suitable for two-photon fluorescence-lifetime recordings. *J Microsc* **225:** 209–213.

Wilms CD, Schmidt H, Eilers J. 2006. Quantitative two-photon Ca^{2+} imaging via fluorescence lifetime analysis. *Cell Calcium* **40:** 73–79.

25 Quantitative Aspects of Calcium Fluorimetry

Erwin Neher

Max Planck Institute for Biophysical Chemistry, D-37077 Göttingen, Germany

ABSTRACT

Ca^{2+} indicator dyes by necessity are Ca^{2+} chelators because it is the binding of Ca^{2+} to dye molecules that induces the change in fluorescence on which the Ca^{2+} signal is based. As chelators, once introduced into a cell, they contribute to cellular Ca^{2+} buffering. It has been a question of much debate to what extent this added Ca^{2+} buffer (exogenous Ca^{2+} buffer) changes Ca^{2+} homeostasis and the signals of interest. I discuss this problem in this chapter, emphasizing the distinction between the influence of the dyes on amplitudes (which may be not so severe) and on the dynamics of Ca^{2+} signals (which may be drastic). Once the Ca^{2+}-buffering action of dyes relative to intrinsic Ca^{2+} buffers is understood for a given preparation, Ca^{2+} dyes can be used as very versatile tools for studying both Ca^{2+} concentrations and Ca^{2+} fluxes. I describe in detail some of my own experiences in calibrating the indicator dye Fura-2. These refer exclusively to experiments in which the dye is loaded into the cell via a patch pipette because acetoxymethyl ester loading introduces problems that very often prohibit precise quantitative conclusions.

THE INFLUENCE OF CA^{2+} DYES ON CA^{2+} SIGNALS

Buffering Capacity of Ca^{2+} Dyes

Ca^{2+} indicators act as Ca^{2+} buffers because they bind Ca^{2+} in a reversible manner. In the following discussion, the term Ca^{2+} buffer is used in the strict sense to designate Ca^{2+}-binding molecules—either endogenous or added by the experimenter. For these arguments, it is important that such Ca^{2+} buffers are not confused with other mechanisms (i.e., Ca^{2+} pumps and exchangers) that sequester or pump Ca^{2+} into organelles or into other compartments. The Ca^{2+}-buffering power of such molecules is most conveniently characterized by the so-called Ca^{2+}-binding ratio, κ (Mathias et al. 1990; Neher 1995):

$$\kappa \equiv \frac{d[\mathrm{CaB}]}{d[\mathrm{Ca}^{2+}]} = \frac{B_\mathrm{T}}{K_\mathrm{D}} \times \frac{1}{(1+[\mathrm{Ca}^{2+}]/K_\mathrm{D})^2}. \quad (1)$$

Here, [CaB] is the concentration of the Ca^{2+}-bound buffer, [Ca^{2+}] is the concentration of the free Ca ions, B_T is the total concentration of the buffer, and K_D is its dissociation constant (for review, see Neher 1995, 1998). One minimum requirement for unperturbed [Ca^{2+}] measurement can readily be written down by postulating that the Ca^{2+}-binding ratio κ_B of the dye (the added Ca^{2+} buffer) should be smaller than the Ca^{2+}-binding ratio κ_s of the cytoplasm,

$$\kappa_B << \kappa_s. \tag{2}$$

Here, the cellular Ca^{2+} buffers, for simplicity, are represented by one hypothetical Ca^{2+}-buffering species, s. κ_s can be measured by a number of techniques and turns out to be in the range of 40–200 for many neuronal cell types (for review, see Neher 1995). In many cases in which it has been studied, its Ca^{2+} dependence has been found to be remarkably constant (Xu et al. 1997), indicating that endogenous Ca^{2+} buffers have low affinity (in that case, Equation 1 reduces to $\kappa = B_T/K_D$). Some types of neurons, however, express specific Ca^{2+}-binding proteins at high concentration. In these cases, κ_s may have a value of 1000 or more (Roberts 1993; Fierro and Llano 1996).

How much indicator dye is tolerable to satisfy Equation 2? Using a κ_s value in the middle of the range indicated above (e.g., 100), assuming that Ca^{2+} signals of ~100 nM are of interest, and inserting an in vivo value for the K_D of Fura-2 (see below) of 238 nM (Zhou and Neher 1993a), then according to Equation 1, $\kappa_s = \kappa_B$ for a total concentration of dye B_T of 48 μM. Thus, Equation 2 is quite restrictive, given the fact that many imaging studies use dye concentrations well above 100 μM. It is, however, borne out by experiments on several types of mammalian central neurons in which it was shown that dye concentrations in the range of 50–100 μM severely distort peak values and time courses of Ca^{2+} signals (Helmchen et al. 1996, 1997). Higher dye concentrations are tolerable for low-affinity dyes, such as Calcium Green-5N, for which, according to Equation 1, the Ca^{2+}-binding ratio is smaller.

Effects of Exogenous Ca^{2+} Buffers on Ca^{2+} Dynamics

In many cases, however, there is another aspect that may impose restrictions much more severe than those of Equation 2. It turns out that most of the endogenous Ca^{2+} buffer in cells with relatively low κ_s is immobile, probably being fixed to organelles, membranes, or cytoskeletal elements (Zhou and Neher 1993a). This means that endogenous buffer in the unperturbed cell retards diffusion of Ca^{2+}, localizing Ca^{2+} changes and slowing down Ca^{2+} redistribution within a cell. Adding a mobile Ca^{2+} buffer, such as a dye, in this situation, dramatically speeds up Ca^{2+} diffusion because the mobile dye competes with the fixed endogenous buffer for Ca^{2+} and will shuttle Ca^{2+} from subcellular regions of high concentration to those of low concentration. Ca^{2+} transport in the presence of both fixed and mobile buffers can be described by an effective diffusion coefficient D_{app} of the form (Wagner and Keizer 1994; Gabso et al. 1997):

$$D_{app} = \left(D_{Ca} + \sum_m \kappa_m D_m\right)\left(1 + \sum_m \kappa_m + \sum_f \kappa_f\right)^{-1}, \tag{3}$$

where κ_m and D_m represent Ca^{2+}-binding ratios and diffusion coefficients of mobile buffer species and κ_f represents Ca^{2+}-binding ratios of fixed buffer species. Unfortunately, κ_m values for some cell types are so small that they cannot be measured. Zhou and Neher (1993a) argued that for adrenal chromaffin cells, the value might be within $2 < \kappa_m < 7$, which, together with reasonable estimates for the other parameters of Equation 3, would give values for D_{app} between 2×10^{-7} and 4×10^{-7} cm^2 sec^{-1}. Adding 50-μM Fura-2 to this background would change D_{app} threefold to fivefold. Fura-2 concentrations less than 5 μM would be required to maintain D_{app} close to its value in an unperturbed cell. A similar conclusion was reached by Gabso et al. (1997), who measured D_{app} in axons of cultured *Aplysia* cells. One must conclude from these considerations that the dynamics of Ca^{2+} signals, inasmuch as they are determined by Ca^{2+} diffusion, are not faithfully represented by Ca^{2+}-imaging studies, basically because no studies are performed at dye concentrations as low as required. The situation is relaxed, however, in cell types that express mobile Ca^{2+}-binding proteins, such as cochlear hair cells (Roberts 1993), cerebellar Purkinje cells (Fierro and Llano 1996), or muscle cells. When studying developmental or activity-induced changes in the expression of Ca^{2+}-binding proteins (for review, see Heizmann and Braun 1995), investigators should be aware that such changes will manifest themselves in Ca^{2+} dynamics only at extremely low dye concentrations (Gabso et al. 1997).

INDICATOR DYES AS TOOLS TO MEASURE CA^{2+} FLUXES

It has been shown that in many cell types, extremely low concentrations of indicator dyes are required, if amplitudes and time courses of Ca^{2+} signals are to be recorded faithfully (see above). The reason is that dyes compete with endogenous Ca^{2+} buffers for Ca^{2+} and that even at concentrations below 100 µM, dyes are quite successful in this competition. This situation can be turned into an advantage if Ca^{2+} fluxes rather than Ca^{2+}-concentration signals are of interest in a research project. Once the level of endogenous Ca^{2+} buffers is known in a given cell type (e.g., let it be equivalent to 50 µM of Fura-2), a moderate increase in dye concentration (such as to 500 µM) will ensure that the majority of Ca^{2+} entering a cell during a short stimulus will bind to the dye. The fluorescence change measured at a Ca^{2+}-sensitive wavelength (such as at 380 nm for Fura-2 or as measured with Calcium Green) will then be directly proportional to the amount of Ca^{2+} entering during the stimulus. It is then relatively straightforward to calibrate a fluorescence setup for Ca^{2+} flux and to measure the contribution of Ca^{2+}-inward current in nonspecific cationic currents (for review, see Neher 1995) or to estimate the amount of Ca^{2+} released from intracellular stores (Ganitkevich 1996).

CALIBRATING A FURA-2 SETUP FOR QUANTITATIVE FLUORIMETRY

Many of the aspects discussed above—particularly estimating endogenous Ca^{2+} buffer strength from the changes in Ca^{2+} signals induced by increasing dye concentration—require a very careful calibration of the fluorescence setup. A variety of procedures has been described previously for such calibrations (Grynkiewicz et al. 1985; Poenie 1990; Williams and Fay 1990). Incomplete hydrolysis of ester-loaded dye or compartmentalization of dye, however, very often prevents a precise determination of calibration constants (Tsien 1989; see Appendix in Zhou and Neher 1993a). Therefore, the following descriptions refer exclusively to experiments in which the indicator dye is loaded into the cell by diffusion from a patch pipette.

For such types of measurements an in vivo calibration, such as described by Almers and Neher (1985), is appropriate. Thereby, the fluorescence ratio is determined in whole-cell recordings, and three different pipette solutions are used, which set $[Ca^{2+}]$ to one of three values: zero (by including an excess of free Ca^{2+} buffer), very high (by including excess Ca^{2+}), and intermediate (by including appropriate amounts of Ca^{2+}-bound and Ca^{2+}-free buffer). From these three ratios, the calibration constants of the equation by Grynkiewicz et al. (1985) (see Equation 4, below) can be calculated. It is advisable to perform at least three to five measurements under each condition such that the procedure is quite laborious. Additionally, recalibration is necessary from time to time because of aging (or after replacement) of the light source or when new batches of dyes are introduced. Thus, calibration may use up a substantial portion of measuring time on a given setup. It is my experience that the calibration procedure is safer and, on the whole, less time-consuming if in addition to the three types of measurement mentioned above, two additional measurements are being performed during a calibration. These, together with the conventional calibration, are described below, following discussion of some of the intrinsic problems of calibration. For further guidelines on calibrating calcium indicators, see Chapter 24.

Problems Encountered in Ca^{2+}-Indicator Calibration

The Dissociation Constant of the Calibrating Buffer and Its pH Dependence

The accuracy of the calibration ultimately depends on the correctness of the dissociation constant of the Ca^{2+} buffer used in the preparation of the calibrating solutions. Values for these parameters can be found in tables (see, e.g., Martell and Smith 1977) for standard conditions (usually 0.1 ionic strength and 20°C or 25°C). Corrections have to be applied to these values to account for ionic strength and temperature of the experiment. An additional complication arises from the fact that the apparent K_D of ethylene glycol tetra-acetic acid (EGTA), the most commonly used calibration buffer, is strongly pH dependent (Miller and Smith 1984). A pH error of 0.1 unit will cause an error in the apparent K_D close to 50%. This is a serious problem because the pH changes when Ca^{2+} binds to EGTA (and replaces H^+) and arises because calibration solutions are usually made in small quantities, which are difficult to titrate correctly. In any case, the pH of the calibration solution should be measured immediately before or after a calibration. Above all, it is advisable to use 1,2-bis-(2-aminophenoxy)ethane-*N,N,N′,N′*-tetra-acetic acid (BAPTA) instead of EGTA for calibration purposes because BAPTA is much less sensitive to pH (above 7.0) than is EGTA. Tsien (1980) gives a K_D for BAPTA of 107 nM, as measured in 100 mM KCl at 22°C. However, this value is quite sensitive to ionic strength and temperature (Harrison and Bers 1987). Under conditions typical for pipette-filling solutions (~150 mM total salt concentration, pH 7.2, 1 mM free Mg^{++}, 22°C) a value of ~225 nM is calculated from the parameters given by Harrison and Bers (1987). Oheim (1995) measured a value of 221 nM by potentiometric titration in a solution that was formulated to allow measurement of Ca^{2+} currents in the whole-cell configuration (120 mM CsCl, 20 mM HEPES buffer, 2 mM BAPTA, 5 mM NaCl, 1 mM $MgCl_2$, pH 7.2). Thus, a value of 225 nM (Zhou and Neher 1993a) seems to be quite appropriate for an in vivo condition. It should be kept in mind that this value is defined in terms of concentration of Ca^{2+} (not activity) and that the concentration of calcium in the cytosol may be quite different from that in the pipette during the calibration measurement, even if the pipette solution is at equilibrium with the cytosol (for a discussion of this problem, see Neher 1995).

Autofluorescence

For typical measurements in adrenal chromaffin cells using 50–100 μM Fura-2, autofluorescence of the cell may be as large as 5%–10% of the total fluorescence after dye loading. This value may be even larger in brain-slice preparations. Autofluorescence can be partially accounted for if its values at the wavelengths used are measured during the cell-attached configuration (this then also corrects for some fluorescence originating from the pipette!) and subtracted from corresponding values after loading. However, two problems arise with such a subtraction.

- Shortly after obtaining the cell-attached configuration, any Fura-2-containing solution that spilled from the pipette during the approach may still be present in the measuring field. Care has to be taken either to wash away such solution by bath perfusion or to allow enough time for it to diffuse away.
- Some of the autofluorescence is mobile, probably representing NADH, and will be lost by washout during whole-cell recording. Both total fluorescence and mobile contribution vary from cell to cell, and more so, from cell type to cell type, depending also on the metabolic state of the cells (Chance et al. 1965). Thus, the procedure to subtract all autofluorescence overcompensates, leading to errors of variable degree in the calibration constants. This is particularly serious when the value R_{max} is determined at high $[Ca^{2+}]$ because Fura-2 fluorescence at long wavelength (385 nm) is very small under these conditions. For very accurate measurements at high $[Ca^{2+}]$, the mobile part of autofluorescence should be measured separately and accounted for. For normal work, it may be sufficient to perform calibration and test measurements under conditions that are as similar as possible, such that these errors cancel. Nevertheless, it is advisable to determine, in a few test measurements, the order of magnitude of autofluorescence, its variability, and its time course during a whole-cell recording without dye, to be able to estimate the order of magnitude of the problem. It has also been found that low autofluorescence (in the excitation range of 360 to 380 nm) is an indicator of healthy cells, and preselecting cells with low autofluorescence is quite helpful for obtaining stable whole-cell patch recordings.

Slow Changes in Fluorescence Properties

The fluorescence properties of cytosol change slowly during prolonged recordings (5–30 min). This may be because of cell swelling (Zhou and Neher 1993b) or because of washout of cellular constituents, with the result that with time the cytosol becomes more similar to the pipette-filling solution. Strictly speaking, a set of calibration parameters is only valid within a certain time window following the establishment of a whole-cell recording. Again, care should be taken to perform critical measurements as similar as possible to the corresponding calibration measurements, that is, at about the same time after break-in (i.e., entry of the pipette into the cell).

Calibration Procedure

The relationship between the fluorescence ratio R and $[Ca^{2+}]$ was given by Grynkiewicz et al. (1985) as

$$[Ca^{2+}] = K_{eff}(R - R_0)/(R_1 - R), \tag{4}$$

where K_{eff}, R_0, and R_1 are three calibration constants that have to be determined for a given setup (see Protocol for details). The three constants are related to K_D, the dissociation constant of the indicator dye, and to a fourth constant, the isocoefficient α (Zhou and Neher 1993a), through the equation

$$K_D = K_{eff}\,(R_0 + \alpha)/(R_1 + \alpha). \tag{5}$$

Therefore, Equation 4 can be written as

$$[Ca^{2+}] = K_D \frac{(R_1 + \alpha)}{(R_0 + \alpha)} \times \frac{(R - R_0)}{(R_1 - R)}, \tag{6}$$

Once K_D is known for a given dye and a given cell type, it is convenient to base a calibration on the measurement of R_0, R_1, and α, which are relatively easy to determine (see below). Alternatively, Equation 5 may be used to calculate K_D from R_0, R_1, K_{eff}, and α. Because K_D is a property of the indicator dye, which should not depend on the specific fluorimeter, such a calculation is a good check for the validity of the procedures used. K_D of Fura-2 was found to be 238 nM in adrenal chromaffin cells (Zhou and Neher 1993a) in calibrations based on BAPTA, when the K_D of BAPTA was assumed to be 225 nM. This value is close to the value of 259 nM determined by titration of Fura-2 in a standard intracellular saline (Oheim 1995). However, this result is quite different from studies on skeletal muscle cells (Konishi et al. 1988) in which the K_D of Fura-2 was found to be higher by a factor of 3–4 in the presence of myoplasmic proteins as compared with the value in simple saline. Two explanations can be given for these discrepancies. First of all, problems of interaction of Fura-2 with intracellular constituents seem to be much more severe in muscle cells than in other cell types. For instance, diffusion of Fura-2 is retarded much less in neuronal axons (Gabso et al. 1997) than in muscle (Baylor and Hollingworth 1988). Second, the calibration buffer used by Zhou and Neher (1993a) is based on BAPTA, a compound much more similar to Fura-2 than EGTA, as used in the studies on muscle. Therefore, the K_D of Fura-2 in cytosol is expected to be quite close to the K_D of BAPTA in cytosol, although both values may be quite different from those in the calibration buffer. Given our calibration procedure, which interprets all concentration values as those in the calibration buffer (at equilibrium with cytosol), the resulting cytosolic K_D of Fura-2 is expected to be also quite similar to the one in the calibration buffer.

As argued above, calibration measurements should be performed in a manner as similar as possible to the corresponding test measurements. The fluorescence of a region of interest (such as an entire cell) should be measured at two appropriate wavelengths after loading cells with the same solution as is used during experiments, except that $[Ca^{2+}]$ is fixed to one of three values (see Protocol) by adding excess buffer or excess $[Ca^{2+}]$. The fluorescence values should be corrected for background, and the fluorescence ratios should be calculated. In all cases, solutions should be titrated to pH 7.2 immediately before use. More details on the procedures used in my laboratory can be found in Neher (1989) and Zhou and Neher (1993a).

Auxiliary Measurements

It has been argued above that it is advisable to complement the calibration by two more auxiliary measurements, which provide checks for consistency. They also can be used to monitor instrumental changes. These are (1) measurement of the isocoefficient and (2) measurement of the fluorescence ratio of a fluorescence standard, such as fluorescent beads.

The isocoefficient: If $F_1(t)$ and $F_2(t)$ denote the background-corrected fluorescence signals from a ratiometric dye measured at wavelengths 1 and 2, respectively, it is possible to find a coefficient α, the so-called isocoefficient, for which the linear combination $F(t)$,

$$F(t) = F_1(t) + \alpha F_2(t), \qquad (7)$$

is independent of $[Ca^{2+}]$ (Zhou and Neher 1993a; Naraghi et al. 1998). Once α has been determined for the pair of wavelengths used, it is possible to calculate a Ca^{2+}-independent fluorescence signal (proportional to the total concentration of dye), even if neither of the two wavelengths is exactly at the isosbestic point. Such a signal is useful for monitoring dye loading and for calculating Ca^{2+} signals from single-wavelength measurements in cases when, during a certain time interval, the total dye concentration and its spatial distribution can be considered as constant (Naraghi et al. 1998). The isocoefficient is readily determined during any standard experiment in which a brief Ca^{2+} excursion can be induced such as by a short depolarization. Such an excursion will show up both in $F_1(t)$ and in $F_2(t)$, and it is straightforward to display on a computer screen $F(t)$ according to Equation 7 with a range of values for the parameter α. The correct isocoefficient can then be chosen as that value of α that minimizes the excursion. A more complicated procedure, appropriate for Ca^{2+} imaging, is given by Naraghi et al. (1998).

Once the isocoefficient has been determined, it is advisable to calculate the K_D of the dye, according to Equation 5, and to verify that this value (which is a property of the dye) is reasonable. As mentioned above, the $K_{D,fura}$ was found to be 238 nM by Zhou and Neher (1993a) in adrenal chromaffin cells. Once this value has been determined for a batch of dye and for a given cell type, it is more convenient to calculate $[Ca^{2+}]$ according to Equation 6, which does not require the knowledge of K_{eff}.

The bead ratio: The most common cause for changes in calibration parameters on a given setup is aging of the lamp. This can be monitored by measuring the fluorescence ratio, R_B, of a suitable fluorescence standard. Fluoresbrite B/B (bright blue) beads (Polysciences, Inc.) are quite useful for this purpose, although their fluorescence properties depend on the ionic composition of the medium (Hoth 1995). If this bead ratio, as measured at the time of calibration, is designated, and a slightly different ratio is measured at some later time, or after a lamp change, then the calibration can be readily updated by multiplying both R_0 and R_1 by the factor R_B^t/R_B^C. K_{eff} should be updated according to Equation 5, based on the knowledge of K_D (as calculated from the original calibration) and the new values of R_0, R_1, and α (the latter to be measured under the new conditions). R_B^C and R_B^t should be measured in the same Ringer-type solution. It is advisable to measure the bead ratio repeatedly, at least once every week of experimentation, and to compare both its value and its fluorescence values at the individual wavelengths with corresponding values during the last calibration. This comparison is particularly important following lamp changes and other major changes on a given setup. Unexpected variations will alert the experimenter to misalignments of the optical path, to deterioration of the light source or the detection device, and to many other problems, which readily can invalidate weeks of experimentation.

Protocol

Measurement of Calibration Constants for Quantitative Calcium Fluorimetry

A careful calibration of the fluorescence setup is essential when using Fura-2 for quantitative fluorimetry. The calibration procedure described here is used for experiments in which the indicator dye is introduced into the cell through a patch-clamp pipette. The fluorescence ratio is determined in whole-cell recordings for three different calibration parameters that reflect different [Ca^{2+}]: very high, intermediate, and zero.

MATERIALS

See the end of the chapter for recipes for reagents marked with <R>.

Reagents

Cells of interest grown in culture
K_{eff} measuring solution <R>
R_0 measuring solution <R>
R_1 measuring solution <R>
Standard internal solution <R>

Equipment

Equipment to perform whole-cell recording
Fluorimeter

EXPERIMENTAL METHOD

Measuring R_0

The measured fluorescence ratio can be used directly as R_0 of Equations 4 and 6.

1. Establish a whole-cell recording, taking care that the access resistance (the resistance between measuring pipette and cell) is <20 MΩ.
2. Measure the limiting fluorescence ratio (derived from two wavelengths, 1 and 2) at low [Ca^{2+}] using a standard internal solution containing 0.1 mM Fura-2 and 10 mM BAPTA-tetrapotassium salt.

   Be sure that the fluorescence ratio is measured at a time when the fluorescence readings at both wavelengths have reached a steady state (which, in adrenal chromaffin cells, is after ~3–4 min).

Measuring R_1

The fluorescence ratio under these conditions is the quantity R_1 of Equations 4 and 6.

3. Measure the limiting fluorescence at high [Ca^{2+}] in the same way as for R_0, except use 10 mM $CaCl_2$ instead of BAPTA.

   Pay attention to the series resistance because pipettes tend to seal off at high [Ca^{2+}], and powerful Ca^{2+}-sequestering mechanisms are at work to reduce cytosolic [Ca^{2+}] (for a discussion of the problem of diffusion between a pipette and the cytosol, see Pusch and Neher 1988; Matthias et al. 1990).

4. Provoke Ca^{2+} influx (e.g., by depolarization in the case of cells expressing Ca^{2+} channels) or Ca^{2+} release from stores to verify that cytosolic [Ca^{2+}] has reached its maximum possible value.

Measuring K_{eff}

Once R_0 and R_1 have been determined, obtain a further measurement at intermediate [Ca^{2+}] to determine K_{eff} from Equation 4. Supplementing the standard internal solution with 0.1 mM Fura-2, 6.6 mM Ca-BAPTA, and 3.3 mM BAPTA-tetrapotassium salt should result in a free [Ca^{2+}] of $2K_{D,BAPTA}$ or 450 nM (see above).

5. Measure the cellular fluorescence ratio R_2, and then calculate K_{eff} (in nanomoles) by rewriting Equation 4 as

$$K_{eff} = 450 \times (R_1 - R_2)/(R_2 - R_0). \quad (8)$$

RECIPES

CAUTION: See Appendix 6 for proper handling of materials marked with <!>.
Recipes for reagents marked with <R> are included in this list.

K_{eff} Measuring Solution

Supplement standard internal solution with 0.1 mM Fura-2, 6.6 mM Ca-BAPTA <!>, and 3.3 mM BAPTA-tetrapotassium salt <!>.

R_0 Measuring Solution

Supplement standard internal solution with 0.1 mM Fura-2 and 10 mM BAPTA-tetrapotassium salt <!> (Sigma-Aldrich).

R_1 Measuring Solution

Supplement standard internal solution with 10 mM $CaCl_2$ <!>.

Standard Internal Solution

Cesium glutamate	145 mM
NaCl	8 mM
$MgCl_2$	1 mM
Mg-ATP	2 mM
HEPES buffer	10 mM
Guanosine triphosphate	0.3 mM

ACKNOWLEDGMENTS

Work in my laboratory related to the subject of this review is supported by the Behrens-Weise-Stiftung and by the Deutsche Forschungsgemeinschaft (SFB 406 and SFB 523), by a grant from the European Community (No. CHRX-CT94-0500), and by the Human Science Frontier Program (RG-4/95B). I thank my colleagues Ralf Schneggenburger and Uri Ashery for helpful suggestions concerning this chapter.

REFERENCES

Almers W, Neher E. 1985. The Ca signal from fura-2 loaded mast cells depends strongly on the method of dye-loading. *FEBS Lett* **192:** 13–18.

Baylor SM, Hollingworth S. 1988. Fura-2 calcium transients in frog skeletal muscle fibres. *J Physiol* **403:** 151–192.

Chance B, Williamson JR, Jamieson D, Schoener B. 1965. Properties and kinetics of reduced pyridine nucleotide fluorescence of the isolated and in vivo rat heart. *Biochem Z* **341:** 357–377.

Fierro L, Llano I. 1996. High endogenous calcium buffering in Purkinje cells from rat cerebellar slices. *J Physiol* **496:** 617–625.

Gabso M, Neher E, Spira M. 1997. Low mobility of the Ca^{2+} buffers in axons of cultured *Aplysia* neurons. *Neuron* **18:** 473–481.

Ganitkevich VY. 1996. The amount of acetylcholine mobilisable Ca^{2+} in single smooth muscle cells measured with the exogenous cytoplasmic Ca^{2+} buffer, Indo-1. *Cell Calcium* **20:** 483–492.

Grynkiewicz G, Poenie M, Tsien RY. 1985. A new generation of Ca^{2+} indicators with greatly improved fluorescence properties. *J Biol Chem* **260:** 3440–3450.

Harrison SM, Bers DM. 1987. The effect of temperature and ionic strength on the apparent Ca-affinity of EGTA and the analogous Ca-chelators BAPTA and dibromo-BAPTA. *Biochim Biophys Acta* **925:** 133–143.

Heizmann CW, Braun K. 1995. *Calcium regulation by calcium-binding proteins in neurodegenerative disorders.* Springer-Verlag, Heidelberg.

Helmchen F, Imoto K, Sakmann B. 1996. Ca^{2+} buffering and action potential-evoked Ca^{2+} signaling in dendrites of pyramidal neurons. *Biophys J* **70:** 1069–1081.

Helmchen F, Borst GG, Sakmann B. 1997. Calcium dynamics associated with a single action potential in a CNS presynaptic terminal. *Biophys J* **72:** 1458–1471.

Hoth M. 1995. Calcium and barium permeation through calcium release-activated calcium (CRAC) channels. *Pflueg Arch Eur J Physiol* **430:** 315–322.

Konishi M, Olson A, Hollingworth S, Baylor SM. 1988. Myoplasmic binding of fura-2 investigated by steady-state fluorescence and absorbance measurements. *Biophys J* **54:** 1089–1104.

Martell AE, Smith RM. 1977. *Critical stability constants.* Plenum, New York.

Mathias RT, Cohen IS, Oliva C. 1990. Limitations of the whole cell patch clamp technique in the control of intracellular concentrations. *Biophys J* **58:** 759–770.

Miller DJ, Smith GL. 1984. EGTA purity and the buffering of calcium ions in physiological solutions. *Am J Physiol* **246:** C160–C166.

Naraghi M, Müller TH, Neher E. 1998. Two-dimensional determination of the cellular Ca^{2+} binding in bovine chromaffin cells. *Biophys J* **75:** 1635–1647.

Neher E. 1989. Combined fura-2 and patch clamp measurements in rat peritoneal mast cells. In *Neuromuscular junction* (ed. LC Sellin et al.), pp. 65–76. Elsevier, Amsterdam.

Neher E. 1995. The use of fura-2 for estimating Ca buffers and Ca fluxes. *Neuropharmacology* **34:** 1423–1442.

Neher E. 1998. Usefulness and limitations of linear approximations to the understanding of Ca^{2+} signals. *Cell Calcium* **24:** 345–357.

Oheim M. 1995. "Methodische Voraussetzungen zur Untersuchung der Calcium Diffusion und Calcium Pufferung im Cytosol lebender Chromaffinzellen." Diploma thesis, University of Göttingen, Germany.

Poenie M. 1990. Alteration of intracellular fura-2 fluorescence by viscosity: Simple correction. *Cell Calcium* **11:** 85–91.

Pusch M, Neher E. 1988. Rates of diffusional exchanges between small cells and a measuring patch pipette. *Pflueg Arch Eur J Physiol* **411:** 204–211.

Roberts WM. 1993. Spatial calcium buffering in saccular hair cells. *Nature* **363:** 74–76.

Tsien R. 1980. New calcium indicators and buffers with high selectivity against magnesium and protons: Design, synthesis, and properties of. *Biochemistry* **19:** 2396–2404.

Tsien R. 1989. Fluorescent probes of cell signaling. *Annu Rev Neurosci* **12:** 227–253.

Wagner J, Keizer J. 1994. Effects of rapid buffers on Ca^{2+} diffusion and Ca^{2+} oscillations. *Biophys J* **67:** 447–456.

Williams D, Fay F. 1990. Intracellular calibration of the fluorescent calcium indicator fura-2. *Cell Calcium* **11:** 75–83.

Xu T, Naraghi M, Kang H, Neher E. 1997. Kinetic studies of Ca^{2+} binding and Ca^{2+} clearance in the cytosol of adrenal chromaffin cells. *Biophys J* **73:** 532–545.

Zhou Z, Neher E. 1993a. Mobile and immobile calcium buffers in bovine adrenal chromaffin cells. *J Physiol* **469:** 245–273.

Zhou Z, Neher E. 1993b. Calcium permeability of nicotinic acetylcholine receptor channels in bovine adrenal chromaffin cells. *Pflueg Arch Eur J Physiol* **425:** 511–517.

26 Genetic Calcium Indicators

Fast Measurements Using Yellow Cameleons

Atsushi Miyawaki,[1] Takeharu Nagai,[2] and Hideaki Mizuno[1]

[1]*Laboratory for Cell Function and Dynamics, Advanced Technology Development Group, Brain Science Institute, RIKEN, 2-1 Hirosawa, Wako-city, Saitama, Japan;* [2]*Laboratory for Nanosystems Physiology, Research Institute for Electronic Science, Hokkaido University, Sapporo, Hokkaido, Japan*

ABSTRACT

Green fluorescent protein (GFP)-based fluorescent indicators for Ca^{2+} offer significant promise for monitoring Ca^{2+} in previously unexplored organisms, tissues, and submicroscopic environments because they are genetically encoded, function without cofactors, can be targeted to any intracellular location, and are bright enough for single-cell imaging. These probes use simple GFP variants, circularly permuted green fluorescent protein (cpGFP), in which the amino and carboxyl portions have been interchanged and reconnected by short spacers between the original termini, or pairs of GFP variants that permit fluorescence resonance energy transfer (FRET). Yellow cameleons (YCs) use FRET between cyan-emitting and yellow-emitting variants of *Aequorea* GFP (cyan fluorescent protein [CFP] and yellow fluorescent protein [YFP], respectively). YCs are composed of a linear combination of CFP, calmodulin (CaM), a glycylglycine linker, the CaM-binding peptide of myosin light-chain kinase (M13), and YFP. Binding of Ca^{2+} to the CaM moiety of the YC initiates an intramolecular interaction between the CaM and the M13 domains, causing the chimeric protein to shift from an extended conformation to a more compact one, thereby increasing the efficiency of FRET from CFP to YFP. This technique is amenable to emission ratioing, which is more quantitative than single-wavelength monitoring. This chapter describes how best to use YCs for fast $[Ca^{2+}]_i$ (intracellular free Ca^{2+} concentration) imaging using laser-scanning confocal microscopy. Further discussion of genetic calcium indicators can be found in Helmchen and Konnerth (2011).

INTRODUCTION

Since the prototype cameleon was released in 1997 (Miyawaki et al. 1997), several improvements have been made so that the cameleon (1) can be shifted to longer wavelengths, (2) is less sensitive to acidic pH, (3) shows more efficient maturity in mammalian cells at 37°C, and (4) shows a significantly wider dynamic range. First, whereas the original version has blue and green mutants of GFP as donor and acceptor, respectively, CFP and YFP have been substituted to make YCs. Subsequently, red cameleons have been constructed by incorporating RFP (red fluorescent protein) (DsRed) as an acceptor (Mizuno et al. 2001). Second, the original version of the YC (YC2.0) has high pH sensitivity because its acceptor, enhanced yellow fluorescent protein (eYFP), is quenched by acidification with a pK_a of 7. The pH sensitivity of YCs has been markedly reduced by introducing the V68L and Q69K mutations into eYFP (eYFP.1) (Miyawaki et al. 1999). The improved YCs, including YC2.1 and YC3.1, permit Ca^{2+} measurement without perturbation by pH changes between pH 6.5 and 8.0. Third, two bright versions of YFP, citrine (Griesbeck et al. 2001) and Venus (Nagai et al. 2002), which mature efficiently at 37°C, have been developed. The rapid maturation of Venus in YC2.12 or YC3.12, for example, allows the immediate detection of $[Ca^{2+}]_i$ transients after gene introduction in freshly prepared brain slices. Fourth, expansion of the Ca^{2+} responses of YCs has been achieved by combining FRET and circular-permutation techniques. To achieve a large Ca^{2+}-dependent change in the relative orientation and distance between the fluorophores of CFP and YFP, Nagai et al. (2004) have used cpYFPs, in which the amino and carboxyl portions were interchanged and were reconnected by a short spacer between the original termini (Baird et al. 1999). Circular permutation was conducted on Venus, and new termini were introduced into surface-exposed loop regions of the β-can. One of the variants, cp173Venus, was given a new amino terminus at Asp-173, which is far removed at the opposite end of the β-can from the original amino terminus, Met1 (Fig. 1A). YC3.60 was generated by replacing Venus in YC3.12 with cp173Venus (Fig. 1B). Compared with YC3.12, YC3.60 is equally bright but shows five- to sixfold larger dynamic range (Fig. 1C). Thus, YC3.60 gives a greatly enhanced signal-to-noise ratio (SNR), thereby enabling Ca^{2+} imaging experiments that were not possible with conventional YCs. In this way, YCs have been improved mainly by optimizing the YFP component.

It has been pointed out that the dynamic range of YCs is damped down in neuronal cell types that have a large amount of CaM and CaM-associated proteins. Thus, the interface between the CaM and the M13 peptide has been redesigned to generate highly specific protein/peptide pairs that are not perturbed by endogenous CaM or CaM-binding proteins (Palmer et al. 2004). A mutant CaM/peptide pair, D1, was cloned between CFP and YFP to yield a reengineered YC. The D1-containing YC is indifferent to a large excess of CaM. Moreover, it displays a low-Ca^{2+} affinity with a K_d value of 60 μM and has proven to be an ideal probe for measuring the Ca^{2+} concentration in the endoplasmic reticulum. The binding interface of CaM and M13 was further reengineered by computationally designing complementary bumps and holes (Palmer et al. 2006). Another attempt to eliminate the sensitivity of genetically encoded Ca^{2+} indicators to endogenous CaM and CaM-binding proteins involved the use of troponin C proteins from skeletal and cardiac muscles (Heim and Griesbeck 2004). Because troponin C is specifically expressed in muscle, these novel probes should not be perturbed by endogenous proteins when expressed in nonmuscle tissues. Generation of transgenic mouse lines expressing variants of the troponin C-based Ca^{2+} indicators, such as CerTN-L15 (Heim et al. 2007) and TN-XXL (Mank et al. 2008), allowed the investigators to perform reliable in vivo two-photon Ca^{2+} imaging of neurons.

Because of their emission of ratiometric responses, YCs are the best choice for observing Ca^{2+} dynamics in motile animals. Calcium transients in neurons have been observed in the process of sensation of gentle touch (Suzuki et al. 2003) or temperature (Clark et al. 2007) of *Caenorhabditis elegans* and during escape behaviors of zebrafish (Higashijima et al. 2003). The ratiometric approach using YCs is also of great advantage for quantitative Ca^{2+} measurements. A version of YC (Synapcam) was localized to the postsynaptic terminals at the *Drosophila* larval neuromuscular junction to selectively and quantitatively measure Ca^{2+} influx through glutamate receptors with single-impulse and single-bouton resolutions (Guerrero et al. 2005). The Ca^{2+} influx signals (i.e., the transmission strength) varied along the axonal branches with a gradient of weak, at the proximal boutons, to strong, at the distal ones. More recently, quantitative in vivo imaging using YC3.60 revealed elevated $[Ca^{2+}]_i$ in neurites near Aβ plaques (Kuchibhotla et al. 2008).

Although FRET-based indicators for Ca^{2+} have been developed, our understanding of the structure-photochemistry relationships of GFP has enabled the development of Ca^{2+} probes based on a single GFP variant. cpGFPs, in which the amino and carboxyl portions had been interchanged

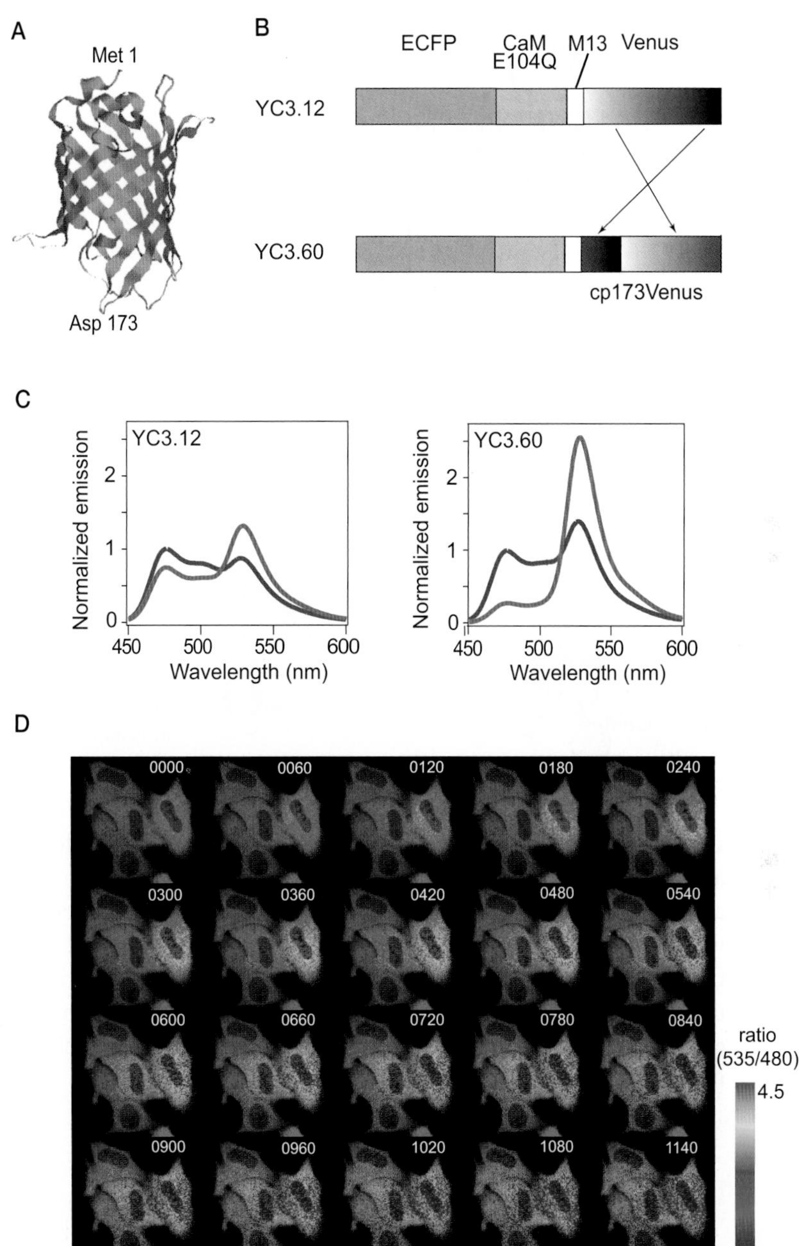

FIGURE 1. Development and performance of YC3.60. (*A*) The three-dimensional structure of GFP with the positions of the original (Met1) and the new (Asp173) amino termini indicated. (*B*) Domain structures of YC3.12 and YC3.60. (CaM) *Xenopus* calmodulin; (E104Q) mutation of the conserved bidentate glutamate (E104) at position 12 of the third Ca^{2+}-binding loop to glutamine. (*C*) Emission spectra of YC3.12 (*left*) and YC3.60 (*right*) (excitation at 435 nm) at zero (blue line) and saturated Ca^{2+} (red line). (*D*) A series of confocal pseudocolored ratio images showing propagation of $[Ca^{2+}]_c$ in HeLa cells that expressed YC3.60.

around position 145 and reconnected by short spacers between the original termini (Baird et al. 1999), were fused to CaM and its target peptide, M13. Chimeric proteins G-CaMP (Nakai et al. 2001) and pericam (Nagai et al. 2001) are fluorescent, and their spectral properties change reversibly with Ca^{2+} concentration, probably caused by the interaction between CaM and M13, leading to an alteration of the environment surrounding the chromophore. G-CaMP is a single-wavelength intensity-modulating probe for Ca^{2+}, which is now widely used. The probe has been iteratively improved in terms of brightness, response linearity, photostability, and calcium affinity. The most updated ver-

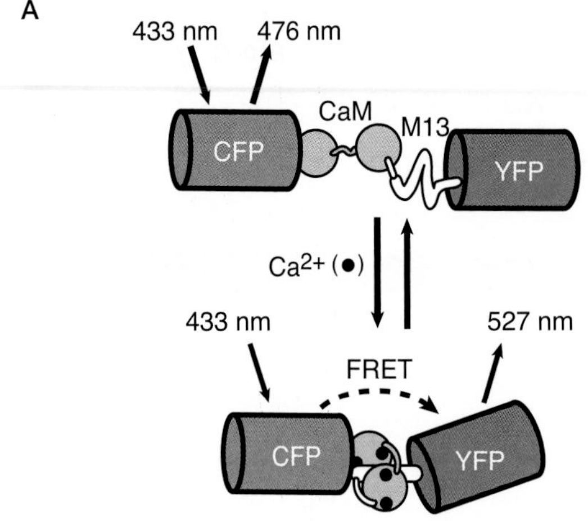

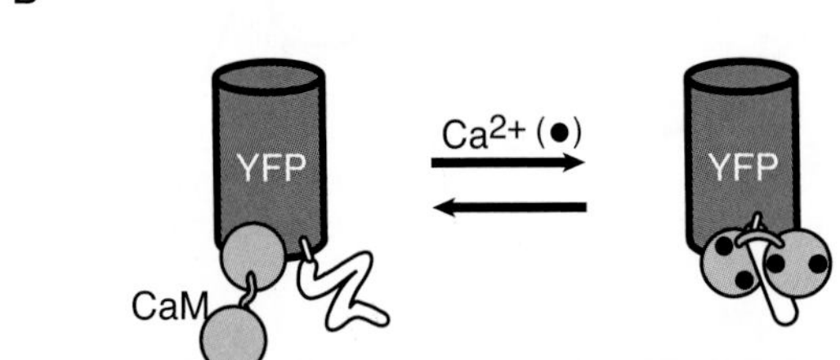

FIGURE 2. Schematics of two genetically encodable indicators for Ca^{2+}. (*A*) Yellow cameleon. (*B*) GCaMP or pericam. (Modified, with permission, from Tsien 2000.)

sion (G-CaMP3) allows for imaging neural activity in multiple model animals, such as worms, flies, and mice (Tian et al. 2009). Three types of pericam have been obtained by mutating several amino acids adjacent to the chromophore. Of these, flash pericam becomes bright with Ca^{2+}-like G-CaMP, whereas inverse pericam dims. On the other hand, ratiometric pericam has an excitation wavelength that changes in a Ca^{2+}-dependent manner, thereby enabling dual excitation ratiometric Ca^{2+} imaging. Ratiometric pericam permits quantitative Ca^{2+} measurement by minimizing the effects of several artifacts that are unrelated to changes in $[Ca^{2+}]_i$. The molecular structures of the genetic calcium indicators are shown in Figure 2.

Protocol

Imaging Intracellular Free Ca^{2+} Concentration Using YCs

YCs are used to image rapid changes in intracellular free Ca^{2+} concentration within HeLa cells. FRET imaging is performed using a laser-scanning confocal microscope.

IMAGING SETUP

For fast and simultaneous acquisition of cpVenus173 and CFP images from HeLa cells expressing YC3.60, a color camera (Hamamatsu Photonics, C7780-22) composed of three charge-coupled device (CCD) chips (RGB: red, green, and blue) and a prism can be used. The cpVenus173 and CFP images are captured by the G and B chips, respectively. In addition, to improve spatial resolution along the *z*-axis, a spinning-disk unit was placed in front of the camera. A confocal image of YC3.60-expressing HeLa cells is shown in Figure 1D. A series of ratio images in pseudocolor acquired at video rate (displayed at 16.7 Hz) shows how the increase in $[Ca^{2+}]_c$ (cytosolic free Ca^{2+} concentration) appeared and propagated within the individual cells after stimulation with histamine.

MATERIALS

CAUTION: See Appendix 6 for proper handling of materials marked with <!>.
See the end of the chapter for recipes for reagents marked with <R>.

Reagents

Ca^{2+}-free medium <R>
Dulbecco's modified Eagle medium (DMEM)
Hanks' balanced salt solution (HBSS) containing 1.26 mM $CaCl_2$ <!>
Heat-inactivated fetal bovine serum (FBS)
HeLa cells
Mammalian expression vector containing the YC3.60 gene; pcDNA3/YC3.60
Transfection reagents such as Superfect (QIAGEN) and Lipofectamine2000 (Invitrogen)

Equipment

3CCD color camera (ORCA-3CCD Ashura, Hamamatsu Photonics)
Diode-pumped solid-state laser, 430-nm (Melles Griot)
Glass-bottomed dishes, 35-mm (Matsunami Glass Ind, Ltd., cat no. D111300)
Image acquisition software: AQUACOSMOS (Hamamatsu Photonics) to control the Ashura 3CCD camera
Image processing software: AQUACOSMOS (Hamamatsu Photonics) or MetaMorph (Molecular Devices) to make pseudocolor ratio images
Incubator for cell culture set at 37°C
Inverted microscope IX-71 (Olympus) equipped with the best chromatically corrected objective UPLSAPO 60x 1.35 numerical aperture (NA) (Olympus)
Spinning-disk confocal unit (CSU21, Yokogawa Electric Corporation)

EXPERIMENTAL METHOD

Transfection

1. Plate HeLa cells on coverslips in a Petri dish, and allow them to grow in DMEM containing 10% heat-inactivated FBS for 12 h.
2. Transfect cells in the dish with 1 µg of cDNA (pcDNA/YC3.60) using Superfect or Lipofectamine2000.

Imaging

3. Between 1 and 3 d posttransfection, image HeLa cells on an inverted microscope. Expose cells to reagents at room temperature in HBSS containing 1.26 mM $CaCl_2$.
4. Choose moderately bright cells in which the fluorescence is uniformly distributed in the cytosolic compartment but excluded from the nucleus, as would be expected for a 74-kDa protein without targeting signals.
5. Define several variables for fast image acquisition (stream mode). They include (a) intensity of the laser, (b) scanning speed, (c) binning, and (d) sensitivity of the camera.

   In our experience, the intensity of the laser should be attenuated greatly using ND filters for Ca^{2+} imaging on a timescale of subseconds. Photobleaching does not happen with a laser that is sufficiently attenuated.
6. Observe fluorescence signals from CFP and cp173Venus by blue and green channels of the 3CCD camera, respectively.
7. Set 4 x 4 binning for high-sensitive fluorescence detection at video rate observation.
8. Adjust the rotational speed of the spinning disk. To remove the artificial streak raster pattern, synchronize spinning with the frame CCD readout.
9. Acquire video rate confocal FRET images (31.5-msec accumulation time/frame) during stimulation such as an application of histamine.
10. At the end of an experiment, convert the fluorescence signal into $[Ca^{2+}]_c$. Obtain R_{max} and R_{min} values as follows.
    1. Saturate intracellular YC3.60 with Ca^{2+} by increasing the extracellular $[Ca^{2+}]$ to 10–20 mM in the presence of 1–5 µM ionomycin. Wait until the fluorescence intensity reaches a plateau.
    2. Deplete the indicator of Ca^{2+} by washing the cells with Ca^{2+}-free medium.
    3. The in situ calibration for $[Ca^{2+}]$ uses the equation

$$[Ca^{2+}] = K'_d \, [(R - R_{min})/(R_{max} - R)]^{(1/n)},$$

       where K'_d is the apparent dissociation constant corresponding to the Ca^{2+} concentration at which R is midway between R_{max} and R_{min} and n is the Hill coefficient. YC3.60 shows a monophasic Ca^{2+} response curve ($K'_d = 0.25$ µM; $n = 1.7$).
11. Take an image without laser illumination to make a background image.

    Background-corrected images for CFP and cpVenus173 were used to make pseudocolor ratio images.

PRACTICAL CONSIDERATIONS

pH

pH-related artifacts were not an issue in the experiments that used HeLa cells because agonist-induced $[Ca^{2+}]_c$ mobilization did not cause any intracellular pH changes detectable by the pH indicator, BCECF (2′,7′-bis-(2-carboxyethyl)-5-(and-6)-carboxyfluorescein) (data not shown). Correspondingly, comparisons of YC2 with YC2.1 and YC3 with YC3.1 showed no major differences in the reported $[Ca^{2+}]_c$ attributable to the difference in pH sensitivity. On the other hand, both YC2 and YC3 expressed in dissociated hippocampal neurons were perturbed by acidification following depolarization or glutamate stimulation. This problem was solved by using YC2.1 or YC3.1 (Miyawaki et al. 1999). YC2.12, YC3.12, and YC3.60 also give reliable Ca^{2+} responses over a physiological range of pH (Nagai et al. 2004).

Concentration of YCs in Cells

The estimation of cameleon concentration in cells is essential for quantifying the trade-off between optical detectability and Ca^{2+} buffering. It is important to consider the ability of various concentrations of YC3.60 to buffer Ca^{2+} transients in HeLa cells. During a 0.1-mM histamine challenge, sharp $[Ca^{2+}]_c$ transients followed by $[Ca^{2+}]_c$ oscillations can be observed in cells expressing <200 μM YC3.60. In contrast, $[Ca^{2+}]_c$ recovers slowly to the baseline over a period of several hundred seconds, and oscillations are never observed when transfected HeLa cells expressed >300 μM YC3.60. Cameleons at cytosolic concentrations below <20 μM are too dim to give favorable SNRs with our current instruments. This limitation cannot be overcome by increasing the intensity of illumination because of YFP photochromism (see below).

Photochromism of YFP (cpVenus173)

If YFP is excited too strongly, its fluorescence will be reduced. This apparent bleaching is actually photochromism because the fluorescence recovers to some extent spontaneously and can be further restored by ultraviolet illumination (Dickson et al. 1997). Intense excitation of YCs also causes photochromism of the YFP moiety, which results in a decrease in the yellow:cyan emission ratio independent of Ca^{2+} change. The extent of photochromism is dependent on excitation power, NA of objective, and exposure time. Therefore, it is necessary to optimize these factors for each cell sample to minimize photochromism while still preserving a high SNR. Because the photochromism is partially reversible, the sampling interval is another factor that should be considered. Illumination at frequent intervals sometimes leads to a decrease in the resting ratio values of YCs. A better solution is to bin pixels at the cost of spatial resolution. The increased SNR permits the decrease in intensity of the excitation light with a neutral-density filter and the observation of $[Ca^{2+}]_i$ oscillations without significant photochromism of the indicators.

Microscope

The dynamic measurement of $[Ca^{2+}]_i$ in space and time is crucial for understanding many important cellular functions. Although common conventional confocal microscopes use a single laser beam to scan a specimen, the Yokogawa Electric Corporation confocal scanning unit (CSU) scans the field of view with approximately 1000 laser beams by means of the microlens-enhanced Nipkow-disk scanning technology. CSU's multibeam scanning technology requires low-light intensity per unit area, thereby significantly reducing photobleaching and phototoxicity in live cells. The most advanced model of the CSU series is the CSU-X1. With an additional camera port, the CSU-X1 makes it easy to image two different emission ranges simultaneously. This is probably best for the fast confocal observation of YC3.60

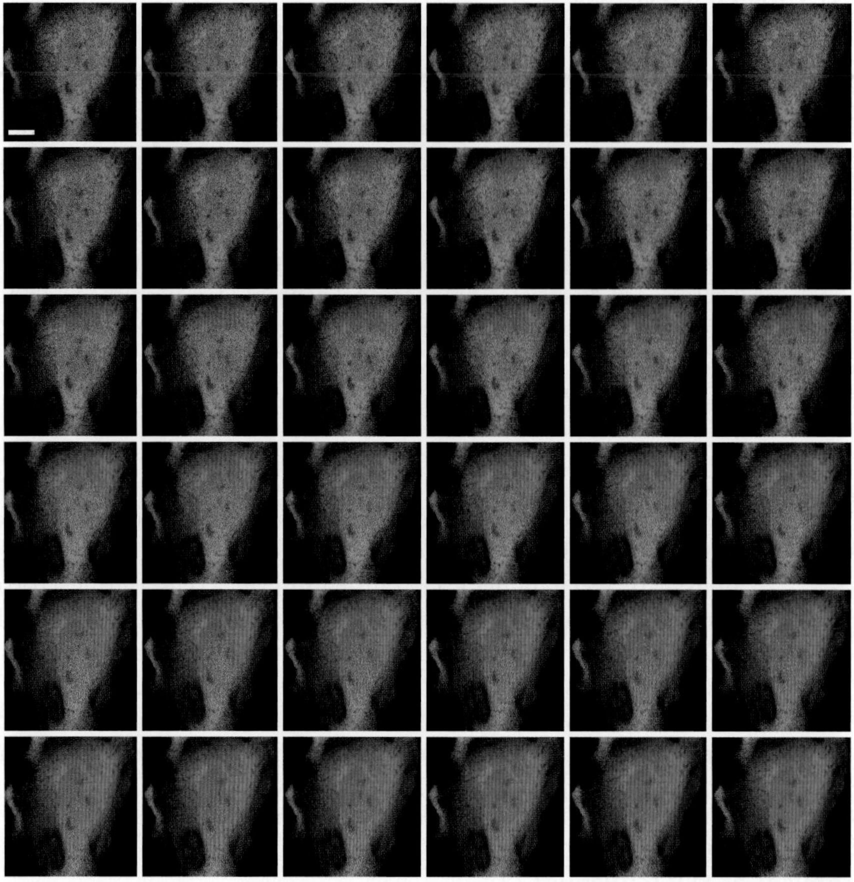

FIGURE 3. Histamine-induced $[Ca^{2+}]_c$ transients in HeLa cells imaged with LSM 5 *LIVE* system equipped with a 440-nm (laser diode) laser line and an objective lens (Plan-Apochromat 63x, 1.4 NA, oil). Images were acquired without averaging at 30 Hz. The pinhole size was set to 28 µm (~2 Airy disk units) for both channels. Scale bar, 10 µm.

or indicators based on FRET between CFP and YFP. The 430-nm laser described in this chapter is no longer available and can be replaced with common solid-state lasers emitting at ~440 nm.

Another option for fast confocal imaging is the LSM 5 *LIVE* (Zeiss). With the use of a coherent line excitation light and a line CCD, this system illuminates 512 pixels along a line in the x direction and scans this line in the y direction with a scan mirror for up to 2048 pixels. This results in the ideal combination of long pixel dwell time and short frame-acquisition time. Unlike the CSU, confocal aperture can be freely adjusted in this system. In addition, scanning speed need not be synchronized with frame CCD readout. Figure 3 shows histamine-induced $[Ca^{2+}]_c$ transients in HeLa cells imaged with the LSM 5 *LIVE* system. The updated system, LSM 7 *LIVE*, is currently available.

Neural activity can be analyzed in dissociated neuronal cultures, slice cultures, acute slices, and intact animals. Gene transfer techniques, including liposome-mediated transfection, various viral vectors, electroporation, and the gene gun, have shown significant progress in recent years. Because there is a trend toward the understanding of brain functions in a physiological context, the ultimate approach is to construct transgenic lines, to which two-photon excitation microscopy is applied. Although two-photon excitation microscopy has been applied to samples loaded with multiple fluorophores for multicolor or FRET imaging, the excitation of one fluorophore relative to the other fluorophores cannot be controlled well in the conventional system that uses a narrowband laser. Recently, a new technology that enables phase modulation of ultrabroadband laser pulses has been reported (Isobe et al. 2009). Versatile control of two-photon excitation of CFP and YFP was achieved with their ratio (CFP/YFP) being varied 100-fold. Selective excitation of CFP over YFP was indeed appreciated in an experiment using YC.

RECIPE

Recipes for reagents marked with <R> are included in this list.

Ca^{2+}-Free Medium

To HBSS that is nominally Ca^{2+} free, add:

Ionomycin	1 µM
EGTA	1 mM
$MgCl_2$	5 mM

REFERENCES

Baird GS, Zacharias DA, Tsien RY. 1999. Circular permutation and receptor insertion within green fluorescent proteins. *Proc Natl Acad Sci* **96:** 11241–11246.

Clark DA, Gabel CV, Gabel H, Samuel ADT. 2007. Temporal activity patterns in thermosensory neurons of freely moving *Caenorhabditis elegans* encode spatial thermal gradients. *J Neurosci* **27:** 6083–6090.

Dickson RM, Cubitt AB, Tsien RY, Moerner WE. 1997. On/off blinking and switching behaviour of single molecules of green fluorescent protein. *Nature* **388:** 355–358.

Griesbeck O, Baird GS, Campbell RE, Zacharias DA, Tsien RY. 2001. Reducing the environmental sensitivity of yellow fluorescent protein. Mechanism and applications. *J Biol Chem* **276:** 29188–29194.

Heim N, Griesbeck O. 2004. Genetically encoded indicators for cellular calcium dynamics based on troponin C and green fluorescent protein. *J Biol Chem* **279:** 14280–14286.

Heim N, Garaschk O, Friedrich MW, Mank M, Milos RI, Kovalchuk Y, Konnerth A, Griesbeck O. 2007. Improved calcium imaging in transgenic mice expressing a troponin C-based biosensor. *Nat Methods* **4:** 127–129.

Helmchen F, Konnerth A, eds. 2011. *Imaging in neuroscience: A laboratory manual.* Cold Spring Harbor Laboratory Press, Cold Spring Harbor, NY (in press).

Higashijima S, Masino MA, Mandel G, Fetcho JR. 2003. Imaging neuronal activity during zebrafish behavior with a genetically encoded calcium indicator. *J Neurophysiol* **90:** 3986–3997.

Isobe K, Suda A, Tanaka M, Kannari F, Kawano H, Mizuno H, Miyawaki A, Midorikawa K. 2009. Multifarious control of two-photon excitation of multiple fluorophores achieved by phase modulation of ultra-broadband laser pulses. *Opt Express* **17:** 13737–13746.

Kuchibhotla KV, Goldman ST, Lattarulo CR, Wu H-Y, Hyman BT, Bacskai BJ. 2008. Aβ plaques lead to aberrant regulation of calcium homeostasis in vivo resulting in structural and functional disruption of neuronal networks. *Neuron* **59:** 214–225.

Mank M, Santos AF, Direnberger S, Mrsic-Flogel TD, Hofer SB, Stein V, Hendel T, Reiff, DF, Levelt C, Borst A, et al. 2008. A genetically encoded calcium indicator for chronic in vivo two-photon imaging. *Nat Methods* **5:** 805–811.

Miyawaki A, Llopis J, Heim R, McCaffery JM, Adams JA, Ikura M, Tsien RY. 1997. Fluorescent indicators for Ca^{2+} based on green fluorescent proteins and calmodulin. *Nature* **388:** 882–887.

Miyawaki A, Griesbeck O, Heim R, Tsien RY. 1999. Dynamic and quantitative Ca^{2+} measurements using improved cameleons. *Proc Natl Acad Sci* **96:** 2135–2140.

Mizuno H, Sawano A, Eli P, Hama H, Miyawaki A. 2001. Red fluorescent protein from Discosoma as a fusion tag and a partner for fluorescence resonance energy transfer. *Biochemistry* **40:** 2502–2510.

Nakai J, Ohkura M, Imoto K. 2001. A high signal-to-noise Ca^{2+} probe composed of a single green fluorescent protein. *Nat Biotechnol* **19:** 137–141.

Nagai T, Ibata K, Park ES, Kubota M, Mikoshiba K, Miyawaki A. 2002. A variant of yellow fluorescent protein with fast and efficient maturation for cell-biological applications. *Nat Biotechnol* **20:** 87–90.

Nagai T, Yamada S, Tominaga T, Ichikawa M, Miyawaki A. 2004. Expanded dynamic range of fluorescent indicators for Ca^{2+} by circularly permuted yellow fluorescent proteins. *Proc Natl Acad Sci* **101:** 10554–10559.

Palmer AE, Jin C, Reed JC, Tsien RY. 2004. Bcl-2-mediated alterations in endoplasmic reticulum Ca^{2+} analyzed with an improved genetically encoded fluorescent sensor. *Proc Natl Acad Sci* **101:** 17404–17409.

Palmer AE, Glacomello M, Kortemme T, Andrew Hires S, Lev-Ram V, Baker D, Tsien RY. 2006. Ca^{2+} indicators based on computationally redesigned calmodulin-peptide pairs. *Chem Biol* **13:** 521–530.

Suzuki H, Kerr R, Bianchi L, Frøkjaer-Jensen C, Slone D, Xue J, Gerstbrein B, Driscoll M, Schafer WR. 2003. In vivo imaging of *C. elegans* mechanosensory neurons demonstrates a specific role for the MEC-4 channel in the process of gentle touch sensation. *Neuron* **39:** 1005–1017.

Tian L, Hires SA, Mao T, Huber D, Chiappe ME, Chalasani SH, Petreanu L, Akerboom J, Mckinney SA, Schreiter ER, et al. 2009. Imaging neural activity in worms, flies, and mice with improved GCaMP calcium indicators. *Nat Methods* **6:** 875–881.

Tsien RY. 2000. Physiological indicators based on fluorescence resonance energy transfer. In *Imaging neurons: A laboratory manual* (ed. Yuste R, et al.), pp. 55.1–55.10. Cold Spring Harbor Laboratory Press, Cold Spring Harbor, NY.

27 Targeted Recombinant Aequorins

Tullio Pozzan and Rosario Rizzuto

Department of Biomedical Sciences, University of Padova and CNR Unit for Study of Biomembranes, Padova, Italy

ABSTRACT

Aequorin is a small protein produced by the genus *Aequorea* that was widely used in the 1960s and 1970s as a probe to measure Ca^{2+} in living cells (Ridgway and Ashley 1967; Allen and Blinks 1978; Cobbold 1980). The invention of the carboxylate Ca^{2+} indicators (Tsien et al. 1982; Grynkiewicz et al. 1985), which are much simpler to load into intact living cells and to calibrate and image at the single-cell level, has led most groups to abandon aequorin. Yet, this latter Ca^{2+} indicator still offers some advantages over the fluorescent probes. In particular, the use of molecular biological techniques for expressing recombinant aequorin in mammalian cells, thus eliminating the need for microinjection, has opened new possibilities for this probe (Brini et al. 1995). Among the new uses of aequorin, one of the most interesting is the potential for targeting it specifically to different cellular locations (Rizzuto et al. 1992; Brini et al. 1993, 1997; Montero et al. 1995; Marsault et al. 1997), thus opening the possibility of monitoring selectively the dynamics of $[Ca^{2+}]$ with unprecedented spatial resolution. This chapter briefly discusses the problems concerned with targeting aequorin to different locations, the advantages and disadvantages offered by the steep dependence of luminescence on $[Ca^{2+}]$, and the instruments needed to obtain reliable measurements.

THE CA^{2+} RESPONSE OF AEQUORIN

Aequorin, as produced by various *Aequorea* species, includes an apoprotein and a covalently bound prosthetic group (coelenterazine). When Ca^{2+} ions bind to three high-affinity sites (EF-hand type), aequorin undergoes an irreversible reaction, in which a photon is emitted and the oxidized coenzyme is released. For $[Ca^{2+}]$ between 10^{-7} M and 10^{-5} M, there is a relationship between the fractional rate of consumption (i.e., L/L_{max}, where L is the actual rate of photon emission and L_{max} is the maximal rate of aequorin discharge at saturating Ca^{2+} concentrations) and $[Ca^{2+}]$ (Allen et al. 1977; Blinks et al. 1978). Because of the cooperativity between the three binding sites, light emission is proportional to the second to third power of $[Ca^{2+}]$. It is important to stress that, unlike commonly used fluorescent indicators, aequorin is consumed on photon emission and, thus, the probe content decreases with time. This decrease in probe concentration depends on the Ca^{2+} concentration to which aequorin is exposed. In addition, because of the steep relationship between [Ca^{2+}] and photon emission, the average signal is biased toward the highest values, if, as often occurs within living cells, the concentration of the cation is inhomogeneous.

Wild-type aequorin is well suited for measuring [Ca^{2+}] between 0.5 μM and 10 μM. However, in some intracellular compartments or regions, the [Ca^{2+}] is much higher (e.g., the lumen of the endoplasmic reticulum [ER] and sarcoplasmic reticulum [SR], near Ca^{2+} channels and pumps). It is possible to generate aequorins with reduced Ca^{2+} affinities by mutating one (or possibly more) of the three EF-hand Ca^{2+}-binding sites (Kendall et al. 1992). A modified aequorin has been used to study [Ca^{2+}] microdomains in presynaptic terminals (Llinás and Sugimori 2005).

The point mutation used by the investigators (D119A; Montero et al. 1995; Brini et al. 1997) affects the second EF-hand domain and produces a mutated aequorin whose affinity for Ca^{2+} is reduced by ~20-fold. The range of Ca^{2+} sensitivity can be expanded further by using divalent cations other than Ca^{2+} or synthetic coelenterazine analogs (e.g., coelenterazine n; Montero et al. 1997).

The advantages of using recombinant aequorin include the following.

1. *Selective intracellular distribution.* Fluorescent probes are usually trapped mainly in the cytoplasm, but they are often in part sequestered, to a variable extent and depending on the cell type, into the lumen of organelles. In addition, slow leakage of the dyes into the extracellular medium also occurs. Recombinantly expressed aequorins, on the other hand, are not released by living cells, and their targeting is extremely selective. For example, <1% of aequorin targeted to the mitochondrial matrix is found in the cytoplasm.
2. *High signal-to-noise ratio.* The background chemiluminescence signal of living cells is intrinsically very low, unlike that of autofluorescence (particularly when low-excitation wavelengths are used). In addition, the ratio between the rate of photon emission by aequorin exposed to resting and saturating Ca^{2+} concentration is >100,000. With fluorescent Ca^{2+} indicators this value is usually 3–10.
3. *Low Ca^{2+}-buffering effect.* The usual concentrations reached intracellularly with fluorescent probes are on the order of 10–100 μM, whereas those achieved with recombinantly expressed chimeric aequorins are in the range of 0.1–1 μM.
4. *Wide dynamic range.* The range of Ca^{2+} concentrations that can be measured with aequorins (wild-type and low-affinity mutants) is between 10^{-7} M and 10^{-3} M, whereas with most commonly used indicators, the usable range is from 10^{-8} M up to concentrations of a few micromolar.
5. *Possibility of coexpression with proteins of interest.* By transiently cotransfecting aequorin and a protein of interest, the Ca^{2+} signal comes only from the cell population expressing the protein investigated, thus bypassing the problem of variability of single-cell response. In the case of fluorescent indicators, this analysis is highly complex and time-consuming.

The disadvantages include the following.

1. *Overestimation of the average increase in cells (or compartments) with inhomogeneous behavior.* As discussed below, the aequorin signal, unlike that of fluorescent indicators, is biased toward the highest Ca^{2+} values. This may lead to substantial overestimations of the mean response of a cell population.
2. *Low-light emission.* The maximum number of photons in theory emittable by aequorin is 1 photon/molecule (in practice, less, ~0.6 photon/molecule). In the case of the fluorescent Ca^{2+} probes, the limit is set by the photobleaching of the dye, and it has been calculated to be more than 10,000 photons/molecule.
3. *Loading procedure.* Loading with fluorescent probes is extremely simple and requires a few tens of minutes. Aequorin must be introduced into cells via transfection or microinjection, and reconstitution with coelenterazine is necessary to generate the functional photoprotein.

4. *Necessity of reconstitution with a coenzyme.* The expressed polypeptide does not emit light if the cells are not supplemented externally with the coenzyme (see also below for the problems of reconstitution with coelenterazine).

STRATEGIES FOR TARGETING AEQUORINS TO SELECTIVE LOCATIONS

Selective targeting of recombinant polypeptides is a major issue of modern cell biology. The machinery and the signals that are responsible for the selective localization of endogenous proteins, in some cases, have been well characterized (e.g., mitochondria [Hartl et al. 1989] and ER [Pelham 1989]), but in others are largely unknown. In the case of aequorin, we have adopted three different strategies for selectively targeting aequorins. Before describing them, it is worth mentioning that fusion of long polypeptides at the amino terminus of aequorin causes no appreciable alterations in the chemiluminescent properties and Ca^{2+} affinity of the probe. To the contrary, even small modifications at the carboxyl terminus result either in irreversible loss of chemiluminescence or in dramatic increases in the rate of Ca^{2+}-independent luminescence, which pose significant problems in quantitation of the signal (Nomura et al. 1991). The following three strategies have been used.

1. *Addition of a known targeting sequence at the amino terminus of aequorin.* This has been the case for aequorins targeted to the mitochondria (Rizzuto et al. 1992) and to the nucleus (Brini et al. 1993).
2. *Fusion of aequorin to the carboxyl terminus of a protein that contains its own targeting strategy.* In this case, it is important that the carboxy-terminal extension (due to the aequorin polypeptide) does not affect the targeting by itself. Using this approach, we have constructed the aequorins targeted to the SR (Brini et al. 1997), the plasma membrane (Marsault et al. 1997), and the intermembrane space of the mitochondria (Rizzuto et al. 1998).
3. *Fusion of aequorin to the carboxyl terminus of a polypeptide that binds firmly to an endogenous protein* (Sitia et al. 1990). This is the strategy adopted for the targeting to the ER lumen (Montero et al. 1995). In this latter case, the leader sequence, the VDJ, and the CH1 region of an IgG2 cDNA have been fused to the 3′ end of the aequorin copy DNA (cDNA). The CH1 domain binds with very high affinity to the endogenous ER lumenal protein Bip (binding immunoglobulin protein), and the chimeric aequorin is thus retained in the ER lumen (Montero et al. 1995).

It should be stressed that efficient and selective targeting is an essential part of the whole measurement, given that the spatial resolution of the method depends not on sophisticated hardware of subcellular imaging but rather on the selective targeting of the probe. Thus, even a small fraction of missorted probe can profoundly affect the measurement, both qualitatively and quantitatively. It is therefore important that the subcellular localization is checked with extreme accuracy. For this purpose, both immunocytochemical approaches (at the optical and electron microscopy level) and functional approaches (subcellular fractionation, specific drugs) have been used.

EXPERIMENTAL PROCEDURE

Transfection

In our laboratory, we currently use three main transfection procedures as follows: calcium phosphate, electroporation, and particle gun. Other groups have used other standard procedures for loading aequorin cDNA into cells, including viral infection. Both transiently transfected and stable clones can be generated. Overall, recombinant expression of aequorin has, so far, proven to be efficient and totally innocuous for the cells, independent of the subcellular localization of the protein. Several different types of cultured cells have been transfected, from stable cell lines (HeLa, COS, 3T3, etc.) to primary cultures (skeletal muscle myoblast, neurons). Typically, the classic calcium phosphate procedure uses 40 μg of cDNA per tissue-culture dish.

Reconstitution

After expression, the recombinant apoprotein must be reconstituted into functional aequorin. This can be accomplished by incubating transfected cells with the chemically synthesized prosthetic group, coelenterazine (now commercially available from Molecular Probes). In our experience, coelenterazine is freely permeable across cell membranes, and reconstitution may occur within all intracellular compartments to which the photoprotein has been targeted. The reconstitution process is relatively slow (optimal reconstitution is usually observed after 1–2 h of incubation with coelenterazine; Rizzuto et al. 1994). In compartments with low Ca^{2+}, the functional aequorin pool gradually increases during the reconstitution, as consumption is negligible. On the other hand, compartments with high Ca^{2+} pose a major problem. In fact, if the rate of discharge is the same as that of reconstitution, no active aequorin will be present at the end of the incubation with the prosthetic group. For this reason, we usually deplete these latter compartments of Ca^{2+} before performing the reconstitution (Montero et al. 1995; Brini et al. 1997).

The concentrations of coelenterazine usually used vary between 1 μm and 10 μm, and the coenzyme is simply added to the medium from a stock solution in methanol. With most chimeric aequorins, coelenterazine is added to the complete growth medium containing 1%–3% of serum. In the case of the aequorins trapped in the lumen of the ER or SR, it is necessary to first reduce the lumenal Ca^{2+} concentration. To achieve this, several protocols have been devised that take advantage of Ca^{2+} ionophores, inhibitors of the Ca^{2+} ATPases, or specific agents leading to opening of intracellular Ca^{2+} channels (e.g., caffeine) (Montero et al. 1995; Brini et al. 1997).

Measurements

Instrumentation

For most of our experiments, we have used a custom-built luminometer (Rizzuto et al. 1995). In this system, the perfusion chamber, which is on top of a hollow cylinder with a thermostat-controlled water jacket, is continuously perfused with buffer via a peristaltic pump. The cells are grown on glass coverslips that are then placed a few millimeters from the surface of a low-noise phototube. The photomultiplier tube (PMT) is kept in a dark, refrigerated box. An amplifier discriminator is built into the photomultiplier housing; the pulses generated by the discriminator are captured by a Thorn EMI photon-counting board, installed in a personal computer. The board allows the storage of data in the computer memory for further analyses. A schematic of the setup is shown in Figure 1.

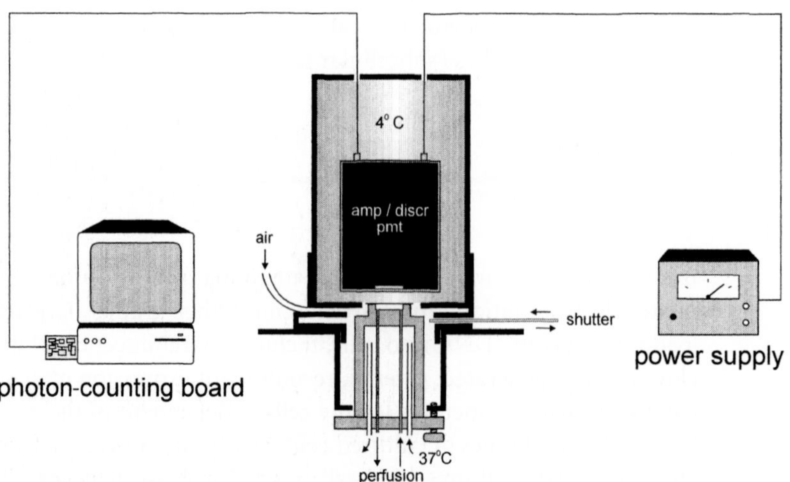

FIGURE 1. Schematic representation of the measuring apparatus. (amp/discr pmt) Amplifier/discriminator photomultiplier tube.

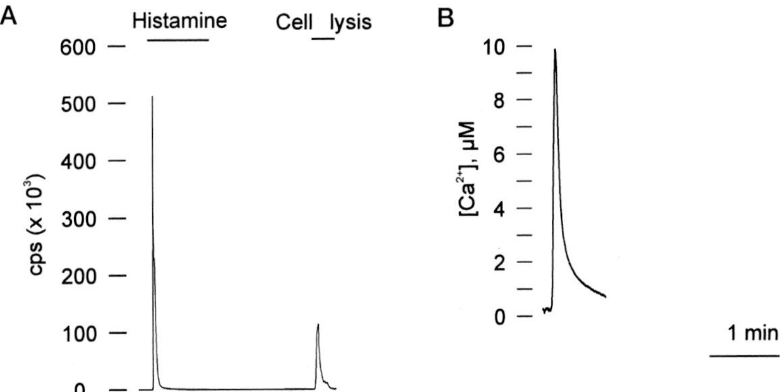

FIGURE 2. Typical experiments performed with HeLa cells that were transiently transfected with the cDNA encoding aequorin targeted to the mitochondrial matrix. (*A*) The trace represents the kinetics of photon emission. (*B*) The trace represents the corresponding calibrated values. Where indicated, the cells were perfused with medium containing 200 mM histamine. "Cell lysis" refers to the perfusion with medium containing 10 mM $CaCl_2$ and 100 μM digitonin. For details, see the original papers by Rizzuto et al. (1994) and Brini et al. (1995).

Calibration

To calibrate the crude luminescent signal in terms of $[Ca^{2+}]$, an algorithm has been developed (based on that of Blinks and coworkers [Allen et al. 1977; Blinks et al. 1978]) that takes into account the instant rate of photon emission and the total number of photons that can be emitted by the aequorin of the sample. To obtain the latter parameter, at the end of each experiment the cells are lysed by perfusing them with a hyposmotic medium containing 10 μm $CaCl_2$ and a detergent (100 μm digitonin) to discharge all the aequorin that was not consumed during the experiment.

Result

Figure 2 shows a typical experiment performed with HeLa cells transiently transfected with aequorin targeted to the mitochondrial matrix. Figure 2A shows the kinetics of photon emission and Figure 2B the calibrated values (using the algorithm described by Brini et al. 1997). As shown in Figure 2B, the addition of histamine causes a very rapid and ample increase in the $[Ca^{2+}]$ of the mitochondrial matrix.

CONCLUSION

Recombinantly expressed aequorins represent a unique tool to investigate cellular Ca^{2+} homeostasis at the subcellular level. The main advantage of this approach with respect to the classic Ca^{2+} indicators resides in the highly selective targeting that can be achieved by appropriate modifications of the encoding cDNA. The techniques for expressing and reconstituting the functional photoproteins in various subcellular compartments are straightforward and can be applied to different model systems in vitro. The development of transgenic animals constitutively expressing the photoprotein will further expand the range of applications of this method.

ACKNOWLEDGMENTS

The experimental work described in this chapter was supported by grants to the authors from Telethon (project no. 845, 850), the Human Frontier Science Program, the Biomed program of the European Union, the Armenise Foundation (Harvard), the Italian University Ministry, and the British Research Council.

REFERENCES

Allen DG, Blinks JR. 1978. Calcium transients in aequorin-injected frog cardiac muscle. *Nature* **273:** 509–513.

Allen DG, Blinks JR, Prendergast FG. 1977. Aequorin luminescence: Relation of light emission to calcium concentration—A calcium-independent component. *Science* **195:** 996–998.

Blinks JR, Allen DG, Prendergast FG, Harrer GC. 1978. Photoproteins as models of drug receptors. *Life Sci* **22:** 1237–1244.

Brini M, Murgia M, Pasti L, Picard D, Pozzan T, Rizzuto R. 1993. Nuclear Ca^{2+} concentration measured with specifically targeted recombinant aequorin. *EMBO J* **12:** 4813–4819.

Brini M, Marsault R, Bastianutto C, Alvarez J, Pozzan T, Rizzuto R. 1995. Transfected aequorin in the measurement of cytosolic Ca^{2+} concentration ($[Ca^{2+}]_c$): A critical evaluation. *J Biol Chem* **270:** 9896–9903.

Brini M, De Giorgi F, Murgia M, Marsault R, Massimino ML, Cantini M, Rizzuto R, Pozzan T. 1997. Subcellular analysis of Ca^{2+} homeostasis in primary cultures of skeletal muscle myotubes. *Mol Biol Cell* **8:** 129–143.

Cobbold PH. 1980. Cytoplasmic free calcium and ameboid movement. *Nature* **285:** 441–446.

Grynkiewicz G, Poenie M, Tsien RY. 1985. A new generation of Ca^{2+} indicators with greatly improved fluorescence properties. *J Biol Chem* **260:** 3440–3450.

Hartl FU, Pfanner N, Nicholson DW, Neupert W. 1989. Mitochondrial protein import. *Biochim Biophys Acta* **988:** 1–45.

Kendall JM, Sala-Newby G, Ghalaut V, Dormer RL, Campbell AK. 1992. Engineering the Ca^{2+}-activated photoprotein aequorin with reduced affinity for calcium. *Biochem Biophys Res Commun* **187:** 1091–1097.

Llinás R, Sugimori M. 2005. Imaging intracellular calcium-concentration microdomains at a chemical synapse. In *Imaging in neuroscience and development: A laboratory manual* (ed. Yuste R, Konnerth A), pp. 325–330. Cold Spring Harbor Laboratory Press, Cold Spring Harbor, NY.

Marsault R, Murgia M, Pozzan T, Rizzuto R. 1997. Domains of high Ca^{2+} beneath the plasma membrane of living A7r5 cells. *EMBO J* **16:** 1575–1581.

Montero M, Brini M, Marsault R, Alvarez J, Sitia R, Pozzan T, Rizzuto R. 1995. Monitoring dynamic changes in free Ca^{2+} concentration in the endoplasmic reticulum of intact cells. *EMBO J* **14:** 5467–5475.

Montero M, Barrero MJ, Alvarez J. 1997. $[Ca^{2+}]$ microdomains control agonist-induced Ca^{2+} release in intact HeLa cells. *FASEB J* **11:** 881–885.

Nomura M, Inouye S, Ohmiya Y, Tsuji FI. 1991. A C-terminal proline is required for bioluminescence of the Ca^{2+}-binding photoprotein, aequorin. *FEBS Lett* **291:** 63–66.

Pelham HRB. 1989. Control of protein exit from the endoplasmic reticulum. *Annu Rev Cell Biol* **5:** 1–23.

Ridgway EB, Ashley CC. 1967. Calcium transients in single muscle fibers. *Biochem Biophys Res Commun* **29:** 229–234.

Rizzuto R, Simpson AWM, Brini M, Pozzan T. 1992. Rapid changes of mitochondrial Ca^{2+} revealed by specifically targeted recombinant aequorin. *Nature* **358:** 325–328.

Rizzuto R, Bastianutto C, Brini M, Murgia M, Pozzan T. 1994. Mitochondrial Ca^{2+} homeostasis in intact cells. *J Cell Biol* **126:** 1183–1194.

Rizzuto R, Brini M, Bastianutto C, Marsault R, Pozzan T. 1995. Photoprotein-mediated measurement of calcium ion concentration in mitochondria of living cells. *Methods Enzymol* **260:** 417–428.

Rizzuto R, Pinton P, Carrington W, Fay FS, Fogarty KE, Lifshitz LM, Tuft RA, Pozzan T. 1998. Close contacts with the endoplasmic reticulum as determinants of mitochondrial Ca^{2+} responses. *Science* **280:** 1763–1766.

Sitia R, Neuberger M, Alberini C, Bet P, Fra A, Valetti C, Williams G, Millstein C. 1990. Developmental regulation of IgM secretion: The role of the carboxy-terminal cysteine. *Cell* **60:** 781–790.

Tsien RY, Pozzan T, Rink TJ. 1982. T-cell mitogens cause early changes in cytoplasmic free Ca^{2+} and membrane potential in lymphocytes. *Nature* **295:** 68–71.

28 Imaging Intracellular Signaling Using Two-Photon Fluorescent Lifetime Imaging Microscopy

Ryohei Yasuda

Department of Neurobiology, Duke University Medical Center, Durham, North Carolina 27710

ABSTRACT

The recent development of Förster resonance energy transfer (FRET) sensors and FRET imaging techniques permits visualization of the dynamics of intracellular signaling events with high spatiotemporal resolution. In particular, fluorescence lifetime imaging in combination with two-photon laser-scanning microscopy (two-photon fluorescence lifetime imaging microscopy [2pFLIM]) is a powerful tool to monitor signaling events in small subcellular compartments in thick tissue. This chapter provides practical guidelines for quantitative imaging of intracellular signaling using 2pFLIM.

INTRODUCTION

FRET is the process of nonradiative energy transfer from an excited donor fluorophore to an acceptor fluorophore through dipole–dipole interactions. The efficiency of FRET strongly depends on the distance between the donor and the acceptor (Förster 1993; Lakowicz 2006), making this phenomenon a useful tool for monitoring interactions between proteins that are fused to fluorophores (Miyawaki 2003). Because FRET increases acceptor fluorescence and decreases donor fluorescence, the ratio between the fluorescence intensities in the donor and the acceptor emission wavelengths is often used to measure FRET (Wallrabe and Periasamy 2005). Alternatively, the fluorescence lifetime of the donor, which is the time between the excitation of the fluorophore and emission of a photon, can be used as a readout of FRET because the fluorescence lifetime shortens as FRET efficiency increases (Wallrabe and Periasamy 2005; Lakowicz 2006; Yasuda 2006). There are several advantages to using fluorescence lifetime imaging microscopy (FLIM) compared with ratiometric imaging. First, the signal is independent of the relative concentration of the donor and the acceptor. Second, the signal is relatively independent of light scattering by the tissue compared with ratiometric imaging, which is affected by the wavelength dependency of light scattering. Third, when multiple populations with different FRET efficiency coexist, the fluorescence decay curve becomes multiexponential, and one can deconvolve each component by fitting the curve (Lakowicz 2006). FLIM has been combined with two-photon laser-scanning microscopy (2pLSM) to image samples in light-scattering tissue. This technique has enabled the measurement of signaling events in small neuronal compartments in brain slices (Yasuda 2006). For other applications and a discussion of time-domain FLIM, see Chapter 41.

TWO-PHOTON FLUORESCENCE LIFETIME IMAGING SETUP

There are several methods for imaging fluorescence lifetime (Lakowicz 2006). The time-correlated single-photon-counting (TCSPC) method in combination with 2pLSM is optimal for imaging small subcellular compartments in light-scattering tissue (Yasuda 2006). TCSPC measures the time that has elapsed between the arrival of a single photon detection pulse and the next laser pulse (Lakowicz 2006). As 2pLSM uses a pulsed laser, it is relatively simple to combine TCSPC with 2pLSM by modifying standard 2pLSM as follows (Yasuda et al. 2006; Murakoshi et al. 2008) (see Fig. 1A).

1. Install TCSPC imaging PCI card (such as PSC-150, Becker & Hickl GmbH).
2. Use a PMT with low transit-time spread (<0.2 nsec) and single photon-counting sensitivity (such as H-7422P, Hamamatsu).
3. Detect the laser pulses reflected by a slide glass using a fast photodiode (such as FDS010, Thorlabs). Some commercial lasers have a built-in fast photodiode output.
4. Feed the photon signal from the PMT, the laser pulse signal from the photodiode and pixels, and line and frame clocks synchronized with scanning into the TCSPC card.

ANALYSIS OF FLUORESCENCE LIFETIME IMAGES

Fitting Fluorescence Lifetime Curve

The fluorescence from a fluorophore following a short laser pulse typically decays exponentially (Fig. 1B). Thus, the fluorescence lifetime of the free donor τ_D can be measured by fitting the fluorescence decay curve with an exponential function convolved with the Gaussian pulse response function of the system (Fig. 1B):

$$H(t,t_0,\tau_D,\tau_G) = \frac{1}{2}\exp\left(\frac{\tau_G^2}{2\,\tau_D} - \frac{t-t_0}{\tau_D}\right)\mathrm{erf}\left(\frac{\tau_G^2 - \tau_D(t-t_0)}{\sqrt{2}\tau_D\tau_G}\right), \tag{1}$$

where τ_G is the width of the Gaussian pulse response function, t_0 is the time offset, and erf is the error function (Fig. 1B). When the free donor and the donor bound to an acceptor coexist, the fluorescence decays according to a double exponential, a mixture of the free donor and the donor bound to the acceptor. Thus, the fraction of the donor bound to the acceptor can be calculated by fitting the fluorescence decay curve with the double exponential curve,

$$F(t) = F_0[P_D \cdot H(t,t_0,\tau_D,\tau_G) \cdot P_{AD} \cdot H(t,t_0,\tau_{AD},\tau_G)], \tag{2}$$

where τ_{AD} is the fluorescence lifetime of the donor bound with the acceptor, and P_D and P_{AD} are the fractions of the free donor and the donor bound with the acceptor ($P_D + P_{AD} = 1$). To reduce the number of parameters to be fitted, τ_D can be fixed to the fluorescence lifetime obtained when the free donor is measured separately. It should be noted that some fluorophores, such as enhanced cyan fluorescent protein (eCFP) and its brighter variant Cerulean, have more than one time constant (Tramier et al. 2002; Yasuda et al. 2006). When using these fluorophores as FLIM donors, calculation of the binding fraction will be more complicated.

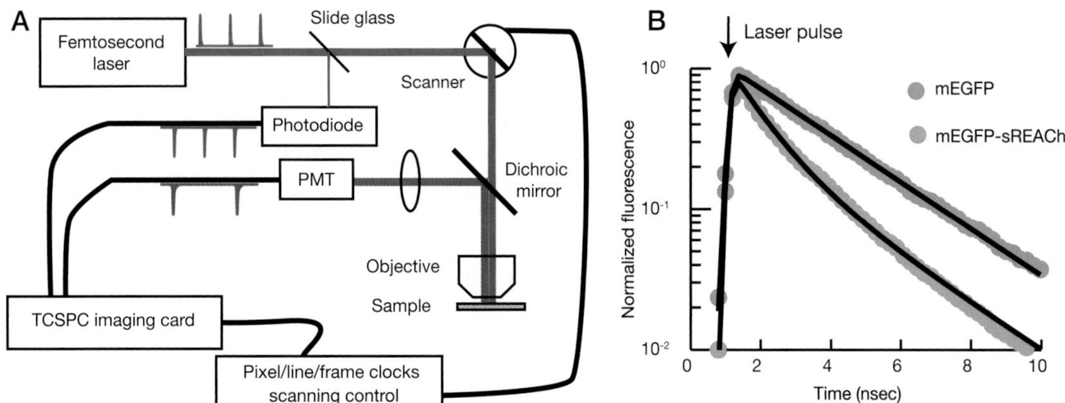

FIGURE 1. Two-photon fluorescence lifetime imaging. (*A*) Schematic of a 2pFLIM setup. TCSPC, time-correlated single-photon counting; PMT, photomultiplier tube. (*B*) Semilog plot of a fluorescence decay curve of monomeric enhanced green fluorescent protein (meGFP) and meGFP-sREACh (superresonance energy-accepting chromoprotein) tandem. meGFP is enhanced green fluorescent protein (eGFP) with the monomeric mutation (A206K) (Zacharias et al. 2002), and sREACh is a variant of nonradiative yellow fluorescent protein (YFP) (Murakoshi et al. 2004). The curves are fitted with Equation 1 for meGFP and Equation 2 for meGFP-sREACh tandem. The curve of meGFP-sREACh shows more than one fluorescence lifetime, presumably corresponding to the fraction of meGFP with folded and unfolded sREACh.

Visualizing Fluorescence Lifetime Images and Binding Fraction

Fluorescence lifetime images are in 3D (x, y, and t) or 4D (x, y, t, and z). Thus, the images need to be folded into 2D to visualize them. One way to visualize fluorescence lifetime images is to code the fluorescence lifetime with color and to code fluorescence intensity with brightness (Yasuda et al. 2006) (Fig. 2). For 4D images, the z-axis can be further folded by the commonly used maximum intensity projection, where the maximum intensity in the z-axis is assigned for each pixel in an x–y image. Curve-fitting the fluorescence lifetime in each pixel is suboptimal because it is computationally time consuming and often does not converge well because the number of photons in each pixel is limited. Instead, fluorescence lifetime images can be produced by calculating the mean fluorescence lifetime τ_m in each pixel as

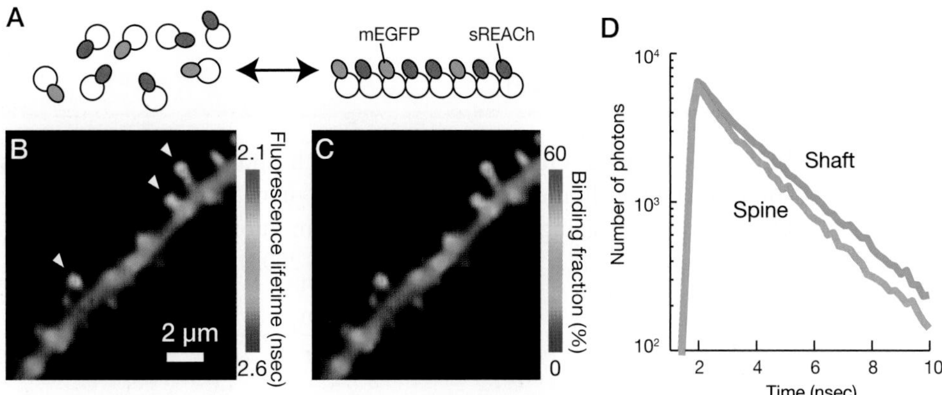

FIGURE 2. Visualization of fluorescence lifetime and binding fraction. (*A*) Schematic of an actin polymerization sensor. meGFP-actin and sREACh-actin are coexpressed (Murakoshi et al. 2004), and FRET occurs when they form polymeric filaments. (*B*) A fluorescence lifetime image of a dendritic segment of a neuron expressing the actin polymerization sensor (*A*) in an organotypic hippocampal slice. The image is produced by using Equation 3. (*C*) A binding fraction image calculated from the fluorescence lifetime image (*B*) using Equation 4. (*D*) The fluorescence decay curves in spines (average of three spines indicated with white arrowheads in *B*) and the parent dendritic shaft.

$$\tau_m = \langle t \rangle - t_0 = \frac{\int dt \cdot tF(t)}{\int dt \cdot F(t)} - t_0, \tag{3}$$

where $F(t)$ is the fluorescence lifetime decay curve (Yasuda et al. 2006; Murakoshi et al. 2008) (Fig. 2B). The offset t_0 depends on the optical path length, the cable length, and so on. Its value can be obtained by fitting the fluorescence lifetime curve integrated over the entire image, using either Equation 1 or Equation 2.

When the free donor and the donor bound to an acceptor coexist (Equation 2), τ_m is given approximately by

$$\tau_m \sim \frac{\int dt \cdot t\left(P_D e^{-t/\tau_D} + P_{AD} e^{-t/\tau_{AD}}\right)}{\int dt \cdot \left(P_D e^{-t/\tau_D} + P_{AD} e^{-t/\tau_{AD}}\right)} = \frac{P_D \tau_D^2 + P_{AD} \tau_A^2}{P_D \tau_D + P_{AD} \tau_{AD}}.$$

Thus, the binding fraction (P_{AD}; see Equation 2) can be obtained by (Murakoshi et al. 2008) (Fig. 2C)

$$P_{AD} = \frac{\tau_D(\tau_D - \tau_m)}{(\tau_D - \tau_{AD})(\tau_D + \tau_{AD} - \tau_m)}, \tag{4}$$

where τ_D and τ_{AD} can be obtained by fitting the fluorescence decay curve integrated over the entire image (Equation 2). A red–green–blue (RGB) pseudocolor image with the color scheme of Figure 2, A and B, can be generated by using the following scheme.

1. Normalize τ_m or P_{AD} so it maps to 0 to 1 range (h).
2. Convert h into RGB colors, such that

R = 0	G = 3h	B = 1	($0 \le h \le 1/3$)
R = 3h – 1	G = 1	B = 2 – 3h	($1/3 < h \le 2/3$)
R = 1	G = 3 – 3h	B = 0	($2/3 < h \le 1$)

3. Normalize the total fluorescence intensity ($\int dt \cdot F(t)$) so that it is in the 0 to 1 range, and multiply it to each color component.

The normalization depends on whether FRET increases or decreases with protein activity (compare Fig. 3B with 3D).

NUMBER OF PIXELS AND SIGNAL LEVEL

In TCSPC, there is some dead time after each photon detection event, so the maximum photon-counting rate is limited to ~10^6 Hz. This may limit the temporal resolution of the system. Quantitatively speaking, the noise of the mean fluorescence lifetime ($\delta\tau_m$) is related to the number of photons counted (N) (Philip and Carlsson 2003) as

$$\frac{\delta\tau_m}{\tau_m} \sim \frac{1}{\sqrt{N}}.$$

Thus, to achieve an ~3% noise level, ~10^3 photons per pixel are required. Thus, with a photon-counting rate of 10^6 Hz, it takes ~1 msec per pixel or ~7 sec per image with a 128 x 128-pixel resolution. Although this temporal resolution is sufficient for many signaling events (Fig. 3), if higher temporal resolution is needed, then the number of pixels must be reduced.

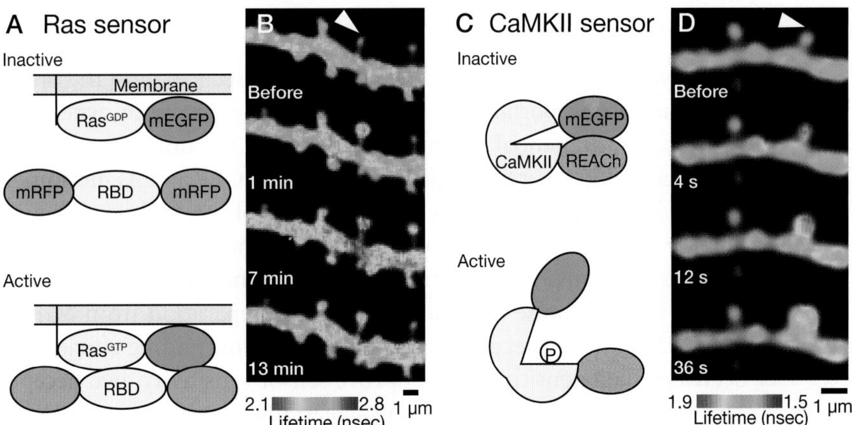

FIGURE 3. Signaling sensors designed for FLIM. (*A*) Schematic of the Ras activity sensor (Yasuda et al. 2006). Ras binds the Ras-binding domain (RBD) of cRaf1 when it is in the active form (GTP-bound form), causing FRET. (*B*) Fluorescence lifetime images of the Ras activity sensor within dendritic spines of pyramidal neurons in organotypic hippocampal slices. The spine marked with the arrowhead is stimulated with two-photon glutamate uncaging at time 0 (4 msec x 5 mW with a 720-nm Ti:sapphire laser, 0.5 Hz, 30 times) in 2.5-mM MNI (4-methoxy-7-nitroindoline)-caged-L-glutamate in zero extracellular Mg^{++} (Harvey et al. 2008). (*C*) Schematic of CaMKII activity sensor (Lee et al. 2009). CaMKII is in a closed conformation (causing FRET) when it is inactive, and opens on activation. Note that a decrease in FRET here denotes activity. (*D*) Fluorescence lifetime images of the CaMKII activity sensor within dendritic spines of pyramidal neurons in organotypic hippocampus slices (Lee et al. 2009). The spine was stimulated with a protocol similar to that used in *B*. The fluorescence lifetime color map has a reversed orientation compared with *B*.

OPTIMIZING SIGNALING SENSORS FOR FLIM

Since the first FRET-based calcium sensors were developed (Miyawaki et al. 1997), a number of genetically encoded FRET sensors of signaling have been created (most of them use FRET between eCFP and eYFP). For example, a class of kinase activity reporters has been produced, which consist of four linked components: a donor, a specific kinase target polypeptide, a phosphoamino-acid-binding domain, and an acceptor (Ni et al. 2006). Phosphorylation of the kinase target polypeptide causes binding between the kinase target polypeptide and the phosphoamino-acid-binding domain, thereby altering the FRET efficiency between the donor and the acceptor. Another class of kinase activity reporter uses fluorescent proteins tagged to both ends of the kinase to measure the conformational change of the kinase that occurs when the protein is activated (Takao et al. 2005; Ni et al. 2006). Most of these sensors are optimized for ratiometric imaging. Because the design criteria for ratiometric measurements and FLIM are different, the sensors must be reoptimized for FLIM (Yasuda 2006; Yasuda et al. 2006). Recently, sensors for Ras and CaMKII activity optimized for FLIM have been developed and were shown to be useful for measuring the activity of these proteins in single dendritic spines in hippocampal slices (Fig. 3). Ras activity was measured by quantifying the binding fraction between meGFP-Ras and Ras-binding domain of downstream protein tagged with two monomeric red fluorescent proteins at both ends (mRFP-RBD-mRFP) (Yasuda et al. 2006). When Ras is activated, mRFP-RBD-mRFP binds to meGFP-Ras, producing FRET between meGFP and mRFP (Fig. 3A,B). The CaMKII activity sensor measures the conformational change of CaMKII associated with the activation of CaMKII by meGFP and resonance energy-accepting chromoprotein (REACh) (nonradiative YFP variant) tagged to both ends of the molecule (Fig. 3C,D) (Lee et al. 2009). These sensors were originally made for the ratiometric method (Mochizuki et al. 2001; Takao et al. 2005) but further optimized for FLIM. To optimize sensors for FLIM, the following parameters can be modified.

Fluorophores

FLIM typically measures only the donor fluorescence; thus, the brightness and the photostability of the donor are important for a good signal-to-noise ratio, whereas the acceptor brightness is not. Also, monoexponential fluorescence decay of the donor is critical for measuring the fraction of donor bound to acceptor (Equations 2 and 3). eGFP and its monomeric mutants meGFP are optimal donors in FLIM because they are bright, are photostable, and have monoexponential fluorescence lifetime decay (Peter et al. 2005; Yasuda 2006; Yasuda et al. 2006).

mRFP and mCherry (a brighter variant of mRFP) (Shaner et al. 2004) have been used as acceptors in FLIM because of their good spectral separation from eGFP and meGFP (Peter et al. 2005; Yasuda 2006; Yasuda et al. 2006). The relatively low folding efficiency of mRFP and mCherry (~50%) decreases their sensitivity. To improve sensor sensitivity, two acceptors can be tagged to a protein. This increases the chance of having at least one mature acceptor (Yasuda et al. 2006) (Fig. 3A,B). The recently developed REACh (nonradiative YFP variant) and sREACh (a REACh variant with improved solubility, environmental insensitivity, and folding) are also reported to be useful as FLIM acceptors (Ganesan et al. 2006; Murakoshi et al. 2008). In particular, the high folding efficiency (~70%) of sREACh allows imaging with a high signal-to-noise ratio (Figs. 2 and 3B). REACh and sREACh, however, do not fluoresce; thus, their concentration cannot be measured easily.

The eCFP–eYFP pair is the most popular for ratiometric imaging. However, this pair is not optimal for FLIM because eCFP is dim and has more than one fluorescence lifetime (Tramier et al. 2002; Yasuda et al. 2006), whereas FLIM does not take advantage of the brightness of eYFP.

Location of Fluorophore and Linker

Because FRET efficiency is sensitive to the distance and the angle between the two fluorophores, it depends on the length and the rigidity of the linkers connecting fluorophores and target proteins as well as the location of the fluorophore within the target protein (Miyawaki 2003). These parameters have to be optimized empirically (Miyawaki 2003). The optimum FRET efficiency is ~50% for a protein-binding assay when using FLIM versus 100% for ratiometric imaging (Yasuda 2006; Yasuda et al. 2006).

Binding Kinetics

When using a sensor to observe interactions between a target protein and a binding partner that specifically binds to the active form of the protein (such as in a Ras sensor; Fig. 3A), the dissociation constants for binding between the two proteins significantly impact the sensitivity and kinetics of the sensor. The optimum dissociation constants are related to the expression level of the binding partner. For example, in the case of our Ras sensor, in which binding between meGFP-Ras and mRFP-RBD-mRFP is used as a readout of Ras activity (Fig. 3A), a typical expression level of mRFP-RBD-mRFP in neurons is in the 1–20-μM range (Murakoshi and Yasuda, unpubl.). Thus, for optimal sensitivity, the dissociation constant between RBD and Ras^{GTP} (active form) must be lower than this range, and the dissociation between RBD and Ras^{GDP} (inactive form) must be higher than these levels. Also, keep in mind that binding of the binding partner may slow down inactivation of the target protein. In the case of the Ras sensor (Fig. 3A), RBD binding to Ras^{GTP} competes with the activity of GTPase-activating proteins (GAPs) (Ras inactivator). More quantitatively, the inactivation time constant τ_{GAP} is a function of [RBD] and K_D,

$$\tau_{GAP} = 1 + \frac{[\mathrm{RBD}]}{K_D}\tau^0_{GAP}, \qquad (5)$$

where τ^0_{GAP} is τ_{GAP} when no RBD is present and K_D is the dissociation constant of RBD and Ras^{GTP} (Yasuda et al. 2006). To minimize slowdown of Ras inactivation by GAPs, K_D has to be higher than [RBD]. However, to obtain an optimal binding signal, K_D has to be lower than [RBD]. Thus, an RBD with a K_D similar to or slightly less than the expression level of RBD (micromolar level) should be chosen. The Ras sensor becomes almost irreversibly active when wild-type cRaf1-RBD ($K_D \sim 0.1$ µM) is used, whereas the sensor is inactivated as fast as endogenous Ras when a low-affinity mutant of Raf-RBD (R59A; $K_D \sim 4$ µM) is used (Yasuda et al. 2006; Harvey et al. 2008).

In cases in which the binding partner and the target protein are in the same polypeptide, such as in the ratiometric Ras sensor (Mochizuki et al. 2001), the optimal dissociation constant must be determined empirically (Ni et al. 2006).

EVALUATION OF THE SENSITIVITY OF SENSORS

One way to evaluate the dynamic range of a sensor is to coexpress the sensor with an activator or an inactivator protein within cells and to measure the sensor signals (Mochizuki et al. 2001). For some proteins, constitutively active or inactive mutants of the target protein could be used to test the dynamic range (Mochizuki et al. 2001; Yasuda et al. 2006). The dynamic range of FLIM sensors can be up to ~50% in terms of the binding fraction or ~25% in the mean fluorescence lifetime (Yasuda et al. 2006; Lee et al. 2009).

EVALUATION OF THE OVEREXPRESSION EFFECTS OF SENSORS

Because the expression of a sensor can affect the kinetics of the sensor signal and can perturb cell signaling, it is necessary to evaluate the effects of overexpression on measured parameters such as time course and spatial spreading (Harvey et al. 2008). To do so, quantify the correlation between the parameters and the overexpression level, which is estimated from measurements of sensor concentration (Harvey et al. 2008). If a significant correlation is found, the value can be extrapolated to zero expression level to estimate the parameter values in the unperturbed system.

Measurements of Sensor Concentration

To measure the concentration of sensors transfected into cells, the fluorescence intensity of the sensor inside a compartment that is larger than the two-photon excitation volume within the cell (such as a primary dendrite in pyramidal neurons) can be measured with 2pLSM (Harvey et al. 2008; Lee et al. 2009). Next, the fluorescence intensity value is converted into concentration by comparing it with the fluorescence of a known concentration of purified fluorescent protein that has been measured separately (Lee et al. 2009). For cells deep within tissue, the effects of light scattering may have to be taken into account. To do so, fluorescent beads are injected near the cells of interest, and the measured fluorescence intensity of the beads is used to normalize the fluorescence intensity from the sensor (Gray et al. 2006; Harvey et al. 2008). That value is then converted into the fluorophore concentration by using the ratio between the fluorescence intensity of the beads in solution and the ratio of purified fluorescent protein with known concentration.

CONCLUSION

2pFLIM has proven to be useful for imaging signaling events in small subcellular compartments, especially in light-scattering tissue (Yasuda 2006; Harvey et al. 2008; Lee et al. 2009). The technique has been used to monitor several key proteins involved in synaptic plasticity, such as Ras (Yasuda 2006; Harvey et al. 2008) and CaMKII (Lee et al. 2009), within single dendritic spines in brain tissue. With continued optimization and development of FLIM sensors, the highly sensitive 2pFLIM will become increasingly applicable for imaging activity of a variety of proteins.

ACKNOWLEDGMENTS

I thank R. Kasliwal, M. Patterson, S.-J. Lee, and H. Murakoshi for discussions.

REFERENCES

Förster VT. 1993. Intermolecular energy migration and fluorescence (Translation of Förster T, 1948). In *Biological physics* (ed. Mielczarek EV, et al.), pp. 183–221. American Institute of Physics, New York.

Ganesan S, Ameer-Beg SM, Ng TT, Vojnovic B, Wouters FS. 2006. A dark yellow fluorescent protein (YFP)-based resonance energy-accepting chromoprotein (REACh) for Förster resonance energy transfer with GFP. *Proc Natl Acad Sci* **103:** 4089–4094.

Gray NW, Weimer RM, Bureau I, Svoboda K. 2006. Rapid redistribution of synaptic PSD-95 in the neocortex in vivo. *PLoS Biol* **4:** e370. doi: 10.1371/journal.pbio.0040370.

Harvey CD, Yasuda R, Zhong H, Svoboda K. 2008. The spread of Ras activity triggered by activation of a single dendritic spine. *Science* **321:** 136–140.

Lakowicz JR. 2006. *Principles of fluorescence spectroscopy*, 3rd ed. Plenum, New York.

Lee SJ, Escobedo-Lozoya Y, Szatmari EM, Yasuda R. 2009. Activation of CaMKII in single dendritic spines during long-term potentiation. *Nature* **458:** 299–304.

Miyawaki A. 2003. Visualization of the spatial and temporal dynamics of intracellular signaling. *Dev Cell* **4:** 295–305.

Miyawaki A, Llopis J, Heim R, McCaffery JM, Adams JA, Ikura M, Tsien RY. 1997. Fluorescence indicators for Ca^{2+} based on green fluorescent proteins and calmodulin. *Nature* **388:** 882–887.

Mochizuki N, Yamashita S, Kurokawa K, Ohba Y, Nagai T, Miyawaki A, Matsuda M. 2001. Spatio-temporal images of growth-factor-induced activation of Ras and Rap1. *Nature* **411:** 1065–1068.

Murakoshi H, Iino R, Kobayashi T, Fujiwara T, Ohshima C, Yoshimura A, Kusumi A. 2004. Single-molecule imaging analysis of Ras activation in living cells. *Proc Natl Acad Sci* **101:** 7317–7322.

Murakoshi H, Lee S-J, Yasuda R. 2008. Highly sensitive and quantitative FRET-FLIM imaging in single dendritic spines using improved non-radiative YFP. *Brain Cell Biol* **36:** 31–42.

Ni Q, Titov DV, Zhang J. 2006. Analyzing protein kinase dynamics in living cells with FRET reporters. *Methods* **40:** 279–286.

Peter M, Ameer-Beg SM, Hughes MK, Keppler MD, Prag S, Marsh M, Vojnovic B, Ng T. 2005. Multiphoton-FLIM quantification of the EGFP-mRFP1 FRET pair for localization of membrane receptor-kinase interactions. *Biophys J* **88:** 1224–1237.

Philip J, Carlsson K. 2003. Theoretical investigation of the signal-to-noise ratio in fluorescence lifetime imaging. *J Opt Soc Am A* **20:** 368–379.

Shaner NC, Campbell RE, Steinbach PA, Giepmans BN, Palmer AE, Tsien RY. 2004. Improved monomeric red, orange and yellow fluorescent proteins derived from *Discosoma sp.* red fluorescent protein. *Nat Biotechnol* **22:** 1567–1572.

Takao K, Okamoto K, Nakagawa T, Neve RL, Nagai T, Miyawaki A, Hashikawa T, Kobayashi S, Hayashi Y. 2005. Visualization of synaptic Ca^{2+}/calmodulin-dependent protein kinase II activity in living neurons. *J Neurosci* **25:** 3107–3112.

Tramier M, Gautier I, Piolot T, Ravalet S, Kemnitz K, Coppey J, Durieux C, Mignotte V, Coppey-Moisan M. 2002. Picosecond-hetero-FRET microscopy to probe protein-protein interactions in live cells. *Biophys J* **83:** 3570–3577.

Wallrabe H, Periasamy A. 2005. Imaging protein molecules using FRET and FLIM microscopy. *Curr Opin Biotechnol* **16:** 19–27.

Yasuda R. 2006. Imaging spatiotemporal dynamics of neuronal signaling using fluorescence resonance energy transfer and fluorescence lifetime imaging microscopy. *Curr Opin Neurobiol* **16:** 551–561.

Yasuda R, Harvey CD, Zhong H, Sobczyk A, van Aelst L, Svoboda K. 2006. Super-sensitive Ras activation in dendrites and spines revealed by two-photon fluorescence lifetime imaging. *Nat Neurosci* **9:** 283–291.

Zacharias DA, Violin JD, Newton AC, Tsien RY. 2002. Partitioning of lipid-modified monomeric GFPs into membrane microdomains of live cells. *Science* **296:** 913–916.

29 Imaging Gene Expression in Live Cells and Tissues

Hao Hong,[1] Yunan Yang,[1] and Weibo Cai[1,2]

[1]*Departments of Radiology and Medical Physics, School of Medicine and Public Health, University of Wisconsin–Madison, Madison, Wisconsin 53705-2275;* [2]*University of Wisconsin Carbone Cancer Center, Madison, Wisconsin 53705*

ABSTRACT

Monitoring gene expression is crucial for studying the responses of gene therapy and clarifying gene function in various environments. Molecular imaging is a powerful tool for noninvasive visualization of gene expression. This chapter will summarize the current status of fluorescence and bioluminescence imaging (BLI) of gene expression in live cells and tissues, with the emphasis mainly on the early studies that pioneered the field. First, we will describe fluorescence imaging of gene expression with a wide variety of fluorescent proteins. Next, we will discuss the strategies for BLI of gene expression. Besides incorporating the reporter gene into the host DNA, mRNA-based BLI of gene expression will also be briefly mentioned. Lastly, the construction of double- and triple-fusion reporter genes will be presented. Because no single imaging modality is perfect and sufficient to obtain all of the necessary information for a given question, combination of multiple molecular imaging modalities can offer synergistic advantages over any modality alone. Noninvasive optical imaging of gene expression has revolutionized biomedical research, and the progress made over the last decade should allow molecular imaging to play a major role in the field of gene therapy. For basic and preclinical research, optical imaging is indispensable for imaging gene expression. However, for clinical imaging of gene expression, positron emission tomography (PET) holds the greatest promise.

INTRODUCTION

With the ability to engineer genes and to create knockin and knockout models of human diseases, it has become clear that gene therapy can play an important role in disease management. Over the last four decades, gene therapy has moved from preclinical to clinical studies for diseases ranging from monogenic recessive disorders (e.g., hemophilia) to more complex diseases, such as cancer, cardiovascular disorders, and human immunodeficiency virus (HIV) infection (Serganova et al. 2008; Gillet et al. 2009). To choose the appropriate gene therapy strategy and to optimize the therapeutic efficacy, an effective monitoring system is needed. One of the most economical and practical approaches is to adopt certain molecular imaging techniques for monitoring gene expression in vivo.

In recent years, noninvasive molecular imaging has emerged as a powerful tool for monitoring cellular and molecular events in vivo (Cai and Chen 2007; Cai et al. 2008a,b). A variety of imaging technologies are being investigated as tools for evaluating gene therapy efficiency and for studying gene expression in living subjects. Noninvasive, longitudinal, and quantitative imaging of gene expression can help in human gene therapy trials and can also facilitate preclinical experimental studies in animal models. Radionuclide-based imaging techniques (i.e., single-photon emission computed tomography [SPECT] and PET) have the greatest clinical potential and offer many advantages for noninvasive imaging of gene expression.

Besides imaging gene vectors to indirectly visualize the gene expression efficiency, PET and SPECT can play a significant role in imaging gene expression using diverse reporter genes and reporter probes (Kang and Chung 2008). If transcription of a reporter gene is induced, translation of the reporter gene mRNA will lead to a protein product that can interact with the imaging reporter probe (administered in trace amounts for PET/SPECT applications; hence the term "tracer"). This interaction may be based on the intracellular enzymatic conversion of the reporter probe with the retention of the metabolite(s), or a receptor-ligand-based interaction.

Examples of intracellular reporters include the herpes simplex virus type 1 thymidine kinase (HSV1-tk) and its mutant gene (HSV1-sr39tk) (Gambhir et al. 1999, 2000; Najjar et al. 2009). Substrates that have been studied to date, as PET reporter probes for HSV1-tk, can be classified into two main categories: pyrimidine nucleoside derivatives (e.g., 2′-fluoro-2′-deoxy-5′-[^{124}I]iodo-1β-D-arabinofuranosyluracil [^{124}I-FIAU]) and acycloguanosine derivatives (e.g., 9-[4-[^{18}F]fluoro-3-(hydroxymethyl) butyl]guanine [^{18}F-FHBG]). A few examples of reporters on the surface of cells include the dopamine 2 receptor (D2R), receptors for the human type 2 somatostatin receptor (hSSTr2), and the sodium iodide symporter (NIS).

Although magnetic resonance imaging (MRI) has relatively low sensitivity, it can be used to image gene expression in vivo (Gilad et al. 2007, 2008). MRI has been used for imaging iron-related gene products (e.g., transferrin), certain enzymes (e.g., β-galactosidase [β-gal]), and chemical exchange saturation transfer–related products (e.g., polylysine-containing proteins) (Gilad et al. 2008).

Whereas PET/SPECT and MRI offer greater clinical potential than many other imaging modalities, optical imaging techniques, such as fluorescence and bioluminescence, have the following advantages: They are more economical, are easier to handle, do not need any radioisotopes, and are quite sensitive in certain scenarios. Therefore, optical imaging has been widely used for monitoring gene expression in live cells and tissues. In the remainder of this chapter, we will summarize the current state of the art in optical imaging of gene expression.

FLUORESCENCE IMAGING OF GENE EXPRESSION

For fluorescence imaging of gene expression, green fluorescent protein (GFP; emission peak at 509 nm) and its variants are the mainstay among fluorescent probes (Fig. 1). Because light at longer wavelengths (i.e., red or near-infrared) penetrates tissues better than light at shorter wavelengths, much effort has been devoted to engineering fluorescent proteins that can emit in the red or near-infrared.

GFP-Based Imaging

From the early 1990s, GFP and its variants have become invaluable markers for monitoring protein localization and gene expression in vivo (Heim and Tsien 1996; Jakobs et al. 2000). One early example of imaging gene expression with GFP was the demonstration of vascular endothelial growth factor (VEGF) promoter activity (Fukumura et al. 1998). The VEGF promoter was linked to the GFP gene, and the expression of GFP was imaged by intravital microscopy. For such intravital studies, in which only limited tissue penetration of the reporter protein is needed, GFP was found to be read-

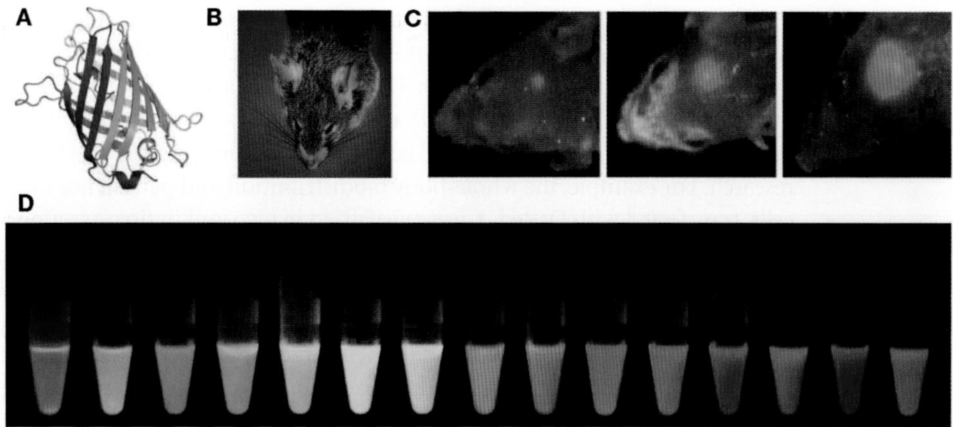

FIGURE 1. Fluorescence imaging of gene expression. (*A*) The structure of GFP. (*B*) A transgenic GFP mouse. (*C*) Real-time whole-body imaging of a GFP-expressing human glioma growing in the brain of a nude mouse at 1 wk (*left*), 3 wk (*center*), and 5 wk (*right*) after surgical orthotopic implantation. (*D*) Monomeric and tandem dimeric fluorescent proteins derived from *Aequorea* GFP or *Discosoma* RFP, expressed in bacteria, and purified. (Adapted, with permission, from Shaner et al. 2004; Hoffman and Yang 2006, ©Macmillan.)

ily detectable, confined to the cell, and stable (with a half-life of up to 24 h in cells). Although GFP penetrates tissue poorly, its expression can be visualized noninvasively in intact animals (Fig. 1) (Yang et al. 2000; Hoffman and Yang 2006). GFP has also been used frequently in stem cell research to monitor gene expression (Meyer et al. 2000; Niyibizi et al. 2004). Modifications of GFP to increase its signal intensity and thermostability, as well as to alter its emission spectrum, have expanded the utility of GFP in gene transfer studies (Welsh and Kay 1997).

GFP is very useful for imaging tumor-related gene expression. Mouse models of metastatic cancer were developed with genetically fluorescent tumors, in which the GFP gene was cloned into cancer cell lines and selected for stable GFP expression, which was imaged in fresh tissue, both in situ and externally (Hoffman 2001). With this model system, tumor location and metastasis were detected and were visualized in situ within host organs down to the single-cell level. Imaging GFP-transfected tumor cells was a fundamental advance in visualizing tumor growth and metastasis in real time in vivo. In addition, tumor growth and metastatic development and inhibition of these processes by certain drugs could now be imaged and quantified for rapid screening of anticancer drugs. Research about imaging tumor-related gene expression has flourished, including the use of a reporter gene system consisting of enhanced green fluorescent protein (eGFP) and wild-type HSV1-tk for both fluorescence and PET imaging (Luker et al. 2002).

GFP can also serve as a powerful tool for imaging other diseases, such as HIV infection. For example, the function of the HIV-1 Tat protein (capable of traversing cell membranes) has been elucidated by fluorescence imaging (Ferrari et al. 2003). With exogenously adding Tat-GFP fusion protein to live HeLa and CHO cells, it was found that the internalization process of full-length Tat, as well as of heterologous proteins fused to the transduction domain of Tat, exploits a caveolar-mediated pathway. With fluorescence imaging, the dynamic movement of individual GFP-tagged, Tat-filled caveolae toward the nucleus was observed directly.

Red Fluorescent Protein–Based Imaging

In 1999, a new red fluorescent protein (RFP) was isolated from tropical corals, which was termed DsRed (Matz et al. 1999). With emission maxima at 509 and 583 nm, respectively, eGFP and DsRed are well suited for virtually crossover-free, dual-color imaging upon simultaneous excitation. Mixed populations of *Escherichia coli* expressing either eGFP or DsRed were imaged by one-photon and two-photon microscopy. Both excitation modes were found to be suitable for imaging cells expressing either of the fluorescent proteins.

The predominant use of DsRed has been for multicolor imaging in plants, together with the GFP variants because its redshifted excitation/emission spectra avoids damaging plant cells and tissues by the excitation light (Heikal et al. 2000; Dietrich and Maiss 2002). Dual gene expression has also been imaged in animal models with GFP and DsRed-2 after transduction with a dual-promoter lentiviral vector (Chen et al. 2004). RFP-based gene expression imaging has also been used extensively in stem cell research. For example, the whole-body biodistribution and persistence of multipotent adult progenitor cells, transfected with DsRed-2, were visualized in vivo and in tissue sections (Tolar et al. 2005). Another important field for RFP-based research is cancer cell imaging. For example, fluorescent pancreatic cancer cells expressing a high level of the DsRed-2 gene were used to establish an orthotopic metastatic pancreatic cancer model (Zhou et al. 2008). The high-level expression of DsRed-2 enabled noninvasive imaging of distant micrometastases in their target organs, even in deep tissue such as the lung.

Wild-type DsRed has several drawbacks including inefficient folding of the protein, extremely slow maturation of the chromophore, and tetramerization even in dilute solutions. Therefore, stepwise evolution of DsRed to a dimer and then to a true monomer, designated as mRFP1, was performed (Campbell et al. 2002). The monomer and DsRed show similar brightness in living cells. In addition, the excitation and emission peaks of mRFP1, 584 and 607 nm, respectively, are ~25-nm redshifted from DsRed, conferring greater tissue penetration and spectral separation from autofluorescence and other fluorescent proteins. Subsequently, a series of monomeric fluorescent proteins have been reported, which further expanded the toolbox for fluorescence imaging of gene expression (Fig. 1) (Shaner et al. 2004; Wang et al. 2004).

Infrared Fluorescent Protein (IFP)

Generally speaking, GFP remains the first choice for fluorescence imaging of gene expression because it is relatively easy to fuse it with other genes and it has enhanced fluorescence characteristics compared with RFP. However, for optimal performance in in vivo imaging, near-infrared fluorescent (NIRF; 700–900 nm) proteins are superior. The NIRF signal can pass through deeper tissues, and the background fluorescence in this part of the spectrum is very low.

Recently, it was reported that a bacteriophytochrome from *Deinococcus radiodurans*, incorporating biliverdin as the chromophore, could be engineered into a monomeric IFP with excitation and emission maxima of 684 and 708 nm, respectively (Shu et al. 2009). The IFPs express well in mammalian cells and mice and spontaneously incorporate biliverdin, which is ubiquitous as the initial intermediate in heme catabolism but has negligible fluorescence by itself. These IFPs provided the basis for further engineering of IFPs with better performance characteristics, which is expected to find broad use in future biomedical research, in particular, those related to gene expression imaging.

β-Galactosidase-Based Imaging

Besides the fusion of fluorescent proteins with target gene products, several other strategies can also be adopted for fluorescence imaging of gene expression. The use of fluorogenic substrates for certain gene translation products (e.g., enzymes and proteins) is one choice. For example, the bacterial *lacZ* gene, which encodes β-gal, is a common reporter gene used in transgenic mice (Watson et al. 2008). However, the absence of fluorogenic substrates usable in live animals greatly hampered the applications of this reporter gene. In 2007, a far-red fluorescent substrate, 9H-(1,3-dichloro-9,9-dimethylacridin-2-one-7-yl) β-D-galactopyranoside, was developed for imaging β-gal expression (Josserand et al. 2007). With β-gal as a reporter of tumor growth, as few as 1000 β-gal-expressing tumor cells located under the skin could be detected with this substrate.

Tomographic Imaging of Gene Expression

Aside from the various strategies for imaging gene expression in live cells and tissues by fluorescence, as mentioned above, there have also been significant advances in imaging instrumentation. To generate a more quantitative fluorescence readout, fluorescence diffuse tomography has been developed recently for small animal imaging (Turchin et al. 2008). The animal is scanned in the transilluminative configuration by a single source and a detector pair, and a reconstruction algorithm was developed to estimate the fluorophore distribution. Another technique, fluorescence-mediated tomography (FMT) can also provide important insights into in vivo gene expression within deep tissues (Ntziachristos et al. 2003).

With the wide variety of tools for fluorescence imaging of gene expression, multiplexing is highly desirable and certainly possible. Using an imaging system capable of spectral unmixing, in which fluorophores with different emission spectra can be readily separated and analyzed simultaneously within the same sample (Cai et al. 2006; Cai and Chen 2008), multiple genes fused to different fluorescent proteins can be imaged at the same time, shedding new light into the biology and the mechanisms of various diseases. With newly developed fluorescent proteins that emit at longer wavelengths and the constantly evolving imaging systems, which can be used clinically to image certain sites of the human body (e.g., tissues close to the skin, accessible by endoscopy, and/or during surgery), fluorescence imaging of gene expression serves as an invaluable research tool for cell- and animal-based studies, as well as for certain potential clinical applications.

BIOLUMINESCENCE IMAGING OF GENE EXPRESSION

Another optical imaging technique, BLI, is an attractive alternative to fluorescence for optical imaging of gene expression. Although BLI has little, if any, clinical potential, it is virtually free of any background signal, an ideal property for sensitive imaging of gene expression in cells and small animals. Bioluminescence is the emission of light from biochemical reactions that occur within a living organism. The luciferases are a family of photoproteins that can be isolated from a variety of insects, marine organisms, and prokaryotes (Hastings 1996). Luciferase has been used as a reporter gene in transgenic mice but, until the instrumentation was created for BLI, the detection of luciferase activity required either sectioning the animal or excising the tissue and homogenizing it to measure the luciferase activity in a luminometer (Sadikot and Blackwell 2008). BLI has proven to be a very powerful method for detecting luciferase activity in intact animal models. It is noninvasive, convenient, and relatively inexpensive, thus making it an excellent method for elucidating the pathobiology of diseases such as inflammation/injury, infection, and cancer in animal models. To date, firefly luciferase (Fluc; Fig. 2A) and *Renilla* luciferase (Rluc; Fig. 2B) are the most widely used reporter proteins for BLI of gene expression in living animals. Another type of luciferase, the bacterial luciferase, is primarily used in the study of infection because it is limited to bacteria that express the reporter gene and produce the substrate (Contag et al. 1995).

Fluc-Based BLI

Light emission (yellow–green color, 557 nm) from the firefly *Photinus pyralis*, which is generally believed to be the most efficient bioluminescence system known to date, makes Fluc an excellent tool for monitoring gene expression (Branchini et al. 2005). Fluc has been used to image gene expression in cells since the early 1990s (Hooper et al. 1990). In 2001, a method was described for repetitively tracking in vivo gene expression of Fluc in skeletal muscles of mice with a cooled CCD camera (Fig. 2C) (Wu et al. 2001). The in vivo bioluminescence signals correlated well with results from in vitro luciferase enzyme assays, which showed the ability of BLI to sensitively and noninvasively track the location, the magnitude, and the persistence of Fluc-related gene expression.

Fluc has broad applications in monitoring disease-related gene expression. For example, the cyclooxygenase-2 (COX-2) gene plays a role in a wide variety of physiologic pathways and is a major target for therapeutic intervention in many pathophysiologic contexts such as pain, fever, inflammation, and cancer. Expression of the COX-2 gene can be induced in a wide range of cells, in response to an ever-increasing number of stimuli. BLI was successfully performed to image the expression of the Fluc gene in tumor xenografts, which were stably transfected with a chimeric gene containing the first kilobase of the murine COX-2 promoter (Nguyen et al. 2003). The imaging data suggested that gene

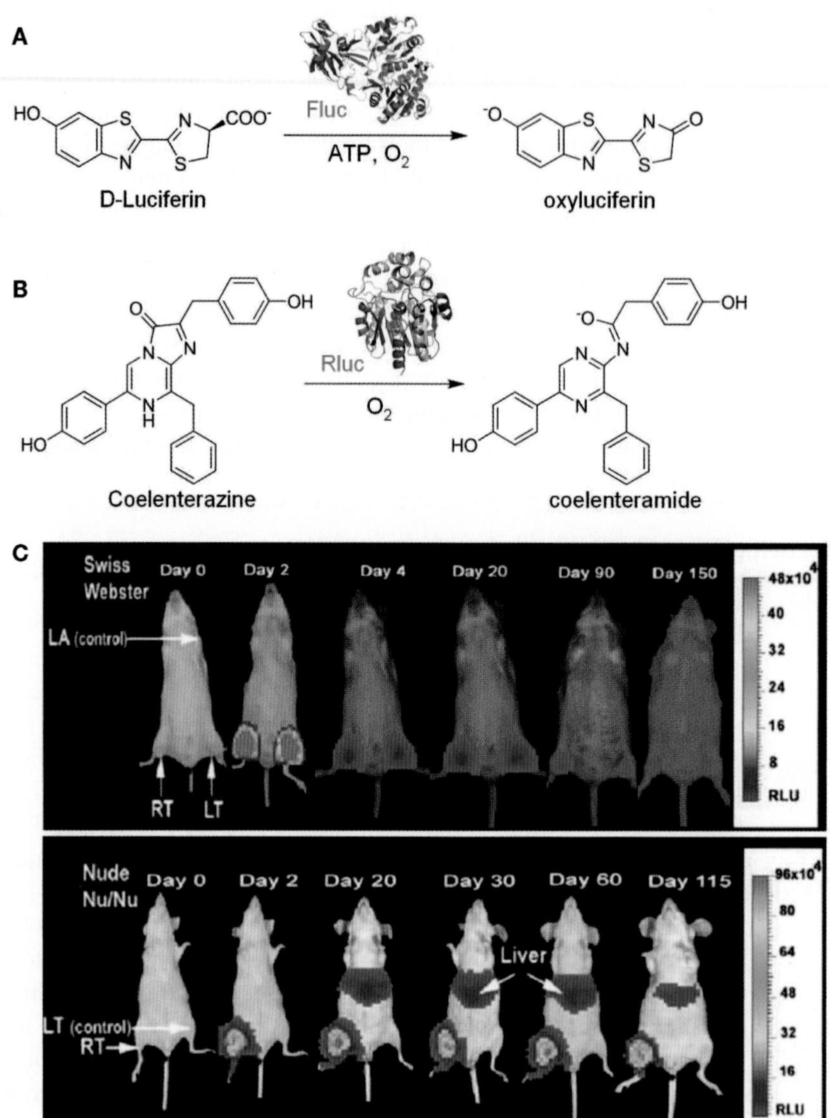

FIGURE 2. Bioluminescence imaging of gene expression. (*A*) The bioluminescent reaction catalyzed by Fluc. (*B*) The bioluminescent reaction catalyzed by Rluc. (*C*) The magnitude and the duration of Fluc gene expression in different host immune systems can be monitored noninvasively and repetitively using a cooled charge-coupled device (CCD) camera. For Swiss Webster mice, the Fluc activity was highest at day 2 and significantly dropped over time. For nude mice, a considerable amount of Fluc activity was seen throughout the study period. (Adapted, with permission, from Wu et al. 2001, ©Macmillan.)

expression from the COX-2 promoter can be easily analyzed in a variety of disease models in which the COX-2 gene is upregulated. BLI of gene expression with Fluc was also tested in larger animals such as rabbits (Li et al. 2005). It was reported that the BLI signal was capable of passing through at least 1 cm of muscle tissue.

BLI with Fluc is also applicable for the evaluation of DNA vaccines. Administration of naked DNA into animals has been used as a research tool for developing DNA vaccines. To monitor the distribution and the duration of gene expression of a DNA vaccine in living subjects, the naked DNA encoding Fluc was used as an imaging reporter gene in a mouse model (Jeon et al. 2006). It was concluded that BLI with Fluc could be useful for monitoring the location, the intensity, and the duration of gene expression of naked DNA vaccines in living animals, both noninvasively and repetitively. Similar to the study of DNA vaccines, BLI with Fluc (which serves as a model gene) can also be used to evaluate the gene delivery efficiency of certain gene delivery systems, especially in tumor models (Hildebrandt et al. 2003; Liang et al. 2004).

One major limitation of Fluc is that the light generated is in the yellow–green range, which has poor tissue penetration. In 2005, a set of red- and green-emitting luciferase mutants were reported (Branchini et al. 2005). The bioluminescence properties of these mutants are suitable for expanding the use of the *P. pyralis* system in dual-color reporter assays, biosensor measurements with internal controls, and imaging. Using a combination of mutagenesis methods, a red-emitting luciferase with a bioluminescence maximum of 615 nm was created, which also has a narrow emission bandwidth and favorable kinetic properties. Studies in animal models showed that these luciferases could detect gene expression at the attomole level, many orders of magnitude more sensitive than Fluc.

Rluc-Based BLI

Rluc is another promising bioluminescence reporter. Distinct from Fluc in terms of its origin, enzyme structure, and substrate requirements (Inouye and Shimomura 1997), Rluc was isolated from the sea pansy (*Renilla reniformis*), which displays blue–green bioluminescence on stimulation. It can catalyze the oxidation of coelenterazine, which leads to a bioluminescence signal (Fig. 2B). Rluc was cloned and sequenced in 1991 (Lorenz et al. 1991) and has been used as a marker for gene expression in bacteria, yeast, plant, and mammalian cells (Lorenz et al. 1996).

BLI of gene expression with Rluc dates back to the early 2000s. In one pioneering report, Rluc was used for noninvasive BLI of advanced human prostate cancer lesions in living mice by a targeted gene transfer vector (Adams et al. 2002). In another early study, the ability to image gene expression based on Rluc bioluminescence was validated by injecting the substrate coelenterazine into living mice (Bhaumik and Gambhir 2002). Cells transiently expressing the Rluc gene, located in the peritoneum, subcutaneous layer, as well as in the liver and the lungs of living mice, were imaged after tail-vein injection of coelenterazine. Importantly, both Rluc and Fluc expressions can be imaged in the same living mouse, although the kinetics of light production is distinct for the two enzymes. The imaging strategy validated in this study has direct application in various studies in which two molecular events need to be tracked, such as the trafficking of two cell populations, two gene therapy vectors, or indirect monitoring of two endogenous genes.

In a follow-up study, the expression in live mice of a novel synthetic Rluc reporter gene (hRluc) was explored, which has previously been reported to be a more sensitive reporter in mammalian cells than the native Rluc (Bhaumik et al. 2004a). It was found that hRluc:coelenterazine yielded stronger signals than the Fluc:D-luciferin in both cell culture and live animal studies.

During the prognosis of cancer treatment, early detection of the tumor and its metastases is crucial. In animal models, Rluc can be used for detecting the expression of diagnostic genes in different stages of tumor development, thus delineating the location of primary and metastatic tumors (Yu et al. 2003). To achieve noninvasive measurement of chemotherapy-induced changes in the expression of genes related to tumor growth, BLI was used to image the alteration in human telomerase reverse transcriptase (hTERT) gene expression in tumor cells before and after 5-fluorouracil treatment (Padmanabhan et al. 2006). Several fusion reporters directed by the hTERT promoter fragments were investigated, which integrate the hRluc (for BLI), the mRFP1 (for fluorescence imaging), and a truncated thymidine kinase (for PET imaging with radiolabeled acycloguanosines), respectively. In vitro studies showed that although all three of the reporter systems can visualize the hTERT promoter activity, BLI was the most powerful one, and it could also provide assistance in choosing the optimal tracers for PET imaging (Wang et al. 2006).

BLI has been widely adopted for stem cell–based studies. The discovery of human embryonic stem cells (hESCs) has dramatically expanded the tools available to scientists and clinicians in the field of regenerative medicine (Thomson et al. 1998). However, direct injection of hESCs, and cells differentiated from hESCs, into living organisms was hampered by significant cell death, teratoma formation, and host immune rejection. Understanding the hESC behavior in vivo after transplantation requires noninvasive imaging techniques, such as BLI, to longitudinally monitor the hESC localization, proliferation, and viability (Wilson et al. 2008). Bioluminescence reporter genes can be transcribed either constitutively or under specific biological or cellular conditions, depending on the type of promoter used. Stably transduced cells that carry the reporter construct within their chromosomal DNA will pass the reporter construct DNA to daughter cells, allowing for longitudinal monitoring of hESC survival and proliferation in vivo with BLI. Because expression of the reporter gene product is required for signal generation, only viable parent and daughter cells will generate a BLI signal, whereas the apoptotic or dead cells cannot, which is the key advantage of reporter gene-based cell tracking over direct cell labeling (Zhang and Wu 2007).

BLI of mRNA

In addition to incorporating Fluc or Rluc into DNA, a strategy for imaging mRNA with BLI was designed using spliceosome-mediated RNA *trans*-splicing (SMaRT) (Bhaumik et al. 2004b; Walls et al. 2008). SMaRT provides an effective means for reprogramming mRNAs and the proteins they encode. It can have a broad range of applications, including RNA repair and molecular imaging, each governed by the nature of the sequences delivered by the pre-*trans*-splicing molecule (PTM). In one groundbreaking study, the ability of SMaRT to optically image the expression of an exogenous gene at the level of pre-mRNA splicing was shown in both cells and living animals (Bhaumik et al. 2004b). Because of the modular design of PTMs, there is potential for SMaRT to be used for imaging the expression of any arbitrary gene of interest in living subjects.

Recently, this strategy was improved (Fig. 3A) (Walls et al. 2008) by using signal amplification and a facile method of delivery for developing a class of generalized probes capable of imaging pre-mRNA in a sequence-specific manner. Incorporating a modular binding domain that confers specificity by base-pair complementarity to the target pre-mRNA, the PTMs were designed to target a chimeric target minigene and *trans*-splice the Rluc gene onto the end of the target. After hydrodynamic delivery of the PTMs and target genes in mice, the efficiency and the specificity of the trans-splicing reaction was found to vary depending on the binding domain length and the structure (Fig.

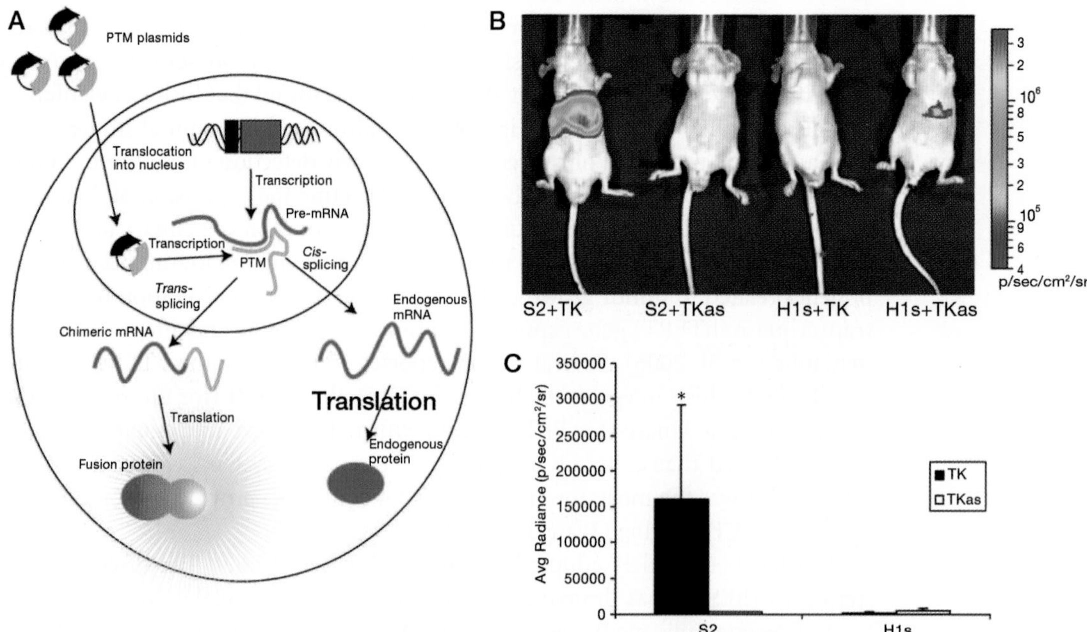

FIGURE 3. BLI of mRNA. (*A*) A schematic of the SMaRT imaging strategy. (*B*) Representative images of nude mice injected hydrodynamically with a combination of PTM and target plasmids, S2+TK:PTM and the target gene. The remaining three mice are negative controls. (C) Average radiance for each mouse (mean ± S.E.M.) in photons/sec/cm²/steradian. (Adapted, with permission, from Walls et al. 2008, ©Society of Nuclear Medicine.)

3). Nonetheless, specific *trans*-splicing was observed in living animals that showed a proof of principle for a generalized imaging probe against RNA, which can amplify the signal on detection and can be delivered with existing gene delivery methodology.

BLI of gene expression has been used in a wide variety of disease models over the last decade, and the above-mentioned studies serve only as representative examples. Theoretically, Fluc or Rluc can be fused to almost any gene of interest, and the expression of the gene can be noninvasively monitored with BLI. One major advantage of BLI is that it is quite quantitative. Therefore, the changes in the bioluminescence signal can reflect the differences in gene expression level, provided that all other related variables remain constant (e.g., the depth of the tissue). Over the years, BLI has been employed in a variety of model systems for applications ranging from stem cell trafficking to cancer therapy to drug screening. Bear in mind, however, that BLI is exclusively a research tool in preclinical studies and cannot be used in humans.

MULTIMODALITY IMAGING OF GENE EXPRESSION

Among all of the molecular imaging modalities, no single modality is sufficient to obtain all of the necessary information for a given question. For example, it is difficult to accurately quantify fluorescence signals in living subjects with fluorescence imaging alone, particularly in deep tissues; BLI is quantitative but is not suitable for clinical studies; radionuclide-based imaging techniques, such as PET, are very sensitive but have relatively poor spatial resolution. A combination of multiple molecular imaging modalities can offer synergistic advantages over any modality alone. Various double-fusion and triple-fusion reporter genes have been reported for multimodality imaging of gene expression. In one early study, a reporter vector encoding a mutant HSV1-sr39tk and Rluc was constructed (Ray et al. 2003). The two genes were joined by a 20-amino-acid spacer sequence. Both PET and BLI were able to delineate tumors stably expressing the fusion gene in live xenograft-bearing mice. Subsequently, several triple-fusion reporter genes that can monitor bioluminescence, fluorescence, and PET imaging have been constructed. A triple-fusion reporter vector consisting of Rluc, RFP, and HSV1-sr39tk (ttk) was found to confer activity of every protein composition in cell culture (Ray et al. 2004). In xenograft mouse models, the lentiviral vector encoding the triple-fusion reporter gene also showed a good correlation of signals from different imaging modalities. Further, the vector expression lasted for 40–50 days in live mice.

To exploit the combined strengths of each imaging technique and to facilitate multimodality imaging, a dual-reporter construct was established in which Fluc was fused in-frame to the amino terminus of a mutant HSV-tk kinetically enhanced for PET (Kesarwala et al. 2006). In addition, a triple-reporter construct was developed in which monster GFP was introduced into the fusion vector downstream of an internal ribosome entry site to allow analysis by fluorescence microscopy or by flow cytometry without compromising the specific activities of the upstream fusion components. In mice, somatic gene transfer of a ubiquitin promoter-driven triple-fusion plasmid showed a >1000-fold increase in liver photon flux and an more than twofold increase in liver retention of ^{18}F-FHBG by micro-PET when compared with mice treated with a control plasmid.

Construction and validation of several triple-fusion genes, each composed of a bioluminescent, a fluorescent, and a PET reporter gene in cell culture and in living subjects, were further explored (Ray et al. 2007). A mutant of a thermostable Fluc, bearing the peroxisome localization signal, was designed to have greater cytoplasmic localization and improved access for its substrate, D-luciferin. This mutant showed several-fold higher activity than Fluc both in vitro and in vivo. The improved version of the triple-fusion vector showed a significantly higher bioluminescence signal than the previous triple-fusion vectors. As the third reporter component of this triple-fusion vector for PET applications, a truncated version of wild-type HSV1-tk also retained a higher expression level than the truncated mutant HSV1-sr39tk. It was suggested that this improved triple-fusion reporter vector should enable high sensitivity detection of lower numbers of cells in living animals using the combined bioluminescence, fluorescence, and PET imaging techniques.

A proof-of-principle study for lentivirus-transduced murine embryonic stem cells (mESCs) that stably express this triple-fusion reporter gene was performed (Cao et al. 2006). It was shown that this

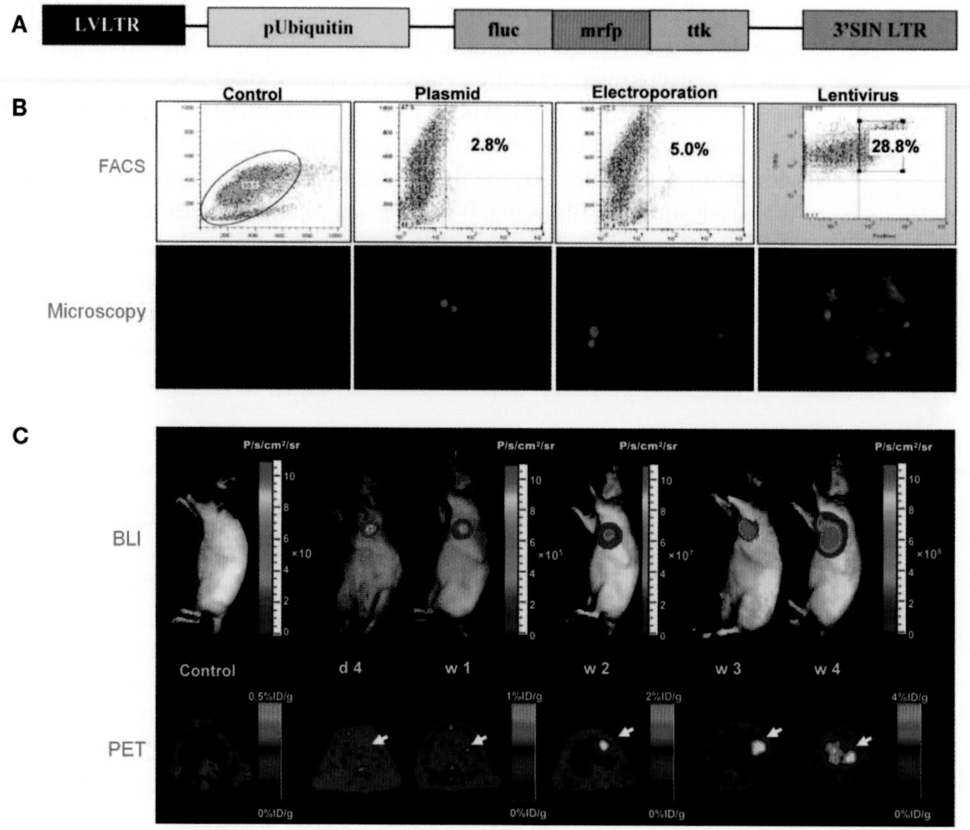

FIGURE 4. Multimodality imaging of mESC survival, proliferation, and migration after cardiac delivery with a triple-fusion reporter gene. (*A*) Schema of the triple-fusion reporter gene. (*B*) FACS histograms of mESCs at 48 h after transduction with plasmid lipofectamine, electroporation, and lentivirus carrying the triple-fusion reporter gene. (*C*) Noninvasive imaging of transplanted mESCs with BLI and PET. (Adapted, with permission, from Cao et al. 2006.)

molecular imaging platform can be used to monitor the kinetics of stem cell survival, proliferation, migration, and ablation of teratoma sites. The fluorescence feature of the reporter gene was used for cell sorting and microscopy studies, whereas both BLI and PET were used for longitudinal noninvasive monitoring of the transplanted mESCs (Fig. 4). Further development of novel imaging techniques will contribute important insights into the biology and the physiology of transplanted stem cells, leading to significant potential clinical applications for years to come.

Radiolabeled luciferase substrates have been explored for multimodality imaging of gene expression. In one report, a few ^{11}C-labeled D-luciferin analogs were synthesized (Wang et al. 2006). PET studies showed a low retention of the ^{11}C label at 45 min postinjection in luciferase-expressing tumors, whereas BLI with the unlabeled substrate D-luciferin and the radiolabeled analogs gave tumor signal within a few minutes of photon counting. Because D-luciferin and the radiolabeled analogs are substrates of Fluc, with a relatively high turnover rate, this is likely responsible for the poor retention of the tracers in the tumor. In addition, the required injected doses for these two imaging modalities also differ by several orders of magnitude. Therefore, although this is an interesting venue for exploring multimodality imaging, implementing it for optimal performance in animal studies is very challenging.

CONCLUSION

Recent advances in molecular imaging technologies have provided the potential for identifying changes at the genetic or molecular level long before they are detectable by conventional diagnostic techniques such as computed tomography or MRI. More importantly, these noninvasive imaging techniques can allow us to interrogate certain biological events (e.g., gene expression) in intact animals that previously could only be determined from in vitro assays of biopsied tissues or body fluids. To date, the vast majority of gene expression imaging studies has been in the preclinical stage. Although there are many gene therapy clinical trials currently ongoing, few of them have incorporated noninvasive imaging of gene expression into the trial.

The progress made over the last decade in the development of noninvasive imaging technologies for monitoring gene expression should allow molecular imaging to play a major role in the field of gene therapy (Min and Gambhir 2004). These tools have been validated in gene therapy models for longitudinal and quantitative monitoring of the location(s), magnitude, and time variation of gene delivery and/or expression. This chapter on optical imaging of gene expression is primarily applicable to preclinical research, and the emphasis has been mainly on the early studies that pioneered the field. The studies published to date clearly indicated that noninvasive imaging can accelerate the validation of preclinical models, which can give important insights toward clinical monitoring of human gene therapy. For clinical imaging of gene expression, PET holds the greatest promise (Yaghoubi et al. 2009).

Noninvasive optical imaging of gene expression has revolutionized biomedical research. It is no surprise that the 2008 Nobel Prize in Chemistry was awarded to three scientists for "the discovery and the development of the GFP." Although both fluorescent and bioluminescent optical imaging have certain inherent disadvantages, such as poor tissue penetration and limitations as quantitative tools, their low cost, convenience, constantly evolving imaging instrumentation, as well as emerging fluorescent proteins (Shaner et al. 2004; Shu et al. 2009) and luciferases (Loening et al. 2006, 2007) with better optical, physical, and biological characteristics, make these optical imaging methods irreplaceable in basic and preclinical research far beyond the field of imaging gene expression.

ACKNOWLEDGMENTS

We acknowledge financial support from the University of Wisconsin (UW) School of Medicine and Public Health's Medical Education and Research Committee through the Wisconsin Partnership Program, the UW Carbone Cancer Center, the National Center for Research Resources 1UL1RR025011, and a Susan G. Komen Postdoctoral Fellowship (to HH).

REFERENCES

Adams JY, Johnson M, Sato M, Berger F, Gambhir SS, Carey M, Iruela-Arispe ML, Wu L. 2002. Visualization of advanced human prostate cancer lesions in living mice by a targeted gene transfer vector and optical imaging. *Nat Med* **8:** 891–897.

Bhaumik S, Gambhir SS. 2002. Optical imaging of *Renilla* luciferase reporter gene expression in living mice. *Proc Natl Acad Sci* **99:** 377–382.

Bhaumik S, Lewis XZ, Gambhir SS. 2004a. Optical imaging of *Renilla* luciferase, synthetic *Renilla* luciferase, and firefly luciferase reporter gene expression in living mice. *J Biomed Opt* **9:** 578–586.

Bhaumik S, Walls Z, Puttaraju M, Mitchell LG, Gambhir SS. 2004b. Molecular imaging of gene expression in living subjects by spliceosome-mediated RNA *trans*-splicing. *Proc Natl Acad Sci* **101:** 8693–8698.

Branchini BR, Southworth TL, Khattak NF, Michelini E, Roda A. 2005. Red- and green-emitting firefly luciferase mutants for bioluminescent reporter applications. *Anal Biochem* **345:** 140–148.

Cai W, Chen X. 2007. Nanoplatforms for targeted molecular imaging in living subjects. *Small* **3:** 1840–1854.

Cai W, Chen X. 2008. Preparation of peptide conjugated quantum dots for tumour vasculature targeted imaging. *Nat Protoc* **3:** 89–96.

Cai W, Shin DW, Chen K, Gheysens O, Cao Q, Wang SX, Gambhir SS, Chen X. 2006. Peptide-labeled near-infrared quantum dots for imaging tumor vasculature in living subjects. *Nano Lett* **6:** 669–676.

Cai W, Niu G, Chen X. 2008a. Imaging of integrins as biomarkers for tumor angiogenesis. *Curr Pharm Des* **14:** 2943–2973.

Cai W, Niu G, Chen X. 2008b. Multimodality imaging of the HER-kinase axis in cancer. *Eur J Nucl Med Mol Imaging* **35:** 186–208.

Campbell RE, Tour O, Palmer AE, Steinbach PA, Baird GS, Zacharias DA, Tsien RY. 2002. A monomeric red fluorescent protein. *Proc Natl Acad Sci* **99:** 7877–7882.

Cao F, Lin S, Xie X, Ray P, Patel M, Zhang X, Drukker M, Dylla SJ, Connolly AJ, Chen X, et al. 2006. In vivo visualization of embryonic stem cell survival, proliferation, and migration after cardiac delivery. *Circulation* **113:** 1005–1014.

Chen HH, Zhan X, Kumar A, Du X, Hammond H, Cheng L, Yang X. 2004. Detection of dual-gene expression in arteries using an optical imaging method. *J Biomed Opt* **9:** 1223–1229.

Contag CH, Contag PR, Mullins JI, Spilman SD, Stevenson DK, Benaron DA. 1995. Photonic detection of bacterial pathogens in living hosts. *Mol Microbiol* **18:** 593–603.

Dietrich C, Maiss E. 2002. Red fluorescent protein DsRed from *Discosoma* sp. as a reporter protein in higher plants. *Biotechniques* **32:** 286, 288–290, 292–283.

Ferrari A, Pellegrini V, Arcangeli C, Fittipaldi A, Giacca M, Beltram F. 2003. Caveolae-mediated internalization of extracellular HIV-1 tat fusion proteins visualized in real time. *Mol Ther* **8:** 284–294.

Fukumura D, Xavier R, Sugiura T, Chen Y, Park EC, Lu N, Selig M, Nielsen G, Taksir T, Jain RK, Seed B. 1998. Tumor induction of VEGF promoter activity in stromal cells. *Cell* **94:** 715–725.

Gambhir SS, Barrio JR, Phelps ME, Iyer M, Namavari M, Satyamurthy N, Wu L, Green LA, Bauer E, MacLaren DC, et al. 1999. Imaging adenoviral-directed reporter gene expression in living animals with positron emission tomography. *Proc Natl Acad Sci* **96:** 2333–2338.

Gambhir SS, Herschman HR, Cherry SR, Barrio JR, Satyamurthy N, Toyokuni T, Phelps ME, Larson SM, Balatoni J, Finn R, et al. 2000. Imaging transgene expression with radionuclide imaging technologies. *Neoplasia* **2:** 118–138.

Gilad AA, Winnard PT Jr, van Zijl PC, Bulte JW. 2007. Developing MR reporter genes: Promises and pitfalls. *NMR Biomed* **20:** 275–290.

Gilad AA, Ziv K, McMahon MT, van Zijl PC, Neeman M, Bulte JW. 2008. MRI reporter genes. *J Nucl Med* **49:** 1905–1908.

Gillet JP, Macadangdang B, Fathke RL, Gottesman MM, Kimchi-Sarfaty C. 2009. The development of gene therapy: From monogenic recessive disorders to complex diseases such as cancer. *Methods Mol Biol* **542:** 5–54.

Hastings JW. 1996. Chemistries and colors of bioluminescent reactions: A review. *Gene* **173:** 5–11.

Heikal AA, Hess ST, Baird GS, Tsien RY, Webb WW. 2000. Molecular spectroscopy and dynamics of intrinsically fluorescent proteins: Coral red (dsRed) and yellow (Citrine). *Proc Natl Acad Sci* **97:** 11996–12001.

Heim R, Tsien RY. 1996. Engineering green fluorescent protein for improved brightness, longer wavelengths and fluorescence resonance energy transfer. *Curr Biol* **6:** 178–182.

Hildebrandt IJ, Iyer M, Wagner E, Gambhir SS. 2003. Optical imaging of transferrin targeted PEI/DNA complexes in living subjects. *Gene Ther* **10:** 758–764.

Hoffman RM. 2001. Visualization of GFP-expressing tumors and metastasis in vivo. *Biotechniques* **30:** 1016–1022, 1024–1016.

Hoffman RM, Yang M. 2006. Whole-body imaging with fluorescent proteins. *Nat Protoc* **1:** 1429–1438.

Hooper CE, Ansorge RE, Browne HM, Tomkins P. 1990. CCD imaging of luciferase gene expression in single mammalian cells. *J Biolumin Chemilumin* **5:** 123–130.

Inouye S, Shimomura O. 1997. The use of *Renilla* luciferase, *Oplophorus* luciferase, and apoaequorin as bioluminescent reporter protein in the presence of coelenterazine analogues as substrate. *Biochem Biophys Res Commun* **233:** 349–353.

Jakobs S, Subramaniam V, Schonle A, Jovin TM, Hell SW. 2000. EFGP and DsRed expressing cultures of *Escherichia coli* imaged by confocal, two-photon and fluorescence lifetime microscopy. *FEBS Lett* **479:** 131–135.

Jeon YH, Choi Y, Kang JH, Chung JK, Lee YJ, Kim CW, Jeong JM, Lee DS, Lee MC. 2006. In vivo monitoring of DNA vaccine gene expression using firefly luciferase as a naked DNA. *Vaccine* **24:** 3057–3062.

Josserand V, Texier-Nogues I, Huber P, Favrot MC, Coll JL. 2007. Non-invasive in vivo optical imaging of the *lacZ* and *luc* gene expression in mice. *Gene Ther* **14:** 1587–1593.

Kang JH, Chung JK. 2008. Molecular-genetic imaging based on reporter gene expression. *J Nucl Med* (suppl 2) **49:** 164S–179S.

Kesarwala AH, Prior JL, Sun J, Harpstrite SE, Sharma V, Piwnica-Worms D. 2006. Second-generation triple reporter for bioluminescence, micro-positron emission tomography, and fluorescence imaging. *Mol Imaging* **5:** 465–474.

Li JZ, Holman D, Li H, Liu AH, Beres B, Hankins GR, Helm GA. 2005. Long-term tracing of adenoviral expression in rat and rabbit using luciferase imaging. *J Gene Med* **7:** 792–802.

Liang Q, Yamamoto M, Curiel DT, Herschman HR. 2004. Noninvasive imaging of transcriptionally restricted transgene expression following intratumoral injection of an adenovirus in which the COX-2 promoter drives a reporter gene. *Mol Imaging Biol* **6:** 395–404.

Loening AM, Fenn TD, Wu AM, Gambhir SS. 2006. Consensus guided mutagenesis of *Renilla luciferase* yields enhanced stability and light output. *Protein Eng Des Sel* **19:** 391–400.

Loening AM, Wu AM, Gambhir SS. 2007. Red-shifted *Renilla* reniformis luciferase variants for imaging in living subjects. *Nat Methods* **4:** 641–643.

Lorenz WW, McCann RO, Longiaru M, Cormier MJ. 1991. Isolation and expression of a cDNA encoding *Renilla reniformis* luciferase. *Proc Natl Acad Sci* **88:** 4438–4442.

Lorenz WW, Cormier MJ, O'Kane DJ, Hua D, Escher AA, Szalay AA. 1996. Expression of the *Renilla reniformis* luciferase gene in mammalian cells. *J Biolumin Chemilumin* **11:** 31–37.

Luker GD, Luker KE, Sharma V, Pica CM, Dahlheimer JL, Ocheskey JA, Fahrner TJ, Milbrandt J, Piwnica-Worms D. 2002. In vitro and in vivo characterization of a dual-function green fluorescent protein—HSV1-thymidine kinase reporter gene driven by the human elongation factor 1 α promoter. *Mol Imaging* **1:** 65–73.

Matz MV, Fradkov AF, Labas YA, Savitsky AP, Zaraisky AG, Markelov ML, Lukyanov SA. 1999. Fluorescent proteins from nonbioluminescent Anthozoa species. *Nat Biotechnol* **17:** 969–973.

Meyer N, Jaconi M, Landopoulou A, Fort P, Puceat M. 2000. A fluorescent reporter gene as a marker for ventricular specification in ES-derived cardiac cells. *FEBS Lett* **478:** 151–158.

Min JJ, Gambhir SS. 2004. Gene therapy progress and prospects: Noninvasive imaging of gene therapy in living subjects. *Gene Ther* **11:** 115–125.

Najjar AM, Nishii R, Maxwell DS, Volgin A, Mukhopadhyay U, Bornmann WG, Tong W, Alauddin M, Gelovani JG. 2009. Molecular-genetic PET imaging using an HSV1-tk mutant reporter gene with enhanced specificity to acycloguanosine nucleoside analogs. *J Nucl Med* **50:** 409–416.

Nguyen JT, Machado H, Herschman HR. 2003. Repetitive, noninvasive imaging of cyclooxygenase-2 gene expression in living mice. *Mol Imaging Biol* **5:** 248–256.

Niyibizi C, Wang S, Mi Z, Robbins PD. 2004. The fate of mesenchymal stem cells transplanted into immunocompetent neonatal mice: Implications for skeletal gene therapy via stem cells. *Mol Ther* **9:** 955–963.

Ntziachristos V, Bremer C, Weissleder R. 2003. Fluorescence imaging with near-infrared light: New technological advances that enable in vivo molecular imaging. *Eur Radiol* **13:** 195–208.

Padmanabhan P, Otero J, Ray P, Paulmurugan R, Hoffman AR, Gambhir SS, Biswal S, Ulaner GA. 2006. Visualization of telomerase reverse transcriptase (hTERT) promoter activity using a trimodality fusion reporter construct. *J Nucl Med* **47:** 270–277.

Ray P, Wu AM, Gambhir SS. 2003. Optical bioluminescence and positron emission tomography imaging of a novel fusion reporter gene in tumor xenografts of living mice. *Cancer Res* **63:** 1160–1165.

Ray P, De A, Min JJ, Tsien RY, Gambhir SS. 2004. Imaging tri-fusion multimodality reporter gene expression in living subjects. *Cancer Res* **64:** 1323–1330.

Ray P, Tsien R, Gambhir SS. 2007. Construction and validation of improved triple fusion reporter gene vectors for molecular imaging of living subjects. *Cancer Res* **67:** 3085–3093.

Sadikot RT, Blackwell TS. 2008. Bioluminescence: Imaging modality for in vitro and in vivo gene expression. *Methods Mol Biol* **477:** 383–394.

Serganova I, Mayer-Kukuck P, Huang R, Blasberg R. 2008. Molecular imaging: Reporter gene imaging. *Handb Exp Pharmacol* **185:** 167–223.

Shaner NC, Campbell RE, Steinbach PA, Giepmans BN, Palmer AE, Tsien RY. 2004. Improved monomeric red, orange and yellow fluorescent proteins derived from *Discosoma* sp. red fluorescent protein. *Nat Biotechnol* **22:** 1567–1572.

Shu X, Royant A, Lin MZ, Aguilera TA, Lev-Ram V, Steinbach PA, Tsien RY. 2009. Mammalian expression of infrared fluorescent proteins engineered from a bacterial phytochrome. *Science* **324:** 804–807.

Thomson JA, Itskovitz-Eldor J, Shapiro SS, Waknitz MA, Swiergiel JJ, Marshall VS, Jones JM. 1998. Embryonic stem cell lines derived from human blastocysts. *Science* **282:** 1145–1147.

Tolar J, Osborn M, Bell S, McElmurry R, Xia L, Riddle M, Panoskaltsis-Mortari A, Jiang Y, McIvor RS, Contag CH, et al. 2005. Real-time in vivo imaging of stem cells following transgenesis by transposition. *Mol Ther* **12:** 42–48.

Turchin IV, Kamensky VA, Plehanov VI, Orlova AG, Kleshnin MS, Fiks II, Shirmanova MV, Meerovich IG, Arslanbaeva LR, Jerdeva VV, et al. 2008. Fluorescence diffuse tomography for detection of red fluorescent protein expressed tumors in small animals. *J Biomed Opt* **13:** 041310.

Walls ZF, Puttaraju M, Temple GF, Gambhir SS. 2008. A generalizable strategy for imaging pre-mRNA levels in living subjects using spliceosome-mediated RNA *trans*-splicing. *J Nucl Med* **49:** 1146–1154.

Wang L, Jackson WC, Steinbach PA, Tsien RY. 2004. Evolution of new nonantibody proteins via iterative somatic hypermutation. *Proc Natl Acad Sci* **101:** 16745–16749.

Wang JQ, Pollok KE, Cai S, Stantz KM, Hutchins GD, Zheng QH. 2006. PET imaging and optical imaging withD-luciferin [11C]methyl ester and D-luciferin [11C]methyl ether of luciferase gene expression in tumor xenografts of living mice. *Bioorg Med Chem Lett* **16:** 331–337.

Watson CM, Trainor PA, Radziewic T, Pelka GJ, Zhou SX, Parameswaran M, Quinlan GA, Gordon M, Sturm K, Tam PP. 2008. Application of *lacZ* transgenic mice to cell lineage studies. *Methods Mol Biol* **461:** 149–164.

Welsh S, Kay SA. 1997. Reporter gene expression for monitoring gene transfer. *Curr Opin Biotechnol* **8:** 617–622.

Wilson K, Yu J, Lee A, Wu JC. 2008. In vitro and in vivo bioluminescence reporter gene imaging of human embryonic stem cells. *J Vis Exp* **14:** 740. doi: 3791/740.

Wu JC, Sundaresan G, Iyer M, and Gambhir SS. 2001. Noninvasive optical imaging of firefly luciferase reporter gene expression in skeletal muscles of living mice. *Mol Ther* **4:** 297–306.

Yaghoubi SS, Jensen MC, Satyamurthy N, Budhiraja S, Paik D, Czernin J, Gambhir SS. 2009. Noninvasive detection of therapeutic cytolytic T cells with ^{18}F-FHBG PET in a patient with glioma. *Nat Clin Pract Oncol* **6:** 53–58.

Yang M, Baranov E, Moossa AR, Penman S, Hoffman RM. 2000. Visualizing gene expression by whole-body fluorescence imaging. *Proc Natl Acad Sci* **97:** 12278–12282.

Yu YA, Timiryasova T, Zhang Q, Beltz R, Szalay AA. 2003. Optical imaging: Bacteria, viruses, and mammalian cells encoding light-emitting proteins reveal the locations of primary tumors and metastases in animals. *Anal Bioanal Chem* **377:** 964–972.

Zhang SJ, Wu JC. 2007. Comparison of imaging techniques for tracking cardiac stem cell therapy. *J Nucl Med* **48:** 1916–1919.

Zhou J, Yu Z, Zhao S, Hu L, Zheng J, Yang D, Bouvet M, Hoffman RM. 2008. Lentivirus-based DsRed-2-transfected pancreatic cancer cells for deep in vivo imaging of metastatic disease. *J Surg Res* **157:** 63–70.

30 Multiphoton Excitation of Fluorescent Probes

Chris Xu and Warren R. Zipfel

Department of Biomedical Engineering, School of Applied and Engineering Physics, Cornell University, Ithaca, New York 14850

ABSTRACT

This chapter reviews the multiphoton excitation cross sections of extrinsic and intrinsic fluorophores, genetically engineered probes, and nanoparticles. We will review the known two-photon excitation cross sections of biological indicators and will discuss several related issues such as how to theoretically estimate and experimentally gauge the two-photon cross section of an indicator. We provide practical guides for experimentally estimating the excitation cross section.

INTRODUCTION

The fluorescence excitation or action cross section is a basic parameter in fluorescence imaging. The action cross section is a combined measure of how strongly a molecule absorbs at a particular excitation wavelength and of how efficiently the fluorophore converts the absorbed light into emitted fluorescence (i.e., the quantum yield). For conventional one-photon imaging, the one-photon absorption cross sections are well documented for a wide range of molecules, as are fluorescence quantum yields. However, significantly less is known about two-photon cross sections of biologically useful fluorophores. Furthermore, it is often difficult to predict multiphoton excitation spectra, especially two-photon spectra, from the one-photon data because of differences in the selection rules and the effects of vibronic coupling. Before the 1990s, two-photon excitation cross-sectional measurements were almost always performed at 694 nm (ruby laser) and 1064 nm (Nd:glass laser) on laser dyes (Smith 1986). Less effort was devoted to accurate quantitative studies of common fluorophores widely used in multiphoton microscopy. In addition, substantial disagreement (sometimes more than one order of magnitude) between published values of two-photon cross sections often exists. The lack of knowledge about two-photon excitation cross sections and spectra for common fluorophores used in biological studies has been a significant obstacle in the use of two-photon laser-scanning microscopy.

The emergence of the mode-locked solid-state femtosecond lasers, most commonly the Ti:sapphire lasers (Spence et al. 1991; Curley et al. 1992), have greatly facilitated the measurement of a multiphoton excitation cross section. When compared with earlier ultrafast lasers (e.g., ultrafast dye lasers), the Ti:sapphire lasers are highly robust and are widely tunable, making femtosecond pulses from ~690 to 1050 nm easily accessible. A large number of fluorescent indicators, intrinsic fluorescent molecules, genetically engineered probes, and some nanoparticles (e.g., quantum dots [QDs]) have since been measured. New measurement techniques have also been developed to further improve the absolute accuracy.

This review on multiphoton excitation cross sections is motivated by the application of multiphoton microscopy as a powerful tool for three-dimensionally resolved fluorescence imaging of biological samples (Denk et al. 1990; Xu et al. 1996a,b; So et al. 2000; Masters 2003; Zipfel et al. 2003b; Masters and So 2004; Helmchen and Denk 2005). We will review the known two-photon excitation (2PE) cross sections of biological indicators and will discuss several related issues such as how to experimentally gauge the two-photon cross section of an indicator. An effort is made to compare cross-sectional values obtained from various research groups in the last 10 years or so. Although we will concentrate on 2PE, three-photon excitation will also be discussed briefly. The remainder of this chapter is divided into three sections. First, we provide simple estimates of multiphoton excitations of dyes and QDs from theory. Second is the methods section, which includes practical guides for experimentally estimating the excitation cross section. The third section is a compilation of two-photon excitation cross sections of extrinsic and intrinsic fluorophores, genetically engineered probes, and nanoparticles.

ESTIMATION OF MULTIPHOTON EXCITATION CROSS SECTIONS

The essence of the theory of multiphoton processes can be represented in perturbation theory. Details of the rigorous derivation can be found elsewhere (Faisal 1987). Here, for the purpose of order of magnitude estimation, a greatly simplified approach is used to describe the multiphoton excitation processes, and only the lowest-order dipole transition will be considered. The single intermediate state approximation (Birge 1983) can be used to give an order of magnitude estimation. The two-photon absorption cross section (σ_2) can be obtained as

$$\sigma_2 = \sigma_{ij}\sigma_{jf}\tau_j, \tag{1}$$

where σ_{ij} and σ_{jf} represent the one-photon absorption cross sections from the initial state (i) to the intermediate state (j) and from the intermediate state (j) to the final state (f), respectively, and τ_j is the intermediate state lifetime. τ_j can be estimated from the uncertainty principle; that is, τ_j must be short enough to avoid violating energy conservation. Thus,

$$\tau_j \approx 1/\Delta\omega = 1/|\omega_{ij} - \omega|, \tag{2}$$

where ω_{ij} and ω are the transition frequency and the incident photon frequency, respectively. For an electronic transition (ω_{ij}) in the visible frequency range and assuming that the intermediate state and the final state are close in energy, then $\tau_j \approx 10^{-15}$–10^{-16} sec. The one-photon absorption cross section of a fluorescent molecule is typically $\sigma_1 \approx 10^{-16}$–$10^{-17}$ cm^2. Hence, the estimated two-photon absorption cross sections should be ~10^{-49} cm^4 sec/photon (Eq. 1), or 10 Göppert-Mayer (GM; 1 GM = 10^{-50} cm^4 sec/photon).

This description and estimation of multiphoton excitation for dye molecules needs to be modified in the case of nanocrystals, such as QDs (Efros and Efros 1982; Alivasatos 1996). A strong quantum confinement occurs when the size of the dot is much smaller than the exciton Bohr radius of the bulk material (i.e., the average physical separation between the electron and the hole in a bulk material). Although such an electron–hole pair (called an exciton) is analogous to a hydrogen atom, it should be emphasized that the charge carriers in a QD are bound by the confining potential of the boundary rather than by the Coulomb potential as in the bulk material (Wise 2000). A zeroth-order model for the description of a QD is a single particle confined in an infinite potential well. A desirable effect of the reduced dimension in a QD is the concentration of the density of state into discrete bands. Because light always interacts with one electron–hole pair regardless of the size of the material, the total or integrated absorption does not change. Thus, the concentration of the density of states significantly increases absorption at certain photon energies, at the expense of reduced absorption at other energies. A rough order of magnitude estimation of the absorption enhancement can be obtained by examining the density of states in a QD. There are approximately 10^4 atoms in a QD of 5-nm radius, resulting in a total integrated number of states of $\sim 10^4$. Assuming all these available states are concentrated in a few discrete bands, a QD will have a peak absorption strength of $\sim 10^4$ of that of a single atom. Using the estimation obtained above for a dye molecule, the two-photon absorption cross section of a QD will be $\sim 10^5$ GM.

MULTIPHOTON EXCITED FLUORESCENCE

For historical reasons, the units for two- and three-photon cross sections are a little confusing. We hope that the following discussion will provide the readers with a practical guide for using the numbers provided above.

The calculation for the experimentally detected two-photon excited fluorescence (F in photon/sec) is quite simple if the sample is much thicker than the confocal length of the focused beam. Such a thick sample limit is usually valid in biological imaging with high-numerical-aperture (high-NA) lenses when the confocal lengths are typically on the order of 1 µm. Using the two-photon cross section (σ_2) in units of GM, the expression for F can be obtained as

$$F \approx 1.28\,\phi\, C\,\eta\,\sigma_2\, n_0 \,\frac{\lambda P^2}{f\tau}, \tag{3}$$

where η is the fluorescence quantum efficiency (QE) of the dye, ϕ is the fluorescence collection efficiency of the measurement system, C (in µM) is the indicator concentration, λ (in µm) is the wavelength of excitation light in a vacuum, n_0 is the refractive index of the sample media, P (in milliwatts) is the average incident power, f is the pulse repetition rate, and τ is the excitation pulse width (full width at half-maximum). For example, if $\eta * \sigma_2 = 1$ GM, $C = 100$ µM, $\tau = 100$ fsec, $\lambda = 1$ µm, $n_0 = 1.3$, and a laser power of $P = 10$ mW at a repetition rate of 100 MHz, then $F \approx \phi \times 1.66 \times 10^9$ photon/sec.

Absolute measurement of two- and three-photon cross sections requires detailed characterization of the spatial and temporal profiles of the excitation beam. Details of the measurement method can be found elsewhere (Xu and Webb 1996, 1997). An effective and simple experimental approach is to compare the generated fluorescence of the specimen with some known two- or three-photon references provided that reliable two- or three-photon standards in the wavelength range of interest exist (Kennedy and Lytle 1986; Jones and Callis 1988). (Note that there was an error in the published Bis-MSB cross section value by Kennedy and Lytle [1986].) We note that the collection efficiencies must be taken into account when comparing the standard with the new indicators. Thus, it is more convenient to compare the indicator with a standard of similar fluorescence emission spectra. For example, for blue-emitting indicators, Cascade Blue (fluorescence peak at 423 nm) may be used as a standard; although for orange-emitting indicators, Rhodamine B (fluorescence peak at 570 nm) is more desirable.

When using such a reference method, the ratio of the experimentally measured fluorescence signals becomes (Albota et al. 1998b)

$$\frac{\langle F(t)\rangle_{\text{cal}}}{\langle F(t)\rangle_{\text{new}}} = \frac{\phi_{\text{cal}}\,\eta_{\text{cal}}\,\sigma_{2\text{cal}}\,C_{\text{cal}}\,\langle P_{\text{cal}}(t)\rangle^2\, n_{\text{cal}}}{\phi_{\text{new}}\,\eta_{\text{new}}\,\sigma_{2\text{new}}\,C_{\text{new}}\,\langle P_{\text{new}}(t)\rangle^2\, n_{\text{new}}}, \tag{4}$$

where the subscripts cal and new indicate the parameters for the calibration and the new fluorophore, respectively. The two-photon excitation action cross section of a new molecular fluorophore is then

TABLE 1. Two-photon absorption cross section of fluorescein in water (pH = 13)

Wavelength (nm)	2PE cross section (GM)	Wavelength (nm)	2PE cross section (GM)
691	16	870	9.0
700	19	880	11
710	17	890	14
720	19	900	16
730	25	910	23
740	30	920	26
750	34	930	23
760	36	940	21
770	37	950	18
780	37	960	16
790	37	970	17
800	36	980	16
810	32	992	13
820	29	1008	5.3
830	19	1020	3.2
840	13	1034	1.0
850	10	1049	0.23
860	8.0		

2PE, two-photon excitation.

related to known experimental wavelength-dependent parameters including the two-photon excitation action cross section of the calibration standard, as described by

$$\sigma_{2\text{new}}(\lambda)\eta_{\text{new}} = \frac{\phi_{\text{cal}}\,\eta_{\text{cal}}\,\sigma_{2\text{cal}}(\lambda)\,C_{\text{cal}}}{\phi_{\text{new}}\,C_{\text{new}}}\,\frac{\langle P_{\text{cal}}(t)\rangle^2}{\langle P_{\text{new}}(t)\rangle^2}\,\frac{\langle F(t)\rangle_{\text{new}}}{\langle F(t)\rangle_{\text{cal}}}\,\frac{n_{\text{cal}}}{n_{\text{new}}}. \tag{5}$$

Table 1 lists our measured two-photon cross-sectional values for fluorescein in the wavelength range of 690–1050 nm (Xu and Webb 1996). These values can be used to calibrate the two-photon cross sections of new indicators.

Although the values of the two-photon cross section appear small, one should still be aware of the possibility that fluorescence excitation can saturate. Detailed calculation of saturation can be found elsewhere (Xu and Webb 1997). As a practical guide, Rhodamine B will be used as an example. Rhodamine B has a 2PE cross section of 210 GM at 840 nm. With a mode-locked Ti:sapphire laser providing 100-fsec pulses at an 80-MHz repetition rate, the average power for the onset of saturation for Rhodamine B at 840 nm is ~8 mW at the specimen assuming a diffraction-limited focus with a 1.3-NA objective lens. Criteria for other indicators and focusing NAs can be extrapolated from this number by using the scaling relationship

$$\text{Saturation power} \approx \sigma 2^{-0.5}\,(\text{NA})^{-2}. \tag{6}$$

Besides 2PE, another way to image living cells is three-photon excitation (Wokosin et al. 1995; Xu et al. 1996b). Three-photon excited fluorescence provides the unique opportunity to excite intrinsic chromophores (such as amino acids, proteins, and neurotransmitters) using relatively benign excitation wavelengths accessible with commercially available near-IR lasers (Maiti et al. 1997). The combination of two- and three-photon excited fluorescence microscopy extends the useful range of nonlinear laser microscopy.

TWO-PHOTON CROSS-SECTIONAL DATA

In this section, we summarize measurements of a variety of dyes, intrinsic molecules, fluorescent proteins, and QDs. The first part of this section will compare the published cross sections and/or the two-photon excitation spectra after 1996. We collected the literature by doing forward citation searches of the papers that we published in cross-sectional measurements. Thus, the literature covered in this section may not include all relevant work.

Xanthene dyes (i.e., Rhodamine B, Rhodamine 6G, and fluorescein) are inexpensive, are widely available, and have good photostability and reasonably low toxicity. These dyes are natural candidates for use as calibration standards. The fact that the most reported cross sections are that of xanthene dyes ensures careful examinations by many independent research groups and further enhances the confidence of using them as standards. In addition to our own measurements (Xu and Webb 1996; Albota et al. 1998b), these dyes have also been measured using both fluorescence and nonlinear transmission methods. Kaatz and Shelton measured the two-photon cross section of Rhodamine B, Rhodamine 6G, and fluorescein at 1064 nm by calibrating two-photon fluorescence to hyper-Rayleigh scattering (Kaatz and Shelton 1999). The results for Rhodamine B are in reasonable agreement with previous data, taking into account various factors such as wavelength difference. Measurements of fluorescein cross sections were repeated by Song et al. (1999). The measured value at 800 nm is ~1.5 times of what we reported in 1996. A two-photon cross section of Rhodamine 6G has been measured by three other groups at 800 nm, using 2PE fluorescence calibrated against a luminance meter (Kapoor et al. 2003) and using nonlinear transmission methods (Sengupta et al. 2000; Tian and Warren 2002). Although the 2PE method obtained results nearly identical to what we published in 1998, values obtained using the nonlinear transmission methods are consistently lower by about a factor of 2. This discrepancy between the fluorescence method and the nonlinear transmission methods has been discussed in the past (Oulianov et al. 2001). By comparing the excited-state methods (i.e., fluorescence and transient spectroscopy following two-photon excitation) and nonlinear transmission methods, it is found that, although the values obtained by these two excited-state methods are comparable and agree with values obtained by the fluorescence method in the past, they differ considerably from the value obtained using the nonlinear transmission method.

The intrinsic fluorescent molecule, flavin mononucleotide (FMN), has also been measured by Blab et al. (2001). The reported value and the spectral shape of FMN are very close to what we reported in 1996. The investigators also reported two-photon action cross sections of fluorescent proteins, including a value of 41 GM for enhanced green fluorescent protein (eGFP). We have measured eGFP using the fluorescence technique and have estimated an action cross-sectional value of ~100 GM at 960 nm (Xu and Webb 1997). An alternative measurement, using an FCS method that essentially measures fluorescence per single molecule, gave a two-photon absorption cross section of eGFP at 180 GM (Schwille et al. 1999), which agreed very well with our measurement assuming a QE of 0.6 for eGFP (Patterson et al. 1997). Two-photon absorption of eGFP has also been measured using a combination of a nonlinear transmission method and an absorption saturation. However, the reported value is ~600,000 GM at 800 nm (Kirkpatrick et al. 2001). Not only is this value four orders of magnitude larger than the values obtained by the fluorescence method, it is also two orders of magnitude larger than any reported two-photon absorption cross section. We are currently not certain about the origin of this large discrepancy.

The importance of fluorescent proteins cannot be overstated. The discovery of green fluorescent protein in the early 1960s catalyzed a new era in biology by enabling investigators to apply molecular cloning methods to fuse a fluorophore moiety to a wide variety of protein and enzyme targets that could be monitored in vivo. There are now multiple mutated forms of the original jellyfish protein with improved functionality available, as well as many new fluorescent proteins from other organisms, such as coral (Tsien 2005; Zacharias and Tsien 2006). Thus, for experiments involving intact tissue or live animals in which multiphoton microscopy has real advantages, measurements of the cross sections of the available fluorescent protein are important.

Below we provide a compilation of the majority of the two-photon action cross sections (two-photon absorption cross sections for fluorescein and Rhodamine B) that we have measured over the past decade. The data are presented in four figures starting with a set of conventional dyes and calcium ion indicators (Figs. 1 and 2), action cross sections of 10 commonly used fluorescent proteins (Fig. 3) (for a detailed review of fluorescent proteins, please see Shaner et al. 2005), QDs (Fig. 3D),

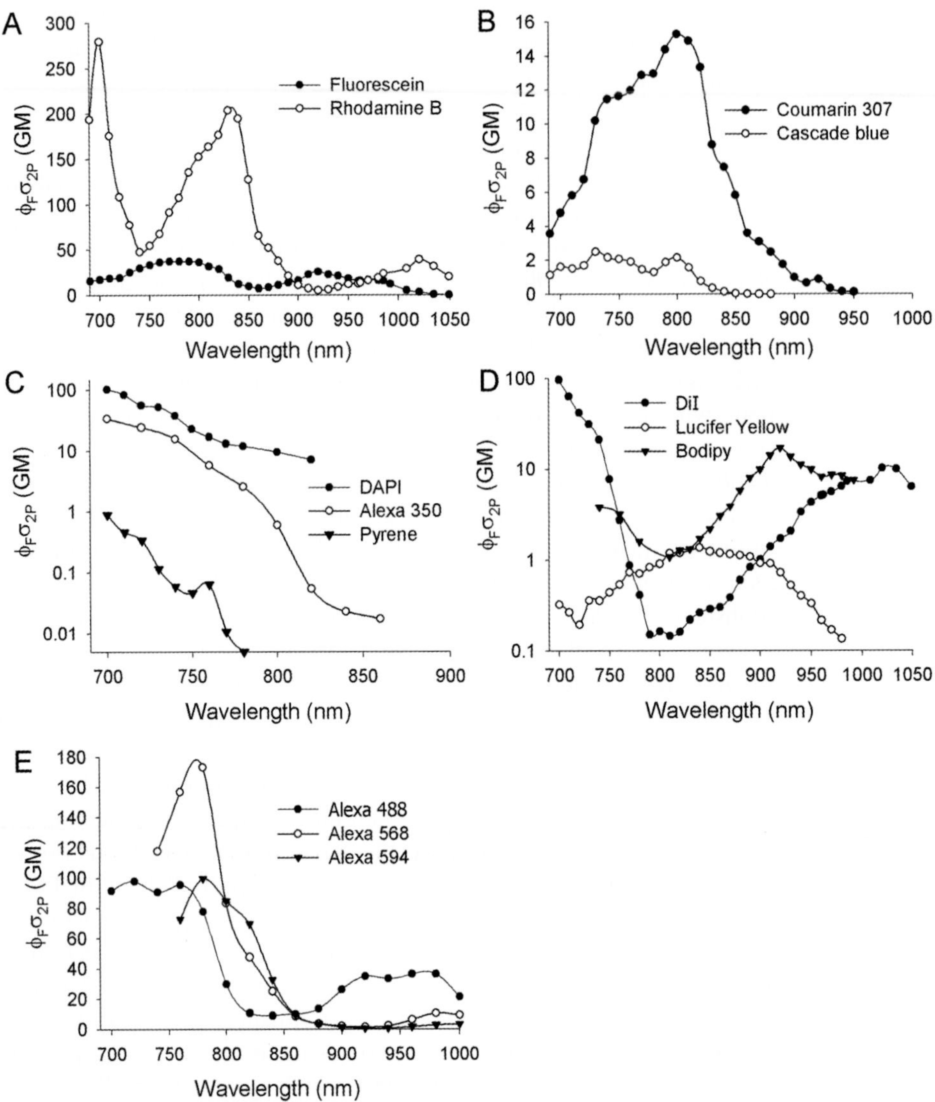

FIGURE 1. Two-photon action cross sections of conventional fluorophores used in multiphoton microscopy. (*A*) Fluorescein (water, pH 11) and Rhodamine B (methanol). (Numbers for fluorescein and Rhodamine B are absorption cross sections calibrated assuming that the one- and two-photon QEs are the same.) (*B*) Cascade Blue (water) and coumarin 307 (methanol). (*C*) DAPI (4′,6-diamino-2-phenylindole; measured in water without DNA, values shown are multiplied by 20 to reflect the known QE enhancement from binding), Alexa 350 hydrazide (water), and pyrene (methanol). (*D*) DiI C-18 (methanol), Lucifer yellow (water), and BODIPY (water); note the logarithmic *y*-axis for *C* and *D*. (*E*) Alexa 488, 568, and 594 hydrazide (water). Units are GM; 1 GM = 10^{-50} cm^4 sec/photon.

and, finally, action cross sections of several intrinsic biological molecules found in cells or in the extracellular matrix (Fig. 4). About 70% of the data has been previously presented (Xu and Webb 1996a,b; Xu et al. 1996; Larson et al. 2003; Zipfel et al. 2003b), with the exception of the Alexa dyes (Fig. 1C,E) and several of the fluorescent proteins. The measured cross sections for QDs are ensemble-averaged values, given the uncertainties caused by blinking and/or nonradiant dark fractions (Yao et al. 2005).

Recently, the advantages of longer wavelength multiphoton imaging for deep tissue penetration have been shown. When compared with imaging at the 800-nm region, multiphoton microscopy at ~1300 nm can penetrate twice as deep in both ex vivo and in vivo brain tissues (Kobat et al. 2009). In addition, an increased ability to image through blood vessels and a greater ability to suppress endogenous fluorescence background with the 1300-nm excitation are also observed. Combined

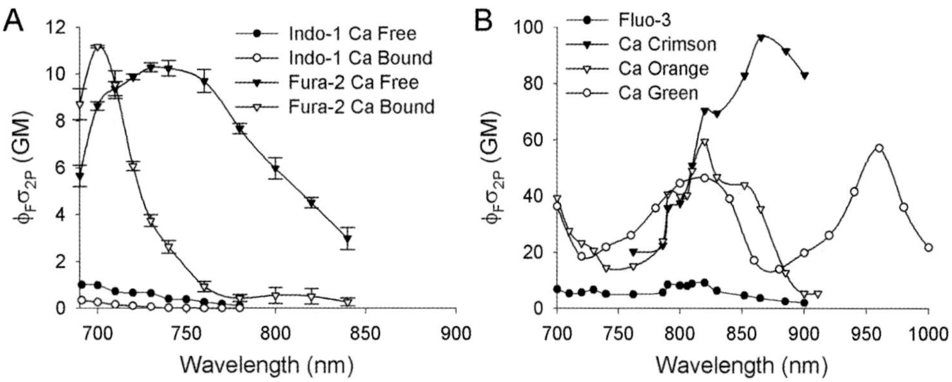

FIGURE 2. Two-photon action cross sections of calcium indicating dyes. (*A*) Indo-1 (±Ca in water) and Fura-2 (±Ca in water). Error bars on Fura-2 represent the standard error of the mean of three independent measurements of the action cross-sectional spectra. (*B*) Calcium bound forms of Fluo-3, Calcium Crimson, Calcium Orange, and Calcium Green-1N. Units are GM; 1 GM = 10^{-50} cm^4 sec/photon.

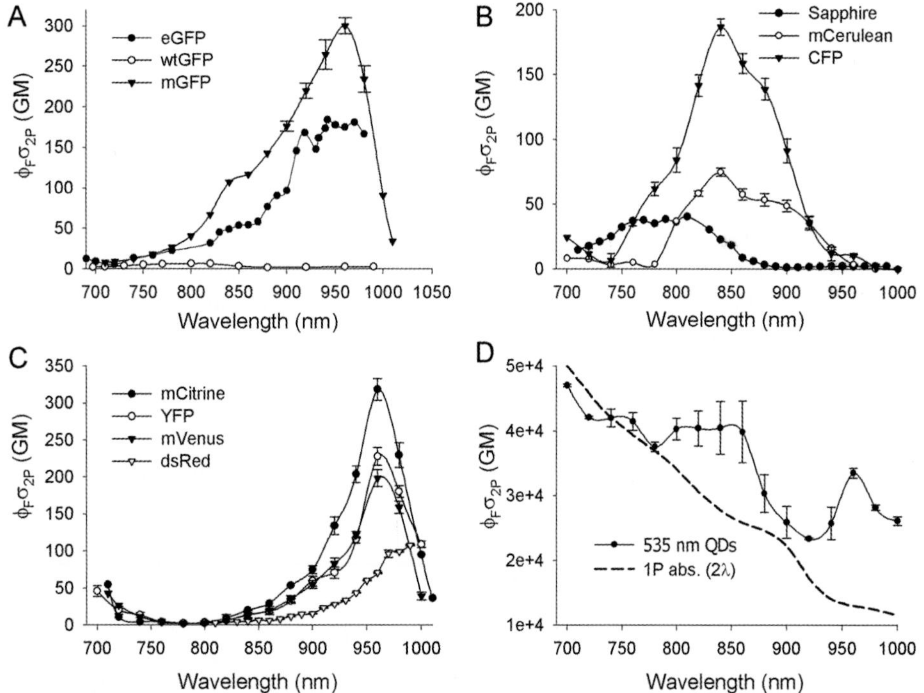

FIGURE 3. Two-photon action cross-sectional spectra of fluorescent proteins and QDs. (*A*) Three variants of green fluorescent protein: eGFP, monomeric enhanced green fluorescent protein (meGFP), and wild-type green fluorescent protein (wtGFP). Error bars for meGFP represent the standard error of the mean for two independent measurements. Concentrations for wtGFP and eGFP were based on measurements of protein concentration; meGFP concentration was measured by fluorescence correlation spectroscopy (FCS) using G(0) of the autocorrelation and a focal volume calibration based on a known concentration of Rhodamine Green. (*B*) Three blue and cyan fluorescent proteins (CFPs): Sapphire, CFP, and a monomeric form of the cerulean protein. Error bars for CFP and mCerulean represent the standard error of the mean for two independent measurements. Concentrations of CFP and mCerulean were measured by FCS; Sapphire concentration was based on measurement of protein concentration. (*C*) Four yellow and red fluorescent proteins: monomeric Citrine (mCIT), yellow fluorescent protein (YFP), monomeric Venus, and dsRed. Error bars for mCIT, YFP, and mVenus represent the standard error of the mean for two independent measurements. Concentrations of mCIT, YFP, and mVenus were measured by FCS; dsRed concentration was based of measurement of protein concentration. (*D*) Two-photon action cross sections of water soluble, high quantum yield (>90%) batch of 535-nm emitting cadmium selenide QDs. The dotted line is the single-photon absorption line shape plotted at double the wavelength for comparison. Error bars represent the standard error of the mean from two independent cross-sectional determinations from the same batch of QDs. The concentration of nanoparticles was determined using FCS. Units are GM; 1 GM = 10^{-50} cm^4 sec/photon.

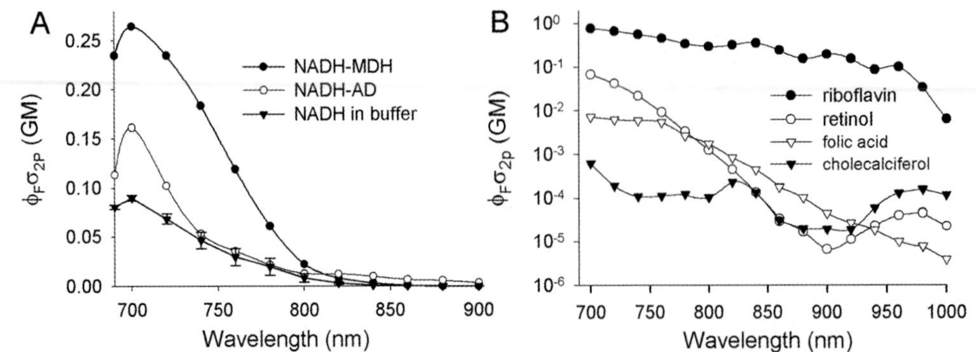

FIGURE 4. Two-photon action cross sections for five intrinsic biological compounds. (*A*) Nicotinamide adenine dinucleotide plus hydrogen (NADH) at pH 7.0 in free and enzyme-bound form. Protein-NADH complexes were prepared by adding excess purified enzyme to free NADH. Measurements of both enzymes free of NADH showed no two-photon generated fluorescence in the 450–530-nm range using excitation in the wavelength range shown. Error bars on free NADH represent the standard deviation from four independent measurements. MDH, malate dehydrogenase. AD or ADH, alcohol dehydrogenase. (*B*) Four intrinsic fluorophores commonly found in cells and tissues measured in phosphate-buffered saline solution (pH 7.0) in free form. As with bound and free NADH, the brightness (QY) of these compounds can vary greatly depending on the environment and the binding. Note: The *y*-axis is logarithmically scaled; units are GM; 1 GM = 10^{-50} cm^4 sec/photon.

with the earlier observations that longer excitation wavelengths also reduce tissue photodamage (Chen et al. 2002), multiphoton microscopy at the 1300-nm spectral window is highly promising for in vivo deep tissue imaging. Figure 5 shows the measured two-photon action cross sections of several commercially available fluorophores from 1220 to 1320 nm (Kobat et al. 2009) following the methodology described in Xu and Webb (1996). Two of these fluorophores (Alexa680 and Cy5.5) have two-photon excitation peaks within the measured wavelength range, and their two-photon action cross sections (50–75-GM range) are comparable with those of widely used shorter wavelength excitable dyes such as Rhodamine B and fluorescein (Xu and Webb 1996).

We have not included measurements of some of the synthesized large cross-sectional molecules (e.g., Albota et al. 1998a) because these dyes have not yet found many actual uses in biological imaging because of their highly lipophilic nature and toxicity. These compounds, however, have cross sections an order of magnitude higher than the conventional dyes and indicators in use today and

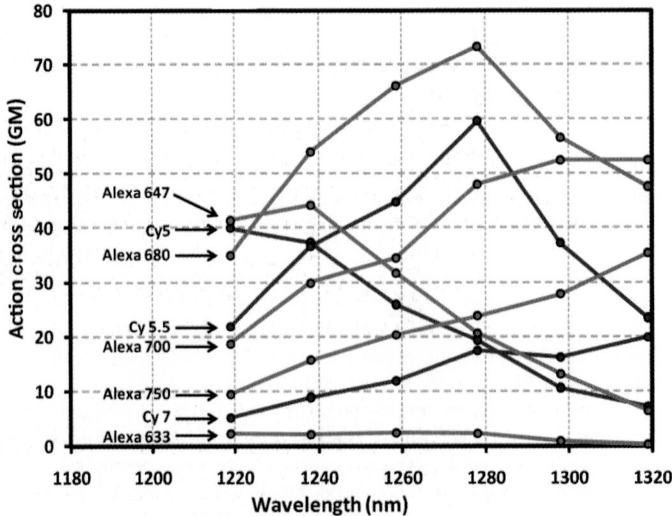

FIGURE 5. Two-photon action cross sections of eight commercial dyes for long-wavelength multiphoton microscopy. Solid lines are guides for the eye.

suggest that two-photon imaging could be further improved by the rational design of fluorophores specifically for nonlinear excitation. In addition, there are currently only a handful of fluorescent indicators for multiphoton microscopy at the longer-wavelength window of ~1300 nm. Significant future effort is required for developing longer-wavelength dyes and fluorescent proteins.

REFERENCES

Albota M, Beljonne D, Bredas JL, Ehrlich JE, Fu JY, Heikal AA, Hess SE, Kogej T, Levin MD, Marder SR, et al. 1998a. Design of organic molecules with large two-photon absorption cross sections. *Science* **281:** 1653–1656.

Albota MA, Xu C, Webb WW. 1998b. Two-photon excitation cross sections of biomolecular probes from 690 to 980 nm. *Appl Opt* **37:** 7352–7356.

Alivisatos 1996. Semiconductor clusters, nanocrystals, and quantum dots. *Science* **271:** 933–937.

Birge RR. 1983. One-photon and two-photon excitation spectroscopy. In *Ultrasensitive laser spectroscopy* (ed. Kliger DS), pp. 109–174. Academic, New York.

Blab GA, Lommerse PHM, Cognet L, Harms GS, Schmidt T. 2001. Two-photon excitation action cross-sections of the autofluorescent proteins. *Chem Phys Lett* **350:** 71–77.

Chen I, Chu S, Sun C, Cheng P, Lin B. 2002. Wavelength dependent damage in biological multi-photon confocal microscopy: A micro-spectroscopic comparison between femtosecond Ti:sapphire and Cr:forsterite laser sources. *Opt Quantum Electron* **34:** 1251–1266.

Cormack BP, Valdivia RH, Falkow S. 1996. FACS-optimized mutants of the green fluorescent protein (GFP). *Gene* **173:** 33–38.

Curley PF, Ferguson AI, White JG, Amos WB. 1992. Application of a femtosecond self-sustaining mode-locked Ti:sapphire laser to the field of laser scanning confocal microscopy. *Opt Quantum Electron* **24:** 851–859.

Denk W, Strickler JH, Webb WW. 1990. Two-photon laser scanning fluorescence microscopy. *Science* **248:** 73–76.

Efros AL, Efros AL. 1982. Interband absorption of light in a semiconductor sphere. *Sov Phys Semicond* **16:** 772–774.

Faisal FHM. 1987. *Theory of multiphoton processes.* Plenum, New York.

Helmchen F, Denk W. 2005. Deep tissue two-photon microscopy. *Nat Methods* **2:** 932–940.

Jones RD, Callis PR. 1988. A power-square sensor for two-photon spectroscopy and dispersion of second-order coherence. *J Appl Phys* **64:** 4301–4305.

Kaatz P, Shelton DP. 1999. Two-photon fluorescence cross-section measurements calibrated with hyper-Rayleigh scattering. *J Opt Soc Am B* **16:** 998–1006.

Kapoor R, Friend CS, Patra A. 2003. Two-photon-excited absolute emission cross section measurements calibrated with a luminance meter. *J Opt Soc Am B* **20:** 1550–1554.

Kennedy SM, Lytle FE. 1986. p-Bis(o-methylstyryl)benzene as a power-square sensor for two-photon absorption measurements between 537 and 694 nm. *Anal Chem* **58:** 2643–2647.

Kirkpatrick SM, Naik RR, Stone MO. 2001. Nonlinear saturation and determination of the two-photon absorption cross section of green fluorescent protein. *J Phys Chem* **105:** 2867–2873.

Kobat D, Durst ME, Nishimura N, Wong AW, Schaffer CB, Xu C. 2009. Deep tissue multiphoton microscopy using longer wavelength excitation. *Opt Express* **17:** 13354–13364.

Larson DR, Zipfel WR, Williams RM, Clark SW, Bruchez MP, Wise FW, Webb WW. 2003. Water soluble quantum dots for multiphoton fluorescence imaging in vivo. *Science* **300:** 1434–1436.

Maiti S, Shear JB, Williams RM, Zipfel WR, Webb WW. 1997. Measuring serotonin distribution in live cells with three-photon excitation. *Science* **275:** 530–532.

Masters BR. 2003. *Selected papers on multiphoton excitation microscopy*. SPIE, Bellingham, MA.

Masters BR, So PTC. 2004. Antecedents of two-photon excitation laser scanning microscopy. *Microsc Res Tech* **63:** 3–11.

Oulianov DA, Tomov IV, Dvornikov AS, Rentzepis PM 2001. Observations on the measurement of two-photon absorption cross-section. *Opt Commun* **191:** 235–243.

Patterson GH, Knobel SM, Sharif WD, Kain SR, Piston DW. 1997. Use of the green fluorescent protein and its mutants in quantitative fluorescence microscopy. *Biophys J* **73:** 2782–2790.

Schwille P, Haupts U, Maiti S, Webb WW. 1999. Molecular dynamics in living cells observed by fluorescence correlation spectroscopy with one- and two-photon excitation. *Biophys J* **77:** 2251–2265.

Sengupta P, Balaji J, Banerjee S, Philip R, Kumar GR, Maiti S. 2000. Sensitive measurement of absolute two-photon absorption cross sections. *J Chem Phys* **112:** 9201–9205.

Shaner NC, Steinbach PA, Tsien RY. 2005. A guide to choosing fluorescent proteins. *Nat Methods* **2:** 905–909.

Smith WL. 1986. Two-photon absorption in condensed media. *Handbook of laser science and technology* (ed. Weber J), pp. 229–258. CRC, Boca Raton, FL.

So PTC, Dong CY, Masters BR, Berland KM. 2000. Two-photon excitation fluorescence microscopy. *Annu Rev Biomed Eng* **2:** 399–429.

Song JM, Inoue T, Kawazumi H, Ogawa T. 1999. Determination of two-photon absorption cross section of fluorescein using a mode-locked titanium sapphire laser. *Anal Sci* **15:** 601–603.

Spence DE, Kean PN, Sibbett W. 1991. 60-fsec pulse generation from a self-mode-locked Ti:sapphire laser. *Opt Lett* **16:** 42.

Tian P, Warren WS. 2002. Ultrafast measurement of two-photon absorption by loss modulation. *Opt Lett* **27:** 1634–1636.

Tsien RY. 2005. Building and breeding molecules to spy on cells and tumors. *FEBS Lett* **579:** 927–932.

Wise FW. 2000. Lead salt quantum dots: The limit of strong quantum confinement. *Acc Chem Res* **33:** 773–780.

Wokosin DL, Centonze VE, Crittenden S, White JG. 1995. Three-photon excitation of blue-emitting fluorophores by laser scanning microscopy. *Mol Biol Cell* **6:** 113a.

Xu C, Webb WW. 1996. Measurement of two-photon excitation cross-sections of molecular fluorophores with data from 690 nm to 1050 nm. *J Opt Soc Am B* **13:** 481–491.

Xu C, Webb WW. 1997. Multiphoton excitation of molecular fluorophores and nonlinear laser microscopy. In *Topics in fluorescence spectroscopy* (ed. Lakowicz J), Vol. 5, pp. 471–540. Plenum, New York.

Xu C, Williams RM, Zipfel W, Webb WW. 1996a. Multiphoton excitation cross sections of molecular fluorophores. *Bioimaging* **4:** 198–207.

Xu C, Zipfel W, Shear JB, Williams RM, Webb WW. 1996b. Multiphoton fluorescence excitation: New spectral windows for biological nonlinear microscopy. *Proc Natl Acad Sci* **93:** 10763–10768.

Yao J, Larson DR, Vishwasrao HD, Zipfel WR, Webb WW. 2005. Blinking and nonradiant dark fraction of water-soluble quantum dots in aqueous solution. *Proc Natl Acad Sci* **102:** 14284–14289.

Zacharias DA, Tsien RY. 2006. Molecular biology and mutation of green fluorescent protein. *Methods Biochem Anal* **47:** 83–120.

Zipfel WR, Williams RM, Christie R, Nikitin AY, Hyman BT, Webb WW. 2003a. Live tissue intrinsic emission microscopy using multiphoton-excited native fluorescence and second harmonic generation. *Proc Natl Acad Sci* **100:** 7075–7080.

Zipfel WR, Williams RM, Webb WW. 2003b. Nonlinear magic: Multiphoton microscopy in the biosciences. *Nat Biotechnol* **21:** 1369–1377.

31 Single-Molecule FRET Using Total Internal Reflection Microscopy

Jeehae Park and Taekjip Ha

Department of Physics and Howard Hughes Medical Institute, University of Illinois at Urbana-Champaign, Urbana, Illinois 61801

ABSTRACT

Single-molecule Förster resonance energy transfer (smFRET) is a fluorescence-based technique used to observe dynamics of single molecules occurring within a range of 30–80 Å. The extreme sensitivity and stability of FRET combined with total internal reflection (TIR) microscopy allows for the observation of hundreds of individual molecules undergoing conformational dynamics with associated distance changes as small as several angstroms. We begin this chapter with a brief description of the theory and the types of questions smFRET can address. These are followed by practical considerations when designing smFRET experiments and constructing a TIR setup. Lastly, data acquisition and analysis methods are described to show how to extract information in more efficient ways. At the end of the chapter, a protocol for preparing a sample chamber is described.

INTRODUCTION

Single-molecule Förster resonance energy transfer (FRET), which was first introduced in 1996 (Ha et al. 1996), has become a popular technique to observe dynamics in many biological systems (Cornish and Ha 2007). FRET is often regarded as a "molecular ruler" with which the distance between two fluorescently labeled positions can be measured. With FRET, distance changes that occur within 30–80 Å can be probed, a distance range that is far below the resolution limit of optical microscopy. FRET occurs between a pair of donor and acceptor fluorophores. When a donor is excited, it emits fluorescence, some of which is transferred to an acceptor (via an induced dipole–dipole interaction) in a distance-dependent manner (Förster 1948). The fluorescence intensities of the two fluorophores change in an anticorrelated manner and the FRET efficiency is typically estimated as the ratio of the acceptor intensity to the total intensities of both fluorophores. The relationship of FRET efficiency to distance (R) is shown in Figure 1, where R_0 is the distance at which half of the energy is transferred (Clegg 1992).

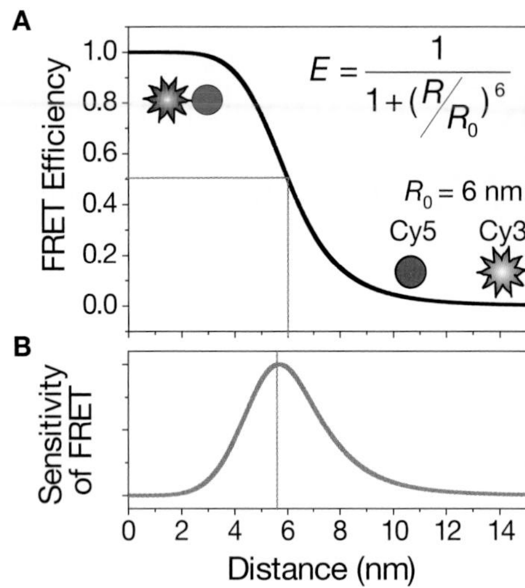

FIGURE 1. FRET. (*A*) Shown is the distance dependence of FRET efficiency. Cy3 (donor, green circle) transfers one-half of its energy to Cy5 (acceptor, red circle) when they are separated by R_0 (6 nm). (*B*) The change of FRET is most prominent around R_0.

Advantages of Single-Molecule FRET

In ensemble FRET experiments, signals from millions of molecules are averaged and it is their net signal that is measured. When populations of molecules in solution are in various states and show nonsynchronized dynamics, it is often impossible to detect the true dynamics of the molecule with ensemble FRET. Through careful design of the surface-tethering scheme, by immobilizing molecules noninvasively, and by illuminating only a small depth above a surface to reject background signals (i.e., total internal reflection microscopy), we can observe the dynamics of single biomolecules in real time. Single-molecule FRET (smFRET) offers the following advantages: (1) distribution of conformers within the population can be identified (McKinney et al. 2003; Lee et al. 2005), (2) short-lived transition states can be detected (Nahas et al. 2004; McKinney et al. 2005), (3) heterogeneous dynamics among molecules can be observed (Zhuang et al. 2002; Okumus et al. 2004), (4) reaction signals can be postsynchronized (Blanchard et al. 2004), and (5) equilibrium dynamics can be observed (Joo et al. 2006).

Comparison of smFRET with Other Techniques

One advantage of smFRET compared with other single-molecule methods (such as optical tweezers, magnetic traps, and fluorescence particle tracking) is that the distance measurement does not require the determination of the absolute positions of the two fluorophores. FRET measures the distance between two fluorescently labeled positions in the center-of-mass frame of the molecule or molecular complex and therefore is less prone to environmental noise or stage drift. smFRET can measure several hundreds of molecules in parallel and therefore has a higher data throughput than many other single-molecule methods. Lastly, as a fluorescent-based method, high distance precision (~0.5 nm) can be achieved with fewer photons (~100) with smFRET when compared with that of fluorescence particle tracking (~1.5 nm precision with ~10,000 photons). However, smFRET is limited to a distance range of ~3–8 nm, and thus is not suitable for observing larger scale dynamics (Cornish and Ha 2007).

DESIGNING FRET EXPERIMENTS

Choosing Fluorophores for the Optimal FRET Pair

The criteria for choosing a suitable FRET pair are as follows.

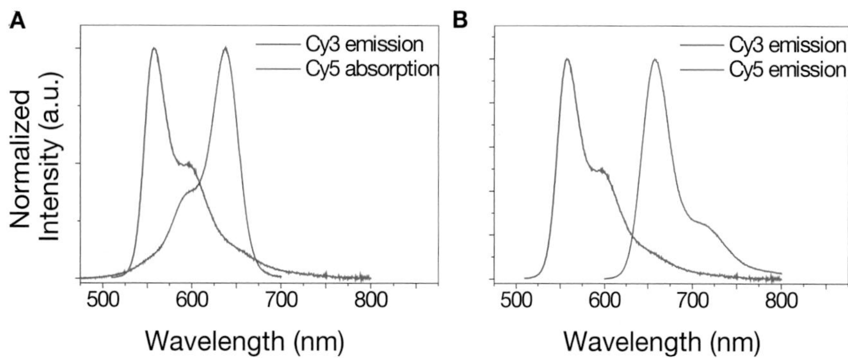

FIGURE 2. Absorption and emission spectrum of Cy3 and Cy5. (*A*) The emission spectrum of Cy3 (donor) overlaps with the absorption spectrum of Cy5 (acceptor). (*B*) The emission spectra of Cy3 and Cy5 are well separated.

1. The donor emission spectrum and acceptor absorption should have significant overlap in order for FRET to occur (Fig. 2A), but at the same time, the maxima of the donor and acceptor emission spectra should be far enough apart that there is minimal signal cross talk (Fig. 2B).
2. The quantum yields of emission for the donor and acceptor should be comparable, so that the anticorrelated changes in their intensities can be observed distinctively.

The Cy3/Cy5 pair is an excellent fluorophore pair for FRET because the two fluorophores have a large spectral separation (~100 nm) with similar quantum yields (~0.25). They are also photostable in oxygen-free conditions and are commercially available in various reactive forms for conjugation.

Determining Where to Attach Fluorescent Labels

When designing biological constructs with fluorescent labels, it is crucial to position the fluorophores so that the signal is maximized for a successful smFRET analysis. For example, the most pronounced FRET changes are observed if the distance between the donor and the acceptor changes around R_0. For three-color FRET experiments, the location of the dye molecules on the biomolecules should be chosen to minimize complications during data analysis. When designing an experiment having one donor and two different acceptors, place the two acceptors far enough apart that FRET between them is minimized. Lastly, when labeling two interacting molecules with a donor and an acceptor, it is useful to attach the acceptor to the surface-immobilized molecule and the donor to its cognate molecule in solution. In such a scheme, the binding and dissociation events can be clearly identified as the appearance and disappearance of fluorescent spots.

Surface Preparation

To observe the dynamics of a single molecule over a period of time, molecules need to be localized within the imaging area (Rasnik et al. 2005). Using a biotin-avidin linkage, molecules can be surface immobilized in a specific manner. Streptavidin (or neutravidin) has multiple biotin binding sites that allow two types of biotin-labeled species to be linked: the reagent passivated over the surface, and the molecule of interest. When the molecule of interest is DNA, biotin-labeled bovine serum albumen (BSA) is used to passivate the glass/quartz slides. Next, neutravidin is applied so that it binds to the biotin-BSA. Finally, the biotin-labeled DNA is applied to the surface to achieve specific binding (>500:1, compared with controls that excluded biotin or neutravidin). It is important to prevent nonspecific interactions between the molecule of interest and the surface. DNA, being negatively charged, is not attracted to other negatively charged species such as glass, BSA, and neutravidin (Fig. 3A). However, we have found that for the study of proteins or protein–DNA interactions, it is essential to passivate the surface with polyethylene glycol (PEG), which forms a polymer brush that prevents nonspecific adsorption of proteins to the underlying surface (Ha et al. 2002). Biotin-

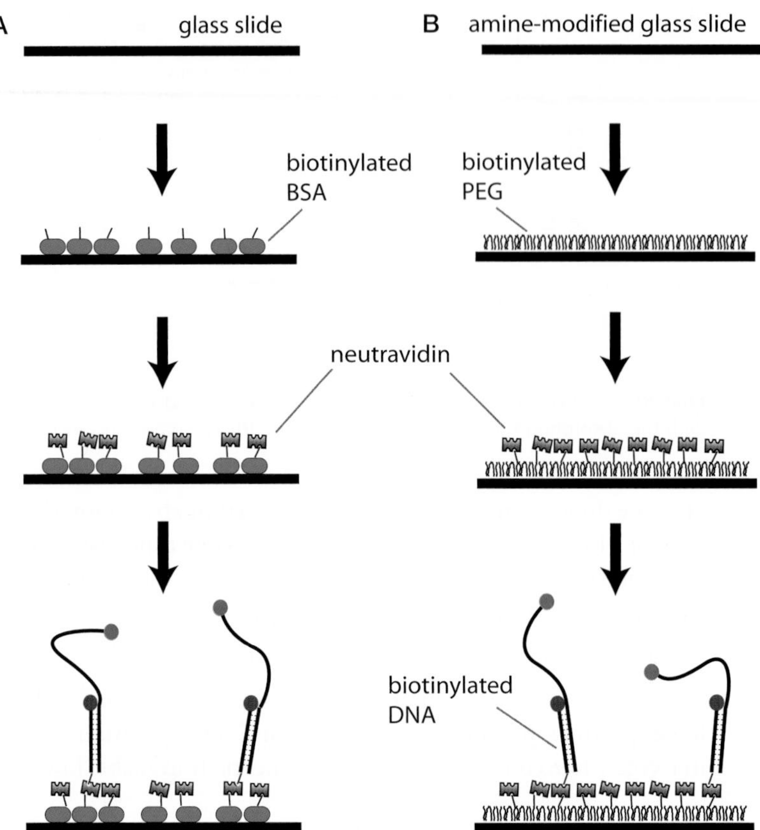

FIGURE 3. Surface preparation for sample chamber assembly. (*A*) Biotinylated BSA is deposited onto an untreated glass slide. Neutravidin is conjugated with the biotinylated BSA. A biotinylated specimen (here DNA labeled with both donor and acceptor) is immobilized via biotin-neutravidin binding. (*B*) On an amine-modified glass slide, the NHS-ester form of PEG is covalently conjugated. Neutravidin specifically binds to a fraction of the PEG molecules that have biotin moieties at one end.

labeled PEG is mixed with unlabeled PEG and applied onto the surface to provide a specific linkage (Fig. 3B). The protocol at the end of this chapter provides details for preparing both BSA-coated and PEG-coated slides.

TOTAL INTERNAL REFLECTION MICROSCOPY SETUP

Our total internal reflection (TIR) microscopy setup is built around a commercial inverted microscope (Olympus). Add-on features of the setup can be divided into excitation and emission components.

Excitation Optics: Prism-Type TIR and Objective-Type TIR

The excitation laser (Nd:YAG [yttrium aluminum garnet] 532 nm) is guided to the microscope so that the beam hits the microscope slide at a large enough angle for TIR. TIR can be achieved by passing the light through a prism placed on top of the slide (prism-type TIR), or by passing it through a high-numerical-aperture (NA) objective lens (objective-type TIR). In prism-type TIR, a microscope slide is sandwiched between an objective lens and the prism. This configuration prevents the prism from being permanently mounted. Therefore, for each experiment it is necessary to optically align the laser to make corrections for the excitation area, a process that can be dificult for inexperienced users. With objective-type TIR, we use a 1.4-NA 100x oil-immersion objective (UPlanSApo, Olympus). Because illumination and detection are through the same objective, it is simple to locate the focal plane. A comparable excitation field of view can be achieved for both types of TIR. The excitation field of view for our setup is 25 x 50 μm. In addition, objective-type TIR has unobstructed access above the sample so that additional manipulations, such as interfacing with microfludics, can be easily performed. However, because objective-type TIR images the coverslip surface, coverslip bowing during flow injection leads to sample defocusing. This can introduce problems, such as making it more difficult to perform nonequilibrium experiments requiring the flow of reaction buffer.

Emission Optics

The emission part of the setup collects the fluorescence signals from the sample, and using a dichroic mirror splits the signal into a donor component and an acceptor component. A long-pass filter, located in the microscope filter turret, screens excitation laser scattering from the signal before the signal exits the side port of the microscope. The separated donor and acceptor signals are sent together and side-by-side to the charge-coupled device (CCD). With this single-detector path design, the need for post-time synchronization between two signals is eliminated. A pair of lenses expands the signal and together with a slit, an optimum sized signal can be generated to fill the CCD view. Figure 4 shows the emission optics for two-color smFRET (Fig. 4A) and three-color smFRET (Fig. 4B).

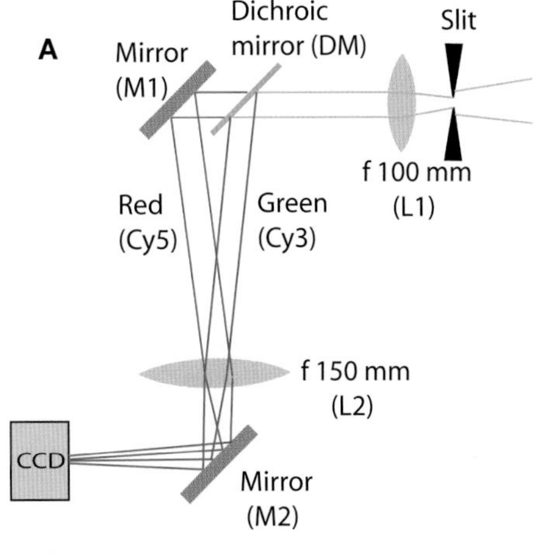

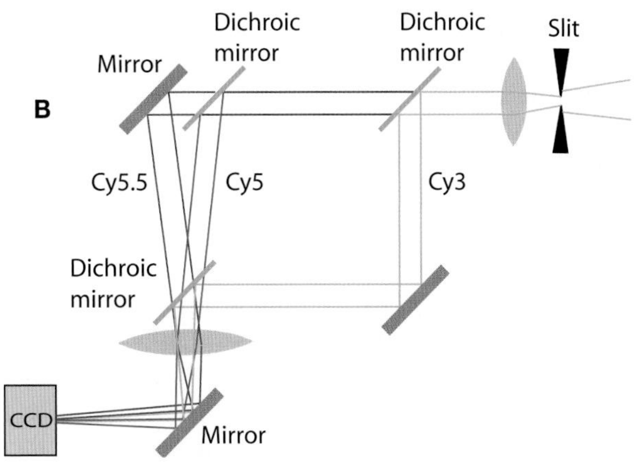

FIGURE 4. Emission optics. (*A*) In the two-color emission optics, the collimated beam goes through a dichroic mirror where the donor signal is reflected. After passing through a lens (L2), the donor and acceptor images are projected onto one-half of the CCD screen each. (*B*) In the three-color setup, the donor signal is separated first, then the other two colors are later split in the same way as in *A*. The donor signal is then combined with the other two by another dichroic mirror and projected onto the CCD screen.

Detection Device

For high-speed CCD imaging, signal amplification is needed. We use an electron-multiplying CCD camera (iXon, Andor Technology) to suppress readout noise. Our current time resolution for a frame that is 512 x 512 pixels is 30 msec without binning and can go down to 8 msec with 2 x 2 binning. Using a 128 x 128-pixel electron-multiplying CCD (EMCCD), time resolution down to 2 msec without binning and 1 msec with binning are possible. When higher time resolution is desired, using a confocal setup with avalanche photodiodes can push the time resolution to the system's photophysical limit.

DATA ACQUISITION AND ANALYSIS

Image Acquisition and Data Extraction

Fluorescence signals collected from CCD camera are recorded as frames of a movie. With our home-written data acquisition software (Visual C++, Microsoft), data from a single pixel is encoded as a single byte (8 bit). A 1-min movie taken with 30-msec time resolution uses ~500 MB. From the image frames, single-molecule peak positions are identified and single-molecule time traces are obtained using scripts written in IDL (Research Systems). To compare donor and acceptor signals and calculate the FRET of each molecule, images from the donor channel and acceptor channel need to be mapped. Because of the optical aberrations and imperfect alignment, the mapping process requires corrections for scaling, rotation, sheer distortion, and lateral offset. We use a calibration sample of sparsely distributed fluorescence beads to obtain the optimal polynomial map.

Calculating FRET

From a typical single movie file, data from 200–400 molecules can be obtained and individual time trajectories visualized and analyzed one by one, using software written in MATLAB. Apparent FRET efficiency can be calculated as

$$E_{app} = I_A / (I_D + I_A),$$

where I_D and I_A are the emission intensity of the donor and acceptor, respectively. What we actually measure are the raw intensities of the donor and acceptor channels, I_D^0 and I_A^0. We then need to correct for the leakage of donor signal into the acceptor channel, which is typically 10%–15% of the donor signal, and can be determined from a donor-only molecule. We subtract a fixed fraction α of I_D^0 from I_A^0 such that the corrected acceptor signal becomes 0 (and $E_{app} = 0$),

$$E_{app} = (I_A^0 - \alpha \times I_D^0) / (I_D^0 + I_A^0).$$

FRET Histogram

FRET data obtained from several movies provides information on more than 1000 molecules. By taking a snapshot of each molecule's FRET efficiency, subpopulations of different FRET values can be identified in an unbiased way. Typically, a FRET histogram is built by averaging each molecule's FRET efficiency over 10 frames.

Time Trajectories

Perhaps the most compelling aspect of the smFRET technique is the ability to obtain real time trajectories of a single molecule that are not blurred by averaging over time or population. In general, time traces are taken with 30-msec resolution for 1 min. For a system that displays relatively slow dynamics, >100-msec resolution can be used and recording for ≥10 min is possible. The total observation time is limited by photobleaching of the fluorophores and therefore is determined by the minimum illumination intensity required to obtain a good signal-to-noise ratio at the desired time resolution.

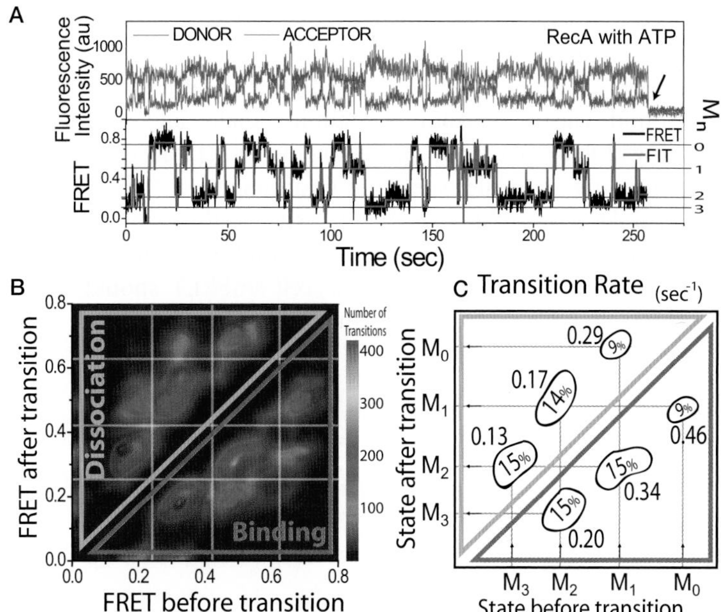

FIGURE 5. HaMMy and transition density plot (TDP). (*A*) When there are more than two states observed (here five are present, including photoblinking of the acceptor), an advanced algorithm, such as HaMMy, is required for reliable data analysis. The fit (green line) over the FRET trace is from HaMMy analysis. (*B*,*C*) TDP. (*B*) This pseudo-3D plot is constructed by adding a Gaussian peak for each transition. There are four states with $E \sim 0.15$, 0.3, 0.5, and 0.7. (*C*) Rates and frequencies (inside circles) of each transition, $M_i \rightarrow M_{i \pm 1}$, are presented in the same scheme. (Reprinted, in part, with permission, from Joo et al. 2006, ©Elsevier.)

General Analysis Method

For a time trajectory that displays two-state dynamics in equilibrium, transition between states can be observed and the transition rate can be calculated by dwell time analysis. When the transitions are fast, making dwell time analysis difficult, a cross-correlation analysis can be performed. A single exponential fit to the cross-correlation function of the donor and acceptor time traces gives the sum of forward and backward transition rates. Together with the population analysis using histograms, both transition rates can be determined (Tan et al. 2003; Joo et al. 2004). If the system is more complex and a molecule visits multiple states, identifying each transition and measuring dwell times become more complicated. Using HaMMy, software based on hidden Markov modeling, each state can be assigned automatically without bias and the transition rates between different states can be determined (McKinney et al. 2006) (Fig. 5A). The data can be further reduced in the form of a transition density plot (TDP), which is a two-dimensional histogram representing the frequency of transitions between beginning and ending FRET states (Fig. 5B). If the experimental system shows dynamics among well-defined states, isolated peaks will appear in the TDP. Each peak represents frequently occurring transitions and by analyzing the data set comprising each peak, the rate of each transition can be calculated (Fig. 5C). As a cautionary note, when the FRET trajectory does not show distinctive states and/or the expected dwell time of the state is short compared with the time resolution, HaMMy should not be used.

SUMMARY AND FUTURE DIRECTIONS

The single-molecule FRET method using TIR provides high-throughput information on individual molecules without temporal and population averaging. This is accomplished by effectively rejecting the background signal and improving the signal from a molecule of interest by enhancing fluorophore stability and using fast, highly sensitive detection devices. Techniques continue to be developed to further increase the power of smFRET. Vesicle encapsulation (Okumus et al. 2004; Cisse et al. 2007) and a zero-mode wave guide have recently been developed, making it possible to perform experiments that require imaging high concentrations of labeled molecules while overcoming background. The use of Trolox, a vitamin E analog and a reducing agent, greatly enhances the photostability of fluorophores by quenching the triplet state (Rasnik et al. 2006). We expect that development of brighter and wider-spectrum fluorophores as well as improvements in detectors will expand the capabilities of smFRET. Techniques such as three-color FRET (Hohng et al. 2004) will become more important for investigating complex, multicomponent systems within living cells. Although many technical difficulties remain, live cell smFRET should become possible in the future, revealing angstrom-scale molecular dynamics in vivo.

Protocol

Sample Chamber Preparation for smFRET Imaging

This protocol describes the preparation of sample chambers with either BSA- or PEG-coated slides to which single molecules can be tethered for use in FRET studies. Fluid access to the sample chamber is provided by holes drilled into the slide. This system has the advantage that many solution conditions can be explored by injecting different solutions into the same sample chamber. The small-diameter holes minimize evaporation during prolonged measurements and reduce oxygen uptake by the solution, thereby reducing photobleaching effects and solution acidification.

MATERIALS

CAUTION: See Appendix 6 for proper handling of materials marked with <!>.

Reagents

Acetone <!>
Alconox <!>
Biotinylated bovine serum albumin (only required for preparing BSA-coated slides)
Biotinylated DNA of interest (50 pM in an appropriate buffer, usually buffer T50 containing 0.1 mg/ml BSA)
Buffer T50 (10 mM Tris-HCl [pH8.0], 50 mM NaCl)
Cy3-labeled DNA (1 nM in buffer T50) for nonspecific binding tests
Neutravidin (ImmunoPure NeutrAvidin Protein; Pierce)
Potassium hydroxide (KOH) <!> (1 M in MilliQ H_2O)

For Preparing PEG-Coated Slides

Acetic acid, glacial <!>
Amino silane <!> (N-(2-aminoethyl)-3-aminopropyltrimethoxysilane; United Chemical Technologies)
Biotin-PEG (MW 5000; Biotin-PEG-SC; hydrolysis half-life >20 min; Laysan Bio)
Cy3-labeled protein (1 nM) for nonspecific binding tests
Methanol <!>
mPEG (MW 5000; mPEG-SC; hydrolysis half-life > 20 min; Laysan Bio)
Sodium bicarbonate

Equipment

Coverslips, 24 x 40-mm (# 1½, rectangular)
Diamond drill bits, 3/4-mm (Kingsley North or UKAM Industrial Superhard Tools)
Double-sided tape, ~100-µm thick (3M)
Epoxy glue, quick-drying
Flask, 100–150-mL (Pyrex; required for preparation of PEG-coated slides)
Needle, 26-gauge, 3/8-in long
Pipette tips, 200-µL

Propane torch <!> (Bernzomatic)
Pump (PHD 22/2000 series syringe pump; Harvard Apparatus) (optional; see Step 28)
Slides, microscope, glass, 1 × 3-in, 1-mm thick (Gold Seal)
Slides, microscope, quartz, 1 × 3-in, 1-mm thick (G. Finkenbeiner) (for prism-type TIR)
Sonicator (Bransonic tabletop ultrasonic cleaner; Branson)
Staining dishes, glass
Syringe, 1-mL
Tubing, 28-gauge (PTFE tubing; Hamilton Company)

EXPERIMENTAL METHOD

Preparing and Cleaning Slides and Coverslips

1. Drill two 0.75-mm-diameter holes into a glass slide to form the inlet and outlet of the chamber that will be made (Fig. 6).
2. Sonicate the slides in a glass staining dish for 20 min in 10% alconox, 5 min in tap H_2O, 15 min in acetone, and 20 min in 1 M KOH. Sonicate the coverslips in another glass staining dish for 20 min in 1 M KOH.
3. Rinse the slides and coverslips with deionized MilliQ H_2O (18.5 MΩ).
4. Burn the surface to be imaged using a propane torch to remove any fluorescent organic molecules. Burn a quartz slide for the prism-type TIR for half a minute. Burn a coverslip for the objective-type TIR for only a few seconds to prevent glass deformation.

   To prepare BSA-coated slides, complete Steps 5–10. To prepare PEG-coated slides and coverslips, skip to Step 11.

Sample Chamber Assembly

Assemble the chamber immediately before each experiment.

5. Attach two pieces of double-sided tape (~100 µm thick) to a cleaned slide such that there is an ~5-mm gap between the pieces of tape. These will serve as spacers. Put a cleaned coverslip over the slide to form a sample chamber with a 10–20-µL volume (Fig. 6).

   When assembling a PEG-coated chamber, be sure to place the PEG-coated side of the slide and the coverslip facing toward each other.

   See *Troubleshooting.*
6. Seal the edges of the sample chamber using 5-min epoxy glue.

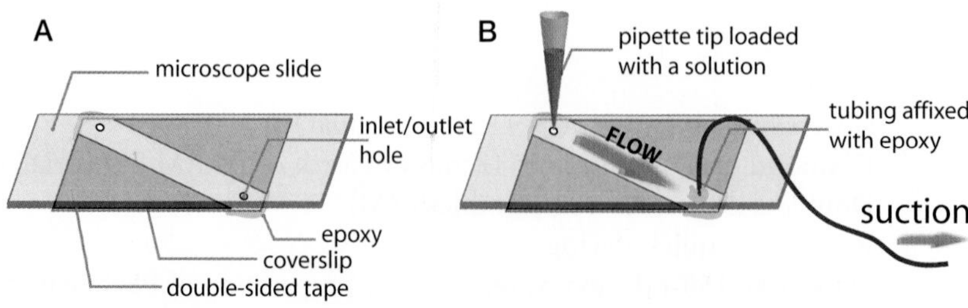

FIGURE 6. Sample chamber design. (*A*) A sample chamber is made by securing a microscope slide and a coverslip together with double-sided tape and sealing with epoxy. The holes on the slide are used for solution exchange. (*B*) A syringe is connected to the chamber by tubing and a pipette tip that contains solution is snugly plugged into an inlet hole. When the syringe is pulled, the solution is introduced into the chamber.

Preparing BSA-Coated Slides

Prepare this surface just before each experiment.

7. Prepare a solution of 1 mg/mL biotinylated BSA in buffer T50. Using a 200-μL pipette tip placed securely into the inlet port of the slide, allow 30 μL of this solution to flow into the chamber. Incubate for 5 min.

   BSA nonspecifically adsorbs to the chamber surfaces.

   See *Troubleshooting*.

8. Wash away the biotinylated BSA solution by pushing 100 μL of buffer T50 through the chamber.

9. Prepare a solution of 0.2 mg/mL neutravidin in buffer T50. Introduce 30 μL of this solution into the chamber. Incubate for 1 min. Wash out the chamber as in Step 8.

10. Add 30 μL of a biotinylated substrate of interest (i.e., ~50 pM of fluorescently labeled DNA in an appropriate buffer) to the chamber.

    This protocol allows the stepwise deposition of reagents without sample drying and typically results in a surface concentration of DNA suitable for single-molecule imaging.

    See *Troubleshooting*.

Amino-Modification of Slides and Coverslips

11. Place the cleaned coverslips and slides (from Step 4) into glass staining dishes containing methanol.

12. Sonicate a Pyrex flask with methanol for 5 min.

13. Discard the methanol from the flask. In the flask, mix 100 mL of methanol, 5 mL of acetic acid, and 1 mL of amino silane.

14. Immediately replace the methanol in the glass staining dishes with the silane solution. Incubate for 10 min, sonicate for 1 min, and incubate for 9 min.

15. Replace the mixture with methanol. Store the slides and coverslips in methanol until the next step.

Preparing PEG-Coated Slides and Coverslips

16. Rinse the slides and coverslips with MilliQ H_2O. Place them on a level surface.

17. Dissolve 0.2 mg of biotin-PEG and 8 mg of mPEG in 64 μL of freshly made 0.1 M sodium bicarbonate (pH 8.5) for each slide. Centrifuge the solution for 1 min at 7200*g* to remove air bubbles.

18. Apply 70 μL of the PEG solution to each slide. Place a coverslip over each slide to spread the solution evenly and to prevent drying. Avoid trapping air bubbles between the slide and coverslip.

    See *Troubleshooting*.

19. Incubate the slide/coverslip sandwiches overnight in a dark humid environment. For example, place them in a box with an open container of H_2O.

20. Disassemble the sandwiches, rinse them with MilliQ H_2O, and store them at –20°C until use.

21. Assemble the sample chamber by completing Steps 5 and 6.

22. Using a 200-μL pipette tip placed securely into the inlet port of the slide, fill the chamber with 30 μL of 0.2 mg/mL neutravidin solution in buffer T50. Incubate for 1 min.

    See *Troubleshooting*.

23. Wash out the neutravidin by injecting 100 µL of buffer T50 through the chamber.
24. Inject 30 µL of a biotinylated substrate of interest (i.e., ~50 pM of fluorescently labeled DNA in an appropriate buffer) into the chamber. Adjust the concentration as necessary.

    DNA sticks nonspecifically to the PEG surface if the pH is <7.4.

    See *Troubleshooting.*
25. For the protein–DNA interaction experiment, add 30 µL of the protein of interest in desired reaction conditions. Choose a protein concentration to use based on the K_d of the protein. However, be cautious when using fluorescently labeled protein at a high concentration (i.e., >5 nM) as it may cause high background fluorescence.

    See *Troubleshooting.*

Solution Exchange during Data Acquisition

Use the following syringe-driven fluid flow system to move fluids into and out of the sample chamber during imaging.

26. In an assembled chamber, attach a piece of 28-gauge tubing to the 0.8-mm-wide hole with epoxy glue.
27. Connect the tubing to a 26-gauge, 3/8-in needle, which is attached to a 1-mL syringe.
28. Draw the desired solutions into a 200-µL pipette tip. Place the pipette tip securely into the inlet port of the slide. Pull the solution into the sample chamber by pulling the syringe either manually or via an automated pump.
29. After imaging is complete, soak the sample chambers overnight in tap H_2O. This makes it easier to remove the double-sided tape and epoxy, thus allowing the slides to be reused.

TROUBLESHOOTING

Problem (Step 5): You forgot which side of the slide/coverslip is coated with PEG.
Solution: Gently flow MilliQ H_2O over the surface. The PEG-coated surface will repel H_2O.

Problem (Step 7 or 22): Epoxy blocks the inlet hole.
Solution: Inject T50 into the ports and sample chamber after assembling the chamber but before applying the epoxy.

Problem (Step 18): Air bubbles become trapped between the coverslip and the slide.
Solution: Gently press a pipette tip across the coverslip. Move the air bubbles toward the edge to remove them.

Problem (Step 10 or 24): Too many (>600) fluorescence spots are observed.
Solution: Check the nonspecific binding of DNA as follows. Prepare an assembled sample chamber using the same procedure as was used for the experimental chamber. Briefly flush with buffer T50. Inject 1 nM of Cy3-labeled DNA in buffer T50. When neutravidin is absent, the DNA should not stick to the surface. On a good PEG-coated surface, fewer than 10–20 molecules should be visible in each 25 x 50-µm imaging area. If nonspecific binding can be ruled out, reduce the DNA concentration and try again.

Problem (Step 10 or 24): Too few (<100) fluorescence spots are observed.
Solution: Check the pH of the buffer T50. High pH (>8.0) prevents biotin binding to the neutravidin. Once the biotin/neutravidin link is formed, then higher-pH buffer can be used if necessary. If the buffer pH is okay, try the experiment with a higher DNA concentration.

Problem (Step 25): Too many fluorescence spots are observed in patches.
Solution: Check the nonspecific binding of proteins by adding 1 nM Cy3-labeled protein to a sample chamber that has been treated with neutravidin and PEG. On a good PEG-coated surface, fewer than 10–20 molecules should be visible in each 25 x 50-µm imaging area. Double PEGylation is an option if nonspecific binding of protein occurs. Using an assembled sample chamber, repeat the application of PEG starting with Step 17.

REFERENCES

Blanchard SC, Gonzalez RL, Kim HD, Chu S, Puglisi JD. 2004. tRNA selection and kinetic proofreading in translation. *Nat Struct Mol Biol* **11:** 1008–1014.

Cisse I, Okumus B, Joo C, Ha TJ. 2007. Fueling protein–DNA interactions inside porous nanocontainers. *Proc Natl Acad Sci* **104:** 12646–12650.

Clegg RM. 1992. Fluorescence resonance energy-transfer and nucleic-acids. *Methods Enzymol* **211:** 353–388.

Cornish PV, Ha T. 2007. A survey of single-molecule techniques in chemical biology. *ACS Chem Biol* **2:** 53–61.

Förster T. 1948. Intermolecular energy migration and fluorescence. *Ann Phys* **2:** 55–75.

Ha T, Enderle T, Ogletree DF, Chemla DS, Selvin PR, Weiss S. 1996. Probing the interaction between two single molecules: Fluorescence resonance energy transfer between a single donor and a single acceptor. *Proc Natl Acad Sci* **93:** 6264–6268.

Ha T, Rasnik I, Cheng W, Babcock HP, Gauss GH, Lohman TM, Chu S. 2002. Initiation and re-initiation of DNA unwinding by the *Escherichia coli* Rep helicase. *Nature* **419:** 638–641.

Hohng S, Joo C, Ha T. 2004. Observing single molecule conformational changes via three-color FRET. *Biophys J* **86:** 326a.

Joo C, McKinney SA, Lilley DM, Ha T. 2004. Exploring rare conformational speciesand ionic effects in DNA Holliday junctions using single-molecule spectroscopy. *J Mol Biol* **341:** 739–751.

Joo C, McKinney SA, Nakamura M, Rasnik I, Myong S, Ha T. 2006. Real-time observation of RecA filament dynamics with single monomer resolution. *Cell* **126:** 515–527

Lee JY, Okumus B, Kim DS, Ha T. 2005. Extreme conformational diversity in human telomeric DNA. *Proc Natl Acad Sci* **102:** 18938–18943.

McKinney SA, Declais AC, Lilley DM, Ha T. 2003. Structural dynamics of individual Holliday junctions. *Nat Struct Biol* **10:** 93–97.

McKinney SA, Freeman AD, Lilley DM, Ha T. 2005. Observing spontaneous branch migration of Holliday junctions one step at a time. *Proc Natl Acad Sci* **102:** 5715–5720.

McKinney SA, Joo C, Ha T. 2006. Analysis of single-molecule FRET trajectories using hidden Markov modeling. *Biophys J* **91:** 1941–1951.

Nahas MK, Wilson TJ, Hohng SC, Jarvie K, Lilley DMJ, Ha T. 2004. Observation of internal cleavage and ligation reactions of a ribozyme. *Nat Struct Mol Biol* **11:** 1107–1113.

Okumus B, Wilson TJ, Lilley DMJ, Ha T. 2004. Vesicle encapsulation studies reveal that single molecule ribozyme heterogeneities are intrinsic. *Biophys J* **87:** 2798–2806.

Rasnik I, McKinney SA, Ha T. 2005. Surfaces and orientations: much to FRET about? *Acc Chem Res* **38:** 542–548.

Rasnik I, Myong S, Ha T. 2006. Unraveling helicase mechanisms one molecule at a time. *Nucleic Acids Res* **34:** 4225–4231.

Tan E, Wilson TJ, Nahas MK, Clegg RM, Lilley DMJ, Ha T. 2003. A four-way junction accelerates hairpin ribozyme folding via a discrete intermediate. *Proc Natl Acad Sci* **100:** 9308–9313.

32 Alternating Laser Excitation for Solution-Based Single-Molecule FRET

Achillefs Kapanidis,[2] Devdoot Majumdar,[1] Mike Heilemann,[3] Eyal Nir,[4] and Shimon Weiss[1]

[1]Department of Chemistry and Biochemistry, Department of Physiology, and the California NanoSystems Institute, University of California, Los Angeles, California 90095; [2]Department of Physics and Biological Physics Research Group, Clarendon Laboratory, University of Oxford, Oxford OX1 3PU, United Kingdom; [3]Department of Physics, Applied Laser Physics and Laser Spectroscopy, Bielefeld University, 33615 Bielefeld, Germany; [4]Department of Chemistry, Ben Gurion University, Beersheba, Israel

ABSTRACT

Single-molecule fluorescence resonance energy transfer (smFRET) has been widely applied to the study of fluorescently labeled biomolecules on surfaces and in solution. Sorting single molecules based on fluorescent dye stoichiometry provides one with further layers of information and also enables "filtering" of unwanted molecules from the analysis. We accomplish this sorting by using alternating laser excitation (ALEX) in combination with smFRET measurements; here we describe the implementation of these methodologies for the study of biomolecules in solution.

INTRODUCTION

The development of single-molecule FRET (smFRET) (Ha et al. 1996) introduced a powerful technique to probe conformational states of single biomolecules. Widespread adoption and application of smFRET in a short period of time by the biological community has helped establish the method as a tool to characterize conformational dynamics (in the 1–10-nm range) of single molecules in very dilute solutions or immobilized on surfaces. Often, this approach has unearthed new and important information about the mechanisms underpinning the function of a broad range of biological machines (Ha 2001, 2004).

Typically, a solution-based smFRET experiment involves the detection of photons from one donor and a nearby acceptor fluorophore, and this information is used to arrive at a ratio for energy transfer (FRET) that reports on the distance between the donor and acceptor dyes. However, if the dyes are far apart, it is impossible to distinguish between a solitary donor (without nearby acceptor)

TABLE 1. Explanation of acronyms and abbreviations

Acronyms/ abbreviations	Explanation
ALEX	Alternating laser excitation
AOM	Acousto-optical modulator
AOTF	Acousto-optical tunable filter
APBS	All-photons burst search
APD	Avalanche photodiode
BSA	Bovine serum albumin
DCBS	Dual-channel burst search
DM	Dichroic mirror
DRLP	Dichroic reflector long pass
E	FRET efficiency
EOM	Electro-optical modulator
FRET	Fluorescence resonance energy transfer
KOH	Potassium hydroxide
OD	Optical density
S	Stoichiometry
smFRET	Single-molecule fluorescence resonance energy transfer
TTL	Transistor–transistor logic

and a donor with a distant companion acceptor (well beyond the range of FRET, ~1–10 nm for common dyes).

To remedy this problem, we developed alternating laser excitation (ALEX) (Kapanidis et al. 2004, 2005a; Laurence et al. 2005; Lee et al. 2005; Nir et al. 2006), which provides additional and direct information on the presence and the state of both donor and acceptor fluorophores in the molecules of interest. The additional information leads to multidimensional histograms of FRET and fluorophore stoichiometry, which, in turn, report on biomolecular structure and stoichiometry, respectively. Initial applications of ALEX focused on studies of gene transcription (Kapanidis et al. 2005b, 2006; Margeat et al. 2006) and protein folding (Jager et al. 2005, 2006; Laurence et al. 2005; Hamadani and Weiss 2008). As with conventional smFRET, ALEX is compatible with studies of both diffusing and immobilized molecules. Diffusing molecules can be studied in solutions, gels, or even porous materials; immobilized molecules can be studied using confocal microscopy and scanning, or total internal reflection (TIR) wide-field microscopy (Margeat et al. 2006).

This chapter describes methods based on ALEX of single molecules in solution; advice is given for building instrumentation for basic analysis of diffusing with general instructions for data analysis. Detailed information is presented on preparing standards for alignment and evaluation of the sensitivity of the setup. See Table 1 for a listing of the acronyms and abbreviations used in this chapter.

THEORY: FRET AND ALEX RATIOS

Single-excitation FRET measures photon counts $f_{D_{ex}}^{D_{em}}$ and $f_{D_{ex}}^{A_{em}}$, where f_X^Y is the photon count for a single molecule on excitation at wavelength X (where D_{ex} is the wavelength of substantial excitation of the FRET donor) and detection at wavelength range Y (where D_{em}, A_{em} are wavelengths of substantial emission of FRET donor and acceptor, respectively). ALEX provides one more nonzero photon count, $f_{A_{ex}}^{A_{em}}$, for acceptor emission on direct acceptor excitation (Kapanidis et al. 2004; Lee et al. 2005). These three-photon counts allow calculation of the proximity ratio E_{PR}^{raw} and the relative fluorescence stoichiometry ratio S^{raw}:

$$E_{PR}^{raw} = \frac{f_{D_{ex}}^{A_{em}}}{f_{D_{ex}}^{A_{em}} + f_{D_{ex}}^{D_{em}}},$$

$$S^{\mathrm{raw}} = \frac{f_{D_{\mathrm{ex}}}}{f_{D_{\mathrm{ex}}} + f_{A_{\mathrm{ex}}}} = \frac{f_{D_{\mathrm{ex}}}^{D_{\mathrm{em}}} + f_{D_{\mathrm{ex}}}^{A_{\mathrm{em}}}}{f_{D_{\mathrm{ex}}}^{D_{\mathrm{em}}} + f_{D_{\mathrm{ex}}}^{A_{\mathrm{em}}} + f_{A_{\mathrm{ex}}}^{A_{\mathrm{em}}}}.$$

This additional information is summarized in two-dimensional histograms (Fig. 1) and allows virtual molecular sorting. Sorting can remove artifacts that complicate FRET (such as the presence of states with inactive FRET donor or acceptor, and the presence of complex fluorophore stoichiometries) while introducing new entities that can be observed, such as an *A*-only population, which is helpful for evaluating biomolecular interactions.

The nature and timescale of the biological question to be addressed, along with the required measurement sensitivity, determine the alternation timescale for ALEX. Basic experiments concerning equilibrium views of a biological system (as well as kinetics occurring in the few-minutes timescale) can be pursued using laser modulation at the microsecond timescale by what we dub as microsecond-ALEX (μs-ALEX), which provides snapshots of diffusing molecules.

As in smFRET experiments, typical ALEX-based methods require use of subnanomolar concentrations of labeled molecules to ensure that single molecules can be detected and analyzed. Moreover, fluorescent probes need to be attached to molecules of interest, and assays need to show that the fluorophores do not significantly perturb their function. Finally, fluorophore photobleaching sets a limit on the observation of immobilized molecules, especially in the absence of oxygen scavengers and chemical additives. For information on bioconjugation chemistries and fluorophores

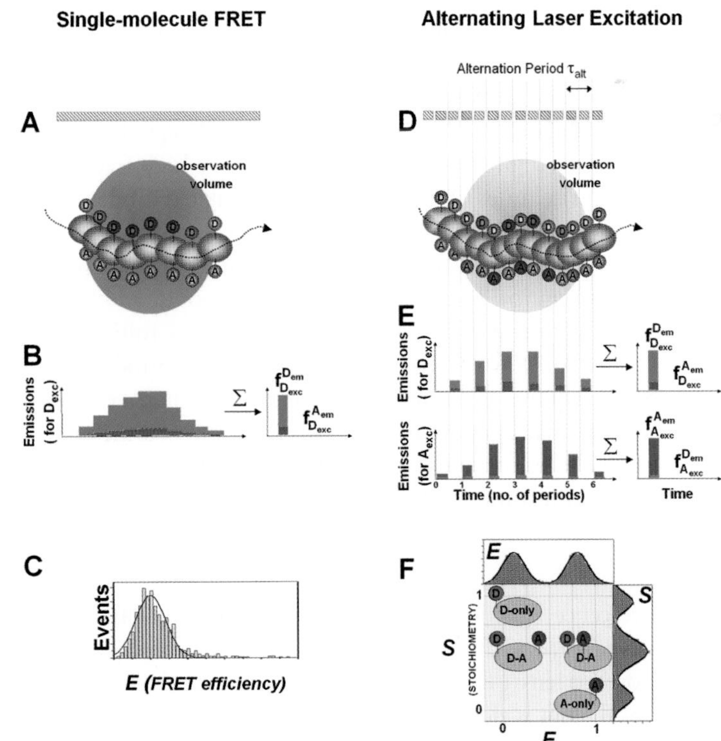

FIGURE 1. Concept of ALEX and comparison with single-excitation single-molecule FRET. (*A–C*) Single-molecule FRET using single-laser excitation; diffusing-molecule example. A fluorescent molecule traverses a focused green-laser beam and emits photons in the donor- and acceptor-emission wavelengths. The photon counts at these two wavelengths are used to generate one-dimensional histograms of FRET efficiency *E*. (*D,E*) Single-molecule FRET using alternating laser excitation. A fluorescent molecule traverses an observation volume illuminated in an alternating fashion using focused green- and red-laser beams. Using the photons emitted in the donor- and acceptor-emission wavelengths for each laser excitation, one can generate a two-dimensional histogram of FRET efficiency *E* and relative fluorophore stoichiometry *S*, enabling molecular sorting (see text for details).

repeat that are compatible with single-molecule fluorescence, see Kapanidis and Weiss (2002) and Jager et al. (2005, 2006).

PRACTICE: DESIGN PRINCIPLES FOR SMFRET SETUP

Excitation

Using confocal optics, single molecules are detected during transit through the confocal volume, wherein a focused laser beam excites the molecules and from which emitted photons are detected. In the "photon-hungry" regime of single-molecule measurements, the confocal microscope enables measurements with negligible background; the low background can be attributed to the pinhole in the optical path. Whereas most details for microscope construction do not at all deviate from traditional confocal setups (also used for fluorescence correlation spectroscopy or imaging), here we discuss some practical implementation specific to smFRET/ALEX-type microscope setups.

To probe both donor (for FRET) and acceptor (for ALEX) dyes, the laser beam must be "chopped" between alternating periods of donor and acceptor excitation. There are three main ways to modulate the laser excitation, depending on the type of laser (diode, solid state, gas) and general requirements (frequency and rise time, flexibility, cost, ease of use).

Acousto-optical modulation: This is easy to align and operate, with frequencies up to 15 MHz and rise times of ~25 nsec (for $PbMnO_4$ as the acousto-optical material). An acousto-optic modulator (AOM) is inexpensive, and operation with all types of linearly polarized continuous wave lasers is possible, thus increasing the flexibility of a setup. Similar to AOMs, acousto-optical tunable filters (AOTFs) can be used, allowing modulation of multiple wavelengths by selective deflection (Nir et al. 2006; Ross et al. 2007).

Direct modulation: This method is applicable to solid-state lasers; modulation is achieved by direct modulation of the laser power via TTL signals (transistor–transistor logic, which refers to a standardized digital signal) which can achieve high frequencies of up to 100 MHz (for diode lasers) and short rise times (2 nsec). Directly modulated diode lasers represent the most convenient option, as no additional optical components are necessary (see Kong et al. 2007). However, the current cost of these lasers and low power output (often 5–10 mW) remains a major drawback to this approach.

Electro-optical modulation: Basic electro-optic modulators (EOMs) can reach frequencies up to ~1 MHz with rise times of ~100 nsec, and thus fulfill the requirements for µs-ALEX; most of the published ALEX work is based on electro-optical modulation. However, this method is rather expensive and requires additional optical elements (quarter-wave plates and half-wave plates) and space.

All modulation devices can easily be addressed through software-controlled TTL or analog signals. All of these options work well for µs-ALEX, as modulation frequencies of 10–100 kHz are used. However, direct modulation of a laser is, if available, the easiest option.

Emission

Fluorescence light leaving the microscope through the base port is directed toward the detection module (Fig. 2). After the base port of the microscope, a pinhole is placed at the focal point of the microscope lens. (This is the "tube lens." The focal length is available from the manufacturer. Usually, the focal point is close to the base port of the microscope.) The pinhole is mounted onto an *x–y* positioner with micrometer screws, allowing easy removal or repositioning. This is helpful for the first alignment, as the detectors can be aligned without a pinhole. Single-molecule experiments will necessitate use of a 50–150-µm pinhole, depending on background of solution and solution viscosity.

After the pinhole, the light is collimated with another lens and spectrally separated with a single-band dichroic beam splitter. In combination with emission filters in front of the detectors, these elements should match the spectral characteristics of the setup and the fluorophores (i.e., eliminating excitation wavelengths and contributions to background signal by Raman scattering), thereby assuring efficient detection of the selected fluorophores. Point source detectors should be chosen for µs-

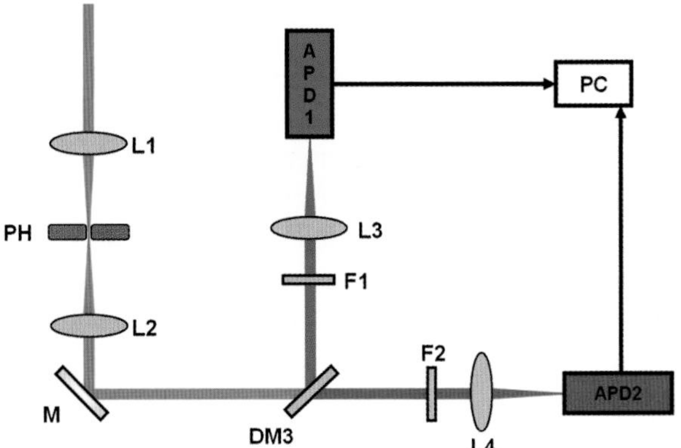

FIGURE 2. Emission module for ALEX. Fluorescence emission collected by the objective and through the DM of the sample holder module is focused with lens L1 onto a pinhole (PH), parallelized by L2, and separated according to wavelength by a dichroic beam splitter DM3 (650 DRLP). Light is directed through filters F1 (585DF70) and F2 (650LP) to select emission spectrum. Lenses L3 and L4 (20-mm focal length) focus onto the active area of the detectors (APD1 and APD2).

ALEX applications. These detectors are most often avalanche photodiodes (APDs; also known as single-photon avalanche diodes or SPADs), showing a low electronic background noise (termed "dark counts," as no photon is actually detected) and high detection efficiency for a single photon.

The arrival of a photon on the active area of the detector generates an electric signal, which is sent to a PCI (Peripheral Component Interconnect) board (e.g., PCI-6602, from National Instruments) and is then processed by acquisition software. It is crucial to shield the whole detection module from external light, using a black box (e.g., made of black cardboard) placed around the module.

DATA ANALYSIS

Burst Search

Because single-molecule measurements are performed using ~10–100 pM solutions of fluorescent molecules, the confocal volume is occupied by a fluorescent molecule for less than ~10% of the time; during the remaining ~90% or more of the time, photons detected are background photons. This background, with intensities of several thousands of photons per second (several kHz), is mostly attributed to Raman scattering of water and buffer molecules, labeled molecules diffusing at the fringe of the confocal spot, and reflections from optics. These contributions are in addition to the APD dark count. During the transit time (~1 msec) of the fluorescent molecule through the confocal volume, a finite number of excitation/emission cycles occur, resulting in a burst of fluorescence photons. This fluorescence burst is the observable result for a single diffusing molecule. When analyzing the stream of photons, one needs to distinguish between background photons and the photons of a burst and to assign the beginning and end of such a burst.

Fixed-Bin Burst Search

A simple way to separate fluorescence from background photons is to use the so-called "Box-Filter" search algorithm. During this search, the photon arrival time record is sectioned into fixed time intervals (bins) (Deniz et al. 1999). Using a photon-count threshold (the minimum number of photons per bin), bins containing fewer photons than the threshold are ignored, whereas bins containing more photons than the threshold are identified and retained for further analysis. Typical search parameters for standard fluorophores are bin durations of 500 µsec and photon-count thresholds of 5–20 photons. This algorithm lacks correlation between fixed bins and stochastically appearing bursts, because a burst (from a single molecule) may be divided into two bins. Moreover, a bin may contain predominantly background photons in addition to the burst of photons. This approach cannot be used efficiently in the dual-channel burst search (see below), because it cannot detect the exact moment in which a donor or acceptor molecule becomes inactive.

Sliding Burst Search

Sliding burst search is an alternative approach introduced by Seidel and co-workers (Eggeling et al. 2001). During this search, a burst is identified if at least L successive photons have at least M neighboring photons within a time window of length T centered on their own arrival time. The first and the last photons satisfying these criteria define the "start" and "stop" photons, respectively. Typical search parameters used are L = 10–50, M = 2–10, and T = 100–1000 µsec. The photons considered in this burst search can be photons detected during donor-excitation periods, acceptor-excitation periods, or all excitation periods ("all-photons" burst search or APBS) (Nir et al. 2006).

ALEX-Based Burst Search

Because the APBS identifies bursts irrespective of the wavelength of laser excitation and fluorescence emission, it may lead to skewed FRET values for molecules that bleach or blink during their transit through the confocal spot. An ALEX-based burst search can eliminate such events; here, the sliding burst search is used separately for (1) photons detected during donor-excitation periods, indicating the time intervals during which the donor is present and is photoactive, and (2) photons detected during acceptor-excitation periods, indicating the time intervals during which the acceptor is present and is photoactive. During this burst search, photons are considered as part of a burst if their time of arrival overlaps with the time intervals (1) and (2). This ensures identification of bursts containing photons detected during time intervals when both fluorophores are photoactive (thus removing any skew in the FRET distribution). The laser-alternation periods are considerably shorter than the sliding window. We dub this search the dual-channel-burst search (DCBS), where "dual-channel" refers to donor excitation versus acceptor excitation (and not donor emission vs. acceptor emission). Typical search parameters are L = 5–25, M = 2–10, and T = 100–1000 µsec. A subset of DCBS monitors solely the presence and state of the acceptor in diffusing molecules (Kapanidis et al. 2005b).

ALEX-Related Histograms and Factors Affecting Them

After burst search and identification, one can calculate two fluorescence ratios: proximity ratio E_{PR}^{raw} and stoichiometry S^{raw} (see above, Theory: FRET and ALEX Ratios, and below, µs-ALEX: Measuring Accurate FRET Efficiencies) for each burst. In this section, we use the general terms E and S to refer to these two ratios and to corrected versions thereof. For a complete treatment, see below, µs-ALEX: Measuring Accurate FRET Efficiencies, and Lee et al. (2005). The two ratios are then plotted in a 2D histogram, as well as two related 1D histograms (the collapse of the 2D histogram onto the proximity-ratio axis or the stoichiometry axis) (Fig. 3). For better representation, the results are "binned," usually by dividing the entire range of E or S (0.00–1.00) in 50–100 bins. To extract reliable information from the 2D histograms, one needs to understand how the histogram is affected by factors such as statistical fluctuations, molecular dynamics, and experimental artifacts. In part, this is achieved using dual-color burst search (Fig. 3).

Shot Noise

In the absence of artifacts (e.g., fluorophore photophysics), the main factors affecting the shape and width of subpopulations in the efficiency versus stoichiometry, or E–S, histogram are the donor to acceptor, or D–A, distance distribution and the shot noise (statistical fluctuation or noise). If the D–A distance is fixed and identical for all molecules, then the histogram shape and width result from the statistical fluctuation of photon counts around an average. For example, if we assume that E is 0.5 and all bursts have 100 photons, then because of statistical noise, the actual number of acceptor photons will vary around 50 (because $E = f_{D_{ex}}^{A_{em}}/(f_{D_{ex}}^{A_{em}} + f_{D_{ex}}^{D_{em}}) = 0.5$ when $f_{D_{ex}}^{A_{em}} = f_{D_{ex}}^{D_{em}} = 50$) but will not always be equal to 50. Because of the stochastic character of the donor and the acceptor emissions, the fluctuation around the average can be described by a binomial law and depends on the size of a burst (i.e., the sum of donor and acceptor photons in a burst) and the probability that a photon in a burst is an acceptor photon. In the absence of background and cross-talk photons (or after

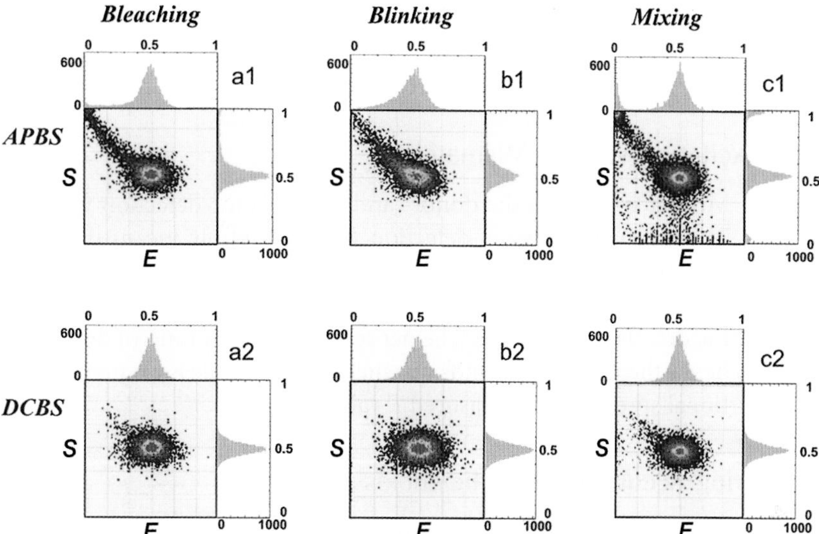

FIGURE 3. Simulated effects of bleaching, blinking, and random coincidence on the 2D *E–S* histogram: Comparison of all-photons burst search (APBS; *upper row*) with the dual-channel-burst-search (DCBS; *lower row*). Bleaching (*left*) and blinking (*middle*): In both cases, APBS reveals a trail of bursts connecting a *D*-only population ($E = 0$, $S = 1$) and the population of interest ($E = S = 0.5$). DCBS eliminates such bursts, yielding symmetric, Gaussian-like *E* and *S* distributions. Random coincidence (*right*): APBS reveals a trail of bursts from the *D–A* population toward the *D*-only subpopulation and toward the *A*-only subpopulation ($S = 0$). Again, DCBS eliminates most of these bursts, resulting in symmetric, Gaussian-like *E* and *S* distributions (see also Nir et al. 2006).

accounting for such contributions), this latter probability is equal to the *E* value (Lee et al. 2005; Nir et al. 2006). Moreover, in a real single-molecule fluorescence experiment, the bursts do not have a fixed size, but rather a distribution of sizes. Because it is difficult to predict such a distribution, the binomial calculation of the shot noise is performed using the empirical burst size distribution. Estimation of the FRET histogram width and shape, and subsequent comparison to the experimental histogram, recovers additional contributions to the FRET histogram width and shape and is necessary to extract the correct *D–A* distance distribution.

Bleaching and Blinking

Because of photophysical or photochemical reactions, a fluorophore may reach permanent (bleaching) or temporary (blinking) nonemitting states (Kong et al. 2007). Figure 3 shows a simulated ALEX histogram of freely diffusing, doubly labeled molecules of $E = 0.5$. In the simulation, the acceptor stochastically bleaches or blinks (Fig. 3, left and middle) once every ~300 cycles of absorption/emission. When APBS is used to analyze bursts during which the acceptor is either bleached or in a temporary dark state, a decrease in the number of acceptor photons and an increase in the number of donor photons are detected. This results in a decrease in the mean *E* value and an increase in the mean *S* value. The exact value depends on the timing and timescale of the bleaching or blinking relative to the burst duration. Because bleaching and blinking are stochastic processes, they occur randomly during the diffusion through the confocal spot, leading to a "tail" of events in the 2D *E–S* histogram bridging the *D–A* subpopulation ($E = 0.5$ and $S = 0.5$) with the *D*-only-like subpopulation ($E = 0$ and $S = 1$). When DCBS is used (lower row), most of the events caused by blinking and bleaching are removed, leaving the histogram with the *D–A* subpopulation.

Random Coincidence of Diffusing Species

As discussed earlier, to achieve single-molecule resolution, the sample concentration is ≤100 pM. This ensures low probability of having two diffusing molecules in the confocal spot at the same time, because this random coincidence will alter the E and S values. If the concentration is >100 pM (Fig. 3, right; 300 pM), "bridges" between subpopulations are formed, altering the shape and the center of the distributions; DCBS minimizes such effects.

Detection and Excitation Volume Mismatch

Mismatch between the donor- and the acceptor-detection volumes broadens the FRET histogram. The mismatch between detection volumes simply means that the ratio between the probability of detecting a donor photon and the probability of detecting an acceptor photon depends on the location of the emitter. This can be owing to chromatic aberrations of the optics or to misalignment of the detectors or pinhole. The net result is that the ratio of donor- to acceptor-detected photons (and hence the E and S) depends on where the molecule is located in the confocal spot, leading to an additional spread of the measured E values.

μs-ALEX: Measuring Accurate FRET Efficiencies

Single-molecule FRET experiments that calculate proximity ratios are sufficient for studying the presence and relative abundance of various FRET states and subpopulations, as well as reporting on the kinetics of transitions between FRET states. However, it is often necessary to determine accurate FRET efficiencies and corresponding distances between fluorophores (e.g., in studies of protein translocation or determination of biomolecular structures). In these cases, it is necessary to perform corrections to remove photon cross talk and instrumental biases (Lee et al. 2005); such correction factors are available to all ALEX methods.

According to Förster theory, the accurate energy transfer efficiency E is related to the distance between the fluorophores by

$$E = \frac{R_0^6}{r^6 + R_0^6},$$

where R_0 represents the Förster radius of the pair of fluorophores used:

$$R_0^6 = \frac{9000(\ln 10)\kappa^2 Q_D}{128\pi^5 N n^4} \int_0^\infty F_D(\lambda)\varepsilon_A(\lambda)\lambda^4\, d\lambda,$$

with $\kappa^2 = (\cos\theta_T - 3\cos\theta_D\cos\theta_A)^2$.

The Förster radius is specifically related to the donor and acceptor fluorophore and can be determined out of experimental values, including the quantum yield of the donor Q_D, the fluorescence spectrum of the donor F_D, the wavelength-dependent extinction coefficient of the acceptor ε_A, the refractive index of the medium n, and Avogadro's number N. The relative orientation of the dye is expressed by κ^2, derived from the angle between the dipole moment of the donor and acceptor with respect to the connecting line (θ_D and θ_A) and relative to each other (θ_T). In case of freely rotating fluorophores, κ^2 can be approximated to 2/3.

CONCLUSION AND FUTURE PROSPECTS

Only a handful of biophysical approaches can be used to directly measure conformational changes, among them nuclear magnetic resonance (NMR) and small angle X-ray scattering (SAXS). These ensemble approaches will often be more informative but necessitate homogeneous and synchronizable conformational changes for proteins at concentrations in the micromolar range. In contrast, smFRET/ALEX permits study of small conformational changes that occur in a subpopulation of measured molecules without the need for synchronization and at picomolar concentrations. The caveat of note, however, is that a FRET only measures a single point-to-point distance that requires time-consuming sample preparation and is susceptible to error owing to changes in dye environment or other unexplained dye-related photophysics.

Nonetheless, smFRET with ALEX brings a new tool to the repertoire of structural biologists with which to probe Angstrom-level structural changes in biomolecules. This has thus far been most successfully applied to nucleic acid–based systems, such as polymerases, helicases, and ribosomes. These systems offer the practical advantage of facile and specific fluorophore labeling, whereas to study structural biology of protein system relies on less robust labeling strategies; nonetheless, several groups have investigated structural changes in protein folding and protein conformational dynamics without issue.

The use of smFRET/ALEX has become increasingly practical over the past 10 years owing to improvements in (1) the brightness of fluorescent dyes, (2) the availability of new and orthogonal biomolecule labeling chemistries, (3) the cost and availability of increasingly higher-precision optical elements, and (4) the cost and availability of lasers spanning the visible spectrum.

Advances in each of these domains will continue, contributing to the overall ease of the approach. Furthermore, groups will be able to probe the more intractable structural questions that require complex biochemical environments by performing smFRET/ALEX inside of cells. With advances in detector technology comes the prospect of multiple-confocal spot excitation and detection, allowing for high-throughput FRET measurements. Finally, the growing commercialization of the technology (by companies such as PicoQuant GmbH) enables one to envision facile smFRET/ALEX measurements at the push of a button by a biologist, akin to a fluorimeter.

Protocol A

Assembling the μs-Alex Setup

This protocol describes the construction of a μs-ALEX using two lasers, a green 532-nm acousto-optically modulated laser and a red 635-nm directly modulated laser. See Table 2 for a listing of suppliers of parts for constructing ALEX microscope modules.

MATERIALS

CAUTION: See Appendix 6 for proper handling of materials marked with <!>.

Equipment

Avalanche photodiodes (APDs)
BNC (Bayonet Neill Concelman) cables
Desktop PC
Dichroic mirror (DM)
Emission beam splitter (650DRLP)
Fiber coupler
Lasers (green 532-nm acousto-optically modulated and red 635-nm directly modulated) <!>
  Extreme caution is required when using lasers.
Microscope with appropriate objectives
Mirrors
Modulator
Optical fiber (single-mode)
Optical table
Paper (white)
Polarizers (λ/4, λ/2)
Slide (glass) (optional; see Step 5)
Spectral filters (585DF70 for the green channel, 650LP for the red channel)

METHOD

1. Arrange lasers, the modulator, and the fiber coupler on an ~50 x 50-cm area on an optical table. Verify that the polarization of the lasers is both linear and vertical. If this is not the case, linearize the polarization using a polarizer, and convert to vertical orientation either by adjusting the position of the laser or by using a combination of λ/2 and λ/4 plates. Use two mirrors to direct the green laser beam into the AOM aperture.
2. Overlap both laser beams using a mirror (M1, Fig. 4) and a DM (DM1, 560DRLP) for the green laser and two mirrors (M2, M3) for the red laser, and couple the combined beam into a single-mode optical fiber.

   Caution: In most cases, lasers used for ALEX applications belong to the class IIIB (underlying the International Laser Safety Standard IEC 60825). It is important to follow laser safety instructions, which include general safety rules and special institutional rules. A laser safety officer should be contacted before any ALEX system is set up and should help with risk assessment and safety guidelines.

TABLE 2. Suppliers of parts used for constructing ALEX microscope modules

Components	Types	Provider
Excitation module		
Lasers	Argon ion	Melles Griot, Newport
	Nd:YAG	Cobolt, Laser2000
	Diode laser	Coherent, Picoquant
	HeNe	Melles Griot
Modulation	EOM	Conoptics, Linos
	AOM	Isomet
	AOTF	AA-Optoelectronics
Optics	Filters, dichroics	Semrock, Omega, Chroma
	Lenses, mirrors	Linos, Thorlabs
	Polarizer, λ/4, λ/2	Newport
	Prism	Melles Griot
Fiber optics	Optical fiber	Thorlabs
	Fiber coupler	Thorlabs, Newport
	Multiplexer	AA-Optoelectronics, Linos
Optomechanics	Posts, mounts, kinematics	Thorlabs, Comar, Linos
	Translation stages	Linos, Newport, New Focus, Thorlabs
Sample holder module		
Microscope	IX71 Inverted microscope	Olympus
	Axiovert 100, 200, 200M	Zeiss
	Leica DM IL	Leica
Objective	PLAPON 60xO/TIRFM 1.45 (oil immersion)	Olympus
	UPLSAPO 60x/1.2 W (water immersion)	Olympus
	Alpha Plan-Fluar 100x/1.45 (oil immersion)	Zeiss
Temperature control	Objective heating collar	Biosciencetools
	Objective cooling collar	Intracel
	Microscope temperature control	Olympus, Zeiss, Leica
Emission module		
Detectors	Avalanche photodiodes	PerkinElmer (SPCM-AQR-14 or 15), Picoquant (PDM 20/50/100CT)
	EMCCD camera	Roper Scientific, Andor
Optomechanics	Translation stages for detectors	New Focus
	Kinematic mounts, filter holders, posts, bases, translation stage for pinholes	Thorlabs, Comar
Optics	Filters, dichroics	Omega, Chroma, Semrock
	Lenses, pinholes	Linos, Comar

The above suppliers are used in the authors' laboratories; many other suppliers are also available, especially for components of the excitation module.

3. Mount the output of the optical fiber on the microscope breadboard, followed by a 10x–20x collimating objective mounted on an *x–y–z*-positioning stage.

   The choice of the collimating objective depends on the diameter of the back aperture of the focusing objective, which should be fully illuminated. With common single-mode optical fibers and a 10x collimating objective, beam diameters of 5–10 mm are achieved.

   For some applications, it may be desirable to underfill the back aperture of the objective to increase the volume of the confocal spot. This in turn increases the diffusion time of molecules and the photons emitted per diffusing molecule.

4. Direct the laser beam to the side port of the microscope. The beam should be centered in the back aperture of the focusing objective as well as in the field of view. Use the *x–y–z*-positioning stage to align the beam position and collimate the beam.

5. Align the detection path by reflecting scattered excitation light through the objective. (Place a glass slide and focus onto the upper side of the slide, or place a mirror on the objective.) The light will pass the excitation dichroic and can be visualized along the detection path using a small piece of white paper. Place the pinhole at the focal point of the microscope lens, onto an

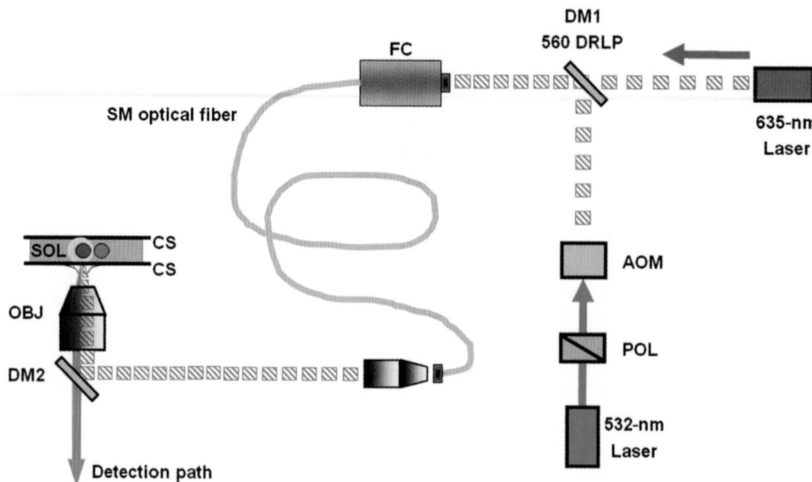

FIGURE 4. Excitation module for ALEX: 635-nm laser (directly modulated and linearly polarized) and 532-nm laser. (POL) Polarizer; (AOM) acousto-optical modulator. Laser beams are overlaid with three mirrors (M1, M2, and M3) and a dichroic beam splitter (DM1; 560DRLP) into a fiber coupler (FC) to which a single-mode (SM) optical fiber is connected; 10x objective serves as fiber output unit; dual-band dichroic beam splitter (DM2) directs light through oil-immersion objective (NA 1.45, 60x or 100x).

x–y or *x–y–z* positioner, followed by a second lens to collimate the light. The parallel light is split by the emission beam splitter on two detectors, split by a dichroic mirror, which is followed by further spectral filters along each beam path. A 20-mm lens focuses the light on the active area of each APD, mounted onto micrometer-precision *x–y–z*-positioning stages.

6. Mount each detector on an *x–y–z*-positioning stage and connect to a counting board in a desktop PC using BNC cables.

Protocol B

Alignment of smFRET/ALEX Setup

To achieve single-molecule sensitivity and thus have the ability to detect single diffusing fluorophores, careful alignment of the setup is crucial. The following protocol describes routine alignment for 2c-ALEX (532 nm/635 nm) with spectral windows $G^{550\text{-}620}R^{650\text{-}750}$.

MATERIALS

CAUTION: See Appendix 6 for proper handling of materials marked with <!>.

Reagents

Alignment sample (e.g., tetramethylrhodamine [TMR; Invitrogen] or Cy3B [Amersham Biosciences])

Equipment

Avalanche photodiodes (APDs)

Camera and closed circuit TV (CCTV) monitor (optional; see Step 3)

Coverslip

Lasers <!>

Extreme caution is required when using lasers.

Microscope with appropriate components

Optical density (OD) filter (flippable) (optional; see Step 3)

METHOD

1. Prepare the alignment sample (concentration of ~10^{-9} M) of a fluorophore that can be excited with 532 nm and shows considerable emission cross talk on the red channel (and thus allows alignment of both detectors). Solutions of tetramethylrhodamine (TMR; Invitrogen) or Cy3B (Amersham Biosciences) are well suited.
2. Set laser power of the green laser to ~50 µW (measured right before entering the microscope). This should yield a considerable signal in the range of 50–100 kHz. If the signal is lower (e.g., owing to setup misalignment), increase the laser power until you reach the desired count rate.
3. Deliver 50 µL of the alignment sample onto a cover slide. Using the eyepiece, find the focus of the glass–water interface, and raise the objective to move the focused beam ~20 µm into the solution.

   Caution: Laser radiation is dangerous for the eyes. Whenever eyepieces are used and laser intensities are >100 µW, the beam should be attenuated before entering the microscope (e.g., with a flippable OD filter). Alternatively, a simple camera mounted on the camera port and connected to a CCTV monitor can be used for focusing.
4. For the first alignment, remove the pinhole and align the detectors using the micrometer screws of an *x–y*-stage. If the lenses in front of the APDs are positioned accurately and the beam is col-

limated, you should observe a plateau for the detection response of the APDs (owing to the small size of the focus compared with the active area of the APD).

5. Insert the pinhole and align its position using the *x–y*-positioning stage. At the first alignment, ideally start with a larger pinhole (200 µm), followed by 100 µm and 50 µm. It is important to ensure that correlated signals on both detectors are observed when the pinhole is repositioned to identify its best position.
6. Align detectors again using the procedure outlined in Step 4.

   It is convenient to establish a range of expected observables after the alignment procedure for a particular setup; for example, this can be the count rate for a standard concentration of an alignment sample (10^{-9} M of a desired fluorophore) and a defined excitation power (50 µW). The important values are the detection ratio between the two detection channels, as well as the count rate of emission.

Protocol C

Sample Preparation and Data Acquisition for µs-ALEX

This protocol describes the preparation of samples and data acquisition for µs-ALEX. Sample preparation requires a dilution that ensures the detection of single events.

MATERIALS

CAUTION: See Appendix 6 for proper handling of materials marked with <!>.

Reagents

Bovine serum albumin (BSA)
Buffer components appropriate for sample (specified "for luminescence spectroscopy," e.g., from Fluka, Merck, or Fisher Scientific)
KOH <!>, 1 M
Sample to be assayed

Equipment

Coverslips
Filters (0.2 µm)
Gaskets (sealable plastic) or enclosable chambers
Lasers <!>

Extreme caution is required when using lasers.

Microscope with appropriate components for data acquisition

METHOD

1. Clean coverslips with 1 M KOH and rinse with H_2O before use.
2. Prepare buffer and filter it through a 0.2-µm sterile filter. Measure the buffer alone with settings identical to those used for sample data acquisition, to decide on purity.

   See *Troubleshooting.*
3. To avoid adsorption of biomolecules to the coverslip (e.g., large proteins that tend to stick to glass), BSA (10–50 µg/mL) may be added to the buffer. Either pretreat surfaces with BSA-containing buffer (the protein will form a layer on the glass surface), or perform measurements using BSA-containing buffer.
4. Dilute the sample in buffer to a concentration of 100 pM.

   Typically, the confocal volume is ~1 fL, and a concentration of 100 pM results in a probability of 0.1 to find a single molecule within the confocal volume.
5. Use sealable plastic gaskets or enclosable chambers to perform measurements. This helps avoid solvent evaporation during data acquisition, which could cause a subsequent increase of sample concentration.

6. Acquire data using appropriate settings and analyze.
   See *Troubleshooting.*

TROUBLESHOOTING

Problem (Step 2): There are impurities in the buffer.

Solution: Buffers for single-molecule experiments need to be free of fluorescence impurities. The number of fluorescence impurities depends strongly on the quality of the chemicals used for buffers. Many of those chemicals are available in high purity and should be used for the preparation of measurement buffers. Buffer purity is best evaluated by the number of bursts observed, which should be considerably lower than the sample itself, and of lower intensity. In general, impurities will appear on the green channel (as impurities showing fluorescence in the red spectral region are rare), and thus can be filtered out during data analysis. In the actual experiment, the fluorescence count rate of impurities should be considerably lower (use less than about one-third as a rough guide) than for the fluorophores used, and the burst frequency (which is proportional to the concentration of impurities) should be <10% compared with a 100 pM solution of labeled molecules.

Problem (Step 6): Photon yields are inadequate.

Solution: The total photon yield of organic fluorophores is limited (~10^6 photons emitted before photobleaching). Depending on the chemical class of fluorophores, oxygen removal (Yildiz et al. 2003) and addition of triplet quenchers to the buffer solution (Rasnik et al. 2006) can increase photon yields significantly.

Problem (Step 6): There is inadequate resolution of subpopulations.

Solution: The ratio of the excitation powers will influence the stoichiometry ratio S (see above, Theory: FRET and ALEX Ratios). Changing this ratio will influence the separation of populations in the 2D E–S histogram and thereby affect the resolution of subpopulations.

Problem (Step 6): Photobleaching occurs.

Solution: High excitation powers will lead to photobleaching of the fluorophores (both donor and acceptor). In FRET experiments using an acceptor excited by 635-nm light, acceptors are more sensitive to photobleaching. In some cases, photobleaching can be reduced by modifying the usual 50%–50% green–red excitation duty cycle into an excitation cycle that allocates 80% to green excitation and 20% to red excitation (Kong et al. 2007). Photobleaching can equally be dealt with using ROXS (reducing and oxidizing system) buffers, at least to a certain extent.

Problem (Step 6): There are insufficient statistics to answer the question at hand.

Solution: µs-ALEX in solution detects single events and derives statistical data out of them. Any experiment should aim at sufficient statistics to answer the questions. Use measurement times from 2 to 30 min. Fast measurements provide a quick view of the sample and allow decisions on whether a longer trace is needed. Proper sealing of the sample, reliable temperature control, use of suitable excitation power (e.g., to allow sufficient photon counts while preventing significant photobleaching), and a physically and optically stable setup are important for ensuring constant conditions for long acquisition times.

Problem (Step 6): Inaccurate results are obtained.

Solution: Record additional experimental variables for proper data analysis. Determine the background on each detector (green, red) during the two excitation cycles (red or green excitation) from buffer-only measurements. At the same time, determine the donor leakage (relative emission of the donor fluorophore on the acceptor channel) and direct excitation of the acceptor by using single-labeled constructs under identical conditions.

Problem (Step 6): Incorrect E values and distances are obtained.

Solution: If experiments aim to determine exact E values (and corresponding distances), determine the correction factor γ for the set of fluorophores used (see above, μs-ALEX: Measuring Accurate FRET Efficiencies).

REFERENCES

Deniz AA, Dahan M, Grunwell JR, Ha T, Faulhaber AE, Chemla DS, Weiss S, Schultz PG. 1999. Single-pair fluorescence resonance energy transfer on freely diffusing molecules: Observation of Förster distance dependence and subpopulations. *Proc Natl Acad Sci* **96:** 3670–3675.

Eggeling C, Berger S, Brand L, Fries JR, Schaffer J, Volkmer A, Seidel CA. 2001. Data registration and selective single-molecule analysis using multi-parameter fluorescence detection. *J Biotechnol* **86:** 163–180.

Ha T. 2001. Single-molecule fluorescence resonance energy transfer. *Methods* **25:** 78–86.

Ha T. 2004. Structural dynamics and processing of nucleic acids revealed by single-molecule spectroscopy. *Biochemistry* **43:** 4055–4063.

Ha T, Enderle T, Ogletree DF, Chemla DS, Selvin PR, Weiss S. 1996. Probing the interaction between two single molecules: Fluorescence resonance energy transfer between a single donor and a single acceptor. *Proc Natl Acad Sci* **93:** 6264–6268.

Hamadani KM, Weiss S. 2008. Nonequilibrium single molecule protein folding in a coaxial mixer. *Biophys J* **95:** 352–365.

Jager M, Michalet X, Weiss S. 2005. Protein–protein interactions as a tool for site-specific labeling of proteins. *Protein Sci* **14:** 2059–2068.

Jager M, Nir E, Weiss S. 2006. Site-specific labeling of proteins for single-molecule FRET by combining chemical and enzymatic modification. *Protein Sci* **15:** 640–646.

Kapanidis AN, Weiss S. 2002. Fluorescent probes and bioconjugation chemistries for single-molecule fluorescence analysis of biomolecules. *J Chem Phys* **117:** 10953–10964.

Kapanidis AN, Lee NK, Laurence TA, Doose S, Margeat E, Weiss S. 2004. Fluorescence-aided molecule sorting: Analysis of structure and interactions by alternating-laser excitation of single molecules. *Proc Natl Acad Sci* **101:** 8936–8941.

Kapanidis AN, Laurence TA, Lee NK, Margeat E, Kong X, Weiss S. 2005a. Alternating-laser excitation of single molecules. *Acc Chem Res* **38:** 523–533.

Kapanidis AN, Margeat E, Laurence TA, Doose S, Ho SO, Mukhopadhyay J, Kortkhonjia E, Mekler V, Ebright RH, Weiss S. 2005b. Retention of transcription initiation factor σ^{70} in transcription elongation: Single-molecule analysis. *Mol Cell* **20:** 347–356.

Kapanidis AN, Margeat E, Ho SO, Kortkhonjia E, Weiss S, Ebright RH. 2006. Initial transcription by RNA polymerase proceeds through a DNA-scrunching mechanism. *Science* **314:** 1144–1147.

Kong X, Nir E, Hamadani K, Weiss S. 2007. Photobleaching pathways in single molecule FRET experiments. *J Am Chem Soc* **129:** 4643–4654.

Laurence TA, Kong XX, Jager M, Weiss S. 2005. Probing structural heterogeneities and fluctuations of nucleic acids and denatured proteins. *Proc Natl Acad Sci* **102:** 17348–17353.

Lee NK, Kapanidis AN, Wang Y, Michalet X, Mukhopadhyay J, Ebright RH, Weiss S. 2005. Accurate FRET measurements within single diffusing biomolecules using alternating-laser excitation. *Biophys J* **88:** 2939–2953.

Margeat E, Kapanidis AN, Tinnefeld P, Wang Y, Mukhopadhyay J, Ebright RH, Weiss S. 2006. Direct observation of abortive initiation and promoter escape within single immobilized transcription complexes. *Biophys J* **90:** 1419–1431.

Nir E, Michalet X, Hamadani KM, Laurence TA, Neuhauser D, Kovchegov Y, Weiss S. 2006. Shot-noise limited single-molecule FRET histograms: Comparison between theory and experiments. *J Phys Chem B* **110:** 22103–22124.

Rasnik I, McKinney SA, Ha T. 2006. Nonblinking and longlasting single-molecule fluorescence imaging. *Nat Methods* **3:** 891–893.

Ross J, Buschkamp P, Fetting D, Donnermeyer A, Roth C, Tinnefeld P. 2007. Multicolor single-molecule spectroscopy with alternating laser excitation for the investigation of interactions and dynamics. *J Phys Chem B* **111:** 321–326.

Yildiz A, Forkey JN, McKinney SA, Ha T, Goldman YE, Selvin PR. 2003. Myosin V walks hand-over-hand: Single fluorophore imaging with 1.5-nm localization. *Science* **300:** 2061–2065.

33 FIONA: Nanometer Fluorescence Imaging

Paul D. Simonson and Paul R. Selvin

Physics Department, University of Illinois at Urbana-Champaign, Urbana, Illinois 61801

ABSTRACT

Fluorescence imaging with one-nanometer accuracy (FIONA) is a technique for localizing a single dye to within a subdiffraction (<250 nm) limit, commonly ~1-nm accuracy. This high degree of precision is technically *not* breaking the diffraction limit, but simply collecting enough photons to obtain this level of accuracy. This is achieved using total internal reflection fluorescence (TIRF) microscopy, deoxygenation reagents, and a high-quantum-yield, low-noise camera. Here we discuss the theory behind FIONA, and applications and variants of FIONA, some of which break the diffraction barrier (i.e., decrease the resolution or measurable distance between two dyes). Examples relevant to molecular motors are discussed. Finally, we provide protocols for doing a simple FIONA experiment.

INTRODUCTION

Fluorescence imaging with one-nanometer accuracy (FIONA) is a simple method for achieving localization of single fluorophores with nanometer accuracy in the *x–y* plane. FIONA can attain ~1-nm accuracy, assuming that the fluorophore emits 5,000–10,000 photons within the time resolution of the experiment (Yildiz et al. 2003). We have attained time resolutions ranging from 500 msec (Yildiz et al. 2003) down to 1 msec (Kural et al. 2005), the latter being important for in vivo measurements of molecular motors. (In vivo, the time resolution must be high because of the native high concentration of ATP, which makes molecular motors "run" instead of simply "walk.") By adding reagents that prevent fast photobleaching, a movie can be recorded that tracks the motion of a single molecule for a good fraction of a minute or longer.

HOW FIONA WORKS

The fundamental idea behind FIONA is presented in Figure 1. In short, a fluorophore is placed on a coverslip and excited with total internal reflection (TIR) laser excitation. The fluorophore is assumed to be essentially stationary during the time resolution of the experiment. Although there are other suitable methods for exciting the fluorophore, TIR produces a small excitation volume, making background fluorescence much lower than with other methods (Yildiz et al. 2003). With wide-field detection using a charge-coupled device (CCD) camera (or better, with an electron-multiplying charge-coupled device [EMCCD] camera), the fluorescence is collected, and the intensity is plotted as a function of x and y. Thompson et al. (2002) have shown that the approximate localization accuracy achievable is given by the width of the point-spread function (which is simply the wide-field diffraction limit of ~250 nm for visible light) divided by the square root of the total number of photons. With 10,000 collected photons, the result is 2.5 nm, or ±1.25 nm. Variables, including the detector and background noise, must also be taken into account, although in reality these can generally be made small enough so that the limiting factor is the number of photons (N) that the fluorophore emits. Taking these variables into account gives the localization as (Thompson et al. 2002)

$$\sigma_{\mu i} = \sqrt{\frac{s_i^2}{N} + \frac{a^2/12}{N} + \frac{8\pi s_i^4 b^2}{a^2 N}}. \tag{1}$$

The positional accuracy σ_μ (i.e., the uncertainty or the standard error of the mean) can be related to the pixel size of the imaging detector (a), the standard deviation of the background (b), which includes background fluorescence and detector noise, and the width of the distribution (standard deviation s_i in direction i, where $i = x$ or y). The pixel size a is equal to the CCD pixel size divided by the total magnification. When the recommended range of 80–120 nm per pixel is used (e.g., a 16-µm CCD pixel divided by 100x objective and 1.6x external magnifier is 100 nm), the effect of the finite size of the pixel a is small. In addition, by not letting a become too small, the effect of polarization is also insignificant (Enderlein et al. 2006). The effect of the background term b is also small when using TIR and a low-noise detector, such as a back-thinned EMCCD. Then, the first term (s_i^2/N), the error due to photon noise, is dominant. If the width is 250 nm (the diffraction limit of light), and 10,000 photons are collected, then $\sigma_\mu = 250/[10^4]^{1/2} = 2.5$ nm, or ±1.25 nm. We find this accuracy experimentally by fitting the image to a two-dimensional Gaussian function (or more computationally intensive Airy function) (Yildiz et al. 2003). When applied to motor proteins labeled with fluorophores, by measuring the position before and after a step, the step size can be determined. Knowing the step size, one can then determine the method of walking: that is, whether one foot bypasses the other foot, walking in a hand-over-hand fashion, or walking by sliding the two feet, moving in an inchworm fashion. This has been performed for myosin V (Yildiz et al. 2003), myosin VI (Yildiz et al. 2004b), kinesin I (Yildiz et al. 2004a), and dynein (Reck-Peterson et al. 2006); the

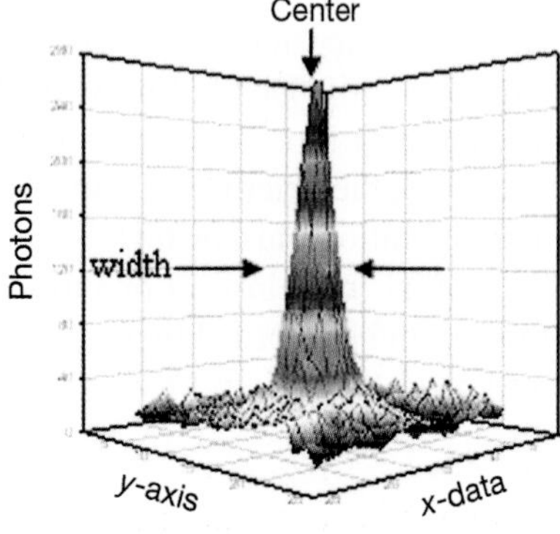

FIGURE 1. CCD image of a single Cy3 molecule. The width is ~250 nm, but the center can be located to within w/(SNR), where SNR is signal to noise ratio (Equation 1).

results are summarized later in this chapter. The motors can be observed for as long as the fluorophores are stable. When using organic fluorophores and by very simple means (Yildiz et al. 2003; Aitken et al. 2008), this can be a minute or longer, and for quantum dots, it can be essentially forever.

VARIANTS AND RELATIVES OF FIONA

FIONA has been applied in many examples, primarily molecular motors. Note that FIONA does *not* break the diffraction limit. The diffraction limit is ~250 nm ($\lambda/2n \sin \theta$ where n is the index of refraction, typically 1.33–1.5, and θ is the angle of collection, with $\sin \theta \approx 1$). The diffraction limit determines resolution—that is, it tells how close two fluorophores can be and still be distinguishable from one another. Variants of FIONA, however, do break the diffraction limit. One such technique is called single-molecule high-resolution imaging with photobleaching (SHRImP) (Gordon et al. 2004). A similar technique was published under the name of nanometer-localized multiple single-molecule (NALMS) fluorescence microscopy (Qu et al. 2004). The techniques resolve two identical dyes that are in close proximity by exciting the two dyes, waiting until one photobleaches, and determining the position of the remaining dye using FIONA. The position of the photobleached fluorophore is then determined by subtracting the image of the unbleached fluorophore from the image taken of the two dyes immediately before photobleaching. The resulting image of the photobleached fluorophore can be fit with FIONA, enabling the position of each dye to be determined with ~10-nm accuracy. NALMS has been applied to distinguish up to five closely spaced fluorophores (Qu et al. 2004).

SHRImP cannot be used if the fluorophores are moving. However, in another variant of FIONA, the distance between two *differently* colored fluorophores can be determined with greater accuracy than the standard far-field diffraction limit (Lacoste et al. 2000; Churchman et al. 2005; Toprak and Selvin 2007). This technique is called single-molecule high-resolution colocalization (SHREC). The two fluorophores are chosen to emit at distinct and separate wavelengths. They are independently imaged using a DualView or similar device. The fluorophores are then individually localized using FIONA. The resolution is 8–10 nm (Churchman et al. 2005; Toprak and Selvin 2007). This technique is good at any timescale and is capable of observing subdiffraction-limit behavior of moving fluorophores, thereby spanning the distance between fluorescence resonance energy transfer (1–10 nm) and standard microscopy (>250 nm).

FIONA has also been extended to localize very large numbers of photoswitchable fluorophores. Two similar techniques, dubbed PALM (photoactivated localization microscopy) (Betzig et al. 2006) or fluorescence PALM (Hess et al. 2006), and stochastic optical reconstruction microscopy (Rust et al. 2006), achieve this by stochastically switching fluorophores on and off. By cycling the fluorophores on and off, and by only switching on a small fraction of the total number of fluorophores at a time—much less than one fluorophore per diffraction-limited area—a high-resolution image can be constructed. Although multiple fluorophores might be physically close together, they are optically separable because they are not fluorescing at the same time. They can thus be localized individually using FIONA, and a composite superresolution image incorporating all of the fluorophores can be constructed.

FIONA does not have to be applied to just fluorophores. In fact, it can also be applied to localizing particles that absorb photons. In this case, the number of photons used to calculate the position of the particles is the number of photons *absorbed*, as opposed to the number *emitted*, as is used in the normal case. Using this principle, we have imaged photon-absorbing objects (specifically, melanosomes), using bright-field microscopy and localized them to nanometer accuracy. We have called this application "bright-field FIONA" (Kural et al. 2007).

In addition to determining the centroid of a fluorophore, the three-dimensional orientation of a fluorophore can be observed using a technique called DOPI (defocused orientation and position imaging) (Toprak et al. 2006). In DOPI, a fluorophore is held rigidly to a protein, for example, by a bis-cysteine reactive dye that is reacted with a second bis-cysteine genetically placed on a protein.

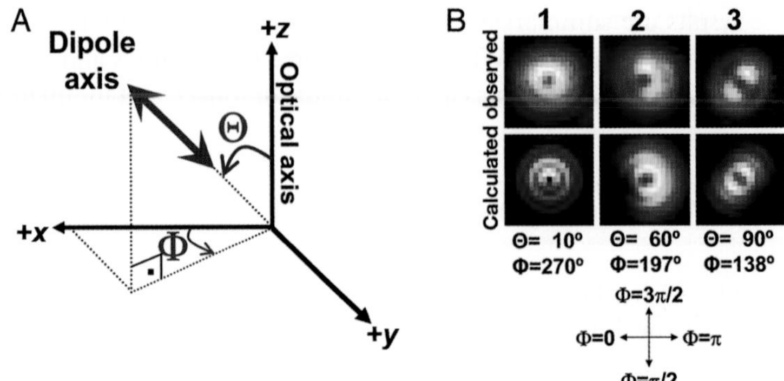

FIGURE 2. Schematic of DOPI obtained with a quantum dot (QS 655, Molecular Probes). (*A*) The definition of angles (Θ, Φ). (*B*) Out-of-focus images of a quantum dot oriented along the *z* direction (*B1*), in the *x*–*y* plane (*B3*), and at a 60° angle with respect to the *z*-axis (*B2*). *B1* is an example of a dipole aligned almost parallel to the optical axis (normal to the specimen plane); a donut pattern is evident. Two symmetric bright lobes appear when the dipole is aligned parallel to the specimen plane (*B3*). Because there is no emission along the dipole axis, a dark "line" appears between the two lobes. The in-plane angle (Θ; as in *A*) can be predicted by looking at this dark line. A dye with an arbitrary out-of-plane angle (*B2*) forms a "Pac-Man-shaped" image that is a composite of *B1* and *B3*.

Whereas FIONA can be used to monitor nanoscale translocations, the transition dipole moment of a fluorophore, via DOPI, can be used in parallel to monitor rotational movements of molecules of interest. By moving the objective lens ~100–500 nm out of focus, the unfocused image shows a characteristic pattern that depends on the orientation of the dye (see Fig. 2). By fitting the resulting patterns to theory, the angular orientation of the fluorophore can be determined to within ~15° in all planes (Toprak et al. 2006). The theory and computational background of DOPI are described by Enderlein et al. (Bohmer and Enderlein 2003; Patra et al. 2004).

TECHNICAL ASPECTS OF PERFORMING FIONA

Equipment for Imaging Single Molecules

Total internal reflection fluorescence microscopy (TIRFM) is typically used to perform FIONA because it gives a very high signal to background ratio that is ideal for observing single fluorophores. Building an objective-type TIRFM setup is relatively straightforward (Selvin et al. 2008). Although care should be taken to ensure "proper alignment," this is a somewhat subjective term. The ultimate litmus test for proper alignment is the quality of the illumination spot and the resulting fluorescence image. Once a satisfactory image is obtained, no further adjustments are necessary. The alignment can always be refined, but a point of diminishing return will be reached, after which further efforts may not result in discernibly improved image quality. The numbers of photons collected is crucial for doing FIONA. High numerical objectives (typically NA = 1.45; for objective-type TIR one needs an NA > 1.33 [Gell et al. 2006]) are optimized for collecting photons. The use of modern CCD array cameras, such as the Andor iXon+ and the Photometrics Cascade EMCCD back-thinned camera, and attention to selecting proper emission filters are also important for maximizing the number of photons collected. (Semrock, Chroma Technology, and Omega Optical manufacture many optimized filters.)

Fluorophore Stability

Stability of the fluorophore is crucial for most applications of FIONA, especially where fluorophore tracking is involved. Fluorophores are inherently unstable; however, the stability can often be dramatically improved by the addition of appropriate reagents. Instability of the fluorophore can usually be categorized as either photobleaching or blinking. Photobleaching is generally thought to be an irreversible process (however, see Hess et al. 2006; Betzig et al. 2006; Rust et al. 2006) caused by a chemical reaction induced by light, and it is apparent that in many cases oxygen dissolved in the solvent plays a key role. The average time before a fluorophore photobleaches is often increased dramatically by introducing oxygen-scavenging systems, such as glucose oxidase/glucose/catalase (Yildiz et al. 2003) or PCA/PCD (Aitken et al. 2008), into the sample solution. The PCA/PCD system has the advantage that the pH does not drop significantly over time. Fluorophore blinking is characterized by the fluorophore going into a dark state that usually lasts from microseconds to seconds. Blinking can often be ameliorated by adding reducing reagents, such as β-mercaptoethanol or Trolox (Rasnik et al. 2006), or a combination of reducing and oxidizing reagents (Vogelsang et al. 2008).

Data Analysis

In our laboratory, analysis of data collected from a FIONA experiment requires several software programs. Using software supplied with our cameras (Andor Technology), movies can be saved as 16-bit TIFF (tagged image file format) files, which can be inspected, cropped, etc., using ImageJ, a free software package available from the National Institutes of Health (http://rsbweb.nih.gov/ij/). ImageJ is also capable of converting TIFF movies to AVI or QuickTime movies.

After selecting images for further analysis, localization and accuracy are determined by fitting a two-dimensional Gaussian function to the point-spread function of the fluorophore at every frame. We use customized software written in IDL (ITT Visual Information Solutions) for this task (see Protocol E for a simple example), but a free National Institutes of Health–sponsored program, Video Spot Tracker, is sufficient for most common analysis needs. The program and instructions can be downloaded from http://cismm.cs.unc.edu/downloads/. Video Spot Tracker can track one or more particles from video files that are compatible with Microsoft DirectShow (e.g., AVI files) and currently runs only on a Windows platform. There are many other commercially available programs that are capable of performing all of the tasks described above, although the price range for these programs is currently ~$10,000. Some of these programs are Andor iQ (Andor Technology), MetaMorph (Universal Imaging Corporation), Image Pro (Media Cybernetics), and Workbench (Intelligent Imaging Innovations, also known as 3i). Additional programs are available through selvin@illinois.edu.

APPLICATIONS OF FIONA TO THE STUDY OF MOLECULAR MOTORS

FIONA techniques lend themselves well to molecular motor studies, because it is often desirable to observe nanometer-scale movements associated with the components of the motors or the cargo they carry. In this section we present some examples of using FIONA in the study of molecular motors.

Using In Vitro FIONA to Analyze Myosin V Movement

We have shown, using FIONA, that myosin Va, myosin VI, and kinesin all walk with a hand-over-hand mechanism (Yildiz et al. 2003, 2004a,b). Myosin V is a homodimer (Fig. 3A). Each monomer consists of a "head" (also called a "hand" or "foot"), which binds actin and binds and hydrolyzes ATP; a neck, which contains six copies of the calmodulin-binding consensus sequence (IQ domain); and a stalk, which holds the two monomers together and facilitates cargo binding.

It is well known by optical trapping techniques that myosin V cargo moves with steps of ~37 nm (Mehta et al. 1999). If myosin walks in a "hand-over-hand" fashion, each "hand" moves past the other with the moving hand taking 74-nm steps and the other hand remaining stationary (thus, the center of mass moves 37 nm; see Fig. 3B). If a fluorophore-labeled calmodulin is bound to myosin V, then the dye moves $(37 - 2x)$ nm, where x is the distance from the center of mass (i.e., the stalk), followed by a $(37 + 2x)$-nm movement. We have used both green fluorescence protein (GFP) bound to the head (Snyder et al. 2004) and calmodulin labeled with rhodamine (Yildiz et al. 2003) to image

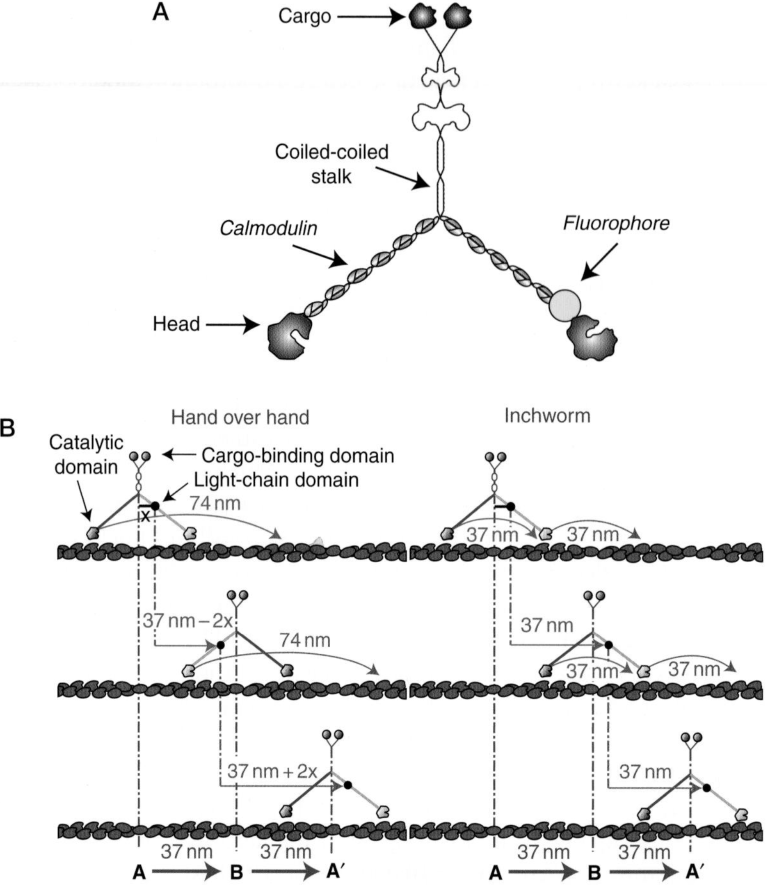

FIGURE 3. (*A*) The structure of myosin V. Each monomer consists of a head, six calmodulin light chains, a coiled-coil domain, and a cargo-binding domain. A calmodulin, isolated separately, is labeled with a fluorophore and exchanged on to the myosin V. (*B*) The exact position of the labeled calmodulin can be calculated from the stepping pattern.

these steps. Figure 4A plots position versus time for several myosin V molecules tracked with rhodamine-labeled calmodulin.

Using In Vivo FIONA to Study Peroxisome Transport

The main difficulty with in vivo FIONA is finding the conditions in which the fluorophore emits enough photons within a short enough period of time to achieve the desired localization accuracy (see Equation 1). Currently, quantum dots and groups of fluorophores (either beads or clusters of GFP) are both bright and long lasting, although obtaining the appropriate binding specificity with quantum dots is an issue (Michalet et al. 2005). In one particular study, we studied peroxisomes filled with a large amount of GFP (Kural et al. 2005). By treating *Drosophila* S2 cells with latrunculin to depolymerize actin, long processes were formed (Fig. 5A) that contained microtubules running the entire length of the cell. The plus ends of the microtubules were located at the termini of the processes, whereas the minus ends were positioned near the nucleus. Note that kinesin shows plus-end-directed motion (away from the nucleus, toward the cell membrane), whereas dynein moves toward the minus end (away from the cell membrane, moving toward the nucleus). Experiments were designed to determine whether cargo is transported by a concerted/cooperative motion of kinesin and dynein or by a competitive "tug of war" between the two. Peroxisomes, like much of the cargo transported by kinesin and dynein, move back and forth in what appears to be a stochastic manner. A "tug of war" would require that both motors pull on the cargo, and the stronger motor

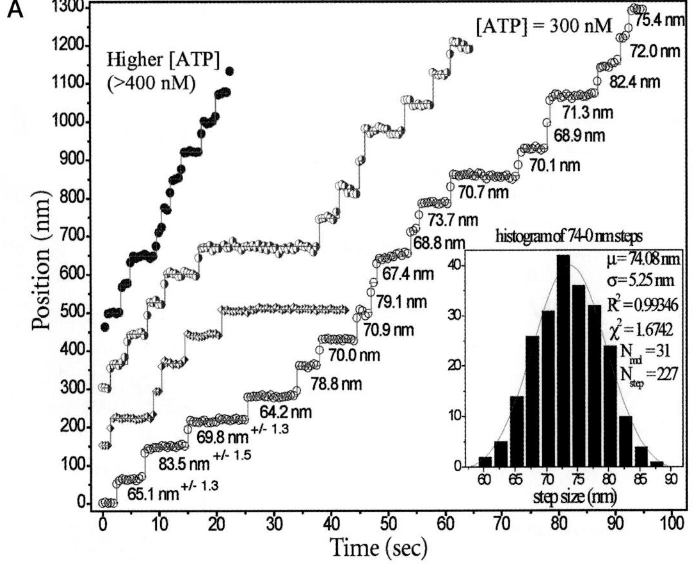

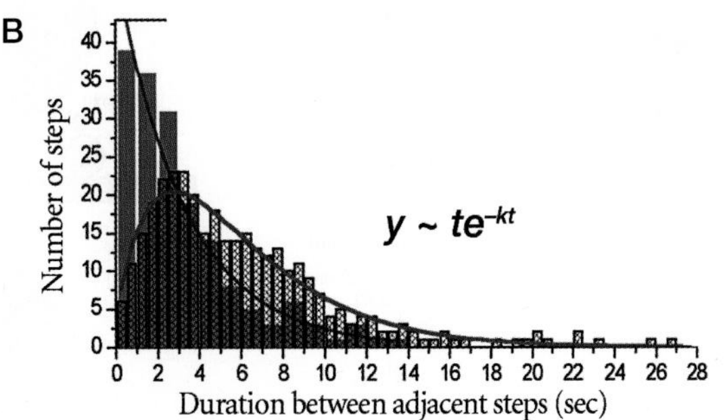

FIGURE 4. The stepping pattern of myosin V. (*A*) Individual traces are shown with a histogram of their steps, giving an average step size of 74.1 ± 5.25 nm (inset). Note that the accuracy of measuring a single step is as small as 1.3–1.5 nm, indicating that the myosin is actually taking different-sized steps. (*B*) A dwell-time histogram of two differently labeled myosin molecules. If the label is on or near to the center stalk of the myosin, then every step will result in movement of the label (33–42 nm, 23–52 nm, etc.). This leads to an exponential dwell-time histogram. On the other hand, if the label exchanges near the head (74.0 nm), it moves with every other step, giving a *t* exp(–*kt*) histogram as expected in the hand-over-hand model.

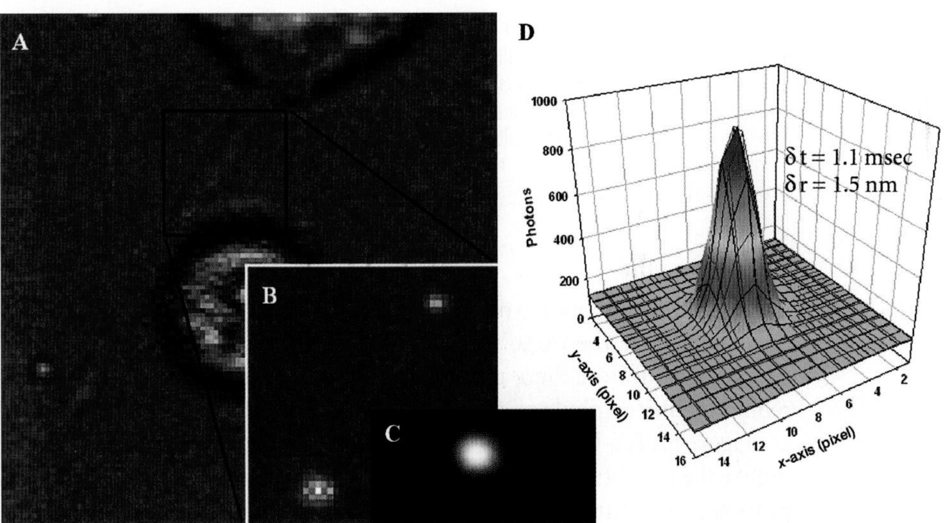

FIGURE 5. In vivo FIONA of *Drosophila* S2 cells. (*A*) A bright-field image of a cell treated with latrunculin, which causes actin depolymerization and the growth of "processes," one of which is located inside the black box. The process has a microtubule running along its length. Kinesin walks on processes away from the cell, and dynein runs toward the cell. If a fluorescent micrograph is taken (*B*) of a GFP-labeled cargo, two such cargoes are seen. Note that the GFP cargo, because it contains many fluorophores, is extremely bright (*C*). Because the multiple GFPs are so bright, excellent time resolution (1.1 msec) can be achieved while still achieving excellent spatial resolution (1.5 nm), which enables 1-msec time resolution to be achieved (*D*).

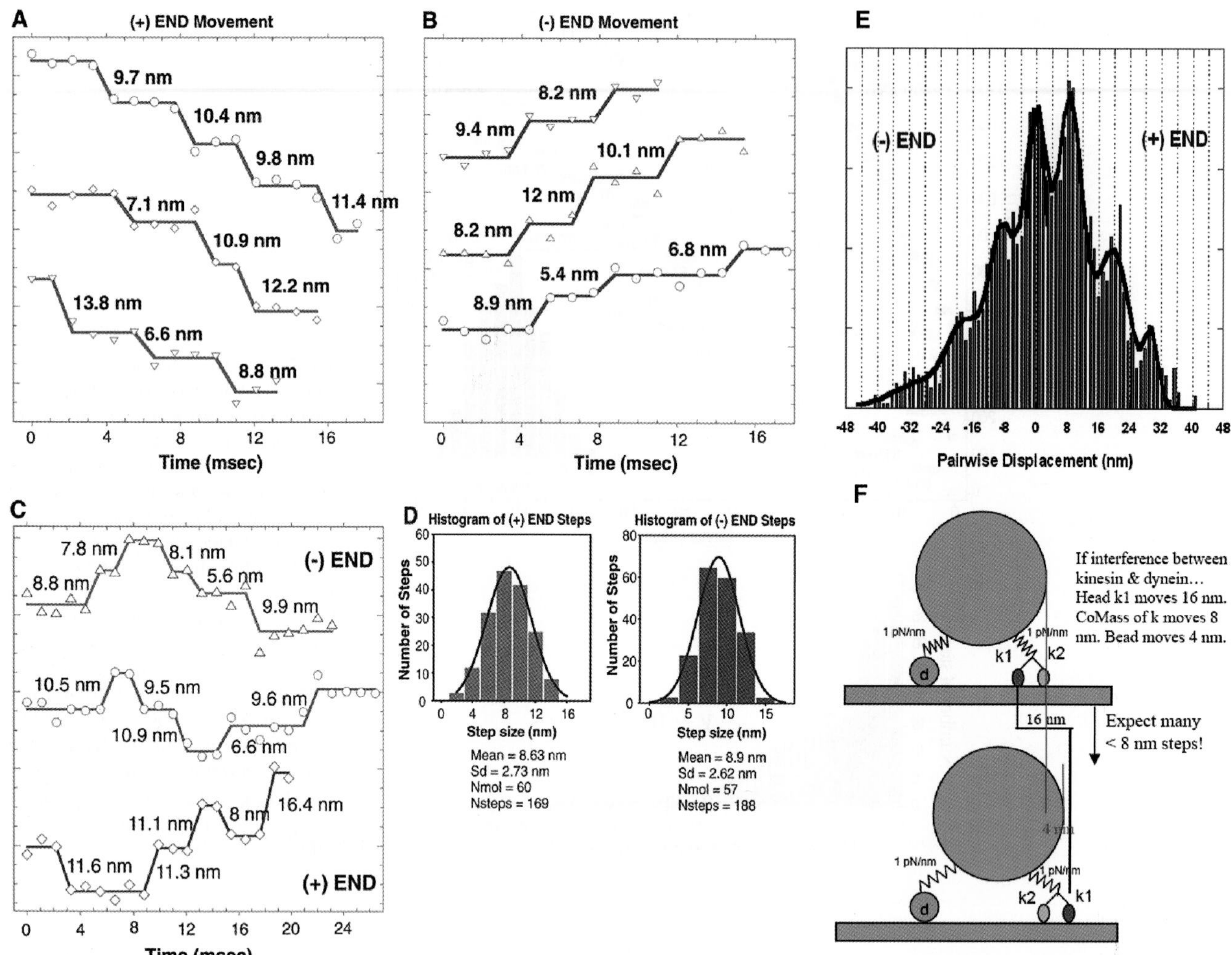

FIGURE 6. Example of in vivo FIONA on GFP peroxisomes in *Drosophila* S2 cells. Individual kinesin (*A*) and dynein (*B*) steps are resolved with millisecond time resolution by tracking peroxisome movement. (*C*) Switching between kinesin-driven movement and dynein-driven movement. (*D*) Histograms of plus- and minus-end-directed steps give an average of ~8 nm, the expected step size of kinesin and dynein. (*E*) A pairwise displacement histogram showing maxima at 0-nm and 8-nm multiples, again revealing an average step size that is the same as for kinesin and dynein in vitro. (*F*) A representation of the tug-of-war model of motion in which kinesin and dynein pull against each other. Note that the average step size should be <8 nm in this model.

would win. A cooperative motion, in which the cargo moves the usual 8.3 nm, requires that kinesin or dynein move the cargo with no drag force being exerted by an opposing motor. Therefore, in cooperative motion, there must be a cellular signal that cues the dynein and kinesin to release and rebind the microtubule and/or the cargo at the appropriate times.

Figure 5A is a bright-field image of a *Drosophila* S2 cell that has been treated with latrunculin to induce the formation of processes. A fluorescence picture (Fig. 5B) shows two bright spots within the process. A quantitative picture that shows one spot's brightness is shown in Figure 5C. If curve fitting to the spot is performed as explained previously (see Fig. 1), the spot is determined to have approximately 5000 photons in a 1.1-msec exposure (Fig. 5D) and fits with an uncertainty of 1.5 nm. This is an approximately 500-fold increase in temporal resolution over the in vitro case shown in Figure 4.

We have analyzed the motion of individual peroxisomes in a *Drosophila* S2 cell and found that individual steps of kinesin and dynein can be resolved (Fig. 6A,B). Switching between kinesin and dynein is also observed (Fig. 6C). Histograms of the kinesin and dynein steps indicate that the two

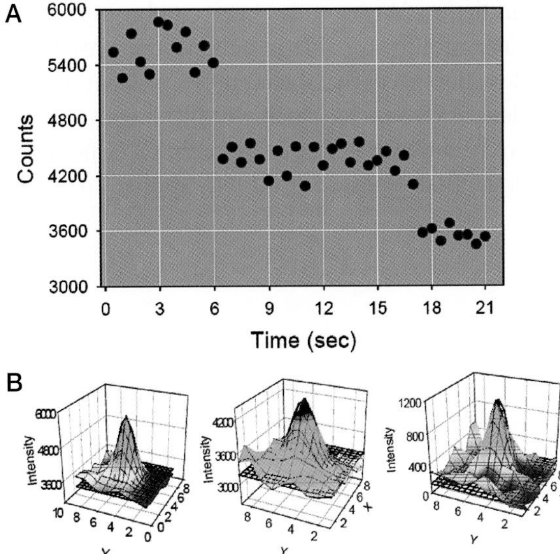

FIGURE 7. SHRImP. (*A*) The number of counts collected from a single spot over time. Note the two-step photobleaching. (*B*) An example of fitting the data. Prephotobleach is a fit to both dyes; postphotobleach is a fit to the dye remaining after the other dye photobleaches, allowing the centroid of that dye to be found. The difference between the two fits then gives the centroid of the photobleached dye.

stepwise histograms (Fig. 6D) are centered at ~8 nm, and the pairwise displacement histogram of kinesin and dynein traces (Fig. 6E) also shows 8-nm steps for both motor proteins. If dynein and kinesin were "fighting" with each other—creating significant drag—a nonsymmetrical distribution would be expected, with many steps being <8 nm and few being >8 nm. This is shown in Figure 6F, where the kinesin walks, with one hand taking the usual 16-nm step, but, because of a finite compliance (shown as springs with 1 pN/nm), the cargo moves <8 nm (shown as 4 nm in Fig. 6F). Hence, we conclude that the motion must be cooperative and involves some unknown regulator.

Using SHRImP to Analyze the Motion of Myosin VI

As noted earlier, SHRImP is a method for imaging the distance between two dyes that are closer than the diffraction limit (10–100 nm) (Gordon et al. 2004). For example, by attaching a fluorophore to each head of myosin VI and imaging the motor proteins bound to actin when in a near-rigor state, it is possible to measure the distance between the heads between steps. Myosin VI molecules that photobleached in two steps (Fig. 7A) were identified, and the spots were fit using the SHRImP technique (Fig. 7B), giving a distance between the heads of 29.3 ± 1.4 nm. This is in agreement with the hand-over-hand model. These results also indicate that there is a relatively constant distance between the heads on stepping, as opposed to an alternating stepping distance of $60 - x, x$ (where x might be significantly different from 30 nm). (For further details of the hand-over-hand and an alternative model, see Balci et al. 2005.)

CONCLUSIONS AND FUTURE DIRECTIONS

FIONA can be used to track nanometer scale movements of proteins. Derivatives of FIONA can be used to break the diffraction limit of optical microscopy, distinguishing and localizing molecules that are as close as 10 nm apart, and to perform superresolution microscopy. Applications of FIONA have yielded insights into the basic functioning of molecular motors, which are crucial for proper transport of cargo within cells.

We have begun expanding the FIONA repertoire, using it to look at other important molecules and processes within cells and even whole animals. FIONA is being used to examine individual ion channels. The results of this work have enormous clinical ramifications for the study of strokes,

Alzheimer's disease, and nicotine addiction. We have also been combining FIONA with optical traps in vivo to understand how molecular motors are able to achieve their high velocities. Recently, we successfully applied FIONA to follow fluorescently labeled molecules within *Caenorhabditis elegans* nerves. GFP- and DENDRA2-labeled ELKS punctae can be localized with sub-10-nm accuracy in ~5 msec (Kural et al. 2009). Our results show that the protein ELKS is occasionally transferred by microtubule-based motors. This is the first example of FIONA applied to a living organism.

Protocol A

Constructing an Objective-Type TIRF Microscopy Setup

Building an objective-type total internal reflection fluorescence (TIRF) microscopy requires all of the components typically used in a normal fluorescence microscope setup (Fig. 8). The key difference, besides the use of high-NA objectives, is the introduction of a lens that deflects the beam off of the optical axis. Deflecting the beam slightly off-axis causes the beam that is exiting the objective to strike the glass–sample interface at an angle that is dependent on the position of the lens. Increasing this angle decreases the penetration depth of the resulting evanescent wave into the sample (Axelrod 1989; Pedrotti and Pedrotti 1993). Because of limitations of the equipment, the angle can only be increased to a certain point before the evanescent wave disappears. It is usually just before this point where the TIRF effect is best for reducing background from the sample. For more detailed discussion, see Selvin et al. (2008).

MATERIALS

CAUTION: See Appendix 6 for proper handling of materials marked with <!>.

Reagents

Immersion oil
Methanol <!>
Rhodamine solution, 1 mM (5-carboxytetramethylrhodamine [Sigma C2734] in methanol)

Equipment

Beam expander, 10x (Thorlabs, Inc.; e.g., BE10M-B or beam expander construction kit)
CCD camera (Andor Technology iXonEM + [DV-897E-CS0])
FIONA sample chamber (see Protocol B)

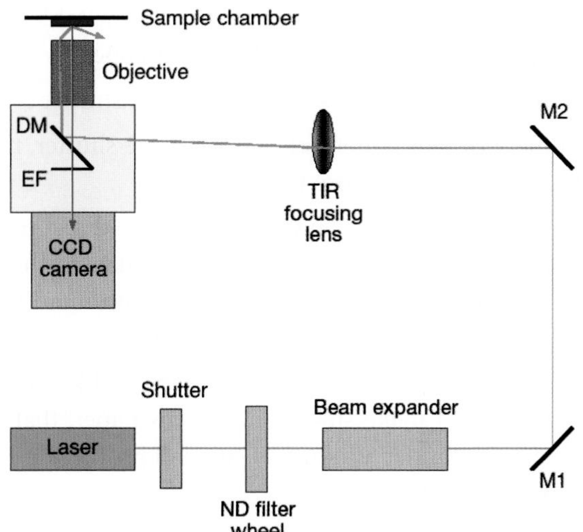

FIGURE 8. Schematic of TIRF microscopy setup. DM, dichroic mirror; EF, emission filter; M1, first silver mirror; M2, second silver mirror.

Inverted microscope (e.g., Olympus IX71)
Kinematic mirror mounts, two (Thorlabs KM200)
Kinematic mount for beam expander (Thorlabs, Inc.)
Laser, 50-mW, 532-nm (CrystaLaser) <!>
Laser shutter
Lens mount (Thorlabs LMR2)
Lens paper (Thorlabs, Inc.)
Long-pass dichroic mirror, 550-nm (Semrock or Chroma Technology)
Long-pass emission filter, 550-nm (Semrock or Chroma Technology)
Objective, 100x 1.45-NA oil immersion (Olympus PlanApo 100x 1.45-NA, /0.17)
Optics table (Newport RS4000)
Optics table supports (I-2000 Lab Legs)
Posts and post holders (Thorlabs, Inc.)
Protected silver mirrors, two, 2-in diameter (Thorlabs PF20-03-P01)
Set of neutral density (ND) filters and filter wheel mounts (Thorlabs, Inc.)
Table clamps (Thorlabs CL5)
Three-axis manual translation stage (Newport 460P-XYZ; also requires three actuators [AJS100-1] and base plate [460P-BA])
TIR focusing lens, 2-in diameter, $f = 400$ mm (Thorlabs LA1725-A)
Two-axis manual translation stage for beam expander (Newport 460P-XY; also requires two actuators [AJS100-1], 90° angle bracket [460P-90BK], and base plate [460P-BA])

EXPERIMENTAL METHOD

Warning: Never look into the laser beam. Be aware of reflective surfaces that might direct beam into eyes. Remove watches and jewelry while aligning lasers. Always use beam stops during intermediate stages of the alignment process, and close the laser shutter when introducing new components in the beam path. When checking the beam path, it is convenient to use a piece of lens paper. Always use the lowest amount of laser power that is reasonable during the alignment process.

1. Set up the optical table and microscope. Secure the microscope with the table clamps, so that it is immobilized on the optics table. Install the CCD camera on the side port of the microscope.
2. Install the dichroic mirror and emission filter in one of the microscope's filter cubes.
3. Mount the laser onto the optics table with the beam parallel to the table and at a height that is approximately the same height as the front port of the microscope.
4. Mount the laser shutter in front of the laser. Make sure that the shutter is closed.
5. Turn on the laser. Using a piece of lens paper held just past the shutter, make sure the laser is passing through the center of the shutter. Close the shutter.
6. Mount the ND filters onto the table just past the shutter. To avoid direct reflection back into the laser, the ND filters should be close but not exactly perpendicular to the beam. Select a filter with a high optical density (e.g., OD 2–3) to reduce the laser power and improve safety during the following steps (later, the OD can be reduced as necessary).
7. Mount the beam expander, using a two-axis translation stage and kinematic mount, just past the ND filter. Open the shutter and make sure, using a lens paper, that the beam is passing through the center of the beam expander and has a Gaussian profile. The beam should be collimated. Make fine adjustments as necessary.
8. Mount one of the silver mirrors so that the beam leaving the beam expander has a 45° incident angle.

9. Mount the second silver mirror past the first mirror and at 45° to the incident beam. Mount the mirror so that the beam is entering the front port of the microscope. Leave ~2–3 ft between the mirror and the microscope.

10. Close the shutter. Make sure the correct filter cube is selected in the microscope. Place a piece of lens paper over the hole where the objective will be mounted. Open the laser shutter and adjust the mirrors using the kinematic mounts until the beam is centered in the hole and passes straight up toward the ceiling, perpendicular to the table. For fine adjustments, the first silver mirror is used to align the laser through the objective-mounting hole, and the second silver mirror is used to align the beam to a point directly above the hole. Through several iterations of very slight overcorrections and undercorrections of the two mirrors, good alignment can be achieved. With practice, this becomes very easy.

    Warning: When the shutter is first opened, the beam will not necessarily be pointing straight up from the microscope. Holding an additional piece of plain paper over the lens paper will protect eyes until it is clear in which direction the beam is pointing.

11. Close the laser shutter. Install the objective in the microscope. Fill a sample chamber with rhodamine solution. Place a drop of immersion oil on the objective. Place the chamber, coverslip side down, above the objective.

12. Open the laser shutter. Look through the eyepiece. Adjust the focus until a bright fluorescent spot is seen. The spot should be somewhat near the center of the field of view. If it is not, close the shutter, remove the chamber and objective (clean the oil off of the objective first using lens paper and methanol), and return to Step 10.

13. Turn on the CCD camera and wait for it to cool to <–40°C. Adjust the camera settings so that there is no gain. Start acquiring images with the camera in continuous acquisition mode. The fluorescent spot should be visible near the center of the field of view (adjust the OD of the ND filter as necessary). Adjust the silver mirrors until the spot is centered in the field of view. If the spot tends to disappear or become faint before reaching the center, iterate slight overcorrection and/or undercorrection of the two silver mirrors in sequence until the spot is bright and centered (if the alignment becomes too bad while trying to adjust the position, return to Step 10).

14. For fine adjustment of the beam, move the objective down (using the microscope focus knob) until the fluorescent spot starts to grow in size (i.e., the focus plane is moving below the sample). This larger spot might be slightly off center and should be corrected using the first silver mirror. Move the objective up until the spot is small again, and center it using the second silver mirror. Repeat this process using slight overcorrections/undercorrections until both the small and enlarged spots are centered in the camera.

Introducing the TIR Focusing Lens

15. Close the shutter. Remove the sample chamber and clean the objective using lens paper and methanol. Remove the objective from the microscope and place a piece of lens paper over the objective mounting hole.

16. Mount the TIR focusing lens (using the three-axis stage) so that the beam passes through the center. Make sure that the flat side of the lens is facing the microscope to reduce spherical aberration. The lens should be mounted so that the beam is focused onto the conjugate plane as imaged through the tube lens of the objective, which amounts to being ~8 in in front of the microscope.

17. Open the shutter. A small, bright spot should appear on the lens paper. Adjust the three-axis stage (using x- and y-axes) until the spot is centered in the mounting hole. Close the shutter.

    Warning: When the shutter is first opened, the beam will not necessarily be pointing straight up from the microscope. Holding an additional piece of plain paper over the lens paper will protect eyes until it is clear in which direction the beam is pointing.

18. Replace the objective. Hold a plain piece of paper above the objective and open the shutter. Adjust the three-axis stage until the beam is pointing straight up (translate the lens in x–y) and collimated (translate the lens along z). Close the shutter.

19. Replace the rhodamine-filled chamber. Open the shutter and check for the fluorescence profile, which should be large and Gaussian. Focus the objective near the coverslip–sample interface. Lower the ND-filter OD and increase the camera gain as necessary. The beam should still be pointing straight up at the ceiling.

20. Choose either the x- or y-axis actuator of the three-axis stage. Turn the actuator until the beam is pointing nearly parallel to the table. While watching the camera image, continue turning the actuator until the fluorescence background starts to decrease dramatically and the beam profile at the glass–sample interface begins to look sharp. Continue turning until the fluorescence suddenly disappears, then reverse the direction until a clear fluorescence image is again visible. The setup is now optimized for TIR.

    Note that the TIR angle is usually optimized each time a new sample is placed on the microscope. For Steps 19 and 20, a chamber with immobilized fluorescent beads attached to the surface can be used in place of rhodamine.

Protocol B

Constructing a Sample Chamber for FIONA

It is not difficult to manufacture an inexpensive sample chamber (or several of them) in the laboratory using readily available components. Figure 9 shows a typical sample chamber used in a FIONA experiment. Once the chamber is constructed, solutions can be introduced into the chamber by inserting pipette tips into the holes on either side of the chamber, from which the solutions will flow through the channel.

MATERIALS

CAUTION: See Appendix 6 for proper handling of materials marked with <!>.

Equipment

Coverslips

The coverslips we use are Fisherbrand 12-544-A, 22x30-1.5 (No. 1.5 is 0.16–0.19 mm thick, which is the correct width for Olympus 1.45-NA objectives). In many cases, it is necessary to clean the coverslips before use (see Protocol C).

Electric drill and 3/4-mm diamond drill bit

Epoxy glue (quick drying, e.g., a "5-min" variety) <!>

Glass slides

The slides we use are "precleaned" Gold Seal Microslides 30103X1, 0.93–1.05 mm. Sources include Becton, Dickinson and Company, and Fisher Scientific.

Lens paper

Pipette tip

Razor blade

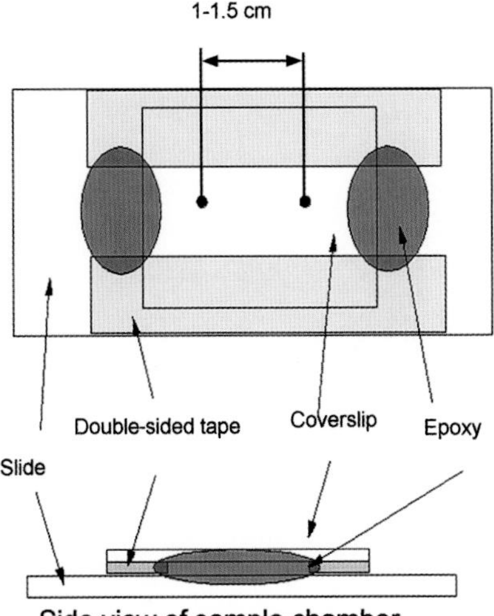

FIGURE 9. Top view and side view of FIONA sample chamber.

Tape, double-sided adhesive
Tissue

EXPERIMENTAL METHOD

1. Find the center of a glass slide and mark two points ~0.75 cm from the center along the long axis of the slide.
2. Drill a hole through the slide at the marked points using a 3/4-mm diamond drill bit.
3. Wash the drilled glass with H_2O.
4. Clean the slide and a coverslip using the method described in Protocol C.
5. Place the drilled and cleaned slide on a piece of lens paper.
6. Apply double-sided tape onto the slide along the long edges, just above and below the drilled holes.
7. Place a clean coverslip on top of the slide. Use a pipette tip to press the coverslip down over the double-sided tape.
8. Seal both ends of the coverslip using epoxy glue. Wait for 15 min for the glue to completely harden.
9. Remove the slide from the paper and use a razor blade to remove excess tape. The chamber volume is 10–20 µl.

Protocol C

Cleaning Slides and Coverslips

Thorough cleaning of coverslips is necessary to remove contaminants that may appear as fluorescent spots during single-molecule experiments. This protocol describes a method for manually cleaning slides and coverslips using sonication. Faster methods exist (Selvin et al. 2008), but this technique uses materials that are common to most laboratories.

MATERIALS

CAUTION: See Appendix 6 for proper handling of materials marked with <!>.

Reagents

Acetone <!>
KOH <!>, 1 M

Equipment

Beaker (glass, large enough to comfortably hold the coverslip holder)
Coverslips
  The coverslips we use are Fisherbrand 12-544-A, 22x30-1.5 (No. 1.5 is 0.16–0.19 mm thick, which is the correct width for Olympus 1.45 NA objectives).
Coverslip holder, Teflon (Invitrogen C-14784)
Glass slides
Nitrogen gas <!>
Nitrocellulose (optional, see Step 13)
Slide holder (glass)
Sonicator
Squirt bottle
Tweezers

EXPERIMENTAL METHOD

1. Rinse the slide holder, coverslip holder, and a glass beaker thoroughly with H_2O at least three times.
2. Fill the slide holder with slides and the coverslip holder with coverslips. Transfer the coverslips, in their holder, into the glass beaker.
3. Rinse the slides and coverslips thoroughly with H_2O.
4. Fill the slide holder (containing the slides) and the beaker containing the coverslips with acetone.
5. Sonicate the materials for 30 min.
6. Remove the coverslip holder and coverslips from the beaker and dispose of the acetone. Discard the acetone from the slide holder.
7. Rinse the slides and coverslips thoroughly with H_2O.

8. Fill the beaker and the slide holder with 1 M KOH. Place the coverslip holder into the filled beaker.
9. Sonicate the materials for 30 min.
10. Remove the coverslip holder and coverslips from the beaker and dispose of the KOH. Discard the KOH from the slide holder.
11. Rinse the slides and coverslips thoroughly with H_2O.
12. Use tweezers to remove the slides and coverslips from their holders. Rinse them with H_2O from a squirt bottle, and dry them with a gentle stream of nitrogen gas, making sure that the tweezers are the last thing the water touches as it flows off each slide or coverslip.
13. If desired, the coverslip can be functionalized at this stage. For instance, coverslips used in myosin assays are usually coated with nitrocellulose; it increases motility, although it also tends to increase the background. Depending on the experiment, other coatings, such as polyethylene glycol or poly-L-lysine, might be applied instead. Apply nitrocellulose to the coverslip as follows.
    i. Pipette 5 mL of nitrocellulose in a line across the bottom of a coverslip.
    ii. Use the coverslip like a paintbrush to drag it across a second coverslip, leaving a continuous film of nitrocellulose on the second coverslip.
    iii. Air-dry the coated coverslip for 15 min.
14. Proceed with the assembly of a flow chamber (Protocol B).

Protocol D

Imaging Immobilized Cy3-DNA under Deoxygenation Conditions

Immobilized Cy3-DNA bound to a coverslip (see Protocol C) is useful as a control to assess the efficacy of the FIONA setup and the ability to attain ~1.5 nm resolution (Yildiz et al. 2003). It is important that the fluorophores remain photostable throughout the experiment. This requires an oxygen-scavenging system (e.g., PCA and PCD) in the medium. Often, Trolox or β-mercaptoethanol is also required to increase the stability and decrease blinking of the fluorophore (Wang et al. 2004; Rasnik et al. 2006).

IMAGING SETUP

A standard TIRF microscopy setup should be used (see Protocol A). Cy3 is excited using 532 nm laser excitation. A proper dichroic mirror (e.g., Chroma Q565LP) and emission filter (e.g., Chroma HQ585/70M) should be used.

MATERIALS

CAUTION: See Appendix 6 for proper handling of materials marked with <!>.
See the end of the chapter for recipes for reagents marked with <R>.

Reagents

BSA-biotin (1 mg/mL; Sigma-Aldrich A-8549) in T50 buffer
Cy3-DNA solution (10 pM) in T50 buffer containing either 0.5–1 mg/mL final concentration of BSA (Sigma-Aldrich A-908) or ~0.5 mg/mL final concentration of casein.
  The double-stranded Cy3-DNA construct is as follows:
  Strand 1: 5′-[biotin]-GCC TCG CTG CCG TCG CCA-3′
  Strand 2: 5′-[Cy3]-TGG CGA CGG CAG CGA GGC TTT-3′
Imaging buffer <R>, made with either Trolox or β-mercaptoethanol
PCA solution <!> <R>, 50 mg/mL (3,4-dihydroxybenzoic acid)
PCD solution <!> <R>, 5 μM (protocatechuate 3,4-dioxygenase)
Streptavidin, 0.2 mg/mL (Molecular Probes S-888) in T50 buffer
T50 buffer (10 mM Tris [pH 8.0] containing 50 mM NaCl)
Trolox solution <!> <R> (saturated; 6-hydroxy-2,5,7,8-tetramethylchromane-2-carboxylic acid; Sigma-Aldrich 238813)

Equipment

Flow chamber (constructed in Protocol B)
Micropipette tips, 200-μL

EXPERIMENTAL METHOD

1. Insert a 200-µL pipette tip into one of the holes in the flow chamber. Introduce 50 µL of 1 mg/mL BSA-biotin in T50 buffer into the flow chamber via the micropipette tip. Incubate the chamber for 10 min at room temperature.
2. Wash the chamber with 100 µL of T50 buffer.
3. Introduce 50 µL of 0.2 mg/mL streptavidin in T50 buffer into the chamber. Incubate the chamber for 5 min at room temperature.
4. Wash the chamber with 100 µL of T50 buffer.
5. Introduce 50 µL of 10 pM Cy3-DNA solution in T50 containing 0.5–1 mg/mL final concentration of BSA or ~0.5 mg/mL final concentration of casein into the chamber. Incubate 5 min; then wash with 100 µL T50 buffer.
6. Introduce 100 µL of imaging buffer into the chamber and image using TIRF microscopy.

DISCUSSION

This setup can be easily adapted for observing molecular motors in vitro. For motility experiments, 40% Trolox is recommended, because at high concentrations, Trolox can interfere with motility. To prepare 40% Trolox, decrease Trolox in the imaging buffer recipe to 40 µL and add 54 µL of T50 buffer or a similar buffer. For myosin assays, use M5 buffer (20 mM HEPES [pH 7.6] containing 2 mM $MgCl_2$, 25 mM KCl, and 1 mM EGTA).

PCA/PCD (Aitken et al. 2008) can be used in place of gloxy and glucose (Yildiz et al. 2003) as a deoxygenation system. Using PCA/PCD can be very useful for observing molecular motors, because the pH does not drop significantly over time.

Protocol E

Analyzing FIONA Data

Single-molecule imaging techniques yield images of single fluorophores that can be fitted to two-dimensional Gaussian functions. This fitting allows one to localize and find the width of the point-spread functions of the fluorophores. By using the values for the widths, the number of photons collected for each fluorophore, the pixel size, and the background noise, the accuracy of the fluorophore localization can be calculated (Thompson et al. 2002; Yildiz et al. 2003) with nanometer accuracy, resulting in FIONA. In this protocol, we use a very basic FIONA fitting program to calculate the position and localization accuracy of a single fluorophore.

MATERIALS

Equipment

IDL version 6.1 or above (ITT Visual Information Solutions)
ImageJ (download from http://rsbweb.nih.gov/ij/)
Images of single fluorophores in a 16-bit TIFF file
Personal computer
"simonson_simple_FIONA.pro" (available at http://www.cshprotocols.org/imaging)

EXPERIMENTAL METHOD

1. Open a TIFF file containing fluorescent spots to be localized using ImageJ. Select one fluorescent spot and crop the image so that it encloses just the spot plus several pixels surrounding it. To do this, use the rectangular selection tool and "Image/Crop" from the menu bar. Save the file as a 16-bit TIFF.
2. Open "simonson_simple_FIONA.pro" using IDL.
3. In IDL, choose "Run/Compile All." This compiles the program and makes it ready to run.
4. In IDL, choose "Run/Run simonson_simple_FIONA."
5. At the dialog box, choose the file you saved in Step 1.
6. IDL will display a message saying, "What is the pixel size (in nm)?" This number is the physical size of the CCD pixel divided by the total magnification. For example, if the size of the image's acquisition camera pixel is 16,000 nm (as is typical of an Andor iXon+ camera) and the total magnification of the image is 100x (for a 100x objective), the pixel size (in the image) is 160 nm. The user would then enter "160" at the command line and press "enter."
7. IDL will display a message saying, "What is the background noise (in terms of counts)?"
   i. To find this number, open the original TIFF file using ImageJ.
   ii. Select a fairly large region of the image that contains no fluorescent spots. In ImageJ, select "Analyze/Set Measurements..." from the menu bar. Check the box for "Standard Deviation" and hit OK.

iii. Select "Analyze/Measure" to measure the standard deviation of the pixel intensity of the selected area in the image. A table will appear, and the value under "StdDev" is the background noise value.

iv. Return to IDL and enter this value at the command line.

8. IDL will display a message saying, "What is the exposure time (in seconds)?" Enter the time required to acquire a single frame in the TIFF file.
9. IDL will display a message saying, "What is the counts-to-photons conversion factor (this is specific to the camera on which the data were acquired)?" This value is the conversion from pixel intensity in the image to actual photons collected by the CCD camera. Contact your camera vendor for the correct conversion value to enter.
10. The program will then display a window with a three-dimensional image of the spot that is being fit. When the program is finished, an output file will be produced. Error messages regarding "floating underflows" typically can be ignored. The accuracy values in the table should be understood as "±" values (i.e., position = Δx).
11. The resulting text file can be opened in any plotting program for further analysis using the calculated positions and localization accuracies.

DISCUSSION

This FIONA analysis protocol is intended to be a starting point for anyone interested in doing FIONA. As mentioned earlier in this chapter, there are several other programs available for doing FIONA, both free and ones that can be purchased. The program "simonson_simple_FIONA.pro" (available at http://www.cshprotocols.org/imaging) is a very simple tool for analyzing FIONA data, but it should provide those interested with a good starting place for writing their own, and inevitably more complex, programs.

RECIPES

CAUTION: See Appendix 6 for proper handling of materials marked with <!>.
Recipes for reagents marked with <R> are included in this list.

Imaging Buffer

To prepare 100 µL of imaging buffer with Trolox:

1. Prepare Trolox <!>, PCA <!>, and PCD <!> solutions.
2. Combine 2 µL of 50 mg/mL PCA, 1 µL of 1 M $MgCl_2$, 93 µL of Trolox solution, and 1 µL of 5 µM PCD.

To prepare 100 µL of imaging buffer with β-mercaptoethanol:

1. Prepare T50, PCA, and PCD solutions.
2. Combine 92 µL of T50 (or similar) buffer, 2 µL of 50 mg/mL PCA, 1 µL of 1 M $MgCl_2$, 1 µm of stock β-mercaptoethanol <!>, and 1 µL of 5 µM PCD.

PCA Solution

To prepare a 50 mg/mL solution:

1. Add 600 mg of PCA <!> (3,4-dihydroxybenzoic acid from Sigma or Fluka) into a 15-mL centrifuge tube.
2. Add ~8.5 mL of H_2O.
3. Add ~2.5 mL of 1 M NaOH <!>.

Most, if not all, of the PCA will be dissolved now.

4. Continue to add NaOH, a few drops at a time, until the pH is ~7.2.
5. Add H_2O to a final solution volume of 12 mL.
6. Divide into 200-µL aliquots and store at –20°C.

PCD Solution

To prepare a 5 µM solution:

1. Resuspend lyophilized powder of PCD (protocatechuate 3,4-dioxygenase [Sigma, P8279]) in a solution of 50 mM KCl, 1 mM EDTA, 100 mM Tris-HCl (pH 8.3), and 50% glycerol.
2. Aliquot the solution and store at –20°C.

Trolox Solution

1. Transfer 5 mg of Trolox <!> (saturated; 6-hydroxy-2,5,7,8-tetramethylchromane-2-carboxylic acid; Sigma-Aldrich 238813) to a 15-mL tube.

2. Add 10 mL of the desired buffer (e.g., M5 for myosin motility experiments) to the tube and mix completely to saturate the buffer with Trolox (which is moderately soluble in H_2O).
3. Filter the mixture through a 0.22-µm syringe filter (e.g., Millex GP; Millipore SLGP033R).
4. Adjust the pH of the buffer to 7.5 with NaOH <!>, because the acidity of Trolox can affect the photostability of the fluorophores.

ACKNOWLEDGMENTS

The authors thank the National Institutes of Health for support.

REFERENCES

Aitken CE, Marshall RA, Puglisi JD. 2008. An oxygen scavenging system for improvement of dye stability in single-molecule fluorescence experiments. *Biophys J* **94:** 1826–1835.

Axelrod D. 1989. Total internal reflection fluorescence microscopy. *Methods Cell Biol* **30:** 245–270.

Balci H, Ha T, Sweeney HL, Selvin PR. 2005. Interhead distance measurements in myosin VI via SHRImP support a simplified hand-over-hand model. *Biophys J* **89:** 413–417.

Betzig E, Patterson GH, Sougrat R, Lindwasser OW, Olenych S, Bonifacino JS, Davidson MW, Lippincott-Schwartz J, Hess HF. 2006. Imaging intracellular fluorescent proteins at nanometer resolution. *Science* **313:** 1642–1645.

Bohmer M, Enderlein J. 2003. Orientation imaging of single molecules by wide-field epifluorescence microscopy. *J Opt Soc Am B* **20:** 554–559.

Churchman LS, Okten Z, Rock RS, Dawson JF, Spudich JA. 2005. Single molecule high-resolution colocalization of Cy3 and Cy5 attached to macromolecules measures intramolecular distances through time. *Proc Natl Acad Sci* **102:** 1419–1423.

Enderlein J, Toprak E, Selvin PR. 2006. Polarization effect on position accuracy of fluorophore localization. *Opt Express* **14:** 8111–8120.

Gell C, Brockwell D, Smith A. 2006. *Handbook of single molecule fluorescence spectroscopy.* Oxford University Press, New York.

Gordon MP, Ha T, Selvin PR. 2004. Single-molecule high-resolution imaging with photobleaching. *Proc Natl Acad Sci* **101:** 6462–6465.

Hess ST, Girirajan TPK, Mason MD. 2006. Ultra-high resolution imaging by fluorescence photoactivation localization microscopy. *Biophys J* **91:** 4258–4272.

Kural C, Kim H, Syed S, Goshima G, Gelfand VI, Selvin PR. 2005. Kinesin and dynein move a peroxisome in vivo: A tug-of-war or coordinated movement? *Science* **308:** 1469–1472.

Kural C, Serpinskaya AS, Chou Y, Goldman RD, Gelfand VI, Selvin PR. 2007. Tracking melanosomes inside a cell to study molecular motors and their interaction. *Proc Natl Acad Sci* **104:** 5378–5382.

Kural C, Nonet ML, Selvin PR. 2009. FIONA on *Caenorhabditis elegans*. *Biochemistry* **48:** 4663–4665.

Lacoste TD, Michalet X, Pinaud F, Chemla DS, Alivisatos AP, Weiss S. 2000. Ultrahigh-resolution multicolor colocalization of single fluorescent probes. *Proc Natl Acad Sci* **97:** 9461–9466.

Mehta AD, Rock RS, Rief M, Spudich JA, Mooseker MS, Cheney RE. 1999. Myosin-V is a processive actin-based motor. *Nature* **400:** 590–593.

Michalet X, Pinaud FF, Bentolila LA, Tsay JM, Doose S, Li JJ, Sundaresan G, Wu AM, Gambhir SS, Weiss S. 2005. Quantum dots for live cells, in vivo imaging, and diagnostics. *Science* **307:** 538–544.

Patra D, Gregor I, Enderlein J. 2004. Image analysis of defocused single-molecule images for three-dimensional molecule orientation studies. *J Phys Chem A* **108:** 6836–6841.

Pedrotti FL, Pedrotti LS. 1993. *Introduction to optics,* 2nd ed. Prentice-Hall, Upper Saddle River, NJ.

Qu X, Wu D, Mets L, Scherer NF. 2004. Nanometer-localized multiple single-molecule fluorescence microscopy. *Proc Natl Acad Sci* **101:** 11298–11303.

Rasnik I, McKinney SA, Ha T. 2006. Nonblinking and long-lasting single-molecule fluorescence imaging. *Nat Meth* **3:** 891–893.

Reck-Peterson SL, Yildiz A, Carter AP, Gennerich A, Zhang N, Vale RD. 2006. Single-molecule analysis of dynein processivity and stepping behavior. *Cell* **126:** 335–348.

Rust MJ, Bates M, Zhuang, X. 2006. Sub-diffraction-limit imaging by stochastic optical reconstruction microscopy (STORM). *Nat Meth* **3:** 793–796.

Selvin PR, Lougheed T, Hoffman MT, Park H, Balci H, Blehm BH, Toprak E. 2008. In vitro and in vivo FIONA and other acronyms for watching molecular motors walk. In *Single-molecule techniques: A laboratory manual* (eds. Selvin PR, Ha T), pp. 37–71. Cold Spring Harbor Laboratory Press, Cold Spring Harbor, NY.

Snyder GE, Sakamoto T, Hammer JA III, Sellers JR, Selvin PR. 2004. Nanometer localization of single green fluorescent proteins: Evidence that myosin V walks hand-over-hand via telemark configuration. *Biophys J* **87:** 1776–1783.

Thompson RE, Larson DR, Webb WW. 2002. Precise nanometer localization analysis for individual fluorescent probes. *Biophys J* **82:** 2775–2783.

Toprak E, Selvin PR. 2007. New fluorescent tools for watching nanometer-scale conformational changes of single molecules. *Annu Rev Biophys Biomol Struct* **36:** 349–369.

Toprak E, Enderlein J, Syed S, McKinney SA, Petschek RG, Ha T, Goldman YE, Selvin PR. 2006. Defocused orientation and position imaging (DOPI) of myosin V. *Proc Natl Acad Sci* **103:** 6495–6499.

Vogelsang J, Kasper R, Steinhauer C, Person B, Heilemann M, Sauer M, Tinnefeld P. 2008. A reducing and oxidizing system minimizes photobleaching and blinking of fluorescent dyes. *Angew Chem Int Ed* **47:** 5465–5469.

Wang CC, Chu CY, Chu KO, Choy KW, Khaw KS, Rogers MS, Pang CP. 2004. Trolox-equivalent antioxidant capacity assay versus oxygen radical absorbance capacity assay in plasma. *Clin Chem* **50:** 952–954.

Yildiz A, Forkey JN, McKinney SA, Ha T, Goldman YE, Selvin PR. 2003. Myosin V walks hand-over-hand: Single fluorophore imaging with 1.5-nm localization. *Science* **300:** 2061–2065.

Yildiz A, Tomishige M, Vale RD, Selvin PR. 2004a. Kinesin walks hand-over-hand. *Science* **303:** 676–678.

Yildiz A, Park H, Safer D, Yang Z, Chen LQ, Selvin PR, Sweeney HL. 2004b. Myosin IV steps via a hand-over-hand mechanism with its lever arm undergoing fluctuations when attached to actin. *J Biol Chem* **279:** 37223–37226.

34 Photoactivated Localization Microscopy (PALM)
An Optical Technique for Achieving ~10-nm Resolution

Haining Zhong

Vollum Institute, Oregon Health Science University, Portland, Oregon 97239

ABSTRACT

The organization of proteins on the scale of a few tens of nanometers, such as in the postsynaptic density and the synaptic vesicles, is a key determinant of their function. However, spatial features on such a fine scale are beyond the resolving power of conventional, diffraction-limited fluorescence light microscopy. In response, several imaging modalities have recently emerged that can surpass the diffraction limit on optically benign samples in which aberration and light scattering are negligible. Although each of these techniques has its own advantages and limitations, photoactivated localization microscopy (PALM) provides the highest shown resolution in biological samples, is the most efficient in utilizing signal photons, and allows for the assessment of individual molecules. This chapter discusses the basic principles of PALM microscopy, its implementation, and the potential applications in neuroscience.

INTRODUCTION

Fluorescence light microscopy is used most often to study the spatial properties of cellular functions because of its unmatchable contrast. Thanks to advances in labeling techniques, such as immunolabeling and fluorescent protein (FP) tagging, fluorescent marking of specific organelles or protein molecules is now routinely achieved with molecular precision. However, diffraction-limited conventional light microscopy cannot resolve objects closer than 200 nm at the focal plane. Many subcellular organelles and groups of proteins, such as the postsynaptic density and the synaptic vesicles, occur on the 10-nm scale. In-depth understanding of cellular physiology, therefore, demands advancements over conventional light microscopy techniques, especially in the area of finer-scale resolution.

Imaging methods that break the diffraction limit have been developed for certain samples in which aberrations and scattering are negligible. At least four superresolution methods have been shown: near-field microscopy (de Lange et al. 2001), stimulated emission depletion microscopy (STED) (Westphal and Hell 2005), structured illumination microscopy (SIM) (Gustafsson 2000), and PALM (Betzig et al.

2006), together with two related techniques—stochastic optical reconstruction microscopy (STORM) (Rust et al. 2006) and fluorescence photoactivated localization microscopy (FPALM) (Hess et al. 2006). Among these techniques, PALM stands out for several reasons: (1) It provides the highest demonstrated resolution in biological samples among these techniques. (2) It is the most efficient in utilizing signal photons. This is particularly important for high-resolution imaging because the photon budget per pixel is limited. (3) It can assess the characteristics (such as diffusion) of single molecules. (4) PALM allows for quantification of the molecular abundance. (5) The design and the optics of PALM are simple and relatively inexpensive (~$150,000 for a complete, do-it-yourself microscope), allowing the technique to be implemented in many biological laboratories. This chapter discusses the principle of PALM and a practical design based on an Olympus Ix81 inverted microscope. The goal is to help biologists, who have a basic understanding of optics, set up a PALM microscope. PALM's potential applications in neuroscience are also discussed.

The Yin and Yang of High-Resolution Imaging

The diffraction limit arises because light emitted by a point source (e.g., a fluorescent molecule) can only be refocused to a distribution called the point-spread function (PSF). Two point sources with significantly overlapping PSFs at the image plane cannot be resolved (Fig. 1A). Such a resolution is

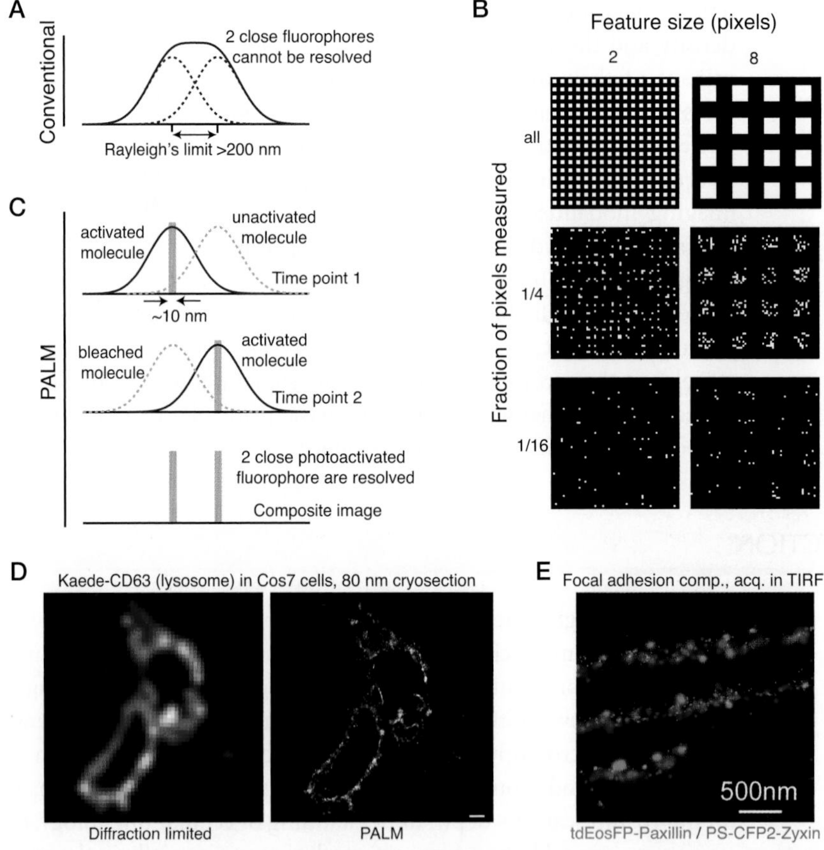

FIGURE 1. The principles of PALM. (*A*) The Rayleigh limit of conventional microscopes: Closely spaced fluorescent molecules having significant overlap in their PSFs cannot be resolved. (*B*) Spatial resolution decreases as the density of imaged molecules decreases. The molecules are represented as pixels here. Finer features (*left column*) require higher molecular density to be recapitulated. (Adapted, with permission, from Shroff et al. 2007, © National Academy of Sciences.) (*C*) PALM breaks the Rayleigh limit while maintaining high-labeling density by using photoactivated fluorophores. At any time point, only a few molecules are activated to become fluorescent, resulting in isolated PSFs. The center of a molecule can be pinpointed with high precision by fitting its PSF. The summation of molecular localization information over many time points (frames) results in a high-resolution image in both a Rayleigh and a Nyquist sense. (*D*) Sample PALM image. Scale bar, 500 nm. (Modified, with permission, from Betzig et al. 2006, ©AAAS.) (*E*) Double-color PALM image. (Modified, with permission, from Shroff et al. 2007, ©National Academy of Sciences.)

defined by the Rayleigh limit $R = 0.6\lambda/\mathrm{NA}$, where λ is the wavelength of light and NA is the numerical aperture of the objective. For visible light, the resolution is typically >200 nm.

Often underappreciated is that spatial information is lost only when the distance between fluorescent molecules is smaller than the Rayleigh limit so that multiple PSFs overlap with one another. If the labeling density is very low (<<25 molecules/μm^2), Gaussian fitting of isolated PSFs can pinpoint the localization of individual fluorophores with nanometer precision. Such a low labeling density, however, is insufficient to reveal detailed structural organization because, by the Nyquist–Shannon theorem, the mean distance between labeled molecules must be more than twice as fine as the desired resolution (Fig. 1B). Although 100 molecules/μm^2 are sufficient to provide a resolution of ~200 nm, 10^4 molecules/μm^2 are required to achieve a Nyquist resolution of 20 nm. Therefore, high-resolution imaging requires two paradoxical feats simultaneously: low-labeling density for precise localization of individual molecules but high-density labeling to reveal spatial features.

THE PRINCIPLE OF PALM

PALM and related techniques solve this paradox by labeling a sample at high density but only visualizing a stochastically selected low-density subset at each imaging frame (Fig. 1C). Overlaying the low-density single-molecule images over time leads to a high-density image in which all of the molecules are precisely localized. The key is utilizing photoactivatable (PA) fluorescent molecules, such as photoactivatable fluorescent proteins (PA-FPs) or organic caged dyes. For example, lysosomes of Cos-7 cells were labeled at high density by expressing the photoconvertible FP Kaede linked to the lysosome membrane protein CD63 (Fig. 1D) (Betzig et al. 2006). Kaede is not fluorescent at the detection wavelength until it is activated by near-ultraviolet (UV) light. The activation efficiency was controlled to be so low that only a sparse subset of Kaede molecules is visible within each frame. After the activated molecules are bleached from imaging, a second subset is activated and is imaged. Such iterations are repeated until all of the molecules have been imaged. Because only a few molecules are imaged within a given frame, individual PSFs do not overlap, allowing the location of single molecules to be determined with nanometer precision. Because of the high molecular density that is sampled over repeated iterations, a final PALM image can be reconstructed at high Nyquist resolution. Dual-color imaging (Fig. 1E) (Bates et al. 2007; Shroff et al. 2007), 3D imaging (Huang et al. 2008; Juette et al. 2008), and live cell PALM imaging (Manley et al. 2008; Shroff et al. 2008a) have also been shown.

MAJOR CONSIDERATIONS FOR PALM

PALM can be set up fairly easily in almost any biological laboratory by coupling two or three properly aligned laser beams to an existing microscope that is equipped with a high-NA objective and a high-sensitivity charge-coupled device (CCD) camera. However, no matter what imaging method is used, superresolution images will only be as good as the labeling technique allows. The labeling method, the choice of fluorophores, and the sample preparation are each crucial for successful results using PALM. We discuss these next and then describe the instrumentation, its setup, and the optimization of components and parameters to yield the best results.

Labeling Method

Fluorescent labels can be introduced either exogenously (e.g., with antibodies) or endogenously (e.g., with FP tagging). Although antibodies exist for many targets, their affinity and specificity often need to be examined for nanometric imaging (Luby-Phelps et al. 2003). Furthermore, the bulky size of antibodies (~25 nm if both primary and secondary antibodies are used) introduces significant localization uncertainty, limits penetration into tissues, and sets an upper limit for the labeling density of a protein target. On the other hand, FPs can be readily attached to most proteins by molecular methods with 100% specificity, and their smaller size (~2 nm) allows for higher packing density and localization precision. Perfect labeling efficiency and quantification of protein abundance are possible if the tagged protein is expressed with genetic gene-replacement techniques (i.e., knockin methodology). However, until knockins become less expensive to generate, expression of FP-tagged proteins will rely mainly on overexpression. Potential mislocalization due to overexpression or interference with the physiological function of the target protein has to be carefully evaluated.

Choice of Fluorophore

The properties of the fluorophore largely determine the resolution and the speed of PALM. An ideal PA fluorophore has an infinite contrast ratio, which is the fluorescence ratio of a molecule between the on and the off states. In practice, infinite contrast ratio never occurs, setting a boundary for the highest Nyquist resolution. For example, the commonly used PA-GFP (Patterson and Lippincott-Schwartz 2002) is weakly fluorescent in the off state and has a contrast ratio of ~40. The accumulated background fluorescence within a diffraction-limited region from the off molecules will be as bright as a single activated PA-GFP when the labeling density approaches ~1000/μm^2 (~40 in a 200 x 200-nm region). This effectively limits the highest Nyquist resolution to be ~70 nm. Currently, the EosFP family of photoswitchable FPs (Wiedenmann et al. 2004) and the PA-mCherry (Subach et al. 2009) have the highest contrast ratio (>2000) among PA proteins. The contrast ratios of most PA organic dyes remain to be determined.

Another important factor is the number of collected photons that a fluorophore emits before bleaching. With some approximation, the precision of localization of a molecule increases in proportion to the square root of the number of photons collected. However, more photons often mean longer bleaching times because there is a limit to the imaging laser power. This leads to longer data acquisition times because PALM requires many iterations of PA and bleaching. Furthermore, other practical factors—such as stage or sample drift (~5 nm; see the sections Drift Correction and Stage and Sample Holder), size of the fluorescent marker (~2 nm with its linker), and limits in the packing density of the target protein—prevent a resolution much beyond 10 nm. EosFPs emit >1000 collected photons per molecule, sufficient to provide localization precision of ~10 nm. This value is also a good balance between speed and resolution.

Additional factors include the brightness and the PA efficacy of the fluorophore. A fluorophore has to be sufficiently bright to stand out from the background fluorescence and the inevitable scattered laser light in the sample and the light path. At a given photon budget per molecule, brighter fluorophores also have shorter bleaching times, leading to higher data acquisition speed. In addition, for many PA-FPs, only a subset of the molecules is photoconvertible, effectively decreasing the labeling density. Again, the investigator favors EosFP, which has ~100% conversion efficacy. Finally, because EosFP PA occurs by shifting the emitting spectrum from green to red, the expression levels and the cell health can be monitored at diffraction-limited resolution in the green wavelength before PALM imaging.

A nearly optimal PA fluorophore for PALM is the recently developed monomeric EosFP variant, mEos2, which inherits all of the advantageous properties for PALM while maturing readily at 37°C with a reduced oligomerization tendency (McKinney et al. 2009). It is, however, sometimes necessary to use other PA fluorophores (e.g., in multicolor imaging). After a screen of existing candidates, Dronpa (Habuchi et al. 2005) and PS-CFP2 (Chudakov et al. 2007) were identified as possible second fluorophores (Shroff et al. 2007) (Fig. 1E). A drawback of these fluorophores is that they can go through many activation/inactivation cycles before they are bleached, preventing quantification of molecular density and slowing down data acquisition. Furthermore, Dronpa emits many fewer photons (~200) than EosFP does, whereas PS-CFP2 has a much inferior contrast ratio between the on and the off states (Shroff et al. 2007). The recently developed PA-mCherry (Subach et al. 2009) is another alternative, but its spectrum overlaps significantly with EosFP.

Maximizing the Signal-to-Noise Ratio

PALM is, by nature, a single-molecule technique. Therefore, every effort has to be made to maximize the signal-to-noise ratio. Filter sets have to be carefully chosen to maximally reject background fluorescence and scattered laser light while efficiently transmitting signal photons. Total internal reflection fluorescence (TIRF) illumination can further reduce out-of-focus fluorescence background because the evanescent excitation wave penetrates <200 nm into the sample (Fig. 1E). However, such a scheme largely limits the application of PALM to cultured cells plated on coverslips and to molecules close to the bottom surface. An alternative method is to section the sample into ultrathin sections (<0.5 µm; e.g., 80-nm sections shown in Fig. 1D).

To achieve TIRF, the microscope objective must have an NA greater than the refractive index of the sample (~1.38 for cellular samples). Practically, only objectives with NA 1.4, 1.45, 1.49, and 1.65 are available, with the 1.65-NA objective requiring expensive high-index sapphire coverslips (these cost ~$30 per coverslip). Even for applications in which TIRF is not required (e.g., when imaging thin sections), higher NA means better photon collection efficiency and smaller PSFs, thus, higher signal-to-background fluorescence ratio. Smaller PSFs also allow for higher density of activated molecules at each imaging frame and, thus, faster data acquisition.

To fully use all the photons collected from the individual molecules, it is necessary to use a cooled electron-multiplying charge-coupled device (EMCCD) camera, which is capable of detecting single photons with low noise. Such a camera uses an on-chip multiplication gain that boosts weak signals above the readout noise of the camera. The ability to chill the CCD chip to –50°C or lower ensures that thermal noise is negligible. These properties allow the camera to detect even single photons. Furthermore, the back-illuminated design (light illuminating at the back of a thinned CCD) of many EMCCD cameras allows for >90% quantum efficiency across the visible spectrum. Finally, the fast (up to 10 MHz) readout rate and the ability to operate in the frame-transfer mode (spooling acquired data to hard drive while simultaneously acquiring new data) of modern EMCCD cameras ensure that they are not the rate-limiting step in PALM imaging.

Effective pixel size is also an important consideration. Magnification of the system has to be carefully chosen to match the desirable effective pixel size of the PSF to the physical pixel size of the EMCCD camera. Pixels that are too large lead to a loss of spatial information, whereas the intensity of individual pixels may not stand out from the background if pixel sizes are too small. It has been estimated that the greatest localization precision comes when the pixel size approximates the standard deviation of the PSF (Thompson et al. 2002), a value slightly smaller than one-half of the full width at half-maximum of the PSF. The PSF of an activated EosFP (emission peaks at 581 nm) has a size of ~240 nm when the 1.49-NA objective is used. An effective pixel size of ~120 nm is, therefore, optimal. For a typical EMCCD camera that has a pixel size of 16 µm, a system magnification of ~133x is required. This can be achieved by combining the objective magnification (60x or 100x) with additional magnification lenses in the detection light path.

Imaging Area and Acquisition Speed

Single-molecule imaging requires a power density high enough to achieve sufficient signal-to-noise ratio. High imaging power density also leads to shorter bleaching time of activated molecules and speeds up data acquisition. However, laser light inevitably deposits heat when going through the imaging oil and the sample. There is a limit on the laser power that can be used before heat builds up so much that individual PSFs start to degrade. With an Olympus 1.49-NA 60x objective, it has been empirically determined that the limit for a 561-nm laser (for imaging activated EosFP) is ~12 mW at the back focal plane of the objective and the limit for a 488-nm laser is ~5 mW. The power limit for the 1.65-NA objective is even lower because the corresponding imaging oil is very volatile. Finally, for live cell imaging, the cells may also have a lower cell type–specific threshold for laser power before their physiology is affected.

To obtain an energy density sufficiently high from limited laser power, the author concentrates the laser to a relatively small area (~35 µm in diameter) at the focal plane to achieve an energy density of ~1 kW/cm^2 (for a 561-nm laser). In the design of a PALM microscope described below, this can be conveniently achieved by adjusting the laser beam size with a beam expander because the illuminating area at the sample is proportional to the beam size. Finally, when TIRF imaging is appropriate, the energy density within the TIRF zone could be higher than when epi-illumination is used.

At a given excitation laser power, the time required for PALM data acquisition largely depends on the highest labeling density within the field of view. Molecules with overlapping PSFs have to be separated by time. Furthermore, because many fluorescent molecules (e.g., EosFP) may blink on and off a few times before they are eventually bleached, the power of the activation light has to be set so low that two activated molecules within a diffraction-limited region are separated by at least a few blank frames to ensure that the first molecule is bleached. The acquisition time is, therefore, determined by the highest number of fluorophores within a diffraction-limited region at a specified photoactivation rate. When depletion of all molecules is not necessary, the desired Nyquist resolution then specifies the number of molecules that need to be accumulated within one diffraction-limited spot. Higher resolution requires higher accumulated molecular density and, thus, slower imaging speed. In applications in which speed is the main consideration, the Nyquist resolution has to be sacrificed. In practice, a rate of ~30 sec per image for a Nyquist resolution of ~60 nm has been achieved (Shroff et al. 2008a). A better analysis algorithm that permits partially overlapping PSFs might be able to improve the data acquisition speed by an order of magnitude.

It is worth noting that within the illumination area, PALM speed is insensitive to the size of the imaging area. This is in contrast to the superresolution techniques that are based on scanning mechanisms, which have to slow down by the square of the size of the field of view to maintain the signal-to-noise ratio at a given laser power.

Drift Correction

A PALM imaging session for high Nyquist resolution often takes >5 min and sometimes >30 min. Every effort has to be made to minimize sample drifting. However, even fairly expensive microscope stages will experience measurable drift over such a long time frame. Furthermore, the high laser power required for PALM inevitably heats up the sample, the coverslip, and the imaging oil, again leading to sample drift. Such drift can be >100 nm. Retrospective correction of the drift by fiducials is, therefore, necessary for most applications. There are 40–100-nm gold beads that can be embedded in the sample, do not bleach, and have fluorescence intensities similar to that of single molecules—these can be used as fiducial markers. After correction, the drifting errors can be minimized to 3–20 nm, depending on the type of the samples and the imaging conditions (personal data).

Sample Preparation

It is not a trivial task to prepare the sample so that both the labeling density and the molecular localization are appropriately maintained for nanometric imaging. PALM can be applied to live cells cultured on coverslips, but it is often difficult to maintain the health of the cell under the harsh imaging conditions required for PALM. TIRF illumination, which is required to reject out-of-focus background fluorescence, also limits the application of PALM to the molecules close to the bottom surface of the cell. For molecules deep inside the cell and for intact brain tissues, fixation and sectioning of the sample are required. However, many fixation protocols for conventional fluorescence microscopy may be unsuitable for examining protein distributions on the nanoscale. For example, a conventional fixation protocol using 2% paraformaldehyde was found to cause alteration of the morphology and the labeling density in cultured cells (Ji et al. 2008). More stringent fixation protocols, similar to those used in electron microscopy (EM), need to be developed. Furthermore, to improve the *z* resolution and to fully eliminate background fluorescence, ultrathin sectioning techniques similar to those used in EM, and more recently used in fluorescence array tomography (Micheva and Smith 2007), are necessary. Practical sample preparation and sectioning methods that preserve both PA-FPs and their localization on the nanoscale are under development.

PALM INSTRUMENTATION

A PALM microscope is composed of a few well-aligned lasers fed into a TIRF-ready microscope with a high-sensitivity EMCCD camera. The major difference from a conventional microscope lies in the excitation light path. Several inexpensive solid-state lasers (~\$10,000 each) for PALM imaging at different wavelengths and a near-UV activation laser are combined before being coupled to a commercial TIRF-ready microscope. At the detection end, the primary difference is an expensive EMCCD camera (~\$35,000). Next, a practical PALM microscope capable of two-color imaging of the fluorophores EosFP and Dronpa is described. Critical part numbers that the investigator uses are also indicated.

Excitation Light Path

In the PALM excitation light path, external lasers are coupled to a commercial microscope to achieve TIRF illumination in a controllable way. Figure 2A illustrates a conceptual design. A relay lens is installed in a commercial TIRF-ready microscope to focus a collimated laser beam to the back focal plane of the objective. The laser beam becomes collimated again after the objective with an angle determined by the lateral position of the laser's focal point at the objective back focal plane. The angle is larger when the focal point is further away from the center of the objective pupil. When the beam angle becomes larger than a critical value, TIRF will take place. To control the lateral position of the beam focus at the back focal plane, a collimating lens (CL) and a focusing lens (FL) are added. The CL forms a telescope with the relay lens so that the back focal plane of the CL is conjugated with the back focal plane of the objective. The FL is used to focus the laser beam to the CL's back focal plane, and the lateral position of the focal point is controlled by a translating mirror (TM).

A Practical Design

Figure 2B shows the realization of the conceptual light path (as described above) on an Olympus IX81 inverted microscope (an IX71 should also suffice). The microscope is equipped with a 1.49-NA 60X objective (Olympus, APON60XOTIRFM), a 1.65-NA 100X objective (AP0100XO-HR-SP), and a 10X objective with necessary differential interference contrast (DIC) imaging accessories. The lower epifluorescence light path is modified by the manufacturer so that only a relay lens (f = 200 mm) is installed. The laser is free-space coupled (i.e., without using an optical fiber) to the light path through a dual port (Olympus, U-DP) that allows for switching light sources between the mercury burner and the external laser beams. To control the lateral position of the laser's focal point at the objective back focal plane, the CL (f = 100 mm) is positioned outside of the dual port (and the scope; see also Fig. 3). The distance between the CL and the relay lens is 300 mm ($f_{relay} + f_{CL}$) to conjugate

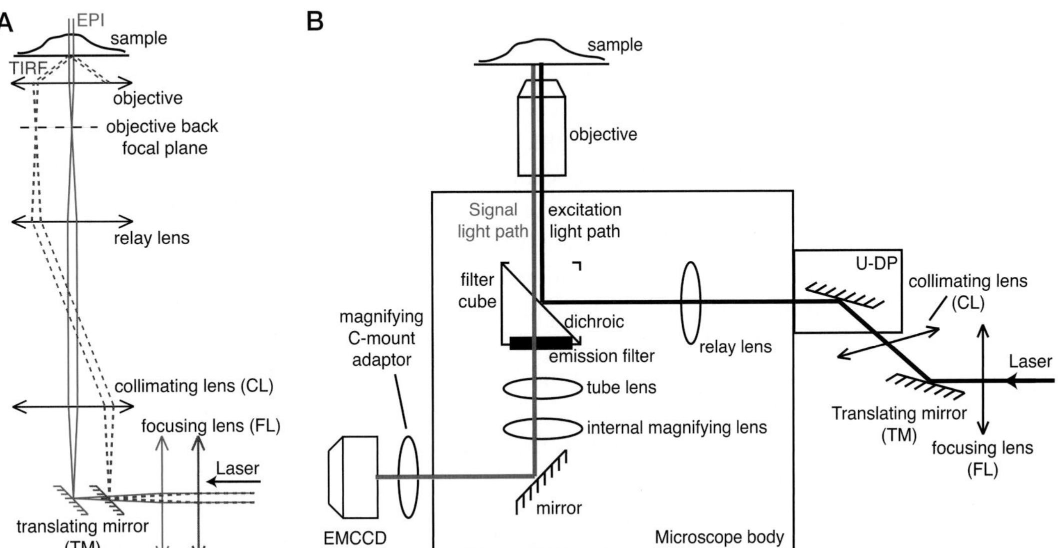

FIGURE 2. Light paths for a PALM microscope. (*A*) Conceptual activation and excitation light paths. (*B*) A practical design of the activation/excitation and detection light paths.

FIGURE 3. A scheme for combining the necessary excitation and activation lights into the microscope for PALM. SH, shutter; BE, beam expander; TX, *x*-direction beam translator; TY, *y*-direction beam translator; FIL, filter group (wavelength, intensity, and polarization); MR, mirror; DC, dichroic beam splitter; AOTF, acousto-optic tunable filter; FL, focusing lens; TM, translating mirror; CL, collimating lens. (Modified, with permission, from Shroff et al. 2007, ©National Academy of Sciences.)

the back focal plane of the CL with the back focal plane of the objective. The FL (f = 50 mm) is positioned at a right angle to the CL with an effective distance of 150 mm ($f_{CL} + f_{FL}$) so that a collimated laser beam is focused to the back focal plane of the CL and, in turn, to the back focal plane of the objective. The TM is fixed on a translational mount (Thorlabs, T12X) to relay the FL with the CL. The translation of the TM changes the position of the laser's focal point at the CL's back focal plane and, thus, at the objective back focal plane. Because the translation at the TM (<2 mm) is much smaller than the focal length of the FL, a compensatory change of the FL's position is unnecessary. However, the FL is still mounted on a *z*-translatable mount (Thorlabs, SM1ZM). This is because the flatness of commercially available dichroics (DCs) (inside the filter cube) varies, leading to a slight change of beam focus when changing filter sets for imaging different colors. Fine-tuning the position of the FL allows for a quick compensation of such variability for all lasers because the different laser beams are combined upstream of the FL.

Figure 3 shows how different laser beams are adjusted and combined. The goal is to combine three solid-state laser beams, a 405-nm laser (Coherent, Inc. 405-50C) for PA EosFP and Dronpa, a 561-nm laser (CrystalLaser, GCL-150-561) for imaging activated EosFP, and a 488-nm laser (Newport Corporation, 488 cyan-50FP) for imaging activated Dronpa, before they are coupled to the microscope. For each beam, it is necessary to fine-tune its *x–y* position, adjust its beam size, and clean up spectral noise. The *x–y* position control is achieved by two optical windows (*x* translating, TX, and *y* translating, TY; CVI Melles Griot, 12-mm thick) mounted on goniometers (Thorlabs, GN05). The beam size is adjusted by a beam expander (BE) with a tunable focus. The tunability of the BE also allows for individual compensation of chromatic aberrations in the light path (e.g., at the FL and the CL). A narrow-band filter is put in the filter groups (FIL) to clean up the spectrum. Finally, because the 405-nm laser has a relatively poor beam profile, it is spatially filtered by inserting a 10-μm pinhole at the focus point of a telescope system inside the FIL. Although ~85% power is lost from such filtering, the remaining power is still more than sufficient to efficiently activate Dronpa and EosFP.

The intensity and the timing of the lasers are controlled by different mechanisms. The 561-nm and 488-nm lasers are controlled by an acousto-optic tunable filter (AOTF; AA Opto-Electronic Company, AA.AOTF.nc), which provides rapid wavelength selection and more than 14 OD intensity controls at submicrosecond temporal resolution. AOTF does not transmit 405 nm efficiently, but the output power of the 405-nm laser can be modulated by software or by an external command signal. Thus, 561-nm and 488-nm laser beams are combined with a DC (Semrock, LM01-503-25) before going through the AOTF. λ/2 plates are added to the respective FIL to match the polarization plane of the laser with the input requirement of AOTF. The 405-nm beam is combined with the other beams after AOTF by another DC (Semrock, LM01-427-25). In a well-fixed sample, the orientations of the fluorophores may not change over time. Fluorophores are best activated by a laser light with a polarization plane in parallel to the fluorophore. A λ/4 plate (unlabeled in Fig. 3) is, therefore, mounted at the back side of the FL at 45° to the polarization plane of the laser beams to make the beams circularly polarized (its polarization plane rotates over time) in order to efficiently excite the maximum number of PA fluorophores positioned at different orientations.

Alignment Procedures to Couple Lasers to the Microscope Light Path

A practical beam alignment procedure is described here.

1. Mount the external light path on a minioptical table (Thorlabs, PBI11106) with the height of the light path adjusted to the same height as the lower light path of the microscope.
2. Bring the imaging objective to focus with respect to a sample mounted on the appropriate sample holder. Remove the sample, and clean the oil objective.
3. Remove the FL and the CL, and visually align the position of the TM with the dual port on the Olympus microscope.
4. Turn on the 561-nm laser, and adjust the corresponding TX, TY, and mirror (MR4) to steer the beam through the AOTF. Adjust the angles of the MR1, the MR2, and the TM to steer the laser beam into the center of the dual port.
5. Project the laser beam onto the ceiling. A ring corresponding to the edge of the relay lens can be seen when the objective is not engaged. Fine-tune the MR1 and the TM until the beam is centered at the relay lens both with and without objective engagement.
6. Put in the FL, and make sure that the laser goes through the center of the lens. Fine-tune the angle of the TM to center the beam with respect to the relay lens when the objective is not engaged.
7. Put in the CL, and use the *x–y* translational mount to position the laser beam through the center of the CL. The beam should still be nearly centered with respect to the relay lens when the objective is engaged.
8. Iteratively adjust the angle of the TM when the objective is not engaged, and the *x–y* position of the CL when the objective is engaged, so that the beam is centered under both conditions.
9. With the objective engaged, adjust the *z*-translation mechanism of the FL to make the laser beam focus on the ceiling (or make the beam size as small as possible). When adjusting the *z* translation of the TM, the angle of the beam coming out of the objective should change in the lateral direction.
10. Turn on the 405-nm laser, and align to the 561-nm laser using the corresponding MR3 and the DC. The criterion is that the 405-nm beam coincides with the 561-nm beam on the ceiling with and without objective engagement. Fine-tune the focus of the corresponding BE when the objective is engaged to make the 405-nm laser also focus on the ceiling.
11. Switch to the filter cube for imaging with the 488-nm laser. Turn on the 488-nm laser, and align it with the 405-nm laser beam using the corresponding TX, TY, and DC.

Excitation, DC, and Emission Filters

At the detection end, filter sets have to be carefully chosen to maximally reject the background fluorescence and the scattered laser light while efficiently allowing the true signal from the fluorophore of interest to reach the camera. Because at least two monochromatic lasers have to reach the specimen simultaneously (e.g., 405 nm for PA and 561 nm for excitation of EosFPs), excitation filters are not used at the filter cube (Fig. 2B). However, a narrow-band excitation filter is often needed at the FIL to minimize the broadband emission noise of the laser. The DC filter should be selected to reflect both activation and excitation wavelengths while transmitting the emitting fluorescence with high efficiency. Finally, emission filters should provide >6-OD suppression of the excitation and activation lights while efficiently transmitting the fluorescence emission. For PALM imaging of the activated EosFP, the investigator uses the DC FF562-Di02-25x36 (Semrock) and the band-pass emission filter FF01-617/73-25 (Semrock). For imaging activated Dronpa, the DC T495lp (Chroma Technology) and the emission filter ET525/50m (Chroma Technology) are used.

Magnification and Signal Detection

Single-molecule imaging would not be possible without a cooled EMCCD camera, which has gain and noise characteristics sufficient for detecting single photons. The back-illumination option results in a quantum efficiency of >90%. The investigator uses an EMCCD camera (Andor Technology, DV887ECS-BV) that offers a 10-MHz pixel readout rate and 512 x 512 pixels with a pixel size of 16 µm. As discussed earlier, a system magnification of ~133x is optimal for such a pixel size. For the 60x objective, the investigator uses a 2x internal magnifying lens (manufacturer installed) and a 1.2x C-mount adaptor (Diagnostic Instruments, DD12NLC) for a total magnification of 144x. When the 100x objective is used, the objective and the 1.2x C-mount adaptor together give a system magnification of 120x. The effective pixel size is, therefore, 111 nm and 133 nm, respectively, depending on the objective. The entire field of view of the camera is 60–70 µm in diameter; but, in practice, 256 x 256 pixels or smaller are often used because the laser beam illuminates only an ~35-µm diameter area.

Stage and Sample Holder

Drifting and vibration must be minimized for nanometric imaging. The microscope body is bolted to a vibration-isolated optical table (Technical Manufacturing Corporation, 4 ft x 5 ft x 12 in), and a mechanically stable rigid stage is bolted to the microscope frame to minimize drift. The investigator uses a motorized stage with a linear encoder (MS-2000 *x–y* stage with a linear encoder; Applied Scientific Instrumentation) that allows marking of sample positions with high precision. To mount the sample, a watertight threaded coverslip holder is firmly attached to a base plate that is, in turn, bolted to the stage (Fig. 4). The coverslip holder allows for application of solutions to keep samples hydrated.

Clean, Clean, and Clean

Debris deposited on the coverslip can be excited by the lasers, which could elevate the overall background or falsely appear as individual fluorescent molecules. It is, therefore, necessary to clean the coverslips thoroughly before use. Among the available methods, the investigator favors washing glass coverslips with 0.5% hydrofluoric acid (HF) for ~1 min followed by a thorough rinse with clean water and air drying them. HF cannot be used to wash the sapphire coverslips used with the 1.65-NA objective because it renders them opaque; however, diluted nitric acid may be an alternative.

Solutions should also be as clean as possible to prevent the introduction of dust that often appears fluorescent. Filtering solutions through 0.22-µm filters eliminates most dust; but note that some commercial filters actually introduce fluorescent particles. In the investigator's laboratory, Fisherbrand syringe filters have been used successfully to filter solutions for PALM.

The sample holder also has to be thoroughly cleaned after each use. The investigator typically sonicates the holder, including the O-ring, in 2% RBS-35 (Pierce, 27950) for 30 min at 50°C before rinsing with water and air drying in a clean environment.

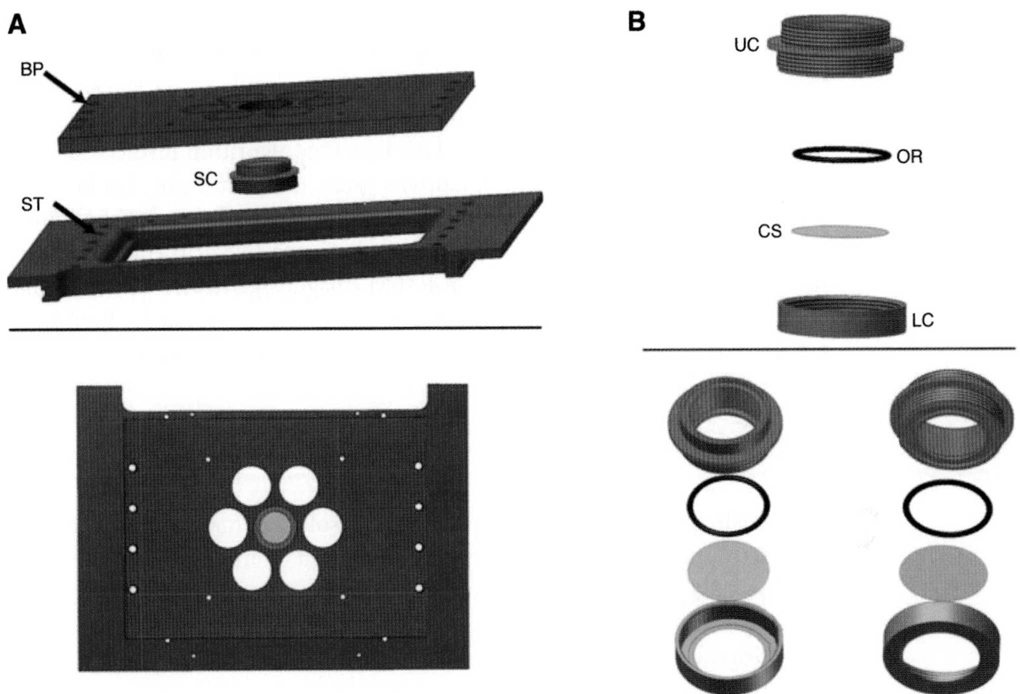

FIGURE 4. Sample holder design. ST, stage; SC, sample holder; BP, base plate; UC, upper steel assembly; OR, O-ring; CS, coverslip; LC, lower steel assembly. (Adapted, with permission, from Shroff et al. 2008b, ©Wiley.)

DATA ACQUISITION

PALM is usually described as consisting of many iterations of PA followed by imaging of single molecules. However, in practice, the two steps can be performed simultaneously and continuously, leading to a significant acceleration of data acquisition. In this regime, the output power of the 405-nm laser is manually increased during data acquisition to maintain a relatively constant density of activated molecules. The fire-out signal from the EMCCD camera that indicates the camera's state is used to shutter the AOTF to block excitation between frames. Image integration of the camera is off for 2 msec between frames while the camera transfers the acquired data from the CCD chip to its internal memory. Although this time is insignificant compared with the image integration time (typically 50 or 100 msec), the investigator has empirically found that shuttering the imaging laser off during this time can lead to an ~20% increase of collected photons per molecule, presumably by interrupting some nonlinear bleaching processes. *It is worth noting that the AOTF driver sometimes backfires a spike that can damage the EMCCD camera.* A buffering circuit is, therefore, necessary to bridge the camera fire-out signal and the AOTF driver.

Computer Hardware and Software

A computer with sufficient random-access memory (RAM), processor power, and hard drive space is essential. The computer needs to meet the necessary requirements for the EMCCD camera (a PCI [peripheral component interconnect] slot for a digital acquisition card, 2.4-GHz Pentium processor, 1-GB RAM, and Microsoft 2000/XP for Andor cameras). An analog output board and a USB (universal serial bus) interface are also useful for controlling the power of activation and excitation lasers. Because PALM data consist of tens of thousands of individual frames, a PALM data file can sometimes exceed 10 GB in size. Ample hard drive space and subsequent long-term storage are necessary. Please also note that the older FAT32 data storage format does not support files larger than 4 GB.

Because PALM experiments can be performed by simultaneously activating and imaging, no sophisticated software is required. The software that comes with the Andor Technology EMCCD camera is sufficient for acquiring PALM single-molecule image stacks, although additional software that controls laser power, laser shuttering, and camera acquisition parameters might be useful.

Data analysis is mostly performed off-line. One algorithm involves identifying molecule peaks on a frame-by-frame basis and summing up the peaks at the same position in adjacent frames. The summed image of a molecule is then fitted by a Gaussian mask to determine the center of the molecule and the error of the center (Thompson et al. 2002). Retrospective drift correction can be performed using the fitted position of fiducials. Finally, a PALM image is generated by rendering molecules as Gaussians whose brightness indicates the probability that the molecule can be found at a given location. Readers are referred to Betzig et al. (2006) for more detailed information on data analysis.

LIMITATIONS OF PALM

It is worth noting that PALM improves the resolution mainly on the *x–y* plane but not along the *z*-axis. Resolution along the *z*-axis can be improved by utilizing TIRF illumination (~200 nm in *z*) or by physically sectioning the sample into ultrathin sections (~50 nm). Although several versions of 3D PALM and related techniques have been reported, they all require very thin samples and sometimes sacrifice lateral resolution. Implementation of these techniques also requires considerable expertise in microscopy.

Although PALM has been applied to live cell imaging, it is limited to the structures within the TIRF zone and only achieves a frame rate of 1 image/30 sec for a Nyquist resolution of 60 nm. Even at such a frame rate, only ~20 frames can be recorded before all PA-FP molecules are depleted. Therefore, PALM is suitable for live cell imaging only for very slow dynamics or when the Nyquist resolution is not required (e.g., single particle tracking). The high laser power that is required with PALM can interfere with normal cellular physiology and needs to be carefully monitored.

PALM'S POTENTIAL IN NEUROSCIENCE

Although PALM and related techniques were developed very recently, investigators have found them useful for studying the organization and the dynamics of the focal adhesion complex in cultured cells as well as for following the diffusion of membrane proteins. It is not a surprise that, in almost all cases, PALM has found its initial application in single-layer cell cultures or bacteria samples in which out-of-focus background fluorescence is minimal. It is, therefore, likely that PALM could be easily adopted to study molecules in cultured neurons plated on coverslips within the TIRF zone or in thin areas with low background fluorescence (e.g., at distal dendrites or axons). Potential applications include the study of synaptic protein organization, molecular diffusion, and vesicle trafficking. Importantly, the latter two studies may not require high Nyquist resolution and could be performed in live neurons, allowing the data to be paired with physiological measurements. PALM is also suitable for investigating in vitro systems reconstituted on a coverslip. The application of PALM to intact tissues, such as the brain, however, is stalled by the lack of a practical procedure to section the tissue while preserving the labeling density of PA fluorophores and the nanometric structure. We are adapting the sample preparation and the resin-embedded sectioning procedures used in EM and fluorescence microscopy. Once established, these sample preparation procedures would make it possible to examine synaptic protein architecture on the nanometer scale within brain tissues. Furthermore, resin-embedded ultrathin sections might also permit correlative PALM-EM studies, allowing the protein localization data to be interpreted within the global cellular context. Such PALM-EM correlation could also help neural tracing in volumetric reconstruction of the neural circuits by assigning colors to individual neuronal processes.

CONCLUSION

PALM has opened up a window for a large number of neuroscience laboratories to peek at the neuronal architectures at 10-nm resolution, thanks to its simple and relatively inexpensive design. PALM is readily applicable to cellular neuroscience questions when neuronal cultures are appropriate. Future developments in sample preparation procedures could further enhance the utility of PALM in revealing the synaptic architecture within intact brain tissues and as a tool in volumetric neuronal reconstruction.

ACKNOWLEDGMENTS

I thank Eric Betzig, Hari Shroff, Na Ji, Thomas Planchon, Jianyong Tang, Tianyi Mao, Craig Jahr, and Nicholas Frost for their comments.

REFERENCES

Bates M, Huang B, Dempsey GT, Zhuang X. 2007. Multicolor super-resolution imaging with photo-switchable fluorescent probes. *Science* **317:** 1749–1753.

Betzig E, Patterson GH, Sougrat R, Lindwasser OW, Olenych S, Bonifacino JS, Davidson MW, Lippincott-Schwartz J, Hess HF. 2006. Imaging intracellular fluorescent proteins at nanometer resolution. *Science* **313:** 1642–1645.

Chudakov DM, Lukyanov S, Lukyanov KA. 2007. Tracking intracellular protein movements using photoswitchable fluorescent proteins PS-CFP2 and Dendra2. *Nat Protoc* **2:** 2024–2032.

de Lange F, Cambi A, Huijbens R, de Bakker B, Rensen W, Garcia-Parajo M, van Hulst N, Figdor CG. 2001. Cell biology beyond the diffraction limit: Near-field scanning optical microscopy. *J Cell Sci* **114:** 4153–4160.

Gustafsson MG. 2000. Surpassing the lateral resolution limit by a factor of two using structured illumination microscopy. *J Microsc* **198:** 82–87.

Habuchi S, Ando R, Dedecker P, Verheijen W, Mizuno H, Miyawaki A, Hofkens J. 2005. Reversible single-molecule photoswitching in the GFP-like fluorescent protein Dronpa. *Proc Natl Acad Sci* **102:** 9511–9516.

Hess ST, Girirajan TP, Mason MD. 2006. Ultra-high resolution imaging by fluorescence photoactivation localization microscopy. *Biophys J* **91:** 4258–4272.

Huang B, Wang W, Bates M, Zhuang X. 2008. Three-dimensional super-resolution imaging by stochastic optical reconstruction microscopy. *Science* **319:** 810–813.

Ji N, Shroff H, Zhong H, Betzig E. 2008. Advances in the speed and resolution of light microscopy. *Curr Opin Neurobiol* **18:** 605–616.

Juette MF, Gould TJ, Lessard MD, Mlodzianoski MJ, Nagpure BS, Bennett BT, Hess ST, Bewersdorf J. 2008. Three-dimensional sub-100 nm resolution fluorescence microscopy of thick samples. *Nat Methods* **5:** 527–529.

Luby-Phelps K, Ning G, Fogerty J, Besharse JC. 2003. Visualization of identified GFP-expressing cells by light and electron microscopy. *J Histochem Cytochem* **51:** 271–274.

Manley S, Gillette JM, Patterson GH, Shroff H, Hess HF, Betzig E, Lippincott-Schwartz J. 2008. High-density mapping of single-molecule trajectories with photoactivated localization microscopy. *Nat Methods* **5:** 155–157.

McKinney SA, Murphy CS, Hazelwood KL, Davidson MW, Looger LL. 2009. A bright and photostable photoconvertible fluorescent protein. *Nat Methods* **6:** 131–133.

Micheva KD, Smith SJ. 2007. Array tomography: A new tool for imaging the molecular architecture and ultrastructure of neural circuits. *Neuron* **55:** 25–36.

Patterson GH, Lippincott-Schwartz J. 2002. A photoactivatable GFP for selective photolabeling of proteins and cells. *Science* **297:** 1873–1877.

Rust MJ, Bates M, Zhuang X. 2006. Sub-diffraction-limit imaging by stochastic optical reconstruction microscopy (STORM). *Nat Methods* **3:** 793–795.

Shroff H, Galbraith CG, Galbraith JA, White H, Gillette J, Olenych S, Davidson MW, Betzig E. 2007. Dual-color super-resolution imaging of genetically expressed probes within individual adhesion complexes. *Proc Natl Acad Sci* **104:** 20308–20313.

Shroff H, Galbraith CG, Galbraith JA, Betzig E. 2008a. Live-cell photoactivated localization microscopy of nanoscale adhesion dynamics. *Nat Methods* **5:** 417–423.

Shroff H, White H, Betzig E. 2008b. Photoactivated localization microscopy (PALM) of adhesion complexes. *Curr Protoc Cell Biol* **Chapter 4:** Unit 4.21.

Subach FV, Patterson GH, Manley S, Gillette JM, Lippincott-Schwartz J, Verkhusha VV. 2009. Photoactivatable mCherry for high-resolution two-color fluorescence microscopy. *Nat Methods* **6:** 153–159.

Thompson RE, Larson DR, Webb WW. 2002. Precise nanometer localization analysis for individual fluorescent probes. *Biophys J* **82:** 2775–2783.

Westphal V, Hell SW. 2005. Nanoscale resolution in the focal plane of an optical microscope. *Phys Rev Lett* **94:** 143903.

Wiedenmann J, Ivanchenko S, Oswald F, Schmitt F, Rocker C, Salih A, Spindler KD, Nienhaus GU. 2004. EosFP, a fluorescent marker protein with UV-inducible green-to-red fluorescence conversion. *Proc Natl Acad Sci* **101:** 15905–15910.

35 Stochastic Optical Reconstruction Microscopy (STORM)

A Method for Superresolution Fluorescence Imaging

Mark Bates,[1,2] Sara A. Jones,[2] and Xiaowei Zhuang[1,2,3]

[1]Howard Hughes Medical Institute, Harvard University, Cambridge, Massachusetts 02138; [2]Department of Chemistry and Chemical Biology, Harvard University, Cambridge, Massachusetts 02138; [3]Department of Physics, Harvard University, Cambridge, Massachusetts 02138

ABSTRACT

Light microscopy is one of the most widely used techniques for the study of cell biology; however, the relatively low-spatial resolution of the optical microscope presents significant limitations for the observation of biological ultrastructure. Subcellular structures and molecular complexes essential for biological function exist on length scales from nanometers to micrometers. When observed with light, however, structural features smaller than ~0.2 µm are blurred and are difficult or impossible to resolve. In this chapter, we describe a method for fluorescence microscopy that extends the resolution of the microscope beyond the classical diffraction limit. This approach is generally applicable to biological imaging and requires relatively simple experimental apparatus; its spatial resolution is theoretically unlimited, and a resolution improvement of an order of magnitude over conventional optical microscopy has been experimentally demonstrated.

INTRODUCTION

As illustrated in the earlier chapters of this manual, fluorescence microscopy is a versatile technique widely used in molecular and cell biology. The noninvasive nature of light allows biological specimens to be imaged with little perturbation, enabling researchers to observe dynamic processes as they occur in living cells and tissues. Fluorescent labeling techniques such as immunofluorescence, fluorescence in situ hybridization, and genetically encoded fluorescent tags (e.g., green fluorescent protein [GFP]) make it possible to label and to image specific biochemical components of a sample (Giepmans et al. 2006; Spector and Goldman 2006). The availability of a broad range of fluorophores enables the simultaneous imaging of multiple targets by multicolor labeling with spectrally distinct fluorescent probes. Furthermore, three-dimensional (3D) images may be generated, for example, by focusing into the sample at different depths.

Despite these advantages, conventional fluorescence microscopy is limited by its spatial resolution, leaving many biological structures too small to be studied in detail. Subcellular structures span a range of length scales from micrometers to nanometers, whereas the light microscope is limited by diffraction to a resolution of ~200 nm in the lateral direction and >500 nm in the axial direction (Pawley 2006). Other imaging techniques such as electron microscopy (EM) have obtained much higher spatial resolutions, and the ability of these methods to visualize biological samples with molecular resolution has had a tremendous impact on our understanding of biology (Koster and Klumperman 2003). To achieve image resolutions comparable to EM but with the labeling specificity and live cell compatibility provided by fluorescence microscopy would create a new opportunity to study the nanoscale structure and dynamics of cells and tissues.

Over the past several years, this goal has been met with the introduction of a number of new methods, which we collectively refer to as superresolution imaging techniques. In this chapter, we present one such technique: stochastic optical reconstruction microscopy (STORM). Here, we review the theoretical basis of this method and focus on practical aspects of its implementation. Requirements for experimental apparatus and data analysis procedures are presented in detail, with the goals of facilitating the implementation of STORM and illustrating its capabilities as a tool for sub-diffraction-limit microscopy. For overviews of the superresolution fluorescence imaging field and descriptions of the various techniques, the reader is referred to several recent review articles (Hell 2007; Bates et al. 2008; Fernandez-Suarez and Ting 2008; Heintzmann and Gustafsson 2009; Hell et al. 2009; Huang et al. 2009; Zhuang 2009).

THEORY AND CONCEPT OF STORM

The Diffraction Limit of Resolution

It was recognized by Abbe in the 19th century that the spatial resolution of an optical microscope is limited by the diffraction of light (Abbe 1873). It is due to diffraction that a point source of light, when imaged through a microscope, appears as a spot with a finite size. The intensity profile of this spot defines the point-spread function (PSF) of the microscope. The full width at half-maximum (FWHM) of the PSF in the lateral (x–y) and axial (z) directions is given approximately by $\Delta x, \Delta y = \lambda/(2\text{NA})$ and $\Delta z = 2\lambda\nu/(\text{NA})^2$, respectively, where λ is the wavelength of the emitted light, ν is the index of refraction of the medium, and NA is the numerical aperture of the objective lens (Pawley 2006).

The diffraction limit of resolution can be understood by considering the image formed by two identical fluorophores spaced some distance d apart in the x–y plane. When $d >> \Delta x$, two well-separated spots are observed in the image plane, each with an intensity profile corresponding to the PSF of the microscope. As the fluorophores are moved closer together, however, the images of the two fluorophores begin to overlap significantly when $d \approx \Delta x$. This is the minimum separation distance at which the two fluorophores are resolvable as separate entities. As the distance between the fluorophores decreases further, the images are fully overlapped, and the two fluorophores are not distinguishable.

The resolution of the microscope is, thus, limited by the width of the PSF. This line of reasoning extends to any sample imaged with an optical microscope: Features smaller than the PSF are blurred and cannot be resolved. Some improvement in resolution can be obtained from reducing the size of the PSF by using shorter-wavelength light or an objective lens with a higher NA. In practice, for visible light ($\lambda \approx 550$ nm) and a high-NA objective (NA = 1.4), the resolution of the conventional light microscope remains limited to ~200 nm laterally and >500 nm along the optical axis.

Nanoscale Localization of Single Fluorophores

The position of an isolated fluorescent emitter, although its image appears as a diffraction-limited spot, can be precisely determined by finding the centroid of its image. For example, shown in

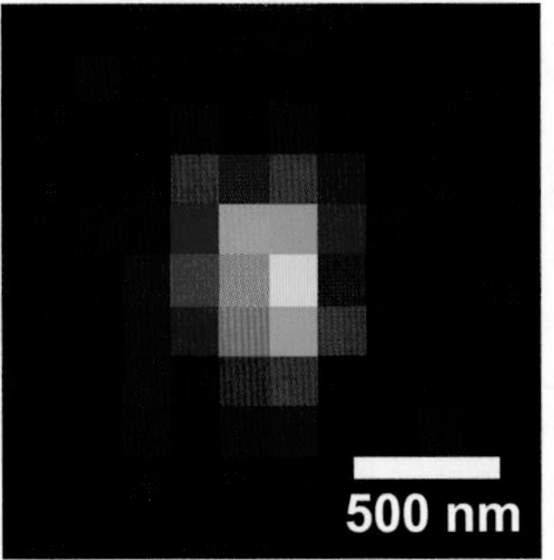

FIGURE 1. Fluorescence image of a single Cy5 fluorophore. The emitted light (wavelength ~670 nm) is imaged onto a CCD camera. Although the fluorophore is only nanometers in size, its image appears as a broad spot with an approximately Gaussian intensity profile. The position of the fluorophore can be determined with nanometer precision by fitting the image to find its centroid.

Figure 1 is the image of a single fluorescent molecule bound to a surface. Despite the nanoscale size of the fluorophore, the image appears as a diffraction-broadened peak several hundred nanometers in width, reflecting the PSF of the microscope. Although the width of the PSF limits the spatial resolution of a conventional optical microscope, it does not prevent the precise determination of the fluorophore position when the image of the fluorophore is isolated from those of the other molecules. The position of the fluorophore corresponds to the center position of the peak, and this can be measured with a high degree of precision. By fitting the image to find the peak centroid, the position of the fluorophore can be determined with a precision limited only by the number of photons collected. The precision of this localization process, expressed in terms of the standard deviation of the position measurement (σ), is given approximately by

$$\sigma \approx \sqrt{\frac{s^2 + (a^2/12)}{N} + \frac{4\sqrt{\pi}s^3b^2}{aN^2}},$$

where s is the standard deviation of the PSF, a is the edge size of the area imaged on each charge-coupled device (CCD) pixel, b is the standard deviation of the image background, and N is the number of photons detected from the fluorophore (Thompson et al. 2002). This concept of high-precision localization has been used to track small particles with nanometer-scale accuracy (Gelles et al. 1988; Ghosh and Webb 1994). Recently it has been shown that, even for a single fluorescent dye molecule under ambient conditions, its position can be determined with a precision as high as ~1 nm (Yildiz et al. 2003).

Nanoscale precision in single-molecule localization does not, however, translate directly into image resolution. When multiple fluorophores are positioned close together such that they are separated by a distance less than the PSF width, their images overlap, and this prevents accurate localization of each of the fluorophores. To distinguish the fluorescence signal from the nearby fluorescent emitters, several approaches have been used, based on differences in the emission wavelength (van Oijen et al. 1998; Lacoste et al. 2000; Churchman et al. 2005), the sequential photobleaching of each fluorophore (Gordon et al. 2004; Qu et al. 2004), or quantum dot blinking (Lidke et al. 2005; Lagerholm et al. 2006). These methods have obtained high accuracy localization for several closely spaced emitters but are difficult to extend to densities of more than a few fluorophores per diffraction-limited area. A fluorescently labeled biological sample, in contrast, may be labeled with hundreds of fluorophores per diffraction-limited region.

Stochastic Optical Reconstruction Microscopy

A solution to the problem of localizing multiple fluorescent molecules positioned in close proximity presents itself in the form of optically switchable fluorophores: molecules that can be switched between a nonfluorescent and a fluorescent state by exposure to light. The fluorescence emission of such molecules can be controlled such that, at any point in time, only a single one, or a sparse subset, of the fluorophores in a given region of the sample is in the fluorescent state. This sparse condi-

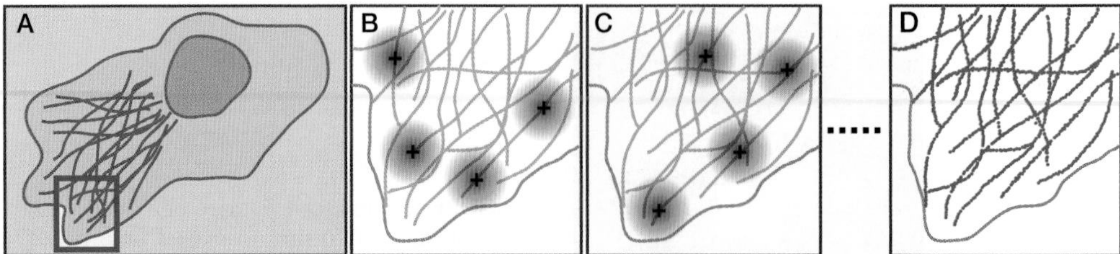

FIGURE 2. The STORM imaging procedure. (*A*) Schematic of a cell in which the structure of interest (gray filaments in this case) are labeled with photoswitchable fluorophores (not shown). All fluorophores are initially in the nonfluorescent state. The red box indicates the area shown in panels *B–D*. (*B*) An activation cycle: A sparse set of fluorophores is activated to the fluorescent state such that their images (large red circles) do not overlap. The image of each fluorophore appears as a diffraction-broadened spot, and the position of each activated fluorophore is determined by fitting to find the centroid of the spot (black crosses). (*C*) A subsequent activation cycle: A different set of fluorophores is activated, and their positions are determined as before. (*D*) After a sufficient number of fluorophores has been localized, a high-resolution image is constructed by plotting the measured positions of the fluorophores (red dots). The resolution of this image is not limited by diffraction but by the accuracy of each fluorophore localization and by the number of fluorophore positions obtained. (Adapted, with permission, from Bates et al. 2008, ©Elsevier.)

tion ensures that the images of the individual molecules are not overlapping, allowing each to be localized with high precision. If the fluorescent molecules are then deactivated to the dark state, and a new subset is activated to the fluorescent state, these newly activated fluorescent molecules can also be localized. By repeating this cycle of activation, localization, and deactivation, the positions of an arbitrary number of closely spaced fluorophores may be determined.

An imaging method based on this principle is illustrated in Figure 2. Beginning with a fluorescently labeled sample, the goal of the method is to determine the positions of the fluorescent labels and to plot them to form an image. The imaging process consists of many cycles during which fluorophores are activated, imaged, and deactivated, as described above. The density of activated molecules is kept low by using a weak activation light intensity such that the images of individual fluorophores do not typically overlap, thereby allowing each fluorophore to be localized with high precision. As this process is repeated, a stochastically different subset of fluorophores is activated in each cycle, enabling the positions of many fluorophores to be determined. After a sufficient number of localizations has been recorded, a high-resolution image is constructed from the measured positions of the fluorophores. The resolution of the final image is not limited by diffraction but by the precision of each localization and the localization density. In this sense, the resolution of the image is limited only by the number of photons collected from the fluorophores and the number of fluorophores labeling the sample. We termed this method "stochastic optical reconstruction microscopy" (STORM). This concept was independently developed by multiple research groups, and it is also referred to as photoactivated localization microscopy (PALM) or fluorescence photoactivation localization microscopy (FPALM) (Betzig et al. 2006; Hess et al. 2006; Rust et al. 2006). For ease of reference, we will refer to the method as STORM in this chapter.

PHOTOSWITCHABLE FLUORESCENT MOLECULES

A fluorescent molecule is termed photoswitchable, photoactivatable, or photochromatic when its fluorescence emission or other spectral characteristics can be converted between two or more distinct states under the control of an external light source. Fluorescence switching was first reported for a variant of the fluorescent protein, YFP (yellow fluorescent protein), which was shown to cycle reversibly between a fluorescent and a dark state by alternating exposure to blue and violet light (Dickson et al. 1997). Similar effects have been shown with synthetic organic fluorophores (Kulzer et al. 1997; Irie et al. 2002). One such example is the red cyanine dye Cy5, which can be reversibly converted between a fluorescent state

TABLE 1. Spectral properties of photoswitchable fluorescent proteins and organic dyes

Fluorophore	Activation (nm)	Before activation: Absorption max (nm)	Before activation: Emission max (nm)	After activation: Absorption max (nm)	After activation: Emission max (nm)	Reference(s)
Fluorescent proteins						
PA-GFP	405	400	515	504	517	Patterson and Lippincott-Schwartz 2002
PS-CFP2	405	400	468	490	511	Chudakov et al. 2004
Dendra-2	405	490	507	553	573	Gurskaya et al. 2006
Kaede	405	508	518	572	582	Ando et al. 2002
EosFP, mEos2	405	506	516	571	581	Wiedenmann et al. 2004; McKinney et al. 2009
mKikGR	405	507	517	583	593	Tsutsui et al. 2005; Habuchi et al. 2008
Dronpa	405	—	—	503	518	Habuchi et al. 2005
Dronpa-2	405	—	—	486	513	Ando et al. 2007
bs-Dronpa	405	—	—	460	504	Andresen et al. 2008
rsFastLime	405	—	—	496	518	Stiel et al. 2007
eYFP	405	—	—	514	529	Dickson et al. 1997; Biteen et al. 2008
PA-mCherry	405	—	—	570	596	Subach et al. 2009
Synthetic dyes						
Cy5	350–570[a]	—	—	649	670	Bates et al. 2005, 2007; Heilemann et al. 2005; Dempsey et al. 2009a
Cy5.5	350–570[a]	—	—	675	694	Bates et al. 2005, 2007; Dempsey et al. 2009a
Cy7	350–570[a]	—	—	747	776	Bates et al. 2005, 2007; Dempsey et al. 2009a
Alexa Fluor 647	350–570[a]	—	—	650	665	Bates et al. 2005, 2007; Dempsey et al. 2009a
Photochromic rhodamine B	375	—	—	565	580	Folling et al. 2007
Rhodamine spiroamides	375			537–591[b]	555–620[b]	Belov et al. 2009
Azido-DCDHF		—	—	570	613	Lord et al. 2009

PA, photoactivatable; GFP, green fluorescent protein; PS, photoswitchable; CFP, cyan fluorescent protein; eYFP, enhanced yellow fluorescent protein; azido-DCDHF, azido-2-dicyanomethylene-3-cyano-2,5-dihydrofuran.

[a]Dependent on the activator dye, if present.

[b]Dependent on the specific compound.

and a nonfluorescent state (Bates et al. 2005; Heilemann et al. 2005). When viewed at the single-molecule level, Cy5 appears to switch its fluorescence on and off. In contrast, the fluorescent protein Kaede undergoes an irreversible spectral change from green fluorescent to red fluorescent emission when exposed to ultraviolet (UV) light (Ando et al. 2002). Photoswitchable fluorophores, such as these, have found increasing application for superresolution imaging and also as optical highlighters in which a specific region of a sample can be marked by photoconverting the fluorophores in a localized area (Elowitz et al. 1997; Marchant et al. 2001). A wide variety of photoswitchable synthetic fluorophores and fluorescent proteins are now available in a range of colors, a selection of which is summarized in Table 1.

Any of the photoswitchable fluorescent proteins and organic dyes listed in Table 1 may be applied to STORM. Some fluorophores are certainly more suited than others, however. Generally, organic dyes tend to be brighter than fluorescent proteins, allowing for higher localization precision. Fluorescent proteins, on the other hand, have the advantage of being genetically encoded, making it easier to label intracellular proteins in living cells. In the following discussion, we will describe the cyanine dyes as a specific example; however, the reader should be aware of the many options available among the fluorescent proteins and other synthetic fluorescent dyes.

Cyanine Dyes

As mentioned above, the red cyanine dye Cy5 can be converted between a fluorescent and a nonfluorescent state under the control of an external light source. This photoswitching behavior is dependent

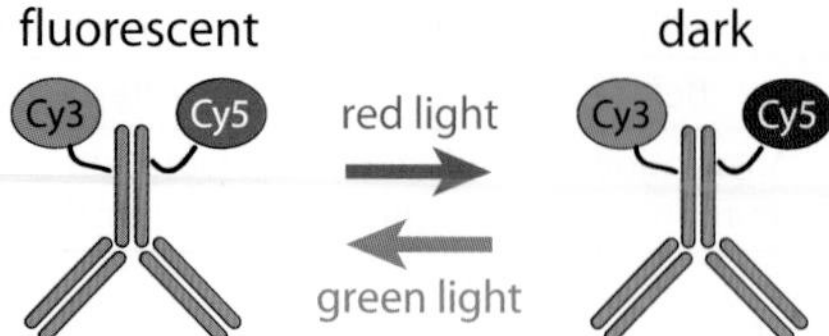

FIGURE 3. A cartoon representation of the photoconversion of Cy5 between a fluorescent and a nonfluorescent state. In this diagram, an antibody molecule is labeled with the green fluorophore Cy3 and the red fluorophore Cy5. When Cy5 is exposed to red light, it emits fluorescence and then switches to a dark nonfluorescent state. Illumination with green light, however, causes Cy5 to revert to its fluorescent state by means of a short-range interaction with the neighboring Cy3. Under illumination conditions allowing single-molecule detection, this single-molecule fluorescent switch can be cycled on and off for hundreds of cycles before permanent photobleaching occurs.

on the presence of a thiol-containing molecule in the imaging solution such as β-mercaptoethanol (βME), β-mercaptoethylamine (MEA), or cysteine (Bates et al. 2005; Heilemann et al. 2005; Dempsey et al. 2009a). When single molecules of Cy5 are exposed to red light, they are initially fluorescent before switching into a stable dark state. If this sample is then exposed to UV light, a significant fraction of the molecules rapidly return to their fluorescent state. This is the hallmark of a photoswitchable fluorophore: that it can be optically converted between distinct fluorescent states. It is worth noting that the photoswitching behavior of Cy5 is observed over a broad range of thiol concentrations, including low concentrations at which living cells remain viable—this feature allows the cyanine dyes to be used with live cell STORM imaging.

The fluorescence reactivation of Cy5 by visible light is made significantly more efficient when Cy5 is paired with a second fluorophore that absorbs visible light such as the green fluorophore Cy3. When Cy3 and Cy5 are positioned in close proximity (e.g., on a doubly labeled antibody as illustrated in Fig. 3), the Cy5 fluorophores switch off as described earlier when exposed to red light in the presence of a thiol-containing reagent. However, when exposed to a low-intensity green excitation light source ($\lambda = 532$ nm), the Cy5 molecules rapidly recover their fluorescence. The rate of fluorescence reactivation by green light is $\sim 10^3$ times greater when Cy5 is paired with Cy3, as compared with an isolated Cy5 molecule. One can think of the pair of dye molecules Cy3 and Cy5 as acting together to form a fluorescent switch. With red illumination, Cy5 emits fluorescence and is converted to the dark state. When exposed to green light, Cy3 absorbs a green photon and causes Cy5 to be reactivated to the fluorescent state through a short-range interaction (Bates et al. 2005). This process is repeatable and is robust, with the dye pair undergoing, on average, more than 100 switching cycles before permanent photobleaching occurs. We refer to the dyes as an activator–reporter pair in which Cy3 is the activator because it serves to activate Cy5 to the fluorescent state and Cy5 is the reporter because its fluorescence emission reports the state of the switch.

Activator–Reporter Dye Pairs for Multicolor Imaging

A useful property of the activator–reporter concept is the ability to create multicolor spectrally distinct probes using different combinations of activator and reporter dyes. In general, the activation wavelength for photoswitching is determined by the absorption spectrum of the activator dye. If a blue fluorescent dye (e.g., Cy2) is substituted in place of Cy3, the Cy2–Cy5 pair shows photoswitching as before, but efficient reactivation occurs only in response to blue light. Similarly, a violet fluorophore (e.g., Alexa Fluor 405) may be used as the activator dye, and, in this case, the Alexa 405–Cy5 construct is reactivated efficiently when exposed to violet light. Each combination represents a spectrally distinct photoswitchable probe that is distinguishable by the wavelength of light required for activation (Bates et al. 2007). This property enables multicolor imaging schemes, as illustrated later in this chapter in the discussion of multicolor STORM.

Characteristics of Photoswitchable Fluorophores

Several properties of photoswitchable fluorophores determine their suitability for superresolution imaging using the STORM method. In particular, the brightness of the fluorophore is crucial for single-molecule detection and precise localization. Also, non-ideal switching characteristics such as residual fluorescence emission from the dark state and spontaneous reactivation from the dark state to the fluorescent state can present significant challenges when imaging densely labeled samples.

Fluorophore Brightness

The precise and accurate measurement of the position of each fluorescent label is an essential part of the STORM method. The precision of each position measurement is primarily determined by the number of photons collected from the activated fluorophore during a single cycle of switching on and off. Photoswitchable fluorescent proteins and organic dyes vary widely in this respect, with the number of photons detected per activation event ranging from a few hundred for some of the relatively dim fluorescent proteins to as high as 6000 for Cy5 and Alexa Fluor 647 (Hess et al. 2006; Bates et al. 2007; Shroff et al. 2007). Among the photoactivatable fluorescent proteins, EosFP is one of the brightest with approximately 1000 photons detected per activation event (Shroff et al. 2007).

Contrast Ratio and Spontaneous Activation

An essential step of the STORM imaging procedure is to activate only sparse subsets of the fluorescent labels at any one time such that each fluorophore is optically resolvable from the rest and can, therefore, be localized with high precision. In most cases, this means that the number of fluorophores in the dark state is many times higher than the number in the fluorescent state. In this scenario, two common characteristics of photoswitchable fluorophores, dark-state fluorescence and spontaneous activation, can have adverse effects on the imaging procedure.

Dark-state fluorescence refers to residual fluorescence emission from fluorophores that are in the deactivated state. This is commonly observed in photoactivatable fluorescent proteins and is characterized by the contrast ratio, defined as the ratio of a fluorophore's brightness in the fluorescent state to that in the dark state (Betzig et al. 2006). In the case of a fluorescent protein with a contrast ratio of 10^3, for example, a single fluorophore in the fluorescent state is equal in brightness to 10^3 fluorophores in the dark state. When the density of fluorophores in the sample is greater than 10^3 per diffraction-limited region, the background fluorescence emitted by the dark-state fluorophores begins to degrade the ability of the analysis procedure to detect and to calculate the location of the individual fluorophores in the fluorescent state.

Spontaneous activation (also called nonspecific activation or blinking) is another effect commonly observed with photoswitchable fluorophores. This refers to the activation of a fluorophore in the absence of specific exposure to the activation light source. In the case of the Cy3–Cy5 dye pair, for example, Cy5 will occasionally be activated when no green light is applied, seemingly blinking on and off on its own. This effect is caused by a relatively small but finite rate of activation of Cy5 by the red imaging laser itself, which is also used to excite fluorescence from the dye and to switch it off. Both the rate of switching off (k_{off}) and the rate of reactivation by the red laser (k_{blink}) are found to be linear in the illumination intensity. Consequentially, at equilibrium under red illumination, a fraction of the fluorophores present on the sample equal to $k_{blink}/(k_{off} + k_{blink})$ will be in the fluorescent state because of spontaneous activation. For Cy5, this fraction has been measured to be ~0.001–0.005 depending on the imaging condition such as the thiol reagent used in the imaging buffer. For densely labeled samples, the population of fluorophores activated due to this effect may violate the requirement for sparse activation and thereby, may interfere with the detection and the localization of individual fluorophores.

STORM MICROSCOPE DESIGN AND IMAGING PROCEDURE

The basic setup for STORM is very similar to that for conventional wide-field fluorescence microscopy, with a few modifications. In this section, the major components of the microscope are described, along with a general outline of the experimental procedure.

Microscope Components

Microscope Frame and Objective Lens

A schematic of the STORM microscope is shown in Figure 4. Although a microscope frame is not an essential component, we have found it convenient to base our microscopes on a standard inverted fluorescence microscope frame. The microscope frame and sample stage should provide a stable mount for the sample and the objective lens, and the focusing mechanism must also be relatively stable. Because the raw data used to generate a STORM image are collected over a period of

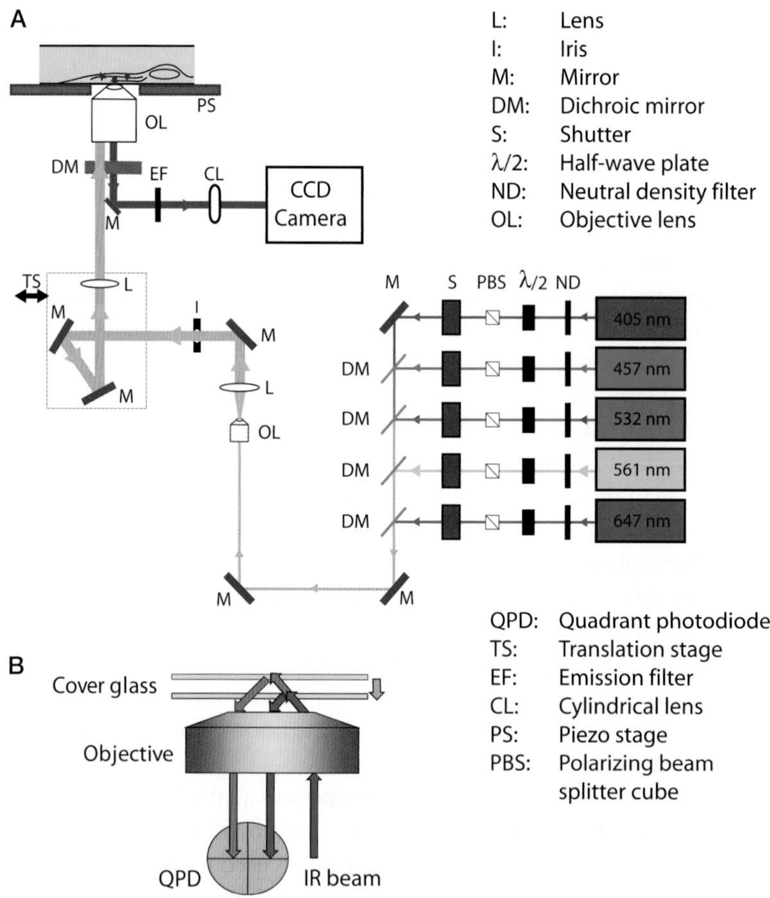

FIGURE 4. A schematic of the STORM microscope. (*A*) The microscope is based on a standard inverted fluorescence microscope. Laser sources are combined into a single beam, which is expanded and focused to the back focal plane of the microscope objective. A translation stage allows adjustment of the incidence angle of the illumination light. The detected light passes through an emission filter and is imaged onto a CCD camera. The specific laser wavelengths shown here are an example of activation and imaging wavelengths that may be useful for STORM. Other wavelengths may be required depending on the choice of photoswitchable dye or fluorescent protein. The cylindrical lens (CL) is used only for astigmatic imaging in 3D STORM applications. (*B*) The geometry of the focus-lock system. An infrared (IR) laser beam is projected through the objective lens from a second dichroic mirror (not shown in panel *A*).The beam reflects off the sample–cover-glass interface and back through the objective to where its position is detected on a quadrant photodiode (QPD). (Adapted from Dempsey et al. 2009b.)

time, sample drift during the measurement degrades the image resolution and must be corrected for in postprocessing. At the nanometer scale, drift of the microscope is inevitable, but any steps taken to minimize drift will improve image quality such as thermal and mechanical isolation of the microscope.

Drift of the focus mechanism is particularly problematic because this effect is difficult to correct for in the analysis procedure. We use an active focus-lock system based on a weak IR laser beam, which passes through the objective lens, is reflected off the sample–cover-glass interface, and is then projected onto a quadrant photodiode (QPD), as illustrated in Figure 4B. The position of the beam on the diode reports any relative motion between the objective lens and the sample, and this signal is fed back to a piezoelectric *z* stage, which corrects for the motion.

The objective lens should be chosen to maximize light collection efficiency and to minimize aberrations and background fluorescence. We typically use an oil-immersion lens such as the UPlanSAPO 100x NA 1.4 microscope objective from Olympus, which allows for both epifluorescence- and objective-type TIRF illumination. Other similar objective lenses may also be used.

Illumination Geometry

For cell imaging, a through-the-objective illumination geometry (Fig. 4) is convenient, allowing cell culture plates to be imaged from underneath such that the cells can remain in solution during imaging. This scheme also allows for the use of short-working-distance high-NA objective lenses, increasing light collection efficiency significantly.

To bring the illumination light to the sample, the imaging and activation lasers are combined into a single path using dichroic mirrors and then are expanded and collimated using a beam expander. This beam is focused to the back focal plane of the objective. A translation stage allows the beam to be shifted toward the edge of the objective so that the light emerging from the objective reaches the sample at a high incident angle near, but not exceeding, the critical angle of the glass–water interface. In this manner, a region of the sample several micrometers in thickness is illuminated by the angled beam. For thinner regions of the sample, adjusting the incident angle allows a narrower depth of illumination and lower background fluorescence (Cui et al. 2007; Tokunaga et al. 2008). For imaging near the cell surface, the incidence angle of the illumination light can be adjusted to greater than the critical angle to allow total internal reflection and to further reduce the background signal.

Detection Optics and CCD Camera

Because STORM relies on the detection of single fluorescent molecules, the optical detection path of the microscope should be optimized for maximum light collection efficiency, and the CCD camera must be suitable for low-light fluorescence imaging. In our experiments, we have used thermoelectrically cooled cameras incorporating electron-multiplying CCD (EMCCD) technology, a form of on-chip signal amplification that raises the signal well above the CCD read noise and dark current levels (Jerram et al. 2001).

The degree of magnification of the image is chosen for optimal fluorophore localization. If the magnification is too low, the image of the fluorophore is smaller than a camera pixel, and no sub-diffraction-limit position information can be determined. If the magnification is too high, the photons collected from the fluorophore are spread over many pixels, and the noise generated as each pixel is read out degrades the localization accuracy. As studied by Thompson and colleagues, the localization process is most precise when the magnification is chosen such that the FWHM of the image of the PSF is equal in size to ~2.4 camera pixels (Thompson et al. 2002).

To relate the STORM data to real distances, the image magnification must be carefully calibrated. For example, the system may be calibrated by imaging a micrometer slide that has a pattern of lines spaced 10 μm apart.

For 3D STORM imaging, a long focal-length cylindrical lens is inserted as the final optical element before the camera (Huang et al. 2008b). This will be discussed further in the sections below.

Light Sources

Laser illumination allows for highly efficient excitation of fluorophores and adds the ability to block the excitation light from entering the detection channel using spectral filters, thereby maximizing detection of the fluorescence signal and minimizing background. For imaging Cy5 or Alexa Fluor 647, a laser source at ~650 nm with a power output >50 mW is sufficient to enable short switching times and rapid data collection. Higher laser powers enable faster fluorophore switching and, hence, faster data collection rates. When imaging a photoactivatable fluorescent protein such as EosFP, a laser source of similar intensity with a wavelength of 561 or 568 nm may be used.

One or more additional lasers may be used for activation of the switchable fluorophores. As described above, multicolor photoswitchable activator–reporter pairs may be activated using various laser wavelengths (e.g., 405, 457, or 532 nm) when the activator dye is Alexa Fluor 405, Cy2, or Cy3, respectively. For photoactivatable fluorescent proteins, 405 nm is the most commonly used activation wavelength.

Finally, the addition of a conventional mercury arc lamp enables the user to quickly image standard fluorophores such as DAPI (4′,6-diamino-2-phenylindole), GFP, or the Cy3 signal from Cy3–Cy5 labeled samples, before starting STORM data collection.

Electronically Controlled Laser Shutters or Acousto-Optical Modulator

A system for switching the laser illumination on and off in synchrony with the data acquisition allows switchable fluorophores to be activated with a predetermined timing. This facilitates multicolor STORM imaging using fluorophores that are activated by distinct wavelengths of light (see the discussion of multicolor STORM methods later in the chapter). If the laser exposure times are synchronized with the frame rate of the camera, the data analysis software may keep track of which frames of the recorded data correspond to which illumination color. In this manner, the time at which a fluorophore switching event appears in the data set may be correlated with a particular activation wavelength. Rapid switching of the laser illumination may be achieved with an acousto-optical modulator (AOM), an acousto-optical tunable filter (AOTF), or mechanical shutters. To synchronize the laser switching with the CCD exposures, the clock signal provided by the CCD camera can be used as the timing source to control the AOM, the AOTF, or the shutters.

Computer Control

When recording STORM data, a computer equipped with a multifunction data acquisition card records image data continuously from the camera, streaming it directly to a disk. The computer also switches the laser modulators/shutters in sync with the camera frames according to a preset laser pulse sequence, and interfaces with the QPD and the piezo-stage controller to maintain the sample focus.

Laser Intensity Control

A means for controlling the laser intensity, in particular, the activation laser intensity, is required for STORM imaging. As discussed above, the fraction of activated fluorophores must be maintained at a low enough level such that the images of individual fluorophores are not typically overlapping. To achieve this condition, the intensity of the activation laser must be adjusted. A neutral-density filter wheel provides a convenient coarse laser intensity adjustment. Fine adjustments can be made by passing the laser through a half-wave plate followed by a polarizing beam-splitter cube. Rotating the wave plate rotates the laser polarization, which, in turn, modulates the amount of light passing through the beam splitter. Intensity control can also be provided by an AOM or an AOTF.

STORM Experimental Procedure

Sample Preparation

As in any fluorescence imaging experiment, the sample must first be fluorescently labeled. There are several approaches to labeling, including immunofluorescence techniques in which the sample is labeled with fluorescent antibodies and fluorescence in situ hybridization in which the probes are fluorescent nucleic acids (Spector and Goldman 2006). Any synthetic fluorophore or genetically encoded fluorescent protein may be used provided they are photoswitchable. The easiest approach to introduce photoswitchable fluorescent proteins into cells is by transfection, although the generation of stable cell lines or transgenic animals expressing fluorescent proteins may also be used and may be preferred because overexpression can adversely affect the sample. The cell or tissue samples may be imaged in either live or fixed conditions.

In general, the staining procedure must be optimized to ensure that the ultrastructure of the sample is not disrupted, to maximize the density of fluorescent labels bound to the structure of interest, and to minimize non-specific labeling. These requirements are more stringent for superresolution fluorescence imaging than they are for conventional microscopy. The high spatial resolution of methods such as STORM must be complemented by a high density of fluorescent labeling to image the sub-diffraction-limit structural features of the sample. Any fixation procedure used must also preserve the ultrastructure of the sample. Fixation protocols previously developed for EM imaging, such as rapid freezing, freeze substitution, and fixation with glutaraldehyde, may be useful in this respect (McIntosh 2007).

Imaging Medium

When using cyanine dyes as photoswitchable probes, certain components are required in the imaging medium. In particular, the imaging buffer should contain a thiol compound, such as βME (~140 mM) or MEA (~100 mM), to enable photoswitching of the dyes. The use of MEA results in a 2x increase in the rate of switching off of Cy5 and a corresponding decrease in both the number of photons collected per switching event and the steady-state fraction of spontaneously activated fluorophores (fluorophores activated by red imaging light, as discussed above). A significantly lower concentration of βME or MEA (~70 mM and ~6 mM, respectively) is used for live cell imaging to keep cells viable under this condition.

An oxygen scavenger system that reduces photobleaching also benefits STORM imaging with photoswitchable fluorescent dyes. A commonly used scavenger system contains glucose oxidase (~0.5 mg/mL), catalase (~40 µg/mL), and glucose (~10% w/v). For live cell imaging, a lower concentration of glucose (~1% w/v) is used to maintain cell viability.

Data Collection

The general procedure for STORM imaging is illustrated in Figure 2, which shows a cell in which the cytoskeleton has been labeled with switchable fluorophores. Initially, the fluorophores are prepared in the dark state (by exposure to the red laser, for example). Imaging then proceeds as follows.

1. *Activate a sparse set of fluorophores.* A brief exposure to the activation light source (e.g., a green 532-nm laser) causes some of the fluorophores on the sample to switch to the fluorescent state. Alternatively, exposure to the imaging light source can serve this purpose (see the discussion of spontaneous activation above).
2. *Image the fluorophores, and switch them off.* The sample is exposed to the imaging light source (e.g., a red 647-nm laser), and the activated fluorophores emit light, as illustrated in Figure 2B. The density of activated fluorophores is low enough that the images of the fluorophores do not overlap. The duration of the imaging light pulse is set to be long enough that most of the activated fluorophores will switch off before the next activation pulse.
3. *Activate a new set of fluorophores, and repeat.* As illustrated in Figure 2C, a different set of fluorophores is stochastically activated by the following activation light pulse. These are imaged, and the process is repeated until a sufficient number of fluorophore positions has been determined to map out the high-resolution structure of the sample.

This simple sequence of activation and imaging is useful for illustrating the basic concept of STORM, but any experimental procedure that allows for sparse subsets of fluorophores to be independently imaged over time may be used. For example, as an alternative to Steps 1–3, the imaging light source and the activation light source may be kept on continually so that imaging and activation are happening throughout the data collection. In this case, the activation light intensity is adjusted to be sufficiently low such that only a small fraction of the fluorophores in the field of view are activated at any moment in time and their images are not overlapping.

In general, the intensity of the imaging laser determines the rate at which the fluorophores switch off. To maximize the signal-to-noise ratio of the data, the camera exposure duration is adjusted such that, on average, the fluorophore on-time is equal to one exposure time of the camera. In this manner, almost all of the light from each fluorophore switching event is collected in a single camera frame, minimizing the noise associated with reading out a CCD image (Bock et al. 2007). In practice, the data collection speed is limited by either one of the maximum camera frame rate or the maximum rate at which the fluorophores can be cycled on and off.

The intensity of the activation laser is adjusted to control the density of fluorophores activated during each cycle. As the experiment proceeds and some fluorophores are permanently photobleached, the activation light intensity may be increased to maintain the activated density at an optimal level.

Spontaneous activation of the fluorophores (e.g., activation by the imaging laser) results in a baseline density of activated fluorophores that depends on the total density of fluorophores bound to the sample (see above). One can take advantage of spontaneous activation to record a STORM data set without the requirement for a separate activation light source—the sample may simply be exposed to the imaging light source, and the switching events may be recorded over time (Zhuang 2009). However, in cases of very concentrated labeling, the density of spontaneously activated fluorophores may be high enough that the images of the activated fluorophores are constantly overlapping, preventing accurate localization. In this situation, it may be necessary to bleach some fraction of the fluorophores by exposure to intense illumination or to reduce the density of labeling during sample preparation.

STORM DATA ANALYSIS

During the STORM experiment, a series of images is recorded showing individual fluorophores switching on and off in time. A STORM image is typically generated from a sequence of between 500 and 100,000 image frames, 256 x 256 pixels each, recorded at a rate of 20–500 Hz. The large volume of data generated in these experiments necessitates efficient automated data analysis software. For this purpose, a custom data analysis and visualization software package was written, and this section describes its principal algorithms. Additional analysis steps for the generation of multicolor or 3D images are described in later sections.

Peak Identification and Fitting

The first task of the analysis software is to identify the fluorescent molecules in each image of the movie and to fit them to determine their precise locations. This task is very similar to the well-studied problem of automated star analysis in astronomical data sets; hence, we can take advantage of algorithms that have already been developed (Stetson 1987). The problem is made simpler because the intensity profile of each fluorophore's image is identical, corresponding to the PSF of the microscope.

Image Filtering and Peak Identification

Each image frame is first convolved with a Gaussian kernel: a unit height Gaussian having the same width as the expected PSF and lowered to have a zero integral (Stetson 1987). This convolution removes high-frequency noise and low-frequency background intensity variations. The image is then thresholded and is searched for local maxima, which are identified as peaks. Figure 5 shows a portion of an image frame taken from a STORM movie. The left panel shows the raw data, and the right panel shows the same image with the identified peaks boxed in white.

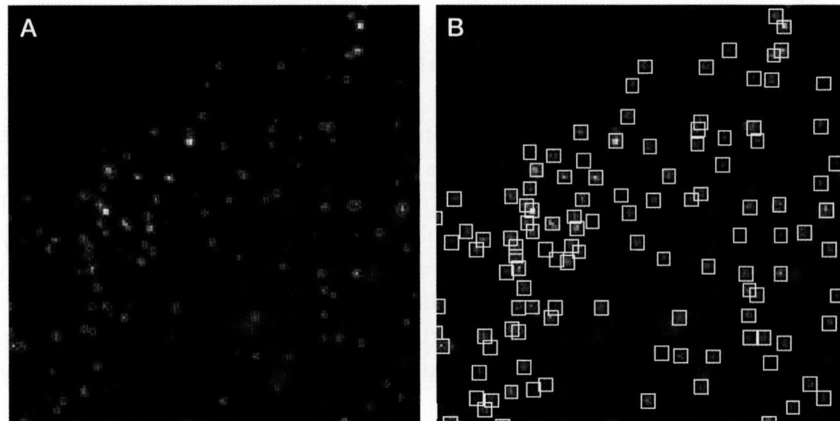

FIGURE 5. A portion of an image frame taken from a STORM movie. (*A*) The bright spots in the image correspond to the fluorescence emission of individual fluorophores. Most of the fluorescent labels are switched off such that the fluorescent molecules are well separated and distinguishable. (*B*) An automated peak finding algorithm is used to identify the peaks. The identified peaks are marked by the white boxes. The areas enclosed by the white boxes indicate the regions of the image that will be fit to determine the positions of the fluorescent molecules that give rise to the peaks.

Peak Fitting

A square region of $n \times n$ pixels ($n = 5$–7) centered on each peak is passed to a nonlinear least-squares fitting algorithm, which attempts to fit a Gaussian function to the data. The data are weighted assuming Poisson noise for the counts in each pixel. Two consecutive fits are performed for each peak. First, the data are fit to a continuous ellipsoidal Gaussian:

$$I(x,y) = A_0 + I_0 \exp\left\{-\frac{1}{2}\left[\left(\frac{x'}{a}\right) + \left(\frac{y'}{b}\right)\right]\right\},$$

where

$$x' = (x - x_0)\cos\theta - (y - y_0)\sin\theta,$$

$$y' = (x - x_0)\sin\theta + (y - y_0)\cos\theta.$$

Here, A_0 is the background fluorescence level, I_0 is the amplitude of the peak, a and b define the widths of the Gaussian distribution along the x and y directions, x_0 and y_0 describe the center coordinates of the peak, and θ is the tilt angle of the ellipse relative to the pixel edges. Based on this seven-parameter fit, we compute a peak ellipticity parameter defined as $\max(a,b)/\min(a,b)$. This measure of the ellipticity is used to identify and to reject peaks corresponding to the overlapping images of two or more closely spaced fluorophores, as the image of a single fluorophore is expected to have circular symmetry (i.e., $a \approx b$). This constraint is relaxed for 3D STORM analysis, as described in the section on 3D STORM.

The second fit step is used to determine the peak center position, the amplitude, and the width. The data are fit to a rotationally symmetric Gaussian function:

$$I(x,y) = A_0 + I_0 \exp\left\{-\frac{1}{2}\left[\left(\frac{x - x_0}{\sigma}\right)^2 + \left(\frac{y - y_0}{\sigma}\right)^2\right]\right\},$$

where σ is the width of the peak. The total number of counts collected in the peak is calculated as $2\pi\sigma_2 I_0$, and this number is then converted to photoelectrons and, thus, the number of photons detected, based on the calibrated camera gain settings used during imaging.

Filtering

A fraction of the peaks identified for peak fitting will correspond to two or more activated fluorophores located in very close proximity such that their image appears as a single peak. This fraction is minimized by activating only a sparse set of fluorophores on the sample at any one time; but, nonetheless, further measures are necessary to avoid including in the final image the incorrect fluorophore positions resulting from fitting peaks corresponding to multiple fluorophores.

For this purpose, several of the fit parameters are stored for each peak and are subsequently used to filter out the erroneous peaks. For each identified peak, the following fit parameters are stored: center position, width, ellipticity, and number of photons detected. These values are compared against a set of filtering criteria, and any peak falling outside the range of acceptable width, ellipticity, and brightness is rejected from further analysis because it is unlikely to yield a precise fluorophore localization.

Trail Generation

The time at which a fluorophore is activated until the time it switches off will often span multiple image frames, generating multiple sets of fit results. Grouping these data together allows a more precise estimate of the fluorophore position. Peaks appearing in consecutive image frames with a relative displacement smaller than one camera pixel are considered to originate from the same fluorescent molecule, and their center positions are grouped into a data structure that we refer to as a trail. The final position of the molecule is calculated as the average of the center positions across the entire trail, weighted by the number of photons detected in each frame.

Drift Correction

An important factor that limits the resolution of a STORM image is sample drift. If the fluorescence image of the sample shifts with respect to the CCD camera pixels during the course of the experiment, the relative positions of fluorophores localized at different points in time are no longer accurate. Any source of mechanical drift in the microscope, such as thermal expansion and contraction of the microscope frame or vibration of the sample stage, can result in such an image shift.

A means of measuring sample drift allows the localized positions of the fluorophores to be corrected during data analysis. We have found two methods to be particularly useful in correcting for mechanical drift. The first method involves adding fiducial markers (e.g., fluorescent beads or gold particles) to the sample, which bind to the glass substrate and are visible in the fluorescence image. These spots can then be tracked for the duration of the movie, and their averaged motion as a function of time is then subtracted from the measured localizations, yielding a drift-corrected image.

The second method involves determining the drift from the STORM data set itself. The data set is divided into a number of equal-time segments, and the STORM image for each time slice is rendered. By calculating the correlation function between the STORM images generated from consecutive time slices, the sample drift during that time window may be determined. After the drift as a function of time has been calculated for all time slices, the drift is then subtracted from each localization to generate the drift-corrected image. This latter approach is useful for fixed samples but not for live samples with time-dependent variations.

STORM Image Rendering

After the data analysis is complete, the final result is a list of fluorophore positions that have passed the peak filtering criteria. The processed data are stored in this form, and STORM images of the data are dynamically generated as the user pans and zooms the image to a given region and magnification level. Dynamic rendering avoids the problem of having to render and to store the entire image area at the highest magnification: a 40 × 40-µm sample region rendered at 1 nm per pixel would yield a 40,000 × 40,000-pixel image, requiring 4.5 Gbyte of computer memory to view.

When the image is generated, each localization is assigned as one point in the STORM image. These points are either represented either by a small marker (e.g., a cross) or rendered as an intensity peak. For example, each peak can be represented as a Gaussian intensity profile with unit volume, and its width can be scaled to correspond to the localization uncertainty (Thompson et al. 2002), based on the number of photons collected for that switching event. In this sense, each intensity peak corresponds to a probability density function for finding the fluorophore at that position in space (Betzig et al. 2006). The summed peaks from all of the localizations then form an intensity image of the sample.

STORM IMAGING OF CULTURED CELLS

Fluorescence imaging of tissue cells in culture is a widely used method for cell and molecular biology (Spector and Goldman 2006), which would benefit significantly from increased spatial resolution. Many parts of the cell have structural features spanning a range of length scales from nanometers to micrometers, and much of this range is beyond the resolution capabilities of conventional microscopy. Moreover, nanometer-scale image resolution would enable precise colocalization analysis for biomolecular complexes.

As an example of fixed cell imaging with STORM, microtubules were labeled by indirect immunofluorescence (Bates et al. 2007). Microtubules are filamentous components of the cytoskeleton of the cell, important for many cellular functions including cell division and intracellular cargo trafficking. Microtubule filaments extend throughout the cell, appearing as long and narrow fibers in fluorescence images. These well-defined structures provide an ideal test sample for demonstrating STORM image resolution. In the following example, Green monkey kidney epithelial (BS-C-1) cells were immunostained with primary antibodies and then with activator–reporter-labeled secondary antibodies. Cy3 was used as the activator and Alexa Fluor 647, a cyanine dye with very similar structural, spectral and photoswitching properties to Cy5, was used as the reporter. The STORM images were obtained by activating and deactivating the Alexa Fluor 647 fluorophores with repeated pulses of green (532 nm) and red light (647 nm).

As illustrated in Figure 6, the STORM image shows a substantial improvement in the resolution of the microtubule network as compared with the conventional fluorescence image. In the regions in which microtubules were densely packed and undefined in the conventional image, individual microtubule filaments were clearly resolved by STORM.

To determine the image resolution more quantitatively, we identified point-like objects in the cell, appearing as small clusters of localizations away from any discernible microtubule filaments. These clusters likely represent individual antibodies not specifically attached to the cell. The width (FWHM) of the localization clusters was 24 ± 1 nm, corresponding to a standard deviation of 10 nm, a measure of the fluorophore localization precision in this experiment. In terms of resolution, therefore, two fluorophores spaced 24 nm apart are resolvable. This represents a resolution improvement of an order of magnitude over the conventional optical microscope.

Another widely used method to fluorescently label cellular structures takes advantage of fluorescent proteins, which can be genetically fused to proteins of interest. For STORM imaging of cellular structures using this labeling approach, the target protein in cells may be labeled with photoswitchable fluorescent proteins (Betzig et al. 2006). Figure 7 shows a comparison of conventional and STORM images of mEos2-labeled vimentin in a cell (Zhuang 2009). The protein mEos2 is a monomeric variant of the Eos fluorescent protein that can be photoactivated from a green-emitting form into a red-emitting form (McKinney et al. 2009). The STORM image was taken by activating mEos2 with a 405-nm laser and imaging the activated mEos2 with a 568-nm laser. Again, a substantial resolution improvement is observed for the STORM image as compared with its corresponding conventional image.

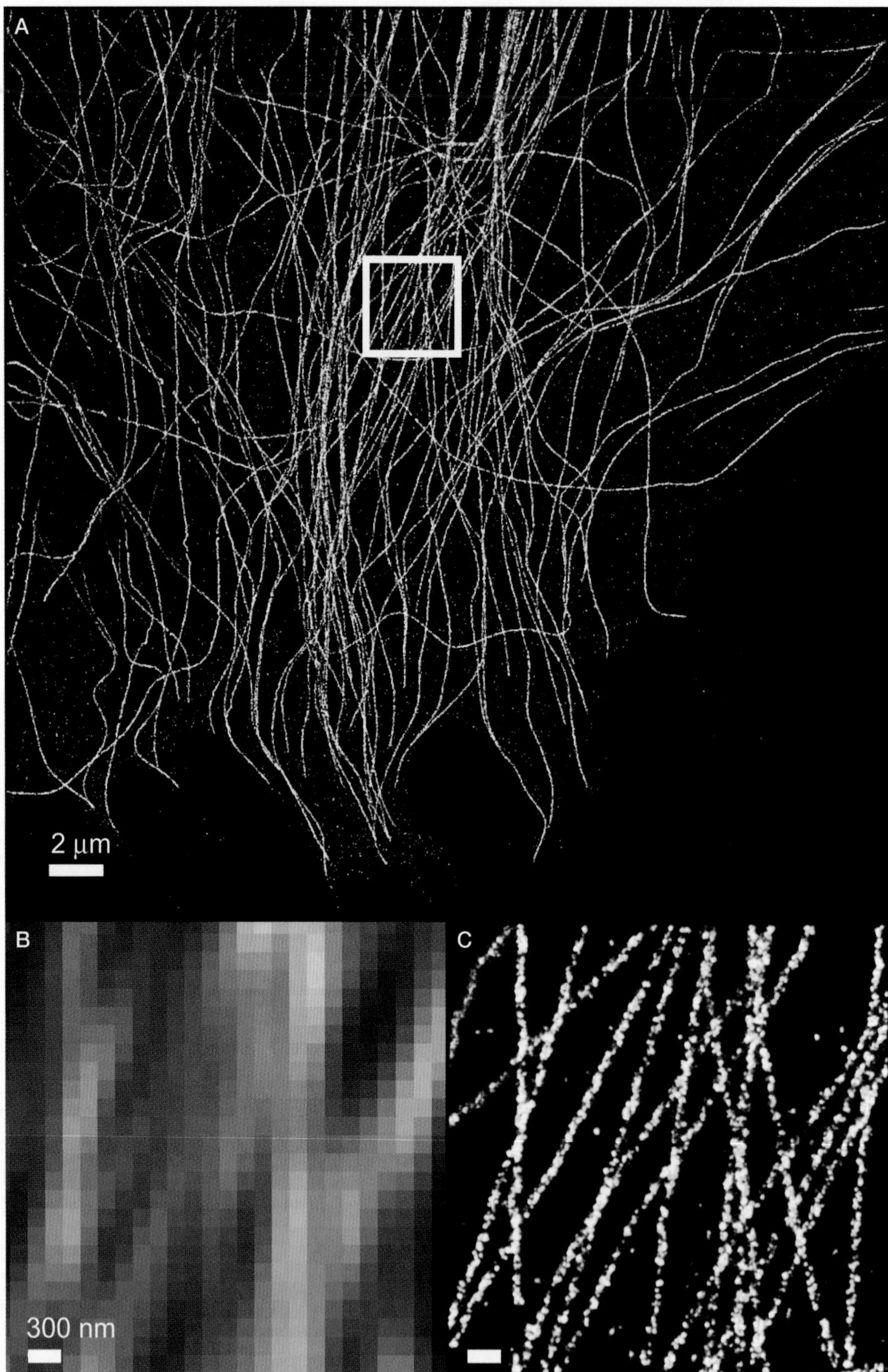

FIGURE 6. STORM image of microtubules in a large area of a BS-C-1 cell. (*A*) The microtubules are immunostained with Cy3-Alexa 647–labeled antibodies. Each fluorophore localization is plotted as a yellow peak with a Gaussian intensity profile and width scaled according to the photon-number-dependent localization precision. The total number of localizations plotted in the image is ~2.4 x 10^6. A magnified view of the boxed region is shown in *B* and *C*. (*B*) Conventional fluorescence image of the boxed region in *A*. To obtain the conventional image, the cell was illuminated with a mercury arc lamp, and the Cy3 channel was imaged directly. (*C*) STORM image of the boxed region in *A*. In comparison with the conventional image in *B*, the STORM image reveals a substantial improvement in spatial resolution.

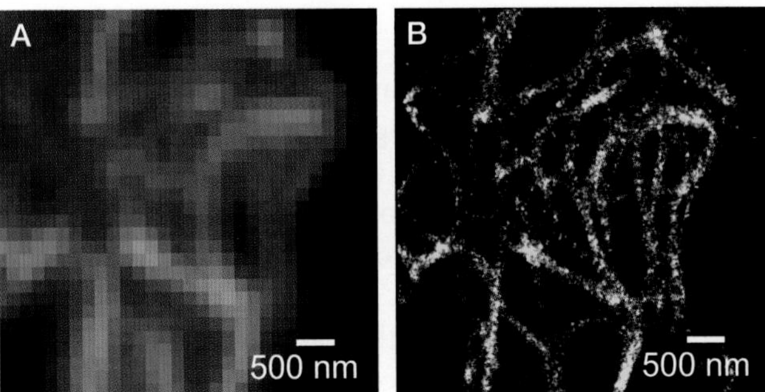

FIGURE 7. STORM imaging of vimentin filaments labeled with photoswitchable fluorescent proteins in a mammalian cell. (*A*) Conventional image of vimentin in a BS-C-1 cell. (*B*) STORM image of the same region. The vimentin filaments are labeled with the photoswitchable fluorescent protein mEos2. (Adapted from Zhuang 2009.)

MULTICOLOR STORM

Conventional multicolor fluorescence imaging relies on the spectral separation of the emission spectrum of each type of fluorophore present in the sample. This approach to multicolor imaging is equally applicable to STORM, using photoswitchable fluorophores with distinct emission wavelengths. For example, among the cyanine dyes, three red-emitting dyes have been shown to display photoswitchable fluorescence: Cy5, Cy5.5, and Cy7 (Bates et al. 2007). Because of the separation of their emission spectra, these dyes can be used for multicolor STORM imaging using a multichannel fluorescence detection scheme. Similarly, other photoswitchable fluorescent dyes and fluorescent proteins are available for multicolor imaging applications (see Table 1). In the case of photoswitchable fluorescent proteins, two-color imaging has been shown using combinations of mEos and Dronpa (Shroff et al. 2007), PA-mCherry and PA-GFP (Subach et al. 2009), and Dronpa with bs-Dronpa (Andresen et al. 2008), for example.

Photoswitchable fluorophores, however, offer a second parameter for multicolor detection: the activation wavelength. The example below shows a multicolor STORM imaging scheme based on the distinct activation wavelengths of different photoswitchable fluorophores. Fluorescent probes with identical emission spectra can be distinguished by the wavelength of the light that causes them to be activated to the fluorescent state.

Experimental Procedure and Data Analysis

When using the activation wavelength as a means to distinguish different photoswitchable fluorophores, the time point in the data set at which the fluorophore is switched on is used to assign a color. For this multicolor imaging scheme, STORM data are acquired by alternately exposing the sample to pulses of activation light and imaging light. In a two-color experiment, for example, the sample may be exposed to the following repeating pattern of illumination pulses: green laser on for a duration of one camera exposure (frame), red laser for three frames, blue laser for one frame, and red laser for three frames. Fluorophores are switched on during the green or blue activation pulses then imaged and switched off when the sample is exposed to red light. If the fluorophore is first detected in an image frame that immediately follows an activation pulse, the fluorophore is assumed to have been activated by that pulse, and a color is assigned accordingly. If detected in a frame not immediately following an activation pulse, the fluorophore may have been activated spontaneously by the imaging laser, and it is impossible to know which activator dye is paired with the reporter. In this case, the color information is not known, but the localization may be assigned to a nonspecific category because its position can still aid in discerning the structure of the sample.

Color Cross Talk

Any incorrect color identification of fluorophores results in cross talk between the color channels. When the activation wavelength is used to distinguish between different types of photoswitchable fluorophores, color cross talk arises mainly because of spontaneous activation of the reporter dye by the imaging laser. Fluorophore activation induced by the imaging laser can be identified if the switching event was first detected in a frame that does not immediately follow an activation pulse. A spontaneous activation event may also occur during the frame immediately following an activation laser pulse, however, and, in this case, the color may be incorrectly assigned.

The frequency of these incorrect assignments will depend on the frequency of nonspecific activation events as compared with color-specific activation events that occur in response to an activation pulse. In practice, this will depend, in part, on the local density of fluorophores in the sample because at higher densities, the activation light intensity must be reduced to maintain the activated fluorophores at a spatially sparse level. In general, this source of cross talk is minimized when the camera frame rate is matched to the average on-time of the dye, and the activation laser intensities are adjusted to maximize the number of specifically activated dyes without increasing the density of activated fluorophores beyond the point in which their images are overlapping. A typical cross-talk level is 10%–30% and can be further corrected statistically in the image-processing stage (Bates et al. 2007).

Multicolor STORM Example

As an example of multicolor STORM imaging, Figure 8 shows a two-color image of microtubules and clathrin-coated pits (CCPs), cellular structures involved in receptor-mediated endocytosis (Bates et al. 2007). Microtubules and clathrin were immunostained with primary antibodies and then with activator–reporter-labeled secondary antibodies. The activator–reporter pairs used were Cy2-Alexa Fluor 647 for microtubules and Cy3-Alexa Fluor 647 for clathrin. Two different laser wavelengths, 457 and 532 nm, were used to selectively activate the two pairs. Cross talk between the two-color channels was subtracted from the image after statistical analysis (Bates et al. 2007).

The green channel (457-nm activation) revealed filamentous structures as expected for microtubules. The red channel shows predominantly circular structures, characteristic of CCPs and vesicles. Interestingly, many of the CCPs appeared to have a higher density of localizations toward the periphery, which is consistent with the two-dimensional projection of a 3D cage structure. The size distribution of CCPs can be measured directly from the image, yielding a mean size of ~180 nm, which agrees quantitatively with results obtained using EM (Heuser and Anderson 1989).

3D STORM

Although the techniques described thus far yield high-resolution information in two dimensions, most organelles and cellular structures have a 3D morphology. Three-dimensional fluorescence imaging is most commonly performed using confocal and multiphoton microscopies, the axial resolution of which is typically in the range of 500–800 nm, two to three times worse than the lateral resolution (Torok and Wilson 1997; Zipfel et al. 2003).

In this section, we introduce a method for 3D STORM imaging. Although the lateral position of a particle can be determined from the centroid of its image, the shape of the image contains information about the particle's axial (z) position. Nanoscale localization precision in the z dimension has been previously achieved by introducing defocusing (van Oijen et al. 1998; Speidel et al. 2003; Prabhat et al. 2006; Toprak and Selvin 2007) or astigmatism (Kao and Verkman 1994; Holtzer et al. 2007) into the image without significantly compromising the lateral position measurement. The STORM technique can be extended to 3D imaging by making use of these methods to determine the 3D coordinates for each fluorescent label with high precision.

The examples given here illustrate applications of the astigmatism imaging approach. Other methods for 3D fluorophore localization with application to nanoscale imaging have also been shown,

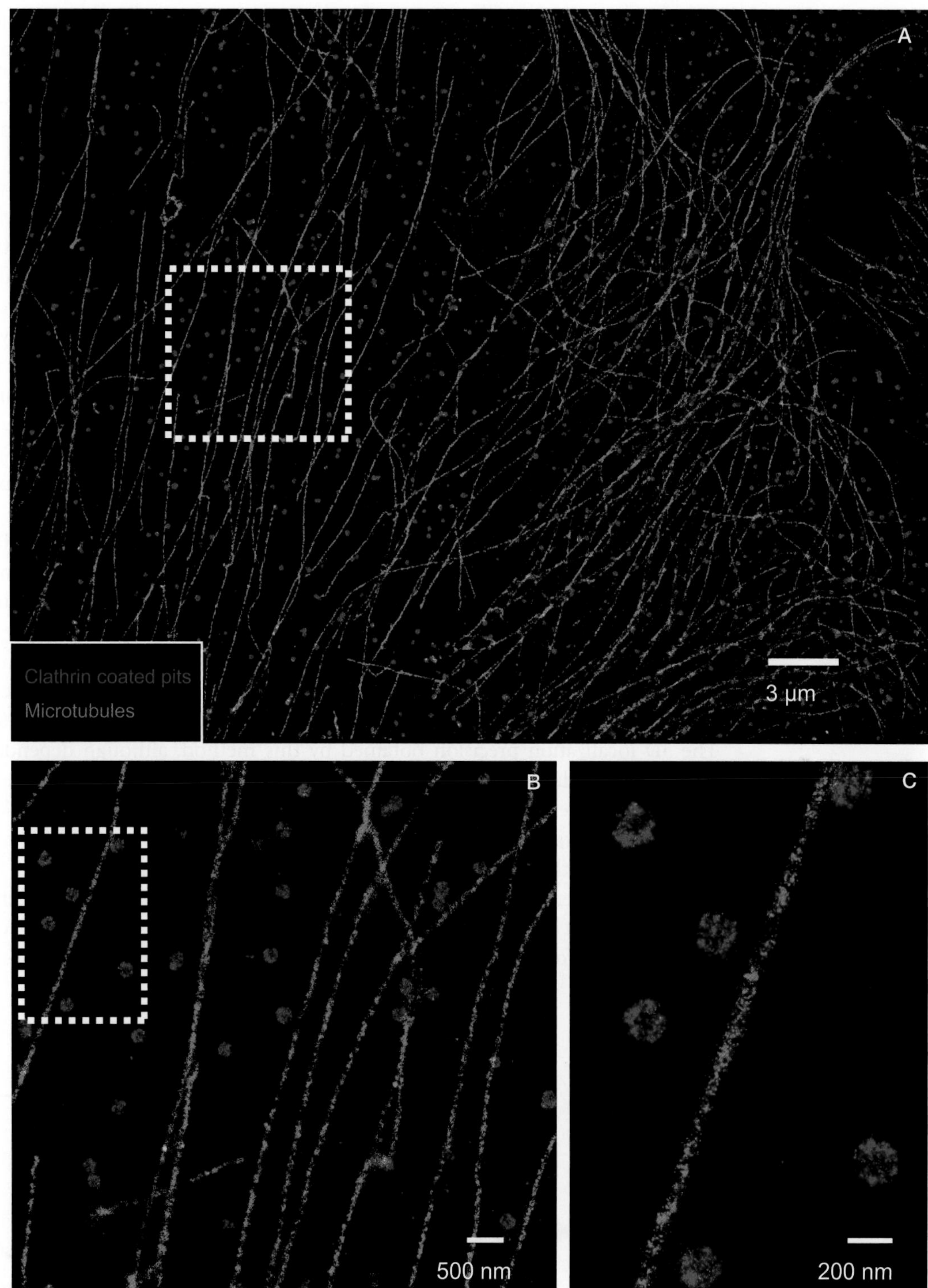

FIGURE 8. Two-color STORM imaging of microtubules and CCPs in a mammalian cell. (*A*) STORM image of a large area of a BS-C-1 cell. The secondary antibodies used for microtubule staining were labeled with Cy2 and Alexa Fluor 647, whereas those for clathrin were labeled with Cy3 and Alexa Fluor 647. Two laser wavelengths, 457 and 532 nm, were used to selectively activate the two pairs. Each localization was falsely colored according to the following code: green for 457-nm activation and red for 532-nm activation. (*B*) STORM image corresponding to the boxed region in *A* shown at a higher magnification. (*C*) A further magnified view of the boxed region in *B*. (Adapted, with permission, from Bates et al. 2007, ©AAAS.)

including multifocal-plane imaging (Juette et al. 2008), PSF engineering to create a double-helix PSF (Pavani et al. 2009), and interferometric methods using two objective lenses (von Middendorff et al. 2008; Shtengel et al. 2009).

Astigmatism Imaging: A Method for 3D Fluorophore Localization

The first implementation of 3D STORM uses the astigmatism imaging method to localize each fluorophore in 3D space (Huang et al. 2008b). This involves the insertion of a weak cylindrical lens into the imaging path to create two slightly different focal planes for the *x* and *y* directions as illustrated in Figure 9A. As a result of this modification to the imaging optics, the images of individual fluorophores typically appear elliptical, and the ellipticity varies as the position of the fluorophore changes in *z*. When the fluorophore is located in the average focal plane approximately halfway between the *x* and *y* focal planes, the image appears round; when the fluorophore is above the average focal plane, its image is more focused in the *y* direction than in the *x* direction and, thus, appears ellipsoidal with its long axis along the *x* axis; conversely, when the fluorophore is below the focal plane, the image appears ellipsoidal with its long axis along the *y* axis. The peak center position corresponds to the *x–y* position of the fluorophore as before, and a measurement of the peak widths along the *x*- and *y*-axes w_x and w_y allows the *z* coordinate of the fluorophore to be unambiguously determined.

The calibration curves of w_x and w_y as a function of *z*, shown in Figure 9B, are experimentally determined for single Alexa Fluor 647 fluorophores bound to a glass surface and scanned in *z* using a piezo-driven sample stage. These curves are used in the data analysis procedure to determine the *z* coordinate of each photoactivated fluorophore by comparing the measured w_x and w_y values of its image with the calibration data. In addition, for samples immersed in aqueous solution on a glass substrate, a small correction is applied to the *z* localization to quantitatively account for the refractive index mismatch between glass and water (Egner and Hell 2006; Huang et al. 2008a,b).

The 3D localization precision obtained by this method, although dependent on many factors unique to an individual microscope setup, is normally sufficient for STORM over a range of ~300 nm above and below the focal plane. This allows high-resolution imaging of samples in a region several hundred nanometers in thickness (Huang et al. 2008b). When combined with *z* stepping of the sample stage relative to the objective, the imaging depth can be increased to several micrometers (Huang et al. 2008a). Another important aspect to consider is the spherical aberration generated by the refractive index mismatch between the sample and the objective. Generally speaking, oil-immersion objectives are preferable for STORM imaging because of their high NA. Oil objectives may pose a problem

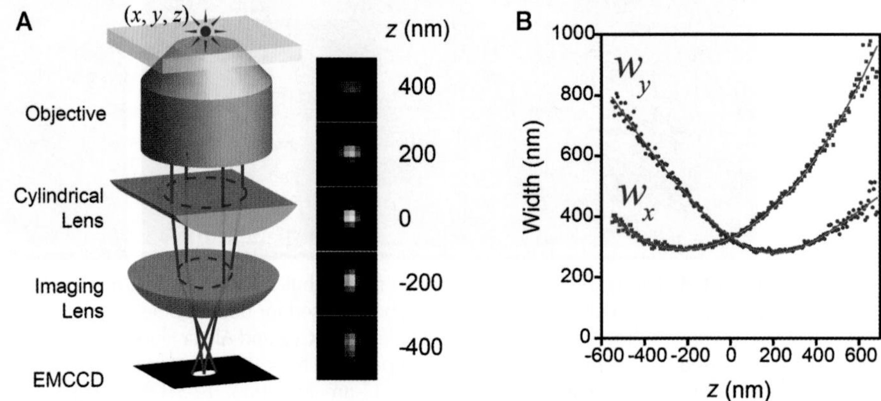

FIGURE 9. Three-dimensional localization of individual fluorophores. (*A*) This simplified optical diagram illustrates the principle of determining the *z* coordinate of a fluorescent object from the ellipticity of its image by introducing a cylindrical lens into the imaging path. The right panel shows the images of a fluorophore at various *z*-positions along the optical axis. (*B*) The calibration curve of the image widths w_x and w_y as a function of *z* obtained from single molecules of Alexa Fluor 647. (Adapted, with permission, from Huang et al. 2008b, ©AAAS.)

when imaging a sample in an aqueous medium, however, as spherical aberration becomes an increasing concern higher above the glass–water interface. Methods have been described to correct imaging aberrations up to ~3 µm into the sample (Huang et al. 2008a). Beyond this limit, it is advisable to switch to a water-immersion objective or to use index-matched media with the oil-immersion objective, despite the other disadvantages these options may present such as decreased light collection efficiency in the former case or incompatibility with live cell imaging in the latter case.

3D STORM Imaging Examples

Three-dimensional STORM images of CCPs provide a clear demonstration of the ability of this method to resolve a sub-diffraction-limit sized cellular structure with a complex 3D morphology. For 3D imaging of CCPs, a direct immunofluorescence strategy was used in which primary antibodies against clathrin were doubly labeled with Cy3 and Alexa Fluor 647. As shown in Figure 10A, when imaged by conventional fluorescence microscopy, all CCPs appeared as nearly diffraction-limited spots with no discernible structure. When imaged using STORM, the 3D structure of the pits becomes apparent (Figure 10B–D). Figure 10B shows an *x–y* cross section of the data, taken from a region near the opening of the pits at the cell surface. The circular ring-like structure of the pit periphery is unambiguously resolved. Moreover, consecutive *x–y* and *x–z* cross sections of the pits (Figure 10C) clearly reveal the half-spherical cage-like morphology of these structures. The spatial resolution of this image was determined to be ~20 nm in the *x–y* directions and ~50 nm in the *z* direction (Huang et al. 2008b).

Whole-Cell 3D STORM Imaging of Mitochondria

The imaging depth of a single 3D STORM image is typically several hundred nanometers in the *z* dimension, as the localization precision of fluorophores at large distances from the focal plane is not sufficient to produce a high-resolution image. However, a typical mammalian cell may span several micrometers, or more, in height. To obtain the STORM image of the entire cell, one can combine the STORM imaging process with *z* stepping of the sample stage relative to the objective lens. Figure 11

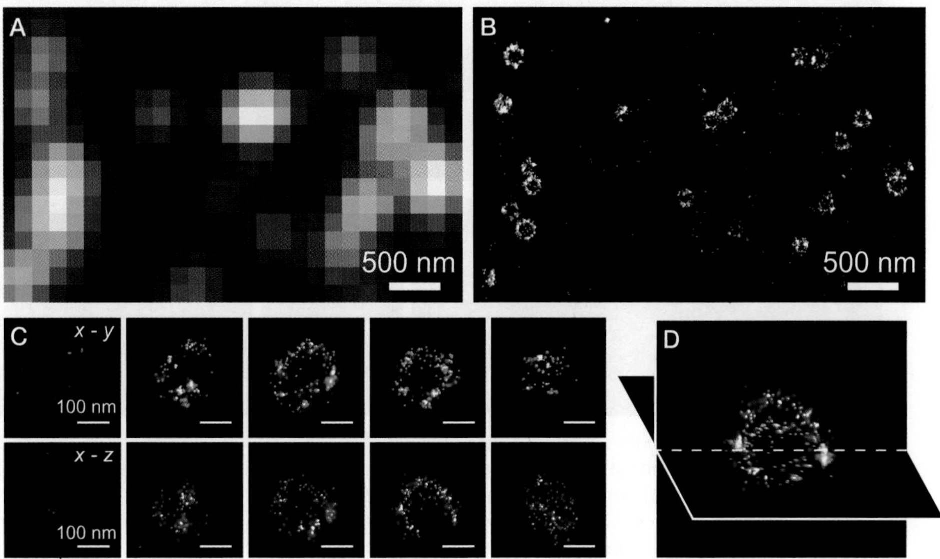

FIGURE 10. Three-dimensional STORM imaging of CCPs in a cell. (*A*) Conventional direct immunofluorescence image of clathrin in a region of a BS-C-1 cell. (*B*) An *x–y* cross section (50-nm thick in *z*) of the 3D STORM image of the same area, showing the ring-like structure of the periphery of the CCPs at the plasma membrane. (*C*) Serial *x–y* cross sections (each 50-nm thick in *z*) and *x–z* cross sections (each 50-nm thick in *y*) of a CCP. (*D*) An *x–y* and *x–z* cross section presented in 3D perspective, showing the half-spherical cage-like structure of the pit. (Adapted, with permission, from Huang et al. 2008b, ©AAAS.)

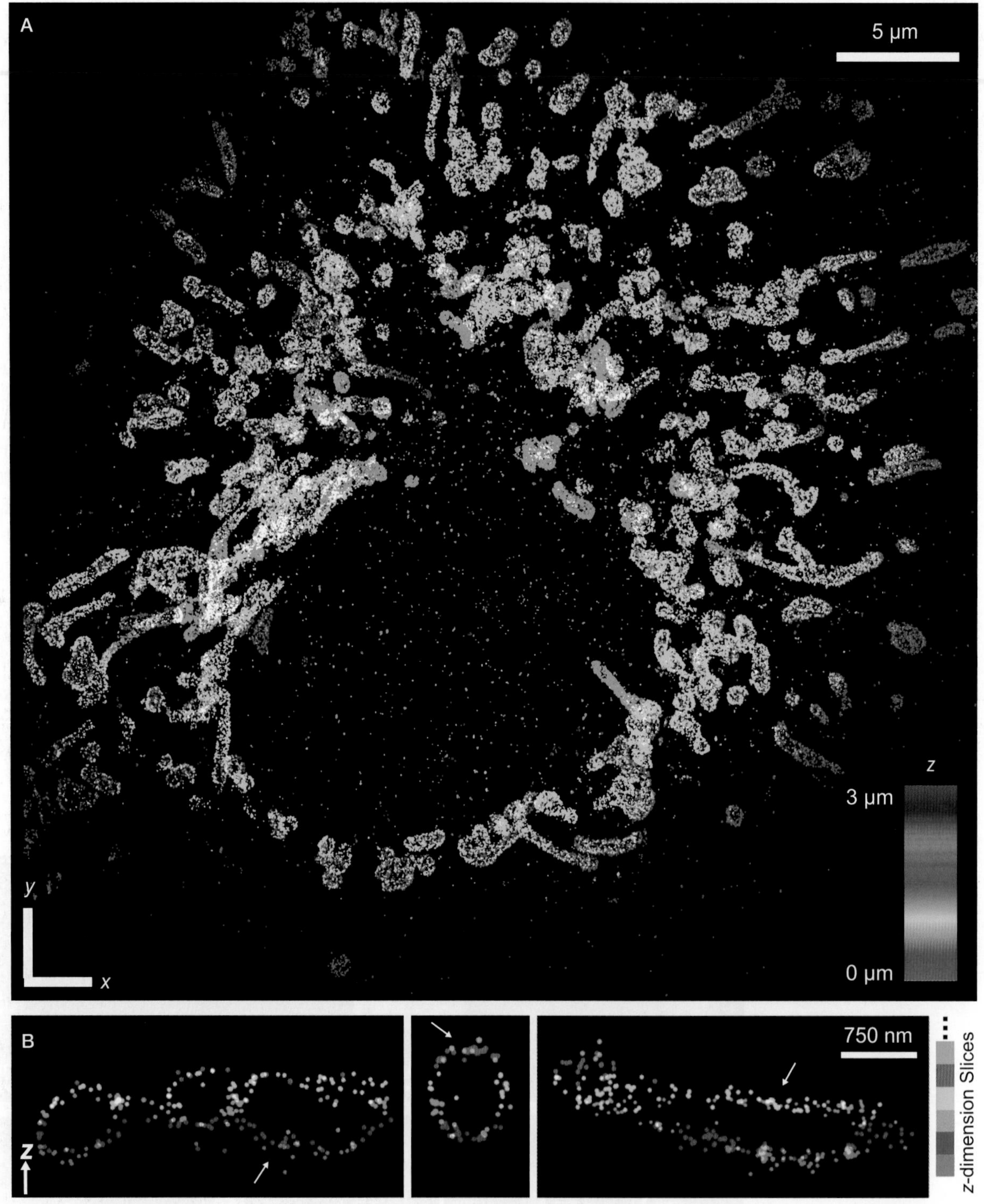

FIGURE 11. Whole-cell 3D STORM imaging of the mitochondrial network in a BS-C-1 cell. (*A*) A 3D STORM image of the mitochondrial outer membrane. Axial stepping of the objective was used to obtain an imaging depth of 3 µm. The *z*-dimension position is encoded according to the inset color bar. (*B*) Magnified vertical cross sections of several mitochondria from the image in *A*. Localizations from each z slice are uniquely colored by the slice in which they were originally recorded with the color index bar shown on the right. The arrows indicate regions of overlap between adjacent z slices of the mitochondrion in which the alignment error in *z* is estimated to be ~18 nm after correction for spherical aberration. (Adapted from Huang et al. 2008a.)

shows a 3D STORM image of the whole mitochondrial network of a mammalian cell, which extends throughout the cell body. Each localization is colored by its z height as indicated by the color bar. The cell was immunostained with a primary antibody against a mitochondrial outer membrane protein Tom20 and a secondary antibody labeled with Alexa Fluor 405 and Cy5. To achieve the imaging depth of 3 µm, the sample was scanned in z with ~300-nm steps, and STORM images were taken at each scan step. As the imaging depth increases, the effect of spherical aberration caused by the refractive index mismatch between the imaging medium and the oil-immersion objective becomes more severe. Only localizations within a range of 350 nm below the focal plane were accepted to minimize the effect of spherical aberration, as the localization precision above the focal plane is subjected to substantially larger degradation because of spherical aberration as compared with that below the focal plane (Huang et al. 2008a). These image slices were then combined to create a 3D image of the entire cell. As shown in Figure 11B, the hollow shape of the outer mitochondrial membrane is clearly visible in the STORM image. The arrows highlight the alignment of regions with significant overlap that were originally recorded in different z slices, illustrating the accuracy of the image alignment and stacking. Other strategies to reduce spherical aberration include the use of a water-immersion objective or an index-matched imaging medium, as discussed above.

SUMMARY

Recent years have witnessed rapid progress in sub-diffraction-limit fluorescence imaging, facilitated by the development of fluorescent probes with novel properties such as photoswitchable fluorescence emission. In this chapter, we have discussed the theory and the implementation of STORM, a method for superresolution imaging based on the high accuracy localization of individual fluorophores. This method yields fluorescence images with spatial resolution an order of magnitude finer than the classical diffraction limit of optical microscopy, and even higher image resolution is possible. STORM requires no specialized apparatus apart from a fluorescence microscope and a sensitive CCD camera. Furthermore, the demonstration of multicolor and 3D STORM illustrates the potential to use this approach for biological imaging applications. Although not discussed in detail here, STORM can also be used to image live cells, enabling the observation of the nanoscale dynamics of cellular structures, with both photoswitchable fluorescent proteins (Biteen et al. 2008; Shroff et al. 2008) and photoswitchable dyes (M. Bates, S.A. Jones, and X. Zhuang, unpubl. data). Superresolution techniques have also found application in prokaryotic biology, in which the small size of bacteria makes these samples difficult to visualize with conventional optical microscopy (Kim et al. 2006; Biteen et al. 2008; Greenfield et al. 2009). Finally, photoswitchable fluorescent probes also facilitate high-density particle tracking measurements in live cells (Hess et al. 2007; Manley et al. 2008). Together with other high-resolution imaging techniques, we expect that superresolution fluorescence microscopy will be broadly applied to biological research and will bring about new insights into life at the nanometer scale.

Protocol A

Transfection of Genetically Encoded Photoswitchable Probes

Determining which photoswitchable fluorescent protein to use from the growing list (Table 1) should be based on several criteria such as brightness, contrast ratio, switching mechanism, and oligomerization state. Most often brightness and contrast ratio are the primary determining factors in selection, as resolution depends critically on them. However, depending on the desired application, one should also consider the emission color of the fluorophore and the mechanism of switching—that is, whether reversible isomerization, allowing for multiple localizations from a single protein, or irreversible backbone break, resulting in only one localization per protein. Finally, many photoswitchable fluorescent proteins are, in their native state, multimers. Mutated variants of most of these proteins have been generated to avoid artificial aggregation of the target protein, but many retain some affinity for self-oligomerization at high concentrations. After selection of an appropriate fluorescent protein, standard molecular biology techniques may be used to generate a plasmid containing the sequence of the photoswitchable protein linked to the gene of interest.

Once the plasmid has been generated and has been verified, it can be introduced into cells via any standard means of gene delivery, such as lipofection or electroporation. Optimal conditions will vary considerably for different cell lines and plasmids. Here, we present an example protocol for the transfection of BS-C-1 cells with an mEos2-vimentin plasmid using the lipid-based reagent FuGENE6.

MATERIALS

Reagents

BS-C-1 cells
Culture medium, free of antibiotics
FuGENE6 (Roche, 11 815 091 001)
mEos2-vimentin plasmid
Minimum Essential Medium (Invitrogen, 51200-038)

Equipment

Culture dish, for example, eight-well chambered cover glass (LabTek II, Nalge Nunc International), 0.5-mL well volume

METHOD

1. Approximately 24 h before transfection, plate BS-C-1 cells in culture medium free of antibiotics such that they are ~70% confluent on transfection.
2. Prepare transfection mixture by diluting 3-µL FuGENE6 in 97-µL Minimum Essential Medium. Mix well, and incubate at room temperature for 5 min.
3. Add 1-µg plasmid to the diluted FuGENE6. Mix well, and incubate at room temperature for 15–20 min.

4. Add the transfection complexes to the cells. Add dropwise and in a circular pattern around the culture well to ensure even distribution. Actual volume added will depend on the size of the culture dish (see manufacturers instructions), but for an eight-well-chambered cover glass, add 25-µL transfection mixture per well.
5. Return cells to the incubator for ~24–30 h.
6. Assay cells for fluorescence and expression level.

DISCUSSION

Expression level is a point of particular care when using genetically encoded fluorescent proteins. In mammalian cells, transfection commonly relies on the overexpression of a given construct without removing the endogenous protein, which may alter the ultrastructure of the cell and induce artifacts in the structures visualized. Careful controls must be performed to ensure that the overall system is not significantly perturbed and that the imaged structure is representative of the endogenous configuration.

Protocol B

Preparation of Photoswitchable Labeled Antibodies

Many synthetic fluorescent dye molecules show photoswitchable fluorescence emission (see the section on Photoswitchable Fluorescent Molecules). In particular, photoswitchable cyanine fluorophores such as Cy5, Alexa 647, Cy7, etc., may be paired with a second fluorophore, which serves as an activator, determining the wavelength of light that re-activates the fluorescence of the photoswitchable molecule (Bates et al. 2007). This protocol describes the preparation of antibodies labeled with one such pairing scheme of synthetic fluorophores, Alexa 405 and Alexa 647. It may easily be adapted for labeling with other fluorophores or for labeling other substrate molecules.

MATERIALS

CAUTION: See Appendix 6 for proper handling of materials marked with <!>.

Reagents

Alexa Fluor 405 dye (Invitrogen, A-30000)
Alexa Fluor 647 dye (Invitrogen, A-20006)
Dimethylsulfoxide (DMSO) <!> (anhydrous)
$NaHCO_3$ solution (0.5 M, pH ~8.5)
Phosphate-buffered saline (PBS)
Secondary antibody (Jackson ImmunoResearch Laboratories, donkey anti-mouse, 1.25-mg/mL stock)

Equipment

Size exclusion column (Illustra NAP5, GE Healthcare), preequilibrated with PBS
Ultraviolet–visible (UV/Vis) spectrophotometer

METHOD

Dye Aliquot Preparation

1. Dissolve one 1.0-mg tube of Alexa Fluor 405 dye in 100 µl of anhydrous DMSO, and divide into 50 aliquots of 0.02 mg each. Dry, and store at –20°C.
2. Dissolve one 1.0-mg tube of Alexa Fluor 647 dye in 100 µl of anhydrous DMSO, and divide into 50 aliquots of 0.02 mg each. Dry, and store at –20°C.

Antibody Labeling

3. Dissolve the Alexa Fluor 647 dye aliquot (from Step 1) in 40 µL of DMSO.
4. Dissolve the Alexa Fluor 405 dye aliquot (from Step 2) in 10 µL of DMSO.
5. Add 40 µL secondary antibody to 10 µL $NaHCO_3$ solution (0.5 M, pH ~8.5).

6. Add 1.0 µL of Alexa 647 and 5 µL of Alexa 405 solution to the reaction, and mix thoroughly.
7. Incubate the reaction for 30 min at room temperature in the dark (with gentle rocking).
8. Add PBS to bring the final volume of the reaction to 200 µL.
9. Load the reaction onto a size exclusion column, preequilibrated with PBS.
10. Collect 300-µL fractions in PBS. The labeled antibody is expected to elute in fraction #3 off the NAP5 column. Verify the peak fraction and the degree of labeling using a UV/Vis spectrophotometer. Store the labeled antibody fraction at 4°C.

DISCUSSION

These reaction conditions should yield a labeling ratio of ~3 Alexa 405 dyes per antibody and 0.7 Alexa 647 dyes per antibody (on average) as determined by UV/Vis absorbance measurements. The labeling ratio can be adjusted by varying the amount of each dye that is added. Typically, we aim for a degree of labeling of 0.1–1.0 photoswitchable reporter dyes (e.g., Alexa 647) and 2.0–4.0 activator dyes (e.g., Alexa 405) per antibody molecule. The final concentration of antibody in the recovered fraction should be ~0.1 mg/mL, and this may be used at a dilution of ~1:200 for immunofluorescence labeling.

ACKNOWLEDGMENTS

We kindly thank M.V. Bujny for her help in preparing Figure 2 and G.T. Dempsey for his assistance in preparing Figures 2, 4, and 7 and also in compiling data for Table 1.

REFERENCES

Abbe E. 1873. Beitrage zur theorie des mikroskops und der mikroskopischen wahrnehmung. *Arch Mikroskop Anat* **9:** 413–420.

Ando R, Hama H, Yamamoto-Hino M, Mizuno H, Miyawaki A. 2002. An optical marker based on the UV-induced green-to-red photoconversion of a fluorescent protein. *Proc Natl Acad Sci* **99:** 12651–12656.

Ando R, Flors C, Mizuno H, Hofkens J, Miyawaki A. 2007. Highlighted generation of fluorescence signals using simultaneous two-color irradiation on Dronpa mutants. *Biophys J* **92:** L97–L99.

Andresen M, Stiel AC, Folling J, Wenzel D, Schonle A, Egner A, Eggeling C, Hell SW, Jakobs S. 2008. Photoswitchable fluorescent proteins enable monochromatic multilabel imaging and dual color fluorescence nanoscopy. *Nat Biotechnol* **26:** 1035–1040.

Bates M, Blosser TR, Zhuang X. 2005. Short-range spectroscopic ruler based on a single-molecule optical switch. *Phys Rev Lett* **94:** 108101.

Bates M, Huang B, Dempsey GT, Zhuang X. 2007. Multicolor super-resolution imaging with photo-switchable fluorescent probes. *Science* **317:** 1749–1753.

Bates M, Huang B, Zhuang X. 2008. Super-resolution microscopy by nanoscale localization of photo-switchable fluorescent probes. *Curr Opin Chem Biol* **12:** 505–514.

Belov VN, Bossi ML, Fölling J, Boyarskiy VP, Hell SW. 2009. Rhodamine spiroamides for multicolor single-molecule switching fluorescent nanoscopy. *Chemistry* **15:** 10762–10776.

Betzig E, Patterson GH, Sougrat R, Lindwasser OW, Olenych S, Bonifacino JS, Davidson MW, Lippincott-Schwartz J, Hess HF. 2006. Imaging intracellular fluorescent proteins at nanometer resolution. *Science* **313:** 1642–1645.

Biteen JS, Thompson MA, Tselentis NK, Bowman GR, Shapiro L, Moerner WE. 2008. Super-resolution imaging in live *Caulobacter crescentus* cells using photoswitchable EYFP. *Nat Methods* **5:** 947–949.

Bock H, Geisler C, Wurm CA, Von Middendorff C, Jakobs S, Schonle A, Egner A, Hell SW, Eggeling C. 2007. Two-color far-field fluorescence nanoscopy based on photoswitchable emitters. *Appl Phys B* **88:** 161–165.

Chudakov DM, Verkhusha VV, Staroverov DB, Souslova EA, Lukyanov S, Lukyanov KA. 2004. Photoswitchable cyan fluorescent protein for protein tracking. *Nat Biotechnol* **22:** 1435–1439.

Churchman LS, Okten Z, Rock RS, Dawson JF, Spudich JA. 2005. Single molecule high-resolution colocalization of Cy3 and Cy5 attached to macromolecules measures intramolecular distances through time. *Proc Natl Acad Sci* **102:** 1419–1423.

Cui BX, Wu CB, Chen L, Ramirez A, Bearer EL, Li WP, Mobley WC, Chu S. 2007. One at a time, live tracking of NGF axonal transport using quantum dots. *Proc Natl Acad Sci* **104:** 13666–13671.

Dempsey GT, Bates M, Kowtoniuk WE, Liu DR, Tsien RY, Zhuang X. 2009a. Photoswitching mechanism of cyanine dyes. *J Am Chem Soc* **131:** 18192–18193.

Dempsey GT, Wang W, Zhuang X. 2009b. Fluorescence imaging at sub-diffraction-limit resolution with stochastic optical reconstruction microscopy. In *Handbook of single-molecule biophysics* (ed. Hinterdorfer P, van Oijen AM), pp. 95–127. Springer Science+Business Media, New York.

Dickson RM, Cubitt AB, Tsien RY, Moerner WE. 1997. On/off blinking and switching behaviour of single molecules of green fluorescent protein. *Nature* **388:** 355–358.

Egner A, Hell SW. 2006. Abberations in confocal and multi-photon fluorescence microscopy induced by refractive index mismatch. In *Handbook of biological confocal microscopy* (ed. Pawley JB), pp. 404–413. Springer, New York.

Elowitz MB, Surette MG, Wolf P-E, Stock J, Leibler S. 1997. Photoactivation turns green fluorescent protein red. *Curr Biol* **7:** 809–812.

Fernandez-Suarez M, Ting AY. 2008. Fluorescent probes for super-resolution imaging in living cells. *Nat Rev Mol Cell Biol* **9:** 929–943.

Folling J, Belov V, Kunetsky R, Medda R, Schonle A, Egner A, Eggeling C, Bossi M, Hell SW. 2007. Photochromic rhodamines provide nanoscopy with optical sectioning. *Angew Chem Int Ed Engl* **46:** 6266–6270.

Gelles J, Schnapp BJ, Sheetz MP. 1988. Tracking kinesin-driven movements with nanometre-scale precision. *Nature* **331:** 450–453.

Ghosh RN, Webb WW. 1994. Automated detection and tracking of individual and clustered cell surface low density lipoprotein receptor molecules. *Biophys J* **66:** 1301–1318.

Giepmans BN, Adams SR, Ellisman MH, Tsien RY. 2006. The fluorescent toolbox for assessing protein location and function. *Science* **312:** 217–224.

Gordon MP, Ha T, Selvin PR. 2004. Single-molecule high-resolution imaging with photobleaching. *Proc Natl Acad Sci* **101:** 6462–6465.

Greenfield D, McEvoy AL, Shroff H, Crooks GE, Wingreen NS, Betzig E, Liphardt J. 2009. Self-organization of the *Escherichia coli* chemotaxis network imaged with super-resolution light microscopy. *PLoS Biol* **7:** e1000137. doi: 10.1371/journal.pbio.1000137.

Gurskaya NG, Verkhusha VV, Shcheglov AS, Staroverov DB, Chepurnykh TV, Fradkov AF, Lukyanov S, Lukyanov KA. 2006. Engineering of a monomeric green-to-red photoactivatable fluorescent protein induced by blue light. *Nat Biotechnol* **24:** 461–465.

Habuchi S, Ando R, Dedecker P, Verheijen W, Mizuno H, Miyawaki A, Hofkens J. 2005. Reversible single-molecule photoswitching in the GFP-like fluorescent protein Dronpa. *Proc Natl Acad Sci* **102:** 9511–9516.

Habuchi S, Tsutsui H, Kochaniak AB, Miyawaki A, van Oijen AM. 2008. mKikGR, a monomeric photoswitchable fluorescent protein. *PLoS One* **3:** e3944. doi: 1371/journal.pone.0003944.

Heilemann M, Margeat E, Kasper R, Sauer M, Tinnefeld P. 2005. Carbocyanine dyes as efficient reversible single-molecule optical switch. *J Am Chem Soc* **127:** 3801–3806.

Heintzmann R, Gustafsson MGL. 2009. Subdiffraction resolution in continuous samples. *Nat Photon* **3:** 362–364.

Hell SW. 2007. Far-field optical nanoscopy. *Science* **316:** 1153–1158.

Hell SW, Schmidt R, Egner A. 2009. Diffraction-unlimited three-dimensional optical nanoscopy with opposing lenses. *Nat Photon* **3:** 381–387.

Hess ST, Girirajan TP, Mason MD. 2006. Ultra-high resolution imaging by fluorescence photoactivation localization microscopy. *Biophys J* **91:** 4258–4272.

Hess ST, Gould TJ, Gudheti MV, Maas SA, Mills KD, Zimmerberg J. 2007. Dynamic clustered distribution of hemagglutinin resolved at 40 nm in living cell membranes discriminates between raft theories. *Proc Natl Acad Sci* **104:** 17370–17375.

Heuser JE, Anderson RGW. 1989. Hypertonic media inhibit receptor-mediated endocytosis by blocking clathrin-coated pit formation. *J Cell Biol* **108:** 389–400.

Holtzer L, Meckel T, Schmidt T. 2007. Nanometric three-dimensional tracking of individual quantum dots in cells. *Appl Phys Lett* **90:** 053902.

Huang B, Jones SA, Brandenburg B, Zhuang X. 2008a. Whole-cell 3D STORM reveals interactions between cellular structures with nanometer-scale resolution. *Nat Methods* **5:** 1047–1052.

Huang B, Wang W, Bates M, Zhuang X. 2008b. Three-dimensional super-resolution imaging by stochastic optical reconstruction microscopy. *Science* **319:** 810–813.

Huang B, Bates M, Zhuang X. 2009. Super-resolution fluorescence microscopy. *Annu Rev Biochem* **78:** 993–1016.

Irie M, Fukaminato T, Sasaki T, Tamai N, Kawai T. 2002. A digital fluorescent molecular photoswitch. *Nature* **420:** 759.

Jerram P, Pool PJ, Bell R, Burt DJ, Bowring S, Spencer S, Hazelwood M, Moody I, Catlett N, Heyes PS. 2001. The LLCCD: Low-light imaging without the need for an intensifier. *Proc SPIE* **4306:** 178–186.

Juette MF, Gould TJ, Lessard MD, Mlodzianoski MJ, Nagpure BS, Bennett BT, Hess ST, Bewersdorf J. 2008. Three-dimensional sub-100 nm resolution fluorescence microscopy of thick samples. *Nat Methods* **5:** 527–529.

Kao HP, Verkman AS. 1994. Tracking of single fluorescent particles in three dimensions: Use of cylindrical optics to encode particle position. *Biophys J* **67:** 1291–1300.

Kim SY, Gitai Z, Kinkhabwala A, Shapiro L, Moerner WE. 2006. Single molecules of the bacterial actin MreB undergo directed treadmilling motion in *Caulobacter crescentus*. *Proc Natl Acad Sci* **103:** 10929–10934.

Koster AJ, Klumperman J. 2003. Electron microscopy in cell biology: Integrating structure and function. *Nat Rev Mol Cell Biol* (suppl.) **2003:** SS6–10.

Kulzer F, Kummer S, Matzke R, Brauchle C, Basche T. 1997. Single-molecule optical switching of terrylene in *p*-terphenyl. *Nature* **387:** 688.

Lacoste TD, Michalet X, Pinaud F, Chemla DS, Alivisatos AP, Weiss S. 2000. Ultrahigh-resolution multicolor colocalization of single fluorescent probes. *Proc Natl Acad Sci* **97:** 9461–9466.

Lagerholm BC, Averett L, Weinreb GE, Jacobson K, Thompson NL. 2006. Analysis method for measuring submicroscopic distances with blinking quantum dots. *Biophys J* **91:** 3050–3060.

Lidke K, Rieger B, Jovin T, Heintzmann R. 2005. Superresolution by localization of quantum dots using blinking statistics. *Opt Express* **13:** 7052–7062.

Lord SJ, Conley NR, Lee HD, Nishimura SY, Pomerantz AK, Willets KA, Lu Z, Wang H, Liu N, Samuel R, et al. 2009. DCDHF fluorophores for single-molecule imaging in cells. *ChemPhysChem* **10:** 55–65.

Manley S, Gillette JM, Patterson GH, Shroff H, Hess HF, Betzig E, Lippincott-Schwartz J. 2008. High-density mapping of single-molecule trajectories with photoactivated localization microscopy. *Nat Methods* **5:** 155–157.

Marchant JS, Stutzmann GE, Leissring MA, LaFerla FM, Parker I. 2001. Multiphoton-evoked color change of DsRed as an optical highlighter for cellular and subcellular labeling. *Nat Biotech* **19:** 645–649.

McIntosh JR. 2007. *Cellular electron microscopy*. Elsevier/Academic Press, Amsterdam/Boston, MA.

McKinney SA, Murphy CS, Hazelwood KL, Davidson MW, Looger LL. 2009. A bright and photostable photoconvertible fluorescent protein. *Nat Methods* **6:** 131–133.

Patterson GH, Lippincott-Schwartz J. 2002. A photoactivatable GFP for selective photolabeling of proteins and cells. *Science* **297:** 1873–1877.

Pavani SRP, Thompson MA, Biteen JS, Lord SJ, Liu N, Twieg RJ, Piestun R, Moerner WE. 2009. Three-dimensional, single-molecule fluorescence imaging beyond the diffraction limit by using a double-helix point spread function. *Proc Natl Acad Sci* **106:** 2995–2999.

Pawley JB, ed. 2006. *Handbook of biological confocal microscopy*. Springer, New York.

Prabhat P, Ram S, Ward ES, Ober RJ. 2006. Simultaneous imaging of several focal planes in fluorescence microscopy for the study of cellular dynamics in 3D. *Proc SPIE* **6090:** 60900–60901.

Qu X, Wu D, Mets L, Scherer NF. 2004. Nanometer-localized multiple single-molecule fluorescence microscopy. *Proc Natl Acad Sci* **101:** 11298–11303.

Rust MJ, Bates M, Zhuang X. 2006. Sub-diffraction-limit imaging by stochastic optical reconstruction microscopy (STORM). *Nat Methods* **3:** 793–795.

Shroff H, Galbraith CG, Galbraith JA, White H, Gillette J, Olenych S, Davidson MW, Betzig E. 2007. Dual-color super-resolution imaging of genetically expressed probes within individual adhesion complexes. *Proc Natl Acad Sci* **104:** 20308–20313.

Shroff H, Galbraith CG, Galbraith JA, Betzig E. 2008. Live-cell photoactivated localization microscopy of nanoscale adhesion dynamics. *Nat Methods* **5:** 417–423.

Shtengel G, Galbraith JA, Galbraith CG, Lippincott-Schwartz J, Gillette JM, Manley S, Sougrat R, Waterman CM, Kanchanawong P, Davidson MW, et al. 2009. Interferometric fluorescent super-resolution microscopy resolves 3D cellular ultrastructure. *Proc Natl Acad Sci* **106:** 3125–3130.

Spector DL, Goldman RD, eds. 2006. *Basic methods in microscopy: Protocols and concepts from* Cells: A laboratory manual. Cold Spring Harbor Laboratory Press, Cold Spring Harbor, NY.

Speidel M, Jonas A, Florin EL. 2003. Three-dimensional tracking of fluorescent nanoparticles with subnanometer precision by use of off-focus imaging. *Opt Lett* **28:** 69–71.

Stetson PB. 1987. DAOPHOT—A computer program for crowded field stellar photometry. *Publ Astron Soc Pac* **99:** 191.

Stiel AC, Trowitzsch S, Weber G, Andresen M, Eggeling C, Hell SW, Jakobs S, Wahl MC. 2007. 1.8 Å bright-state structure of the reversibly switchable fluorescent protein Dronpa guides the generation of fast switching variants. *Biochem J* **402:** 35–42.

Subach FV, Patterson GH, Manley S, Gillette JM, Lippincott-Schwartz J, Verkhusha VV. 2009. Photoactivatable mCherry for high-resolution two-color fluorescence microscopy. *Nat Methods* **6:** 153–159.

Thompson RE, Larson DR, Webb WW. 2002. Precise nanometer localization analysis for individual fluorescent probes. *Biophys J* **82:** 2775–2783.

Tokunaga M, Imamoto N, Sakata-Sogawa K. 2008. Highly inclined thin illumination enables clear single-molecule imaging in cells. *Nat Methods* **5:** 159–161.

Toprak E, Selvin PR. 2007. New fluorescent tools for watching nanometer-scale conformational changes of single molecules. *Annu Rev Biophys Biomol Struct* **36:** 349–369.

Torok P, Wilson T. 1997. Rigorous theory for axial resolution in confocal microscopes. *Opt Commun* **137:** 1270135.

Tsutsui H, Karasawa S, Shimizu H, Nukina N, Miyawaki A. 2005. Semi-rational engineering of a coral fluorescent protein into an efficient highlighter. *EMBO Rep* **6:** 233–238.

van Oijen AM, Kohler J, Schmidt J, Muller M, Brakenhoff GJ. 1998. 3-dimensional super-resolution by spectrally selective imaging. *Chem Phys Lett* **292:** 183–187.

von Middendorff C, Egner A, Geisler C, Hell SW, Schönle A. 2008. Isotropic 3D nanoscopy based on single emitter switching. *Opt Express* **16:** 20774–20788.

Wiedenmann J, Ivanchenko S, Oswald F, Schmitt F, Rocker C, Salih A, Spindler KD, Nienhaus GU. 2004. EosFP, a fluorescent marker protein with UV-inducible green-to-red fluorescence conversion. *Proc Natl Acad Sci* **101:** 15905–15910.

Yildiz A, Forkey JN, McKinney SA, Ha T, Goldman YE, Selvin PR. 2003. Myosin V walks hand-over-hand: Single fluorophore imaging with 1.5-nm localization. *Science* **300:** 2061–2065.

Zhuang X. 2009. Nano-imaging with STORM. *Nat Photon* **3:** 365–367.

Zipfel WR, Williams RM, Webb WW. 2003. Nonlinear magic: Multiphoton microscopy in the biosciences. *Nat Biotechnol* **21:** 1369–1377.

36 Imaging Live Cells Using Quantum Dots

Jyoti K. Jaiswal[1] and Sanford M. Simon[2]

[1]Childen's National Medical Center, Washington, D.C. 20010; [2]The Rockefeller University, New York, New York 10065

ABSTRACT

Quantum dots (QDs) are nanoparticles with fluorescent properties that offer advantages over organic fluorophores. As a result, QDs have found wide application in biological imaging. In this chapter we discuss the approaches for using QDs for labeling and imaging individual cells and cellular processes in live cells both in vivo and in culture.

INTRODUCTION

Quantum dots (QDs) are nanometer-scale particles that, like other fluorescent molecules, absorb photons of light at one wavelength and emit photons at a different wavelength. However, the physicochemical characteristic of QDs is very different from conventional fluorophores. A QD consists of a core made of two or more semiconductors with several layers of coating (usually zinc sulfide). The semiconductor materials are elements paired from groups II and VI, III and V, or IV and VI in the chemical periodic table and frequently include cadmium, selenium, or tellurium. QDs are semiconductors whose electronic structure is closely related to the size and shape of the individual crystal because of quantum confinement. Simply put, quantum confinement describes the condition in which, because of small size of the QD crystal, the distance an electron can move on being excited is smaller than the Bohr radii. Photoabsorption results in promotion of an electron to an excited state, which initially undergoes rapid relaxation, losing some energy. At longer timescales, the excited state decays back to the ground state by emitting a photon (fluorescence). The photon energies required for absorption and emission are critically dependent on the size of the crystals produced, so it is possible to have very precise control over the fluorescent properties of the material.

The material difference between QDs and conventional fluorophores results in different physical and fluorescent properties. Some of these properties make QDs a better choice for certain biological applications, whereas other properties limit their usefulness compared with conventional fluorophores for different biological applications. The following section outlines some properties that distinguish QDs from other fluorophores. Whether a particular feature is advantageous will often depend on the intended use.

- *Brightness:* Each individual QD is several orders of magnitude brighter than most individual organic fluorophores (Wu et al. 2003).
- *Spectral characteristics:* QDs differ from conventional fluorophores in both their emission spectra and their excitation spectra. The emission spectra of QDs are narrow compared with conventional fluorophores (full width at half-maximal intensity of <30 nm vs. ~100 nm) (Jaiswal and Simon 2007). This is advantageous when simultaneously monitoring the emission of multiple fluorophores. The excitation spectrum of a QD is considerably broader than that of conventional fluorophores. As a result, it is possible to use an excitation wavelength that is well separated from the emission wavelength. Also, one excitation wavelength can be used to excite multiple QDs having different emission spectra (Jaiswal and Simon 2007). In some circumstances—specifically, when there is a need to use a minimal part of the spectrum for excitation to maximize the wavelengths available for collecting emission data—this feature of QDs is advantageous. In other circumstances, such as when there is a need to use separate excitation lines to selectively excite different fluorophores, using organic fluorophores is advantageous.
- *Photostability and resistance to metabolic degradation:* QDs are orders of magnitude more photostable than conventional fluorophores (Jaiswal et al. 2003). Additionally, because they are inorganic, they are resistant to metabolic degradation for periods ranging from weeks to months. This is a significant advantage when tracking cells. However, for some other applications, such as FRAP (fluorescence recovery after photobleaching) and superresolution techniques that depend on the ability to photobleach the fluorophores, this feature is disadvantageous.
- *Universal approaches for conjugating biomolecules:* QDs differ from each other with respect to the size of their core. However, their surface chemistry is the same, which allows biomolecules of interest to be conjugated to any QD using the same approach (Michalet et al. 2005). This can be advantageous when there is a need to change the fluorophore conjugated to a protein of interest without having to be concerned about altering the specificity. However, this interferes with the ability to simultaneously tag different biomolecules each with a different fluorophore.
- *Size:* The core of a QD ranges from 3 to 10 nm in size. Although QDs that emit light below 585 nm are round, those emitting at higher wavelengths appear rod-shaped (Deerinck et al. 2007). However, the coatings that are applied to make QDs biofunctional yield commercial QDs all being >20 nm in size. This size is not a limitation if the goal is to tag tissues (e.g., sentinel lymph nodes; Kim et al. 2004) or track cells (e.g., detect tumor cells in tissue; Voura et al. 2004) in vivo. However, if the goal is to label individual biomolecules, then the size of the QDs becomes a significant impediment.
- *Valency:* It has been difficult to develop QD surface chemistry so that the QDs can be linked monovalently to biomolecules. Multivalency simplifies the construction of a fluorescent probe having many molecules on the surface of the QD, such as a biosensor; but multivalency is clearly a problem if the goal is to label single molecules.
- *Delivery into cells:* The size and inorganic nature of QDs makes it difficult to deliver them into the cytosol. The use of "cell penetrating peptides" leaves a large percentage of the QDs in the endocytic system. Additionally, QDs tend to aggregate in the cytosol, limiting their use for various live cell–based applications.

In the sections below we present approaches for using QDs for imaging live cells and discuss considerations in using QDs for these purposes.

APPLICATION 1: LABELING LIVE CELLS WITH QDs

Labeling Methods

QDs are particularly useful for long-term imaging of live cells in situ because of their photostability, the large spectral shift between excitation and emission (which helps reduce the contribution of autofluorescence), and their large two-photon action cross section, which greatly facilitates imaging with multiphoton excitation. There are multiple methods for labeling live cells with QDs by directly delivering them into cells. Many of these approaches rely on the endocytic ability of the cells. Labeled cells can remain fluorescent for weeks, and the label does not produce any detectable adverse effects on the physiology of the cell (Jaiswal et al. 2003). Other labeling approaches involve breaching the cell membrane to enable delivery of QDs directly into the cytoplasm.

Direct QD Endocytosis

Cells are incubated for a few hours with water-soluble QDs in the appropriate growth medium. Excess QDs are washed away with growth medium or an appropriate buffer. The remaining QDs enter the cells inside endosomes and remain there for extended periods of time (Jaiswal et al. 2003).

Inclusion of Cationic Lipid-Based Reagents

Like direct QD endocytosis, this approach also allows efficient and rapid labeling of cells utilizing the endocytic ability of the cell. Anionic QDs (e.g., with COOH groups at the surface) are incubated in serum-free medium containing a lipid-based transfection reagent, such as Lipofectamine 2000 (Invitrogen) or FuGENE 6 (Roche). This method works well for labeling tumor cells. Labeled cells show no obvious differences in their physiology compared with unlabeled cells (Voura et al. 2004).

Inclusion of Carrier Peptides

Peptides are included to enhance endocytic uptake of QDs. One such peptide is Pep-1, which enhances transport of protein molecules into the cell (Morris et al. 2001). Cells are incubated for 1 h in serum-free medium containing a preformed Pep1-QD complex (Jaiswal et al. 2004). Other carrier peptides include polyarginine peptide and HIV-TAT peptide.

Microinjection

Glass capillaries having submicron-sized tips are used to deliver QDs locally and directly into the cell of interest, either in vitro or in situ. This requires the use of very small amount of QDs (1–10 pmol) and is reported to have no effect on cell physiology and development (Dubertret et al. 2002).

Scrape-Loading

Instead of microinjecting QDs into individual cells, cells growing on a substrate are scraped in the presence of 10–100 pmol of QDs (Uyeda et al. 2005). The cells are then transferred to a fresh cell culture dish, allowing cells to reseal the damage to their cell membranes and recover. Compared with microinjection, scrape-loading cannot be used for in situ cell labeling and a greater number of cells die during the process, but it does allow cytoplasmic delivery of QDs into a larger population of cells.

Considerations

As with any live cell reporter, a key requirement for the use of QDs for live cell imaging is ensuring that the labeling process has little to no impact on the cells or molecules being monitored. Features of QDs that need to be considered in this respect are the following.

Surface Coating

For cell biological applications QDs must be stable in aqueous medium. The QD core itself is not stable in water, so QDs are coated with polyacrylic acid polymer, phospholipid micelles, or similar reagents to make them water stable (Michalet et al. 2005). These coatings contribute significantly to the final size of the QD, and reducing its thickness often increases nonspecific interactions of QDs with cells and other QDs, as well as reducing quantum yield (Pinaud et al. 2004). Many of the properties of water-stable QDs in solution and inside cells depend on the nature of the surface coating. Certain coatings can cause the QDs to reduce cell viability, whereas other coatings can cause QDs to aggregate or to bind nonspecifically to other biomolecules. QDs should be carefully evaluated for their toxicity to cells and the efficiency of labeling. Two commercial suppliers, Invitrogen and Evident Technology, offer water-stable QDs with their patented, standardized surface coatings.

Effect of QD Labeling on the Physiology and Viability of Cells

A large number of studies using QDs report that under typical cell growth conditions, QDs are inert and safe to use for studying QD-labeled live cells over long periods of time (Jaiswal and Simon 2004). However, some studies have also reported deleterious effects of QD labeling on specific cell types or cellular properties. Whenever QDs are used to label live cells, establish that the QD delivery method and extended labeling with QDs have no effect on cell growth and other physiological parameters even after prolonged periods of time (i.e., at least 1 d) (Jaiswal and Simon 2007).

Excitation Light

QDs can be excited by any wavelength of light that is lower than the emission wavelength, with the efficiency of excitation increasing with lower wavelengths. Thus for many in vitro experiments UV light is used for exciting QDs and some cellular studies have also been performed using UV excitation. Moreover many suppliers provide excitation filters for use with QDs that transmit 350–400-nm light. Aside from regular UV-induced damage to live cells, UV excitation also causes increased degradation of QDs, causing leaching of cadmium from the QD core, which then causes additional damage to cells. Thus, despite the reduced effectiveness of QD excitation with visible light (compared with UV light), light above 400 nm should be used for imaging QD-labeled live cells. With multiphoton excitation, it is possible to access states similar to those directly populated with UV light, but with much less photodamage to the cells (Larson et al. 2003).

APPLICATION 2: SPECIFIC LABELING OF PROTEINS

Labeling Methods

In addition to making QDs stable in aqueous medium, surface coatings also provide functional groups for conjugating biomolecules to QDs.

Labeling the Cell Surface

In this approach QDs are conjugated to biomolecules such as lectin for targeting glycoproteins or to streptavidin for binding to proteins present on the cell surface that can be conjugated to biotin. Streptavidin-QD conjugates are available commercially, and lectins can be conjugated to commercially available QDs that have been designed for that purpose. Suppliers of such QDs provide optimized protocols for conjugating lectins and other biomolecules to their QDs.

Labeling Specific Cell Proteins

Bioconjugated QDs can be used to label specific cellular proteins (Jaiswal et al. 2003) or subsets of cells using either of two different methods. In one approach, QDs are conjugated with target-protein-specific ligand or antibody, and then incubated with the cells (Fig. 1A). In the second approach, cells are incubated with a biotinylated primary antibody or ligand of interest and then, after washing, avidin-conjugated QDs are allowed to bind these molecules (Fig. 1B). Both approaches can be used

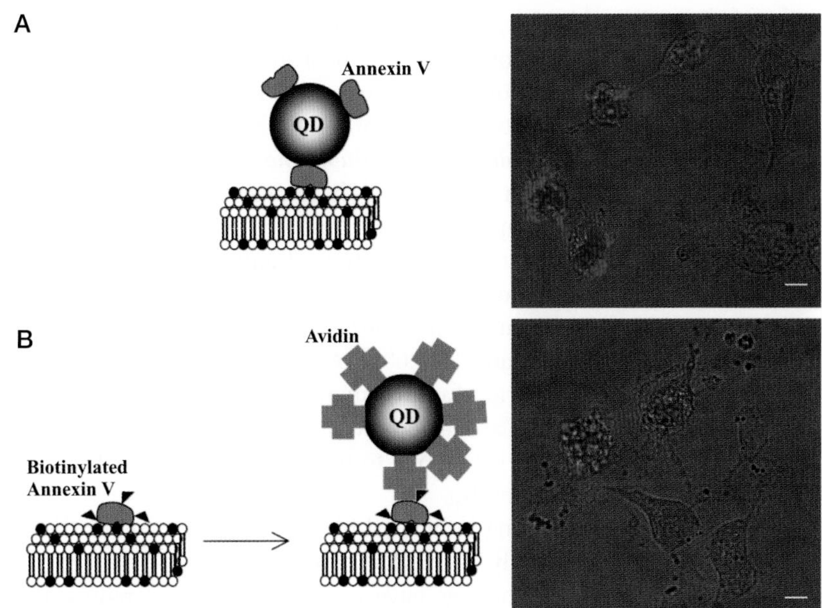

FIGURE 1. Labeling apoptotic cells using quantum dots. Phosphatidylserine (PS) externalized in cells triggered to apoptose by staurosporine treatment is labeled using (*A*) AnV-QD655 conjugate or (*B*) AnV-biotin followed by streptavidin-QD655 conjugate. The *left* panels show this schematically and the *right* panels show corresponding images of cells. In *A*, cells were incubated with 4 nM AnV-QD655 and, in *B*, cells were incubated with 4 nM QD655 conjugated to streptavidin. The images show an overlay of bright field (gray), Hoechst (blue), and QD (red) channels. Scale bar, 10 μm. For further details, see Koeppel et al. (2007).

in vitro to label specific intracellular and cell surface proteins (Jaiswal and Simon 2004; Koeppel et al. 2007). In live cells, these methods permit labeling of cell surface proteins only.

Labeling Single Molecules

Several QD features, including brightness, high photobleaching threshold, and high Stoke's shift, facilitate using QDs to track single fluorophores at a very high signal-to-noise ratio, often for extended periods of time (Dahan et al. 2003). To label single molecules, the approaches described in the previous paragraph are followed, except that lower QD concentrations should be used to minimize QD aggregation and nonspecific labeling.

Considerations

Reducing Nonspecific Binding

In addition to making QDs hydrophilic, the molecules used to coat the QD surface often include additional polymers, such as PEG (polyethylene glycol), that further suppress the tendency of QDs to bind nonspecifically to other QDs and to proteins in cell culture medium or on the cell surface. This reduces aggregation of QDs in solution and nonspecific binding on or inside the cell (Uyeda et al. 2005).

Labeling Single Molecules

Monovalent labeling: Commercially available streptavidin- or antibody-functionalized QDs contain up to 10 functional molecules per QD, causing multivalent binding of biotinylated biomolecules to a single QD (Fig. 1) (Jaiswal and Simon 2004). This is a big impediment to using QD to study single molecules, because QD multivalency can cause crosslinking of surface proteins, which is known to alter normal cellular physiology by activating unwanted signaling pathways, and reducing mobility

(Howarth et al. 2008) and normal interactions of the protein thus labeled with QD (Saxton and Jacobson 1997). Use of the F_{ab} fragment in place of antibodies can reduce the QD valency. Using a recently developed monomeric streptavidin or adopting a scheme for electrophoretic purification of monovalent QDs are two other approaches that can reduce the valency of QDs (Howarth et al. 2008).

Establishing singularity of the fluorophores: Many commonly used approaches to establish that fluorescence is from a single fluorophore cannot be used with QDs; for example, single-step photobleaching cannot be applied to QDs. Another diagnostic approach, fluorescence blinking from a single fluorophore under continuous wave illumination, is hindered by the sensitivity of QD blinking to changes in excitation intensity (Kagan et al. 1996), temperature (Banin et al. 1999), and the surrounding environment (Wang 2001). These factors alter QD blinking or under certain conditions eliminate it (Hohng and Ha 2004). As the cellular environment is highly reducing, it could make blinking a poor criterion for identifying single QDs in live cells. Similarly, it has been reported that for commercially available avidin-conjugated QDs, this interval could be as much as 100 sec (Hohng and Ha 2004), which is too long to track single QD-labeled molecules in the cytoplasm of live cells where they diffuse rapidly.

Size: When labeling individual biomolecules, the size of the biofunctional QDs (>20 nm) can contribute significantly to—perhaps even dominate—the observed properties of the molecule being studied. Improved coatings have made it possible to reduce the size of the biofunctional QDs (Howarth et al. 2008), and use of such QDs should be considered for single-molecule imaging.

DISCUSSION AND FUTURE DEVELOPMENTS

The utility and use of QDs have been steadily increasing. As described above, certain limitations remain in the use of QDs for live cell imaging. Thus, it is advisable to consider the strengths and weaknesses of QDs versus other fluorescent probes for your desired application. Despite the significant advantages offered by QDs, several improvements would increase their utility as probes within living cells. Important among these are the following:

- improved methods for conjugating QDs to biomolecules, particularly in a monomeric fashion (this would help minimize the impact of QD labeling on the functioning of the labeled protein)
- improved approaches to allow efficient delivery of QDs into the cytosol of live cells
- modifications of QD coatings that would reduce QD aggregation in the cytosol

The latter two developments would enable more widespread use of QDs for imaging intracellular molecules in live cells. There are also ongoing efforts to use QDs to assess the function of labeled biomolecules—for example, using QDs as FRET (Förster resonance energy transfer)-based reporters of enzyme activity. These developments together with the advantages that QDs already offer for live cell imaging will continue to increase the utility of QDs for live cell imaging applications.

REFERENCES

Banin U, Bruchez M, Alivisatos AP, Ha T, Weiss S, Chemla DS. 1999. Evidence for a thermal contribution to emission intermittency in single CdSe/CdS core/shell nanocrystals. *J Chem Phys* **110:** 1195–1201.

Dahan M, Levi S, Luccardini C, Rostaing P, Riveau B, Triller A. 2003. Diffusion dynamics of glycine receptors revealed by single-quantum dot tracking. *Science* **302:** 442–445.

Deerinck TJ, Giepmans BN, Smarr BL, Martone ME, Ellisman MH. 2007. Light and electron microscopic localization of multiple proteins using quantum dots. *Methods Mol Biol* **374:** 43–53.

Dubertret B, Skourides P, Norris DJ, Noireaux V, Brivanlou AH, Libchaber A. 2002. In vivo imaging of quantum dots encapsulated in phospholipid micelles. *Science* **298:** 1759–1762.

Hohng S, Ha T. 2004. Near-complete suppression of quantum dot blinking in ambient conditions. *J Am Chem Soc* **126:** 1324–1325.

Howarth M, Liu W, Puthenveetil S, Zheng Y, Marshall LF, Schmidt MM, Wittrup KD, Bawendi MG, Ting AY. 2008. Monovalent, reduced-size quantum dots for imaging receptors on living cells. *Nat Methods* **5:** 397–399.

Jaiswal JK, Simon SM. 2004. Potentials and pitfalls of fluorescent quantum dots for biological imaging. *Trends Cell Biol* **14:** 497–504.

Jaiswal JK, Simon SM. 2007. Optical monitoring of single cells using quantum dots. *Methods Mol Biol* **374:** 93–104.

Jaiswal JK, Mattoussi H, Mauro JM, Simon SM. 2003. Long-term multiple color imaging of live cells using quantum dot bioconjugates. *Nat Biotechnol* **21:** 47–51.

Jaiswal JK, Goldman ER, Mattoussi H, Simon SM. 2004. Use of quantum dots for live cell imaging. *Nat Methods* **1:** 73–78.

Kagan CR, Murray CB, Nirmal M, Bawendi MG. 1996. Electronic energy transfer in CdSe quantum dot solids. *Phys Rev Lett* **76:** 1517–1520.

Kim S, Lim YT, Soltesz EG, De Grand AM, Lee J, Nakayama A, Parker JA, Mihaljevic T, Laurence RG, Dor DM, et al. 2004. Near-infrared fluorescent type II quantum dots for sentinel lymph node mapping. *Nat Biotechnol* **22:** 93–97.

Koeppel F, Jaiswal JK, Simon SM. 2007. Quantum dot–based sensor for improved detection of apoptotic cells. *Nanomed* **2:** 71–78.

Larson DR, Zipfel WR, Williams RM, Clark SW, Bruchez MP, Wise FW, Webb WW. 2003. Water-soluble quantum dots for multiphoton fluorescence imaging in vivo. *Science* **300:** 1434–1436.

Michalet X, Pinaud FF, Bentolila LA, Tsay JM, Doose S, Li JJ, Sundaresan G, Wu AM, Gambhir SS, Weiss S. 2005. Quantum dots for live cells, in vivo imaging, and diagnostics. *Science* **307:** 538–544.

Morris MC, Depollier J, Mery J, Heitz F, Divita G. 2001. A peptide carrier for the delivery of biologically active proteins into mammalian cells. *Nat Biotechnol* **19:** 1173–1176.

Pinaud F, King D, Moore HP, Weiss S. 2004. Bioactivation and cell targeting of semiconductor CdSe/ZnS nanocrystals with phytochelatin-related peptides. *J Am Chem Soc* **126:** 6115–6123.

Saxton MJ, Jacobson K. 1997. Single-particle tracking: Applications to membrane dynamics. *Annu Rev Biophys Biomol Struct* **26:** 373–399.

Uyeda HT, Medintz IL, Jaiswal JK, Simon SM, Mattoussi H. 2005. Synthesis of compact multidentate ligands to prepare stable hydrophilic quantum dot fluorophores. *J Am Chem Soc* **127:** 3870–3878.

Voura EB, Jaiswal JK, Mattoussi H, Simon SM. 2004. Tracking metastatic tumor cell extravasation with quantum dot nanocrystals and fluorescence emission-scanning microscopy. *Nat Med* **10:** 993–998.

Wang LW. 2001. Calculating the influence of external charges on the photoluminescence of a CdSe quantum dot. *J Phys Chem B* **105:** 2360–2364.

Wu X, Liu H, Liu J, Haley KN, Treadway JA, Larson JP, Ge N, Peale F, Bruchez MP. 2003. Immunofluorescent labeling of cancer marker Her2 and other cellular targets with semiconductor quantum dots. *Nat Biotechnol* **21:** 41–46.

37 Imaging Biological Samples with Atomic Force Microscopy

Pedro J. de Pablo[1] and Mariano Carrión-Vázquez[2]

[1]*Departamento de Física de la Materia Condensada, C-III, Facultad de Ciencias, Universidad Autónoma de Madrid, 28049 Madrid, Spain;* [2]*Instituto Cajal (Consejo Superior de Investigaciones Científicas), Centro de Investigación Biomédica en Red sobre Enfermedades Neurodegenerativas (CIBERNED) and IMDEA Nanosciencia, E-28002 Madrid, Spain*

ABSTRACT

Atomic force microscopy (AFM) is an invaluable tool both for obtaining high-resolution topographical images and for determining the values of mechanical and structural properties of specimens adsorbed onto a surface. AFM is useful in an array of fields and applications, from materials science to biology. It is an extremely versatile technique that can be applied to almost any surface-mounted sample and can be operated in ambient air, ultrahigh vacuum, and, most importantly for biology, liquids. AFM can be used to explore samples ranging in size from atoms to molecules, molecular aggregates, and cells. Individual biomolecules can be viewed and manipulated at the nanoscale, providing fundamental biological information. In particular, the study of the mechanical properties of biomolecular aggregates at the nanoscale constitutes an important source of data to elaborate mechanochemical structure/function models of single-particle biomachines, expanding and complementing the information obtained from bulk experiments.

BASICS OF ATOMIC FORCE MICROSCOPY

The first thing that comes to mind on hearing the word "microscope" is an optical device that manipulates light to obtain a magnified image of a sample. As a consequence, the first question that is usually posed when seeing an atomic force microscope for the first time is the following: Where do I have to look to see the specimen? A microscope is generally considered a machine in which a source emits particles such as photons or electrons that are used as probes directed onto or emitted from the specimen. These particles are registered by a detector and subsequently analyzed, yielding information about the sample. There are two kinds of microscopes that fit readily into this source-specimen-detector-analyzer scheme. The first is the optical microscope, where the photons emitted from an incandescent lamp are manipulated by a system of lenses and mirrors located both before and after interacting with the specimen, and arriving at the eyepiece, where the detector (i.e., the eyes) collects the information. A typical optical microscope can reach a resolution of $\lambda/2$ ~200 nm

(λ being the wavelength of the light). Antonie van Leeuwenhoek (1632–1723) is credited with bringing the optical microscope to the attention of biologists, even though simple magnifying lenses were already being produced in the 1500s.

The electron microscope (EM) is a little more complicated. In this case the particles that act as probes are not photons but electrons produced by thermionic emission from an incandescent wire. Here, electromagnetic lenses are used to manipulate and focus the electron beam to provoke the right interaction with the specimen. The electrons are then collected by a screen, which is conveniently monitored. The first prototype electron microscope was built in 1931 by the German engineers Ernst Ruska and Max Knoll. Two years later, Ruska constructed an electron microscope that exceeded the resolution possible with an optical microscope, reaching ~1 nm.

In scanning probe microscopy, a sharp tip a few nanometers in diameter, which can be considered a probe, approaches the surface of the sample. The first member of this family of microscopes was the scanning tunneling microscope (STM) invented by Binnig and Rohrer (Binnig and Rohrer 1982), who received the Nobel Prize for Physics with Ruska in 1986. This system is based on a quantum effect (tunneling) that occurs when a sharp metallic tip is brought to a distance (z) of <1 nm from a conductive surface. This effect involves the flow of an electronic current (I) between the surface and the tip according to the formula $I \propto \exp(-\sqrt{\phi\zeta})$, where ϕ is the work function of the metallic surface (Chen 1993). The strong dependence of the current on the tip-to-surface distance can be used to obtain topographic and electronic maps of the sample by moving (i.e., scanning) the tip on the surface, while keeping the tip–sample distance constant through a feedback algorithm. Although this tool provides true atomic resolution in ultrahigh vacuum (UHV) conditions, a prerequisite is that both the tip and sample should be conductive. Therefore, it follows that STM is not suitable for biological samples because these are mainly insulators, which would need to be covered with a metallic layer (Baró et al. 1985).

In 1986, Binnig, Quate, and Gerber (Binnig et al. 1986) invented the atomic force microscope, combining the principles of both the STM and the so-called stylus profilometer (Schmalz 1929). In an atomic force microscope, a sharp stylus (appproximately tenths of a nanometer in diameter) attached to the end of a cantilever is brought close to the surface. As a consequence of this interaction, a force appears between the tip and surface that can be attractive or repulsive (see below), causing the cantilever to bend. When this bending is controlled with a feedback algorithm, it is possible to obtain a topographic map by scanning the surface in a plane perpendicular to the tip. In the original paper (Binnig et al. 1986) the topographic profiles of a ceramic sample (an insulator) were shown. This is one of the main advantages of AFM: Both tip and sample may be insulators. This property greatly expands the range of possibilities for scanning probe microscopy, making it possible to study biological samples (e.g., proteins, membranes, and whole cells).

AFM IMPLEMENTATION: TECHNICAL ISSUES

Although there are a variety of ways to control the deflection of the cantilever in AFM, here we will focus on the beam deflection method (Meyer and Amer 1988) because it is commonly used when working with biological samples. The beam deflection system involves focusing a laser beam on the end of the cantilever and collecting the reflected light with a photodiode. As a consequence, any bending of the cantilever will affect the position of the reflected laser spot on the photodiode. A normal bending generates a normal force F_N on the photodiode sectors, whereas a lateral torsion will result in a lateral force F_l. The core of an atomic force microscope is the head (Fig. 1A) in which the beam deflection system is integrated along with the piezoelectric tube that moves the sample in three directions (x, y, and z). In Figure 1A, an atomic force microscope head configuration is shown in which the tip is fixed and a piezo tube moves the sample. In another configuration, known as "stand alone," the sample remains stationary while the tip scans the sample surface. The electronic components receive the signals coming from the photodiode, mainly F_N and F_l, and provide high voltages (~240 V) to move the piezo tube to which the sample is attached. A computer manages the data and calculates all of the parameters required to move the piezo tube.

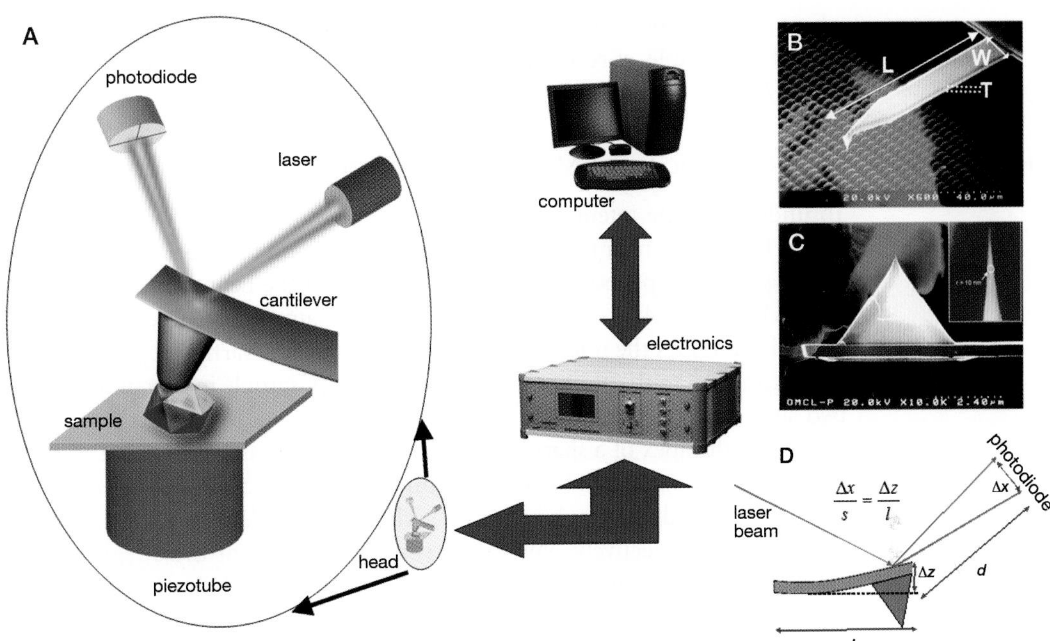

FIGURE 1. Principles of an atomic force microscope. The software running on the computer (*A*) controls the electronics, which communicates with the atomic force microscope head (inside the oval). (*B–D*) The cantilever, which is the atomic force microscope sensor. (*B*) A rectangular cantilever attached to a chip. (*C*) A typical tip at its final diameter. (*D*) The vertical resolution of the bending cantilever is explained as a function of several geometric parameters. (*B,C,* Adapted, with permission, from www.olympus.com.)

Integrated tip and cantilever assemblies can be fabricated from silicon or silicon nitride using photolithographic techniques. More than 1000 tip and cantilever assemblies can be produced on a single silicon wafer. The cantilevers can be rectangular (see Fig. 1B) or V-shaped and they typically range from 60 to 200 μm in length, 10 to 40 μm in width, and 0.3 to 2 μm in thickness. The typical tip radius is ~30 nm, although sometimes smaller diameters can be obtained (Fig. 1C). The cantilever spring constant k ranges between 0.02 and 40 N/m, and it strongly depends on the cantilever's dimensions. For example, for a rectangular cantilever

$$k = \frac{EW}{4}\left(\frac{T}{L}\right)^3,$$

where E is the Young modulus of the cantilever, and W, T, and L are the cantilever's width, thickness, and length, respectively (Fig. 1B). Whereas W and L can be fairly precisely known, T is always difficult to measure. As a consequence, manufacturers normally provide the cantilever spring constant with an error of 10%–30% and users should calibrate each cantilever (Sader et al. 1999). The spring constant k is used to calculate the Hookean force applied on the cantilever as a function of bending Δz, that is $F = k \times \Delta z$ (Fig. 1D).

The vertical resolution of the cantilever, Δz, strongly depends on the noise of the photodiode, because it defines the minimum significant displacement, Δx, on the detector (see Fig. 1D). A typical cantilever of 100 μm length has ~0.1 Å resolution, assuming a signal-to-noise ratio of ~1 (Meyer and Amer 1988).

Another parameter that can influence the vertical resolution is thermal noise (Butt et al. 1995). The cantilever oscillates at its resonance frequency with an amplitude

$$\Delta z = \sqrt{\frac{k_B T}{k}},$$

where k_B is the Boltzman constant, T is the absolute temperature, and k is the cantilever spring constant. For example, at room temperature, there is a noise of ~5 Å for a cantilever of spring constant 0.02 N/m.

Interaction Between the Atomic Force Microscope Tip and the Sample Surface

To understand the interaction between the tip and sample, we shall refer to potentials rather than forces. Physicists prefer this as potentials are scalar, and therefore easier to deal with, than vectors such as forces. For this purpose, let us graphically depict the Lennard-Jones potential (see dotted graph in Fig. 2A), $U(r) = -A/r^6 + B/r^{12}$, where r is the tip–sample distance, $A = 10^{-77}$ Jm^6, and $B = 10^{-134}$ Jm^{12}.

This describes the interaction of all the atoms in a particular solid (Israelachvili 2002). A rough approximation to the atomic force microscope tip–sample interaction is to consider the approach of two such atoms where, as depicted in the inset in Figure 2A, the lower part is a solid surface and above it is the apex of a sharp tip that moves downward. We can find an interaction force such that $F(r) = -dU/dr$, and it can be seen that this force is attractive ($F < 0$) when $r > r_0$ or repulsive ($F > 0$) when $r < r_0$ (solid line graph in Fig. 2A). These regions define the attractive and the repulsive regimes of operation, respectively. Now let us consider the role of these regimes in a force versus distance (F–z) AFM experiment that involves the tip approaching the surface (Fig. 2B). The experiment starts with the tip situated far from the surface in the attractive regime. As the tip is approaching and as

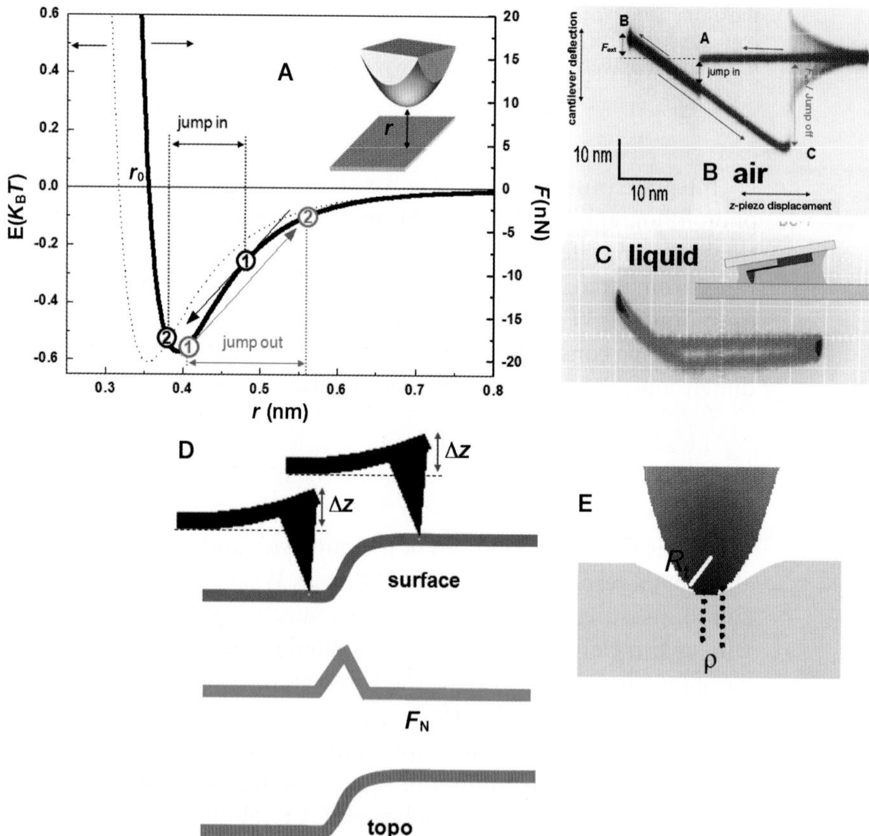

FIGURE 2. Force curves, feedback, and lateral resolution. (*A*) The potential of the tip–sample interaction is shown. The dotted line represents a Lennard-Jones potential as a function of the distance between atoms, *r*, mimicking the tip–surface interaction (represented in the *inset*). The solid line is the interaction force obtained from the potential. The numbers inside the circles indicate the cantilever instabilities (see text). (*B*) A force vs. z-piezo displacement curve in air. (*C*) A force vs. z-piezo displacement curve in liquid. (*D*) Contact mode showing the variation of F_N (green), and z-piezo tube voltage (topography, pink) as a function of a step (blue). (*E*) The geometric features of the tip–surface contact.

soon as the gradient (i.e., the slope) of the force equals the cantilever spring constant, the tip jumps to the surface from black point 1 to black point 2 (both connected by the slope). This is seen in the *F–z* of Figure 2B at point A like a sudden jump of the cantilever deflection (vertical scale). Thus the tip establishes mechanical contact with the surface and it rapidly enters the repulsive regime ($F_N >$ 0). The *z*-piezo (horizontal scale) is elongated until a given F_N or deflection value is reached and then stops (point B of Fig. 2B). The external loading force F_{ext} can be calculated as the difference between the zero deflection position (i.e., before the jump to contact in A) and the deflection at point B. Because the vertical scale is ~10 nm per division and the cantilever spring constant $k = 0.1$ N/m, then F_{ext} ~0.5 div × 10 nm/div × 0.1 nN/nm = 0.5 nN. Subsequently, the *z*-piezo retrace cycle starts, and the tip is released from the surface at C (i.e., once more where the derivative of the tip–surface force equals the cantilever spring constant), jumping from the red point 1 to 2 following the red arrow. The cantilever deflection jumps off to zero with a dampened oscillation. This jump-off is known as the adhesion force F_{adh} (in this case F_{adh} ~ 2 div × 10 nm/div × 0.1 nN/nm = 2 nN). The total force at point B is the sum of F_{ext} and F_{adh} (i.e., 2.5 nN). It is interesting to note that no matter how small F_{ext} is, the total force applied to the surface will always be at least F_{adh}.

Contact Mode

Contact mode is the simplest AFM operational method and it was the first to be developed (Binnig et al. 1986). Here the tip is brought into contact with the surface until a given deflection in the cantilever (F_N) is reached and the tip then scans a square area of the surface to obtain a topographic map. By elongating or retracting the *z*-piezo, the feedback algorithm tries to maintain a constant cantilever deflection by comparing the F_N signal with a set-point reference value established by the user. The topographic data are obtained by recording the *z*-piezo voltage that the feedback algorithm is applying to correct the cantilever deflection at each position on the surface. Because the *z*-piezo is calibrated, the voltages are transformed into heights and a topographic map is obtained. Let us consider a simple example in which the tip is scanning a step (blue line in Fig. 2D) with $F_N = k \times \Delta z$. When the cantilever moves to the upper part of the step, it undergoes a deflection greater than Δz. Therefore the feedback algorithm retracts the *z*-piezo to achieve the same deflection Δz as when the tip was in the lower part of the step. As a consequence, a topographic profile of the step is obtained (pink line in Fig. 2D). On the other hand, F_N varies at the step that is corrected by the feedback, which can be observed as a peak in the deflection signal (green line in Fig. 2D). The latter is known as the constant deflection mode or the constant height mode, because the *z*-piezo is not modified and a map of the changes in F_N is obtained. The reader is encouraged to reproduce Figure 2D when the tip goes down the step.

Let us now consider the lateral resolution that can be achieved in contact mode. We can estimate this parameter by applying the Hertz theory (Johnson 1985), which accounts for the deformation of solids in contact. Once the tip is in contact with the sample, the radius ρ of the tip–surface contact area is given by (Fig. 2E)

$$\rho = \left(\frac{3 \times F \times R}{4E^*} \right)^{1/3},$$

where F is the applied force F_N and E^* is the effective Young modulus expressed by

$$\frac{1}{E^*} = \frac{1-\nu_t^2}{E_t} + \frac{1-\nu_s^2}{E_s},$$

with (E_t, ν_t) and (E_s, ν_s) being the Young modulus and Poisson ratio for the tip and sample, respectively. R is the effective radius, expressed as a combination of the tip radius R_t and sample radius R_s:

$$\frac{1}{R} = \frac{1}{R_{tip}} + \frac{1}{R_{sample}}.$$

In the case of metals, $E = 100$ GPa, $\nu \sim 0.5$, and the mechanical thermal noise of the cantilever is 10 pN. With $R_t \sim 20$ nm, the radius of contact ρ is ~0.15 nm, which implies atomic resolution. However, the adhesion force in air (see below) is ~5 nN, increasing ρ to ~1 nm. Atomic resolution can be achieved by either working in liquids (Ohnesorge and Binnig 1993) or in UHV conditions (Giessibl 1995), where even individual atoms can be chemically identified (Sugimoto et al. 2007).

Geometrical Dilation

When the surface asperities are comparable to the tip radius (which is common in AFM experiments), the size of the tip plays an important role. The tip distorts the image owing to the dilation of certain features of the image by the finite tip size (Villarrubia 1997). Such dilation effects occur in all AFM operational modes. An example can be seen in Figure 3, A and B, in which single-walled carbon nanotubes have been imaged. These are graphene in the form of a cylinder a few nanometers in diameter. In Figure 3C two profiles of similar carbon nanotubes are compared, showing that the nanotubes of Figure 3B are wider than those in Figure 3A. The geometric dilation effect can be seen in Figure 3C, in which scanning of the carbon nanotube by the tip results in the red profile, because the tip cannot get closer to the tube than the tip radius r_t. Therefore, based on geometrical considerations in Figure 3D, the tip, radius r_t can be calculated as $r_t = b^2/2h$. The topographies of Figure 3, A and B, have been obtained using tips with a radius of 15 nm and 70 nm, respectively. It is evident that the sharper the tip, the better it is for imaging.

A very popular and impressive experiment used to teach AFM is to image graphite (in particular, highly oriented pyrolitic graphite) in air conditions (Marti et al. 1988), where the elastic defor-

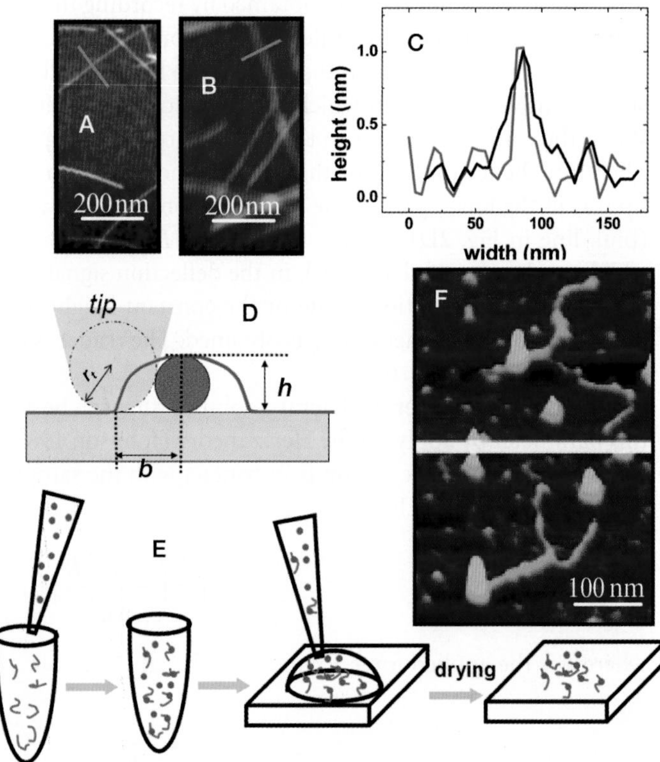

FIGURE 3. Geometrical dilation and sample preparation. (*A,B*) Atomic force microscope images of the same samples of carbon nanotubes adsorbed on silicon oxide. (*C*) The topographic profiles obtained along the green lines in *A* and *B* show different widths due to dilation. (*D*) The geometric parameters in the dilation process. The red line depicts the dilated section of the carbon nanotube section (gray circle). (*E*) Illustration of sample preparation for an experiment to study a DNA–protein complex. (*F*) The DNA–protein complexes imaged by AFM in air (see text). (Adapted from Dame et al. 2002.)

mation of the tip–sample contact plus the dilation effects result in atomic corrugation. Although the image seems to provide atomic resolution, atomic defects are not visualized.

Dynamic Modes

Dynamic modes (DMs) (Martin et al. 1987) are those in which the cantilever is made to oscillate near or at its resonance frequency ($\omega_0 \propto \sqrt{E}\ T/L^2$ for a rectangular cantilever). As the tip approaches the sample, the oscillating amplitude decreases until it establishes contact with the sample, following a similar cycle as that in Figure 2B, but now with oscillation. Therefore the feedback loop involves the amplitude rather than the F_N, and by keeping the oscillation amplitude constant, a topographical map can be obtained. The amplitude is reduced because the resonance frequency ($\Delta\omega$) changes with the tip–sample distance z and with the tip–sample interaction force F_{ts} according to

$$\frac{\Delta\omega}{\omega_0} \propto \left(\frac{1}{2k}\right)\frac{dF_{ts}}{dr},$$

thereby decreasing with attractive forces. The new resonance frequency ω is positioned to the left of ω_0 and because the cantilever is still oscillating at ω_0, the cantilever amplitude decreases.

The very high lateral forces that are applied to the surface in the contact mode (Carpick et al. 1997) can damage the sample. This is especially problematic for single biomolecules adsorbed onto a surface, because these are delicate samples from which to obtain images. However, when operated in noncontact mode (Garcia and Perez 2002), DM does not apply large dragging forces and so it is commonly used to image molecules weakly attached to surfaces in air. Maps other than topographical maps can also be obtained in DM, such as a phase map (the time difference between the excitation and the response of the cantilever), which in air carries information on the composition of the sample (see specific details in Garcia and Perez 2002).

When DM is used in liquid, the landscape completely changes, as the viscosity of water reduces the resonance frequency approximately fourfold and the quality factor (a measure of the cantilever damping) of the oscillation is reduced from ~100 to 10. When oscillating the cantilever in liquid, a mechanical contact between tip and sample is established, resulting in the application of lateral and normal forces that may damage the specimen (Legleiter et al. 2006).

Jumping or Pulse Force Mode

Jumping or pulse force mode (JM) (Miyatani et al. 1997; de Pablo et al. 1998) is a contact mode in which lateral tip displacement occurs when the tip and sample are not in mechanical contact, thereby avoiding shear forces and the corresponding damage to the tip–sample system. An *F–z* curve (Fig. 2B) is obtained at every point of the image, moving the tip to the next point at the end of each cycle when the tip and sample lose contact. Feedback is engaged at point B of Figure 2B, moving the *z*-piezo so that a constant deflection or loading force F_{ext} is maintained. Adhesion force maps can provide compositional or geometrical information about the surface (de Pablo et al. 1999). The adhesion force between the tip and the surface can be described as $F_{adh} = 4\pi R\gamma_L \cos\theta + 1.5\pi\gamma\Delta\gamma R$, where γ_L is the water surface tension and θ is the angle of the water meniscus present between the tip and surface, R is the effective radius (described above), and $\Delta\gamma$ is the tip–surface energy difference. Although the effective radius R is present in both terms, the first one provides information mainly about the hydrophobicity of the sample and as such a rough estimate in air at room temperature results in ~7 nN for an R_t of ~20 nm. In air, the first term is the main contribution to F_{adh} whereas the second one depends mainly on the tip–surface geometry. The importance of the first term can be appreciated by comparing the *F–z* curves taken from glass in both air (Fig. 2B) and liquid (Fig. 2C). In liquid the adhesion force is almost absent, because there is no water meniscus between the tip and sample, although some hysteresis appears in the *F–z* owing to the dragging of water on the cantilever.

USING AFM TO IMAGE BIOLOGICAL SAMPLES

Imaging Biological Samples in Air

AFM is used in dynamic mode for imaging DNA on mica. As DNA and mica are both negatively charged, $MgCl_2$ is added to the DNA solution. The Mg^{++} ions become sandwiched between DNA and mica, allowing the DNA molecules to adsorb onto the surface. The sample is then dried and DM-AFM is used to image the DNA molecules on the surface. DNA itself has been the focus of much research using AFM (Hansma et al. 1992), and more recently, AFM has been used to investigate the binding of proteins to DNA (Lyubchenko et al. 1995; Dame et al. 2003; Janicijevic et al. 2003). In these kinds of single-molecule experiments, the researcher is not interested in the average result of the bulk reaction but rather in the action of individual proteins on DNA. Hence, once the protein is pipetted into the DNA solution (Fig. 3E) the reaction starts. The DNA–protein complex is then adsorbed onto the mica at the desired time points where it is air dried (Fig. 3E). AFM thus provides a snapshot of the process that is taking place between the proteins and DNA, and the topographic map that is generated provides single-molecule information about the protein–DNA complex. For example, this technique has been used to study how the *Escherichia coli* H-NS enzyme interferes with DNA polymerase activity (Dame et al. 2002). AFM images revealed that the DNA polymerase (the big round blob on the DNA filament) becomes trapped between two pieces of DNA that are bridged by the H-NS enzyme (Fig. 3F). This entrapment appears to be sufficient to stop transcription initiation.

Imaging Biological Samples in Aqueous Media

When imaging samples in buffer or other aqueous media, proper attachment of the biomolecule to the supporting surface is required to achieve good resolution during the imaging process. Although biomolecules can be covalently linked to a chemically modified surface (Wagner et al. 1995), covalent modifications could potentially damage the biomolecules. Fortunately, physisorption is usually sufficient and thus, specimens in a physiological buffer can be directly adsorbed to the desired surface. The physisorption process is driven by van der Waals forces, the electrostatic double-layer force (EDL force), and the hydrophobic effect (Muller et al. 1997). The EDL force depends strongly on the concentration and valence of charged solutes, as well as the surface charge density of both surface and specimen. The EDL force between two equally charged surfaces is repulsive and hence opposite to the van der Waals attraction (Muller et al. 2002). Contact mode AFM has been used extensively to image two-dimensional protein crystals in liquid, such as membranes. Biological macromolecules become attached to the surface (e.g., mica, silicon, gold, or glass) when there is a net attractive force between the macromolecules and the surface. The Derjaguin–Landau–Verwey–Overbeek force (F_{DLVO}) can be estimated (Israelachvili 2002) as the sum of the electrostatic force between surface and molecule F_{el}, and the van der Waals interaction F_{vdW},

$$F_{DLVO} = F_{el}(z) + F_{vdW}(z) = \frac{2\sigma_{surf}\sigma_{sample}}{\varepsilon_e \varepsilon_0} e^{-z/\lambda_D} - \frac{H_a}{6\pi z^3},$$

where z is the distance between the surface and specimen; σ_{surf} and σ_{sample} are the charge densities of surface and specimen, respectively; ε_e and ε_0 are the dielectric constants of the electrolyte and the vacuum, respectively; λ_D is the Debye length, which depends on the electrolyte valence (Muller et al. 1997); and H_a is the Hamaker constant.

The adsorption of a sample onto freshly cleaved mica (atomically flat) can be manipulated by adjusting both the ion content and the pH of the buffer solution. An estimate of the F_{DLVO} between a purple membrane (a two-dimensional crystal lattice formed by bacteriorhodopsin) and mica is shown in Figure 4A, highlighting the strong influence of the electrolyte concentration (Muller et al. 1997). Figure 4B shows a region of purple membrane adsorbed onto mica (Muller et al. 1995). Interestingly, in this type of setup, the cantilever can be used to apply forces that trigger conformational changes in single proteins, such as GroEL (Viani et al. 2000).

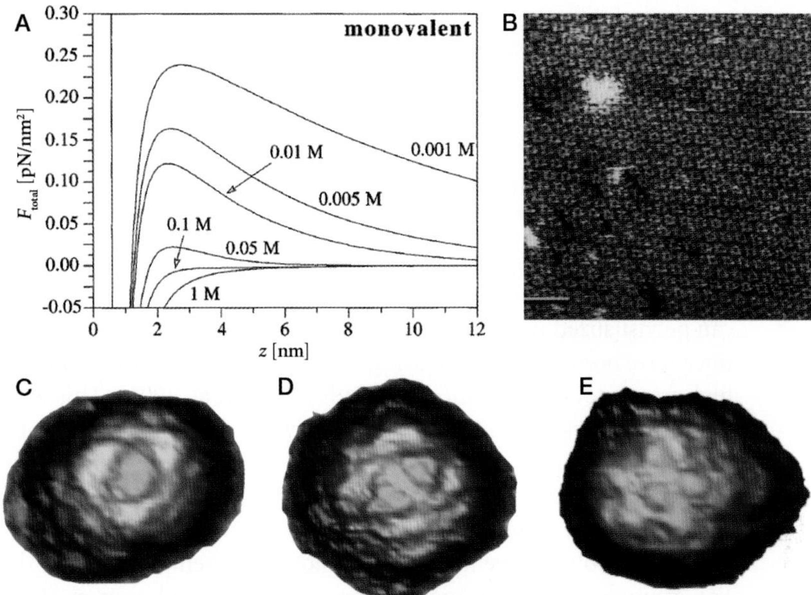

FIGURE 4. Imaging the purple membrane and single viruses. (*A*) Establishing the buffer conditions for imaging. The interaction force between the purple membrane and the mica surface is depicted as a function of the distance and the electrolyte concentrations. (*B*) A typical atomic force microscope image of a purple membrane in liquid with bacteriorhodopsin, a light-absorbing membrane protein (Reprinted, with permission, from Muller et al. 1997 and Muller et al. 1995, ©Elsevier). (*C–E*) Three MVM particles showing threefold, twofold, and fivefold symmetry, respectively. (Adapted from Carrasco et al. 2006.)

AFM can also be used to visualize single proteins at work. To do so, AFM is used in dynamic mode and in liquid (Moreno-Herrero et al. 2004). Maximum peak forces of a few nanoNewtons are applied, relative to the stiffness of the sample, for very short periods of time (Legleiter et al. 2006; Xu et al. 2008), to avoid damaging the sample. For example, the force application period could be 10% of an oscillating period (i.e., for a cantilever with a resonance frequency of 10 kHz in liquid, the forces are applied for 10 µsec every 100 µsec). Using this method, it has been possible to visualize the activity of RNA polymerase on DNA (Kasas et al. 1997), and the conformational changes in a DNA-repair complex on binding DNA (Moreno-Herrero et al. 2005).

Imaging Viruses Using AFM

Structural and chemicophysical characterization has been critical for our understanding of the biology of viruses. X-ray crystallography and EM (cryo-EM/IR) techniques have traditionally been used. Although they provide direct three-dimensional structural information and allow the interior and the surface of the virus to be visualized, these are averaging ("bulk") techniques, and thus they present an average time and space model of the entire population of particles found in the crystal or on the EM grid. These techniques provide limited information about the characteristics of individuals within a population of viruses that distinguishes them from the average. For this reason, the beautifully symmetrical models of larger viruses derived from these techniques may be somewhat deceptive and not fully representative of every individual virus particle within a population (Plomp et al. 2002).

Because viruses are individual particles that adsorb weakly to typical surfaces, they are prone to destruction by lateral forces when imaged in AFM contact mode. It is preferable to use the jumping mode because loading forces can be accurately controlled to avoid the application of lateral forces. Figure 4, C–E, shows single viral particles of the minute virus of mice (MVM) adsorbed in threefold, twofold, and fivefold symmetry, respectively. By making nanoindentations with *F–z* curves on single viruses, it has been shown that single-stranded DNA within the MVM virus contributes to the overall mechanical stiffness of the virus particles, which could be important for viral stability during the extracellular cycle (Carrasco et al. 2006). Moreover, it is possible to selectively disrupt these DNA–protein interactions and thereby engineer a virus with altered mechanical properties (Carrasco et al. 2008).

SUMMARY AND FUTURE DIRECTIONS

A typical AFM image takes minutes to acquire, making it difficult to view dynamic biological processes. Work is underway to increase the rate of AFM image capture, so that biological processes can be visualized in real time. For example, conformational changes of single myosin V proteins on mica have been observed by using high video rates (80 msec/frame) and very soft cantilevers with a high frequency of resonance in liquids (Ando et al. 2001). This was achieved by decreasing the cantilever thickness and reducing the cantilever width and length proportionally. In addition to obtaining topographical images to understand the structure and dynamics of a system, AFM can also extract information about mechanical properties (e.g., stiffness) by using the phase-in dynamic modes in liquid (Melcher et al. 2009).

Finally, noninvasive imaging techniques are being developed that minimize sample destruction. For example, frequency modulation AFM (Hoogenboom et al. 2006) is a dynamic technique in which forces on the order of tens of picoNewtons can be applied to the surface. This promising technique is based on the use of three simultaneous feedback systems. A phase lock loop ensures that the cantilever is always in resonance, whereas a second feedback process (working over the phase lock loop) changes the tip–sample gap to keep a set point frequency, such that its output gives the topography. Finally, a third feedback component is used to maintain the oscillation amplitude constant by changing the amplitude of the cantilever driving signal, which results in more stable operation.

REFERENCES

Ando T, Kodera N, Takai E, Maruyama D, Saito K, Toda A. 2001. A high-speed atomic force microscope for studying biological macromolecules. *Proc Natl Acad Sci* **98:** 12468–12472.

Baró AM, Miranda R, Alamán J, García N, Binnig G, Rohrer H, Gerber C, Carrascosa JL. 1985. Determination of surface-topography of biological specimens at high-resolution by scanning tunnelling microscopy. *Nature* **315:** 253–254.

Binnig G, Rohrer H. 1982. Scanning tunneling microscopy. *Helv Phys Acta* **55:** 726–735.

Binnig G, Quate CF, Gerber C. 1986. Atomic force microscope. *Phys Rev Lett* **56:** 930–933.

Butt HJ, Jaschke M. 1995. Calculation of thermal noise in atomic-force microscopy. *Nanotechnology* **6:** 1–7.

Carpick RW, Ogletree DF, Salmeron M. 1997. Lateral stiffness: A new nanomechanical measurement for the determination of shear strengths with friction force microscopy. *Appl Phys Lett* **70:** 1548–1550.

Carrasco C, Carreira A, Schaap IA, Serena PA, Gómez-Herrero J, Mateu MG, de Pablo PJ. 2006. DNA-mediated anisotropic mechanical reinforcement of a virus. *Proc Natl Acad Sci* **103:** 13706–13711.

Carrasco C, Castellanos M, de Pablo PJ, Mateu MG. 2008. Manipulation of the mechanical properties of a virus by protein engineering. *Proc Natl Acad Sci* **105:** 4150–4155.

Chen CJ. 1993. *Introduction to scanning tunneling microscopy*. Oxford University Press, Oxford.

Dame RT, Wyman C, Wurm R, Wagner R, Goosen N. 2002. Structural basis for H-NS-mediated trapping of RNA polymerase in the open initiation complex at the rrnB P1. *J Biol Chem* **277:** 2146–2150.

Dame RT, Wyman C, Goosen N. 2003. Insights into the regulation of transcription by scanning force microscopy. *J Microsc* **212:** 244–253.

de Pablo PJ, Colchero J, Gomez-Herrero J, Baró AM. 1998. Jumping mode scanning force microscopy. *Appl Phys Lett* **73:** 3300–3302.

de Pablo PJ, Colchero J, Gomez-Herrero J, Baró AM, Schaefer DM, Howell S, Walsh B, Reifenberger R. 1999. Adhesion maps using scanning force microscopy techniques. *J Adhes* **71:** 339–356.

Garcia R, Perez R. 2002. Dynamic atomic force microscopy methods. *Surf Sci Rep* **47:** 197–301.

Giessibl FJ. 1995. Atomic-resolution of the silicon (111)–(7x7) surface by atomic-force microscopy. *Science* **267:** 68–71.

Hansma HG, Sinsheimer RL, Li MQ, Hansma PK. 1992. Atomic force microscopy of single-stranded and double-stranded DNA. *Nucleic Acids Res* **20:** 3585–3590.

Hoogenboom BW, Hug HJ, Pellmont Y, Martin S, Frederix PLTM, Fotiadis D, Engel A. 2006. Quantitative dynamic-mode scanning force microscopy in liquid. *Appl Phys Lett* **88:** 193109.

Israelachvili J. 2002. *Intermolecular and surface forces.* Academic, London.

Janicijevic A, Ristic D, Wyman C. 2003. The molecular machines of DNA repair: Scanning force microscopy analysis of their architecture. *J Microsc* **212:** 264–272.

Johnson KL. 1985. *Contact mechanics.* Cambridge University Press, Cambridge.

Kasas S, Thomson NH, Smith BL, Hansma HG, Zhu X, Guthold M, Bustamante C, Kool ET, Kashlev M, Hansma PK. 1997. *Escherichia coli* RNA polymerase activity observed using atomic force microscopy. *Biochemistry* **36:** 461–468.

Legleiter J, Park M, Cusick B, Kowalewski T. 2006. Scanning probe acceleration microscopy (SPAM) in fluids: Mapping mechanical properties of surfaces at the nanoscale. *Proc Natl Acad Sci* **103:** 4813–4818.

Lyubchenko YL, Jacobs BL, Lindsay S, Stasiak A. 1995. Atomic-force microscopy of nucleoprotein complexes. *Scanning Microsc* **9:** 705–727.

Marti O, Drake B, Gould S, Hansma PK. 1988. Atomic resolution atomic force microscopy of graphite and the native oxide on silicon. *J Vac Sci Technol A* **6:** 287–290.

Martin Y, Williams CC, Wickramasinghe HK. 1987. Atomic force microscope force mapping and profiling on a sub 100-a scale. *J Appl Phys* **61:** 4723–4729.

Melcher JC, Xu X, Carrascosa JL, Gómez-Herrero J, de Pablo PJ, Raman A. 2009. Origins of phase contrast in the atomic forces microscopy in liquids. *Proc Natl Acad Sci* **106:** 13655–13660.

Meyer G, Amer NM. 1988. Novel optical approach to atomic force microscopy. *Appl Phys Lett* **53:** 1045–1047.

Miyatani T, Horii M, Rosa A, Fujihira M, Marti O. 1997. Mapping of electrical double-layer force between tip and sample surfaces in water with pulsed-force-mode atomic force microscopy. *Appl Phys Lett* **71:** 2632–2634.

Moreno-Herrero F, Colchero J, Gomez-Herrero J, Baró AM. 2004. Atomic force microscopy contact, tapping, and jumping modes for imaging biological samples in liquids. *Phys Rev E* **69:** 031915.

Moreno-Herrero F, De Jager M, Dekker NH, Kanaar R, Wyman C, Dekker C. 2005. Mesoscale conformational changes in the DNA-repair complex Rad50/Mre11/Nbs1 upon binding DNA. *Nature* **437:** 440–443.

Muller DJ, Schabert FA, Buldt G, Engel A. 1995. Imaging purple membranes in aqueous-solutions at subnanometer resolution by atomic-force microscopy. *Biophys J* **68:** 1681–1686.

Muller DJ, Amrein M, Engel A. 1997. Adsorption of biological molecules to a solid support for scanning probe microscopy. *J Struct Biol* **119:** 172–188.

Muller DJ, Janovjak H, Lehto T, Kuerschner L, Anderson K. 2002. Observing structure, function and assembly of single proteins by AFM. *Prog Biophys Mol Biol* **79:** 1–43.

Ohnesorge F, Binnig G. 1993. True atomic-resolution by atomic force microscopy through repulsive and attractive forces. *Science* **260:** 1451–1456.

Plomp M, Rice MK, Wagner EK, Mcpherson A, Malkin AJ. 2002. Rapid visualization at high resolution of pathogens by atomic force microscopy—Structural studies of herpes simplex virus-1. *Am J Pathol* **160:** 1959–1966.

Sader JE, Chon JWM, Mulvaney P. 1999. Calibration of rectangular atomic force microscope cantilevers. *Rev Sci Instrum* **70:** 3967–3969.

Schmalz G. 1929. Uber Glatte und Ebenheit als physikalisches und physiologishes Problem. *Verein Deutscher Ingenieure*: 1661–1467.

Sugimoto Y, Pou P, Abe M, Jelinek P, Pérez R, Morita S, Custance O. 2007. Chemical identification of individual surface atoms by atomic force microscopy. *Nature* **446:** 64–67.

Viani MB, Pietrasanta LI, Thompson JB, Chand A, Gebeshuber IC, Kindt JH, Richter M, Hansma HG, Hansma PK. 2000. Probing protein–protein interactions in real time. *Nat Struct Biol* **7:** 644–647.

Villarrubia JS. 1997. Algorithms for scanned probe microscope image simulation, surface reconstruction, and tip estimation. *J Res Natl Inst Stand* **102:** 425–454.

Wagner P, Hegner M, Guntherodt HJ, Semenza G. 1995. Formation and in-situ modification of monolayers chemisorbed on ultraflat template-stripped gold surfaces. *Langmuir* **11:** 3867–3875.

Xu X, Carrasco C, de Pablo PJ, Gomez-Herrero J, Raman A. 2008. Unmasking imaging forces on soft biological samples in liquids when using dynamic atomic force microscopy: A case study on viral capsids. *Biophys J* **95:** 2520–2528.

WWW RESOURCE

www.olympus.com Olympus Corporation homepage.

38 Total Internal Reflection Fluorescence Microscopy

Ahmet Yildiz[1] and Ronald D. Vale[2]

[1]*Department of Physics, University of California Berkeley, Berkeley, California 94720;*
[2]*Department of Cellular and Molecular Pharmacology, University of California San Francisco, San Francisco, California 94158*

ABSTRACT

The goal in fluorescence microscopy is to detect the signal of fluorescently labeled molecules with great sensitivity and minimal background noise. In epifluorescence microscopy, it is difficult to observe weak signals along the optical axis, owing to the overpowering signal from the out-of-focus particles. Confocal microscopy uses a small pinhole to produce thin optical sections (~500 nm), but the pinhole rejects some of the in-focus photons as well. Total internal reflection fluorescence microscopy (TIRFM) is a wide-field illumination technique that illuminates only the molecules near the glass coverslip. It has become widely used in biological imaging because it has a significantly reduced background and high temporal resolution capability. TIRFM has been used to study proteins in vitro as well as signaling cascades by hormones and neurotransmitters, intracellular cargo transport, actin dynamics near the plasma membrane, and focal adhesions in living cells. Because TIRF illumination is restricted to the glass–water interface and does not penetrate the specimen, it is well suited for studying the interaction of molecules within or near the cell membrane in living cells.

INTRODUCTION

Principles of Total Internal Reflection

Total internal reflection fluorescence microscopy (TIRFM) relies on the refraction and reflection of light from a planar surface. When the excitation beam travels from a high index of refraction medium (n_1 = 1.518 for glass) to a low index of refraction medium (n_2 = 1.33 for water), the angle of the beam changes in accordance with Snell's law. At a specific critical angle, the incident beam is totally reflected back from the glass surface, rather than passing through the water (Fig. 1). TIR is achieved when the incident angle is ~2° higher than the critical angle (θ_c); $\sin(\theta_c) = n_2/n_1$, where θ_c = 61.5° for a glass–water interface.

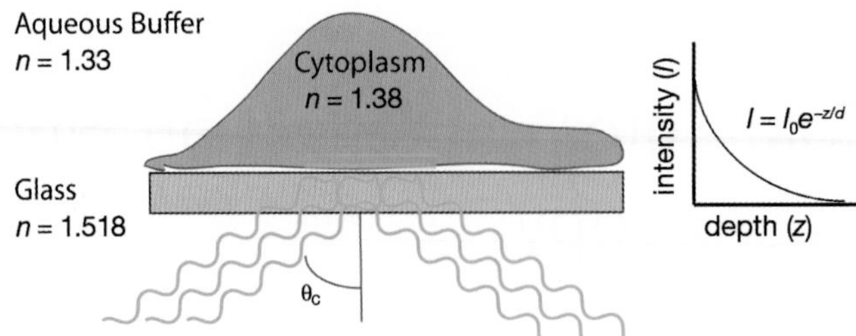

FIGURE 1. Total internal reflection. The angle of the incident light above the critical angle is totally internally reflected back from the coverslip surface. The reflected beam generates an evanescent field that decays exponentially into the sample. TIR illuminates molecules near (~100 nm) the glass surface. In live cell imaging, TIR only excites molecules in or near the cell membrane and significantly reduces background fluorescence.

In TIR, the reflected beam partially emerges from the glass and travels a subwavelength distance in water before returning back to the glass medium. Therefore, TIR generates a very thin electromagnetic field in the aqueous medium, which is called an evanescent field. The evanescent field decays exponentially with increasing distance from the glass–water boundary. The penetration depth (d) is a function of the wavelength (λ) and incident angle (θ) of the excitation beam and the refractive indices of the two media. The characteristic penetration depth is given by

$$d = \frac{\lambda}{4\pi\sqrt{([n]_1 \sin\theta)^2 - n_2^2}}.$$

The penetration depth usually ranges between 80 and 200 nm for the visible spectrum and decreases as the incident angle grows larger. Beyond this distance, the electric field is insufficient to excite fluorophores. Exponential decay of the evanescent wave intensity can be used to study the distance of molecules from the glass surface. In most live cell experiments, the TIRF interface is not as clearly defined as the glass–water boundary. The index of refraction of the cellular membrane (n = 1.38) is slightly higher than the aqueous medium, which results in higher penetration depth. Overall TIRFM generates a much thinner optical section and achieves better signal-to-noise ratio images than does confocal microscopy. However, because TIRF occurs only at the interface, it is not suitable for generating optical sections for three-dimensional imaging, as in the case of confocal microscopy. Daniel Axelrod has greatly stimulated the use of TIRFM for biological applications (Axelrod et al. 1984). Yanagida and co-workers have used the dramatic reduction in background fluorescence to image single molecules under TIRFM (Tokunaga et al. 1997).

Evanescent wave intensity at the surface is a function of the polarization of the incident beam. The plane of incidence is defined as the x–z plane. Two independent polarization directions (perpendicular [P] and parallel [S]) are possible with respect to the microscope's image plane (x–z plane). S-polarized light is perpendicular to the plane of incidence and generates polarization in the y-axis. P-polarized light remains in the plane of incidence and generates elliptical polarization in the x–z plane. Therefore, P-polarized light leads to slightly higher evanescent wave intensity with the same decay constant. To excite all possible orientations of fluorophores, it is suitable to introduce a quarter-wave plate into an optical pathway to generate a circularly polarized excitation beam.

Thin metallic layers are also used to enhance the fluorescence signal in TIRF. In these three-layer systems, light travels through glass, enters a metallic intermediate layer, and reflects back from the metal–water boundary. As the penetration depth changes in regard to the equations given above, the evanescent wave intensity at the surface is enhanced by surface plasmon resonance. This resonance is achieved through excitation of the surface plasmon mode of the metal, which increases the intensity of the signal by an order of magnitude. One interesting feature of metallic films is that they do not require a collimated beam for TIRF. Intensity is only enhanced for a narrow band of angles. Beams with higher or lower incident angles than the surface-plasmon angle produce only negligible fields. Thin films also lead to a highly *P*-polarized evanescent wave, regardless of the polarization angle of the incident beam. One consequence of a metallic coating is that particles closer than 10-nm proximity to the metal–water interface have their fluorescence quenched. To prevent fluorescence quenching, the surface of the metal can be coated with a transparent insulator material.

TIRFM Geometry

TIRFM can be achieved by two primary geometries that differ in their illumination pathways. In a prism-type TIRFM system, the laser beam is guided onto a surface through a prism to achieve the correct angle. In an objective-type TIRFM system, the laser beam travels through the objective before illuminating the sample. Figure 2 shows illumination pathways of both TIRFM types for an inverted microscope. A sample chamber is usually assembled from a clean coverslip and slide that have been attached together by double-sided tape. In the prism-type, the prism and slide are optically coupled with immersion oil that matches their index of refraction. The angle of incidence is adjusted by moving the focusing lens via an *x–y–z*-translation stage. The size of illumination can be adjusted by the beam diameter and the focal length of the lens used (50–100-mm focal length lenses are commonly used). TIR is generated at the slide–water boundary and the fluorescence signal passes through aqueous medium, coverslip, and immersion medium to reach the objective. Because the excitation beam passes through a thick microscope slide, quartz slides and prisms are used to minimize autofluorescence. Light is collected by a high-numerical-aperture (NA = 1.3) water-immersion objective. To prevent the mismatch of refractive index owing to coverslip thickness, microscope companies provide adjustment collar for various coverslip thicknesses. Even though oil objectives can achieve higher NA, they suffer spherical aberration that is caused by the mismatch of the aqueous buffer and the immersion oil. The mismatch becomes negligible if the flow chamber is thinner than 15 µm and oil objectives can be used under these conditions.

Using high-NA objectives, it is possible to send the excitation beam through the objective below the critical angle for TIR. In this method, a collimated laser beam is focused onto the outer edge of

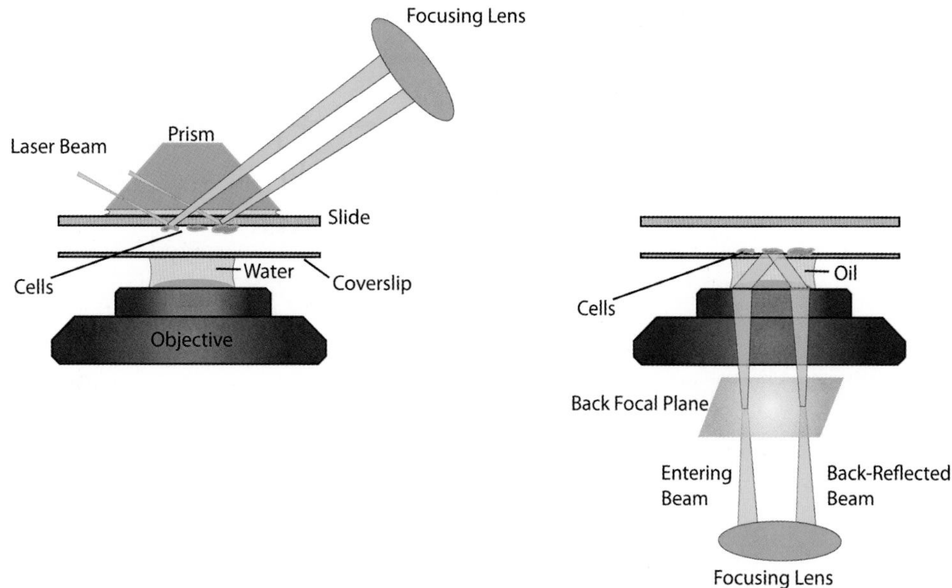

FIGURE 2. Geometries of TIRF illumination. (*Left*) In prism-type TIRF, the angle of incidence is adjusted by the focusing lens. Prism and quartz slide are optically coupled with matching immersion oil. TIR is generated on the surface of the slide and emitted photons are collected by a water-immersion objective through the coverslip. (*Right*) Using high-NA oil objectives, it is possible to guide the light through the sample above the critical angle to achieve TIR. The laser beam is focused to the objective back focal plane by a TIR lens. The lens is moved along the perpendicular axis to adjust the angle of incidence. Cells immobilized on a coverslip are illuminated by TIRF.

the back focal plane of the microscope objective to obtain parallel beams exiting the objective's pupil with a high incidence angle. The need for a high-NA objective is illustrated by Snell's law, in which the critical angle is 61.5° for a glass–water interface (n_{water} = 1.33). A 1.4-NA oil-immersion objective makes it barely possible to achieve TIR. Recently, major microscope companies have begun to provide 1.45–1.49-NA oil-immersion objectives, resulting in a maximum illumination angle of 72°–75°. TIR alignment with these objectives is easier than with 1.4-NA objectives. Investment in a 1.45–1.49-NA objective is important for successful objective-type TIRFM.

Prism-type TIRF is easier to set up and provides more uniform illumination, which results in a lower background compared with the objective-type. However, this method blocks the upper surface of the flow chamber, which is commonly used for buffer exchange, electrophysiology, and microinjection of the sample in life sciences. In addition, the prism has to be removed to exchange the sample, which necessitates realignment of the prism for each experiment. Even though objective-type TIRF leads to increased levels of light scattering within the objective, it is clearly the method of choice for most laboratories. Because the emitted fluorescence travels through a coverslip, an objective collar readily corrects refractive index mismatch, and minimum spherical aberration is achieved.

TIRFM IMAGING SETUP

Most inverted microscopes can easily be converted into an objective-type TIRFM. Figure 3 shows the layout of an objective-type TRIFM system equipped with two lasers. To properly align a TIRFM system, do the following.

1. Adjust the two laser beams to the height of the microscope port using periscopes and combine them using a dichroic mirror placed on a kinematic mount. To ensure that the beams travel along the same optical path, use a set of two irises adjusted to the same height along the optical table.

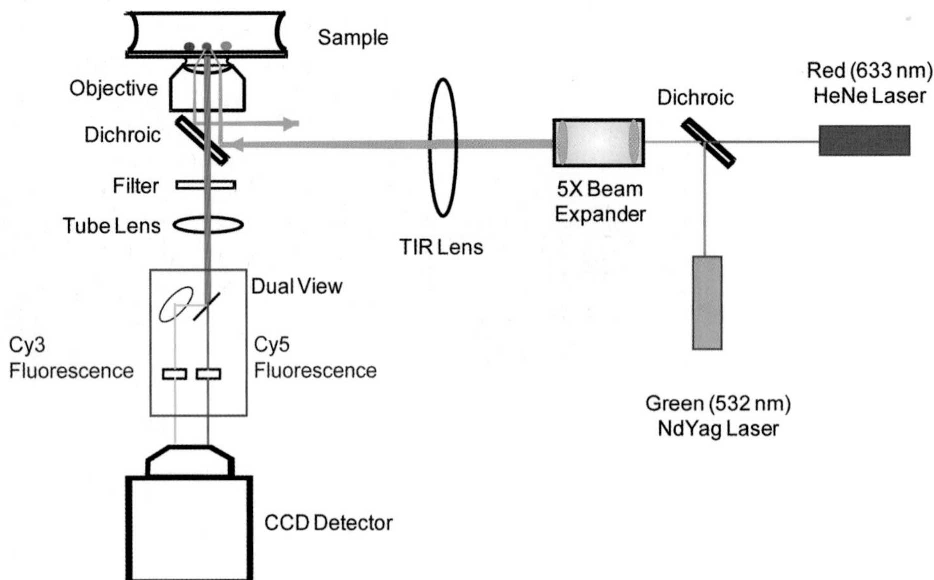

FIGURE 3. Schematics of a multicolor TIRFM. To simultaneously image Cy3 and Cy5 fluorescence, an objective-type TIRF microscope is equipped with 532-nm and 633-nm lasers. The laser beams are combined with a dichroic mirror and expanded through a Gaussian beam expander. The laser beams are focused onto the back focal plane of the microscope objective with an achromatic doublet lens (f = 500 mm). A set of multiband dichroic and emission filters are used to reflect the 532-nm and 633-nm laser beams onto the objective and to transmit Cy3 and Cy5 fluorescence simultaneously. The fluorescence is then separated by a Dual View instrument (Optical Insights Inc.), which is composed of a dichroic mirror to split Cy3 and Cy5 fluorescence, and band-pass emission filters to further reduce the cross talk between the two channels.

2. Place a 5× beam expander on an *x–y*-translation stage along the optical pathway to obtain a large (~60 µm diameter) illumination spot. If the two lasers have a significantly different beam diameter, then a combination of a 10× beam expander and an adjustable iris set at a 1-cm-diameter opening can be used to obtain the same illumination area for both lasers. To minimize chromatic aberrations in the beam expander, achromatic doublet lenses should be used.
3. (Optional) Send the beam to a kinematic mirror placed on a linear translation stage. The light reflects off at 90° at this mirror and is sent to the back illumination port of an inverted microscope.
4. Place an appropriate filter set in the filter turret of the microscope, and a circular field diaphragm onto the nosepiece. Adjust the mirror position to allow the center of the beam to coincide with the center of the object plane.
5. Place an achromatic doublet TIR lens with an *x–y–z*-translation stage along the optical path. The focal length of this lens determines the size of the illumination area, and can be 25–50 cm. First, adjust the lens position to allow the beam to pass through the center of the lens. Second, adjust its *z*-position to focus the laser beam to the center of the microscope nosepiece.
6. Remove the circular diaphragm from the nosepiece and place a high-NA (>1.4) objective lens. Readjust the *x–y*-position of the TIR lens to project the laser beam onto the ceiling, and the *z*-position to obtain smallest possible diameter of the laser beam. Because of the chromatic aberrations in most objectives, it is not possible to properly focus multiple laser beams simultaneously. Therefore, an intermediate position should be found or the *z*-position of the TIR lens should be adjusted before switching to another laser beam.
7. Add matching immersion oil to the objective. Prepare a flow chamber with surface immobilized fluorescent microspheres on a coverslip and place it on the microscope stage. Focus the objective lens on the sample by using binoculars to find the immobilized beads on the coverslip surface.
8. Adjust the *x*- or *y*-position of the TIR lens to move the focus spot within the back focal of the objective. The beam will be sent to the objective at off axis, which will tilt the direction of the beam at the objective exit pupil. When the critical angle is reached, parallel light traveling along the glass coverslip will be observed. Tilting the beam ~2° more will generate TIR. A secondary spot corresponding to the back-reflected beam should be observed on a TIR lens along the direction of tilting.
9. Acquire images with a charge-coupled device (CCD) camera located on a plane conjugate to the object plane.
10. Adjustment of either the *x*- or *y*-position of the TIR lens may result in shifting the illumination area away from the center of the CCD chip. Readjust the *x*- or *y*-position of the 5× beam expander (or the mirror, if used) and then the TIR lens, until the center of the illumination area matches well with the center of the CCD chip.

EQUIPMENT

Light Sources

Noncoherent light sources are not recommended for TIRFM, because only a fraction of light exits the objective at higher angles than the critical angle. Argon-ion, krypton, and helium-neon are the common gas lasers used to excite the sample in the visible region of light. Recently, diode-pumped solid-state (DPSS) lasers have become popular owing to their small size, high stability, and low maintenance and power requirements. Major peaks of gas lasers (405, 488, 514, 532, 547, and 647 nm) are matched in commercial DPSS lasers to easily implement them into existing experimental setups. For multicolor TIRFM setups, different wavelength laser beams must be combined with the proper set of dichroic mirrors to attain collinear illumination. The laser beam can then be coupled to an optical fiber to purify nonuniform modes, and to obtain the same beam size for each laser beam. To perform single molecule polarization assays in TIRFM, polarization-maintaining fibers should be used.

In multicolor time-lapse imaging assays, shuttering and changing the laser power may be required. DPSS lasers can be readily modulated by electronic signals at a megahertz frequency. Excitation of other lasers can be controlled in time series on a millisecond timescale by mechanical shutters. An acousto-optic tunable filter (AOTF) is also commonly used to rapidly (a few microseconds) modulate the transmission wavelength and intensity of the laser excitation. In AOTFs, there is a miniscule leakage of the blocked beams, which may lead to increased background noise. Therefore, in these systems, all possible laser wavelengths in the illumination pathway should be carefully blocked by respective emission filters to obtain a high signal-to-noise ratio.

In TIRFM, it is ideal to fully illuminate the field of view of the CCD cameras. Image magnification should fulfill the Nyquist criterion, which requires that the actual pixel size of the camera be less than half of the minimum resolvable distance (~250 nm for visible light). Most camera companies provide CCD chips with 8–24-µm pixels. TIRF microscopes mostly use 100x objectives, and an additional 1.5x magnifier (built-in optivar in many commercial microscopes) can be used to meet this criterion if needed. We have found that the optimum pixel size of the CCD is between 80 and 160 nm for high spatial resolution. Therefore, an ~40–60 µm diameter circle (0.003 mm^2) should be illuminated with a laser light to fully illuminate a 512 x 512 camera chip. We have also found that 100–300 mW/mm^2 of laser power incident on the sample is sufficient to saturate most organic dyes. Therefore, 1 mW of incident beam on the sample is required to perform the measurements. TIRFM users should take into account the losses in the optical pathway before making a decision on which laser to purchase. In custom TIRFM setups, at least 50% of the light is lost in dichroic mirrors, lenses, filters, and the objective. In fiber optic–coupled systems, an additional 50%–70% of light is lost in the fiber couplers and the AOTF.

Filters

The proper choice of dichroic mirrors and emission filters can significantly improve the performance of the TIRFM system. Because the excitation beam is reflected back from the glass surface and travels back along the detection pathway, blocking the excitation beam is essential for detecting the fluorescence signal by a camera. Assuming that the excitation laser power is roughly a million times more powerful than the fluorescence signal, even miniscule leakage of the laser light into the detection pathway can dominate the signal. Because the entering beam is physically separated from the returning beam, two small mirrors can be used instead of a dichroic mirror to fully reflect the returning beam out of the fluorescence detection path.

For each set of fluorescent probes, a separate filter cube is assigned. The first step is to clean up the laser beam by using an excitation filter that has a narrow (~10 nm) band pass. This step is required if the laser emits multiple wavelengths. Using an excitation filter also reduces the background even though the sample is illuminated with spectrally pure laser light. This is because excitation filters are designed based on the principles of thin film interference at normal incident angles and they significantly block the off-axis light. In the absence of an excitation filter, a fraction of the off-axis light can pass through an emission filter and reach the camera chip.

The dichroic mirror should have the highest possible transmission for the entire emission spectrum of the fluorescent probe used, and reflection of the excitation wavelength. Because the emission spectrum often starts near the absorption maxima, a steep transition from reflection to transmission band is anticipated. Razor edge dichroic mirrors (Semrock) have been designed for most of the common excitation wavelengths to allow measurement of signals very close (2 nm) to the blocked laser light. Recent advances allow for ultrabright (up to 99% transmission), durable, and easy-to-clean dichroic mirrors that can be safely used at high laser powers without burning out. It is extremely important to ensure that the dichroic surface is flat in TIRFM. Any curvature leads to astigmatism in the beam profile, which would result in light entering the sample at various angles and increasing the background fluorescence. Bending of dichroic surfaces may be caused by differences in the intrinsic stress between the hard glass coatings and the substrate. In addition, overtightening the springs in the filter cube can easily bend the filter surface and cause higher-order aberrations. Care should be taken when assembling each filter cube.

For multipurpose TIRFM setups, we have often found it difficult to align the optical pathway for each dichroic mirror used in the microscope. Slight angular differences between dichroic mirrors shift the incident angle of the laser beam and change the location of the illumination area. For these purposes, we recommend a single multi-band-pass dichroic mirror that can block each excitation wavelength used in the setup and transmit the other wavelengths between them. With such mirrors, up to four lasers emitting at different wavelengths can be blocked by a single dichroic. Emission filters that match well with the spectrum of these dichroics can be used so that a single filter cube works for all combinations of excitation wavelengths and fluorescent dyes. This design significantly improves the temporal resolution of multicolor experiments, because it does not require rotating the filter wheel for each laser and eliminates the alignment problems caused by the tilt in the dichroic orientation.

Dichroic mirrors can only reflect 99% of the excitation light. The remaining light is blocked by the emission filter that can transmit down to $\sim 10^{-7}$ of the laser light. Current filters can achieve this performance even if the laser wavelength is only a few nanometers away from the transmission band. Recently, transmission efficiency of these filters has been optimized to >95%, so that there are minimal losses within the detection pathway. Long-pass emission filters can be used to detect a redshifted tail of the fluorescence emission for bright samples. However, at low-light levels, Raman scattering of water can become significant and band-pass emission filters are recommended to block this background noise. The Raman peak is redshifted to the excitation wavelength by 3400 cm^{-1} for pure water. For example, excitation of a Cy3- or rhodamine-stained sample with a 532-nm laser line results in Raman scattering at 649 nm. Using a band-pass filter that collects fluorescence emission between 540 and 620 nm is likely to provide the highest signal to noise.

Objectives

To achieve TIR, a specimen must be illuminated with a numerical aperture greater than the refraction index of the media (n = 1.38 for cells). It is therefore possible to use 1.4-NA oil-immersion objectives, but only a very small fraction (0.02) of the lens numerical aperture can be used. Major microscope vendors now offer up to 1.49 NA for oil-immersion objectives that can send the beam to the sample at higher incidence angles. Alignment of TIR with these objectives is therefore much easier. We use Plan Apochromatic objective lenses in TIRFM because they are well corrected for chromatic aberration and for flatness of field, resulting in high-quality images.

Olympus also offers a 100X 1.65-NA oil Apochromatic objective to achieve a very steep angle in illumination. Increasing the angle of incidence generates a thinner evanescent field and further reduces the background. The high NA of the objective also increases the resolving power of the microscope, which is limited by the diffraction limit (λ/2 NA) of light. However, this objective requires a sapphire coverslip (n = 1.78) with matching volatile immersion liquid, a costly and impractical solution for everyday usage. The 1.65-NA objective is usually used only for very weak signals.

Fluorescence Detection and Image Acquisition

Because TIRFM is a wide-field illumination technique, the image of the object is formed on the surface of a camera. CCD cameras are used in TIRFM to detect the fluorescence signal. Image acquisition speed of the camera, which is a function of camera readout rate and pixel size, should be compatible with the dynamics of the biological system under study. For example, tracking the diffusion of transmembrane proteins, which is on the order of 1 μm^2/sec, requires a millisecond temporal resolution. High acquisition speeds can be achieved by either selecting a small region of interest or by increasing the readout rate of the camera. For weak signals (i.e., single-particle tracking), the system noise is mainly determined by the readout noise of the camera. Intensified CCD cameras are used to multiply the incoming photons before they reach the CCD chip. Therefore, the effect of the readout noise is minimized. However, the image intensifier causes a spread of photons to neighboring pixels, which results in wider point spread functions and blurry images. Although intensified CCDs can achieve high gain and fast acquisition speeds, they are relatively expensive and susceptible to photodamage.

Electron-multiplying (EM) CCD cameras now dominate the market for imaging weak signals. In an EMCCD, photons are multiplied after they reach the chip by an on-chip gain registry. By applying high clock voltages, secondary electrons can be obtained by a process called impact ionization.

Even though this process is stochastic and probability of secondary electron generation is low (0.013), transferring the charges in CCD registry multiple times can achieve a 1000-fold gain in the signal. Therefore, EMCCDs are capable of detecting single photon events by eliminating the readout noise of the output amplifier. However, the stochastic nature of the impact ionization process adds an uncertainty to the number of counts per pixel. This is called an *excess noise factor*, which multiplies photon shot noise by 1.4. The high clock voltages may also produce a secondary electron even when no primary electron is present for transfer. Therefore, the current state of EM is beneficial for low-light-level imaging. At higher signals, excess noise factors in EM registry reduce the signal-to-noise ratio. Andor Technology, Hamamatsu, Roper Scientific, and QImaging offer various chip sizes of these cameras, which can also be operated as a conventional CCD for high signal levels. By using back-illuminated chip design, 92% of the incoming photons can be detected in the visible region of light. To perform photon acquisition and readout simultaneously, interline or frame-transfer chip design are commonly used. In short, these cameras can detect the signal from single fluorophores with high sensitivity by generating minimal detection noise. Currently, 2-msec temporal resolution can be achieved with 10-MHz readout of a small-chip (128 x 128) EMCCD. New cameras with higher readout rates and lower excess noise factors are anticipated in the near future.

Commercial TIRF Systems

Objective-type TIRFM setups have been commercialized recently by major microscope vendors (Agilent Technologies, Leica, Nikon, Carl Zeiss, and Olympus). Commercial TIRFM systems are equipped with multiple laser sources coupled to an optical fiber and controlled through an AOTF. The fiber is collimated in front of a TIRF illuminator that consists of a focusing lens and a motorized *x–y–z* stage to adjust the illumination angle. The illuminator spreads out the light over the full field of view within the binoculars and minimizes the amount of scattered light from the objective. High power lasers are recommended for commercial systems, because a large portion of light (50%) is lost through the fiber coupling, and increasing the field of view decreases the laser power per area. These systems are easier to use and do not require much experience with optics to align them. Many cell biology laboratories have started to use commercial systems for live cell imaging applications.

TIRFM APPLICATIONS

TIRFM is well suited for studying single molecule dynamics for time-resolution demanding applications. Early studies were performed on purified proteins immobilized onto a glass coverslip. In the initial reports of TIRFM for single molecule imaging, Funatsu and colleagues showed binding and dissociation of fluorescently labeled nucleotides to single myosin motors (Funatsu et al. 1995). By monitoring the intensities of diffraction limited spots corresponding to single molecules, diffusional movement of rhodamine-labeled membrane proteins was observed within a lipid bilayer (Schmidt et al. 1996). The TIRFM pathway can be modified to perform fluorescence polarization or fluorescence resonance energy transfer assays to measure conformational changes of motors in vitro (Sosa et al. 2001; Mori et al. 2007). In addition, superior image quality and high temporal resolution of TIRFM allows for high-precision tracking of the diffraction-limited image of single molecules. The center of the fluorescent spot is localized within 1 nm precision using a Gaussian algorithm to study the movement of motor proteins (Yildiz et al. 2003); a method that will be discussed in Chapter 33.

TIRFM provides a significant advantage for exploring protein dynamics on a plasma membrane in live cells. The evanescent field eliminates the high autofluorescence of the cytoplasm and allows tracking of individual membrane proteins. By measuring the intensity and the diffusion constant of GFP-tagged membrane proteins, cluster formation of focal-adhesion molecules (Iino et al. 2001), ion channels (Harms et al. 2001), and anchoring proteins (Murakoshi et al. 2004) has been shown experimentally. Our laboratory has used glass-adsorbed lipid bilayers to introduce immune recognition proteins found in antigen-presenting cells (e.g., integrins, such as ICAM1) to T cells. As first shown by Dustin and colleagues, these bilayers can mimic the antigen-presenting cells and create a

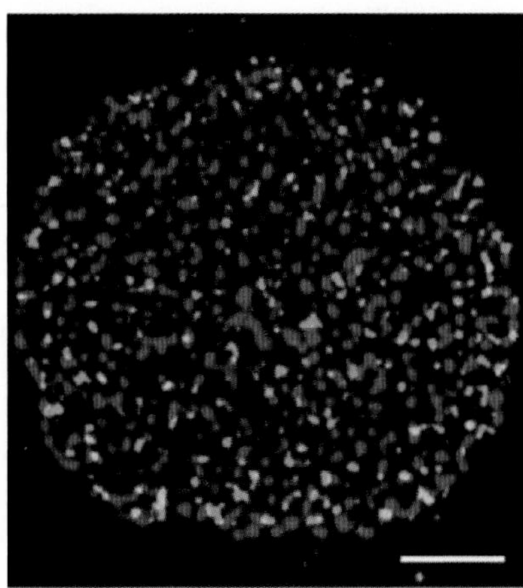

FIGURE 4. Multicolor tracking of single molecules in a T cell membrane. Superimposed images of single molecules of green fluorescent protein (GFP)-tagged Lck kinase (green) and a confocal image (multiple molecules imaged) of the membrane receptor CD2 (red) in Jurkat T cells. Diffusion of individual kinase molecules was tracked relative to CD2-containing microdomains by TIRFM. Scale bar, 5 µm. (Reprinted from Douglass and Vale 2005.)

complex signaling structure called the "immunological synapse" (Grakoui et al. 1999). Using multicolor TIRFM, we have imaged fluorescently labeled proteins on the bilayer and the T-cell membrane simultaneously, and have observed the formation of a microdomain of signaling proteins during T cell signaling (Fig. 4) (Douglass and Vale 2005).

Cellular components close to the membrane also can be observed by TIRF, because the characteristic depth of the evanescent wave is higher in live cells. We have used multicolor TIRFM to image myosin and microtubules at the cortex during the process of cytokinesis (Fig. 5) (Vale et al. 2009). To facilitate the use of TIRFM for such applications, cells can be spread and flattened on the coverslip by using adhesion molecules (e.g., concanavalin A).

Recent advances in TIRFM imaging have made a huge impact in understanding the mechanism and function of cellular components including membrane proteins, signaling cascades, and molecular motors. The relative ease of alignment and use of TIRF combined with high sensitivity in single-molecule detection makes it an extremely powerful technique to tackle a wide array of cellular questions. The availability of the commercial systems is now allowing much wider applications of TIRFM in cell biology laboratories.

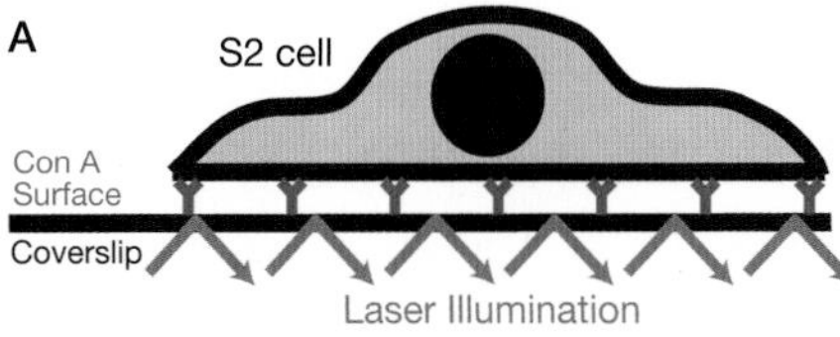

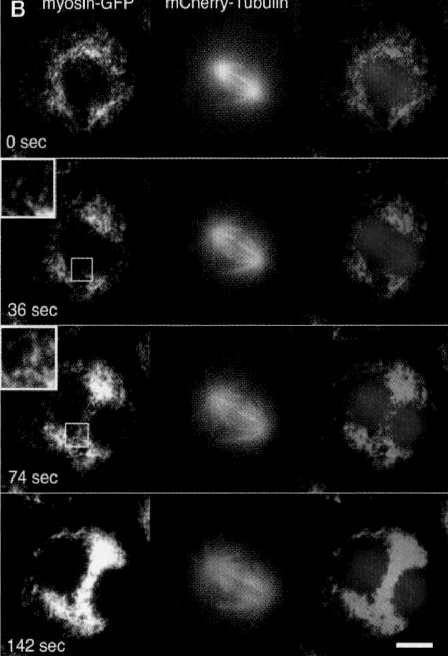

FIGURE 5. TIRFM Imaging of myosin proteins during cell division. (*A*) *Drosophila* S2 cells were spread over a concanavalin A–coated glass surface and imaged under TIRFM. (*B*) Sequence of TIRFM images of GFP-myosin (*left*) superimposed onto wide-field images of mCherry-tubulin (*right*) shows the redistribution of myosin during anaphase. Scale bar, 10 µm. (Reprinted from Vale et al. 2009.)

Protocol

Tracking Movements of the Microtubule Motors Kinesin and Dynein

The principles of TIRFM are illustrated in this protocol, in which the movements of motor proteins are imaged as they move along microtubules within live axonemes.

MATERIALS

CAUTION: See Appendix 6 for proper handling of materials marked with <!>.
See the end of the chapter for recipes for reagents marked with <R>.

Reagents

β-mercaptoethanol (BME) <!>
Biotinylated motor protein (either dynein or kinesin)
BRB12 buffer <R> for kinesin
Casein (blocking agent) solution prepared from lyophilized powder
Do not use dry milk because it contains massive amounts of biotin.
Dynein loading buffer (DLB) <R> for dynein motility
Glucose, 40% (w/v)
Mg^{++}-ATP, 100 mM
Oxygen-scavenging agents (glucose oxidase <!> [100 mg/mL] and catalase) (Yildiz et al. 2003)
Quantum dots, streptavidin-coated
Sea urchin axonemes

Equipment

Coverslip
Microscope slide
Nail polish
Tape, double-sided
TIRFM setup equipped with 488-nm laser, quantum dot filter set, and an EMCCD camera

EXPERIMENTAL METHOD

Sample Preparation

1. Prepare a flow chamber by sandwiching a slide and a clean coverslip with two pieces of double-sided tape. Solutions flow through the two open ends of the chamber, which holds ~10 µL of aqueous solution.
2. Flow 10 µL of sea urchin axonemes (1 mg/mL) into the chamber. Axonemes stick nonspecifically to the chamber surface and are stable for several hours at room temperature.
3. Remove unbound axonemes with 50 µL of either BLB12 or DLB buffer.
4. Block the surface with buffer containing 1 mg/mL casein (buffer+casein).
5. Add 40 µL of 100 pM biotinylated motor protein diluted in buffer+casein.

6. Remove unbound motor protein with two washes of 100 µL buffer+casein.
7. Add 10 µL of 1 µM quantum dots and incubate the sample for 5 min to allow streptavidin-biotin attachment.

   Labeling the microtubule-bound motors with quantum dots is performed to prevent crosslinking of multiple motors to single quantum dots. 1 mg/mL of casein is absolutely required to prevent the quantum dots from sticking to the chamber surface.

8. Remove excess quantum dots with two washes of 100-µL buffer+casein.
9. Add 20 µL of buffer+casein containing 1 mM ATP, 140 mM of BME, and 1 mg/mL casein and oxygen scavenging mixture (1 µL glucose oxidase, 1 µL catalase, and 1 mL glucose). BME enhances the brightness of the quantum dots by reducing the rates of fluorescent blinking (Rasnik et al. 2006).
10. Seal the sample with clean nail polish.

Image Acquisition and Data Analysis

11. Excite the samples with 20 mW of a 488-nm laser with TIRFM.
12. Find the coverslip surface by locating the back-reflected beam on TIRF lens next to the illumination spot and by observing the quantum dot signal through microscope binoculars.
13. Direct the fluorescence signal to the camera port.
14. Acquire images at a 10-MHz readout rate in a frame transfer mode. EM cameras with 128 x 128 pixels and 512 x 512 pixels can acquire 500 frames and 30 frames per second, respectively.
15. Use 60x EM gain to reduce the readout noise rate to below 1 electron rms.
16. Collect 2,000–20,000 images per movie.
17. Use the program of choice (e.g., IDL, MATLAB, C++) to find fluorescent spots with Gaussian masking and track the spots throughout the movie. Using a two-dimensional Gaussian fitting, determine the position of the quantum dots in each image.
18. The resulting position versus time plot can be fitted to a stepping algorithm to determine the sizes of the nanometer-range steps taken by the molecular motors.

RECIPES

Recipes for reagents marked with <R> are included in this list.

BRB12 Buffer

PIPES (pH 6.8)	12 mM
EGTA	1 mM
$MgCl_2$	2 mM

DLB Buffer

HEPES (pH 7.2)	25 mM
EGTA	1 mM
$MgCl_2$	2 mM

ACKNOWLEDGMENTS

We thank Dr. Nico Stuurman for his comments and Dr. Jigar Bandaria for carefully reviewing the manuscript.

REFERENCES

Axelrod D, Burghardt TP, Thompson NL. 1984. Total internal reflection fluorescence. *Annu Rev Biophys Bioeng* **13:** 247–268.

Douglass AD, Vale RD. 2005. Single-molecule microscopy reveals plasma membrane microdomains created by protein–protein networks that exclude or trap signaling molecules in T cells. *Cell* **121:** 937–950.

Funatsu T, Harada Y, Tokunaga M, Saito K, Yanagida T. 1995. Imaging of single fluorescent molecules and individual ATP turnovers by single myosin molecules in aqueous solution. *Nature* **374:** 555–559.

Grakoui A, Bromley SK, Sumen C, Davis MM, Shaw AS, Allen PM, Dustin ML. 1999. The immunological synapse: A molecular machine controlling T cell activation. *Science* **285:** 221–227.

Harms GS, Cognet L, Lommerse PH, Blab GA, Kahr H, Gamsjäger R, Spaink HP, Soldatov NM, Romanin C, Schmidt T. 2001. Single-molecule imaging of l-type Ca^{2+} channels in live cells. *Biophys J* **81:** 2639–2646.

Iino R, Koyama I, Kusumi A. 2001. Single molecule imaging of green fluorescent proteins in living cells: E-cadherin forms oligomers on the free cell surface. *Biophys J* **80:** 2667–2677.

Mori T, Vale RD, Tomishige M. 2007. How kinesin waits between steps. *Nature* **450:** 750–754.

Murakoshi H, Iino R, Kobayashi T, Fujiwara T, Ohshima C, Yoshimura A, Kusumi A. 2004. Single-molecule imaging analysis of Ras activation in living cells. *Proc Natl Acad Sci* **101:** 317–322.

Rasnik I, McKinney SA, Ha T. 2006. Nonblinking and long-lasting single-molecule fluorescence imaging. *Nat Methods* **3:** 891–893.

Schmidt T, Schütz GJ, Baumgartner W, Gruber HJ, Schindler H. 1996. Imaging of single molecule diffusion. *Proc Natl Acad Sci* **93:** 2926–2929.

Sosa H, Peterman EJ, Moerner WE, Goldstein LS. 2001. ADP-induced rocking of the kinesin motor domain revealed by single-molecule fluorescence polarization microscopy. *Nat Struct Biol* **8:** 540–544.

Tokunaga M, Kitamura K, Saito K, Iwane AH, Yanagida T. 1997. Single molecule imaging of fluorophores and enzymatic reactions achieved by objective-type total internal reflection fluorescence microscopy. *Biochem Biophys Res Commun* **235:** 47–53.

Vale RD, Spudich JA, Griffis ER. 2009. Dynamics of myosin, microtubules, and Kinesin-6 at the cortex during cytokinesis in *Drosophila* S2 cells. *J Cell Biol* **186:** 727–738.

Yildiz A, Forkey JN, McKinney SA, Ha T, Goldman YE, Selvin PR. 2003. Myosin V walks hand-over-hand: Single fluorophore imaging with 1.5-nm localization. *Science* **300:** 2061–2065.

39 Fluorescence Correlation Spectroscopy

Principles and Applications

Kirsten Bacia,[1] Elke Haustein,[2] and Petra Schwille[2]

[1]*HALOmem, Martin-Luther-Universtität Halle-Wittenberg, D-06120 Halle, Germany;*
[2]*BioTec, Technische Universität Dresden, D-01307 Dresden, Germany*

ABSTRACT

Fluorescence correlation spectroscopy (FCS) is used to study the movements and the interactions of biomolecules at extremely dilute concentrations, yielding results with good spatial and temporal resolutions. Utilizing a number of technical developments, FCS has become a versatile technique that can be used to study a variety of sample types and can be advantageously combined with other methods. Unlike other fluorescence-based techniques, the analysis of FCS data is not based on the average intensity of the fluorescence emission but examines the minute intensity fluctuations caused by spontaneous deviations from the mean at thermal equilibrium. These fluctuations can result from variations in local concentrations due to molecular mobility or from characteristic intermolecular or intramolecular reactions of fluorescently labeled biomolecules present at low concentrations. In this chapter, we provide a basic introduction to FCS, including its technical development and theoretical basis, experimental setup of an FCS system, adjustment of a setup, data acquisition, and analysis of FCS measurements. Finally, the application of FCS to the study of lipid bilayer membranes and to living cells is discussed. For an introduction to the imaging analog of this technique, see Chapter 40.

INTRODUCTION

A steadily increasing demand for better minimally invasive analytical tools to answer specific biological questions has boosted the development of fluorescence-based techniques. FCS is a well-established method for analyzing solutions of biomolecules at low concentrations, ranging from nanomolar to low micromolar. One of the advantages of FCS is that it requires only small sample volumes. In a basic type of experiment, the optically delimited, femtoliter-sized FCS detection volume is placed inside a drop of a few microliters of a homogeneous sample solution. In a more sophisticated experiment, the FCS detection volume can be positioned inside or on the membrane of a single eukaryotic cell, on a small biosynthetic container (such as a giant liposome), or inside a microfluidic structure. FCS is a rather flexible method with regard to the sample preparation and the optical setup and can be readily combined with other microscopic techniques such as confocal microscopy or atomic force microscopy.

FCS is often referred to as a single-molecule technique because the fluctuations in fluorescence intensity arise from fluctuations in the number or in the brightness of single fluorescent particles. However, each particle is not followed and analyzed individually, which is why FCS may be more accurately described as a single-molecule-sensitive technique.

All fluctuating physical parameters influencing the fluorescence signal are, in principle, accessible by FCS. These comprise local concentrations, mobility coefficients (translational diffusion, active transport, rotational diffusion), and rate constants of intramolecular and photophysical reactions (triplet blinking and photon antibunching). Moreover, processes that indirectly influence these parameters can be analyzed, such as a molecular binding that increases the size and reduces the diffusion coefficient of a molecule. In addition, dual-color fluorescence cross-correlation spectroscopy (dcFCCS) can be implemented to analyze the correlated motion of two distinct fluorescent labels. In this way, cleavage and binding reactions as well as dynamic colocalization during intracellular transport can be studied.

TECHNIQUE DEVELOPMENT

The theoretical foundation of FCS is based on the laws of molecular diffusion, formulated from Brown's observations of random particle motion, and their analysis by Einstein and Smoluchowski at the beginning of the 20th century (Smoluchowski 1906). FCS as a technique was developed in the early 1970s (Magde et al. 1972, 1974; Elson and Magde 1974) as a "miniaturization" of dynamic light scattering. The novelty of FCS lay in the analysis of fluctuations in the fluorescence emission of the sample molecules, induced by spontaneous fluctuations of physical parameters. FCS was first used to measure the diffusion and chemical kinetics of ethidium bromide intercalation into DNA (Magde et al. 1972) and rotational diffusion (Ehrenberg and Rigler 1974). Following these proof-of-principle measurements, a variety of studies were devoted to the investigation of particle concentration and mobility: for example, three-dimensional and two-dimensional (2D) diffusion (Aragón and Pecora 1976; Fahey et al. 1977) or laminar flow (Magde et al. 1978). In an attempt to restrict the system under investigation to small molecular numbers and, thus, to enhance both detection sensitivity and background suppression, Rigler and coworkers combined FCS with confocal detection (Rigler and Widengren 1990; Rigler et al. 1993).

The analytical potential of FCS for the life sciences has been shown in numerous applications since the original work of Rigler and coworkers (Eigen and Rigler 1994; Schwille et al. 1997b). FCS was successfully used to study association and dissociation of nucleic acids (Kinjo and Rigler 1995) and proteins (Rauer et al. 1996). Moreover, a multitude of environmental effects inducing fluctuations in the fluorescence yield of single dye molecules could be studied, including reversible protonation (Haupts et al. 1998), electron transfer, and oxygen and ion concentrations.

By introducing a dual-color cross-correlation scheme for the simultaneous observation of different fluorescent species (Schwille et al. 1997a), the detection specificity in bimolecular association and dissociation processes was significantly enhanced. Combined with extremely short data acquisition times, this even allowed for real-time investigation of enzyme kinetics (Kettling et al. 1998). Cellular measurements were also reported at a very early stage (Elson et al. 1976). However, in turbid media, especially (although not exclusively) inside the cell, the signal-to-noise ratio is influenced by autofluorescence and scattering. In addition to this, photobleaching may irreversibly deplete the limited supply of labeled molecules within the cell, thus, restricting the ability of FCS to measure within small intracellular compartments or small types of cells.

The first commercial FCS instrument was released in 1996 by Carl Zeiss (ConfoCor), permitting the first turnkey FCS applications and triggering the evolution of FCS into a standard technique (Weisshart et al. 2004). Subsequent models added laser-scanning modules and were able to perform dual-color cross correlation. Many cell biologists use intracellular FCS applications to investigate in situ dynamics of fluorescent probes. Moreover, the complexity of many biological systems of interest, including developing tissue and organisms, is fostering innovation and improvements in FCS and related optical methods.

Various extensions and variations of FCS have been introduced. Because of its inherent depth discrimination, two-photon excitation (TPE) was proposed to reduce the problem of out-of-focus bleaching (Denk et al. 1990). The first two-photon–FCS experiments performed in cells labeled with fluorescent beads were reported by Gratton and colleagues (Berland et al. 1995). Four years later, FCS was sensitive enough to detect single molecules (Schwille et al. 1999). Two-photon-induced transitions to the excited state show selection rules that are different from those of their corresponding one-photon equivalent. This makes it possible to accomplish simultaneous excitation of spectrally distinct dyes (Heinze et al. 2000, 2002, 2004).

Optical configurations other than the standard confocal setup are being explored with the goal of reducing the size of the detection volume. These include total internal reflection (TIR) (Starr and Thompson 2001); zero-mode waveguides, which consist of subwavelength apertures within a metal film (Levene et al. 2003); and the stimulated emission/depletion approach for obtaining a detection volume below the conventional diffraction limit (Kastrup et al. 2005; Eggeling et al. 2009).

Additional information can be obtained from FCS experiments through the use of pulsed excitation with time-resolved detection, which can distinguish fluorophores based on their lifetimes (Felekyan et al. 2005; Kapusta et al. 2007). The use of alternating (pulsed interleaved) excitation with a matching detection scheme in dcFCCS eliminates spectral cross-talk artifacts, which can cause a false-positive or increased cross-correlation amplitude, and can complicate the interpretation of dcFCCS data (Muller et al. 2005; Thews et al. 2005).

Conventional FCS rapidly provides dynamical information about molecules at one spot within a sample, but it is very slow at providing spatial information because FCS measurements at different locations have to be acquired sequentially. In contrast, conventional laser-scanning microscopy provides spatial information about the time-averaged fluorescence intensity, but is not as well suited for picking up temporal changes. One variant of FCS that can provide spatial information is parallel multispot FCS, which originally used two parallel detection volumes (Brinkmeier et al. 1999) and has more recently been extended by using charge-coupled device (CCD) detectors (Burkhardt and Schwille 2006; Kannan et al. 2006). Another modification to FCS that makes it possible to obtain both spatial and temporal information is to scan the beam (or the sample) in a linear, a circular, or a random fashion (scanning fluorescence correlation spectroscopy [sFCS]). sFCS is valuable for measuring very slow diffusion phenomena, which are often encountered with membrane probes (Ruan et al. 2004; Ries and Schwille 2006). Raster image correlation spectroscopy (RICS) was recently developed as a type of sFCS that can be performed using conventional laser-scanning microscopes (Digman et al. 2005).

SELECTING A FLUORESCENT LABEL

The outstanding sensitivity and selectivity of FCS is achieved by using strong fluorescent tags for detection because most naturally occurring fluorophores in biological systems are inherently dim. The most crucial decision is, therefore, the selection of the proper fluorescent tag. A good fluorophore undergoes, on average, about 10^6 excitation cycles before being irreversibly destroyed by photobleaching. To achieve good results, it is important to choose a photostable fluorophore within the proper wavelength range. Moreover, biologically relevant phenomena must be distinguished from potential dye-induced artifacts through control experiments. During the past decade, along with the growing impact of fluorescence techniques on the life sciences, techniques and tools for protein labeling have greatly improved.

Autofluorescent proteins, for example, can be incorporated into proteins by recombinant expression, thus allowing for good labeling efficiency and inherent biocompatibility. Novel photostable synthetic chromophores are available from various manufacturers in a wide range of colors from ultraviolet (UV) to near-infrared and with different functional groups. In addition, new developments in mesoscopic physics have made it possible to exploit completely novel fluorescent systems: semiconductor nanocrystals, also known as quantum dots.

The choice of fluorescent dyes may be limited by the available laser lines and the corresponding filter systems. For one-color applications, a wide range of emission wavelengths may be used, although when venturing toward the UV or infrared, spectral region limitations due to detector sensitivity and the transmission curves of relevant optical components have to be considered. Multicolor applications require that chromophores be spectrally distinguishable (i.e., have minimal spectral overlap). This requirement depends on the form of the emission spectra and the chosen filters, but generally, the larger the separation between the emission peaks, the better. A typical synthetic dye combination is Alexa 488 and Cy5 (or Alexa 647). The green fluorophore is very hydrophilic and, thus, ideal for solution measurements, whereas Cy5 is rather hydrophobic and also displays a quite complicated photophysical behavior.

Synthetic Fluorophores

Synthetic fluorophores offer the flexibility to choose a label having suitable spectral and photophysical characteristics. The label does, however, have to be attached to the biomolecule of interest. Proteins are most often labeled on lysine or cysteine residues of purified proteins. Labeling lysine residues typically yields a mixed population of protein molecules carrying varying numbers of fluorophores. Although the resulting heterogeneity in the fluorescence brightness does not affect measurements of the diffusion coefficient, it does affect the correlation amplitudes and is an issue in the quantitative interpretation of dcFCCS measurements (Kim et al. 2005). Therefore, labeling a single engineered cysteine side chain is often the method of choice for obtaining a homogeneously labeled preparation. Alternatively, approaches such as expressed protein ligation (Becker et al. 2006) can be used to obtain stoichiometric coupling of the desired fluorophore to the protein of interest.

The small synthetic fluorophores (<1 kDa) most commonly used in FCS are rhodamine and cyanine dyes and the more hydrophilic derivatives of common dyes such as Alexa dyes (Molecular Probes/Invitrogen). BODIPY dyes (Molecular Probes/Invitrogen) have been used as labels for small ligand molecules (Briddon et al. 2004) or on lipid chains (Molecular Probes/Invitrogen). Although labeling a large protein with a small synthetic fluorophore often has only a minimal effect on the activity of the protein (depending on the site of modification), fluorescence labeling of lipids can alter their properties dramatically, as seen from studies of lipid phase preference of a labeled cholesterol (Ariola et al. 2009) and a labeled GM1 ganglioside (Bacia et al. 2005).

Autofluorescent Proteins

Originally isolated from the light-emitting organ of the jellyfish *Aequorea victoria* by Shimomura et al. (Shimomura et al. 1962) and cloned by Prasher et al. (Prasher et al. 1992), the green fluorescent protein (GFP) has one distinct advantage compared with synthetic chromophores: This protein and—even more important—chimeric GFP-fusion proteins can be expressed in situ in cells. The wild-type protein (wtGFP) shows absorption and emission maxima at 397 and 504 nm, respectively. Altering the structure can influence the excitation and the emission properties, so that now, a variety of different wavelengths is accessible. The most commonly used GFP variant contains an S65T mutation and has its absorption peak at 489 nm, which is convenient for excitation with the 488-nm line of an argon laser (S65T-GFP; also F64L/S65T-GPF, formerly available as EGFP from Clontech; Fig. 1).

Proteins fluorescing more toward the red spectral region are of special interest for measurements in vivo because most cells display reduced autofluorescence at longer wavelengths. Additionally, these proteins are also extremely useful for multicolor binding assays. Matz et al. (Matz et al. 1999) reported the discovery of novel GFP-like proteins from Anthozoa (corals). One of them, originating from the coral *Discosoma* sp. and now known as DsRed, has its emission maximum at 583 nm. Unfortunately, DsRed oligomerizes (Baird et al. 2000), but newer monomeric red fluorescent proteins, such as mRFP1, mStrawberry, and mCherry, are now used for FCS (Hendrix et al. 2008).

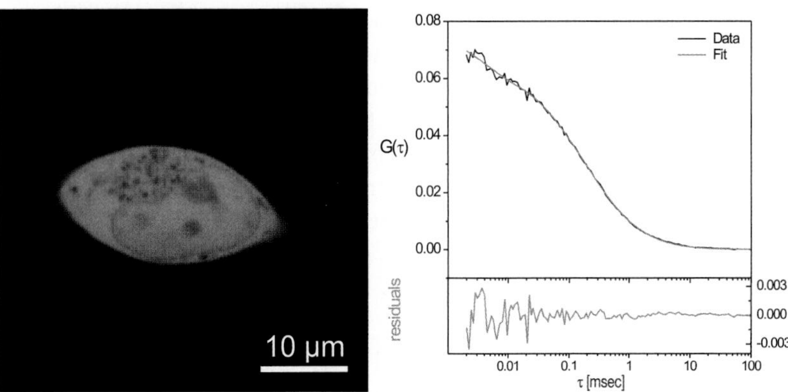

FIGURE 1. FCS in live cells. *Left*: Confocal laser-scanning image of cells expressing eGFP, excited at 488 nm. *Right*: One-photon excitation autocorrelation curve recorded for eGFP freely diffusing in the cytosol. Fitting a model curve (marked Fit) to the data (marked Data) yields a diffusion time of ~0.25 msec, corresponding to a diffusion coefficient of about $D = 3 \times 10^{-11}$ m²/sec. The residuals, which represent the deviations of the data from the fit curve, show that a good fit was obtained. See Figure 3 for more examples on fitting.

One drawback with fluorescent proteins is the time needed for their maturation, which can result in mixed populations. Another drawback are pH-dependent (Haupts et al. 1998) and excitation-dependent conversions to a nonfluorescent state (Schwille et al. 2000; Malvezzi-Campeggi et al. 2001), causing a flickering in the fluorescence signal. If the timescales of the flickering and the diffusion processes are sufficiently separated, the flickering can be accounted for in the analysis of the FCS curves. However, flickering at longer timescales can severely limit the usability of red fluorescent proteins (Hendrix et al. 2008).

THEORETICAL OUTLINE OF FCS

Autocorrelation

Tiny fluctuations in the fluorescence signal from excited molecules within the focal volume are continuously occurring at ambient temperatures. They are generally observed as (unwanted) noise patterns of the measured signal. These fluctuations can be quantified with respect to their strength and duration by temporal autocorrelation of the recorded intensity signal. Autocorrelation analysis provides a measure of the self-similarity of a time series signal and, therefore, describes the persistence of the information that it carries. Essential information about processes governing molecular dynamics can, thus, be derived from the temporal pattern by which fluorescence fluctuations arise and decay. Thus, the normalized autocorrelation function for the fluorescence fluctuations $\delta F(t)$ of the signal $F(t)$ is defined as

$$G(\tau) = \frac{\langle F(t) \cdot F(t+\tau) \rangle}{\langle F(t) \rangle^2} = \frac{\langle \delta F(t) \cdot \delta F(t+\tau) \rangle}{\langle F(t) \rangle^2} + 1, \qquad (1)$$

with

$$\delta F(t) = F(t) - \langle F(t) \rangle \quad \text{and} \quad \langle F(t) \rangle = \frac{1}{T} \int_0^T F(t)\, dt. \qquad (2)$$

It is a temporal average of the signal F at time t multiplied by the signal F at a later time $t + \tau$, normalized by the square of the average of F over the acquisition time T. This autocorrelation routine provides a measure of the self-similarity (i.e., the memory) of a time signal after the lag time τ and highlights characteristic time constants of the underlying processes.

Because the relative fluctuations become smaller with increasing numbers of measured particles, it is important to have a small number of molecules in the focal volume. Roughly, the temporal average of the particle number should be between 0.1 and 1000, which corresponds to concentrations between subnanomolar (~10^{-10} M) and micromolar (~10^{-6} M) for a diffraction-limited focal volume (<1 fL). The most common model for the observation volume is a simple three-dimensional Gaussian profile, which was introduced by Aragón and Pecora (1975) and was found to be a good approximation for a confocal setup (Rigler et al. 1993). In the radial direction, the $1/e^2$ radius is given by r_0; in the axial direction, it is given by z_0. The effective focal volume V_{eff} can be defined as

$$V_{eff} = \pi^{3/2} \cdot r_0^{\ 2} \cdot z_0.$$

Note that the parameters r_0 and z_0 serve to characterize the size and shape of the detection volume. The detection volume depends not only on the illumination beam but also on the degree of overfilling of the back aperture of the objective and on the size of the pinhole.

Cross Correlation

When performing an autocorrelation analysis, a measured signal is compared with itself, and hidden temporal patterns are highlighted. The process of looking for common features of two independently measured signals F_i and F_j by cross correlation not only removes unwanted artifacts introduced by the detector (e.g., the so-called afterpulsing of an avalanche photodiode [APD]), but also results in much higher detection specificity. Generalizing Equation 1, the normalized cross correlation is given by

$$G_\times(\tau) = \frac{\langle \delta F_i(t) \cdot \delta F_j(t+\tau) \rangle}{\langle F_i(t) \rangle \cdot \langle F_j(t) \rangle} + 1. \qquad (3)$$

For dcFCCS, two spectrally distinct dyes are excited within the same detection element (Schwille et al. 1997a) and used to probe interactions between different molecular species. Assuming ideal conditions, in which both channels have the same effective volume element V_{eff}, fully separable emission spectra, and a negligible emission-absorption overlap integral, the following correlation curves can be derived:

Autocorrelation: $$G_{i,j}(\tau) = \frac{\left(\langle C_{i,j} \rangle M_i(\tau) + \langle C_{ij} \rangle M_{ij}(\tau)\right)}{V_{eff}\left(\langle C_{i,j} \rangle + \langle C_{ij} \rangle\right)^2}, \qquad (4)$$

Cross correlation: $$G_\times(\tau) = \frac{\langle C_{ij} \rangle M_{ij}(\tau)}{V_{eff}\left(\langle C_i \rangle + \langle C_{ij} \rangle\right)\left(\langle C_j \rangle + \langle C_{ij} \rangle\right)}, \qquad (5)$$

where $M_{ij}(\tau)$ is the motion-related part of the correlation function, $C_{i,j}(\vec{r}, t)$ are the concentrations for the single-labeled species, i, j, and $C_{ij}(\vec{r}, t)$ is the concentration of the double-labeled species. More complicated situations are discussed, for example, in Weidemann et al. (2002), Hess and Webb (2002), and Kim et al. (2005). The amplitude of the cross-correlation function is directly proportional to the concentration of double-labeled particles:

$$\langle C_{ij} \rangle = \frac{G_\times(0)}{G_i(0) \cdot G_j(0) \cdot V_{eff}}. \qquad (6)$$

Cross talk (and, thus, incomplete separation of the detection channels), which originates mainly from long-wavelength emission of the dye emitting at a shorter wavelength, leads to a nonzero amplitude of the cross-correlation curve, even in the absence of double-labeled molecules. To correct for the cross-talk artifact, the relative fluorescence signals in both detection channels (from the actual measurement) and the cross-talk ratio ("bleed-through," determined from a calibration measurement) are used (Bacia and Schwille 2007).

Fit Functions

Considering only translational diffusion for a single species and a three-dimensional Gaussian volume element, the autocorrelation function is

$$G(\tau)=\underbrace{\frac{1}{\langle C\rangle\cdot V_{\text{eff}}}}_{=\frac{1}{\langle N\rangle}}\cdot\frac{1}{1+\tau/\tau_D}\cdot\frac{1}{\sqrt{1+\left(r_0^2/z_0^2\right)\left(\tau/\tau_D\right)}}. \tag{7}$$

First, the lateral diffusion time τ_D that a molecule stays in the focal volume depends on the dimension r_0 and the setup-independent diffusion coefficient D:

$$\tau_D=\frac{r_0^2}{4\cdot D}\text{ (for one-photon excitation)}. \tag{8}$$

For known particle properties, the local viscosity can be derived from the diffusion coefficient, which scales inversely proportional to the viscosity η_V of the medium:

$$D_i=\frac{kT}{6\pi\eta_V R_{h,i}}. \tag{9}$$

$R_{h,i}$ is the hydrodynamic radius of the particle, T is the temperature, and k is the Boltzmann constant. For constant viscosity and temperature, the diffusion coefficient is inversely proportional to the hydrodynamic radius.

The term (z_0/r_0) is known as the form factor or structure parameter S, and the first term in Equation 7 represents the inverse of the average particle number in the focal volume. Therefore, provided the dimensions r_0 and z_0 are known from calibration, the local concentration of fluorescent molecules can be determined from the amplitude $G(0)$ of the autocorrelation curve.

The autocorrelation function for a 2D Gaussian observation volume, which is encountered, for example, when molecules are confined to diffusing within the plane of a lipid bilayer, is obtained from Equation 7 by taking the limit $z_0 \to \infty$:

$$G_{2D}(\tau)=\frac{1}{\langle N\rangle}\cdot\frac{1}{1+\tau/\tau_D}. \tag{10}$$

If there is more than one species of noninteracting purely diffusing particles, the autocorrelation function for multiple species can be expressed as the weighted sum of all independent contributions,

$$G(\tau)=\sum_{i=0}^{n}\frac{\left(\eta_i\cdot\langle C_i\rangle\right)^2}{\left(\sum_{i=0}^{n}\eta_i\cdot\langle C_i\rangle\right)^2}\cdot G_{ii}^{\text{diff}}(D_i,\tau), \tag{11}$$

with η_i being the molecular brightness of species i and $G_{ii}^{\text{diff}}(D_i, \tau)$ being the diffusion autocorrelation function of species i:

$$G_{ii}^{\text{diff}}(D_i,\tau)=\frac{1}{\langle C_i\rangle\cdot V_{\text{eff}}}\cdot\frac{1}{1+\left(4\cdot D_i\cdot\tau\right)/r_0^2}\cdot\frac{1}{\sqrt{1+\left(4\cdot D_i\cdot\tau\right)/z_0^2}}. \tag{12}$$

Note how the weighting takes place with the square of the brightness. This is important to consider when describing the cross correlation in the presence of spectral cross talk or when changes in brightness occur on binding.

It was hitherto assumed that the chromophore fluorescence properties are not changing while it is traversing the laser focus. Unfortunately, this assumption often does not hold for real dyes and higher excitation powers. The most common cause for a flickering in the fluorescence intensity is the transition of the dye into the first excited triplet state. As decay from this state to the ground electronic state is forbidden by quantum mechanics, the chromophore needs a comparably long time to relax back to the ground state. During these intervals, the dye cannot emit any fluorescence photons and appears dark. Indeed, one can imagine the intersystem crossing as a series of dark intervals interrupting the otherwise continuous fluorescence emission of the molecule on its path through the illuminated region. Typically, the triplet blinking mentioned above is described by a simple exponential decay:

$$G_{\text{unimol}}(\tau)=G_{\text{diff}}(\tau)\cdot G_{\text{triplet}}(\tau)=G_{\text{diff}}(\tau)\cdot\left(1+\frac{T}{1-T}\cdot e^{-\tau/\tau_{\text{triplet}}}\right). \quad (13)$$

This process appears as an additional shoulder in the measured curves at short timescales. The pH-dependent flickering of GFP and the isomerization-dependent flickering of cyanine dyes can also be described by such an exponential term, albeit with different time constants (Haupts et al. 1998; Widengren and Schwille 2000).

EXPERIMENTAL SETUP

A custom-built FCS setup can be made with standard optical parts (see Fig. 2), preferably built around an inverted microscope using one of the side ports for FCS detection. FCS instruments are also commercially available, ranging from stand-alone versions for autocorrelation measurements in solution (e.g., Hamamatsu C9413 spectrometer) to microscopes capable of bright-field and fluorescence microscopy, laser-scanning microscopy (LSM), and FCS. A confocal LSM can be modified by using add-on modules (e.g., PicoQuant's LSM upgrade kit or ISS's Alba FCS) or an integrated FCS/LSM combination microscope system can be purchased (e.g., the LSM 710 ConfoCor 3 by Zeiss, or Leica's TCS SP5 SMD FCS), which combines imaging, FCS, and dcFCCS analysis. The integrated systems come as complete packages with the necessary hardware and software, supporting both acquisition and data analyses.

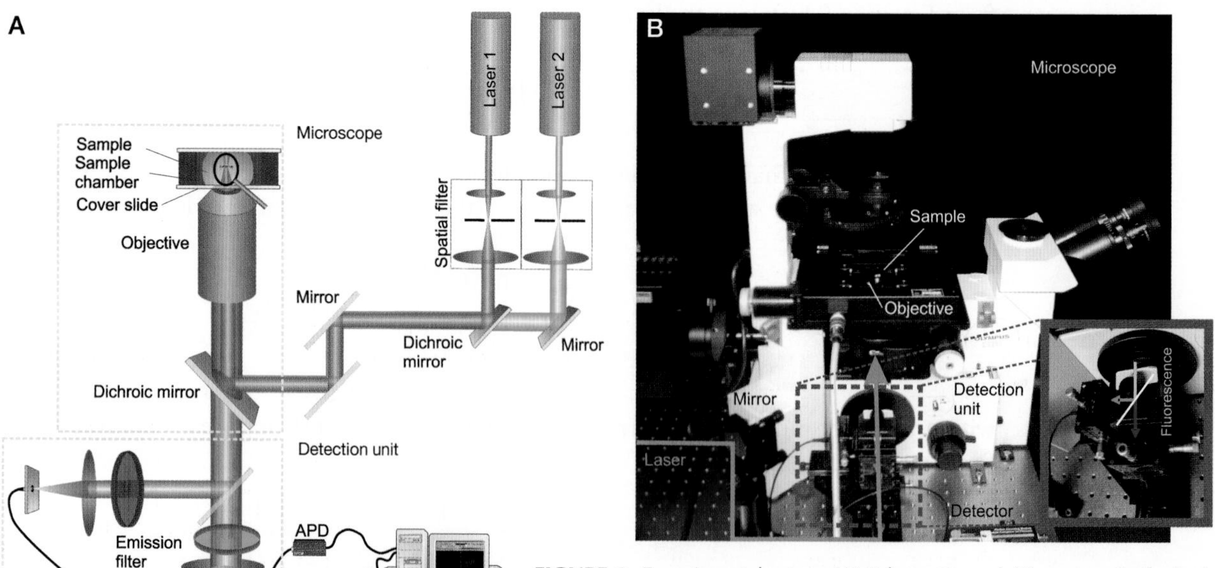

FIGURE 2. Experimental setup. (*A*) Schematic and (*B*) custom-built dual-channel FCS setup. Parallel laser light from one or two lasers is projected onto the back aperture of the objective. Overfilling the back aperture results in a diffraction-limited excitation volume in the sample. The Stokes-shifted fluorescence light from the sample passes through the main dichroic mirror, is focused onto a pinhole (for optical sectioning), and is detected by an avalanche photodiode (APD). For dcFCCS, the fluorescence emissions from the two fluorophores in the sample are separated by a secondary dichroic mirror (*B*, *inset*) and detected by two separate APDs.

Components of a Custom-Built Setup

Different kinds of single-mode lasers are used for FCS. The intensity can be regulated by neutral-density filters or acousto-optical tunable filters. For continuous-wave (cw) excitation of common dye systems, argon or argon–krypton ion lasers or solid-state lasers can be used. Inexpensive alternatives include helium–neon single-line tubes or laser diodes. In principle, parallel laser beams with Gaussian beam profiles (single-mode laser in TEM_{00} [fundamental transverse electromagnetic mode]) of one or more colors are directed via a dichroic mirror onto the back aperture of a water-immersion objective with high numerical aperture (NA; preferably >0.9, typically 1.2). Although oil-immersion objectives with higher NA values are available, the difference in refractive index between the immersion medium and the sample can lead to severe distortions of the volume elements when focusing more than a few micrometers into aqueous solution. The degree of overfilling or underfilling of the back aperture influences the size of the detection volume. Overfilling decreases the size of the focus but augments deviations from the three-dimensional Gaussian approximation (Hess and Webb 2002).

The laser beam is focused into the sample, and the redshifted fluorescence is collected by the same objective and is transmitted by the dichroic and the emission filters, which are adapted to the emission properties of the dye. A confocal pinhole of variable diameter (typically 30–100 μm) in the image plane (field aperture) ensures axial resolution. Optimization of the pinhole size is discussed in detail by Rigler et al. (1993) and Hess and Webb (2002). For convenience, the pinhole may be replaced by the entrance aperture of a multimode optical fiber with suitable core diameter. For TPE, no additional pinholes are required because of the inherent axial sectioning.

To detect the fluorescence light, photomultiplier tubes or APDs with single-photon sensitivity can be used. Traditionally, the best options were single-photon avalanche diodes from PerkinElmer. Other companies (e.g., Micro Photon Devices) now offer similar detectors with a lower quantum efficiency in the red spectral region but improved instrument response times, which may be viable alternatives for fluorescence lifetime measurements (Michalet et al. 2008).

The detector signal is correlated either by using a multiple-τ hardware correlator (e.g., from ALV GmbH or correlator.com) or by using software correlation (Magatti and Ferri 2001, 2003; Weisshart et al. 2004). Data evaluation is performed by Levenberg–Marquardt nonlinear least-squares fitting of the correlation curves. In the commercial systems, special data evaluation routines are already implemented. As a rule of thumb, data acquisition times should be at least about three orders of magnitude longer than the slowest time component to be resolved.

Adjustment

An FCS setup must be carefully adjusted before each session to ensure a detection distribution in the focal volume that is as close as possible to the theoretical model and, thus, may be approximated by a three-dimensional Gaussian. If this assumption does not hold, the parameters obtained from fitting to the model may not have the presumed meaning.

Once the appropriate chromophore is chosen, the exciting laser and, subsequently, a suitable filter system and objective lens are selected. The filter system is crucial: For one-color applications, a dichroic mirror is required with high reflectance at the wavelength of the laser and high transmission for the fluorescence. To further separate excitation and emission, an additional detection filter is required. Band-pass filters are preferred because they can also be chosen to suppress the Raman scattering at higher wavelengths. Using Alexa 488 dye and the 488-nm line of an argon-ion laser, a 495dc dichroic as a beam splitter can be combined with a D525/50 band-pass emission filter. For a good introduction to optical filter properties, see Reichman (2007).

After mounting the optical components, the laser beam must be coupled to the microscope so that it is reflected by the dichroic mirror and hits perpendicularly in the center of the back aperture of the objective lens. In commercial instruments, this is already taken care of. In a custom-built rig, the laser must pass at least two adjustable mirrors before reaching the microscope. When a fluorescent sample (either a concentrated dye solution or a colored plastic slide) is placed onto the objective, a fluorescent spot becomes visible. This must be centered in the field of view. Depending on the experimental conditions, various techniques can achieve this aim. The easiest one consists of adjusting the mirrors alternately in a procedure termed beam walking. First, the brightest position is found using the last mirror. One direction is selected, and the bright spot is moved toward the center until it is barely visible. The second mirror is used to find the brightest position again. The process is repeated until the focal spot is brightest in the center and becomes dimmer equally fast when moved to either side. Then, the pinhole—or fiber entrance—must be adjusted. In both commercial and home-built systems, the confocal pinhole is adjusted repetitively in the x and y directions first, finding the position for which the detector returns maximum intensity. After that, the same is performed for the z position, although, here, the optimum position may be less evident because of the more extended intensity distribution. The procedure is then repeated (x–y–z) until the global maximum of the fluorescence intensity is found. The position of the correction ring should be checked for final adjusting. The adjustment is then tested by recording the autocorrelation curve of a dye with known properties (see below). The adjustment is completed if the curve can be fitted well with the correct model, the structure parameter is between 4 and 6 (for one-photon excitation), and the molecular brightness is sufficiently high. Generally, it is a good idea to adjust the system using the same buffer solutions that will be used during the experiment. Also, for the detection volumes to remain unchanged, the calibration should be performed at the same laser power as will be used in the experiment.

The most basic version of an FCS setup consists of one excitation wavelength and a single detection channel. Nevertheless, the recorded data reveal a wealth of information about local concentrations, mobility parameters, and even photophysical or chemical reactions. By using a second detection channel, significantly more can be achieved than merely reducing detector artifacts. dcFCCS allows a comparison of two spectrally distinct chromophores, whereas spatial FCCS highlights similarities between two different regions in the sample. For dual-color or multicolor applications, laser beams can be combined using dichroic mirrors, or a multiline laser may be used. It can be hard to obtain a stable and accurate superposition of two confocal volumes using separate laser lines, but the precision and the stability of laser beam superposition have greatly improved with the newest generation of commercial setups. Alternatively, TPE can be exploited to excite two distinct dyes using just a single laser line.

The detection channels are also separated by a dichroic mirror and specific detection filters. In principle, adjustment is the same as for one channel, starting, however, with excellent overlap of the laser beams. The excitation volumes are then centered, and the overlap is checked by eye for a custom-built setup. On the detection side, first, the short-wavelength laser is used to adjust the shorter-wavelength detection channel. The spectral cross talk of the shorter-wavelength chromophore is then used to adjust the longer-wavelength detection channel for maximum overlap: Both the autocorrelation and the cross-correlation curves should overlap completely. When the longer-wavelength laser is switched on, the adjustment of the corresponding detection channel should not need to be altered. The quality of the adjustment can be assessed through the use of reference samples (e.g., a double-labeled stable nucleic acid sample from IBA GmbH Göttingen) and by scanning the detection volume overlap with a very small or very thin sample (e.g., a double-labeled lipid bilayer [Bacia and Schwille 2007]).

Integrated Commercial Setups

Standard FCS applications are most easily performed on commercial systems that rely on the extensive use of motorized instrumentation and software control. With this type of system, it is a simple matter to quickly move between filter configurations and to restore the optical alignment, making these setups convenient for work on perishable biological samples and for long-term projects in a multiuser environment.

The selection of the objective, laser line(s), dichroic(s), and filter(s) is essentially the same for both commercial and custom-built instruments, although the number of available options may differ. However, in the case of a commercial instrument, alignment of the incident laser beam(s) is normally performed by the manufacturer. On a properly adjusted FCS system, the user only needs to adjust the correction collar to account for the coverslip thickness and the refractive index of the medium. Depending on the type of setup, the user may also be able to check and to adjust the pinhole position(s) or the incident beam path.

Combining confocal imaging with FCS capabilities is especially useful for measurements in live cells because it allows three-dimensional imaging of the cells and precise positioning of the confocal volume for FCS measurements. Because the photomultiplier detectors normally used with confocal scanning are less sensitive than the APDs used for FCS, confocal images taken on samples with fluorophore concentrations suited to FCS either are very noisy or require long acquisition times. To overcome this problem, with newer instruments, the user can choose the APD detectors for high sensitivity imaging.

DATA ANALYSIS

Interpretation of FCS data generally requires fitting the curves to a mathematical model. The type of model that is applicable and the parameters that can be extracted depend on the system under investigation. For freely diffusing dye molecules in aqueous solution, a model containing one diffusing species and a term describing the (almost) omnipresent triplet blinking is sufficient (see Equation 13). Most commercial systems provide their own analysis software, but any mathematics program capable of nonlinear least-squares (Levenberg–Marquardt) fitting routines can be used (e.g., Origin, Igor, or MATLAB). To assess the quality of the fit, both χ^2 returned by the fit routine and the residuals must be taken into account. Smaller values of χ^2 typically reflect a better fit, but the shape of the residuals (i.e., the difference between the calculated fit curve and the experimental values) is a more visual criterion. The residuals should be distributed symmetrically around 0, reflecting small random deviations without any residual structure (see Fig. 3).

Calibration

The diffusion coefficient D is derived from the characteristic decay time of the correlation function τ_D as mentioned above (see Equation 8). However, the values that will be obtained for the diffusion coefficient and the concentration depend critically on the geometry of the observation volume (i.e.,

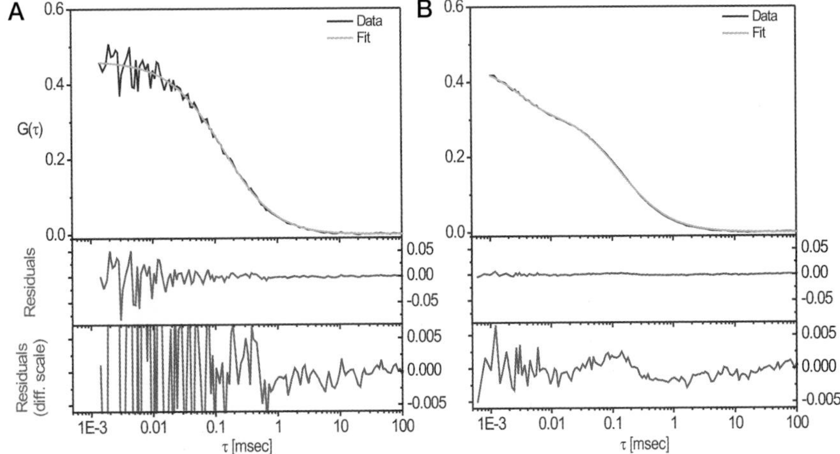

FIGURE 3. Fitting an FCS curve. Different autocorrelation curves of Alexa 488 in aqueous solution. *Upper panels* show the measured data and the corresponding fits; *lower panels* show the residuals of the fits. (*A*) The curve shows a good fit with mostly random deviations, as seen from visual inspection of the residuals. However, the signal-to-noise ratio of the FCS curve is not very good. (*B*) The signal-to-noise ratio of this FCS curve obtained at higher laser power is much better, and the deviations of the fitted curve are smaller. However, systematic (nonrandom) deviations are now discernible when the residuals are plotted on a magnified scale (*bottom panel*). Systematic deviations indicate that the model does not perfectly describe the data.

r_0 and z_0, which are difficult to determine directly) (Enderlein et al. 2005). The most common way to determine these values is to use calibration measurements of a dye with a known diffusion coefficient. From the fit to the FCS curve, the structure parameter and the diffusion time can be obtained and, hence, the axial and lateral dimensions of the effective observation volume can be obtained. For one-photon excitation, the equations are

$$r_0 = 2 \cdot \sqrt{D \cdot \tau_D}, \tag{14a}$$

$$z_0 = \omega \cdot r_0, \tag{14b}$$

$$V_{\text{eff}} = \pi^{3/2} \cdot \omega \cdot r_0^3 = 8 \cdot (\pi \cdot D \cdot \tau_D)^{3/2}. \tag{14c}$$

Rhodamine 6G is widely used as a calibration standard with a diffusion coefficient of 2.8×10^{-10} m^2/sec, although the accuracy of this value is only ~25% (Magde et al. 1974). More recent independent measurements have yielded a slightly higher value (e.g., $D = 4.14 \times 10^{-10}$ m^2/sec for Rhodamine 6G [Culbertson et al. 2002], $D = 4.26 \times 10^{-10}$ m^2/sec for Atto655, which is of similar size [~500 Da] [Dertinger et al. 2007]). For multichannel measurements, the calibration must be performed for each color separately.

Determining the Photobleaching Threshold

To determine the optimal laser power for a particular fluorescently labeled biomolecule under specific environmental conditions, it is good practice to perform a power series before starting the actual measurements. The laser power is increased from very low values to beyond the bleaching threshold, and for each setting, an FCS curve is recorded. The data are then analyzed to determine both the average intensity and the number of molecules in the focal volume and, hence, the count rate per single particle (i.e., the molecular brightness, η). By plotting these values versus the excitation intensity, as shown in Figure 4, one gets an initially linear increase in molecular brightness with

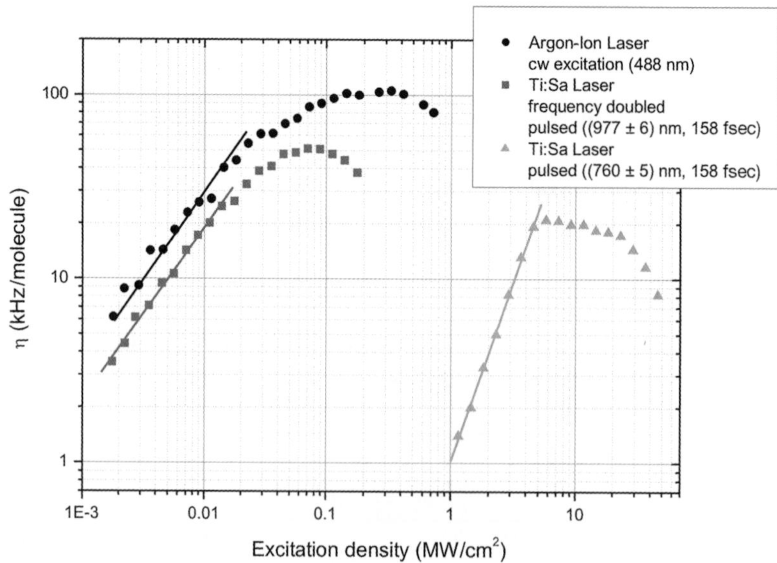

FIGURE 4. Saturation and photobleaching. Comparison of the count rate per single molecule obtained from autocorrelation measurements of Alexa 488 in water. Different modes of excitation were used. One-photon excitation (OPE) continuous wave (cw) using the 488-nm line of an argon-ion laser (closed circles), one-photon pulsed using the frequency doubled Ti:sapphire laser at (977 ± 6) nm, and pulse width of ~200 fsec (gray squares) and two-photon excitation (TPE) at (760 ± 5) nm (light gray triangles). Whereas for OPE (slope 1, *left side*), the brightness depends linearly on the excitation power, for TPE (slope 2, *right side*), it should be proportional to the square of the intensity. The saturation level is highest for cw OPE.

increasing laser power on a double-logarithmic plot. The initial slope is expected to be close to 1 for one-photon excitation. In TPE, the slope should be 2 because the excitation depends on the square of the laser intensity. For higher excitation powers, this curve levels off and drops again because of saturation and photobleaching.

It is advisable to measure at laser powers for which the brightness increase is still linear; the safety margin depends mainly on the system under investigation. If an increase in diffusion time is expected (e.g., because of binding or potential changes in local viscosity for in vivo measurements), this must be taken into account because the photobleaching probability will increase for longer residence times in the focal volume. Moreover, correlation curves may start to be affected at lower excitation intensities than expected from the double-logarithmic plot (Wu and Berland 2007). Because the excitation intensity is nonuniform, saturation and photobleaching are more likely to occur in the center of the focus, resulting in a change of the effective shape of the detection volume when going to higher laser powers.

Interpretation of the Data

Although FCS curves are easy to record, interpreting them may not be straightforward. Many free parameters in the fitting procedure will return seemingly good fit results, even if the values for these parameters fail to reflect the sample properties or have no physical meaning at all. Prior knowledge about the biological system can, thus, help to eliminate incorrect conclusions. Input substances should be characterized as carefully as possible before starting the actual FCS measurements.

The number of potential components for the fit routine is restricted by the ability of the mathematical algorithm to unequivocally resolve the individual contributions. In practice, this means that two or three diffusion components plus the triplet blinking term are about the upper limit. Furthermore, changes in molecular mass due to binding have to be sufficiently large to have a resolvable influence on the diffusion coefficient. Because the diffusion coefficient is inversely proportional to the hydrodynamic radius, which, in turn, depends on the cube root of the molecular mass for spherical particles, an increase in mass by a factor of 8 will only halve the diffusion coefficient. This effect is hardly visible on the logarithmic timescale used for FCS curves. In contrast, binding of a small fluorescently labeled molecule to a large unlabeled molecule or to a slowly diffusing receptor in a membrane will be easily discernable.

Assuming 100% labeling efficiency, one term for translational diffusion and one blinking term will be used to describe the curve (see Equations 12 and 13):

$$G(D,\tau) = \frac{1}{\langle N \rangle} \cdot \frac{1}{1+\tau/\tau_D} \cdot \frac{1}{\sqrt{1+\tau/(S^2 \cdot \tau_D)}} \cdot \left(1 + \frac{T}{1-T} \cdot e^{-\tau/\tau_{\text{triplet}}}\right). \qquad (15)$$

If free dye is still present in the sample, a second diffusion term must be added for this faster component based on Equation 11. If the two diffusing species are difficult to distinguish, it is possible to perform a one-component reference measurement of the free dye in solution. However, additional purification of the labeled biomolecules to remove the free dye is preferable.

Any changes in molecular shape or size on binding or cleavage that affect the hydrodynamic radius of the particle are also reflected in the diffusion coefficient and, thus, in the average diffusion time through the observation volume. For binding studies, in which only the smaller ligand is labeled, at least two differently diffusing species must be accounted for.

Fortunately, dcFCCS facilitates this kind of analysis. If both differently labeled reaction partners are of comparable size, they may be treated as one effective component in the autocorrelation analysis. From the amplitudes of the autocorrelation curves, only the total concentrations are determined. A simple one-component fit to the cross correlation then yields the amplitude of the cross-correlation curve so that the concentration of double-labeled particles can be determined according to Equation 6.

Considering only the amplitudes of the cross correlation and one of the autocorrelation curves, $G_x(0)$ and $G_j(0)$, Equation 6 can be used to obtain the fraction of double-labeled molecules:

$$\frac{\langle N_{ij} \rangle}{\langle N_i \rangle} = \frac{G_x(0)}{G_j(0)}. \tag{16}$$

This means that, for constant autocorrelation amplitudes, an increase in the cross-correlation amplitude directly reflects an increase in the number of particles carrying both colors.

APPLICATIONS

Live Cell Measurements

Measurements in living cells are generally more challenging than measurements in homogeneous solutions. Because of the complex and individual nature of the cells, which can, for example, be at different stages of the cell cycle and have different levels of expression, obtaining meaningful data from intracellular measurements requires looking at the distribution of parameters in numerous cells and in different preparations.

Another difficulty with live cell measurements is that a number of vital substances inside the cells show weak fluorescence. Among them are NADH (the reduced form of nicotinamide adenine dinucleotide), FAD (flavin adenine dinucleotide), and fluorescent amino acids such as tryptophan. Taken together at high concentrations, they form the so-called autofluorescence background, which varies from cell to cell and especially between different cell lines. Choosing a cell line with intrinsically low autofluorescence may significantly increase the signal-to-noise ratio as will using a cell culture medium without added indicator dye. For most cell lines, autofluorescence is most prominent in the green–yellow spectral region. Orange or red chromophores may, therefore, be easier to detect and to distinguish from the background. Prebleaching the autofluorescent material (i.e., exposing the cells to intense radiation for a prolonged period of time before taking measurements) can be hazardous to the cells and is not recommended. However, if strong photobleaching is observed, it may be justified to wait for the initial strong decay in the fluorescence signal to level off before starting actual measurements. This corresponds to a selection of the more mobile species that are not bleached while traversing the focal volume. Alternatively, scanning FCS may be applied to analyze more slowly diffusing molecules.

It is rather difficult to correctly adjust an FCS setup for intracellular measurements because of the differences in refractive index between the cytosol and the surrounding medium. However, because cells are rather thin, the deformations of the confocal volume are small, although they deteriorate rapidly for multiple cell layers or tissues. The procedure of FCS adjustments for intracellular measurements, therefore, does not deviate significantly from that for buffer measurements, and free three-dimensional diffusion is a common model. Two-dimensional diffusion is assumed for membrane measurements. However, depending on the actual position of the observation volume inside the cell and the biological system to be monitored, the correlation function can also take different forms. For example, local confinement of the diffusion to volumes comparable to that of the focal volume element (e.g., when measuring in small organelles) requires more sophisticated models (Gennerich and Schild 2000, 2002).

Membrane Measurements

FCS and other optical techniques have been very useful in elucidating the dynamic structure of biological membranes (Groves et al. 2008), which is difficult to assess with biochemical and genetic techniques. FCS measurements of the diffusion rates of fluorescent molecules can be performed in both native and artificial membranes (Bacia et al. 2004). Studies on supported lipid bilayers and on the free-standing lipid bilayers of giant unilamellar vesicles have yielded insight into the composi-

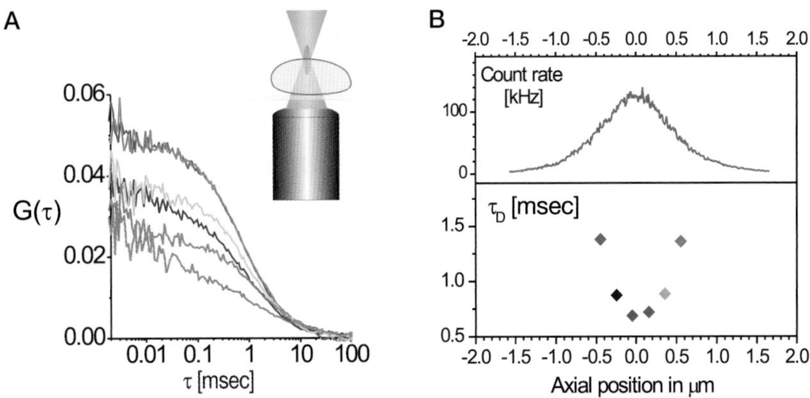

FIGURE 5. FCS on membranes. (*A*) FCS curves obtained on a giant unilamellar vesicle. The membrane contains the long-chain cyanine dye $DiIC_{18}$, which acts as a fluorescent lipid analog and diffuses in the plane of the bilayer. The focal volume is positioned on the upper membrane of the GUV, as shown schematically in the *inset*. (*B*) An axial fluorescence intensity scan (*top*) serves to optimally position the detection volume. When the count rate is at its maximum, the diffusion time reaches a minimum (*bottom*), and the amplitude of the correlation curves reaches a maximum (*A*, red curves).

tion-dependent packing and phase behavior of various combinations of lipids. Furthermore, FCS has been used to measure the diffusion of integral and membrane-bound proteins and to study the binding of ligands to membrane receptors (Briddon et al. 2004). A comprehensive review of FCS studies on membranes can be found in Chiantia et al. (2009).

For membrane measurements, the focal detection volume is normally centered on a horizontal membrane by using an axial fluorescence intensity scan. Accurate positioning is important: If the focal volume deviates from the position of maximal fluorescence, this will lead to a reduced correlation amplitude and an increased apparent diffusion time (Fig. 5). If the membrane does not rest stably in the center of focus, but shows positional fluctuations (undulations), an additional decay may be seen in the correlation curve, which can be easily mistaken as another slow component (Petrov and Schwille 2008). Membrane undulations as well as photobleaching can, therefore, preclude measuring slowly diffusing molecules in membranes.

A very useful approach for circumventing these problems with slow membrane diffusion is to repeatedly scan the FCS focus (or two foci) sideways through a vertical lipid bilayer and to analyze only the fluorescence bursts that are obtained when the foci cross the bilayer (Ries and Schwille 2006). In this way, positional fluctuations can be excluded from the analysis. In addition, photobleaching is reduced because the illumination of the membrane by the scanning beam is not continuous but effectively pulsed.

CONCLUSIONS

Although data interpretation may be initially less intuitive with FCS than with imaging techniques, FCS is a method worth considering for a variety of biological and physicochemical questions. FCS has great advantages for quantitatively analyzing biomolecular mobilities and interactions in situ. Precise values of physical parameters such as diffusion coefficients are determined relatively easily and quickly because data acquisition takes only minutes and the correlation is typically performed online. Many types of FCS analyses can be performed on well-established standard setups. Moreover, new variations of the FCS technique have been worked out to overcome specific experimental problems in some applications. As novel optics, electronics, and analytical concepts are being developed, the scope of FCS applications is expected to continue to grow.

REFERENCES

Aragón SR, Pecora R. 1975. Fluorescence correlation spectroscopy and Brownian rotational diffusion. *Biopolymers* **14:** 119–138.

Aragón SR, Pecora R. 1976. Fluorescence correlation spectroscopy as a probe of molecular dynamics. *J Chem Phys* **64:** 1791–1803.

Ariola FS, Li Z, Cornejo C, Bittman R, Heikal AA. 2009. Membrane fluidity and lipid order in ternary giant unilamellar vesicles using a new bodipy-cholesterol derivative. *Biophys J* **96:** 2696–2708.

Bacia K, Schwille P. 2007. Practical guidelines for dual-color fluorescence cross-correlation spectroscopy. *Nat Protoc* **2:** 2842–2856.

Bacia K, Scherfeld D, Kahya N, Schwille P. 2004. Fluorescence correlation spectroscopy relates rafts in model and native membranes. *Biophys J* **87:** 1034–1043.

Bacia K, Schwille P, Kurzchalia T. 2005. Sterol structure determines the separation of phases and the curvature of the liquid-ordered phase in model membranes. *Proc Natl Acad Sci* **102:** 3272–3277.

Baird GS, Zacharias DA, Tsien RY. 2000. Biochemistry, mutagenesis, and oligomerization of DsRed, a red fluorescent protein from coral. *Proc Natl Acad Sci* **97:** 11984–11989.

Becker CF, Seidel R, Jahnz M, Bacia K, Niederhausen T, Alexandrov K, Schwille P, Goody RS, Engelhard M. 2006. C-terminal fluorescence labeling of proteins for interaction studies on the single-molecule level. *ChemBioChem* **7:** 891–895.

Berland KM, So PTC, Gratton E. 1995. Two-photon fluorescence correlation spectroscopy: Method and application to the intracellular environment. *Biophys J* **68:** 694–701.

Briddon SJ, Middleton RJ, Cordeaux Y, Flavin FM, Weinstein JA, George MW, Kellam B, Hill SJ. 2004. Quantitative analysis of the formation and diffusion of A1-adenosine receptor-antagonist complexes in single living cells. *Proc Natl Acad Sci* **101:** 4673–4678.

Brinkmeier M, Dorre K, Stephan J, Eigen M. 1999. Two beam cross correlation: A method to characterize transport phenomena in micrometer-sized structures. *Anal Chem* **71:** 609–616.

Burkhardt M, Schwille P. 2006. Electron multiplying CCD based detection for spatially resolved fluorescence correlation spectroscopy. *Opt Express* **14:** 5013–5020.

Chiantia S, Ries J, Schwille P. 2009. Fluorescence correlation spectroscopy in membrane structure elucidation. *Biochim Biophys Acta* **1788:** 225–233.

Culbertson CT, Jacobson SC, Ramsey JM. 2002. Diffusion coefficient measurements in microfluidic devices. *Talanta* **56:** 365–373.

Denk W, Strickler JH, Webb WW. 1990. Two-photon laser scanning fluorescence microscopy. *Science* **248:** 73–76.

Dertinger T, Pacheco V, von der Hocht I, Hartmann R, Gregor I, Enderlein J. 2007. Two-focus fluorescence correlation spectroscopy: A new tool for accurate and absolute diffusion measurements. *ChemPhysChem* **8:** 433–443.

Digman MA, Brown CM, Sengupta P, Wiseman PW, Horwitz AR, Gratton E. 2005. Measuring fast dynamics in solutions and cells with a laser scanning microscope. *Biophys J* **89:** 1317–1327.

Eggeling C, Ringemann C, Medda R, Schwarzmann G, Sandhoff K, Polyakova S, Belov VN, Hein B, von Middendorff C, Schonle A, et al. 2009. Direct observation of the nanoscale dynamics of membrane lipids in a living cell. *Nature* **457:** 1159–1162.

Ehrenberg M, Rigler R. 1974. Rotational Brownian motion and fluorescence intensity fluctuations. *Chem Phys* **4:** 390–401.

Eigen M, Rigler R. 1994. Sorting single molecules: Application to diagnostics and evolutionary biotechnology. *Proc Natl Acad Sci* **91:** 5740–5747.

Elson EL, Magde D. 1974. Fluorescence correlation spectroscopy. I. Conceptual basis and theory. *Biopolymers* **13:** 1–27.

Elson EL, Schlessinger J, Koppel DE, Axelrod D, Webb WW. 1976. Measurement of lateral transport on cell surfaces. *Prog Clin Biol Res* **9:** 137–147.

Enderlein J, Gregor I, Patra D, Dertinger T, Kaupp UB. 2005. Performance of fluorescence correlation spectroscopy for measuring diffusion and concentration. *ChemPhysChem* **6:** 2324–2336.

Fahey PF, Koppel DE, Barak LS, Wolf DE, Elson EL, Webb WW. 1977. Lateral diffusion in planar lipid bilayers. *Science* **195:** 305–306.

Felekyan S, Kuhnemuth R, Kudryavtsev V, Sandhagen C, Becker W, Seidel CAM. 2005. Full correlation from picoseconds to seconds by time-resolved and time-correlated single photon detection. *Rev Sci Instrum* **76:** 083104.

Gennerich A, Schild D. 2000. Fluorescence correlation spectroscopy in small cytosolic compartments depends critically on the diffusion model used. *Biophys J* **79:** 3294–3306.

Gennerich A, Schild D. 2002. Anisotropic diffusion in mitral cell dendrites revealed by fluorescence correlation spectroscopy. *Biophys J* **83:** 510–522.

Groves JT, Parthasarathy R, Forstner MB. 2008. Fluorescence imaging of membrane dynamics. *Annu Rev Biomed Eng* **10:** 311–338.

Haupts U, Maiti S, Schwille P, Webb WW. 1998. Dynamics of fluorescence fluctuations in green fluorescent protein observed by fluorescence correlation spectroscopy. *Proc Natl Acad Sci* **95:** 13573–13578.

Heinze KG, Koltermann A, Schwille P. 2000. Simultaneous two-photon excitation of distinct labels for dual-color fluorescence crosscorrelation analysis. *Proc Natl Acad Sci* **97:** 10377–10382.

Heinze KG, Rarbach M, Jahnz M, Schwille P. 2002. Two-photon fluorescence coincidence analysis: Rapid measurements of enzyme kinetics. *Biophys J* **83:** 1671–1681.

Heinze KG, Jahnz M, Schwille P. 2004. Triple-color coincidence analysis: One step further in following higher order molecular complex formation. *Biophys J* **86:** 506–516.

Hendrix J, Flors C, Dedecker P, Hofkens J, Engelborghs Y. 2008. Dark states in monomeric red fluorescent proteins studied by fluorescence correlation and single molecule spectroscopy. *Biophys J* **94:** 4103–4113.

Hess ST, Webb WW. 2002. Focal volume optics and experimental artifacts in confocal fluorescence correlation spectroscopy. *Biophys J* **83:** 2300–2317.

Kannan B, Har JY, Liu P, Maruyama I, Ding JL, Wohland T. 2006. Electron multiplying charge-coupled device camera based fluorescence correlation spectroscopy. *Anal Chem* **78:** 3444–3451.

Kapusta P, Wahl M, Benda A, Hof M, Enderlein J. 2007. Fluorescence lifetime correlation spectroscopy. *J Fluoresc* **17:** 43–48.

Kastrup L, Blom H, Eggeling C, Hell SW. 2005. Fluorescence fluctuation spectroscopy in subdiffraction focal volumes. *Phys Rev Lett* **94:** 178104.

Kettling U, Koltermann A, Schwille P, Eigen M. 1998. Real-time enzyme kinetics monitored by dual-color fluorescence cross-correlation spectroscopy. *Proc Natl Acad Sci* **95:** 1416–1420.

Kim SA, Heinze KG, Bacia K, Waxham MN, Schwille P. 2005. Two-photon cross-correlation analysis of intracellular reactions with variable stoichiometry. *Biophys J* **88:** 4319–4336.

Kinjo M, Rigler R. 1995. Ultrasensitive hybridization analysis using fluorescence correlation spectroscopy. *Nucleic Acids Res* **23:** 1795–1799.

Levene MJ, Korlach J, Turner SW, Foquet M, Craighead HG, Webb WW. 2003. Zero-mode waveguides for single-molecule analysis at high concentrations. *Science* **299:** 682–686.

Magatti D, Ferri F. 2001. Fast multi-τ real-time software correlator for dynamic light scattering. *Appl Opt* **40:** 4011–4021.

Magatti D, Ferri F. 2003. 25 ns software correlator for photon and fluorescence correlation spectroscopy. *Rev Sci Instrum* **74:** 1135–1144.

Magde D, Elson EL, Webb WW. 1972. Thermodynamic fluctuations in a reacting system—Measurement by fluorescence correlation spectroscopy. *Phys Rev Lett* **29:** 705–708.

Magde D, Elson EL, Webb WW. 1974. Fluorescence correlation spectroscopy. II. An experimental realization. *Biopolymers* **13:** 29–61.

Magde D, Webb WW, Elson EL. 1978. Fluorescence correlation spectroscopy. III. Uniform translation and laminar flow. *Biopolymers* **17:** 361–376.

Malvezzi-Campeggi F, Jahnz M, Heinze KG, Dittrich P, Schwille P. 2001. Light-induced flickering of DsRed provides evidence for distinct and interconvertible fluorescent states. *Biophys J* **81:** 1776–1785.

Matz MV, Fradkov AF, Labas YA, Savitsky AP, Zaraisky AG, Markelov ML, Lukyanov SA. 1999. Fluorescent proteins from nonbioluminescent Anthozoa species. *Nat Biotechnol* **17:** 969–973.

Michalet X, Cheng A, Antelman J, Suyama M, Arisaka K, Weiss S. 2008. Hybrid photodetector for single-molecule spectroscopy and microscopy. *Proc SPIE* **6862:** 68620F.

Muller BK, Zaychikov E, Brauchle C, Lamb DC. 2005. Pulsed interleaved excitation. *Biophys J* **89:** 3508–3522.

Petrov E, Schwille P. 2008. State of the art and novel trends in fluorescence correlation spectroscopy. In *Standardization and quality assurance in fluorescent measurements, I* (ed. Resch-Genger U), pp. 145–197. Springer, Berlin.

Prasher DC, Eckenrode VK, Ward WW, Prendergast FG, Cormier MJ. 1992. Primary structure of the *Aequorea victoria* green-fluorescent protein. *Gene* **111:** 229–233.

Rauer B, Neumann E, Widengren J, Rigler R. 1996. Fluorescence correlation spectrometry of the interaction kinetics of tetramethylrhodamin α-bungarotoxin with *Torpedo california* acetylcholine receptor. *Biophys Chem* **58:** 3–12.

Reichman J. 2007. *Handbook of optical filters for fluorescence microscopy*, pp. 1–30. Chroma Technology Corporation, Brattleboro, VT.

Ries J, Schwille P. 2006. Studying slow membrane dynamics with continuous wave scanning fluorescence correlation spectroscopy. *Biophys J* **91:** 1915–1924.

Rigler R, Widengren J. 1990. Ultrasensitive detection of single molecules by fluorescence correlation spectroscopy. *BioScience* **3:** 180–183.

Rigler R, Mets U, Widengren J, Kask P. 1993. Fluorescence correlation spectroscopy with high count rate and low-background: Analysis of translational diffusion. *Eur Biophys J* **22:** 169–175.

Ruan Q, Cheng MA, Levi M, Gratton E, Mantulin WW. 2004. Spatial-temporal studies of membrane dynamics: Scanning fluorescence correlation spectroscopy (SFCS). *Biophys J* **87:** 1260–1267.

Schwille P, Meyer-Almes FJ, Rigler R. 1997a. Dual-color fluorescence cross-correlation spectroscopy for multicomponent diffusional analysis in solution. *Biophys J* **72:** 1878–1886.

Schwille P, Bieschke J, Oehlenschläger F. 1997b. Kinetic investigations by fluorescence correlation spectroscopy: The analytical and diagnostic potential of diffusion studies. *Biophys Chem* **66:** 211–228.

Schwille P, Haupts U, Maiti S, Webb WW. 1999. Molecular dynamics in living cells observed by fluorescence correlation spectroscopy with one- and two-photon excitation. *Biophys J* **77:** 2251–2265.

Schwille P, Kummer S, Heikal AA, Moerner WE, Webb WW. 2000. Fluorescence correlation spectroscopy reveals fast optical excitation-driven intramolecular dynamics of yellow fluorescent proteins. *Proc Natl Acad Sci* **97:** 151–156.

Shimomura O, Johnson FH, Saiga Y. 1962. Extraction, purification and properties of aequorin, a bioluminescent protein from the luminous hydromedusan, *Aequorea. J Cell Comp Physiol* **59:** 223–239.

Smoluchowski M. 1906. Zur kinetischen Theorie der Brownschen Molekularbewegung und der Suspensionen. *Ann Phys* **21:** 756–780.

Starr TE, Thompson NL. 2001. Total internal reflection with fluorescence correlation spectroscopy: Combined surface reaction and solution diffusion. *Biophys J* **80:** 1575–1584.

Thews E, Gerken M, Eckert R, Zapfel J, Tietz C, Wrachtrup J. 2005. Cross talk free fluorescence cross-correlation spectroscopy in live cells. *Biophys J* **89:** 2069–2076.

Weidemann T, Wachsmuth M, Tewes M, Rippe K, Langowski J. 2002. Analysis of ligand binding by two-colour fluorescence cross-correlation spectroscopy. *Single Mol* **3:** 49–61.

Weisshart K, Jungel V, Briddon SJ. 2004. The LSM 510 Meta-ConfoCor2 system: An integrated imaging and spectroscopic platform for single-molecule detection. *Curr Pharm Biotechnol* **5:** 135–154.

Widengren J, Schwille P. 2000. Characterization of photoinduced isomerization and back-isomerization of the cyanine dye Cy5 by fluorescence correlation spectroscopy. *J Phys Chem* **104:** 6416–6428.

Wu J, Berland K. 2007. Fluorescence intensity is a poor predictor of saturation effects in two-photon microscopy: Artifacts in fluorescence correlation spectroscopy. *Microsc Res Tech* **70:** 682–686.

40 Image Correlation Spectroscopy
Principles and Applications

Paul W. Wiseman

Departments of Physics and Chemistry, McGill University, Montreal, Quebec H3A 2T8, Canada

ABSTRACT

Image correlation spectroscopy (ICS) was developed as the imaging analog of fluorescence correlation spectroscopy. Using standard fluorescence microscopy image series as input, different versions of ICS can be used to extract parameters on the molecular transport properties (diffusion and flow) and oligomerization state for fluorescently labeled species in cells. This chapter introduces the various forms of spatial and temporal ICS and discusses application of these methods to reveal properties of the biomolecules that can be measured from standard fluorescence image time series sampled from cells and neurons. Fluorescence correlation spectroscopy and its applications are described in Chapter 39.

INTRODUCTION TO IMAGE CORRELATION SPECTROSCOPY

Background

Image correlation spectroscopy (ICS) was originally introduced as the imaging analog of the temporal fluctuation method called fluorescence correlation spectroscopy (FCS; see Chapter 39) as a method to measure membrane receptor densities and oligomerization states in the plasma membranes of cells (Petersen et al. 1993). In its original guise, ICS was entirely a spatial domain technique; however, the base method has been extended for measurements in the spatial and temporal domains via fluorescence microscopy image time series. The method has also been extended to analysis in k-space and time. This chapter introduces the background theory of ICS and provides a guide for the types of measurements that can be made with the variants of these methods for applications on cells.

Principles of ICS

All forms of ICS are based on the analysis of fluctuations in intensity from pixels that compose fluorescence microscopy images. However, it is important to understand that the intensity recorded in a given pixel (or voxel) represents an integrated light intensity from a focal volume defined by the laser beam focus for a laser scanning microscope (LSM) or the effective optical resolution element

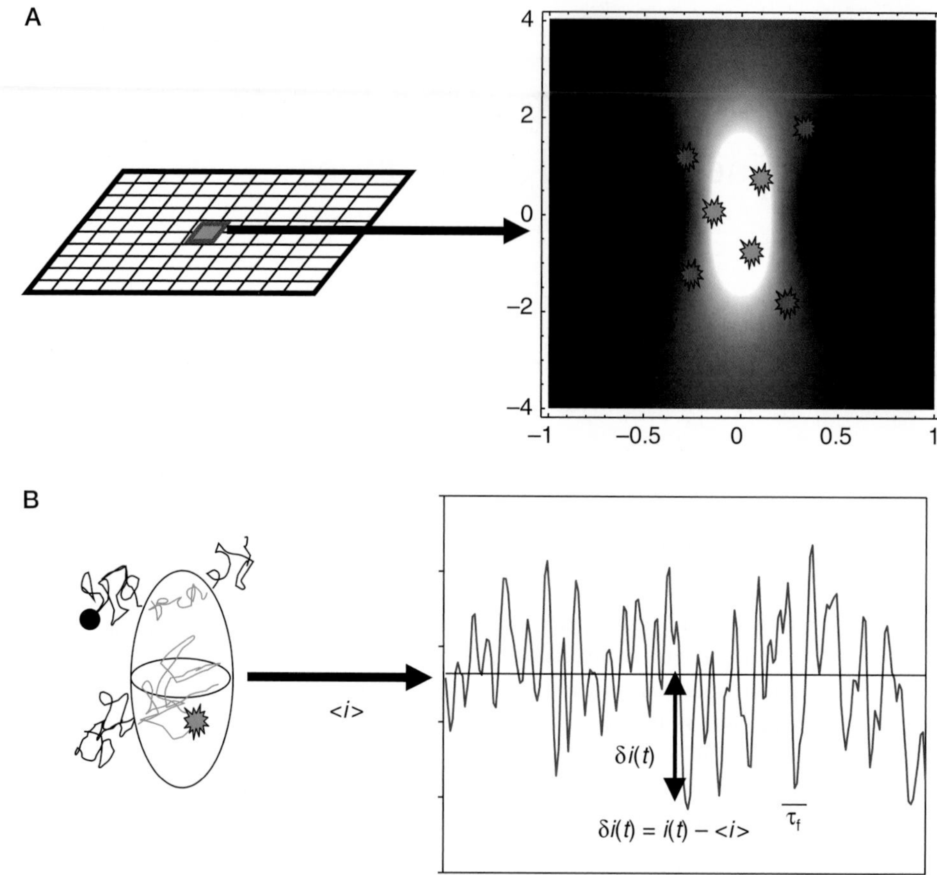

FIGURE 1. Temporal fluorescence fluctuations. (*A*) Schematic depicting that the intensity in any pixel in the image is an integration of fluorescence photons emitted by fluorophores within an optical focal volume (beam focus). The volume of the focal volume can be on the order of a femtoliter. (*B*) Schematic showing that transport of fluorescent molecules in and out of the focal volume leads to changes in occupation number, which are recorded as intensity fluctuations. The fluctuation magnitude (δi) and the characteristic fluctuation time (τ) are the important measureable observables for fluctuation spectroscopy.

or point-spread function (PSF) for a fluorescence microscope in general (see Figs. 1 and 2). Temporal domain fluctuation methods like FCS rely on the measurement and analysis of fluorescence fluctuations as a function of time. These temporal fluctuations arise from spontaneous changes in the number of fluorescent molecules within the focus of a stationary excitation laser beam as a result of molecular transport or chemical reactions (Fig. 1). Both the amplitude of the fluctuation (size of the deviation from the mean) and the duration of the fluctuation in time contain important information about the fluorescently tagged macromolecules.

In contrast to FCS, ICS relies on the characterization of fluorescence fluctuations as a function of space across an image (Fig. 2) or space and time from a fluorescence microscopy time series (Fig. 3). In any form of correlation spectroscopy, it is assumed that the fluorescence intensity is proportional to the number of fluorophores within the focal volume contributing to the integrated signal intensity. We must define the pixel intensity fluctuation as this is the basic input datum for all forms of ICS, and we will assume an *x–y–t* image time series. Any given pixel in the image time series will have an intensity value $i_a(p, q, s)$, where "a" represents the detection channel (e.g., green channel) for the image series, p and q represent the discrete spatial coordinates in *x–y* space, and s is the discrete time point (image number) when the value was recorded. Following convention, we define the fluorescence intensity fluctuation at this pixel as simply the difference between the pixel intensity value and the mean intensity:

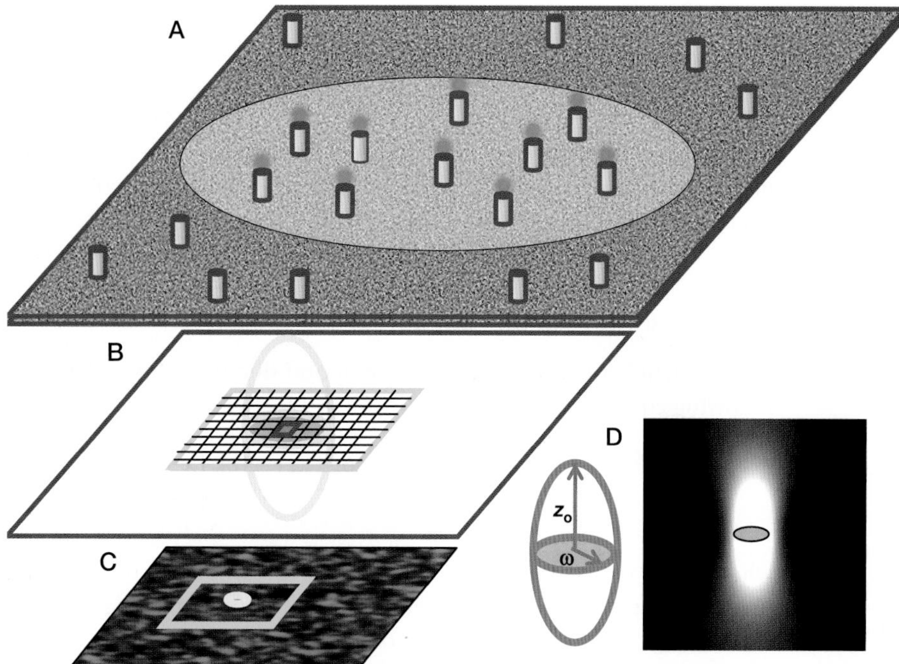

FIGURE 2. Spatial fluorescence intensity fluctuations. (*A*) Schematic showing the molecular basis for spatial fluctuations is the distribution of fluorescent molecules in space. This case depicts a distribution of fluorescently tagged proteins in the membrane with some found in the beam focal area where they are excited and emit fluorescence. (*B*) This shows that the optical focal volume from which fluorescence intensity is integrated is larger than the pixel dimensions in the image. This leads to spatial correlation between adjacent pixels in an image that are exploited for spatial ICS. (*C*) Image level showing the region of interest and approximate focal area size scale outlined by the circle. (*D*) The optical focal volume (defined by the laser beam focus for an LSM) dimensions define the system in which correlations are measured in space and time.

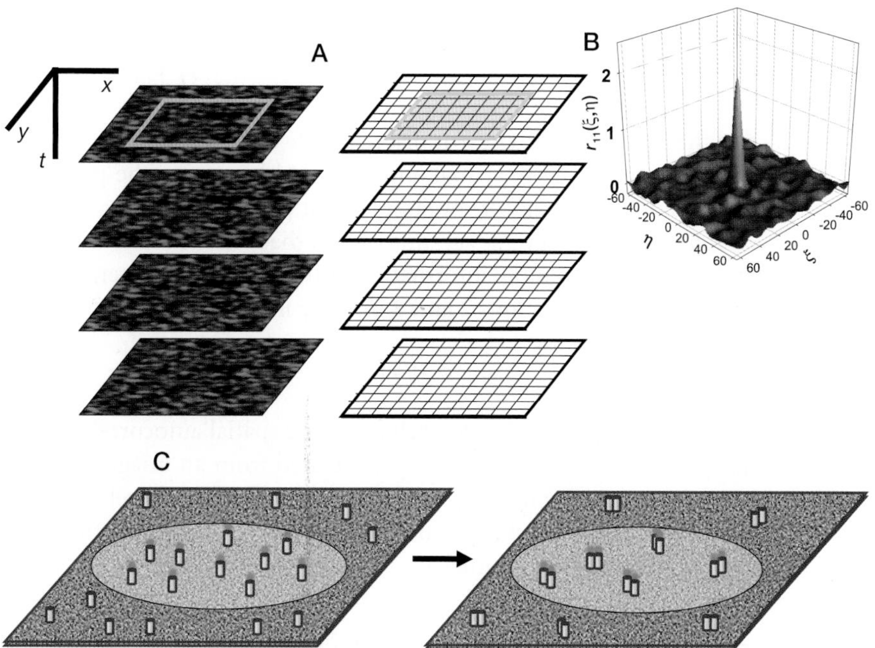

FIGURE 3. Spatial ICS. (*A*) Schematic of an image series showing that for spatial ICS, a region of interest (ROI) in *x*–*y* space in a single image is selected, and a spatial correlation function is calculated from the chosen pixels. (*B*) A spatial correlation function calculated in spatial ICS showing the peak central amplitude at zero spatial lags. This amplitude is inversely proportional to the mean number of independent fluorescent entities in the focal volume/area. (*C*) Schematic of a molecular aggregation event that changes the mean number of independent fluorescent entities in the focal area that is measurable by spatial ICS.

$$\delta i_a(p, q, s) = i_a(p, q, s) - \langle i_a \rangle. \tag{1}$$

The mean intensity can be calculated in several different ways depending on which variant of image correlation is applied, as will be outlined below. Given an *x–y–t* image time series as an input data set from a fluorescence microscope, it is possible to calculate a corresponding matrix of fluctuations using Equation 1, which can then be analyzed by one of the variants of ICS.

Generalized Spatiotemporal Correlation Function

We can define a fully general spatiotemporal fluctuation correlation function, $r(\xi, \eta, \tau)$ using the following equation:

$$r_{ab}(\xi,\eta,\tau) = \frac{\left\langle\left\langle \delta i_a(x,y,t)\,\delta i_b(x+\xi, y+\eta, t+\tau)\right\rangle_{xy}\right\rangle_t}{\langle i_a\rangle_t \langle i_b\rangle_{t+\tau}}, \tag{2}$$

where ξ and η are spatial lag variables that represent shifts of the image in *x–y* space for calculation of spatial correlation $\langle\cdots\rangle_{xy}$, whereas *t* is a temporal lag variable representing shifts in time in the *x–y–t* image series for calculation of the temporal correlation $\langle\cdots\rangle_t$. The angular brackets in the denominator simply represent calculation of a mean intensity, and the subscripts a and b stand for detection channels a and b for image series collected for two fluorophores of different emission wavelengths. Equation 2 represents a general spatiotemporal fluctuation cross-correlation function when a and b are different (two-color collection), whereas it becomes a spatiotemporal autocorrelation if a and b are identical (single-detection-channel image series). In addition, it represents a correlation function that is continuous in the spatial and temporal lag variables. In fact, the image series data sets are discrete pixels and represent a sampling of the underlying molecular distribution by the diffraction limited focus modeled by the microscope's PSF (see Fig. 2). So, in practice, we actually calculate a discrete approximation to this correlation function from the input image series, where the symbols are the same as defined in Equations 1 and 2:

$$r_{ab}(\Delta p,\Delta q,\Delta s) = \frac{\left\langle\left\langle \delta i_a(p,q,s)\,\delta i_b(p+\Delta p, q+\Delta q, s+\Delta s)\right\rangle_{xy}\right\rangle_t}{\langle i_a\rangle_s \langle i_b\rangle_{s+\Delta s}}, \tag{3}$$

and the lag variables represent discrete (integer) pixel shifts in *x*, *y*, or *t*. The discrete pixel shifts can always be converted to spatial and temporal lags using the pixel spatial dimension ($\delta p = \delta q$) and image frame time step (δt) by simple multiplication (e.g., $\xi = \Delta x = \Delta p \delta p$, and $\tau = \Delta s \delta t$).

There are now several variants of ICS that differ in how the image data set or sets are analyzed and consequently on the information that they provide in terms of output. Most can be considered as variations of Equations 2 and 3 in some limit, and these will be described in turn.

Spatial ICS

The original form of ICS was exclusively the spatial autocorrelation variant (Petersen et al. 1993). A spatial autocorrelation function is calculated from an image or region of interest (ROI) within an image by correlating the ROI with itself as a function of pixel shifts in the *x* and *y* directions (Fig. 3). This is equivalent to taking Equation 2 or 3 in the limit that the time lag or shift variable goes to 0 for a single detection channel a:

$$r_{aa}(\xi,\eta,0) = \frac{\left\langle \delta i_a(x,y,t)\,\delta i_a(x+\xi, y+\eta, t)\right\rangle_{xy}}{\langle i_a\rangle_t^2}. \tag{4}$$

An equivalent, and more computationally efficient, way to calculate the discrete spatial correlation function is using fast Fourier transforms (FFTs):

$$r_{aa}(\xi, \eta, 0) = \mathrm{FFT}^{-1}\{\mathrm{FFT}[\mathrm{ImageROI}] \cdot \mathrm{FFT}^{*}[\mathrm{ImageROI}]\}, \tag{5}$$

where FFT indicates the forward fast Fourier transform operation, FFT^{-1} indicates the inverse fast Fourier transform, and the * indicates the complex conjugate operation. In practice, Equation 5 is usually preferred over the brute force calculation of Equation 4 as it is much faster computationally.

Once the raw spatial autocorrelation function is calculated from the image ROI, a Gaussian function is then fit to it by nonlinear least squares:

$$r_{aa}(\xi, \eta, 0) = \mathbf{g_{aa}}(\mathbf{0}, \mathbf{0}, \mathbf{0}) \exp\left\{\frac{\xi^2 + \eta^2}{\boldsymbol{\omega}^2}\right\} + \mathbf{g}_{\boldsymbol{\infty}}, \tag{6}$$

where the fitting parameters **are shown in bold** and represent the zero-lags amplitude of the correlation function $\mathbf{g_{aa}}(\mathbf{0}, \mathbf{0}, \mathbf{0})$; the correlation radius $\boldsymbol{\omega}$ that is related to the e^{-2} radius of the laser beam focus; and $\mathbf{g}_{\boldsymbol{\infty}}$, which is an offset at large spatial lags to take into account incomplete decay of the correlation function.

The laser beam has a Gaussian intensity profile that acts as the correlator in spatial ICS. As the beam excites fluorescence from fluorophores throughout the illumination region, the integrated intensity reflects the underlying distribution of fluorescent entities, and there will be intensity correlations between nearby pixels in the image as the beam area is larger than the pixel area (Fig. 2). The zero-lags amplitude is equal to the inverse of the mean number of independent fluorescent entities, $\langle N \rangle$, within the beam focal area:

$$g_{aa}(0, 0, 0) = \frac{1}{\langle N \rangle}. \tag{7}$$

If the beam radius is known (through calibration measurement), it is possible to calculate the beam area and hence the surface density of fluorescent molecules from the zero-lags autocorrelation amplitude. If the fluorescently tagged molecules change their oligomerization state (see Fig. 3), the number of independent entities in the beam focus changes, and this is measurable as a change in the amplitude of the correlation function (Wiseman and Petersen 1999).

Temporal ICS

Temporal ICS (TICS) is the variant of ICS that is closest to FCS. Intensity fluctuations are correlated in time through the image series, and a time correlation function is calculated from the time series. The rate of decay of the correlation function reflects the average decay time of the fluctuations as fluorescent entities move in and out of the area defined by the beam focus (Fig. 4). Any process that contributes to fluorescence fluctuations on the timescale of the sampling will contribute to the decay of a correlation function, and each dynamic process will have a mean characteristic fluctuation time. To measure transport with TICS, the labeled macromolecules must have a characteristic fluctuation time that is longer than the image frame time. In other words, the fluorophores have to be within the same general beam focal area when the raster scan returns the beam to the same position to acquire the next frame in order to preserve molecular correlations. With standard imaging rates ~1 Hz, the slower transport of membranes proteins is measurable, but the faster diffusion of cytoplasmic components is not (Wiseman et al. 2000).

The temporal autocorrelation function is calculated from the image series using an equation that follows from the general correlation function, Equation 2 (continuous) or Equation 3 (discrete pixels), evaluated with zero spatial lags variables or zero pixel shift variables for a single channel:

$$r_{aa}(0, 0, \tau) = \frac{\left\langle \left\langle \delta i_a(x, y, t)\, \delta i_a(x, y, t+\tau) \right\rangle_{xy} \right\rangle_t}{\langle i_a \rangle_t \langle i_a \rangle_{t+\tau}}, \tag{8}$$

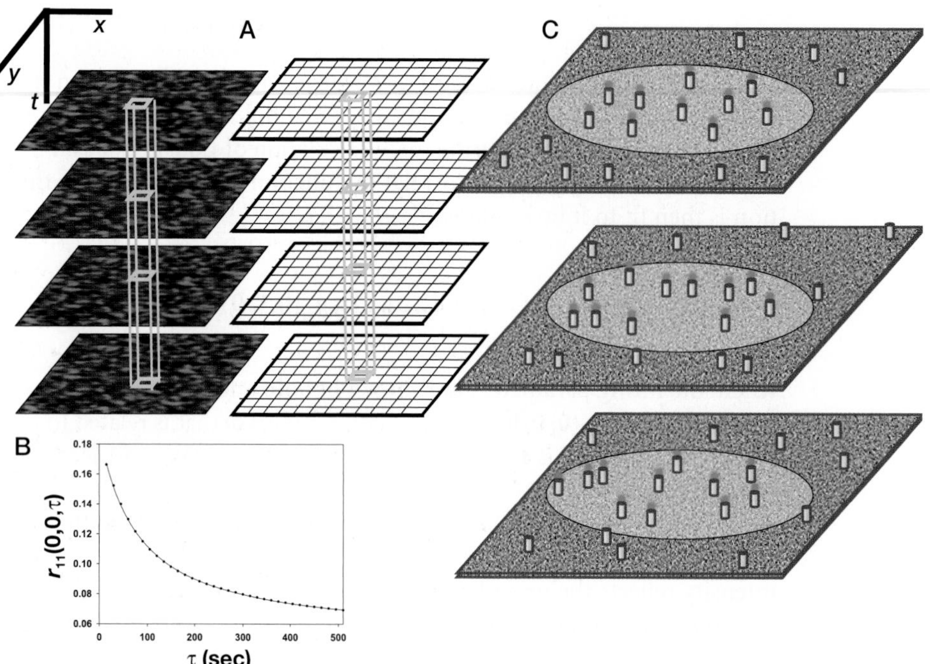

FIGURE 4. Temporal ICS (TICS). (*A*) Schematic of an image series showing that for TICS, pixels are correlated in time. (*B*) A temporal correlation function calculated by TICS. The decay shape and rate reflect the transport dynamics of the underlying molecules. (*C*) Schematic of molecular diffusion that changes the mean number of independent fluorescent entities in the focal area in time that is measureable by TICS.

$$r_{aa}(0,0,\Delta s) = \frac{\left\langle \left\langle \delta i_a(p,q,s)\,\delta i_a(p,q,s+\Delta s)\right\rangle_{xy} \right\rangle_t}{\left\langle i_a \right\rangle_s \left\langle i_a \right\rangle_{s+\Delta s}}. \tag{9}$$

With temporal correlation, we are presented with a choice in terms of order of correlation in space and time. Although Equations 8 and 9 show spatial correlation (at zero lags) being performed in time, it is also possible to perform temporal correlation on a given pixel location first and then average the correlation functions for each pixel position. For the former, the average intensity in the denominator is the mean intensity for the entire image region of interest. In the case of the latter (temporal correlation at given pixel positions), the average intensity is the mean intensity calculated in time for the pixel stack at $i(x, y, t)$. In practice, we usually choose the first approach of spatial correlation in time because it tends to converge more quickly for limited time stacks inherent in imaging and is also sensitive to immobile populations of the labeled species (see below).

The decay of the temporal correlation function essentially records the average time decay of fluctuations for any dynamic process that contributes to changes in the number of fluorescent species in the beam volume in time (Fig. 4). However, in order to extract molecular transport or kinetic properties for the system, the correlation function must be fit with an appropriate physical decay model for the process(es) that causes the fluctuations in occupation number in the focus. For each dynamic process, there will be an associated characteristic fluctuation time that represents an average lifetime of the fluorescent species within the beam focus. Transport parameters can then be calculated from the characteristic fluctuation time obtained from the decay model best fit and the beam focal volume radii in x–y and z (obtained by calibration measurement of the microscope system PSF usually using fluorescent microspheres).

For a system with 3D diffusion of the fluorescent species, the correlation function can be fit using a decay model that assumes a laser beam with a Gaussian intensity profile in x–y and z (Aragon and Pecora 1976):

$$r_{aa}(0,0,\tau) = \frac{g_{aa}(0,0,0)}{\left(1+\frac{\tau}{\tau_d}\right)\left(1+\left[\frac{\omega^2}{z_o^2}\right]\frac{\tau}{\tau_d}\right)^{1/2}} + g_\infty, \quad (10)$$

where $\mathbf{g_{aa}(0, 0, 0)}$ is the zero time lag best-fit amplitude, $\boldsymbol{\tau_d}$ is the best-fit characteristic diffusion time, and $\mathbf{g_\infty}$ is a fitting offset parameter that accounts for cases in which the correlation function does not decay to 0 at longer lag times because of the presence of an immobile population (if spatial correlation is performed before temporal correlation). The e^{-2} beam radii in x–y and z are ω and z_0, respectively, and are fixed as constants for the fitting using values obtained from a beam focus calibration measurement. The $g_{aa}(0, 0, 0)$ is the inverse of the mean number of independent fluorescent entities in the focal volume, as was the case for spatial correlation (Equation 7).

If the system is restricted to 2D diffusion, as is the case for membrane proteins, the fit model above reduces to the following hyperbolic decay assuming a Gaussian intensity beam profile in 2D:

$$r_{aa}(0,0,\tau) = \frac{g_{aa}(0,0,0)}{\left(1+\frac{\tau}{\tau_d}\right)} + g_\infty. \quad (11)$$

The diffusion coefficient can then be calculated from the best-fit characteristic diffusion time knowing the laser beam radius at focus:

$$D = \frac{\omega^2}{4\tau_d}. \quad (12)$$

Calibration for TICS diffusion measurements may be done using fluorescent microspheres of known radius (r) suspended in a fluid of known viscosity (η) such as glycerol or concentrated sucrose solutions in water using the Stokes–Einstein relationship (Wiseman et al. 2000):

$$D_{\text{theory}} = \frac{kT}{6\pi\eta r}, \quad (13)$$

where k is Boltzmann's constant and T is the absolute temperature in kelvins.

For systems in which there is flow, the appropriate fit mode for the temporal correlation function is a Gaussian, as has been shown for FCS (Magde et al. 1978):

$$r_{aa}(0,0,\tau) = g_{aa}(0,0,0)\exp\left\{\left(\frac{|v|\tau}{\omega}\right)^2\right\} + g_\infty, \quad (14)$$

where $\mathbf{g_{aa}(0, 0, 0)}$ and $\mathbf{g_\infty}$ are the same fit parameters as described above for the diffusion case and $|\boldsymbol{v}|$ is the best-fit speed of the flowing population. The ratio of $|v|$ to ω is simply the inverse of the characteristic flow time: $1/\tau_f$.

A single population undergoing flow with superimposed diffusion is fit by a combined model that is the product of each dynamic contribution:

$$r_{aa}(0,0,\tau) = \frac{g_{aa}(0,0,0)}{\left(1+\frac{\tau}{\tau_d}\right)}\exp\left\{-\left(\frac{|v|\tau}{\omega}\right)\left(1+\frac{\tau}{\tau_d}\right)^{-1}\right\} + g_\infty. \quad (15)$$

If there is a mixture of two fluorescent populations having different dynamics, then the decay fit model is a linear combination of each dynamic contribution:

$$r_{aa}(0,0,\tau) = A\left(1+\frac{\tau}{\tau_d}\right) + B\ \exp\left\{-\left(\frac{|v|\tau}{\omega}\right)^2\right\} + g_\infty . \tag{16}$$

It is important to note that TICS analysis is sensitive to flow speed; that is, magnitude but not direction as the spatial correlation is calculated only for zero lags (no pixel shift). So the method is able to measure how quickly the fluorescent species traverse the focal volume but is blind to the direction of flow. An extension of TICS to full spatial and temporal correlation permits measurement of true flow velocities, as is outlined in the next section.

Spatiotemporal ICS

Spatiotemporal image correlation spectroscopy (STICS) (Hebert et al. 2005) effectively calculates the generalized autocorrelation limit of Equation 2 for a single detection channel:

$$r_{aa}(\xi,\eta,\tau) = \frac{\left\langle\left\langle \delta i_a(x,y,t)\,\delta i_a(x+\xi,y+\eta,t+\tau)\right\rangle_{xy}\right\rangle_t}{\langle i_a\rangle_t \langle i_a\rangle_{t+\tau}} . \tag{17}$$

If there is a flowing fluorescent population in the sample, it may be revealed by spatial correlation as a function of time lag (Fig. 5). Spatial correlation at zero time lag effectively averages the spatial correlation of each image in the series with itself so there can be no movement; hence, $r_{aa}(\xi, \eta, 0)$ maps as a Gaussian spatial correlation function with its peak centered at the origin ($\xi = 0, \eta = 0$). At longer time lags, the mobile fluorescent species will move, and this will affect the shape and decay of the spatiotemporal correlation function. If the population is diffusing, the particles undergo Brownian motion, which is isotropic. In this case, the Gaussian spatial correlation peak will remain centered at the origin and decay in amplitude and increase in width at a rate that depends on D in

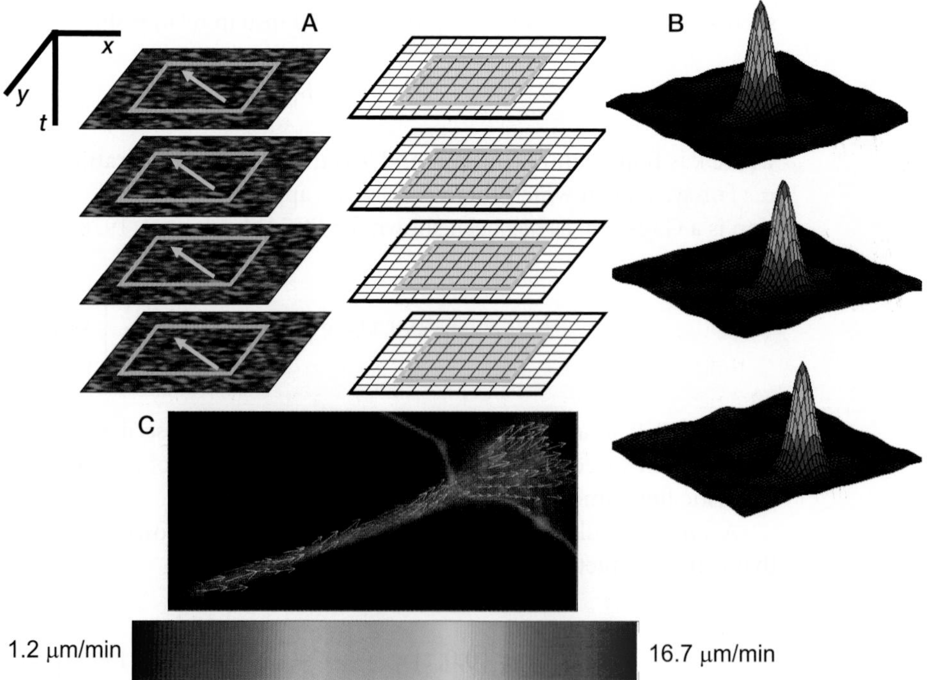

FIGURE 5. Spatiotemporal ICS (STICS). (*A*) Schematic of an image series showing that for STICS, image ROIs are spatially correlated in time. (*B*) The time evolution of the spatial correlation function for a system where there is flow showing directed movement of the correlation peak. (*C*) Example of a cellular vector map obtained by STICS measurement on a chick dorsal root ganglion with fluorescently labeled microtubule tips. (Image series courtesy of the Fournier Lab, Montreal Neurological Institute.)

accordance with the laws of diffusion. For a single uniformly flowing population, the fluorescent species will spatially correlate in time in a direction determined by the flow. Thus at longer time lags, the spatiotemporal correlation function can exhibit a Gaussian peak that moves from the central origin at a uniform rate in a specific direction (see Fig. 5). This correlation peak due to flow can be fit and tracked in time to calculate a flow vector (magnitude and direction).

In practice, the presence of an immobile or slowly diffusing population complicates the measurement of flow at short time lags (when the peak due to flow is in the vicinity of the origin). It is possible to filter the immobile population to fully reveal the dynamic populations. The immobile population contribution to the spatiotemporal correlation function can be removed by Fourier filtering in frequency space the DC component for every pixel trace in time before running the space–time correlation analysis. For each given pixel location, the corrected intensities are given by

$$i_a'(x, y, t) = F_f^{-1}\{F_t\{i_a(x, y, t)\} \times H_{1/T}(f)\}, \tag{18}$$

where T is the total acquisition time of the image series; $H_{1/T}(f)$ is the Heaviside function, which is 0 for $f < 1/T$ and 1 for $f > 1/T$; F_i^{-1} denotes the (inverse) Fourier transform with respect to variable i; and f is the pixel temporal frequency variable. It is also possible to apply a moving average filter to remove dynamics on a range of time scales.

Edge boundaries will also perturb spatial correlation analysis because of the sharp discontinuity in intensity at the periphery of the cell. This issue can be especially problematic for dendritic morphologies characteristic of neurons. However, there is a way to circumvent this issue if we remember that we are correlating fluctuations and that fluctuations are simply the difference between an intensity value and the average intensity. If we measure the on-cell average intensity for a region near an edge that we wish to measure, we can pad this average value in image pixels for areas outside the cell and effectively create a carpet of zero fluctuations that masks the boundary (Comeau et al. 2008). This requires application of image processing routines to identify cell boundaries; however, spatiotemporal correlation can then be carried out across the boundary (see Fig. 5 for an example on a growth cone).

Image Cross-Correlation Spectroscopy

The previous sections dealt with applications of image correlation in which a single protein species was labeled with a fluorophore and spatial autocorrelation was carried out to reveal number densities and dynamics. It is also possible to extend these approaches to measurements on cells in which two protein species are labeled with two different fluorophores. It is simply a matter of collecting the fluorescence image data in detection channels a and b and performing image cross-correlation analysis between fluctuations in each image channel. The equations are simply obtained using the fully general correlation functions defined in Equations 2 and 3 with a ≠ b (which by definition is cross-correlation). The cross correlation is sensitive to the presence of interacting species (hetero-oligomers) that carry both color fluorophores and will be nonzero when such species are present in the sample (see Fig. 6).

In the ideal case, the focal volumes for the lasers used to excite both fluorophores should be the same size and overlap in space. For single-photon excitation, this is difficult to achieve without care-

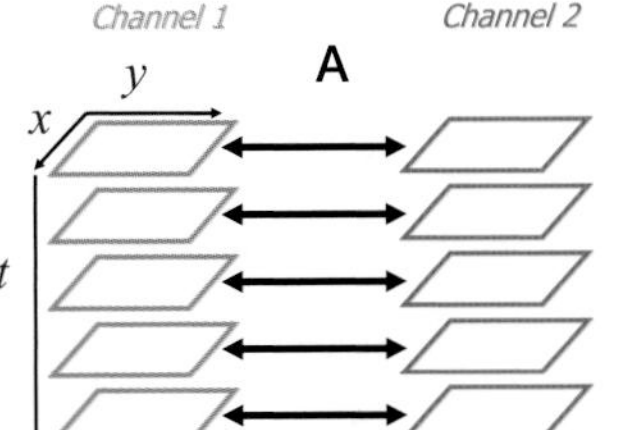

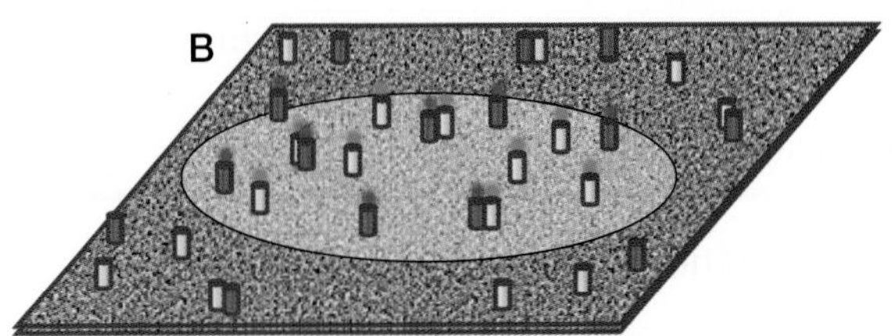

FIGURE 6. Image cross-correlation spectroscopy (ICCS). (*A*) Schematic of a two-channel image series showing that for ICCS, cross correlation is performed between the two detection channels in space and/or time to detect interacting species. (*B*) Schematic of the molecular distribution of two species labeled with different fluorophores showing interacting heterodimers in the focal area that would be detectable by ICCS.

ful alignment and specialized adjustment of the optics. For two-photon excitation, a single wavelength can often be used to excite multiple fluorophores to avoid this problem. However, it is also possible to apply a focal area mismatch correction for 2D systems (membrane proteins) for regular single-photon excitation. The average number of interacting particles $\langle N \rangle_{ab}$ in the focal area for a 2D system is given by the cross-correlation function fit zero-lags amplitude normalized by the two autocorrelation function fit zero-lags amplitudes multiplied by a ratio of the focal areas for the two channels to correct for small mismatches in focal overlap (Comeau et al. 2006):

$$\langle N \rangle_{ab} = \frac{g_{ab}(0,0,0)}{g_{aa}(0,0,0)\, g_{bb}(0,0,0)} \frac{A_b}{A_a}, \tag{19}$$

where $A_i = \pi\omega_i^2$ is the beam area for channel i. The fraction of interacting particles to total number of particles with a given fluorophore within a focal area may also be calculated from the ratio of the best-fit cross-correlation amplitude to the best-fit autocorrelation amplitude:

$$\frac{\langle N \rangle_{ab}}{\langle N \rangle_{aa}} = \frac{g_{ab}(0,0,0)}{g_{bb}(0,0,0)} \text{ and } \frac{\langle N \rangle_{ab}}{\langle N \rangle_{bb}} = \frac{g_{ab}(0,0,0)}{g_{aa}(0,0,0)}. \tag{20}$$

Transport information (diffusion coefficients, flow speeds, and vectors) on the interacting species may be obtained by temporal or spatiotemporal cross-correlation calculations using the two channel variants of the generalized correlation functions (Equations 2 or 3).

Other Variants of ICS

The temporal resolution for the measurement of transport dynamics using TICS or STICS is set by the frame time between images. For standard confocal imaging this is typically ~1 sec and is too slow to measure the diffusion of free molecules in the cytoplasm of cells. An extension that takes into account the actual sampling of the raster scan permits measurement of the faster dynamics used on standard laser scanning microscopes (confocal or two photon). The method is called raster image correlation spectroscopy (RICS) and essentially breaks the analysis down in terms of spatial correlations along the fast scan direction of the raster scan (microsecond pixel dwell times) and slow scan direction (microsecond line times) (Digman et al. 2005). With RICS, image correlation measurements can be performed on freely diffusing molecules in solution just as is the case for FCS.

Another extension of ICS has been introduced as *k*-space image correlation spectroscopy (kICS) (Kolin et al. 2006b). This reciprocal space variant relies on calculation of a time correlation function after each image in the time series has been converted to *k*-space by a 2D FFT. The kICS approach allows measurement of diffusion coefficients and flow directions, but offers the advantage of separating time-dependent photophysics fluctuations (such as fluorophore blinking and bleaching) from the transport fluctuations that are space/time-dependent. This allows measurement of transport coefficients that are not biased by blinking or photobleaching, as well as allowing separate measurement of the photophysics fluctuations. Another advantage of kICS is that it does not require calibration of the beam radii for transport measurements as the beam parameters separate out because of the mathematics of the transforms. Further details on both RICS and kICS can be found in a recent review (Kolin and Wiseman 2007).

INSTRUMENTATION AND ANALYSIS

Instrumentation for ICS

ICS analysis can be applied to fluorescence images obtained using a variety of microscopes. The method has been applied to images obtained using laser scanning microscopes, both with two-photon excitation (Wiseman et al. 2000; Hebert et al. 2005) and conventional single-photon confocal microscopy (Wiseman and Petersen 1999; Hebert et al. 2005). Image series collected using evanescent wave excitation by total internal reflection fluorescence (TIRF) microscopy have also been analyzed using ICS (Brown et al. 2006; Comeau et al. 2008).

Sampling, Signal-to-Noise Ratio, and Photobleaching

The signal-to-noise ratio in any fluorescence fluctuation measurement will depend on several variables including the molecular brightness of the species of interest (number of photons collected per sampling time), the density or concentration in the beam volume, and the number of characteristic fluctuations sampled for the dynamic process of interest.

As image correlation is usually performed using analog detection, we will assume that the user has a sample that is bright enough to be imaged above background. In practice, it is useful to define an ROI that is off cell and calculate the average intensity of this region to define a background. This mean background value is then subtracted from every pixel in the image before carrying out image correlation analysis. This assumes that the background is relatively constant in space and time so a control measurement may be performed to confirm this assumption.

The relative magnitude of a fluctuation varies inversely with the square root of the number of labeled particles in the beam volume or area. This sets an upper limit on concentrations that are accessible to ICS. If the density or concentration is too high, then the fluctuations become too small to be measured (uniform intensity in the image), and ICS cannot be performed. Simulation results show that a density of around 100 particles per beam focal area still yields fluctuations of reasonable magnitude for such measurements. Fortunately, expression levels in cells often result in densities that are on the order of 1 to 10 particles per beam focal area for many membrane proteins. At the other extreme, the concentration has to be high enough such that the ROI actually contains a sampling of the species of interest.

If the signal-to-noise ratio is sufficiently high for imaging and the density of the species of interest falls in the proper range, then ICS measurements can be performed with an uncertainty that is established by the sampling. The uncertainty for any fluctuation method will vary inversely with the square root of the number of characteristic fluctuations sampled. For example, if the characteristic diffusion time for a membrane protein is 1 sec, then sampling one point for 100 sec would yield 100 fluctuations in the measurement, and the uncertainty would be 10% (assuming sufficient intensity). The ability to sample in both space and time provides ICS methods with a sampling advantage because limited time sampling can be compensated by spatial sampling in the image ROI. The number of independent spatial fluctuations sampled is given by the image ROI area divided by the beam area of the focus ($A = \pi\omega^2$). So a user can adjust the ROI size and the size of the time window to optimize measurements and map processes across a cell. Typical values used for TICS would be an ROI of 32 x 32 pixels and 50 to 100 images in the time series, and for STICS an ROI of 16 x 16 pixels and 10 to 50 images in the time series (assuming a 1-sec image frame time). These values are a rough guide only and, of course, depend on the rate of the transport process of interest and the sampling time for the imaging in a given experiment.

Photobleaching of the fluorophore is a major perturbation for most forms of ICS (but not for kICS). As photobleaching changes the number of detectable species on the focal volume over time, it will affect both spatial and temporal ICS measurements. For TICS, it is possible to determine the decay form of the photobleaching from the image series. If the characteristic time scale for the bleaching is longer than that for molecular transport, then it can be neglected. If the decay form of the photobleaching can be fit by an exponential or bi-exponential, then the perturbation of the photobleaching can be corrected in the TICS analysis (Kolin et al. 2006a). Photobleaching will introduce a significant systematic error to any measurement of number densities as a function of time by spatial correlation of each image. It is best to adjust the imaging conditions to minimize bleaching while still obtaining sufficient signal.

Software for Image Correlation

Several stand-alone graphical user interface ICS programs for the PC have been developed by our research group at McGill University. These programs are available for download from the Wiseman

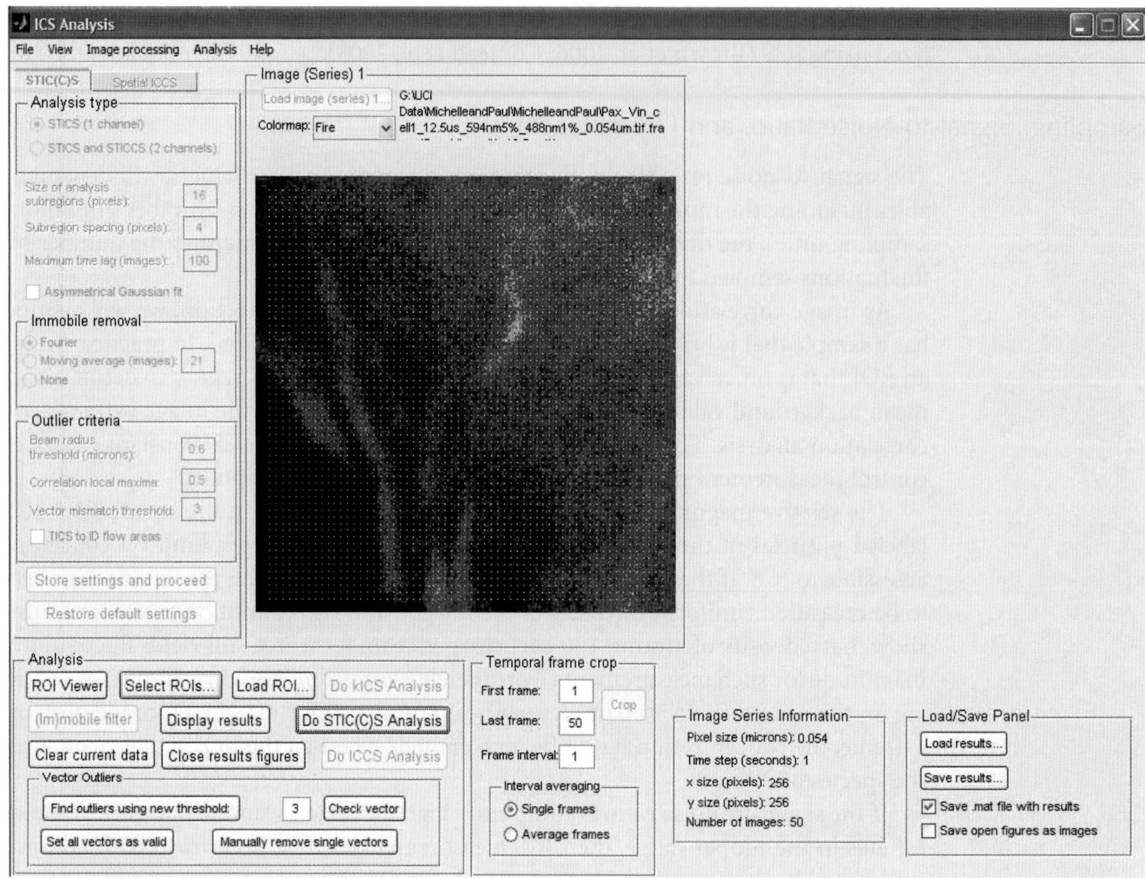

FIGURE 7. The graphical user interface (GUI) for the STICS analysis program showing a loaded image data set of a cell and some of the range of operations accessible in the program.

Research group (chose the software link) and are continually extended (http://wiseman-group.mcgill.ca/). The programs allow users to open and load image data sets, select regions of interest to analyze in the image frame, perform filtering operations, run various forms of ICS analysis, and save and output the results. A screen capture of the interface for STICS is shown in Figure 7. In addition, Enrico Gratton of the Laboratory for Fluorescence Dynamics (LFD) has developed an extensive SimFCS software package that includes RICS analysis and is available for download from the LFD website (http://www.lfd.uci.edu/).

CONCLUSION

Image correlation methods offer a new window to analyze images obtained using standard forms of fluorescence microscopy. As such, they can reveal information about biological molecule transport, oligomerization, and interactions that are entirely accessible in standard measurements performed in many biomedical labs. New developments and graphical user interface (GUI)-based analysis programs are extending the range of applications, making them accessible to nonexperts.

REFERENCES

Aragon SR, Pecora R. 1976. Fluorescence correlation spectroscopy as a probe of molecular dynamics. *J Chem Phys* **64:** 1791–1803.

Brown CM, Hebert B, Kolin DL, Zareno J, Whitmore L, Horwitz AR, Wiseman PW. 2006. Probing the integrin-actin linkage using high-resolution protein velocity mapping. *J Cell Sci* **119:** 5204–5214.

Comeau JW, Costantino S, Wiseman PW. 2006. A guide to accurate fluorescence microscopy colocalization measurements. *Biophys J* **91:** 4611–4622.

Comeau JW, Kolin DL, Wiseman PW. 2008. Accurate measurements of protein interactions in cells via improved spatial image cross-correlation spectroscopy. *Mol Biosyst* **4:** 672–685.

Digman MA, Brown CM, Sengupta P, Wiseman PW, Horwitz AR, Gratton E. 2005. Measuring fast dynamics in solutions and cells with a laser scanning microscope. *Biophys J* **89:** 1317–1327.

Hebert B, Costantino S, Wiseman PW. 2005. Spatiotemporal image correlation spectroscopy (STICS) theory, verification, and application to protein velocity mapping in living CHO cells. *Biophys J* **88:** 3601–3614.

Kolin DL, Wiseman PW. 2007. Advances in image correlation spectroscopy: Measuring number densities, aggregation states, and dynamics of fluorescently labeled macromolecules in cells. *Cell Biochem Biophys* **49:** 141–164.

Kolin DL, Costantino S, Wiseman PW. 2006a. Sampling effects, noise, and photobleaching in temporal image correlation spectroscopy. *Biophys J* **90:** 628–639.

Kolin DL, Ronis D, Wiseman PW. 2006b. *k*-space image correlation spectroscopy: A method for accurate transport measurements independent of fluorophore photophysics. *Biophys J* **91:** 3061–3075.

Magde D, Webb WW, Elson EL. 1978. Fluorescence correlation spectroscopy. III. Uniform translation and laminar flow. *Biopolymers* **17:** 361–376.

Petersen NO, Höddelius PL, Wiseman PW, Seger O, Magnusson KE. 1993. Quantitation of membrane receptor distributions by image correlation spectroscopy: Concept and application. *Biophys J* **65:** 1135–1146.

Wiseman PW, Petersen NO. 1999. Image correlation spectroscopy. II. Optimization for ultrasensitive detection of preexisting platelet-derived growth factor-β receptor oligomers on intact cells. *Biophys J* **76:** 963–977.

Wiseman PW, Squier JA, Ellisman MH, Wilson KR. 2000. Two-photon image correlation spectroscopy and image cross-correlation spectroscopy. *J Microsc* **200:** 14–25.

41 Time-Domain Fluorescence Lifetime Imaging Microscopy

A Quantitative Method to Follow Transient Protein–Protein Interactions in Living Cells

Sergi Padilla-Parra, Nicolas Audugé, Marc Tramier, and Maïté Coppey-Moisan

Department of Cell Biology, Institut Jacques Monod, UMR 7592, CNRS, University Paris-Diderot, 75205 Paris Cedex 13, France

ABSTRACT

Quantitative analysis in Förster resonance energy transfer (FRET) imaging studies of protein–protein interactions within live cells is still a challenging issue. Many cellular biology applications aim at the determination of the space and time variations of the relative amount of interacting fluorescently tagged proteins occurring in cells. This relevant quantitative parameter can be, at least partially, obtained at a pixel-level resolution by using fluorescence lifetime imaging microscopy (FLIM). Indeed, fluorescence decay analysis of a two-component system (FRET and no FRET donor species), leads to the intrinsic FRET efficiency value (E) and the fraction of the donor-tagged protein that undergoes FRET (f_D). To simultaneously obtain f_D and E values from a two-exponential fit, data must be acquired with a high number of photons, so that the statistics are robust enough to reduce fitting ambiguities. This is a time-consuming procedure. However, when fast-FLIM acquisitions are used to monitor dynamic changes in protein–protein interactions at high spatial and temporal resolutions in living cells, photon statistics and time resolution are limited. In this case, fitting procedures are unreliable, even for single lifetime donors. We introduce the concept of a minimal fraction of donor molecules involved in FRET (mf_D), obtained from the mathematical minimization of f_D. Here, we discuss different FLIM techniques and the compromises that must be made between precision and time invested in acquiring FLIM measurements. We show that mf_D constitutes an interesting quantitative parameter for fast FLIM because it gives quantitative information about transient interactions in live cells.

INTRODUCTION

During the last 40 years, FRET has been used to understand a great variety of molecular interactions, both in vitro and in vivo. Advances in different photonic imaging techniques and the development of fluorescent probes, and particularly fluorescent proteins (FPs) (Tsien 1998; Shaner et al. 2005), have made FRET microscopy an extremely useful methodology. Protein–protein interactions in living cells can be directly monitored using FRET. This aspect is critical to improve our understanding of different processes occurring in vivo (biochemical protein cascades) and, if it is performed quantitatively, to build or to improve biological mathematical models (Tuszynski et al. 2006). A quantitative parameter of FRET is the quantum yield of the energy transfer process (E). Donor fluorescence lifetime decreases because of energy transfer in the excited state, and the percentage of the decrease is equal to E. The determination of FRET efficiency by fluorescence lifetime measurements is advantageous in living cell studies because the fluorescence lifetime is independent of the fluorophore concentration and the excitation light path—parameters that are unknown in cells under the microscope. Other quantitative FRET techniques based on steady-state intensity allow determination of the apparent FRET efficiency, E_{app} (Gordon et al. 1998; Hoppe et al. 2002; Berney and Danuser 2003; Elangovan et al. 2003; Zal and Gascoigne 2004; Stockholm et al. 2004; van Rheenen et al. 2004; Wlodarczyk et al. 2008). E_{app} depends directly on the product of two parameters: the intrinsic FRET efficiency value (E) and the fraction of the donor that undergoes FRET (f_D) (Hoppe et al. 2002). Steady-state intensity-based approaches are not able to obtain f_D out of E_{app} (Neher and Neher 2004). To determine f_D, the intrinsic FRET efficiency (E) must be calculated independently. FLIM is a well-established technique to determine the fluorescence kinetics of the donor emission for FRET measurements (Verveer et al. 2000; Emiliani et al. 2003; Peter et al. 2005). In FLIM, using the mean lifetime does not require a high number of measured photons. However, to simultaneously obtain f_D and E values from the fit, fluorescence decay analysis must be performed with two or more components, and data have to be acquired with the highest number of photon counts so that statistics are robust enough to reduce fitting ambiguities. Using time-correlated single-photon counting (TCSPC), a high temporal resolution of the fluorescence kinetics (few tens of picoseconds) can be achieved. Several minutes of data acquisition are necessary, however, to obtain sufficient photon statistics per pixel. Such long acquisition times are incompatible with high spatiotemporal resolution of quantitative FRET images. A fast-gated charge-coupled device (CCD) camera combined with a pulsed, wide-field, or pseudo-wide-field excitation (TriM-FLIM system) can be used to acquire fluorescence decays faster than with the TCSPC method but at the expense of a smaller temporal resolution. To acquire images as fast as possible, a small number of time-gated images together with a low value of photon counts are required. Note, however, that under these conditions, the quality of a double-exponential fit is far from optimal. However, if we consider a two-component system with a narrow distribution of E, the concept of a minimal percentage of donor molecule involved in FRET (mf_D) can be calculated without fitting directly from the mean fluorescence lifetime of the donor (in the absence and in the presence of the acceptor). mf_D is an interesting approach because it provides information about a known threshold of interacting donor protein and is related to the relative concentration of the interacting protein.

PRINCIPLES OF FRET QUANTIFICATION BY TIME-DOMAIN FLIM

Theory of FRET

Förster resonance energy transfer (or, more correctly, Förster-type resonance energy transfer; Förster 1948) (FRET) is a nonradiative process that occurs between the excited state of a donor and the ground state of an acceptor.

The Transfer Rate

The rate of energy transfer ($k_t(r)$) from the donor to the acceptor is defined by the following equations (Valeur 2002; Lakowicz 2006):

$$k_t(r) = 1/\tau_D(R_0^6 / r^6), \tag{1}$$

where R_0 is the Förster radius, r is the distance between donor and acceptor, and τ_D is fluorescence lifetime of the donor in the absence of the acceptor. Because the Förster radius is defined as the distance at which FRET is 50% efficient, if $R_0 = r$, the rate transfer will be the same as the decay rate of the donor. R_0 can be defined as

$$R_0 = 0.211 \cdot [\kappa^2 n^{-4} Q_D J]^{1/6} \, (\text{Å}), \tag{2}$$

with

$$J = \int \varepsilon_A(\lambda) \cdot f_D(\lambda) \cdot \lambda^4 \cdot d\lambda / \int f_D(\lambda) \cdot d\lambda, \tag{3}$$

and κ^2 is derived from

$$\kappa^2 = (\cos\theta - 3\cos\theta_D \cos\theta_A)^2, \tag{4}$$

where n is the refractive index, Q_D is the fluorescence quantum yield of the donor, κ is the parameter related to the orientation of donor and acceptor (Dale et al. 1979; Cheung 1991), $\varepsilon_A(\lambda)$ is the acceptor absorption spectrum, and $f_D(\lambda)$ is the donor emission spectrum. Note that θ is the angle between the emission transition dipole of the donor and the absorption transition dipole of the acceptor, and θ_D and θ_A are the angles between the two dipoles and the vector that goes from the donor to the acceptor (Lakowicz 2006). Förster distances are reported in literature for a value of $\kappa^2 = 2/3$.

The Efficiency of Energy Transfer

We can define the efficiency by calculating the proportion of photons absorbed in the donor versus the excitation transferred to the acceptor:

$$E = k_T/(k_T + k_D), \tag{5}$$

where k_D is the sum of the all of the relaxation pathways (Fig. 1) of the excited donor other than FRET.

Experimentally, the transfer efficiency is calculated by using the lifetime of the donor alone or in the presence of the acceptor, applying a lifetime methodology:

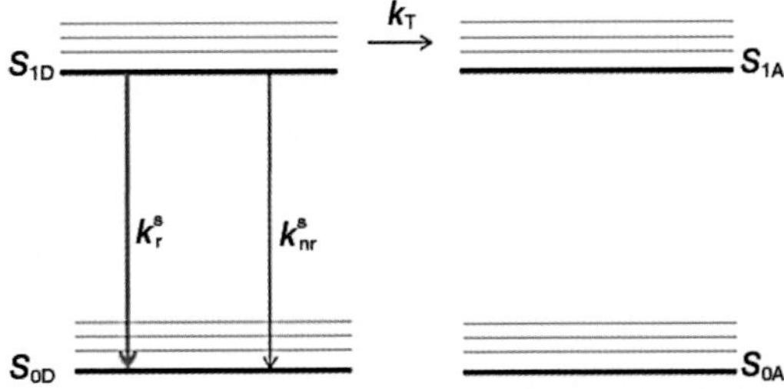

FIGURE 1. Modified Perrin–Jablonski diagrams showing photonic processes occurring in the donor and the FRET process. This diagram takes into consideration the process of FRET in which the donor transfer is indicated with a horizontal black arrow (k_T). One can see the competition between two pathways of relaxation (k_r, the radiative process, and k_{nr}, the nonradiative energy dissipation).

$$E = 1 - \frac{\tau_F}{\tau_D}, \quad (6)$$

in which τ_F and τ_D are the lifetimes of the donor in the presence of the acceptor and the donor alone, respectively. Observe that τ_F and τ_D are defined on the basis of the following rate constants:

$$\tau_F = \frac{1}{k_r + k_{nr} + k_T}, \quad (7)$$

$$\tau_D = \frac{1}{k_r + k_{nr}}. \quad (8)$$

Fluorescence Decay of the Donor

Fluorescence is a radiative process (rate constant k_r), which takes place when molecules excited by light absorption revert to their original state by light emission. Fluorescence is a very brief and transient phenomenon (typically occurring in the picosecond to nanosecond range), in which the molecules in the excited state S_1 release energy radiatively as they return to their unexcited state S_0 (Fig. 1). The kinetics of the fluorescence decay (rate constant $k = k_{nr} + k_r$) depends on the relative proportion of the various pathways for returning to the ground state. In time-domain (td) methods, a pulsed excitation source is used to excite the sample. The aim is to acquire the fluorescence decay after the laser pulse. For a single fluorophore in a homogeneous environment, the fluorescence decay can be written as

$$i(t) = (k_r)[M^*] = k_r\,[M^*]_0 \exp(-t/\tau), \quad (9)$$

where $[M^*]$ is the concentration of molecules in the excited state and τ is the lifetime of the excited state. Equation 9 defines the fluorescence decay profile of the donor in the absence of the acceptor.

Now, if we consider a donor–acceptor interaction in which not all of the donor is engaged in the process of FRET, then

$$i(t) = (k_r)[M^*] = k_r([D]_0 \exp(-\,t/\tau_D) + [DA]_0 \exp(-\,t/\tau_F)) \quad (10)$$

applies. In the above expression, the process of energy transfer is taken into account, and a discrete double exponential describes the fluorescence decay of the donor in the presence of the acceptor. Equation 10 assumes that there is only one orientation that enables the energy transfer to occur. If we normalize the amplitudes of both pre-exponential factors to 1, we can introduce the concept of the fraction of the interacting donor (f_D), and the last equation simplifies to

$$i(t) = (1 - f_D) \exp(-\,t/\tau_D) + (f_D) \exp(-t/\tau_F), \quad (11)$$

where f_D is a parameter that is particularly interesting in the study of the interactions of a related protein in cellular biology. td-FLIM allows for the simultaneous calculation of the fluorescence lifetime of the donor and the other important parameters that describe the fluorescence decay (pre-exponential factors and, hence, f_D) pixel by pixel. This makes this technique extremely useful because it retrieves information about the location and the extent of the interaction under study.

Time-Domain Picosecond FLIM: TCSPC-FLIM

This method is based on TCSPC detection. For each acquired photon, the time between the excitation pulse and the detection of the photon is measured and is used to sample the detected photons (Fig. 2).

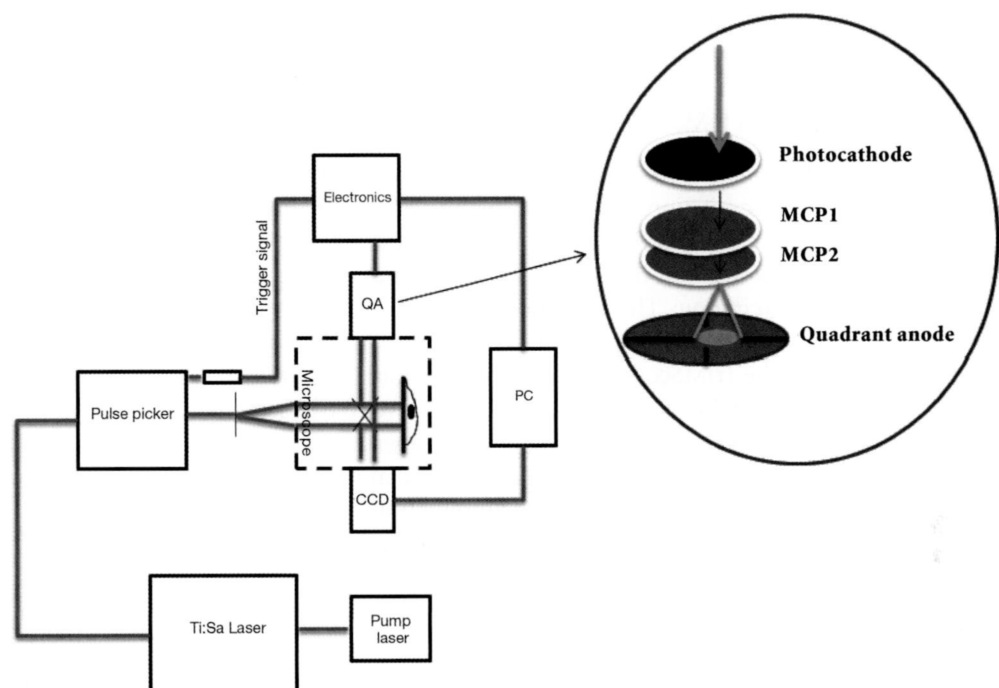

FIGURE 2. Time-domain picosecond TCSPC-FLIM setup. The *inset* on the right shows the components of the quadrant-anode detector. QA, Quadrant anode; PC, personal computer; CCD, charge-coupled device; MCP1, microchannel plate1; MCP2, microchannel plate 2.

Instrumentation Setup

Excitation: Mode-locked titanium:sapphire lasers can deliver picosecond (1 psec) or femtosecond (100 fsec) pulses and can be tuned between 760 and 980 nm. They can be used for two-photon excitation or for one-photon excitation after frequency doubling. Pulsed laser diodes (1 psec) are now also available for one-photon excitation at 440-, 470-, and 635-nm and longer wavelengths. Another option for pulsed excitation is supercontinuum sources.

Microscope and detection: Conventional confocal microscopes have an external port to couple an infrared or visible-light pulsed laser. The laser beam is scanned through the sample, and the single photons are detected by a photomultiplier tube with a time resolution corresponding to 10 psec. Avalanche photodiodes are also used because they are more sensitive in the red part of the spectrum; however, they provide a lower time resolution for the corresponding fluorescence decay. A computer card (e.g., Becker & Hickl GmbH or Picoquant GmbH) integrates delay time measurements (between emitted photons and excitation pulses) and scan position. Wide-field microscopy can also be used (an approach usually used in our laboratory). In this case, the laser beam of a titanium:sapphire laser (after passing through a frequency doubler and a pulse picker, which provide a reduced repetition rate of 4 MHz) is coupled to an inverted microscope. A quadrant-anode detector (Europhoton GmbH) provides information relative to the temporal and spatial correlation times of each single photon counted. Fluorescence decays are acquired by counting and sampling single emitted photons according to both the time delay between their arrival and the laser pulse and their *x–y* coordinates.

Note that TCSPC is a statistical method that requires high photon counts to analyze the fluorescence decay: The faster the count rate, the lower the acquisition time. The detector count rate is the limiting step for the acquisition time. Indeed, in the case of two photons emitted in a very short time (high intensity of the laser), only the first one will be processed, which will induce artificially faster fluorescence kinetics (shorter lifetime). The benefit of using a very low excitation intensity level (currently 20 nW at the sample) is to avoid photobleaching and/or photoconversion of the FPs (Tramier et al. 2006). Moreover, because of the high sensitivity, very dim cells can be analyzed, avoiding undesirable overexpression effects. Using a high-numerical-aperture (high-NA) 60x objective, a typical acquisition of an enhanced green fluorescent protein (eGFP)-labeled cell takes ~2–3 min.

Multiphoton Multifocal FLIM: TriM-FLIM

The objective of multifocal multiphoton FLIM (TriM-FLIM) is to combine pseudo-wide-field pulsed excitation (here in two-photon excitation) and a fast-gated CCD camera (Fig. 3; for an example of an application of two-photon FLIM see also Chapter 28). The principle is to open the intensifier in front of the camera for short times (during the fluorescence decay) after the laser pulse.

Excitation: The excitation source for the TriM-FLIM is a femtosecond-pulsed mode-locked Ti:sapphire laser (Spectra-Physics, France) that is tunable from 700 to 950 nm. The laser is directed into the TriMScope (LaVision Biotec GmbH) box, which contains a 50/50 beam splitter and mirrors to divide the incoming laser into from two to 64 beams. The set of beams passes through a 2000-Hz scanner before illuminating the back aperture of a 60x NA 1.2 infrared water-immersion objective (Olympus). A line of foci is then created at the focal plane, which can be scanned across the sample, producing pseudo-wide-field illumination.

Detection: A filter wheel of different spectral filters is used to select the fluorescence imaged onto a fast-gated light intensifier connected to a CCD camera (PicoStar, LaVision Biotec GmbH). The gate of the intensifier (adjusted to 1 or 2 nsec, depending on the experiment) is triggered by an electronic signal coming from the laser and a programmable delay box, which together are used to acquire a stack of time-correlated images. The number of gates and their time width determine the time resolution of the fluorescence kinetics. The acquisition time for each gate determines the signal-to-noise ratio. The time-correlated gated images are acquired sequentially. Thus, use caution to avoid photobleaching during acquisition, which can artificially shorten the fluorescence decay.

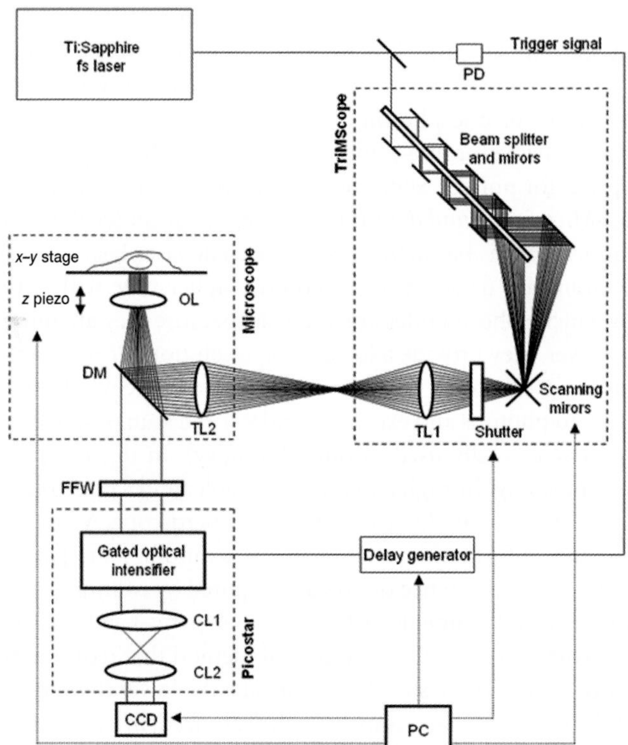

FIGURE 3. Time-domain TriM-FLIM setup for fast acquisitions. PD, Photodiode; OL, objective lens; DM, dichroic mirror; TL1, transfer lens 1; TL2, transfer lens 2; FFW, fast filter wheel; CL1, camera lens 1; CL2, camera lens 2; CCD, charge-coupled device; PC, personal computer.

ANALYSIS OF THE FLUORESCENCE DECAYS OF THE DONOR

The biological information that can be obtained from the fluorescence decay depends on the analysis that is performed. Mainly, the fluorescence decay can be fitted with different models (one or two discrete exponentials or more complex models) or the mean fluorescence lifetime τ can be obtained without fitting procedures to obtain raw FLIM data. If, in either case, FRET is detected, the biological information will be different in terms of quantification and spatial and temporal resolutions. To obtain quantitative measurements about the fraction of donor molecules interacting with the acceptor, it is necessary to perform a fit of the donor fluorescence decay obtained with a high signal-to-noise ratio, and hence a high photon count, using a two-exponential model. This fitting process requires an acquisition method, which is time consuming (TCSPC method) and impedes the measurement of transient interactions at high spatial resolution. A compromise can be found by using a fast-gated CCD, exploiting the mean fluorescence lifetime of the donor by calculating the minimal fraction of interacting donor mf_D pixel by pixel. Interestingly, the mf_D spatiotemporal variations within a single cell are similar to those of f_D obtained from the fit of the fluorescence decay using a double-exponential model (Padilla-Parra et al. 2008).

Fitting Data from a TCSPC-FLIM System

The fluorescence decay obtained by using the TCSPC method shows a high photon count (binning the pixels over large area) with picosecond time resolution. The least-squares method is a valid way to find mechanistic parameters of the system, such as the fluorescence lifetimes and amplitudes, by using Globals (University of California at Irvine) with a Levenberg–Marquardt algorithm (LMA). Nonlinear LMA for parameter estimation should take into consideration the following hypotheses: (i) Data uncertainties should be related to the fluorescence intensity (photon counts); (ii) fluorescence intensity uncertainties should be Gaussian distributed; (iii) there should not be systematic uncertainties related to all data; (iv) the model used (single exponential [in the case of eGFP alone expressed in live cells]; double exponential, stretched exponential, or Gaussian exponential in all cases) should be a good description of the experimental data; and (v) enough photon counts should be acquired to obtain a good fit.

Among the different models that can be applied to describe the fluorescence intensity decay profile of a related fluorophore, we use a two-species model in which two populations are taken into consideration (an interacting fraction corresponding to a population that relaxes through FRET (f_D) and a noninteracting fraction in which the donor lifetime remains undisturbed $[1 - f_D]$). The donor lifetime obtained out of the single-exponential fit from cells expressing the donor alone is assigned and is fixed into the noninteracting fraction of the double-exponential model in the corresponding cotransfected cell in which a given protein interaction is under study (Emiliani et al. 2003),

$$i(t) = (1 - f_D)e^{-t/\tau_D} + f_D e^{-t/\tau_F}, \tag{12}$$

where f_D stands for the fraction of the interacting donor, τ_D is the fixed donor lifetime, and τ_F is the discrete FRET lifetime. In this type of analysis, the total intensity (I_0) is normalized to 1; and, therefore, both pre-exponential factors, and particularly f_D, vary from 0 to 1. As an example, Figure 4 depicts the fluorescence decay of eGFP fused to the double bromodomain ([BD]-eGFP) of the transcription factor TAFIID 250, together with the corresponding fits (black line) expressed in a live HeLa cell (green line) in the absence (Fig. 4A) and the presence (Fig. 4B) of mCherry fused to histone H4 (mCherry-H4).

Data Treatment Coming from the TriM-FLIM System

With the TriM-FLIM system, the mean fluorescence lifetime is determined for each pixel of the image.

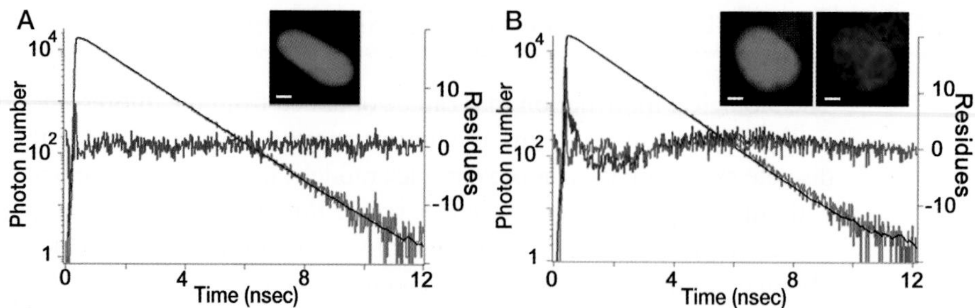

FIGURE 4. eGFP-BD interaction with acetylated mCherry-H4 in the nucleus of HEK293 live cells using the TCSPC method. (*A*) Steady-state intensity image of eGFP-BD expressed alone in HEK293 cells (green). The corresponding eGFP-BD fluorescence decay (green curve) extracted from the whole nucleus is fitted by a single exponential (black line, $\tau_D = 2.59$ nsec), and the residues are presented (blue curve). (*B*) Steady-state intensity images of eGFP-BD (green) and mCherry-H4 (red) coexpressed in HEK293 cells. The corresponding eGFP-BD fluorescence decay (green curve) extracted from the whole nucleus is fitted by a single model ($\tau = 2.46$ nsec) and by a biexponential model ($\tau_D = 2.59$ nsec fixed and $\tau_F = 0.65$ nsec) and residues are presented (blue curve and red curve, respectively). Note that the fluorescence decay is better fitted with a biexponential model as shown by residues. Scale bar, 2 μm.

Considering a fluorescence decay $i(t)$, the mean fluorescence (τ) is defined as

$$<\tau> = \int t \cdot i(t) \cdot dt / \int i(t) \cdot dt, \tag{13}$$

in which t is the time. To analyze data coming from a discrete sampling, the mean lifetime is directly calculated by applying Equation 13. For a time-gated stack of images, we have

$$<\tau> = \Sigma \Delta t_i \cdot I_i / \Sigma I_i, \tag{14}$$

where Δt_i corresponds to the time delay after the laser pulse of the ith image acquired and I_i corresponds to the pixel intensity map in the ith image (Fig. 5). The map of the mean fluorescence lifetime can be calculated and displayed.

If enough time-gated images are acquired (up to 11), the fluorescence decay can be fitted with an LMA, applying, for example, a discrete two-exponential model as in the case of the TCSPC method, and τ_D and τ_F will be determined together with f_D.

The fraction of donor that is involved in FRET can be calculated directly from the mean lifetime of the donor τ_D and τ_F:

$$\tau = [(1 - f_D) \cdot (\tau_D^2 + f_D \cdot \tau_F^2] / [(1 - f_D) \cdot \tau_D + f_D \cdot \tau_F]. \tag{15}$$

Isolating f_D and normalizing the last expression by dividing by τ_D, we find an expression that accounts for the fraction of interacting donor:

$$f_D = [1 - (<\tau>/\tau_D)] / [1 - (\tau/\tau_D) - (\tau_F/\tau_D)^2 + (\tau/\tau_D) \cdot (\tau_F/\tau_D)]. \tag{16}$$

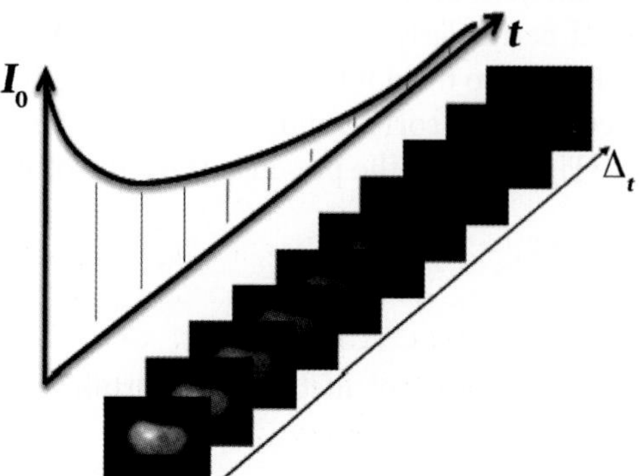

FIGURE 5. Cartoon representing a sequential acquisition of images from the time-domain TriM-FLIM setup.

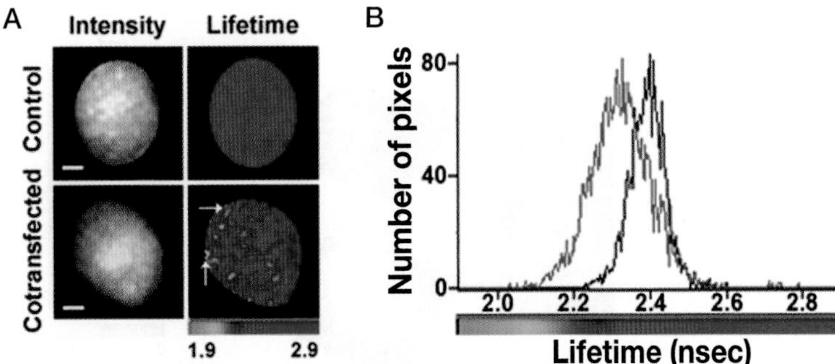

FIGURE 6. Intensity and FLIM images of eGFP-BD expressed alone as the control or with mCherry-H4 as the cotransfection in HEK293 live cells using a TriM-FLIM system and 11 time-gated images. (*A*) Intensity images were obtained by summing the time-gated stack. FLIM images were obtained by using Equation 14 in a pixel-by-pixel manner. White arrows show two chromatin domains in which eGFP-BD mean lifetime decreases significantly. (*B*) Histograms of the mean fluorescence lifetime for the control (black line) and the cotransfected cell (red line). The mean lifetime averaged throughout the nucleus decreased from 2.41 nsec for the control to 2.34 nsec for the cotransfection. Scale bar, 2 μm.

Maps of τ and of f_D (assuming the same τ_F value for each pixel) are obtained as shown in Figure 6, in which the interaction between BD-eGFP and mCherry-H4 is visualized in the nucleus of living cells by the decrease of the fluorescence lifetime of eGFP-BD compared with that of eGFP-BD in the absence of mCherry-H4 (monotransfected cells, the mean lifetime averaged throughout the nucleus decreased from 2.41 nsec for the control to 2.34 nsec for the cotransfection). The fraction of eGFP-BD interacting with mCherry-H4 can be determined and mapped within the nucleus.

The Minimal Fraction of Interacting Donor

The minimal fraction of interacting donor, mf_D (Padilla-Parra et al. 2008), is calculated directly from the values of the mean fluorescence lifetime of the donor in the absence and in the presence of the acceptor.

Mathematically, f_D depends on two variables (τ_F and τ) and can be minimized following τ_F:

$$mf_D = [1 - (\tau/\tau_D)]/[(\tau/2 \cdot \tau_D) - 1]^2. \tag{17}$$

mf_D provides instantaneous knowledge about the minimal extent of the interaction under study. This is particularly relevant in biology because, without knowing the intrinsic transfer efficiency (because mf_D does not require previous knowledge of τ_F), quantitative data related to the relative concentration are immediately at hand.

The impact of fast acquisitions of f_D and mf_D with the same biological example (eGFP-BD and mCherry-H4) was investigated by using the TriM-FLIM system (Fig. 7). On the same cells, two windowing schemes were chosen, one using only five time-gated images (2-nsec gate width) for very fast acquisition times (~3 sec) and the other using 11 time-gated images (1-nsec gate width) for longer acquisition times (~30 sec).

F_D was calculated from the value of τ_F determined from the biexponential fit of the fluorescence decay of the donor in the presence of the acceptor from the stack of 11 time-gated images ($\tau_F = 0.65 \pm 0.59$ nsec, $n = 33$). The donor lifetime was fixed to $\tau_D = 2.40$ nsec, coming from the monoexponential fit of the donor alone. A 3D representation of both f_D and mf_D images using a threshold limit given by the control is included to reveal the differences between control and FRET images (Fig. 7). Both f_D and mf_D images are very similar and present the same pattern when FRET occurs with a very close mean value of f_D (0.13) and mf_D (0.11). The two histograms are also superimposed in Figure 7 (right panel). The underestimation of the amount of donor that undergoes FRET when taking mf_D relative to f_D is only 15% in this example.

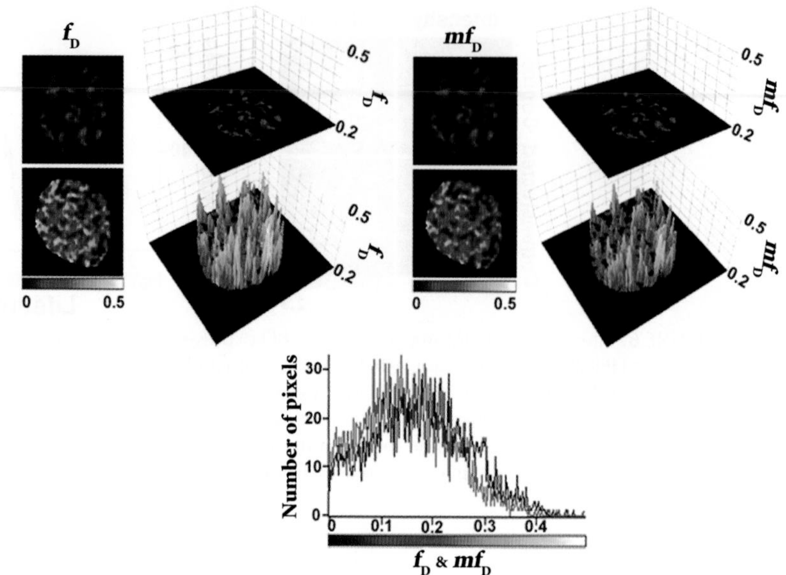

FIGURE 7. (*Top*) f_D and mf_D images obtained by using Equation 16 (with $\tau_D = 2.41$ nsec and $\tau_F = 0.65$ nsec) and Equation 17 (with $\tau_D = 2.41$ nsec), respectively. Three-dimensional representation of the corresponding f_D and mf_D images using a threshold limit given by the control (0.2) are also presented. (*Bottom*) The comparison between the histogram coming from f_D and mf_D clearly show that, for this example, both values behave similarly for all pixels of the image. Note that this is the same cell showed in Figure 6 and the dimensions of the micrograph are 12 x 12 μm.

Number of Interacting Particles and Relative Concentration

When we are interested in following the spatiotemporal changes of a highly localized protein, it is convenient to use a different approach related to the relative concentration: the number of interacting particles (NP). Assuming a two-species model and single-exponential behavior of the donor, fluorescence intensity as a function of time can be defined as

$$i(t) = I_0 f_D \cdot e^{-t/\tau_F} + I_0(1 - f_D) \cdot e^{-t/\tau_D}, \tag{18}$$

where I_0 is the fluorescence intensity at time 0. All of the other parameters have already been defined. Note that here, $i(t)$ is not normalized to 1, and its amplitude corresponds to the relative concentration of fluorophores:

$$i(t) = k_r([DA] \cdot e^{-t/\tau_F} + [D] \cdot e^{-t/\tau_D}), \tag{19}$$

where k_r is the rate constant of the radiative process, $[DA]$ is the concentration of the interacting complex, and $[D]$ is the concentration of the noninteracting donor population. In Equation 19, $[DA]$ is proportional to $(I_0 \times f_D)$. Note that we can approximate f_D by using the mf_D image and I_0 by using the corresponding first gated image. We present a biological example corresponding to the activation of Rac followed by FRET occurring between eGFP-Rac and PBD-mCherry (which is the binding domain of an effector of Rac fused to mCherry) and applying the number of particles in Figure 8 (see Movie 41.1, which shows a time lapse of the spatiotemporal activation of Rac [relative NP = $mf_D \times I_0$], and Movie 41.2, which shows a time lapse of the corresponding eGFP-Rac fluorescence intensity).

PRACTICAL CONSIDERATIONS

Limitations and Use of mf_D

The use of mf_D described here is applicable if we consider a two-component system, the unbound donor and donor involved in FRET with the acceptor with a narrow distribution of E. Although one

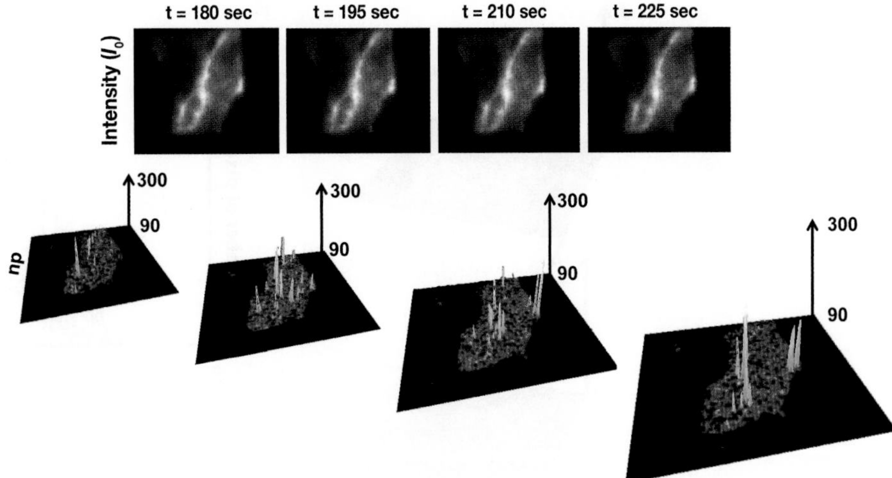

Threshold given by the negative control

FIGURE 8. Time lapse of Rac activation followed by fast FLIM. A cotransfected cell with eGFP-Rac and Pbd-mCherry shows a transient activation localized preferentially in subcellular domains at the periphery of the cell. (*Top*) The first time-gated intensity image of a sequential acquisition of five time-gated images (2 nsec each) is presented (see Movie 41.2). (*Bottom*) the corresponding NP images are shown with a threshold given by the negative control (eGFP-Rac expressed alone coming from a cell with the same signal-to-noise ratio) (see Movie 41.1). The NP is determined by the product of the first gated image and the mf_D value, as described in the text. The Rac activation (*bottom*) is independent of the eGFP-Rac fluorescence intensity (*top*). The dimensions of the micrograph are 16 x 15 µm.

of the constraints mentioned above deals with the single-exponential behavior of the donor (e.g., eGFP), the mf_D concept can be extended to multiexponential donors, such as cyan fluorescent protein (CFP) (Padilla-Parra et al. 2008).

Choice of the Best FRET Couple

The most important characteristics of good FRET–FLIM donors are that (i) their fluorescence decay profile must be fit best with a single-exponential model, (ii) their fluorescence intensity as a function of time should remain constant under different illumination conditions (i.e., they have good photostability), and (iii) they should not be susceptible to undergoing photoconversion.

mTFP1 is a monomeric cyan protein whose fluorescence decay is best described with a single-exponential model (Ai et al. 2006; Padilla-Parra et al. 2009) as is eGFP (Tramier et al. 2006). Different FRET pairs were tested by linking two different FPs with a polypeptide chain (tandem). By using TCSPC-FLIM, we compared different tandems (Table 1). The relatively low f_D percentages could come from a possible spectroscopic heterogeneity of the acceptor population, which is partially caused by different maturation rates for the donor and the acceptor (Padilla-Parra et al. 2009).

TABLE 1. Intrinsic FRET efficiency (*E*) and the fraction of interacting donor (f_D) for a set of different tandem constructs using a two-species model with a double-exponential model fixing the donor to a previously calculated value $N = 5$

Tandem	f_D	E
mRFP1-eGFP	0.26 ± 0.08	0.56 ±0.02
mStrawberry-eGFP	0.37 ± 0.07	0.58 ±0.02
mCherry-eGFP	0.45 ± 0.02	0.58 ±0.03
mTFP1-mOrange	0.37 ± 0.01	0.68 ±0.02
mTFP1-eYFP	0.71 ± 0.01	0.61 ±0.08

RFP, red fluorescent protein; eGFP, enhanced green fluorescent protein; eYFP, enhanced yellow fluorescent protein.

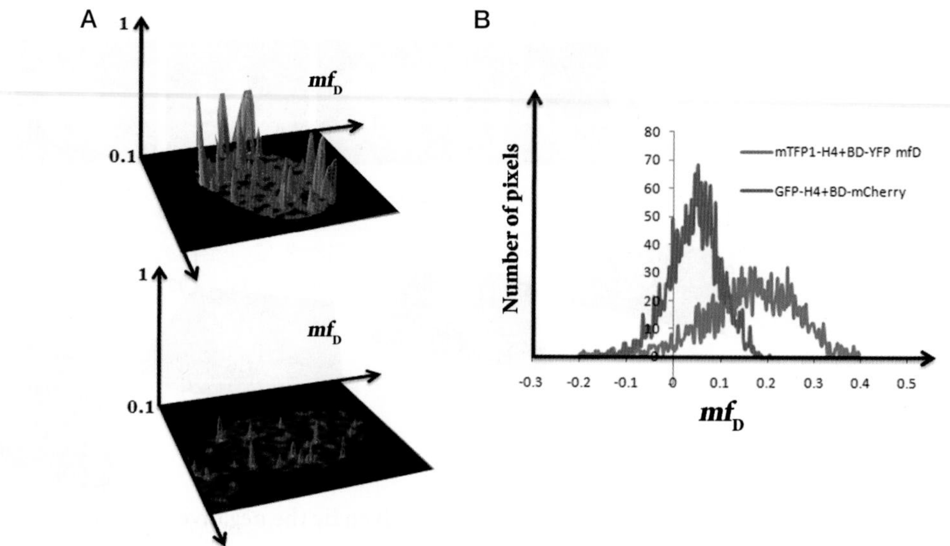

FIGURE 9. mf_D comparison between the FRET couples mTFP1/eYFP and eGFP/mCherry from fast-FLIM acquisitions. (*A*) Three-dimensional maps of mf_D for mTFP1-H4 coexpressed with eYFP-BD (*upper panel*) and eGFP-H4 coexpressed with mCherry-BD (*lower panel*). (*B*) The overlapping mf_D distribution for both examples is shown as well as the higher average mf_D for the mTFP-H4+ eYFP-BD couple. The dimensions of both micrographs are 14 x 14 µm.

The usefulness of mTFP1-eYFP (enhanced yellow fluorescent protein) as a FRET couple is supported by the twofold increase of the minimal fraction of interacting histone H4 with the double BD of TAFII 250, when using mTFP1-H4/eYFP-BD instead of eGFP-H4/mCherry-BD using the TriM-FLIM (Fig. 9). Note that mTFP1 is fairly resistant to photobleaching. However, attention should be paid to the effect of photobleaching when performing time-lapse TriM-FLIM acquisition with fusion proteins using mTFP1 in biological systems in which the proteins are highly immobile, as is the case for histone H4 incorporated into chromatin.

Effect of Photobleaching on FRET Determination: False FRET Signals

The existence of photobleaching and/or photoconversion before or during the acquisition can induce false-positive FRET determination. In steady-state measurements, especially with the CFP/YFP (yellow fluorescent protein) couple, donor photobleaching (CFP is more sensitive to excitation light than YFP) or photoconversion of YFP into a CFP-like species (with the acceptor photobleaching method) results in a ratio of CFP and YFP intensity similar to what can be obtained in a FRET situation (Valentin et al. 2006). The excitation light intensity is much lower with the TCSPC method; and thus this method is less prone to photobleaching. However, photobleaching can occur before the acquisition when observing and choosing the cells. Importantly, when using fast acquisitions (e.g., TriM-FLIM or any other sequential method for the fluorescence lifetime measurement; Grant et al. 2007), the occurrence of photobleaching on single-exponential fluorophores during the acquisition of FLIM has a drastic effect on the mean fluorescence lifetime, which is not the case with TCSPC (Tramier et al. 2006; Padilla-Parra et al. 2009). When acquiring images very quickly (e.g., by taking five time-gated images), if photobleaching occurs, the intensity of the fifth image can be too weak to have a significant value. This phenomenon shortens the mean lifetime determination as in the FRET situation, causing a bias in the interpretation of results.

Reasons for the Absence of a FRET Signal

The occurrence of FRET with a particular FP couple relies on the close proximity of the donor and the acceptor ($R < 80$ Å) and on the correct orientation of their dipoles (they cannot be perpendicular to each other). In addition to these basic requirements, FRET cannot be detected in certain situations even if the corresponding endogenous proteins interact to some extent. The classic reason for the absence of FRET is the localization of the fluorescent tag to a position of the protein sequence that impedes the interaction with its partners. More often encountered is the situation in which the amount of interacting donor per pixel is very weak in front of the noninteracting donor; therefore, it is difficult to get a significant FRET signal. We observe this situation, for example, in polymeric protein structures, such as actin filaments or microtubules. Hetero-FRET detection arises from the close proximity of a donor-tagged and an acceptor-tagged monomer within the polymer. The occurrence of this situation is statistically weak, and no FRET is detected. A related situation, in which it is difficult to unveil FRET, occurs when the interactions between the proteins of interest are transient and subject to spatiotemporal fluctuations, as we saw in the examples above. In this situation, fast time-lapse FLIM has to be performed to avoid the average of the FRET signals in space and time. The existence of a big proportion of immature red acceptor (up to 60%) artificially decreases the amount of bound donor. All these considerations show the difficulties (and likely the impossibility) of obtaining true quantitative measurements for the amount of donor protein that is interacting when we carry out live cell studies. This state of affairs reinforces the use of the concept of the minimal fraction of interacting donor protein mf_D. When using bioprobes based on intramolecular FRET (such as Raichu or other tandem probes used to monitor changes in biochemical activity), the steady-state intensity ratio measurements can be competitive in front of the fast FRET–FLIM method. However, for protein–protein interaction measurements (intermolecular FRET with no control on the amount of each partner within live cells), determining mf_D by using fast FRET–FLIM (time or frequency domain) is the best method because it is independent of the local concentration of protein.

CONCLUSIONS

The classical analysis FLIM technique in which the fluorescent decay is fitted to each pixel of a related image remains challenging when it is used to monitor protein–protein interactions in live cells. Compromises must be made between the precision and the time invested in the measurement. We have shown a simple way to analyze data derived from setups based on TCSPC detection, and we have also provided an alternative way to quantify transient protein interactions with faster systems based on sequential acquisitions. In that sense, the mf_D approach is an original method to quantify protein interactions for very fast acquisitions in which low photon counts and time points are required. This method is capable of providing information related to transient interactions in live cells.

ACKNOWLEDGMENTS

We thank Jean Claude Mevel, Dr. Marie Jo Masse, Dr. Allison Marty, and Dr. Guy Van Tran Nhieu for technical assistance with plasmid constructs. Work described in this chapter was performed at the Imaging Center (ImagoSeine) of the Institut Jacques Monod and was supported by the Fondation pour la Recherche Medicale, the Region Ile de France (Soutien aux Equipes Scientifiques pour l'Acquisition de Moyens Expérimentaux), the Centre National de la Recherche Scientifique (Action Concertee Incitative Biologie Cellulaire, Moleculaire et Structurale), the Association pour la Recherche sur le Cancer, and the Association Nationale pour le Recherche. SP-P is a recipient of a European Union predoctoral fellowship (Marie-Curie Grant No. MRTN-CT 2005-019481).

REFERENCES

Ai HW, Henderson JN, Remington SJ, Campbell RE. 2006. Directed evolution of a monomeric, bright and photostable version of Clavularia cyan fluorescent protein: Structural characterization and applications in fluorescence imaging. *Biochem J* **400:** 531–540.

Berney C, Danuser G. 2003. FRET or no FRET: A quantitative comparison. *Biophys J* **84:** 3992–4010.

Cheung HC. 1991. Resonance energy transfer. In *Topics of fluorescence microscopy, Volume 2, Principles* (ed. Lakowicz JL), Chap. 3, pp. 123–176. Plenum, New York.

Dale RE, Eisinger J, Blumberg WE. 1979. The orientational freedom of molecular probes: The orientation factor in intramolecular energy transfer. *Biophys J* **26:** 161–194. Errata: **30:** 365.

Elangovan M, Wallrabe H, Che Y, Day RN, Barroso M, Periasamy A. 2003. Characterization of one- and two-photon excitation fluorescence resonance energy transfer microscopy. *Methods* **29:** 58–73.

Emiliani V, Sanvitto D, Tramier M, Piolot T, Petrasek Z, Kemnitz K, Durieux C, Coppey-Moisan M. 2003. Low intensity two-dimensional imaging of fluorescence lifetimes in living cells. *Appl Phys Lett* **83:** 2471–2473.

Förster T. 1948. Intermolecular energy migration and fluorescence. *Ann Phys* **6:** 55–75.

Gordon GW, Berry G, Liang XH, Levine B, Herman B. 1998. Quantitative fluorescence resonance energy transfer measurements using fluorescence microscopy. *Biophys J* **74:** 2702–2713.

Grant DM, McGinty J, McGhee EJ, Bunney TD, Owen DM, Talbot CB, Zhang W, Kumar S, Munro I, Lanigan PMP, et al. 2007. High speed optically sectioned fluorescence lifetime imaging permits study of live cell signaling events. *Opt Express* **15:** 15656–15673.

Hoppe A, Christensen K, Swanson JA. 2002. Fluorescence resonance energy transfer-based stoichiometry in living cells. *Biophys J* **83:** 3652–3664.

Lakowicz JR. 2006. *Principles of fluorescence spectroscopy*, 3rd ed. Springer, New York.

Neher RA, Neher E. 2004. Applying spectral fingerprinting to the analysis of FRET images. *Microsc Res Tech* **64:** 185–195.

Padilla-Parra S, Audugé N, Coppey-Moisan M, Tramier M. 2008. Quantitative FRET analysis by fast acquisition time domain FLIM at high spatial resolution in living cells. *Biophys J* **95:** 2976–2988.

Padilla-Parra S, Audugé N, Coppey-Moisan M, Tramier M. 2009. Quantitative comparison of different fluorescent protein couples for fast FRET-FLIM acquisition. *Biophys J* **97:** 2368–2376.

Peter M, Ameer-Beg SM, Hughes MKY, Keppler MD, Prag S, Marsh M, Vojnovic B, Ng T. 2005. Multiphoton-FLIM quantification of the eGFP-mRFP1 FRET pair for localization of membrane receptor-kinase interactions. *Biophys J* **88:** 1224–1237.

Shaner NC, Steinbach PA, Tsien RY. 2005. A guide to choosing fluorescent proteins. *Nat Methods* **2:** 905–909.

Stockholm D, Bartoli M, Sillon G, Bourg N, Davoust J, Richard I. 2004. Imaging calpain protease activity by multiphoton FRET in living mice. *J Mol Biol* **346:** 215–222.

Tramier M, Zahid M, Mevel JC, Masse MJ, Coppey-Moisan M. 2006. Sensitivity of CFP/YFP and GFP/mCherry pairs to donor photobleaching on FRET determination by fluorescence lifetime imaging microscopy in living cells. *Microsc Res Tech* **11:** 933–942.

Tsien RY. 1998. The green fluorescence protein. *Annu Rev Biochem* **67:** 509–524.

Tuszynski J, Portet S, Dixon J. 2006. Nonlinear assembly kinetics and mechanical properties of biopolymers. *Nonlinear Anal* **63:** 915–925.

Valentin G, Verheggen C, Piolot T, Neel H, Coppey-Moisan M, Bertrand E. 2006. Photoconversion of YFP into a CFP-like species during acceptor photobleaching FRET experiments. *Nat Methods* **3:** 491–492.

Valeur B. 2002. *Molecular fluorescence: Principles and applications.* Wiley-VCH, Weinheim, Germany.

van Rheenen J, Langeslag M, Jalink K. 2004. Correcting confocal acquisition to optimize imaging of fluorescence resonance energy transfer by sensitized emission. *Biophys J* **86:** 2517–2529.

Verveer PJ, Wouters FS, Reynolds AR, Bastiaens PIH. 2000. Quantitative imaging of lateral ErbB1 receptor signal propagation in the plasma membrane. *Science* **290:** 1567–1570.

Wlodarczyk J, Woehler A, Kobe F, Ponimaskin E, Zeug A, Neher E. 2008. Analysis of FRET signals in the presence of free donors and acceptors. *Biophys J* **94:** 986–1000.

Zal T, Gascoigne NRJ. 2004. Photobleaching-corrected FRET efficiency imaging of live cells. *Biophys J* **86:** 3923–3939.

MOVIE LEGENDS

Movies are freely available online at www.cshprotocols.org/imaging.

MOVIE 41.1. Time lapse of Rac activation followed by fast fluorescence lifetime imaging microscopy (FLIM). Rac activation is visualized by the relative number of interacting particles (NP = $I_0 \times mf_D$) of eGFP-Rac that undergoes Förster resonance energy transfer (FRET) with PBD-mCherry as described in Figure 8. I_0 and mf_D were obtained with the TriM-FLIM (five time-gated images of 2 nsec each; 5-sec acquisition time per NP image; time-lapse duration: 100 sec). The dimensions of the micrograph are 16 x 15 μm.

MOVIE 41.2. Time lapse of eGFP-Rac fluorescence intensity image (I_0 images) corresponding to the time lapse of relative NP of Movie 41.1.

42 Single- and Two-Photon Fluorescence Recovery after Photobleaching

Kelley D. Sullivan,[1] Ania K. Majewska,[2] and Edward B. Brown[3]

[1]Department of Physics and Astronomy, University of Rochester, Rochester, New York 14627; [2]Department of Neurobiology and Anatomy, University of Rochester, Rochester, New York 14642; [3]Department of Biomedical Engineering, University of Rochester, Rochester, New York 14627

ABSTRACT

Fluorescence recovery after photobleaching (FRAP) is a microscopy technique for measuring the kinetics of fluorescently labeled molecules and can be applied both in vitro and in vivo for two- and three-dimensional systems. This chapter discusses the three basic FRAP methods: traditional FRAP, multiphoton FRAP(MPFRAP), and FRAP with spatial Fourier analysis (SFA-FRAP). Each discussion is accompanied by a description of the mathematical analysis appropriate for situations in which the recovery kinetics is dictated by free diffusion. In some experiments, the recovery kinetics is dictated by the boundary conditions of the system, and FRAP is then used to quantify the connectivity of various compartments. Because the appropriate mathematical analysis is independent of the bleaching method, the analysis of compartmental connectivity is discussed last, in a separate section.

INTRODUCTION

Since its introduction in the 1970s (Peters et al. 1974; Axelrod et al. 1976), FRAP has been used to measure the diffusion coefficient (or analogous transport parameters) of labeled molecules in both two-dimensional systems such as cell membranes, small regions of cells, and lamellipodia (Feder et al. 1996; Braga et al. 2007) and three-dimensional systems such as tumor tissues or cell bodies (Chary and Jain 1989; Berk et al. 1997; Pluen et al. 2001; Stroh et al. 2004; Chauhan et al. 2009). Each of the three FRAP techniques is performed by first photobleaching a small region of interest within a sample, and then monitoring the region as fluorescent molecules from outside the region diffuse in to replace the photobleached molecules. The original spot FRAP technique has undergone a variety of modifications to accommodate different photobleaching methods, including patterned (Abney et al. 1992), continuous (Wedekind et al. 1996), line (Braeckmans et al. 2007), and disk-shaped (Mazza et al. 2008) photobleaching. Modifications to the recovery analysis have also expanded FRAP as a tool to analyze binding kinetics (Kaufmann and Jain 1991; Berk et al. 1997;

Schulmeister et al. 2008), to quantify the connectivity of compartments (Majewska et al. 2000; Cardarelli et al. 2007), and to investigate polymer structure-property relationships (Li et al. 2010).

FRAP AND MPFRAP

In a FRAP experiment, a focused laser beam bleaches a region of fluorescently labeled molecules in a thin sample of tissue (Axelrod et al. 1976). The same laser beam, greatly attenuated, then generates a fluorescence signal from that region as unbleached fluorophores diffuse in. A photomultiplier tube, or a similar detector, records the recovery in fluorescence signal, producing a fluorescence-versus-time curve. In a conventional (one-photon) FRAP experiment, simple analytical formulas can be fit to the fluorescence recovery curve to generate the two-dimensional diffusion coefficient of the fluorescent molecule but only if the sample is sufficiently thin (see FRAP Diffusion Analysis). If the sample is not thin enough for the analytical solution to hold, the diffusion coefficient can be estimated by comparing the recovery time with that of molecules with known diffusion coefficients in samples of identical thickness.

In an MPFRAP experiment, a focused beam from a mode-locked laser provides both bleaching and monitoring, generating fluorescence and photobleaching via multiphoton excitation (Brown et al. 1999). Most current MPFRAP experiments are designed for two-photon fluorescence and photobleaching; however, single- and multiple (three or higher)-photon excitation can also be achieved. Therefore, in this chapter, we will refer to these experiments generally as "multiphoton." The intrinsic spatial confinement of multiphoton excitation means that the bleaching/monitoring volume is three-dimensionally resolved (Denk et al. 1990); consequently, there is no upper limit on the sample thickness. Simple analytical formulas can be applied to the fluorescence recovery curve to generate the three-dimensional diffusion coefficient of the fluorescent molecule.

FRAP Instrumentation

The primary instrumentation of one-photon FRAP consists of a laser source, an acousto-optic modulator (AOM), a dichroic mirror, an objective lens, a gated photomultiplier tube (PMT), and a data recording system such as an analog-to-digital (A/D) board or scaler (photon-counting device) (Fig. 1A). The laser source is directed through the AOM to the dichroic mirror and the objective lens and into the fluorescent sample.

The laser is typically an argon-ion laser operating in the lowest-order transverse electromagnetic (TEM_{00}) mode to produce a Gaussian transverse intensity profile, suitable for analysis of recovery curves (see FRAP Diffusion Analysis). The laser must be modulated on a much faster timescale than the diffusive recovery time of the system, often requiring modulation times of fractions of a millisecond. This necessitates the use of an AOM as the beam-modulation device because of its fast response time. To generate significant variation in transmitted intensity, the first diffraction maximum of the AOM should be used, not the primary transmitted beam.

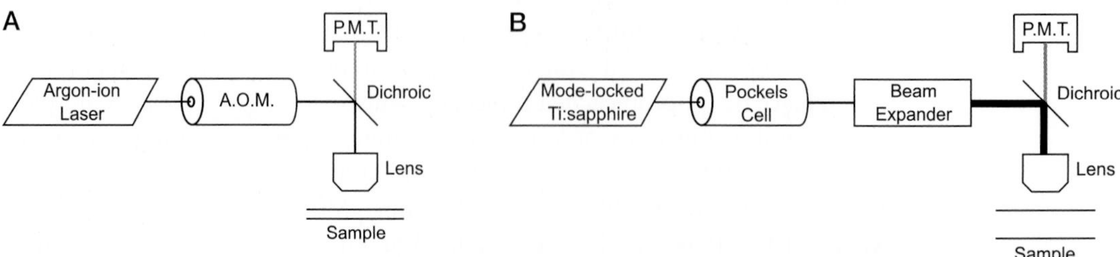

FIGURE 1. (*A*) Equipment for fluorescence recovery after photobleaching (FRAP). (*B*) Equipment for multiphoton FRAP (MPFRAP). Laser light is passed through the AOM or the Pockels cell/beam expander and directed by the dichroic to the objective lens, which focuses the light within the sample. Fluorescence from the sample then passes back through the lens and the dichroic to the PMT.

MPFRAP Instrumentation

The primary instrumentation of MPFRAP consists of a laser source, a Pockels cell, a beam expander, a dichroic mirror, an objective lens, a gated PMT, and a data recording system (Fig. 1B). The laser source is directed through the Pockels cell to the beam expander, the dichroic mirror, and the objective lens and into the fluorescent sample.

The laser is typically a mode-locked (100-fsec pulses) Ti:sapphire laser. This beam is expanded to overfill the objective lens (Zipfel et al. 2003), thereby producing a uniformly illuminated back aperture, resulting in the formation of the highest resolution spot in the plane of the sample (Pawley 2006). The intrinsic spatial confinement of multiphoton excitation produces a three-dimensionally defined bleach volume, whose size depends on the numerical aperture (NA) and the wavelength of the excitation light and is typically ~0.5 x 0.5 x 1 µm. This extremely small bleached volume dissipates rapidly (hundreds of microseconds for smaller fluorescently labeled molecules such as fluorescein isothiocyanate [FITC]-bovine serum albumin [BSA] or green fluorescent protein [GFP]). Consequently, MPFRAP requires a beam-modulation system with response times as fast as 1 µsec. The AOM traditionally used in one-photon FRAP relies on diffraction of the laser beam to achieve intensity modulation, whereas 100-fsec pulses, typical of a mode-locked Ti:sapphire laser used in multiphoton FRAP, have a bandwidth of 15 nm. Different wavelengths of light will diffract in different directions; therefore, the nondiffractive Pockels cell is often used for MPFRAP beam modulation instead of an AOM. A Pockels cell operates by passing a beam through a crystal, across which a voltage is applied. Varying the voltage rotates the plane of polarization of the incident light. Before exiting the Pockels cell, the beam passes through a polarizer, which converts the rotation in the plane of polarization to a variation in intensity.

FRAP or MPFRAP Procedure

1. Set the laser modulator (AOM or Pockels cell) to a low transmission state, producing the monitor beam and generating fluorescence from the sample, which is collected by the objective lens and is detected by the PMT. The scaler monitors the output of the PMT for a short duration (tens of microseconds to milliseconds), recording the prebleach signal.
2. Switch the laser power modulator to a high transmission level, producing the bleach beam, which photobleaches a fraction of the fluorophores within the sample. The modulator is returned rapidly to the low monitoring state (after a total bleach time that depends on the sample dynamics). If possible, gate the PMT during the bleach pulse to avoid damage to the PMT.
3. Continuously record the fluorescence generated by the monitor beam. As unbleached fluorophores diffuse into the region excited by the laser beam, the fluorescence signal recovers to equilibrium levels.
4. Analyze the fluorescence recovery curve to yield diffusion coefficients or analogous parameters as well as the fraction of immobile fluorophores (Fig. 2).

Considerations when Determining FRAP and MPFRAP Diffusion Parameters

A number of steps must be taken to ensure accurate determination of the diffusion parameters.

1. The power of the monitoring beam must be high enough to generate a sufficient fluorescence signal but not so high as to cause significant photobleaching. The generation of photobleaching by the monitoring beam can be detected easily by photon counting over an integration time much larger than the expected recovery time for a range of typical monitoring powers. For MPFRAP with two-photon excitation, a reduction from a slope of 2 on a log–log plot of photon counts as a function of power indicates the presence of photobleaching. For FRAP, a deviation from a slope of 1 on a linear plot of photon counts as a function of power indicates the presence of photobleaching. This is because fluorescence scales as intensity to the nth power, where n is the number of photons absorbed during excitation. In a one-photon FRAP experi-

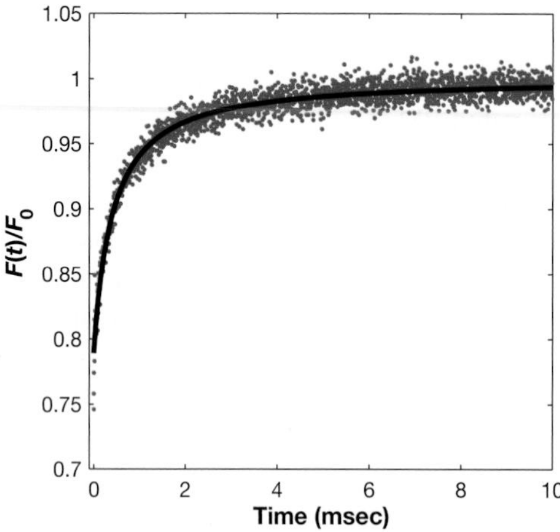

FIGURE 2. MPFRAP recovery curve. MPFRAP was performed on FITC-BSA in free solution. (Solid line) Least-χ^2 fit, producing a diffusion coefficient at 22°C of 52.2 μm²/sec.

ment, if the resultant monitoring power is too low to allow a sufficient signal-to-noise ratio, the monitoring beam can be repeatedly cycled between 0 and a high power that causes some limited photobleaching, allowing intermittent recording of the fluorescence recovery at higher signal rates, while limiting the total photobleaching by the monitoring beam (Waharte et al. 2005).

2. The duration of the bleaching flash must be short enough that no significant diffusion (i.e., recovery in the fluorescent signal) occurs during the bleach pulse. A rule of thumb is that the bleach pulse should be less than one-tenth of the half-recovery time of the subsequent recovery curve.

3. If the acquisition of the fluorescence recovery curve does not occur for a long enough period, overestimation of the immobile fraction and underestimation of the diffusion recovery time can result. A rule of thumb is that the recovery curve should be visibly flat (i.e., any systematic change in signal is less than shot noise) for the latter half of the recording time.

4a. In FRAP and MPFRAP, the fluorescence excitation rate is assumed to scale as rate $\sim\sigma\langle I^b\rangle$, where σ is the absorption cross section (units of cm² for one-photon excitation and cm⁴·sec for two-photon excitation), I is the intensity of the bleach beam, b is the number of photons absorbed in a bleaching event, and $\langle\ \rangle$ denotes a time average. There is an upper limit to the excitation rate of a fluorescent molecule, however, because fluorescent molecules have excited-state lifetimes of $t_L \approx 10$ nsec and, hence, cannot be excited at a faster rate than $1/t_L$. Furthermore, the pulsed lasers used in MPFRAP have a duty cycle of $t_D = 12.5$ nsec. Consequently, when the excitation rate of a fluorophores during the bleaching pulse approaches a significant fraction of $1/t_L$ or $1/t_D$, the rate of excitation will deviate from $\sim\sigma\langle I^b\rangle$ and will asymptotically approach a limiting value, which depends on $1/t_L$ or $1/t_D$. This phenomenon is known as excitation saturation. FRAP or MPFRAP curves generated in the saturation regime in which the photobleaching rate does not scale as $\sim\sigma\langle I^b\rangle$, will produce erroneously low diffusion coefficients.

4b. To avoid excitation saturation, a series of FRAP or MPFRAP curves must be generated at increasing bleach powers. The curves are then analyzed (see Diffusion Analysis below) using the bleach depth parameter and the diffusive recovery time as the fitting parameters. For MPFRAP with two-photon excitation, a reduction from a slope of 2 on a log–log plot of bleach depth parameter as a function of power indicates the presence of excitation saturation. For FRAP, a deviation from a slope of 1 on a linear plot of bleach depth parameter as a function of power indicates the presence of excitation saturation.

5. To convert a measured diffusive recovery time to a diffusion coefficient, the characteristic size of the bleached region must be known. In both FRAP and MPFRAP, the excitation probability as a function of position transverse to the beam axis at the focal spot (i.e., the transverse beam intensity profile to the nth power) can be well represented by a Gaussian function (see below),

whose characteristic half-width at e^{-2} must, therefore, be determined to convert recovery times to diffusion coefficients. In the case of one-photon FRAP, the e^{-2} half-width of the excitation probability must be measured transverse to the beam axis only, whereas in MPFRAP, the e^{-2} half-width in both the transverse and the axial directions must be measured. This is typically accomplished by scanning the focus of the laser beam across a subresolution (~10 nm or less) fluorescent bead and recording the fluorescent signal versus the position of the bead. Unless the excitation beam of a one-photon FRAP system is provided by a confocal laser-scanning microscope, there is no mechanism for easily altering the position of the laser focus in the sample plane, so a simple method to measure the excitation probability is to scan a subresolution bead transversely across the stationary beam focus with a stepper motor or a piezoelectric motor (Schneider and Webb 1981). In an MPFRAP system, the laser position is usually governed by galvanometers and stepper motors as part of a multiphoton laser-scanning microscope system. Consequently, it is relatively easy to scan the laser across a stationary subresolution fluorescent bead to determine the transverse e^{-2} half-width, whereas the axial e^{-2} half-width can be determined by scanning the bead across the focus, using the focus stepper motor that accompanies most laser-scanning microscope systems.

Diffusion Analysis: Conventional (One-Photon) FRAP

If the sample thickness in a FRAP experiment is sufficiently thin, the complex three-dimensional hourglass shape of the focused bleaching beam (Pawley 2006) can be ignored. This is because bleaching only occurs in a thin slice at the focus of the beam, and the postbleach recovery kinetics occurs laterally in a two-dimensional system. If the excitation laser is operating in the TEM_{00} mode and significantly underfills the objective lens (i.e., the beam is significantly smaller than the back aperture of the objective lens), then the transverse intensity profile is a simple Gaussian one, and an analytical formula for the fluorescence recovery curve can be derived. For a Gaussian laser beam, the fluorescence recovery curve describing free diffusion in a two-dimensional system is given by (Axelrod et al. 1976)

$$F(t) = F_\infty \sum_{n=0}^{\infty} \frac{(-\beta)^n}{n!} \frac{1}{(1+n+2nt/\tau_D)},$$

where F_∞ is the $t = \infty$ fluorescence signal, β is the bleach depth parameter, and τ_D is the two-dimensional diffusion recovery time. The fraction of immobile fluorophores in the sample is given by $(F_0 - F_\infty)/F_\infty$, where F_0 is the prebleach fluorescence signal and immobile is defined as having a diffusive mobility significantly slower than the timescale of the experiment. The diffusion coefficient is given by $D = w^2/4\tau_D$, where w is the transverse e^{-2} half-width of the laser beam at the sample.

FRAP can be extended to thicker samples, but analytic derivations of the diffusion coefficient become problematic because of the complex nature of the hourglass-shaped laser focus distribution. Furthermore, the fluorescence recovery time becomes dependent on the thickness of the sample, which may or may not be known. In these cases, the FRAP technique is often limited to a simple comparison of recovery times between samples of unknown diffusion coefficients and samples with known diffusion coefficients that have been measured with analytical techniques such as those described above.

Diffusion Analysis: MPFRAP

In an MPFRAP experiment, the highest spatial resolution is achieved by overfilling the objective lens, producing a diffraction-limited intensity distribution at the beam focus. The square (or higher power) of this intensity profile is well approximated by a Gaussian distribution, both transverse to and along the optical axis, although the half-width at e^{-2} is typically longer in the axial dimension than in the transverse dimension (Pawley 2006). For an overfilled objective lens inducing two-photon fluorescence and photobleaching, the fluorescence recovery curve describing free diffusion in a three-dimensional system is given by Brown et al. (1999) as

$$F(t) = F_\infty \sum_{n=0}^{\infty} \frac{(-\beta)^n}{n!} \frac{1}{(1+n+2nt/\tau_D)} \frac{1}{\sqrt{1+n+2nt/(R\tau_D)}},$$

where F_∞ is the $t = \infty$ fluorescence signal, β is the bleach depth parameter, and τ_D is the three-dimensional diffusion recovery time. Because the signal is limited by restrictions on the bleaching and the monitoring powers (as described in the section, Considerations when Determining FRAP and MPFRAP Diffusion Parameters, above), MPFRAP experiments are generally performed with a series of bleach/monitor sequences at a location of interest, and the resultant curves averaged together to improve the signal-to-noise ratio. Therefore, MPFRAP is insensitive to immobile fluorophores, as they are bleached out during repeated flashes and do not contribute to recovery kinetics. The diffusion coefficient is given by $D = w_r^2/8\tau_D$, where w_r is the transverse e^{-2} half-width of the laser beam at the sample and R is the square of the ratio of the axial (w_z) to the radial (w_r) e^{-2} half-width (Fig. 2).

In some experiments, it may be the case that the mobility of the fluorophore of interest is influenced by convective flow as well as diffusion. In this case, the shape of the fluorescence recovery curve changes and so must the mathematical model describing the recovery. By introducing a coordinate shift into the derivation of the equation immediately above, the new derivation yields (Sullivan et al. 2009)

$$F(t) = F_0 \sum_{n=0}^{\infty} \frac{(-\beta)^n}{n!} \frac{\exp\left[\frac{4nt(1/\tau_{v_x}^2 + 1/\tau_{v_y}^2)}{1+n+2nt/\tau_D}\right] \exp\left[\frac{4nt/\tau_{v_z}^2}{1+n+2nt/R\tau_D}\right]}{(1+n+2nt/\tau_D)(1+n+2nt/R\tau_D)^{1/2}},$$

where $\tau_{v_x} = w_r/v_x$, $\tau_{v_y} = w_r/v_y$, and $\tau_{v_z} = w_z/v_z$; and $v^2 = v_x^2 + v_y^2 + v_z^2$ is the speed of the convective flow. All other variables are defined as above. With this diffusion-convection model, accurate values for the diffusion coefficient can be obtained even under the influence of moderate flows, defined by a dimensionless speed parameter, $v_s = v(w_r/8D) \leq 3$ (Fig. 3) (Sullivan et al. 2009).

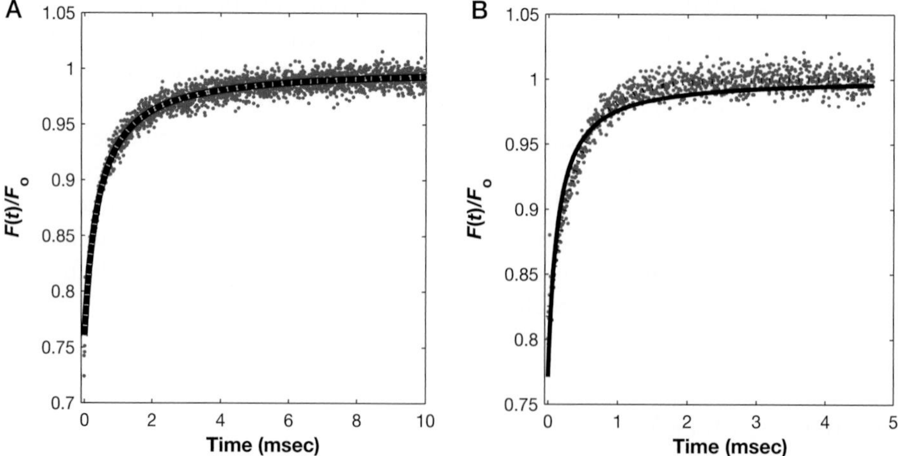

FIGURE 3. MPFRAP recovery curves. In both *A* and *B*, MPFRAP was performed on FITC-BSA in free solution, and the data were fit to both the conventional diffusion-only (solid line) and the diffusion-convection (dashed line) models. (*A*) Solution shows no convective flow. Both fits yield a diffusion coefficient at 22°C of 52.9 μm²/sec. (*B*) Solution shows convective flow with a scaled speed of 0.8. The diffusion-only model fits the data poorly and yields a diffusion coefficient at 22°C of 160 μm²/sec. The diffusion-convection model fits the data well and yields a diffusion coefficient of 51.9 μm²/sec, in good agreement with the value determined in the absence of convective flow.

SFA-FRAP

In an SFA-FRAP experiment, a focused laser beam is used to bleach a region of fluorophore, as in FRAP (described above). Unlike conventional FRAP, however, the evolution of the bleached region is imaged repeatedly using a charge-coupled device (CCD) camera with wide-field illumination provided by a mercury lamp. The sequential images of the recovery of the bleached spot are Fourier transformed, and the decay of selected spatial frequency components produces the diffusion coefficient of the diffusing fluorophore, without requiring knowledge of the details of the bleaching distribution or the sample thickness (Tsay and Jacobson 1991; Berk et al. 1993). Consequently, SFA-FRAP can be performed in thick samples, although its reliance on epifluorescence means that the diffusion coefficients it measures are averages over the entire depth of view of the microscope and are not three-dimensionally resolved.

SFA-FRAP Instrumentation

The primary instrumentation of conventional SFA-FRAP consists of a laser source, a dichroic mirror, two fast shutters, a galvanometer-driven movable mirror, an objective lens, a CCD camera and mercury arc lamp, and an image recording system such as a frame grabber card (Fig. 4). The laser source is directed through one shutter and dichroic mirror to the objective lens and into the fluorescent sample, whereas the lamp is directed by the movable mirror to be collinear with the laser.

The laser is typically an argon-ion laser as in FRAP. SFA-FRAP is traditionally used in thick (many hundreds of micrometers) tissues, and the imaged bleach spot is several tens of micrometers wide. Consequently, the diffusive recovery times can be relatively slow (several tens of milliseconds to many minutes), and the laser modulation rate does not have to be as rapid as in FRAP. Furthermore, the laser modulation is binary, with a bright bleaching flash and zero power being the only required states, and no intermediate power monitoring beam is needed. Therefore, a fast shutter is generally sufficient for SFA-FRAP, instead of a rapid (and more expensive) analog device such as the AOM or Pockels cell.

SFA-FRAP Procedure

1. Shutter the laser so that the movable mirror directs the mercury lamp illumination into the sample. The CCD records a few prebleach images.
2. Shutter the CCD to avoid damage, and move the movable mirror out of the laser path. Open the laser shutter, allowing the bleaching flash through. Close the laser shutter again, ending the bleaching flash, and unshutter the CCD.
3. The movable mirror periodically shifts back into the light path, directing the mercury arc lamp illumination into the objective lens and allowing epifluorescence images to be captured as the bleached distribution recovers to equilibrium levels and the bleached spot disappears.

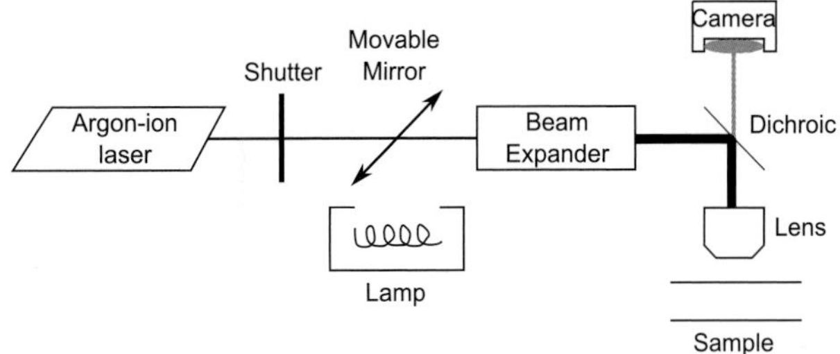

FIGURE 4. Equipment for FRAP with spatial Fourier analysis. Laser and lamp light are passed collinearly, but independently, through the beam expander and are directed by the dichroic to the objective lens, which focuses the light within the sample. The presence of the laser light is modulated by the shutter, whereas the movable mirror controls the presence of lamp light. Fluorescence from the sample is passed back through the lens and the dichroic to the CCD camera.

4. The series of images of the evolution of the bleached spot are Fourier transformed, and six of the lowest spatial frequencies are plotted. The exponential decay of the spatial frequencies yields the diffusion coefficient of the labeled molecule as well as the fraction of immobile fluorophores.

Considerations when Determining SFA-FRAP Diffusion Parameters

A number of steps must be taken to ensure accurate determination of the diffusion parameters.

1. The epifluorescence lamp must be bright enough to generate a sufficient signal, but not so bright as to cause significant photobleaching. Photobleaching caused by the lamp can be easily detected by performing a test SFA-FRAP experiment with no bleach pulse. If the fluorescence signal decays significantly over the course of the image series, the epifluorescence lamp must be attenuated. Additionally, the duty cycle of the monitoring pulses can be altered, thereby reducing the total exposure time of the sample during the recovery period.
2. The duration of the bleaching flash must be short enough that no significant diffusion (i.e., recovery in fluorescent signal) occurs during the bleach pulse. A rule of thumb is that the bleach pulse must be less than one-tenth of the half-recovery time of the subsequent spatial frequency decay curve.
3. If the acquisition of the fluorescence recovery images does not occur for a long enough period, overestimation of the immobile fraction and underestimation of the diffusion recovery time can result. A rule of thumb is that the decay curve of the spatial frequencies (see below) should be visibly flat (any change is less than shot noise) for the latter half of the recording time.

Diffusion Analysis

After a photobleaching pulse, the concentration distribution of unbleached fluorophores $c(x, y, t)$ evolves according to the diffusion equation. If the concentration distribution is first Fourier transformed with respect to x and y, the solution to this differential equation is a simple exponential (Berk et al. 1993),

$$C(u, v, t) = C(u, v, 0)e^{-4\pi^2(u^2 + v^2)Dt},$$

where u and v are the spatial frequencies and D is the diffusion coefficient. The relationship between dye concentration at the sample and intensity at the corresponding location in the CCD image may vary in space but is expected to be constant in time. Therefore, the Fourier transform of the CCD images also decays with a simple exponential in which the exponential decay time of a given pair of spatial frequencies is $1/(4\pi^2(u^2 + v^2)D)$. To analyze SFA-FRAP data, the series of CCD images of the evolving dye distribution are each Fourier transformed, and the exponential decay of a number (typically six) of selected spatial frequency pairs is analyzed to produce the exponential decay time, which directly yields the diffusion coefficient. Unlike FRAP or MPFRAP, no knowledge of the initial spatial distribution of photobleaching is required. Note that this analysis ignores diffusion along the optical axis and, furthermore, uses epifluorescence images of the diffusing system, therefore generating a diffusion coefficient that is an average over the visible depth of the system. In other words, any depth-dependent differences in the diffusion coefficient will be averaged out to a single value (Berk et al. 1993).

COMPARTMENTALIZATION ANALYSIS

Single-photon and multiphoton FRAP can also be used to measure the diffusional coupling between two connected compartments. A characteristic time course for the diffusion or the transport of different fluorescent molecules can be obtained by bleaching one compartment and by monitoring the fluorescence recovery curve as it is refilled with a fluorophore from the unbleached compartment. This information can then be used to determine parameters such as resistivity and pore size of the separating barrier. MPFRAP has been used to examine diffusion of dyes between the excitatory synapse (dendritic spine) and its parent dendrite (Svoboda et al. 1996; Majewska et al. 2000; Sobczyk et al. 2005) and between plant plastids (Kohler et al. 1997). FRAP has examined the cell cytoplasm and the nucleus (Wei et al. 2003) as well as the turnover of fluorescently tagged actin filaments between the spine and the dendrite (Star et al. 2002).

Instrumentation for Compartmentalization Analysis

A FRAP or MPFRAP instrument (described above) can be used for compartmentalization analysis. Because diffusion between compartments tends to be slower than diffusion within a compartment, these experiments can often be performed in the line-scan mode on a laser-scanning microscope. In the line-scan mode, the excitation beam is scanned repeatedly along a single line that intersects an object of interest, and a position-versus-time curve of the fluorescence is generated. The line-scan mode uses the acquisition electronics of the laser-scanning microscope and obviates the need to purchase a separate photon-counting device. This mode can limit the acquisition speed, however, depending on the design of the microscope.

Compartmentalization Analysis Procedure

1. Choose one of the compartments for bleaching and monitoring (typically the smaller compartment). Set the laser modulator (AOM or Pockels cell) to a low transmission state, producing the monitor beam and generating fluorescence from a region of interest (ROI) within the sample, which is collected by the objective lens and is detected by the PMT as a measure of the prebleach fluorescence.
2. Gate the PMT (i.e., the dynode voltage is set to 0) to avoid damage, and switch the laser modulator to a high transmission level, producing the bleach beam, which photobleaches a fraction of the fluorophores within the ROI. The modulator rapidly returns to the low monitor state (after a total bleach time that depends on the sample dynamics, as described above), and the PMT is subsequently ungated.
3. Continuously record the fluorescence generated in the ROI by the monitor beam. As unbleached fluorophores diffuse in from the unbleached compartment, the fluorescence signal recovers back to equilibrium levels.
4. Analyze the fluorescence recovery curve to yield the characteristic coupling time as well as the fraction of immobile fluorophores.

Considerations when Determining Compartmentalization Analysis Diffusion Parameters

1. Bleaching during the monitoring phase, the bleach duration, and the total acquisition time must be evaluated as in points 1–3 in the section, Considerations when Determining FRAP and MPFRAP Diffusion Parameters, described above.
2. Diffusional coupling is typically studied between two well-mixed compartments (i.e., compartments in which the timescale of diffusional equilibrium within the compartments is much faster than between compartments). This can be verified by spot bleaching within each of the compartments to determine the diffusion characteristics for fluorophores in each of the compartments.
3. Because the communicating compartments are well mixed, the initial spatial distribution of bleached molecules is irrelevant because the bleached compartment undergoes diffusive mixing before significant communication with other compartments can occur. Consequently, neither the bleach spot profile nor the excitation saturation is a significant concern. However, it is

important to determine that bleaching occurs in only one compartment. This can be accomplished by performing a line scan that intersects both compartments while restricting the bleaching pulse to a single compartment. In this case, both compartments can be monitored to ensure that bleaching is spatially restricted.

Compartmentalization Analysis

The fluorescence recovery curve in a well-mixed photobleached compartment diffusionally coupled to larger well-mixed compartments is given by (Svoboda et al. 1996; Majewska et al. 2000)

$$F(t) = F(\infty) - \Delta F_0 e^{-t/\tau},$$

where $F(\infty)$ is the fluorescence at $t = \infty$, ΔF_0 is the change in fluorescence level following the bleach pulse, and τ is the timescale of diffusion between the two compartments. The timescale τ of recovery between compartments provides insight into the characteristic resistivity of the coupling pathway, the number of coupling pathways, the diffusion coefficient of the tracer, etc., depending on the geometry of the system (Majewska et al. 2000; Wei et al. 2003).

SUMMARY

FRAP is a valuable tool for measuring the diffusion coefficient (or analogous transport parameters) in biological samples. The three FRAP techniques described here vary in resolution and depth penetration. FRAP requires thin samples (~1 μm) and yields an effective two-dimensional diffusion coefficient. SFA-FRAP is not limited by sample size but offers a diffusion coefficient averaged over the axial extent of the light cone of the objective. In addition, MPFRAP offers three-dimensional resolution and greatly improved depth penetration. Another common technique for measuring transport, fluorescence correlation spectroscopy (FCS), also comes in one-photon and two-photon varieties (Madge et al. 1974; Berland et al. 1995) and offers analogous resolution and depth penetration to the FRAP family. FRAP and FCS form a strong complement because FCS requires low fluorophore concentrations and is highly sensitive to background noise, whereas FRAP requires high fluorophore concentrations and is, therefore, less sensitive to background noise.

The relevance of two-photon FRAP, in particular, appears to grow as biomedical research delves further in vivo. As described above, MPFRAP was recently expanded to allow for the accurate measurement of diffusion in the presence of convective flow, making it a strong candidate for use in leaky physiological systems. Because of its superior spatial confinement, MPFRAP may also prove effective at measuring diffusion in the presence of bounding geometries, allowing it to be taken inside cell compartments, for example. There is also great promise for MPFRAP to take the leap into clinical applications via endoscopy.

REFERENCES

Abney JR, Scalettar BA, Thompson NL. 1992. Evanescent interference patterns for fluorescence microscopy. *Biophys J* **61:** 542–552.

Axelrod D, Koppel DE, Schlessinger J, Elson E, Webb WW. 1976. Mobility measurement by analysis of fluorescence photobleaching recovery kinetics. *Biophys J* **16:** 1055–1069.

Berk DA, Yuan F, Leunig M, Jain RK. 1993. Fluorescence photobleaching with spatial Fourier analysis: Measurement of diffusion in light-scattering media. *Biophys J* **65:** 2428–2436.

Berk DA, Yuan F, Leunig M, Jain RK. 1997. Direct in vivo measurement of targeted binding in a human tumor xenograft. *Proc Natl Acad Sci* **94:** 1785–1790.

Berland KM, So PT, Gratton E. 1995. Two-photon fluorescence correlation spectroscopy: Method and application to the intracellular environment. *Biophys J* **68:** 694–701.

Braeckmans K, Remaut K, Vandenbroucke RE, Lucas B, De Smedt SC, Demeester J. 2007. Line FRAP with the confocal laser scanning microscope for diffusion measurements in small regions of 3-D samples. *Biophys J* **92:** 2172–2183.

Braga J, McNally JG, Carmo-Fonseca M. 2007. A reaction-diffusion model to study RNA motion by quantitative fluorescence recovery after photobleaching. *Biophys J* **92:** 2694–2703.

Brown EB, Wu ES, Zipfel W, Webb WW. 1999. Measurement of molecular diffusion in solution by multiphoton fluorscence photobleaching recovery. *Biophys J* **77:** 2837–2849.

Cardarelli F, Serresi M, Bizzarri R, Giacca M, Beltram F. 2007. In vivo study of HIV-1 Tat arginine-rich motif unveils its transport properties. *Mol Ther* **15:** 1313–1322.

Chary SR, Jain RK. 1989. Direct measurement of interstitial convection and diffusion of albumin in normal and neoplastic tissues by fluorescence photobleaching. *Proc Natl Acad Sci* **86:** 5385–5389.

Chauhan VP, Lanning RM, Diop-Frimpong B, Mok W, Brown EB, Padera TP, Boucher Y, Jain RK. 2009. Multiscale measurements distinguish cellular and interstitial hindrances to diffusion in vivo. *Biophys J* **97:** 330–336.

Denk W, Strickler JH, Webb WW. 1990. Two-photon laser scanning microscopy. *Science* **248:** 73–76.

Feder TJ, Brust-Mascher I, Slattery JP, Baird B, Webb WW. 1996. Constrained diffusion or immobile fraction on cell surfaces: A new interpretation. *Biophys J* **70:** 2767–2773.

Kaufmann EN, Jain RK. 1991. Measurement of mass transport and reaction parameters in bulk solution using photobleaching. Reaction limited binding regime. *Biophys J* **60:** 596–610.

Kohler RH, Cao J, Zipfel WR, Webb WW, Hanson MR. 1997. Exchange of protein molecules through connections between higher plant plastids. *Science* **276:** 2039–2042.

Li J, Sullivan KD, Brown EB, Anthamatten M. 2010. Thermally activated diffusion in reversibly associating polymers. *Soft Matter* **6:** 235–238.

Madge D, Elson EL, Webb WW. 1974. Fluorescence correlation spectroscopy. II. An experimental realization. *Biopolymers* **13:** 29–61.

Majewska A, Brown E, Ross J, Yuste R. 2000. Mechanisms of calcium decay kinetics in hippocampal spines: Role of spine calcium pumps and calcium diffusion through the spine neck in biochemical compartmentalization. *J Neurosci* **20:** 1722–1734.

Mazza D, Braeckmans K, Cella F, Testa I, Vercauteren D, Demeester J, De Smedt SS, Diaspro A. 2008. A new FRAP/FRAPa method for three-dimensional diffusion measurements based on multiphoton excitation microscopy. *Biophys J* **95:** 3457–3469.

Pawley J, ed. 2006. *Handbook of biological confocal microscopy.* Springer, New York.

Peters R, Peters J, Tews K, Bahr W. 1974. Microfluorimetric study of translational diffusion of proteins in erythrocyte membranes. *Biochim Biophys Acta* **367:** 282–294.

Pluen A, Boucher Y, Ramanujan S, McKee TD, Gohongi T, di Tomaso E, Brown EB, Izumi Y, Campbell RB, Berk DA, Jain RK. 2001. Role of tumor–host interactions in interstitial diffusion of macromolecules: Cranial vs. subcutaneous tumors. *Proc Natl Acad Sci* **98:** 4628–4633.

Schneider MB, Webb WW. 1981. Measurement of submicron laser beam radii. *Appl Opt* **20:** 1382–1388.

Schulmeister S, Ruttorf M, Thiem S, Kentner D, Lebiedz D, Sourjik V. 2008. Protein exchange dynamics at chemoreceptor clusters in *Escherichia coli. Proc Natl Acad Sci* **105:** 6403–6408.

Sobczyk A, Scheuss V, Svoboda K. 2005. NMDA receptor subunit-dependent [Ca^{2+}] signaling in individual hippocampal dendritic spines. *J Neurosci* **25:** 6037–6046.

Star EN, Kwiatkowski DJ, Murthy VN. 2002. Rapid turnover of actin in dendritic spines and its regulation by activity. *Nat Neurosci* **5:** 239–246.

Stroh M, Zipfel WR, Williams RM, Ma SC, Webb WW, Saltzman WM. 2004. Multiphoton microscopy guides neurotrophin modification with poly(ethylene glycol) to enhance interstitial diffusion. *Nat Mater* **3:** 489–494.

Sullivan KD, Sipprell WH III, Brown EB Jr, Brown EB III. 2009. Improved model of fluorescence recovery expands the application of multiphoton fluorescence recovery after photobleaching in vivo. *Biophys J* **96:** 5082–5094.

Svoboda K, Tank DW, Denk W. 1996. Direct measurement of coupling between dendritic spines and shafts. *Science* **272:** 716–719.

Tsay TT, Jacobson KA. 1991. Spatial Fourier analysis of video photobleaching measurements. Principles and optimization. *Biophys J* **60:** 360–368.

Waharte F, Brown CM, Coscoy S, Coudrier E, Amblard F. 2005. A two-photon FRAP analysis of the cytoskeleton dynamics in the microvilli of intestinal cells. *Biophys J* **88:** 1467–1478.

Wedekind P, Kubitscheck U, Heinrich O, Peters R. 1996. Line-scanning microphotolysis for diffraction-limited measurements of lateral diffusion. *Biophys J* **71:** 1621–1632.

Wei X, Henke VG, Strubing C, Brown EB, Clapham DE. 2003. Real-time imaging of nuclear permeation by EGFP in single intact cells. *Biophys J* **84:** 1317–1327.

Zipfel WR, Williams RM, Christie R, Nikitin AY, Hyman BT, Webb WW. 2003. Live tissue intrinsic emission microscopy using multiphoton-excited native fluorescence and second harmonic generation. *Proc Natl Acad Sci* **100:** 7075–7080.

43 Fluorescent Speckle Microscopy

Lisa A. Cameron,[1] Benjamin R. Houghtaling,[2] and Ge Yang[3]

[1]*Confocal and Light Microscopy Core Facility, Dana-Farber Cancer Institute, Boston, Massachusetts 02115;* [2]*Laboratory of Chemistry and Cell Biology, Rockefeller University, New York, New York 10021;* [3]*Department of Biomedical Engineering and Lane Center for Computational Biology, Carnegie Mellon University, Pittsburgh, Pennsylvania 15213*

ABSTRACT

Fluorescent speckle microscopy (FSM) is a live imaging and quantitative measurement technique used for analyzing motion and turnover of macromolecular assemblies in vivo and in vitro. It differs from related imaging techniques such as photobleaching and photoactivation in its use of substantially lower concentrations of fluorescently labeled assembly subunits. When small numbers of labeled subunits and large numbers of unlabeled subunits become randomly incorporated together into a macromolecular structure, the random distribution of fluorophores generates nonuniform fluorescence intensity patterns that appear as distinct puncta against low background fluorescence. These puncta, called speckles, serve as fiduciary markers so that motion and turnover of the structure are visualized. Computational analysis of speckle image data transforms FSM into a powerful tool for high-resolution quantitative analysis of macromolecular assembly dynamics.

Successful application of FSM depends on the ability to reliably generate and image speckles, which are characterized by their weak emission signals, and to effectively extract quantitative information through computational analysis of speckle image data, which are characterized by their stochastic fluctuations, low signal-to-noise ratios, and high spatiotemporal complexity. This chapter aims to provide a practical introduction to basic principles, experimental implementation, and computational data analysis of FSM. Examples are used to show the application of FSM in analyzing the dynamic organization and assembly/disassembly of cytoskeletal filament networks, an area in which FSM analysis has found great success. A brief comparison of FSM with other imaging techniques and an outlook into the future of FSM are given at the end.

INTRODUCTION

Fluorescent speckle microscopy (FSM) is a method developed for the imaging, measurement, and analysis of motion and turnover of macromolecular assemblies in vivo and in vitro. It originated from an accidental discovery that microinjection of low concentrations of fluorescently labeled tubulin into live epithelial cells generated nonuniform fluorescence intensity patterns along microtubules that appeared as distinct puncta against a low-level fluorescent background (Danuser and Waterman-Storer 2006). These puncta, called speckles, resulted from the random distribution of fluorophores after random incorporation of small numbers of labeled tubulin with large numbers of unlabeled tubulin into polymerizing microtubules (Waterman-Storer and Salmon 1998) (Fig. 1A). Using speckles as fiduciary markers, FSM provides a powerful approach to visualize dynamics of microtubules and other macromolecular structures (see, e.g., Movie 43.1). The use of low concentrations of labeled subunits is critical to FSM in generating separated speckles and low background fluorescence. In comparison, related imaging techniques such as photobleaching and photoactivation use substantially higher labeled subunit concentrations and offer much lower spatial and temporal resolutions because of more continuous and uniform fluorescence signals.

Since the initial discovery of FSM, significant progress has been made in several important directions. Following its initial implementation in wide-field epifluorescence microscopy (Waterman-Storer et al. 1998), FSM has now been performed using spinning-disk confocal microscopy (Adams et al. 2003) and total internal reflection fluorescence (TIRF) microscopy (Adams et al. 2004). Speckle imaging at the single-fluorophore (SF) level has been achieved (Watanabe and Mitchison 2002; Yang et al. 2007). Besides microtubules, FSM has also been used to analyze a variety of other macromolecular structures (Danuser and Waterman-Storer 2006). Most importantly, computational techniques and software developed for processing speckle image data have transformed FSM from a tool of visualization into a tool of high-resolution quantitative measurement and analysis.

This chapter presents basic principles, experimental implementation, and computational data analysis of FSM from a practical perspective. After a brief review of basic principles of FSM, important practical considerations in generating and imaging speckles, which are characterized by their weak emission signals, are discussed. The general procedure and basic techniques of computational analysis of speckle image data, which are characterized by their stochastic fluctuations, low signal-to-noise ratios, and high spatiotemporal complexity, are introduced. Two FSM applications are presented, including the dynamic architecture of microtubule networks in metaphase spindles and the polymerization/depolymerization dynamics of F-actin networks at the leading edge of migrating epithelial cells. The chapter concludes with a brief comparison of FSM with other related imaging techniques and a look into the future of FSM.

BASIC PRINCIPLES OF FSM

Speckle formation is the end result of two processes. First, random incorporation of a small number of labeled subunits into a macromolecular structure produces a random distribution of fluorophores. This process also stabilizes positions of fluorophores of incorporated subunits so that sufficiently long exposure time can be used to collect their weak emission signals without being affected by their position drift (Danuser and Waterman-Storer 2006). Second, the spatial distribution of fluorophores is recorded in images through a light microscope, which effectively performs a convolution of its point-spread function with fluorophore signals. The imaging process, however, can only resolve separated fluorophores up to the resolution limit of light microscopy, which can be characterized using the Rayleigh diffraction limit. Here, we use speckle formation on cytoskeletal filaments as an example to explain basic principles of FSM.

We first consider speckle formation on single microtubules (Fig. 1A). A microtubule is an assembly of α/β-tubulin heterodimers, which organize longitudinally into 13 protofilaments that jointly form the microtubule wall. Because each tubulin dimer is 8 nm in length, there are a total of 1625 dimers within 1 µm of microtubule. The probability of k labeled dimers being incorporated among

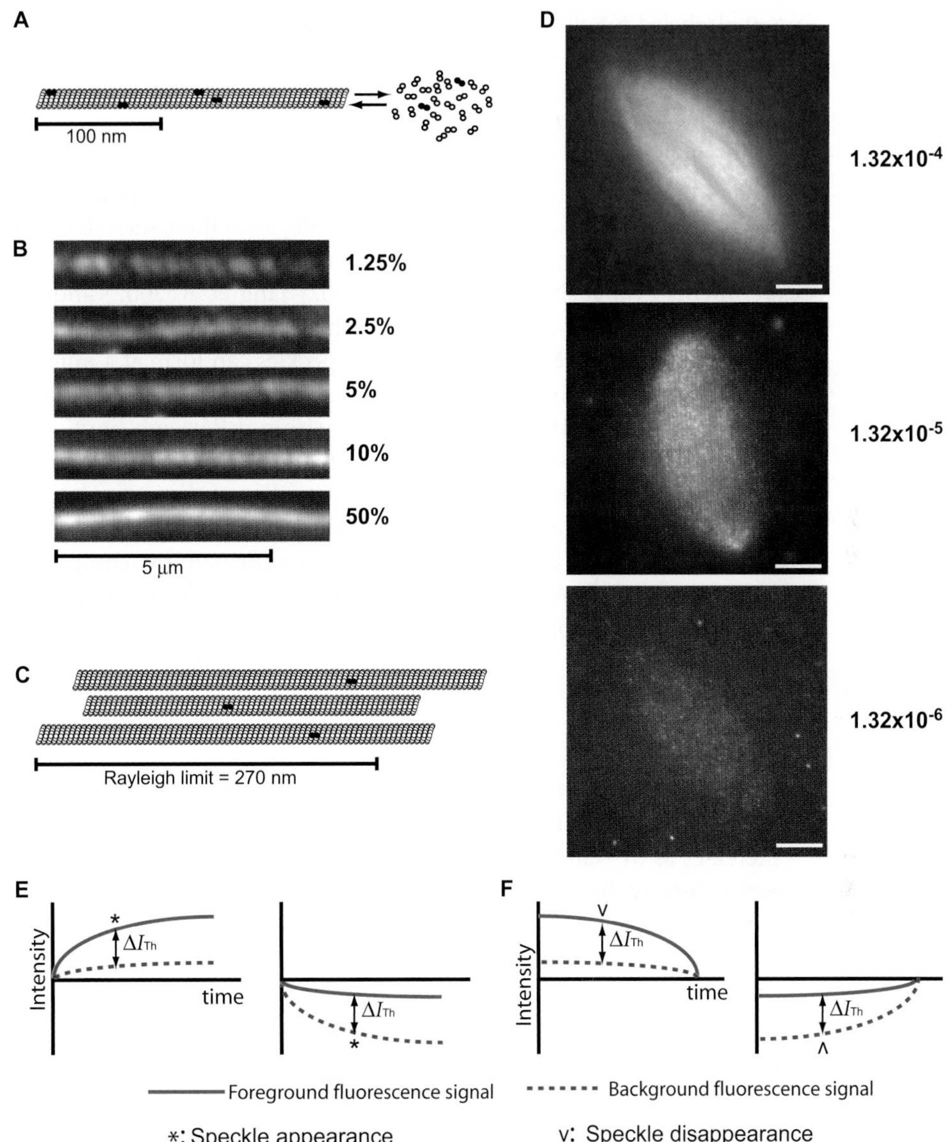

FIGURE 1. Basic principles of speckle formation. (*A*) Random incorporation of small numbers of labeled tubulin (solid color) into a polymerizing microtubule generates a random distribution of fluorophores. (*B*) Speckle formation is determined by local image contrast, which can be controlled by changing the fraction of labeled tubulin (listed to the *right* of the images). (*C*) Speckle formation depends on the spatial organization of microtubules because multiple microtubules can reside in a Rayleigh limit. (*D*) Speckle formation in dense microtubule networks of *Xenopus* extract spindles under different fractions of labeled tubulin (listed to the *right* of images). Compared with the case of single microtubules in *B*, the fraction of labeled tubulin required for speckle formation in microtubule networks is substantially lower. Scale bars, 10 μm. (*E*) Two cases that lead to speckle appearance when differences between foreground and background fluorescence intensities become higher than a threshold ΔI_{Th}, determined by the statistical distribution of foreground and background intensities. (*F*) Two cases that lead to speckle disappearance when differences between foreground and background intensities drop below ΔI_{Th}. (*B*, Reprinted, with permission, from Danuser and Waterman-Storer 2006. *D*, Reprinted, with permission, from Yang et at. 2007, ©Macmillan.)

N dimers follows a Poisson distribution $P\{n = k\} = (v^k e^{-v})/k!$ with $v = N \cdot r$ as its mean and $\sigma = \sqrt{Nr}$ as its standard deviation, where r is the fraction of labeled tubulin in the tubulin pool. Assuming that tubulin is labeled with X-rhodamine fluorophores, whose emission wavelength l is at ~620 nm, and that the numerical aperture (NA) of the microscope objective lens is 1.4, the Rayleigh limit is $(0.61 \cdot l)/\text{NA}$ = 270 nm, which contains 440 tubulin dimers. The distance d between consecutively incorporated labeled tubulin subunits follows an exponential distribution $p(d) = \lambda e^{-\lambda d}$ (Ross 2006), whose mean and standard deviation both are $1/\lambda$, which equals $1000/(1625 \cdot r)$ in nanometers.

These models predict that as the fraction of labeled tubulin decreases, the average number of fluorophores within the Rayleigh limit decreases while the average distance between sequentially incorporated fluorophores increases. Under the previously assumed condition, when $r < 1/440$, the average number of fluorophores would be <1, whereas the spacing between sequentially incorporated fluorophores would be larger than the Rayleigh limit such that individual X-rhodamine fluorophores would appear as individual speckles. This predicted condition has been achieved in experiments (Watanabe and Mitchison 2002; Yang et al. 2007) and is referred to as SF-FSM. Its implementation will be discussed later in the chapter.

Experiments, however, also show that speckles can form at a much higher labeled tubulin fraction than what is required for SF-FSM (Fig. 1B). At $r = 1.25\%$, for example, the mean and standard deviation of the number of fluorophores within a Rayleigh limit are calculated as 5.5 and 2.35, respectively. The average distance between two consecutively incorporated fluorophores is ~50 nm, much smaller than the Rayleigh limit. Appearance of distinct speckles under this condition indicates that the predominant cause of speckle formation is the significant local contrast of fluorescence signals. This conclusion has been confirmed by computer simulation and imaging experiments (Waterman-Storer and Salmon 1998). The local speckle intensity contrast can be quantified using a score,

$$c = \frac{\sqrt{Nr}}{Nr} = \frac{1}{\sqrt{Nr}},$$

which is the ratio between the standard deviation and the mean of the number of fluorophores within the Rayleigh limit (Waterman-Storer and Salmon 1998). For $r = 1.25\%$, the score is 0.42. As r increases to 10% and 50%, the scores drop significantly to 0.15 and 0.05, respectively, under which speckles largely disappear (Fig. 1B). Overall, speckle formation is determined by whether the foreground fluorescence signal (i.e., the speckle signal) is significantly higher than its local background fluorescence signal. Separation of individual fluorophores minimizes background fluorescence but is not necessarily required for speckle formation.

In comparison to the case of single microtubules, speckle formation in a three-dimensional (3D) cytoskeletal filament network is more complicated because it depends on the spatial organization of filaments. This dependence can be illustrated using a simple conceptual example. The small diameter of microtubules makes it possible for a bundle of microtubules to fit within a Rayleigh limit (Fig. 1C). Assuming a bundle of 10 microtubules, the speckle contrast score under $r = 1.25\%$ would decrease to ~0.13 and would result in nearly uniform fluorescence distributions similar to those under $r = 10\%$ for single microtubules (Fig. 1B). This prediction has been confirmed in experiments under similar conditions (Fig. 1D).

Because of the dependence of speckle formation on the spatial organization of the filaments in networks, detailed analysis of speckle imaging condition requires computer simulation. For actin networks, computer simulation confirms that speckle formation follows the same principle as for single microtubules and is determined by local intensity contrast (Ponti et al. 2003). However, different from the case under single microtubules, foreground and background fluorescence signals now both depend on network assembly/disassembly (Ponti et al. 2003). A speckle may appear either because of polymerization in the foreground, when its intensity undergoes a significant increase relative to its local background, or because of depolymerization in the local background, when its intensity undergoes a significant decrease relative to the foreground (Fig. 1E). Conversely, a speckle may disappear because of either depolymerization in the foreground or polymerization in the local background (Fig. 1F). Statistical tests are required to rigorously classify the different scenarios of speckle formation and annihilation (Ponti et al. 2003). However, a general and rigorous interpretation of a speckle is that it is the diffraction-limited image of fluorophores whose intensity is significantly higher than that of their local background.

IMPLEMENTATION: PRACTICAL CONSIDERATIONS

Preparation of Labeled Protein

Successful FSM requires both proper sample preparation and proper microscope configuration for efficient collection of speckle emission light. The protein of interest must be tagged with a fluorescent label and incorporated into the normal cellular structure in a living cell. Purified proteins that are chemically labeled and microinjected into cells by researchers in their own laboratories often produce the best speckle imaging and provide the freedom to choose different fluorescent tags depending on the application. Alternatively, fluorescently labeled proteins are commercially available from suppliers, such as Cytoskeleton Inc., but are less economical and flexible if FSM experiments will be ongoing. For labeling cytoskeletal proteins, such as tubulin and actin, succinimidyl ester derivatives have historically produced the best functional proteins, as they covalently bind to lysine residues on the surface of the protein allowing polymerization and depolymerization. Higher dye-to-protein ratios are possible because there is more than one lysine available for reaction. Other chemical labelings are possible. For instance, 5-iodoacetamide reacts with cysteine residue 374 on actin (Wang and Taylor 1980), but this produces a lower dye-to-protein labeling ratio. Alexa dyes (especially those in the 500–650-nm range) are also a good choice as they have high photostability. For tubulin fluorophore labeling protocols, refer to Hyman et al. (1991), Waterman-Storer (2002), and the website of Dr. Timothy Mitchison's laboratory (http://mitchison.med. harvard.edu/protocols. html). For actin purification, refer to Pardee and Spudich (1982), Turnacioglu et al. (1998), and Waterman-Storer (2002).

Fluorophore Selection

When choosing the excitation and emission spectra of a fluorophore label for the protein of interest, it is important to consider the cellular environment, namely, phototoxicity effects and possible cellular autofluorescence. The investigators have found that the cytoskeletal polymer subunits, tubulin and actin, produce the best speckles when labeled with X-rhodamine via succinimidyl ester (Molecular Probes). Alexa 568 also provides a good option. Far red (such as Alexa 647) works well too but has the added difficulty of not being able to view the sample with the FSM fluorophore because the human eye cannot detect far-red wavelengths. If one chooses to use a far-red label, it is recommended that an additional fluorophore in another color be used for a similarly localized protein so that the sample and the appropriate focal plane can be chosen by viewing through the eyepieces of the microscope. Short wavelengths are very damaging to cells and can result in cell edge retracting, rounding up of cells, and cell death. Therefore, fluorophores that are excited by <450 nm should be avoided. Even those in the 450–500-nm range can produce phototoxicity. Most cellular autofluorescence is caused by flavoproteins that are excited in the ultraviolet and blue wavelengths and emit in blue and green wavelengths. This autofluorescence contributes to the background when using fluorescein and similar dyes and reduces speckle contrast. Thus, the best choices to reduce phototoxicity and autofluorescence are the yellow/orange dyes excited in the 570–600-nm range.

As there are many cell lines expressing GFP-tagged proteins, it is attractive to try FSM with cells containing the protein of interest with a GFP tag. Currently, however, this is not optimal in most cases because the typical level of expression of a GFP-tagged protein is higher than what is required for speckle imaging. For instance, cells expressing GFP-tubulin usually have microtubules that are uniformly labeled with GFP. However, it is possible to use GFP-tagged proteins if the expression level is very low and/or if expression is controlled with a leaky promoter that results in lower nonuniform expression to produce a nonuniform labeling. For example, FSM has been successfully performed in cell lines expressing GFP-tubulin, such as *Drosophila* S2 cells, but at a less than optimal level, by putting the expression under the control of a leaky inducible metallothionein promoter. Without induction, low levels of GFP-α-tubulin are expressed, and cells can be found to have microtubules with a speckled appearance (Matos et al. 2009). Also, GFP fusions to focal adhesion proteins, including vinculin, talin, paxilin, α-actinin, zyxin, or integrin, have been expressed in epithelial cells from crippled promoters to achieve very low levels to give these focal adhesion proteins a speckled appearance (Danuser and Waterman-Storer 2006).

Microscope System Configuration

FSM has now been implemented using wide-field epifluorescence microscopy, spinning-disk confocal microscopy, and TIRF microscopy. Laser-scanning confocal microscopy has not been used successfully for FSM, mostly because of strong photobleaching and the higher noise and limited dynamic range of photomultiplier detectors compared with charge-coupled device (CCD) cameras.

FSM requires a stable biological research-grade microscope, preferably an inverted microscope for optimal live cell imaging. A useful development in microscope technology is the availability of automated focus-lock systems, which maintain the in-focus position using a closed-loop feedback control system that maintains a user-defined in-focus position over a long period of time.

Illumination

For wide-field epifluorescence, use a high-quality mercury lamp or a metal-halide lamp with sufficient controls for proper alignment. A manual field diaphragm enables the investigator to control the field-of-view dimensions and reduces photodamage. To further reduce photobleaching, electronically controlled shutters should be mounted in the light path after the lamp house to block illumination when the camera is not capturing an image. For FSM imaging via spinning-disk confocal microscopy or TIRF microscopy when laser illumination is used, software control of the shutters and neutral-density filter wheels or acousto-optic tunable filters are required.

Objective Lens Selection

For fluorescence microscopy, image resolution is determined by the Rayleigh limit $(0.61 \cdot \lambda)/\mathrm{NA}$, where NA is the numerical aperture of the objective lens and λ is the emission wavelength. The power of a microscope to collect speckle emission is proportional to NA^4/M^2, where M is the magnification. Choosing an objective lens with high NA is, therefore, critical for fluorescent speckle microscopy to resolve speckles and to reduce the exposure time. High magnification at 100x is often chosen for high spatial sampling, which is required for high-resolution computational analysis of speckle image signals.

Camera Selection

To image fluorescent speckles, a scientific-grade cooled CCD camera is required for high-efficiency collection of speckle emission light. Such cameras can have a quantum efficiency (QE) of >90% if the chip is back-thinned. For FSM, a cooled CCD with a QE of at least 60% is required. Refer to Waterman-Storer (2002) for a detailed overview of various factors in selecting cameras for FSM. An important recent development in CCD camera technology is the production of electron-multiplying CCD (EMCCD) cameras, which have the sensitivity to detect single photons and thus greatly facilitate the implementation of SF speckle imaging. The high sensitivity of an EMCCD camera permits shorter exposure times and a longer imaging time before photobleaching occurs. One limitation, however, of an EMCCD camera is its relatively large pixel sizes, currently at 13 × 13 µm or 16 × 16 µm. Improving the spatial sampling by using a higher magnification will come at the expense of lower signal intensities. If spatial resolution is of the utmost importance or when extremely high sensitivity is not required, a more traditional cooled scientific-grade CCD camera may suffice. An excellent introduction to the technology, performance, and selection of CCD cameras, including EMCCD cameras, is given in Pawley (2006).

Filter Selection

The detection of speckles requires a sensitive imaging system. Filter sets must be precisely matched with the excitation and emission wavelengths of the fluorophore of choice. Off-peak filter sets will result in longer exposure times, an increased photobleaching rate, and compromised intensities, preventing efficient collection of emission light.

Single-Fluorophore FSM

The decision to pursue SF versus multifluorophore FSM is dependent on the biological questions to be addressed (Danuser and Waterman-Storer 2006). Here, we describe experimental considerations for SF-FSM, as it is the more technically challenging form of FSM.

Following selection of an appropriate fluorophore for SF-FSM, the fluorophore must be coupled to exogenous protein subunits in a manner that ensures most subunits do not contain more than one fluorophore. Amine-reactive fluorophores (e.g., succinimidyl esters) are commonly used for this purpose. These fluorophores react with nonprotonated aliphatic amine groups, including lysine side chains and the amino-terminal amine of peptides, to conjugate to purified protein subunits. Consequently, the number of reactive groups within any target protein will vary, as well as the reactivity of the fluorophores. To obtain a substoichiometric labeling ratio of fluorophore to protein subunit, optimization of the labeling protocol is required, typically involving varying the molar ratio of fluorophore to protein, as well as the reaction incubation time. A substoichiometric ratio ensures that most, but not all, subunits have zero or one fluorophore. The closer the molar ratio of fluorophore to protein subunit is to 1, the higher the probability of having a significant fraction of single subunits labeled with multiple fluorophores.

To perform SF-FSM, an estimate of the required fraction of unlabeled subunits in the experimental system can be computed (Yang et al. 2007). However, the actual amount of labeled subunits to add must be determined empirically, with the expected concentration likely to be several fold lower than that of multifluorophore imaging. Experimental titrations of labeled subunits should be combined with rigorous tests (Fig. 2A–D) to determine fluorophore numbers before analyzing or interpreting any SF-FSM data.

DATA ANALYSIS

Speckle image data are characterized by two basic properties. First, speckle signals are weak and noisy. Computational and statistical analyses become essential for reliable information extraction. Second, FSM images often show high spatiotemporal complexity. For complex cellular structures such as the metaphase spindle, it is common for an FSM image to contain thousands of speckles (Fig. 3A), while the total number of speckles in a time-lapse movie can easily approach one million. Automated data analysis is essential for managing the complexity of FSM data.

Kymograph Analysis of FSM Data

Kymograph analysis is a commonly used approach for manual analysis of FSM data (Waterman-Storer 2002). It provides a convenient way to visualize movement of speckles within a user-defined thin slice of image region (Fig. 3B). Simple estimation of speckle velocities can be made from kymographs by approximating speckle trajectories using straight lines and then calculating their slopes (Fig. 3B). However, this approach suffers from several basic limitations. First, kymograph analysis is only possible when the speckles are easily identified by eye. This requires high signal-to-noise ratios that are often not available in FSM. For noisy speckle image data, computation-based image analysis often is the only tool for reliable information extraction. Second, kymograph analysis tends to be biased. Long and distinct speckle trajectories are more likely to be selected by users from kymographs for analysis. Approximating speckle trajectories using straight lines also makes it impossible to detect

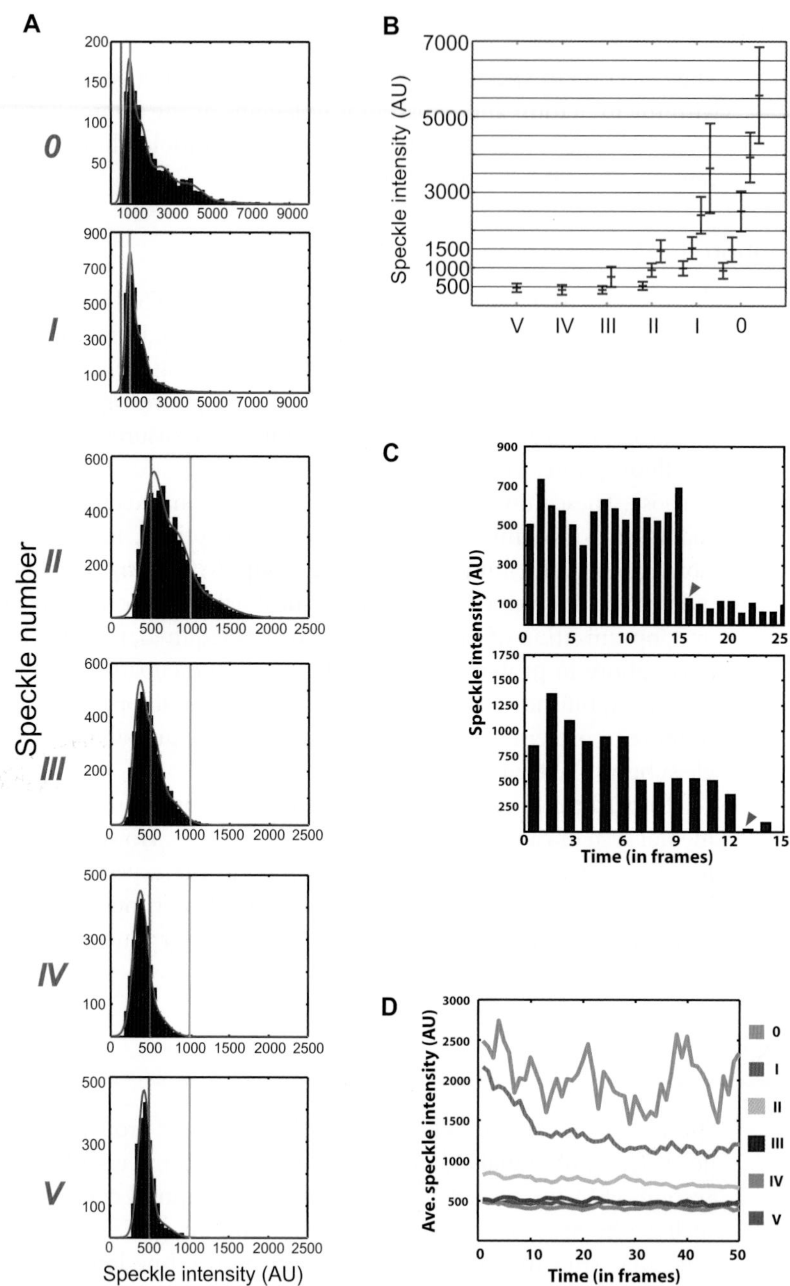

FIGURE 2. Establishing an SF-FSM condition. (*A*) Intensity distributions of tubulin speckles in *Xenopus* extract spindles under different fractions of labeled tubulin (*0*: 1.32E-4; *I*: 4.40E-5 *II*: 1.32E-5 *III*: 4.40E-6, *IV*: 1.32E-6, *V*: 4.40E-7). Each speckle intensity distribution is fitted with a mixture of normal distributions (red line), each representing an intensity data cluster (Fraley and Raftery 2002). Green and brown lines mark 500 AU and 1000 AU, respectively. (*B*) Speckle intensity clusters under conditions in *A* follow approximately multiples of 500 AU. (*C*) (*Upper panel*) A single step decrease of speckle intensity from ~500 AU to the background, reflecting SF photobleaching or blinking. (*Lower panel*) A two-step decrease of speckle intensity: first from ~1000 AU to ~500 AU, and then to the background. Red arrows indicate when speckles could no longer be detected. (*D*) Changes over time of average speckle intensities in each frame. The average speckle intensity remained at ~500 AU under the three lowest labeled tubulin fractions, under which an SF-FSM condition was achieved. In comparison, the average speckle intensity under higher tubulin concentrations fluctuated over time because of photobleaching of multifluorophore speckles. Overall, three different tests are recommended for verification of a SF-FSM condition. First, average intensities under higher fractions of labeled subunits should be multiples of the average intensity under the lowest fraction of labeled subunit. Second, intensities of speckles should decrease by steps that are multiples of the average intensity under the lowest fraction of labeled subunit. Third, the average speckle intensity should remain approximately unchanged over time under an SF condition. (*A–D*, Reprinted, with permission, from Yang et al. 2007, ©Macmillan.)

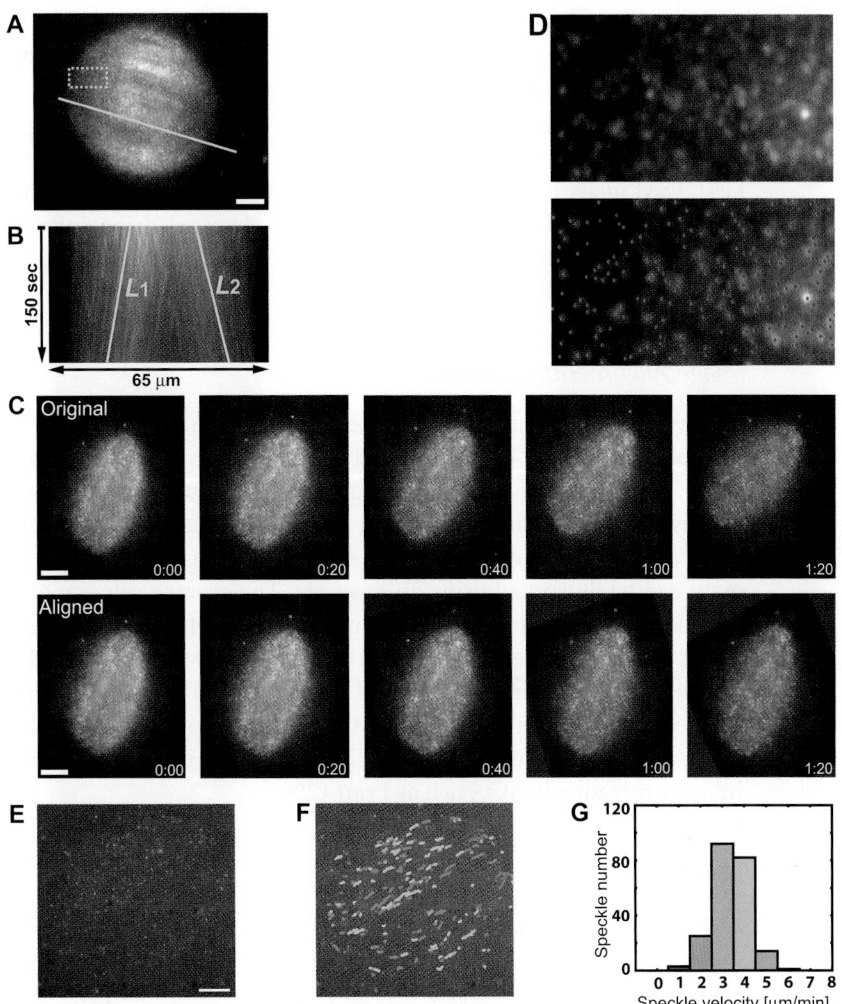

FIGURE 3. Different steps in FSM data analysis. (*A*) X-rhodamine tubulin speckles in a *Xenopus* extract spindle (total number of speckles in this frame is 4264). Scale bar, 10 μm. (*B*) A kymograph generated along the solid line in *A*. Velocities of two selected speckles, whose trajectories were approximated by lines L_1 and L_2, were estimated to be 2.26 μm/min and 2.91 μm/min, respectively. (*C*) Image alignment eliminates translational and rotational drift of a spindle. Scale bars, 10 μm. (*D*) Detection of speckles. (*Upper panel*) A magnified image of the region marked by the dotted box in *A*. (*Lower panel*) Detected speckles marked by red dots overlaid onto the original image. The number of detected speckles in this region is 195. (*E*) SF-FSM image of a *Xenopus* extract spindle. Scale bar, 10 μm. (*F*,*G*) Speckle trajectories within the spindle in *E* recovered by software were color coded according to the average instantaneous velocity histogram in *G* and overlaid on the image in *E*. (*E–G*, Reprinted, with permission, from Yang et al. 2007, ©Macmillan.)

fine speckle movement (Fig. 3B). Third, manual kymograph analysis provides only limited information because it can only be applied in a small number of user-defined regions on a small number of user-selected trajectories. It cannot provide a reliable lifetime measurement because speckles may disappear from the kymographs when they move outside the regions selected for analysis. Overall, kymograph analysis should only be used as a supplement to computer-based image analysis tools.

General Procedure for Computational Analysis of FSM Data

Conceptually, computational analysis of FSM image data consists of two steps. In the first step, the goal is to recover complete speckle trajectories (i.e., time histories of speckle position and intensity over time). The second part is strongly dependent on the biological question to be answered. The goal is to transform low-level speckle trajectory data into higher-level information about the structure, function, and/or regulation of the macromolecular assembly being studied. Here, we focus on the first step of the process and, later, show the second step using specific examples (see Applications of FSM).

Overall, speckle signal analysis should follow general principles of fluorescent data analysis described in Dorn et al. (2008). Specifically, recovery of speckle trajectories requires three substeps. First, FSM images are aligned to eliminate any image drift that may cause bias in subsequent analysis. Second, speckles within each frame are detected using software. Third, detected speckles are tracked automatically using software to recover their trajectories. Next, these steps are discussed in greater detail.

Image Alignment/Registration

Image drift is a commonly encountered issue in live cell imaging that is typically caused by either drift of the specimen being observed (Fig. 3C, upper panels) and/or thermal drift of the microscope stage. The latter often can be eliminated using an automated focus-lock system. However, specimen drift is difficult to eliminate. Image alignment/registration, the removal of image drift, is a classic application of computational image analysis (Fig. 3C, lower panels). The importance of image alignment can be illustrated using an example. The average velocity of tubulin speckles in a metaphase spindle is ~2 µm/min. Assuming a magnification of 100x, a CCD camera pixel size of 6.7 µm, and a frame rate of 4 frames per second, the average movement of speckles with each frame is ~2 pixels. Without image alignment, a drift of the spindle by one pixel per frame would cause an ~50% bias in speckle velocity measurement. A detailed introduction to image alignment/registration techniques, including the correlation alignment technique used in Figure 3D, can be found in many image-processing books, such as Pratt (2007).

Speckle Detection

Automated detection of speckles (Fig. 3D) is based on their definition: A speckle is a local intensity maximum whose intensity is, in a statistical sense, significantly higher than that of its background. Speckle software locates local intensity maxima and then performs a statistical significance test of their intensity against their background (Ponti et al. 2003). Weak speckles can be included or excluded in detection by adjusting the level-of-significance threshold.

When the spatial sampling of imaging is sufficiently high (e.g., when the effective pixel size is less than one-third of the Rayleigh limit), it is possible to use curve fitting to break the Rayleigh limit and to achieve subpixel level resolution in speckle position detection (Dorn et al. 2008).

Speckle Tracking

A general review of particle tracking techniques is given in Dorn et al. (2008). For speckle tracking, the global nearest-neighbor algorithm is frequently used when the numbers of speckles to be tracked in each frame are very high (Yang et al. 2008) (Fig. 3E–G). For thick 3D structures such as the *Xenopus* extract spindle, it is common to encounter speckle trajectories of short lifetime, which are caused by speckles moving transiently in and out of the microscope focal plane. These trajectories can be excluded by setting a threshold for the minimum trajectory lifetime.

Software

Image alignment, speckle detection, and speckle tracking are basic techniques shared by nearly all FSM data analysis applications. MATLAB-based software packages providing these functions are available on request from the Laboratory for Computational Cell Biology (http://lccb.hms.harvard.edu/) at Harvard Medical School and the Computational Cell Dynamics Laboratory at Carnegie Mellon University (http://ccdl.compbio.cmu.edu). Many publicly available programs, such as Imaris and ImageJ, also provide some of the required analysis functions. Higher-level analysis of FSM data after recovery of speckle trajectories is application specific. Related software is often customized and may be requested from their developers.

APPLICATIONS OF FSM

As a live cell imaging and quantitative measurement technique, FSM has been developed to analyze the dynamic organization and assembly/disassembly of macromolecular structures. In particular, it has been applied with great success in investigating the structure, the function, and the regulation of the cytoskeleton. Here, we show applications of FSM in two studies of the cytoskeleton.

FSM Analysis of Metaphase Spindle Architecture

The metaphase spindle of higher eukaryotic cells is a highly complex and dynamic molecular machine whose primary responsibility is to reliably segregate replicated chromosomes. The spindle structure is composed primarily of microtubules and their associated proteins. Remarkably, tubulin subunits of spindle microtubules undergo constant poleward movement in a process called microtubule flux, whose direct visualization was achieved as a result of the discovery of FSM (Waterman-Storer et al. 1998). For the first time, it became possible to visualize the architecture of the highly dense and dynamic spindle microtubule network.

Using FSM, we generated high-resolution quantitative maps of microtubule flux in *Xenopus* egg extract meiotic spindles, an in vitro metaphase spindle model (Yang et al. 2008) (Fig. 4A–C). Automated extraction of individual speckle trajectories using particle detection and tracking allowed us to fully map the movement of individual speckles within the spindle (Fig. 4B). Color coding the speckle trajectories based on their velocities revealed regional velocity decreases near the spindle poles (Fig. 4B). To quantify this regional heterogeneity of speckle movement, we divided each spindle spatially into 23 regions (Fig. 4D) and examined speckle velocities within each region. We found that average speckle velocity decreased by ~20% near the spindle poles compared with the spindle midzone (Fig. 4E,F). We then conducted statistical cluster analysis (Fraley and Raftery 2002) of speckle velocities within the entire spindle and found that speckles move at two distinct velocity modes (Fig. 4G).

To investigate the possible molecular mechanism of the regional variation and distinct modes of speckle movement, we combined molecular perturbations with FSM analysis. Separate and concurrent inhibition of molecular motors, dynein/dynactin and kinesin-5, which play critical roles in maintaining the dynamic spindle architecture, removes the regional heterogeneity and reduces the number of velocity modes to 1. Together, the results led to an architectural model of spindle microtubules in which different speckle velocity modes were associated with two architecturally distinct yet spatially overlapping and dynamically crosslinked arrays of microtubules: Focused polar microtubule arrays of a uniform polarity and slower flux velocities were interconnected by a dense barrel-like microtubule array of antiparallel polarities and faster flux velocities (Fig. 4H). This proposed spindle architecture model is experimentally confirmed in a follow-up study (Houghtaling et al. 2009). Overall, this study shows the strength of FSM in providing high-resolution quantitative characterization of the motion and dynamic architecture of macromolecular structures and the strength of using computational and statistical analyses to investigate the spatial heterogeneity and the velocity distribution of subunit movement. Further analysis using SF-FSM has made it possible to analyze local spindle architecture at the level of individual microtubules (Yang et al. 2007).

FSM Analysis of Polymerization/Depolymerization of F-Actin Networks

We explained previously that speckle appearance and disappearance within a cytoskeletal filament network are directly associated with filament polymerization and depolymerization (Fig. 1E,F). This provides a way for quantifying polymerization/depolymerization by analyzing speckle signals.

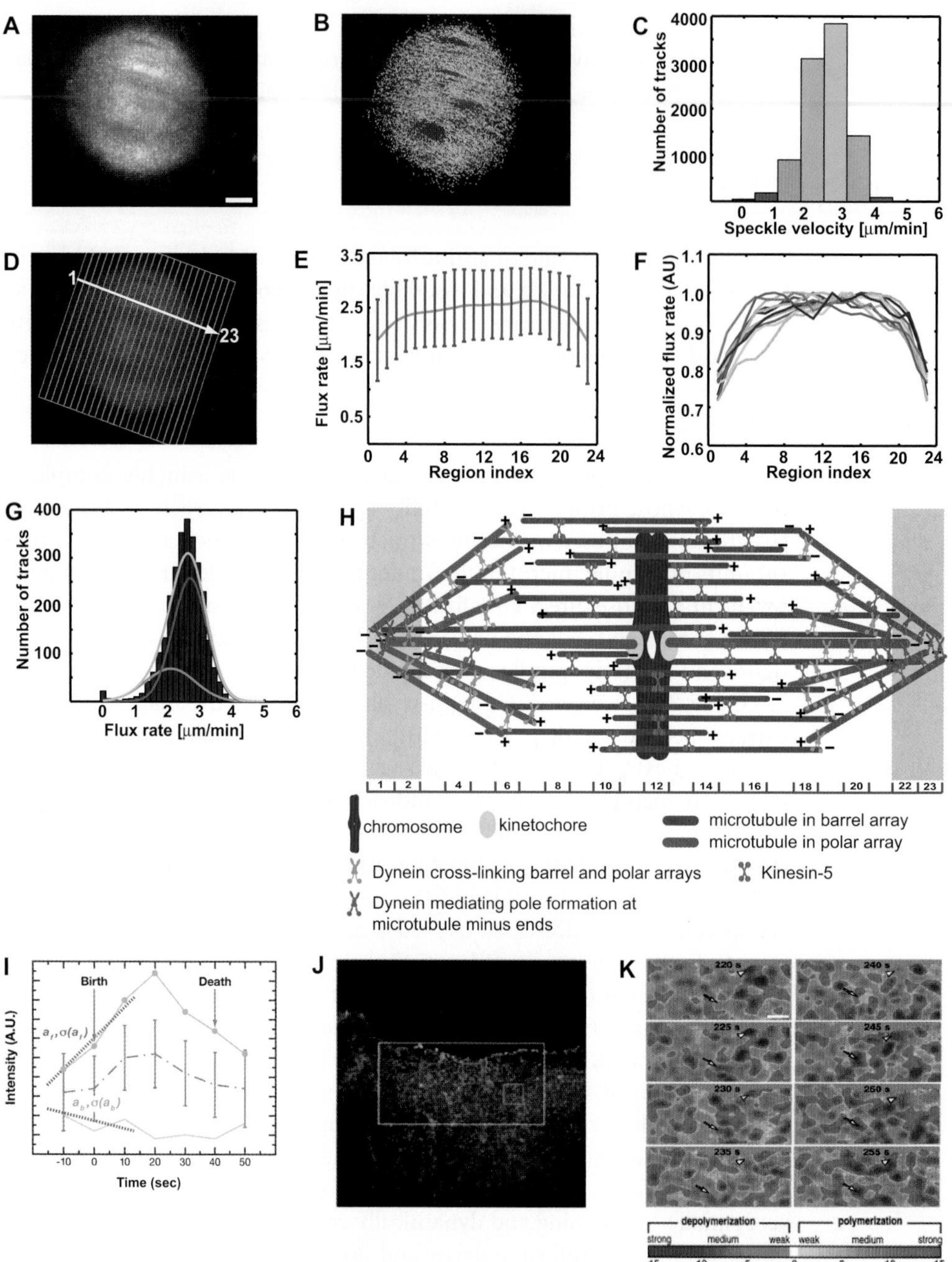

FIGURE 4. Applications of FSM in analyzing motion and turnover of the cytoskeleton. (*A*) The same spindle as in Fig. 3A. Scale bar, 10 μm. (*B,C*) Speckle trajectories (n = 9562) recovered by particle tracking are color coded according to their average instantaneous velocity histogram in *C* and are overlaid on the image in *A*. (*D*) Each spindle is divided into multiple (n = 23) regions for analyzing spatial heterogeneity of speckle velocities. (*E*) Distribution of the mean and standard deviation of speckle velocities in regions defined in *D*. Reduction of speckle velocities is observed at spindle poles. (*F*) Spatial distributions of normalized speckle velocities from 11 spindles further confirm the reduction of speckle velocities at spindle poles. (*G*) Statistical clustering of speckle velocities within the spindle in *A* shows that speckles move at either a fast (red) or a slow (green) velocity mode. (*H*) Regional variations and different modes of speckle velocities suggest a metaphase spindle architecture model. (*I*) Actin network polymerization/depolymerization can be quantified through analyzing foreground (i.e., speckle) and background intensity changes. (Red) Foreground intensity; (blue) background intensity. (Dotted line) calculated threshold intensity over time (refer to Fig. 1*E,F*). (*J*) An F-actin speckle image of the lamellipodium of a newt lung epithelial cell. (*K*) Computed polymerization/depolymerization scores over time within the window (white) shown in *J*. (*A–H*, Reprinted, with permission, from Yang et al. 2008. *I*, Reprinted, with permission, from Danuser and Waterman-Storer 2006. *J,K*, Reprinted with permission from Ponti et al. 2003, ©Elsevier.)

Here, we show this analysis using the speckle appearance event shown in Figure 4I. After such an event is detected from the speckle trajectory, a polymerization/depolymerization score can be calculated (Ponti et al. 2003). First, the foreground intensity (i.e., speckle intensity) and the background intensity are fitted by linear regression. This determines the foreground intensity change rate (slope) α_f with its variance $\sigma(\alpha_f)$ and background intensity change rate (slope) α_b with its variance $\sigma(\alpha_b)$. Then, both the foreground and the background rates are tested for statistical significance based on their p values. Finally, events in which both foreground and background rates are nonsignificant are excluded from the analysis. If there is one significant rate for an event, it is reported as the polymerization/depolymerization score. If both rates are significant, the more significant one (i.e., the one with the lower p value) is reported as the polymerization/depolymerization score. After the raw scores are computed spatially, spatiotemporal smoothing is used to further reduce the influence of noise. Overall, the score provides a relative measure of polymerization and depolymerization of the network. This approach was used to map the assembly dynamics of the F-actin network in the lamellipodium of a migrating newt lung epithelial cell (Fig. 4J,K).

In the previous analysis, quantifying polymerization/depolymerization is complex because the speckles are at multiple-fluorophore levels such that their formation and annihilation depend on both foreground and background intensities (see Fig. 1E,F). Indeed, a large percentage of events (~60%) could not be conclusively classified for the actin network shown in Figure 4J (Ponti et al. 2003). This complexity is avoided in SF-FSM (Watanabe and Mitchison 2002). However, as discussed before, implementation of SF-FSM is more challenging, and its spatial resolution is lower than FSM because of the sparse spatial coverage by lower numbers of speckles.

Independent of the fluorophore level adopted, two issues must be addressed when quantifying polymerization/depolymerization by FSM. First, the appearance and disappearance of speckles must not be caused by their movement in and out of the focal plane. This is why SF 2D FSM data of thick and large extract spindles (Yang et al. 2007) cannot be used to interpret their polymerization/depolymerization dynamics. Overcoming this difficulty requires either that the network to be analyzed resides fully within the focal plane of the microscope or that the speckle movement be fully imaged and computationally tracked in 3D, so that speckle appearance and disappearance can be classified without ambiguity. Currently, however, 3D FSM is not feasible because of technical barriers in both imaging and computational analysis. Second, speckle disappearance can also result from photobleaching. The influence of photobleaching must be corrected based on calibration when calculating polymerization/depolymerization rates (Ponti et al. 2003).

SUMMARY AND OUTLOOK

FSM is a powerful technique for live cell imaging and quantitative measurement that is most suitable for analyzing motion and turnover of macromolecular assemblies in vitro and in vivo. So far, it has found great success in studies of the cytoskeleton, although extensions to other macromolecular structures are possible (Danuser and Waterman-Storer 2006). Its application is more specific when compared with more general techniques for analyzing protein dynamics such as photoactivation and photobleaching. On the other hand, because each speckle is a particle, FSM can achieve much higher spatial and temporal resolutions than can photoactivation and photobleaching techniques and fluorescence correlation techniques. FSM, however, is more sensitive to the influence of random noise. Furthermore, FSM is capable of mapping polymerization/depolymerization dynamics of macromolecular assemblies in vivo. In terms of implementation, FSM has been performed using wide-field epifluorescence microscopy, spinning-disk confocal microscopy, and TIRF microscopy but has not been successfully used with laser-scanning confocal microscopy, mostly because of strong photobleaching and the higher noise and more limited dynamic range of photomultiplier detectors compared with CCD cameras.

Despite its success in analyzing cytoskeletal structure and assembly/disassembly, FSM has yet to become a routine technique for macromolecular assembly analysis. In particular, the need for microinjection seems to have limited the adoption of FSM. The requirement for customized data analysis and software to address specific biological questions also poses some technical challenges. Future development of FSM will overcome these obstacles. Currently, to avoid the need for microinjection, stable cell lines expressing low fractions of labeled proteins have been developed; but, so far, speckle signal quality is marginal. This problem will be solved using improved fluorescent proteins and molecular biology techniques. The expansion of FSM applications will likely increase the number of analytical techniques and related software.

It is expected that FSM will be applied to the study of more macromolecular assemblies. Development of microscopy techniques will make it more convenient to implement simultaneous speckle imaging at multiple wavelengths, at different fluorophore levels, and into 3D. Development of computational techniques will permit the analysis of complex 3D speckle image data and high-resolution correlative analysis of multiple molecular assemblies simultaneously. These developments will further establish FSM as the method of choice for the study of macromolecular assembly dynamics.

ACKNOWLEDGMENTS

GY receives startup fund support from Carnegie Mellon University. We gratefully acknowledge Ted Salmon, Clare Waterman-Storer, and Gaudenz Danuser for their pioneering work in FSM and their support.

REFERENCES

Adams MC, Salmon WC, Gupton SL, Cohan CS, Wittmann T, Prigozhina N, Waterman-Storer CM. 2003. A high-speed multispectral spinning-disk confocal microscope system for fluorescent speckle microscopy of living cells. *Methods* **29:** 29–41.

Adams MC, Matov A, Yarar D, Gupton SL, Danuser G, Waterman-Storer CM. 2004. Signal analysis of total internal reflection fluorescent speckle microscopy (TIR-FSM) and wide-field epi-fluorescence FSM of the actin cytoskeleton and focal adhesions in living cells. *J Microsc* **216:** 138–152.

Danuser G, Waterman-Storer CM. 2006. Quantitative fluorescent speckle microcopy of cytoskeleton dynamics. *Annu Rev Biophys Biomol Struct* **35:** 361–387.

Dorn JF, Danuser G, Yang G. 2008. Computational processing and analysis of dynamic fluorescence image data. In *Methods in cell biology*, Vol. 85, pp. 497–538. Academic, New York.

Fraley C, Raftery AE. 2002. Model-based clustering, discriminant analysis, and density estimation. *J Am Stat Assoc* **97:** 611–631.

Houghtaling BR, Yang G, Matov A, Danuser G, Kapoor TM. 2009. Op18 reveals the contribution of non-kinetochore microtubules to the dynamic organization of the vertebrate meiotic spindle. *Proc Natl Acad Sci* **106:** 15338–15343.

Hyman A, Drechsel D, Kellogg D, Salser S, Sawin K, Steffen P, Wordeman L, Mitchison T. 1991. Preparation of modified tubulin. *Methods Enzymol* **196:** 478–485.

Matos I, Pereira AJ, Lince-Faria M, Cameron LA, Salmon ED, Maiato H. 2009. Synchronizing chromosome segregation by flux-dependent force equalization at kinetochores. *J Cell Biol* **186:** 11–26.

Pardee JD, Spudich JA. 1982. Purification of muscle actin. *Methods Enzymol* **85:** 164–181.

Pawley JB. 2006. *Handbook of biological confocal microscopy*, 3rd. ed. Springer, New York.

Ponti A, Vallotton P, Salmon WC, Waterman-Storer CM, Danuser G. 2003. Computational analysis of F-actin turnover in cortical actin meshworks using fluorescent speckle microscopy. *Biophys J* **84:** 3336–3352.

Pratt WK. 2007. *Digital image processing*. Wiley-Interscience, New York.

Ross SM. 2006. *Introduction to probability models*. Academic, New York.

Turnacioglu KT, Sanger JW, Sanger JM. 1998. Sites of monomeric actin incorporation in living PTK2 and REF-52 cells. *Cell Motil Cytoskeleton* **40:** 59–70.

Wang Y, Taylor D. 1980. Preparation and characterization of a new molecular cytochemical probe: 5-iodoacetamidofluorescein-labeled actin. *J Histochem Cytochem* **28:** 1198–1206.

Watanabe N, Mitchison TJ. 2002. Single-molecule speckle analysis of actin filament turnover in lamellipodia. *Science* **295:** 1083–1086.

Waterman-Storer CM. 2002. Fluorescent speckle microscopy (FSM) of microtubule and actin in living cells. In *Current protocols in cell biology* (ed. Bonifacino JS, et al.), Vol. 1, pp. 4.10.1–4.10.26. Wiley, New York.

Waterman-Storer CM, Salmon ED. 1998. How microtubules get fluorescent speckles. *Biophys J* **75:** 2059–2069.

Waterman-Storer CM, Desai A, Bulinski JC, Salmon ED. 1998. Fluorescent speckle microscopy, a method to visualize the dynamics of protein assemblies in living cells. *Curr Biol* **8:** 1227–1230, S1.

Yang G, Houghtaling BR, Gaetz J, Liu JZ, Danuser G, Kapoor TM. 2007. Architectural dynamics of the meiotic spindle revealed by single-fluorophore imaging. *Nat Cell Biol* **9:** 1233–1242.

Yang G, Cameron LA, Maddox PS, Salmon ED, Danuser G. 2008. Regional variation of microtubule flux reveals microtubule organization in the metaphase meiotic spindle. *J Cell Biol* **182:** 631–639.

MOVIE LEGEND

Movies are freely available online at www.cshprotocols.org/imaging.

MOVIE 43.1. Fluorescent speckle microscopy (FSM) of poleward tubulin movement (flux) in a control *Xenopus laevis* egg extract spindle. (*Left panel*) Individual tubulin speckles undergoing antiparallel poleward movement. (*Right panel*) Trajectories of individual speckles are recovered by software for quantitative analysis. Red and green squares mark speckles moving to the left and the right poles, respectively. In each frame, the center of each square denotes the current position of the corresponding speckle being followed. Only those speckles whose trajectories last at least four frames are displayed using squares. Frame rate: 5 sec/frame. Display rate: 15 frames/sec. Field of view: width x height, 83 x 66 μm.

44 Polarized Light Microscopy
Principles and Practice

Rudolf Oldenbourg

Marine Biological Laboratory, Woods Hole, Massachusetts 02543, and Physics Department, Brown University, Providence, Rhode Island 02912

ABSTRACT

Polarized light microscopy provides unique opportunities for analyzing the molecular order in heterogeneous systems, such as living cells and tissues, without using exogenous dyes or labels. This chapter briefly discusses the theory of polarized light microscopy and elaborates on its practice using a traditional polarized light microscope and more specialized polarization microscopes such as the LC-Polscope, Oosight, or Abrio. The microscope components specific to analyzing the polarization of light, such as polarizer and compensator, are introduced, and quantitative techniques for measuring the birefringence of the specimen point by point using a traditional polarizing microscope are discussed. The new LC-PolScope greatly improves the analytic power of the technique, providing quantitative birefringence data simultaneously for every image point, thereby revealing molecular order with unprecedented sensitivity and at the highest resolution of the light microscope. Practical aspects discussed include the choice of optics, sample preparation, and combining polarized light with differential interference contrast and fluorescence microscopy. A glossary of polarization optical terms is included at the end of the chapter to facilitate the discussion of observations made with a polarized light microscope.

INTRODUCTION

Polarized light microscopy probes the local anisotropy of a specimen's optical properties such as refraction and absorption. Anisotropy of the refractive index is called birefringence, whereas anisotropy of the absorption coefficient is called dichroism. Specimens with either property exhibit a characteristic variation of intensity as they are rotated between crossed linear polarizing filters. In this short introduction to the principles and practice of polarized light microscopy we will describe the basic components of a polarizing microscope and its use for identifying specimen anisotropy and for measuring the birefringence of anisotropic structures that typically occur in biological tissues and living cells.

Optical anisotropy is a consequence of molecular order, such as is found in crystals. In materials with molecular order, the absorption, refraction, and scattering of light typically become dependent on the orientation of the material with respect to the polarization of the light. Polarized light microscopy exploits this dependency and provides a sensitive tool to analyze the alignment of molecular bonds or fine structural form in a specimen. By its very nature, polarizing microscopy provides structural information at a submicroscopic level, including the alignment of polymers in plastics and filaments in biological tissues and cells. Neuronal processes like axons and dendrites are birefringent because microtubules and other filaments align parallel to the extended processes. Hence, the polarizing microscope gives us information not only about where and when certain structures form and change inside organisms, tissues, and cells, but also about some of the submicroscopic features of these structures. Most importantly, the polarizing microscope can perform live cell imaging dynamically, repeatedly in short time intervals and over long time periods, at the highest resolution of the light microscope, and with only minimal interference to the physiological conditions required for live specimens.

This chapter gives a brief discussion of the theory of polarized light microscopy and its practice using a traditional polarized light microscope and more specialized polarization microscopes such as the LC-Polscope (Cambridge Research and Instrumentation, Inc., Woburn, MA, cri-inc.com), Oosight, or Abrio. For lucid discussions of polarized light microscopy and its application to biophysical inquiries inside living cells, the reader is referred to articles and books by Shinya Inoué, including the recent publication of his collected works (Inoué 1986, 2002, 2007, 2008; Inoue and Oldenbourg 1998).

TRADITIONAL POLARIZED LIGHT MICROSCOPY

Basic Setup

The polarized light microscope (also called polarizing microscope or polarization microscope; Fig. 1) generally differs from a standard trans-illuminating microscope by the addition of a polarizer before the condenser; a compensator slot and analyzer behind the objective lens; strain-free optics;

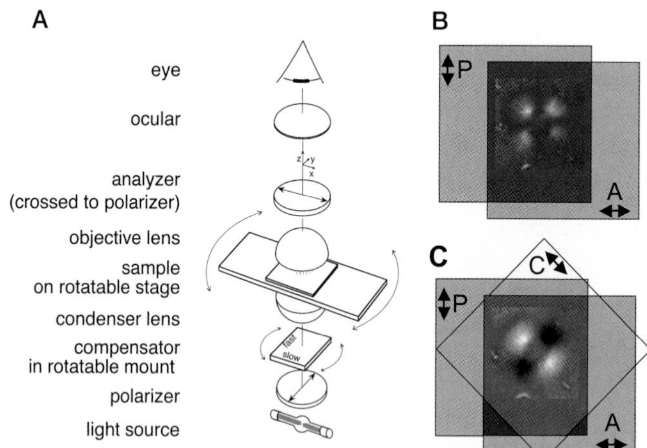

FIGURE 1. Traditional polarized light microscope schematic and image cartoons. (*A*) Schematic of optical arrangement of a conventional polarizing microscope. (*B*) Cartoon depicting at its center the image of an aster as it appears when located between a crossed polarizer P and analyzer A. The arrows on the polarizer and analyzer sheet indicate their transmission directions. An aster is made of birefringent microtubule (MT) arrays radiating from a centrosome (aster diameter = 15 µm). MTs that run diagonal to the polarizer and analyzer appear bright, whereas MTs that run parallel to the polarizer or analyzer appear dark. (*C*) Cartoon of an aster as it appears when located between polarizer, analyzer, and a compensator C, which is made of a uniformly birefringent plate. The arrow in the compensator plate indicates its slow axis direction. Microtubules that are nearly parallel to the slow axis of the compensator appear bright, whereas those that are more perpendicular to the slow axis are dark. Therefore, the birefringence of microtubules has a slow axis that is parallel to the polymer axis, as is the case for many biopolymers.

a graduated, revolving stage; centerable lens mounts; crosshairs in the ocular aligned parallel or at 45° to the polarizer axes; and a focusable Bertrand lens that can be inserted for conoscopic observation of interference patterns in the back aperture of the objective lens.

Polarizers

Most light sources (halogen bulb, arc burner, light-emitting diode) generate unpolarized light; hence, the first polarizer located before the condenser optics polarizes the light that illuminates the specimen. The second polarizer serves to analyze the polarization of the light after it passed through the specimen; therefore, it is called the analyzer. Usually the polarizer and analyzer are in crossed orientation, so that the analyzer blocks (absorbs) most if not all of the light that has passed through the specimen. Therefore, the image of the specimen looks mostly dark, except for structures that are birefringent or otherwise optically anisotropic and appear bright against the dark background. When the specimen is rotated on a revolving stage (around the axis of the microscope), the birefringent parts change brightness, changing from dark to bright and back to dark four times during a full 360° rotation. A uniformly birefringent specimen part appears darkest when its optical axes are parallel to polarizer and analyzer. This is called the extinction orientation (or extinction position). Rotating the specimen by 45° away from the extinction orientation makes the birefringent part appear brightest. Not all birefringent parts in the field of view will turn dark at the same time, because in general each part has different axis orientations. In summary, by rotating the specimen between crossed polarizers, one can recognize birefringent components and determine their axis orientations.

Compensator

Although not absolutely necessary for some basic observations, especially of highly birefringent objects, the compensator (1) can significantly improve the detection and visibility of weakly birefringent objects, (2) is required to determine the slow and fast axis of specimen birefringence, and (3) is an indispensable tool for the quantitative measurement of object birefringence. There are several types of compensators; most of them are named for their original inventors. For the observation of weak birefringence, typically encountered in biological specimens, the Brace–Köhler compensator is most widely used. It consists of a thin birefringent plate, often made from mica, with a retardance of 1/10 to 1/30 of a wavelength ($\lambda/10$ to $\lambda/30$; for definition of retardance, see below). The birefringent plate is placed in a graduated rotatable mount and inserted either between the polarizer and condenser, as in Figure 1, or between the objective lens and the analyzer. The location varies between microscope manufacturers and specific microscope types. In either location, the effect of the Brace–Köhler compensator on the observed image is the same and its standard usage is independent of these locations. In general, the birefringence of the compensator, when inserted into the optical path, causes the image background to become brighter, whereas birefringent specimen parts can turn either brighter or darker than the background, depending on their orientation with respect to the compensator. If the birefringent structure becomes brighter, its slow axis aligns more parallel to the compensator slow axis, whereas if the structure turns darker, then its slow axis aligns more perpendicular to the compensator slow axis. Hence, the compensator with known slow axis orientation can be used to determine the slow axis of birefringent parts in the specimen. The compensator can also enhance specimen contrast, and it is used to quantify specimen birefringence by measuring its retardance, as discussed next.

Birefringence, Retardance, and Slow Axis

Birefringence provides the unique opportunity to measure and analyze the structural order of specimens without having to treat them with exogenous dyes, fluorescence labels, or stains, or to resort to destructive and expensive electron microscopy. Birefringence occurs when there is a molecular order, that is, when the average molecular orientation is nonrandom, as in crystals or in aligned polymeric materials. Molecular order usually gives the material two orthogonal optical axes, with the index of refraction along one axis being different from that along the other axis. The difference between the two indices of refraction is called birefringence and is an intrinsic property of the material:

$$\text{birefringence} = \Delta n = n_{\|} - n_{\perp},$$

where $n_{\|}$ and $n_{\perp}$ are the refractive indices for light polarized parallel or perpendicular to the two optical axes.

Light polarized parallel to one axis travels at a different speed through the sample than does light polarized parallel to the orthogonal axis. As a result, these two light components, which were in phase when they entered the sample, are retarded at a different rate and exit the sample out of phase. Using the polarizing microscope, one can measure this differential retardation, also called retardance, and thereby quantify the magnitude and orientation of molecular order in the specimen. Retardance is an extrinsic property of the material and a product of the birefringence and the path length l through the material:

$$R = \Delta n \cdot l.$$

In addition to retardance, birefringence has an orientation associated with it. The orientation refers to the specimen's optical axes: One is called the fast axis and the other is called the slow axis. Light polarized parallel to the slow axis experiences a higher refractive index and travels more slowly than light polarized parallel to the fast axis. In materials that are built from aligned filamentous molecules, the slow axis is typically parallel to the average orientation of the filaments. Birefringence orientation always correlates with molecular orientation, which usually changes from point to point in the specimen (see the aster in Fig. 1B,C).

Quantitative Analysis of Specimen Retardance

Birefringence and retardance are optical properties that can be directly related to molecular order, such as the alignment of polymeric material. Using the traditional polarizing microscope, Sato, Ellis and Inoué, for example, have measured the retardance of mitotic spindles in living cells and definitively concluded that the birefringence is caused by the array of aligned spindle microtubules (Sato et al. 1975). Their measurements were made possible by the careful analysis of specimen birefringence using a traditional compensator in addition to a polarizer and analyzer in the microscope optical train. Using a Brace–Köhler compensator, the retardance of a resolved image point or uniformly birefringent area in the field of view can be measured by carrying out the following steps.

1. With the specimen in place and viewed through the eyepiece (or on the monitor of a video camera), make sure that polarizer and analyzer are crossed and the compensator is in the extinction position. Given those settings, background areas that have no birefringence appear dark, and birefringent structures change from dark to bright four times when rotating the specimen on the rotatable stage by 360°.
2. Again, using the rotatable stage, orient the birefringent structure of interest so that it appears darkest (extinction position). Note the orientation and then rotate the specimen 45° away from the extinction position. At that orientation the structure appears brightest. (By eye, it is usually easier to determine the orientation that leads to the lowest intensity than the highest one. In any case, the two orientations are 45° apart.)
3. Now rotate the compensator either clockwise or counterclockwise, until the birefringent structure of interest appears dark. When rotating the compensator, any background area becomes brighter, whereas birefringent structures become either brighter or darker. For structures that turn darker than the background, the birefringence of the compensator is said to "compensate" the birefringence of the sample structure. Take as an example the aster image in the cartoon of Figure 1C. The radial array of microtubules shows two bright quadrants and two dark quadrants. In the bright quadrants, the microtubules' slow axis runs more parallel to the compensator slow axis, whereas in the dark quadrants the slow axis of the microtubule birefringence is oriented nearly perpendicular to the slow axis of the compensator.

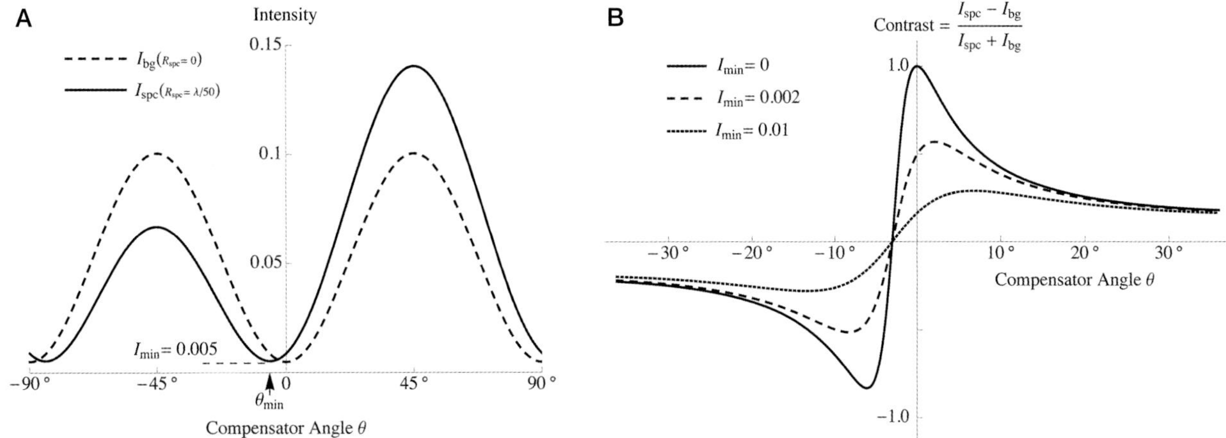

FIGURE 2. (*A*) The graph shows intensities detected in a single image point or uniform area of the specimen versus the rotation angle of a Brace–Köhler compensator. The dashed curve shows the background intensity I_{bg} of a specimen area that exhibits no birefringence. The solid curve shows the intensity I_{spc} of a specimen point that exhibits a small retardance (l/50). The transmitted intensity is given as a fraction of the amount of light that has passed through the first polarizer. I_{min} represents the spurious intensity that is detected when the compensator is in the extinction position. (*B*) The graph shows the contrast of a birefringent specimen point (R_{spc} = l/50) as a function of compensator angle, for three different values of I_{min}. I_{min} is affected by the quality of the polarizers used and the polarization distortions introduced by the intervening optics (e.g., condenser and objective lens). Both graphs were generated computationally using the Jones calculus and assuming a retardance of l/10 for the birefringent crystal plate of the Brace–Köhler compensator and a specimen retardance of l/50 with slow axis oriented at 45° to polarizer and analyzer.

As the compensator is rotated from its extinction position, the specimen turns darkest at a characteristic angle θ_{min}. Given the known compensator retardance R_{cmp}, the angle θ_{min} is a direct measure of the specimen retardance R_{spc}:

$$R_{spc} = R_{cmp}\sin(2\theta_{min}).$$

Figure 2A shows a graph of the intensity calculated for a uniformly birefringent specimen area (or single specimen point) as a function of the rotation angle θ of a Brace–Köhler compensator. At the rotation angle θ_{min} the intensity is a minimum. For other rotation angles the intensity varies according to a complex expression of trigonometric functions that can be derived using the Jones calculus. This expression can also be used to calculate the expected contrast of a birefringent object against its nonbirefringent background. The graph in Figure 2B shows the computed contrast of the birefringent specimen area versus θ. The contrast is defined as the intensity difference between the specimen area and the background, divided by their sum:

$$\text{contrast} = \frac{I_{spc} - I_{bg}}{I_{spc} + I_{bg}}.$$

As is apparent from Figure 2B, the highest contrast and therefore the best visibility of the birefringent specimen is achieved when adjusting for a compensator angle of $\sim\pm\theta_{min}$.

Figure 2B further illustrates the importance of using so-called high extinction optics when observing a specimen with low birefringence. Figure 2B introduces the parameter I_{min}, which represents the spurious intensity observed when polarizer and analyzer are crossed. The better the quality of the polarizers employed and of the intervening optical components (including condenser and objective lens), the lower I_{min} will be and the higher the contrast that can be achieved for a specimen structure of given retardance. High extinction optics will produce high contrast images that also result in images with high signal-to-noise ratio (Oldenbourg 1999).

THE LC-POLSCOPE

The LC-PolScope is a birefringence imaging technique that was first developed in the author's laboratory at the MBL and is commercially available from Cambridge Research and Instrumentation (CRi), Inc., Woburn, MA (www.cri-inc.com) under the trade names Abrio and Oosight. The optical design of the LC-PolScope builds on the traditional polarized light microscope, introducing two essential modifications: The specimen is illuminated with nearly circularly polarized light and the traditional compensator is replaced by a liquid crystal–based universal compensator. The LC-PolScope also requires the use of narrow bandwidth (≤40 nm) or monochromatic light. In Figure 3, the schematic of the optical train shows the universal compensator located between the monochromatic light source (arc lamp with interference filter) and the condenser lens. The analyzer for circularly polarized light is placed after the objective lens. The universal compensator is built from two variable retarder plates and a linear polarizer. The variable retarder plates are implemented as electro-optical devices made from two liquid crystal plates. Each liquid crystal plate has a uniform retardance that depends on the voltage applied to the device. A computer controlled electronic circuitry supplies the voltage for each plate. The computer is also connected to the electronic camera, typically a CCD (charge-coupled device) camera for recording the specimen images projected onto the camera by the microscope optics. Specialized software synchronizes the image acquisition process with the liquid crystal settings and implements image processing algorithms. These algorithms compute images that represent the retardance and slow axis orientation at each resolved image point.

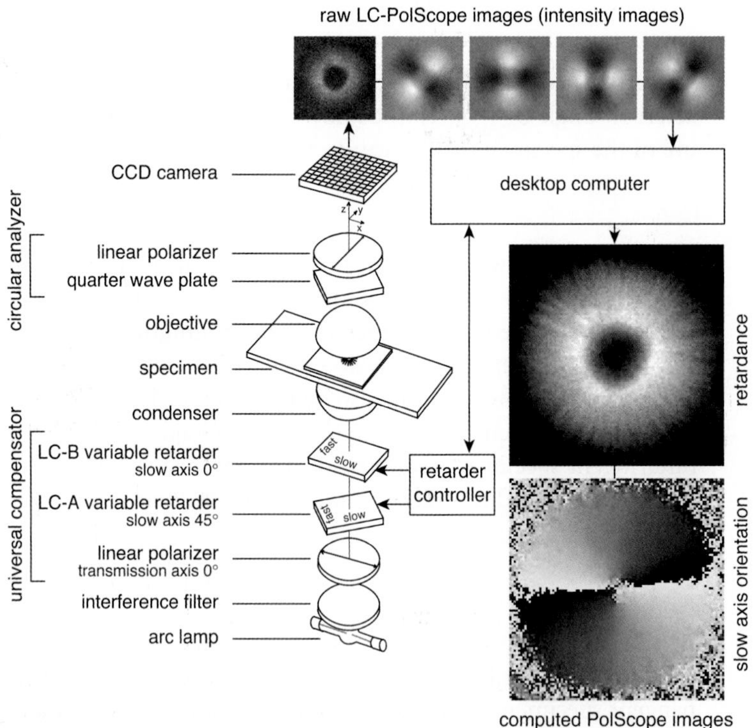

FIGURE 3. Schematic of the LC-PolScope. The optical design (*left*) builds on the traditional polarized light microscope with the conventional compensator replaced by two variable retarders LC-A and LC-B. The polarization analyzer passes circularly polarized light and is typically built from a linear polarizer and a quarter wave plate. Images of the specimen (*top row*, aster isolated from surf clam egg) are captured at five predetermined retarder settings, which cause the specimen to be illuminated with circularly polarized light (*top row, leftmost* image) and with elliptically polarized light of different axis orientations (*top row, second to fifth* images). Based on the raw PolScope images, the computer calculates the retardance image and the slow axis orientation or azimuth image shown on the *lower right*.

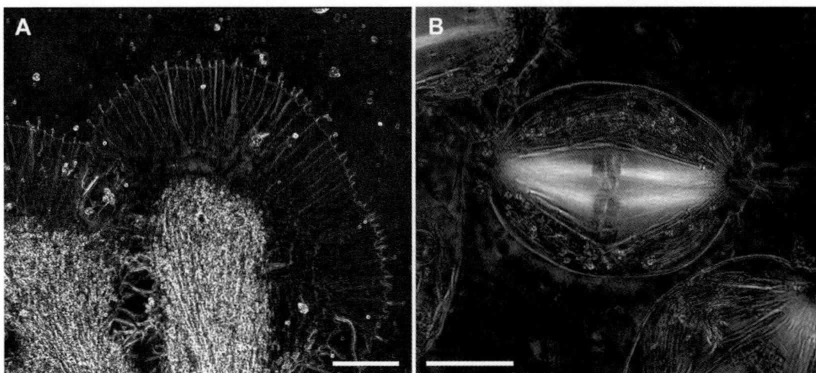

FIGURE 4. Retardance images recorded with the LC-PolScope. (*A*) Living growth cone of an *Aplysia* bag cell neuron. The peripheral lamellar domain contains radially aligned fibers composed of 15–40 actin filaments. A detailed description of the birefringent fine structure including a time-lapse movie of the growth cone dynamics is published in Katoh et al. (1999). Image brightness represents measured retardance between 0 nm (black) and 0.5 nm or larger (white). Scale bar, 10 µm. (Reprinted, with permission, from Katoh et al. 1999, ©American Society of Cell Biology.) (*B*) Meiotic spindle in living spermatocyte of crane fly *Nephrotoma suturalis*. At metaphase, chromosomes have congressed to the spindle equator and kinetochore fibers, each composed of approximately 60 microtubules, connect each bivalent to opposite poles. A cage of birefringent mitochondria surrounds the spindle. (Reprinted from LaFountain and Oldenbourg 2004, ©American Society of Cell Biology.) (Time-lapse movies of meiosis I and II recorded with the LC-PolScope are available at http://cellimages.ascb.org/.) Image brightness represents measured retardance between 0 nm (black) and 2.0 nm or larger (white). Scale bar, 10 µm.

The commercial system, developed and distributed by CRi Inc., is available as an accessory to microscope stands of all major microscope manufacturers (Leica, Nikon, Olympus, Zeiss). It usually includes the universal compensator, circular polarizer, a camera with control electronics, and a computer with software for image acquisition and processing. Three slightly differing versions are available, each optimized for research in the life sciences (Abrio LS), for research in industrial metrology (Abrio IM), and for in vitro fertilization and related laboratory techniques (Oosight).

Figure 4 shows retardance images recorded with the LC-PolScope, illustrating the clarity, resolution, and analytic potential of analyzing the birefringent fine structure in living cells.

Practical Considerations

Choice of Optics

As indicated earlier, the polarization distortions introduced by the objective and condenser lenses limit the extinction that can be achieved in a polarizing microscope. Most microscope manufacturers offer lenses that are designated “Pol” to indicate low polarization distortions, which can arise from a number of factors, including stress or crystalline inclusions in the lens glass and the type of antireflection coatings used on lens surfaces. Some lens types are available only with “DIC” designation. “DIC” lenses do not meet the more stringent Pol requirements but pass for use in differential interference contrast and can also be used with the LC-PolScope (without the Wollaston or Nomarski prisms specific to DIC, of course).

The polarization performance of the most highly corrected lenses, so-called Plan-Apochromat objectives, is often compromised by a large number of lens elements, special antireflection coatings, and, in some cases, special types of glass used to construct the lenses (Oldenbourg and Shribak 2010). For some applications, these high-quality lenses, which provide a large, highly corrected viewing field over a wide spectral range, might not be required. If the objective lens is to be used only with the PolScope that requires monochromatic light and typically acquires images from a region near the center of the viewing field, a less stringent correction might well suffice. To find out what works best for a particular imaging situation, several lenses should be tested. It is also helpful to be able to select the best performing combination of condenser and objective lens from a batch of the same lens types.

Whenever possible or practical, oil-immersion lenses should be used. This is because the transition of a light ray between two media of different refractive index (n_{air} = 1.00, n_{glass} = 1.52) introduces polarization distortions, especially for high-numerical-aperture (high-NA) lenses. The peripheral rays leaving the condenser front lens are highly tilted to the slide surface and their polarization is typically rotated when traversing the air–glass interface. Oil and, to a lesser degree, water and other immersion liquids greatly reduce polarization aberrations caused by air–glass interfaces between the specimen and the lenses. For a discussion of the origin of polarization distortions in lenses and optical systems, and of ways to reduce them, see (Shribak et al. 2002). This publication also discusses the various conoscopic images that can be observed in a polarizing microscope equipped with crossed, or nearly crossed, polarizers and a Bertrand lens.

Lenses and other optical elements that are part of the optical train but not placed between the crossed polarizers do not affect the polarization performance of the microscope setup. Filter cubes for epi-illumination, for example, should be placed outside the polarization optical train. (The polarization optical train is defined as the stretch between the crossed polarizers.) In a dissecting microscope, the polarizing elements and compensator should be placed in front of the objective lens instead of behind it. The objective of a dissecting scope has very low NA and the image-forming rays have a small angle of divergence (in contrast to high-NA lenses.) Rays with a small tilt angle with respect to the normal of the polarizer and compensator plates do not appreciably affect the performance of these devices.

Specimen Preparation

The following are recommendations for preparing living cells for observation with a polarizing microscope. Other resources for polarized light microscopy of living cells include Inoué (2002) and Oldenbourg (1999), which have sections on suitable cell types and their preparation.

- If cells have to be kept in plastic Petri dishes during observations, only use dishes with glass coverslip bottoms. The strong birefringence of a plastic bottom ruins the polarization of the light.
- Prepare specimens such that cells or structures of interest are as close to the coverslip as possible. When using high-NA (>0.5) oil-immersion or dry optics, a layer of >10 μm of aqueous medium introduces enough spherical aberration to noticeably reduce the resolution of cell images. The use of water-immersion optics alleviates this problem.
- For observations of structures near the cell surface, it might be necessary to increase the medium's refractive index to match the refractive index of the cytoplasm and reduce the effect of edge birefringence (Oldenbourg 1991). Adding polyethyleneglycol, polyvinylpyrrolidone, or some other harmless polymer or protein substitute to the medium will increase its refractive index to match that of the cytoplasm inside the cell. The concentration of substitute required might vary from a few percent to >10%, depending on the average refractive index of the cytoplasm of the cells of interest. It is best to test various media containing different polymer concentrations by suspending cells in them and examining the refractive index mismatch in a slide and coverslip preparation with DIC or polarized light. At the optimal polymer concentration, no distinct cell boundary is visible, giving the impression that the organelles and cytoskeleton are freely suspended. After determining the optimal concentration for imaging, the medium should be tested for compatibility with growing cells. Many cells grow and develop normally in media with inert additives. The molecular weight of the polymer needs to be ~40 kD or more to prevent its uptake into cells.
- For observations with the LC-PolScope, prepare the specimen so that, when mounted in the microscope, at least one clear area without cells or birefringent material can be identified and moved into the viewing field when required. A clear area is needed for calibrating the PolScope and for recording background images. The background images are used to remove spurious background retardance from specimen images, allowing precise measurement and imaging of the cells and structures of interest. Many cell cultures and cell-free systems can be prepared

without special attention to this requirement. For example, a free area can often be found around sparsely plated cells. Cultures of free-swimming cells might have to be diluted before mounting a small drop of the suspension between slide and coverslip to observe free areas. Sometimes it is helpful to add a tiny drop of oil or other nontoxic, immiscible liquid to the preparation. This clear drop can provide an area for calibrating the instrument and taking background images in a preparation that is otherwise dense with birefringent structures.

Combining Polarized Light with DIC and Fluorescence Imaging

For differential interference contrast (DIC), two matching prisms are added to the polarization optical train (for the discussion of a standard DIC arrangement, see Salmon and Tran 1998; Oldenbourg and Shribak 2010). The two prisms, called Wollaston or Nomarski prisms, are specially designed to fit in specific positions, one before the condenser and the other after the objective lens, and to match specific lens combinations. The prisms in their regular positions can also be combined with the LC-PolScope setup, in which case the universal compensator and circular analyzer take the places of the linear polarizer and analyzer in a standard DIC arrangement. In fact, the same DIC prisms function equally well using either linearly or circularly polarized light. Note, however, that Carl Zeiss has recently combined linear polarizers with their condenser DIC prisms, thus preventing their use with circularly polarized light.

For best results, the LC-PolScope is first calibrated without prisms to find the extinction setting for the universal compensator. With the compensator set to extinction, the prisms are entered into the optical path. Typically, one of the two prisms has a mechanism for fine-adjusting its position, which in turn affects the brightness and contrast of the DIC image. In a PolScope setup, the mechanical fine adjustment or the universal compensator can be used to add a bias retardance to change the brightness and contrast of the DIC image. When using the universal compensator, retardance must be added or subtracted from the extinction setting of either the LC-A or LC-B retarder (Fig. 3), depending on the orientation of the shear direction of the prisms.

For live cell imaging, it is very useful to be able to switch easily between polarized light and DIC imaging. DIC provides good contrast of many morphological features in cells using direct viewing through the eyepiece or by video imaging. When viewing specimens that have low polarization contrast, it is more effective to align the optics, including the visualization of the specimen, using DIC. After the optics have been aligned and the specimen is in focus, the DIC prisms are removed from the optical path and the PolScope specific adjustments completed.

Fluorescence imaging can be combined with the PolScope in several ways. Fluorescence is commonly observed using epi-illumination, which requires a filter cube in the imaging path. The filter cube includes a dichromatic mirror and interference filters for separating the excitation and emission wavelengths. For best results the filter cube should be removed for PolScope observations to avoid the polarization distortions caused by the dichromatic mirror. For observing fluorescence, on the other hand, the PolScope analyzer should be removed, because it attenuates the fluorescence emission by at least 50%. To meet both requirements, the PolScope analyzer can be mounted in an otherwise empty filter cube holder, which is moved into the optical path as the fluorescence cube is moved out, and vice versa.

If, however, the option of removing the fluorescence cube is unavailable, the cube can remain in the optical path for observations with polarized light and the LC-PolScope. In this case, the following points should be considered.

- For the light source of the PolScope, choose a wavelength that is compatible with the emission wavelength filter in the fluorescence cube. For example, fluorescein isothiocyanate (FITC) requires excitation with blue light (485 nm) and fluoresces in green. The FITC dichromatic beam splitter and a barrier filter passing green fluorescence light with wavelength >510 nm would be compatible with 546-nm light for polarized light observations using a mercury arc burner for the transmission light path.

- The light source for one imaging mode must be blocked while observing with the other imaging mode.
- Removing the polarization analyzer while observing fluorescence more than doubles the fluorescence intensity.
- The LC-PolScope must be calibrated with the fluorescence filter cube in place. The polarization distortions caused by the dichromatic mirror are partially counteracted by the calibrated settings of the universal compensator.

CONCLUSION AND FUTURE PROSPECTS

Polarized light microscopy provides complementary information to the many imaging techniques discussed in this series of manuals. In contrast to fluorescence microscopy, imaging with polarized light reveals information about the organization of the endogenous molecules that built the complex and highly dynamic architecture of cells and tissues. Although fluorescence identifies the chemical nature of tagged structures, such as f-actin and microtubules, it does not provide information about their submicroscopic architecture. Polarized light microscopy, on the other hand, is not specific to the chemical nature but to the structural nature of macromolecular assemblies such as the submicroscopic alignment of molecular bonds and the architectural fine structure of the specimen.

Recently, the chemical specificity of fluorescence tagging and the structural information available through polarization analysis were combined in a study to reveal septin organization in yeast bud necks (Vrabioiu and Mitchison 2006, 2007). The study took advantage of the polarized florescence emitted by every single GFP fluorophore. GFP was fused to the protein of interest, in this case septin, with a rigid linker that limits the GFP's ability to rotate relative to the septin molecule. When septin assembled into an ordered structure, a consistent orientation of the GFP dipoles was detected throughout the structure. In the future we can expect more studies of this kind that highlight specific structures through fluorescence tagging and analyze their organization by polarized fluorescence.

From an optical point of view, polarized light microscopy is related to differential interference contrast (DIC) and phase contrast microscopy, because all of them were designed to highlight changes of the refractive index of the specimen, affecting the phase of the transmitted light. Phase contrast and DIC microscopy highlight changes of refractive index that occur from point to point in the specimen, whereas polarized light microscopy highlights changes in refractive index that occur when changing the polarization of light. The change of refractive index due to polarization is called birefringence and it turns that it can be measured very sensitively using relatively simple optical means such as two polarizers and a compensator. The measured quantity is retardance, which can be used to highlight and analyze the molecular architecture, including the estimation of the number of filaments in a bundle or the alignment of molecular bonds in a complex structure. It is this submicroscopic information, gleaned from live, actively functioning cells, tissues, and whole organisms that makes polarized light microscopy an indispensable tool in the array of analytic and quantitative experimental techniques that provide us a glimpse into the secrets of life itself.

GLOSSARY OF POLARIZATION OPTICAL TERMS

The following is a brief introduction to terms that are relevant for observations with a polarized light microscope. A more detailed explanation of these and other terms that describe physical phenomena or optical devices can be found in the following references (Shurcliff 1962; Born and Wolf 1980; Chipman 1995; Hecht 1998).

Analyzer

An analyzer is a polarizer that is used to analyze the polarization state of light (see Polarizer).

Azimuth

The azimuth is an angle that refers to the orientation of the slow axis of a uniformly birefringent region. The azimuth image refers to the array of azimuth values of a birefringent specimen imaged with the LC-PolScope. The azimuth is typically measured from the horizontal orientation with values increasing for counterclockwise rotation. Angles range between 0° and 180°, where both end points indicate horizontal orientation.

Birefringence

Birefringence is a material property that can occur when there is molecular order, that is, when the average molecular orientation is nonrandom, as in crystals or in aligned polymeric materials such as those comprising the mitotic spindle in a dividing cell. Molecular order usually renders the material optically anisotropic, leading to a refractive index that, in general, changes with the polarization of the light. For example, assume that a beam of light passes through a mitotic spindle in a direction perpendicular to the spindle axis. The light that is polarized parallel to the spindle axis experiences a higher refractive index than the light that is polarized perpendicular to the spindle axis (i.e., the mitotic spindle is birefringent). The birefringence originates from the parallel alignment of microtubules (Sato et al. 1975). The difference between the two indices of refraction is called birefringence:

$$\text{birefringence} = = \Delta n = n_{\parallel} - n_{\perp},$$

where $n_{\parallel}$ and $n_{\perp}$ are the refractive indices for light polarized parallel and perpendicular to one of the principal axes.

Compensator

A compensator is an optical device that includes one or more retarder plates and is commonly used to analyze the birefringence of a specimen. For a traditional polarizing microscope, several types of compensators exist that typically use a single fixed retarder plate mounted in a mechanical rotation stage. With the help of a compensator it is possible to distinguish between the slow and fast axis direction and to measure the retardance of a birefringent object after orienting it at 45° with respect to the polarizers of the microscope.

The LC-PolScope employs a universal compensator that includes two electro-optically controlled, variable retarder plates. Using the universal compensator it is possible to measure the retardance and slow axis orientation of birefringent objects that have any orientation in the plane of focus.

Dichroism

Dichroism is a material property that can occur in absorbing materials in which the light-absorbing molecules are arranged in a nonrandom orientation. Dichroism refers to the difference in the absorption coefficients for light polarized parallel and perpendicular to the principal axis of alignment.

The measurement of optical anisotropy by the LC-PolScope is affected by the dichroism of absorbing materials. In nonabsorbing, clear specimens, however, dichroism vanishes and birefringence is the dominant optical anisotropy measured by the LC-PolScope. Like absorption, dichroism is strongly wavelength dependent, whereas birefringence only weakly depends on wavelength.

Extinction

Extinction is defined as the ratio of maximum to minimum transmission of a beam of light that passes through a polarization optical train. Given a pair of linear polarizers, for example, the extinction is the ratio of intensities measured for parallel versus perpendicular orientation of the transmission axes of the polarizers (extinction = $I_{\parallel}/I_{\perp}$). In addition to the polarizers, the polarization optical train can also include other optical components, which usually affect the extinction of the complete train. In a polarizing microscope, the objective and condenser lens are located between the polarizers and significantly reduce the extinction of the whole setup.

Fast Axis

The fast axis describes an orientation in a birefringent material. For a given propagation direction, light that is polarized parallel to the fast axis experiences the lowest refractive index, and hence travels the fastest in the material (see also Slow Axis).

Optic Axis

The optic axis refers to a direction in a birefringent material. Light propagating along the optic axis does not change its polarization; hence, for light propagating along the optic axis, the birefringent material behaves as if it were optically isotropic.

Polarized Light

A beam of light is said to be polarized when its electric field is distributed nonrandomly in the plane perpendicular to the beam axis. In unpolarized light, the orientation of the electric field is random and unpredictable. In partially polarized light, some fraction of the light is polarized, whereas the remaining fraction is unpolarized. Most natural light is unpolarized (sun, incandescent light), but can become partially or fully polarized by scattering, reflection, or interaction with optically anisotropic materials. These phenomena are used to build devices to produce polarized light (see Polarizer).

Linearly Polarized Light

In a linearly polarized light beam, the electric field is oriented along a single axis in the plane perpendicular to the propagation direction.

Circularly Polarized Light

In circularly polarized light, the electric field direction rotates either clockwise (right-circularly) or counterclockwise (left-circularly) when looking toward the source. Although the field direction rotates, the field strength remains constant. Hence, the end point of the field vector describes a circle.

Elliptically Polarized Light

In elliptically polarized light, as in circularly polarized light, the electric field direction rotates either clockwise or counterclockwise when looking toward the source. However, although the field direction rotates, the field strength varies in such a way that the end point of the field vector describes an ellipse. The ellipse has a long and short principal axis that are orthogonal to each other and have fixed orientation. Any type of polarization (linear, circular, or elliptical) can be transformed into any other type of polarization by means of polarizers and retarders.

Polarizer

A polarizer, sometimes called a polar, is a device that produces polarized light of a certain kind. The most common polar is a linear polarizer made from dichroic material (e.g., a plastic film with small, embedded iodine crystals that have been aligned by stretching the plastic), which transmits light of one electric field direction while absorbing the orthogonal field direction. Crystal polarizers are made of birefringent crystals that split the light beam into orthogonal linear polarization components. A polarizer that produces circularly polarized light, a circular polarizer, is typically built from a linear polarizer followed by a quarter-wave plate.

The LC-PolScope employs a universal compensator that also serves as a universal polarizer in that it converts linear polarization into any other type of polarization by means of two variable retarders.

Retardance

Retardance is a measure of the relative optical path difference, or phase change, suffered by two orthogonal polarization components of light that has interacted with an optically anisotropic material. Retardance is the primary quantity measured by the LC-PolScope. Assume a nearly collimated beam of light traversing a birefringent material. The light component that is polarized parallel to the high refractive index axis travels at a slower speed through the birefringent material than the component polarized perpendicular to that axis. As a result, the two components, which were in phase when they entered the material, exit the material out of phase. The relative phase difference, expressed as the distance between the respective wave fronts, is called the retardance:

$$\text{retardance} = R = (n_{\|} - n_{\perp}) \cdot l = \Delta n \cdot l,$$

where l is the physical path length or thickness of the birefringent material. Hence, retardance has the dimension of a distance and is often expressed in nanometers. Sometimes it is convenient to express that distance as a fraction of the wavelength λ, such as $\lambda/4$ or $\lambda/2$. Retardance can also be defined as a differential phase angle, in which case $\lambda/4$ corresponds to 90° and $\lambda/2$ to 180° phase difference.

As a practical example consider a mitotic spindle observed in a microscope that is equipped with low-NA lenses (NA $\leq$ 0.5). When the spindle axis is contained in the focal plane, the illuminating and imaging beams run nearly perpendicular to the spindle axis. Under those conditions, the retardance measured in the center of the spindle is proportional to the average birefringence induced by the dense array of aligned spindle microtubules. To determine Δn, it is possible to estimate the thickness, l, either by focusing on spindle fibers located on top and bottom of the spindle and noting the distance between the two focus positions or by measuring the lateral extent of the spindle when focusing through its center. The latter approach assumes a rotationally symmetric shape of the spindle. Typical values for the spindle retardance of crane fly spermatocytes (Fig. 4B) and of other cells is 3–5 nm and the spindle diameter is ~30–40 µm, leading to an average birefringence of around 10^{-4}. It has been found that the retardance value of the spindle is largely independent of the NA for imaging systems using NA $\leq$ 0.5 (Sato et al. 1975).

On the other hand, when using an imaging setup that employs high-NA optics (NA > 0.5) for illuminating and imaging the sample, the measured retardance takes on a somewhat different context. For example, the retardance measured in the center of a microtubule image recorded with a LC-PolScope equipped with a high-NA objective and condenser lens is 0.07 nm. A detailed study showed that the peak retardance decreased inversely with the NA of the lenses. However, the retardance integrated over the cross section of the microtubule image was independent of the NA (Oldenbourg et al. 1998). Although a conceptual understanding of the measured retardance of submicroscopic filaments has been worked out in the aforementioned publication, a detailed theory of these and other findings about the retardance measured with high-NA optics has yet to materialize.

Retarder

A retarder, or wave plate, is an optical device that is typically made of a birefringent plate. The retardance of the plate is the product of the birefringence of the material and the thickness of the plate. Fixed retarder plates are either cut from crystalline materials such as quartz, calcite, or mica or they are made of aligned polymeric material. If the retardance of the plate is $\lambda/4$, for example, the retarder is called a quarter-wave plate.

A variable retarder can be made from a liquid crystal device. A thin layer of highly birefringent liquid crystal material is sandwiched between two glass windows, each bearing a transparent electrode. A voltage applied between the electrodes produces an electric field across the liquid crystal layer that reorients the liquid crystal molecules. This reorientation changes the birefringence of the layer without affecting its thickness or the direction of its slow axis.

Slow Axis

The slow axis describes an orientation in a birefringent material. For a given propagation direction, light polarized parallel to the slow axis experiences the highest refractive index and hence travels the slowest in the material (see also Fast Axis).

Wave Plate

See Retarder.

ACKNOWLEDGMENTS

I gratefully acknowledge many years of illuminating discussions on polarized light microscopy with Shinya Inoué and Michael Shribak of the MBL. This work was supported by funds from the National Institute of Biomedical Imaging and Bioengineering (grant EB002045).

REFERENCES

Born M, Wolf E. 1980. *Principles of optics: Electromagnetic theory of propagation, interference and diffraction of light.* Pergamon, Elmsford, NY.

Chipman RA. 1995. Polarimetry. In *Handbook of optics* (ed. Bass M), pp. 22.21–22.37. McGraw-Hill, New York.

Hecht E. 1998. *Optics.* Addison-Wesley, Reading, MA.

Inoué S. 1986. *Video microscopy.* Plenum, New York.

Inoué S. 2002. Polarization microscopy. In *Current protocols in cell biology* (ed. Bonifacino JS, et al.), pp. 4.9.1–4.9.27. Wiley, New York.

Inoué S. 2007. Exploring living cells and molecular dynamics with polarized light microscopy. In *Optical imaging and microscopy* (ed. Török P, Kao FJ), pp. 3–20. Springer-Verlag, Berlin.

Inoué S. 2008. *Collected works of Shinya Inoué.* World Scientific, Singapore.

Inoué S, Oldenbourg R. 1998. Microtubule dynamics in mitotic spindle displayed by polarized light microscopy. *Mol Biol Cell* **9:** 1603–1607.

Katoh K, Hammar K, Smith PJ, Oldenbourg R. 1999. Birefringence imaging directly reveals architectural dynamics of filamentous actin in living growth cones. *Mol Biol Cell* **10:** 197–210.

LaFountain JR Jr, Oldenbourg R. 2004. Maloriented bivalents have metaphase positions at the spindle equator with more kinetochore microtubules to one pole than to the other. *Mol Biol Cell* **15:** 5346–5355.

Oldenbourg R. 1991. Analysis of edge birefringence. *Biophys J* **60:** 629–641.

Oldenbourg R. 1999. Polarized light microscopy of spindles. *Methods Cell Biol* **61:** 175–208.

Oldenbourg R, Shribak M. 2010. Microscopes. In *Handbook of optics* (ed. Bass M), pp. 28.1–28.62. McGraw-Hill, New York.

Oldenbourg R, Salmon ED, Tran PT. 1998. Birefringence of single and bundled microtubules. *Biophys J* **74:** 645–654.

Salmon ED, Tran PT. 1998. High-resolution video-enhanced differential interference contrast (VE-DIC) light microscopy. *Methods Cell Biol* **56:** 153–184.

Sato H, Ellis GW, Inoué S. 1975. Microtubular origin of mitotic spindle form birefringence. Demonstration of the applicability of Wiener's equation. *J Cell Biol* **67:** 501–517.

Shribak M, Inoué S, Oldenbourg R. 2002. Polarization aberrations caused by differential transmission and phase shift in high NA lenses: Theory, measurement and rectification. *Opt Eng* **41:** 943–954.

Shurcliff WA. 1962. *Polarized light, production and use.* Harvard University Press, Cambridge, MA.

Vrabioiu AM, Mitchison TJ. 2006. Structural insights into yeast septin organization from polarized fluorescence microscopy. *Nature* **443:** 466–469.

Vrabioiu AM, Mitchison TJ. 2007. Symmetry of septin hourglass and ring structures. *J Mol Biol* **372:** 37–49.

45 Array Tomography
High-Resolution Three-Dimensional Immunofluorescence

Kristina D. Micheva, Nancy O'Rourke, Brad Busse, and Stephen J Smith

Department of Molecular and Cellular Physiology, Stanford University School of Medicine, Stanford, California 94305

ABSTRACT

Array tomography is a volumetric microscopy method based on physical serial sectioning. Ultrathin sections of a plastic-embedded tissue specimen are cut using an ultramicrotome, bonded in ordered array to a glass coverslip, stained as desired, and then imaged. The resulting two-dimensional image tiles can then be computationally reconstructed into three-dimensional volume images for visualization and quantitative analysis. The minimal thickness of individual sections provides for high-quality, rapid staining and imaging, whereas the array format provides for reliable and convenient section handling, staining, and automated imaging. In addition, the array's physical stability permits the acquisition and registration of images from repeated cycles of staining, imaging, and stain elution and from imaging by multiple modalities (e.g., fluorescence and electron microscopy). Array tomography offers high resolution, depth invariance, and molecular discrimination, which justify the relatively difficult tomography array fabrication procedures. With array tomography it is possible to visualize and quantify previously inaccessible features of tissue structure and molecular architecture. This chapter will describe one simple implementation of fluorescence array tomography and provide protocols for array tomography specimen preparation, image acquisition, and image reconstruction.

INTRODUCTION

Our understanding of tissue function is constrained by incomplete knowledge of tissue structure and molecular architecture. Genetics, physiology, and cell biology make it overwhelmingly clear that all cell and tissue function depends critically on the composition and precise three-dimensional configuration of subcellular organelles and supramolecular complexes, and that such structures may consist of very large numbers of distinct molecular species. Unfortunately, the intricacies of tissue molecular architecture badly outstrip the analytical capability of all presently known tissue imaging methods.

Array tomography is a new high-resolution, three-dimensional microscopy method based on constructing and imaging two-dimensional arrays of ultrathin (70–200 nm thickness) specimen sections on solid substrates. (The word "tomography" derives from the Greek words *tomos*, to cut or section, and *graphein*, to write: The moniker "array tomography" thus simply connotes the "writing" of a volume image from an array of slices.) Array tomography allows immunofluorescence imaging of tissue samples with resolution, quantitative reliability, and antibody multiplexing capacity that is greatly superior to previous tissue immunofluorescence methods (Micheva and Smith 2007). Array tomography was developed with neuroscience applications in mind (e.g., Smith 2007; Stephens et al. 2007; Koffie et al. 2009), and the following description will be illustrated with examples from neuroscience and particularly from studies of synapses and circuits in rodent brain.

ARRAY TOMOGRAPHY PROCEDURES

A sequence of eight steps for a very basic array tomography protocol is illustrated in Figure 1. Array tomography begins with (Step 1) the chemical fixation of the specimen, followed by (Step 2) dissection and embedding in resin (LR White). Resin-embedded specimen blocks are then (Step 3) mounted in an ultramicrotome chuck, trimmed, and prepared for ultrathin sectioning. Block preparation includes careful trimming of the block edges and application of a tacky adhesive to the top and bottom block edges. As shown in the magnified detail of Step 3, this adhesive causes the spontaneous formation of a stable splice between successive serial sections as they are cut by the ultramicrotome's diamond knife blade. The automated cycling of a standard ultramicrotome produces automatically a ribbon up to 45 mm in length, which may consist of more than 100 serial sections held on a water surface. Ribbons are then manually transferred to the surface of a specially coated glass coverslip (Step 4). The resulting array can be stained using antibodies or any other desired reagents (Step 5). After immunostaining, arrays can be imaged using fluorescence microscopy (Step 6). The minimal thickness of array sections promotes very rapid and excellent staining and imaging, whereas the array format promotes convenient and reliable handling of large numbers of serial sections. The individual two-dimensional section images are then computationally stitched and aligned into volumetric image stacks (Step 7) to provide for three-dimensional image visualization and analysis (Fig. 2). The volumetric image stacks are stored electronically for analysis and archiving

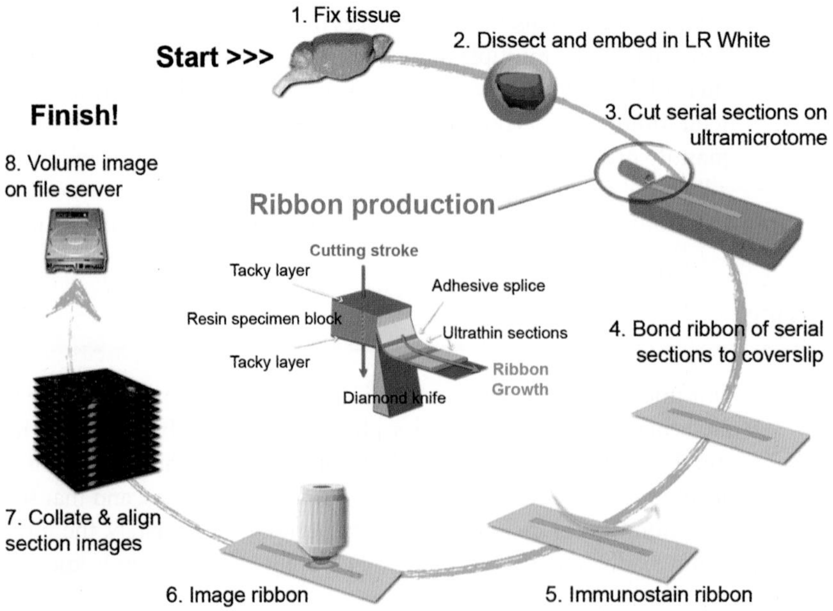

FIGURE 1. The sequence of steps for a basic immunofluorescence array tomography process.

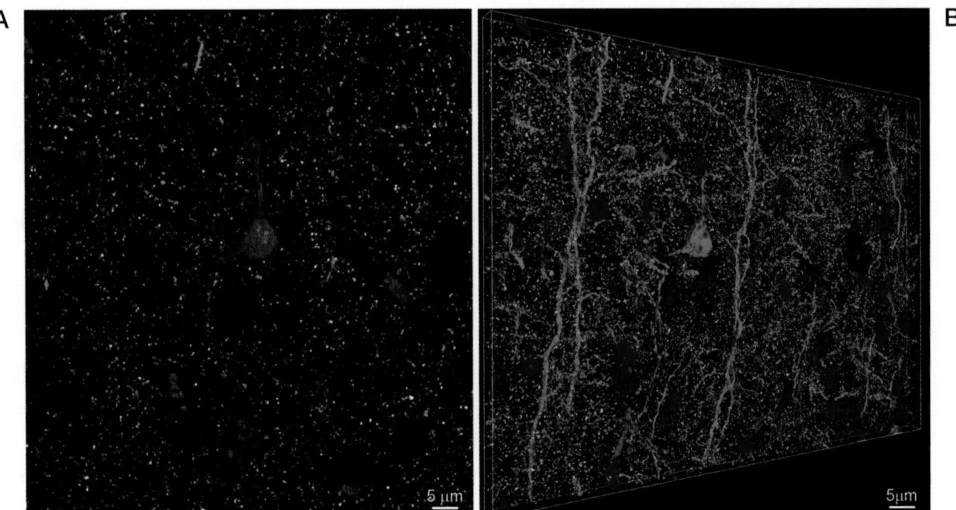

FIGURE 2. Array tomographic images of layer 5 neuropil, barrel cortex of YFP-H Thy-1 transgenic mouse (Feng et al. 2000). Yellow fluorescent protein (YFP) expression in a subset of pyramidal cells (green), Synapsin 1 immunostaining (white), PSD95 (red), DAPI staining of nuclear DNA (blue). (*A*) Four-color fluorescence image of a single, ultrathin section (200 nm). (*B*) Volume rendering of a stack of 30 sections after computational alignment as described in this chapter.

(Step 8). Although array tomography procedures are at present relatively complex and demanding in comparison to many other imaging methods, each of the steps lends itself potentially to automated and highly parallel implementations, and for many applications the advantages outlined below can easily justify this extra effort.

Resolution

The volumetric resolution of fluorescence array tomography compares very favorably with the best optical sectioning microscopy methods. The axial resolution limit for array tomography is simply the physical section thickness (typically 70 nm). For a confocal microscope, the *z*-axis resolution is limited by diffraction to ~700 nm. The confocal's limiting *z*-axis resolution is usually worsened, however, by spherical aberration when a high-numerical-aperture (high-NA) objective is focused more deeply than a few micrometers into any tissue specimen. Array tomography physical sectioning thus improves on ideal confocal optical sectioning by at least an order of magnitude. Spherical aberrations also adversely impact the lateral resolution of confocal microscopes as they are focused into a tissue depth. Array tomography avoids this problem, because the high-NA objective is always used at its design condition (immediate contact between specimen and coverslip), with no chance of focus depth aberration. The degradation of lateral resolution that occurs at focus depths of just a few micrometers can easily exceed a factor of 2 (see http://www.microscopy.fsu.edu/), so a very conservative approximation would imply that array tomography using ordinary high-NA, diffraction-limited optics would improve volumetric resolution (the product of improvements in *x*-, *y*-, and *z*-axes) by a factor of 40 (= 2 x 2 x 10). The improved volumetric resolution realized by array tomography can be very significant. For instance, individual synapses in situ within mammalian cortex generally cannot be resolved optically from their nearest neighbors by confocal microscopy but can be resolved quite reliably by array tomography (Micheva and Smith 2007).

Depth Invariance

The major limitation to quantitative interpretation of whole-mount tissue immunofluorescence images arises from reductions in both immunostaining and imaging efficiencies as focal plane depth increases. Diffusion and binding regimes typically limit the penetration of labeling antibodies to the first few micrometers below the surface of a tissue, even after multiday incubations. Imaging efficiency likewise decreases with depth, as increasing spherical aberration and light scattering reduce signals profoundly with focal plane depths of just a few micrometers. These staining and imaging efficiency gradients make any quantitative comparison of specimen features at different depths with whole-mount (e.g., confocal) volume microscopy difficult and unreliable. Array tomography completely circumvents depth dependence issues, because each specimen volume element is stained identically owing to minimal section thickness, and imaged identically because every section is bonded directly to the coverslip surface.

Multiplexity

Traditional multicolor immunofluorescence techniques have provided compelling evidence for the localization of multiple molecular species at individual subcellular complexes. For example, because there is a very large number and a great diversity of distinct molecules at individual synapses, there is a pressing need for imaging techniques that can simultaneously discriminate many more than the three or four species that can be distinguished by standard multicolor immunofluorescence. Attempts have been made in the past to improve the multiplexity of immunofluorescence microscopy by repeated cycles of staining, imaging, and stain elution, but the results have been disappointing owing to the tendency of antibody elution treatments to destroy samples. In array tomography, specimens are stabilized by the embedding resin matrix and by tight attachment to the coverslip substrate. An example of multiplexed staining with array tomography is shown in Figure 3. We have shown as many as nine cycles of staining, imaging, and elution thus far (Micheva and Smith 2007). With four fluorescence "colors" per cycle, this would mean that 36 or more antigens could be probed in one specimen. We now routinely acquire four colors in each of three cycles for a total of 12 marker channels. Although 12–36 markers may still fall short of the degree of multiplexing needed to fully probe the many and diverse molecules composing a synapse, it is a substantial advance in comparison to traditional multicolor immunofluorescence methods.

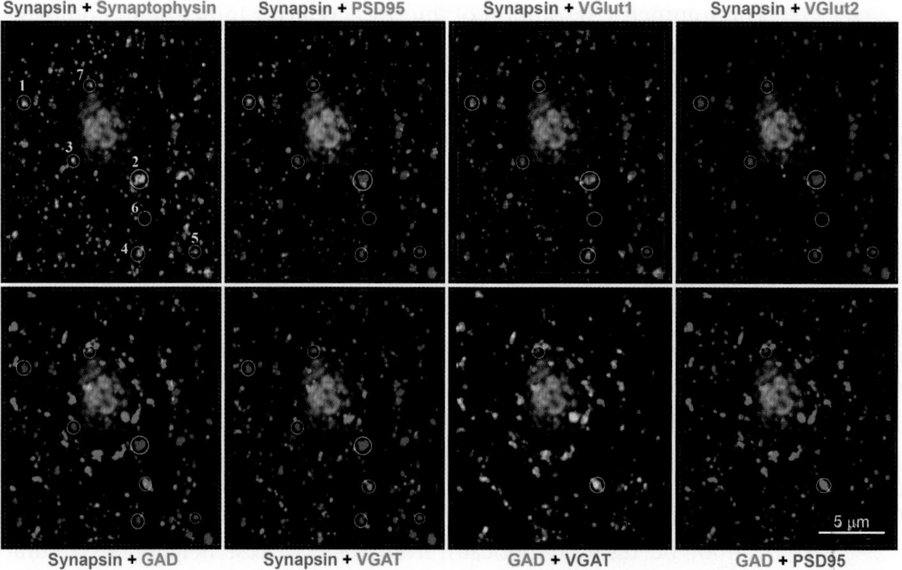

FIGURE 3. Multiplexed staining for seven synaptic proteins in mouse cerebral cortex (layer 2/3, barrel cortex) using five cycles of staining and elution. This volume of 18 x 16 x 1.3 µm was reconstructed from 19 serial sections (70 nm each). Individual synapsin puncta 1, 2, and 3 colocalize with synaptophysin and VGlut1 and are closely apposed to PSD95 and thus appear to be excitatory synapses. Synapsin puncta 4–7 colocalize with synaptophysin, but do not have adjacent PSD95 puncta. Puncta 6 and 7 also colocalize with GAD and VGAT and thus have the characteristics of inhibitory synapses.

Volume Field of View

In principle, array tomography offers unique potential for the acquisition of high-resolution volume images that extend "seamlessly" over very large tissue volumes. The depth invariance of array tomography noted above eliminates any fundamental limit to imaging in depth, whereas the availability of excellent automated image mosaic acquisition, alignment, and stitching algorithms allows tiling over arbitrarily large array areas. Ultimate limits to the continuous arrayable volume will be imposed by difficulties in tissue fixation, processing, and embedding (owing to diffusion limitations) as thicker volumes are encountered, and by mechanical issues of ultramicrotome and diamond knife engineering as block face dimensions increase. Successful array tomography has already been shown for volumes with millimeter minimum dimensions, and it seems likely that volumes with minimum dimensions of several millimeters (e.g., an entire mouse brain) may be manageable eventually.

In practice, the size of seamless array tomography volumes is limited by the requirement that numerous steps in the fabrication, staining, and imaging of arrays be performed through many iterations without failure. At present, the most error-prone steps are those involved in array fabrication, whereas the most time-consuming are those involved in image acquisition. Ongoing engineering of array fabrication materials and processes will advance present limits to the error-free production of large arrays, whereas image acquisition times will be readily reducible by dividing large arrays across multiple substrates and imaging those subarrays on multiple microscopes.

The following protocols describe one simple implementation of immunofluorescence array tomography suitable for any laboratory with standard equipment and some expertise in basic fluorescence microscopy and ultrathin sectioning. In addition, algorithms designed to fully automate the acquisition of array images are described for the benefit of any laboratory having or planning to acquire the appropriate automated fluorescence microscopy hardware and software.

Protocol A

Rodent Brain Tissue Fixation and Embedding

Careful preparation of the tissue is essential for successful array tomography. These steps take time to complete and require some practice to perfect.

MATERIALS

CAUTION: See Appendix 6 for proper handling of materials marked with <!>.
See the end of the chapter for recipes for reagents marked with <R>.

Reagents

Ethanol <!>, 4°C
Fixative <R>
Isoflurane <!> (VWR International)
LR White resin <!> (medium grade, SPI Supplies 2646 or Electron Microscopy Sciences 14381)
Mice
Wash buffer <R>, 4°C

Equipment

Capsule mold (Electron Microscopy Sciences 70160)
Dissection instruments: handling forceps, small scissors, bone rongeur, forceps #5, small spatula, scalpel
Gelatin capsules, size 00 (Electron Microscopy Sciences 70100)
Guillotine
Microscope, dissection
Microwave tissue processor system (PELCO with a ColdSpot set at 12°C; Ted Pella, Inc.) (optional)
Oven (set at 51°–53°C)
Paintbrush, fine
Petri dishes, 35-mm
Scintillation vials, glass, 20-mL

EXPERIMENTAL METHOD

Dissecting and Fixing Tissue

1. Anesthetize the rodent with isoflurane.
2. Remove head using the guillotine.
3. In a hood, using the dissection tools quickly remove the brain and plunge it into a 35-mm Petri dish filled with fixative (room temperature). Remove the tissue region of interest.
4. Transfer tissue to a scintillation vial with fixative solution. Use ~1 mL of fixative per vial, or just enough to cover the tissue; excessive liquid volume will cause overheating in the microwave.

5. Microwave the tissue in the fixative using a cycle of 1 min on/1 min off/1 min on at 100–150 W. After this and each subsequent cycle feel the glass vial to check for overheating. If solutions are getting too warm (>37°C), decrease the amount of liquid added.
6. Microwave using a cycle of 20 sec on/20 sec off/20 sec on at 350–400 W. Repeat three times.
7. Leave the tissue at room temperature for ~1 h.

   If a microwave is unavailable, fix the samples at room temperature for up to 3 h or overnight at 4°C. Tissue can also be fixed by perfusion.
8. Prepare ethanol dilutions: 50%, 70%, 95%, and 100% in ultrapure H_2O. Keep at 4°C.
9. Wash the tissue in wash buffer (4°C) twice for 5 min each.
10. Transfer the tissue to a 100-mm Petri dish, cover with wash buffer, and under a dissecting microscope dissect the tissue into smaller pieces (<1 mm in at least one dimension).
11. Return the samples to scintillation vials and rinse them twice with wash buffer for 15 min each at 4°C.
12. Change to 50% ethanol (4°C) and microwave the samples for 30 sec at 350 W. Use just enough liquid to cover the tissue; excessive liquid volume will cause overheating.

    If a microwave processor is unavailable, Steps 12–20 can be performed for 5 min per step on the bench.
13. Change to 70% ethanol (4°C) and microwave the samples for 30 sec at 350 W.

Processing Samples that Contain Fluorescent Proteins

If processing samples with fluorescent proteins, then complete Steps 14–16. If samples do not contain fluorescent proteins, then skip Steps 14–16, and instead continue with Step 17.

14. Change one more time to 70% ethanol and microwave for 30 sec at 350 W.
15. Change to a mixture of 70% ethanol and LR White (1:3; if it turns cloudy add 1–2 extra drops of LR White) and microwave for 30 sec at 350 W.
16. Go to Step 20.

Processing Samples that *Do Not* Contain Fluorescent Proteins

17. Change to 95% ethanol (4°C) and microwave for 30 sec at 350 W.
18. Change to 100% ethanol (4°C) and microwave for 30 sec at 350 W. Repeat once.
19. Change to 100% ethanol and LR White resin (1:1 mixture, 4°C) and microwave for 30 sec at 350 W.

Embedding Brain Tissue

20. Change to 100% LR White (4°C) for 30 sec at 350 W. Repeat two more times.
21. Change to fresh LR White (4°C) and leave either overnight at 4°C or 3 h at room temperature.
22. Using a fine paintbrush, place the tissue pieces at the bottom of gelatin capsules (paper labels can also be added inside the capsule) and fill to the rim with LR White.

    See *Troubleshooting*.
23. Close the capsules well and put in the capsule mold.

    Gelatin capsules are used because they exclude air that inhibits LR White polymerization. The little bubble of air that will remain at the top of the capsule will not interfere with the polymerization.
24. Put the mold with capsules in the oven set at 51°–53°C. Leave overnight (~18–24 h).

TROUBLESHOOTING

Problem (Step 22): It is difficult to orient the tissue.

Solution: If tissue orientation is important, it should be dissected in a shape that will make it naturally sink in the resin the desired way—for example, for mouse cerebral cortex, a 300-μm coronal slice can be cut and trimmed to a rectangle, ~1 × 2 mm, that includes all of the cortical layers. Alternately, if the tissue is elongated and has to be cut perpendicular to the long axis, the capsules can be positioned on the side, instead of standing up in the mold.

Protocol B

Production of Arrays

Once the tissue has been embedded, the arrays are prepared. This protocol requires familiarity with ultramicrotome sectioning for electron microscopy.

MATERIALS

CAUTION: See Appendix 6 for proper handling of materials marked with <!>.
See the end of the chapter for recipes for reagents marked with <R>.

Reagents

Borax
Contact cement (DAP Weldwood)
Subbing solution <R>
Tissue, fixed and embedded as in Protocol A
Toluidine blue
Xylene <!>

Equipment

Coverslips (for routine staining: VWR International Micro Cover Glasses, 24 x 60-mm, No.1.5, 48393-252; for quantitative comparison between different arrays: Bioscience Tools High Precision Glass Coverslips CSHP-No1.5-24 x 60)
Diamond knife (Cryotrim 45; Diatome) (optional)
Diamond knife (Histo Jumbo; Diatome)
Eyelash tool
Marker
Razor blades
Paintbrush, fine
Slide warmer set at 60°C
Staining rack (Pacific Southwest Lab Equipment, Inc. 37-4470 and 4456)
Syringe
Transfer pipettes, extra fine-tip polyethylene (Fisher Scientific 13-711-31)
Ultramicrotome (e.g., Leica EM UC6)

EXPERIMENTAL METHOD

1. Prepare subbed coverslips. They can be prepared in advance and stored in coverslip boxes until needed.
   i. Put clean coverslips into the staining rack.
   ii. Immerse the rack in the subbing solution and remove bubbles formed at the surface of the coverslips using a transfer pipette.

iii. After 30–60 sec, lift out and drain off excess liquid. Leave the coverslips in a dust-free place until they are dry.

2. Using a razor blade, trim the block around the tissue. A blockface ~2 mm wide and 0.5–1 mm high works best.
3. Using a glass knife or an old diamond knife cut semithin sections until you reach the tissue. Mount a couple of the semithin sections on a glass slide and stain with 1% toluidine blue in 0.5% borax. View the stained sections under a microscope to determine whether they contain the region of interest and decide how to trim the block.
4. Trim the block again, to ensure that the blockface is not too big and the leading and trailing edges of the blockface are parallel. The Cryotrim 45 diamond knife works well for this purpose.
5. Using a paintbrush, apply contact cement diluted with xylene (1:2) to the leading and trailing sides of the block pyramid. Blot the extra glue using a tissue.
6. Insert a subbed coverslip into the knife boat of the Histo Jumbo diamond knife. You may need to push it down and wet it using the eyelash tool. Make sure that the knife angle is set at 0°.
7. Carefully align the block face with the edge of the diamond knife. If the block starts cutting at an angle, the leading and trailing edge of the block face will no longer be parallel.
8. Start cutting ribbons of serial sections (70–200 nm) with the diamond knife. In general, thinner sections stick better to the glass.

   See *Troubleshooting*.
9. When the desired length of the ribbon is achieved, carefully detach it from the edge of the knife by running an eyelash along the outer edge of the knife. Then use the eyelash to gently push the ribbon toward the coverslip, so that the edge of the ribbon touches the coverslip at the interface of the glass and the water. The edge of the ribbon will stick to the glass.
10. Using a syringe, slowly lower the water level in the knife boat until the entire ribbon sticks to the glass.
11. Remove the coverslip from the water and label it on one edge. Also, mark the position of the ribbon by circling it with a marker on the backside of the coverslip.

    This allows you to keep track of the samples and provides a way to tell which side of the coverslip the ribbon is mounted on (without a label, after the ribbon dries, it is not possible to tell which side it is on).
12. Let the ribbon dry at room temperature and place the coverslip on the slide warmer (~60°C) for 30 min. The slides can be stored at room temperature for at least 6 mo.

TROUBLESHOOTING

Problem (Step 8): The ribbons curve.
Solution: Sometimes, even when the leading and trailing edges of the blockface are parallel, the ribbons are curved. This can happen when there is more resin around the tissue on one side of the block than the other. As the section comes in contact with water it expands, however, the resin and tissue expand to different degrees, causing curving of the ribbon. Thus, make sure that the extra resin is trimmed on either side of the block.

Problem (Step 8): The ribbons break.
Solution: Trim the block using a very sharp razor blade or, even better, the Cryotrim diamond knife. Make sure that the blockface is at least twice as wide as it is high. Apply glue again and take care to align the block so the edge of the blockface is parallel to the knife edge.

Protocol C

Immunostaining and Antibody Elution

The tissue arrays are prepared for imaging by binding primary antibodies against specific cellular targets followed by secondary fluorescent antibodies. Alternatively, fluorescent proteins can be used that have been introduced into the tissue before dissection.

MATERIALS

CAUTION: See Appendix 6 for proper handling of materials marked with <!>.
See the end of the chapter for recipes for reagents marked with <R>.

Reagents

Alternative antibody dilution solution with normal goat serum (NGS) <R>
Alternative blocking solution with NGS <R>
Blocking solution with bovine serum albumin (BSA) <R>
Elution solution <R>
Glycine
Mounting medium: SlowFade Gold antifade reagent with DAPI <!> (Invitrogen S36939) or without DAPI (Invitrogen S36937)
Primary antibodies, see Table 1

A detailed list of antibodies that have been tested for array tomography is available from www.smithlab.stanford.edu.

Secondary antibodies: for example, the appropriate species of Alexa Fluor 488, 594, and 647, IgG (H+L), highly cross-adsorbed (Invitrogen)
Tissue sectioned as in Protocol B
Tris buffered saline tablets (Sigma-Aldrich T5030)

Equipment

Microcentrifuge
Microscope slides (precleaned Gold Seal Rite-On micro slides; Fisher Scientific 12-518-103)
PAP pen (ImmEdge Pen, Vector Laboratories H-4000)
Petri dishes, 100-mm diameter

TABLE 1. Primary antibodies used with array tomography

Antibody	Source	Supplier	Dilution
Synapsin I	Rabbit	Millipore AB1543P	1:100
PSD95	Mouse	NeuroMabs 75-028	1:100
VGluT1	Guinea pig	Millipore AB5905	1:1000
GAD	Rabbit	Millipore AB1511	1:300
Gephyrin	Mouse	BD Biosciences 612632	1:100
Tubulin	Rabbit	Abcam ab18251	1:200
Tubulin	Mouse	Sigma-Aldrich T6793	1:200
Neurofilament 200	Rabbit	Sigma-Aldrich N4142	1:100

Slide warmer set at 60°C
Transfer pipettes, extra fine-tip polyethylene (Fisher Scientific 13-711-31)

EXPERIMENTAL METHOD

1. Encircle the ribbon of sectioned tissue with a PAP pen.
2. Place the coverslip into a humidified 100-mm Petri dish and treat the sections with 50 mM glycine in Tris buffer for 5 min.
3. Apply blocking solution with BSA for 5 min.

   If there is a problem with high background staining, see the alternate blocking and staining protocol beginning with Step 21.
4. Dilute the primary antibodies in blocking solution with BSA. Approximately 150 µL of solution will suffice to cover a 30-mm-long ribbon.
5. Centrifuge the antibody solution at 13,000 revolutions per minute (rpm) for 2 min before applying it to the sections.
6. Incubate the sections in primary antibodies either overnight at 4°C or for 2 h at room temperature.

   Primary antibodies are diluted to 10 µg/mL, although the best concentration will need to be determined for each antibody solution.
7. Rinse the sections three to four times with Tris buffer for a total of ~20 min. Wash the sections using a manual "perfusion" method, simultaneously adding Tris buffer on one end and removing if from another with plastic transfer pipettes.
8. Dilute the appropriate secondary antibodies in blocking solution with BSA (1:150 for Alexa secondaries).
9. Centrifuge secondary antibody solution at 13,000 rpm for 2 min.
10. Incubate the sections in secondary antibodies for 30 min at room temperature in the dark.
11. Rinse the sections three to four times with Tris buffer for ~5 min each.
12. Wash the coverslip thoroughly with filtered ultrapure H_2O to remove any dust or debris, leaving some H_2O on the sections so that they do not dry out.
13. Mount the sections on a clean, dust-free microscope slide with SlowFade Gold Antifade containing DAPI.
14. Image the sections as soon as possible after immunostaining, or at least the same day. If you are planning to restain the sections with additional antibodies, elute the antibodies (Steps 15–19) as soon as possible after imaging.

Elute Antibodies Before Restaining

15. Add filtered ultrapure H_2O around the edge of the coverslip to help slide it off the microscope slide.

    Wash the coverslip gently with filtered ultrapure H_2O to rinse off the mounting medium.
16. Apply elution solution for 20 min.
17. Gently rinse the coverslips twice with Tris, allowing them to sit for 10 min with each rinse.
18. Rinse the coverslips with filtered ultrapure H_2O and let them air dry completely.
19. Bake the coverslip on a slide warmer set to 60°C for 30 min.

Staining the Sections Multiple Times

20. Restain using the Steps 2–13 above or store array at room temperature until needed.

    See *Troubleshooting*.

Alternative Staining Method to Reduce Background

21. Proceed through Steps 1 and 2 of the staining protocol above.
22. Incubate the sections for 30 min with alternative blocking solution with NGS.

    If secondary antibodies are made in donkey, use normal donkey serum; if secondary antibodies are made in horse, use normal horse serum, etc. This protocol can only be used if all of the secondary antibodies are made in the same animal.
23. Dilute the primary and secondary antibodies in alternative antibody dilution solution with NGS.
24. Follow the rest of the staining protocol above, using the solutions with NGS.

TROUBLESHOOTING

Problem (Step 20): There is incomplete elution of antibodies.

Solution: To check for incomplete elution, which could interfere with subsequent antibody staining, perform the following control experiment. Stain with the antibody of interest and image a region that you can relocate later. Elute and apply the secondary antibody again. Image the same region as before, using the same exposure time; this will give an estimate of how much primary antibody was left after the elution. Increase the exposure time to determine if longer exposure times reveal the initial pattern of antibody staining. If the first antibody was not eluted sufficiently, try longer elution times. Some antibodies elute poorly (e.g., rabbit synapsin or tubulin) and, if followed by a weaker antibody, may still be detectable after the elution. In such cases, begin the experiment with the weaker antibodies.

Protocol D

Imaging Stained Arrays

Tissue arrays are imaged using a conventional wide-field fluorescence microscopy. Images can be captured manually or, with the appropriate software and hardware, the process can be automated.

MATERIALS

Reagents

Immunostained brain sections prepared as in Protocol C

Equipment

Digital camera (Axiocam HR, Carl Zeiss)
Fluorescence filters sets (all from Semrock) YFP, 2427A; GFP, 3035B; CFP, 2432A; Texas Red, 4040B; DAPI, 1160A; FITC, 3540B; and Cy5, 4040A
Illuminator series 120 (X-Cite)
Objective (Zeiss Immersol 514 F Fluorescence Immersion Oil)
Piezo Automated Stage (Zeiss)
10x Plan-Apochromat 0.45 NA
63x Plan-Apochromat 1.4 NA oil objective
Software (e.g., Zeiss Axiovision with Interactive Measurement Module, Automeasure Plus Module and Array Tomography Toolbar; the toolbar can be downloaded from http://www.stanford.edu/~bbusse/work/downloads.html)
Upright microscope (Zeiss Axio Imager.Z1)

EXPERIMENTAL METHOD

Manual Image Acquisition

1. Focus on your sample using the 10x objective. Find the ribbon by focusing on the DAPI label or another bright label that is not prone to bleaching. Once you have found the right general area of the sample, switch to the 63x objective.

   See *Troubleshooting*.

2. Find the exact area of the sample that you want to image. Choose a landmark that you can use to find the same spot in the next section. A useful landmark should not change dramatically from one section to the next (e.g., a DAPI-stained nucleus or blood vessel). Because the sections are 70–200 nm thick we can often follow the same nucleus through the entire length of a long array. Line up your landmark with a crosshair in the middle of the field.
3. Set the correct exposure for each of your fluorescence channels.
4. Beginning with the first section, collect an image of your area of interest.
5. Manually, move to the same area of the next section. The glue on the edge of each section is autofluorescent, so you can tell when you have moved to the next section. Align your landmark carefully in each section to assure that your image alignment will run smoothly.

   See *Troubleshooting*.

6. Continue to the end of the ribbon, collecting an image from each section. Align your stack of images using Protocol E.

Automated Image Acquisition

Although we have developed our automated tools to work with Zeiss Axiovision software, any microscopy software suite (such as Micro-Manager) controlling an automated stage should be adaptable to this approach. Some steps may be altered or eliminated, depending on your framework and implementation.

7. With the 10x objective, find the ribbon by focusing on the DAPI label or another bright label that is not prone to bleaching.

   See *Troubleshooting*.

8. Acquire a mosaic image of the entire ribbon with the MosaicX Axiovision module, using a bright label that does not vary much between sections, such as DAPI.
9. Find the top left and bottom right corners of the ribbon and use them to define the limits of the mosaic in the Mosaic Setup dialog.
10. Set three to four focus positions along the length of the ribbon and enable focus correction.
11. Collect the mosaic image. Convert the mosaic to a single image with the "Convert Tile Images" dialog, setting the Zoom factor to 1 so that the resulting image is the same size.

    See *Troubleshooting*.

12. Choose a point of interest to be imaged in the ribbon. Place a marker on that point via Measure → Marker. Place another marker at the same spot in the next consecutive section. Create a table of the *x* and *y* coordinates of the markers, "DataTable," via Measure → Create Table, with the "list" option. This allows Axiovision's Visual Basic scripts to read the marker locations.

    See *Troubleshooting*.

13. With the large, stitched image selected, call "PrepImage" and "MarkLoop" from the Array Tomography toolbar.
14. The preceding step will create a file (csv) with a list of the coordinates for the same position in each section, which will be automatically saved in the same folder as the mosaic and with the same name as the stitched image. To load the position list, go to Microscope → Mark and Find, click the "New" icon, and then the "Import Position list" button. In the Mark and Find dialog, switch to the "Positions" tab which will let you review or edit the calculated positions by double-clicking on any position.
15. Collect one field of view at each point via Multidimensional Acquisition with the "position list" checkbox set. We recommend using a bright label that is present throughout the field as the first channel, setting it to autofocus at each position. Review your images at the end to make sure they are all in focus.

    See *Troubleshooting*.

TROUBLESHOOTING

Problem (Steps 1 and 7): Sections cannot be found under the microscope.

Solution: Use DAPI in the mounting medium—it will stain the nuclei brightly and make it easy to find the sections with the 10x objective. Make sure the coverslip has been mounted with the sections on the same side as the mounting medium and that there are no bubbles in the immersion oil.

Problem (Steps 5 and 15): Sections are wrinkled.

Solution: Section wrinkling can occur at several steps in the procedure. First, it can occur during array preparation if the coverslip is put on the slide warmer while the ribbon is still wet. Make sure that the sections are dry before putting them on the slide warmer. It can also occur if the blockface is too big (>1 x 2 mm) or sections are too thick (>200 nm). Second, wrinkles can be caused by improper subbing of the coverslips. The gelatin must be 300 Bloom (measure of stickiness, higher number indicates stickier) and should not be heated above 60°C during solution preparation. Third, sections can wrinkle if the ribbon is stored with the mounting solution for >2 d. Finally, wrinkling can occur after antibody elution, especially with sections 200 nm thick. Make sure that the solutions are applied gently during the elution and the array is completely dry before putting it on the slide warmer.

Problem (Steps 5 and 15): There is no staining or fluorescent signal.

Solution: Use a high-power, high-NA objective—ideally a 63x oil objective. Only immunofluorescence with antibodies against abundant antigens (e.g., tubulin, neurofilament) will be visible with a low-power objective. Also, check if there are two coverslips stuck to each other; this will make it impossible to focus at higher magnification.

Problem (Steps 5 and 15): Punctate staining is seen with a seemingly random distribution.

Solution: Immunostaining with thin array sections (≤200 nm) looks different from staining on thicker cryosections or vibratome sections. Because a very thin layer of tissue is probed, many stains that appear continuous on thicker sections will appear punctate with array tomography. A 3D reconstruction of a short ribbon (10–20 sections) can be helpful for comparison. You may also need to test antibody performance. First, compare the antibody staining pattern to that of different antibodies against the same antigen or a different antigen with a similar distribution. For example, a presynaptic marker should be adjacent to a postsynaptic marker. Other common controls for immunostaining can be used, such as omitting primary antibodies, staining a tissue that does not contain the antigen, etc. Second, specific controls for array tomography include comparison of the antibody staining patterns from adjacent sections or from consecutive stains (i.e., stain → image → elute → stain with the same antibody → image the same region → compare). Not all antibodies that work well for other applications will work for array tomography.

Problem (Steps 5 and 15): There is high background fluorescence.

Solution: Background fluorescence can have many causes. Often, there is high autofluorescence when using the low-power (but not high-power) objectives. If the autofluorescence levels are high with the 63x objective, try the following. First, check whether the immersion oil is designed to be used with fluorescence. Second, labeling marks on the back of the coverslip can dissolve in the immersion oil causing autofluorescence—wipe labels off with ethanol before imaging. Third, use high-quality fluorescence filter sets. Fourth, try a longer fluorescence quenching step (glycine treatment in Protocol C, Step 2), the alternative staining method (Protocol C, Step 21), or introduce an additional quenching step with 1% sodium borohydride in Tris buffer for 5 min.

Problem (Steps 5 and 15): Green fluorescent protein (GFP)/YFP fluorescence is lost.

Solution: First, confirm that the tissue was dehydrated only to 70% ethanol (Protocol A, Step 14). Second, make sure you are using a high-power, high-NA objective. To check for GFP fluorescence use a short array with ultrathin sections (<200 nm). Let it sit for 5–10 min or more with Tris-glycine (50 mM glycine in Tris), mount over a glass slide and look with the 63x objective. GFP can bleach very fast, so work quickly to find the region with GFP fluorescence. For acquiring images, select the region of interest with another stain (e.g., Alexa 594) and focus. Do not use the DAPI stain for this purpose, because it can cause DAPI to bleed into the GFP channel. In cases of weak GFP fluorescence, GFP antibodies may help identify GFP-positive cell bodies and large processes, but are generally not useful for thinner processes. GFP antibodies for array tomography include Roche 11814460001 (mouse), MBL 70 (rabbit), Invitrogen A11122 (rabbit), NeuroMabs 75-131 (mouse), GeneTex GTX13970 (chicken). All of these antibodies should be used at 1:100 dilution.

Problem (Step 11): The "Convert Tile Images" step keeps downsampling the stitched image.
Solution: In the Tools → Options → Acquisition menu, change the Mx. MosaicX image size to the maximum allowed: 1000000000 pixel.

Problem (Step 12): The microscopy software is not designed for array tomography.
Solution: We have developed an algorithm that automates position finding in the arrays by using simple extrapolation to estimate the neighborhood of an unknown point and then refining the estimate with an autocorrelation search. Given two known points Pn and Pn−1, we find the next point Pn+1 such that Pn+1 = Pn + (Pn − [Pn−1]) (Fig. 4). This does not take into account ribbon curvature or changes in section width, but gives a rough approximation of the unknown point's locale. Pn+1 becomes the center of an autocorrelation search to find the point's true position. The size of the search varies with the width of the sections; larger sections will have larger warping and curvature effects, and any miscalculation in the estimate of Pn+1 will be magnified.

To conduct the search, the algorithm compares the area centered at Pn+1 with a Kalman-filtered image of recently processed points. Although our fiducial labels (DAPI and tubulin immunostaining) have minor variations from section to section, it does not disrupt the accuracy of the correlation search. To make the Kalman-filtered image at each iteration, use the area around the current Pn, *newSample*, to update the image using the following pseudocode: image = 0.3 x image + 0.7 x *newSample*. The purpose of using the Kalman filter, when *newSample* alone would do, is to add a measure of robustness to the algorithm. If the ribbon is damaged or has aberrant staining on a single section, using *newSample* alone may result in the algorithm going off course. With a running average of previous iterations to compare with, a defect in a single section has a good chance of being ignored. This process continues until one end of the ribbon is reached, then starts in the other direction.

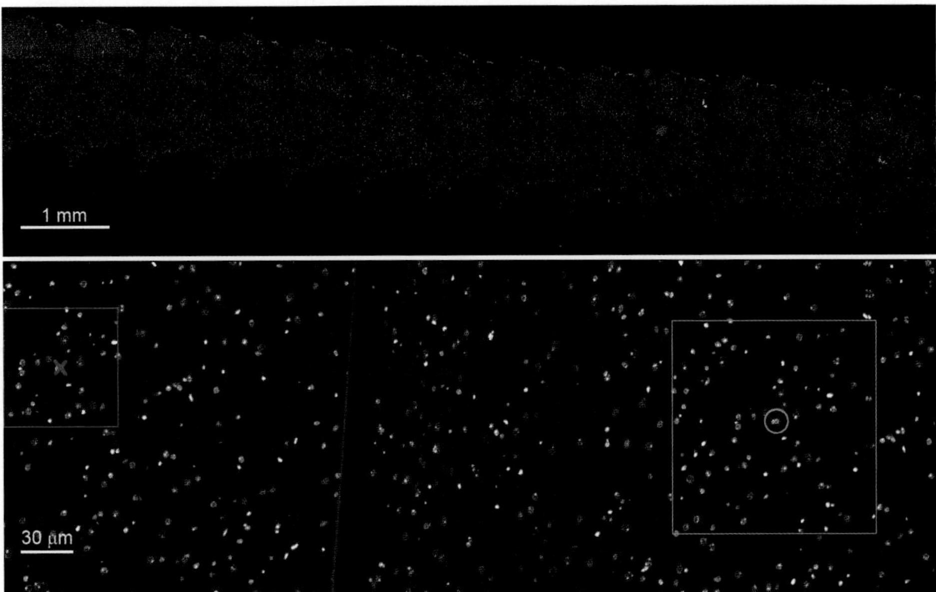

FIGURE 4. (*Top*) A fragment of an array tomography ribbon stained with DAPI. (*Bottom*) A closer view of two sections in the ribbon showing a single iteration of the position-finding algorithm. An established field (red x) is used to maintain a reference patch (red square) for a correlation-based search (green square) to find the next point (green circle).

We developed an implementation of this algorithm in Visual Basic script for Zeiss Axiovision, available from http://www.stanford.edu/~bbusse/work/downloads.html, and would welcome any ports to other microscopy software.

Problem (Step 15): Autofocus does not work using Axiovision.
Solution: The autofocus does not work every time. Typically, ~5% of the images collected with autofocus may be out of focus. In that case, you can move to the positions on the ribbon with bad focus, focus by hand, and collect individual images. Replace the out-of-focus images with the newly focused ones in the stack before to alignment. If 10% or more of the images are out of focus, you can try using the autofocus with a different channel. Pick a channel with antibody staining that is bright, and present throughout the field of view. Using a channel with dim or sparse immunostaining will not work well.

Problem (Step 15): Autofocus is grayed out.
Solution: In the Tools → Options → Acquisition menu, check the box marked either "Use calibration-free Autofocus" or "Enable new Autofocus."

Protocol E

Semiautomated Image Alignment

Successful array tomography requires that the captured images be properly stacked and aligned. Software to achieve these ends is freely available.

MATERIALS

Software

Fiji can be obtained at http://pacific.mpi-cbg.de/wiki/index.php/Main_Page
MultiStackReg is available at http://www.stanford.edu/~bbusse/work/downloads.html

EXPERIMENTAL METHOD

1. Load your images into Fiji. If using Axiovision, Fiji's Bio-Formats Importer plugin can read .zvi files directly.
2. Pick a channel that is relatively invariant from one section to the next (e.g., DAPI or tubulin), and select a slice near the middle of the ribbon.
3. Align the sections of that channel using "affine" in MultiStackReg (Fiji), but do not save over the misaligned stack. Save the resulting transformation matrix. This is the intrasession matrix. See *Troubleshooting*.
4. Using MultiStackReg, apply that matrix to the other channels of the same imaging session.
5. For each subsequent imaging session, choose the same channel. Align the new (misaligned) channel to the old (misaligned) channel, saving the matrix. This is the intersession matrix.
6. For each channel in that imaging session, first apply the intersession matrix from Step 5 and then the intrasession matrix from Step 3.
7. Repeat until all imaging sessions have been registered.

TROUBLESHOOTING

Problem (Step 3): The alignment steps are not working properly.
Solution: Detailed instructions with graphical illustration, compiled by Andrew Olson, are available at http://nisms.stanford.edu/UsingOurServices/Training.html. If an "affine" transformation does not align the images well, try either the "rigid body" then "affine" or try "rigid body" alone. For each registration step, save the transformation matrix and apply it to the other channels in sequence.

MultiStackReg is an extension of the StackReg ImageJ plugin, which is dependent on TurboReg (Thévenaz et al. 1998). TurboReg aligns a single pair of images using a pyramid registration scheme. StackReg aligns an entire stack by calling TurboReg on each pair of consecutive slices in the stack, propagating the alignment to later slices. The two principle changes added by

MultiStackReg are the ability (1) to load and save transformation matrices and (2) to align one stack to another by registering each pair of corresponding sections independently. MultiStackReg can process TurboReg alignment files in the same manner as the files it generates for itself, so if your alignment is failing owing to a single section, it is possible to manually align that section in TurboReg, apply that transform to a copy of the stack, and splice the two together.

CONCLUSION AND FUTURE DIRECTIONS

One important application of array tomography in the field of neuroscience is the analysis of synapse populations. With this method it is possible to resolve individual synapses in situ within brain tissue specimens. Because 10 or more antibodies can be used on an individual sample, the molecular signature of each synapse can be defined with unprecedented detail. The throughput of the technique is inherently high, approaching the imaging of one million synapses per hour. Compared with 3D reconstruction at the electron microscopic level, array tomography can image much larger volumes and provide information about the presence of a much larger number of molecules, but cannot presently provide the fine ultrastructure of electron microscopy. On the other hand, the amount of effort involved in array tomography may not be warranted for all studies. If it is not considered critical to resolve individual synapses, immunostaining of vibratome sections or cryosections and confocal microscopy imaging may be sufficient.

Currently, we are focused on developing array tomography in three directions. First, we are refining current staining and imaging approaches to image larger and larger tissue volumes with more antibodies. Second, we are combining light and electron microscopic imaging to visualize both immunofluorescence and ultrastructure on the same tissue sections. Finally, we are applying advanced computational methods for data analysis, in particular with the goal to both count and classify millions of synapses on a routine basis.

RECIPES

CAUTION: See Appendix 6 for proper handling of materials marked with <!>.
Recipes for reagents marked with <R> are included in this list.

Alternative Antibody Dilution Solution with NGS (1 mL)

Reagent	Quantity	Final concentration
Tween (1%) (make the stock solution using Tween-20 [Electron Microscopy Sciences 25564])	100 µL	0.1%
NGS (Invitrogen PCN5000)	30 µL	3%
Tris buffer	870 µL	

Prepare on the same day it is used. NGS can be kept frozen in aliquots for several months.

Alternative Blocking Solution with NGS (1 mL)

Reagent	Quantity	Final concentration
Tween (1%) (make the stock solution using Tween-20 [Electron Microscopy Sciences 25564])	100 µL	0.1%
NGS (Invitrogen PCN5000)	100 µL	10%
Tris buffer	800 µL	

Prepare on the same day it is used. NGS can be kept frozen in aliquots for several months.

Blocking Solution with BSA (1 mL)

Reagent	Quantity	Final concentration
Tween (1%) (make the stock solution using Tween-20 [Electron Microscopy Sciences 25564])	50 µL	0.05%
BSA (10%) (AURION BSA C [acetylated BSA], Electron Microscopy Sciences 25557)	10 µL	0.1%
Tris buffer	940 µL	

Prepare the same day. The 1% Tween stock (10 µL Tween in 1 mL of H_2O) and the 10% BSA stock can be kept at 4°C for several months.

Elution Solution (10 mL)

Reagent	Quantity	Final concentration
NaOH <!>, 10 N	200 µL	0.2 N
SDS <!> (20%)	10 µL	0.02%
Distilled H_2O	10 mL	

Can be prepared in advance and stored at room temperature for several months.

Fixative (4 mL)

Reagent	Quantity	Final concentration
Paraformaldehyde <!> (8%, EM grade; Electron Microscopy Sciences 157-8)	2 mL	4%
PBS, 0.02 M (use PBS powder, pH 7.4 [Sigma-Aldrich P3813])	2 mL	0.01 M
Sucrose	0.1 gm	2.5%

Prepare the same day as it will be used.

Subbing Solution (300 mL)

Reagent	Quantity	Final concentration
Gelatin from porcine skin, 300 Bloom (Sigma-Aldrich G1890)	1.5 g	0.5%
Chromium potassium sulfate (Sigma-Aldrich 243361)	0.15 g	0.05%
Distilled H_2O	300 mL	

Prepare the same day. Dissolve the gelatin in 290 mL of distilled H_2O by heating to <60°C. Dissolve 0.15 gm of chromium potassium sulfate in 10 mL of H_2O. When the gelatin solution cools down to ~37°C, combine the two solutions, filter, and pour into the staining tank. Use fresh.

Wash Buffer (50 mL)

Reagent	Quantity	Final concentration
Glycine	187.5 mg	50 mM
Sucrose	1.75 g	3.5%
PBS, 0.02 M	25 mL	0.01 M
Distilled H_2O	25 mL	

Can be prepared in advance and stored at 4°C for up to 1 mo; discard if it appears cloudy.

ACKNOWLEDGMENTS

We thank JoAnn Buchanan and Nafisa Ghori for their help in refining the methods. This work was supported by grants from McKnight Endowment Fund for the Neurosciences, the National Institutes of Health (NS 063210), The Gatsby Charitable Foundation, and the Howard Hughes Medical Institute.

REFERENCES

Feng G, Mellor RH, Bernstein M, Keller-Peck C, Nguyen QT, Wallace M, Nerbonne JM, Lichtman JW, Sanes JR. 2000. Imaging neuronal subsets in transgenic mice expressing multiple spectral variants of GFP. *Neuron* **28:** 41–51.

Koffie RM, Meyer-Luehmann M, Hashimoto T, Adams KW, Mielke ML, Garcia-Alloza M, Micheva KD, Smith SJ, Kim ML, Lee VM, et al. 2009. Oligomeric amyloid β associates with postsynaptic densities and correlates with excitatory synapse loss near senile plaques. *Proc Natl Acad Sci* **106:** 4012–4017.

Micheva KD, Smith SJ. 2007. Array tomography: A new tool for imaging the molecular architecture and ultrastructure of neural circuits. *Neuron* **55:** 25–36.

Smith SJ. 2007 Circuit reconstruction tools today. *Curr Opin Neurobiol* **17:** 601–608.

Stevens B, Allen NJ, Vazquez LE, Howell GR, Christopherson KS, Nouri N, Micheva KD, Mehalow A, Huberman AD, Stafford B, et al. 2007. The classical complement cascade mediates CNS synapse elimination. *Cell* **131:** 1164–1178.

Thévenaz P, Ruttimann UE, Unser M. 1998. A pyramid approach to subpixel registration based on intensity. *IEEE Trans Image Process* **7:** 27–41.

46 Monitoring Membrane Potential with Second-Harmonic Generation

Stacy A. Wilson,[1] Andrew Millard,[1] Aaron Lewis,[2] and Leslie M. Loew[1]

[1]*Richard D. Berlin Center for Cell Analysis and Modeling, University of Connecticut Health Center, Farmington, Connecticut 06030-1507;* [2]*Division of Applied Physics, Hebrew University of Jerusalem, Jerusalem 91904, Israel*

ABSTRACT

This chapter describes the nonlinear optical phenomenon known as second-harmonic generation (SHG) and discusses its special attributes for imaging membrane-potential changes in single cells and multicellular preparations. Styryl and naphthylstyryl dyes, also known as hemicyanines, are a class of electrochromic membrane-staining probes that have been used to monitor membrane potential by fluorescence; they also produce SHG images of cell membranes with SHG intensities that are sensitive to voltage. The voltage sensitivity of this technique can be significantly greater than the voltage sensitivity of fluorescence for many of the dyes that have been tested. Furthermore, the ability of SHG to image only the plasma membrane and not intracellular organelles is especially useful for experiments in which dye is applied internally to an individual cell. Thus, second-harmonic imaging of membrane potential has great promise for spatiotemporal mapping of electrical activity in neurons.

INTRODUCTION

The work of Lawrence Cohen and colleagues in the mid-1970s led to the establishment of optical methods as a way to measure the electrical activity of cells in situations in which traditional microelectrode methods are not possible or are too limiting (Cohen et al. 1974). The authors' laboratory soon joined the effort to develop potentiometric dyes by applying rational design methods based on molecular orbital calculations of the dye chromophores and characterization of their binding and orientations in membranes (Loew et al. 1978, 1979a). Several important general-purpose hemicyanine dyes have emerged from this effort, including di-5-ASP (Loew et al. 1979b), di-4-ANEPPS (Fluhler et al. 1985; Loew et al. 1992), and di-8-ANEPPS (Bedlack et al. 1992; Loew 1994). The fluorescence signals from the naphthylstyryl ANEP dyes have been particularly effective in studies aimed at mapping the activity of excitable cells in complex preparations (Wu et al. 1998; Obaid et al. 1999; Antic et al. 2000; Zochowski et al. 2000; Loew 2001, 2010a). An important and advantageous attribute of the hemicyanine dyes is their strong fluorescence when bound to membranes and the almost insignificant fluorescence from dye in the bathing aqueous medium. This not only increases the sensitivity of membrane-potential measurements, but also has led to a completely different application for these dyes as indicators of synaptic vesicle release and recycling; the "FM" series of dyes are styryl dyes that have been optimized for such measurements (Betz and Bewick 1992; Betz et al. 1992).

In the mid-1980s, investigators realized that the large charge redistribution that occurs upon absorption of a photon by the ANEP chromophores, which makes these dyes electrochromic, should also make them promising materials for second-harmonic generation (SHG) (Huang et al. 1988). SHG is a nonlinear optical process that can take place at the focus of an ultrafast near-infrared laser. As in the case of two-photon excitation of fluorescence (2PF), the probability of SHG is proportional to the square of the incident light intensity, so that three-dimensional (3D) optical sectioning is a natural benefit of scanning microscopy with either of these nonlinear optical modalities. The physics behind these phenomena are, however, quite distinct. Whereas 2PF involves the near-simultaneous absorption of two photons to excite a fluorophore, followed by relaxation and noncoherent emission, SHG is a near-instantaneous process in which two photons are converted into a single photon of precisely twice the energy. The SHG light propagates coherently in the forward direction. The intensity of the SHG signal depends on the molecular hyperpolarizability of the array of molecules that experience the intense laser field. The molecular hyperpolarizability, in turn, can be resonance-enhanced when the incident laser wavelength is close to twice the wavelength of an absorption band of the molecules; this resonance enhancement also depends on the sensitivity of this absorption band to electric fields (i.e., their degree of electrochromism). The full theory of SHG and its application to biological systems is beyond the scope of this chapter, but has been thoroughly reviewed elsewhere (Moreaux et al. 2000; Loew et al. 2002; Campagnola and Loew 2003; Millard et al. 2003a; Pons et al. 2003; Jiang et al. 2007; Loew 2010b). One key condition for SHG, in addition to a large molecular hyperpolarizability, is that the molecules producing an SHG signal be organized in a noncentrosymmetric array—a condition that is nicely met when the dye molecules stain one side of a cell membrane.

The relationship between SHG and electrochromism also prompted investigation of whether SHG from membranes stained with ANEP dyes could be sensitive to membrane potential (Bouevitch et al. 1993; Ben-Oren et al. 1996; Campagnola et al. 1999). Many of these earlier studies provided indications that the SHG sensitivity to membrane potential could be much larger than the typical best sensitivity of ~10% fluorescence change per 100 mV. However, it has been only recently that SHG has been measured in stained cells that are simultaneously voltage-clamped via whole-cell patch clamp (Millard et al. 2003b, 2004, 2005a,b; Dombeck et al. 2004, 2005; Araya et al. 2006; Nuriya et al. 2006; Teisseyre et al. 2007). These experiments have allowed for the precise characterization of the voltage sensitivity of SHG and identification of the optimal wavelength for the incident laser fundamental light. The details of these measurements are presented in this chapter along with a discussion of the prospects of this new technology for imaging membrane potential changes in neuronal preparations.

Protocol

Second-Harmonic Imaging of Membrane Potential

As a model cellular system for membrane electrophysiology, we use undifferentiated N1E-115 mouse neuroblastoma cells. Voltage-sensitive dyes are prepared using procedures adapted from Hassner et al. (1984) and Wuskell et al. Dyes may be purchased from Invitrogen, and small samples of non-commercially available dyes may be obtained from the authors' laboratory upon request. The 1PF properties of the dyes are characterized using a hemispherical lipid bilayer (HLB) apparatus (Loew et al. 1979a; Loew and Simpson 1981).

IMAGING SETUP

We have adapted a FluoView scanning confocal imaging system (Olympus) with an inverted Axiovert 100TV microscope (Carl Zeiss) for nonlinear optical imaging, as shown in Figure 1. A Mira 900 Ti:sapphire (Ti:S) ultrafast laser (Coherent) is pumped by a 10-W Verdi doubled solid-state laser (Coherent) and purged with nitrogen gas to make wavelengths >930 nm accessible. A second excitation source, a FemtoPower fiber laser (Fianium) operating at a fixed wavelength of 1064 nm, provides a robust, turnkey source of long-wavelength excitation. ER4 protected gold coating mirrors (Newport) are used to direct the beam through the various optical components, through the FluoView scan head, and finally into the microscope. The Ti:S beam is first passed through a Faraday isolator (Electro-Optics Technology, Inc.) to prevent back-propagating reflections from knocking

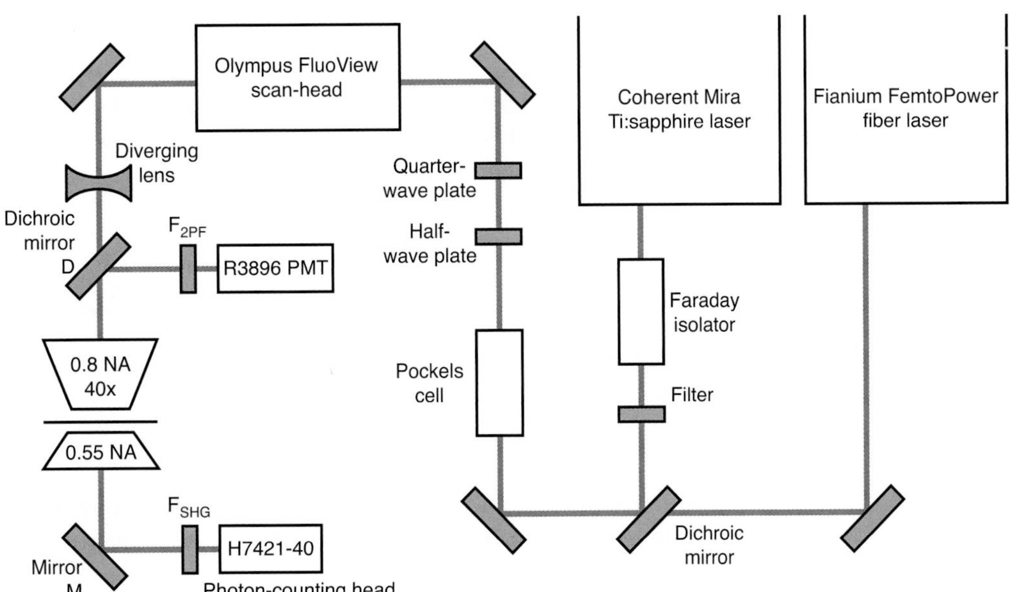

FIGURE 1. Schematic of our nonlinear imaging system. Excitation light leaves the scan head and is directed through a diverging lens into the microscope. The light passes through the dichroic (D) to enter the objective. 2PF is collected back through the objective, and the dichroic reflects it through filters (F_{2PF}) into the photomultiplier tube. The condenser collects all transmitted light, directing it to the mirror (M), which selectively reflects SHG through filters (F_{SHG}) into the photon-counting head.

TABLE 1. Filters used for SHG detection with various excitation wavelengths

Wavelengths	Filters, F_{SHG}
830–870 nm	Chroma Technology D425/40 and CVI Laser SPF-450, passing 425 ± 10 nm
890–910 nm	Chroma Technology D460/50 and Oriel Instruments 57530, passing 450 ± 10 nm
930–970 nm	Chroma Technology D470/40 and 475 ± 50 nm band pass, passing 470 ± 15 nm
1064 nm	Semrock FF01-530/11, passing 530 ± 11 nm

the Mira out of mode-lock. A 700-nm (CVI Laser) or 880-nm (Chroma) long-pass filter then removes residual pump light from the Ti:sapphire or the fiber laser, respectively. We use a half-wave plate and a Glan Thompson polarizer to modulate the beam intensity and then half- and quarter-wave plates (CVI Laser) to produce circularly polarized light at the sample; this effectively eliminates distorting contrast patterns produced by direct excitation with the linearly polarized laser light. The scanning beam enters the microscope from below, passing through a plano-concave BK7 lens (f = –150 mm) (Newport) to colocalize the bright-field focus and the focus for nonlinear excitation.

For nonlinear (simultaneous SHG and 2PF) imaging mode, an infinity-corrected 40x 0.8 NA water-immersion IR-Achroplan objective (Zeiss) focuses the ultrafast beam into the sample. 2PF is collected by the water-immersion objective, reflected by a dichroic mirror, and passed through filters to select emission wavelengths of interest. For imaging of bluer styryl dyes, a 770-nm long-pass dichroic mirror (Chroma; D in Fig. 1) is used with a 750-nm short-pass filter (CVI Laser) in combination with either a 540-nm band-pass filter (Chroma) or a 675-nm band-pass filter (Chroma). This setup covers wavelength ranges to either side of the emission spectrum maximum wavelength (i.e., ~615 nm for di-4-ANEPPS); by restricting the emission range to one side of the spectrum, the voltage sensitivity can reinforce the sensitivity associated with the corresponding wing of the excitation spectrum (Fluhler et al. 1985). For more recently synthesized longer-wavelength dyes, an 880-nm dichroic is used with an 850-nm short-pass filter and either a 640-nm band pass or a 750-nm long-pass filter, allowing access to dyes with redder wavelength maxima. The filtered 2PF is directly detected by a photomultiplier tube (Hamamatsu R3896) that is connected to one of the FluoView channel inputs via a PMT amplifier board (Olympus). Second-harmonic light is produced in the forward direction and is collected using a 0.55-NA condenser (Zeiss), then reflected from a broadband dielectric mirror (Thorlabs; M in Fig. 1) to focus through filters that are appropriate for the second-harmonic wavelength (see Table 1) onto a GaAsP photon-counting head (Hamamatsu H7421-40) connected to the other FluoView channel input.

MATERIALS

CAUTION: See Appendix 6 for proper handling of materials marked with <!>.
See the end of the chapter for recipes for reagents marked with <R>.

Reagents

Carboxyethyl-γ-cyclodextrin, 20 mM (Cyclodextrin Technologies Development)
Dulbecco's modified Eagle's medium (DMEM) with 10% fetal bovine serum and 1% antibiotic-antimycotic (Invitrogen)
Earle's balanced salt solution (EBSS, Sigma)
Ethanol <!>, 100%
External patch clamp buffer <R>
Internal patch pipette buffer
N1E-115 mouse neuroblastoma cells
Styryl and/or naphthylstyryl dyes <!> (Invitrogen)

Equipment

Argon gas <!>
Borosilicate glass capillary tubes, 1.5-mm outer diameter, 0.86-mm inner diameter (Sutter Instruments)
Clampex software (Axon Instruments)
Culture dishes, 50-mm, glass bottom (Matek Corporation)
Electrophsiology rig, including a patch-clamp amplifier (Axon Instruments)
Imaging system (see Imaging Setup)
Incubator for cell culture set at 37°C and 5% CO_2
Micropipette puller (P2000, Sutter Instruments)
Rotary vacuum evaporator (Savant Instruments)

EXPERIMENTAL METHOD

Growing Cells

1. In preparation for patching and subsequent imaging, grow N1E-115 mouse neuroblastoma cells in DMEM with 10% fetal bovine serum and 1% antibiotic-antimycotic, maintaining them at 37°C with 5% CO_2 in 50-mm glass-bottomed culture dishes.

Preparation of Dyes

Preparation of Hydrophobic Dyes

Aqueous solutions of the more hydrophobic dyes are complexed with cyclodextrin to facilitate and accelerate staining (Bullen and Loew 2001). The complexed form of the dye is used because it provides a more efficient method of staining than methods that employ surfactants (Lojewska and Loew 1987) such as Pluronic F-127 (Molecular Probes).

2. Prepare a 4 mM stock solution of the hydrophobic dye in 100% ethanol.
3. Dilute the dye by a factor of 20 with 20 mM carboxyethy-γ-cyclodextrin in dH_2O.
4. Dry small aliquots (0.5 mL) of this mixture in a rotary vacuum evaporator and store them at 4°C.
5. Just prior to use, reconstitute an aliquot of the dye with EBSS to make 100 μM aqueous dye solution. This solution can be applied directly to the cells (see Step 12).

Preparation of Dyes That Are Soluble in Water

6. Prepare a 4 mM stock solution of the dye in 100% ethanol. Store it at 4°C.
7. Just before use, take small aliquots of the dye stock solution and dry them under argon gas until all of the ethanol has evaporated.
8. Dissolve the dried dye in EBSS to make a 100 μM dye solution. This solution can be applied directly to the cells (see Step 12).

   Long-term storage of dilute aqueous dye solutions is not recommended.

Electrophysiology Preparation

9. Prepare patch pipettes on a micropipette puller from 1.5-mm outer diameter, 0.86-mm inner diameter borosilicate glass. The pipettes should have a resistance of ~5 MΩ when filled with internal buffer.
10. Set up the electrophysiology rig. The authors control a patch-clamp amplifier by a computer

running Clampex. The software provides diagnostics such as seal resistance during patching and synchronization of voltage-clamping operations with nonlinear imaging.

11. Remove the DMEM from the cells, gently rinse with 2 mL external buffer, and then add 3 mL of the buffer.

Staining, Patching, and Imaging Cells

12. Stain the cells by adding aqueous dye solution (from Step 5 or 8) to the glass-bottom dish and swirling it immediately to promote prompt mixing and to avoid prolonged exposure to locally high dye concentrations. An overall dye concentration of 3–4 μM is typically used for staining.
13. Using the air objective and bright-field imaging, select a cell for patching to a gigaohm seal (Penner 1995), and then switch to the water immersion objective. After forming the whole cell patch, carefully switch the microscope configuration from bright-field to nonlinear imaging.

    If this presents difficulty, the configuration of the microscope can be switched after gigaohm seal formation but before breaking in to establish the more delicate whole cell patch.
14. Perform imaging. The authors use FluoView v.4.3, Olympus's imaging software designed for use with the FluoView scanning confocal imaging system. The vertical sync trigger from FluoView (pin 5 of the FluoView output cable) is used to trigger Clampex as appropriate. Combined SHG and 2PF images are typically recorded in a time series and analyzed within FluoView or using ImageJ software (developed by Wayne Rasband; http://rsb.info.nih.gov/ij/).

EXAMPLE OF APPLICATION

To determine the voltage sensitivity of SHG from styryl dyes, we typically select single cells having no physical contact with other cells. Once a cell is patched and stained, we take a series of 27 images with the clamp voltage switched back and forth between 0 mV and a test voltage *V* after every three image frames. A total intensity value for each image is obtained by summing the intensity values of the pixels associated with the cell membrane; that is, a total intensity value $I(t_i)$ is obtained for frame *i*. Fifteen total intensity values, $I_0(t_i)$, correspond to the 0-mV reference voltage, and 12 values, $I_V(t_i)$, correspond to the test voltage *V*. In addition to scatter, the $I_0(t_i)$ and $I_V(t_i)$ drift a little over time, because of a small but continuous incorporation of dye into the membrane during the course of the experiment. Thus, a normalization process that successfully corrects for the drift has been developed as follows: We fit a second-order polynomial $A_0(t_i)$ to the 15 $I_0(t_i)$ values, and then the normalized total intensity values $N(t_i)$ are just $I(t_i)/A_0(t_i)$ for each *i*. The average relative signal change *C* for the test voltage is $\langle N_V(t_i)\rangle - \langle N_0(t_i)\rangle$. Each image series yields one value for *C*, which is considered a single experimental measurement, although each measurement obtained in this way actually represents data collected over an entire cell membrane and averaged over 15 0-mV images and 12 test voltage images.

To illustrate the voltage sensitivity visually, an SHG image series can be processed to produce a montage such as that shown in Figure 2. Although on-the-fly averaging is not generally used while obtaining image series, each image in Figure 2 is a Kalman average of three acquisitions to yield a clearer visual presentation. Modulation of SHG intensity by membrane potential is apparent. The 81 acquisitions of the original image series, each corresponding to an ~2.5-sec acquisition time, took place without any major degradation in SHG intensity and with a stable response to the step changes in membrane potential. Such stability appears to be dependent on excitation wavelength, with greater degradation of both SHG and 2PF intensity at wavelengths <830 nm (e.g., attempts to image at 780 nm typically result in rapid, readily visible damage [Campagnola et al. 1999]) along with gradual loss of cell viability with increased light exposure.

To explore the kinetics of observed SHG changes, measurements are obtained via a separate set of electrophysiology protocols. Cells are selected using the same criteria as those chosen for the voltage sensitivity measurements detailed above, with additional emphasis on roundness. After the cell is stained and the voltage clamped, 22 images are acquired. The clamp voltage is left unchanged for even-numbered images, whereas odd-numbered images have a step waveform triggered halfway down the page scan, as shown in Figure 3. This results in a series of images with frames alternating between switched and unswitched voltage clamp. The intensities of the switched images were normalized by an average of the nearest unswitched images (e.g., frames 2 and 4 are averaged, and then that average is used to normalize frame 3; frames 4 and 6 are averaged to normalize frame 5; etc.). Such normalization corrects for slow drifts in laser power and staining level during the acquisition time. The 10 resulting normalized images are then analyzed to obtain an average intensity value for each row containing nonzero pixels.

Despite the large SHG sensitivities determined for a series of ANEP dyes (Millard et al. 2003b, 2004), several of the dyes exhibit slow kinetics with time constants of tens of milliseconds. The slower speed of the second-harmonic generation response for these dyes precludes their use as a practical tool for measuring electrical activity in neuronal systems, which will require millisecond temporal resolution. Such noninstantaneous responses suggest that the mechanism by which SHG is sensitive to membrane potential is not purely electrochromic and may involve a reorientation or change in molecular conformation. More recently developed dyes have fast SHG voltage responses (Teisseyre et al. 2007). Figure 4 illustrates kinetics for the SHG responses of two hemicyanine dyes. In Figure 4A, di-4-ANEPPS exhibits a slow kinetic response to a –75-mV change in voltage that can be fit to a single exponential with a time constant of 70.6 msec. A newer dye, PY1261, shown in Figure 4B, displays a response to a –50-mV voltage step that is instantaneous on the 5-msec timescale of our measurements.

We postulate that SHG probes exhibiting fast or slow kinetics likely share the same underlying non-electro-optic mechanism. Evidence for this can be seen by comparing the responses for several hemicyanine dyes developed in our laboratory containing aminothiophene moieties. PY1280, PY1261, and PY1278 all contain the same chromophore and differ only by head group (trimethylammonium, tri-

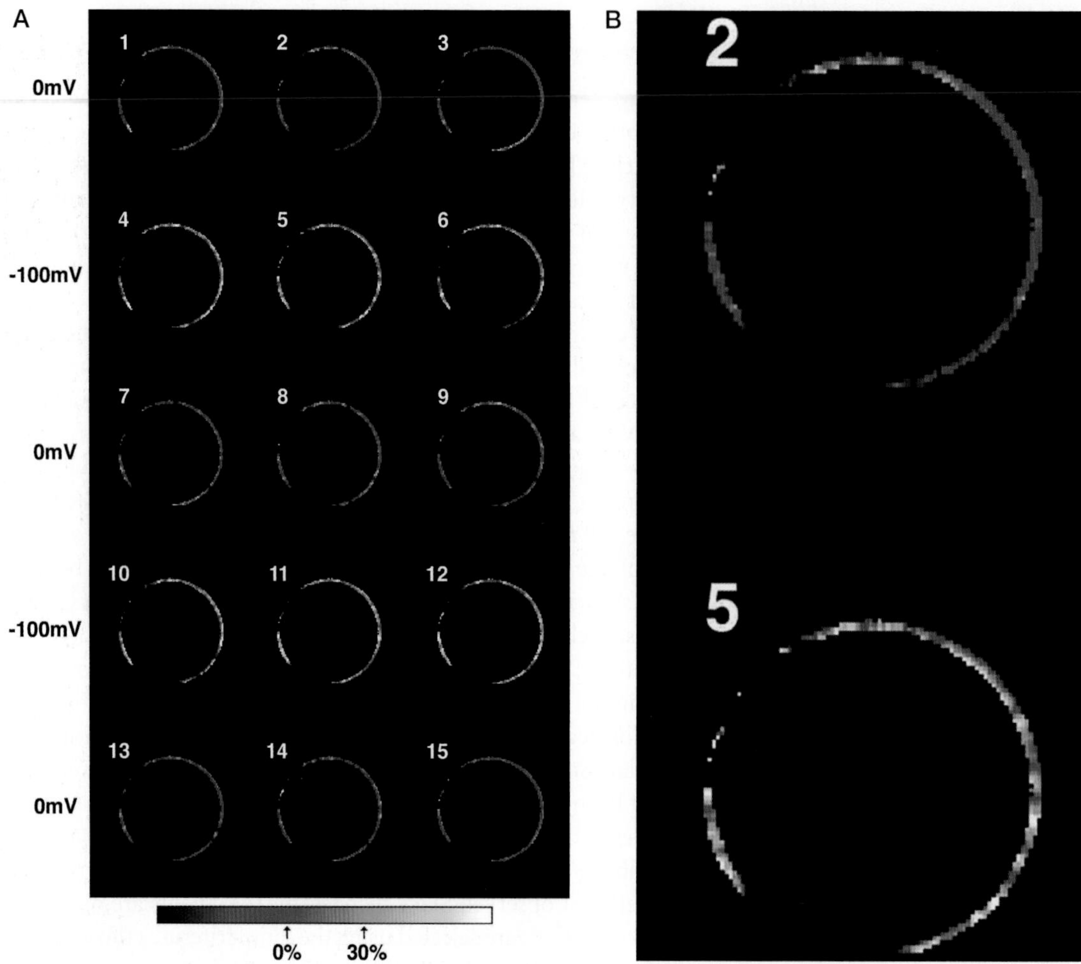

FIGURE 2. (*A*) Montage of SHG images of an N1E-115 mouse neuroblastoma cell stained with ~3 μM di-4-ANEPPS and excited at 910 nm. Each image is a Kalman average of three acquisitions. This cell is patched, with the pipette entering the field of view from the lower-left corner and resulting in a reduction in membrane in the lower-left quadrant. A 2PF image of the cell (not shown) includes a significant number of filopodia in the upper-left quadrant of the cell; they are sufficiently small that opposing membranes are within the optical coherence length of ~λ/10 (Campagnola et al. 2002), violating the noncentrosymmetric constraint, and hence the filopodia do not appear in the SHG images, likely reducing the membrane SHG in the upper-left quadrant of the cell. The images are from a time series, with time increasing from *left* to *right* and from *top* to *bottom*. The first, third, and fifth rows were all acquired at 0 mV, while the second and fourth rows were acquired at –100 mV. The color bar shows levels within the "Fire" look-up table corresponding to the normalized reference intensity (labeled 0%) and to a normalized intensity 30% above the reference. The average relative signal change for this image series is 27.0% ± 2.6%. The images were processed using code written in Perl (http://www.perl.com/) and using the Netpbm package (http://netpbm.sourceforge.net/). The frames were first nudged, at most by a few pixels, to correct for any slight drift in the position of the cell during imaging. With the frames aligned, a mask corresponding to the cell membrane was generated. Within each frame, the mask was applied to select the bright membrane pixels, and a nearest-neighbor mean filter was applied, with the filter constrained to use only pixels within the mask. Just as the total intensity values were normalized as described in the Example of Application section, so were the intensity values of each pixel normalized across the frames by a running average interpolated from the values of that pixel in the frames corresponding to the reference voltage. (*B*) The second and fifth time points are zoomed up for better visualization of the contrast between the two applied voltages. (Adapted, with permission, from Millard et al. 2004.)

ethylammonium, and tripropylammonium, respectively). Although all three dyes exhibit approximately the same SHG sensitivity [~20%/(100 mV)] on patch-clamped neuroblastoma cells, there is a marked difference in the speed of this response, as shown in Table 2. PY1261 and PY1278 both have instantaneous (<5 msec) SHG kinetics, possibly consistent with an electro-optic mechanism, in contrast to the significantly slower response of PY1280 (109 msec). But a purely electro-optic mechanism of voltage response would not be expected to be sensitive to head-group change since the head group

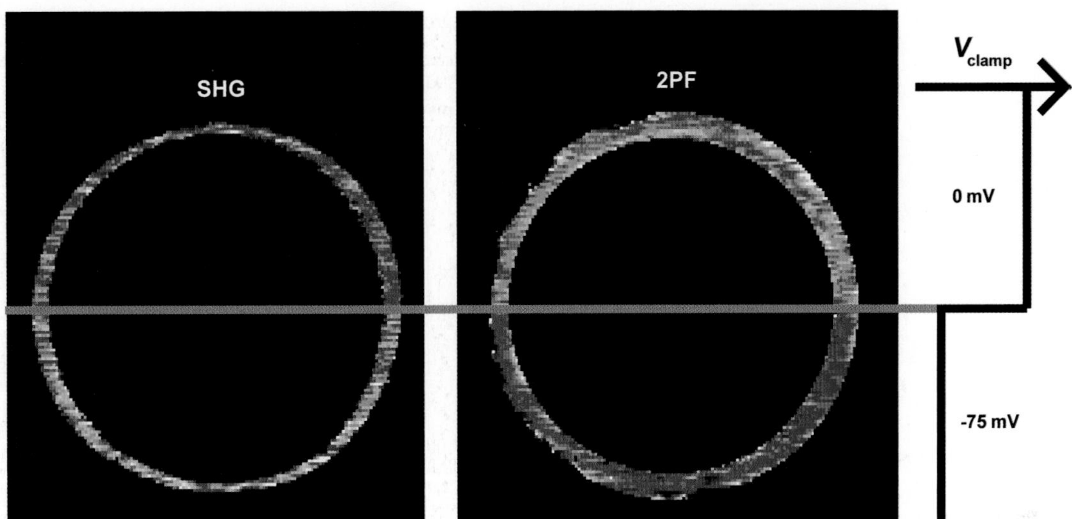

FIGURE 3. SHG and 2PF images of an N1E-115 mouse neuroblastoma cell stained with ~3 μM di-4-ANEPPS and excited at 910 nm. To determine the kinetics of the observed signal changes, the voltage clamp is synchronized to switch from the reference voltage to a test voltage approximately halfway through the page scan. Each row of the image takes 5 msec to acquire. The colors are mapped such that violet and blue represent lower signal levels, and orange and yellow represent higher signal levels. These images were processed with custom Perl code, as described for Figure 2. (Adapted, with permission, from Millard et al. 2005.)

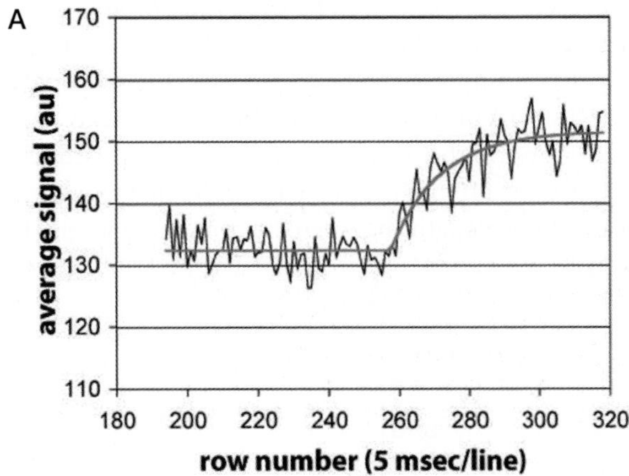

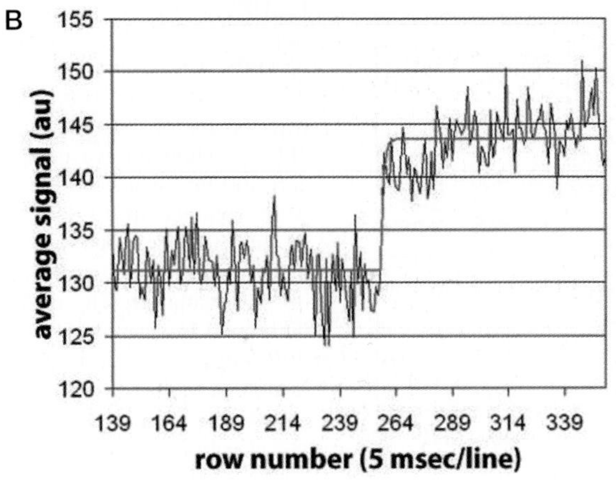

FIGURE 4. Kinetic responses for two hemicyanine dyes on patch-clamped stained neuroblastoma cells. The voltage is changed at the midpoint of the trace. (*A*) The SHG signal from di-4-ANEPPS excited at 916 nm for a –75-mV voltage change. The black line is data from about 167 averaged images; the gray line shows a fit ($R = 0.94$) based on an exponential rise to a final value, with $\tau = 70.6$ msec. (*B*) The change in SHG for a –50-mV voltage switch for PY1261 excited at 1064 nm. Here about 100 images have been averaged, and the data have been fit to a simple step function. (Adapted, with permission, from Millard et al. 2005 and Teisseyre et al. 2007.)

TABLE 2. Comparison of the effect of head group on SHG sensitivity and kinetic response for a series of aminothiophene dyes

Dye	SHG change [%/(50 mV)]	SHG kinetics (msec)	Structure
PY1261	9.56 ± 0.42	<5	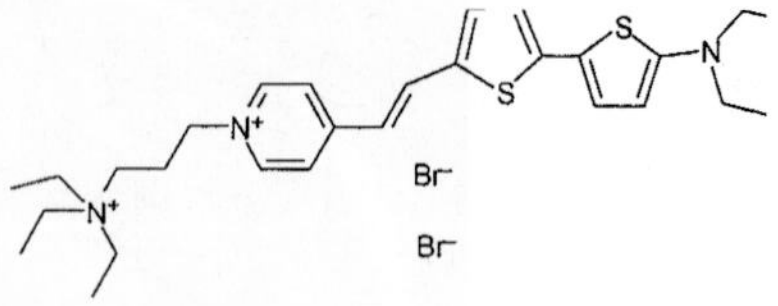
PY1278	10.70 ± 0.59	<5	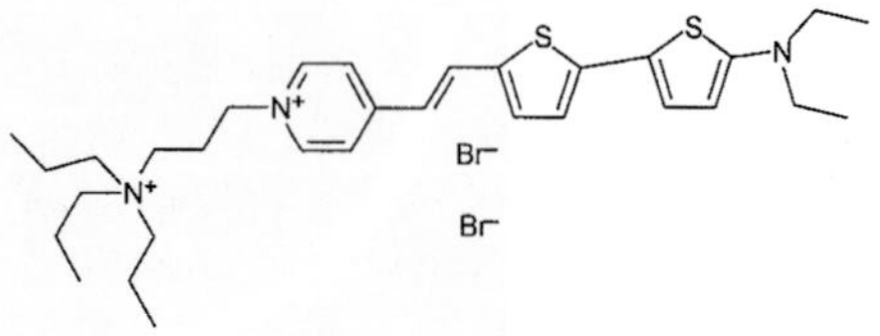
PY1280	10.13 ± 0.88	109 ± 30	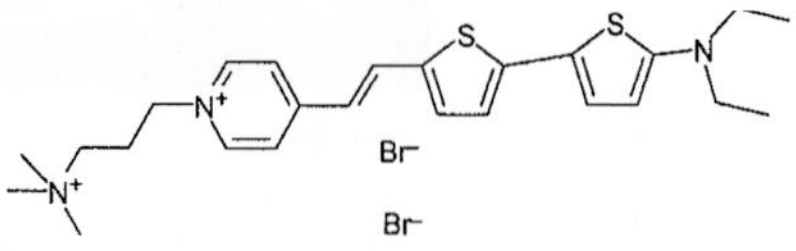

is not part of the chromophore. It is very unlikely that the mechanism would change entirely upon a switch of methyl to ethyl substituents. However, the kinetics of a non-electro-optic mechanism could very well be quite sensitive to the nature of the substituents on the chromophore.

Figure 5 shows relative signal changes in SHG and 2PF versus clamp voltage for di-4-ANEPPS at 850 nm and at 910 nm. All four sets of data have been fit with straight lines going through the origin. We define the voltage sensitivity, either for SHG or 2PF (or 1PF), as the slope of the linear fit for signal change relative to the 0-mV intensity; we shall express a voltage sensitivity calculated in this way as a percentage change per 100 mV. From the linear fits, the voltage sensitivity for SHG is (−18.5% ± 0.5%)/(100 mV) at 850 nm (on the basis of a total of 38 image sequences using about 30 cells) and (−26.3% ± 0.8%)/(100 mV) at 910 nm (on the basis of 52 image sequences using about 40 cells). For 2PF detection at emission wavelengths between 490 nm and 560 nm, the voltage sensitivities are ~6.6%/(100 mV) and ~4.8%/(100 mV), respectively. The best voltage sensitivity under 1PF

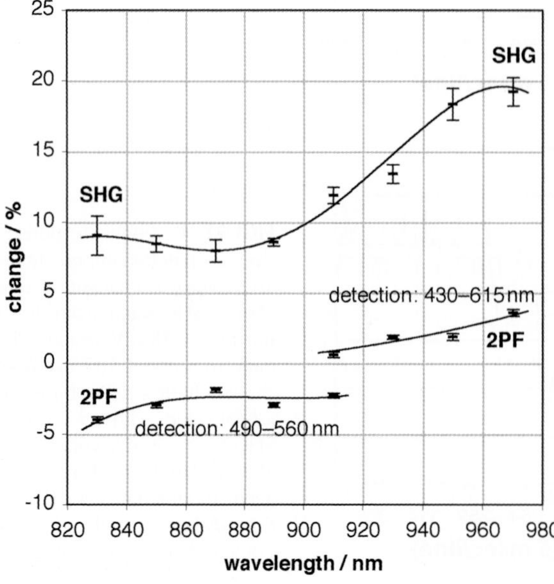

FIGURE 5. Signal changes for a −50-mV voltage step versus wavelength for SHG and 2PF from di-4-ANEPPS. The arbitrary fits are intended for trend visualization only. The 2PF data for excitation wavelengths of 830–910 nm were obtained from 2PF detection for wavelengths between 490 nm and 560 nm, while the 2PF data for 910–970 nm were obtained from 2PF detection between 430 nm and 615 nm. (Although this wavelength range includes the SHG wavelength, there is essentially no backscattered SHG on the 2PF channel.) The difference in sign for 2PF changes from the two wavelength ranges indicates that the 2PF emission exhibits a biphasic response similar to that of 1PF emission (Fluhler et al. 1985), a characteristic consistent with an electrochromic mechanism for voltage sensitivity.

is ~10%/(100 mV) at optimal excitation and emission wavelengths. The data in Figure 5 show that the voltage sensitivity of SHG from di-4-ANEPPS is increased by ~42% between 850 nm and 910 nm, whereas the voltage sensitivity of 2PF is reduced by ~27%.

Advantages and Limitations

The properties of SHG offer several advantages over fluorescence-based techniques for live cell imaging and optical mapping of membrane potential. Clearly, the fourfold increase in sensitivity to membrane potential of the SHG signal is a major advantage. Because SHG does not involve promotion of molecules to an excited state, the bleaching and photodynamic damage that accompany fluorescence may also be minimized (however, resonance-enhanced SHG is usually accompanied by collateral 2PF with its attendant photodamage). Relatively minor enhancements to a 2PF microscope can enable second-harmonic imaging, and the fact that these two modalities use separate light paths permits both 2PF and SHG to be acquired from the same specimen simultaneously. A key advantage of SHG for recording from internally stained cells is that it displays little or no background from internal organelle membranes, such as that of the endoplasmic reticulum, because the subresolution membrane structure results in an effectively centrosymmetric distribution of the chromophore on the scale of the coherence length of the laser light.

Despite the superior sensitivity of SHG voltage responses over fluorescence-based techniques, the low intensity of the SHG signal still presents a major challenge for its routine application as a method for imaging neuronal electrical activity. The key factor is that there are typically two orders of magnitude fewer SHG photons produced than fluorescent photons. Thus, the number of photons collected in a frame scan during an action potential is insufficient to surmount the shot noise inherent in the quantal nature of light. One clear approach to overcome these limits is to develop improved dyes with greater voltage sensitivities and hyperpolarizabilities. Some preliminary promise has been displayed by a radically different approach to dye development in which available electrochromic chromophores, such as ANEP, are linked to silver or gold nanoparticles (Peleg et al. 1996, 1999; Clark et al. 2000); the metal particles locally focus the laser electromagnetic field via a plasmon resonance effect, thereby enhancing SHG from neighboring dye molecules.

Most common experimental neuronal preparations, including brain slices, do not have significant SHG background signal. Therefore, the background actually is much lower than the typical contribution of autofluorescence to 2PF. However, since the SHG signal is most effectively collected via the transmitted light path, in vivo imaging presents a significant challenge. The transmitted light path has been shown to be fully viable for brain slice preparations if enough laser power is available. SHG from Purkinje cells has been collected in the forward direction through 200-µm acute cerebellar slices using 15–20 mW of 1064-nm laser power (Sacconi 2008). Transmitted SHG of brain tissue can provide micrometer-scale resolution at depths up to 400 µm (Dombeck et al. 2003). Additionally, significant progress has been made in using backscattered SHG to produce images of collagen from thick tissue specimens (Zoumi et al. 2002), and this technology should be adaptable to second-harmonic imaging.

The large sensitivity to membrane potential of SHG from styryl membrane probes has motivated further exploration of second-harmonic imaging to measure electrical activity in brain slices in the laboratories of Watt Webb, Rafael Yuste, and Ken Eisenthal. An early successful SHG measurement of action potentials (APs) in intact mammalian neural systems involved intracellular application of FM4-64 to neurons in hippocampal slices (Dombeck et al. 2005). FM4-64 is a hemicyanine dye composed of an aniline donor and a pyridinium acceptor with a trienyl linker moiety. Fifty-five averaged line scans were needed to obtain optically recorded APs because of the poor signal-to-noise ratio of each single line scan. Second-harmonic imaging of membrane potential in individual spines and adjacent dendrites was performed on FM4-64 filled pyramidal neurons (Araya et al. 2006). These measurements were obtained at a frame rate of 1–5 sec/frame and were repeated five to 10 times. For optical mapping of electrical activity in more complex neuronal preparations via SHG, recent investigations have attempted to further limit the number of spatial points sampled by the laser scanning microscope and instead collect as many photons as possible from specific points of interest. This approach has been used for "point-scan" imaging of FM4-64-stained pyramidal neurons in brain slices (Araya et al. 2006; Nuriya et al. 2006) and random access laser excitation of clusters of Purkinje cells in cerebellar slices bulk-loaded with the same dye (Sacconi et al. 2008).

As both instrumentation and available dyes continue to make dramatic advances, the use of SHG-based optical recording for measuring membrane potential in neurons will become more accessible and common. Efforts continue toward the design and synthesis of improved SHG-based voltage sensors, which ideally will have increased efficiency and low collateral photobleaching.

RECIPES

CAUTION: See Appendix 6 for proper handling of materials marked with <!>.
Recipes for reagents marked with <R> are included in this list.

External Patch Clamp Buffer

20 mM HEPES (Merck Biosciences)
Earle's balanced salt solution (EBSS, Sigma-Aldrich)

Adjust to pH 7.35 with stock KOH <!> and HCl <!> before use.

Internal Patch Pipette Buffer

Potassium aspartate	140 mM
NaCl	5 mM
HEPES	10 mM

Adjust to pH 7.35 with stock KOH <!> and HCl <!> before use.

ACKNOWLEDGMENTS

We thank Thomas Teisseyre, Lei Jin, Joseph Wuskell, Ping Yan, and Mei-De Wei for their contributions to this work and Paul Campagnola for his advice. We gratefully acknowledge financial support from the National Institutes of Health, National Institute of Biomedical Imaging and Bioengineering grant R01EB00196.

REFERENCES

Antic S, Wuskell JP, Loew LM, Zecevic D. 2000. Functional profile of the giant metacerebral neuron of *Helix aspersa*: Temporal and spatial dynamics of electrical activity in situ. *J Physiol* **527:** 55–69.

Araya R, Jiang J, Eisenthal KB, Yuste R. 2006. The spine neck filters membrane potentials. *Proc Natl Acad Sci* **103:** 17961–17966.

Bedlack RS, Wei M-D, Loew LM. 1992. Localized membrane depolarizations and localized intracellular calcium influx during electric-field-guided neurite growth. *Neuron* **9:** 393–403.

Ben-Oren I, Peleg G, Lewis A, Minke B, Loew LM. 1996. Infrared nonlinear optical measurements of membrane potential in photoreceptor cells. *Biophys J* **71:** 1616–1620.

Betz WJ, Bewick GS. 1992. Optical analysis of synaptic vesicle recycling at the frog neuromuscular junction. *Science* **255:** 200–203.

Betz WJ, Mao F, Bewick GS. 1992. Activity-dependent fluorescent staining and destaining of living vertebrate motor nerve terminals. *J Neurosci* **12:** 363–375.

Bouevitch O, Lewis A, Pinevsky I, Wuskell JP, Loew LM. 1993. Probing membrane potential with nonlinear optics.

Biophys J **65:** 672–679.

Bullen A, Loew LM. 2001. Solubility and intracellular delivery of hydrophobic voltage-sensitive dyes with chemically-modified cyclodextrins. *Biophys J* **80:** 168a.

Campagnola PJ, Loew LM. 2003. Second harmonic imaging microscopy for visualizing biomolecular arrays in cells, tissues and organisms. *Nat Biotechnol* **21:** 1356–1360.

Campagnola PJ, Wei M-D, Lewis A, Loew LM. 1999. High resolution nonlinear optical microscopy of living cells by second harmonic generation. *Biophys J* **77:** 3341–3349.

Campagnola PJ, Millard AC, Terasaki M, Hoppe PE, Malone CJ, Mohler W. 2002. Three dimensional high resolution second harmonic generation imaging of endogenous structural proteins in biological tissues. *Biophys J* **81:** 493–508.

Clark HA, Campagnola PJ, Wuskell JP, Lewis A, Loew LM. 2000. Second harmonic generation properties of fluorescent polymer encapsulated gold nanoparticles. *J Am Chem Soc* **122:** 10234–10235.

Cohen LB, Salzberg BM, Davila HV, Ross WN, Landowne D, Waggoner AS, Wang CH. 1974. Changes in axon fluorescence during activity: Molecular probes of membrane potential. *J Membr Biol* **19:** 1–36.

Dombeck DA, Kasischke KA, Vishwasrao HD, Ingelsson M, Hyman BT, Webb WW. 2003. Uniform polarity microtubule assemblies imaged in native brain tissue by second-harmonic generation microscopy. *Proc Natl Acad Sci* **100:** 7081–7086.

Dombeck DA, Blanchard-Desce M, Webb WW. 2004. Optical recording of action potentials with second-harmonic generation microscopy. *J Neurosci* **24:** 999–1003.

Dombeck DA, Sacconi L, Blanchard-Desce M, Webb WW. 2005. Optical recording of fast neuronal membrane potential transients in acute mammalian brain slices by second-harmonic generation microscopy. *J Neurophysiol* **94:** 3628–3636.

Fluhler E, Burnham VG, Loew LM. 1985. Spectra, membrane binding and potentiometric responses of new charge-shift probes. *Biochemistry* **24:** 5749–5755.

Hassner A, Birnbaum D, Loew LM. 1984. Charge-shift probes of membrane potential: Synthesis. *J Org Chem* **49:** 2546–2551.

Huang YA, Lewis A, Loew LM. 1988. Nonlinear optical properties of potential sensitive styryl dyes. *Biophys J* **53:** 665–670.

Jiang J, Eisenthal KB, Yuste R. 2007. Second harmonic generation in neurons: Electro-optic mechanism of membrane potential sensitivity. *Biophys J* **93:** L26–L28.

Loew LM. 1994. Voltage-sensitive dyes and imaging neuronal activity. *Neuroprotocols* **5:** 72–79.

Loew LM. 2001. Mechanisms and principles of voltage-sensitive fluorescence. In *Optical mapping of cardiac excitation and arrhythmias* (ed. Rosenbaum DS, Jalife J), pp. 33–46. Futura Publishing, Armonk, NY.

Loew LM. 2010a. Design and use of organic voltage sensitive dyes. In *The voltage imaging book* (ed. Zecevic D, Canepari M). Springer, New York (in press).

Loew LM. 2010b. Second harmonic imaging of membrane potential. In *The voltage imaging book* (ed. Zecevic D, Canepari M). Springer, New York (in press).

Loew LM, Simpson L. 1981. Charge shift probes of membrane potential: A probable electrochromic mechanism for asp probes on a hemispherical lipid bilayer. *Biophys J* **34:** 353–365.

Loew LM, Bonneville GW, Surow J. 1978. Charge-shift optical probes of membrane potential: Theory. *Biochemistry* **17:** 4065–4071.

Loew LM, Scully S, Simpson L, Waggoner AS. 1979a. Evidence for a charge-shift electrochromic mechanism in a probe of membrane potential. *Nature* **281:** 497–499.

Loew LM, Simpson L, Hassner A, Alexanian V. 1979b. An unexpected blue shift caused by differential solvation of a chromophore oriented in a lipid bilayer. *J Am Chem Soc* **101:** 5439–5440.

Loew LM, Cohen LB, Dix J, Fluhler EN, Montana V, Salama G, Wu JY. 1992. A naphthyl analog of the aminostyryl pyridinium class of potentiometric membrane dyes shows consistent sensitivity in a variety of tissue, cell and model membrane preparations. *J Membr Biol* **130:** 1–10.

Loew LM, Campagnola PJ, Lewis A, Wuskell JP. 2002. Confocal and nonlinear optical imaging of potentiometric dyes. *Methods Cell Biol* **70:** 429–452.

Lojewska Z, Loew LM. 1987. Insertion of amphiphilic molecules into membranes is catalyzed by a high molecular weight non-ionic surfactant. *Biochim Biophys Acta* **899:** 104–112.

Millard AC, Campagnola PJ, Mohler W, Lewis A, Loew LM. 2003a. Second harmonic imaging microscopy. *Methods Enzymol* **361:** 47–69.

Millard AC, Jin L, Lewis A, Loew LM. 2003b. Direct measurement of the voltage sensitivity of second-harmonic generation from a membrane dye in patch-clamped cells. *Opt Lett* **28:** 1221–1223.

Millard AC, Jin L, Wei M-D, Wuskell JP, Lewis A, Loew LM. 2004. Sensitivity of second harmonic generation from styryl dyes to trans-membrane potential. *Biophys J* **86:** 1169–1176.

Millard AC, Jin L, Wuskell JP, Boudreau DM, Lewis A, Loew LM. 2005a. Wavelength- and time-dependence of potentiometric nonlinear optical signals from styryl dyes. *J Membr Biol* **208:** 103–111.

Millard AC, Lewis A, Loew LM. 2005b. Second harmonic imaging of membrane potential. In *Imaging in neuroscience and development* (ed. Yuste R, Konnerth A), pp. 463–474. Cold Spring Harbor Laboratory Press, Cold Spring Harbor, NY.

Moreaux L, Sandre O, Mertz J. 2000. Membrane imaging by second harmonic generation microscopy. *J Opt Soc Am B* **17:** 1685–1694.

Nuriya M, Jiang J, Nemet B, Eisenthal KB, Yuste R. 2006. Imaging membrane potential in dendritic spines. *Proc Natl Acad Sci* **103:** 786–790.

Obaid AL, Koyano T, Lindstrom J, Sakai T, Salzberg BM. 1999. Spatiotemporal patterns of activity in the intact mammalian network with single-cell resolution: Optical studies of nicotinic activity in an enteric plexus. *J Neurosci* **19:** 3073–3093.

Peleg G, Lewis A, Bouevitch O, Loew LM, Parnas D, Linial M. 1996. Gigantic optical nonlinearities from nanoparticle enhanced molecular probes with potential for selectively imaging the structure and physiology of nanometric regions in cellular systems. *Bioimaging* **4:** 215–224.

Peleg G, Lewis A, Linial M, Loew LM. 1999. Nonlinear optical measurement of membrane potential around single molecules at selected cellular sites. *Proc Natl Acad Sci* **96:** 6700–6704.

Penner R. 1995. A practical guide to patch clamping. In *Single-channel recording* (ed. Sakmann B, Neher E), pp. 3–30. Plenum, New York.

Pons T, Moreaux L, Mongin O, Blanchard-Desce M, Mertz J. 2003. Mechanisms of membrane potential sensing with second harmonic generation microscopy. *J Biomed Opt* **8:** 428–431.

Sacconi L, Mapelli J, Gandolfi D, Lotti J, O'Connor RP, D'Angelo E, Pavone FS. 2008. Optical recording of electrical activity in intact neuronal networks with random access second-harmonic generation microscopy. *Opt Express* **16:** 14910–14921.

Teisseyre TZ, Millard AC, Yan P, Wuskell JP, Wei M-D, Lewis A, Loew LM. 2007. Nonlinear optical potentiometric dyes optimized for imaging with 1064 nm light. *J Biomed Opt* **12:** 044001–044008.

Wu J-Y, Lam Y-W, Falk CX, Cohen LB, Fang J, Loew LM, Prechtl JC, Kleinfeld D, Tsau Y. 1998. Voltage-sensitive dyes for monitoring multi-neuronal activity in the intact central nervous system. *Histochem J* **30:** 169–187.

Wuskell JP, Boudreau D, Wei MD, Jin L, Engl R, Chebolu R, Bullen A, Hoffacker KD, Kerimo J, Cohen LB, et al. 2006. Synthesis, spectra, delivery and potentiometric responses of new styryl dyes with extended spectral ranges. *J Neurosci Meth* **151:** 200–215.

Zochowski M, Wachowiak M, Falk CX, Cohen LB, Lam Y-W, Antic S, Zecevic D. 2000. Imaging membrane potential with voltage-sensitive dyes. *Biol Bull* **198:** 1–21.

Zoumi A, Yeh A, Tromberg BJ. 2002. Imaging cells and extracellular matrix in vivo by using second-harmonic generation and two-photon excited fluorescence. *Proc Natl Acad Sci* **99:** 11014–11019.

47 Grating Imager Systems for Fluorescence Optical-Sectioning Microscopy

Frederick Lanni

Department of Biological Sciences, Carnegie Mellon University, Pittsburgh, Pennsylvania 15213

ABSTRACT

In fluorescence microscopy, optical sectioning is defined as the attenuation or removal of out-of-focus features from an image, and it is a prerequisite for quantitative analysis of three-dimensional structure or function within the specimen. Optical sectioning is most commonly performed by confocal scanning fluorescence microscopy or two-photon scanning fluorescence microscopy. However, structured illumination can be used in conventional fluorescence microscopes to obtain optical sectioning performance, and, in advanced systems, 3D superresolution. The simplest structured-illumination system uses a Ronchi grating as a mask to project parallel stripes within the sharp depth-of-focus of the objective to encode in-focus specimen features differently from out-of-focus features. By shifting the grating, the in-focus image component can be discriminated and separated by elementary image processing operations. This implementation of structured illumination, the fluorescence grating imager, uses a conventional light source, is compatible with all high-quality fluorescence filter sets, and provides high optical-sectioning performance when used to image specimens in which (1) the out-of-focus image component is not much brighter than the in-focus features, and (2) there is no significant movement in the specimen during the grating shift and image capture process.

OPTICAL SECTIONING

A single optical section is a view of a slice within the specimen centered axially on the plane of focus. A through-focus stack of optical sections therefore constitutes a 3D representation of the specimen. For the past 25 years, optical sectioning has been driven by the development of the laser confocal scanning fluorescence microscope (CSFM), the two-photon scanning fluorescence microscope (TPSFM), and deconvolution algorithms. In 1997, Wilson and colleagues described both a principle and an optical system to achieve nearly direct optical sectioning in a conventionally illuminated reflectance microscope (Wilson et al. 1997, 1998a,b,c; Neil et al. 1997; Juskaitis et al. 1998). The essential idea was the use of "structured light" or striped illumination projected with sharply defined depth of focus to achieve axial selectivity. The extension to fluorescence was straightforward (Neil et al. 1998; Wilson et al. 1998b; Ben-Levy and Peleg 1999; Lanni and Wilson 2000; Neil et al. 2000; Lagerholm et al. 2003; Vanni et al. 2003), and there are now instruments available from Carl Zeiss (ApoTome; www.zeiss.com/) and from

Qioptiq LINOS (OptiGrid; www.qioptiqlinos.com/) that incorporate the concept. In addition, much more complex coherent structured-illumination microscopy (SIM) systems for optical sectioning and superresolution have been developed which are beyond the scope of this chapter (Lanni 1986; Lanni et al. 1986; Bailey et al. 1993; Frohn et al. 2000; Gustafsson 2000; Gustafsson et al. 2008; Shao et al. 2008; Schermelleh et al. 2008). For the purposes of the present discussion, we refer to the basic instrument as a fluorescence grating imager (FGI). The attractive characteristics of the FGI are its versatility and performance, given its simplicity and relatively low cost.

Operating Principle and Instrumentation

In a grating imager, only a single essential modification is made to a conventional research-grade fluorescence microscope. A movable Ronchi grating mask is placed in the field iris plane of the incident-light illuminator, thus producing in the specimen striped illumination that is coincident with the geometric focus plane and within the depth of focus of the objective (Fig. 1). The actual focus range of the stripes can be made very sharp if a fine grating is used with an objective of high numerical aperture (NA). Therefore, the image formed by the microscope will consist of striped in-focus features superposed with uniformly illuminated out-of-focus features (background). The period of the projected stripes is the actual grating period divided by the demagnification ratio between the illuminator field iris plane and the specimen, which is objective dependent (and generally not equal

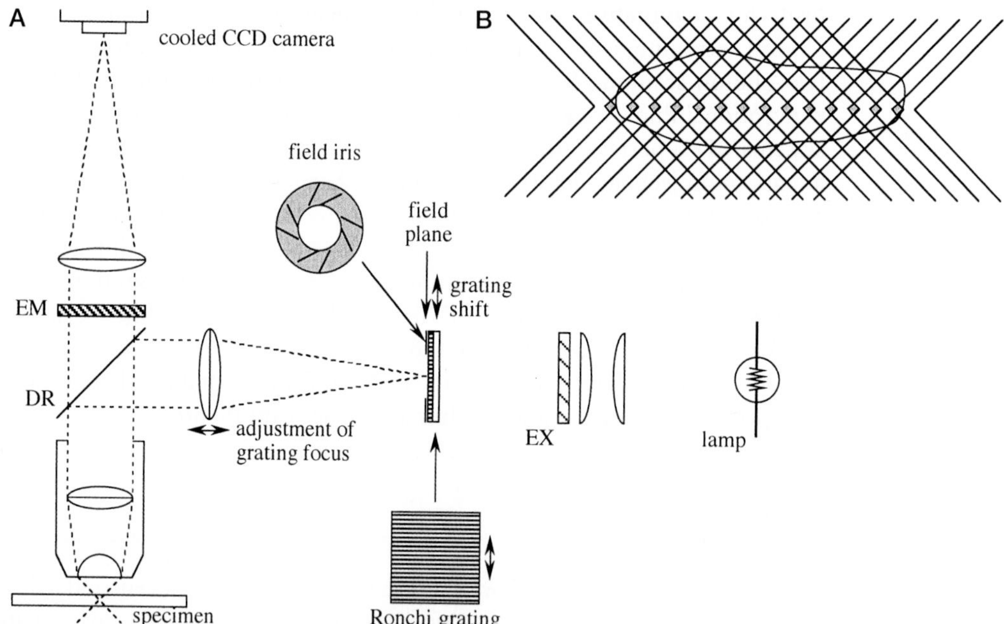

FIGURE 1. Schematic of fluorescence grating imager. (*A*) Optics layout showing movable grating mask inserted at location of illuminator field iris. Ronchi gratings (equal-width square-wave pattern) in the range of 8–50 line pairs per mm (LP/mm) are used, depending on the demagnification ratio determined by the objective and the focal length of the illuminator tube lens. A means for bringing the grating to a sharp focus in the plane of focus of the objective is necessary, here shown as an axial adjustment of the illuminator tube lens. To minimize aberration in the grating image, the excitation filter (EX) is shown removed from its usual position to precede the grating. This is conveniently performed by use of an external filter wheel between lamp collector lens and microscope stand. Additionally, the dichroic reflector (DR) must be flat and strain-free. Actuators for grating movement can be piezoelectric or electromagnetic, or the grating may be fixed but refractively shifted by a separate optic. Shift increments are generally one-third or one-quarter of the grating period, so they will be in the range 5.00–42.0 µm. Computer control synchronizes discrete grating movement and CCD (charge-coupled device) readout between periods of light exposure. A fast light shutter (not shown) is essential for precision exposure timing and minimization of unnecessary exposure. (EM) Emission filter. (*B*) Schematic showing finite grating depth-of-focus within a 3D specimen. Geometrically, it can be seen that both grating spatial frequency and objective NA will affect the depth-of-focus of the pattern.

to the specified objective magnification). When the grating is shifted so as to move the projected stripes transversely across the object, the fluorescence detected in any pixel element in the image will consist of (1) a steady or DC component owing to the out-of-focus background, plus (2) an oscillating (AC) component owing to the shifting of the stripes across in-focus structures, and (3) shot noise owing to both the DC and AC signal components. In brief, the amplitude of the AC component is the in-focus signal level for that pixel. By shifting the grating between a minimum of three defined positions to obtain three images, a very simple digital image processing operation can subtract away the background and demodulate the AC component to produce an optical section.

Microscope optical system quality has a definite effect on FGI performance. First, proper immersion is essential to avoid spherical aberration in both the projected grating and the fluorescence image. Indirect or direct water-immersion must be used when optically sectioning living cells, or other low-index specimens. Oil immersion must be used for specimens mounted in high-index medium. Second, the illuminator optics must provide diffraction-limited imaging of the field iris plane onto the specimen, so that the grating is projected at the full resolution of the objective. Although excitation filters are generally of excellent quality in terms of pass band and blocking, they may not be optically flat, thus introducing aberration into the projected grating pattern. Mounting the excitation filter between the lamp and the grating mask (as shown in Fig. 1) obviates this problem. For the same reason, the dichroic reflector must be strain-free and flat. Third, the full illumination numerical aperture (INA) of the incident light optics must be used to minimize the depth of focus of the projected grating. Fourth, if the instrument is used over a significant range in the UV–visible spectrum (e.g., with DAPI and Cy3 dyes), chromatic shift of focus could necessitate axial adjustment of the grating when filter sets are interchanged. This is analogous to the differential adjustment of pinhole focus in a confocal scanner, and is minimized by use of apochromatic optics. In the ApoTome and Optigrid, this correction can be set and automated. Lastly, the grating shift mechanism must be precise and accurate, and must hold the mask motionless during each image exposure.

Computation of the Optical Section

The simplest projected grating image is a sine wave pattern, which, when translated across the specimen, produces a sinusoidal modulation of the fluorescence. Translation of the grating is denoted by the phase shift (ϕ), where, for example, a shift of 90° is equivalent to movement equal to one-quarter of the grating period. In any pixel, the fluorescence signal can be expressed as a periodic function of ϕ, plus a constant term due to out-of-focus background:

$$f(\phi) = \text{DC term} + \text{AC term} = a_0/2 + (a_1 \cos \phi + b_1 \sin \phi). \qquad (1)$$

The in-focus part of the total fluorescence signal in that pixel is the AC amplitude, which is given by the Pythagorean sum of the sine and cosine coefficients: $\sqrt{(a_1^2 + b_1^2)}$. It can be shown generally that the AC amplitude can be computed on a pixel-by-pixel basis from three images made with the grating shifted between three distinct positions. Three images are necessary because there are three unknown coefficients in Equation 1. The most commonly used shift sequences are given in Table 1, along with corresponding formulas for computing the optical section. The out-of-focus image com-

TABLE 1. FGI image processing formulas

Grating shift sequence	Computation of optical section	Uniform exposure
One-third period, three images 0°-120°-240°	$(2^{1/2}/3)[(i_0 - i_{120})^2 + (i_{120} - i_{240})^2 + (i_{240} - i_0)^2]^{1/2}$	Yes
One-quarter period, four images 0°-90°-180°-270°	$(1/2)[(i_0 - i_{180})^2 + (i_{90} - i_{270})^2]^{1/2}$	Yes
One-quarter period, three images 0°-90°-180°	$(2^{-1/2})[(i_0 - i_{90})^2 + (i_{90} - i_{180})^2]^{1/2}$	No

ponent, which is the same in all images in the set, is removed by computing differences between image pairs (Fig. 2). The resulting bipolar, striped images are then demodulated by algebraic operations that effectively form the Pythagorean sum.

A real projected grating will include spatial harmonics. For a perfect Ronchi (square wave) grating, only odd harmonics occur: $f(\phi) = \text{DC} + (a_1 \cos \phi + b_1 \sin \phi) + (a_3 \cos 3\phi + b_3 \sin 3\phi) + (a_5 \cos 5\phi + b_5 \sin 5\phi) + \cdots$. The amplitude of the AC term is still the in-focus part of the total fluorescence signal in that pixel. However, now it can be seen that the simple image processing formulas will not give $\sqrt{(a_1^2 + b_1^2)}$ exactly. The harmonics occur as error terms, and may appear as fine stripes in the optical sections. Two factors can minimize or eliminate this problem. (1) The shift sequence and algorithm used can suppress one or more harmonics. As originally pointed out by Wilson et al. (1997), the one-third-period shift sequence exactly compensates the third harmonic, therefore the first error term is due to the fifth harmonic. Because, for a Ronchi grating, the amplitudes drop off as $1/n$, this error is small. (2) Because the incoherent modulation transfer function of the microscope decreases with spatial frequency (to zero at the inverse of Abbe's resolution limit), the microscope attenuates the higher harmonics that are projected into the specimen—further reducing the error terms. In principle, harmonic error can be eliminated altogether by choosing the Ronchi so that its fifth harmonic matches or exceeds Abbe's resolution limit for the objective. This is not a severe restriction. With the period of the fifth harmonic set equal to Abbe's resolution limit (λ/2NA), the fundamental period will equal 5x(λ/2NA). This is 2.5 times more coarse than the optimal period (λ/NA; see Equation 3 below), but it is projected with high contrast. Alternatively, the grating could be chosen so that the third harmonic was equal to Abbe's limit, in which case the fundamental period would be 3x(λ/2NA). This is close to the optimal period (λ/NA), but it will have lower contrast than with fifth-harmonic suppression.

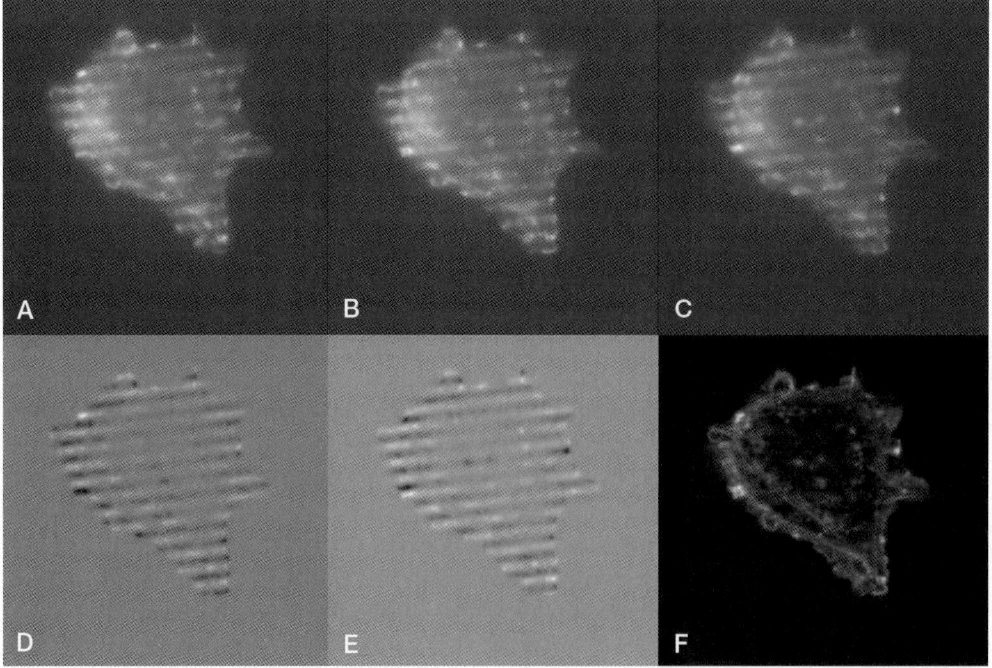

FIGURE 2. Optical section computation from image data showing the actin cytoskeleton in a 3T3 fibroblast. Cells were grown on a cover glass under standard incubator conditions, fixed, permeabilized, and stained with rhodamine-phalloidin to show F-actin. *(A,B,C)* i_0, i_{90}, and i_{180} with focus set close to the adherent basal region of the cell. *(D,E)* difference images *(A – B)* and *(B – C)* as in Table 1, formula 3. In *D* and *E*, the gray scale is bipolar with 0 at midrange gray. *(F)* Optical section computed by Pythagorean summation of *D* and *E*. Field of view, 39 μm.

In general, a wide variety of grating shift sequences can be formulated, all of which, in principle, provide the basis for computation of an optical section. Sequences of more than three images per focus plane can be used to compute the contribution of higher harmonics, or, for example, to eliminate second-harmonic error. However, this is a trade-off in terms of speed and light exposure. The particular shift sequence used, the number of images recorded, and the processing formula all affect various performance characteristics of the instrument: For example, as noted, the 0°-120°-240° shift sequence eliminates third-harmonic error. It also uniformly exposes the plane of focus, which minimizes patterned photobleaching (Table 1). In general, light exposure of the specimen in the grating imager is comparable to conventional fluorescence microscopy. Because the Ronchi passes 50% of the incident light, the three exposures required for one optical section deliver 1.5 times the light dose needed for a conventional image in which the in-focus features are equally bright (but superposed with out-of-focus features).

Sharpness of the Optical Section

Based on an analysis of the optical transfer function of the FGI (Lagerholm et al. 2003), it can be shown that the optical section thickness, δ, is related to the NA and projected grating period, L, by the formula

$$\delta = (\lambda/2)/\{[n^2 - (\mathrm{NA} - \lambda/L)^2]^{1/2} - [n^2 - \mathrm{NA}^2\,]^{1/2}\} \quad \text{for } L > \lambda/2\mathrm{NA}. \tag{2}$$

The combination of a high-NA objective and a fine grating can produce sharp sections that, in principle, match the optical-sectioning performance of a confocal scanner. Ideally, the sharpest sectioning is obtained when the projected grating period is twice Abbe's resolution limit:

$$\delta_{\min} = (\lambda/2)/\{n - [n^2 - \mathrm{NA}^2\,]^{1/2}\} \quad \text{for } L = \lambda/\mathrm{NA}. \tag{3}$$

In practice, a twofold to fourfold coarser grating generally gives better performance owing to increased stripe contrast. As in any form of fluorescence microscopy, FGI signal per pixel is reduced as sectioning is sharpened.

Signal-to-Noise Ratio

Relative to a confocal scanner, performance of the FGI is limited ultimately by photon counting noise due to out-of-focus (background) features in the object. Unlike a confocal scanner, in which the pinhole blocks most background light before detection, background light in an FGI is detected in each image, and then removed by digital subtraction (Table 1 formulas). Both signal (S) and background (B) photocount levels can be treated as Poisson-distributed variables, with means N_S and N_B. The subtraction steps remove the background mean, N_B, but leave a residual noise with a level roughly equal to the Poisson root-mean-square (RMS) value, $\sqrt{N_B}$. For example, if the background amounts to 10^4 counts, and the in-focus signal is 10^3, then $\sqrt{N_B} = \sqrt{10^4} = 100$, therefore a signal-to-noise ratio (SNR) of $\sim 10^3/100 = 10$ is expected. If the in-focus signal were 100 in a background of 10^4, it could barely be distinguished from the noise. A more detailed analysis shows the RMS additivity of noise in the image processing steps leading to the optical section. The intrinsic FGI noise level can be estimated from the ideal case of removal of a perfectly defocused background in a "0°-120°-240°" image set with $S = 0$ (blank specimen). By definition, N_B is the same for all three images, and the RMS noise level in the resulting blank optical section is 1.155 $\sqrt{N_B}$, only 15% greater than the noise level in a single image. Therefore, a simple SNR approximation is the ratio of S to the RMS sum of the Poisson noise levels in the signal ($\sqrt{S + 2B}$) and the background ($\sqrt{B}$): SNR $S/\sqrt{S + 2B}$.

Sampling in the Image Plane

What sampling is required for a camera to pick up all of the detail in the optical image field that is formed by a microscope? This is set by Nyquist's sampling theorem, when noise is not a limiting fac-

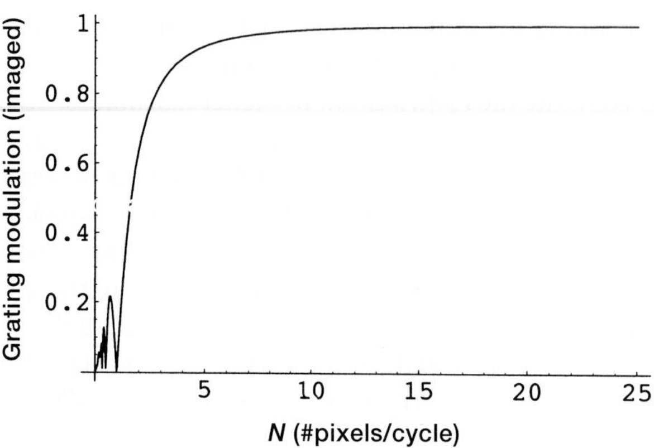

FIGURE 3. Effect of finite camera CCD element size on image contrast. Graph shows the contrast of a projected sine wave grating versus number of contiguous pixels per cycle (N). The plotted function is $|\sin(\pi/N)/(\pi/N)|$. As N is increased to more than 10 pixels per cycle, the contrast of the projected grating approaches 1 (100%).

tor: The density of sample points must exceed the inverse of the signal bandwidth. In the case of a microscope, Abbe's resolution limit, $\lambda/2\text{NA}$, is the period of the finest grating that can be resolved, and sets the signal bandwidth. The spatial frequency of that grating is $2\text{NA}/\lambda$ (cycles/μm), and the bandwidth is defined as the frequency interval from $-2\text{NA}/\lambda$ to $2\text{NA}/\lambda$, or $4\text{NA}/\lambda$. Therefore, the Nyquist sampling interval in nonmagnified coordinates is 1/bandwidth or $\lambda/4\text{NA}$; that is, the camera pixel spacing divided by total magnification must be $\leq\lambda/4\text{NA}$ (Lanni and Baxter 1992). For typical high-NA microscopes, $\lambda/4\text{NA}$ is generally close to 0.1 μm. When using a camera having, for example, 6.45-μm charge-coupled device (CCD) elements, a total magnification ≥64.5x is required to achieve Nyquist sampling. In real cameras, finite pixel size also causes spatial averaging of the fine details in the image field (Lanni and Baxter 1992). The effect of finite CCD pixel size in the grating imager is a reduction in the percent modulation in an image of the grating projected on a uniformly fluorescent specimen. Clearly, if the grating period matched the pixel size, no grating would be seen. A graph of percent modulation versus number of camera pixels across one grating line pair (Fig. 3) shows that modulation is preserved to a high degree when at least 10 pixels span a projected line pair. As an example, a grating chosen to set $L = 5\lambda/2\text{NA}$ (see Computation of the Optical Section) would also make $L/10 = \lambda/4\text{NA}$, the Nyquist interval.

Performance

A very basic measure of performance for any optical-sectioning fluorescence microscope is its axial response, defined as the graph of pixel brightness versus focus increment for a uniform planar specimen. In practice, a stable fluorescent film that is thin compared with the smaller of (1) the wave-optical depth of field ($\sim 2n\lambda/\text{NA}^2$) or (2) the optical-sectioning limit must be used to obtain a meaningful measurement. Figure 4A shows graphically the axial response of a grating imager based on a Zeiss Axiovert 200M operated with a 1.30-NA objective (Zeiss 100x Plan-Neofluar) and 1.33-μm projected grating period, as measured using a 0.05-μm rhodamine-labeled thin film specimen. The optical section thickness, defined by the graphical full width at half-maximum (FWHM), was 0.65 μm. In comparison, Equation 2 gives a value $\delta = 0.623$ μm for NA = 1.3, $L = 1.33$ μm, and $\lambda =$ 570 nm (strictly speaking, Equation 2 should distinguish between the excitation and emission wavelengths, but these are usually close; in this example they are 540 nm and 605 nm, respectively). The experimental FWHM is comparable to a confocal scanner operated with the same objective. Furthermore, the FGI axial response could be further sharpened by use of a finer grating because, in this example, L exceeded λ/NA by a factor of 3 (see Equation 3). In Figure 4B showing the perinuclear actin cytoskeleton of a fibroblast, a focus increment of 0.6 μm (encoded in magenta/green pseudocolor) clearly discriminates axial structure in the FGI optical sections.

In cell biological applications, where out-of-focus image features are generally less than an order of magnitude greater in brightness than in-focus features, grating imagers generally produce fluorescence optical sections comparable to high-quality confocal scans. Figure 5 shows the quantitative effect of optical sectioning in a specimen of chick embryo fibroblasts grown in a 3D collagen gel,

A

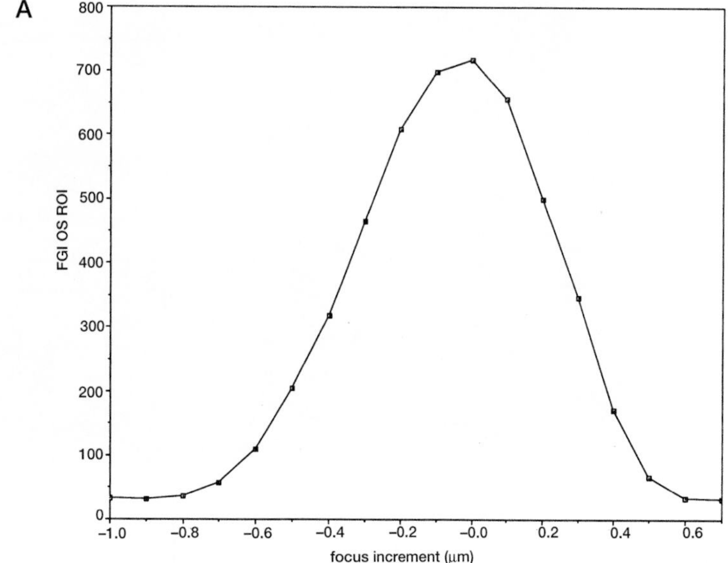

B

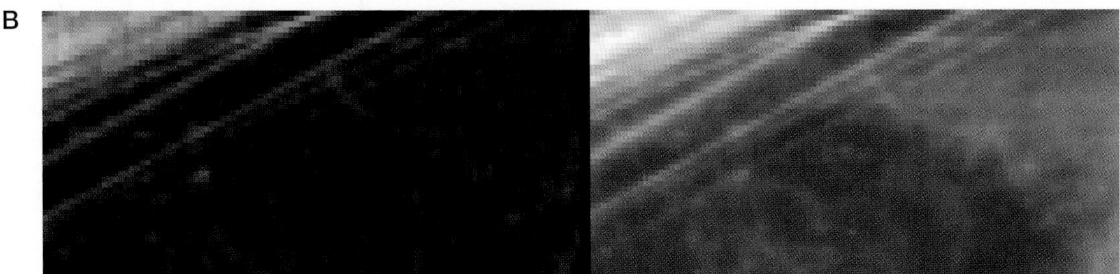

FIGURE 4. (*A*) Grating imager axial response. The graph shows optical section signal level versus focus drive movement in a small region-of-interest (ROI) in the center of an image of a thin film. For the test reported here, poly-[methyl methacrylate] covalently labeled with rhodamine (Rh-PMMA) was spin-cast from chlorobenzene onto a coverglass to form a thin film. A simple reflectance interferometer was used to determine the film thickness (50 nm) based on fringe shift and a refractive index equal to 1.49 for PMMA. The coverglass was then mounted with the film in contact with optical cement (Norland Optical Adhesive #60, index = 1.56) on a slide, and UV light was used to polymerize the cement. It was previously determined that Rh-PMMA is insoluble in NOA#60. The advantage of this type of standard specimen is its simple refractive structure and similar refractive indexes (glass 1.52, PMMA 1.49, NOA 1.56). The full width at half-maximum (FWHM) defines the sharpness of the optical section, here 0.65 µm. Instrument: Zeiss Axiovert 200M with Queensgate piezoelectric grating shifter, 100x 1.30 NA oil-immersion objective, 50 LP/mm Ronchi grating projected to 1.33-µm period, Hamamatsu Orca-ER CCD camera with 6.45-µm pixels, and QED Imaging InVivo instrument control software. (*B*) Pseudocolor overlay showing F-actin structures in the perinuclear region of a fibroblast. Field-of-view width, 10.0 µm. The basal plane of focus (magenta) shows small actin fibers below the nucleus. A 0.6-µm focus increment brings stress fibers into view (green). In the FGI optical sections (*right*), the structures are readily discriminated. Overlay of the corresponding pseudocolor conventional images (*left*) shows that most structures are not discriminated by focus alone.

then fixed and stained for F-actin with rhodamine phalloidin. The left panel in the figure shows a fluorescence image in the absence of optical sectioning, and the right panel shows the FGI optical section at the same plane of focus. A pixel-row profile across the images between the guidelines shows the large effect that out-of focus light would have on quantification of actin in the cytoskeleton, and the degree of improvement obtained with the grating imager.

As in any form of high-resolution microscopy, a number of use-related factors have been identified that have a significant effect on performance. Table 2 lists those known to date. These can be classified generally as (1) reduction in aberration and grating defocus (which reduce projected grating contrast

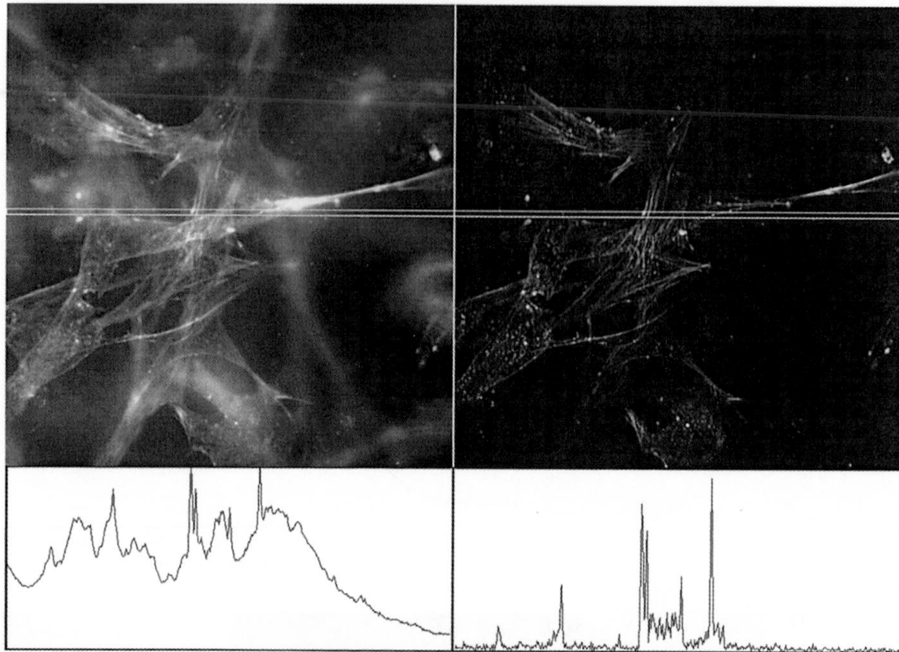

FIGURE 5. Quantification of actin cytoskeletal structures in a grating imager optical section (*upper right*) versus the corresponding conventional image (*upper left*). (*Lower panels*) Pixel-row profile between guidelines shows fluorescence from out-of-focus features mixed with in-focus features and how attenuation of the background in the optical section makes quantification possible. Gray scale in optical section profile (*lower right*) is boosted threefold. The field of view (width and height) of each square image is 165 μm.

and image contrast), (2) optimization of the light source and minimization of light exposure, and (3) optimization of sampling. In comparison to confocal scanning, the FGI has a number of advantages and disadvantages, which are listed in Table 3. An outstanding advantage is that the instrument is immediately compatible with almost any standard high-quality fluorescence filter set. The only wavelength range limitation so far encountered is longitudinal chromatic aberration when using UV excitation. Finally, although the grating imager has been developed as a fast and compact means to obtain optical sections, it is possible to apply 3D deconvolution methods to the unprocessed serial-focus image sets. Although this is computationally much more demanding, such processing can be performed offline while still obtaining fast optical sections using one of the image-processing formulas in Table 1.

TABLE 2. Use factors affecting fluorescence grating imager performance

Minimization of aberration
Proper immersion/dial-in spherical aberration correction
Proper parfocalization of CCD (at ±0 diopter ocular setting)
Imaging-grade excitation filters (or use external filter wheel)
Strain-free dichroic reflector
Alignment of light source
Use of full illumination numerical aperture (INA)
Stability of light source
Precision of exposure timing
Shift sequence: Cancellation of harmonics and uniformity of summed exposure in plane of focus
Grating parfocus
Axial color correction
Speed vs. movement in specimen
Nyquist sampling of image field in camera (pixel spacing $\leq\lambda/4NA$)
Oversampling of grating image ($N > 10$ pixels/line pair)
Camera linearity, >10-bit digitization
Minimization of photobleaching

TABLE 3. Comparison of FGI to other optical-sectioning methods

Advantages
Simple accessory to microscope—all other modes unaffected
Fast compared with point-scan CSFM
Low light exposure (3 x 50%)
Obtain optical section from single focus plane—image stack not required
Deconvolution not required—processing is noniterative and fast
Conventional light source
Standard filter sets
Axial bandwidth matches CSFM
Low cost
Disadvantages
Slower than Yokogawa-type spinning-disk multipoint CSFM
Background rejection not as high as CSFM
Photobleaching or specimen movement gives stripe artifact

Future Outlooks

In the past 15 years, a number of multiplex optical sectioning systems have come into general use in addition to FGIs. These would include the Yokogawa-type pinhole array scanning confocal microscope and various slit-scan confocal systems. Most recently, a pinhole aperture-correlation confocal scanner (Juskaitis et al. 1996; Wilson et al. 1996) with highly efficient use of the light source has become available to biologists. The FGI, however, remains the simplest accessory modification to a fluorescence microscope giving confocal-quality optical sectioning.

ACKNOWLEDGMENTS

The author gratefully acknowledges past support from the National Science Foundation IID and STC Programs, and from Carl Zeiss Inc., Thornwood, New York. The author thanks David Pane (Carnegie Mellon University) and Michel Nederlof (QED Imaging) for much good advice and for the development of instrument control software. Finally, the author especially thanks Winfried Denk for communicating his deep insight into the problem of optical-sectioning microscopy, which inspired much of the past and current work.

REFERENCES

Bailey B, Farkas DL, Taylor DL, Lanni F. 1993. Enhancement of axial resolution in fluorescence microscopy by standing-wave excitation. *Nature* **366:** 44–48.

Ben-Levy M, Peleg E. 1999. Imaging measurement system. U.S. Patent #5,867,604.

Frohn JT, Knapp HF, Stemmer A. 2000. True optical resolution beyond the Rayleigh limit achieved by standing wave illumination. *Proc Natl Acad Sci* **97:** 7232–7236.

Gustafsson MGL. 2000. Surpassing the lateral resolution limit by a factor of two using structured illumination microscopy. *J Microsc* **198:** 82–87.

Gustafsson MG, Shao L, Carlton PM, Wang CJ, Golubovskaya IN, Cande WZ, Agard DA, Sedat JW. 2008. Three-dimensional resolution doubling in wide-field fluorescence microscopy by structured illumination. *Biophys J* **94:** 4957–4970.

Juskaitis R, Wilson T, Neil MA, Kozubek M. 1996. Efficient real-time confocal microscopy with white light sources. *Nature* **383:** 804–806.

Juskaitis R, Neil MAA, Wilson T. 1998. Microscopy using an interference pattern as illumination source. *Proc SPIE* **3261:** 27–28.

Lagerholm BC, Vanni S, Taylor DL, Lanni F. 2003. Cytomechanics applications of optical sectioning microscopy. *Methods Enzymol* **361:** 175–197.

Lanni F. 1986. Standing wave fluorescence microscopy. In *Applications of fluorescence in the biomedical sciences* (ed. Taylor DL, et al.), pp. 505–521. A.R. Liss, New York.

Lanni F, Baxter GJ. 1992. Sampling theorem for square-pixel image data. *Proc SPIE* **1660:** 140–147.

Lanni F, Wilson T. 2000. Grating image systems for optical sectioning fluorescence microscopy of cells, tissues, and small organisms. In *Imaging neurons: A laboratory manual* (ed. Yuste R, et al.), pp. 8.1–8.9. Cold Spring Harbor Laboratory Press, Cold Spring Harbor, NY.

Lanni F, Waggoner AS, Taylor DL. 1986. Standing-wave luminescence microscopy. U.S. Patent #4,621,911.

Neil MAA, Juskaitis R, Wilson T. 1997. Method of obtaining optical sectioning by using structured light in a conventional microscope. *Opt Lett* **22:** 1905–1907.

Neil MAA, Juskaitis R, Wilson T. 1998. Real time 3D fluorescence microscopy by two beam interference illumination. *Opt Commun* **153:** 1–4.

Neil MAA, Squire A, Juskaitis R, Bastiaens PIH, Wilson T. 2000. Wide-field optically sectioning fluorescence microscopy with laser illumination. *J Microsc* **197:** 1–4.

Schermelleh L, Carlton PM, Haase S, Shao L, Winoto L, Kner P, Burke B, Cardoso MC, Agard DA, Gustafsson MGL, et al. 2008. Subdiffraction multicolor imaging of the nuclear periphery with 3D structured illumination microscopy. *Science* **320:** 1332–1336.

Shao L, Isaac B, Uzawa S, Agard DA, Sedat JW, Gustafsson MG. 2008. I5S: wide-field light microscopy with 100-nm-scale resolution in three dimensions. *Biophys J* **94:** 4971–4983.

Vanni S, Lagerholm BC, Otey C, Taylor DL, Lanni F. 2003. Internet-based image analysis quantifies contractile behavior of individual fibroblasts inside model tissue. *Biophys J* **84:** 2715–2727.

Wilson T, Juskaitis R, Neil MA, Kozubek M. 1996. Confocal microscopy by aperture correlation. *Opt Lett* **21:** 1879–1981.

Wilson T, Juskaitis R, Neil MAA. 1997. A new approach to three-dimensional imaging in microscopy. *Cell Vis J Anal Morphol* **4:** 231.

Wilson T, Neil MAA, Juskaitis R. 1998a. Real-time three-dimensional imaging of macroscopic structures. *J Microsc* **191:** 116–118.

Wilson T, Neil MAA, Juskaitis R. 1998b. Optically sectioned images in widefield fluorescence microscopy. *Proc SPIE* **3261:** 4–6.

Wilson T, Neil MAA, Juskaitis R. 1998c. Microscopy imaging apparatus and method. International Patent #WO 98/45745.

48 Coherent Raman Tissue Imaging in the Brain

Brian G. Saar,[1,9] Christian W. Freudiger,[1,2,9] Xiaoyin Xu,[3] Anita Huttner,[4,7] Santosh Kesari,[5,8] Geoffrey Young,[3,6] and X. Sunney Xie[1]

[1]*Department of Chemistry and Chemical Biology, Harvard University, Cambridge, Massachusetts 02138;* [2]*Department of Physics, Harvard University, Cambridge, Massachusetts 02138;* [3]*Department of Radiology, Brigham and Women's Hospital, Boston, Massachusetts 02115;* [4]*Department of Pathology, Brigham and Women's Hospital, Boston, Massachusetts 02115;* [5]*Dana-Farber Cancer Institute and Department of Neurology, Brigham and Women's Hospital, Boston, Massachusetts 02115;* [6]*Department of Radiology, Dana-Farber Cancer Institute, Boston, Massachusetts 02115*

ABSTRACT

Imaging in neuroscience has been dramatically impacted by the advent of multiphoton microscopy (Denk et al. 1990). Multiphoton-excited fluorescence (MPF) in combination with endogenous fluorophores or labeling by fluorescent molecules has proven to be particularly powerful. However, endogenous fluorescence is limited to relatively few molecular species, and practical labeling schemes do not exist for many classes of molecules. Coherent Raman scattering (CRS) techniques, including coherent anti-Stokes Raman scattering (CARS) (Zumbusch et al. 1999; Evans and Xie 2008) and stimulated Raman scattering (SRS) (Freudiger et al. 2008), allow imaging without the need for staining or fluorescent labeling. Such label-free imaging is desirable in biomedical research, because labeling often perturbs the function of small metabolite and drug molecules and may be too toxic to use in vivo. CRS techniques have similar imaging parameters to MPF, making use of pulsed near-infrared lasers to deliver high-sensitivity, high spatial resolution in three dimensions and rapid image acquisition. In this chapter, we will discuss the basic principles of CRS imaging, present the instrumentation requirements for high-speed CRS imaging, and show an example of imaging brain tumors and healthy tissue based on their intrinsic vibrational signatures. This chapter is intended to introduce the benefits and tradeoffs associated with different CRS techniques and show one example of the powerful capabilities of label-free chemical imaging.

[7]Present address: Department of Pathology, Yale University School of Medicine, New Haven, Connecticut 06520.
[8]Present address: Moores Cancer Center, Department of Neurosciences, University of California San Diego, La Jolla, California 92093.
[9]Equal contributors.

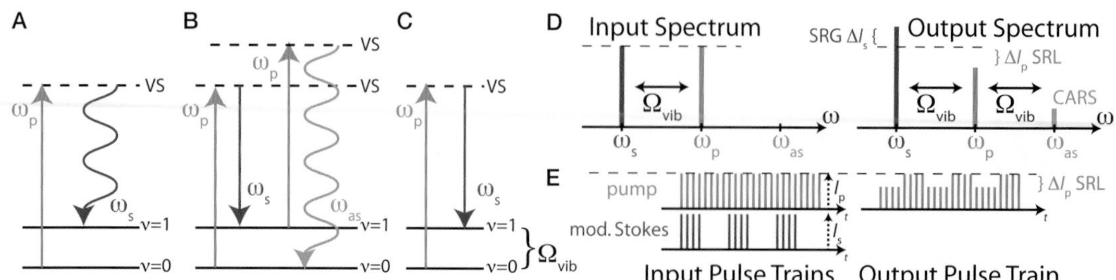

FIGURE 1. Schematic depiction of Raman processes. (*A*) Spontaneous Raman scattering: Incident light at ω_p is scattered to create new frequencies, shown as ω_s. Because $\omega_p - \omega_s = \Omega_{vib}$, and Ω_{vib} is characteristic of molecular structure, Raman scattering can be used to probe the chemical composition of a sample without the use of labels. The Raman scattering efficiency is low, making rapid imaging at high resolution impossible. The individual vibrational states are labeled as $n = 0$ (ground state) and $n = 1$ (first vibrational excited state). The scattering process is mediated by one or more virtual states (VSs). (*B*) Coherent anti-Stokes Raman scattering (CARS): Lasers at two wavelengths, ω_p and ω_s, chosen such that $\omega_p - \omega_s$ matches Ω_{vib} of interest, are applied to the sample. This results in a four-wave mixing process, and the sample emits a new frequency, $\omega_{as} = 2\omega_p - \omega_s$. The coherent nature of the process results in a strong signal, making rapid imaging possible by detecting the intensity at ω_{as} as a function of position in the sample. (*C*) Stimulated Raman scattering (SRS): As in CARS, two wavelengths, ω_p and ω_s, chosen such that $\omega_p - \omega_s = \Omega_{vib}$ of interest, are applied to the sample. Intensity transfer from the pump beam (stimulated Raman loss; SRL) to the Stokes beam (stimulated Raman gain, SRG) is detected. (*D*) Input and output spectra of the CARS and SRS processes, which occur simultaneously at the laser focus. Initially, two laser beams at ω_s and ω_p are present. After passing through the sample containing a species resonant at Ω_{vib}, intensity is transferred from the pump beam (SRL) to the Stokes beam (SRG), and light appears at ω_{as}. The schematic intensity changes are not to scale. SRL and SRG can be as large as one part in 10^3 of the laser beam, and the light at ω_{as} can be as large as one part in 10^8 of the laser intensity. (*E*) Schematic of the SRL detection scheme. The pump beam and modulated Stokes beam are applied to the sample. Because SRL occurs when the Stokes beam is present, and not when it is absent, a modulation in the detected pump beam can be observed owing to the SRL process. By blocking the Stokes beam with an optical filter and detecting only the amplitude-modulated component of the pump beam, SRS can be measured with high sensitivity.

COHERENT RAMAN TECHNIQUES: CARS AND SRS

CRS techniques derive their contrast from the inelastic scattering of light by molecules, known as the Raman effect (Raman 1928). In spontaneous Raman scattering spectroscopy, a narrowband laser beam of frequency ω_p is focused into a sample and the spectrum of the scattered light is measured. This spectrum contains frequency components shifted from the incident frequency because of inelastic scattering from molecular vibrations. This scattering can result either in red ("Stokes") shifting of the incident light, in which case a quantum of vibrational energy is deposited in the sample by the light field, or blue ("anti-Stokes") shifting, in which case a quantum of vibrational energy is removed from the sample by the light field. Figure 1A shows the energy diagram of Stokes-shifted spontaneous Raman scattering. The frequency shift Ω_{vib}, known as the Raman shift, is diagnostic of particular chemical groups in the sample. For example, C-H bonds scatter light with a different frequency shift (2845 cm^{-1} or ~85 THz) than O-H bonds (3250 cm^{-1} or ~97 THz) (Socrates 2001). In principle, Raman scattering can be used to image biological samples by scanning the excitation spot over the sample and recording the scattering spectrum as a function of position. However, the Raman effect is intrinsically weak, and typical biological imaging requires pixel dwell times of 100 msec to 1 sec or longer (Uzunbajakava et al. 2003; Gierlinger and Schwanninger 2006). Acquiring a 256 x 256-pixel image at 0.1 sec per pixel would require ~2 h, meaning that fast dynamic processes cannot be monitored and the sample must remain static for the (potentially very long) imaging time.

Coherent Raman techniques solve this problem by offering contrast based on the intrinsic Raman response of molecules, but with imaging speeds that are four to seven orders of magnitude faster, as fast as 100 nsec/pixel. This means that images based on the intrinsic chemical contrast can be produced on a similar timescale to typical multiphoton-excited fluorescence (MPF) images, up to video rate (30 frames/sec). This makes applications to high-resolution imaging of dynamic, biological systems possible. This dramatic enhancement in the signal size is obtained by exploiting multiphoton processes to coherently drive the molecular vibrations of the sample in phase with each other.

In CRS, two laser beams are used to excite the sample, the "pump" beam, at frequency ω_p, and "Stokes" beam, at frequency ω_s. The frequency difference $\Delta\omega = \omega_p - \omega_s$ can be chosen to be on resonance with a particular vibrational frequency of interest, Ω_{vib}. This combination of two frequencies creates a beating frequency in the sample at the difference frequency $\Delta\omega$. In the case of CARS (Fig. 1B), the beating creates a coherence between the ground and first vibrational excited state, which is probed by a third interaction with the laser field, resulting in the generation of a strong signal at the new anti-Stokes frequency $\omega_{as} = 2\omega_p + \Delta\omega = 2\omega_p - \omega_s$ (Mukamel 1995). The CARS process is dramatically enhanced when $\Delta\omega$ is tuned into resonance with a chemical species in the sample at Ω_{vib}.

CARS was exploited for spectroscopic mapping of flames and other samples as early as the 1960s (Maker and Terhune 1965), but its renaissance as a microscopic technique dates to 1999 (Zumbusch et al. 1999), when it was implemented using colinear pump and Stokes beams and a high-numerical-aperture microscope objective. This approach offers high spatial resolution, high signal levels, and experimental convenience. CARS microscopy offers a number of major benefits, including reported pixel dwell times as short as 100 nsec (but more typically in the 1–10-µsec range), the ability to detect the signal even in the presence of a strong one photon-excited fluorescence background, and a strong backward-directed signal in tissue, which is key for diagnostic applications in thick samples in which the transmitted light is not detectable. However, it has a few major drawbacks. The most important is the presence of an unwanted nonresonant background, which is a signal at ω_{as} even if $\Delta\omega$ is tuned off resonance or no resonant molecules are present in the sample. This background can overwhelm weak resonant signals and results in limited sensitivity, image artifacts, and distorted Raman spectra, which makes image interpretation based on the wealth of Raman spectroscopy literature difficult. In addition, the concentration dependence of CARS is nonlinear (Li et al. 2005). Together these features make quantitative imaging with CARS challenging. Various methods to suppress the CARS nonresonant background have been devised (Cheng et al. 2001; Evans et al. 2004; Ganikhanov et al. 2006b; Jurna et al. 2009; Saar et al. 2009), but they often require complex instrumentation, sacrifice sensitivity, or are not robust enough to implement in tissue samples.

Stimulated Raman scattering (SRS) is closely related to CARS, and both effects happen simultaneously in a CRS imaging instrument. In SRS, rather than detecting the newly generated light at ω_{as}, a small intensity change is detected: either a small intensity decrease of the pump beam ΔI_p (stimulated Raman loss) or a small intensity increase of the Stokes beam ΔI_S (stimulated Raman gain) associated with the excitation of molecular vibrations of the sample when $\Delta\omega$ is in resonance with Ω_{vib} (Fig. 1C,D) (Ploetz et al. 2007; Freudiger et al. 2008). These intensity changes are extremely small when lasers with biocompatible excitation conditions are used (less than one part in 10^3 of the laser intensity), but can be detected by implementation of a high-frequency phase-sensitive detection scheme. The amplitude of either the pump or Stokes beam is modulated at a known frequency and the transfer of that modulation due to SRS to the other beam is detected using a lock-in amplifier, which is a narrowband electrical amplifier that provides high gain at a precise frequency of interest while suppressing noise at other frequencies. If the modulation frequency is chosen to be faster than the laser noise (>1 MHz), high detection sensitivity necessary for imaging of biological samples can be achieved (Freudiger et al. 2008).

Unlike CARS, the SRS energy transfer can only occur when the difference frequency $\Delta\omega$ matches a molecular vibration of the sample. This means that SRS is free of the nonresonant background that limits CARS. It also offers a linear concentration dependence and high sensitivity (Freudiger et al. 2008) and spectral response identical to the spontaneous Raman spectrum, making quantitative or semiquantitative analysis much more straightforward than CARS. Despite its relatively recent application to microscopy, SRS has already proven to be able to image a wider range of vibrational frequencies including those in the low-wave-number region in which the spectral congestion and nonresonant background make CARS imaging problematic. Compared with CARS, SRS offers pixel dwell times that are slightly longer (in the 30–100-µsec/pixel range) and offers weaker backward-detected (epi) signal; thus application to thick tissue specimens remains challenging.

It is important to emphasize that both CARS and SRS happen simultaneously in the focal volume. Although they require different electrical and optical detection hardware, the two modalities can coexist on one imaging system and images can be acquired simultaneously using both techniques and the same excitation laser system. The additional costs and complexity of upgrading an existing CARS system to also perform SRS measurements are minimal and worthwhile for the improved sensitivity and more easily interpreted images.

COHERENT RAMAN INSTRUMENTATION

There are two major components to a CRS imaging system: the laser system and the scanning microscope. Both components can be purchased independently and combined to form a CRS imaging system. In the future, commercial systems combining all of the necessary elements into a single package will become available, opening up CRS techniques to a wider audience by making the setup and operation more user-friendly. However, to optimize a system for particular measurements and retain flexibility for future upgrades, many groups have designed custom modifications of the instrumentation described here, thus trading ease of use for increased flexibility.

The laser system for CRS imaging is the biggest difference between a CRS system and an MPF system. The main reason for this is because two wavelengths, at least one of which must be tunable, are required to excite the CRS process. In addition, because the CRS techniques have an overall nonlinear dependence on the incident intensity ($I_{SRS} \propto I_p \cdot I_s$; $I_{CARS} \propto I_p^2 \cdot I_s$), pulsed lasers are used to maximize the peak intensity while maintaining moderate average powers to minimize sample damage. In CRS, in contrast to MPF, picosecond pulses are preferred for rapid, high-selectivity imaging. To excite the process, the pulses must arrive at the sample at the same instant in time. Because pulse repetition rates on the order of 100 MHz derived from mode-locked oscillators are used, some method of synchronizing the pump and Stokes beams is required. Although feedback electronics to lock two independent lasers (e.g., Ti:sapphire oscillators) are available (Jones et al. 2002), these electronics are extremely sensitive to mechanical vibrations, laser alignment, and other subtle factors. For this reason, synchronously pumped optical systems are preferred, in which one pulsed laser pumps another oscillator and two pulses are automatically optically synchronized. Details of this scheme have been presented in a recent review article (Evans and Xie 2008). Typically it makes use of a passively mode-locked, diode-pumped solid state laser that produces a 76-MHz pulse train of 5–7-psec pulses from an Nd:YVO_4 gain medium. A portion of this output at 1064 nm can be used directly as the Stokes beam for CARS or SRS. The remainder of the beam is frequency-doubled to produce 532 nm light, which synchronously pumps an optical parametric oscillator (OPO). An OPO is a resonant wavelength conversion device that makes use of optical parametric generation in a specially designed nonlinear crystal to provide tunable output pulses at two additional colors (in this case, one from 680 to 1010 nm, called the "signal," and one from 1120 to 2500 nm, called the "idler"). The signal wave can be used as the pump beam for the CRS process. Typically the idler wave is not used, although CARS imaging using the signal and idler waves simultaneously has been shown (Ganikhanov et al. 2006a). However, this offers lower spatial resolution in CRS imaging because of the longer wavelengths, and, in the case of CARS, also requires detection of longer anti-Stokes wavelengths (typically >730 nm), at which the available photomultiplier detectors have low quantum efficiency.

To perform CARS imaging, the pump and Stokes beams must be overlapped in space using a dichroic mirror and overlapped in time by using a passive delay stage in one of the beam lines (Fig. 2A). Once overlapped, the two beams are steered to the laser-scanning microscope, discussed below. For SRS, an additional amplitude modulator must be placed in one of the beam paths (Fig. 2B). Using an acousto- or electro-optic modulator (AOM or EOM, respectively) allows one beam to be amplitude-modulated at high frequency (>1 MHz). AOMs are cheaper and require lower radio frequency (RF) voltages, minimizing electrical noise pickup in the detection circuitry. In a combined CARS/SRS system, the amplitude modulator can be switched off when only CARS imaging is required, or on when SRS is needed. Modulation of the Stokes beam is preferred because then the detected pump beam is shorter in wavelength and optimally detected using silicon photodiodes, as

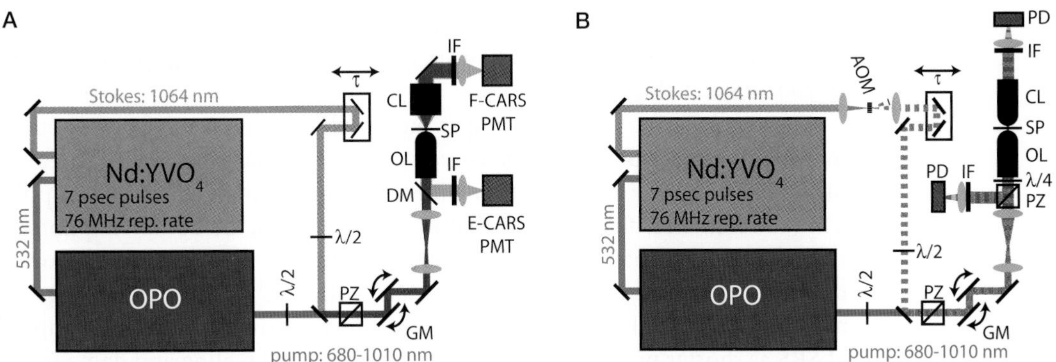

FIGURE 2. Schematic of the CARS and SRS laser systems. A passively mode-locked Nd:YVO_4 laser producing 7-psec pulses at 1064 nm is used as the Stokes beam. A portion of the beam is also frequency doubled to 532 nm and used to synchronously pump an OPO, which produces tunable output pulses from 680 to 1010 nm. The two beams are combined using a passive delay stage, *t*. Half wave plates in both the pump and Stokes beams (λ/2) are combined with a polarizer (PZ) to independently adjust the two power levels. The combined beams are passed into a laser-scanning microscope, which consists of galvanometer mirrors (GM) that scan the beam in two dimensions, an objective lens (OL) to focus the beam, a specimen (SP), a condenser lens (CL) to collect the transmitted beam, an interference filter (IF) to block undesired laser light, and detectors. In the case of CARS, shown in *A*, the anti-Stokes signal is spectrally separated from the pump and Stokes signals by means of a filter in the forward direction and a dichroic mirror (DM) and filter in the EPI direction. In both cases, a high-sensitivity photomultiplier tube (PMT) detects the anti-Stokes signal. In the case of SRS (*B*), the Stokes beam is amplitude-modulated by an acousto-optic modulator (AOM) in zeroth-order mode before combination with the pump beam. After passing through the specimen and being collected using a high-NA condenser, the modulated Stokes beam is blocked and a photodiode (PD) detects the amplitude modulation transferred to the pump beam by the SRS process. In the EPI direction, light passes through a quarter wave plate (λ/4), reflects off the sample, and passes again through the quarter wave plate, so that the reflected light is rotated 90° with respect to the incident light. A polarizing beam splitter then reflects the light through an interference filter and onto the photodiode in the EPI direction.

described below. Once the two beams are combined in space and time and the Stokes beam is passed through the AOM, they are ready to enter the microscope.

A number of groups have published papers that report variations on the laser system described above using 50–200-fsec pulse widths rather than the 2–6-psec pulses (Murugkar et al. 2007; Chen et al. 2009; Pegoraro et al. 2009). Although CRS processes may be excited by femtosecond pulses, this comes at the cost of decreased signal levels, limited tunability, loss of spectral selectivity, and an increased nonresonant background in CARS. There are two main reasons for this. First, the typical width of Raman spectral features is ~15 cm^{-1}. Near 800 nm, this corresponds to a bandwidth of ~1 nm. For any laser system, the inverse relationship between laser pulse width and laser spectral bandwidth means that for a given spectral bandwidth there is a fundamental limit to the shortest pulse that can be achieved. In the case of a 1-nm bandwidth, this pulse width is ~0.95 psec, assuming a Gaussian pulse shape. Shortening the pulse below this value will result in an increase in the nonresonant background/resonant signal ratio in CARS, degrading the contrast and reducing image quality.

In SRS, the result will simply be that no additional signal will be generated even though the peak power (and therefore the nonlinear photodamage) will increase because the frequency components that are not on resonance with the Raman-active transition will not contribute signal. In addition, if two nearby resonances occur, the broader bandwidth will mean that the spectral resolution will be lower and the obtained images will be contaminated by signal from both resonances. For the typical 8-nm bandwidth obtained from the commercial mode-locked femtosecond Ti:sapphire lasers used in multiphoton microscopy, this means that only about one-eighth of the laser energy applied to the sample is productively used by the CRS process. In contrast, for pulses of a few picoseconds, all the laser intensity is concentrated into narrower frequency bands that are completely matched to the Raman resonance, which can be well resolved. Although spectrally resolved detection with broadband femtosecond lasers can recover the CARS or SRS spectrum at high resolution, it typically requires a multielement detector such as a charge-coupled device camera that has very long (>10 msec) readout time at each pixel, severely limiting the imaging speed (Cheng et al. 2002; Wurpel et al. 2002; Lim et al. 2007).

A second feature of slightly longer pulses with higher average power but reduced peak power is that the nonlinear photodamage is reduced. This actually has the practical benefit of allowing for more total SRS signal by excitation with 6-psec pulses than with 150-fsec pulses, even in the case of a broad resonance. The reason is that in many samples, the nonlinear photodamage increases faster than the signal of interest with decreasing laser pulse width (e.g., in work on imaging calcium activity, it was determined that the damage scaled with $1/(\tau^{1.5})$; Hopt and Neher 2001). Since the SRS signal scales with $1/\tau$, the photodamage is clearly expected to increase more rapidly than the signal level of SRS as shorter pulses are used. Of course, the actual scaling and damage threshold is highly sample dependent, so it is difficult to make absolute statements about safe power levels. However, to test this experimentally, we imaged a collection of bacterial cells that had incorporated deuterium from deuterated carbon sources in their growth medium. The carbon-deuterium stretching resonance is exceptionally broad—it has a width of >100 cm^{-1}, and thus even femtosecond pulses may be used to excite it with good efficiency. We then compared the maximum average power that could be applied to bacterial cells without obvious morphological changes in the sample owing to the optical power. This is an extreme test of the sample damage limit, but gives some insight when applied using similar laser parameters with only the pulse width varying. At 180 fsec, an average power of 25 mW was found to be usable. At 1 psec the power level was 80 mW, and at 6 psec the power level was 275 mW without obviously burning the sample. In all cases, the laser repetition rate was 76 MHz, the Stokes wavelength was between 1040 and 1064 nm, and the pump wavelength was between 800 and 816 nm. The same beam parameters and microscope optics were used in all cases. At the maximum power at 6 psec, the absolute SRS signal is higher than the signal at the maximum power at 150 fsec by a factor of 4, even for a broad resonance. In the case of a narrow resonance, the difference is expected to be a factor of 30–40 because of the combination of the damage limits and the spectral widths. Because the major advantage of coherent Raman techniques is the imaging speed and sensitivity, this large factor in signal level obtainable is important and the proper choice of laser pulses in the 1–10-psec range (as opposed to 100–200 fsec) is critical to obtaining the best results. Within this range, the exact choice makes only a small difference in the obtainable signal (less than a factor of 2) while maintaining good spectral resolution.

The microscope imaging system can be readily modified from a commercial MPF laser-scanning microscope. CARS detection can be performed in either the forward or EPI direction, which offer different contrast mechanisms (Volkmer et al. 2001). With the addition of some commercially available optical filters and dichroic mirrors (which can typically be replaced by the user), the two-photon microscope is ready for CARS imaging. The beams should be steered down the center of the optical axis of the microscope, in the same alignment as the MPF laser. Unlike MPF, color correction in the objective lens is critical to obtaining good performance because the pump and Stokes beams must be focused to the same position in space to obtain strong signal. Typically high-numerical-aperture (NA) lenses with NAs of 1–1.2 and magnifications from 40x–63x are used. The choice of objective lens is critical and difficult to predict beforehand, so testing the individual objective lenses is necessary. High-sensitivity photomultipliers in the forward and EPI direction are needed to collect the signal. In CARS, as in MPF, good blocking filters with high optical density throughout the near infrared spectrum and good (ideally >90%) transmission at the anti-Stokes wavelength of interest are needed.

As in MPF, the nonlinear excitation means that the technique is intrinsically depth resolved: Signal is only obtained from the tightest part of the focus. Thus, nondescanned detectors that are close-coupled to the back aperture of the objective lens or the condenser lens offer the most efficient collection of the CARS signal, particularly in scattering samples. A typical series of wavelengths to address the CH_2 stretching resonance in brain tissue (Fig. 3A) would be 1064 nm for the Stokes beam, 816.7 nm for the pump beam, and 660 nm for the anti-Stokes beam, meaning that multialkali photomultiplier tubes can be used.

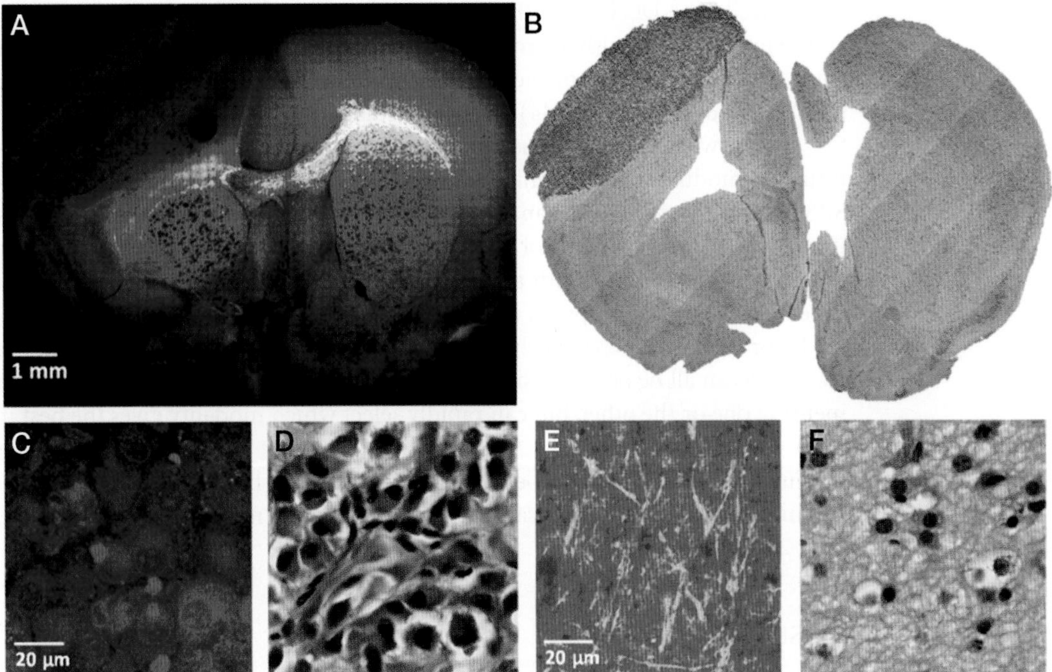

FIGURE 3. Rapid CARS imaging allows acquisition of large area images of whole organs with cell- and tissue-level resolution. As such it can be used for virtual histology in situ to identify tumors. (*A*) CARS mosaic image of a slice through a whole mouse brain with a human stem cell–derived glioma tuned into the CH_2 stretching mode at 2845 cm^{-1}. The tumor can be seen as a dark area in the *upper left corner* above the bright *corpus callosum*, because of the difference in lipid density between tumor and healthy tissue. (*B*) The identification of the tumor is confirmed by conventional histopathology with an H&E-stained micrograph, showing the tumor in purple owing to the hematoxylin-stained nucleic acids. (*C–F*) Subsections from the mosaic. Tumor pathology is characterized by enlarged cell bodies and nuclei (*C*), which are also seen in the traditional histology image (*D*). Healthy white matter tissue is rich in strongly myelinated axons (*E*) and large nuclei are not seen, as confirmed in the traditional histology image (*F*). The comparison of the CARS images (lipid contrast) with traditional histology confirms that similar diagnostic pathological features can be highlighted with CRS imaging.

For SRS detection, the best results so far have been obtained in the forward direction. Forward SRS detection requires a high-NA condenser, ideally greater than the excitation NA. Oil-immersion condensers with NA of 1.4 are available commercially and work well. The use of a high-NA condenser has been shown to reduce spurious background signal owing to other nonlinear processes, such as cross-phase modulation, and to allow for the maximum SRS sensitivity (Freudiger et al. 2008). In addition, epi-SRS detection has also been shown by using a polarizing beam splitter in the epi direction instead of a dichroic mirror, and a quarter wave plate below the objective lens (Freudiger et al. 2008). In this way, light that passes through the quarter wave plate and objective and is then reflected backward is rotated in polarization by 90° and then directed to the epi-SRS detector.

The high-frequency phase-sensitive detection scheme for SRS (Fig. 1E) involves applying an amplitude modulation waveform at a known frequency (e.g., 10 MHz) to the Stokes beam and then detecting the amplitude modulation transferred to the pump beam. For this reason, a filter that blocks the modulated Stokes beam with an optical density of at least 8 is required, so residual leakage of the Stokes beam does not contaminate the signal. Once the Stokes beam is blocked, the pump beam can be detected by a silicon PiN photodiode. These photodiodes can operate at high speed (>10 MHz bandwidth) with a large active area (>100 mm^2) if properly biased. The photodiode output, terminated in 50 Ω, can be directly fed into a lock-in amplifier, which is also referenced to the modulation frequency. The lock-in measures the amplitude and phase of the 10 MHz signal as a function of time and uses that to provide the intensity of a pixel in the image. A passive inductor/capacitor bandpass filter is used to avoid overloading the input amplifier of the lock-in.

Typically, the DC level of the laser impinging on the photodiode is several volts. The SRS signal at maximum power with a strongly resonant sample can be as large as 5 mV. With a 100-µsec time constant, the noise floor of the lock-in amplifier can be as low as 100 nV. One commercial lock-in (the Stanford Research SR844) offers adjustable time constants as short as 10 µsec using the "no filter" option, which allows the user to balance between rapid imaging and signal-to-noise ratio. In the future, demodulation electronics with shorter time constants may be developed to allow even faster SRS imaging. Currently, the analog output of the lock-in amplifier is digitized to produce the images using the microscope control software, which also controls the beam scanning. For this reason, obtaining a microscope with an analog input channel that can be used to feed in the lock-in (or other demodulator) output is important in implementing SRS imaging.

By properly choosing a dichroic mirror in the forward direction, epi- and forward CARS and forward SRS can all be obtained simultaneously. Thus the user does not need to choose between implementing one or the other, but can rapidly select which modality gives the best images for a particular task.

Finally, we note that good-quality MPF images can be obtained with the picosecond laser system, meaning that this platform is capable of multimodality nonlinear optical imaging beyond CRS (Evans et al. 2005).

IMAGING BRAIN TISSUE

As an example of the type of resolution and contrast available from fresh animal tissue we show lipid imaging in brain tissue using both CARS (similar to Evans et al. 2007) and SRS. As discussed above, the major advantages of CARS include the strong epi-detected signal and short pixel dwell times. This contrast is particularly strong for lipid in brain tissue imaged at the CH_2 stretching resonance, because the myelin sheathes in brain tissue strongly generate CARS signal (Evans et al. 2007). This leads to a contrast mechanism that highlights white matter, and individual axons can easily be observed with 1-µsec pixel dwell times.

The rapid pixel dwell times that can be achieved in CARS imaging allow for the acquisition of very large scan areas, up to 1 x 1 cm in ~1 h, with diffraction-limited ~400-nm resolution. For these large maps, imaging is performed at two length scales. The individual field of view of the microscope in beam scanning mode is ~350 x 350 µm. To obtain the larger view, the individual field is scanned, and then an automated stage tiling program moves the stage to the next field. The computer software automatically stitches the images together to form the mosaic. Such tile scanning systems are commercially available from many microscope manufacturers and are ideal for viewing large areas at high resolution. Figure 3A shows such an image of fresh mouse brain tissue without fixing, freezing, sectioning, or staining. More intriguingly, a large dark area in the brain tissue in Figure 3A in the top left corresponds to a glioma implanted into the mouse brain. Comparison to histology (Fig. 3B) shows that the same region is identified using a hematoxylin and eosin stain, the workhorse for diagnostic neuropathology. Figure 3, C and E, shows individual frames of the large area overview in Figure 3A, showing tumor versus healthy tissue, respectively. In the healthy tissue, strongly myelinated neurons are visible and in the tumor, fast growing tumor cells with enlarged nuclei and lipid droplets can be seen. The comparison with the H&E-stained micrographs from comparable regions (Fig. 3D,F) highlights similar microscopic tissue- and cell-level details that can be used by a pathologist for diagnosis. In contrast to traditional pathology, CRS imaging is possible in fresh brain tissue or even in vivo without the lengthy and perturbative fixing, freezing, sectioning, staining, and analysis protocols required for traditional histology.

The motivation for obtaining label-free contrast with diagnostic potential is to allow for imaging of brain structures in vivo. Currently, during surgery to remove a brain tumor, a major challenge is to precisely delineate where the cancerous tissue ends and the healthy tissue begins. The consequences of a mistake—either removing healthy tissue and damaging the patient's neurological function, or incompletely excising the tumor, which can cause a relapse—are potentially catastrophic. The current

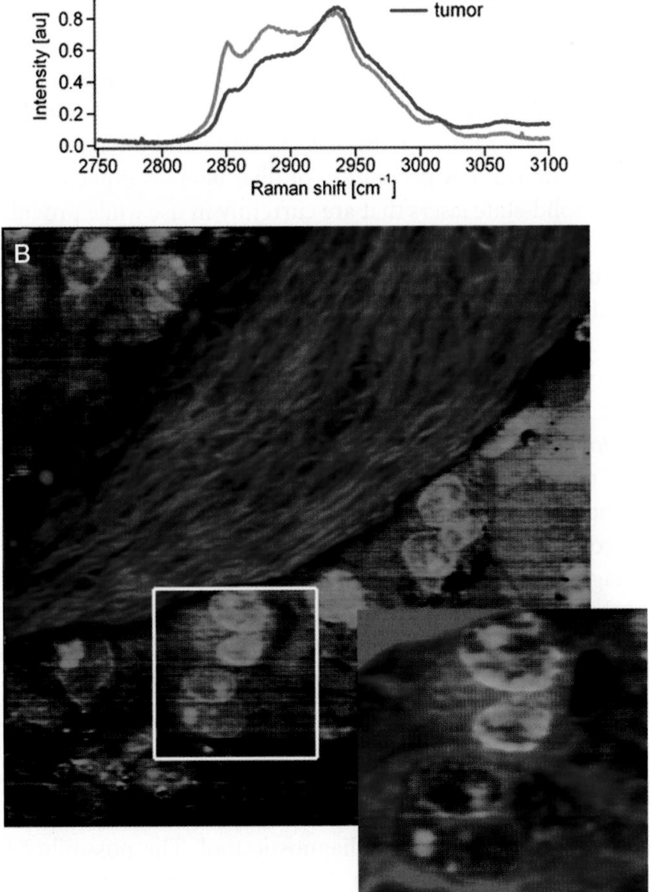

FIGURE 4. SRS microscopy preserves Raman spectra and thus enables multicolor imaging for better efficacy. (*A*) Raman spectroscopy provides a basis for distinguishing healthy tissue from brain tumor tissue based on the different ratios of the lipid (2845 cm^{-1}) and protein (2950 cm^{-1}) bands. (*B*) Diagnostic multicolor contrast can be achieved in SRS based on the ratio of CH_2 (red) to CH_3 bonds (green), which is related to the ratio of lipids to proteins. The CH_3 vibrational counterstain allows for highlighting of the nucleus that normally appears dark in the CH_2-contrasted SRS images. The *inset* shows that it is possible to image the nuclear membrane and subnuclear features.

practice involves the surgeon removing a small portion of tissue near the margin of the excision and sending it for pathological diagnosis. This tissue is frozen, sliced, stained, and viewed under a microscope by a trained neuropathologist, who then provides a diagnosis back to the surgeon. The entire process can take 15 min or more and often concludes with a verdict of "nondiagnostic sample." This feedback process is slow and uncertain, adding time to the surgical procedure and increasing patient morbidity and mortality. Addressing this and related problems with label-free, all-optical contrast methods is a major goal in biomedical optics. Coherent Raman microscopy techniques are uniquely suited to this challenge because they offer subcellular resolution in tissue with chemical contrast. Those two properties are the key to making histological diagnoses, which involve looking at subtle features of the tissue architecture such as patterns in the shape and size of cell nuclei or the presence of various types of cells involved in the tumor response (e.g., macrophages and lymphocytes).

Multicolor contrast has turned out to be essential in traditional pathology. In CRS, an analogous effect using different Raman shifts can be used to generate different colors for different cellular features within an image. By comparing the contrast available at two Raman shifts in the CH band (Fig. 4A) with SRS, nuclei can be highlighted easily using simple image subtraction, because of the differences in contrast between lipids and proteins in the cell body and nucleus. The cell body can also be highlighted by using the sum of the two images, producing an image that is comparable to the histological stains, but is obtained using only light applied to the tissue (Fig. 4B). This approach holds promise as an in vivo diagnostic method for tumor excision, which will allow for more rapid and precise delineation of the tumor margin in patients.

OUTLOOK

Over the past ten years dramatic progress has been made in CRS imaging, especially in the areas of instrumentation, understanding of the contrast mechanism, and detection schemes, culminating in the recent development of SRS imaging. Nonetheless, many challenges remain and the field is continuing to expand by adding laboratories with new expertise and interests.

In terms of instrumentation, a new generation of fiber lasers is expected to compete in performance with diode-pumped solid-state lasers that are currently in use while providing major advantages in cost and ease-of-use. Thus far, fiber sources of high-power, fixed-wavelength picosecond pulse trains are available (Kieu et al. 2009), but further work on fiber frequency shifting (Andresen et al. 2007) is needed to produce the high-power tunable second color for CRS in optical-fiber-format device. In addition, manufacturers of laser systems are currently in the process of developing turnkey, one-box systems of the two-color pulse-train needed for CARS (Huber et al. 2009). This is expected to open the field to groups without ultrafast laser expertise in much the same way that MPF microscopy became dramatically easier with the development of closed, one-box Ti:sapphire lasers.

For applications in neuroscience, significant effort has gone into improving the penetration depth of MPF microscopy to 1000 μm (Theer et al. 2003; Rueckel et al. 2006). So far, the imaging depth in CRS microscopy in brain tissue has been limited to <100 μm (Evans et al. 2007). Fundamentally, the imaging parameters used in CRS imaging, in terms of the excitation and detection wavelengths and the laser energies, are quite similar to MPF microscopy. For this reason, it is expected that using adaptive optics (Wright et al. 2007), amplified laser systems, and/or highly optimized lenses may allow for far greater imaging depths than have been achieved so far in CRS imaging. This is a challenging area and one in which exchange of ideas between lens manufacturers and end users could greatly improve performance.

Finally, understanding the contrast mechanism in more detail and gaining experience with various models of disease in animal and human tissue are the key areas for research aimed at obtaining widespread acceptance of CRS imaging as a diagnostic tool. The possibility for real time, in vivo optical biopsies without removing tissue from the patient is intriguing and will motivate research into biomedical applications of CRS imaging for years to come.

ACKNOWLEDGMENTS

We thank Dr. Conor L. Evans and Dr. Wei Min for early work and helpful discussions. Funding for this work was provided by the NIH Director's Pioneer Award and by the U.S. National Science Foundation (grant DBI-0649892).

REFERENCES

Andresen ER, Nielsen CK, Thøgersen J, Keiding SR. 2007. Fiber laser-based light source for coherent anti-Stokes Raman scattering microspectroscopy. *Opt Exp* **15:** 4848–4856.

Chen H, Wang H, Slipchenko MN, Jung Y, Shi Y, Zhu J, Buhman KK, Cheng J-X. 2009. A multimodal platform for nonlinear optical microscopy and microspectroscopy. *Opt Exp* **17:** 1282–1290.

Cheng J-X, Book LD, Xie XS. 2001. Polarization coherent anti-Stokes Raman scattering microscopy. *Opt Lett* **26:** 1341–1343.

Cheng J-X, Volkmer A, Book LD, Xie X S. 2002. Multiplex coherent anti-Stokes Raman scattering microspectroscopy and study of lipid vesicles. *J Phys Chem B* **106:** 8493–8498.

Denk W, Strickler JH, Webb WW. 1990. Two-photon laser scanning fluorescence microscopy. *Science* **248:** 73–76.

Evans CL, Xie XS. 2008. Coherent anti-Stokes Raman scattering microscopy: Chemical imaging for biology and medicine. *Annu Rev Anal Chem* **1:** 883–909.

Evans CL, Potma EO, Xie XS. 2004. Coherent anti-Stokes Raman scattering spectral interferometry: Determination of the real and imaginary components of nonlinear susceptibility $\chi^{(3)}$ for vibrational microscopy. *Opt Lett* **29:** 2923–2925.

Evans CL, Potma EO, Puoris'haag M, Côté D, Lin CP, Xie XS. 2005. Chemical imaging of tissue in vivo with video-rate coherent anti-Stokes Raman scattering microscopy. *Proc Natl Acad Sci* **102:** 16807–16812.

Evans CL, Xu X, Kesari S, Xie XS, Wong STC, Young GS. 2007. Chemically-selective imaging of brain structures with CARS microscopy. *Opt Exp* **15:** 12076–12087.

Freudiger CW, Min W, Saar BG, Lu S, Holtom GR, He C, Tsai JC, Kang JX, Xie XS. 2008. Label-free biomedical imaging with high sensitivity by stimulated Raman scattering microscopy. *Science* **322:** 1857–1861.

Ganikhanov F, Carrasco S, Xie XS, Katz M, Seitz W, Kopf D. 2006a. Broadly tunable dual-wavelength light source for coherent anti-Stokes Raman scattering microscopy. *Opt Lett* **31:** 1292–1294.

Ganikhanov F, Evans CL, Saar BG, Xie XS. 2006b. High-sensitivity vibrational imaging with frequency modulation coherent anti-Stokes Raman scattering (FM CARS) microscopy. *Opt Lett* **31:** 1872–1874.

Gierlinger N, Schwanninger M. 2006. Chemical imaging of poplar wood cell walls by confocal Raman microscopy. *Plant Phys* **140:** 1246–1254.

Hopt A, Neher E. 2001. Highly nonlinear photodamage in two-photon fluorescence microscopy. *Biophys J* **80:** 2029–2036.

Huber HP, Zoppel S, Rimke I. 2009. Ultrafast lasers: CARS ultrafast light source is hands-free. *Laser Focus World* **45:** 47–51.

Jones DJ, Potma EO, Cheng J-X, Burfeindt B, Pang Y, Ye J, Xie XS. 2002. Synchronization of two passively mode-locked, picosecond lasers within 20 fs for coherent anti-Stokes Raman scattering microscopy. *Rev Sci Instrum* **73:** 2843–2848.

Jurna M, Korterik JP, Otto C, Herek JL, Offerhaus HL. 2009. Vibrational phase contrast microscopy by use of coherent anti-Stokes Raman scattering. *Phys Rev Lett* **103:** 043905.

Kieu K, Saar BG, Holtom GR, Xie XS, Wise FW. 2009. High-power picosecond fiber source for coherent Raman microscopy. *Opt Lett* **34:** 2051–2053.

Li L, Wang H, Cheng J-X. 2005. Quantitative coherent anti-Stokes Raman scattering imaging of lipid distribution in coexisting domains. *Biophys J* **89:** 3480–3490.

Lim S-H, Caster AG, Leone SR. 2007. Fourier transform spectral interferometric coherent anti-Stokes Raman scattering (FTSI-CARS) spectroscopy. *Opt Lett* **32:** 1332–1334.

Maker PD, Terhune RW. 1965. Study of optical effects due to an induced polarization third order in the electric field strength. *Phys Rev* **137:** A801–A818.

Mukamel S. 1995. *Principles of nonlinear optical spectroscopy.* Oxford University Press, New York.

Murugkar S, Brideau C, Ridsdale A, Naji M, Stys PK, Anis H. 2007. Coherent anti-Stokes Raman scattering microscopy using photonic crystal fiber with two closely lying zero dispersion wavelengths. *Opt Express* **15:** 14028–14037.

Pegoraro AF, Ridsdale A, Moffatt DJ, Jia Y, Pezacki JP, Stolow A. 2009. Optimally chirped multimodal CARS microscopy based on a single Ti:sapphire oscillator. *Opt Exp* **17:** 2984–2996.

Ploetz E, Laimgruber S, Berner S, Zinth W, Gilch P. 2007. Femtosecond stimulated Raman microscopy. *Appl Phys B* **87:** 389–393.

Raman CV. 1928. A change of wave-length in light scattering. *Nature* **121:** 619.

Rueckel M, Mack-Bucher JA, Denk W. 2006. Adaptive wavefront correction in two-photon microscopy using coherence-gated wavefront sensing. *Proc Natl Acad Sci* **103:** 17137–17142.

Saar BG, Holtom GR, Freudiger CW, Ackermann C, Hill W, Xie XS. 2009. Intracavity wavelength modulation of an optical parametric oscillator for coherent Raman microscopy. *Opt Exp* **17:** 12532–12539.

Socrates G. 2001. *Infrared and Raman characteristic group frequencies: Tables and charts.* Wiley, New York.

Theer P, Hasan MT, Denk W. 2003. Two-photon imaging to a depth of 1000 µm in living brains by use of a Ti:Al_2O_3 regenerative amplifier. *Opt Lett* **28:** 1022–1024.

Uzunbajakava N, Lenferink A, Kraan Y, Volokhina E, Vrensen G, Greve J, Otto C. 2003. Nonresonant confocal Raman imaging of DNA and protein distribution in apoptotic cells. *Biophys J* **84:** 3968–3981.

Volkmer A, Cheng J-X, Xie XS. 2001. Vibrational imaging with high sensitivity via epi-detected coherent anti-Stokes Raman scattering microscopy. *Phys Rev Lett* **87:** 023901.

Wright AJ, Poland SP, Girkin JM, Freudiger CW, Evans CL, Xie XS. 2007. Adaptive optics for enhanced signal in CARS microscopy. *Opt Exp* **15:** 18209–18219.

Wurpel GWH, Schins JM, Müller M. 2002. Chemical specificity in three-dimensional imaging with multiplex coherent anti-Stokes Raman scattering microscopy. *Opt Lett* **27:** 1093–1095.

Zumbusch A, Holtom GR, Xie XS. 1999. Three-dimensional vibrational imaging by coherent anti-Stokes Raman scattering. *Phys Rev Lett* **82:** 4142.

49 Ultramicroscopy

Light-Sheet-Based Microscopy for Imaging Centimeter-Sized Objects with Micrometer Resolution

Klaus Becker,[1,2] Nina Jährling,[1,2,3] Saiedeh Saghafi,[1,2] and Hans-Ulrich Dodt[1,2]

[1]*Vienna University of Technology, Institute of Solid State Electronics, Department of Bioelectronics,1040 Vienna, Austria;* [2]*Center for Brain Research, Medical University of Vienna, Section Bioelectronics, 1090 Vienna, Austria;* [3]*University of Oldenburg, Department of Neurobiology, 26129 Oldenburg, Germany*

ABSTRACT

This chapter presents an overview of light-sheet-based microscopy and describes the underlying physics of light sheet generation. The assembly of an "ultramicroscope" for investigating fixed chemically cleared tissue is described in detail, and the functions of the essential components, such as mechanics, camera, and objectives, are discussed. We present some protocols for specimen preparation and clearing for investigations in autofluorescent light or using immunostaining, lectin-labeling, or fluorescent protein expression. Further possible advantages in ultramicroscopy are discussed.

INTRODUCTION

Detailed 3D anatomical reconstruction of intact specimens (e.g., whole mouse brains and embryos), which range in size from several millimeters to centimeters, is extremely difficult using conventional light microscopy. Alternatives such as micromagnetic resonance imaging (μMRI) are currently not capable of achieving the required resolution (Tyszka et al. 2005) and cannot be combined with fluorescence labeling (Tyszka et al. 2005). Consequently, until quite recently, techniques involving mechanical slicing procedures, despite being labor intensive, were the only methods available for overcoming these problems, although the results obtained were often unsatisfactory. Some problems, such as mechanical tissue distortion by the microtome knife, are unavoidable, and the complexity of reference point mapping associated with accurately and precisely registering hundreds of single sections to create a reliable 3D model force the investigator to compromise on the desired results (Weninger and Mohuhn 2002).

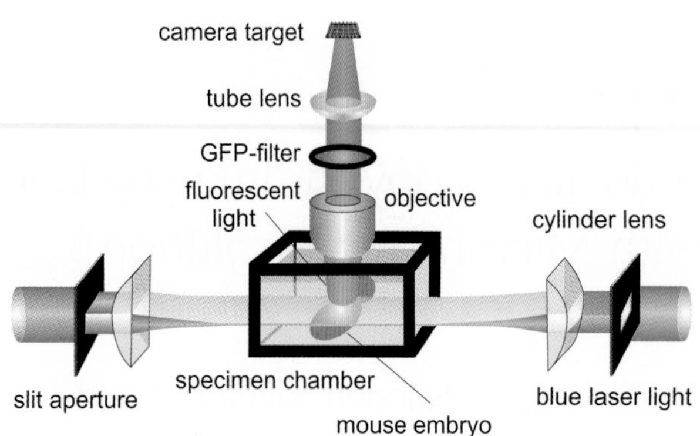

FIGURE 1. Principle of UM. The sample is illuminated from the side by two thin counterpropagating sheets of laser light. Fluorescent light is thus emitted only from a thin optical section and collected by the objective. Hence, photobleaching in out-of-focus regions is avoided. (Adapted, with permission, from Becker et al. 2008.)

Ultramiscroscopy (UM) is a microscopy technique in which the specimen is illuminated perpendicular to the observation pathway by two thin counterpropagating sheets of laser light, which are formed by one or more optical components (e.g., cylindrical lenses; Fig. 1). Thus, only the in-focus "area" of the specimen is illuminated. Hence, out-of-focus "areas" cannot contribute to stray-light generation and image blurring. The two-sided illumination is especially useful for large specimens, such as whole mouse brains, because it counterbalances the loss of light intensity due to light absorption by the tissue to a certain degree. To further minimize the generation of stray light, typically specimens are chemically cleared before investigation.

UM achieves precise and accurate 3D reconstructions of macroscopic specimens with micrometer resolution, without requiring microtome slicing (Dodt et al. 2007; Becker et al. 2008; Jährling et al. 2008). UM is closely related to a growing family of comparable microscopy approaches based on light sheet illumination developed in recent years. Among these techniques are single-plane illumination microscopy (SPIM) (Huisken et al. 2004; also see Chapter 51), orthogonal-plane fluorescence optical sectioning (OPFOS) (Voie et al. 1993), and a number of other methods utilizing the same principle (Huisken and Stainier 2009). Light sheet illumination has been used for more than a century and can be traced to an invention by H. Siedentopf and R. Zsigmondy (1903), in which the term ultramicroscopy was originally coined. Using UM, it is possible to generate a three-dimensional reconstruction of a whole mouse embryo in <1 h. The technique can be applied to immunolabeled specimens, XFP-expressing transgenic animals, and autofluorescence investigations. Speed and simplicity make UM ideally suited for high-throughput phenotype screening of transgenic mutants of mice, flies, and the large number of other specimens generated by biomedical research.

PRINCIPLE OF ULTRAMICROSCOPY

Imaging intact macroscopic specimens requires large fields of view, which necessitates the use of objectives with a low numerical aperture (NA). Unfortunately, these normally have poor axial resolution, thus restricting 3D reconstructions using either confocal or two-photon microscopy to specimen sizes that are less than a few hundred micrometers (Helmchen and Denk 2005). Efforts have been made to overcome this limitation using computational techniques such as wide-field deconvolution (Shaw 1995). Although such an approach has become more applicable with increases in computing power and memory, it still can be applied in real time. Furthermore, the results are severely limited by the presence of noise (Shaw 1995). In UM, the illumination and the observation pathway of the microscope are separated to achieve an uncoupling of lateral and axial resolution. One consequence is that the axial resolution is not limited by the NA of the objective, but by the thickness of the light sheet. In UM, optical sectioning is achieved by stepping the specimen chamber vertically through the light sheet in small increments using a computer-controlled jack.

Using light sheet illumination, no photobleaching is generated in the out-of-focus planes of the specimen, because these are always in the dark. Critically, when working with highly bleachable fluorophores, this is an important advantage compared with standard confocal microscopy.

Theory of Light Sheet Generation by a Single Cylinder Lens

The spot size of a symmetrical ideal Gaussian beam along the propagation axis is generally given by

$$w(z) = w_0\sqrt{1+\left(\frac{\lambda \cdot z}{\pi \cdot w_0^2}\right)^2}. \tag{1}$$

In practice, most lasers do not produce diffraction-limited Gaussian beams. The output intensity distribution of a real laser beam is related to an ideal Gaussian beam (TEM_{00} mode) through a standard parameter (ISO Standard 11146), the M^2-beam propagation factor (Siegman 1998). M^2 is usually provided by the laser manufacturer; if not, there are several standard methods for measuring this factor experimentally (ISO Standard 11146; Sasnett and Johnson 1991). Considering M^2, Equation 1 becomes, for the x direction,

$$w_x^2 = w_{0x}^2 + M_x^4 \times \left(\frac{\lambda}{\pi w_{0x}}\right)^2 (z - z_{0x})^2, \tag{2}$$

and, similarly for the y direction,

$$w_y^2 = w_{0y}^2 + M_y^4 \times \left(\frac{\lambda}{\pi w_{0y}}\right)^2 (z - z_{0y})^2 \tag{3}$$

(Siegman 1998).

If a cylindrical lens of focal length f is illuminated by a Gaussian beam, the beam width and curvature change in one direction, while both remain approximately unchanged in the direction perpendicular to the focused direction (Siegman 1986) (Fig. 2A).

The time-averaged intensity distribution of this beam of wavelength λ along the z-axis can be described by

$$I(r,z) = I_0\left(\frac{w_{0x}w_{0y}}{w_x(z)w_y(z)}\right)\exp\left(-2\left(\frac{x^2}{w_x^2(z)} + \frac{y^2}{w_y^2(z)}\right)\right) \tag{4}$$

(Fig. 2A). In Equation 4, x and y are the distance from the center of the beam in Cartesian coordinates. w_{0x} and w_{0y} are the beam waist (minimum spot size value along the z-axis) at x and y direction, respectively. Similarly, $w_x(z)$ and $w_y(z)$ are the beam radius along the propagation axis z at any arbitrary point (e.g., at $1/e^2$ cutoff values), and I_0 is the maximum intensity at the beam center. When the laser beam is focused, a new beam waist (constructed waist) is produced (Fig. 2A). The minimal half-width w_{0f} of this new constructed waist near the focal plane approximately is

$$w_{0f} \cong M^2 \frac{\lambda \cdot f}{\pi \cdot w_D}. \tag{5}$$

In Equation 5, w_D is the half-width of the lens-illuminating laser beam in the focused direction. The beam width and intensity, at any arbitrary point along the z-direction, can be obtained using Equations 2–5 by replacing the beam waist with the constructed waist in the related direction.

Effects of the Medium in the Specimen Chamber on the Beam Propagation

The width w_D of the beam illuminating the cylindrical lenses is controlled by removable slit apertures, which are placed directly before the cylindrical lenses (Fig. 1). Using different slit apertures (e.g., 2, 4, 6, 8, or 10 mm with cylindrical lenses of focal length f = 80 mm) the shape of the light sheet can be adjusted.

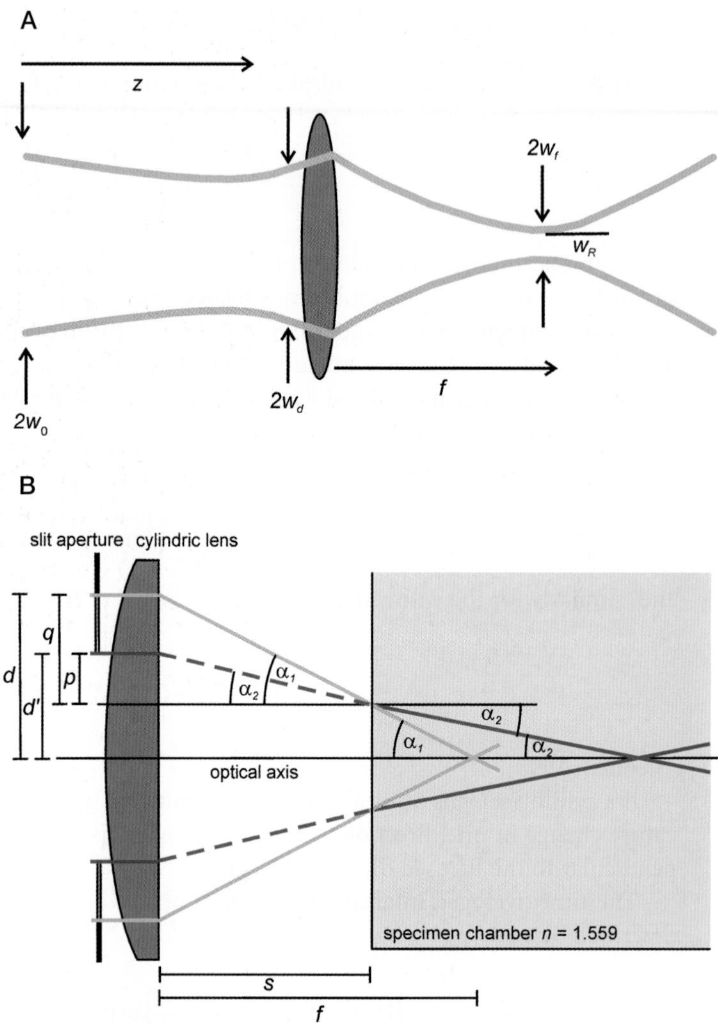

FIGURE 2. (*A*) Refraction of a real Gaussian beam ($M^2 > 1$) at a single cylindrical lens. When the laser beam is focused, a new waist of the laser beam (constructed waist) is produced. Symbols: *z*, propagation axis; w_0, initial beam radius; w_D, half-width of the lens-illuminating laser beam; w_f, constructed beam waist; w_R, Raleigh range. (*B*) Because of the refraction at the specimen chamber, the effective width of the slit aperture *d′* is smaller than the width of the physical slit aperture *d* by a certain factor γ. This factor depends on the distance between cylindrical lens and specimen chamber, the focal length of the cylindrical lens, and on the refractive index of the medium in the specimen chamber. (*B*, Modified, with permission, from Jährling et al. 2008, ©Elsevier GmbH.)

As the illumination beam is refracted at the wall of the specimen chamber, owing to the changing refractive index, the numerical values *d* for the width of the slit apertures have to be corrected by a constant arbitrary number γ, because the effective slit aperture is smaller than the physical aperture *d* (Fig. 2B):

$$d' = \gamma \cdot d. \qquad (6)$$

γ can be approximately determined from geometrical optics:

$$\gamma = 1 - \frac{s}{f} + \frac{s}{\sqrt{n^2\left(f^2 + d^2\right) - d^2}} \qquad (7)$$

(Jähring et al. 2008).

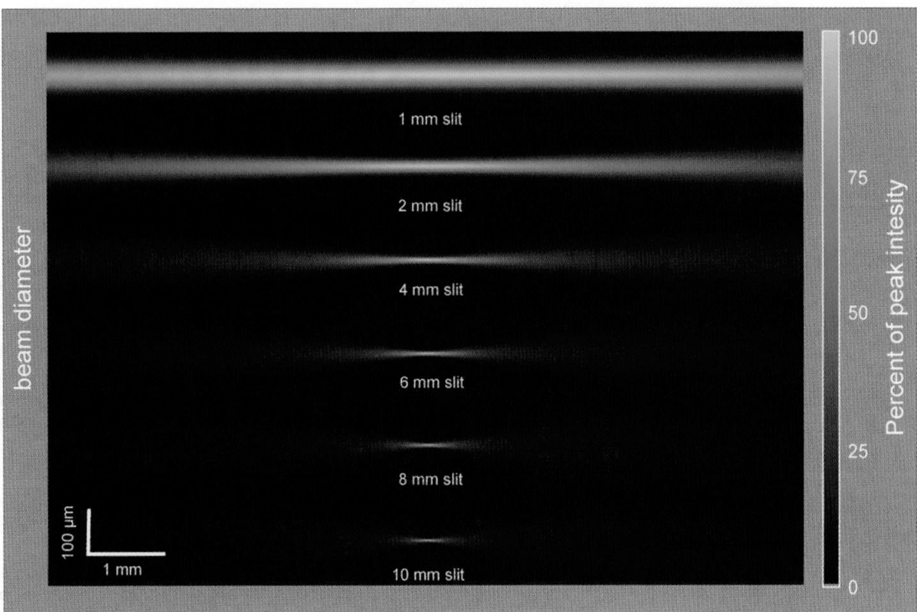

FIGURE 3. Spatial light intensity profiles for different values of the slit aperture width *d*, calculated according to Equations 4–7. The waist of the light sheet becomes thinner with increasing slit widths, but it strongly expands with increasing distance from the focus (simulation parameters: laser wavelength λ = 488 nm, focal length of cylindrical lens f = 80 mm, distance between cylindrical lens and specimen chamber s = 60 mm, refractive index of specimen chamber medium n = 1.559). (Modified, with permission, from Jährling et al. 2008, ©Elsevier GmbH.)

Figure 3 shows a plot of the simulated beam shapes calculated according to Equations 4–7 for different slit apertures ranging from 1 mm up to 10 mm width. Clearly, the waist w_f of the light sheet becomes thinner with increasing slit widths; conversely, the Raleigh range z_R increases (Fig. 2A). Consequently, the width of the light sheet has to be individually modified for each specimen, depending on its size.

Specimen Dehydration and Clearing

Because UM requires the excitation light sheet to travel throughout the entire horizontal width of the specimen, specimens usually have to be rendered transparent before microscope inspection. Such clearing can be obtained by a technique first applied by the German anatomist W. Spalteholz in his studies of heart vascularization (Spalteholz 1914). It relies on incubation of the specimen in an oily medium having almost the same refractive index as protein. Thus, light scattering is strongly reduced, and light traverses the specimen with minimal diffraction. Additionally, if the light absorption by dyes is low, the specimens will become almost completely transparent (Fig. 4). Detailed clearing procedures depend on the type and size of the specimen. Protocols have been established for *Drosophila*, mouse embryos, mouse brains, and isolated mouse hippocampi, and are presented at the end of this chapter.

Image Enhancement by Deconvolution

The quality of ultramicroscopic 3D reconstructions can be significantly improved by deconvolution (Fig. 5). Commercial software packages like Huygens (Scientific Volume Imaging) or AutoQuant (Media Cybernetics) developed for confocal microscopy have been found suitable for this purpose. With Huygens the best results are obtained in the "confocal deconvolution" mode, using an experimentally determined point-spread function (PSF) (Jährling et al. 2008). Figure 5C shows a reconstruction obtained from mouse nerve fibers in the hindlimb of an E12.5 mouse. Figure 5D, obtained

FIGURE 4. A mouse embryo E12.5 (light brown) placed in the specimen chamber filled with clearing solution (BABB). A Siemens star allows for an estimation of the degree of transparency. (Modified, with permission, from Jährling et al. 2008, ©Elsevier GmbH.)

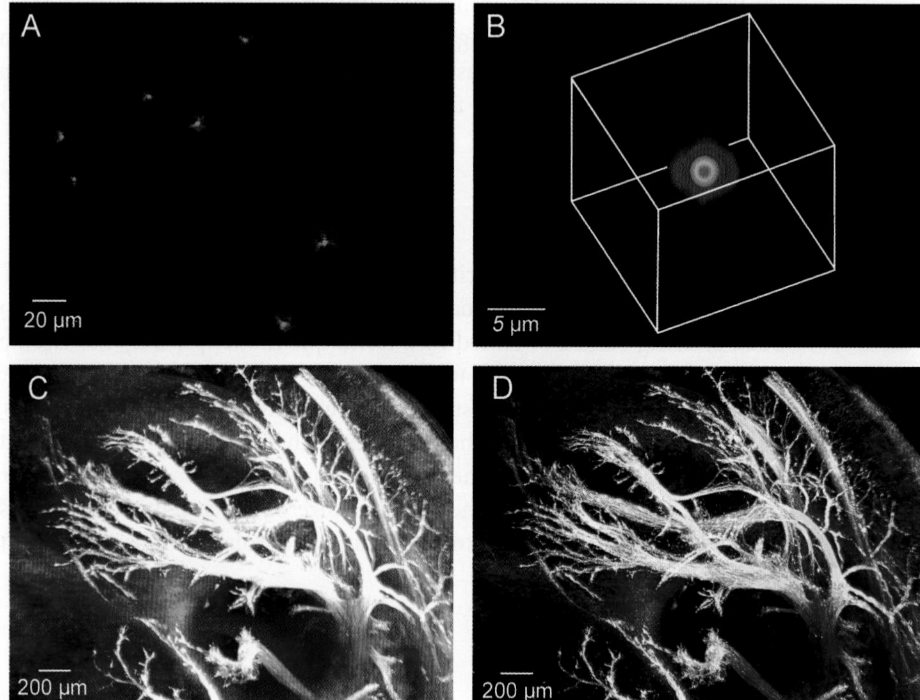

FIGURE 5. Quality improvement of UM reconstructions by non-blind deconvolution. (*A*) Resin-embedded latex beads for measuring the point-spread function (FluoSpheres 1 µm, 488 nm; Invitrogen). Objective: Olympus Fluar 10x, NA 0.3 with 4x postmagnification. (*B*) PSF determined by averaging the bead recordings shown in *A*. (*C*) 3D reconstruction of a part of the right hindlimb of a mouse embryo E12.5. (*D*) Deconvolution of *C* using a measured PSF. In contrast to *C* even fine nerve branches are distinguishable. (*C,D*, Reprinted, with permission, from Jährling et al. 2008, ©Elsevier GmbH.)

from the same stack of data, has been deconvolved using Huygens. As a result, fine details, like small fiber branches, become more pronounced and background intensity is strongly reduced (Fig. 5D).

Measuring the PSF of an UM setup can be performed with fluorescent beads that are 0.5–1 µm in diameter (e.g., FluoSpheres, 1 µm, 488 nm; Invitrogen). Because the beads are delivered in an aqueous solution, which is not miscible with the solution in the specimen chamber, 1 µL of the bead suspension is dried overnight in an open tube in a desiccator, and then mixed with 1 mL of polyester resin (Polyester-Klarharz, Carl Roth GmbH, Germany), and 20 µL of the included accelerator. Curing is performed overnight at room temperature in a small cube-shaped rubber form of ~1 x 1 x 1-cm feed size. The cured resin has approximately the same refractive index as BABB ($n \approx 1.565$). A stack of 150–200 images is recorded from the resin-embedded beads (Fig. 5A). To obtain sufficient resolution, either a twofold or fourfold postmagnification can be used. From the obtained image stack, a 3D model of the PSF can be calculated (e.g., by the "bead distiller" provided with Huygens; Fig. 5B). The PSF measurements have to be performed separately for different combinations of objectives and slit apertures. Before starting a certain deconvolution, the appropriate PSF data have to be loaded into Huygens. Because the deconvolution process is computationally very intensive and consumes an extensive amount of memory, the computer should be sufficiently powerful. For deconvolving overnight a large stack containing ~2048 x 2048 x 800 voxels, a multiprocessor system (e.g., two Intel Quad Core processors) equipped with at least 24 GB of main memory is recommended.

ULTRAMICROSCOPY SETUP

A complete ultramicroscope can be configured at relatively low cost using standard components from various optics and microscope distributors. To avoid artifacts from mechanical vibrations, the whole UM setup should be assembled on a vibration-cushioned optical breadboard (e.g., 150 x 90 cm in size) (Fig. 6).

Mechanical Components

The microscope stand (e.g., Leica), the jack for vertically moving the specimen chamber, and two linear stages (e.g., miCos GmbH) for fine focus adjustment of the cylindrical lenses are the essential

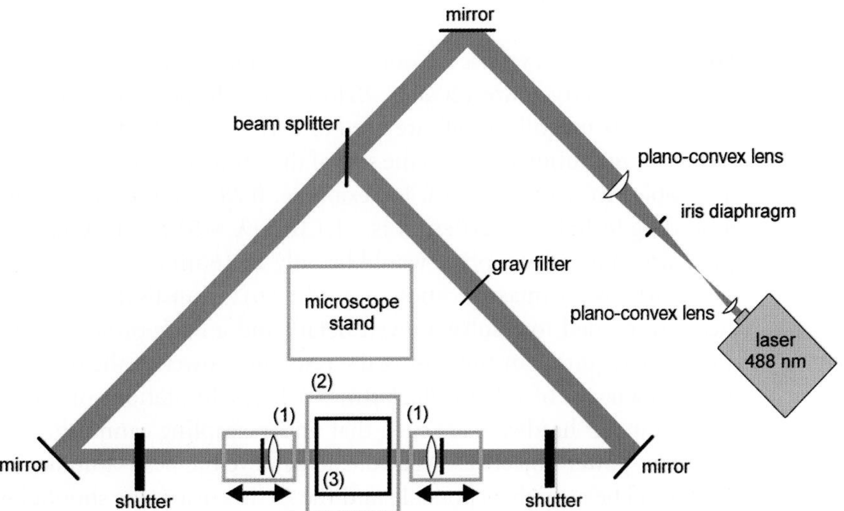

FIGURE 6. Schematic drawing of the UM setup. (1) Motor-driven linear stages with cylindrical lenses and slit apertures. (2) Computer-controlled jack. (3) Specimen chamber filled with clearing medium. (Modified, with permission, from Jährling et al. 2010.)

mechanical components of an UM setup. These components are operated by computer-controlled stepping motors, allowing precise and vibration-free finetuning in the submicrometer range. In our laboratory, the optical components for light sheet generation, like lenses and mirrors, are assembled using the "microbench" optical construction system from LINOS Photonics. Some minor elements (e.g., the holders for the 45° tilted mirrors) are custom made (Fig. 6).

Optical Components

The microscope consists of a binocular eyepiece (e.g., Olympus) with integrated camera interface, mounted on a custom-made intermediate tube equipped with a thread at its bottom side for aligning the objectives. The tube fits into the holding ring of the microscope stand and has a collateral slit for inserting a variety of optical detection filters.

For projecting the microscopic image onto the camera target, we use standard fluorescence objectives (XL Fluor 2x NA 0.14, XL Fluor 4x NA 0.28, or Fluor 10x NA 0.30; Olympus). The objectives are encased by custom-made plastic caps, having an ~0.5-mm-thick circular glass window at their lower surface. The caps allow for the immersion of the objective into the specimen chamber medium without damage. As the objectives are calculated for air, image quality suffers from a certain degree of spherical aberration, when viewing specimens embedded in a highly refractive medium, such as BABB. The caps ensure that the optical pathway through the medium between the light sheet and the front lens of the objective always remains constant, while stepping the specimen chamber vertically through the light sheet. Thus, the degree of spherical aberration does not depend on how deeply one focuses into the specimen.

As a light source, we use a 488-nm optically pumped semiconductor laser of 500 mW power (sapphire; Coherent), generating a monomode laser beam with an approximately Gaussian intensity profile (TEM_{00}, $M^2 = 1.1$), and a beam diameter of 0.7 ± 0.05 mm. The beam is expanded and homogenized by two convex lenses of 6-mm and 200-mm focal length to a diameter of ~15 mm (Fig. 6).

A system of mirrors and a beam splitter is used to guide the laser beam into two cylindrical lenses of 80-mm focal length, placed at the left and right side of the specimen chamber. Custom-made holders are mounted directly behind the lenses, which allow for the insertion of different slit apertures, ranging from 2 to 16 mm in width. As discussed before, these slit apertures aid in shaping the light sheet to accommodate different specimen sizes (Fig. 6).

Camera and Software

For data sampling, we use a CoolSNAP K4 cooled charge-coupled device (CCD) camera with 15.16 x 15.16-mm target size (2048 x 2048 pixels) (Roper Scientific). In general, low-power objectives (e.g., 2x–20x magnification) are used in UM. Consequently, when selecting a camera, possible limitations of resolution owing to the size of the camera target have to be taken into consideration. Using a 4x objective with a NA of, for example, 0.28, the theoretical lateral resolution of the objective according to Raleigh's criterion is ~1.1 µm ($\lambda = 510$ nm). On the other hand, the back-projected pixel size in the object plane would be only 15.16 mm/(2048 x 4) = 1.85 µm, in this example, without using any postmagnification. According to Nyquist's theorem, at least 2 x 2 pixels (more, in practice) are needed to resolve a given detail; and a back-projected pixel size of (1.1/2) x (1.1/2) µm would be required for fully using the resolving power of the objective. This results in an undersampling by a factor of 1.85 µm/0.55 µm = 3.4. The limitations of today's scientific grade CCD cameras and computer hardware are such that undersampling cannot be completely avoided when working with low-power objectives. As a rule, to achieve the best sampling rates, the number of camera pixels should be as high as possible and the CCD sensor size should be small. Camera operation, jack, shutters, and the two linear stages are controlled by custom-made software programmed in Visual Studio 2008 (Microsoft), allowing for the automatic recording of an image stack once all optical components are correctly adjusted.

PRACTICAL APPLICATION

UM was developed for specimens in the size range of ~1–15 mm, such as whole mouse brains, mouse embryos, mouse organs, and *Drosophila melanogaster*. Specimens for UM need to be sufficiently transparent. This requires chemical clearing in most cases. Therefore, the specimens have to be able to withstand dehydration and treatment with chemical clearing agents. When working with XFP-expressing or immunolabeled material, the presence of autofluorescence can be a problem. Therefore, blood contained in mouse brains or organs has to be removed by careful transcardical perfusion before fixation. Specimens should not contain structures that strongly absorb light. In certain cases, pigments can be removed by bleaching with hydrogen peroxide (Dent et al. 1992). UM allows for discriminating between XFP-expressing neuronal somata and dendrites in the hippocampus through the overlying neocortex. In excised mouse hippocampi, dendritic networks of pyramidal neurons, up to the level of single spines, can be visualized (Dodt et al. 2007).

Imaging Autofluorescence

The simplest UM reconstruction technique is autofluorescence imaging. Because of the presence of endogenous chromophores, such as elastin, collagen, tryptophan, NADH, porphyrins, and flavins, with absorption spectra ranging from blue to ultraviolet, most biological tissues show a distinct autofluorescence throughout the visible to the near-infrared (Billinton and Knight 2001). Using a 488-nm laser for excitation in combination with a 520 nm ± 50 nm detection filter, the exterior and interior anatomy of *Drosophila* can be clearly imaged (Becker et al. 2008). This includes the direct and indirect flight muscles and structures of the nervous system, such as the optic lobes and the thoracic ganglia (Fig. 7A). Major blood vessels in mouse embryos and fiber tracts and the barrel cortex in isolated mouse brains can also be directly imaged using autofluorescence (Dodt et al. 2007).

Immunostaining Mouse Embryos

In combination with fluorescein isothiocyanate (FITC) immunostaining, UM allows visualization of somatic motor and sensorial nerve fibers in whole mouse embryos. Even the fine branches of the sensomotoric fibers can be visualized over a distance of up to several millimeters (Figs. 5C,D and 7B) (Becker et al. 2008; Jährling et al. 2008). An immunostaining protocol is provided at the end of this chapter (Protocol E).

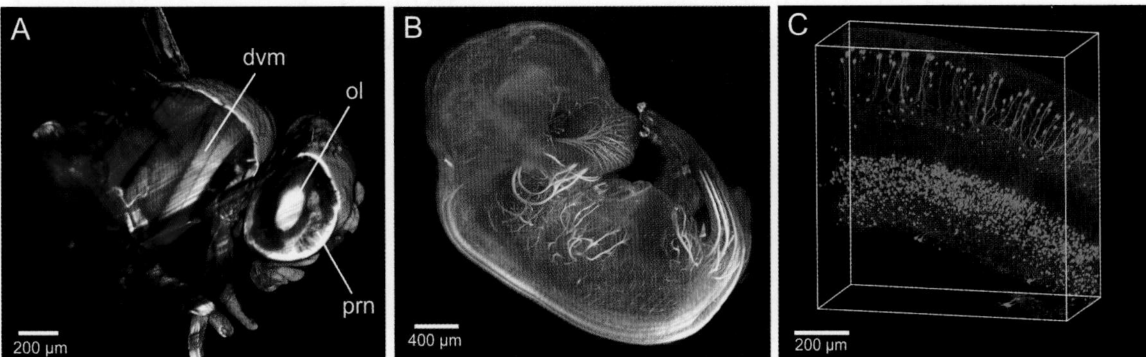

FIGURE 7. (*A*) View into the body of a *Drosophila* male, reconstructed from 238 optical sections. Easily recognized are some dorsoventral indirect flight muscles (ligamentous structures in the thorax, dvm), and one of the optic lobes as a part of the insect brain. (*B*) Images obtained from a mouse embryo E12.5 immunostained with neurofilament 160, reconstructed from 4 x 413 optical sections. The thinnest sensory nerve fibers innervating the vibrissae are detectable. (*C*) 3D reconstruction of a part of a mouse hippocampus using 132 optical sections. The complex dendritic network of pyramidal cells becomes apparent in the reconstruction. (*A,B*, Reprinted, with permission, from Becker et al. 2008. *C*, Modified, with permission, from Dodt et al. 2007.)

Lectin-Labeled Mouse Organs

The vascular system can be labeled by infiltration of ~10 mL of fluorescent conjugated lectins during transcardial perfusion (e.g., 10 µg lectin-FITC from *Lycopersicon esculentum* per mL [L0401, Sigma-Aldrich]). This allows studies of vascular networks in mouse organs, like brains, spinal cord, or heart (Jährling et al. 2009).

XFP-Expressing Transgenic Mice

Green fluorescent protein (GFP)-labeled neurons can be imaged by UM in isolated hippocampi, as well as in whole brains of Thy-1 GFP-M (c57/Bl6) mice (Feng et al. 2000). The complex dendritic network of pyramidal neurons can be imaged using a 5x objective with NA 0.25 (Fig. 7C). By using a 20x objective providing a NA of 0.4 and deconvolution, spines can be identified in isolated mouse hippocampi (Dodt et al. 2007). For XFP-expressing specimens, it is essential that fluorescence be maintained during the clearing procedure. Thus clearing time should be kept as short as possible (e.g., 6 h for isolated mouse hippocampi and ~2 d for whole mouse brains).

CONCLUSION AND FUTURE DIRECTIONS

Ultramicroscopy has proven to be a powerful tool for the study of specimens that are too large to be imaged in a single pass by confocal microscopy but are too small to be accessible to macrophotography. Thus, UM bridges a gap in which previously no adequate imaging techniques were available at the required resolution. UM is capable of providing micrometer-level resolution in specimens that are in the centimeter size range. This opens up numerous novel applications, especially in neuroscience and developmental studies.

UM provides potential for further advances. Currently, commercial CCD cameras do not provide enough resolution to fully take advantage of low-power objectives with high numerical apertures. Furthermore, 3D reconstructions of very highly resolved image stacks are presently limited by available computing power. Some specimens show a distinct autofluorescence, which reduces the visualization depth for small details. To improve imaging depth, XFP fluorescence could be distinguished from autofluorescence by spectral imaging (Dickinson et al. 2001). Because low-power objectives corrected for use in oil are currently not commercially available, correction of spherical aberration by custom-made correction optics may improve resolution and imaging depth. By using more sophisticated optics for light sheet generation, thinner and more homogenous light sheets may be obtained. With further improvements, a UM-based microscope providing a resolution of 1 µm throughout should make it possible to generate a whole mouse brain 3D reconstruction.

Protocol A

Dehydration and Clearing of Adult *Drosophila* for Ultramicroscopy

Following fixation, *Drosophila melanogaster* adults are dehydrated in ethanol and then cleared in a solution of benzyl alcohol and benzyl benzoate (Becker et al. 2008).

MATERIALS

CAUTION: See Appendix 6 for proper handling of materials marked with <!>.

Reagents

BABB clearing solution (two parts benzyl benzoate <!> and one part benzyl alcohol <!>)
Drosophila melanogaster adults
Ethanol <!> (50%, 70%, 80%, 96%, and 100% for dehydration series)
Ether <!>
Paraformaldehyde <!>
Phosphate-buffered saline (PBS)

Equipment

Vials

EXPERIMENTAL METHOD

Perform all of the steps at room temperature. Slowly agitate the vials containing the specimens during each incubation step.

1. Anesthetize adult *Drosophila* with ether and fix them in 4% paraformaldehyde (three times for 1 h each).
2. Wash them twice for 10 min in PBS.
3. Dehydrate the flies in an ascending ethanol series (50%, 70%, 80%, 96%, 3x 100%, in 1-h steps, last step overnight).
4. Incubate the flies in BABB clearing solution for 3 d.

Protocol B

Dehydration and Clearing of Mouse Embryos for Ultramicroscopy

Mouse embryos are dehydrated in ethanol and then cleared in benzyl benzoate/benzyl alcohol solution (Becker et al. 2008).

MATERIALS

CAUTION: See Appendix 6 for proper handling of materials marked with <!>.

Reagents

BABB clearing solution (two parts benzyl benzoate <!> and one part benzyl alcohol <!>)
Ethanol <!> (50%, 70%, 80%, 96%, and 100% for dehydration series)
Mouse embryos, immunostained and fixed

Equipment

Vials

EXPERIMENTAL METHOD

Perform all of the steps at room temperature. Slowly agitate the vials containing the specimens during each incubation step.

1. Dehydrate immunostained and fixed embryos in an ascending ethanol series (E12.5 embryos: 50%, 70%, 80%, 96%, 3x 100%, 1 h each, last step overnight). For older embryos, the incubation times may have to be prolonged.
2. Transfer the embryos into BABB and incubate them for at least 2 d (Fig. 4).

Protocol C

Dehydration and Clearing of Whole Mouse Brains for Ultramicroscopy

Mouse brains are carefully dissected and then dehydrated in ethanol followed by hexane. The dehydrated brains are cleared in BABB solution in preparation for ultramicroscopy (Dodt et al. 2007).

MATERIALS

CAUTION: See Appendix 6 for proper handling of materials marked with <!>.

Reagents

Anesthesia <!>
BABB clearing solution (two parts benzyl benzoate <!> and one part benzyl alcohol <!>)
Ethanol <!> (30%, 50%, 70%, 80%, 96%, and 100% for dehydration series)
Heparin <!>
Mice, Thy-1 GFP-M (C57/Bl6)
Paraformaldehyde <!>, 4% in 0.1 M ice-cold PBS
PBS, 0.1 M ice cold (pH 7.4)
Pentobarbital <!>

Equipment

Dissection instruments
Vials

EXPERIMENTAL METHOD

Obtain whole mouse brains from Thy-1 GFP-M (C57/Bl6) mice (Feng et al. 2000). The vials containing the brains are slowly agitated during each incubation step.

1. Deeply anesthetize the mice.
2. Kill the mice by intraperitoneal injection of pentobarbital (10 mg/kg).
3. Transcardically perfuse them with 40 mL of ice-cold 0.1 M PBS (pH 7.4) containing 1000 units/mL heparin, followed by 80 mL of 4% paraformaldehyde in 0.1 M ice-cold PBS.
4. Remove the brains and place them in fixative for 1 h at 4°C.
5. Rinse them two times in PBS, and store the brains in PBS at 4°C.
6. Dehydrate the brains in an ascending ethanol series (30%, 50%, 70%, 80%, 96%, and 2x 100%) for 1 d each at room temperature.
7. Incubate the brains in 100% hexane for 1 h. This is a crucial step to achieve maximal dehydration.
8. Transfer them into BABB for at least 2 d at room temperature.

Protocol D

Dehydration and Clearing of Dissected Mouse Hippocampi for Ultramicroscopy

Mouse hippocampi are carefully removed from dissected mouse brains and then dehydrated in ethanol. The dehydrated brains are cleared in a solution of benzyl benzoate and benzyl alcohol in preparation for ultramicroscopy (Dodt et al. 2007).

MATERIALS

CAUTION: See Appendix 6 for proper handling of materials marked with <!>.

Reagents

Anesthesia
BABB clearing solution (two parts benzyl benzoate <!> and one part benzyl alcohol <!>)
Ethanol <!> (50%, 80%, 96%, and 100% for dehydration series)
Heparin <!>
Mice, Thy-1 GFP-M (C57/Bl6)
Paraformaldehyde <!>, 4% in 0.1 M ice-cold PBS
PBS, 0.1 M ice cold (pH 7.4)
Pentobarbital <!>

Equipment

Dissection instruments
Vials

EXPERIMENTAL METHOD

The vials containing the hippocampi are slowly agitated during each incubation step.

1. Prepare the mouse brains as described in Protocol C, Steps 1–5.
2. Dissect the hippocampi under red light.
3. Dehydrate the brains in an ascending ethanol series (50%, 80%, and 96%) for 1 h each at room temperature.
4. Leave them in 100% ethanol overnight.
5. Incubate the brains in BABB for 6 h at room temperature.

Protocol E

Immunostaining Mouse Embryos

In combination with fluorescein isothiocyanate (FITC) immunostaining, UM allows visualization of somatic motor and sensorial nerve fibers in whole mouse embryos. Even the fine branches of the sensomotoric fibers can be visualized over a distance of up to several millimeters (Figs. 5C,D and 7B) (Becker et al. 2008; Jährling et al. 2008).

MATERIALS

CAUTION: See Appendix 6 for proper handling of materials marked with <!>.
See the end of the chapter for recipes for reagents marked with <R>.

Reagents

Blocking serum (four parts calf serum and one part DMSO <!>)
DENT's fix (one part dimethylsulfoxide [DMSO] and four parts methanol <!>; Dent et al. 1992)
Goat anti-mouse conjugated to Alexa488 (Invitrogen)
Hydrogen peroxide (H_2O_2) <!>, 30%
Primary antibody (monoclonal anti-neurofilament 160 clone NN18, Sigma-Aldrich)
TBS <R>

Equipment

Vials or other containers

EXPERIMENTAL METHOD

1. Fix embryos overnight in DENT's fix.
2. Bleach the embryos overnight in one part 30% H_2O_2 and two parts DENT's fix.
3. Wash the embryos three times in TBS for 30 min.
4. Incubate them for 2 d at room temperature in the primary antibody solution (monoclonal anti-neurofilament 160 clone NN18 diluted 1:200 in blocking serum).
5. Wash the embryos three times for 1 h in TBS.
6. Incubate them for 2 d at room temperature in the secondary antibody solution (goat anti-mouse conjugated to Alexa488 diluted 1:200 in blocking serum).
7. Wash the embryos at least five times in TBS (1 h each).
8. Dehydrate and chemically clear the embryos as described in Protocol B.

RECIPE

CAUTION: See Appendix 6 for proper handling of materials marked with <!>.
Recipes for reagents marked with <R> are included in this list.

TBS (10x)

Reagent	Amount to add
NaCl	87.7 g
Tris-Cl <!>, 1 M (pH 8.0)	100 mL
H_2O	to 1 L

Dilute to 1x before use.

REFERENCES

Becker K, Jährling N, Kramer ER, Schnorrer F, Dodt H-U. 2008. Ultramicroscopy: 3D-reconstruction of large microscopic specimens. *J Biophotonics* **1:** 36–42.

Billinton N, Knight A. 2001. Seeing the wood through the trees: A review of techniques for distinguishing green fluorescent protein from endogenous autofluorescence. *Anal Biochem* **291:** 175–197.

Dent JA, Cary RB, Bachant JB, Domingo A, Klymkowski MW. 1992. Host cell factors controlling vimentin organization in the *Xenopus* oocyte. *J Cell Biol* **119:** 855–866.

Dickinson ME, Bearman G, Tille S, Lansoford R, Fraser SE. 2001. Multi-spectral imaging and linear unmixing add a whole new dimension to laser scanning fluorescence microscopy. *BioTechniques* **31:** 1272–1278.

Dodt H-U, Leischner U, Schierloh A, Jährling N, Mauch CP, Deininger K, Deussing JM, Eder M, Zieglgänsberger W, Becker K. 2007. Ultramicroscopy: Three-dimensional visualization of neural networks in the whole mouse brain. *Nat Methods* **4:** 331–336.

Feng G, Mellor RH, Bernstein M, Keller-Peck C, Nguyen QT, Wallace M, Nerbonne JM, Lichtman JW, Sanes JR. 2000. Imaging neuronal subsets in transgenic mice expressing multiple spectral variants of GFP. *Neuron* **28:** 41–51.

Helmchen F, Denk W. 2005. Deep tissue two-photon microscopy. *Nat Methods* **2:** 932–940.

Huisken J, Stainier DY. 2009. Selective plane illumination microscopy techniques in developmental biology. *Development* **136:** 1963–1975.

Huisken J, Swoger J, del Bene F, Wittbrodt J, Stelzer EH. 2004. Optical sectioning deep inside live embryos by selective plane illumination microscopy. *Science* **305:** 1007–1009.

Jährling N, Becker K, Kramer ER, Dodt H-U. 2008. 3D-visualization of nerve fiber bundles by ultramicroscopy. *Med Laser Appl* **23:** 209–215.

Jährling N, Becker K, Dodt H-U. 2009. 3D-reconstruction of blood vessels by ultramicroscopy. *Organogenesis* **5:** 50–54.

Jährling N, Becker K, Schönbauer C, Schnorrer F, Dodt H-U. 2010. Three-dimensional reconstruction and segmentation of intact *Drosophila* by ultramicroscopy. *Front Syst Neurosci* **4:** 1–6.

Sasnett MW, Johnston TF Jr. 1991. Beam characterization and measurement of propagation attributes. *Proc SPIE* **1414:** 21–32.

Shaw PJ. 1995. Comparison of wide-field/deconvolution and confocal microscopy for 3D imaging. In *Handbook of biological confocal microscopy*, 2nd ed. (ed. Pawley JB), pp. 373–387. Plenum, New York.

Siedentopf H, Zsigmondy R. 1903. Visualization and size measurement of ultramicroscopic particles, with special application to gold-colored ruby glass. *Ann Phys* **10:** 1–39.

Siegman AE. 1986. Wave optics and Gaussian beams, and physical properties of Gaussian beams. In *Lasers*, pp. 626–662 and 663–697. University Science Books, Sausalito, CA.

Siegman AE. 1998. How to (maybe) measure laser beam quality. In *Diode pumped solid state (DPSS) lasers: Applications and issues* (OSA Trends in Optics and Phottonics series) (ed. Dowley MW), Vol. 17. Optical Society of America, Washington, DC.

Spalteholz W. 1914. *Über das Durchsichtigmachen von menschlichen und tierischen Präparaten.* S. Hierzel, Leipzig, Germany.

Tyszka JM, Fraser SE, Jacobs RE. 2005. Magnetic resonance microscopy: Recent advantages and applications. *Curr Opin Biotechnol* **26:** 93–99.

Voie AH, Burns DH, Spelman FA. 1993. Orthogonal-plane fluorescence optical sectioning: Three dimensional imaging of macroscopic biological specimens. *J Microsc* **170:** 229–236.

Weninger WJ, Mohun T. 2002. Phenotyping transgenic embryos: A rapid 3D screening method based on episcopic fluorescence image capturing. *Nat Genet* **30:** 59–65.

50 In Vivo Optical Microendoscopy for Imaging Cells Lying Deep within Live Tissue

Robert P.J. Barretto[1] and Mark J. Schnitzer[1,2]

[1]*James H. Clark Center for Biomedical Engineering & Sciences, Stanford, California 94305;* [2]*Howard Hughes Medical Institute, Stanford University, Stanford, California 94305*

ABSTRACT

Although in vivo microscopy has been pivotal in enabling studies of neuronal structure and function in the intact mammalian brain, conventional intravital microscopy has generally been limited to superficial brain areas such as the olfactory bulb, the neocortex, or the cerebellar cortex. For imaging cells in deeper areas, this chapter presents in vivo optical microendoscopy using gradient refractive index (GRIN) microlenses that can be inserted into tissue. Our general methodology is broadly applicable to many deep brain regions and areas of the body. Microendoscopes are available in a wide variety of optical designs, allowing imaging across a range of spatial scales and with spatial resolution that can now closely approach that offered by standard water-immersion microscope objectives. The incorporation of microendoscope probes into portable miniaturized microscopes allows imaging in freely behaving animals. When combined with the broad sets of available fluorescent markers, animal preparations, and genetically modified mice, the methods described here enable sophisticated experimental designs for probing how cellular characteristics may underlie or reflect animal behavior and life experience, in healthy animals and animal models of disease.

INTRODUCTION

Recent strides in intravital light microscopy have enabled seminal studies of both neuronal structure and dynamics in the intact mammalian brain (Gobel and Helmchen 2007; Kerr and Denk 2008; Rochefort et al. 2008; Holtmaat et al. 2009; Holtmaat and Svoboda 2009; Wilt et al. 2009). Applications of two-photon microscopy in awake but head-restrained animals have even permitted Ca^{2+}-imaging studies during active animal behavior (Dombeck et al. 2007; Mukamel et al. 2009; Nimmerjahn et al. 2009). However, photon scattering limits the optical penetration of light microscopy into tissue, restricting the utility of conventional intravital microscopy to superficial tissue areas such as the olfactory bulb, the neocortex, and the cerebellar cortex (Helmchen and Denk 2005; Wilt et al. 2009). Penetration depths are typically limited to ~50–100 µm with epifluorescence microscopy and ~500–700 µm with conventional two-photon microscopy.

To extend the microscope's penetration depth into tissue, a range of innovative optical strategies has been experimentally explored in the last few years (Helmchen and Denk 2005; Wilt et al. 2009). Here, we describe one of these approaches: optical microendoscopy (Jung and Schnitzer 2003; Jung et al. 2004; Levene et al. 2004), which can penetrate the furthest of these and reach >1 cm into tissue (Llewellyn et al. 2008) via the use of needle-like micro-optical probes. These probes typically act like an optical relay and can be inserted into tissue. Subject to some optical constraints discussed below, the length of the probe can be tailored to the anatomical depth of the tissue under examination. Optical microendoscopy provides spatial resolution that can approach that of a conventional water-immersion objective lens (Barretto et al. 2009); is compatible for use with multiple contrast modalities including epifluorescence, two-photon excited fluorescence, and second-harmonic generation (Mehta et al. 2004; Flusberg et al. 2005); and has been used in both live mice and humans (Llewellyn et al. 2008; Wilt et al. 2009). In this protocol, we present optical considerations in the choice of a microendoscope probe, modifications to the upright light microscope that facilitate microendoscopy, and methods for imaging cellular characteristics in the intact mammalian brain. The methodology is also compatible for use with portable miniaturized microscopes (Gobel et al. 2004; Engelbrecht et al. 2008; Flusberg et al. 2008) that are based on micro-optics and enable imaging in freely behaving mice (Flusberg et al. 2008).

IMAGING SETUP

Microscope Body

Nearly any upright microscope that has infinity optics and has already been adapted for in vivo imaging (e.g., Ultima IV, Prairie Technologies, Inc.) can readily be used for microendoscopy. There are two main options for how the microendoscope probe can be held (Barretto et al. 2009), one of which requires custom modifications to the microscope.

In the simpler approach, the microendoscope probe is held by its insertion into the animal subject, instead of being coupled mechanically to the body of the microscope. When the animal and the microendoscope probe are positioned correctly, the microendoscope relays the focal plane of the microscope objective into deep tissue, with a demagnification or magnification that depends on the optical details of the probe. This approach has the advantage of not requiring any alterations to the microscope but the disadvantage that any fine adjustments of the microendoscope relative to the tissue are not automatically referenced to the optical axis.

In an alternative approach, the microendoscope probe is mounted on the microscope's focusing unit, which is modified to permit two modes of fine focal adjustment (Fig. 1A,C). The first mode adjusts the position of the microscope objective lens relative to the microendoscope probe. The second mode moves the objective lens and the microendoscope probe in tandem, permitting the microendoscope to be inserted into tissue without affecting the optical coupling to the objective lens. Both modes can be motorized. To grip the microendoscope probe on its sides, we use a two-pronged pincer holder (e.g., Thorlabs, Inc., Micro-V-Clamp) (Fig. 1C). This holder is attached to a miniature probe clamp (e.g., Siskiyou, Inc., MXC-2.5) that can be rotated about its long axis and swung in and out of the optical pathway. By adjusting the two angular degrees of freedom of the probe clamp, we align the microendoscope with the optical axis (Barretto et al. 2009). Adjustments in the axial position of the objective lens while keeping the microendoscope fixed are performed using a stepper motor (Sutter Instruments, MP-285) mounted on the microscope's nosepiece. These adjustments modify the intermediate plane at which the illumination is focused above the microlens, leading to corresponding focal adjustments in the specimen. The microendoscope and objective lens are moved in tandem using the microscope's normal focusing actuator (Fig. 1C).

Microendoscope Probes

Microendoscope probes can be customized for specific applications (Fig. 1D), and distinct values of the probes' basic optical parameters are preferred in different situations. For example, some optical

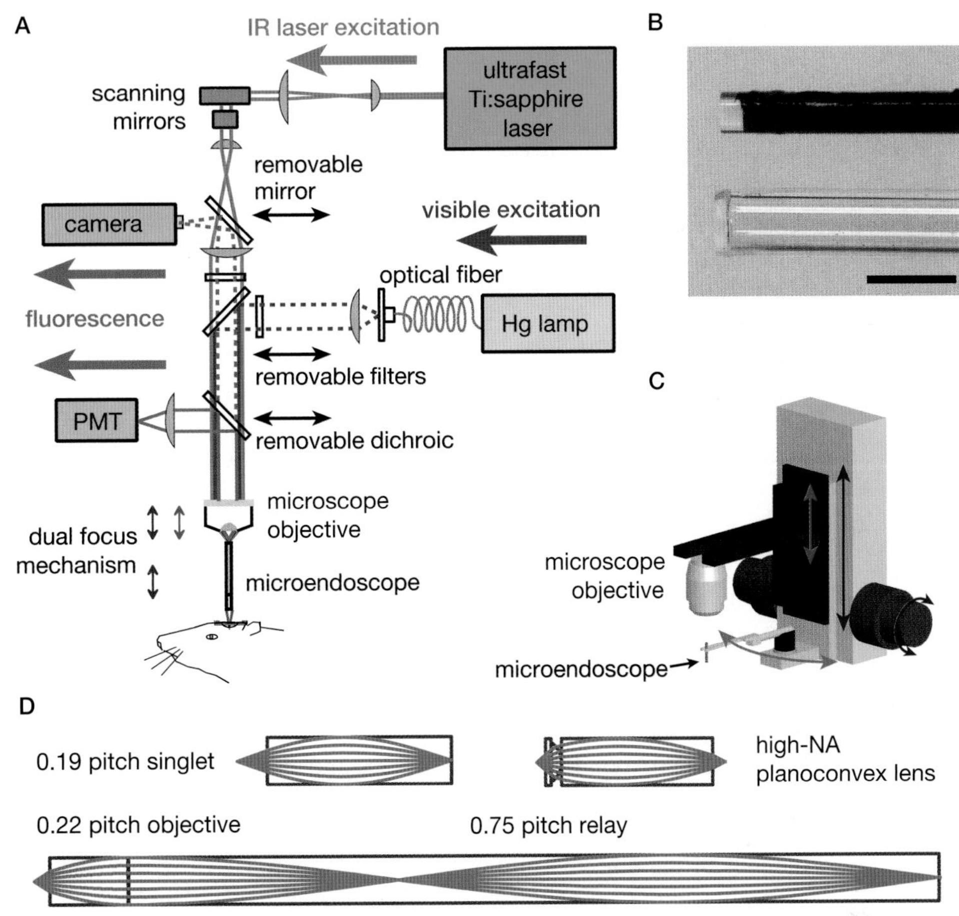

FIGURE 1. Methodologies for in vivo optical microendoscopy. (*A*) Optical schematic of an upright microscope modified to permit both one- and two-photon fluorescence microendoscopy. For two-photon imaging, the beam from an ultrashort-pulsed infrared (IR) Ti:sapphire laser is scanned within the focal plane of the microscope objective. By adjusting the axial separation between the objective and the microendoscope (red arrow of the dual-focus mechanism; see also C), this focal plane of the microscope objective is also set to the microendoscope's back focal plane. Another focal adjustment (blue arrows of the dual mechanism) is used to lower the objective and the microendoscope in tandem toward the animal. For one-photon imaging, a mercury (Hg) arc lamp provides illumination. In both imaging modes, fluorescence emissions route back through the microendoscope and to either a camera or a photomultiplier tube (PMT) for one- or two-photon imaging, respectively. (*B*) Photographs of the tips of a 0.5-mm-diameter microendoscope of doublet design (*top*) and a 0.8-mm-outer-diameter glass capillary guide tube (*bottom*) into which this microendoscope can be inserted. The relay of the microendoscope is coated black. A glass coverslip is attached to the tip of the guide tube. The guide tube facilitates the rapid exchange of microendoscopes without perturbation to the underlying tissue. Scale bar, 1 mm. (C) The microscope objective and the microendoscope probe are mounted on a pair of cascaded focusing actuators that provide dual-focus capability. This allows the objective to be moved either alone (red arrow) or together with the microendoscope (blue arrow). The microendoscope can also be swung out of the optical axis (green arrow) to permit conventional microscopy. (*D*) Optical ray diagrams for sample microendoscopes of the singlet GRIN (*top left*), compound plano-convex and GRIN (*top right*), and GRIN doublet (*bottom*) types. Scale bar, 1 mm.

designs are better suited for examining subcellular features such as dendritic structures, whereas other designs are preferred for wide-field Ca^{2+} imaging of neuronal dynamics. In our own work, we have explored three main types of optical designs. The first design involves a single GRIN lens (Fig. 1D, top left) that provides low magnification and a large field of view. The second design involves a GRIN lens attached in series to a high-numerical-aperture (NA) (~0.65–0.82) plano-convex microlens (Barretto et al. 2009) (Fig. 1D, top right); this combination can provide superior light collection and diffraction-limited resolution but has a smaller field of view. The third design has a

TABLE 1. Characteristics of sample microendoscope probes

Microendoscope type	Diameter (mm)	Length (mm)	Usable field of view (μm)	Lateral magnification) (×)	Two-photon lateral resolution (FWHM, μm)	NA
Doublet (0.75/0.21 pitch)	1.0	20.6	275	2.52	0.9	0.49
Doublet (1.25/0.19 pitch)	0.5	16.4	130	2.48	1.0	0.48
Doublet (1.75/0.16 pitch)	0.35	15.8	75	2.69	1.2	0.45
Singlet (0.46 pitch)	1.0	4.4	700	0.97	0.9	0.49
Singlet (0.94 pitch)	0.5	4.3	350	0.92	1.0	0.47
GRIN/plano-convex doublet (BK7 plano-convex lens)	1.0	3.7	120	1.41	0.8	0.65
GRIN/plano-convex doublet (LaSFN9 plano-convex lens)	1.0	4.0	75	1.86	0.6	0.82

To facilitate comparisons between parameter values, each microendoscope listed has an optical working distance of 250 μm. The lateral resolution of two-photon imaging is given for 920-nm illumination.

FWHM, full width at half-maximum; NA, numerical aperture; GRIN, gradient refractive index.

GRIN relay lens coupled to a GRIN objective (Fig. 1D, bottom), allowing longer probe designs (>5 mm) and intermediate-sized fields of view (Jung et al. 2004; Levene et al. 2004).

Microendoscope probes of all three types (Fig. 1D, Table 1) can generally be conceptualized as consisting of two optical components in series: an infinity micro-objective that focuses illumination to the specimen and collects emission photons, combined with a micro-optical relay lens that receives focused illumination from the microscope and also focuses the sample's emissions to the front focal plane of the upright microscope's objective lens (Fig. 1A,C). In singlet GRIN lenses, both of these functions occur within a single optical element; in GRIN doublet probes or in high-resolution probes, the jobs of the objective and the relay are accomplished by two micro-optical entities attached in series. In epifluorescence microendoscopy, the relay microlens projects a real image of the sample to the microscope objective's focal plane. Table 1 presents optical parameters for some microendoscopes, with each of the three major types represented. Below, we consider these parameters in further detail. For mathematical formulas to guide optical design, see Jung et al. (2004); with these equations researchers can design probes to custom specifications and have them fabricated commercially.

Microendoscope Diameter

Microendoscope probes with diameters ranging from 0.35 to 2.8 mm are commercially available (e.g., from GRINTECH GmbH); our laboratory most commonly uses 0.35-, 0.5-, and 1.0-mm sizes. For a given NA value, the smaller diameter probes (e.g., 0.35 or 0.5 mm) offer resolution and magnification values comparable to those of wider diameter probes. However, the wider probes of the same NA will generally have longer working distances to the sample and broader fields of view. In addition, smaller diameter probes are more fragile. Encasing these probes in thin-walled stainless-steel hypodermic sheaths will make them more robust. Probes as thin as 0.35 mm in diameter have been successfully applied for high resolution in vivo laser-scanning imaging, including in humans (Llewellyn et al. 2008).

Microendoscope Length

The lengths of microendoscope probes are typically designed to meet the mechanical constraints posed by the depth of the tissue under examination and the surgical preparation. The probe length should be sufficient to guide photons from the specimen plane lying deep within the tissue to an unobstructed intermediate focal plane that is also the focal plane of the microscope objective (Fig. 1A). It is the length of the relay microlens that is typically adjusted in applications requiring imaging at substantial tissue depths.

Valid lengths for the relay microlens are calculated by first determining the pitch length of its glass GRIN substrate (Jung et al. 2004). Within a paraxial approximation, light rays propagate down the optical axis of the microendoscope along a trajectory for which the rays' radial distance from the axis varies as a sinusoidal function of the distance propagated axially (Fig. 1D). One pitch length is defined as the length of the GRIN substrate within which a ray will propagate a full sinusoidal cycle. This length depends on the radially varying refractive index profile of the GRIN material. Cylindrical rods of this material can then be cut to various lengths measured in units of the pitch length. GRIN lenses of integral or half-integral pitch—that is, 1/2, 1, 3/2 pitch, etc.—refocus light rays emanating from a single focus on one side of the lens to another focal spot on the opposite side of the lens. By comparison, 1/4-pitch lenses—or 3/4-pitch lenses, 5/4-pitch lenses, etc.—are infinity lenses that focus collimated rays entering one side of the microlens to a focal spot on the lens's opposing side (see the 0.75-pitch relay in Fig. 1D, bottom). Longer microendoscope probes can be designed by adding multiple 1/2-pitch lengths to the relay lens as necessary. Such additions extend the probe's length without altering the NA, the magnification, the field of view, or the working distance. However, probes of longer length often suffer from poorer optical resolution caused by the accumulation of spherical aberrations over multiple half-pitch lengths of the GRIN substrate.

Optical Working Distance to the Specimen

The working distance to the specimen is set for a GRIN objective lens by the degree to which the objective is slightly shorter than a 1/4-pitch design (see the 0.19-pitch singlet and the 0.22-pitch objective in Fig. 1D). An objective of shorter pitch has a longer working distance, but the objective's NA is reduced. Typical values of working distance range from 0 µm to 800 µm. In one-photon fluorescence imaging, light scattering precludes efficient imaging beyond ~100 µm from the tip of the endoscope (Flusberg et al. 2008), so the working distance will be relatively short. By comparison, in two-photon imaging, microendoscopy can be performed up to ~650 µm into tissue beyond the probe tip, which generally necessitates a design of longer working distance.

Although the focal plane can be adjusted to a depth other than the working distance, microendoscopes are often designed to have minimal optical aberrations at their specified working distance. In particular, the high-resolution GRIN/plano-convex compound lenses (Fig. 1D, top right) are designed so that aberrations from the objective component are compensated at a specific working distance by an appropriate choice of the GRIN relay's radial refractive index profile (Barretto et al. 2009). High-resolution experiments should, thus, be performed with the tissue of interest located at the designed working distance. However, for imaging experiments that permit modest degradation in resolution, it is convenient to design the optical working distance to be a few hundred micrometers longer than what will be used for the experiment. This choice ensures that neither the plane of laser scanning in two-photon imaging nor the intermediate real image in one-photon imaging is located at external glass surfaces of the microendoscope, where surface imperfections can degrade image quality.

Microscope Objectives for Optical Coupling to the Microendoscope Probe

Microscope Objective Magnification

In applications requiring large fields of view, the magnification of the microscope objective should suffice to permit imaging of the entire top surface of the microendoscope probe. For example, a typical 10x objective has a sufficient field of view to image the entire aperture of a 1-mm-diameter microendoscope probe. Other parameters of the entire optical system must also suffice to image this entire aperture. For example, in one-photon microendoscopy, the camera chip must be sufficiently wide; and in two-photon microendoscopy, the range of laser scanning must be sufficiently broad to sample the entire face of a 1-mm-diameter microendoscope probe.

Microscope Objective NA

To achieve high-resolution imaging, the NA of the microscope objective should be higher than that of the microendoscope probe's relay lens. In one-photon imaging, this condition ensures that the microscope objective captures the full NA of fluorescence emissions exiting the microendoscope's relay lens, thereby preserving signal power as well as image resolution. In two-photon imaging, this condition ensures that the laser illumination fills the back aperture of the probe's objective lens, typically located at the boundary between the micro-objective and the relay (Fig. 1D) and thereby uses the full NA of the microendoscope's objective in focusing the laser beam at the specimen plane. A portion of the laser illumination will be lost, however, because the NA of the beam striking the relay lens is higher than the NA that the relay can accept.

Imaging Parameters

One-Photon Imaging

Excitation filter: Approximately 470/40 nm for fluorescein-conjugated dextrans (for blood-flow imaging), green fluorescent protein (GFP) and yellow fluorescent protein (YFP).

Emission filters: Approximately 525/50 nm for fluorescein-conjugated dextrans, GFP, and YFP.

Images/frame rate: 512 x 512 pixels at 100 Hz with a high-speed electron-multiplying charge-coupled device (CCD) camera (e.g., iXon DU-897E, Andor Technology), or 1392 x 1040 pixels with a cooled CCD camera (e.g., Coolsnap HQ, Roper Scientific).

Recording duration: Typically 30–40 sec for a given field of view.

Two-Photon Imaging

Excitation wavelength: Approximately 800 nm for fluorescein-conjugated dextrans in vascular imaging, ~920 nm for GFP and YFP.

Excitation power at sample surface: Always <25 mW for tissues proximal to the microendoscope, more distal tissues require greater power.

Images/dwell times: 512 x 512 pixels (typically ~0.8–4 µsec per pixel or as permitted by tissue motion). For high-resolution imaging, multiple images can be acquired and can be averaged, after motion correction, to produce an improved image.

Section/stack (3D imaging): Approximately 5–10-µm axial spacing for GRIN singlets and doublets that have ~10-µm axial resolution, 1–2.5-µm axial spacing for high-resolution microlenses.

Recording duration: Typically 5–10 min for a given field of view.

Protocol

In Vivo Microendoscopy of the Hippocampus

Microendoscopic probes can be used to investigate deep tissues within the brain and other parts of the body.

MATERIALS

CAUTION: See Appendix 6 for proper handling of materials marked with <!>.
See the end of the chapter for recipes for reagents marked with <R>.

Reagents

Agarose, Type III-A (Sigma-Aldrich)
Analgesic (e.g., buprenorphine)
Anti-inflammatory (e.g., carprofen <!>, dexamethasone <!>)
Artificial cerebral spinal fluid (ACSF; e.g., from Harvard Apparatus)
Dental acrylic (e.g., Ortho-Jet, Lang Dental Mfg Co., Inc.)
Eye ointment (e.g., Puralube Vet Ointment, PharmaDerm Nycomed US)
Local anesthetic (e.g., 1% lidocaine)
Physiologic saline or lactated Ringer's solution (e.g., from Electron Microscopy Sciences)
Skin disinfectant (e.g., betadine, Baxter)
Tissue adhesive (e.g., Vetbond, 3M)

Equipment

Guide Tube Preparation

Coverslips (#0 thickness; e.g., from Electron Microscopy Sciences)
Capillary tubing (e.g., thin-walled glass 1.0–2.5-mm inner diameter, Vitrocom, Inc.)
Curing light (e.g., COLTOLUX 75, Coltène Whaledent)
Diamond-scribing tool (e.g., from Electron Microscopy Sciences)
Glass polisher (e.g., ULTRAPOL, ULTRA TEC Manufacturing, Inc.)
Microdrill (e.g., Osada, Inc. EXL-M40)
Optical adhesive (e.g., NOA 81; Norland Products, Inc.)
Sandpaper, fine-500 grit (e.g., 3M)
Sonicator (e.g., Model 1510, Branson Ultrasonics Corp.)
Stereomicroscope (e.g., MZ12.5, Leica)

Surgery

Aseptic instruments/surgical tools (e.g., from Fine Science Tools)
Balance (for weighing animals; e.g., Mettler Toledo International, Inc. PG503-S)
Cold light source (e.g., KL 1500, SCHOTT North America, Inc.)
DC temperature regulation system (e.g., FHC Inc. 40-90-8; 40-90-5; 40-90-2-07)
Glass bead sterilizer (e.g., model BS-500, Dent-EQ)

Laboratory animal anesthesia system (e.g., VetEquip Inc. item #901806), anesthetic gas (e.g., isoflurane, Southmedic, Inc.), carrier gas tank (e.g., medigrade oxygen from Praxair); recommended: waste anesthetic gas system (e.g., VetEquip Inc. item #933101)
Mounting post (custom-made; aluminum 15 x 3 x 2-mm bar with 2.7-mm through hole on end)
Mounting-post holder (custom-made; aluminum bar with M2 tapped hole)
Standard microwave (for agarose gel preparation)
Stereotaxic apparatus (custom-made)
Surgical eye spears (e.g., 1556455, Henry Schein Medical)
Waste liquid suction line (custom-made)

EXPERIMENTAL METHOD

Glass Guide Tube Construction (~25 min)

An optically transparent guide tube (Fig. 1B) is often used to assist in delivering the microendoscope to the tissue of interest. Because the tube is sealed at the tip with a small cover glass that permits optical but not physical access to the tissue, microendoscopes can be quickly delivered and can be interchanged with minimal mechanical disturbance to the field of view under inspection.

1. Choose a thin-walled capillary glass of appropriate diameter. Typical inner diameters safely exceed the microendoscope diameter by 10%–15%.
2. Cut the thin-walled capillary glass to the desired length. Use a microdrill to uniformly thin the circumference of the glass at the location of the cut. Snap the glass at the thinned portion, and coarsely smooth with the microdrill or sandpaper.
3. Polish one end of the guide tube. Use a fiber-optic polisher or a fine grit sandpaper. Inspect the guide tube end under a stereomicroscope, and ensure flatness. Repolish as necessary.
4. Cut circular pieces of #0-thickness cover glass with diameters matching the outer diameter of the guide tube. Using a diamond scribe, score circular patterns onto the cover glass, and break with the forceps. Tolerances for the cover-glass dimensions are set by the inner and the outer diameters of the guide tube.
5. Clean all glass pieces by sonication while immersed in the cleanser, and store in ethanol until assembly. In subsequent steps, use gloves, and work in a dust-free area.
6. Apply a thin layer of ultraviolet-curing optical adhesive to the polished end of the guide tube. Using a high-magnification stereomicroscope, orient the guide tube toward the objective, and use a fine 30-gauge needle to apply the adhesive onto the guide tube.
7. Attach the circular coverslips to the guide tube. Use forceps to hold the cover glass, and gently drop the coverslip onto the guide tube. Ensure that glue does not enter the central area of the guide tube and that an epoxy seal is formed around the entire circumference of the guide tube. Set the epoxy using an ultraviolet light source.
8. Store guide tubes in clean containers until use (e.g., sterile culture dishes). If possible, allow at least 12 h for the optical epoxy to cure before use. Rinse with saline solution before implantation.

Surgery (~1 h)

The following animal procedures are outlined for the examination of the dorsal hippocampus in adult mice but are applicable to other regions. All procedures were approved by the Stanford Administrative Panel on Laboratory Animal Care (APLAC). Consultation with those overseeing institutional guidelines for animal surgery care and anesthesia is recommended.

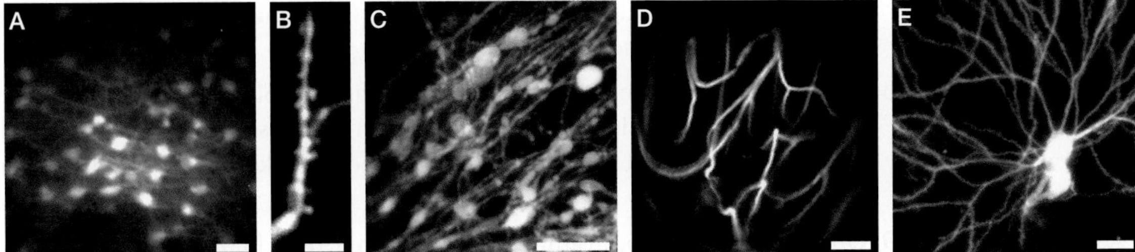

FIGURE 2. Images acquired by fluorescence microendoscopy in live mice. (*A*) GFP-labeled pyramidal neurons in CA1 hippocampus imaged with a 1-mm singlet probe. Scale bar, 50 µm. (*B*) High-resolution image of CA1 hippocampal dendritic spines acquired using an LaSFN9 high-resolution probe. Scale bar, 5 µm. (*C*) GFP-labeled neurons in the brain stem's external cuneate nucleus imaged with a 1-mm doublet probe of 20-mm length and a 0.75-pitch relay. Scale bar, 50 µm. (*D*) Fluorescein-labeled vasculature in CA1 hippocampus imaged with a 0.5-mm singlet probe. Scale bar, 50 µm. (*E*) GFP-labeled pyramidal neurons in CA1 hippocampus imaged with a 1-mm singlet probe. Scale bar, 50 µm. *A–C* and *E* are 2D projections of 3D stacks acquired by two-photon microendoscopy. These stacks were composed of 108 image slices acquired at 2-µm axial separation between adjacent slices for *A*; nine images with 1.6-µm axial separation for *B*; 50 images with 0.43-µm axial separation for *C*; four slices taken at 4.2-µm axial separation for *E*. *D* was obtained by one-photon microendoscopy and shows the standard deviation image of a high-speed video sequence of blood flow, which is a postprocessed image that highlights blood vessels.

9. Deeply anesthetize mice with isoflurane gas (2.0%–2.5%; mixed with 2-L/min oxygen) or interperitoneal injection of ketamine (75 mg/kg) and xylazine (15 mg/kg). Assess depth of anesthesia by monitoring toe pinch withdrawal, eyelid reflex, and respiration rate.
10. (Optional) Administer dexamethasone (2-mg/kg intramuscular) and carprofen (5-mg/kg subcutaneous) to minimize tissue swelling and inflammation.
11. Secure the animal in a stereotaxic frame. Maintain body temperature at 37°C with a heating blanket. Apply ophthalmic ointment to the eyes.
12. Trim or shave the fur from the top of the head, and disinfect the exposed skin with alternating washes of 70% ethanol and betadine. The use of a bead sterilizer to disinfect surgical instruments is recommended.
13. Expose the cranium in the vicinity dorsal to the brain structure of interest. Remove the periosteum using a probe or a scalpel, and rinse with 0.9% saline solution. After rinsing, use a cotton swab to dry the exposed skull. Subcutaneous delivery of lidocaine or other local anesthetic may be administered before exposing the cranium, as necessary.
14. Apply a thin layer of cyanoacrylate (e.g., Vetbond) to the regions of exposed skull outside of the expected craniotomy site. Use a fine applicator (e.g., hypodermic needle) to spread the cyanoacrylate over the boundaries of the exposed cranium to seal the skin cut sites. Allow the cyanoacrylate to dry for 5 min.
15. Drill a round craniotomy centered over the stereotaxic coordinates of interest (e.g., 2.0-mm posterior and 2.0-mm lateral of the bregma in the hippocampus). A trephine is helpful in marking craniotomy dimensions matched to the microendoscope diameter. Remove the dura with forceps.
16. Perform blunt dissection and aspiration to gradually remove a cylindrical column of neocortical brain tissue with a 27-gauge blunt needle. Continuously irrigate the applied area with sterile ACSF or Ringer's solution. Bleeding from disrupted vasculature is normal, increase irrigation rates to maintain visibility within the column.
17. As the desired imaging area is approached, aspiration with a fine 29-gauge blunt needle can be used to expose the imaging area. Under optimal conditions, a thin layer of tissue remains overlying the cells of interest, to minimize direct mechanical tissue damage from aspiration. In the

hippocampal preparation, the overlying corpus callosum can be readily identified by its stereotyped white matter tract patterns.

18. Minimize bleeding from the sides of the aspirated column. This is performed by following applications of saline irrigation and aspiration with 5-sec pause intervals to allow clot formation. Gel foam may be applied to control bleeding. Take care not to allow a clot to form over the imaging area.
19. Optionally, at this step, an animal may be examined for fluorescence labeling, using a low-magnification long working distance objective. (See Imaging below.)
20. Gradually insert a closed-end glass guide tube into the aspirated column. Lower the guide tube until it is in contact with the distal tissue regions. Check that neither air pockets nor bleeding regions are present under the guide tube. If necessary, irrigate with buffer, and repeat guide tube insertion. The tissue should be visible on inspection through the guide tube with a stereomicroscope.
21. Suction any liquids that are present on the cyanoacrylate layer.
22. Apply melted agarose (~1.5%) to the sides of the guide tube, filling gaps between skull and the guide tube. Allow agarose to harden. Excess agarose can be removed by dicing with a scalpel blade.
23. Apply a layer of dental acrylic over all of the exposed skull and sides of the guide tube. Affix a metal connection bar approximately parallel to the plane of the guide tube surface. The distal end of the bar must be at least 1 cm away from the guide tube to prevent obstruction during imaging. Wait 10 min for the acrylic to harden.
24. Affix a piece of flexible tape or adhesive dressing over the guide tube. This will prevent dirt from entering the tube.
25. Allow the animal to recover from anesthesia. Return mouse to a clean home cage, and maintain heating until righting reflex is shown. Analgesics (e.g., buprenorphine or carprofen) can be administered as necessary.

Imaging Session (>30 min)

26. Anesthetize mice with isoflurane gas (2.0%–2.5%; mixed with 2-L/min oxygen) or interperitoneal injection of ketamine (75 mg/kg) and xylazine (15 mg/kg). Assess depth of anesthesia by monitoring toe pinch withdrawal, eyelid reflex, and respiration rate.
27. Secure animal into a position suitable for imaging. Use appropriate adaptors to clamp the metal connection bar. Maintain body temperature at 37°C with a heating blanket. Apply ophthalmic ointment to the eyes as necessary.
28. Insert the microendoscope probe into the guide tube. Remove protective tape to expose the guide tube. Examine the guide tube for any dirt particles. If necessary, deliver H_2O into the guide tube, and rinse. Using air suction through a 25–29-gauge blunt needle, remove all fluid from the guide tube. Take care not to damage the bottom face of the guide tube with excess pressure.
29. Using an eyepiece and bright-field illumination, focus the microscope objective onto the proximal microendoscope surface. Align the microendoscope to the optical axis of the microscope by adjusting the clamp orientation. Under bright-field illumination, a well-aligned microendoscope will appear circular, not elliptical (which would indicate tilt relative to the optical axis).
30. If available, use one-photon fluorescence imaging to locate the desired tissue region. Use the minimal intensity of light necessary to illuminate the tissue. Typically, one gradually adjusts the focal plane of the microscope objective upward (i.e., away from the specimen), assuming that the tissue plane of interest is located closer to the face of the micro-optical objective than to the

microendoscope probe's design working distance. Optionally, switch to the two-photon fluorescence mode.

See *Troubleshooting*.

31. Microendoscope probes may be interchanged without displacing the animal by using suction to remove the microendoscope from the glass guide tube.

    See *Troubleshooting*.

32. At the end of the experiment, microendoscopes should be cleaned by rinsing and gentle scrubbing with H_2O and lens paper. Sacrifice animal as appropriate for post hoc histological examination.

TROUBLESHOOTING

Problem (Step 30): Excessive tissue motion during imaging.
Solution: Most commonly observed tissue motions are caused by breathing rhythms. First, check the depth of anesthesia during imaging. Second, adjust the head position relative to the animal's trunk to facilitate unconstrained breathing while providing modest mechanical decoupling of the head from motions of the trunk. Another common cause of tissue motion is an excess gap between the tissue and the end of the guide tube; this is the fault of either an improper guide tube placement during the surgery or any swelling that occurred then and later subsided. As the brain tissue stabilizes, the guide tube may no longer be optimally positioned for the desired imaging experiment. Reducing the overall duration of surgery, adjusting the dosage of anti-inflammatory agents, and decreasing the potential heating of the tissue during skull drilling all generally improve experimental quality.

Problem (Step 31): Image quality degrades during image acquisition.
Solution: Clean and inspect the microendoscope, and replace it as necessary. Excessive laser power focused to surfaces of the microendoscope can result in damage to the glass. When this occurs, background photon levels in the image typically increase. Inspection of the microendoscope with an epifluorescence microscope will reveal autofluorescent patterns in which laser scanning occurred on the glass surface.

Alternatively, image degradation may be an indication of cellular damage. During imaging of subcellular structures such as dendrites or axons, blebbing may appear as well as general fading of fluorescence in the scanned regions across the imaging sessions. In such cases, use lower intensity illumination. As an alternative to acquiring a single image at a higher illumination power, averaging of multiple images each taken at a faster acquisition speed and lower power may also improve image quality.

DISCUSSION

Optical microendoscopy is suited for cellular level imaging deep within tissue in live animals or humans. Researchers can choose among a wide variety of microendoscope probe designs to select those best matched to their needs. For the combined acquisition of high-speed videos and 3D image stacks from the same specimen, it is useful to have a microscope that allows online toggling between one-photon fluorescence and laser-scanning imaging (Jung et al. 2004) (Fig. 1A). Laser-scanning second-harmonic generation microendoscopy can generally be performed on any microscope intended for intravital two-photon imaging by an appropriate choice of emission filter (Llewellyn et al. 2008). Overall, microendoscopy is a flexible technique that can be used with multiple modes of contrast generation, at different tissue depths, and with a wide variety of imaging parameters. In the brain, this flexibility has enabled the examination of intracellular calcium dynamics, microcirculatory flow, and neuronal morphology. Because the microendoscope is conceptually, at core, an optical relay, any fluorescent marker that performs well under conventional one- or two-photon fluorescence microscopy will generally perform comparably well under microendoscopy in similar optical conditions.

Comparison to Other Strategies for Imaging Deep Tissues

Some deep structures may be accessed by conventional microscope optics. In one strategy, more invasive aspiration of the tissue allows direct access to the tissue of interest (Mizrahi et al. 2004). A wide column of tissue must be removed to prevent blocking light to and from the specimen if imaging with a high NA is to be achieved. The applicability of this technique seems limited because deeper structures require surgery and aspiration that are substantially more invasive.

A second strategy for deep imaging extends the penetration depth of conventional two-photon microscopy to tissues as deep as 1 mm below the surface, as reviewed in Wilt et al. (2009). To achieve this, several methods exist to improve fluorescence generation, including the use of illumination sources with higher pulse energies and longer wavelengths and adaptive optics to improve the focusing of light in the tissue. In addition to providing a relatively noninvasive means of imaging structures at intermediate depths, such as the infragranular layers of the neocortex, these improvements are also compatible with microendoscopy. However, because of the exponential increase with depth of a photon's probability of being scattered, these methods for extending the reach of conventional light microscopy are unlikely to reach the tissue depths of several millimeters to ~1 cm that have already been demonstrated by microendoscopy.

CONCLUSION

In conclusion, microendoscopy is a useful technique for expanding the range of tissues accessible to cellular level imaging in live animals or humans. Microendoscope probe designs can be customized to accommodate a wide range of optical requirements. Overall, microendoscopy opens a broad set of possibilities for imaging cells in brain areas outside the reach of conventional light microscopy, for basic research purposes, studies of animal disease models, or testing of new therapeutics.

ACKNOWLEDGMENTS

This work was supported by the Stanford Biophysics training grant to RPJB from the U.S. National Institutes of Health and research funding provided to M.J.S. under the National Institute on Drug Abuse Cutting-Edge Basic Research Awards (NIDA CEBRA) DA017895, the National Institute of Neurological Disorders and Stroke (NINDS) R01NS050533, and the National Cancer Institute (NCI) P50CA114747. We thank our collaborators Bernhard Messerschmidt of Grintech GmbH and

Tony Ko, Juergen C. Jung, Alessio Attardo, Yaniv Ziv, Michael Llewellyn, Scott Delp, George Capps, Alison Waters, Tammy J. Wang, and Lawrence Recht of Stanford University for their contributions to the methodologies summarized here.

REFERENCES

Barretto RP, Messerschmidt B, Schnitzer MJ. 2009. In vivo fluorescence imaging with high-resolution microlenses. *Nat Methods* **6:** 511–512.

Dombeck DA, Khabbaz AN, Collman F, Adelman TL, Tank DW. 2007. Imaging large-scale neural activity with cellular resolution in awake, mobile mice. *Neuron* **56:** 43–57.

Engelbrecht CJ, Johnston RS, Seibel EJ, Helmchen F. 2008. Ultra-compact fiberoptic two-photon microscope for functional fluorescence imaging in vivo. *Opt Express* **16:** 5556–5564.

Flusberg BA, Cocker ED, Piyawattanametha W, Jung JC, Cheung EL, Schnitzer MJ. 2005. Fiber-optic fluorescence imaging. *Nat Methods* **2:** 941–950.

Flusberg BA, Nimmerjahn A, Cocker ED, Mukamel EA, Barretto RP, Ko TH, Burns LD, Jung JC, Schnitzer MJ. 2008. High-speed, miniaturized fluorescence microscopy in freely moving mice. *Nat Methods* **5:** 935–938.

Gobel W, Helmchen F. 2007. In vivo calcium imaging of neural network function. *Physiology* **22:** 358–365.

Gobel W, Kerr JN, Nimmerjahn A, Helmchen F. 2004. Miniaturized two-photon microscope based on a flexible coherent fiber bundle and a gradient-index lens objective. *Opt Lett* **29:** 2521–2523.

Helmchen F, Denk W. 2005. Deep tissue two-photon microscopy. *Nat Methods* **2:** 932–940.

Holtmaat A, Svoboda K. 2009. Experience-dependent structural synaptic plasticity in the mammalian brain. *Nat Rev Neurosci* **10:** 647–658.

Holtmaat A, Bonhoeffer T, Chow DK, Chuckowree J, De Paola V, Hofer SB, Hubener M, Keck T, Knott G, Lee WC, et al. 2009. Long-term, high-resolution imaging in the mouse neocortex through a chronic cranial window. *Nat Protoc* **4:** 1128–1144.

Jung JC, Schnitzer MJ. 2003. Multiphoton endoscopy. *Opt Lett* **28:** 902–904.

Jung JC, Mehta AD, Aksay E, Stepnoski R, Schnitzer MJ. 2004. In vivo mammalian brain imaging using one- and two-photon fluorescence microendoscopy. *J Neurophysiol* **92:** 3121–3133.

Kerr JN, Denk W. 2008. Imaging in vivo: Watching the brain in action. *Nat Rev Neurosci* **9:** 195–205.

Levene MJ, Dombeck DA, Kasischke KA, Molloy RP, Webb WW. 2004. In vivo multiphoton microscopy of deep brain tissue. *J Neurophysiol* **91:** 1908–1912.

Llewellyn ME, Barretto RP, Delp SL, Schnitzer MJ. 2008. Minimally invasive highspeed imaging of sarcomere contractile dynamics in mice and humans. *Nature* **454:** 784–788.

Mehta AD, Jung JC, Flusberg BA, Schnitzer MJ. 2004. Fiber optic in vivo imaging in the mammalian nervous system. *Curr Opin Neurobiol* **14:** 617–628.

Mizrahi A, Crowley JC, Shtoyerman E, Katz LC. 2004. High-resolution in vivo imaging of hippocampal dendrites and spines. *J Neurosci* **24:** 3147–3151.

Mukamel EA, Nimmerjahn A, Schnitzer MJ. 2009. Automated analysis of cellular signals from large-scale calcium imaging data. *Neuron* **63:** 747–760.

Nimmerjahn A, Mukamel EA, Schnitzer MJ. 2009. Motor behavior activates Bergmann glial networks. *Neuron* **62:** 400–412.

Rochefort NL, Jia H, Konnerth A. 2008. Calcium imaging in the living brain: Prospects for molecular medicine. *Trends Mol Med* **14:** 389–399.

Wilt BA, Burns LD, Wei Ho ET, Ghosh KK, Mukamel EA, Schnitzer MJ. 2009. Advances in light microscopy for neuroscience. *Annu Rev Neurosci* **32:** 435–506.

51 Light-Sheet-Based Fluorescence Microscopy for Three-Dimensional Imaging of Biological Samples

Jim Swoger,[1] Francesco Pampaloni,[2] and Ernst H.K. Stelzer[2]

[1]EMBL/CRG Systems Biology Research Unit, Centre for Genomic Regulation (CRG), Barcelona, Spain; [2]Cell Biology & Biophysics Unit, European Molecular Biology Laboratory (EMBL), Heidelberg, Germany

ABSTRACT

In modern biology, most optical imaging technologies are applied to two-dimensional cell culture systems; that is, they are used in a cellular context that is defined by hard and flat surfaces. However, a physiological context is not found in single cells cultivated on coverslips. It requires the complex three-dimensional (3D) relationship of cells cultivated in extracellular matrix gels, tissue sections, or in naturally developing organisms. In fact, the number of applications of 3D cell cultures in basic research as well as in drug discovery and toxicity testing has been increasing over the past few years. Unfortunately, the imaging of highly scattering multicellular specimens is still challenging. The main issues are the limited optical penetration depth, the phototoxicity, and the fluorophore bleaching. Light-sheet-based fluorescence microscopy (LSFM) overcomes many drawbacks of conventional fluorescence microscopy by using an orthogonal/azimuthal fluorescence arrangement with independent sets of lenses for illumination and detection. The basic idea is to illuminate the specimen from the side with a thin light sheet that overlaps with the focal plane of a wide-field fluorescence microscope (Huisken et al. 2004). Optical sectioning and minimal phototoxic damage or photobleaching outside a small volume close to the focal plane are intrinsic properties of LSFM. We discuss the basic principles of LSFM and the basic protocols for the preparation, the embedding, and the imaging of 3D specimens used in the life sciences in an implementation of LSFM known as the single (or selective) plane illumination microscope (SPIM).

INTRODUCTION

This chapter describes the technique of light-sheet-based fluorescence microscopy (LSFM) and one of the microscopes that is based on its principles, the single (or selective) plane illumination microscope (SPIM). The introduction covers the limitations of traditional optical imaging techniques, which helps to illuminate the advantages of LSFM and the reasons why it is a valuable method for studying developmental biology and the biology of three-dimensional (3D) cell cultures. The chapter also describes the optical setup and concepts underlying the SPIM, and several applications are provided to illustrate its utility. The chapter ends with several protocols for the preparation, embedding, and imaging of multicellular systems using the SPIM.

TRADITIONAL OPTICAL IMAGING TECHNIQUES

An understanding of modern developmental biology requires us to monitor the complex 3D relationship of cells, whether cultivated (e.g., in an extracellular matrix [ECM]-based gel), in developing embryos, or in tissue sections (Pampaloni et al. 2007). However, the observation and the optical manipulation of multicellular biological specimens remains a challenge, principally for two reasons: (1) Such specimens are optically dense. They scatter and absorb light; thus, the delivery of the probing light and the collection of the signal light tend to become inefficient. (2) In addition to the fluorophores, many endogenous biochemical compounds absorb light and suffer phototoxicity, which can induce damage or can kill the specimen (Pampaloni et al. 2007). The situation is particularly dramatic in confocal fluorescence microscopy. Even when only a single plane is observed, the entire specimen is illuminated. Recording stacks of images along the optical *z*-axis thus illuminates the entire specimen once for each plane. Hence, cultured cells are illuminated 10–20 times, and fish embryos are illuminated 100–300 times more often than they are observed. As we will see in the next section, this can be avoided by a simple change to the optical arrangement.

Traditionally, biological imaging is dominated by wide-field imaging techniques, in which the full field of view is captured in a single exposure. Although invaluable for two-dimensional applications, conventional wide-field fluorescence microscopy provides no optical sectioning. This means that any image recorded in a particular plane in a thick specimen will contain both the in-focus and the out-of-focus signals from the fluorophores above and below the plane of focus. This is the fundamental limitation of conventional wide-field fluorescence light microscopy and makes the construction of quantitative 3D models even of moderately thick specimens difficult, if not impossible, without some form of further innovation.

One obvious way to avoid the problem of out-of-focus light in wide-field fluorescence microscopy is to physically section the sample so that material that would be a source of out-of-focus light is physically removed. Traditionally, the sample is mechanically cut into thin slices, which are treated as two-dimensional specimens, whose images can be computationally assembled into a 3D reconstruction. However, because it is destructive, the technique is not useful for the live time-lapse imaging that is essential in the modern life sciences. In addition, physical sectioning is a low-throughput technique that is not suitable for automated scanning of many specimens.

Over the last few decades, confocal fluorescence microscopy (Pawley 2006) has become a standard tool for biological imaging in which 3D data sets are required. This is a technique whereby, through the use of a confocal pinhole, images containing mainly in-focus information are acquired. Therefore, an optical rather than a physical sectioning of the specimen is achieved. By using confocal fluorescence microscopes to localize fluorophore distributions in three dimensions, studies of complex biological structures can be performed with fixed and live samples. Although confocal fluorescence microscopy does provide 3D voxel data sets, the resolution along the axial direction of a confocal fluorescence microscope is always lower than along the lateral directions (Pawley 2006). This is because of the physics of diffraction. This resolution anisotropy can be reduced by working with high-numerical-aperture (NA) objective lenses. However, in practical lens design, a very large NA is accompanied by a small working distance (generally substantially less than a millimeter). In addition to the mechanical limitations imposed by the working distance of the lens, the image quality in confocal data stacks degenerates rapidly with depth because of the high optical scattering caused by the refractive index heterogeneity of thick biological specimens. Taken together, these properties limit the suitability of confocal fluorescence microscopy for imaging thick 3D samples.

An alternative to confocal fluorescence microscopy that provides optically sectioned data sets through a different physical mechanism is multiphoton microscopy (Denk et al. 1990; Konig 2000). Instead of using a pinhole to discriminate against out-of-focus light, nonlinear interactions between the excitation light and the fluorophores are used to ensure that only molecules exposed to very high illumination intensities (i.e., those located directly in the focal plane) are excited. The resulting optically sectioned images produced by multiphoton microscopy are, in many respects, similar to those recorded with confocal fluorescence microscopy. However, because of the longer wavelengths used, the resolution is considerably worse along all three dimensions than in a confocal or a conventional fluorescence microscope (Stelzer and Lindek 1994).

Several other imaging methods have been used successfully in developmental biology. Optical coherence tomography, a technique based on interferometric detection of low-coherence illumination, has received considerable attention in the past decade as a technique for noninvasive biological imaging. Magnetic resonance imaging (MRI; Bamforth et al. 2011; Metscher 2011; Ruffins and Jacobs 2011), ultrasound imaging (Foster and Brown 2011), and microcomputed tomography (μCT; Quintana and Sharpe 2011) all allow imaging of relatively large samples and are, therefore, suitable for imaging the later stages of development of specimens such as mouse embryos. Although, in many ways, these are ideal methods for imaging a range of samples, a common drawback is that they cannot produce images using the wide range of highly specific chemical and genetically encoded fluorophores that have become invaluable in light microscopy (Lichtman and Conchello 2005). Optical projection tomography (OPT; Gu et al. 2011; Quintana and Sharpe 2011) is one of the few optical imaging techniques suitable for developmental biology that is capable of operating with both fluorescence and absorption contrasts.

PRINCIPLES OF LSFM

The basic idea of LSFM is to use a light sheet that illuminates the specimen from the side and that overlaps with the focal plane of a wide-field fluorescence microscope (Huisken et al. 2004). In contrast to an epifluorescence arrangement, which uses the same lens, a light-sheet-based orthogonal/azimuthal fluorescence arrangement uses two independent lenses for illumination and detection. An LSFM system is able to provide optical sectioning without phototoxic damage or photobleaching outside of a small volume located close to the focal plane. LSFM takes advantage of modern charge-coupled device (CCD) camera technologies to generate images with a signal-to-noise ratio (SNR) that, for a given fluorophore, objective lens, and specimen, is at least one order of magnitude better than that of a confocal fluorescence microscope (Keller et al. 2008). (Note that good-quality confocal fluorescence images have an SNR between 10 and 40 [given by the square root of the number of detected photons; see the Scientific Volume Imaging-Wiki, http://support.svi.nl/wiki/WikiWikiWeb]. CCD cameras used in fluorescence microscopy have an SNR between 100 and 1000 [defined as the square root of the full-well capacity]. For example, the ORCA-AG [Hamamatsu] has a full-well capacity of 18,000 electrons, which gives an SNR of ~130.)

The idea of using light-sheet illumination for microscopy has occurred to various groups at different times and has been implemented in a number of configurations. These are summarized in Table 1. All of these techniques intrinsically provide optical sectioning; and, except for the original implementation of ultramicroscopy (Siedentopf and Zsigmondy 1903), which works with incoherent illumination and scattered light contrast, all rely on laser illumination focused in one dimension to excite fluorescence. In this chapter, we refer to an instrument known as a single (or selective) plane illumination microscope (SPIM). The optical properties of LSFM can be calculated by following the approach in Engelbrecht et al. (2007) and Stelzer and Lindek (1994). In SPIM, the sample is illuminated by a sheet of laser light. A laser array, in combination with dichroic mirrors and an acousto-optical tunable filter (AOTF), is used to provide a variety of wavelengths for the excitation of different fluorescent dyes (Fig. 1A). Cylindrical optics (Huisken et al. 2004; Greger et al. 2007) or a scanning mechanism in the digital scanned laser light-sheet fluorescence microscopy (DSLM)

TABLE 1. Some methods that use LSFM

Name	Distinctions/features	Reference(s)
Scattering ultramicroscopy[a]	Developed to study subwavelength colloidal particles(these are illuminated from the side by an incoherent illumination)	Siedentopf and Zsigmondy 1903
Confocal θ fluorescence microscopy[a]	Point-scanning device using an orthogonal or, in general, azimuthal optical arrangement	Stelzer and Lindek 1994
Fluorescence ultramicroscopy	Applied to fixed chemically cleared samples	Dodt et al. 2007
Orthogonal-plane fluorescence optical sectioning device	Developed for imaging very large and chemically cleared organs, such as the guinea pig cochlea	Voie et al. 1993; Buytaert and Dirckx 2009
Thin laser light-sheet microscopy	Used in microbial oceanography	Fuchs et al. 2002
SPIM, scanned light-sheet microscopy (DSLM)	Developed for live, fast, multiple-view, and multiple-dye imaging of embryos, cell clusters, and single cells	Huisken et al. 2004; Keller et al. 2008
Objective-coupled planar illumination microscopy	Illumination optics mechanically coupled to detection objective lens	Holekamp et al. 2008; Turaga and Holy 2008

[a]The first two examples (original ultramicroscopy and confocal θ fluorescence microscopy) are forerunner techniques exploiting the idea of side illumination of the specimen.

(Keller et al. 2008) is used to focus a collimated laser beam along one dimension, thereby creating a single plane of light. The specimen is embedded in a cylinder of agarose or other mounting gel, which is generally immersed in an aqueous medium. The fluorescence emitted from the narrow volume of the specimen that is excited by the light sheet is imaged onto a camera using standard fluorescence microscopy optics: an objective lens, a filter wheel, and a tube lens (Fig. 1A). The sample can be moved along the detection axis (i.e., perpendicular to the plane of the light sheet) and imaged in a stepwise fashion to generate a 3D data set. The SPIM setup usually includes a mechanism for rotating the specimen, so that its orientation can be optimized for imaging in a particular experiment or for acquiring image stacks from multiple directions (Fig. 1C).

With a SPIM, optical sectioning is obtained in a direct and efficient way. The sectioning capability is defined by the thickness of the light sheet (typically 2–6 µm, depending on the extent of the field of view). By illuminating only the plane of interest, no out-of-focus light is created, and there is no need for a confocal pinhole for the discrimination of this background (Fig. 1Bi). Therefore, photobleaching and photodamage are less than with confocal or wide-field fluorescence microscopy, in which the complete specimen is illuminated even when observing only a single plane (Fig. 1Bii). Objective lenses with a long working distance and a relatively low NA can be used for detection when imaging large specimens (a few millimeters). The lateral resolution is defined by the NA of the detection objective lens, whereas the axial resolution of the system is determined by the light-sheet thickness, when it is thinner than the axial detection resolution. The combination of optical sectioning, long working distances, and minimal photodamage make LSFM ideal for imaging of samples typically of interest in developmental and 3D multicellular biology. A schematic of an LSFM setup can be found in Keller et al. (2011).

One of the advantages of a SPIM is that the sample can be rotated and can be observed from multiple orientations (Fig. 1C). Three-dimensional stacks can be acquired from different directions, and the images can be computationally fused into a single representation of the sample (Swoger et al. 2007; Verveer et al. 2007). This has the advantage of improving the resolution in small transparent specimens and of improving the uniformity of image quality in larger specimens for which absorption and/or scattering are significant limitations to imaging. Examples are provided in the next section.

SPIM APPLICATIONS

Figure 2 illustrates the effect of multiple-view image fusion on a grain of paper mulberry pollen. The same grain of pollen has been imaged along 18 directions, two of which are shown in Figure 2, A and

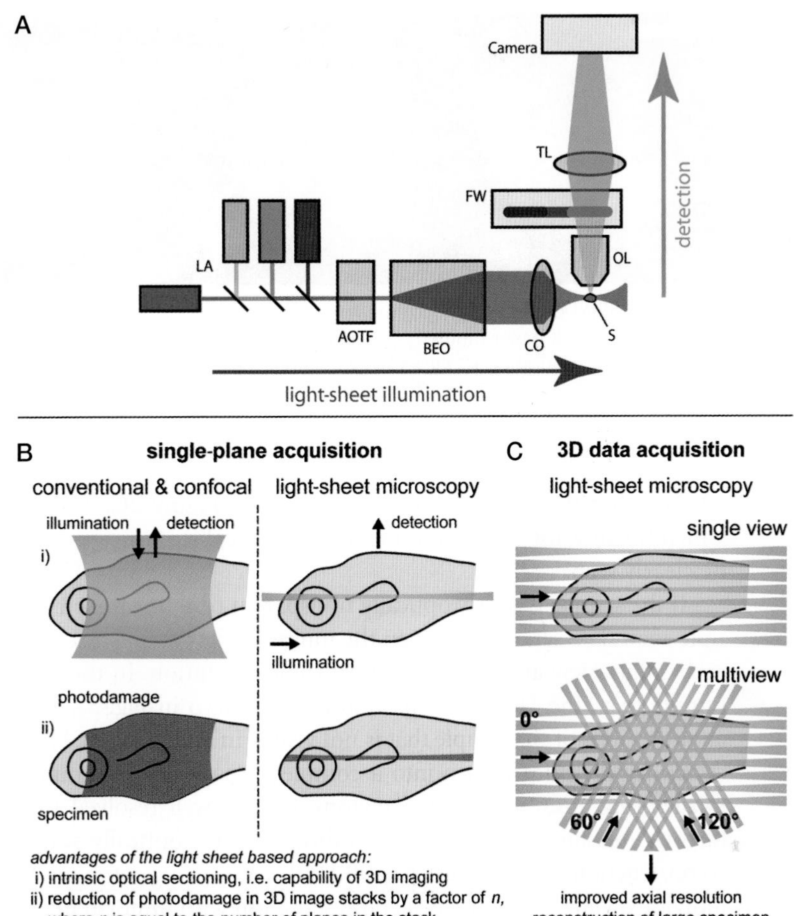

FIGURE 1. Principles of light-sheet-based microscopy. (*A*) The overall layout of a SPIM. The emission from a laser array (LA) of various wavelengths for exciting fluorescence is combined into a single beam. An acousto-optic tunable filter (AOTF) is used to select the wavelengths and the intensity of the excitation beam. The beam expanding and cylindrical optics (BEO and CO) create the light sheet that illuminates the sample (S). Detection is through a wide-field fluorescence microscope consisting of an objective lens (OL), a filter wheel (FW) that contains filters for rejection of scattered excitation light and background fluorescence, a tube lens (TL), and a camera. (*B*) Comparison of sample illumination and fluorescence detection in conventional/confocal microscopy and in LSFM. (*i*) The entire region of interest in the specimen is illuminated in conventional/confocal microscopy, although only a single plane in the specimen is being observed. (*ii*) In contrast, no photodamage is inflicted outside the in-focus plane of the detection system in the LSFM. (*C*) Three-dimensional imaging in LSFM is performed by moving the specimen step by step through the light sheet while recording two-dimensional images. In multiple-view imaging, the same volume inside the specimen or even inside the entire specimen is recorded along several angles. The resulting multiple-view information is combined into a single image stack by a fusion algorithm. (*B,C*, Adapted, with permission, from Keller and Stelzer 2008.)

B. Note that although the light-sheet illumination improves the axial resolution over wide-field or confocal fluorescence microscopy with the same detection objective lens, the resolution is anisotropic. This is indicated schematically by the red ellipses: These point-spread functions (PSFs) are narrowest along the direction with the highest resolution. In the image shown in Figure 2A, the resolution is good vertically but poor horizontally; in Figure 2B, which was acquired with the sample oriented at 90° to the orientation in Figure 2A, the situation is the reverse.

By recording 18 three-dimensional data sets (i.e., a rotation of the sample in 20° steps), aligning them, and taking the average, the image shown in Figure 2C is obtained. It can be seen that the resolution (the PSF is represented by the red circle) is quite high along all directions and approximately equal to the resolution along the best axis in either Figure 2, A or B. A further improvement in resolution and contrast can be obtained by using a multiple-view image deconvolution (Verveer et al.

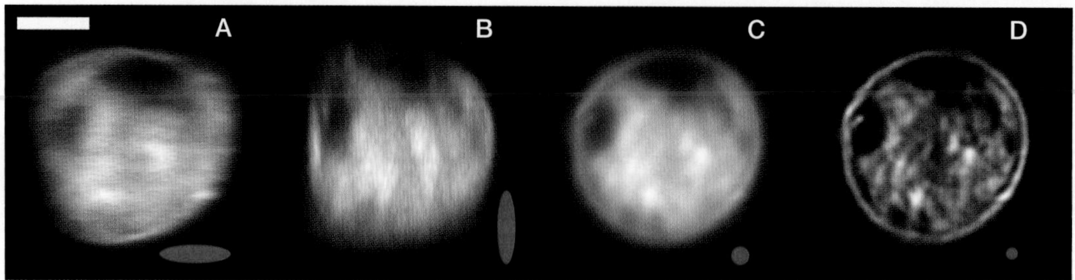

FIGURE 2. SPIM of paper mulberry pollen, multiview reconstruction. Slices from volumetric data stacks of an autofluorescent grain of paper mulberry pollen are imaged with a SPIM. (*A,B*) Single-view images along orthogonal directions. (*C*) Combination of 18 views by averaging. (*D*) Combination of the same 18 views as in *C* by multiple-view deconvolution. Scale bar, 5 µm. (Image acquisition by Klaus Greger, EMBL, Heidelberg.)

2007) rather than the simple average used in Figure 2C. The result is illustrated in Figure 2D, in which the interior details and the quasispherical surface layer or sporoderm of the pollen grain can be clearly seen.

If the sample is nearly optically transparent (e.g., the pollen grain in Fig. 2) and the different multiple-view data sets overlap, then the multiple-view reconstruction compensates for the anisotropy in each single view and provides an isotropic resolution. In the case of a more translucent or opaque sample (e.g., the *Drosophila melanogaster* embryo in Fig. 3), additional views provide information about regions of the sample that is not visible in a single view. Here, the multiple-view reconstruction combines the information into a complete image of the sample. It can be seen in the single-view images (Fig. 3A,C) that some regions are not well resolved (arrowheads) or are entirely missing (arrows) because they were imaged through thick optically scattering tissue. When a multiple-view reconstruction is performed (Fig. 3B,D) the processing combines the best-resolved features from each individual view into a single 3D representation with uniform high-resolution coverage of the sample.

Imaging 3D Cell Biology Specimens with a SPIM

Increasingly, cell biologists appreciate that conventional two-dimensional cell cultures do not provide a physiological environment for the cell (Pampaloni et al. 2007; Mazzoleni et al. 2009). The surfaces of plastic and glass substrates are flat and rigid. In contrast, a tissue is a 3D soft-matter

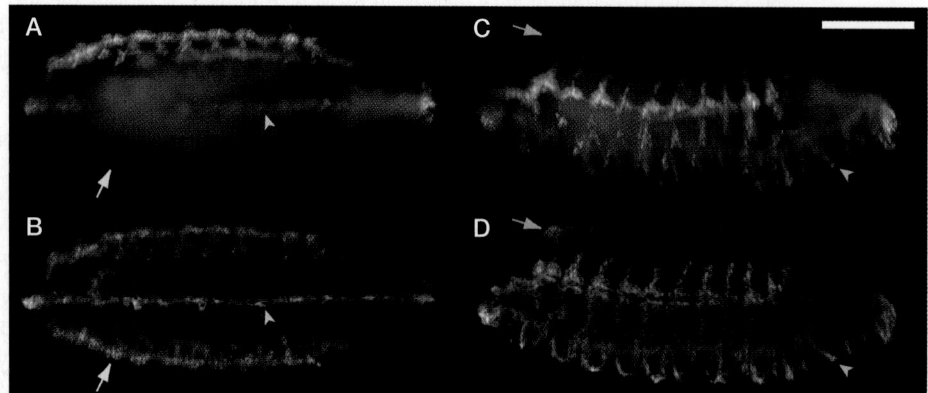

FIGURE 3. SPIM of a *D. melanogaster* embryo. Maximum-value projections through volumetric data stacks of a fruit fly embryo with the trachea labeled with green fluorescent protein. (*A*) Single-view data set, projected along the dorsoventral axis. (*B*) Multiple-view fusion of 12 views similar to that shown in *A* but taken with different orientations of the sample. (*C,D*) Projections corresponding to those in *A* and *B* but along the lateral axis of the embryo. Scale bar, 100 µm. (Sample courtesy of Ferenc Jankovics and Damian Brunner, EMBL, Heidelberg; image acquisition by Klaus Greger, EMBL, Heidelberg.)

structure. There are many instances in which cells have drastically different phenotypes in two dimensions than in three dimensions. For example, fibroblasts cultured in two dimensions have a flat shape dissimilar from the bipolar/stellate shape found in tissue. Strikingly, culturing fibroblasts in 3D collagen induces their tissue-specific phenotype (Beningo et al. 2004; Rhee et al. 2007; Rhee and Grinnell 2007). It is currently acknowledged that establishing 3D cell–cell interactions and using ECMs or 3D gels reduces the gap between cell culture and real tissue. There are strong indications that toxicity assays based on 3D cultures of human cells provide more accurate results than two-dimensional cultures (Pampaloni et al. 2009). Three-dimensional cultures are promising for toxicity screening of chemicals and to sort out toxic substances at early stages of drug discovery. Imaging of fluorophore-tagged targets provides the specificity that is required to perform detailed studies at the molecular level in live 3D cultures. Novel advanced fluorescence imaging technologies are required to study the behavior of living 3D cultures. LSFM provides an excellent tool to study the live behavior of 3D cellular systems with high spatial resolution and for long periods of time. Light-sheet illumination provides optical sectioning deep inside large 3D tissue. The extremely low illumination level induces minimum photobleaching and photodamage.

Embedding in agarose is a well-established approach to stably mount specimens for imaging with a SPIM. The agarose concentration used is 0.5%–1% (w/v) in buffer or culture medium. Agarose is a nearly ideal embedding medium because it is optically clear and introduces very limited aberrations. Agarose is nontoxic and, in most cases, does not interfere with the physiology of a living specimen. Next, we provide detailed protocols for embedding and imaging, with a SPIM, two typical specimens used in 3D cell biology, namely, cellular spheroids and MDCK (Madin–Darby canine kidney) cysts grown in ECM hydrogels.

Protocol A

Imaging Cellular Spheroids with a SPIM

Cellular spheroids are aggregations of hundreds to thousands of cells into spheres with diameters of 100–800 µm (Friedrich et al. 2009) (Fig. 4A). Spheroids were probably the first 3D cell system used in clinical pharmacology (Mueller-Klieser 1997). Many common cell lines form spheroids, including MCF-10a, Caco-2, and HepG2 (Kelm et al. 2003). Spheroids from the human teratocarcinoma cell line Ntera-2 are a useful model system for biomedical studies and toxicity assays of the nervous system (Podrygajlo et al. 2009). Cellular spheroids obtained by the aggregation of primary cells, tumor, or nontumor cell lines are often used in tissue engineering, regenerative medicine, oncology, and the development of more reliable drug toxicity assays (Pampaloni et al. 2009). Cell aggregation is induced by physical forces (buoyancy or stirring) and mostly occurs without having to add exogenous scaffolds or ECM proteins. Buoyancy is exploited in the hanging-drop method, in which droplets of culture medium containing trypsinated cells are suspended on the surface of a Petri dish lid. Compact spheroids are harvested from the droplets within 3–20 d of growth (Kelm et al. 2003; Timmins et al. 2004). Spheroids can also be obtained by seeding cells on nonadhesive surfaces such as 3D alginate porous scaffolds (Glicklis et al. 2000) or from cell cultures in rotating well vessels (Bilodeau and Mantovani 2006). Standardized protocols for the reproducible and reliable culture, handling, and screening of cellular spheroids for pharmacology and toxicology have been published recently (Friedrich et al. 2009).

IMAGING SETUP

Live spheroids can be stained with a variety of fluorescent dyes. Members of the CellTracker family (Invitrogen) are suitable dyes for identifying and tracking a population of cells inside the spheroid. Nuclear morphology can be investigated with DAPI (4′,6-diamino-2-phenylindole) (excitation 345

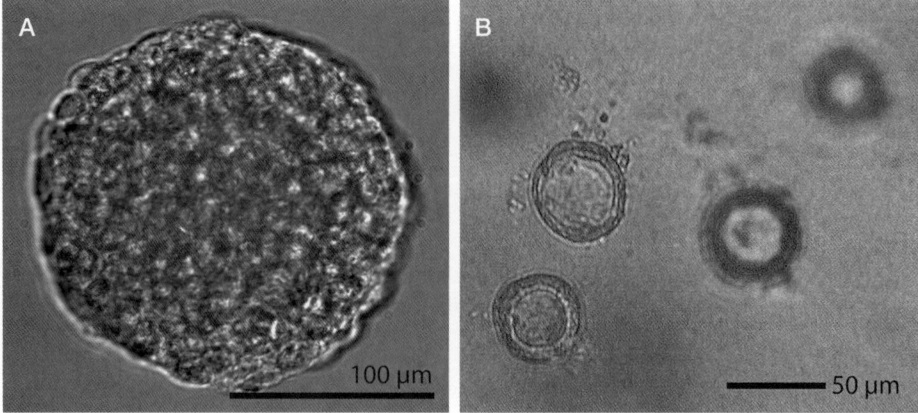

FIGURE 4. Typical 3D cell biology specimens. (*A*) This spheroid derives from a BxCP-3 pancreas adenocarcinoma cell line and has been obtained with the hanging-drop method. The spheroid is embedded in agarose and is composed of about 300 cells. Transmitted light imaging obtained on the SPIM setup, objective lens CZ Plan Apochromat 40x/NA 0.8. (*B*) MDCK cysts cultured in 3D hydrated collagen type I for ~1 wk. MDCK cells embedded in 3D collagen I or reconstituted basement membrane (commercial name is Matrigel) form hollow spherical monolayers (called cysts) within a few days. MDCK cysts are a well-established model of epithelial development and epithelial cell polarization. The specimen was extracted from the original 3D collagen matrix by using collagenase and subsequently embedded in agarose. Transmitted light imaging obtained on the SPIM setup, objective lens Zeiss Plan Apochromat 40x/NA 0.8.

nm; emission 458 nm) or DRAQ5 (Biostatus Limited, excitation 647 nm; emission 665 nm). Of course, cells tagged with fluorescent proteins are, in many cases, more convenient than using dyes.

To image ~100-μm spheroids at subcellular resolution, use a SPIM with (1) a 40x/NA 0.75–0.80 water-dipping objective lens (e.g., Zeiss Achroplan 40x/0.8 W) and (2) a CCD camera suitable for fluorescence microscopy (e.g., Hamamatsu ORCA-AG, 1344 x 1024 pixels, cell size 6.45 x 6.45 μm). With this combination of objective lens and camera, the field of view is 217 x 165 μm. A CCD camera with a larger chip (e.g., the PCO.2000 camera [PCO AG imaging] chip size 2112 x 2072 pixels, cell size 7.40 x 7.40 μm) could be used with a 63x objective lens, such as a Zeiss Achroplan 63x/1.0 W. Objective lenses with lower magnification and NA (e.g., Zeiss Achroplan 10x/0.3 W or Achroplan 20x/0.5 W) are recommended when subcellular resolution is not the main issue.

MATERIALS

CAUTION: See Appendix for proper handling of materials marked with <!>.

Reagents

Agarose, low-gelling 1% (w/v) in phosphate-buffered saline (PBS)
  Prepare aliquots of 1% agarose in 1.5-mL microfuge tubes.
Cellular spheroids stained or tagged with fluorescent probes (see Imaging Setup)
Dulbecco's modified Eagle's medium (DMEM), 1x without phenol red
Ethanol <!> (Troubleshooting)
PBS (pH 7.4)

Equipment

Beaker, glass
Dissection stereomicroscope with transmitted light illumination and incident light illumination with fiber-optic cold-light source (e.g., Stemi 2000, Carl Zeiss MicroImaging, Inc.)
Electrical wire with diameter 1–1.8 mm (e.g., 10/0.1-mm tinned copper wire 01-1535 or 7/0.2-mm wire 01-2246; Rapid Electronics)
Embryo dishes or watch glasses
  These are molded clear glass dishes for viewing free-floating specimens (e.g., Agar Scientific L4161). Alternatively, glass well/depression slides can be used.
Glass cutter, diamond-tipped
Heating blocks set at 37°C and 65°C
Micropipettes, glass, 100-μL or 200-μL (e.g., BLAUBRAND intraMARK 708744, inner diameter [ID] 1.0 mm, outer diameter [OD] 1.7 mm, or intraMARK 708757, ID 1.45 mm, OD 2.25 mm, BRAND GMBH & Co.)
Molding dough (e.g., Nakiplast, Pelikan)
Nail polish
Pipette tips, 200-μL yellow tips with cut ends and intact 10-μL tips
Scalpel or razor blade
SPIM (see Imaging Setup)
SPIM imaging chamber
SPIM sample holder

EXPERIMENTAL METHOD

Mounting Spheroids for Imaging

1. Transfer several 1.5-mL aliquots of 1% low-gelling-temperature agarose into a 65°C heating block.
2. Once the agarose has melted, transfer the tubes to a 37°C heating block or water bath.

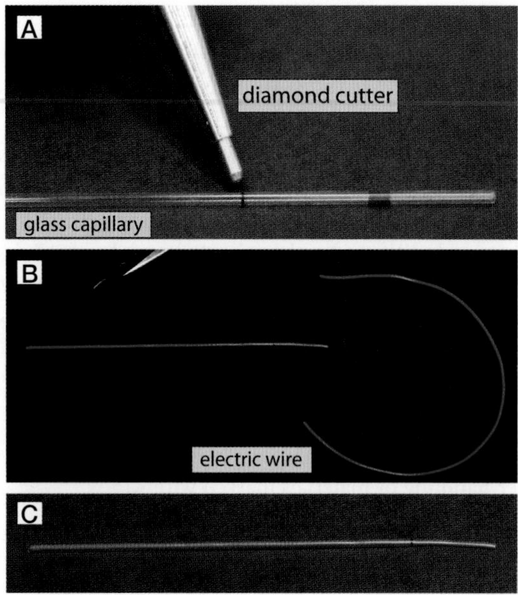

FIGURE 5. Preparation of glass capillary holder. (*A*) A 100-µL glass disposable micropipette (BLAUBRAND intraMARK, ID 1.0 mm, OD 1.7 mm) is cut to the desired length with a diamond-tip cutter. (*B*) A slightly longer electrical wire is inserted into the capillary as a plunger. (*C*) Close-up of the complete capillary-wire system. The length of the cut glass capillary is 12.5 cm (cut to 9 cm).

3. Prepare sample holders by cutting glass micropipettes to the desired length with a diamond-tip cutter (Fig. 5A).
4. Insert a section of electrical wire into the glass capillaries to serve as a plunger. The wire should be slightly longer than the capillary (Fig. 5B,C).
5. Using yellow pipette tips with cut ends, harvest one to five spheroids from the culture, and deposit them in a watch glass.

   Cut tip ends minimize shear flow and mechanical damage to the spheroids.
6. Using a dissection stereomicroscope, carefully remove as much medium as possible from the watch glass using a 10-µL pipette tip. Carefully replace the medium with a generous amount of 1% agarose solution.
7. Pick up the spheroids one by one by sucking them into a glass capillary (Fig. 6). Practice is required to efficiently perform this step. Work quickly, picking up all of the spheroids before the agarose sets. For multiple-view imaging, the spheroids should be placed in the center of the

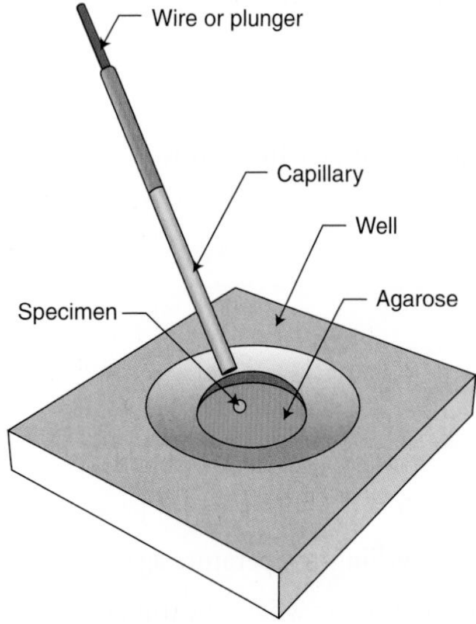

FIGURE 6. Embedding the specimens in agarose. A specimen (e.g., an MDCK cyst or a cellular spheroid) is suspended in a droplet of liquid low-gelling agarose and deposited in a shallow well. A few microliters of agarose containing the specimen are drawn into a thin glass capillary by using, for example, an electrical wire of suitable diameter as a plunger. The agarose sets within 5 min with the specimen embedded.

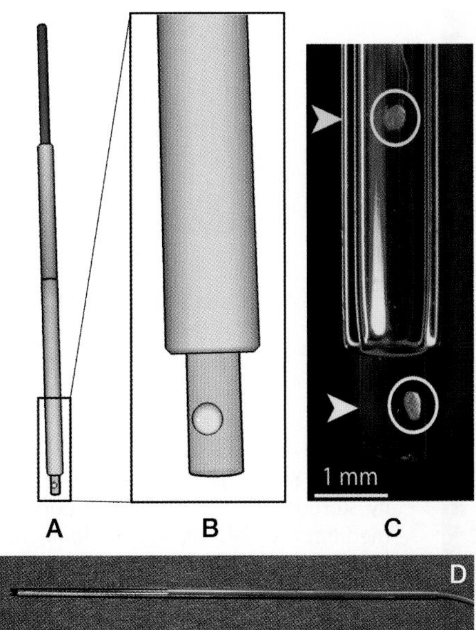

FIGURE 7. Preparation of the specimen for imaging. (A,B) On polymerization, the solid agarose tip with the specimen is pushed out of the capillary. The specimen is ready for imaging. (C) Photograph of a specimen. The bottom arrow shows an agarose-embedded cellular spheroid outside of the capillary. A second embedded spheroid is still inside the capillary (top arrow). (D) Photograph of the complete setup comprising glass capillary, wire, and agarose. The capillary length is 9 cm.

capillary. In practice, however, the spheroid's position within the agarose is hard to control. Thus, preparing several samples is recommended.

See *Troubleshooting*.

8. Once the spheroids are picked, wait 10–15 min while the agarose inside the capillary fully sets. Then, push the agarose cylinder with the spheroid out of the capillary by sliding the plunger (Fig. 7).
9. Remove excess agarose below the spheroid with a scalpel or a razor blade. Only a short segment (1–2 mm) of agarose should be exposed outside the capillary (Fig. 7). A short agarose segment is more mechanically stable than a long one. Agarose cylinders longer than 1–2 mm may produce blurred or offset images.
10. For short-term storage of the specimens, keep the capillaries in a laboratory beaker filled with culture medium or PBS. Use a small slab of soft molding dough attached to the beaker's rim to keep the specimens in place (Fig. 8). To avoid drifting of the agarose cylinder during imaging, glue the wire at the top of the capillary with nail polish (Fig. 8).
11. Insert the capillary into the SPIM specimen holder (examples of holders used at EMBL are shown in Fig. 9), and place it into the imaging chamber by firmly connecting it with the *x–y–z–Φ* motorized stage (Fig. 10).

Imaging Cellular Spheroids with a SPIM

12. Image small spheroids (diameter of ~100 μm) at subcellular resolution using a 40X water-dipping objective lens (NA 0.75–0.80) and a CCD camera suitable for fluorescence microscopy (see Imaging Setup above for details on instrumentation).
13. Adjust the illuminating laser light sheet to obtain the best compromise between light-sheet thickness and homogeneity of axial resolution. For a 40X objective, this corresponds to a light-sheet thickness of 3.5 μm full width at half-maximum.
14. Acquire stacks of lateral (*x–y*) images using a step size of 1.8 μm along the *z*-axis to obtain a reasonable axial (*z*) sampling. Typical recording parameters for a spheroid with a diameter of ~100 μm are listed in Table 2.

    Given the nearly perfect spherical shape of the spheroids, multiple-view imaging can be performed quite easily (Figs. 11 and 12).

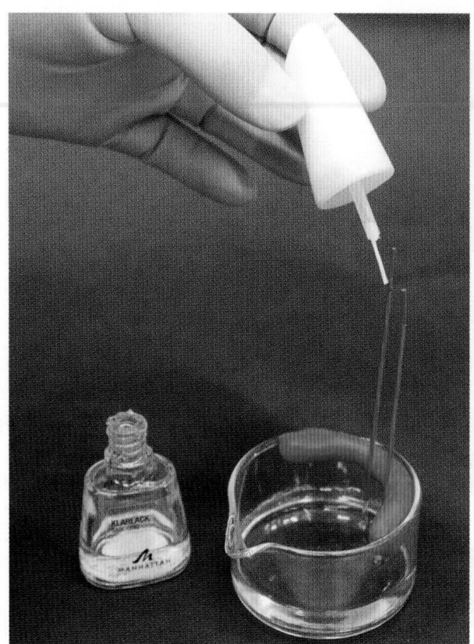

FIGURE 8. Storage of the specimen. The specimen is temporarily stored within a glass laboratory beaker filled with PBS or medium. A slab of soft molding dough is attached to the beaker's rim and is used to keep the glass capillary in place. After pushing out the specimen, the wire is fixed in its final position with nail polish. This avoids a further drift of the specimen during imaging with a SPIM.

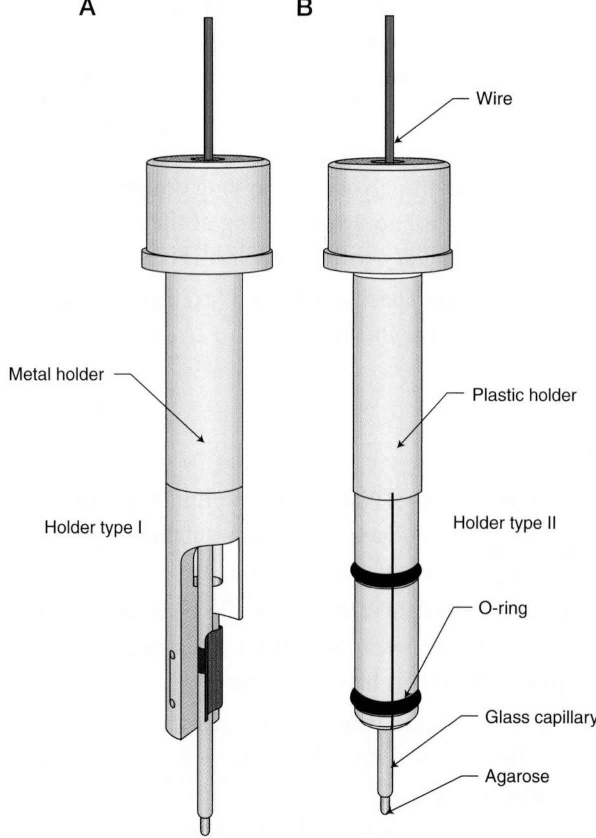

FIGURE 9. Capillary holders for SPIM. Two types of glass capillary holders developed at EMBL. (*A*) Aluminum holder. A metal spring keeps the glass capillary firmly in place. (*B*) Plastic holder. In this simplified holder type, two rubber O-rings keep a tight hold on the capillary.

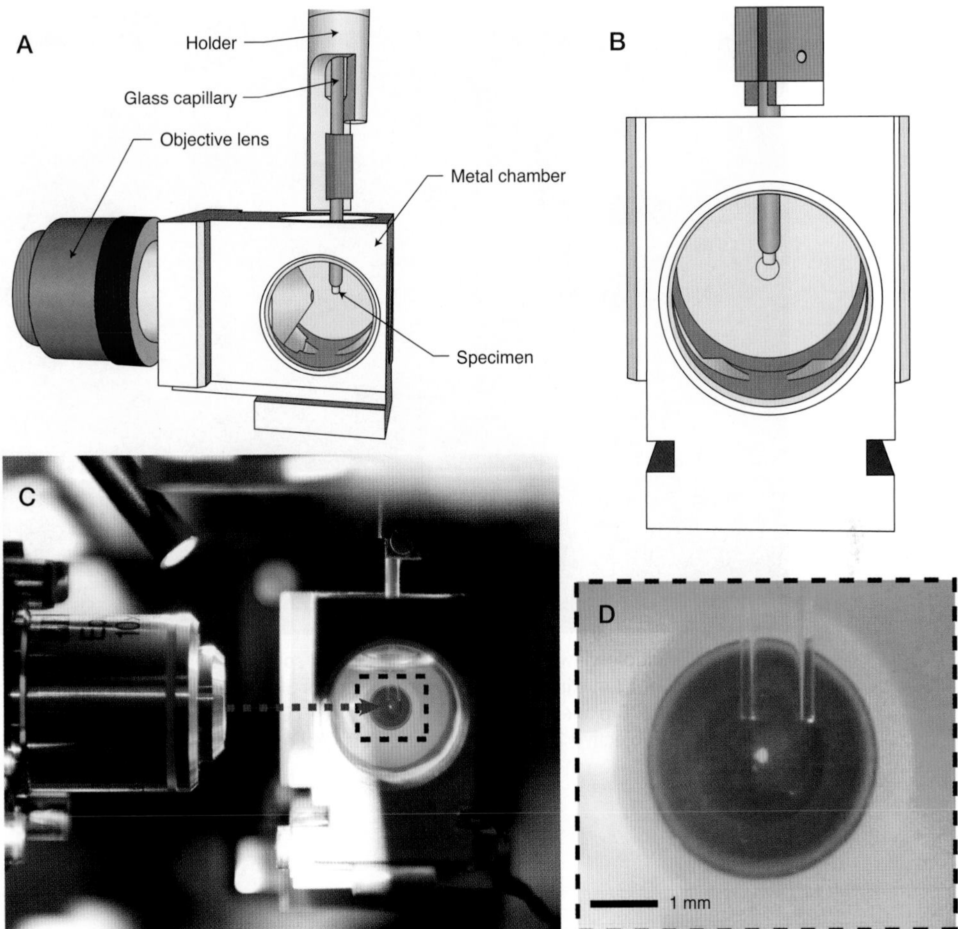

FIGURE 10. SPIM imaging setup. (*A,B*) Schematic of the SPIM imaging setup. The specimen mounted in the glass capillary is immersed in a liquid medium (e.g., DMEM without phenol red or PBS). The medium is contained in a chamber with glass windows on the side and on the front. The glass windows are glued to the chamber by using nail polish or a silicon-based glue and can be easily replaced. The detection objective lens is directly inserted into the chamber. A spring-loaded radial shaft seal ensures that the junction of the objective lens and the chamber is watertight. (*C*) Photograph of the real system. On the *left*, the illumination lens is visible. The light sheet illuminates the sample from the side as indicated by the blue arrow. (*D*) A close-up of the specimen, a cellular spheroid with a diameter of ~100 µm.

TABLE 2. Typical SPIM recording parameters

Specimen	Cellular spheroid (diameter ~100µm)
Objective lens	Zeiss Plan Apochromat 63↔/NA 1.0
Fluorescence labeling	CyTRAK dye (nuclei + cytoplasm)
Excitation wavelength	488 nm
Light-sheet thickness	2.6 µm
Measured laser power on the specimen	0.540 mW (30 nW/µm^2)[a]
Exposure time	100 msec
CCD camera type	Hamamatsu ORCA-AG
CCD chip cell size	6.45 µm
CCD chip total size	1340 x 1024 pixels
z-stack (total travel)	130 µm
Voxel size	102 x 102 x 819 nm
Number of *x*–*y* planes	159
Total recording time	18 sec

[a]Specimen, cellular spheroid with a diameter of ~150 µm. The laser power was measured by placing the detector at the position of the specimen.

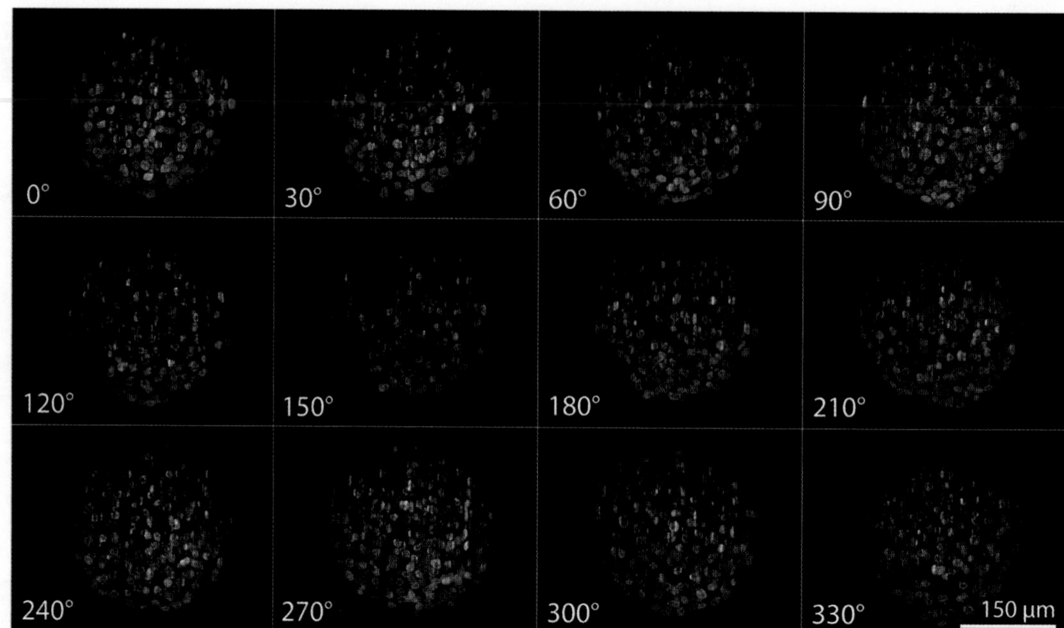

FIGURE 11. Multiple-view angular imaging series of a spheroid derived from BxCP-3 pancreas adenocarcinoma cells. The figure shows 12 maximum-intensity projections of SPIM stacks recorded along 12 different directions. Each stack is composed of 278 planes with 0.5-μm spacing. The cell nuclei are stained with DRAQ5 (excitation 633 nm; emission 665 nm). Objective lens water-dipping Zeiss Achroplan 40X/NA 0.8. The CCD camera used is a Hamamatsu ORCA-AG.

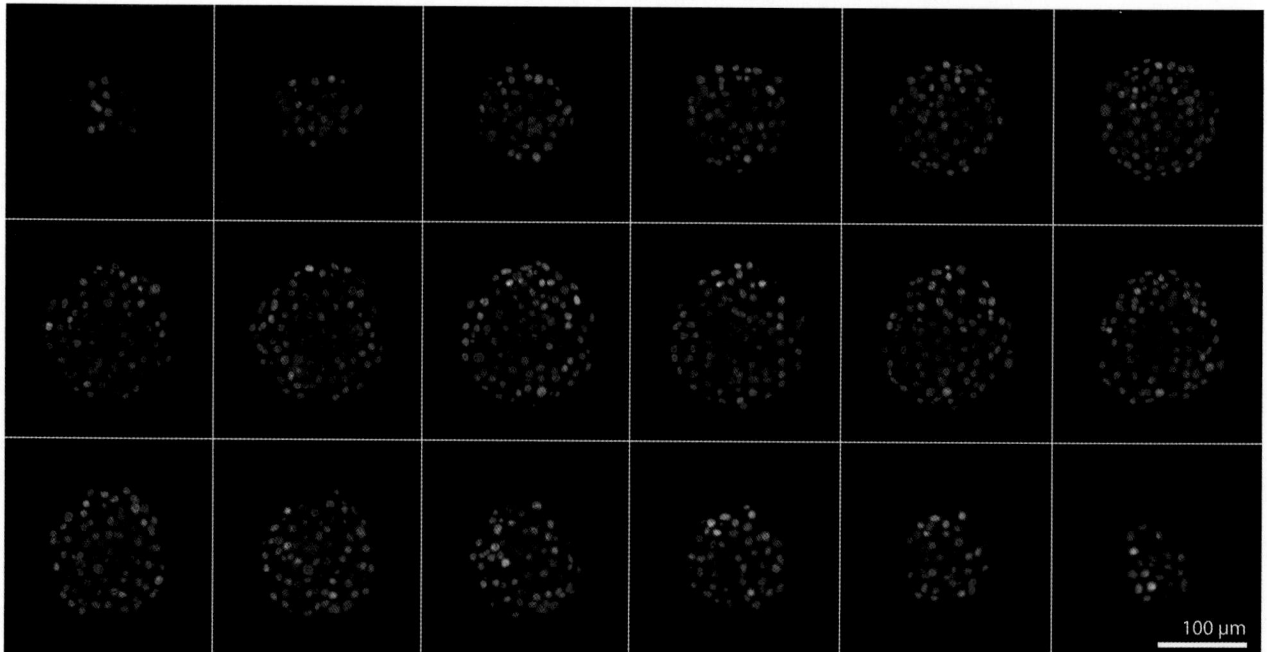

FIGURE 12. Images of a spheroid derived from BxCP-3 pancreas adenocarcinoma cells recorded with a SPIM. Selected single frames from a single-view SPIM stack. The frames are spaced ~10 μm along the z-axis.

TROUBLESHOOTING

Problem (Step 7): Spheroids stick to the watch glass bottom.

Solution: Clean the glass with ethanol before adding spheroids.

Protocol B

Imaging MDCK Cysts with a SPIM

The MDCK cell line was originally isolated from a dog kidney's distal tubules (Herzlinger et al. 1982). MDCK cells form hollow spherical monolayers (called cysts) when grown within 3D type I collagen or reconstituted basement membrane (Matrigel). MDCK cysts show the cellular polarization of kidney tubules, with the apical side facing a fluid-filled internal lumen and the basal part facing the ECM. MDCK cyst formation is a well-established model of epithelial morphogenesis in vitro (Herzlinger et al. 1982; O'Brien et al. 2002). Live and fixed MDCK cysts can be fluorescently stained by following the protocols described in Elia and Lippincott-Schwartz (2009). SPIM instrumentation details are in Imaging Setup in Protocol A.

MATERIALS

See the end of the chapter for recipes for reagents marked with <R>.

Reagents

Bicarbonate buffer (pH 9.0) <R>
Collagen, bovine type I hydrated (e.g., AteloCell I AC-30 or Nutragen)
Collagenase type 1A, 2 mg/mL in PBS
DMEM, 1x without phenol red (e.g., D2902 from Sigma-Aldrich) supplemented with 2.2 g/L $NaHCO_3$ and 10% fetal bovine serum (FBS)
DMEM, 10x without phenol red
Matrigel without phenol red (BD Biosciences 356237)
MDCK type II cells stained or tagged with fluorescent probes (CCL-34, American Type Culture Collection) (Elia and Lippincott-Schwartz 2009)
PBS (pH 7.4)

Equipment

Beaker, glass
Dissection stereomicroscope with transmitted light illumination and incident light illumination with fiber-optic cold-light source (e.g., Stemi 2000, Carl Zeiss MicroImaging, Inc.)
Electrical wire with diameter 1–1.8 mm (e.g., 10/0.1-mm tinned copper wire 01-1535 or 7/0.2-mm wire 01-2246; Rapid Electronics)
Glass cutter, diamond-tipped
Incubator, tissue culture set at 37°C and 5% CO_2
Micropipettes, glass, 100- or 200-µL (e.g., BLAUBRAND intraMARK 708744, ID 1.0 mm, OD 1.7 mm, or intraMARK 708757, ID 1.45 mm, OD 2.25 mm)
Molding dough (e.g., Nakiplast, Pelikan)
Nail polish
Petri dishes, plastic, 55-mm
Pipette tips, 200-µL yellow tips with cut ends
Plates, 96-well (e.g., Nunc 269620)
Scalpel or razor blade
Scissors, fine
SPIM (see Imaging Setup in Protocol A)

SPIM imaging chamber
SPIM sample holder
Tubes, 10-mL

METHOD 1: PREPARING AND IMAGING MDCK CYSTS IN COLLAGEN

Culturing MDCK Cysts in 3D Collagen

1. Prepare 5.6 mL of collagen-medium mixture in a 10-mL tube on ice as follows:

   0.1 mL 1x DMEM with $NaHCO_3$ and FBS
   0.2 mL bicarbonate buffer
   0.6 mL 10x DMEM
   4.7 mL bovine type I hydrated collagen

   The final collagen concentration is 2.5 mg/mL.

2. Add 2.8 mL of MDCK cell suspension (4×10^5 cells/mL) in DMEM to the collagen-medium mixture.
3. Warm the cell-gel mixture for 3–4 min at room temperature by gently shaking the tube.
4. Aliquot the cell-gel mixture into two 55-mm Petri dishes (~2 mL of cell-gel mixture per dish). Incubate the dishes for 15 min at 37°C in a CO_2 incubator until collagen polymerization is complete.
5. Add 9 mL of DMEM supplemented with 10% FBS to each dish.
6. Keep the collagen-embedded MDCK cells in a humidified cell culture incubator at 37°C and 5% CO_2 for long-term culture.
7. Exchange the culture medium daily. Following 9 d of growth in 3D collagen gel, MDCK cells form spherical cysts with a diameter of ~50 µm, consisting of about 40–50 cells.

Imaging MDCK Cysts Isolated from Collagen

8. Remove the collagen slab containing the cysts from the Petri dish, and cut it into small fragments with a pair of fine scissors.
9. Transfer the fragments into a 10-mL tube, and mix it with 5 mL of collagenase type 1A solution. Incubate the tube for ~15 min at 37°C to completely dissolve the collagen.
10. Using the dissecting microscope, pick up isolated cysts, and put them into a 55-mm Petri dish containing DMEM without phenol red.
11. Mount the cysts into glass capillaries (see Protocol A).
12. Transfer the mounted cysts to a sample holder, place into the SPIM imaging chamber containing phenol red–free DMEM or PBS, and image the isolated cysts.

    An example of an isolated MDCK cyst imaged with a SPIM is shown in Figure 13.

Imaging Collagen-Embedded MDCK Cysts

13. Cut the collagen slab with a pair of fine scissors, and transfer the fragments into a 55-mm Petri dish containing DMEM without phenol red.
14. Using the dissecting microscope, cut the fragments so that they are smaller than the diameter of the glass capillary.

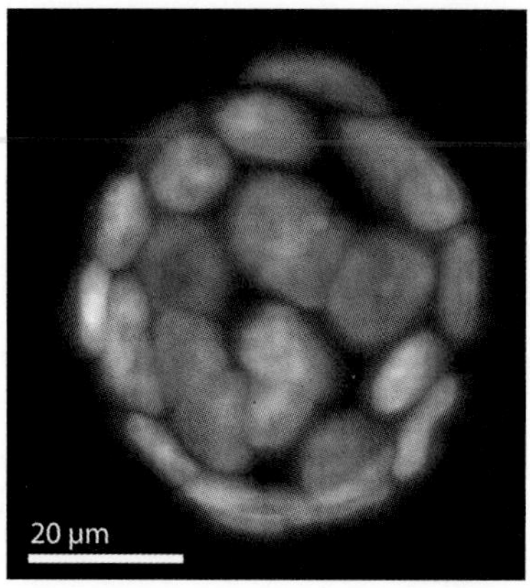

FIGURE 13. Image of an isolated MDCK cyst recorded with a SPIM. The image is a maximum-intensity projection of a stack of about 50 planes spaced 0.8 μm along the z-axis. The stack was recorded with a water-dipping objective lens Zeiss Achroplan 40x, NA 0.8. The cell nuclei are stained with the dye DRAQ5, which selectively labels the nuclear DNA. The excitation laser wavelength is 633 nm; the emission is ~665 nm.

15. Insert the collagen fragments into the glass capillary (see Protocol A) (Fig. 14).
16. Transfer the mounted cysts to a sample holder, place into the SPIM imaging chamber, and image the embedded cysts.

METHOD 2: PREPARING AND IMAGING MDCK CYSTS IN MATRIGEL

Culturing MDCK Cells in Matrigel

1. In a 10-mL tube on ice, combine two parts Matrigel and one part MDCK cells suspended in phenol red–free DMEM (4 x 10^5 cells/mL).
2. Warm the cell-gel mixture for 3–4 min at room temperature while gently shaking the tube.
3. Aliquot 200 μL of the cell-gel mixture into each well of a 96-well plate. Incubate the plates for 15 min at 37°C in a CO_2 incubator until Matrigel polymerization is complete.
4. Add 150 μL of DMEM supplemented with 10% FBS to each well.
5. Culture the Matrigel-embedded MDCK cells in a humidified cell culture incubator at 37°C and 5% CO_2.
6. Exchange the culture medium daily. Within 4–6 d of growth in Matrigel, MDCK cells form spherical cysts with a diameter of ~50 μm.

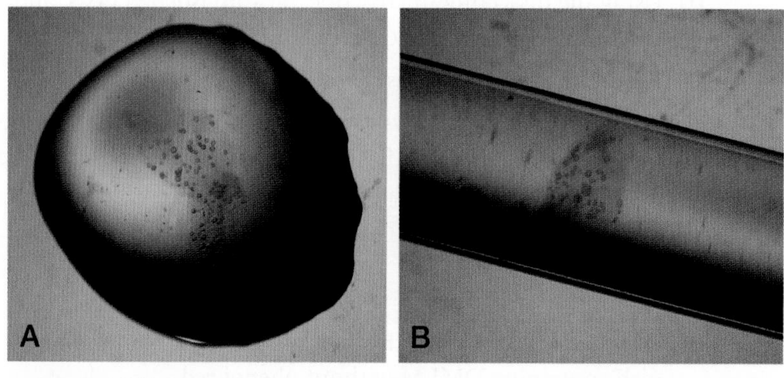

FIGURE 14. Image of collagen-embedded MDCK cysts in agarose. (*A*) A fragment of collagen containing MDCK cysts, suspended in liquid low-gelling agarose. (*B*) The same specimen embedded in agarose inside the glass capillary. Both images are recorded with a Stemi 2000 from Carl Zeiss MicroImaging, Inc. (magnification 50x).

Imaging Matrigel-Cultured MDCK Cysts

7. Using a pipette, remove the culture medium from the wells.
8. Pipette Matrigel containing the cysts from the wells with a 200-μL yellow tip with a cut end.
   Carefully avoid the formation of air bubbles in the gel.
9. Pipette the Matrigel into a 55-mm Petri dish containing 2 mL of phenol red–free DMEM. Slowly pipette the material up and down to fragment the Matrigel.
10. Mount the small Matrigel fragments into glass capillaries (see Protocol A).
11. Transfer the mounted cysts to a sample holder, place into the SPIM imaging chamber containing phenol red–free DMEM or PBS, and image the embedded cysts.

SUMMARY

We have described the SPIM, a microscope that uses the principles of LSFM for imaging 3D specimens, such as are commonly studied in developmental biology and in 3D cell culturing. The principal advantages of SPIM over techniques such as wide-field or confocal microscopy are as follows:

- reduced photobleaching and phototoxicity
- the ability to achieve high resolution with long working distance objectives, which allows imaging of thick 3D specimens
- the option of achieving high isotropic resolution via multiple-view reconstructions

Autofluorescent pollen grains and genetically labeled *D. melanogaster* embryos were used to show the advantages of multiview reconstructions, which (to the best of our knowledge) is not a technique available in systems other than SPIM. In addition, we have described, in detail, the necessary materials and protocols for imaging two types of 3D cell cultures with SPIM: cellular spheroids derived from BxCP-3 pancreas adenocarcinoma cells or other cell lines and MDCK cell cysts. The ability of SPIM to generate high-resolution images of live samples as they develop over time will lead to novel types of experiments in developmental biology and 3D cell cultures.

RECIPE

CAUTION: See Appendix 6 for proper handling of materials marked with <!>.
Recipes for reagents marked with <R> are included in this list.

Bicarbonate Buffer

Mix 7.5 g $NaHCO_3$ and 7.5 g Na_2CO_3<!> per 100 mL H_2O. The pH should be 9.0.

REFERENCES

Bamforth SD, Schneider JE, Bhattacharya S. 2011. High-throughput analysis of mouse embryos by magnetic resonance imaging. In *Imaging in developmental biology: A laboratory manual* (ed. Sharpe J, Wong RO). Cold Spring Harbor Laboratory Press, Cold Spring Harbor, NY (in press).

Beningo KA, Dembo M, Wang YL. 2004. Responses of fibroblasts to anchorage of dorsal extracellular matrix receptors. *Proc Natl Acad Sci* **101:** 18024–18029.

Bilodeau K, Mantovani D. 2006. Bioreactors for tissue engineering: Focus on mechanical constraints. A comparative review. *Tissue Eng* **12:** 2367–2383.

Buytaert JAN, Dirckx JJJ. 2009. Tomographic imaging of macroscopic biomedical objects in high resolution and three dimensions using orthogonal-plane fluorescence optical sectioning. *Appl Opt* **48:** 941–948.

Denk W, Strickler JH, Webb WW. 1990. Two-photon laser scanning fluorescence microscopy. *Science* **248:** 73–76.

Dodt HU, Leischner U, Schierloh A, Jahrling N, Mauch CP, Deininger K, Deussing JM, Eder M, Zieglgansberger W, Becker K. 2007. Ultramicroscopy: Three-dimensional visualization of neuronal networks in the whole mouse brain. *Nat Methods* **4:** 331–336.

Elia N, Lippincott-Schwartz J. 2009. Culturing MDCK cells in three dimensions for analyzing intracellular dynamics. *Curr Protoc Cell Biol* **4:** 22.

Engelbrecht CJ, Greger K, Reynaud EG, Krzic U, Colombelli J, Stelzer EH. 2007. Three-dimensional laser microsurgery in light-sheet based microscopy (SPIM). *Opt Express* **15:** 6420–6430.

Foster SF, Brown AS. 2011. Micro-ultrasound and its application to longitudinal studies of mouse eye development and disease. In *Imaging in developmental biology: A laboratory manual* (ed. Sharpe J, Wong RO). Cold Spring Harbor Laboratory Press, Cold Spring Harbor, NY (in press).

Friedrich J, Seidel C, Ebner R, Kunz-Schughart LA. 2009. Spheroid-based drug screen: Considerations and practical approach. *Nat Protoc* **4:** 309–324.

Fuchs E, Jaffe J, Long R, Azam F. 2002. Thin laser light sheet microscope for microbial oceanography. *Opt Express* **10:** 145–154.

Glicklis R, Shapiro L, Agbaria R, Merchuk JC, Cohen S. 2000. Hepatocyte behavior within three-dimensional porous alginate scaffolds. *Biotechnol Bioeng* **67:** 344–353.

Greger K, Swoger J, Stelzer EH. 2007. Basic building units and properties of a fluorescence single plane illumination microscope. *Rev Sci Instrum* **78:** 023705.

Gu S, Jenkins MW, Watanabe M, Rollins AM. 2011. High-speed optical coherence tomography (OCT) imaging of the beating avian embryonic heart. In *Imaging in developmental biology: A laboratory manual* (ed. Sharpe J, Wong RO). Cold Spring Harbor Laboratory Press, Cold Spring Harbor, NY (in press).

Herzlinger DA, Easton TG, Ojakian GK. 1982. The MDCK epithelial cell line expresses a cell surface antigen of the kidney distal tubule. *J Cell Biol* **93:** 269–277.

Holekamp TF, Turaga D, Holy TE. 2008. Fast three-dimensional fluorescence imaging of activity in neural populations by objective-coupled planar illumination microscopy. *Neuron* **57:** 661–672.

Huisken J, Swoger J, Del Bene F, Wittbrodt J, Stelzer EH. 2004. Optical sectioning deep inside live embryos by selective plane illumination microscopy. *Science* **305:** 1007–1009.

Keller PJ, Stelzer EH. 2008. Quantitative in vivo imaging of entire embryos with digital scanned laser light sheet fluorescence microscopy. *Curr Opin Neurobiol* **18:** 624–632.

Keller PJ, Schmidt AD, Wittbrodt J, Stelzer EH. 2008. Reconstruction of zebrafish early embryonic development by scanned light sheet microscopy. *Science* **322:** 1065–1069.

Keller PJ, Schmidt AD, Wittbrodt J, Stelzer EHK. 2011. Imaging the development of entire zebrafish and *Drosophila* embryos with digital scanned laser light-sheet fluorescence microscopy (DSLM). In *Imaging in developmental biology: A laboratory manual* (ed. Sharpe J, Wong RO). Cold Spring Harbor Laboratory Press, Cold Spring Harbor, NY (in press).

Kelm JM, Timmins NE, Brown CJ, Fussenegger M, Nielsen LK. 2003. Method for generation of homogeneous multicellular tumor spheroids applicable to a wide variety of cell types. *Biotechnol Bioeng* **83:** 173–180.

Konig K. 2000, Multiphoton microscopy in life sciences. *J Microsc* **200:** 83–104.

Lichtman JW, Conchello JA. 2005. Fluorescence microscopy. *Nat Methods* **2:** 910–919.

Mazzoleni G, Di Lorenzo D, Steimberg N. 2009. Modelling tissues in 3D: The next future of pharmaco-toxicology and food research? *Genes Nutr* **4:** 13–22.

Metscher BD. 2011. X-ray microtomography (microCT) imaging of vertebrate embryos. In *Imaging in developmental biology: A laboratory manual* (ed. Sharpe J, Wong RO). Cold Spring Harbor Laboratory Press, Cold Spring Harbor, NY (in press).

Mueller-Klieser W. 1997. Three-dimensional cell cultures: From molecular mechanisms to clinical applications. *Am J Physiol* **273:** C1109–1123.

O'Brien LE, Zegers MM, Mostov KE. 2002. Opinion: Building epithelial architecture: Insights from three-dimensional culture models. *Nat Rev Mol Cell Biol* **3:** 531–537.

Pampaloni F, Reynaud EG, Stelzer EH. 2007. The third dimension bridges the gap between cell culture and live tissue. *Nat Rev Mol Cell Biol* **8:** 839–845.

Pampaloni F, Stelzer EH, Masotti A. 2009. Three-dimensional tissue models for drug discovery and toxicology. *Recent Pat Biotechnol* **3:** 103–117.

Pawley JB. 2006. *Handbook of biological confocal microscopy*. Springer, New York.

Podrygajlo G, Tegenge MA, Gierse A, Paquet-Durand F, Tan S, Bicker G, Stern M. 2009. Cellular phenotypes of human model neurons (NT2) after differentiation in aggregate culture. *Cell Tissue Res* **336:** 439–452.

Quintana L, Sharpe J. 2011. Optical projection tomography of vertebrate embryo development. In *Imaging in developmental biology: A laboratory manual* (ed. Sharpe J, Wong RO). Cold Spring Harbor Laboratory Press, Cold Spring Harbor, NY (in press).

Rhee S, Grinnell F. 2007. Fibroblast mechanics in 3D collagen matrices. *Adv Drug Deliv Rev* **59:** 1299–1305.

Rhee S, Jiang H, Ho CH, Grinnell F. 2007. Microtubule function in fibroblast spreading is modulated according to the tension state of cell-matrix interactions. *Proc Natl Acad Sci* **104:** 5425–5430.

Ruffins SW, Jacobs RE. 2011. MRI in developmental biology and the construction of developmental atlases. In *Imaging in developmental biology: A laboratory manual* (ed. Sharpe J, Wong RO). Cold Spring Harbor Laboratory Press, Cold Spring Harbor, NY (in press).

Siedentopf H, Zsigmondy R. 1903. Über Sichtbarmachung und Größenbestimmung ultramikroskopischer Teilchen, mit besonderer Anwendung auf Goldrubingläser. *Ann Phys* **10:** 1–39.

Stelzer EHK, Lindek S. 1994. Fundamental reduction of the observation volume in far-field light-microscopy by detection orthogonal to the illumination axis: Confocal θ microscopy. *Opt Commun* **111:** 536–547.

Swoger J, Verveer P, Greger K, Huisken J, Stelzer EHK. 2007. Multi-view image fusion improves resolution in three-dimensional microscopy. *Opt Express* **15:** 8029–8042.

Timmins NE, Dietmair S, Nielsen LK. 2004. Hanging-drop multicellular spheroids as a model of tumour angiogenesis. *Angiogenesis* **7:** 97–103.

Turaga D, Holy TE. 2008. Miniaturization and defocus correction for objective-coupled planar illumination microscopy. *Opt Lett* **33:** 2302–2304.

Verveer PJ, Swoger J, Pampaloni F, Greger K, Marcello M, Stelzer EHK. 2007. High-resolution three-dimensional imaging of large specimens with light sheet-based microscopy. *Nat Methods* **4:** 311–313.

Voie AH, Burns DH, Spelman FA. 1993. Orthogonal-plane fluorescence optical sectioning: 3-dimensional imaging of macroscopic biological specimens. *J Microsc* **170:** 229–236.

52 Photoacoustic Imaging

Yin Zhang,[1] Hao Hong,[2] and Weibo Cai[1–3]

[1]*Department of Medical Physics, School of Medicine and Public Health, University of Wisconsin-Madison, Madison, Wisconsin 53705;* [2]*Department of Radiology, School of Medicine and Public Health, University of Wisconsin-Madison, Madison, Wisconsin 53705;* [3]*University of Wisconsin Carbone Cancer Center, Madison, Wisconsin 53705*

ABSTRACT

Photoacoustic imaging, which is based on the photoacoustic effect, has developed extensively over the last decade. Possessing many attractive characteristics, such as the use of nonionizing electromagnetic waves, good resolution and contrast, portable instrumention, and the ability to partially quantitate the signal, photoacoustic techniques have been applied to the imaging of cancer, wound healing, disorders in the brain, and gene expression, among others. As a promising structural, functional, and molecular imaging modality for a wide range of biomedical applications, photoacoustic-imaging systems can be briefly categorized into two types: photoacoustic tomography (PAT), which is the focus of this chapter, and photoacoustic microscopy (PAM). We first briefly describe the endogenous (e.g., hemoglobin and melanin) and the exogenous (e.g., indocyanine green [ICG], various gold nanoparticles, single-walled carbon nanotubes [SWNTs], quantum dots [QDs], and fluorescent proteins) contrast agents for photoacoustic imaging. Next, we discuss in detail the applications of nontargeted photoacoustic imaging. Recently, molecular photoacoustic (MPA) imaging has gained significant interest, and a few proof-of-principle studies have been reported. We summarize the current state of the art of MPA imaging, including the imaging of gene expression and the combination of photoacoustic imaging with other imaging modalities. Last, we point out obstacles facing photoacoustic imaging. Although photoacoustic imaging will likely continue to be a highly vibrant research field for years to come, the key question of whether MPA imaging could provide significant advantages over nontargeted photoacoustic imaging remains to be shown in the future.

INTRODUCTION

More than a century ago, Alexander G. Bell first observed the photoacoustic effect. He found that absorption of electromagnetic waveforms, such as radio frequency (rf) or optical waves, can generate transient acoustic signals in media. Such absorption leads to local heating and thermoelastic expansion, which can produce megahertz ultrasonic waves in materials. Because different biological tissues have different absorption coefficients, by measuring the acoustic signals with ultrasonic transducers, it is possible to rebuild the distribution of optical energy deposition and ultimately obtain images of the biological tissues.

Photoacoustic imaging, based on the photoacoustic effect, is a noninvasive imaging modality, which has come a long way over the last decade (Wang 2008). Optical and rf waves, instead of electromagnetic waves at other wavelengths, are used in photoacoustic imaging because of their desirable physical properties such as deeper tissue penetration and better absorption by contrast agents. The combination of high ultrasonic resolution with good image contrast because of differential optical/rf absorption is quite advantageous for imaging purposes (Xu and Wang 2006). When compared with fluorescence imaging, in which the scattering in tissues limits the spatial resolution with increasing depth, photoacoustic imaging has higher spatial resolution and deeper imaging depth because scattering of the ultrasonic signal in tissue is much weaker. When compared with ultrasound imaging, in which the contrast is limited because of the mechanical properties of biological tissues, photoacoustic imaging has better tissue contrast, which is related to the optical properties of different tissues. In addition, the absence of ionizing radiation also makes photoacoustic imaging safer than other imaging techniques such as computed tomography and radionuclide-based imaging techniques.

Typically, the spatial resolution provided by current photoacoustic-imaging systems is ~100 µm, which may be further improved in the future (Debbage and Jaschke 2008). Given that optical absorption can reveal various physiological parameters such as hemoglobin/melanin/water/ion concentration and oxygen saturation in living subjects, photoacoustic imaging is a promising structural, functional, and molecular imaging modality for a wide range of biomedical applications (Wang et al. 2003b; Li et al. 2008; Wang 2008). Photoacoustic-imaging systems can be briefly categorized into two types: photoacoustic tomography (PAT; also referred to as optoacoustic tomography or thermoacoustic tomography) and photoacoustic microscopy (PAM) (Fig. 1). Although PAM has gained significant attention over the last several years (Maslov et al. 2005; Zhang et al. 2007; Maslov et al. 2008; Hu et al. 2009; Stein et al. 2009a; Xie et al. 2009), we will primarily focus on PAT in this chapter. Both nontargeted and molecularly targeted photoacoustic imaging will be discussed.

CONTRAST AGENTS FOR PHOTOACOUSTIC IMAGING

An ideal scenario for photoacoustic imaging would be that light absorption of normal tissue should be low for deeper signal penetration, whereas the absorption for the object of interest should be high for optimal image contrast (Xiang et al. 2009). The contrast agents used for photoacoustic imaging can be categorized into two types: endogenous and exogenous contrast agents. Certain endogenous molecules, such as hemoglobin and melanin (Fig. 2), have much stronger light absorption than normal tissue in both the visible and the near-infrared (700–900-nm) regions. Two of the biggest advantages of using endogenous contrast agents for imaging applications are safety and the possibility of revealing the true physiological condition because the physiological parameters do not change during image acquisition if a relatively slow biological process is imaged.

In many scenarios such as the detection of early stage tumors, an endogenous contrast agent alone is insufficient to provide useful information. Because the intensity of a photoacoustic signal in biological tissue is proportional to optical energy absorption, which is proportional to the amount of the contrast agent (Rajian et al. 2009), exogenous contrast agents are frequently needed to provide better signal/contrast for photoacoustic imaging. Commonly used contrast agents for photoacoustic imaging include ICG, various gold nanoparticles (Yang et al. 2007), SWNTs (Hong et al. 2009; Pramanik et al. 2009a), QDs (Shashkov et al. 2008), and fluorescent proteins (Fig. 2).

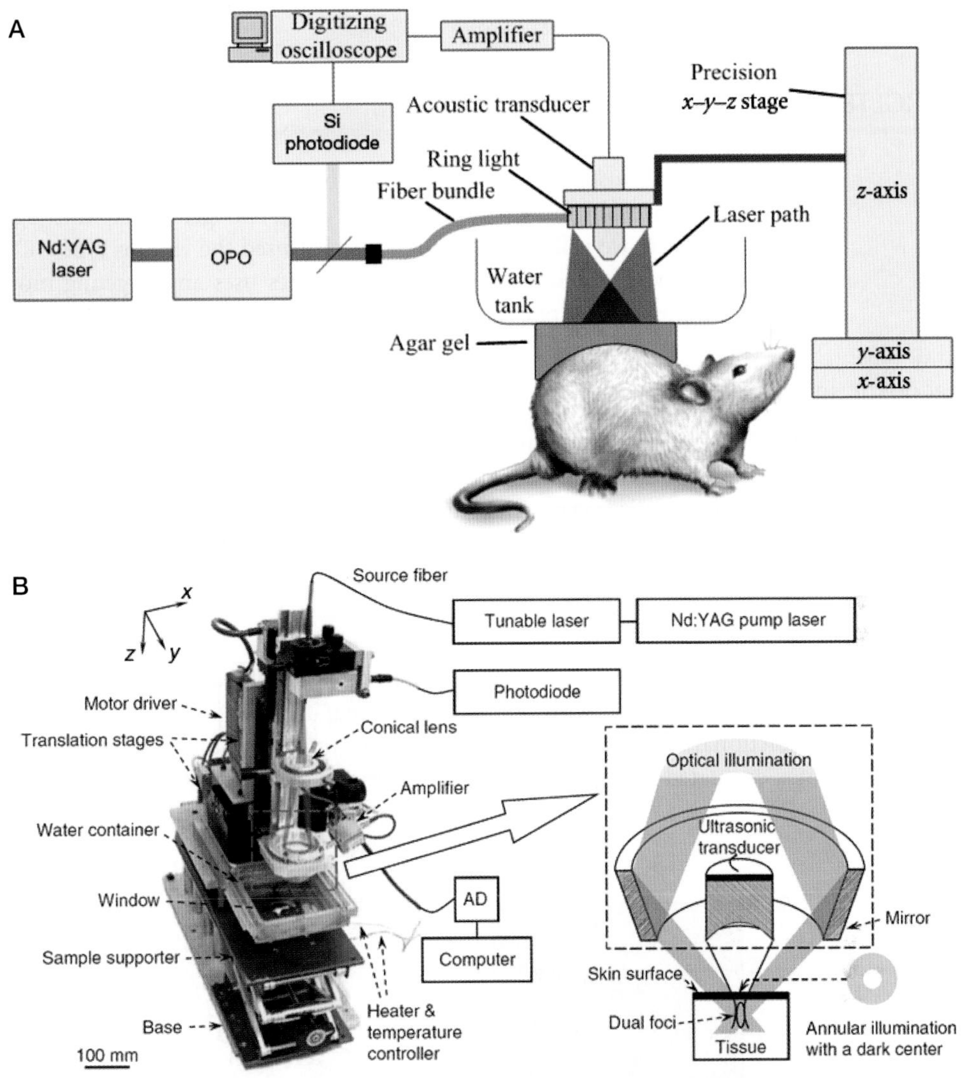

FIGURE 1. Typical setups for photoacoustic tomography (*A*) and photoacoustic microscopy (*B*). (Reprinted, with permission, from Zhang et al. 2006, © Macmillan, and from De la Zerda et al. 2008, © Macmillan.)

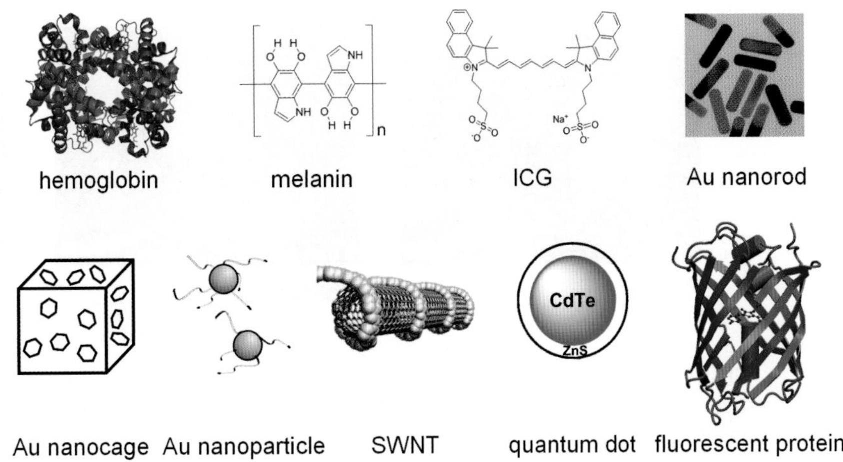

FIGURE 2. A wide variety of contrast agents can be used for photoacoustic-imaging applications.

NONTARGETED PHOTOACOUSTIC IMAGING

To date, photoacoustic imaging has generally been used in preclinical research and animal studies. Here, we will first give an overview of the studies using endogenous contrast agents (i.e., hemoglobin and melanin). Then, we will discuss the reports that used the various types of exogenous contrast agents mentioned above. Although none of these studies uses any targeting moiety, with the assistance of proper contrast agents, photoacoustic imaging could be used in many aspects of biomedical research.

Photoacoustic Imaging with Hemoglobin and/or Melanin

Hemoglobin and melanin are the two most important naturally occurring contrast agents for enhanced photoacoustic imaging. As an endogenous contrast agent, hemoglobin has been explored for photoacoustic imaging in a number of experimental scenarios such as visualizing brain structure and lesions (Wang et al. 2003a), delineating tumor vasculature (Siphanto et al. 2005), monitoring hemodynamics (Wang et al. 2006), imaging small animals (Kruger et al. 2003), and measuring microvascular blood flow (Fang et al. 2007).

Photoacoustic imaging with hemoglobin can greatly facilitate brain-and-blood-dynamics-related research (Wang et al. 2003a,b; Kolkman et al. 2004). In 2003, structures of the rat brain were accurately mapped with photoacoustic imaging (Fig. 3A) (Wang et al. 2003b). Not only were the functional cerebral hemodynamic changes in cortical blood vessels around the whisker-barrel cortex evaluated in response to whisker stimulation, hyperoxia-induced and hypoxia-induced cerebral

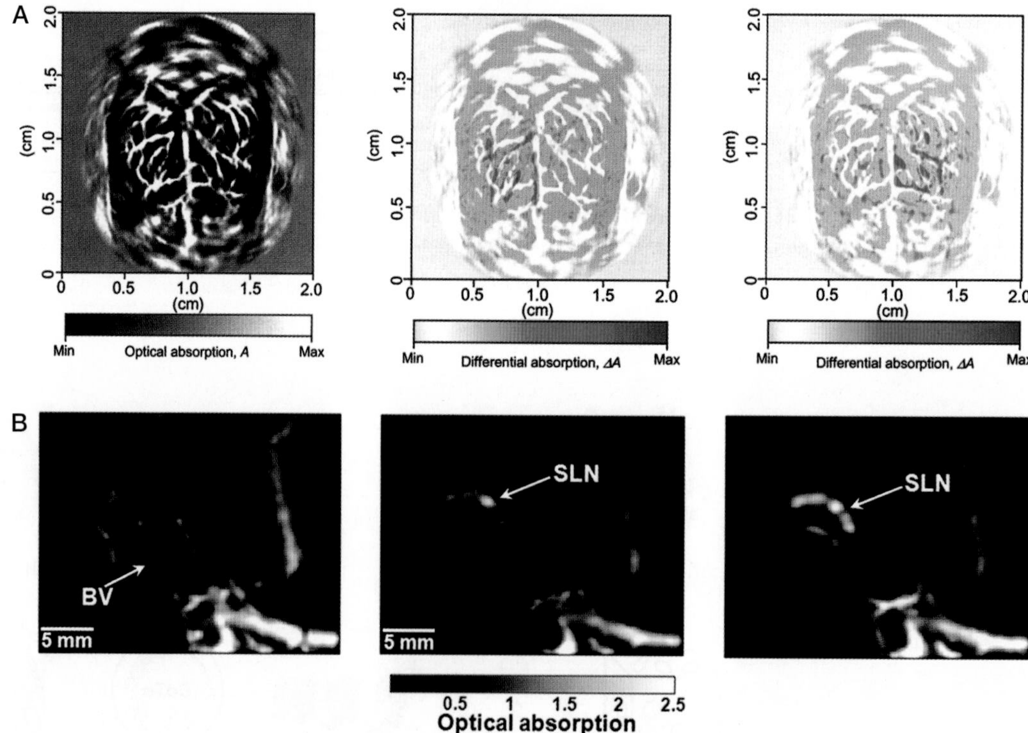

FIGURE 3. Nontargeted photoacoustic imaging. (*A*) Functional PAT imaging of cerebral hemodynamic changes in response to whisker stimulation. (*Left*) A PAT image of the vascular pattern in the superficial layer of the rat cortex acquired with the skin and the skull intact. (*Middle*) Noninvasive functional PAT images corresponding to left-side whisker stimulation. (*Right*) Noninvasive functional PAT images corresponding to right-side whisker stimulation. (*B*) Photoacoustic images acquired before (*left*), 5 min after (*middle*), and 140 min after (*right*) gold nanocage injection for sentinel lymph node (SLN) mapping in live animals. BV, blood vessel. (Adapted, with permission, from Wang et al. 2003b, © Macmillan, and Song et al. 2009, © American Chemical Society.)

hemodynamic changes were also imaged. This pioneering study represented an important milestone for photoacoustic imaging, which suggested that this imaging modality holds promise for a number of applications in neurophysiology, neuropathology, and neurotherapy.

Subsequently, the pattern of optical absorption in a mouse brain was also imaged (Wang et al. 2003a). The intrinsic tissue contrast revealed not only blood vessels but also other brain structures such as the cerebellum, hippocampus, and ventriculi lateralis, which corresponded well with brain histology. Recently, noninvasive, high-resolution imaging of mouse brain activity has been reported (Stein et al. 2009b). With endogenous hemoglobin as the contrast agent, a contrast-to-noise ratio of 25 dB was achieved. Such noninvasive visualization of the brain's vascular system could be of great importance for studying the function of the brain, diagnosing possible disorders, and providing clinically translatable insights into the progression of various human neurological diseases.

Hemoglobin-based photoacoustic imaging can also facilitate the monitoring of many other biological processes such as burn recovery. For example, a two-wavelength photoacoustic-imaging technique was used to discriminate coagulated and noncoagulated blood in a dermal burn phantom (Talbert et al. 2007). Different optical absorption spectra of coagulated and noncoagulated blood produced different ratios of the peak photoacoustic amplitude, which could be analyzed to identify the different blood samples. Recently, multiwavelength photoacoustic measurement was performed to monitor the wound-healing process of extensive deep dermal burns in rats (Aizawa et al. 2008). The peak of the photoacoustic signal at 532 nm, an isosbestic point for oxyhemoglobin and deoxyhemoglobin, was found to shift to a shallower region of the injured skin tissue over time. Histological analysis showed that the photoacoustic signals reflected angiogenesis (the formation of new blood vessels) in the wound, indicating that multiwavelength photoacoustic measurement could be useful for monitoring the changes in local hemodynamics during wound healing.

Tumor growth is dependent on the formation of new blood vessels, and noninvasive imaging of tumor angiogenesis could play a pivotal role in many aspects of cancer patient management (Cai and Chen 2008; Cai et al. 2008b). Imaging of tumor angiogenesis is another important application for hemoglobin-based photoacoustic imaging (Ku et al. 2005). Photoacoustic imaging was used more than a decade ago to image blood vessels in highly scattering samples (Hoelen et al. 1998). It was shown that the sensitivity of this technique could reach single red blood cell detection on a glass plate. In another study, photoacoustic images of tumor neovascularization were obtained over a 10-day period after subcutaneous inoculation of pancreatic tumor cells in rats (Siphanto et al. 2005). Three-dimensional data sets were acquired to visualize the development and to quantify the extent of individual blood vessels around the growing tumor as well as to measure the blood concentration changes inside the tumor.

Photoacoustic imaging with melanin is mainly used for the diagnosis, prognosis, and treatment planning of melanotic melanoma (>90% of all melanomas) (Oh et al. 2006; Wang 2008). The combination of hemoglobin and melanin can be used for more accurate detection of melanomas because more parameters can be measured. Dual-wavelength reflection-mode photoacoustic imaging has been used to noninvasively obtain three-dimensional images of subcutaneous melanomas and their surrounding vasculature in living nude mice (Oh et al. 2006). At the two wavelengths used for hemoglobin and melanin, 584 nm and 764 nm, respectively, the absorption coefficients of blood and melanin-pigmented melanomas vary greatly relative to each other. Therefore, in vivo photoacoustic imaging with a 764-nm light source was able to image the three-dimensional melanin distribution inside the skin to a maximum thickness of 0.5 mm melanoma. In another study, photoacoustic imaging was reported to be capable of detecting melanoma cells within the human circulation system (Weight et al. 2006).

Photoacoustic imaging may also be potentially useful in visualizing certain breast cancers based on intrinsic optical absorption contrast, and proof-of-principle studies have been reported (Manohar et al. 2007; Pramanik et al. 2008). However, despite all of the progress made to date by photoacoustic imaging with hemoglobin and/or melanin, the use of exogenous contrast agents would undoubtedly benefit this imaging modality by enhancing the image quality and contrast, particularly in cancer research. To date, the majority of contrast agents for photoacoustic imaging are not molecularly targeted.

ICG-Based Photoacoustic Imaging

ICG, a tricarbocyanine dye with peak absorption at ~800 nm, has been used as a diagnostic aid for measuring blood volume, cardiac output, or hepatic function. Noninvasive photoacoustic angiography of animal brains has been reported with ICG (Wang et al. 2004). When it is injected into the circulation system of a rat, ICG significantly enhances the absorption contrast between the blood vessels and the background tissues. Because near-infrared light can penetrate deep into brain tissues through the skin and the skull, the vascular distribution in the rat brain was successfully reconstructed from the photoacoustic signals with high spatial resolution and low background. Subsequently, it was reported that PAT could image objects embedded at depths of up to 5 cm at a resolution of <1 mm, with a sensitivity of <10 pM of ICG in the blood (Ku and Wang 2005). The resolution was found to deteriorate slowly with increasing imaging depth.

Gold Nanoparticle–Based Photoacoustic Imaging

Gold nanoparticles (e.g., nanorods and nanocages) have been used for photoacoustic imaging (Cai and Chen 2007; Cai et al. 2008a). Initially, gold nanorods were proposed for photoacoustic flow measurements by the use of laser-induced shape transitions (Li et al. 2005). A series of studies were performed to investigate the shape dependence of the optical absorption of gold nanorods as well as the shape transition induced by pulsed laser irradiation (Wei et al. 2006; Wei et al. 2007). It was found that photon-induced shape transition of gold nanorods involves mainly a rod-to-sphere conversion and a shift in the peak optical absorption wavelength. The application of laser pulses will induce shape changes in gold nanorods as they flow through a region of interest, and the quantitative flow information can be derived based on the photoacoustic signals (measured as a function of time) from the irradiated gold nanorods.

The plasmon resonant absorption and scattering of gold nanorods in the near-infrared region makes them attractive for in vivo imaging applications (Tong et al. 2009). In one study, it was reported that 25 µL of gold nanorod solution with a concentration of 1.25 pM, when injected into nude mice, could be detected by a single-channel acoustic transducer (Eghtedari et al. 2007).

Current sentinel lymph node (SLN) mapping methods, based on blue dye and/or nanometer-sized radioactive colloid injections, are intraoperative because of the need for visual detection of the blue dye and low spatial resolution of Geiger counters in detecting radioactive colloids. Compared with these techniques, photoacoustic mapping of SLNs with gold nanocages could have certain attractive features: noninvasiveness, strong optical absorption in the near-infrared region for deeper tissue penetration, and high concentration accumulation of the gold nanocages. Recently, it was shown that gold nanocages could be used for noninvasive photoacoustic imaging of SLNs (Fig. 3B) (Song et al. 2009). In an animal model, gold nanocage–containing SLNs, as deep as 33 mm below the skin surface, could be detected with good contrast. Potentially, these gold nanocages could also be conjugated with certain targeting moieties (e.g., antibodies or peptides) for molecular-imaging applications.

Photoacoustic Imaging with Other Nanoparticles

In addition to gold nanoparticles, two other types of nanoparticles could also be promising contrast agents for in vivo photoacoustic imaging: SWNTs and QDs. When compared with blood in phantom studies, SWNTs were found to show significant signal enhancement for PAT at a 1064-nm wavelength (Pramanik et al. 2009b). Recently, noninvasive SWNT-enhanced photoacoustic identification of SLNs in a rat model was reported (Pramanik et al. 2009a). The SLNs were successfully imaged in vivo with high contrast-to-noise ratio (>80) and good resolution (~500 µm). Because SWNTs have optical absorption over a wide excitation wavelength range, the imaging depth could be maximized by varying the incident light wavelength to the near-infrared region where biological tissues (e.g., hemoglobin, tissue pigments, lipids, and water) have low light absorption.

Over the last decade, QDs have become one of the fastest growing areas of research in nanotechnology (Cai et al. 2007b; Li et al. 2007b). The vast majority of QD-based research uses their fluorescent properties. Interestingly, the use of QDs for photoacoustic imaging has been described with a nanosecond pulse laser excitation (Shashkov et al. 2008). The laser-induced photoacoustic phenomena were studied with an advanced multifunctional microscope. It was shown that QDs could be used as photoacoustic contrast agents and sensitizers, thereby providing an opportunity for multimodal (photoacoustic and fluorescence) imaging and potentially also photothermal therapy.

MOLECULAR PHOTOACOUSTIC IMAGING

The field of molecular imaging—the visualization, characterization, and measurement of biological processes at the molecular and cellular levels in humans and other living systems (Mankoff 2007)—has expanded tremendously over the last decade. In general, molecular imaging modalities include molecular magnetic resonance imaging (MRI), magnetic resonance spectroscopy, optical bioluminescence, optical fluorescence, targeted ultrasound, single-photon emission computed tomography, and positron emission tomography (Massoud and Gambhir 2003). Continued development and wider availability of scanners dedicated to small animal imaging studies, which can provide a similar in vivo imaging capability in mice, primates, and humans, can enable the smooth transfer of knowledge and molecular measurements between species, thereby facilitating clinical translation.

Molecular imaging takes advantage of the traditional diagnostic imaging techniques and introduces molecular-imaging probes to measure the expression of indicative molecular markers at different stages of diseases. Noninvasive detection of various molecular markers of diseases can allow for much earlier diagnosis, earlier treatment, and better prognosis that will eventually lead to personalized medicine (Cai et al. 2006a, 2008b,c). Recently, MPA imaging has gained significant interest, and a few proof-of-principle studies have been reported. By incorporating a targeting moiety into the above-mentioned contrast agents, photoacoustic imaging can be used to obtain the molecular signatures of various diseases, in particular, cancer.

ICG-Based MPA Imaging

ICG-embedded nanoparticles (~100 nm in diameter), using modified silicate as a matrix, have been developed as contrast agents for photoacoustic imaging (Kim et al. 2007). These ICG-embedded nanoparticles showed improved stability in aqueous solution when compared with the free dye. When conjugated with anti-HER-2 antibody, the nanoparticles showed high contrast and high efficiency for binding to prostate cancer cells in vitro. No in vivo MPA imaging with these ICG-embedded nanoparticles has been reported yet. The fact that ICG can also be used as a photosensitizer for photodynamic therapy may expand the future use of this agent, which can potentially combine both diagnostic and therapeutic functions in a single entity.

Gold Nanoparticle–Based MPA Imaging

Targeted gold nanoparticles have been used mainly for molecular imaging and therapy of cancer (Cai et al. 2008a). Recently, it was reported that antibody-conjugated, epidermal growth factor receptor (EGFR)-targeted spherical gold nanoparticles undergo molecular specific aggregation when they bind to EGFR on the cell surface, thereby leading to a redshift in their plasmon resonance frequency (Fig. 4A) (Mallidi et al. 2009). Capitalizing on this effect, the efficacy of MPA imaging was shown using subcutaneous tumor-mimicking gelatin implants in excised mouse tissue, suggesting that selective and sensitive detection of cancer cells is potentially possible with multiwavelength photoacoustic imaging and molecular-specific gold nanoparticles.

The potential of gold nanoparticle–based MPA imaging in the monitoring of antitumor necrosis factor (anti-TNF) therapy has been evaluated (David et al. 2008). In this study, the contrast agent is composed of gold nanorods conjugated with Etanercept molecules, a fusion protein of the soluble TNF receptor and the Fc component of human immunoglobulin G1 (Cao et al. 2007). In ex vivo

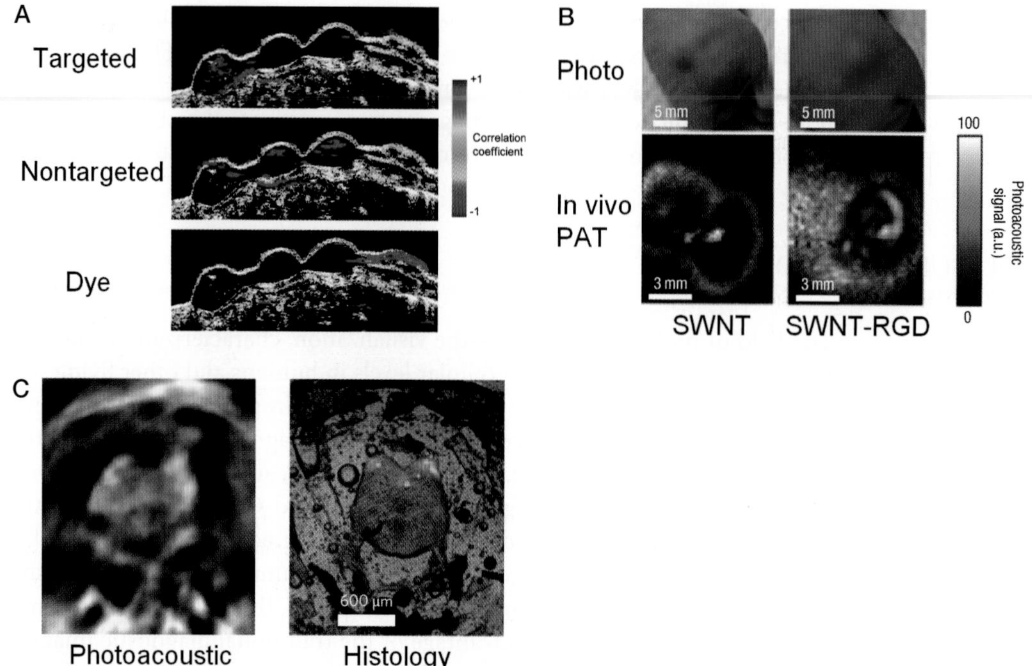

FIGURE 4. Molecular photoacoustic imaging. (*A*) Correlation coefficient images, obtained by comparing multiwavelength photoacoustic images with optical spectra of targeted gold nanoparticles, nontargeted gold nanoparticles, and a dye, overlaid on ultrasound images of the subcutaneous gelatin implants in mouse tissue ex vivo. Only correlation coefficient values >0.75 were displayed in the images. Images measure 44 mm laterally and 9.1 mm axially. (*B*) Photographs of the tumors in mice and the corresponding photoacoustic subtraction images (green) shown as horizontal slices through the tumors. (*C*) (*Left*) Photoacoustic image of the zebrafish brain with fluorescent protein expression shown in color. (*Right*) Corresponding histology of a dissected fish. (Adapted, with permission, from De la Zerda et al. 2008, ©Macmillan, from Mallidi et al. 2009, ©American Chemical Society, and from Razansky et al. 2009, ©Macmillan.)

studies, it was found that gold nanorods with a concentration down to 1 pM in phantoms or 10 pM in biological tissues could be imaged with good signal-to-noise ratio and high spatial resolution (David et al. 2008). Further investigations are warranted to test this agent for in vivo MPA imaging.

One of the attractive features of a gold nanorod is its tunable optical absorption property, which is dependent on its size and its aspect ratio (i.e., the length divided by the width). The use of gold nanorods for MPA imaging with simultaneous multiple targeting has been reported (Li et al. 2008). Two types of nanorods with different peak absorption wavelengths were each conjugated with anti-EGFR or anti-HER-2 antibody, respectively. By simply switching the wavelength of the excitation laser, multiple molecular signatures could be obtained both in vitro and in vivo. Future studies with more gold nanorods/nanoparticles with different optical properties may potentially allow for simultaneous visualization of even more cancer-related molecular targets.

SWNT-Based MPA Imaging

The SWNT is perhaps the most successful contrast agent that has been used for MPA imaging. Integrin $\alpha_v\beta_3$ and arginine-glycine-aspartic acid (RGD), a potent integrin $\alpha_v\beta_3$ antagonist (Cai and Chen 2006), is one of the most extensively studied and validated receptor-ligand pairs (Cai and Chen 2008; Cai et al. 2008b). In a recent study, SWNT conjugated with cyclic RGD peptides was used as a contrast agent for MPA imaging of tumors in living mice (Fig. 4B) (De la Zerda et al. 2008). Intravenous administration of the SWNT-RGD conjugate to tumor-bearing mice showed an eight times greater photoacoustic signal in the tumor than the photoacoustic signal showed in tumor-bearing mice injected with nontargeted SWNTs. Taking advantage of the intrinsic Raman signal of the SWNT, the in vivo MPA imaging results were further validated ex vivo with Raman microscopy.

With regard to sensitivity, a concentration of 50 nM of SWNTs was found to produce a photoacoustic signal equivalent to the mouse tissue (the background signal) in this study (De la Zerda et al. 2008). However, the minimum detectable concentration of SWNTs is likely less than that because the photoacoustic images were acquired before and after the administration of the contrast agent, which makes it possible to separate the signal of the SWNT from the background.

Targeted SWNTs can be used for both MPA imaging and photoacoustic therapy. Recently, the large photoacoustic effect of SWNTs was explored for targeting and selective destruction of cancer cells (Bin et al. 2009). Under the irradiation of a 1064-nm Q-switched millisecond pulsed laser, SWNTs showed a large photoacoustic effect in suspension, which could trigger an explosion at the nanoscale. By conjugating the SWNTs with folic acid, which can bind to cancer cells overexpressing the folate receptor on the membrane, the laser power used for cancer cell killing could be reduced 150–1500 times. This discovery has opened up new perspectives for exploring the photoacoustic properties of SWNTs in cancer therapy.

MPA Imaging of Gene Expression

In one interesting study, the first demonstration of PAT for reporter gene imaging was reported (Li et al. 2007a). Rats inoculated with 9L/lacZ gliosarcoma tumor cells were imaged with PAT before and after injection of X-gal, an optically transparent lactose-like substrate, which yields a stable dark blue product following the cleavage of the glycosidic linkage by β-galactosidase (the protein encoded by *lacZ*). A spatial resolution of ~400 μm as well as a sensitivity of ~0.5 μM were achieved. With the future development of new absorption-based reporter gene systems, MPA imaging is expected to become a valuable technique in molecular imaging research.

Fluorescent proteins are widely used in cell biology and in the study of transgenic animals because they can be genetically targeted to a specific molecule of interest (Giepmans et al. 2006). One major advantage of using fluorescent proteins as the contrast agent for photoacoustic imaging is that they overcome the limitations of conventional fluorescence microscopy and allow imaging of gene expression in much deeper tissues (Burgholzer et al. 2009).

Recently, it was reported that multispectral PAT can achieve tissue penetration of several millimeters (potentially centimeters) with a resolution of 20–100 μm, which remains constant as a function of depth and depends only on the ultrasonic detector characteristics (Razansky et al. 2009). This technique is capable of visualizing fluorescent proteins in small living organisms (Fig. 4C). Whole-body imaging of *Drosophila melanogaster* pupae and adult zebrafish revealed that tissue-specific expression of different fluorescent proteins could be resolved for precise morphological and functional observations in vivo. This report represents an important step forward in high-resolution imaging of fluorescent proteins expressed in genetically manipulated organisms.

Multimodality Molecular Imaging

Among all of the molecular imaging modalities, no single modality is perfect and sufficient to obtain all of the necessary information for a given question. For example, it is difficult to accurately quantify fluorescence signals in living subjects with fluorescence imaging alone, particularly in deep tissues; and MRI has exquisite soft tissue contrast yet it suffers from very low sensitivity. Combining imaging modalities can offer synergistic advantages more than any modality alone. In several reports, photoacoustic imaging has been used together with other imaging modalities to obtain complementary information.

Because of the overwhelming scattering of light in biological tissues, the spatial resolution and imaging depth of conventional fluorescence imaging is unsatisfactory. Dual-modality imaging, with both fluorescence and PAT, was found to be beneficial in an animal tumor model (Wang et al. 2005). An integrin $\alpha_v\beta_3$-targeted peptide, conjugated to ICG, was used as the molecular probe for tumor detection in nude mice inoculated with M21 human melanoma cell lines (integrin $\alpha_v\beta_3$-positive). PAT was able to provide noninvasive images of both the brain structure and the angiogenesis associated with tumor growth, whereas fluorescence imaging offered high sensitivity for tumor detection. Further, co-registration of the PAT and fluorescence images enabled simultaneous visualization of the tumor location, angiogenesis, and brain structure.

Another recent report explored the combination of PAT with MRI (Bouchard et al. 2009). The probe used in this study consists of ferromagnetic cobalt particles coated with gold, which renders both biocompatibility and a unique shape that enables optical absorption over a broad range of frequencies. This dual-modality agent was shown to be useful for detecting trace amounts of the nanoparticles in biological tissues, in which MRI provides volume detection and PAT performs edge detection. Modification of this dual-modality imaging agent with a targeting ligand should be pursued in future studies.

CONCLUSION

Photoacoustic imaging possesses many attractive characteristics: the use of nonionizing electromagnetic waves, good resolution and contrast, portable instrumentation, and the capability of quantitating the signal to a certain extent. Big strides have been made over the last decade, and many diseases can be investigated with photoacoustic imaging, such as cancer, wound healing, and disorders in the brain, among others. PAT compares very favorably to other imaging modalities with its precise depth information, submillimeter resolution, and nanomolar sensitivity. With further improvement in background reduction and hardware/software, as well as the use of lasers with high repetition rates, it is likely that PAT will find wide uses in the future in both basic research and clinical care. While the photoacoustic research community continues to discover new phenomena and to invent new technologies, several companies are actively commercializing the instrument for PAT. Therefore, PAT will continue to be a highly vibrant research field in the years to come.

A significant portion of the contrast agents used for photoacoustic imaging is based on certain nanoparticles. Although nanoparticles offer many advantages over conventional small molecule-based imaging/therapeutic agents such as multifunctionality and flexibility, many barriers exist for in vivo applications in preclinical animal models and future clinical translation of these nanoparticles, among which are biocompatibility, in vivo kinetics, (tumor) targeting efficacy, acute/chronic toxicity, the ability to escape the reticuloendothelial system, and cost-effectiveness.

In terms of molecularly targeted photoacoustic imaging, (tumor) vasculature targeting is the best bet for nanoparticle-based contrast agents because many of these nanoparticles are too large to extravasate (Cai et al. 2006b; Cai et al. 2007a). A circulation half-life of several hours may be sufficient to allow for efficient (tumor) vasculature targeting. Close partnerships among scientists in various disciplines (e.g., engineering, chemistry, oncology, physics, and biology, just to name a few) are needed to move the field forward in a timely manner. Although several proof-of-principle studies have been reported, whether MPA imaging could provide significant advantages over nontargeted photoacoustic imaging remains to be shown.

ACKNOWLEDGMENTS

We acknowledge financial support from the University of Wisconsin (UW) School of Medicine and Public Health's Medical Education and Research Committee through the Wisconsin Partnership Program, the UW Carbone Cancer Center, NCRR 1UL1RR025011, and a Susan G. Komen Postdoctoral Fellowship (to H. Hong).

REFERENCES

Aizawa K, Sato S, Saitoh D, Ashida H, Obara M. 2008. Photoacoustic monitoring of burn healing process in rats. *J Biomed Opt* **13:** 064020.

Bin K, Decai Y, Yaodong D, Shuquan C, Da C, Yitao D. 2009. Cancer-cell targeting and photoacoustic therapy using carbon nanotubes as "bomb" agents. *Small* **5:** 1292–1301.

Bouchard LS, Anwar MS, Liu GL, Hann B, Xie ZH, Gray JW, Wang X, Pines A, Chen FF. 2009. Picomolar sensitivity MRI and photoacoustic imaging of cobalt nanoparticles. *Proc Natl Acad Sci* **106:** 4085–4089.

Burgholzer P, Grun H, Sonnleitner A. 2009. Photoacoustic tomography: Sounding out fluorescent proteins. *Nat Photon* **3:** 378–379.

Cai W, Chen X. 2006. Anti-angiogenic cancer therapy based on integrin $\alpha_v\beta_3$ antagonism. *Anti-Cancer Agents Med Chem* **6:** 407–428.

Cai W, Chen X. 2007. Nanoplatforms for targeted molecular imaging in living subjects. *Small* **3:** 1840–1854.

Cai W, Chen X. 2008. Multimodality molecular imaging of tumor angiogenesis. *J Nucl Med* (suppl 2) **49:** 113S–128S.

Cai W, Rao J, Gambhir SS, Chen X. 2006a. How molecular imaging is speeding up anti-angiogenic drug development. *Mol Cancer Ther* **5:** 2624–2633.

Cai W, Shin DW, Chen K, Gheysens O, Cao Q, Wang SX, Gambhir SS, Chen X. 2006b. Peptide-labeled near-infrared quantum dots for imaging tumor vasculature in living subjects. *Nano Lett* **6:** 669–676.

Cai W, Chen K, Li ZB, Gambhir SS, Chen X. 2007a. Dual-function probe for PET and near-infrared fluorescence imaging of tumor vasculature. *J Nucl Med* **48:** 1862–1870.

Cai W, Hsu AR, Li ZB, Chen X. 2007b. Are quantum dots ready for in vivo imaging in human subjects? *Nanoscale Res Lett* **2:** 265–281.

Cai W, Gao T, Hong H, Sun J. 2008a. Applications of gold nanoparticles in cancer nanotechnology. *Nanotechnol Sci Appl* **1:** 17–32.

Cai W, Niu G, Chen X. 2008b. Imaging of integrins as biomarkers for tumor angiogenesis. *Curr Pharm Des* **14:** 2943–2973.

Cai W, Niu G, Chen X. 2008c. Multimodality imaging of the HER-kinase axis in cancer. *Eur J Nucl Med Mol Imaging* **35:** 186–208.

Cao Q, Cai W, Li ZB, Chen K, He L, Li HC, Hui M, Chen X. 2007. PET imaging of acute and chronic inflammation in living mice. *Eur J Nucl Med Mol Imaging* **34:** 1832–1842.

Chamberland DL, Agarwal A, Kotov N, Fowlkes JB, Carson PL, Xueding W. 2008. Photoacoustic tomography of joints aided by an Etanercept-conjugated gold nanoparticle contrast agent—An ex vivo preliminary rat study. *Nanotechnology* **19:** 095101.

De la Zerda A, Zavaleta C, Keren S, Vaithilingam S, Bodapati S, Liu Z, Levi J, Smith BR, Ma TJ, Oralkan O, et al. 2008. Carbon nanotubes as photoacoustic molecular imaging agents in living mice. *Nat Nanotechnol* **3:** 557–562.

Debbage P, Jaschke W. 2008. Molecular imaging with nanoparticles: Giant roles for dwarf actors. *Histochem Cell Biol* **130:** 845–875.

Eghtedari M, Oraevsky A, Copland JA, Kotov NA, Conjusteau A, Motamedi M. 2007. High sensitivity of in vivo detection of gold nanorods using a laser optoacoustic imaging system. *Nano Lett* **7:** 1914–1918.

Fang H, Maslov K, Wang LV. 2007. Photoacoustic Doppler effect from flowing small light-absorbing particles. *Phys Rev Lett* **99:** 184501.

Giepmans BN, Adams SR, Ellisman MH, Tsien RY. 2006. The fluorescent toolbox for assessing protein location and function. *Science* **312:** 217–224.

Hoelen CG, de Mul FF, Pongers R, Dekker A. 1998. Three-dimensional photoacoustic imaging of blood vessels in tissue. *Opt Lett* **23:** 648–650.

Hong H, Gao T, Cai W. 2009. Molecular imaging with single-walled carbon nanotubes. *Nano Today* **4:** 252–261.

Hu S, Maslov K, Wang LV. 2009. Noninvasive label-free imaging of microhemodynamics by optical-resolution photoacoustic microscopy. *Opt Express* **17:** 7688–7693.

Kim G, Huang SW, Day KC, O'Donnell M, Agayan RR, Day MA, Kopelman R, Ashkenazi S. 2007. Indocyanine-green-embedded PEBBLEs as a contrast agent for photoacoustic imaging. *J Biomed Opt* **12:** 044020.

Kolkman RG, Klaessens JH, Hondebrink E, Hopman JC, de Mul FF, Steenbergen W, Thijssen JM, van Leeuwen TG. 2004. Photoacoustic determination of blood vessel diameter. *Phys Med Biol* **49:** 4745–4756.

Kruger RA, Kiser JWL, Reinecke DR, Kruger GA. 2003. Thermoacoustic computed tomography using a conventional linear transducer array. *Med Phys* **30:** 856–860.

Ku G, Wang LV. 2005. Deeply penetrating photoacoustic tomography in biological tissues enhanced with an optical contrast agent. *Opt Lett* **30:** 507–509.

Ku G, Wang X, Xie X, Stoica G, Wang LV. 2005. Imaging of tumor angiogenesis in rat brains in vivo by photoacoustic tomography. *Appl Opt* **44:** 770–775.

Li PC, Huang SW, Wei CW, Chiou YC, Chen CD, Wang CR. 2005. Photoacoustic flow measurements by use of laser-induced shape transitions of gold nanorods. *Opt Lett* **30:** 3341–3343.

Li L, Zemp RJ, Lungu G, Stoica G, Wang LV. 2007a. Photoacoustic imaging of lacZ gene expression in vivo. *J Biomed Opt* **12:** 020504.

Li ZB, Cai W, Chen X. 2007b. Semiconductor quantum dots for in vivo imaging. *J Nanosci Nanotechnol* **7:** 2567-2581.

Li PC, Wang CR, Shieh DB, Wei CW, Liao CK, Poe C, Jhan S, Ding AA, Wu YN. 2008. In vivo photoacoustic molecular imaging with simultaneous multiple selective targeting using antibody-conjugated gold nanorods. *Opt Express* **16:** 18605–18615.

Mallidi S, Larson T, Tam J, Joshi PP, Karpiouk A, Sokolov K, Emelianov S. 2009. Multiwavelength photoacoustic imaging and plasmon resonance coupling of gold nanoparticles for selective detection of cancer. *Nano Lett* **9:** 2825–2831

Mankoff DA. 2007. A definition of molecular imaging. *J Nucl Med* **48:** 18N, 21N.

Manohar S, Vaartjes SE, van Hespen JC, Klaase JM, van den Engh FM, Steenbergen W, van Leeuwen TG. 2007. Initial results of in vivo non-invasive cancer imaging in the human breast using near-infrared photoacoustics. *Opt Express* **15:** 12277–12285.

Maslov K, Stoica G, Wang LV. 2005. In vivo dark-field reflection-mode photoacoustic microscopy. *Opt Lett* **30:** 625–627.

Maslov K, Zhang HF, Hu S, Wang LV. 2008. Optical-resolution photoacoustic microscopy for in vivo imaging of single capillaries. *Opt Lett* **33:** 929–931.

Massoud TF, Gambhir SS. 2003. Molecular imaging in living subjects: Seeing fundamental biological processes in a new light. *Genes Dev* **17:** 545–580.

Oh JT, Li ML, Zhang HF, Maslov K, Stoica G, Wang LV. 2006. Three-dimensional imaging of skin melanoma in vivo by dual-wavelength photoacoustic microscopy. *J Biomed Opt* **11:** 34032.

Pramanik M, Ku G, Li C, Wang LV. 2008. Design and evaluation of a novel breast cancer detection system combining both thermoacoustic (TA) and photoacoustic (PA) tomography. *Med Phys* **35:** 2218–2223.

Pramanik M, Song KH, Swierczewska M, Green D, Sitharaman B, Wang LV. 2009a. In vivo carbon nanotube-enhanced non-invasive photoacoustic mapping of the sentinel lymph node. *Phys Med Biol* **54:** 3291–3301.

Pramanik M, Swierczewska M, Green D, Sitharaman B, Wang LV. 2009b. Single-walled carbon nanotubes as a multimodal-thermoacoustic and photoacoustic-contrast agent. *J Biomed Opt* **14:** 034018.

Rajian JR, Carson PL, Wang X. 2009. Quantitative photoacoustic measurement of tissue optical absorption spectrum aided by an optical contrast agent. *Opt Express* **17:** 4879–4889.

Razansky D, Distel M, Vinegoni C, Ma R, Perrimon N, Koster RW, Ntziachristos V. 2009. Multispectral opto-acoustic tomography of deep-seated fluorescent proteins in vivo. *Nat Photon* **3:** 412–417.

Shashkov EV, Everts M, Galanzha EI, Zharov VP. 2008. Quantum dots as multimodal photoacoustic and photothermal contrast agents. *Nano Lett* **8:** 3953–3958.

Siphanto RI, Thumma KK, Kolkman RG, van Leeuwen TG, de Mul FF, van Neck JW, van Adrichem LN, Steenbergen W. 2005. Serial noninvasive photoacoustic imaging of neovascularization in tumor angiogenesis. *Opt Express* **13:** 89–95.

Song KH, Kim C, Cobley CM, Xia Y, Wang LV. 2009. Near-infrared gold nanocages as a new class of tracers for photoacoustic sentinel lymph node mapping on a rat model. *Nano Lett* **9:** 183–188.

Stein EW, Maslov K, Wang LV. 2009a. Noninvasive, in vivo imaging of blood-oxygenation dynamics within the mouse brain using photoacoustic microscopy. *J Biomed Opt* **14:** 020502.

Stein EW, Maslov K, Wang LV. 2009b. Noninvasive, in vivo imaging of the mouse brain using photoacoustic microscopy. *J Appl Phys* **105:** 102027.

Talbert RJ, Holan SH, Viator JA. 2007. Photoacoustic discrimination of viable and thermally coagulated blood using a two-wavelength method for burn injury monitoring. *Phys Med Biol* **52:** 1815–1829.

Tong L, Wei Q, Wei A, Cheng JX. 2009. Gold nanorods as contrast agents for biological imaging: Optical properties, surface conjugation and photothermal effects. *Photochem Photobiol* **85:** 21–32.

Wang LV. 2008. Prospects of photoacoustic tomography. *Med Phys* **35:** 5758–5767.

Wang X, Pang Y, Ku G, Stoica G, Wang LV. 2003a. Three-dimensional laser-induced photoacoustic tomography of mouse brain with the skin and skull intact. *Opt Lett* **28:** 1739–1741.

Wang X, Pang Y, Ku G, Xie X, Stoica G, Wang LV. 2003b. Noninvasive laser-induced photoacoustic tomography for structural and functional in vivo imaging of the brain. *Nat Biotechnol* **21:** 803–806.

Wang X, Ku G, Wegiel MA, Bornhop DJ, Stoica G, Wang LV. 2004. Noninvasive photoacoustic angiography of animal brains in vivo with near-infrared light and an optical contrast agent. *Opt Lett* **29:** 730–732.

Wang L, Xie X, Oh JT, Li ML, Ku G, Ke S, Similache S, Li C, Stoica G. 2005. Combined photoacoustic and molecular fluorescence imaging in vivo. *Conf Proc IEEE Eng Med Biol Soc* **1:** 190–192.

Wang X, Xie X, Ku G, Wang LV, Stoica G. 2006. Noninvasive imaging of hemoglobin concentration and oxygenation in the rat brain using high-resolution photoacoustic tomography. *J Biomed Opt* **11:** 024015.

Wei CW, Liao CK, Tseng HC, Lin YP, Chen CC, Li PC. 2006. Photoacoustic flow measurements with gold nanoparticles. *IEEE Trans Ultrason Ferroelectr Freq Control* **53:** 1955–1959.

Wei CW, Huang SW, Wang CR, Li PC. 2007. Photoacoustic flow measurements based on wash-in analysis of gold nanorods. *IEEE Trans Ultrason Ferroelectr Freq Control* **54:** 1131–1141.

Weight RM, Viator JA, Dale PS, Caldwell CW, Lisle AE. 2006. Photoacoustic detection of metastatic melanoma cells in the human circulatory system. *Opt Lett* **31:** 2998–3000.

Xiang L, Yuan Y, Xing D, Ou Z, Yang S, Zhou F. 2009. Photoacoustic molecular imaging with antibody-functionalized single-walled carbon nanotubes for early diagnosis of tumor. *J Biomed Opt* **14:** 021008.

Xie Z, Jiao S, Zhang HF, Puliafito CA. 2009. Laser-scanning optical-resolution photoacoustic microscopy. *Opt Lett* **34:** 1771–1773.

Xu M, Wang LV. 2006. Photoacoustic imaging in biomedicine. *Rev Sci Instrum* **77:** 041101.

Yang X, Skrabalak SE, Li Z-Y, Xia Y, Wang LV. 2007. Photoacoustic tomography of a rat cerebral cortex in vivo with Au nanocages as an optical contrast agent. *Nano Lett* **7:** 3798–3802.

Zhang HF, Maslov K, Stoica G, Wang LV. 2006. Functional photoacoustic microscopy for high-resolution and noninvasive in vivo imaging. *Nat Biotechnol* **24:** 848–851.

Zhang HF, Maslov K, Wang LV. 2007. In vivo imaging of subcutaneous structures using functional photoacoustic microscopy. *Nat Protoc* **2:** 797–804.

53 Imaging the Dynamics of Biological Processes via Fast Confocal Microscopy and Image Processing

Michael Liebling

Department of Electrical & Computer Engineering, University of California, Santa Barbara, California 93106

ABSTRACT

Confocal microscopy offers the ability to optically section biological samples and to build three-dimensional (3D) volumes that can be digitally rendered, viewed from arbitrary directions, and quantitatively analyzed. Capturing, in addition to their 3D structure, the dynamics of living cells and organisms further requires that image acquisition be performed rapidly to avoid measurements that are corrupted by sample movement. Several variations of the confocal principle have been incorporated into microscopes that can achieve fine optical sectioning, high frame rate, and high signal-to-noise ratio (SNR) while minimally perturbing biological samples. In this chapter, we present some of the challenges and the requirements that are tied to imaging live samples and discuss how these are addressed by three optical versions of fast confocal microscopes: single-beam point scanning, spinning-disk, and slit-scanning confocal microscopes. Finally, we show that, for samples undergoing periodic motions, dynamic 3D image representations with virtually identical frame rates as in the case of two-dimensional imaging can be achieved via postacquisition image synchronization. These tools and techniques open new avenues for imaging fast dynamic processes in biology.

INTRODUCTION

Development and commercialization of confocal microscopes have permitted imaging of fixed and live fluorescent biological samples on a routine basis. The key feature of confocal microscopy is its ability to optically (and, therefore, to a large extent, nondestructively) section samples and to localize fluorescent features in 3D within samples at depths up to 200 μm. Optical sectioning is achieved by measuring fluorescence emission in a confined region within the sample (Fig. 1A). When light is focused into the sample, fluorophores that are illuminated by the collimated light, predominantly a small spot in the focal region, emit fluorescence photons. However, because the illumination pattern has an hourglass shape, fluorescent molecules in sections above and below the section of interest are also excited and contribute to the emitted signal. To collect only the fluorescence signal from

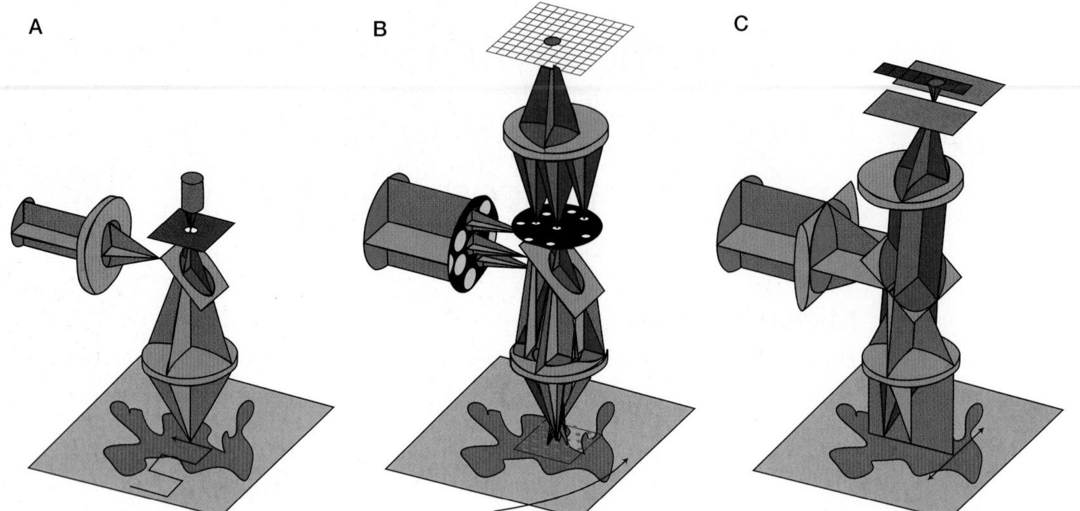

FIGURE 1. Confocal microscopy rejects out-of-focus light via a pinhole aperture that is confocal with the imaging plane. (*A*) Point-scanning confocal microscopes require raster scanning of the focused light over the entire sample to acquire an image. (*B*) Parallel scanning of multiple beams (with confocal pinholes) arranged, for example, on a spinning disk makes it possible to increase the frame rate while preserving the dwell time. (*C*) A special case of parallel scanning in which the illumination light is shaped into a line and a slit is used instead of a pinhole. Scanning is only required in a single direction.

the focal region while rejecting out-of-focus light, a pinhole is placed in front of the detector located in the image plane, which is confocal with the plane of interest within the object. The pinhole blocks light that is defocused as it reaches the pinhole, that is, light that originated away from the focal plane.

The above fluorescence emission measurement procedure is limited to a single point and, to form an image, it must be repeated sequentially at every location within a section or a volume. In early confocal microscopy implementations, the illumination beam remained static while the sample was raster scanned (Minsky 1988), a procedure that would result in acquisition times of several minutes. Even with the advent of faster beam-scanning technology, second-to-minute-long acquisition times for a single frame are common. For fixed samples, slow image acquisition rates are not an issue. If, however, during the time the image is raster scanned (by scanning the beam across lines from top to bottom), the sample changes its shape or moves, the resulting image is distorted and no longer accurately represents the object (Fig. 2A–A″).

In addition to being used for display, data collected via live imaging are increasingly used for analysis. Analysis includes tracking migrating cells during embryo development or intracellular trafficking of organelles and molecules, measuring changes in cell or organ morphology, quantifying cell or organ physiology, or determining the precise colocalization of gene expression within cells or tissues. The precision and the ability to automate quantitative analysis techniques rely on input data that have high signal-to-noise ratios and frame rates and faithfully reproduce morphological features without distortion. As the observation scale decreases, capturing data of sufficient quality is particularly challenging as even processes occurring at seemingly slow velocities may require imaging at surprisingly high frame rates. For example, while it may take from hours to months for complex organisms to develop, the motion of a cell over the field of view when observed at high magnification can occur in minutes, seconds, or a fraction thereof.

In this chapter, we review the requirements and the caveats of fast imaging and discuss three types of fast confocal microscopes as well as a method for imaging periodic motions that allows extending hardware limitations of fast confocal microscopes to achieve dynamic 3D imaging.

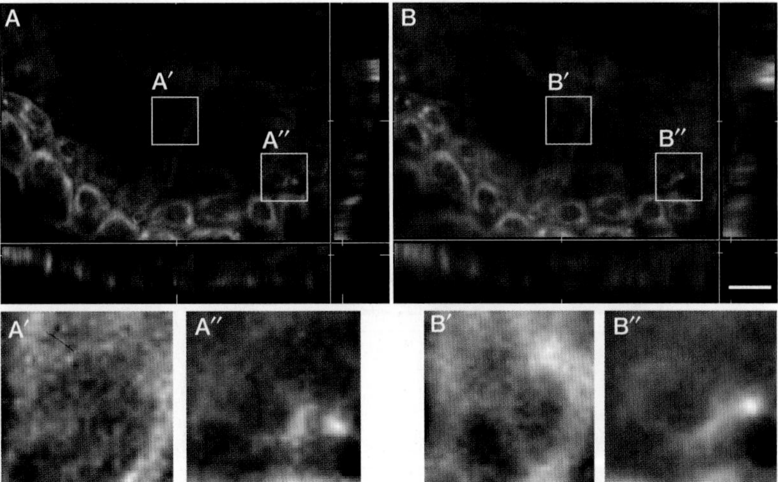

FIGURE 2. *(A,A′,A″)* Single point-scanning confocal microscopes provide high spatial resolution and axial selectivity *(A,A″)*, yet fast moving structures, such as red blood cells in a 30-h-postfertilization (hpf) zebrafish embryo that was soaked in BODIPY FL fluorescent dye, appear distorted *(A′)*. *(B,B′,B″)* Line-scanning confocal microscopes have 10–100-fold faster frame rates, and the actual shape of fast moving objects is preserved in the acquired images *(B′)* while also preserving spatial resolution and optical sectioning ability *(B,B′)*. Scale bar, 20 μm.

CAVEATS OF DYNAMIC IMAGING IN BIOLOGY

When imaging fixed samples, resolution is typically limited by diffraction to a few hundreds of nanometers in both the axial and the lateral directions, depending on the numerical aperture of the microscope objective and the imaging wavelength. The ability to resolve dynamic biological processes in space and time (e.g., cell migration, division, and growth, or the dynamics of the cardiovascular system in developing embryos) does not exclusively depend on the optical features of the microscope: When imaging dynamic samples, resolution also depends on the frame rate at which images are acquired. For example, if acquiring one image takes 1/100 of a second, a blood cell that moves at a speed of 1 mm/sec will cover a distance of 10 μm during the time it takes to capture one image, resulting in a corresponding uncertainty about the blood cell shape or position. The image may appear blurred or distorted. When imaging a moving sample, the point-spread function (PSF)—the image produced by a (theoretical) static point source, which characterizes the optical performance of the microscope—must take that motion into account. A modified PSF is obtained by combining the contributions from shifted (static) PSFs. This is illustrated in Figure 3A, in which two nearby point sources that move along the direction that joins them (but whose relative distance is fixed, assuming, specifically, a uniform motion of velocity v) yield PSFs that depend on the time over which acquisition is performed. Resolution, the smallest distance between the point sources that yield PSFs that can be distinguished from one another, thereby directly depends on the product of sample motion v and the integration time T, which corresponds to the additional width introduced by their compounded effect.

Dynamic processes in molecular, cellular, and developmental biology span a broad range of velocities and scales (Vermot et al. 2008). Although the 1–2 frames per second (fps) frame rate of a conventional confocal microscope may be sufficient to monitor the slower movements, the swifter ones (or those for which a high spatial resolution is required) can only be captured with microscopes that are several orders of magnitude faster. In Figure 4, given the sample velocity and the target spatial resolution, the required frame rate is indicated by diagonal lines. Interestingly, the amount of blur Tv introduced by motion can very rapidly exceed, by several orders of magnitude, the purely optical effects of the microscope. This highlights the importance of being able to capture images at a high frame rate when sample dynamics plays a major role in the studied process.

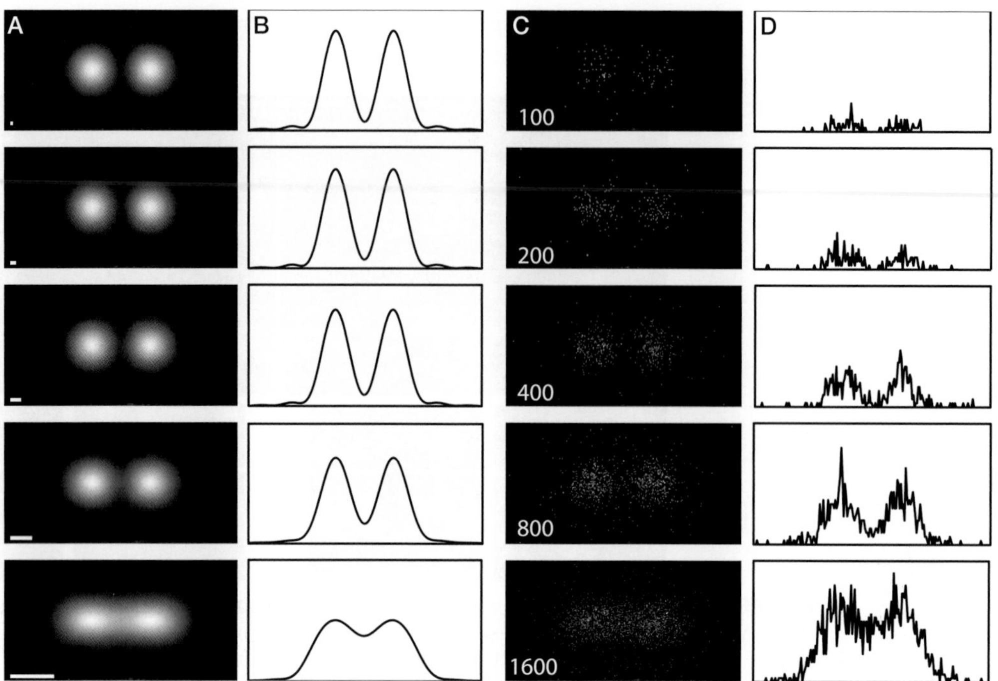

FIGURE 3. Short integration time reduces motion blur but also limits photon count. (*A*) Simulated two-dimensional (2D) point spread functions (PSFs) generated by two point sources that move at constant velocity along the horizontal direction. As the integration time increases, the distance traveled by the two point sources (marked by white lines) increases and results in motion-blurred PSFs that blend into a single spot past a certain integration time. From *top* to *bottom*, integration time doubles for each experiment resulting in a 16-fold increase between the *top* and the *bottom* integration times. (*B*) PSFs in *A* projected along the vertical direction. (*C*) Simulated photon counts assuming the point sources emit, on average, 100 photons during the integration time in the *top* row (and increasing proportionally to the integration time in *A* up to 1600 in the *bottom* row). Resolution is both dependent on well-separated PSFs and dependent on sufficient photon count (which specify SNR) implying a trade-off when the frame rate is adjusted. (*D*) Photon counts projected along the vertical direction.

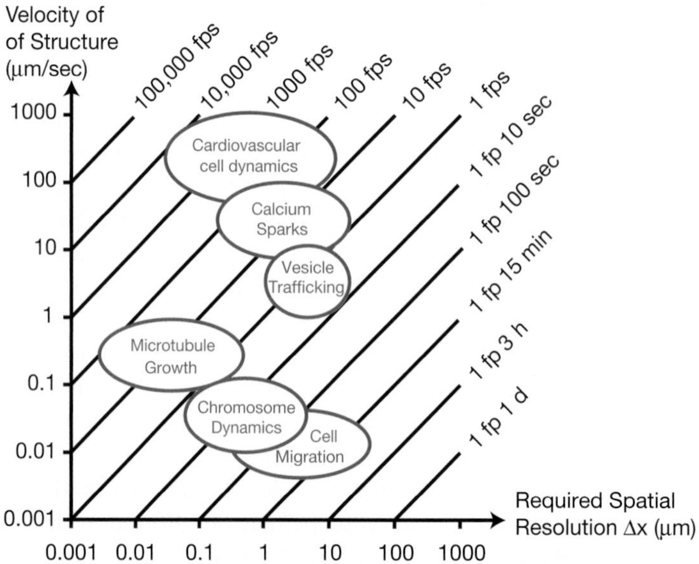

FIGURE 4. Sample motion and required spatial resolution specify target frame rate. Observable biological events are arranged according to the spatial resolution required to study them and the velocity of the involved structures with oblique lines representing the target frame rate required to resolve the events properly. A structure moving at a velocity v and imaged with a device requiring a time T (equal to the inverse of the frame rate) to capture a frame will introduce a blur width Tv. To achieve the desired resolution, the frame rate must be adjusted to be $<Tv$. For example, to reach a resolution of 1 μm for a sample that moves at a velocity of $v = 10$ μm/sec, one requires $T \times 10$ μm/sec < 1 μm; and, therefore, $T < 1/10$ sec, corresponding to a frame rate of 10 fps. (Adapted, with permission, from Vermot et al. 2008.)

Although imaging at high frame rates is clearly desirable, two fundamental aspects of imaging play against each other. On the one hand, as described above, images should be acquired with as short an integration time as possible to freeze motion and avoid blurring. For a scanning confocal microscope, this means scanning over the sample at the highest possible speed. On the other hand, to get a crisp image with good contrast and a low noise level, a high number of photons must reach the detector. As integration time is reduced, fewer photons contribute to the image, and the SNR is reduced. Bright fluorescent molecules, higher fluorophore concentrations, and higher illumination levels may, in some cases, increase the photon count on the detector. Beyond a certain point, however, these are poor alternatives because they lead to fluorophore quenching or bleaching. For a given sample and illumination intensity, increasing the number of photons contributing to the image is, therefore, synonymous to increasing the integration time, that is, the dwell time—the time the laser beam spends over a single point during the scanning process. During the time the illumination beam remains at a single location, the detector gathers photons, gradually improving the SNR. Indeed, photon-limited imaging is best modeled by a Poisson process for which the variance of the detected photons (the noise) is given by the square root of the average number of photons detected (specified by the brightness of the fluorophore, the microscope efficiency at collecting emitted photons, and the time over which photons are gathered) with the SNR increasing as the photon count increases. We must, therefore, include the effect of low photon numbers to accurately model motion and integration-time-dependent PSFs. Figure 3C,D illustrates this process. In a setting in which imaging is not limited by photon count, the PSF is given by an Airy pattern (Fig. 3A,B). If the point source is moving, the effect of motion is compounded, and the PSF is elongated along the direction of motion by a distance corresponding to the product of the sample velocity and the integration time. In a photon-limited case, the PSF represents the spatial probability function for a photon emitted by the point source to hit the detector in any given spatial position. The temporal emission dynamics of the point source can be modeled by a Poisson process with the mean number of photons emitted by the point source in a set amount of time as its parameter. When evaluating resolution by taking two closely positioned point sources (that we assume are moving with equal velocity and direction along the direction joining both points), one strives to keep blurring, that is, the spatial extent of the PSF, as small as possible while still ensuring that sufficient photons are captured to allow distinguishing any peaks at all. Finally, just as in the case of static samples, sampling and quantization of the PSF play similar roles as in the static case (a detailed description of these effects for static samples is given in Stelzer 1998).

FAST CONFOCAL MICROSCOPES

We turn to a discussion of how several implementations of the basic confocal principle lead to fast microscopes that ally high frame rate and optical sectioning capabilities.

Early implementations of the confocal microscopy principle (Minsky 1988) required scanning the sample through the light beam. Point-scanning microscopes (Wilson and Sheppard 1984; see Fig. 1A) are typically capable of acquiring a 512 x 512-sized image in <1 sec. Fast scanning across the *x* and the *y* directions can be achieved using mirrors mounted on galvanometers, acousto-optic tunable filters (Goldstein et al. 1990), and resonant scanning mirrors, or a combination of different methods (Rietdorf and Stelzer 2006). With fast scanning, tens of frames per second can be acquired. Although the frame rate increases with scanning speed, the dwell time decreases, which affects the SNR. Because increasing the illumination power and changing the dye concentrations are usually not viable solutions to counter the shorter dwell time, microscopes that aim at achieving high frame rates with point scanners are particularly dependent on limiting any photon losses along the acquisition path.

Because decreasing the dwell time negatively affects image brightness (and, therefore, the resolution), alternative solutions are desirable. One general solution is to simultaneously scan multiple beams (instead of a single beam) over the sample. Such a parallel illumination scheme, combined with a parallel acquisition procedure, is implemented in spinning-disk confocal microscopes (Petráň et al. 1968; Tanaami et al. 2002). Multiple beams arranged with a spiral pattern on a disk cover the entire field of view as the disk revolves around its axis (Fig. 1B). At any given time, about 1000 beams illuminate the sample. Light emitted after illumination by one of the beams is passed by a corresponding pinhole located on a second rotating disk. Images are captured on a fast digital camera, such as an electron-multiplying charge-coupled device. To limit the losses when shaping the multiple beams, microlenses are usually used on the first rotating disks. Another design for simultaneously producing and scanning multiple beams combines illumination of a microlens array with an oscillating mirror that scans and descans the beams onto a pinhole array (Pawley 2006). A central limitation of systems that includes multipinhole arrays is that the size of the individual pinholes cannot be adjusted. On single-beam scanners, in which the pinhole size is adjustable, such flexibility allows trading brightness for optical sectioning capabilities (as the pinhole is opened, images are brighter while the optical section thickness increases). Reduced photodamage is one of the benefits of imaging with multiple beams in parallel, rather than with a single beam. Indeed, scanning low-intensity beams multiple times over the field of view rather than one scan of a single higher intensity beam appears to be less damaging to samples (Tadrous 2000; Egner et al. 2002; Graf et al. 2005).

Finally, we consider a special case of parallelized confocal microscopy. Instead of illuminating the sample pointwise with an adjustable pinhole in a confocal position, the illumination beam is shaped to a blade, and an adjustable conjugated slit acts as the pinhole. This situation is a special case of parallelization via multiple beams and conjugated pinholes because it corresponds, in essence, to aligning all of the pinholes in a row (Fig. 1B). With advances in beam shaping, fast line detectors, and efficient hardware, such systems allow for both thin optical sectioning and fast frame rates while retaining the advantages of a variable-size pinhole (Wolleschensky et al. 2006). The simultaneous acquisition of an entire line of pixels reduces the scanning dimensionality to one direction, which makes it possible to acquire 2D image sequences (512 × 512 pixels) at frame rates of up to 120 fps without reducing the dwell time.

For all systems that involve scanning multiple beams (or an entire line), optical sectioning capabilities are limited by cross talk. Although a single pinhole blocks most of the fluorescence photons emitted outside of the focal region corresponding to the confocal position in the sample, these photons may be accepted by pinholes (or, in the case of a slit-scanning microscope, by other pixels) corresponding to regions other than the confocal ones. This phenomenon is particularly problematic when imaging through thick samples that show high fluorescence contributions from sections above and below the section of interest. This is illustrated in Figure 1.

DIGITAL POSTPROCESSING FOR FAST 3D IMAGING OF PERIODICALLY MOVING SAMPLES

Although the newer generation of fast confocal microscopes can acquire 2D images at high frame rates, imaging full 3D volumes, that is, sequences of optical sections at successive depths, remains limited by the mechanical constraints of adjusting the focus position rapidly as well as the electronic and digital challenges of transferring and recording 3D volumes of data at high rates. The usual sequence in which 3D volumes are acquired over time is x–y–z–t: An x–y slice is acquired, the stage is moved axially (along z), the next x–y slice is acquired, and so on, until the full z-stack is acquired to complete one time point, and this entire procedure is repeated for subsequent time points. Even with fast z-positioning procedures or remote scanning (Botcherby et al. 2008), recording full 3D data remains one to three orders of magnitude slower than acquiring image sequences of a single section. To avoid deformation or blurring artifacts during 3D imaging, the motion of the sample during the acquisition of one volume must be low, similar to the 2D case. For highly dynamic samples, this usually represents a problem, and the number of z-slices must be limited.

However, when the motion of the imaged sample is periodic (for example, the motion of the beating heart in developing embryos), data volumes at high frame rates can still be assembled (Liebling et al. 2005, 2006; Forouhar et al. 2006). The procedure is as follows. At a fixed axial position, an image sequence corresponding to a single optical slice is acquired at a high frame rate for the duration of at

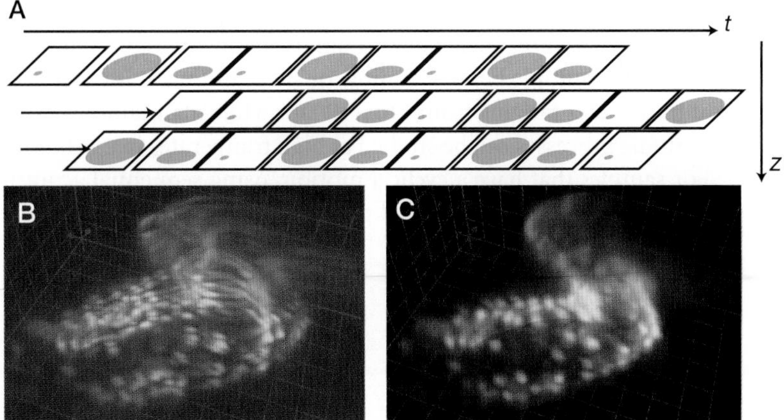

FIGURE 5. A periodically moving structure can be imaged at high speed, requiring postprocessing for temporal image alignment. (*A*) Image sequences of a beating embryonic heart in a transgenic zebrafish (Tg(*cmlc2*:GFP); Huang et al. 2003), acquired at high speed and fixed axial positions (one high-speed sequence at each axial position), are temporally registered to produce a 3D model of the beating heart. (*B*) Direct reconstruction of raw (nonsynchronized) data and (*C*) proper heart shape is revealed after synchronization. Grid spacing, 20 µm.

least one period of the motion. Next, the axial position of the sample (or the objective focus) is adjusted to a new position. When the new axial focus position is reached, rapid imaging is resumed, and another image series of one optical section is acquired over at least one period. Focus change and fast acquisition are alternated until the sample has been imaged in its entirety. Changing the focus position rapidly is not critical because imaging is only performed once a new focus position has been reached, and periodicity ensures that the same motion can be captured in any cycle. Therefore, no fast device to adjust the focus is required. If the acquisitions are not triggered at any particular time during the periodic motion (in practice, gating the acquisition to the motion can be difficult to carry out on biological samples), the image sequences at each axial position cannot be assembled directly: Sections of the sample in arbitrary positions assembled into volumes would result in a severely distorted shape (Fig. 5B). Instead of being assembled directly, the sequences must first be registered along the temporal direction—that is, synchronized. This operation may be performed manually by cropping each sequence such that the start and the end cover a full period. Use of automatic digital pixel-based registration methods are preferable because they are faster, less time consuming, less error prone, and do not suffer from interoperator variability (Liebling et al. 2005). Also, some synchronization methods permit compensating for slight variations in the temporal pattern such as minor arrhythmias when imaging the heart (Liebling et al. 2006). Automatic synchronization methods operate by recursively synchronizing pairs of sequences such that the sum of the absolute value of the pixel intensity differences across image pairs from neighboring sections is minimized. This ensures that, as one sequence is temporally shifted with respect to its axial neighbor, pairs of axially neighboring images should be approximately similar (Fig. 5A). This implies that the optical sections must overlap to some extent, so neighboring sections share similar image content and must translate to a practical requirement when adjusting the optical slice thickness (by adjusting the pinhole or slit size). Synchronized slice sequences can then be assembled into a sequence of 3D volumes that accurately portray the shape of the imaged sample. This is illustrated in Figure 5C, in which an embryonic zebrafish heart is expressing green fluorescent protein (GFP).

SUMMARY

Complex dynamic movements at a microscopic scale in living biological samples can be studied by acquiring, by visualizing, and by analyzing a 3D time series (4D data). Point-scanning microscopes allow optical sectioning of the heart but are usually slow because of the scanning nature of the image acquisition process. Acquiring images at a high frame rate, a high SNR, and with thin optical sections is, therefore, challenging. Using versions of rapid confocal microscopes based on fast point scanning, multiple beam scanning, or slit scanning, it is possible to follow many biological processes at frame rates ranging from 1 to ~100 fps while maintaining good optical sectioning capabilities. Fast periodic processes can be studied at high frame rates in 2D, but imaging in 3D remains much slower. For samples that have a cyclical motion, using sequential acquisition of 2D optical sections at high frame rates followed by digital postprocessing permits 3D models to be built without compromising frame rate.

REFERENCES

Botcherby EJ, Juškaitis R, Booth MJ, Wilson T. 2008. An optical technique for remote focusing in microscopy. *Opt Commun* **281:** 880–887.

Egner A, Andresen V, Hell SW. 2002. Comparison of the axial resolution of practical Nipkow-disk confocal fluorescence microscopy with that of multifocal multiphoton microscopy: Theory and experiment. *J Microsc* **206:** 24–32. Part 1.

Forouhar AS, Liebling M, Hickerson A, Nasiraei-Moghaddam A, Tsai H-J, Hove JR, Fraser SE, Dickinson ME, Gharib M. 2006. The embryonic vertebrate heart tube is a dynamic suction pump. *Science* **312:** 751–753.

Goldstein SR, Hubin T, Rosenthal S, Washburn C. 1990. A confocal video-rate laser-beam scanning reflected-light microscope with no moving parts. *J Microsc* **157:** 29–38.

Gräf R, Rietdorf J, Zimmermann T. 2005. Live cell spinning disk microscopy. *Adv Biochem Eng Biotechnol* **95:** 57–75.

Huang C-J, Tu C-T, Hsiao C-D, Hsieh F-J, Tsai H-J. 2003. Germ-line transmission of a myocardium-specific GFP transgene reveals critical regulatory elements in the cardiac myosin light chain 2 promoter of zebrafish. *Dev Dyn* **228:** 30–40.

Liebling M, Forouhar AS, Gharib M, Fraser SE, Dickinson ME. 2005. Four-dimensional cardiac imaging in living embryos via postacquisition synchronization of nongated slice sequences. *J Biomed Opt* **10:** 054001.

Liebling M, Forouhar AS, Wolleschensky R, Zimmerman B, Ankerhold R, Fraser SE, Gharib M, Dickinson ME. 2006. Rapid three-dimensional imaging and analysis of the beating embryonic heart reveals functional changes during development. *Dev Dynam* **235:** 2940–2948.

Minsky M. 1988. Memoir on inventing the confocal scanning microscope. *Scanning* **10:** 128–138.

Pawley JB, ed. 2006. *Handbook of biological confocal microscopy*, 3rd ed. Springer, New York.

Petráň M, Hadravský M, Egger MD, Galambos R. 1968. Tandem-scanning reflected-light microscope. *J Opt Soc Am* **58:** 661–664.

Rietdorf J, Stelzer EHK. 2006. Special optical elements. In *Handbook of biological confocal microscopy* (ed. Pawley JB), Chap. 3, pp. 43–58. Springer, New York.

Stelzer EHK. 1998. Contrast, resolution, pixelation, dynamic range and signal-to-noise ratio: Fundamental limits to resolution in fluorescence light microscopy. *J Microsc* **189:** 15–24. Part 1.

Tadrous PJ. 2000. Methods for imaging the structure and function of living tissues and cells: 3. Confocal microscopy and micro-radiology. *J Pathol* **191:** 345–354.

Tanaami T, Otsuki S, Tomosada N, Kosugi Y, Shimizu M, Ishida H. 2002. High-speed 1-frame/ms scanning confocal microscope with a microlens and Nipkow disks. *Appl Opt* **41:** 4704–4708.

Vermot J, Fraser SE, Liebling M. 2008. Fast fluorescence microscopy for imaging the dynamics of embryonic development. *HFSP J* **2:** 143–155.

Wilson T, Sheppard C. 1984. *Theory and practice of scanning optical microscopy*. Academic, London.

Wolleschensky R, Zimmermann B, Kempe M. 2006. High speed confocal fluorescence imaging with a novel line scanning microscope. *J Biomed Opt* **11:** 064011 (1–14).

54 High-Speed Two-Photon Imaging

Gaddum Duemani Reddy and Peter Saggau
Department of Neuroscience, Baylor College of Medicine, Houston, Texas 77030

ABSTRACT

The small size of neuronal dendrites and spines combined with the high speed of neurophysiological signals, such as transients in membrane potential or ion concentration, necessitates that any functional study of these structures uses recording methods with both high spatial and high temporal resolutions. In this regard, conventional two-photon microscopy, in combination with fluorescent indicators sensitive to physiological parameters, has proven to be only a partial solution by providing near-diffraction-limited spatial resolution even when imaging structures deep inside light-scattering tissue. This is because the relatively slow beam-scanning methods used in most conventional two-photon microscopes severely limit the extent to which functional data can be recorded. Here, we detail developments to create high-speed two-photon imaging systems that overcome this limitation and discuss important considerations that must be taken into account when attempting to construct such systems.

INTRODUCTION

Techniques that increase the temporal resolution of optical sectioning imaging modalities such as two-photon microscopy are becoming increasingly necessary in experimental biology. Applications include tracking cellular and subcellular movements, monitoring for rare events, and recording fast temporal signals from small structures such as neuronal dendrites and dendritic spines. In this last application, the need for increased temporal resolution can be understood by the temporal dynamics of the signals of interest, such as back-propagating action potentials, dendritic spikes, or ionic transients, which occur on the order of milliseconds. Similarly, the micrometer and submicrometer sizes of these fine structures and their locations deep inside studied tissue require the use of high spatial resolution optical sectioning techniques such as two-photon microscopy.

Most high-speed two-photon imaging systems closely resemble, in their optical design, conventional two-photon systems, containing four primary components: the light source, the scanner, a base microscope, and a detector. However, the optimal choice for these components is primarily dependent on the scanning technique used. In this chapter, we will review several of these techniques, some of which are detailed elsewhere in this book series, such as scanless spatial light modulator (SLM)-based microscopy (see Chapter 56), and we will discuss the considerations in choosing the additional elements needed to optimize them for effective use.

HIGH-SPEED TWO-PHOTON IMAGING SETUP

The four main components of most high-speed two-photon microscopes, whether they are designed for fast two-dimensional (2D) or three-dimensional (3D) imaging, are the near-infrared (NIR) laser light source, the laser scanner, an adaptable light microscope, and the photodetector (Fig. 1). Each of these elements is usually coupled to the next element in the optical path using relay optics, which could be a series of 1:1 telescopes. However, if the beam size needs to be adjusted to fill an aperture (such as the objective back focal aperture) or the scan angles need to be increased to fill the effective field of view, magnification or demagnification telescopes are used accordingly.

Light Source

Although two-photon excitation can be achieved utilizing a high-powered continuous-wave laser (Hell et al. 1998), such a scheme is not usually conducive to biological experimentation. As a result, modern two-photon microscopes use pulsed laser sources, which maintain moderate average power levels while ensuring sufficient peak powers for significant two-photon excitation. In this regard, since its introduction in 1986, titanium-doped sapphire (Moulton 1986), which has a very high gain bandwidth that allows for very short pulses as well as a large range of wavelength tunability, has become the preeminent gain medium for most two-photon laser sources.

However, the optimal laser for a two-photon microscope is not always the one with the shortest pulses. Indeed, the broad spectral profiles of transform-limited short pulses when combined with wavelength-sensitive refractive objects, such as lenses and optical fibers, or diffractive objects, such as gratings or acousto-optic devices (AODs), lead to two forms of distortion: spatial dispersion and temporal dispersion. These forms of dispersion can have detrimental effects on both image quality and efficient power use, unless their effects are minimized.

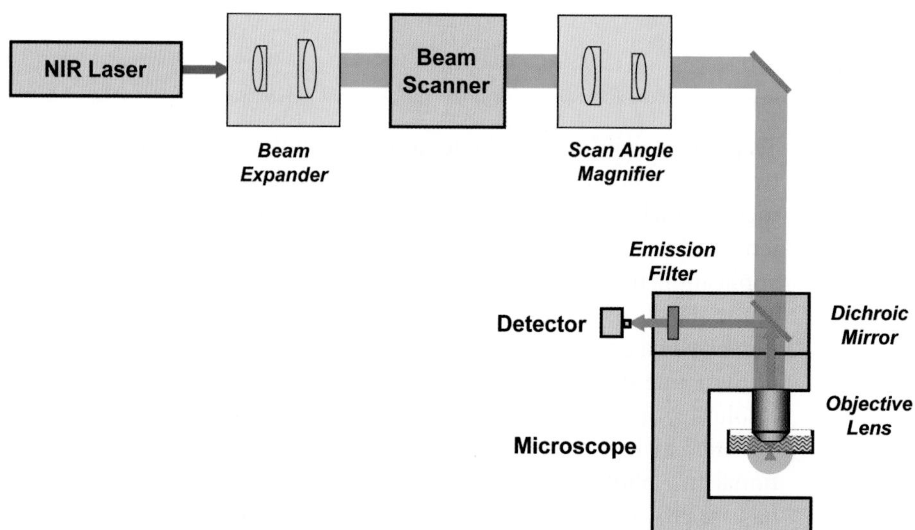

FIGURE 1. Basic high-speed two-photon imaging system. Primary system components are a pulsed near-infrared (NIR) laser, various configurations of a beam scanner, an adaptable epifluorescence microscope, and a sensitive photodetector. Important optical path components include two telescopes acting as beam expander and scan angle magnifier, and a dichroic mirror to pass excitation light to the objective lens and direct fluorescence to the emission filter, which selects for the desired fluorescence band.

Spatial dispersion results because both refraction and diffraction are wavelength dependent. For the multiple wavelengths inherently present in a short pulse, this means that different wavelengths within the pulse will travel different paths, which leads to a less compact focus and a worsened spatial resolution. Because shorter pulses have larger spectral profiles, the degree to which the spatial resolution is reduced increases with the shortening of the temporal profile of the pulse.

Temporal dispersion results from the velocity differences between differing wavelengths of light when traveling through an optical medium, which result in pulse broadening and thus lower peak powers. In general, short pulses broaden to a greater extent than longer pulses, and in many optical layouts, a longer transform-limited pulse at the exit of the laser can lead to a shorter pulse at the image plane. This is increasingly true with long optical paths through material having a high index of refraction.

Although the amount of spatial or temporal dispersion introduced by conventional mirror-based two-photon microscopes is small, the value is significant for some high-speed two-photon imaging systems, particularly ones that use AOD-based scanning. In addition, although methods for both spatial and temporal dispersion compensations are available (Treacy 1969; Martinez et al. 1984; Iyer et al. 2003; Zeng et al. 2006), often the most practical and resource-efficient means for reducing the degree to which these dispersive mechanisms affect imaging is by selecting laser sources with slightly longer pulse widths and, hence, smaller bandwidths. This, in turn, will reduce both forms of dispersion, often negating the need for placing additional elements in the optical path for compensation. Similarly, when using 3D AOD-based scanning, in which spatial dispersion is inherently compensated by the scanning mechanism (Reddy and Saggau 2005), a resource-efficient solution can be to use a laser with higher average power to compensate for the reduction in peak power resulting from the anticipated temporal broadening of pulses.

The Scanning Mechanism

The crucial difference between various high-speed two-photon imaging devices is the scanning mechanism that is used (see Fig. 2). As mentioned above, in conventional two-photon microscopes, the primary limitation in temporal resolution is caused by the laser beam-scanning mechanisms, which use galvanometer-driven rotating mirrors (Fig. 2A). Given the inherent inertia of both the mirror and the galvanometer, there is a fundamental physical restriction on the highest speed at which these beam-scanning mechanisms can operate. Although higher speeds can be achieved by driving the galvanometers at their resonant frequency to minimize the rate-limiting inertia of the system (Fan et al. 1999), the increase in speed requires restricting the scan to linear or simple patterns. In general, restricting scan patterns to single scan lines allows for acquisition rates on the order of hundreds of images per second. However, when imaging highly irregular and sparse structures, such as neuronal dendrites, a line scan is highly inefficient because much time is spent visiting areas without relevant structures.

2D Scanning Systems

Structural Imaging Systems

Several systems have been developed that increase the speed of conventional two-photon microscopes to video rate. In some of these systems, scanning an individual focal point is replaced with scanning entire lines of two-photon excitation (Fig. 2B) (Brakenhoff et al. 1996). Although these systems offer higher temporal resolution scanning, they suffer from a loss of resolution as well as a decrease in excitation efficiency. Other methods use a rotating polygonal mirror in place of one of the galvanometers (Fig. 2C) (Kim et al. 1999). This approach preserves spatial resolution; however, it restricts zooming capabilities. Still other popular approaches that resemble the Nipkow spinning-disk system in confocal microscopy, deterministically separate the input laser beam and use multiple foci of two-photon excitation to simultaneously excite several locations in a region of interest (Fig. 2D) (Bewersdorf et al. 1998). By dividing the excitation light into several beams, these systems suffer from reduced power for excitation at each individual focus and also require the use of an imaging

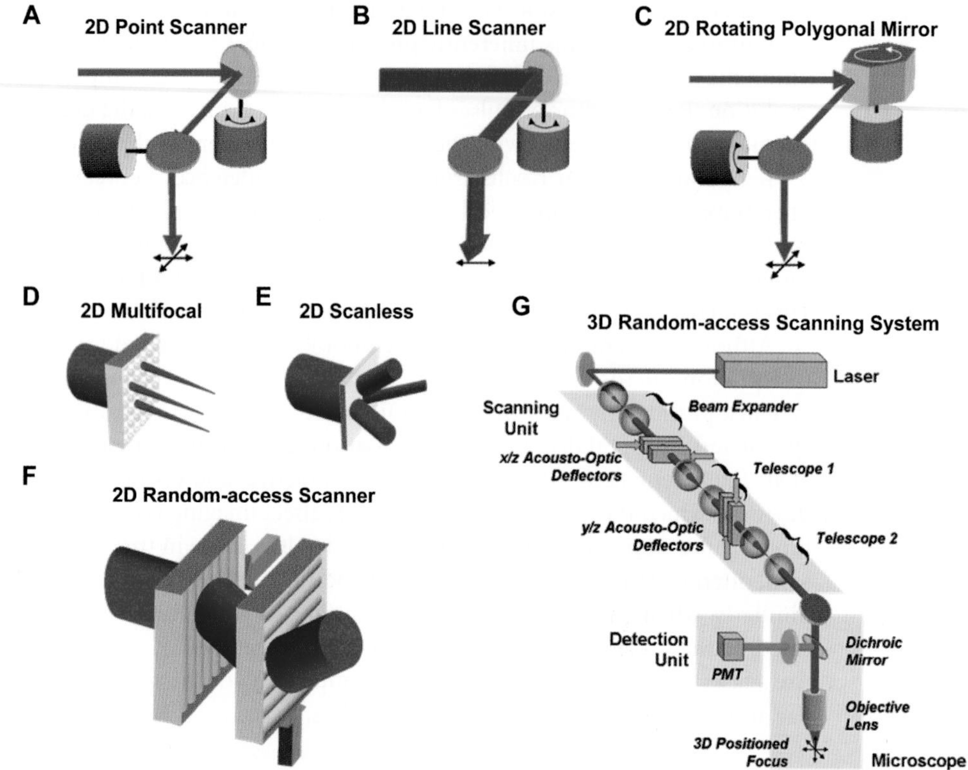

FIGURE 2. Beam-scanning schemes. (*A–F*) Two-dimensional scanners: (*A*) Point scanner uses two galvanometer-driven mirrors; (*B*) line scanner sweeps laterally to spread out the laser beam with a single galvanometer-operated mirror; (*C*) rotating polygonal mirrors are used to increase the scan speed in one dimension; (*D*) multifocal scanners use a microlens array to achieve multiple scan points; (*E*) scanless beam positioning uses an SLM to control one beam or multiple beams; (*F*) random-access scanner uses two acousto-optical deflectors to direct a single beam. (*G*) 3D random-access scanning system uses four AODs to position a laser focus in a volume without moving the objective lens.

detector (e.g., a sensitive high-speed camera) to separate the emission light from each excited focus. All these systems are designed primarily for structural imaging, in which raster scans are used by necessity. Thus, although they increase the temporal resolution over galvanometer-based systems, they do not overcome the primary limitation imposed by them. Therefore, similar to such systems, they are forced into simple linear-based scans that are highly time inefficient when functionally imaging fast signals on sparse structures.

Functional Imaging Systems

For applications in which functional imaging is of primary importance, scanless or random-access systems can be used. As discussed in Chapter 56, scanless systems use SLMs to create several excitation foci (Fig. 2E) (Nikolenko et al. 2008), similar to the multifocal systems described above. However, unlike those systems, the SLMs in scanless systems allow the locations of the foci to be arbitrarily chosen by the user to match a complex structure such as a dendrite. These systems suffer from similar drawbacks to multifocal systems, including diminished excitation power at each foci as well as the need for a fast imaging detector to spatially separate the individual foci from each other.

Random-access scanning systems operate by using either a single AOD for fast single-axis imaging (Lechleiter et al. 2002) or dual AODs for *x–y* imaging (Iyer et al. 2006; Lv et al. 2006; Salome et al. 2006). AODs work by utilizing an acoustic wave traveling through a medium to set up a virtual diffraction grating, thereby diffracting the incident light by an angle θ, which is proportional to the frequency of the incident acoustic wave by

$$\theta = \frac{\lambda f}{v}, \tag{1}$$

where λ is the wavelength of the incident light and f and v are the frequencies and the velocities of the propagating acoustic waves, respectively. The advantage of such systems comes from the ability to change quickly from one acoustic frequency to a noncontiguous frequency. This makes it possible to use scan patterns that are no longer confined to raster, line, or trajectory scans. However, as mentioned above, utilizing AODs generates significantly higher levels of both spatial and temporal dispersions. As a result, either compensating elements have to placed in the optical path (Iyer et al. 2003; Salome et al. 2006), or longer pulses, which reduce two-photon excitation efficiency, must be used (Iyer et al. 2006).

3D Scanning Systems

The systems mentioned above use high-speed scanning methods that were designed for fast lateral, or *x–y*, scans. To selectively move to different axial planes, such systems typically rely on moving an objective lens via a stepper motor or moving the imaging detector. Although certain techniques manage to synchronize the movements of the lateral scanning mechanism and the stepper motor to create a scan pattern trajectory that visits most user-defined locations with sufficient temporal resolution to record physiological data (Gobel et al. 2007), all such systems are still limited by the inertia of the objective lens and its actuator.

One relatively inexpensive method to circumvent this problem is to use an axicon, which extends the depth of focus at the image plane (Dufour et al. 2006). This expands the axial range at which points of interest can be excited by a laser beam focused at a particular *x–y* location. Although this technique does offer the ability to visit several locations in different axial planes, it suffers from reduced available excitation energy at sites of interest because the power is spread over the increased depth of field. In addition, with this method, it is not possible to differentiate sites of interest that are located directly above or below each other.

Another technique that avoids the need for an objective lens actuator is 3D random-access imaging (Duemani Reddy et al. 2008). As with 2D random-access systems, this technique uses AODs to create scan patterns that can quickly change between several regions of interest. However, 3D random-access imaging uses four AODs to create a user-defined scan pattern in 3D (Fig. 2G). In such cases, two AODs are arranged such that chirped acoustic waves (which are defined by a continuously changing frequency over time) propagating within them are in opposing directions (Reddy and Saggau 2005). This arrangement allows for the generation of a cylindrical focus with a focal length, f_{AOL} proportional to the degree α at which the acoustic wave is being ramped and is determined by

$$f_{AOL} = \frac{v^2}{2\alpha\lambda}, \tag{2}$$

where λ and v are the wavelength and the velocity of the propagating acoustic waves, respectively. By utilizing two orthogonal pairs of deflectors, a full spherical focus can be generated. Lateral scanning is maintained in the *x–y* plane by manipulating the differences in the absolute frequencies between the two deflectors organized in the same direction. Although these systems show a greater temporal dispersion than 2D random-access systems, they are self-compensating with respect to spatial dispersion (Reddy and Saggau 2005). An example of dispersion-compensated fast imaging with 3D random-access scanning is shown in Figure 3 (Y. Liang, A.R. Faustov, and P. Saggau, unpubl.).

Using an SLM for 3D scanning without moving the objective lens would also be possible. However, its multibeam feature (Nikolenko et al. 2007) could not be straightforwardly utilized, because the imaging detector necessary for this scanning mechanism might not be able to separate fluorescence received from different axial distances. Therefore, SLM-based 3D imaging will most likely be limited to single-beam scanning, where AODs have been superior because of their approximately 10 times higher beam positioning speed.

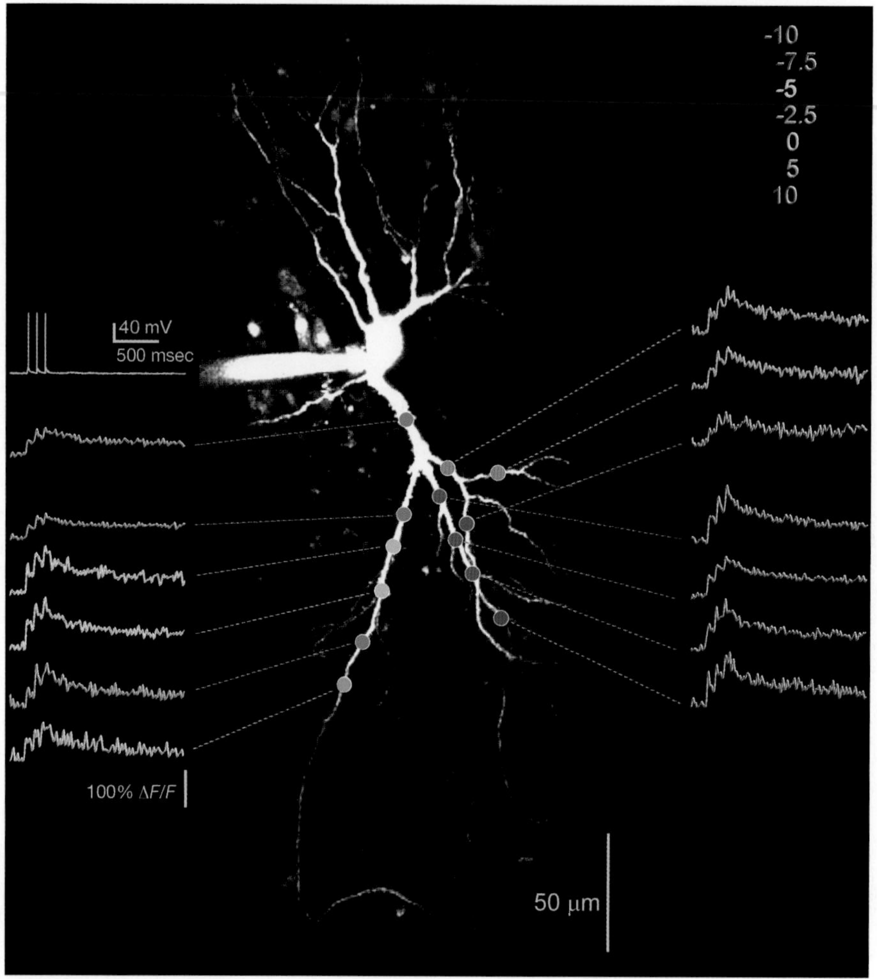

FIGURE 3. Fast 3D imaging. Hippocampal CA3 neuron in organotoypic culture, filled by patch pipette with two fluorescent dyes: structural label Alexa 594 (50 μM) and calcium indicator Oregon Green BAPTA-1 (100 μM). Current injection into the soma elicits three back-propagating action potentials. Calcium transients at user-selected sites measured simultaneously over an axial extension of 20 μm. Note color coding in micrometers relative to the physical focal plane of the objective lens.

Other methods for fast focusing, such as deformable mirrors (Zhu et al. 1999), fluid-filled lenses (Zhang et al. 2003), and electro-optic lenses (Shibaguchi and Funato 1992), have all been shown and are possibilities for offering fast 3D scanning when combined with either galvanometers or other 2D scanning methods. However, to our knowledge, none has yet been shown in two-photon imaging systems.

Microscopes

Regardless of the scanning method used, most two-photon imaging systems are constructed around adapted research microscopes. The type of microscope chosen depends on the biological task to be accomplished. For example, for optically studying electrophysiology, an upright microscope is used because it allows for easy patching of cells to be filled with the appropriate optical indicators. By adding an additional port to the microscope, the excitation and emission light for two-photon imaging can be separated from the wide-field illumination usually required for other gross visualization of tissues of interest.

Detectors

The final element to any two-photon imaging system is the photodetector. The first concern about what type of detector is optimal is whether an imaging detector is needed or not. For single-beam laser scanning, such as that utilizing galvanometers or AODs, an imaging detector is unnecessary and either photomultiplier tubes or photodiodes can be used to detect the emission light. The benefits of each method are beyond the scope of this chapter. For scanning methods that require multiple beams, such as multifocal or scanless two-photon microscopy, or for those that image with a line instead of a point, an imaging detector, often a high-speed charge-coupled device camera, is a necessity.

CONCLUSION AND FUTURE PROSPECTS

High-speed two-photon imaging systems are becoming increasingly important in experimental biology. Nowhere is this more obvious than in the functional study of neuronal dendrites and their spines. Although most such systems are typically built with the same four components, optimizing these systems relies on choosing elements depending on the scanning methods used. Indeed, it is advances in high-speed beam scanning or the successful implementation of fast and effective scanless methods that will continue to advance the field of high-speed two-photon imaging.

REFERENCES

Bewersdorf J, Pick R, Hell SW. 1998. Multifocal multiphoton microscopy. *Opt Lett* **23:** 655–657.

Brakenhoff GJ, Squier J, Norris T, Bliton AC, Wade MH, Athey B. 1996. Real-time two-photon confocal microscopy using a femtosecond, amplified Ti:sapphire system. *J Microsc* **181:** 253–259.

Duemani Reddy G, Kelleher K, Fink R, Saggau P. 2008. Three-dimensional random access multiphoton microscopy for functional imaging of neuronal activity. *Nat Neurosci* **11:** 713–720.

Dufour P, Piché M, De Koninck Y, McCarthy N. 2006. Two-photon excitation fluorescence microscopy with a high depth of field using an axicon. *Appl Opt* **45:** 9246–9252.

Fan GY, Fujisaki H, Miyawaki A, Tsay RK, Tsien RY, Ellisman MH. 1999. Video-rate scanning two-photon excitation fluorescence microscopy and ratio imaging with cameleons. *Biophys J* **76:** 2412–2420.

Gobel W, Kampa BM, Helmchen F. 2007. Imaging cellular network dynamics in three dimensions using fast 3D laser scanning. *Nat Methods* **4:** 73–79.

Hell SW, Booth M, Wilms S, Schnetter CM, Kirsch AK, Arndt-Jovin DJ, Jovin TM. 1998. Two-photon near- and far-field fluorescence microscopy with continuous-wave excitation. *Opt Lett* **23:** 1238–1240.

Iyer V, Hoogland TM, Saggau P. 2006. Fast functional imaging of single neurons using random-access multiphoton (RAMP) microscopy. *J Neurophysiol* **95:** 535–545.

Iyer V, Losavio BE, Saggau P. 2003. Compensation of spatial and temporal dispersion for acousto-optic multiphoton laser-scanning microscopy. *J Biomed Opt* **8:** 460–471.

Kim KH, Buehler C, So PT. 1999. High-speed, two-photon scanning microscope. *Appl Opt* **38:** 6004–6009.

Lechleiter JD, Lin DT, Sieneart I. 2002. Multi-photon laser scanning microscopy using an acoustic optical deflector. *Biophys J* **83:** 2292–2299.

Lv X, Zhan C, Zeng S, Chen WR, Luo Q. 2006. Construction of multiphoton laser scanning microscope based on dual-axis acousto-optic deflector. *Rev Sci Instrum* **77:** 046101.

Martinez OE, Gordon JP, Fork RL. 1984. Negative group-velocity dispersion using refraction. *J Opt Soc Am A* **1:** 1003–1006.

Moulton PF. 1986. Spectroscopic and laser characteristics of Ti-Al2O3. *J Opt Soc Am B* **3:** 125–133.

Nikolenko V, Watson BO, Araya R, Woodruff A, Peterka DS, Yuste R. 2008. SLM microscopy: Scanless two-photon imaging and photostimulation with spatial light modulators. *Front Neural Circuits* **2:** 5.

Reddy GD, Saggau P. 2005. Fast three-dimensional laser scanning scheme using acousto-optic deflectors. *J Biomed Opt* **10:** 064038.

Salome R, Kremer Y, Dieudonné S, Léger JF, Krichevsky O, Wyart C, Chatenay D, Bourdieu L. 2006. Ultrafast random-access scanning in two-photon microscopy using acousto-optic deflectors. *J Neurosci Methods* **154:** 161–174.

Shibaguchi T, Funato H. 1992. Lead-lanthanum zirconate-titanate (PLZT) electrooptic variable focal-length lens with stripe electrodes. *Jpn J Appl Phys* **31:** 3196–3200.

Treacy EB. 1969. Optical pulse compression with diffraction gratings. *IEEE J Quantum Electron* **5:** 454–458.

Zeng S, Lv X, Zhan C, Chen WR, Xiong W, Jacques SL, Luo Q. 2006. Simultaneous compensation for spatial and temporal dispersion of acousto-optical deflectors for two-dimensional scanning with a single prism. *Opt Lett* **31:** 1091–1093.

Zhang DY, Lien V, Berdichevsky Y, Choi J, Lo YH. 2003. Fluidic adaptive lens with high focal length tunability. *Appl Phys Lett* **82:** 3171–3172.

Zhu L, Sun PC, Fainman Y. 1999. Aberration-free dynamic focusing with a multichannel micromachined membrane deformable mirror. *Appl Opt* **38:** 5350–5354.

55 Digital Micromirror Devices

Principles and Applications in Imaging

Vivek Bansal[1] and Peter Saggau[2]

[1]*Department of Radiology, Baylor College of Medicine, Houston, Texas 77030;* [2]*Department of Neuroscience, Baylor College of Medicine, Houston, Texas 77030*

ABSTRACT

A digital micromirror device (DMD) is an array of individually switchable mirrors that can be used in many advanced optical systems as a rapid spatial light modulator. With a DMD, several implementations of confocal microscopy, hyperspectral imaging, and fluorescence lifetime imaging can be realized. The DMD can also be used as a real-time optical processor for applications such as the programmable array microscope and compressive sensing. Advantages and disadvantages of the DMD for these applications as well as methods to overcome some of the limitations will be discussed in this chapter. Practical considerations when designing with the DMD and sample optical layouts of a completely DMD-based imaging system and one in which acousto-optic deflectors (AODs) are used in the illumination pathway are also provided.

INTRODUCTION

Advanced microscopy techniques are often used to optimize the spatial and/or the temporal resolution of image acquisition, compared with wide-field microscopy, for applications involving structural and functional imaging. This is especially true when imaging in optically thick light-scattering tissues. These imaging techniques often rely on some form of spatial light modulator to accomplish spatial filtering, wavelength selection, or other more advanced forms of real-time optical processing. For applications in which the spatial light modulator can act in a binary fashion (i.e., on/off), an array of individual rapidly switchable mirrors, such as the digital micromirror device (DMD), can be used.

DIGITAL MICROMIRROR DEVICE

The DMD is a microlithographed micro-opto-electromechanical system made by the Digital Light Processing (DLP) group at Texas Instruments. It is most commonly encountered in widely available

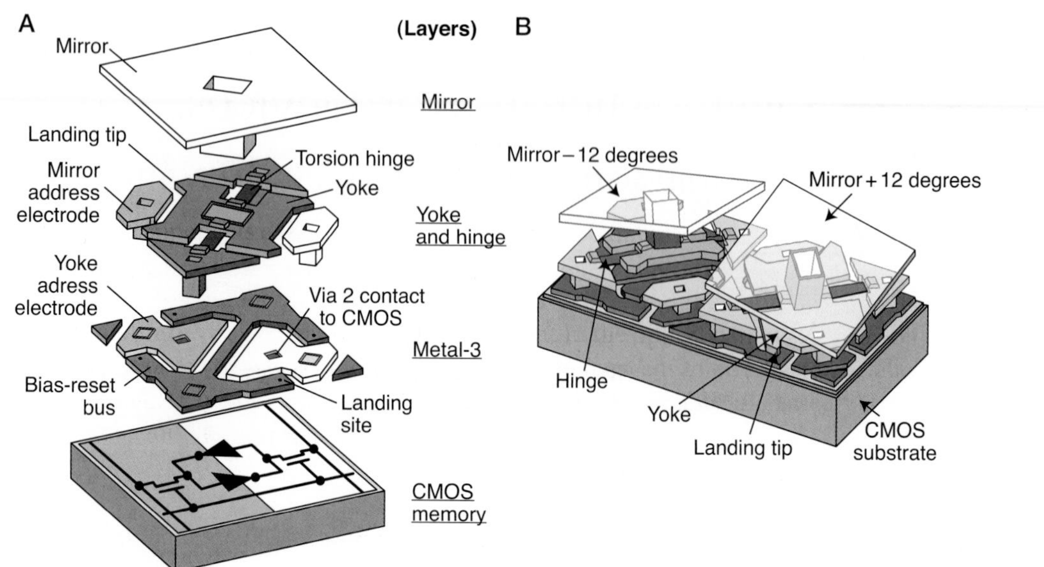

FIGURE 1. DMD microstructure. (*A*) The DMD is composed of a square mirror on top of a diagonal hinge. A complementary metal-oxide semiconductor (CMOS) memory controls the landing pads that determine the orientation of the mirror. (*B*) The mirrors are mechanically constrained to tilt angles of ±12°. The DMD acts as a binary light switch. (Images from Hornbeck 1997.)

DLP video projectors and televisions. The semiconductor chip has an array of aluminum mirrors on its surface, which can be positioned into one of two orientations (+12° or –12°) by tilting on a hinge running diagonally across the bottom of each mirror (Fig. 1A). Because of the way the DMD is fabricated, the hinge obeys the rules of thin films rather than the rules of bulk metals, and thus is not subject to the usual wear from repetitive bending, as has been verified through exhaustive stress testing (Douglass and McMurray 1997; Douglass 2005).

The two free corners of each mirror hover over address electrodes, which are controlled by a CMOS static random-access memory (SRAM). After the SRAM activates one of the electrodes, the corresponding mirror corner is pulled down by an electrostatic force. Once the yoke underlying the mirror makes full contact with the landing site next to the electrode, the mirror is 12° from the normal (Fig. 1B). The logic state of the SRAM controls which address electrode is activated and determines whether the mirror tilts in the +12° or the –12° direction. Because the yoke under the mirror corner is mechanically constrained by the landing site, there is no way for a mirror to settle at a tilt angle >12°. Additionally, because the yoke must physically rest on the landing site, there is no provision to tilt a mirror <12°. Therefore, under normal operation, each mirror must sit in one of the two orientations and must act as a binary light switch. The mirrors can only return to the 0° position if the entire DMD chip is put in the park mode. Note that the earlier generation of the DMD only had a ±10° tilt angle. The increased tilt angle in the current DMDs allows for an improved contrast ratio.

Although each DMD mirror can flip orientations in ~4 µsec, it takes ~14–18 µsec for a mirror to fully settle when switching from +12° to –12°. If multiple mirrors are switched simultaneously, they can all settle at the same time, and the total settling time remains ~14–18 µsec for one full frame.

Multiple array sizes exist for the DMD including 1024 × 768 extended graphics array (XGA), 1400 × 1050 superextended graphics array plus (SXGA+), and 1920 × 1080 high-definition television (HDTV). The XGA chip is available with a mirror pitch of 13.7 µm or 10.8 µm. The DMD can be optimized for different electromagnetic spectra by changing the material used for the front window, which protects the chip. All of the available DMD matrix sizes are available with a window for use in the visible spectrum. The 13.7-µm pitch 1024 × 768 chip and the 1920 × 1080 chip have window options that allow them to be used in the ultraviolet spectrum. Additionally, the 13.7-µm pitch 1024 × 768 chip has a window that allows it to be used in the near-infrared spectrum.

DMD CONTROL ELECTRONICS

Several commercially available control boards are available to simplify the interface between a computer and the DMD. The most widely available control boards are in the DMD Discovery series developed by Texas Instruments and their partners. There are currently two models available—the DLP Discovery 3000 and the DLP Discovery 4000.

The DLP Discovery 3000 works with the 13.7-µm pitch XGA and SXGA+ DMDs and is capable of transferring data to the DMD at 12.8 Gb/sec. This translates to a frame rate of 16,300 frames/sec for the XGA chip and a frame rate of 8600 frames/sec for the SXGA+ chip. The DLP Discovery 4000 works with the 10.8-µm and the 13.7-µm pitch XGA DMDs as well as the 1920 × 1080 DMD. It can transfer data to the XGA chips at 25.6 Gb/sec corresponding to a frame rate of 52,550 frames per second. The data transfer speed to the 1920 × 1080 chip is 51.2 Gb/sec resulting in a frame rate of 24,690 frames/sec.

For increased functionality and special applications, there are several daughter boards available designed to interface with the Discovery series control boards such as the accessory light-modulator package board. Added functions include communication with the host computer through a USB 2.0 port and increased onboard memory. By increasing the memory available locally to the DMD, more pattern data can be preloaded to eliminate any latency caused by computer communication. This is useful for applications that need many different mirror patterns displayed on the DMD at a high speed. There are also daughter boards with field-programmable gate arrays (FPGAs) that allow programming of special functions and direct computations locally at the DMD to further minimize communication time with the host computer for applications in which the data patterns cannot be determined a priori.

Basic functions of the Discovery series control boards and the daughter boards can be controlled via software provided by the manufacturers. Access to advanced features as well as development of custom applications can be accomplished using a graphical programming language such as LabVIEW (National Instruments) or in a programming language such as C/C++ with a provided software developer's kit.

DMD-BASED IMAGING SYSTEMS

In addition to the video projector applications for which the DMD was originally designed, a DMD can be placed in the illumination and/or the detection path of any imaging system to act as a binary spatial light modulator. This is accomplished by designating one mirror orientation as "on" to direct light toward the object in the illumination path and/or toward a photodetector/camera in the detection path and the other orientation as "off" to block light.

When all of the DMD mirrors are flipped to the on position, the DMD effectively acts as a full mirror in the optical path; however, there is ~30%–40% loss of light intensity in the visible spectrum because of the mirror fill factor, the limited reflectivity, and the inherent diffraction (Hornbeck 1997). Losses are similar in the ultraviolet and the near-infrared spectra.

Confocal Imaging

A DMD located in an image plane of a microscope can be used to perform confocal fluorescence or reflectance imaging with epi-illumination (Liang et al. 1997; Sorensen et al. 2001). This is accomplished by flooding the DMD with collimated excitation light and then using a small group of micromirrors in the on orientation as both the point source for illumination and the pinhole in the detection path; this allows light to go to and from the preparation. The remaining micromirrors are set in the off orientation, which blocks light from reaching the preparation and diverts out-of-focus and scattered fluorescence/reflectance away from the detector (Fig. 2). The spatial resolution can be set by changing the number of mirrors that define one site of interest (e.g., 1 × 1, 2 × 2, or 4 × 4 micromirrors for a single pinhole).

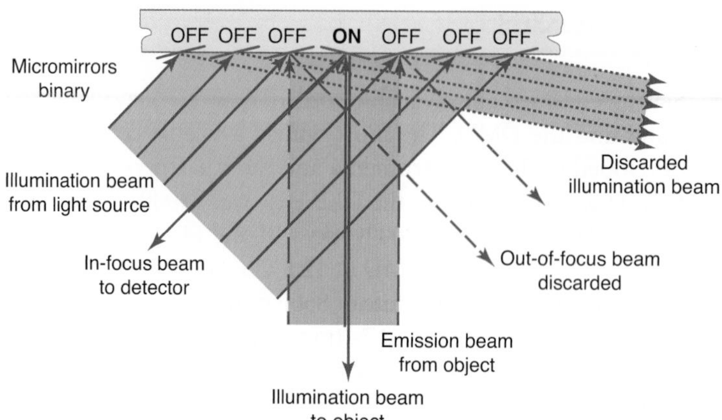

FIGURE 2. Light paths for a DMD-based confocal microscope. The illumination beam (blue) is set up to flood the face of the DMD. By turning on one mirror or one group of mirrors, excitation light can be directed to the specimen. The returning fluorescence (red) will strike the same group of DMD mirrors. The out-of-focus halo will be rejected, while the central in-focus spot will be sent to the detector.

There are several advantages to this scheme compared with commercially available confocal microscopes, which use galvanometer-driven mirrors or a combination of a galvanometer-driven mirror and an acousto-optic deflector (AOD) to position the excitation beam. One limitation of galvanometer-based systems is the amount of time necessary to raster scan a complete field of view to generate a single optical section. Current systems scan on the order of 1–2 frames/sec. For functional imaging applications in which increased temporal resolution is needed, such as tracking shifts in cellular ion concentrations using chelating fluorescent indicators, the scan speed is often increased by limiting the field of view to a single scan line, which is aligned with a few sites of interest. In this mode, the line can be scanned approximately 1000 times per second. An advantage of the confocal systems utilizing two galvanometer-driven scanning mirrors is that collected fluorescence is easily descanned and a stationary pinhole can be used for spatial filtering.

Systems replacing one galvanometer-driven mirror with an AOD can decrease the overall raster scan time for a full frame. AODs rely on diffraction of light, unlike traditional scanning mirrors, which use reflection. Acoustic waves created by a piezoelectric transducer travel through an acousto-optic medium as a series of compressions and rarefactions, which change the refractive index of the medium in a periodic manner. Thus, the acoustic wave effectively creates a diffraction grating whose slit spacing (i.e., the grating constant) is inversely proportional to the frequency of the acoustic wave. Light traveling perpendicular to the sound wave will be diffracted by an angle that is proportional to the acoustic frequency. The frequency of the acoustic wave can be rapidly updated, which allows for much faster positioning of the point excitation compared with scanning mirrors; however, because AODs rely on diffraction rather than on reflection, they cannot be efficiently used for descanning.

To overcome the descanning limitation in confocal systems that use one AOD, a slit aperture is used instead of a pinhole, and descanning is only performed in the mirror-scanned axis. This allows for improved spatial resolution compared with wide-field imaging, but it is not truly confocal. Furthermore, the AOD axis cannot be directly used in a line-scan scheme because of the lack of descanning, so functional imaging with this configuration has limitations that are similar to a fully galvanometer-based system.

When using the DMD in a confocal scheme, the mirrors in the on orientation form both the point illumination and the point detection at a particular site. By controlling the location of this group of micromirrors, a virtual addressable pinhole is created. This ensures that the criteria for confocal microscopy are met at each site of interest and that the detection pinhole is automatically in register with the illuminated point, regardless of the address of the site being probed. Because the

DMD is freely addressable and not inertia limited, sites of interest can be accessed anywhere on the specimen in any sequence and need not lie on a single scan line. Furthermore, the dwell time at a given site of interest can be adaptively changed to increase photon collection and, subsequently, the signal-to-noise ratio. This allows for true confocal imaging at user-selected sites of interest anywhere on a specimen with a higher signal-to-noise ratio than can be achieved with other systems. The DMD can be incorporated into a microscope scan head (Sorensen et al. 1999; Bansal et al. 2006) or into an endoscopic system (Rector et al. 2003).

The major limitation when using the DMD for both point illumination and point detection is its low utilization of excitation light. If a square 768 × 768 mirror region on the DMD is mapped to the field of view, then for pinholes composed of 2 × 2 mirrors, there are approximately 147,500 potential sites that can be independently addressed. This means that even if all of the excitation light could be directed to the mirrors, each site would receive <0.0007% of the total power. Additionally, this does not take into account the 35% loss of excitation light resulting from flooding the square DMD face with a round beam nor does it account for previously described inherent DMD losses.

To overcome the decreased efficiency, there are several techniques that can be used to optimize the use of excitation light for full-field confocal imaging and/or functional imaging at sites of interest. Increasing the intensity of excitation light serves to increase the signal-to-noise ratio of measured parameters.

The simplest method to better use excitation light is to sacrifice spatial resolution by increasing the number of mirrors that are used to define one pinhole. An 8 × 8 group of mirrors transmits 16 times more light than a 2 × 2 group of mirrors and results in four times worse spatial resolution in each dimension of the *x–y* plane. The larger pinhole also limits the optical sectioning ability and results in a decrease in axial resolution. Furthermore, the modest increase in light utilization may still not permit functional imaging of weakly fluorescent specimens.

Another method for full-field imaging is to use several sufficiently spaced mirror groups in the on position to create an array of multiple synchronous pinholes (Liang et al. 1997). With this scheme, the light at each site does not increase, but the overall utilization is improved, thus decreasing the time needed to collect each image and increasing the frame rate. The achievable spacing between additional pinholes depends on the degree of light scattering present in the target specimen. If the spacing is too small, then optical sectioning ability will be compromised. Also, because multiple sites are probed simultaneously, an imaging detector must be used instead of a higher-sensitivity detector such as a photodiode or a photomultiplier tube (PMT). With most biological specimens, the number of pinholes that can be used without sacrificing confocality is on the order of 25–100, which still results in a utilization of excitation light of <0.1%.

One method to markedly increase the utilization of light in a multipoint confocal system involves using closely spaced apertures that are opened and closed to form random patterns on the preparation (Juskaitis et al. 1996). By ensuring that the state of an aperture is completely uncorrelated with the state of its neighbors, an image can be made that consists of a confocal image superimposed on a conventional image. The confocal image can then be isolated by mathematically subtracting the conventional image in software. The DMD is well suited to act as the spatial light modulator for this scheme (Verveer et al. 1998; Hanley et al. 1999). This programmable array microscope (PAM) uses 50% of the incident light by setting one-half of the DMD mirrors to the on orientation. By rapidly cycling through several aperture patterns, it can collect confocal images at a frame rate of 60 Hz, which allows for video-rate full-field dynamic imaging.

For applications that require even greater temporal resolution, such as measuring intracellular ion fluxes with fluorescent chelating indicators or changes in membrane potential with voltage-sensitive dyes, the intensity of excitation light must be dramatically increased to allow for an appropriate decrease in the dwell time while maintaining an appropriate signal-to-noise ratio. One method of accomplishing this is to decouple the point excitation and the point detection needed for confocal imaging by using two AODs to position the point illumination and the DMD for point detection (Bansal et al. 2006). This scheme bypasses the AODs in the detection pathway to overcome the inability of AODs to perform descanning. The AODs allow for focusing of the full power of the excitation source at each site of interest on the specimen while the DMD acts as an addressable spatial filter that can track the excitation point. Both devices have no macroscopic moving parts and can be positioned to any new site address in ~20 μsec. The available scan time can then be adaptively allotted to the signals of interest (e.g., two to five sites for kilohertz signals or 20–40 sites for 500–1000 Hz signals). The number of sites probed can also be adapted to maximize the signal-to-noise ratio at each site.

Compressive Sensing

Another application in which the DMD has been used to increase light efficiency is compressive sensing (Duarte et al. 2008). A full description of the mathematics involved in this technique is beyond the scope of this chapter. Briefly, compressive sensing uses the fact that many sparse signals can be exactly recovered after a lossless compression in which some data is effectively thrown out. If the way that the signal is sampled is optimized, then the data that would have otherwise been thrown out need not be sampled in the first place, thus saving a significant amount of scan time (Sun et al. 2008).

Compressive sensing uses a similar scheme as the PAM in displaying a random pattern on the DMD with half of the micromirrors in the on orientation; however, it uses a high-sensitivity detector such as a photodiode or a PMT, rather than an imaging camera, to combine all of the captured light to a single point. Because of the compressive nature, fewer measurements than needed to perform a raster scan are required to mathematically reconstruct the collected data into a complete real-space image. For the standard compressive-sensing camera, the DMD is only used in the detection pathway; however, the random patterns used for compressed sensing can also be imaged onto the specimen in the illumination pathway, effectively creating a compressive-sensing confocal system.

Hyperspectral and Fluorescence Lifetime Imaging

The DMD has also been used for hyperspectral imaging (Zuzak et al. 2008). To accomplish this, a grating is used to project a dispersed spectrum of light onto the DMD. By selecting different groups of micromirrors, the bandwidth of light projected onto a specimen can be rapidly cycled, and a three-dimensional data set composed of a spectroscopic chemical signature at each pixel in the image plane can be recorded. Similarly, the DMD can be used for sequential illumination of sites of interest with wide-field fluorescence collection to create an imaging system that can record fluorescence lifetime data at each pixel in the image plane. In these schemes, the DMD is used only in the illumination pathway.

Other Applications

For a review of several other nonbiological applications of the DMD, refer to Dudley et al. (2003).

OPTICAL PATHWAYS

There are many possible layouts that can be used for a DMD-based imaging system. Several issues must be considered during the optical design to maximize efficiency for a given application. The most basic step is to determine if the DMD will be used in the illumination pathway, the detection pathway, or both.

DMD-Based Illumination Pathway

When using the DMD for illumination, it is important to use a light source that can be collimated to the aperture size of the DMD. If possible, coherence from the light source should be removed to decrease the amount of diffraction that occurs because of the periodicity of mirrors on the DMD surface. For schemes in which the DMD is exclusively used for illumination, the emission light is

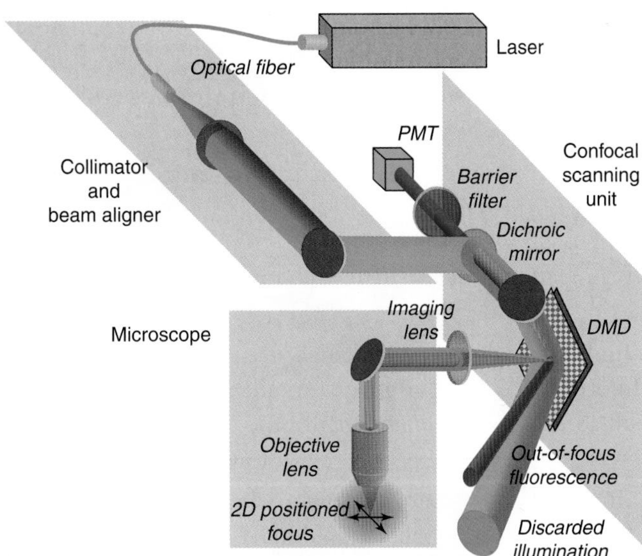

FIGURE 3. Layout of a DMD-based confocal microscope. Self-registering scheme with a single DMD for both point scanning and point detection. The DMD is rotated 45° so that optical paths to and from the specimen are perpendicular to the optical table.

directly collected by a photodetector in either an epi-illumination or a transillumination configuration before it is allowed to propagate back to the DMD as it does in the confocal setup.

DMD-Based Illumination and Detection Pathways

In a basic confocal fluorescence imaging system, the DMD should be used in both pathways. A sample optical layout is shown in Figure 3. To accommodate standard optical bench hardware, we have found that it is easiest to rotate the entire DMD by 45° so that the hinges of the individual mirrors are parallel to the optical table. This allows the mirrors to direct light up or down on an axis that is perpendicular to the optical table rather than at a 45° angle. It also makes it easier to eliminate free space light propagation and to enable relay telescope imaging throughout the system.

To align the illumination light, the central pinhole can be turned to the on position, and beam steering mirrors and telescopes can be used to ensure that the image of this spot is centered in the microscope specimen plane in all three dimensions.

In a fluorescence-based system, emission light from the specimen will be directed back to the last mirror used before the DMD in the illumination path and can be isolated from the excitation light using a standard dichroic mirror. The emission light can then be directed to either an imaging detector or a high-sensitivity point detector such as a photodiode or PMT.

DMD-Based Detection Pathway

For applications in which the DMD is used only in the detection pathway, such as when using AODs for the point illumination in a high-speed confocal system, the dichroic mirror used to separate excitation and emission light should be located between the DMD and the microscope specimen plane (Fig. 4). The emission light can still be collected after spatial filtering by the DMD in a similar fashion to the DMD-only system with either a camera or a single photodetector. In this scheme, it is important to ensure registration between the point illumination and the point detection to achieve confocality.

The DMD can also be used exclusively in the detection pathway for applications such as the nonconfocal version of the single-pixel camera based on compressive sensing. In this layout, the object

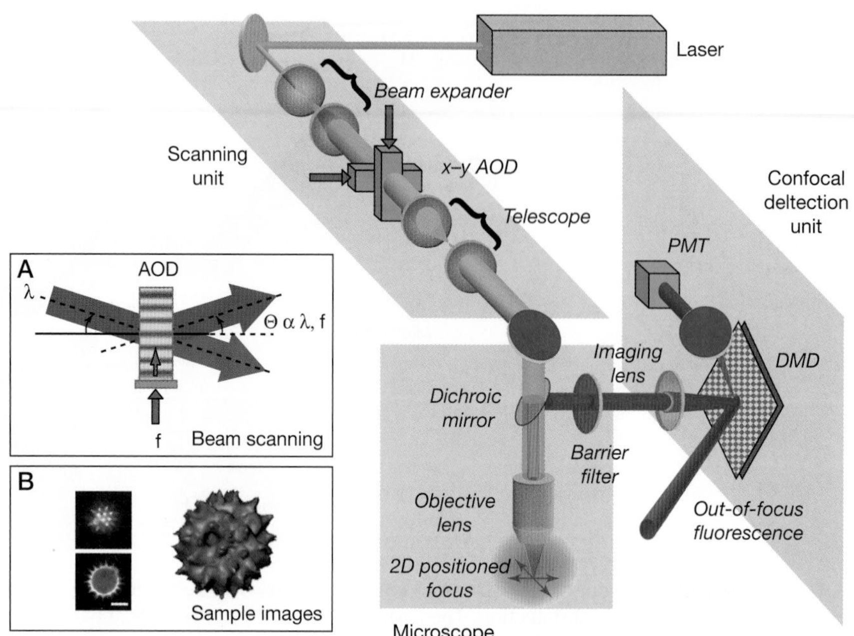

FIGURE 4. Confocal microscope using AODs for point illumination and a DMD for point detection. A pair of AODs is used to scan the beam, while the DMD is used only for spatial filtering. AODs and DMD must be kept in register. (*A*) Beam deflection in an AOD is determined by the acoustic frequency used as input. (*B*) Optical sections and surface-rendered 3D reconstruction of a pollen grain. Scale bar, 10 µm.

of interest is illuminated with ambient light or via wide-field transillumination. The collected light from the DMD is then directed to a single photodetector without the need for a dichroic mirror.

COMPARISON WITH OTHER TECHNIQUES

Standard Confocal Microscopy

As described above, the main advantage of DMD-based implementations of confocal imaging systems compared with standard systems is the increase in temporal resolution as well as the ability to adaptively increase the signal-to-noise ratio at any given site of interest. The only disadvantage is the increased complexity of the optical system.

Multiphoton Microscopy

Compared with multiphoton microscopes, DMD-based confocal microscopes allow for the use of many more fluorescent indicators because of the ability to use light in the spectrum extending from ultraviolet light to near-infrared light. As with standard confocal systems, the DMD-based confocal microscope can achieve a higher spatial resolution than multiphoton systems and can be used in both fluorescent and reflective modes. The disadvantages are also similar to standard confocal microscopes and include less penetration depth into thick tissue as well as increased photobleaching in planes above and below the site of interest.

Patch-Clamp Recording

Although membrane potential measurements can be made with high temporal resolution using patch-clamp pipettes, it is technically very challenging to probe more than one or two sites simultaneously as is possible with optical imaging techniques. Furthermore, the physical size of patch-clamp pipettes limits the evaluation of the smallest cellular structures such as fine dendrites of neurons.

CONCLUSION

There are currently several imaging schemes utilizing DMDs that are suitable for biological imaging. Choosing an optimal scheme will depend on the particular application (structural or functional imaging) as well as the desirable trade-off between spatial and temporal resolutions. In general, the high degree of spatiotemporal flexibility for creating optical patterns with DMDs makes them an attractive building block for advanced imaging systems.

REFERENCES

Bansal V, Patel S, Saggau P. 2006. High-speed addressable confocal microscopy for functional imaging of cellular activity. *J Biomed Opt* **11:** 34003.

Douglass MR. 2005. *Lifetime estimates and unique failure mechanisms of the digital micromirror device (DMD).* Texas Instruments white papers. Texas Instruments, Dallas, TX.

Douglass MR, McMurray IS. 1997. *Why is the Texas Instruments digital micromirror device so reliable?* Texas Instruments white papers. Texas Instruments, Dallas, TX.

Duarte MF, Davenport MA, Tahar D, Laska JN, Sun T, Kelly KF, Baraniuk RG. 2008. Single-pixel imaging via compressive sampling. *IEEE Signal Process Mag* **25(2):** 83–91.

Dudley D, Duncan W, Slaughter J. 2003. Emerging digital micromirror device (DMD) applications. *Proc SPIE* **4985:** 14–25.

Hanley QS, Verveer PJ, Gemkow MJ, Arndt-Jovin D, Jovin TM. 1999. An optical sectioning programmable array microscope implemented with a digital micromirror device. *J Microsc* **196:** 317–331.

Hornbeck LJ. 1997. Digital light processing for high-brightness, high-resolution applications. Projection displays III. Electronic imaging. *Proc SPIE* **3013:** 27–40.

Juskaitis R, Wilson T, Neil MA, Kozubek M. 1996. Efficient real-time confocal microscopy with white light sources. *Nature* **383:** 804–806.

Liang MH, Stehr RL, Krause AW. 1997. Confocal pattern period in multiple-aperture confocal imaging systems with coherent illumination. *Opt Lett* **22:** 751–753.

Rector DM, Ranken DM, George JS. 2003. High-performance confocal system for microscopic or endoscopic applications. *Methods* **30:** 16–27.

Sorensen ME, Patel SS, Saggau P. 1999. A novel direct-view confocal microscope system. *J Gen Physiol* **114:** 7a.

Sun T, Takhar D, Duarte M, Bansal V, Baraniuk R, Kelly K. 2008. Realization of confocal and hyperspectral microscopy via compressive sensing. Abstract in the Bulletin of the American Physical Society March Annual Meeting, New Orleans, LA. http://meetings.aps.org/link/BAPS.2008.MAR.U36.8.

Verveer PJ, Hanley QS, Verbeek PW, vanVliet LJ, Jovin TM. 1998. Theory of confocal fluorescence imaging in the programmable array microscope (PAM). *J Microsc* **189:** 192–198.

Zuzak K, Francis R, Smith J, Tracy C, Cadeddu J, Livingston E. 2008. Novel hyperspectral imager aids surgeons. *SPIE Newsroom (Biomed Opt Med Imaging).* doi: 10.1117/2.1200812.1394.

56 Spatial Light Modulator Microscopy

Volodymyr Nikolenko, Darcy S. Peterka, Roberto Araya, Alan Woodruff, and Rafael Yuste

Department of Biological Sciences, Howard Hughes Medical Institute (HHMI), Columbia University, New York, New York 10027

ABSTRACT

The use of spatial light modulators (SLMs) for two-photon laser microscopy is described. SLM phase modulation can be used to generate nearly any spatiotemporal pattern of light, enabling simultaneous illumination of any number of selected regions of interest. We take advantage of this flexibility to perform fast two-photon imaging or uncaging experiments on dendritic spines and neocortical neurons. By operating in the spatial Fourier plane, an SLM can effectively mimic any arbitrary optical transfer function and thus replace, in software, many of the functions provided by hardware in standard microscopes, such as focusing, magnification, and aberration correction.

INTRODUCTION

Laser-scanning microscopy, such as confocal or two-photon fluorescence, has revolutionized the life sciences by making it possible to image biological tissues with high resolution under physiological conditions. Laser scanning is most often performed either with a pair of galvanometer mirrors, which steer the beam by rotating mirrors, or by acousto-optic modulators, which deflect the beam by diffraction caused by sound-induced changes of the refractive index (Chapter 5). In these scanning modes, the image is reconstructed by sequentially moving the laser beam point by point and line by line across the sample in raster fashion. The drawback of this approach is that the temporal resolution is limited by the finite response time of the scanners. Even if it were possible to increase the scanning speed of the devices, a more fundamental limitation arises. Specifically, to maintain adequate fluorescence signal with the shorter dwell times per pixel (i.e., the time during which the beam remains at a certain point in the sample and collects optical signals from that point), it is usually necessary to increase the laser intensity. However, the rate of signal acquisition is limited by the number of chromophore molecules present and how often they can be excited (i.e., the duty cycle per lifetime in the case of fluorescence). This implies that even in the complete absence of photodamage, excitation intensity cannot be continually increased to achieve faster scanning or shorter dwell times because, regardless of the excitation power, the chromophores or fluorophores cannot produce more than a certain number of excitation-emission cycles per unit time. Thus, the signal cannot be made stronger by increasing the power because it is effectively saturated (Koester et al. 1999; Hopt and Neher 2001).

A logical way to overcome this second limitation is to parallelize the excitation process and use a scheme that can excite and acquire signals from multiple points of the sample at the same time. Traditional wide-field illumination does exactly that (although it is subject to other problems such as limited spatial resolution and blurring caused by cross talk between adjacent pixels). However, wide-field illumination is not a practical option for nonlinear optical methods such as two-photon fluorescent microscopy because existing ultrafast pulsed laser sources do not provide enough power to excite the whole field of view simultaneously.

Diffractive Spatial Light Modulators

Although ultrafast lasers cannot illuminate the entire field, they are powerful enough to illuminate many points of interest at the same time. The difficulty is efficiently redistributing the light to only the areas that are of interest. Phase-only SLMs are ideally suited for this task, and they can dynamically adjust the number and location of active beamlets that can be used for imaging and photostimulation. Phase-only SLMs usually use a matrix of nematic liquid crystals similar to those used in multimedia projectors. However, in contrast to creating an image by masking certain pixels, phase-only SLMs take advantage of the wave properties of light and essentially work as a computer-controlled diffraction grating in which each pixel introduces different phase delays instead of modulating the intensity of passing light. That, in turn, leads to the creation of an image in the far field in a manner similar to classical Fraunhofer diffraction. The power of this approach is that almost any arbitrary pattern of intensity distribution can be created with minimal loss of power. This differs from the situation in which pixels are simply masked such as with digital micromirror devices (DMDs) (Wang et al. 2007; Chapter 55). It is important to emphasize that if intensity modulators, such as DMDs, create illumination patterns by removing light, phase-only SLMs work by redistributing it. This redistribution of light leaves practically all of the power available, making it possible for nonlinear imaging modalities, such as two-photon absorption or second-harmonic imaging, to operate.

IMAGING SETUP

Adding a diffractive SLM to an existing microscope is a straightforward process. The SLM is a single small element, which is placed in the optical path, and can be positioned at almost any point before the objective lens, but ideally it should be on the plane that is optically conjugated to the back aperture of the objective. Existing laser-scanning systems can be easily modified to work with a diffractive SLM by placing the SLM on the plane that is conjugated to the existing scanners (i.e., the galvanometer mirrors) via a simple telescope.

Figure 1 illustrates the optical design of the SLM microscope used in our laboratory. A full description of the components used in the system can be found in Nikolenko et al. (2008). Here the design and logic of the SLM microscope are summarized. The elements of the optical pathway are listed approximately in the functional order of signal propagation. The numbered items below correspond to the numbers in Figure 1. Individual mirrors are not numbered, and unless otherwise noted, EO3 dialectical mirrors from Thorlabs, Inc. were used. These were optimized for the near-infrared (NIR) region (700–1200 nm) and did not introduce noticeable pulse broadening.

1. The source of illumination is an ultrafast pulsed (mode-locked) laser, the Chameleon Ultra from Coherent, Inc.

A

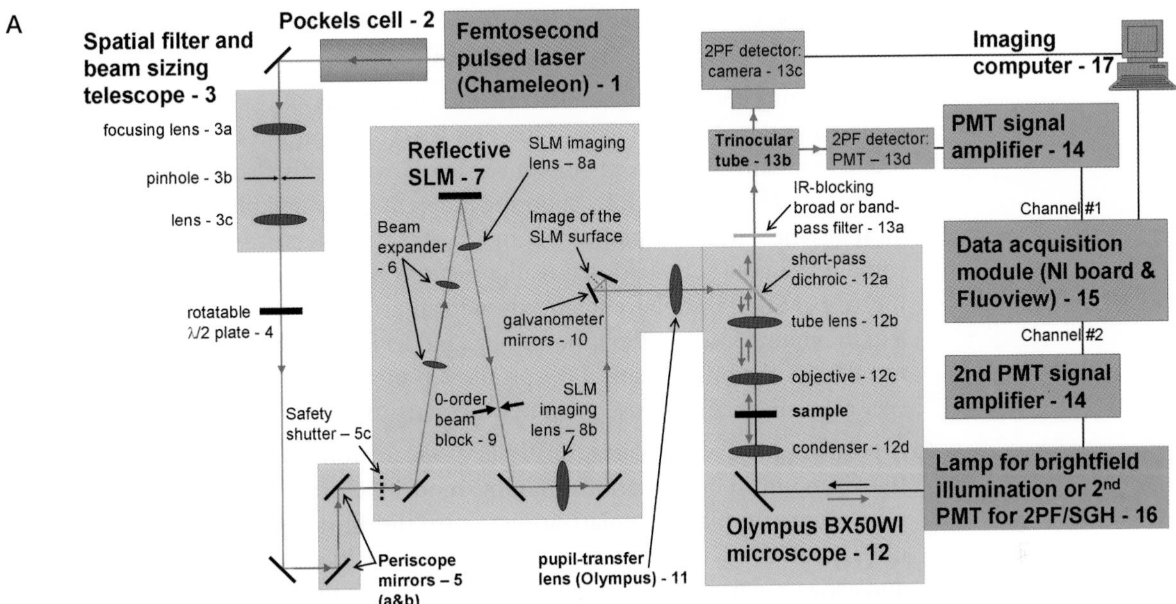

B

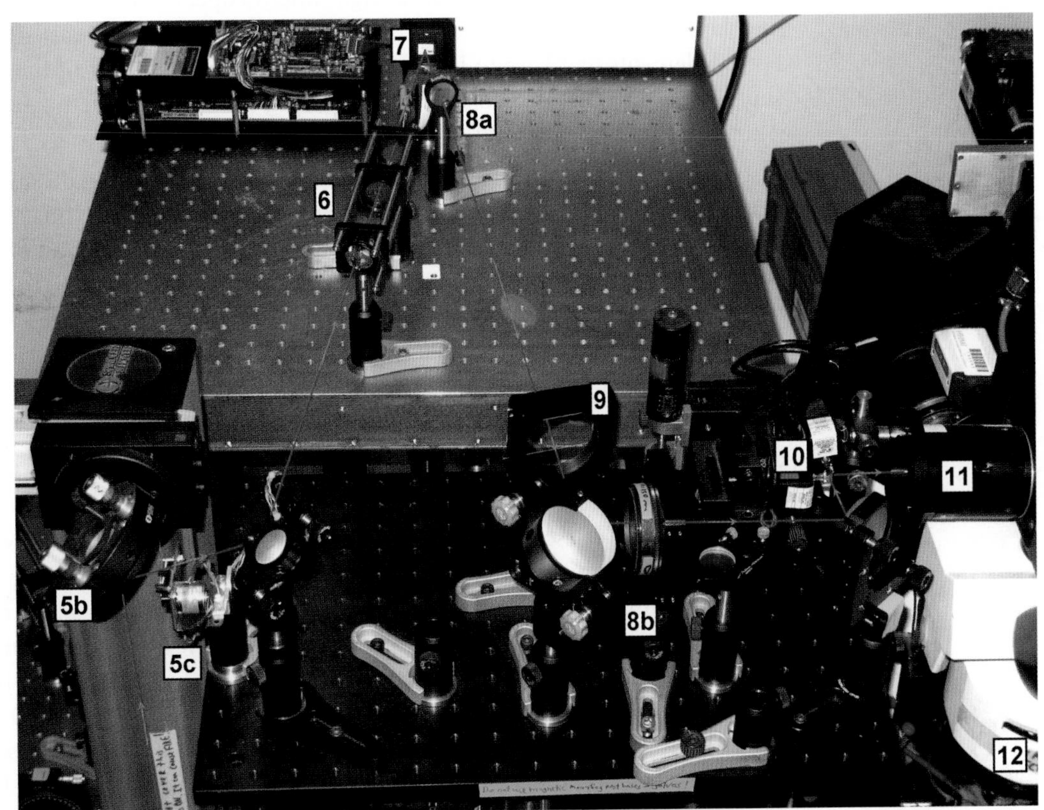

FIGURE 1. Optical design of an SLM microscope. (*A*) Optical diagram of the SLM microscopy system used in our laboratory. (*B*) Photograph of a portion of our SLM microscopy system. Red lines illustrate the laser excitation pathway. (Modified, with permission, from Nikolenko et al. 2008.)

2. The Pockels cell (Conoptics model 350-160) is controlled by a data acquisition board through a high-voltage driver (275 linear amplifier; Conoptics, Inc.).
3. A beam sizing and reshaping telescope that also works as a spatial filter if a pinhole (item 3b) is placed at the plane of focus of the first lens (3a), and the second lens (3c) recollimates the beam: Standard BK7 thin plano-convex lenses with antireflection coating optimized for NIR were used (Thorlabs, Inc.). By choosing different lenses and placing them at the corresponding focal distances from the pinhole (item 3b), it is possible to change the size of the output beam without the need for additional realignment. It is convenient to use a lens kit from Thorlabs, Inc. (e.g., LSB01-B) to have the freedom to adjust the size of the beam easily. It is also convenient to mount lenses on flip mounts (e.g., model FM90; Thorlabs, Inc.), which makes it easy to reconfigure the optical path. Low-profile flip mounts (e.g., model 9891; New Focus) are also very convenient, and we use them in other parts of the optical path.
4. A polarizing half-wave plate (AHWP05M-950 achromatic $\lambda/2$ plate, 690–1200 nm; Thorlabs, Inc.) is mounted on a rotational mount (model PRM1; Thorlabs, Inc.). Rotating the half-wave plate rotates the plane of polarization to turn the diffraction of the SLM "on" or "off" (our liquid-crystal SLM is fully sensitive to polarization). When off, the SLM works as a passive mirror and allows regular scan mode imaging using galvanometer scanners (for high-resolution calibration images).
5. Periscope mirrors: When using an upright microscope, it is convenient to bring the light from the plane of the optical table up to the "second floor"—that is, a raised breadboard with other optical elements that have to be in the vicinity of the input port of the upright microscope. A shutter (item 5c) is used to block the laser light when there is no scanning of the sample. Alternatively, the Pockels cell or the SLM could block the beam.
6. A secondary beam-resizing telescope, which is similar to item 3 and is made from a pair of thin plano-convex lenses: The main function of this telescope is to make the laser beam large enough to fill the aperture of the SLM (0.7-in chip). This enables all of the available pixels to be used and spreads the power across a larger area to avoid damage to the SLM by the laser. This telescope is not absolutely required, although it is convenient to have, because its function can be fulfilled by the telescope in item 3.
7. The diffractive SLM used is a model HEO 1080 P (HOLOEYE Photonics), configured to work in the reflection mode. It provides high-definition television resolution (1920 x 1080-pixels) and comes with simple software to compute the phase mask. To avoid distortions, it is important to minimize the angle of reflection for the SLM.
8. The second SLM telescope is arranged such that the surface of the SLM is conjugate to an intermediate image, which is itself conjugated to the back aperture of the microscope objective through the microscope's pupil transfer and tube lens. This same plane is also occupied by the galvanometer scanning mirrors (item 10), which are left from the original Olympus FluoView system. The first lens (item 8a) is a model LA1906-B, F = 500 mm (1-in diameter; Thorlabs, Inc.). The second lens (item 8b) is larger (model LA1417-B, F = 150 mm [2-in diameter]; Thorlabs, Inc.) to accommodate the full range of scanning angles necessary for the full field of view. To save space, the mirror (also 2-in diameter) is placed between the lenses. The ratio of the telescope (~1:3) shrinks the beam and increases the deflection angles to match the range of angles expected by the scan lens of the microscope imaging port. The relative distances are important for matching the optical planes; thus, in its current configuration, the distance between the SLM (item 7) and the first lens (item 8a) is 90 mm, the total distance between lenses (8a) and (8b) is 650 mm (the sum of focal distances for telescope configuration), and the total distance between the second lens (8b) and the plane of galvanometers (10) is ~190 mm.
9. The zeroth-order beam block allows only the diffracted (first-order diffraction) beam to reach the sample. The block is made from a small piece of metal foil glued to a thin glass cover slide.

The element is mounted onto a flip mount (model FM90; Thorlabs, Inc.) for quick reconfiguration between actively using the SLM for multiplexed excitation, or using it as a passive mirror, when the diffraction is turned off by the half-wave plate (for high-resolution standard imaging).

10. Galvanometer scanning mirrors (Olympus FV200 system): Although separate mechanical scanners are not necessary for an SLM microscope, they provide additional flexibility to our system and were already present in our microscope (described in Chapter 8). Standard Olympus FluoView software is used for slow high-resolution imaging, which is used for calibration purposes.
11. The scan (or pupil transfer) lens is a standard part of the Olympus FluoView system (FVX-PL-IBX50/T). In combination with the microscope tube lens (item 12b), it forms a telescope and images the plane of galvanometers (and therefore, also the plane of the SLM chip) onto the back aperture of the microscope objective.
12. The Olympus BX50WI upright microscope is used without significant modifications. A dichroic mirror (item 12a; Chroma Technology) reflects excitation (NIR) light toward the sample and transmits visible fluorescence from the sample to the detector.

    The emission path consists of the following components.

13. A short-pass (IR-block) filter or a combination of an IR-block and a band-pass filter (Chroma Technology) is used to reject scattered excitation light and to detect the signal within a chosen spectral region. The trinocular tube (item 12b; Olympus FV3-LVTWI) allows switching between two imaging ports, either multibeam SLM imaging with a camera (item 13c; Hamamatsu Orca C9100-12 cooled electron-multiplying charge coupled device [EMCCD] camera) or single-beam whole-field-of-view-scanning imaging using a photomultiplier tube (PMT) (item 13d; Hamamatsu H7422-40P cooled GaAs PMTs).
14. The signal amplifier (model PE 5113 preamplifier; Signal Recovery AMETEK Advanced Measurement Technology), used in combination with a current-to-voltage converter (a passive 5-kΩ load resistor in the simplest case), converts signals into a convenient range of voltages for digitizing.
15. The data acquisition system is controlled by Olympus FluoView scanning software. The signal from the PMT is digitized by the FV 200 data acquisition module. In certain situations, generic data acquisition cards (e.g., PCI 6259; National Instruments) and custom software are used.
16. Alternatively, the optical signal can be detected in a transmissive configuration. In our setup, a separate PMT is installed after the microscope condenser and detects either the second channel in two-photon fluorescence (different color) or the second-harmonic generation (SHG) signals. It is possible to install a camera in this pathway for multibeam imaging of transmissive SHG signals.
17. The computer receives images from the camera and/or digitizes PMT signals. The computer is also used to control excitation intensity via the Pockels cell. Our setup uses three computers with their software synchronized by through-the-lens triggers.

Phase-Mask Computation

Most algorithms for phase-mask computation are based on the fact that the phase-mask and far-field diffraction patterns are related through the Fourier transform (see Fig. 2 for a detailed explanation). Commercial software (e.g., from HOLOEYE Photonics) is available for computing phase masks, but there are circumstances in which it is valuable to write custom software that computes the phase pattern using generic algorithms based on iterative-adaptive algorithms (Fig. 2).

To show the ability of SLM microscopy to form arbitrary complex patterns, a series of tests were performed (Fig. 3). Binary (Fig. 3A, left panel) and grayscale patterns (Fig. 3B, left panel) were used

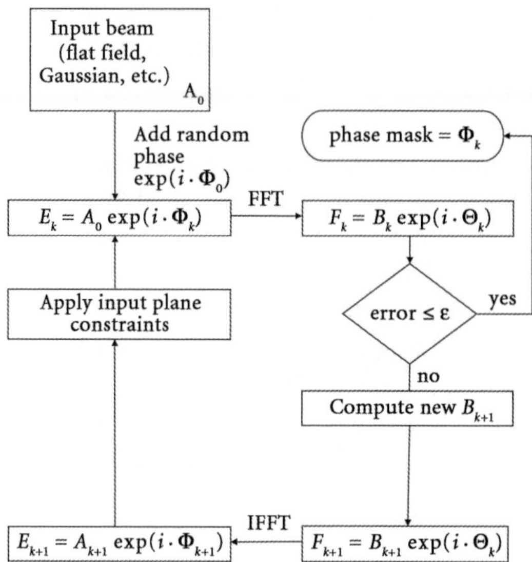

FIGURE 2. SLM phase-mask formation. The algorithm starts with the known intensity distribution of the laser and then adds a random phase (which speeds convergence), generating E_k. It then computes the spatial fast Fourier transform (FFT), F_k, and compares the computed image with the desired image. If the error exceeds a threshold, the amplitude, but not the phase, is modified to match the desired image better. An inverse fast Fourier transform (IFFT) is then performed, and constraints applied, such as phase quantization, giving rise to a new input field, and the cycle begins again. This description is deliberately nonspecific about the comparison process and modification because the optimal implementation critically depends on the physical parameters of the system (such as SLM resolution and desired deflection angles). More complete information on the variety of algorithms can be found in Kuznetsova (1988) and Bauschke et al. (2002). (Reprinted, with permission, from Nikolenko et al. 2008.)

as test images to be formed by diffractive SLM. The algorithm calculated the corresponding phase masks to be uploaded to the SLM (Fig. 3A,B, middle panels). Grayscale tones here have a meaning different from the imaging plane: White corresponds to 0 phase delay at that pixel, and black corresponds to a 2π phase shift (the whole wave) of the wave front at that point. These phase masks form diffraction images on the sample, which are shown in the right panels of Figure 3, A and B. The optical signal in these images is actually the two-photon fluorescence that was generated in the sample of agarose gel infused with a fluorescent dye. These images were acquired by a camera and show the power of the nonlinear imaging approach with its inherent three-dimensional (3D) sectioning property. The relatively thick sample (i.e., fluorescent gel) represents one of the most challenging imaging scenarios with respect to contrast because every point of the sample is capable of generating fluorescence. With linear (one-photon) excitation, the image would be very blurry because linear excitation causes points of the sample both above and below the point of interest (on the focal plane) to fluoresce. To make the image usable, either deconvolution or confocal detection would be necessary.

Focusing with an SLM

Figure 3, C and D, shows another useful property of the SLM: 3D focusing. The SLM works as a universal modulator of the phase of light waves, so it can change the axial position of patterns by convolving phase masks with corresponding lens function. In fact, lenses simply convert plane waves into a spherical one, in a manner analogous to a physical thin lens (Fig. 3C, two middle panels). By varying the optical strength of the lens function, it is possible to shift the excitation pattern in the axial dimension in the sample (Fig. 3D). Moreover, any SLM that changes the phase of light can be used as a universal motionless scanner because, in principle, almost any complicated 3D pattern of illumination can be created by an appropriate phase mask.

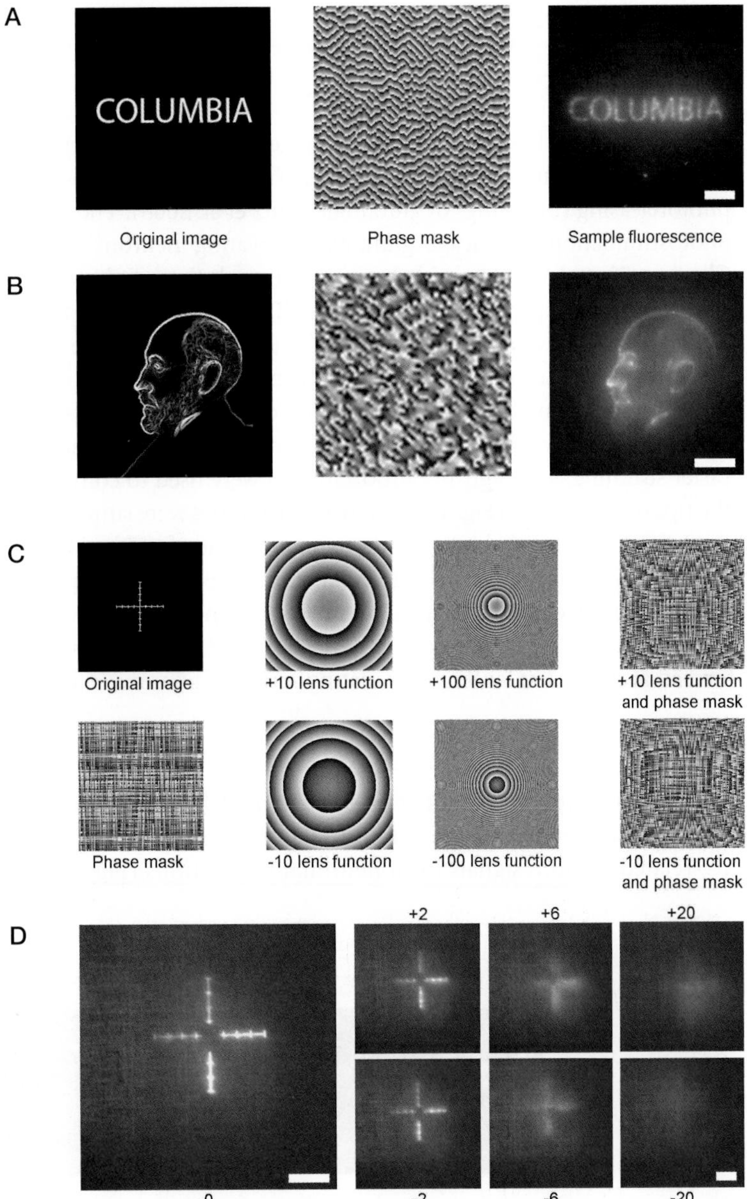

FIGURE 3. SLM light patterning and depth focusing. Imaging samples from an agarose gel saturated with Alexa Fluor 488 fluorescence indicator to test the efficiency of two-photon excitation. Images were acquired using a 60x 0.9-numerical-aperture (NA) objective. Scale bar, 20 μm. (*A*) A simple binary bitmap pattern (COLUMBIA) was uploaded into the SLM software, and the phase mask obtained is shown in the *middle* panel. Grayscale corresponds to a phase shift from 0 to 2π. The resulting two-photon fluorescence image of the sample acquired with the CCD is shown in the *right* panel. These data also show that liquid-crystal-based diffractive SLMs can withstand illumination by a powerful pulsed mode-locked ultrafast laser and can be used effectively for structured nonlinear illumination. (*B*) Complex grayscale patterns can be used to program the SLM. A stylized picture of Santiago Ramón y Cajal, based on a historical photograph, was used. The panels are similar to those in *A*. (*C*) Focusing with an SLM. The SLM software allows additional optical functions to be applied on top of the phase mask. In this example, a lens function was used to shift the focus of excitation in the axial dimension. The original image and its corresponding phase mask as well as the lens phase function alone and added to the original phase mask are shown. The –10, –100, +10, and +100 are arbitrary units used by the software to indicate correspondingly a diverging or converging lens function and its relative optical strength. Note that increasing the optical strength of the lens function created by the SLM corresponds to a faster change of phase from the center to the edges of the SLM (rings of 2π phase reversal are spaced close together). (*D*) Two-photon fluorescence image of the test pattern acquired with the CCD camera. The virtual focus plane is moved away in both directions from the camera's imaging plane using a lens function of corresponding strength. A 40x 0.8-NA objective was used. Scale bars, 50 μm. These data illustrate that SLMs can be used as universal scanners that do not require moving parts. (Modified, with permission, from Nikolenko et al. 2008.)

APPLICATIONS OF SLM MICROSCOPY

Two-Photon Activation of Multiple Dendritic Spines

SLM microscopy has been used to study neurons and their dendritic spines. This work complements (and is in excellent agreement with) experiments using diffractive SLM microscopy for one-photon photoreleasing (uncaging) of glutamate (Lutz et al. 2008). The system described here uses two-photon photostimulation to uncage glutamate-activating neurons and dendritic spines (see Yuste 2011; Chapters 59 and 60) to provide high spatial resolution of photostimulation. In an extension of single-spine uncaging experiments (Araya et al. 2006b, 2007), a two-photon laser is used here to simultaneously stimulate several dendritic spines by uncaging glutamate near their heads (Fig. 4A). Whole-cell electrodes were used to record the somatic membrane potential of pyramidal neurons in slices bathed with 2.5-mM 4-methoxy-7-nitroindolinyl-caged-L-glutamate (MNI-glutamate). The basal dendrites and selected arrays of dendritic spines from these pyramidal neurons were imaged with conventional raster scanning. The high-resolution images were used to compute phase masks targeting spines near the tips of their heads (Fig. 4). Next, five to 15 spines were simultaneously activated with the SLM phase masks, generating reliable uncaging potentials with fast kinetics (Nikolenko et al. 2008). Using SLM microscopy to simultaneously stimulate multiple individual spines or dendritic locations could help address fundamental problems in dendritic biophysics such as integration of inputs (Cash and Yuste 1998; Araya et al. 2006a; Losonczy and Magee 2006). In principle, similar experiments could be performed at the circuit level by stimulating arrays of neurons (Nikolenko et al. 2007, 2008).

Two-Photon Imaging of Multiple Neurons

SLM microscopy can also be used for fast multifocal two-photon imaging, providing the potential for rapid deep-tissue imaging of multiple sites of interest. This method has been used to image activity-related calcium signals from neuronal populations (Fig. 5). Multiple neurons were identified as

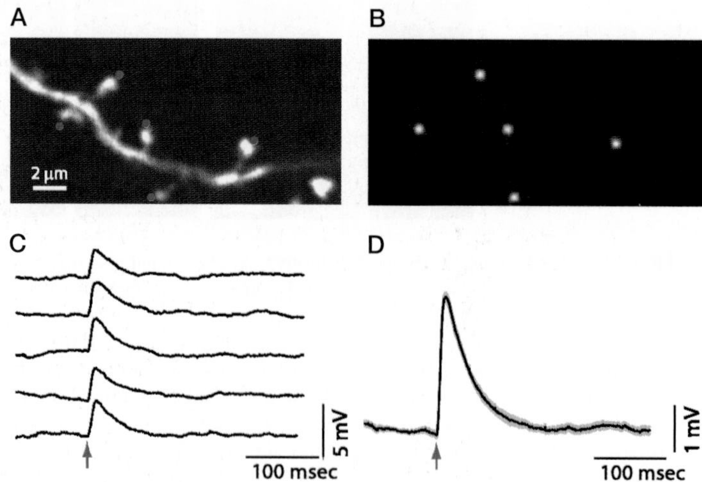

FIGURE 4. Simultaneous glutamate uncaging on multiple dendritic spines using an SLM. (*A*) Basal dendrite from a layer-5 pyramidal neuron, loaded with Alexa Fluor 488, in a mouse neocortical slice bathed in 2.5-mM MNI-glutamate. The red spots indicate the sites of simultaneous two-photon uncaging of glutamate using an SLM. First, an image (*A*), of the dendritic spines selected to be activated, was acquired with galvanometer raster scanning. Second, a bitmap file (*B*) was generated with the uncaging locations selected in *A*. Next, a Fourier transform of the image was set as the command to generate a phase mask and the desired diffraction pattern, in this case, five uncaging spots next to spine heads (*A*). The voltage responses triggered after uncaging glutamate right next to the spine heads were recorded with a whole-cell patch-clamp recording electrode from the cell somata (*C*). Five (out of fifteen) representative uncaging potentials are depicted here. These were generated after simultaneously uncaging glutamate right next to the spine heads shown in *A*. (*D*) Average of 15 uncaging potentials (including the ones in *C*). The black trace is the average uncaging potential as shown in *D*. Light gray in *D* is ±SEM. (Modified, with permission, from Nikolenko et al. 2008.)

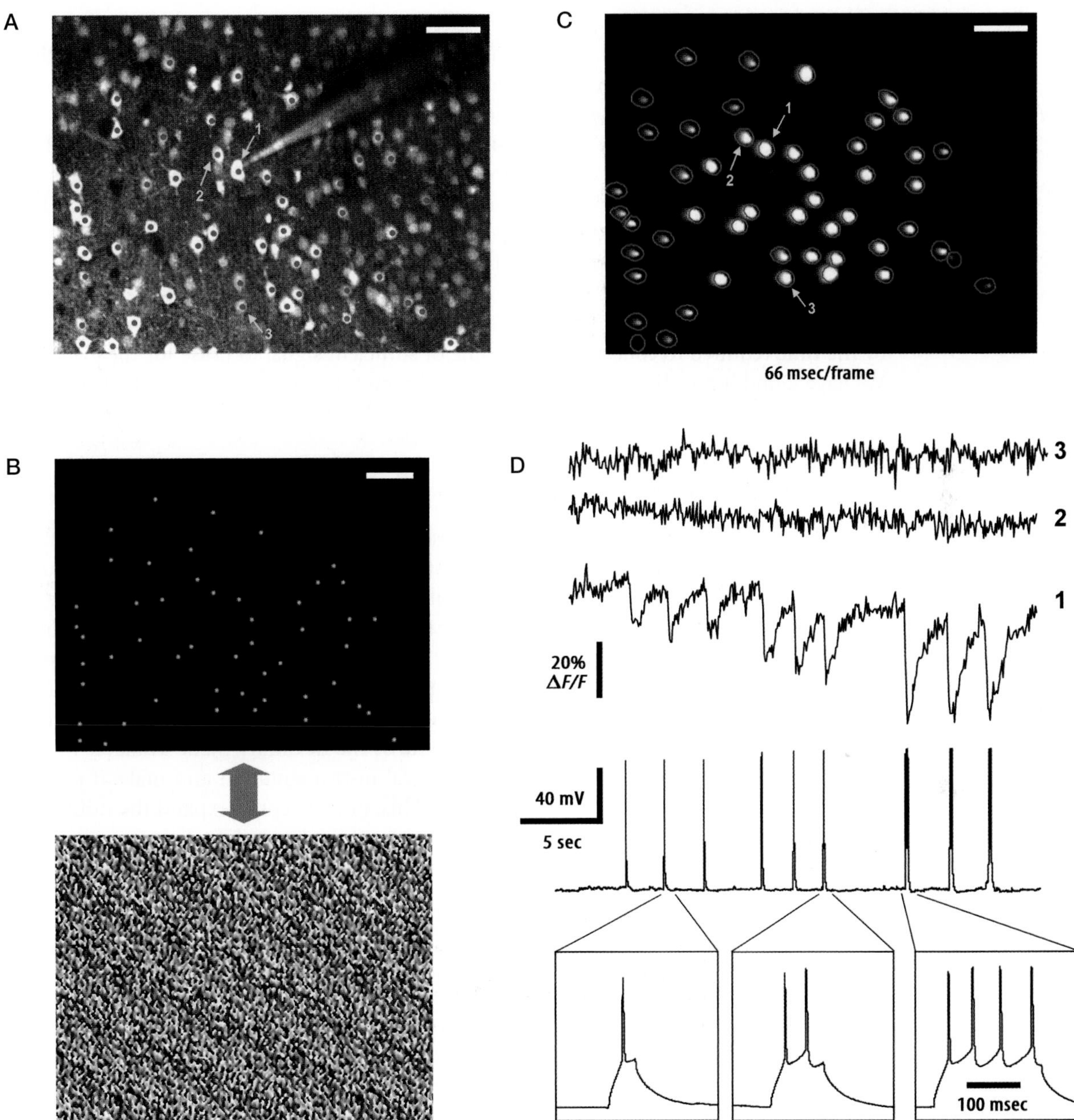

FIGURE 5. SLM multibeam imaging. (*A*) A neocortical slice (L2/3, area S1, P15 mouse) was bulk loaded with a Ca^{2+} indicator (a 10:1 mixture of Fura-2AM and mag-Indo-1AM [Nikolenko et al. 2007]). The image shown was taken using a standard two-photon raster imaging mode (790-nm excitation). Fifty neurons were targeted for imaging using a diffractive SLM (red spots). One of the neurons (labeled "1") was targeted for patch-clamp recording to trigger action potentials using current injection. The intracellular solution contained 50-μM Fura-2AM pentapotassium salt, a concentration that corresponds roughly to the intracellular concentration of Fura-2AM achieved by bulk loading (Peterlin et al. 2000). The pipette also contained 10-μM Alexa 594 for localization of the patched neuron using a different emission filter. (*B*) Command image file for SLM software and corresponding phase mask. (*C*) Image of two-photon fluorescence from multiple locations obtained with a camera. The diffractive SLM splits the laser beam to continuously illuminate spatially different locations with a static pattern (~4.4 mW of average excitation power per spot on the sample plane). Red contours were detected using custom software to quantify time-lapsed signals from different cells. Scale bar, 50 μm. (*D*) Calcium signals recorded from cells identified in *A* and *C* and corresponding to different numbers of action potentials elicited in cell 1 (nine current pulses that triggered triplets of one, two, and four action potentials, respectively, are shown). Even individual spikes can be detected with a good signal-to-noise ratio. Neurons 2 and 3 were not stimulated and did not show any change in fluorescent signals. Imaging was performed with ~15-frames/sec temporal resolution (66 msec/frame). No noticeable photobleaching or photodamage was observed over the course of the experiment (several minutes of continuous illumination). (Modified, with permission, from Nikolenko et al. 2008.)

imaging targets (Fig. 5A) within brain slices loaded with the calcium indicator Fura-2-acetoxymethyl ester (Fura-2AM). The phase pattern generated from the coordinates of these cellular locations (Fig. 5B) was used to create the diffractive pattern of excitation light. The resultant fluorescence was imaged using an EMCCD camera (Fig. 5C).

In the experiment shown, one neuron was whole-cell-patched with 50-µM Fura-2 to approximate the dye concentration in other neurons, and varying numbers of action potentials were evoked in the patched neuron. Action potential–related fluorescence transients were faithfully reported, with single spikes clearly detectable (Fig. 5D).

It should be pointed out that there is no scanning in this imaging mode because all points of interest are continuously illuminated. Thus the maximum possible temporal resolution of the microscope is defined only by the frame rate of the detector. The practical temporal resolution is set by the time required to collect a single frame of sufficient signal to noise for the given experiment. We have found it possible to obtain good two-photon Ca^{2+} signals at rates as high as 60 frames/sec.

CONCLUSION

A novel design of an optical microscope is presented based on an SLM that changes only the phase of light and redistributes the total available power among the points of interest. The utility of this approach is shown for simultaneous stimulation of multiple dendritic spines and neurons (Nikolenko et al. 2008) using two-photon uncaging of excitatory neurotransmitters. In addition, a new imaging paradigm has been designed that takes advantage of the selective illumination of the points of interest and combines it with fast imaging using a camera as the detector. This makes it possible to achieve high rates of signal acquisition not easily available with other nonlinear imaging techniques. The SLM microscope highlights some of the flexibility made available by this "universal" optic, and which could greatly simplify optical instrumentation and make it possible to construct miniaturized lightweight optical devices. This, in turn, could expand the utility of nonlinear imaging and photostimulation methods for a variety of in vivo and medical uses.

ACKNOWLEDGMENTS

We thank the National Eye Institute, the HHMI, and the Kavli Institute for funding, and we also thank the members of the Yuste laboratory for support and help.

REFERENCES

Araya R, Eisenthal KB, Yuste R. 2006a. Dendritic spines linearize the summation of excitatory potentials. *Proc Natl Acad Sci* **103:** 18799–18804.

Araya R, Jiang J, Eisenthal KB, Yuste R. 2006b. The spine neck filters membrane potentials. *Proc Natl Acad Sci* **103:** 17961–17966.

Araya R, Nikolenko V, Eisenthal KB, Yuste R. 2007. Sodium channels amplify spine potentials. *Proc Natl Acad Sci* **104:** 12347–12352.

Bauschke HH, Combettes PL, Luke DR. 2002. Phase retrieval, error reduction algorithm, and Fienup variants: A view from convex optimization. *J Opt Soc Am A* **19:** 1334–1345.

Cash S, Yuste R. 1998. Input summation by cultured pyramidal neurons is linear and position-independent. *J Neurosci* **18:** 10–15.

Hopt A, Neher E. 2001. Highly nonlinear photodamage in two-photon fluorescence microscopy. *Biophys J* **80:** 2029–2036.

Koester HJ, Baur D, Uhl R, Hell SW. 1999. Ca^{2+} fluorescence imaging with pico- and femtosecond two-photon excitation: Signal and photodamage. *Biophys J* **77:** 2226–2236.

Kuznetsova TI. 1988. On the phase retrieval problem in optics. *Sov Phys Usp* **31:** 364–371.

Losonczy A, Magee JC. 2006. Integrative properties of radial oblique dendrites in hippocampal CA1 pyramidal neurons. *Neuron* **50:** 291–307.

Lutz C, Otis TS, DeSars V, Charpak S, DiGregorio DA, Emiliani V. 2008. Holographic photolysis of caged neurotransmitters. *Nat Methods* **5:** 821–827.

Nikolenko V, Poskanzer KE, Yuste R. 2007. Two-photon photostimulation and imaging of neural circuits. *Nat Methods* **4:** 943–950.

Nikolenko V, Watson BO, Araya R, Woodruff A, Peterka DS, Yuste R. 2008. SLM microscopy: Scanless two-photon imaging and photostimulation with spatial light modulators. *Front Neural Circuits* **2:** 1–14.

Peterlin ZA, Kozloski J, Mao BQ, Tsiola A, Yuste R. 2000. Optical probing of neuronal circuits with calcium indicators. *Proc Natl Acad Sci* **97:** 3619–3624.

Wang S, Szobota S, Wang Y, Volgraf M, Liu Z, Sun C, Trauner D, Isacoff EY, Zhang X. 2007. All optical interface for parallel, remote, and spatiotemporal control of neuronal activity. *Nano Lett* **7:** 3859–3863.

Yuste R. 2011. Circuit mapping with two-photon uncaging. In *Imaging in neuroscience: A laboratory manual* (ed. Helmchen F, Konnerth A). Cold Spring Harbor Laboratory Press, Cold Spring Harbor, NY (in press).

57 Temporal Focusing Microscopy

Dan Oron and Yaron Silberberg

Department of Physics of Complex Systems, Weizmann Institute of Science, Rehovot 76100, Israel

ABSTRACT

Axial localization of multiphoton excitation to a single plane is achieved by temporal focusing of an ultrafast pulsed excitation. We take advantage of geometrical dispersion in an extremely simple experimental setup, where an ultrashort pulse is temporally stretched and hence its peak intensity is lowered outside the focal plane of the microscope. Using this, out-of-focus multiphoton excitation is dramatically reduced, and the achieved axial resolution is comparable to line-scanning multiphoton microscopy for wide-field excitation and to point-scanning multiphoton microscopy for line excitation.

A detailed description of the considerations in choosing the experimental parameters, as well as the alignment of a temporal focusing add-on to a multiphoton microscope, is provided. We also review current advances and applications for this technique.

INTRODUCTION

The ability to perform optical sectioning is one of the great advantages of laser-scanning multiphoton microscopy and photoactivation systems. This introduces, however, a number of difficulties caused by the scanning process, such as lower frame rates caused by the serial acquisition process. Whenever these need to be circumvented, temporal focusing multiphoton microscopy makes it possible to confine multiphoton excitation to a single plane completely without scanning. The method relies on utilization of geometrical dispersion to compress the ultrashort excitation pulse as it propagates through the sample, reaching its shortest duration at the focal plane, before stretching again beyond it. This chapter outlines the requirements and the limitations of this technique and presents a brief summary of recent applications, including scanningless depth-resolved microscopy, improved depth resolution in line-scanning multiphoton microscopy, and confined photoactivation within a single two-dimensional plane.

AXIALLY RESOLVED MICROSCOPY

The confocal microscope, invented by Minsky more than 40 years ago (Minsky 1961), marked the dawn of an era of significant departure from the traditional principles of optical microscopy. In this new generation of microscopes, unlike their more traditional counterparts, an optical image of the sample is formed by scanning the sample point by point with a tightly focused illumination beam. The scattered laser light, or, more commonly, fluorescence induced by it, is detected through a confocal pinhole and collected to form an image. The main advantage of confocal microscopes is their optical sectioning capability. The introduction of multiphoton optical processes as imaging modalities in laser-scanning microscopy offers another mechanism for obtaining optical sections (Sheetz and Squier 2009, and references therein). Because of the nonlinear dependence of the signal on the illuminating electric field, multiphoton scattering is observed only from regions in which the excitation field peaks (i.e., from the focal volume). This inherent mechanism for rejection of out-of-focus scattering eliminates the need for a confocal pinhole in the detection path. To date, a variety of multiphoton processes has been used for imaging, including both incoherent processes such as two-photon and three-photon fluorescence and coherent processes such as second-harmonic and third-harmonic generation and coherent Raman processes (see Chapters 7, 46, and 48). More recently, multiphoton photoactivation has become an important tool in biosciences (Nikolenko et al. 2007), enabling either controlled uncaging of a biologically active molecule in a particular spot within the specimen or as a tool for switching fluorescent molecules between an emissive and a non-emissive state. In the proper configuration, the latter method can enable subdiffraction-limited localization of the fluorophores.

In both confocal microscopes and multiphoton ones, however, optical sectioning capability comes at a cost—not all points in an image are simultaneously illuminated. For imaging purposes, this results in longer image acquisition times: in the range of a few tens of milliseconds per section with current commercial systems. When photoactivating caged molecules, serial excitation limits the number of points at which uncaging can occur within a given time window. This is an inherent limitation of laser-scanning microscopy that significantly limits its utility for imaging and control of fast dynamics.

Most methods developed to increase the image acquisition rate in multiphoton microscopes rely on illumination of more than one point at a time or "space multiplexing." As long as the excited points are sufficiently far from each other, depth resolution will not be significantly affected. The simplest example is the use of single-axis scanning combined with line illumination (Brakenhoff et al. 1995). The primary disadvantage of such a simple implementation is degradation in depth resolution. Rotation of a lenslet array is a more complex implementation of single-axis scanning that, when using a properly designed optical element, does not compromise depth resolution (Bewersdorf et al. 1998). Multiple-point illumination can also be accomplished using beam-splitter arrays (Fittinghof et al. 2000; Nielsen et al. 2001) and digital holography (Sacconi et al. 2003).

TEMPORAL FOCUSING: THE WORKING PRINCIPLE

The use of a pulsed excitation field as in multiphoton microscopy offers yet another degree of freedom for improving the depth resolution in space-multiplexed microscopy—the temporal degree of freedom.

Taking into account the temporal shape (more specifically, the duration) of the excitation pulse, it is, in principle, possible to imagine a multiphoton microscopy scheme in which time and space exchange roles: The sample is illuminated by a pulsed plane wave whose duration becomes shorter as it propagates, reaches a temporal focus at the focal plane of the temporal lens, and then increases by further propagation. This scheme is shown in Figure 1A along with the conventional spatial focusing technique. Such a scheme can be used to achieve axially resolved excitation even for simultaneous illumination of the entire plane (Oron et al. 2005).

Optical Setup for Temporal Focusing

The experimental apparatus functions as a lens in which time and space exchange roles. A perfect lens focuses light along the directions perpendicular to the propagation direction without affecting

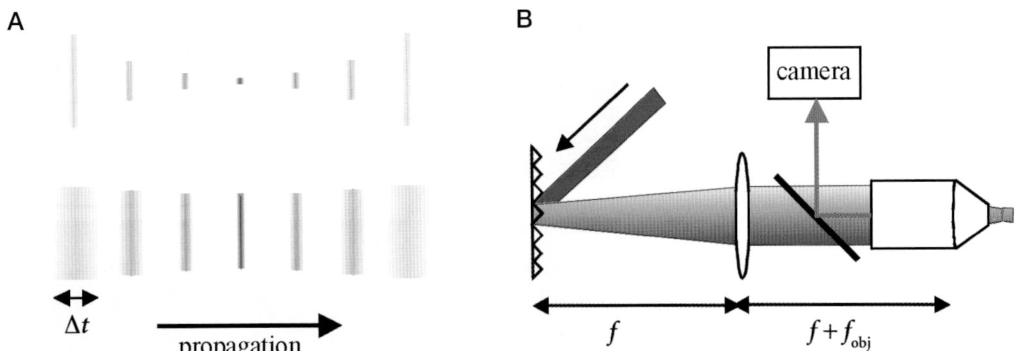

FIGURE 1. (*A*) The concept of temporal focusing. In standard point-scanning multiphoton microscopy, the beam is focused to a point (*top*), where the intensity is high enough to generate nonlinear scattering. In a temporally focused beam (*bottom*), the pulse reaches its shortest duration at the focal plane, so that the multiphoton signal is generated preferentially at the entire focal plane. (*B*) The optical setup for temporal focusing microscopy. A short-pulse laser beam is scattered by a grating and is demagnified by a telescope into the specimen.

its temporal profile (i.e., the duration of a light pulse is maintained along the propagation). By analogy, a pulsed optical field passing through the time lens is temporally compressed as it propagates, reaching its shortest duration (and highest peak intensity) at the temporal focal plane, before stretching again beyond it. This should be performed while in effect maintaining the spatial profile of the pulse. For this to be practical, the depth of the temporal focus should be as short as several micrometers, implying that dispersion has to be of geometrical origin. The setup for this, shown in Figure 1B, is closely related to the 4f grating-based pulse compressor, a common device in ultrafast optics applications (Oron et al. 2005).

The setup is composed of a grating imaged onto the sample by a high magnification telescope made with an achromatic lens and a microscope objective. The grating is illuminated by an ultrafast laser beam at an angle that enables the laser's central frequency to be scattered toward the optical axis of the telescope. At the back focal plane of the objective, the pulse is separated into its constituent frequency components (colors). After the objective lens, each frequency propagates at a different angle, which introduces a relative phase shift between them because of propagation. It is only at the focal plane of the objective that all frequency components are in phase, reconstructing the original short pulse. Where it is out of focus, the relative phase is quadratic in frequency, corresponding to group velocity dispersion (GVD). Hence, outside the Rayleigh range of the objective lens, the pulse is temporally stretched, leading to weaker multiphoton excitation. For microscopy applications, the two-photon fluorescence image can be epi-detected and imaged onto a camera. An example of such a set of axially resolved images of a *Drosophila* egg chamber, obtained without any scanning, is given in Figure 2.

Axial Resolution in Temporal Focusing

Temporal focusing has been shown to be equivalent, in terms of the axial response, to line focusing (i.e., spatial focusing along a single spatial dimension). For both coherent and incoherent multiphoton microscopies, this shows a full width at half-maximum (FWHM) axial resolution worse by only less than a factor of 2 but with a significantly increased signal at the tails. This applies, however, only to the case of a uniformly illuminated grating.

Improvement of the axial response can be achieved by combining temporal focusing with spatial focusing along one axis (Tal et al. 2005; Zhu et al. 2005). The end result, in this case, is an illuminated line at the image plane while the axial resolution of point-scanning microscopy is maintained. In practice, the simplest implementation is by illuminating the grating with a single line perpendicular to the grating grooves. Alternatively, the achromat of Figure 1B can be replaced with a cylindrical lens of a similar focal length.

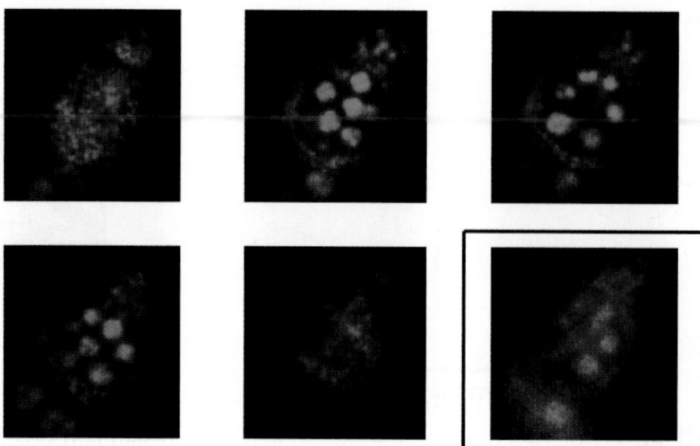

FIGURE 2. Scanningless depth-resolved images of a *Drosophila* egg chamber stained with DAPI (4′,6-diamino-2-phenylindole), a fluorescent DNA-binding probe. The images are optical sections of a *Drosophila* egg chamber containing 15 nurse cells and a single oocyte and wrapped by a layer of follicle cells. The images move spatially from the bottom of the egg chamber (*upper left* image) to its top (*lower middle* image). The area of each image is 140 x 140 µm. Images are separated by 7.5 µm. The *lower right* image shows a standard wide-field two-photon excited fluorescence image, to be compared with the temporally focused image directly above it, both focused at the same depth. Although some detail can be seen in the standard image, strong out-of-focus background dominates; a background that is practically eliminated in the temporally focused image.

In general, the axial response depends on the details of the illuminated pattern on the grating. For example, if the illumination pattern consists of several parallel lines, the axial resolution varies between that of point scanning and that of line scanning, depending on the distance between the lines. A full analysis of the effect of the excitation pattern on the axial resolution can be found in Papagiakoumou et al. (2009).

Choice of System Parameters for Temporal Focusing

To implement temporal focusing, care has to be taken to choose the right combination of grating, lens, and objective to fit the laser pulse parameters. Because, as in standard laser-scanning microscopy, the axial resolution stems from the spread of illumination angles at the focal plane, it is essential that the excitation pulse spectrum fill the back aperture of the objective, having a diameter d_{obj}. This dictates a relation between the pulse bandwidth $\Delta\lambda$, the groove density of the grating, k_g, and the focal length of the achromat, f:

$$\frac{f \cdot \Delta\lambda \cdot k_g}{d_{obj}} \approx 1.$$

For a typical 100-fsec Ti:sapphire oscillator, for which $\Delta\lambda \sim 10$ nm, and assuming $k_g = 1200$ lines/mm and $d_{obj} = 6$ mm, this corresponds to $f = 50$ cm. Because the pulse spectrum varies smoothly, the axial resolution only weakly depends on small deviations from the value of unity.

Once the focal length of the achromatic lens has been set, one needs to determine the required illuminated area on the grating. Because the system presented in Figure 1B is essentially a magnifying telescope, the illuminated area is simply smaller by a magnification ratio $M = f/f_{obj}$ than that on the grating. Again, for the above system parameters, with $f = 50$ cm, this ratio corresponds to about three times the objective magnification. The maximal illuminated area is determined by the total available laser power and the required excitation power per diffraction-limited spot. For typical values—1 mW average power per spot and 1 W of laser power—about 1000 points, equivalent to a 30 x 30 array, can be simultaneously illuminated without inducing a significant increase in image acquisition times. For larger illuminated areas, the required acquisition time will scale quadratically with the imaged area.

One convenient outcome for the temporal focusing setup is that dispersion precompensation of the excitation pulse can be performed using the same setup. Because the entire system is an asymmetric pulse compressor, a shift in the axial position of the grating introduces GVD, axially shifting the position of the temporal focus. For optimal imaging, the temporal focal plane has to coincide with the focal plane of the objective. To test this, a thin fluorescent sample should be axially scanned through the focal plane. When the grating is placed in the optimal position, the axial fluorescence response will be symmetric to small deviations from this position.

Line-Scanning Temporal Focusing

In line-scanning temporal focusing, the axial resolution is comparable with standard point-scanning multiphoton microscopes. This is an extremely useful and simple method for covering large areas. As discussed above, for typical laser parameters, the aspect ratio of the illuminated line can readily be as high as 1000, which, when combined with scanning along the orthogonal direction, covers a significant portion of the objective field of view. As discussed above, illumination of a single line can be achieved by replacing the achromat with a cylindrical lens (Zhu et al. 2005). For scanning, however, it is more beneficial to illuminate the grating with a line and to scan its position on the grating. This is readily achieved by a $2f$ combination of a galvanometric scanner and a cylindrical lens (Tal et al. 2005). The focal length of the cylindrical lens f_{cyl} is determined by the ratio of the laser beam diameter d_{laser} and the objective back aperture, such that

$$f_{cyl} = \frac{f \cdot d_{laser}}{d_{obj}},$$

ensuring that the objective back aperture is filled along both axes. In line-scanning temporal focusing, it is essential that the spatial focus and the temporal focus spatially overlap for optimal axial resolution. This may require more delicate alignment. In addition, achievement of optimal axial resolution may require external dispersion precompensation, unlike the case of plane illumination. Alternatively, the cylindrical lens should be slightly axially shifted to overlap the two foci. Using this setup along with a sufficiently strong fluorescent signal, video-rate axially resolved imaging can be readily achieved.

Generation of Arbitrary 2D Excitation Patterns

For a variety of applications, particularly photoactivation of caged compounds, it is useful to illuminate an arbitrarily shaped region in an axially resolved manner for the purpose of spatially confining multiphoton excitation to a single cell (Lutz et al. 2008). Because, in temporal focusing, the illumination pattern on the grating is practically imaged (and magnified) in an axially resolved manner onto the sample, the task is practically reduced to that of generating the corresponding magnified image on the grating. This requires conversion of the near-Gaussian spatial mode of the laser to an arbitrary shape (Papagiakoumou et al. 2008).

The problem of conversion from one intensity pattern to another is a classical one in Fourier optics (Goodman 2005). The best known solution is to place a 2D phase spatial light modulator (SLM) in the back focal plane and the grating in the front focal plane of a lens and to use an iterative algorithm (such as Gerchberg–Saxton) to generate the required pattern. Alternatively, a 2D amplitude SLM can be imaged onto the grating. The latter approach results in some speckle pattern in the image as well as slightly reduced axial resolution. The former approach, depending on the required shape, wastes more laser excitation power. If the presence of speckle interferes with the application, it can be smoothed by the introduction of a rotating diffuser placed in proximity to the grating.

Image Detection in Temporal Focusing Microscopy

Temporal focusing is an imaging modality that can be used in conjunction with any multiphoton microscopy technique, either coherent or incoherent. By far, the most common imaging modality is two-photon fluorescence. In this case, fluorescence from the excited plane is most easily collected by epi-detection and a camera. The choice of camera should be determined by the nature of the measurement. Electron-multiplying charge-coupled devices (CCDs) are optimal for imaging fast dynamics at low light levels. For slower imaging or higher levels of signal, cooled CCDs are probably the best choice. Because the signal is detected by an imaging detector, it should be noted that, unlike with point-scanning multiphoton microscopy, the image quality deteriorates on scattering in the sample. Hence, temporal focusing is inappropriate for imaging deep into scattering tissue.

For coherent multiphoton imaging modalities, such as harmonic generation or coherent Raman scattering, the generated signal is mostly directed in the forward direction. Hence, the sample should be placed between two objective lenses. Epidetection of coherent signals usually selects only for coherent scattering from subwavelength objects.

Troubleshooting and Practical Limitations

Aligning a temporal focusing setup is relatively simple and forgiving of small alignment errors. Care should be taken to align the grating perpendicular to the optical axis of the telescope because the temporal focal plane is parallel to the grating surface. Care should also be taken when implementing temporal focusing with very short laser sources (<20 fsec). In this case, higher-order dispersion (i.e., third and above) should be precompensated, as it results in broadening of the axial response. Because the grating surface is illuminated at an angle, a circular laser beam profile results in an elliptic illumination pattern on the grating. Hence, a cylindrical telescope (before the temporal focusing setup) should be used to generate circular illumination patterns in the sample plane. Generally, the axial response should be equivalent to that of line-scanning multiphoton microscopy for wide-field temporal focusing and to that of point scanning for line-scanning temporal focusing. A broader axial response is usually accompanied by asymmetry and is caused by incorrect positioning of the grating along the telescope axis or is caused by an axial shift between the spatial focus and the temporal focus for line-scanning temporal focusing microscopy.

Other Applications of Temporal Focusing

Temporal focusing has recently been used in several novel imaging situations. One is rapid axial scanning. A shift in the axial position of the temporal focus corresponds to the introduction of GVD to the excitation pulse. In a recent experiment, an acousto-optic deflector was used to control the GVD, inducing a shift of the axial position of the temporal focal plane within 10 µsec (Rui et al. 2009), which is orders of magnitude faster than standard mechanical scanning devices such as direct current motors and piezoelectric crystals. Using the GVD dependence of the focal position, axial scanning was recently shown also through a fiber delivery system (Durst et al. 2006).

Selective photoactivation in a single plane has also been used to perform 3D subdiffraction-limited imaging via photoactivated light microscopy (PALM). The standard geometry for PALM limits the photoactivation volume using TIRF (total internal reflection fluorescence) excitation. This is because out-of-focus background from a thick specimen usually overwhelms the fluorescent signal from the objective focal plane. By using temporally focused two-photon photoactivation, out-of-focus background can be suppressed by orders of magnitude, enabling practical multilayer PALM (Vaziri et al. 2008).

The combination of temporal focusing with Fourier-domain spectral-pulse-shaping techniques has been used to improve the axial resolution of temporal focusing–based imaging (Oron and Silberberg 2005). This is based on a lock-in technique, relying on the fact that out-of-focus background is only weakly dependent on the pulse shape, whereas the signal from the focal plane strongly depends on it. By subtracting signals obtained with two different pulse shapes, it is possible to eliminate out-of-focus background and to achieve some narrowing of the FWHM axial response.

CONCLUSION

Temporal focusing is a versatile technique for localizing multiphoton excitation to a single plane and, as such, can replace many alternative space-multiplexing techniques in multiphoton microscopy, such as line-scanning, microlens arrays, beam-splitter arrays, and holographic beam splitters. Currently, it is the only available wide-field axially resolved multiphoton technique. Various ways to implement multiphoton imaging and photoactivation have been shown, but the possible range of applications is clearly broader, including, for example, multiphoton lithography and space-multiplexed fluorescence correlation spectroscopy. The alignment of a temporal focusing system is relatively simple, and the required components are available in practically any laboratory utilizing ultrafast laser excitation. The technique can be practically implemented with any short-pulse (~<200 fsec) laser source.

REFERENCES

Bewersdorf J, Pick R, Hell SW. 1998. Multifocal multiphoton microscopy. *Opt Lett* **23:** 655–657.

Brakenhoff GJ, Squier J, Norris T, Bliton AC, Wade MH, Athey B. 1995. Real-time two-photon confocal microscopy using a femtosecond, amplified, Ti:sapphire system. *J Microsc* **181:** 253–259.

Durst ME, Zhu G, Xu C. 2006. Simultaneous spatial and temporal focusing for axial scanning. *Opt Express* **14:** 12243–12254.

Fittinghoff DN, Wiseman PW, Squier JA. 2000. Widefield multiphoton and temporally decorrelated multifocal multiphoton microscopy. *Opt Express* **7:** 273–279.

Goodman WJ. 2005. *Introduction to Fourier optics*, 3rd ed. Roberts and Company, Greenwood Village, CO.

Lutz C, Otis TS, de Sars V, Charpak S, DiGregorio DA, Emiliani V. 2008. Holographic photolysis of caged neurotransmitters. *Nat Methods* **5:** 821–827.

Minsky M. 1961. Microscopy apparatus. U.S. Patent #3,013,467.

Nielsen T, Fricke M, Hellweg D, Andresen P. 2001. High efficiency beam splitter for multifocal multiphoton microscopy. *J Microsc* **201:** 368–376.

Nikolenko V, Poskanzer KE, Yuste R. 2007. Two-photon photostimulation and imaging of neural circuits. *Nat Methods* **4:** 943–950.

Oron D, Silberberg Y. 2005. Spatiotemporal coherent control using shaped, temporally focused pulses. *Opt Express* **13:** 9903–9908.

Oron D, Tal E, Silberberg Y. 2005. Scanningless depth resolved microscopy. *Opt Express* **13:** 1468–1476.

Papagiakoumou E, de Sars V, Oron D, Emiliani V. 2008. Patterned two-photon illumination by spatiotemporal shaping of ultrashort pulses. *Opt Express* **16:** 22039–22047.

Papagiakoumou E, de Sars V, Emiliani V, Oron D. 2009. Temporal focusing with spatially modulated excitation. *Opt Express* **17:** 5391–5401.

Rui D, Kun B, Shaoqun Z, Derong L, Songchao X, Qingming L. 2009. Analysis of fast axial scanning scheme using temporal focusing with acousto-optic deflectors. *J Mod Opt* **56:** 81–84.

Sacconi L, Froner E, Antolini R, Taghizadeh MR, Choudhury A, Pavone FS. 2003. Multiphoton multifocal microscopy exploiting a diffractive optical element. *Opt Lett* **28:** 1918–1920.

Sheetz KE, Squier J. 2009. Ultrafast optics: Imaging and manipulating biological systems. *J Appl Phys* **105:** 051101.

Tal E, Oron D, Silberberg Y. 2005. Improved depth resolution in video-rate line-scanning multiphoton microscopy using temporal focusing. *Opt Lett* **30:** 1686–1688.

Vaziri A, Tang J, Shroff H, Shank CV. 2008. Multilayer three-dimensional super resolution imaging of thick biological samples. *Proc Natl Acad Sci* **105:** 20221–20226.

Zhu G, van Howe J, Durst M, Zipfel W, Xu C. 2005. Simultaneous spatial and temporal focusing of femtosecond pulses. *Opt Express* **13:** 2153–2159.

58 Caged Neurotransmitters and Other Caged Compounds
Design and Application

George P. Hess, Ryan W. Lewis, and Yongli Chen
Department of Molecular Biology & Genetics, Cornell University, Ithaca, New York 14853-2703

ABSTRACT

The approaches using caged neurotransmitters described here enable transient kinetic investigations to be made with membrane-bound proteins (receptors) on a cell surface with the same time resolution as was previously possible only with proteins in solution. Caged neurotransmitters also provide good spatial resolution to identify/locate receptors in cells, tissue slices, and organisms as described elsewhere in this series.

INTRODUCTION

Elegant statistical techniques, including recording single-channel currents (Sakmann and Neher 1983; Sakmann and Neher 1995), exist for investigating receptor-mediated reactions on cell surfaces and provide valuable information about the ion specificity, conductance, and lifetime of the open channel. In single-channel current measurements the receptors and ligands are in a *quasi* equilibrium. However, there remain interesting questions that can be answered if, before initiating receptor-mediated reactions, one can equilibrate ligands with the receptors on cell surfaces in times short compared to channel opening and desensitization, thus greatly improving the temporal resolution of the experiments. This goal can be accomplished by using photolabile, inert precursors of neurotransmitters ("caged" neurotransmitters) that can be equilibrated with cell-surface receptors before photolytically releasing the neurotransmitter, thus avoiding diffusional barriers. Once equilibrated, the caged neurotransmitter can be rapidly cleaved in the microsecond time region by a pulse of light of the appropriate wavelength and energy, thus releasing free neurotransmitter. Caged compounds can also provide spatial resolution (Li et al. 1997), depending on the area illuminated and the duration of illumination.

Photocleavable protecting groups for biologically important compounds have many uses. This is particularly true of cases in which access of a compound to its reaction partner is slow, but the induced reaction is fast (Kaplan et al. 1978; McCray and Trentham 1989; for reviews, see Adams and Tsien 1993; Corrie and Trentham 1993; Hess 1993; Nerbonne 1996). Several common caging groups that are used in biological assays include 2-methoxy-5-nitrophenyl (MNP) esters (Ramesh et al.

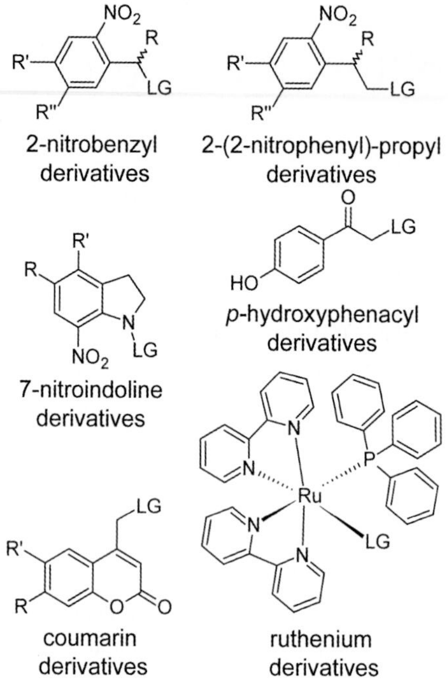

FIGURE 1. A noninclusive list of several generic caging groups that have been used in biological assays. The most commonly used caging group, or derivative thereof, is the 2-nitrobenzyl group. Both 7-nitroindoline and coumarin derivates are also widely used. (LG) Leaving group; (R) unspecified functional group.

1993; Niu et al. 1996c), *p*-hydroxyphenacyl derivatives (Park and Givens 1997), desyl-based compounds (Givens et al. 1998), coumarin esters (Bendig et al. 1997; Furuta and Iwamura 1998), and ruthenium complexes (Rial Verde et al. 2008; also see Chapter 60). The basic structure of several of these caging groups can be found in Figure 1.

EXPERIMENTAL CONSIDERATIONS

Photolabile, Biologically Inert Neurotransmitter Precursors (Caged Neurotransmitters)

Many photolabile protecting groups have been identified and studied for their synthetic properties (Bochet 2002), and several of these groups have become excellent tools for the study of biological systems (Mayer and Heckel 2006). The most frequently used protecting group is the 2-nitrobenzyl group, with various substituents (De Mayo 1960; Barltrop et al. 1966; Patchornik et al. 1970; McCray et al. 1980; Corrie and Trentham 1993; also see Chapter 59). The use of this protecting group was pioneered by Engels and Schlaeger (1977) and for biologically important phosphates by Kaplan et al. (1978) and McCray and Trentham (1989), and led to widespread use of "caging groups" with many other biological molecules, including neurotransmitters, enzyme substrates, cofactors, nucleic acids, oligonucleotides such as aptamers and siRNAs, specific residues of peptides and proteins, Ca^{2+}, phospholipids, steroids, hormones, and many others (for review, see Mayer and Heckel 2006).

The caging group and the functional group of the neurotransmitter to which it is attached, the photolysis characteristics, and the by-products of photolysis all play a role in determining whether a caged compound is satisfactory for a particular purpose. This is not, so far, predictable and must be determined experimentally. As an example, the αCNB-caged GABA (γ-aminobutyric acid) and other caged GABA molecules that are satisfactory for use with α1β2γ2L $GABA_A$ receptors inhibit α1β2δ $GABA_A$ receptors at the same concentration (K.P. Eagen, G.P. Hess, unpubl. data).

A systematic approach to the development of a caged compound and to its use in answering interesting biological questions is recommended.

1. The quantum yield and rate of photolysis of the caged compound at the wavelength to be used must be known. The quantum yield determines the maximum amount of neurotransmitter that can be released by photolysis. The photolysis rate determines how fast a reaction can be measured.
2. A functional assay must be used to determine whether the caged compound (before photolysis) or its photoproducts other than the desired compound are biologically inert in the system one wishes to use.
3. A method for calibrating the concentration of compound released upon photolysis is generally necessary.

Caged neurotransmitters must meet several important criteria.

1. They must be soluble in aqueous solutions and sufficiently stable at physiological pH before photolysis.
2. They must be photolyzed at a wavelength >335 nm to avoid cell damage.
3. Neither the caged compound nor the photolysis products, with the exception of the liberated neurotransmitter, should modify the receptor-mediated reaction being studied.

For use in transient kinetic experiments, caged neurotransmitters must also meet the following criteria.

4. They must be photolyzed in the microsecond time region so that photolysis is not rate limiting.
5. They must photolytically release the neurotransmitter with sufficient quantum yield to allow kinetic investigations to be made over a wide range of neurotransmitter concentration.

The quantum yield of a caged compound can be measured with several techniques, the simplest of which is photolysis of a sample of a caged molecule with pulses of light of known energy while monitoring spectroscopic changes in the absorption or fluorescence spectrum (Milburn et al. 1989). If spectroscopic changes are not observed, an alternative method is to analytically separate and quantify the caged and uncaged molecules by techniques such as high performance liquid chromatography (HPLC) (Milburn et al. 1989).

The rate of photolysis can be approximated from the data obtained in a quantum yield determination if the duration of a single light pulse is known. It can also be determined by measuring the rate of change of any observed transient absorption that occurs during photolysis, a method that is described in the literature (Walker et al. 1986, 2002; Milburn et al. 1989). All these criteria were kept in mind when we initiated the synthesis of photolabile inert precursors of neurotransmitters. We tried using derivatives of the 2-nitrobenzyl group to cage carbamoylcholine (Walker et al. 1986), a stable and well-characterized analog of acetylcholine with an amino group. The initial compounds were not suitable for rapid kinetic investigations because they were not biologically inert (Walker et al. 1986) or they photolyzed too slowly. However, when we introduced the use of the α-carboxy-2-nitrobenzyl group (α-CNB) (Milburn et al. 1989) to protect the carboxyl group of neurotransmitters, we obtained compounds that meet all the criteria listed above for transient kinetic investigations of the acetylcholine, glutamate (Wieboldt et al. 1994), kainate (Niu et al. 1996a), and GABA (Gee et al. 1994) receptors. In the case of the neurotransmitter glycine, we used the 2-methoxy–5-nitrophenol protecting group (MNP), creating a derivative that is photolyzed in the 1-μsec time region but that is not stable in aqueous solution (Patchornik et al. 1970; Ramesh et al. 1993). So we turned to β-alanine, which also activates the glycine receptor (Choquet and Korn 1988), to make an MNP-caged β-alanine that has all the desired properties for transient kinetic investigations of the glycine receptor (Niu et al. 1996c). It is important to note that if the ligand to be "caged" absorbs light at the same wavelength as the caging group, problems may be encountered (Breitinger et al. 2000).

Visible light–sensitive photolabile neurotransmitters have several advantages compared to their UV-sensitive counterparts because visible light is less damaging to the cells/receptors and because visible light flash lamps are more affordable than lasers. An additional advantage is that caged compounds photolyzable with visible light can be used for two-photon microscopy (Denk 1994). The neurotransmitters glutamic acid and glycine were initially caged (Shembekar et al. 2005, 2007) with a coumarin derivative that had been used previously to cage cAMP and cGMP (Hagen et al. 2001). The coumarin-caged glutamic acid and glycine can be photolyzed in the microsecond time region by visible light with a good quantum yield and are suitable for transient kinetic investigations (Shembekar et al. 2005, 2007). Several other visible light–sensitive compounds have been reported using various caging groups, including ruthenium complexes (Rial Verde et al. 2008), coumarin derivatives (Fan et al. 2009), and a 2-nitrobenzyl derivative (Banerjee et al. 2003).

Purification and Storage of Caged Compounds

Contamination of a caged compound by a small amount of the uncaged compound is one of the most frequently encountered problems in their use. Small amounts of uncaged neurotransmitter can desensitize the receptors during equilibration with the caged compound before photolysis. To avoid these problems, every caged neurotransmitter must be tested for purity (even if obtained from a commercial source), purified if necessary, and stored appropriately. The inertness of caged neurotransmitters we use are tested with the cell-flow technique (Udgaonkar and Hess 1987) described below. Testing the caged neurotransmitter in this manner easily identifies if a solution of caged compound is contaminated by free neurotransmitter that will activate the receptor. We find that this method is much more sensitive for detecting low concentrations of neurotransmitter than can be measured by separation techniques such as TLC, HPLC, and the like. If necessary, the caged compounds are purified using various chromatography techniques. Purification details are given in the pertinent references for each compound.

To avoid degradation of caged compounds and the generation of free neurotransmitter, precautions must be taken to protect the compounds from light, and they should be stored over a desiccant at –20°C to –80°C. Occasionally it is found necessary to work under controlled lighting conditions, choosing the lighting relative to the caged compound (e.g., red lamps for compounds sensitive to visible light). We store caged compounds protected from light in brown vials or amber Eppendorf tubes, wrapped in black electrical tape or aluminum foil. To avoid releasing free neurotransmitter as a result of hydrolysis, solutions of caged compounds should be prepared immediately before use. Thermal stability is also an issue for some compounds, in which case the solutions should be kept on ice until used.

Cell-Flow Technique: For Testing Biological Inertness of a Caged Neurotransmitter, Calibrating Concentration of Free Neurotransmitter Released by Photolysis, and Checking Cell Viability

The cell-flow technique (Hess et al. 1987; Udgaonkar and Hess 1987) consists of (i) the whole-cell current-recording technique (Hamill et al. 1981; Marty and Neher 1995), which allows one to determine the current arising from open receptor channels on the cell surface at constant voltage; (ii) a U-tube device (Krishtal and Pidoplichko 1980; Udgaonkar and Hess 1987) that allows solutions containing neurotransmitter to flow over a cell; and (iii) when necessary, a method to correct the current amplitude for receptor desensitization that occurs while the neurotransmitter equilibrates with the receptors (Hess et al. 1987; Udgaonkar and Hess 1987).

The cell-flow method is used for several purposes.

1. The first is to determine that the caged neurotransmitter itself is not an agonist or inhibitor of the receptor and that the receptor is not modulated by the protecting group after photolysis. This is done by determining the whole-cell current in (i) cell-flow measurements using a standard concentration of neurotransmitter, (ii) the same measurement in the presence of a large excess of caged neurotransmitter, and (iii) the same measurements after the cell has been exposed to caged neurotransmitter for several seconds.
2. During photolysis experiments the stability of the whole-cell patch may begin to deteriorate, causing a change in the current response of a cell to a standard agonist concentration. As a con-

trol measurement to determine the integrity of the whole-cell patch, a standard neurotransmitter concentration, typically a saturating concentration, is applied to cells before and after photolysis measurements. Equivalent responses to the standard solution verify that the receptors and whole-cell patch have not changed over the course of an experiment. If a change is detected, all the measurements after the initial control measurement are discarded and a new cell is used. Whole-cell patches in cell-flow experiments are typically stable enough that three independent cells may be sufficient to construct a full dose-dependent response curve covering a wide range of neurotransmitter concentration (Fig. 2D).

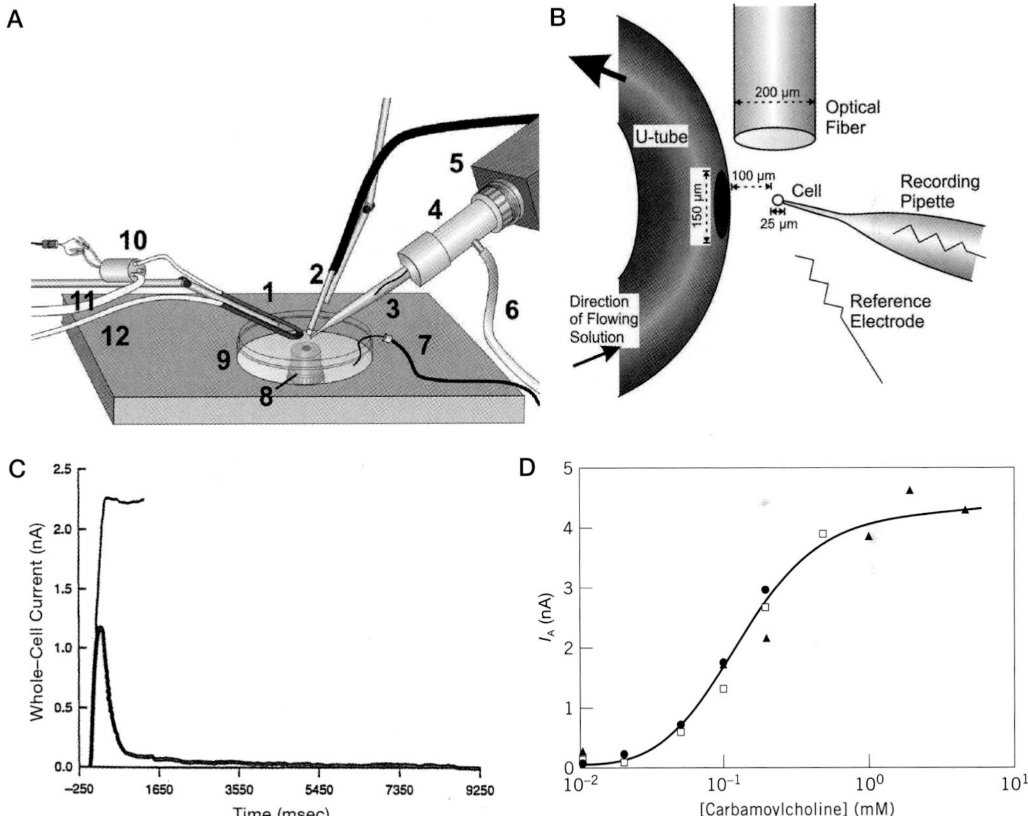

FIGURE 2. (*A*) A schematic drawing of the device used for the flash/laser-pulse photolysis technique. The components shown include the stainless steel U-tube (1), the fiber optic cable (2), a borosilicate recording pipette containing intracellular buffer and the recording electrode (3), the pipette holder (4), the head stage (5), the suction/vacuum tube (6), the reference electrode (7), the microscope (objective) for viewing cells and aligning the U-tube (8), a cell culture plate with cells expressing the receptor of interest (9), a three-port solenoid valve (10), 0.38-mm or 0.42-mm inner-diameter peristaltic tubing drawing solution away from the U-tube (11), and 0.25-mm or 0.5-mm inner-diameter peristaltic tubing with solution flowing toward the U-tube (12). The cell-flow technique requires all of the same components, omitting only the optical fiber. (*B*) A zoomed-in diagram depicting the alignment of components needed for the flash/laser-pulse photolysis technique. While the solenoid valve is open, solution is actively drawn away from the U-tube at a higher rate than that at which solution flows into the U-tube. The U-tube draws extracellular buffer in through the porthole from the dish, preventing any leakage or diffusion of ligand solution over the cell. When the solenoid valve closes, solution being pumped to the U-tube is forced out of the U-tube porthole and over the surface of the cell. Linear flow rates of 1–4 cm/sec are typically used. (*C*) Cell-flow measurement with a BC_3H1 muscle cell containing nicotinic acetylcholine receptors, pH 7.4, 23°C, –60 mV transmembrane potential. A 200-μm acetylcholine solution emerged from the cell-flow device (*A*) at a rate of 1 cm/sec. (Thick solid line) The observed current; (thinner line) the calculated current corrected for receptor desensitization. (Reprinted from Hess et al. 1987.) (*D*) Concentration dependence of the current amplitude corrected for receptor desensitization, I_A. BC_3H1 muscle cells, pH 7.4, 22°C–23°C, and –60 mV transmembrane potential. Data are from Udgaonkar and Hess (1987) (●, single–channel current recordings; ▲, cell-flow measurements) and Matsubara et al. (1992) (□, laser-pulse photolysis). The line indicating the concentration of open receptor channels was calculated from the constants pertaining to the channel-opening process determined in laser-pulse photolysis experiments (Matsubara et al. 1992). (Reprinted from Hess and Grewer 1998.) (*Continued on next page.*)

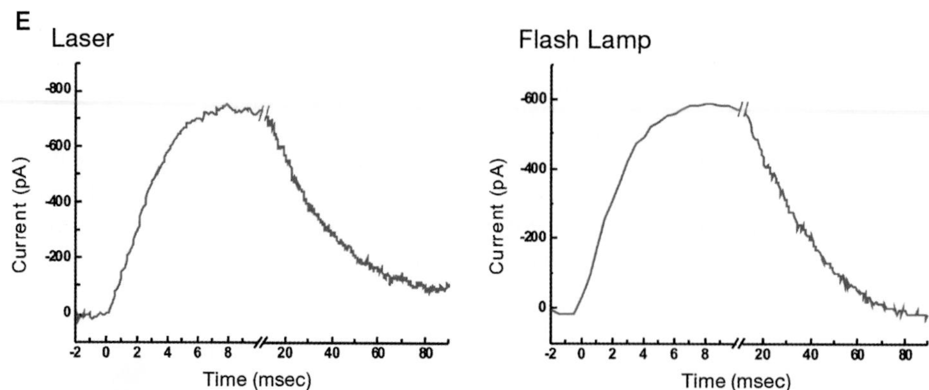

FIGURE 2 *(Continued)*. *(E)* Whole-cell currents induced by the photolytic release of ~100-μm glutamate from 2 mM coumarin-caged glutamate using a laser *(left)* or a flash lamp *(right)* were recorded from HEK 293 cells stably transfected with cDNA encoding GluR6 kainate receptors (Y. Chen, G.P. Hess, unpubl. data). The bath buffer contained 150 mM NaCl, 10 mM HEPES, and 1 mM $CaCl_2$. The pipette buffer contained 120 mM CsCl, 10 mM HEPES, and 10 mM EGTA. The buffers were adjusted to pH 7.4 using NaOH and CsOH, respectively. The experiments were performed at room temperature and a clamped voltage of –60 mV.

3. Neurotransmitter dose-dependent curves obtained with the cell-flow technique are needed for calibrating the concentration of neurotransmitter generated by photolysis of a caged precursor in measurements using the laser-pulse photolysis technique (Milburn et al. 1989; Matsubara et al. 1992; Hess and Grewer 1998). These estimates are obtained by comparing current responses from a known concentration of free neurotransmitter and the amplitude of the current evoked upon the photolysis of the caged compound, together with a dose–response curve.

WHOLE-CELL CURRENT-RECORDING TECHNIQUE

This variant of the patch-clamp technique is described in great detail by Hamill et al. (1981) and by Marty and Neher (1995).

Cell-Flow Device—U-Tube

Various flow devices have been used to flow neurotransmitter solutions over cells containing receptors (Krishtal and Pidoplichko 1980; Trussell and Fischbach 1989; Vyklicky et al. 1990; Franke et al. 1993; Edmonds et al. 1995). Problems are encountered with some devices, such as piezo-electric translators and theta tubes, and multi-barreled tubes. With many of the devices, the orientation between the porthole of the flow device and the cell is not maintained absolutely constant. Orientation is critical for reproducibility in kinetic experiments (Hess et al. 1987; Udgaonkar and Hess 1987; Niu et al. 1996b).

To be able to change the composition of the flowing solution during an experiment and still maintain a constant orientation, we modified (Niu et al. 1996b) the design of Krishtal and Pidoplichko (1980) and use stainless steel tubing. A diagram of the U-tube cell-flow device is shown in Figure 2, A and B. The porthole has a diameter of ~150 μm and is placed ~100–200 μm from a cell suspended from the recording electrode. In brief, the neurotransmitter solution emerges from the porthole of the device at a linear velocity of 1–3 cm/sec. A more rapid flow is a disadvantage because the integrity of the whole-cell seal between the cell and the electrode tends to deteriorate, and fewer measurements can be made with each cell.

One must also use a laminar flow of solution and avoid turbulence. Accordingly, the cell must be suspended in the flowing solution and must be nearly spherical. This is accomplished after making the whole-cell seal between the recording electrode and a cell by lifting the cell from the bottom of the culture dish so that the cell is suspended from the recording electrode (Udgaonkar and Hess 1987).

Alternatively, vesicles or patches of ~10-μm diameter can be pulled from the cell. Patch formation is described in detail in Sakmann and Neher (1995). The method for obtaining vesicles has been described by Walstrom and Hess (1994) and is similar to that described by Sather et al. (1992). In brief, a vesicle is obtained from a cell body by first making a whole-cell seal (Hamill et al. 1981) and then gently lifting the recording pipette until the membrane pinches off from the cell body, thus forming a vesicle. The vesicles typically have a diameter of ~10 μm and a capacitance of ~1–3 pF. A disadvantage of membrane patches is that the receptor concentration is considerably lower than in a whole cell, and consequently measurements must be made with more patches to achieve satisfactory statistics.

Correcting the Observed Whole-Cell Current for Receptor Desensitization

If the receptor being studied desensitizes during the current rise time, the amplitude of the maximum current recorded can be corrected to take into account the desensitization (Udgaonkar 1986; Hess et al. 1987). The correction is based on theories of solution flow over submerged spherical objects (Landau and Lifshitz 1959; Levich 1962) and on the observation that many cells become spherical when detached from the substratum. At the flow rates we use, 1–4 cm/sec, the rate-limiting step in equilibration of ligand with the cell surface is the velocity of a layer of the solution (which becomes the diffusion boundary ~2 μm above the cell surface [Levich 1962]) emerging from the flow device and from which the ligand diffuses to the cell surface. The buildup in ligand concentration on the cell surface below this solution layer is rapid (2–5 msec), depending on the flow rates used (Udgaonkar 1986; Hess et al. 1987). Knowing the ligand concentration on the cell surface allows one to correct the observed current during the current rise for receptor desensitization that occurs during the measurement (Udgaonkar 1986; Hess et al. 1987). Receptor desensitization is characterized by the rate coefficient α and can be measured independently in each experiment (Fenwick et al. 1982; Clapham and Neher 1984). The corrected current, I_A, is defined as the amplitude of the current arising from receptors on the cell surface in the absence of desensitization and at a definite ligand concentration (Udgaonkar 1986; Hess et al. 1987). To obtain the value of I_A from measurements of the observed current I_{obs}, we divide the current time course into constant (~5-msec) time intervals to take into account the equilibration time of small segments of the cell surface with ligand as the solution flows from the U-tube over the cell. The current is then corrected for the desensitization occurring during each time interval Δt. After n constant time intervals ($n\Delta t = t_n$), during each of which the current $(I_{obs})\Delta t$ is measured, the corrected current is given by (Udgaonkar 1986; Hess et al. 1987)

$$I_A = (e^{\alpha \Delta t} - 1)\Sigma(I_{obs})\Delta t_i + (I_{obs})\Delta t_{in}, \qquad \text{(A)}$$

where $(I_{obs})\Delta t_i$ is the observed current during the ith time interval and t_n is equal to or greater than the current rise time (Udgaonkar 1986; Hess et al. 1987). The value of I_A was found to be independent of the solution velocities used in the cell-flow method and can be determined with good precision (±10%) (Udgaonkar 1986; Hess et al. 1987). The observed current in a cell-flow experiment (solid line) and the current corrected for desensitization (thin line) are shown in Figure 2C.

The relationship between I_A and the concentration of receptors in the open-channel form in the absence of desensitization is given by Equation 1B in Table 1. I_M represents the current produced by 1 mol of open receptor channels, R_M the moles of receptors on the cell surface, and $(\overline{AL}_2)_o$ the fraction of receptors present that are in the open-channel form. In terms of the scheme at the head of Table 1 (Equation 1), $(\overline{AL}_2)_o$ is given (Cash and Hess 1980) by Equation 1A in Table 1.

Using the nicotinic acetylcholine receptor in BC3H1 cells as an example, Figure 2D shows the dependence of I_A [~$(AL_2)_o$] on carbamoylcholine concentration over a 500-fold range. The solid triangles represent results obtained by the cell-flow technique after the current was corrected for desensitization. The circles show the dependence of $(AL_2)_o$ on the carbamoylcholine concentration when determined by an entirely different approach and methodology. The probability P_0 that the channel is open while the receptor is in a nondesensitized state (Neher 1983; Ogden and Colquhoun 1985)

TABLE 1. Equations used in analysis of the mechanism of the nicotinic acetylcholine receptor in BC3H1 cells

$$A+L \underset{}{\overset{K_1}{\rightleftharpoons}} AL_2 \underset{k_{cl}}{\overset{k_{op}}{\rightleftharpoons}} \overline{AL}_2 \tag{1}$$

$$(\overline{AL}_2)_o = \frac{\overline{AL}_2}{A+AL+\overline{AL}_2} = \frac{L^2}{(L+K_1)^2\Phi + L^2} = P_o. \tag{1A}$$

In Equation (1A), A and AL_2 represent the receptor in the closed-channel form, and $(\overline{AL}_2)_o$ is the fraction of nondesensitized receptor molecules in the open-channel form. $\Phi^{-1} = k_{op}/k_{cl}$ is the channel opening equilibrium constant; k_{op} and k_{cl} are the rate constants for channel opening and closing, respectively. L represents the molar concentration of activating ligand, and K_1 is the dissociation constant of the neurotransmitter from the sites controlling channel opening. P_o is the conditional probability, determined in single-channel current recordings, that the receptor is in the open-channel form (Udgaonkar and Hess 1987).

$$I_A = I_M R_M(\overline{AL}_2)_o. \tag{1B}$$

In Equation (1B), I_A is the current due to open receptor channels in the cell membrane corrected for receptor desensitization, I_M is the current due to 1 mol of open receptor channels, and R_M represents the number of moles of receptor in the cell membrane. Equation (1C) is a linear version of Equation 1B (Hess et al. 1983):

$$[I_M R_M (I_A)^{-1} - 1]^{1/2} = \Phi^{1/2} + \Phi^{1/2} K_1 [L_1]_{-1}. \tag{1C}$$

The desensitization phase of the observed current is described by Equation (2A):

$$(I_{obs})_t - (I_{obs})_{t=\infty} = [(I_{obs})_{t=0} - (I_{obs})_{t=\infty}]e^{-\alpha t}. \tag{2A}$$

Here α represents the rate coefficient for receptor desensitization obtained from the falling phase of the current in cell-flow experiments (Udgaonkar and Hess 1987). I_{obs} represents the current during the falling phase. The subscripts t and ($t = 0$) refer to the time of measurement; ($t = \infty$) refers to the time when an equilibrium between active and desensitized forms has been reached. The exponential parameter α is described as

$$\alpha = \Phi\left[\frac{L k_{43} + 2K_2 k_{21}}{(L+2K_2)} + \frac{(L^2 k_{34} + 2K_1 k_{12})L}{L^2(1+\Phi) + 2K_1 L\Phi + K_1^2\Phi}\right]. \tag{2B}$$

Here k_{12}, k_{34} and k_{21}, k_{43} represent the rate constants for desensitization and resensitization, respectively. When k_{34} is the dominant rate constant, a simplified equation is obtained (Udgaonkar and Hess 1987):

$$\alpha = \frac{k_{43}\Phi L^2}{(L+K_1)^2\Phi + L^2}. \tag{2C}$$

The dissociation constant of the inhibitor from the nondesensitized receptor can be determined by both cell-flow and photolysis measurements. To simplify the equations, we use a ratio method, I_A/I_A', where I_A and I_A' represent the current maxima corrected for receptor desensitization in the absence and presence of inhibitor, respectively. The relationship between the observed inhibitor dissociation constant K and the inhibitor dissociation constant for the A, AL, AL_2, and $\overline{AL}_2$ receptor forms F_A, F_{AL}, F_{AL_2}, and $F_{\overline{AL}_2}$, represent the fraction of receptors in forms A, AL, AL_2, and $\overline{AL}_2$. I_0 and II_0 represent the concentrations of two different inhibitors, and K_I and K_{II} are the observed dissociation constants.

$$I_A/I_A' = 1 + I_0/K_1, \tag{3A}$$

$$\frac{1}{K_1} = \frac{F_A}{(K_1)_1} + \frac{F_{AL}}{(K_1)_2} + \frac{F_{AL_2}}{(K_1)_3} + \frac{F_{\overline{AL}_2}}{(K_1)_4}, \tag{3A'}$$

$$\frac{I_A}{I_A'} = 1 + \frac{I_0}{K_1}\frac{(2K_1 L + K_1^2)\Phi}{(L+K_1)^2\Phi + L^2}. \tag{3B}$$

Equation (3B): For a competitive inhibitor, $1/K_I$ is multiplied by the fraction of receptor molecules in the A form and the AL form.

$$\frac{I_A}{I_A'} = 1 + \frac{I_0}{K_1}(\overline{AL}_2)_o. \tag{3C}$$

Equation (3C): An inhibitor binding only to the open-channel form.

$$\frac{I_A}{I_A'} = 1 + \frac{I_0}{K_I}\frac{II_0}{K_{II}}. \tag{3D}$$

TABLE 1. *(Continued)*

Equation (3D): Two inhibitors, I_0 and II_0, binding to the same receptor site.

$$\frac{I_A}{I_A'} = 1 + \frac{I_0}{K_I} + \frac{II_0}{K_{II}} + \frac{I_0}{K_I}\frac{II_0}{K_{II}} = 1 + \frac{I_0}{K_I} + \frac{II_0}{K_{II}}\left(\frac{K_I + I_0}{K_I}\right). \tag{3E}$$

Equation (3E): Two inhibitors, I_0 and II_0, binding to two different receptor sites.

$$I_t = I_{max}[1 - \exp(-k_{obs}t)]. \tag{4A}$$

Equation (4A): In the laser-pulse photolysis experiments with BC_3H1 cells containing nicotinic acetylcholine receptors, the current rise time was observed to follow a single, exponential rate law over 85% of the reaction (Matsubara et al. 1992). In this equation, I_t represents the observed current at time t and I_{max} the maximum current. k_{obs} is the first-order rate constant for the current rise.

$$k_{obs} = k_{cl} + k_{op}\left(\frac{L}{L+K_1}\right)^2. \tag{4B}$$

Equation (4B): The relationship between the observed rate constant for the current rise k_{obs}, and k_{op}, k_{cl}, k_1 of the reaction scheme (see the beginning of this table).

$$k_{obs} = k_{cl}\frac{K_1}{K_1 + I_0} + k_{op}\left(\frac{L}{L+K_1}\right)^2. \tag{4C}$$

Equation (4C): k_{obs} in the presence of an inhibitor that binds only to the open–channel form of the receptor.

$$k_{obs} = k_{cl} + k\left(\frac{L}{L+K_1}\right)^2\frac{K_1}{K_1 + I_0}. \tag{4D}$$

Equation (4D): k_{obs} in the presence of an inhibitor that binds only to the closed–channel form of the receptor. If the inhibitor binds both to the open- and closed-channel forms, a combination of Equations 4C and 4D is obtained.

Reprinted from Hess and Grewer 1998.

The equations are based on the assumption that the binding of two ligand molecules (acetylcholine or carbamoylcholine) is required before the opening of the transmembrane channel (Katz and Thesleff 1957; Reynolds and Karlin 1978; Hess et al. 1983). A, AL, and AL_2 represent receptor forms with none, one, or two ligand molecules bound, and $\overline{AL}_2$ represents the open-channel form of the receptor (with two ligand molecules bound).

was determined at three carbamoylcholine concentrations. The P_0 values were obtained from single-channel current measurements (Neher and Sakmann 1976) and represent the fraction of time the channel is open while the receptor is in a nondesensitized state (Equation 1A in Table 1). The open squares represent results obtained by the laser-pulse photolysis technique (see below) (Milburn et al. 1989; Matsubara et al. 1992; Hess and Grewer 1998), the time resolution of which is sufficient so that the observed current obtained using the whole-cell current-recording technique does not have to be corrected for desensitization. The agreement between the results obtained using the three different techniques confirms the validity of the approach.

SETTING UP FOR WHOLE-CELL PATCH-CLAMPING

Brief Overview of Whole-Cell Patch-Clamping

Since the development of the patch-clamp technique, many variations and improvements have been made to improve electrophysiological measurements for the study of ion channels. Whole-cell patch-clamping procedures are thoroughly explained in Hamill et al. (1981), Sakmann and Neher (1995), and Waltz et al. (2002). Here, we give a brief description of how we measure whole-cell currents.

Cells expressing the receptor of interest are cultured and grown as required for the cell type used. We have used primary cells, the PC12 cell line with sympathetic ganglionic nicotinic acetylcholine receptors, the BC_3H1 cell line with endogenous muscle–type nicotinic acetylcholine receptors, and HEK293T cells transiently transfected with cDNAs encoding receptor subunits of interest. In the case of HEK293T cells transiently expressing recombinant receptors, we often cotransfect a plasmid with the cDNA for green fluorescent protein (GFP) as a transfection marker.

When cells are to be used in an experiment, the growth medium is removed and replaced with extracellular buffer. A borosilicate recording pipette is backfilled with intracellular buffer and inserted into the pipette holder by sliding it carefully over the recording electrode. A gigaohm seal between the pipette and the membrane of a cell is achieved by positioning the pipette on the surface of the cell membrane and applying a small amount of vacuum on the inside of the pipette through a tube connected to the pipette holder. The cell-attached state is easily turned into a whole-cell configuration by breaking the membrane between the cytosol and the intracellular buffer of the pipette with a brief and sharp increase in vacuum or a brief voltage transient (referred to as "zapping" the cell). The whole-cell configuration is then voltage-clamped with the patch-clamp amplifier to the desired potential across the cell membrane. The resulting whole-cell conformation is then used in both cell-flow and flash/laser-pulse photolysis techniques.

Preparing the Reagents and Electrodes

Cultured Cells

The cells used for whole-cell current recordings either must express the receptor of interest or be transfected with cDNAs leading to the expression of the receptor of interest. The cells are typically seeded and grown for at least 24 h on 35-mm cell culture dishes under the optimal conditions for the particular cells. The health of the cells is critical for successful whole-cell patch-clamping and current recording. Only the healthiest-looking cells should be used. The health of the cells is typically assessed by several criteria, including a "normal" cell morphology and a sharp/clearly defined cellular membrane.

Intracellular and Extracellular Buffers

There are no standardized buffer compositions for electrophysiology, or even for particular receptors (Walz et al. 2002). Buffer compositions have several roles; they must mimic the osmotic composition of the cytosol or extracellular environment, provide the ions needed for studying the receptor of interest (e.g., chloride ions in the case of $GABA_A$ receptors), and limit the conductance of ion channels that may be present in the cell membrane but that are not being studied. The pH of the solutions should be ~7, and the buffering capacity should be adequate if photolysis is likely to cause changes in the value. Deionized, distilled water should be used for all solutions. Buffers should be passed through sterile 0.22-µm filters and stored in sterile containers.

Recording and Reference Electrodes

Recording and reference electrodes are predominately made of silver wire that has been chloride-coated by either electroplating or chemical (bleach) treatment (Sakmann and Neher 1995). Both methods seem to work equally well, and the coating helps to stabilize the open electrode potential.

Recording pipettes are made from borosilicate capillaries (World Precision Instruments); the dimensions used are dependent on the currents to be measured. For instance, for $GABA_A$ receptors we use capillaries with a 1.5-mm outer diameter and a 1.12-mm internal diameter. We use a two-stage vertical pipette puller, finding this preferable to a horizontal puller, but either will work. The tips of the pulled pipettes are heat-polished (fire-polished) on a microforge. Heat-polishing is not required, but we find that it increases the likelihood of forming a stable gigaohm seal with the cell.

Setting Up the Equipment

An inverted microscope with good working distance above the stage is set on a vibration-resistant table and is surrounded by a Faraday cage to diminish background electromagnetic noise. A copper cage is ideal, but aluminum window screens can be used to make an inexpensive cage. If possible, place power supplies and lamps outside the Faraday cage to reduce noise and turn them off when not needed during current recording. A computer (with electrophysiological software for data acquisition, e.g., Clampex), digitizer, and oscilloscope (optional) are connected to a patch-clamp amplifier. A head stage connected to the amplifier is the connection point for the grounding electrode and the recording electrode/pipette holder. The pipette holder is also connected to a vacuum tube with a closing valve. Inclusion of a small 25–50-mL flask in the vacuum tube line increases the control over the degree of vacuum one can draw on the pipette. Vacuum is applied to the line either by mouth or by a syringe, depending on the preference of the experimenter. The entire head stage is mounted onto a micromanipulator, allowing for precise moment and positioning of the recording pipette when creating a membrane gigaohm seal. It is critical that all the equipment is electrically grounded. Electrically grounding the Faraday cage, microscope, and other equipment to a single ground source helps to eliminate ground loops and minimizes background noise.

Protocol

Cell-Flow Technique

In the cell-flow technique for ligand-gated ion channels, a peristaltic pump, a stainless steel U-tube, two different sizes of peristaltic tubing, and a solenoid valve are used to create a simple solution exchange system that can rapidly apply and remove solutions over the surface of a cell in tens of milliseconds. This system allows one to test multiple conditions on a cell containing the receptor of interest while constantly "washing" the cell with extracellular buffer solution between experimental applications. The use of the solenoid valve allows for the application of solutions to be precisely timed and controlled by a computer during electrophysiological current recording.

MATERIALS

Reagents

(See "Preparing the Reagents and Electrodes," above.)
Cultured cells expressing the receptor of interest
Extracellular buffer
Intracellular buffer

Equipment

(See "Setting Up the Equipment," above.)
Analog-to-digital converter ("digitizer"; Molecular Devices, Digidata 1322A)
Antivibration, floating table (table or table top)
Borosilicate recording pipettes of suitable dimensions

Pipettes can be made using a pipette puller (HEKA, PIP5), which can be programmed to pull pipettes of different geometries.

Cell culture dishes (35-mm; Corning)
Color-coded peristaltic pump tubing (Elkay, Krackler Scientific, or Cole Parmer); for example, 0.25-mm inner diameter (orange-blue) for the application of the solution combined with 0.38-mm inner diameter (orange-green) for the suction side, or tubing with comparable size ratios
Computer with appropriate sampling/acquisition software (Clampex)
Faraday cage
Gilson Minpuls 3 peristaltic pump, or similar
Headstage (Molecular Devices)
Inverted microscope
Microforge (Narishige, MF-830)
Micromanipulators (Narishige)
Oscilloscope (optional)
Patch-clamp amplifier (Molecular Devices, Axopatch 200B)
Pipette holder
Recording electrode
Reference electrode
Solenoid valve (Lee valve; Lee Co.)
Stainless steel Hamilton syringe needle tubing (outer diameter 300–400 μm; inner diameter

200–300 μm) for creation and assembly of the U-tube (for details, see Step 1)
Stand to hold manipulator and U-tube arm
U-tube arm with a clamp

METHOD

Making the U-Tube and Setting Up the Peristaltic Pump

1. Create a U-tube by bending the stainless steel Hamilton syringe needle tubing into a U-shape with a distance between the arms of ~5 mm. Drill a port hole with a diameter of 150 μm at the apex of the tube for solution to flow through (Fig. 1).

   Stainless steel tubing is used because the internal surface is more uniform than that of plastic or glass tubing.

2. Clamp the U-tube to a U-tube arm made of Plexiglas [poly(methyl 2-methylpropenoate)] to minimize electrical noise. To allow adjustments in the position of the U-tube during testing, insert the arm into a coarse manipulator that fastens either to the microscope or to the vibration-resistant table.

3. Set up the connections (a connection diagram is shown in Fig. 2A).

   i. Connect the inlet of the U-tube to the narrower peristaltic tubing with an internal diameter of 0.25 mm.

   ii. Connect the outlet of the U-tube by a very short (3–6-cm) tubing, also with an internal diameter of 0.25 mm, to the inlet of the solenoid valve.

   iii. Connect the outlet of the solenoid valve to the wider peristaltic tubing with a large internal diameter of 0.38 mm.

   This larger diameter causes a larger volume of solution to be drawn from the U-tube than the volume sent to the U-tube, thus causing extracellular buffer to be drawn through the porthole into the U-tube and preventing any experimental solutions from leaking onto the cell.

4. Place both the inlet and outlet peristaltic tubes onto the peristaltic pump head such that the inlet (narrower) and the outlet (wider) tubing flow solution to and from the U-tube, respectively. Using a syringe, fill the tubing with water, clamping the tubes on the peristaltic pump head, and insert the inlet tubing into the solution desired to run through the U-tube and the outlet tubing into a waste container.

5. Place a 35-mm dish containing water on the microscope stage, submerge the U-tube, and turn on the peristaltic pump. The flow of solutions should be as smooth as possible. Test the flow by allowing a small air bubble into the inlet tubing. The bubble should move at a smooth and constant rate, and the best place to look for this is in the tubing between the peristaltic pump head and the U-tube, where the solution is under slight back pressure.

   Fluctuations in the flow rate are usually caused by improper tightening of the peristaltic pump clamps. Overtightening not only causes poor and uneven flow of testing solutions out of the U-tube porthole, but it also decreases the lifetime of the peristaltic tubing.

6. Calculate and/or measure the solution flow rate, and adjust the rate to achieve a balance between rapid solution exchange and a gentle flow that will not damage a cell being tested.

7. Connect the solenoid valve with a BNC cable to a digital output on the digitizer. The solenoid valve can be controlled through the use of data acquisition software such as Clampex. Design an episodic protocol in the software that will switch the solenoid valve on for a given number of seconds (e.g., 1–4 sec), causing the water to flow out of the U-tube porthole, and then switch it off again.

   This protocol is used to test that the U-tube system is working correctly (as in Steps 8–14) and for the application of experimental solutions to a cell.

Testing the Flow of Solution

Now that the U-tube flow system is set up, it is important to test that the solution flowing through the U-tube and out of the porthole exchanges rapidly and is consistent.

8. Turn on the pump and draw distilled, deionized water through the inlet tubing.
9. Submerge the U-tube in a dish of extracellular buffer (replacing the dish of water), and bring the porthole of the U-tube into focus under the microscope.
10. Backfill a recording pipette with intracellular buffer, and insert the filled pipette into the pipette holder on the head stage.
11. Lower the tip of the pipette into the extracellular buffer in the dish, and bring it into focus using the coarse manipulator holding the head stage, making sure the electrode is in the buffer.
12. Using the fine manipulator, center the tip of the open-ended pipette ~100–200 μm in front of the U-tube porthole. Release any back pressure within the pipette holder that may be pushing solution out of the pipette tip by opening and then closing the valve used for applying a vacuum. Make sure the reference (grounding) electrode is also in the dish.
13. Set (but do not yet turn on) the holding potential of the amplifier to a moderate voltage, such as –60 mV. While in the voltage-clamp mode and metering current, adjust the resting conductance to 0 on the amplifier. Now switch on the negative holding potential; there should be a large negative conductance.
14. Run the acquisition software protocol created in Step 7 to close the solenoid valve for a short period while recording the current.

    If the flow is ideal, the current should sharply increase toward 0 pA, followed by a flat conductance level and then a rapid decrease in conductance back to the original conductance level at times corresponding to the protocol. This change in conductance is due to the flowing of non-conductive, pure deionized water out of the U-tube porthole and over the tip of the recording pipette. Ideally, the rising phase of the conductance change from 10% to 90% should only span 10 to 20 msec. If the flow is constant and without pulsing or pausing, the current observed should be perfectly flat during the application time.

15. If the flow looks good, start the flow of experimental solutions over a cell. Turn off the pump, and change the solutions to those to be used for measurements.

    To maintain and clean the U-tube system, it is a good practice to wash the tubing with distilled/deionized water after conducting experiments. This prevents the buildup of salt deposits and the growth of microbes within the tubing and the solenoid valve that may lead to irregularities in the flow of solutions.

Cell-Flow Measurements

16. Place a 35-mm culture dish of adherent cells in extracellular buffer on the microscope stage. Select a cell, establish a whole-cell patch-clamp configuration with the cell (see the section Brief Overview of Whole-Cell Patch-Clamping), and voltage-clamp the cell at the desired membrane potential.
17. While the cell is in the whole-cell configuration, gently lift the cell from the surface of the cell culture dish using the micromanipulator.

    This may require a great deal of patience for some cell types.

18. Lower the U-tube into the extracellular buffer in the dish, and turn on the peristaltic pump to start the flow of ligand solution through the U-tube.
19. Using the microscope and micromanipulators, bring into focus both the U-tube porthole and the cell at the end of the recording pipette. When both of these are in focus, the U-tube and the cell are aligned on the same vertical plane. Center the cell ~100–200 μm in front of the porthole of the U-tube.
20. Run a desired protocol in the acquisition software to apply the ligand solution to the cell suspended from the recording pipette.

FURTHER APPLICATIONS AND DISCUSSION

Kinetic measurements and localization studies can be carried out by using the photolysis technique. The apparatus used to perform these measurements is depicted in Figure 2A and includes the whole-cell current-recording instrumentation described in the protocol above, a fiber-optic cable, and a light source. Photolysis can be carried out using either a laser or a suitable light source.

Laser-Pulse Photolysis Technique

A variety of different lasers can be used as light sources for photolysis, depending on which models are available. The advantage of a dye laser is the relatively long laser pulse length (500 msec); experimentally we find that the cells are less stable when pulse lengths shorter than 600 nsec are used. Whatever the light source, the light beam is introduced from an optical fiber of 200–400-μm (core) diameter (Newport) depending on the amount of energy needed to be delivered. We typically adjust the amount of energy of a single pulse of laser light emerging from the fiber to ~500 μJ, as determined with a Joule meter (model ED-200, Gentec, Palo Alto, CA).

Light from a tungsten source (Newport 780) can also be projected through the optical fiber. Although not essential, this allows one to position the fiber, which also carries the laser light, so that the cell is in the center of the beam of laser light. The irradiated area is adjusted by positioning the optical fiber above the cell surface to have a light diameter between 300 and 400 μm. This area is sufficient so that the concentration of photoliberated neurotransmitter is constant during the time interval of the kinetic (current) measurements.

The cell attached to the current-recording electrode is equilibrated with caged compound. At time 0 a single pulse of laser light photolyzes the caged compound in the microsecond time region. The liberated neurotransmitter binds to the receptors on the cell surface and initiates the formation of transmembrane channels. The time resolution of the technique allows one to observe three distinct phases of the reaction: a rising phase of the current reflecting the opening of acetylcholine receptor channels, maximum current amplitude, a measure of the concentration of open receptor channels, and on a different and slower time scale, the falling phase of the current reflecting receptor desensitization. In experiments with the excitatory acetylcholine, glutamate, kainate receptors, and the inhibitory GABA and glycine receptors (Milburn et al. 1989; Matsubara et al. 1992; Ramesh et al. 1993; Gee et al. 1994; Wieboldt et al. 1994; Niu et al. 1996c), the rise time of the current followed a single exponential rate equation over 85% of the reaction.

The dependence of the first-order rate constant for the current rise time is plotted versus carbamoylcholine concentration according to Equation 4B in Table 1. The ordinate intercept of the line gives the value of the channel-closing rate constant k_{cl}, and the slope allows evaluation of the channel-opening rate constant k_{op}. From the dependence of the maximum current amplitude on the concentration of carbamoylcholine, one can obtain the value of the channel-opening equilibrium constant Φ ($= k_{op}/k_{cl}$), and the value of K_1, the dissociation constant of the receptor site controlling channel opening, by using Equation 1C in Table 1. The values of K_1 and Φ obtained from the effect of ligand concentration on the current amplitude can be compared to the values of K_1, k_{cl}, and k_{op} obtained from the effect of ligand concentration on the observed rate constant k_{obs} for the rise time of the current (Table 1, Eq. 4B). The falling phase of the current gives information about the rate of desensitization. When present, receptor desensitization is slow compared to channel opening, occurs in a different time scale, and is investigated more conveniently by the cell-flow method (Udgaonkar and Hess 1987).

The laser-pulse photolysis technique allows one to determine the rate constant of the current rise indicative of channel opening and, therefore, to study the effects of inhibitors on k_{cl} (Table 1, Eq. 4C) and k_{op} (Table 1, Eq. 4D) independently of one another.

The time resolution of the kinetic methods just described allows one to determine the maximum current amplitudes and, therefore, the concentration of open receptor channels before desensitization occurs. This, in turn, allows one to determine the effects of inhibitors or potentiators (or mutations that modulate receptor current) on the current amplitudes at low concentrations of neurotransmitter when the receptor is mainly in the closed-channel form, and at high concentrations when the receptor is mainly in the open-channel form (Niu and Hess 1993; Niu et al. 1995; Ferster et al. 1996; Hess and Grewer 1998). These measurements allow one to differentiate between noncompetitive and competitive inhibitors (Table 1, Eqs. 3A and 3B) and to determine whether two noncompetitive inhibitors bind to two different sites (Table 1, Eq. 3E) or to the same site (Table 1, Eq. 3D).

Flash-Lamp Photolysis with Visible Light

Originally, suitable caged neurotransmitters were photolyzed in the near-UV region using light from a laser (for review, see Marriott 1998; Goeldner and Givens 2005). In an effort to make transient pre–steady state kinetic studies of reactions mediated by ligand-gated ion channels more widely accessible and for use in multiphoton microscopy (Matsuzaki et al. 2001; Smith et al. 2003; Trigo et al. 2009), we embarked on the development of caged neurotransmitters that can be photolyzed by visible light.

A light source appropriate for such studies must meet certain requirements. Light of a suitable wavelength, delivered from the exit of an optical fiber to a cell in the whole-cell current-recording mode, must be of sufficient energy in a *single* pulse to release free neurotransmitter but not of such energy as to damage the protein or cell. For transient kinetic studies, the pulse duration must be within the microsecond domain. A variety of visible light sources were considered. Engert et al. (1996) built a low-cost nitrogen laser with a 5-nsec pulse for flash photolysis of caged compounds. However, although the entire pulse energy was ~250 μJ, only 20 μJ reached the sample, which is too low to photo-release sufficient concentrations of neurotransmitter to activate receptors in transient kinetic studies. A light-emitting diode (LED) has been used to study intracellular calcium homeostasis (Bernardinelli et al. 2005). However, with a 100–1000-msec pulse, it is not suitable for studying receptor activation in the microsecond–millisecond time domain. Xenon flash lamps have been used with caged compounds to map functional neuronal connections (Matsuzaki et al. 2001), but in those cases, the goal was high spatial resolution; the temporal resolution was not sufficient for transient kinetic studies.

The use of flash lamps for photolysis of caged compounds has been reviewed in detail (Rapp 1998). An XF-10 xenon flash lamp (Hi-Tech Scientific, UK) has been used to photo-release Ca^{2+} from DM-nitrophen and from nitr-5–caged Ca^{2+} (Hardie 1995), but it has a pulse duration of 200 msec, which is too long for transient kinetic measurements. Canepari et al. (2001) used light at 290–370 nm with a pulse length of 1 msec from a Rapp xenon flash lamp, a duration suitable for kinetic measurements, to photo-release glutamate, glycine, and GABA from a 7-nitroindolinyl (NI)–caged precursor or 4-methoxy-7-nitroindolinyl (MNI)–caged glutamate. The wavelength range of the Rapp xenon flash lamp extends into the visible wavelength region, making it suitable for transient kinetic investigations.

An SP 450 385–450-nm band-pass filter is used with the flash lamp. The coumarin-caged compounds do not absorb light >500 nm. Other wavelengths can be used by changing the filter combination. The wide (200–1100-nm) spectrum of the lamp allows many filter combinations to be used to achieve a desired wavelength range.

The pulse length can be adjusted from 2 to 400 μsec by changing the lamp capacitors. A single 220-μsec pulse provides sufficient energy (~350 μJ) to photo-release free neurotransmitter. A Joule meter (Molectron) is used to measure the energy of the light pulse emerging from the quartz fiber. The available short pulse length (2–400 μsec) and tunable high pulse energy (up to 150 J/pulse) make the SP-20 flash lamp system a good light source for use in transient kinetic measurements of reactions mediated by neurotransmitter receptors. A separate shutter system is not needed because the data acquisition software used, pClamp, controls the flash-lamp system operation directly via the built-in external trigger of the SP-20 flash lamp. In any case, the shutter systems available are too slow for use in transient kinetic measurements.

To minimize light exposure before photolysis is initiated, the Faraday cage surrounding the current recording instrumentation is covered with aluminum foil, and the tubing that is used to deliver the caged-glutamate solution to the cells via a peristaltic pump is also wrapped with foil. Overhead lights should be turned off. If necessary, a red light can be used to illuminate the room. Measurements are made as described in the Laser-Pulse Photolysis Technique section.

REFERENCES

Adams RS, Tsien RY. 1993. Controlling cell chemistry with caged compounds. *Annu Rev Physiol* **55:** 755–784.

Banerjee A, Grewer C, Ramakrishnan L, Jäger J, Gameiro A, Breitinger H-GA, Gee KR, Carpenter BK, Hess GP. 2003. Toward the development of new photolabile protecting groups that can rapidly release bioactive compounds upon photolysis with visible light. *J Org Chem* **68:** 8361–8367.

Barltrop JA, Plant PJ, Schofield P. 1966. Photosensitive protecting group. *J Chem Soc Chem Commun* 822–823.

Bendig J, Helm S, Hagen V. 1997. (Coumarin-4-yl)methyl ester of cGMP and 8-Br-cGMP: Photochemical fluorescence enhancement. *J Fluorescence* **7:** 357–361.

Bernardinelli Y, Haeberli C, Chatton JY. 2005. Flash photolysis using a light emitting diode: An efficient, compact, and affordable solution. *Cell Calcium* **37:** 565–572.

Bochet CG. 2002. Photolabile protecting groups and linkers. *J Chem Soc Perkin Trans* **1:** 125–142.

Breitinger H-GA, Wieboldt R, Ramesh D, Carpenter BK, Hess GP. 2000. Synthesis and characterization of photolabile derivatives of serotonin for chemical kinetic investigations of the serotonin 5-HT3 receptor. *Biochemistry* **39:** 5500–5508.

Canepari M, Nelson L, Papageorgiou G, Corrie JE, Ogden D. 2001. Photochemical and pharmacological evaluation of 7-nitroindolinyl and 4-methoxy-7-nitroindolinyl–amino acids as novel, fast caged neurotransmitters. *J Neurosci Methods* **112:** 29–42.

Casey JP, Blidner RA, Monroe WT. 2009. Caged siRNAs for spatiotemporal control of gene silencing. *Mol Pharm* **6:** 669–685.

Cash DJ, Hess GP. 1980. Molecular mechanism of acetylcholine receptor–controlled ion translocation across cell membranes. *Proc Natl Acad Sci* **77:** 842–846.

Choquet D, Korn H. 1988. Does β-alanine activate more than one chloride channel associated receptor? *Neurosci Lett* **84:** 329–334.

Clapham DE, Neher E. 1984. Substance P reduces acetylcholine-induced currents in isolated bovine chromaffin cells. *J Physiol* **347:** 255–277.

Corrie JET, Trentham DR. 1993. Caged nucleotides and neurotransmitters. In *Bioorganic photochemistry*, Vol. 2: *Biological applications of photochemical switches* (ed. Morrison H), pp. 243–305. Wiley, New York.

De Mayo P. 1960. Ultraviolet photochemistry of simple unsaturated systems. *Adv Org Chem* **2:** 367–425.

Denk W. 1994. Two-photon scanning photochemical microscopy: Mapping ligand-gated ion channel distributions. *Proc Natl Acad Sci* **91:** 6629–6633.

Edmonds B, Gibb AJ, Colquhoun D. 1995. Mechanisms of activation of muscle nicotinic acetylcholine receptors and the time course of endplate currents. *Annu Rev Physiol* **57:** 469–493.

Engels J, Schlaeger EJ. 1977. Synthesis, structure, and reactivity of adenosine cyclic 3′,5′-phosphate benzyl triesters. *J Med Chem* **20:** 907–911.

Engert F, Paulus GG, Bonhoeffer T. 1996. A low-cost UV laser for flash photolysis of caged compounds. *J Neurosci Methods* **66:** 47–54.

Fan L, Lewis RW, Hess GP, Ganem B. 2009. A new synthesis of caged GABA compounds for studying $GABA_A$ receptors. *Bioorg Med Chem Lett* **19:** 3932–3933.

Fenwick E, Marty A, Neher E. 1982. A patch-clamp study of bovine chromaffin cells and of their sensitivity to acetylcholine. *J Physiol* **331:** 577–597.

Ferster D, Chung S, Wheat H. 1996. Orientation selectivity of thalamic input to simple cells of cat visual cortex. *Nature* **380:** 249–252.

Franke C, Parnas H, Hovac G, Dudel J. 1993. A molecular scheme for the reaction between acetylcholine and nicotinic channels. *Biophys J* **64:** 339–356.

Furuta T, Iwamura M. 1998. New caged groups: 7-substituted coumarinylmethyl phosphate esters. *Methods Enzymol* **291:** 50–63.

Gee KR, Wieboldt R, Hess GP. 1994. Synthesis and photochemistry of a new photolabile derivative of GABA. Neurotransmitter release and receptor activation in the microsecond time region. *J Am Chem Soc* **116:** 8366–8367.

Givens RS, Weber JF, Jung AH, Park CH. 1998. New photoprotecting groups: Desyl and *p*-hydroxyphenacyl phosphate and carboxylate esters. *Methods Enzymol* **291:** 1–29.

Goeldner M, Givens RS, eds. 2005. *Photoswitches, and caged biomolecules*. Wiley-VCH, Weinheim, Germany.

Hagen V, Bendig J, Frings S, Eckardt T, Helm S, Reuter D, Kaupp UB. 2001. Highly efficient and ultrafast phototriggers for cAMP and cGMP by using long-wavelength UV/Vis-activation. *Angew Chem Int Ed Engl* **40:** 1045–1048.

Hamill OP, Marty A, Neher E, Sakmann B, Sigworth FJ. 1981. Improved patch–clamp techniques for high–resolution current recording from cells and cell–free membrane patches. *Pflügers Arch* **391:** 85–100.

Hardie RC. 1995. Photolysis of caged Ca^{2+} facilitates and inactivates but does not directly excite light–sensitive channels in *Drosophila* photoreceptors. *J Neurosci* **15:** 889–902.

Hess GP. 1993. Determination of the chemical mechanism of neurotransmitter receptor–mediated reactions by rapid chemical kinetic techniques. *Biochemistry* **32:** 989–1000.

Hess GP, Grewer C. 1998. Development and application of caged ligands for neurotransmitter receptors in transient kinetic and neuronal circuit mapping studies. *Methods Enzymol* **291:** 443–473.

Hess GP, Cash DJ, Aoshima H. 1983. Acetylcholine receptor–controlled ion translocation: Chemical kinetic investigations of the mechanism. *Annu Rev Biophys Bioeng* **12:** 443–473.

Hess GP, Udgaonkar JB, Olbricht WL. 1987. Chemical kinetic measurements of transmembrane processes using rapid reaction techniques: Acetylcholine receptor. *Annu Rev Biophys Biophys Chem* **16:** 507–534.

Kaplan JH, Forbush B, Hoffman JF. 1978. Rapid photolytic release of adenosine 5′-triphosphate from a protected analogue: Utilization by the Na:K pump of human red blood cell ghosts. *Biochemistry* **17:** 1929–1935.

Katz B, Thesleff S. 1957. A study of the desensitization produced by acetylcholine at the motor end-plate. *J Physiol* **138:** 63–80.

Krishtal OA, Pidoplichko VI. 1980. A receptor for protons in the nerve cell membrane. *Neuroscience* **5:** 2325–2327.

Landau VH, Lifshitz EM. 1959. *Fluid mechanics.* Pergamon Press, Oxford.

Levich AG. 1962. *Physicochemical hydrodynamics.* Prentice-Hall, Upper Saddle River, NJ.

Li H, Avery L, Denk W, Hess GP. 1997. Identification of chemical synapses in the pharynx of *Caenorhabditis elegans. Proc Natl Acad Sci* **94:** 5912–5916.

Marriott G, ed. 1998. *Caged compounds. Methods in enzymology*, Vol. 291. Academic Press, New York.

Marty G, Neher E. 1995. Tight-seal whole-cell recording. In *Single-channel recording*, 2nd ed. (ed. Sakmann B, Neher E), pp. 31–52. Plenum Press, New York.

Matsubara N, Billington AP, Hess GP. 1992. How fast does an acetylcholine receptor channel open? Laser-pulse photolysis of an inactive precursor of carbamoylcholine in the microsecond time region with BC3H1 cells. *Biochemistry* **31:** 5507–5514.

Matsuzaki M, Ellis-Davies GC, Nemoto T, Miyashita Y, Iino M, Kasai H. 2001. Dendritic spine geometry is critical for AMPA receptor expression in hippocampal CA1 pyramidal neurons. *Nat Neurosci* **4:** 1086–1092.

Mayer G, Heckel A. 2006. Biologically active molecules with a "light switch." *Angew Chem Int Ed Engl* **45:** 4900–4921.

McCray JA, Trentham DR. 1989. Properties and uses of photoreactive caged compounds. *Annu Rev Biophys Biophys Chem* **18:** 239–270.

McCray JA, Herbette L, Kihara T, Trentham DR. 1980. A new approach to time-resolved studies of ATP-requiring biological systems; laser flash photolysis of caged ATP. *Proc Natl Acad Sci* **77:** 7237–7241.

Milburn T, Matsubara N, Billington AP, Udgaonkar JB, Walker JW, Carpenter BK, Webb WW, Marque J, Denk W, McCray JA, Hess GP. 1989. Synthesis, photochemistry, and biological activity of a caged photolabile acetylcholine receptor ligand. *Biochemistry* **28:** 49–55.

Neher E. 1983. The charge carried by single-channel currents of rat cultured muscle cells in the presence of local anaesthetics. *J Physiol* **339:** 663–678.

Neher E, Sakmann B. 1976. Single-channel currents recorded from membrane of denervated frog muscle fibres. *Nature* **260:** 799–802.

Nerbonne JM. 1996. Caged compounds: Tools for illuminating neuronal responses and connections. *Curr Opin Neurobiol* **6:** 379–386.

Niu L, Hess GP. 1993. An acetylcholine receptor regulatory site in BC3H1 cells: Characterized by laser-pulse photolysis in the microsecond-to-millisecond time region. *Biochemistry* **32:** 3831–3835.

Niu L, Abood LG, Hess GP. 1995. Cocaine: Mechanism of inhibition of a muscle acetylcholine receptor studied by a laser-pulse photolysis technique. *Proc Natl Acad Sci* **92:** 12008–12012.

Niu L, Gee KR, Schaper K, Hess GP. 1996a. Synthesis and photochemical properties of a kainate precursor and activation of kainate and AMPA receptor channels on a microsecond time scale. *Biochemistry* **35:** 2030–2036.

Niu L, Grewer C, Hess GP. 1996b. Chemical kinetic investigations of neurotransmitter receptors on a cell surface in the ms time region. *Tech Protein Chem* **7:** 139–149.

Niu L, Wieboldt R, Ramesh D, Carpenter BK, Hess GP. 1996c. Synthesis and characterization of a caged receptor ligand suitable for chemical kinetic investigations of the glycine receptor in the 3-microseconds time domain. *Biochemistry* **35:** 8136–8142.

Ogden DC, Colquhoun D. 1985. Ion channel block by acetylcholine, carbachol and suberyldicholine at the frog neuromuscular junction. *Proc R Soc Lond B Biol Sci* **225:** 329–355.

Park C-H, Givens RS. 1997. New photoactivated protecting groups. 6. *p*-hydroxyphenacyl: A phototrigger for chemical and biochemical probes. *J Am Chem Soc* **119:** 2453–2463.

Patchornik A, Amit B, Woodward RB. 1970. Photosensitive protecting groups. *J Am Chem Soc* **92:** 6333–6335.

Ramesh D, Wieboldt R, Niu L, Carpenter BK, Hess GP. 1993. Photolysis of a protecting group for the carboxyl function of neurotransmitters within 3 microseconds and with product quantum yield of 0.2. *Proc Natl Acad Sci* **90:** 11074–11078.

Rapp G. 1998. Flash lamp–based irradiation of caged compounds. *Methods Enzymol* **291:** 202–222.

Reynolds JA, Karlin A. 1978. Molecular weight in detergent solution of acetylcholine receptor from *Torpedo californica*. *Biochemistry* **17:** 2035–2038.

Rial Verde EM, Zayat L, Etchenique R, Yuste R. 2008. Photorelease of GABA with visible light using an inorganic caging group. *Front Neural Circuits* **2:** 2. doi: 10.3389/neuro.04.002.2008.

Sakmann B, Neher E, eds. 1983. *Single-channel recording*. Plenum Press, New York.

Sakmann B, Neher E, eds. 1995. *Single-channel recording*, 2nd ed. Plenum Press, New York.

Sather W, Dieudonne S, MacDonald JF, Ascher P. 1992. Activation and desensitization of *N*-methyl-D–aspartate receptors in nucleated outside-out patches from mouse neurones. *J Physiol* **450:** 643–672.

Shembekar VR, Chen Y, Carpenter BK, Hess GP. 2005. A protecting group for carboxylic acids that can be photolyzed by visible light. *Biochemistry* **44:** 7107–7114.

Shembekar VR, Chen Y, Carpenter BK, Hess GP. 2007. Coumarin-caged glycine that can be photolyzed within 3 microseconds by visible light. *Biochemistry* **46:** 5479–5484.

Smith MA, Ellis-Davies GC, Magee JC. 2003. Mechanism of the distance-dependent scaling of Schaffer collateral synapses in rat CA1 pyramidal neurons. *J Physiol* **548:** 245–258.

Trigo FF, Corrie JE, Ogden D. 2009. Laser photolysis of caged compounds at 405 nm: Photochemical advantages, localisation, phototoxicity and methods for calibration. *J Neurosci Methods* **180:** 9–21.

Trussell LO, Fischbach GD. 1989. Glutamate receptor desensitization and its role in synaptic transmission. *Neuron* **3:** 209–218.

Udgaonkar JB. 1986. "Kinetic studies of the acetylcholine receptor using rapid flow and current recording techniques." Ph.D. thesis, Cornell University, Ithaca, NY.

Udgaonkar JB, Hess GP. 1987. Chemical kinetic measurements of a mammalian acetylcholine receptor by a fast-reaction technique. *Proc Natl Acad Sci* **84:** 8758–8762.

Vyklicky L, Benveniste M, Meyer ML. 1990. Modulation of *N*-methyl-D-aspartic acid receptor desensitization by glycine in mouse cultured hippocampal neurones. *J Physiol* **428:** 313–331.

Walker JW, McCray JA, Hess GP. 1986. Photolabile protecting groups for an acetylcholine receptor ligand. Synthesis and photochemistry of a new class of *o*-nitrobenzyl derivatives and their effects on receptor function. *Biochemistry* **25:** 1799–1805.

Walker JW, Reid GP, McCray JA, Trentham DR. 2002. Photolabile 1-(2-nitrophenyl)ethyl phosphate esters of adenine nucleotide analogs. Synthesis and mechanism of photolysis. *J Am Chem Soc* **110:** 7170–7177.

Walstrom KM, Hess GP. 1994. Mechanism for the channel-opening reaction of strychnine-sensitive glycine receptors on cultured embryonic mouse spinal cord cells. *Biochemistry* **33:** 7718–7730.

Walz W, Boulton AA, Baker GB, eds. 2002. Patch-clamp analysis: Advanced techniques. In *Neuromethods* (series eds., Boulton AA, Baker GB), Vol. 35. Humana Press, Totowa, NJ.

Wieboldt R, Gee KR, Niu L, Ramesh D, Carpenter BK, Hess GP. 1994. Photolabile precursors of glutamate: Synthesis, photochemical properties, and activation of glutamate receptors on a microsecond time scale. *Proc Natl Acad Sci* **91:** 8752–8756.

59 Nitrophenyl-Based Caged Neurotransmitters

Graham C.R. Ellis-Davies

Department of Neurosceince, Mt. Sinai School of Medicine, New York, New York 10029

ABSTRACT

Nitroaromatic photochemical protecting groups were developed for organic synthesis in 1966. Since the early 1990s, this type of chromophore has been used by neuroscientists to liberate a wide variety of amino acid neurotransmitters such as ACh, glutamate, GABA, and glycine, among others. Since 2001, several laboratories have used two-photon excitation of nitroaromatic cages for highly localized uncaging of glutamate in acute brain slices.

INTRODUCTION

What Are Caged Neurotransmitters?

From a chemical point of view, caged neurotransmitters are synthetically modified derivates of amino acids or nucleotides in which a photochemical protecting group has typically been attached to the acid or amino functionality (see Chapter 58) (Mayer and Heckel 2006; Ellis-Davies 2007, 2008). Irradiation breaks the covalent bond that connects the organic chromophore to the neurotransmitter, liberating the caged compound. The most widely used caged compounds all use the same photochemistry, namely, nitroaromatic intramolecular photo-redox reactions. There are two principal reasons for this: historical and practical. The first caged compounds used by biologists (Kaplan and Somlyo 1989) were *ortho*-nitrobenzyl derivatives of ATP (Kaplan et al. 1978) and c-AMP (Engels and Schlaeger 1977). The striking success of these pioneering experiments led to the development of several other caged compounds (caged calcium and caged neurotransmitters) using the same photochemistry. The *ortho*-nitrobenzyl caging chromophore was originally developed by organic chemists (Baltrop et al. 1966) for general use in complex synthetic organic sequences to protect reactive functionalities (Dorman and Prestwich 2000). The *ortho*-nitrobenzyl cage is the only caging chromophore that photo-cleaves all organic functionalities (e.g., ether, thioether, phosphate, thiophosphate, carboxylate, diol, amine, amide, carbamate, and carbonate). Thus, it is eminently practical and has consequently been applied to the widest possible array of substrates for biological sciences (Mayer and Heckel 2006; Ellis-Davies 2007). This is in contrast to other photochemical pro-

TABLE 1. Summary of some of the chemical and pharmacological properties of important mitrophenyl caged neurotransmitters

Cage	Quantum yield	Extinction coefficient (M^{-1} cm^{-1})	$f \cdot e$	Rate (sec^{-1})	Stability (pH 7.4)	Solubility	Pharmacology	2PE
NPE-ATP	0.63	430	271	83	High	High		ND
CNB-CC	0.8	500	400	17,000	High	High	Inert to nicotinic	ND
DMNPE-ATP	0.07	5000	350	18	ND	High	ND	low
CNB-Glu	500	0.14	70	40,000	Moderate	Good	Inert to AMPA	ND
CNB-GABA	500	0.16	80	36,000	Moderate	Good	Mild $GABA_A$ antagonist at 200 µM	ND
CNO	0.05	5000	250	>200	High	Good	Inert	ND
NI-Glu	0.026	1400	36	1000	High	High	Inert to AMPA	ND
MNI-Glu	0.085	4200	357	100,000	High	High	Inert to AMPA	0.06 GM
MNI-D-Asp	0.1	4200	420	ND	High	High	Inert to NMDA	ND
CmNB-anadamide	0.06	2000	120	50,000	High	ND	Inert to eCB-R	ND
CDNI-Glu	0.5	6400	3200	ND	Moderate	High	Inert to AMPA	Like MNI
CNI-GABA	0.1	4200	420	ND	High	High	No $GABA_A$ antagonism at 0.3 mM	ND
CDNI-GABA	0.6	6400	3840	ND	Moderate	High	Mild $GABA_A$ antagonist at 0.4 mM	ND

Abbreviations: ND, no datum; *f*, quantum yield; *e*, extinction coefficient; 2PE, two-photon excitation.

The two-photon cross section MNI-Glu has been reported (Matsuzaki et al. 2001), and CDNI-Glu has similar properties (Ellis-Davies et al. 2007). By analogy, MNI-D-Asp, CNI-GABA, and CDNI-GABA will have similar two-photon cross sections. The other nitrobenzyl caged neurotransmitters probably have much lower cross sections. MNI-Glu and CNDI-Glu have strong $GABA_A$ receptor antagonism at 10 mM.

tecting groups such as coumarins (Furuta et al. 1995), nitroindolinyl (Pass et al. 1981), biyridyl-ruthenium (Zayat et al. 2003), *para*-hydroxyphenacyl (Givens and Park 1996), or desyl (Sheehan and Wilson 1964) chromophores, which are more specialized cages that are only applicable to a limited number of functionalities (Pelliccioli and Wirz 2002; Mayer and Heckel 2006).

For caged transmitters to be useful for neuroscience, they must also possess several important chemical properties. The probes should be chemically stable and highly soluble at physiological pH. Some chemical bonds that attach organic chromophores to transmitters are quite sensitive to aqueous hydrolysis above pH 4, necessitating that fresh solutions of the caged compound be made daily, whereas others are stable for many months in frozen aqueous solution (Table 1). Adding a chromophore containing one or more aromatic rings to natural products, such as glutamate or serotonin, reduces their aqueous solubility considerably, often requiring the addition of solubilizing substituents onto the caging chromophore. Finally, the caged transmitter must be sufficiently photochemically active for useful amounts of neurotransmitter to be released without phototoxicity to the living tissue under study.

From a biological point of view, caged neurotransmitters must have several characteristics for use with living cells such as glia or neurons. First, the *sine qua non* of a caged neurotransmitter is that it be biologically inert. In other words, if the caged compound activates the target receptor before irradiation, it is not "caged," but merely "photolabile." Second, the caged neurotransmitter should not block receptors. This "antagonist criterion" is much harder to fulfill, especially for $GABA_A$ receptors. For example, most caged glutamate probes seem to be inert toward AMPA and NMDA receptors. However, since the initial report that CNB-GABA is antagonistic toward $GABA_A$ receptors (Molnar and Nadler 2000), other nitroaromatic caged transmitters based on acylindole have been reported to be strong antagonists to this diverse family of receptors (Canepari et al. 2001). More recently, several reports on the pharmacology of nitroaromatic caged glutamate and GABA have appeared. Specifically, the widely used MNI-Glu was reported to be highly antagonistic toward $GABA_A$ receptors (Fino et al. 2009; Matsuzaki et al. 2010). This result was quite surprising because MNI-Glu had been reported to be inert toward APMA receptors (Canepari et al. 2001; Matsuzaki et al. 2001; Andrasfalvy et al. 2003; Ngo-Anh et al. 2005; Beique et al. 2006) (but see Fino et al. 2009, wherein small perturbations of AMPA receptors were reported). Nitroindolinyl caged GABA probes bearing pendant negative charges on the 4-methoxy have been developed with the idea of reducing the antagonism of nitroaromatic caged transmitters. Monocarboxylate derivatives of GABA (called CNI-GABA) at low concentrations (bath application of 0.325 mM) are inert (Alvina et al. 2008), whereas puffer application (0.4 mM) of another monocarboxylate derivative (CDNI-GABA) was mildly antagonistic. Finally, bath application of a bisphosphate derivative of GABA (Trigo et al. 2009b) revealed an IC_{50} = 0.5 mM. This "promiscuity" is also reflected by the wide range of structure–activity relationships seen for drugs targeted toward GABA receptors (Mohler et al. 1990). Thus, development of a completely biologically inert caged GABA probably remains the outstanding challenge in the area of caged neurotransmitters. However, antagonism toward $GABA_A$ receptors of nitroaromatic caged GABA compounds does not prevent these probes from being used to good effect in many studies of synaptic inhibition (Eder et al. 2004; Trigo et al. 2009b; Matsuzaki et al. 2010).

Why Use Caged Neurotransmitters?

Photochemical uncaging of neurotransmitters is especially useful when studying neurons and astrocytes in acutely isolated brain slices (Callaway and Katz 1993; Callaway and Yuste 2002). Such intact tissue is the preparation of choice for many neuroscientists because it preserves much of the three-dimensional (3D) architecture of cellular connectivity that exists in vivo, yet, being ex vivo, permits relatively easy access for multi-electrode stimulation and recording, along with pharmacological intervention. However, the very complexity of the brain slice preparation can restrict the reductionist scientific drive to control and understand smaller structures such as single synapses. For example, each hippocampal CA1 neuron has approximately 10,000 synapses, so that the stimulation of a single synapse by traditional electrical means is at best haphazard, or virtually impossible. Uncaging provides an answer to this quandary (Eder et al. 2004; Judkewitz et al. 2006). As light can be focused and directed to any point across a neuron or astrocyte, it can be used to uncage known amounts of neurotransmitter at visually designated positions for defined periods, permitting controlled receptor activation of single synapses. Thus, the photolysis beam is a "magic wand" for synaptic physiology.

Uncaging does not replace traditional electrode stimulation; rather, it is a useful complement to it for several reasons.

1. A single transmitter is normally photo-released. Electrode stimulation of afferent fibers in brain slice preparations typically evokes secretion of many transmitters, requiring pharmacology to isolate the effects of individual transmitters. Uncaging allows one to release a single transmitter.
2. Voltage-gated ion channels are not required for synaptic stimulation. Because the presynaptic machinery is bypassed by uncaging, drugs to inactivate ion channels can be co-applied to delineate the biological role(s) of such channels.
3. Many synapses can be activated simultaneously according to the area (or volume) of illumination. Because light can be easily focused, one may activate single synapses in a complex 3D network using two-photon uncaging, or globally stimulate many neurons by bathing a preparation in near-UV light.
4. Unnatural amino acids or drugs can be photo-released. Because virtually any organic molecule can be caged, photolysis can be used to rapidly release unnatural products. By definition, this can only be accomplished by application of exogenous probes.
5. Subquantal or supraquantal neurotransmitter release is feasible. Uncaging allows one to produce stereotypical, reproducible stimulation of neurons and astrocytes. By controlling the light intensity and probe, activation of single synapses in brain slices can be "tuned" to mimic quantal release at a visually selected synapse. Such stimulation by two-photon uncaging of glutamate has been widely used in the past few years; logically it also allows subquantal or supraquantal stimulation as well.

Comparison of Transmitter Uncaging with Other Photostimulation Techniques

In the past five years, two new optical methods to control neuronal membrane potential have been developed (Gorostiza and Isacoff 2008). One uses *cis–trans* isomerization of azobenzene chemical probes to block/unblock ion channels (Banghart et al. 2004; Volgraf et al. 2006). The second uses light-activated channels isolated from the unicellular green alga *Chlamydomonas reinhardtii*, called channelrhodopsin-2 (ChR2) and halorhodopsin (NpHR), to control membrane potential (Han and Boyden 2007; Zhang et al. 2007). Fortuitously, NpHR has a different action spectrum from ChR2—the former absorption maximum is at 580 nm, and the latter is at 430 nm. Even though NpHR is not optically transparent at short wavelengths, ChR2 and NpHR can be coexpressed in neurons so that blue light can depolarize and yellow light can hyperpolarize the membrane potential with excellent temporal fidelity. The azobenzene approach can fire action potentials (illumination at 380 nm) and extinguish the evoked current (illumination at 500 nm reverts the azobenzene probe to its inactive state). Spike rates of 50 Hz are possible without probe fatigue (Volgraf et al. 2006). Initially, the designers of the azobenzene method reasoned that it would be important to tether their probe to a specific point near the mouth of the channel using site-directed mutagenesis, in order to ensure specificity. Recently, however, it has been discovered that native membrane channels can be cleanly modified with reactive probe or that covalent modification is not required at all for the azobenzene method to work effectively (Fortin et al. 2008). This powerful new method is still really in its infancy, and so its full potential will only be apparent with further use. In contrast, ChR2 is already starting to have considerable impact as an optical method to stimulate neurons, probably because it *only* requires DNA in the appropriate neurons for expression. These approaches are covered in detail in Weissman et al. (2011).

What are the advantages of caged compounds and azobenzenes compared to optogenetic methods for neuronal photostimulation? Both of the former methods, unlike the latter, require the addition of exogenous chemical probes to control membrane potential or cell physiology. Thus, because a simple chemical probe is used, caged compounds or azobenzenes can be applied to any neural tissue without prior application of DNA to encode the seven-transmembrane receptors necessary for the channelrhodopsin family. A second advantage of chemical probes is that they control membrane potential by directly targeting native membrane receptors, whereas channelrhodopsin requires light activation of a non-native ion channel. Finally, two-photon excitation (Denk et al. 1991) produces much improved 3D resolution when compared with regular UV-visible light photolysis (Rial Verde et al. 2008; Trigo et al. 2009a). Modern technology (lasers and caged compounds) has enabled two-photon uncaging of neurotransmitters to be effected very close to the diffraction limit, enabling highly localized, subcellular stimulation of dendritic and axonal cholinergic, AMPA and GABA receptors (Denk 1994; Matsuzaki et al. 2001, 2010). Thus, I would argue that optogenetics and caged compounds are highly complementary methods for optical stimulation of synapses, neurons, and neuronal circuits. Optogenetics is a uniquely powerful method for controlling neurons in freely moving mice, allowing one to connect behavior to circuit activation (Gradinaru et al. 2007). Application of caged compounds is particularly powerful when addressing classical questions in cellular neuroscience, because it allows fine subcellular optical control of membrane potential. In some cases, it is even possible to combine the two optical methods for pipette-less all-optical induction of LTP (Zhang et al. 2008).

EXPERIMENTAL USE OF CAGED NEUROTRANSMITTERS

Cellular Application of Caged Neurotransmitters

Two perfusion methods have been used for the application of caged neurotransmitters to brain slices. The simpler method is to bath the entire brain slice with the probe dissolved in the aCSF perfusate (Alvina et al. 2008). The advantage of this method is that a known concentration of probe is applied evenly to all neurons and astrocytes. The disadvantage is that it is quite costly in terms of caged compound, so typically small volumes (~10 mL) of aCSF are recirculated for many hours (Carter and Sabatini 2004; Matsuzaki et al. 2008). Alternatively, several laboratories have used local perfusion of caged neurotransmitters for two-photon photolysis, while allowing normal aCSF to wash over the brain slice (Matsuzaki et al. 2001; Smith et al. 2003). This method has been criticized for delivering an imprecise concentration of probe to the cells. However, in skilled hands, highly reproducible results have been reported. Local application does permit the simple and rapid exchange of applied probes to the same neuron (Tanaka et al. 2007).

Light Sources for Uncaging

Many different light sources have been used for uncaging neurotransmitters. Most of the caged compounds listed in Table 1 are photorelease following excitation with light from 300 to 420 nm and have maximum absorption near 350 nm. There is a wide variety of near-UV laser sources supplying this region, but most modern confocal microscopes have a 405-nm laser for DAPI imaging. These small lasers are very reliable and can be used for efficient photolysis of caged transmitters like MNI-Glu. A detailed description of how 405-nm lasers can be added to standard microscope has been published (Trigo et al. 2009a). Uncaging of neurotransmitters in the redder part of the visible spectrum with 473-nm lasers has also been reported (Rial Verde et al. 2008). This is the same laser wavelength that is used for ChR2 activation (Gradinaru et al. 2007). Flash lamps are good sources of energy in the 300–400-nm range and are especially useful for illumination of a large area, but typically do not provide diffraction-limited uncaging. Two-photon lasers (mode-locked, Ti:sapphire lasers) are now one-box, turnkey systems, allowing uncaging of neurotransmitters in a small focal volume, because of the nonlinear nature of excitation in the axial dimension. Typically, the laser is tuned to 720 nm (Matsuzaki et al. 2001, 2008; Carter and Sabatini 2004), as shorter wavelengths (<700 nm) seem quite phototoxic to cells.

Handling Caged Neurotransmitters

"Will room lights photolyze caged compounds?" is an important practical question. We have bath-applied MNI-Glu (1 mM) to brain slices and can detect no current when fluorescent room lights are turned on for brief periods; however, for long-term exposure we cover room lights with yellow-filtered lights (Roscolux 10 filter). Most of the caged transmitters listed in Table 1 are quite stable long term in frozen solution. A solution of MNI-D-Asp can be stored for 4 d at 4°C without detectable hydrolysis as measured by activation of astrocytic transporter currents (Huang et al. 2005).

Software for the Control of Photolysis of Caged Neurotransmitters

Modern commercial two-photon microscopes now have very sophisticated software packages for the control of many devices, including lasers and patch-clamp amplifiers. As a cheaper alternative, several laboratories have invested considerable effort in the development of comparable software suites. These are freely distributed and are evolving projects that are useful resources for the neuroscience community (Pologruto et al. 2003; Nguyen et al. 2006).

CRITICAL ASSESSMENT OF COMMERCIAL CAGED COMPOUNDS

The commercial availability of caged compounds is crucial for use by the biomedical research community. The reason for this is simple: Caged transmitters are used in large quantities, so cannot be supplied in useful amounts by the laboratories that originally develop the probes to all who ask for samples. In contrast, DNA-based probes are somewhat easier to distribute (see footnote 31 in Tsai et al. 2009). Table 1 shows the properties of the commercially available nitroaromatic caged neurotransmitters plus a few historically important or recently developed caged transmitters.

Two of the earliest caged compounds continue to be commercially available (NPE-ATP and CNB-carbamoylcholine; both from Invitrogen). In general, both have many highly desirable properties that should characterize all caged neurotransmitters. NPE-caged phosphates (viz., ATP) are released slowly but are otherwise excellent caged compounds. Unfortunately, faster caged ATP probes are not available commercially. CNB-carbamoylcholine, the first really effective caged neurotransmitter, has only one real shortcoming: It is insensitive to two-photon photolysis. CNB-Glu and CNB-GABA (both from Invitrogen) have also been widely used in many important experiments for UV uncaging. These caged compounds are reasonably stable in solution, but the more recently developed nitroindolinyl caged glutamates ("NI"-Glu [Sigma-Aldrich] and MNI-Glu [Tocris]) are much more stable at pH 7.4, and MNI-Glu is reasonably sensitive to two-photon photolysis, so they are now preferred to CNB caged transmitters. MNI-D-Asp and CmNB-anadamide are not commercially available but are excellent examples of very useful caged compounds, as both have near-ideal ranges of properties for use in neurological experiments. There has been a recent flurry of activity to develop a good caged GABA probe. BC204, CNI-GABA, and CDNI-GABA are all effective for UV uncaging, with relatively low antagonism toward $GABA_A$ receptors. Unfortunately, none of these probes are commercially available yet. For a discussion of RuBiGABA and other bipyridyl-ruthenium caged compounds, see Chapter 60 in this volume.

SUMMARY AND FUTURE PROSPECTS

The advantages of caged compounds became clear when the first biological experiments were performed with them in the late 1970s and early 1980s. The development of a wide variety of other caged compounds (e.g., neurotransmitters and calcium) in the 1990s enabled a wide variety of problems to be tackled by cell biologists, physiologists, and neuroscientists. UV-visible and two-photon uncaging of neurotransmitters remains a powerful method to address classical questions in the field of neuroscience. The strengths of this method nicely complement the recently developed "optogenetic" method for photocontrol of cell signaling.

REFERENCES

Alvina K, Walter JT, Kohn A, Ellis-Davies G, Khodakhah K. 2008. Questioning the role of rebound firing in the cerebellum. *Nat Neurosci* **11:** 1256–1258.

Andrasfalvy BK, Smith MA, Borchardt T, Sprengel R, Magee JC. 2003. Impaired regulation of synaptic strength in hippocampal neurons from glur1-deficient mice. *J Physiol* **552:** 35–45.

Baltrop JA, Plant PJ, Schofield P. 1966. Photosensitive protecting groups. *J Chem Soc Chem Commun* 822–823.

Banghart M, Borges K, Isacoff E, Trauner D, Kramer RH. 2004. Light-activated ion channels for remote control of neuronal firing. *Nat Neurosci* **7:** 1381–1386.

Beique JC, Lin DT, Kang MG, Aizawa H, Takamiya K, Huganir RL. 2006. Synapse-specific regulation of AMPA receptor function by PSD-95. *Proc Natl Acad Sci* **103:** 19535–19540.

Callaway EM, Katz LC. 1993. Photostimulation using caged glutamate reveals functional circuitry in living brain slices. *Proc Natl Acad Sci* **90:** 7661–7665.

Callaway EM, Yuste R. 2002. Stimulating neurons with light. *Curr Opin Neurobiol* **12:** 587–592.

Canepari M, Nelson L, Papageorgiou G, Corrie JE, Ogden D. 2001. Photochemical and pharmacological evaluation of 7-nitroindolinyl and 4-methoxy-7-nitroindolinyl-amino acids as novel, fast caged neurotransmitters. *J Neurosci Methods* **112:** 29–42.

Carter AG, Sabatini BL. 2004. State-dependent calcium signaling in dendritic spines of striatal medium spiny neurons. *Neuron* **44:** 483–493.

Denk W. 1994. Two-photon scanning photochemical microscopy: Mapping ligand-gated ion channel distributions. *Proc Natl Acad Sci* **91:** 6629–6633.

Dorman G, Prestwich GD. 2000. Using photolabile ligands in drug discovery and development. *Trends Biotechnol* **18:** 64–77.

Eder M, Zieglgansberger W, Dodt HU. 2004. Shining light on neurons—Elucidation of neuronal functions by photostimulation. *Rev Neurosci* **15:** 167–183.

Ellis-Davies GC. 2007. Caged compounds: Photorelease technology for control of cellular chemistry and physiology. *Nat Methods* **4:** 619–628.

Ellis-Davies GCR. 2008. Photolysis of caged glutamate for use in the CNS. In *Encyclopedia of neuroscience* (ed. Squire LR), pp. 639–644. Academic Press, New York.

Ellis-Davies GCR, Matsuzaki M, Paukert M, Kasia H, Bergles DE. 2007. 4-carboxymethoxy-5,7-dinitroindolinyl-Glu: An improved caged glutamate for expeditious ultraviolet and 2-photon photolysis in brain slices. *J Neurosci* **27:** 6601–6604.

Engels J, Schlaeger EJ. 1977. Synthesis, structure, and reactivity of adenosine cyclic 3′,5′-phosphate benzyl triesters. *J Med Chem* **20:** 907–911.

Fino E, Araya R, Peterka DS, Salierno M, Etchenique R, Yuste R. 2009. RuBi-glutamate: Two-photon and visible-light photoactivation of neurons and dendritic spines. *Front Neural Circuits* **3:** 2. doi: 10.3389/neuro.04.002.2009.

Fortin DL, Banghart MR, Dunn TW, Borges K, Wagenaar DA, Gaudry Q, Karakossian MH, Otis TS, Kristan WB, Trauner D, Kramer RH. 2008. Photochemical control of endogenous ion channels and cellular excitability. *Nat Methods* **5:** 331–338.

Furuta T, Torigai H, Sugimoto M, Iwamura M. 1995. Photochemical properties of new photolabile camp derivatives in a physiological saline solution. *J Org Chem* **60:** 3953–3956.

Givens RS, Park CH. 1996. *p*-hydroxyphenacyl ATP: A new phototrigger. *Tetrahedron Lett* **37:** 6259–6262.

Gorostiza P, Isacoff EY. 2008. Optical switches for remote and noninvasive control of cell signaling. *Science* **322:** 395–399.

Gradinaru V, Thompson KR, Zhang F, Mogri M, Kay K, Schneider MB, Deisseroth K. 2007. Targeting and readout strategies for fast optical neural control in vitro and in vivo. *J Neurosci* **27:** 14231–14238.

Han X, Boyden ES. 2007. Multiple-color optical activation, silencing, and desynchronization of neural activity, with single-spike temporal resolution. *PLoS One* **2:** e299. doi: 10.1371/journal.pone.0000299.

Huang YHH, Sinha SR, Fedoryak OD, Ellis-Davies GCR, Bergles DE. 2005. Synthesis and characterization of 4-methoxy-7-nitroindolinyl-D-aspartate, a caged compound for selective activation of glutamate transporters and *N*-methyl-D-aspartate receptors in brain tissue. *Biochemistry* **44:** 3316–3326.

Judkewitz B, Roth A, Hausser M. 2006. Dendritic enlightenment: Using patterned two-photon uncaging to reveal the secrets of the brain's smallest dendrites. *Neuron* **50:** 180–183.

Kaplan JH, Somlyo AP. 1989. Flash photolysis of caged compounds: New tools for cellular physiology. *Trends Neurosci* **12:** 54–59.

Kaplan JH, Forbush B III, Hoffman JF. 1978. Rapid photolytic release of adenosine 5′-triphosphate from a protected analogue: Utilization by the Na:K pump of human red blood cell ghosts. *Biochemistry* **17:** 1929–1935.

Matsuzaki M, Ellis-Davies GC, Nemoto T, Miyashita Y, Iino M, Kasai H. 2001. Dendritic spine geometry is critical for AMPA receptor expression in hippocampal CA1 pyramidal neurons. *Nat Neurosci* **4:** 1086–1092.

Matsuzaki M, Ellis-Davies GC, Kasai H. 2008. Three-dimensional mapping of unitary synaptic connections by two-photon macro photolysis of caged glutamate. *J Neurophysiol* **99:** 1535–1544.

Matsuzaki M, Hayama T, Kasai H, Ellis-Davies GCR. 2010. Two-photon uncaging of γ-aminobutyric acid in intact brain tissue. *Nat Chem Biol* **6:** 255-257.

Mayer G, Heckel A. 2006. Biologically active molecules with a "light switch." *Angew Chemie Int Ed Engl* **45:** 4900–4921.

Mohler H, Malherbe P, Draguhn A, Sigel E, Sequier JM, Persohn E, Richards JG. 1990. $GABA_A$-receptor subunits: Functional expression and gene localisation. *Adv Biochem Psychopharmacol* **46:** 23–34.

Molnar P, Nadler JV. 2000. γ-Aminobutyrate, α-carboxy-2-nitrobenzyl ester selectively blocks inhibitory synaptic transmission in rat dentate gyrus. *Eur J Pharmacol* **391:** 255–262.

Ngo-Anh TJ, Bloodgood BL, Lin M, Sabatini BL, Maylie J, Adelman JP. 2005. SK channels and NMDA receptors form a Ca^{2+}-mediated feedback loop in dendritic spines. *Nat Neurosci* **8:** 642–649.

Nguyen QT, Tsai PS, Kleinfeld D. 2006. MPScope: A versatile software suite for multiphoton microscopy. *J Neurosci Methods* **156:** 351–359.

Pass S, Amit B, Patchornik A. 1981. Racemization-free photochemical coupling of peptide segments. *J Am Chem Soc* **103:** 7674–7675.

Pelliccioli AP, Wirz J. 2002. Photoremovable protecting groups: Reaction mechanisms and applications. *Photochem Photobiol Sci* **1:** 441–458.

Pologruto TA, Sabatini BL, Svoboda K. 2003. ScanImage: Flexible software for operating laser scanning microscopes. *Biomed Eng Online* **2:** 13. doi: 10.1186/1475-925X-2-13.

Rial Verde EM, Zayat L, Etchenique R, Yuste R. 2008. Photorelease of GABA with visible light using an inorganic caging group. *Front Neural Circuits* **2:** 2. doi: 10.3389/neuro.04.002.2008.

Sheehan JC, Wilson RM. 1964. Photolysis of desyl compounds. A new photolytic cyclization. *J Am Chem Soc* **86:** 5277–5281.

Smith MA, Ellis-Davies GC, Magee JC. 2003. Mechanism of the distance-dependent scaling of Schaffer collateral synapses in rat ca1 pyramidal neurons. *J Physiol* **548:** 245–258.

Tanaka K, Khiroug L, Santamaria F, Doi T, Ogasawara H, Ellis-Davies GC, Kawato M, Augustine GJ. 2007. Ca^{2+} requirements for cerebellar long-term synaptic depression: Role for a postsynaptic leaky integrator. *Neuron* **54:** 787–800.

Trigo FF, Corrie JE, Ogden D. 2009a. Laser photolysis of caged compounds at 405 nm: Photochemical advantages, localisation, phototoxicity and methods for calibration. *J Neurosci Methods* **180:** 9–21.

Trigo FF, Papageorgiou G, Corrie JE, Ogden D. 2009b. Laser photolysis of DPNI-GABA, a tool for investigating the properties and distribution of GABA receptors and for silencing neurons in situ. *J Neurosci Methods* **181:** 159–169.

Tsai HC, Zhang F, Adamantidis A, Stuber GD, Bonci A, de Lecea L, Deisseroth K. 2009. Phasic firing in dopaminergic neurons is sufficient for behavioral conditioning. *Science* **324:** 1080–1084.

Volgraf M, Gorostiza P, Numano R, Kramer RH, Isacoff EY, Trauner D. 2006. Allosteric control of an ionotropic glutamate receptor with an optical switch. *Nat Chem Biol* **2:** 47–52.

Weissman TA, Sanes JR, Lichtman JW, Livet J. 2011. Generating and imaging multicolor Brainbow mice. In *Imaging in neuroscience: A laboratory manual* (ed. Helmchen F, Konnerth K). Cold Spring Harbor Laboratory Press, Cold Spring Harbor, NY (in press).

Zayat L, Calero C, Albores P, Baraldo L, Etchenique R. 2003. A new strategy for neurochemical photodelivery: Metal-ligand heterolytic cleavage. *J Am Chem Soc* **125:** 882–883.

Zhang F, Wang LP, Brauner M, Liewald JF, Kay K, Watzke N, Wood PG, Bamberg E, Nagel G, Gottschalk A, Deisseroth K. 2007. Multimodal fast optical interrogation of neural circuitry. *Nature* **446:** 633–639.

Zhang YP, Holbro N, Oertner TG. 2008. Optical induction of plasticity at single synapses reveals input-specific accumulation of αCaMKII. *Proc Natl Acad Sci* **105:** 12039–12044.

60 Uncaging with Visible Light
Inorganic Caged Compounds

Leonardo Zayat, Luis M. Baraldo, and Roberto Etchenique

Departamento de Química Inorgánica, Analítica y Química Física, INQUIMAE, Facultad de Ciencias Exactas y Naturales, Universidad de Buenos Aires, Buenos Aires C1428EHA, Argentina

ABSTRACT

In this chapter, we present a new strategy to uncage biologically active molecules based on the photochemistry of metal coordination complexes. Complexes containing a ruthenium-bipyridine core are efficient photoactive cages for amines, including several neurotransmitters and their analogs such as γ-aminobutyric acid (GABA), glutamate, serotonin, nicotine, and 4-aminopyridine. The uncaging process occurs very rapidly following absorption of one photon of visible light ($<$532 nm) or two photons of infrared light (~800 nm).

INTRODUCTION

The chemistry of coordination compounds (i.e., inorganic substances involving a metallic ion chemically bonded to nonmetallic atoms of ligand molecules) is such that a wide range of photochemical reactions are used in applications ranging from photography and lithography to light harvesting in solar cells. Despite this fact, they have not been exploited for uncaging biomolecules.

Of the 90 stable or quasistable elements in the periodic table, approximately 60 are metals or present metal-like behavior. These elements have very rich chemistry. Most metals can form complexes with several ligands, which often present interesting photochemical properties. The photochemistry of coordination compounds of transition metals has been exhaustively studied (Balzani and Carassitti 1970). The main difference between the photochemical properties of metal complexes and organic compounds is that although most organics photodecompose only in response to ultraviolet (UV) photons, many coordination complexes can also respond to visible light. Because of the interaction between the transition ion and the ligands, the spectroscopic properties of the coordination compounds can be tuned, thus changing the properties of the ligands. This chapter presents a brief survey of metal-based caged compounds and their advantages arising from their chemical and physical properties.

Ruthenium-Bipyridine Complexes

The use of Ru(II) complexes as photosensitizer dyes in solar cells was pioneered by Grätzel (Desilvestro et al. 1985; O'Regan and Grätzel 1991). Modified 2,2′-bipyridines (bpy) were used as ligands to produce highly efficient light-absorbing complexes. Upon irradiation, a charge transfer occurs between ruthenium (Ru) and the bpy ligand, generating an excited state that can lead to decomposition. One of the first reports of photodecomposition in a Ru-bpy complex was for $[Ru(bpy)_2(py)_2]^{2+}$ (Bosnich and Dwyer 1966), which, in aqueous solution, releases one of the pyridines with a quantum yield of 0.26 (Pinnick and Durham 1984).

The mechanism for the release of a ligand from a metal complex is usually different from that observed in the breaking of a covalent bond in the release from an organic molecule. Organic molecules tend to decompose through a series of steps, generating reactive intermediates or radicals, which decay into the final biomolecule through a dark process. Photoinduced reactions of inorganic complexes usually involve the heterolytic release of a ligand and its substitution in the coordination sphere by a solvent molecule within a single photochemical step. The stability of the resulting complex and the absence of subsequent dark reactions are the main reason why many inorganic photochemical processes do not present side reactions.

The fact that visible light ($\lambda > 400$ nm) can be used to release a ligand is a great advantage for the uncaging of compounds in living tissue, as it is much less harmful and much more penetrating than UV light. The metallic ions usually present multiple coordination positions (frequently up to 6); and, hence, nonphotolabile ligands could also be present in the cage. Their chemical derivatization opens up the possibility for tuning the absorption wavelength, hydrophilicity, redox potential, steric hindrance, and fluorescence properties of these compounds. Another advantage of metal caged compounds is that they can be irradiated without requiring expensive optics. Good-quality objectives, suitable for normal light microscopy, can be used without the energy losses caused by glass absorption. Although most organic caged compounds require excitation in the 300–350-nm range, the absorption wavelength of two coordination caged compounds can be differentially tuned, providing two different visible irradiation ranges for the release of two different neurochemicals in different locations or at different times.

The first metal-based caged compound capable of releasing organic biomolecules was $[Ru(bpy)_2(4AP)_2]^{2+}$, in which 4AP is 4-aminopyridine, a K^+ channel blocker, depicted in Figure 1A (Zayat et al. 2003). The chloride salt of this complex is very soluble in water, and its solutions present a strong metal-to-ligand charge transfer (MLCT) band centered at 490 nm. Figure 2 shows the states diagram of Ru-bpy complexes. Irradiation at the MLCT band promotes a transition to an excited state in which some electronic density is shifted from the Ru center to one of the bipyridines, formally a $[Ru^{3+}bpy^{-}]$ state. This state can populate a d-d dissociative state, which leads to decomposition, yielding a free 4AP and the aquo-complex $[Ru(bpy)_2(4AP)(H_2O)]^{2+}$ as the two unique photoproducts. Almost all Ru-bpy complexes present similar photochemistry. The complex of 4AP is remarkably stable (no dark decomposition is apparent in aqueous solution after 24 h at ambient temperature), and it is nontoxic when applied in extracellular media. Its quantum yield at 473 nm is ~0.03, which is significant for visible-light decomposition. The caged complex fluoresces at ~590 nm, whereas the uncaged form does not present emission. This difference can be useful for in situ quantification of photodecomposition. Biological assays using this complex have been made in *Hirudo medicinalis* leech ganglia and mouse brain cortical slices. The complex is also useful in a two-photon (2P) regime, in which the simultaneous absorption of two photons of infrared light (~800 nm) leads to the release of 4AP (Nikolenko et al. 2005).

Other complexes based in the same Ru-bpy fragment have been synthesized. Caged serotonin ($[Ru(bpy)_2(5HT)_2]^{2+}$) has properties similar to those of $[Ru(bpy)_2(4AP)_2]^{2+}$ with the absorption maxima shifted to slightly longer wavelengths. Other amines can also be caged in this way (Zayat et al. 2006). Although acetylcholine, being a quaternary amine, cannot be caged in this way, its agonist nicotine (Nic) can be caged using the Ru-bpy system. The complex $[Ru(bpy)_2(Nic)_2]^{2+}$ presents a very high quantum yield of photodecomposition ($\phi_{PD} = 0.28$, $\lambda = 473$ nm) and can be used effectively with light up to 532 nm. (O. Filevich, M. Salierno, and R. Etchenique, in prep.). The structures for these complexes are depicted in Figure 1, B and C.

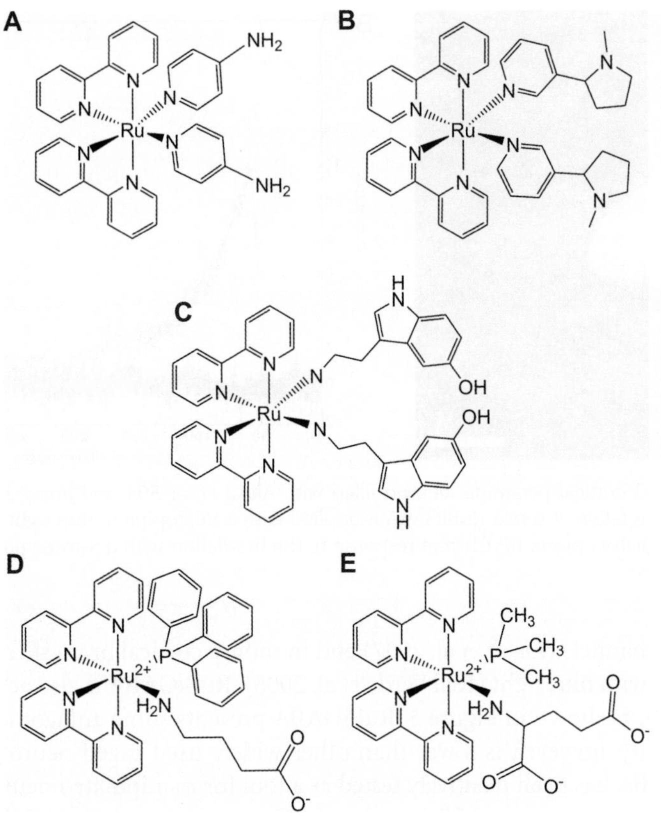

FIGURE 1. Structures of the photoactive complexes based on the Ru-bpy core. (*A*) $[Ru(bpy)_2(4AP)_2]^{2+}$, (*B*) $[Ru(bpy)_2(Nic)_2]^{2+}$, (*C*) $[Ru(bpy)_2(5HT)_2]^{2+}$, (*D*) $[Ru(bpy)_2(PPh_3)(GABA)]^{+}$, and (*E*) $[Ru(bpy)_2(PMe_3)(Glu)]$.

Use of Phosphines as Auxiliary Ligands

Heteroleptic complexes, bearing two bipyridines and two different monodentate ligands, offer a new path to enhance the chemical and photochemical properties of the cages. The use of phosphines as an auxiliary ligand provides a simple path to tune the quantum yield and the hydrophilicity of the cages.

It is easy to synthesize the complex $[Ru(bpy)_2(GABA)_2]$, which releases the inhibitory neurotransmitter GABA ($\phi_{PD} = 0.036$) (Zayat et al. 2006). Unfortunately, the complex presents activity in the absence of light, possibly because of some ligand exchange between GABA and the surrounding water. The replacement of one of the GABA ligands by triphenylphosphine (PPh_3) yields the mixed complex $[Ru(bpy)_2(PPh_3)(GABA)]^+$ (RuBiGABA; see Fig. 1D), which presents a high obscure stability and an increased quantum yield of $\phi_{PD} = 0.21$. This compound has been used in oocytes

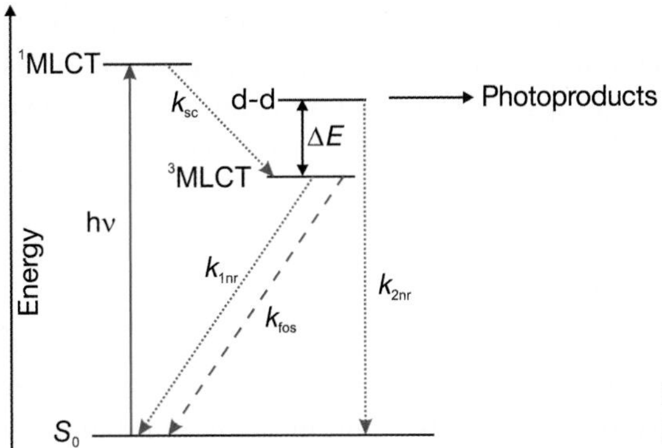

FIGURE 2. States diagram of the complexes $[Ru(bpy)_2XY]^{n+}$. Irradiation at the MLCT band finally leads to heterolytic decomposition through a d-d nonligand state.

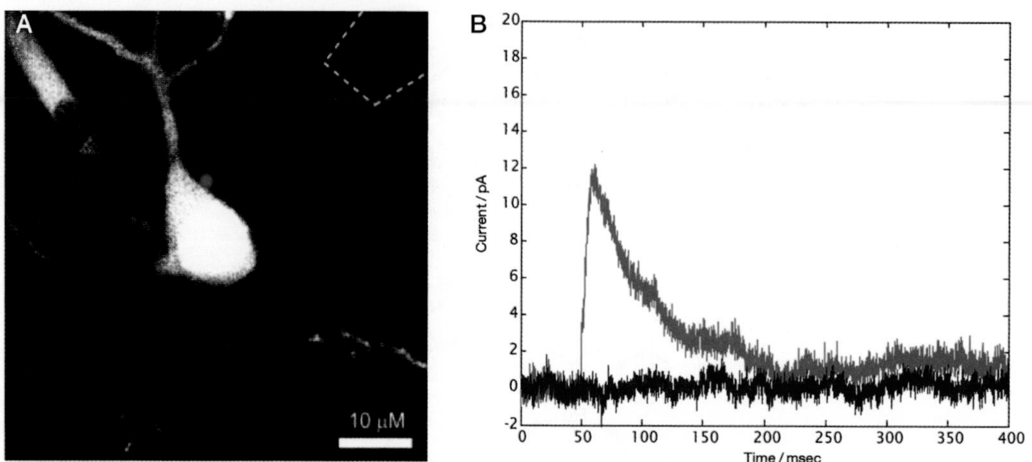

FIGURE 3. *(A)* Layer 2/3 cortical pyramidal neuron filled with Alexa Fluor 594 and imaged in two-photon fluorescence. After the image is taken, 1.6 mM RuBiGABA is applied with a micropipette (top right, dotted line). The small circle indicates the irradiation place. *(B)* Current response to the irradiation with a 5-msec pulse of 800-nm light.

expressing $GABA_C$ channels (Zayat et al. 2007) and in mouse cortical brain slices, being active at the one-photon regime with blue light (Rial Verde et al. 2008). RuBiGABA is also active in the 2P regime (Peterka et al. 2010), as shown in Figure 3. RuBiGABA presents some antagonist activity in $GABA_A$ channels. This activity, however, is lower than other widely used caged neurotransmitters such as MNI-Glut. RuBiGABA has been positively tested as a tool for manipulated neural circuits and stopping epileptic seizures by means of LED irradiation (Yang et al. 2009). RuBiGABA is the first significantly caged GABA known to be active and usable in the 2P regime.

Glutamate, the major excitatory neurotransmitter in mammals' central nervous systems, can also be effectively caged using Ru-bpy-phosphine. The complex $[Ru(bpy)_2(PMe_3)(Glu)]$ (PMe_3, trimethylphosphine) is very soluble in water and in saline solutions and uncages glutamate with ($\phi_{PD} = 0.12$) at 450 nm. Its structure can be seen in Figure 1E (RuBiGlutamate). This caged glutamate can also be effectively used in a 2P regime, having an effective action cross section of ~0.14 GM, more than twice that of MNI-Glut (Salierno et al. 2010). Its stability is remarkable; it can be stored as a solution for months without any perceptible change in its properties. Figure 4 shows the effect of irradiation in the 2P regime, which elicits action potentials in a reliable way. Its $GABA_A$ antagonist activity is much lower than MNI-Glut when used in the concentration needed to induce physiological responses (Fino et al. 2009). RuBiGlutamate and RuBiGABA are now commerically available from Tocris and Ascent.

The use of phosphines as auxiliary ligands in the design of caged compounds opens a wide spectrum of possible cages. For example, the quantum yield of the molecules can be tuned by changing the substituents in the phosphine ligand. A series of Ru-bpy GABA phototriggers has been devised in which the phosphine can be PMe_3, PMe_2Ph, $PMePh_2$, or PPh_3. The quantum yield of GABA photouncaging increases through the series from ~0.12 to 0.30. On the other hand, each complex in the series has increasingly lower solubility in water and higher membrane affinity because of the increasing number of lipophilic phenyl groups. In the same way, complexes with phosphines bearing specific chemical groups can be synthesized without loss of their main photochemical capabilities.

WORKING WITH INORGANIC CAGED BIOMOLECULES

In addition to all of the general precautions regarding purity, stability, and biocompatibility associated with conventional caged compounds, visible-light uncaging substances also require that all of

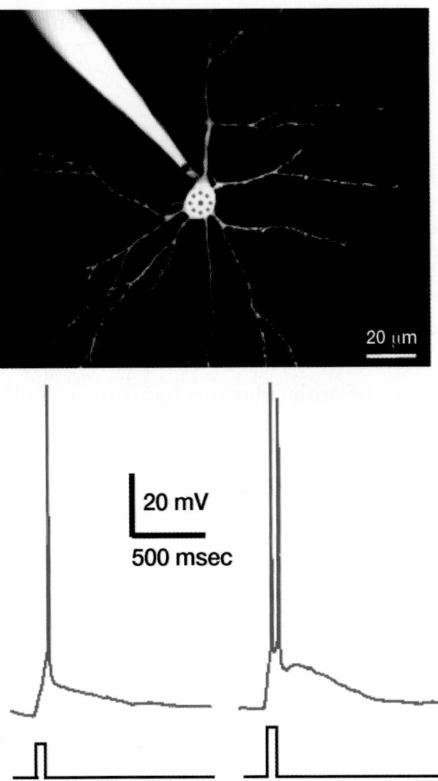

FIGURE 4. Action potentials triggered by RuBiGlutamate (300 μM) in a layer 2/3 pyramidal neuron using two-photon excitation. In this example, one action potential was evoked with a power-on sample of 180 mW, and two action potentials were evoked with 220 mW at 800 nm.

the manipulations be performed either in the dark or by using a filter that blocks the uncaging portion of the spectrum. Diminished tungsten light can be enough for manipulation, but fluorescent light—with a high content of short-wavelength light—should be avoided. Yellow or red LED (light-emitting diode) lamps are the best sources to prevent undesired uncaging. The same precautions have to be taken into account when synthesizing these compounds.

In principle, any molecule having a donor nitrogen can be cageable, making it possible to create a Ru-bpy complex using standard synthesis techniques (Zayat et al. 2006, 2007). In practice, however, steric hindrance can prevent the caging of some ligands. Deactivation through nonradiative vibrational processes can compete with photocleavage, particularly in cases in which the excited state has several ways to decay by interacting with the surrounding solvent. Some sulfur and phosphorus compounds can also be caged in a similar manner. RuBiGlutamate, RuBiGABA, and other widely used caged neurochemicals are now offered by Tocris Bioscience.

Most visible-light sources can be used for uncaging. Xenon-pulsed lamps are cheap and are easy to use in the microsecond-to-millisecond range, but focusing is very difficult. This kind of source can be used for kinetic investigation of fast channels. High-power blue LEDs are now inexpensive and reliable and can be electronically strobed and dimmed with microsecond control. Several lasers are available in the 400–500-nm range. Argon (488-nm) lasers are suitable but are big and expensive. A better choice is the new blue diode-pumped solid-state Nd:YAG laser at 473 nm. Alternatively, a Nd:YAG laser at 532 nm can be used, but the efficiency of photorelease is rather poor because of the low extinction coefficient of the complexes at such a long wavelength. A new 405-nm solid-state laser is becoming popular because of Blue-Ray technology. Although its wavelength is slightly shorter than the optimum for most Ru-bpy caged compounds, its ease of use, availability, and very low cost configure a good compromise, particularly in cases in which low size and weight become important (i.e., in vivo uncaging). Many Ru-bpy caged compounds are also active in the 2P regime around 800 nm, which is usually achieved using Ti:sapphire lasers. Ru-bpy caged compounds also absorb strongly in the 330–380-nm range, making them compatible with the near-UV lasers used for organic caged compounds, although at the cost of their main convenience: the ability to provide mild irradiation.

CONCLUSION

Ru-based caged compounds present several advantages over organic phototriggers: the ability to uncage with visible light instead of UV, easy derivatization, wavelength tunability, intrinsic fluorescence, chemical tunability, and high stability. Most of the caged compounds are also active in a 2P regime. Several important caged neurochemicals have already been prepared, including GABA, glutamate, serotonin, 4AP, and nicotine. Given their distinctive properties, this family of caged compounds should be useful for the exploration of molecular mechanisms in biological systems.

REFERENCES

Balzani V, Carassitti V. 1970. *Photochemistry of coordination compounds*. Academic, London.

Bosnich B, Dwyer FP. 1966. Bis-1,10-phenanthroline complexes of divalent ruthenium. *Aust J Chem* **19:** 2229–2233.

Desilvestro J, Grätzel M, Kavan L, Moser J, Augustynski J. 1985. Highly efficient sensitization of titanium dioxide. *J Am Chem Soc* **107:** 2988–2990.

Fino E, Araya R, Peterka DS, Salierno M, Etchenique R, Yuste R. 2009. RuBi-Glutamate: Two-photon and visible-light photoactivation of neurons and dendritic spines. *Front Neural Circuits* **3:** 2. doi:10.3389/neuro.04.002.2009.

Nikolenko V, Yuste R, Zayat L, Baraldo LM, Etchenique R. 2005. Two-photon uncaging of neurochemicals using inorganic metal complexes. *Chem Commun* **2005:** 1752–1754. doi: 10.1039/b418572b.

O'Regan B, Grätzel M. 1991. A low cost, high efficiency solar cell based on dye sensitized colloidal Tio2 films. *Nature* **353:** 737–740.

Peterka DS, Nikolenko V, Fino E, Araya R, Etchenique R, Yuste R. 2010. Fast two-photon and neuronal imaging and control using a spatial light modulator and ruthenium compounds. *Proc SPIE* **7548:** 75484P. doi:10.1117/12.842606.

Pinnick DV, Durham B. 1984. Photosubstitution reactions of $Ru(bpy)_2XY^{n+}$ complexes. *Inorg Chem* **23:** 1440–1445.

Rial Verde EM, Zayat L, Etchenique R, Yuste R. 2008. Photorelease of GABA with visible light using an inorganic caging group. *Front Neural Circuits* **2:** 2. doi: 10.3389/neuro.04.002.2008.

Salierno M, Marceca E, Peterka DS, Yuste R, Etchenique R. 2009. A fast ruthenium polypyridine cage complex photoreleases glutamate with visible or IR light in one and two photon regimes. *J Inorg Biochem* **104:** 418–422.

Yang X, Peterka DS, Yuste R, Rothman SM. 2009. Anticonvulsant effect of photorelease of GABA using a visible light emitting diode and a novel caged compound in rat brain slices. 63rd Annual Meeting of the American Epilepsy Society, December 2009, Boston, MA.

Zayat L, Calero C, Albores P, Baraldo L, Etchenique R. 2003. A new strategy for neurochemical photodelivery: Metal ligand heterolytic cleavage. *J Am Chem Soc* **125:** 882–883.

Zayat L, Salierno M, Etchenique R. 2006. Ruthenium(II) bipyridyl complexes as photolabile caging groups for amines. *Inorg Chem* **45:** 1728–1731.

Zayat L, Noval MG, Campi J, Calero CI, Calvo DJ, Etchenique R. 2007. A new inorganic photolabile protecting group for highly efficient visible light GABA uncaging. *ChemBioChem* **8:** 2035–2038.

Appendix 1

Electromagnetic Spectrum

Marilu Hoeppner

The Electromagnetic Spectrum

The different forms of electromagnetic radiation include radio waves, infrared (IR) rays, visible light, ultraviolet (UV) light, X-rays, and γ rays. All electromagnetic rays exhibit the properties of waves, each having a characteristic wavelength (λ) and frequency (ν). Electromagnetic waves travel with the speed of light c, which in vacuum is $\sim 3 \times 10^8$ m/sec. The relationship between the wavelength, the frequency, and the speed of light is given by $c = \lambda\nu$. Thus, it is possible to refer to electromagnetic radiation in terms of wavelength or frequency. Electromagnetic radiation may also be considered in terms of quanta or photons. The relationship between the energy of the electromagnetic wave E and the frequency is given by $E = h\nu$, where h is Planck's constant ($h = 6.63 \times 10^{-34}$ J sec). The higher the frequency of the electromagnetic wave (i.e., the shorter the wavelength), the greater the amount of energy carried by a photon.

Abbreviations

Photon energy: eV, electron volt (1 eV = 1.60219×10^{-19} joule); keV, kiloelectron volt (10^3 eV); MeV, megaelectron volt (10^6 eV).

Frequency: Hz, hertz (cycles per second); kHz, kilohertz (10^3 Hz); MHz, megahertz (10^6 Hz); GHz, gigahertz (10^9 Hz); THz, terahertz (10^{12} Hz); PHz, petahertz (10^{15} Hz); EHz, exahertz (10^{18} Hz).

Wavelength: m, meter; km, kilometer (10^3 m); Mm, megameter (10^6 m); mm, millimeter (10^{-3} m); µm, micrometer (10^{-6} m); nm, nanometer (10^{-9} m); pm, picometer (10^{-12} m); fm, femtometer (10^{-15} m).

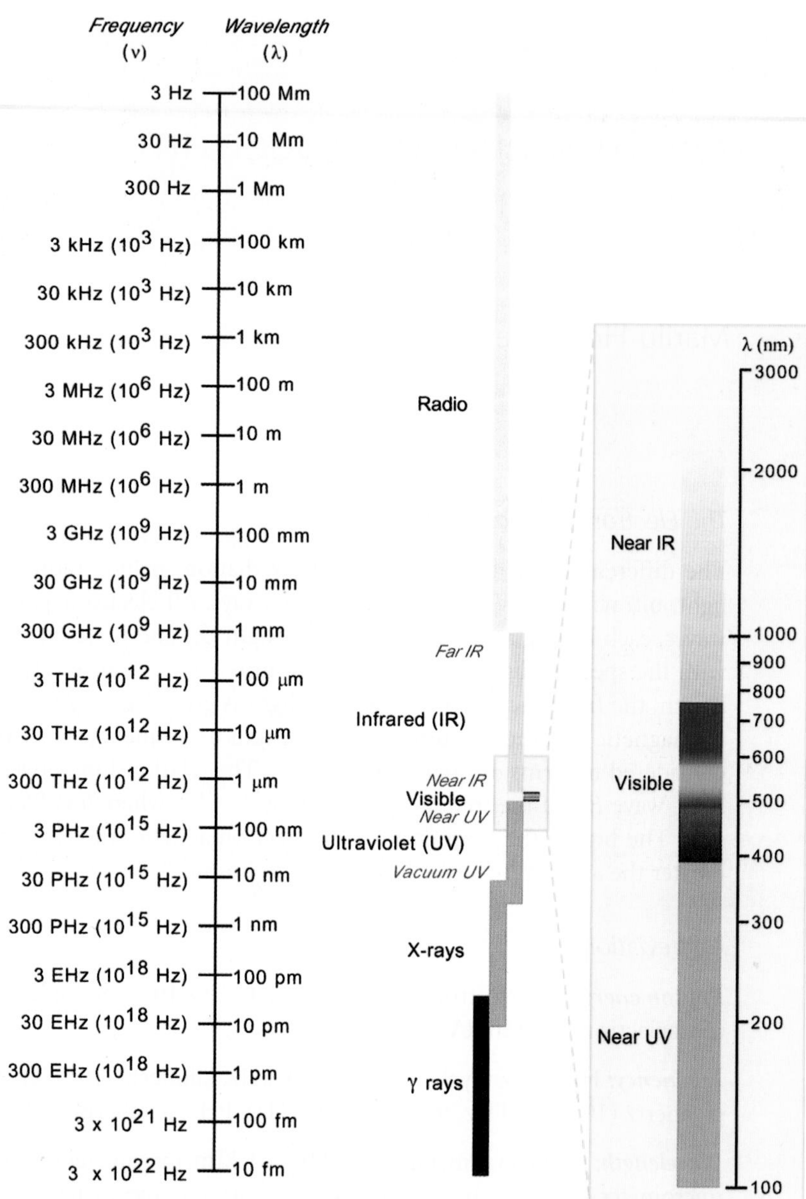

FIGURE 1. The electromagnetic spectrum. (Provided by Marilu Hoeppner.)

Appendix 2 Fluorescence Microscopy Filters and Excitation/Emission Spectra

FILTERS USED IN FLUORESCENCE MICROSCOPY

The primary filtering element in the epifluorescence microscope is the set of three filters housed in the fluorescence **filter cube**, also called the **filter block**: the **excitation filter**, the **emission filter**, and the **dichroic beam splitter**. A typical filter cube is illustrated schematically in Figure 1.

- The **excitation filter** (also called the **exciter**) is a color filter that transmits only those wavelengths of the illumination light that efficiently excite a specific dye. Common filter blocks are named after the type of excitation filter: UV or U (ultraviolet) for exciting DAPI, Indo-1, etc.; B (blue) for exciting FITC; and G (green) for exciting TRITC, Texas Red, etc. Although short-pass filter designs were used in the past, band-pass filter designs are now used.
- The **emission filter** (also called the **barrier filter** or **emitter**) is a color filter that attenuates all of the light transmitted by the excitation filter and very efficiently transmits any fluorescence emitted by the specimen. This light is always of longer wavelength (more to the red) than the excitation color.

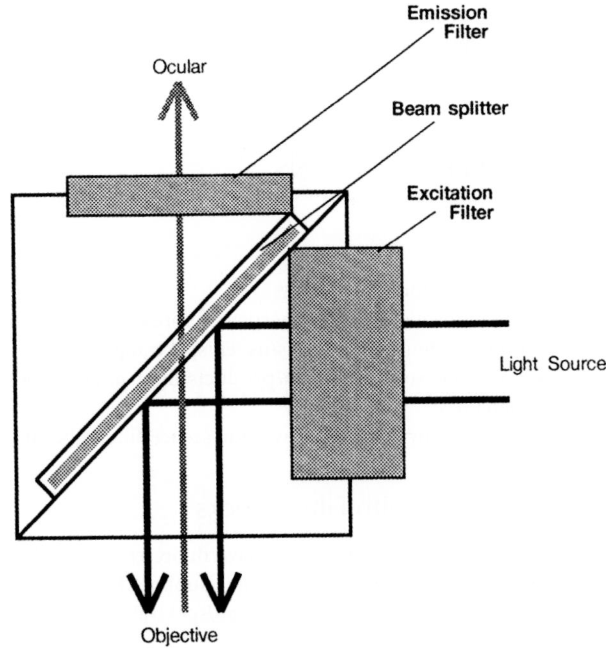

FIGURE 1. Schematic of a fluorescence filter cube.

These can be either band-pass filters or long-pass filters. Common barrier filter colors are blue or pale-yellow in the U-block; green or deep-yellow in the B-block; and orange or red in the G-block.

- The **dichroic beam splitter** (also called the **dichroic mirror** or **dichromatic beam splitter**) is a thin piece of specially coated glass (the substrate) set at an angle 45° to the optical path of the microscope. This coating has the unique ability to reflect one color (the excitation light) but transmit another color (the emitted fluorescence). Current dichroic beam splitters achieve this with great efficiency—that is, with >90% reflectivity of the excitation along with ~90% transmission of the emission. This is a great improvement over the traditional gray half-silvered mirror, which reflects only 50% and transmits only 50%, giving only ~25% efficiency.

Most microscopes have a slider or turret that can hold from two to four individual filter cubes. It must be noted that the filters in each cube are a matched set, and mixing filters and beam splitters should be avoided unless the complete spectral characteristics of each filter component are known.

Other optical filters can also be found in fluorescence microscopes.

- A heat filter (also called a hot mirror) is incorporated into the illuminator collector optics of most but not all microscopes. It attenuates infrared light (typically wavelengths >800 nm) but transmits most of the visible light.
- Neutral-density filters, usually housed in a filter slider or filter wheel between the collector and the aperture diaphragm, are used to control the intensity of illumination.
- Filters used for techniques other than fluorescence, such as color filters for transmitted-light microscopy and linear polarizing filters for polarized-light microscopy, are sometimes installed. (Information courtesy of Chroma Technology Corp.)

REFERENCES

Bieber F. 1994. Microscope and image analysis. In *Current protocols in human genetics* (ed. Dracopoli NC, et al.), p. 4.4.5. John Wiley, New York.

Spector DL, Goldman RD, Leinwand LA, eds. 1998. Microscopy: Lenses, filters, and emission/excitation spectra. In *Cells: A laboratory manual*, Vol. 3. *Subcellular localization of genes and their products*, pp. A3.1–A3.10. Cold Spring Harbor Laboratory Press, Cold Spring Harbor, NY.

WWW RESOURCES: FILTER SPECIFICATIONS

Filter Sets

http://chroma.com/products/catalog/
http://www.olympusmicro.com/primer/java/fluorescence/matchingfilters/index.html
https://www.micro-shop.zeiss.com/us/us_en/spektral.php?f=fi
http://www.leica-microsystems.com/products/light-microscopes/accessories/filtercubes-and-fluochromes/find-your-filtercubes/
http://www.microscopyu.com/articles/fluorescence/filtercubes/filterindex.html

Useful Tools for Matching Spectra with Filter Blocks

http://www.olympusmicro.com/primer/java/fluorescence/matchingfilters/index.html
http://www.invitrogen.com/site/us/en/home/support/Research-Tools/Fluorescence-SpectraViewer.html
https://www.micro-shop.zeiss.com/us/us_en/spektral.php?f=fa

Two-Photon Action Cross Sections

http://www.drbio.cornell.edu/cross_sections.html

TABLE 1. Excitation (Exc) and emission (Em) maxima for typical fluorochromes

	Exc (nm)	Em (nm)
4-methylumbelliferone	360	450
6-amino-quinoline	360	443
7-amino-actinomycin D	555	655
Acid Fuchsin	540	630
Acridine orange + DNA	502	526
Acridine orange + RNA	460	650
Acridine yellow	470	550
Acriflavin-Feulgen	480	550–600
Alexa Fluor 350	346	442
Alexa Fluor 405	401	421
Alexa Fluor 430	433	541
Alexa Fluor 488	496	519
Alexa Fluor 532	532	553
Alexa Fluor 546	556	573
Alexa Fluor 555	555	565
Alexa Fluor 568	578	603
Alexa Fluor 594	590	617
Alexa Fluor 610	612	628
Alexa Fluor 633	632	647
Alexa Fluor 635	633	647
Alexa Fluor 647	650	665
Alexa Fluor 660	663	690
Alexa Fluor 680	679	702
Alexa Fluor 700	702	723
Alexa Fluor 750	749	775
Alexa Fluor 790	784	814
Alizarin complex	530–560	580
Allphycocyanin	630	660
AMCA	345	425
Amino-methylcumarin	354	441
Auramine	460	550
BCECF	430/480	520
Berberine sulfate	430	550
BFP (Blue Fluorescent Protein)	382	448
BOBO-1	462	481
BOBO-3	570	602
BODIPY	503	512
BODIPY 581/591 phalloidin	584	592
Calcein	495	500–550
Calcein Blue	375	420–450
Calcium Crimson	590	615
Calcium Green	505	535
Calcium Orange	549	576
Cascade Blue	376/399	423
Catecholamine	410	470
Cerulean	433	475
Chromomycin	436–460	470
Coriphosphine	460	575
Coumarin maleimide	385	471
Cyanine Cy3.5	578	691
Cyanine Cy2	489	505
Cyanine Cy3	549	562
Cyanine Cy5	640	664
Cyanine Cy5.5	673	692
Dansyl chloride	380	475
DAPI + DNA	359	461
DiBAC4	439	516
DIDS thiocyanatostilbene	342	418
Di-2 ANEPEQ (JPW 1114)	517	721
Di-8 ANEPPS	469	630
DiI	557	565
DiO	551	569
DsRed	558	583
DTAF	495	528
dTomato	554	581
Ethidium bromide + DNA	510	595
Euchrysin	430	540
Evans Blue	550	611
eYFP	514	527
Feulgen	480	560
FITC	495	517
Fluo-3	506	526
Fluo-4	494	516
Fluorescein	494	521
Fluoro Gold	350–395	530–600
FluoSpheres (blue)	360	415
FluoSpheres (crimson)	625	645
FluoSpheres (dark red)	650	690
FluoSpheres (red)	580	605
FluoSpheres (yellow-green)	490	515
FM1-43	479	598
FM4-64	508	751
Fura Red	420/480	640/665
Fura-2	340/380	500/530
GFP (enhanced)	489	509
Hoechst 33258, 33342	352	461
Indo-1	328/352	403/520
Lissamine Rhodamine B	575	595
Lucifer Yellow	428	540
Magdala Red	524	600
Mag-indo-1	331	415
mBanana	540	553
mCFP	433	475
mCherry	587	610
Merocyanine	555	578
mHoneydew	487/504	537/562
Mithramycin	420	575
mOrange	548	562
mPlum	590	649
mRaspberry	598	625
mRFP1	584	607
mStrawberry	574	596
mTangerine	568	585
NBD–amine	460–485	534–542
NBD–chloride	480	510–545
Nile Red	559	637
Olivomycin	350–480	470–630
Oregon Green 488	498	526
Pararosaniline-Feulgen	560	625
PBFI	340/380	420
Phosphine	465	565
Phycoerythrin	565	575
POPO-1	434	456
POPO-3	534	570
Primuline	410	550
Propidium iodide	536	617
Pyrene	343	380–400
Pyronine	410	540
Quin-2	340/365	490
Quinacrine mustard	440	510

Continued on following page.

TABLE 1. *(Continued)*

	Exc (nm)	Em (nm)		Exc (nm)	Em (nm)
Resorufin	571	585	Tetracycline	390	560
Rhodamine	551	573	Tetramethylrhodamine	540	566
Rhodamine 123	507	529	Texas red	595	620
Rhodol green	500	525	Thiazine red	510	580
R-phycoerythrin	565	578	Thioflavin 5	430	550
SBFI	340	555	TOTO-1 + DNA	509	533
SITA	365	460	TOTO-3 + DNA	642	661
SNARF	563	639	TRITC	557	580
Stilbene	335	440	XRITC	582	601
Sulfaflavin	380–470	470–580	Xylenol orange	377	610
Sulforhodamine 101	579	594	YOYO-1	491	509
tdTomato	554	581	YOYO-3	612	631

Provided by Adam Packer, Columbia University.

Appendix 3 Safe Operation of a Fluorescence Microscope

George McNamara

Fluorescence microscopes are used by thousands of researchers every day with few publicized reports of major accidents. This may be a matter of luck! Here we highlight several safety issues.

The standard illuminator for fluorescence microscopy is the high-pressure mercury arc lamp (sometimes called an Hg bulb or Hg burner), which can explode, spraying shards of glass. The glass can ruin the mirror and collector lens and injure people in the vicinity. Explosions typically occur either during startup or when someone is trying to change the bulb position by adjusting one or more controls on the lamphouse. Safety recommendations include the following.

1. New users should be trained by an experienced user in the laboratory or by the microscope sales representative.
2. Never operate an arc lamp without its housing.
3. If the lamphouse does not incorporate an explosion trap, then replace it with one that does.

A famous cartoon in many laser laboratories shows a scientist in a lab coat with an eye patch and the caption, "Do not look at laser with remaining eye." Commercial laser microscope systems should have one or more safety interlocks to prevent laser light from illuminating the specimen if the condenser arm is raised back. This is especially a concern for total internal reflection fluorescence (TIRF) microscopes, whose safety interlocks (if ever present) can be defeated by the user to enable shining the light up to the ceiling for inverted systems or down to the microscope stand on upright systems (if the condenser has been removed). Illuminating the ceiling may facilitate visualizing that the TIRF critical angle has been exceeded (the illumination disappears because it is trapped in the cover glass), but it is a major safety issue. Standard confocal and TIRF microscope lasers range in power from a modest 1 mW to a fairly powerful 100 mW up to >3 W for the average power of a multiphoton laser. Almost all multiphoton lasers emit ~100 femtosecond (100×10^{-15} sec) pulses at an 80-MHz pulse frequency (80×10^{6} times per second). The total pulse time is thus 8×10^{-6} sec, and the peak power is the product of the inverse of this number times the average power. For 1 W, the peak power is 125,000 W, which is significantly greater than the threshold required for permanent eye damage. When operating any high-power laser (i.e., >10 mW for a CW laser and for ultrafast lasers of any average power), observe the following.

1. Do not remove the safety interlocks.
2. Do not insert reflective objects into the laser light path during operation—especially metal or glass at the output of the objective lens.
3. Operate the laser behind blackout curtains.
4. Do not design the optical light path such that anyone's eyes are going to be at laser output height.

This latter caution is especially important for multiphoton laser systems in which the laser beam is guided around the optical table in an open-path design. Redesign these systems so that they are enclosed.

Most modern arc lamps have the high-voltage starter unit integrated into the lamphouse. With many older arc lamp systems, however, the starter unit is in the power supply. Starting an older unit requires sending an electromagnetic pulse (EMP) down the cable from the power supply to the lamphouse. This EMP can cross over from its cable to other electrical cables, especially if they are aligned in parallel to the arc lamp cable. The EMP can destroy electronics attached to the other cables, such as the computer, filter wheel and its controller, or other peripherals attached to your instrument. Therefore, do not bundle the arc lamp cable with other cables. In addition, arc lamps should not be attached to power line conditioners or to standard uninterruptible power supplies. The arc lamp power supply should not have an automatic restart, because power outages (especially multiple cycles) can trigger a lamp explosion. There should be no cables, paper notes, or walls near an arc lamp and there should be free air circulation under and above the lamp. The use of a liquid light guide or fiber optic is a good investment (see below).

The newest generation of fluorescence illuminators may be safer, as well as having longer lasting bulbs and (hopefully) much more stable light output. Interesting designs include LED illuminators such as the Colibri (Zeiss) or precisExcite (CoolLED), and metal halide lamps such as the X-Cite 120 (EXFO), PhotoFluor II (Chroma Technology or 89North.com), or Lambda XL plasma lamp (Sutter Instruments). Xenon arc lamps, such as the DG-4 or Lambda LS300 (Sutter Instruments), may be preferable because they do not use high pressure, may have more favorable spectral output, and may have longer bulb lifetimes than do mercury arc lamps. All lamps can—and ideally should—be attached to the microscope by a liquid light guide or fiber-optic scrambler, which scrambles the lamp image generating much more uniform illumination than is provided by a standard arc lamp attached directly to the microscope. Protect the light guide or fiber-optic scrambler from heat by including IR (infrared)-cutoff filters in the lamphouse, and protect it from users, because a kink could ruin a $1000 light guide. Each liquid light guide should be regarded as a consumable, because even if well maintained, it will become dimmer over time (1–3 yr).

Many laboratories will keep their standard mercury arc lamp fluorescence illuminators for a long time. The standard 100-W Hg arc lamp bulb (~$125/bulb) is rated for 200–300 h. In our laboratory, the first user of the day turns on the lamp and the last user or last person to leave the laboratory turns off the arc lamp at the end of the workday. When using our Zeiss FluoArc lamphouse/power supplies, we turn the power down to 20% between sessions. When replacing a bulb, wear gloves to prevent finger oils from contaminating the glass envelope, and gently rub the glass envelope with an ethanol-moistened Kimwipe to remove any chemical residues picked up during manufacture or in storage (tip from J. Zhang, Leica Microsystems, via a Leica customer). This cleaning and handling procedure can triple the useful life of the Hg bulb.

Appendix 4

Microscope Objective Lenses

Adapted from Inoué 1986 by A. Packer (Columbia University)

CARE AND CLEANING OF OPTICAL EQUIPMENT

Purpose of Cleanliness

Microscopes provide optimal images only if lens surfaces are kept clean. Cleanliness requirements are most stringent for polarizing and interference microscopes, but a fingerprint on any lens of a microscope can spoil the image. Similarly, it is important to use clean, unscratched, high-quality slides and coverslips and to keep fingerprints and all other dirt from their surfaces.

How to Clean an Optical Surface

1. Use only high-quality lens paper.
2. Do not touch a lens surface except as a last resort. Dust can often be removed from the surface of a lens by using a small bulb to puff air lightly over the surface. Do not try to blow off the surface with your mouth as you are likely to deposit small droplets of spit. Never rub a lens with lens paper or other items such as Q-tips, commercial facial tissue, or Kimwipes. Small dust particles are always present in the air and, if rubbed into a microscope lens, these dust particles can cause microscopic scratches in the lenses and the image will be degraded. Do not clean lenses with commercial facial tissue or bathroom tissue, which contain diatom frustules (glass) as filler. Lens coatings must be treated with respect.
3. If deposits have dried onto the surface of the lens, first attempt to clean the lens with distilled water by placing a drop on the optical surface and sliding a pristine piece of high-quality lens paper across the surface. If repeated attempts of this procedure do not produce a clean optical surface, then try alcohol as an alternative solvent. Be particularly cautious with filters as they are coated with antireflective coatings—read the cleaning instructions that come with the filter (e.g., http://chroma.com/support/filter-support).
4. Inspect the lens using an inverted ocular from the microscope as a magnifier. Observe the objective lens in the reflected light from a lamp (generally in the ceiling). To observe microscopic imperfections and dirt, place the lens to be inspected and cleaned under a dissecting microscope and focus on the lens. The lens will need to be illuminated from above.

Keeping Lenses Clean

1. Only use oil on lenses designed to be used with oil, as it can be time-consuming to clean a lens that has accidentally been immersed in oil. Clean oil off a lens after each session.

2. Keep objectives that are not in use in the screw-top containers in which they came. Always inspect lens surfaces with an inverted ocular before putting them away.
3. Some microscopists find it useful not to clean the lens between slides if they are continually using an oil-immersion lens with a series of slides. Instead, wipe off excess oil (especially important in inverted microscopes) with lens paper and place new slide with a bit of new oil onto the old oil, being careful to avoid air bubbles.
4. If using seawater, be especially careful to remove it from all metallic and lens surfaces as it will corrode metals and etch away the antireflective coating of some lenses.
5. Keep condensers, compensators, etc., in boxes, plastic "baggies," or Petri dishes until used.
6. Prohibit all smoking in microscopy laboratories because tobacco tar accumulates on lens surfaces and reduces their efficiency.

Personal Safety

See Appendix 3 for additional safety information.

1. Never open an ether bottle if there is a flame in the room.
2. Observe the no smoking regulation.
3. Avoid damage to the retinas of your eyes by protecting them from unfiltered arc sources (e.g., mercury, xenon), strobe lamps, and especially lasers.
4. If a high-pressure mercury arc lamp should explode or even stop working, *evacuate the room without delay*. If possible, open a window as you exit and then notify your institution's safety office.

REFERENCES

Inoué S. 1986. *Video microscopy*. Plenum Press, New York.

Spector DL, Goldman RD, Leinwand LA, eds. 1998. Microscopy: Lenses, filters, and emission/excitation spectra. In *Cells: A laboratory manual*, Vol. 3. *Subcellular localization of genes and their products*, pp. A3.1–A3.10. Cold Spring Harbor Laboratory Press, Cold Spring Harbor, NY.

WWW RESOURCES: OBJECTIVE LENSES

http://www.olympusamerica.com/seg_section/uis2/seg_uis2.asp The Olympus website is a searchable database of their complete line of objectives. Suitable imaging modalities are suggested for each objective, and a resource table and a transmittance curve are provided for each objective. A brochure in pdf format with a comprehensive resource table can be accessed from the website.

https://www.micro-shop.zeiss.com/us/us_en/objektive.php?cp_sid= The Carl Zeiss website is a searchable database of their full line of objectives. Search returns are flexible, allowing comparisons among user-selected objectives and providing detailed resource information including transmittance curves.

http://www.leica-microsystems.com/products/light-microscopes/accessories/objectives/ The Leica Microsystems website is a searchable database of their complete line of objectives. Detailed resource information is provided for each objective.

http://www.nikoninstruments.com/Products/Optics-Objectives The Nikon Instruments website provides information on their complete line of objectives, suggested uses, a selection guide, and a brochure in pdf format that contains a comprehensive resource table.

Appendix 5

Glossary

Absorption: The attenuation of light as it passes through a material, generally due to its conversion to other energy forms (typically heat). At the molecular level, absorption is the process by which a quantum of energy is transferred from the radiation field to the bound electrons of the molecule. In general, absorption is a resonant process, in which the frequency of the light must match a difference frequency of oscillation between the initial and final quantum states of the molecule. In the radiation field, each energy quantum is carried by a photon, one of which is annihilated for every absorption event. *See also* Extinction coefficient, Multiphoton absorption, Photon.

Absorption spectrum: A graph showing the frequency dependence of the efficiency or rate of absorption of light in a material, on a relative or per-molecule basis. Because photon energy is proportional to frequency (given by Planck's formula; $E = h\nu$), and frequency is inversely proportional to wavelength ($\nu = c/\lambda$, in vacuo), the graph may be plotted versus frequency (Hz), energy (eV), wavelength (nm), or wave number ($1/\lambda$). In biological applications, by far the most common plot is extinction coefficient versus wavelength. *See also* Emission spectrum, Excitation spectrum.

Access resistance (or series resistance): The electrical resistance (usually in megaohms) of the connection between a measurement device and a biological cell. Typically, the device is a micropipette electrode.

Achromatic: Free of color; absence of primary chromatic aberration. In a simple lens, or singlet, the focal length and magnification vary slightly with wavelength, mainly because of slight variation in refractive index with wavelength. This leads to color-dependent sharpness of focus or color-dependent magnification in the image, known as longitudinal and lateral chromatic aberration, respectively. An Achromat is a multi-element lens that is designed to have a null point in chromatic aberration in the mid-visible spectrum and improved performance across the spectrum. An Apochromat is more broadly corrected, with significantly improved performance over an achromat.

Acousto-optic tunable filter (AOTF): An electro-optic device used as a tunable band-pass filter. An AOTF is composed of a birefringent crystal to which is coupled one or more piezoelectric transducers. Driving the transducers with a radio frequency electronic signal produces an acoustic wave train in the crystal that causes it to act as a diffraction grating with a band-pass characteristic. The center wavelength of the filter can be shifted (on the microsecond timescale) over much of the visible spectrum by variation of the radio frequency in the 40–68-MHz range. The bandwidth of the filter can be adjusted electronically in advanced designs.

Afocal: An optical system in which two lenses are positioned on a common axis with a separation equal to the sum of the two focal lengths. In this configuration, entering parallel rays are brought to a focus by the first lens, and then recollimated by the second. The combination therefore has no finite focal length, in the usual sense. However, for an object located in the front focal plane of one lens, a real image is formed in the back focal plane of the other.

Airy disk: In an aberration-free circularly symmetric optical system, the central bright peak in the in-focus image of a point source of light. The central zone of the Airy pattern.

Amplifier: An electronic or optical device that actively enlarges or strengthens a signal without significant distortion. Power amplifiers drive output devices such as motors or loudspeakers; isolation amplifiers buffer a weak signal source against drainage of power by a measuring device. The operational amplifier (Op-Amp) is an integrated differential amplifier with high gain and high input impedance. Op-Amps used with simple feedback networks form the basis for many instrumentation systems such as voltage and current clamps.

Amplitude: The strength of a wave field at a particular location and instant. For light, which is an electromagnetic wave, amplitude usually refers to the electric field strength of the wave. Because this is a vector quantity, the amplitude can be resolved into directional, or polarized, components. In general, the intensity of light is proportional to the (time-averaged) square of the amplitude.

Analog: A device or record in which a signal is represented by a proportional voltage, current, brightness, magnetization, or other physical variable. A continuous (nondigital) representation.

Analog-to-digital (A/D) converter: A device that converts an analog signal, that is, a signal in the form of a continuous variable voltage or current, to a digital signal, in the form of bits.

Antireflection coating: A thin layer of material applied to a lens surface to reduce reflected light. Ideally, the index of refraction of the film should be equal to the geometric mean of the refractive indices of the materials on either side of the coating. The ideal thickness for a single layer coating is one-quarter of the wavelength at which reflectance is to be minimized. For the most common case (glass–air interface), magnesium fluoride (MgF_2) is used.

Aperture: A measure of the angular range over which an optical system can accept entering rays. In photography, aperture is specified by f-number (f/#). In microscopy, it is specified by numerical aperture (NA). For an objective lens, NA = $n \sin \theta_{max}$, where n is the refractive index of the specimen and immersion medium, and θ_{max} is the largest angle at which light may enter the objective. The illuminating numerical aperture (INA) is defined similarly, for rays exiting the condenser into the specimen.

Aperture stop: An iris or circular mask at a location in an optical system where the diameter of the iris sets the maximum angle at which light may enter or leave the objective. *See also* Field stop.

Aplanatic: A lens system that is free from spherical aberration and coma.

Arc lamp: A light source in which emission occurs from an ionized plasma in an electric discharge between two electrodes. The electrodes are contained in a glass or fused quartz bulb containing a specific gas or vapor. For fluorescence microscopy, intense mercury-vapor and xenon lamps are most common. The 100-W mercury direct-current short arc is the brightest available nonlaser light source that is commonly available for a microscope.

Axial resolution: The sharpness with which an optical system can discriminate between features that are superimposed in an image (i.e., that differ only in best focus setting). Axial resolution is related to depth of field and to the degree to which out-of-focus features are blurred or filtered out of the image. Strictly, axial resolution is defined by the shape of the optical transfer function.

Back focal plane (of an objective): The distance from the last optical element (of an objective) to the rear focal point.

Ban-pass filter: A filter with high transmissivity in a particular wavelength interval and high blocking power outside of this range.

Beam: (1) A bundle of light rays that may be parallel, converging, or diverging. (2) A concentrated, unidirectional stream of particles. (3) A concentrated, unidirectional flow of electromagnetic waves.

Beam expander: A system of optical components designed to increase the diameter of a laser beam. Usually an afocal system composed of a low-power microscope objective used in reverse plus a conventional achromatic collimating lens.

Beam splitter: An optical element for dividing a beam into two or more separate beams. Usually, a beam splitter is a semitransparent plate on which an incident beam is partially reflected and partially transmitted. Beam splitters may be polarization or wavelength selective. *See also* Dichroism.

Binning: Detector readout and digitization process in which pixels are grouped prior to measuring signal level. For example, a 512 x 512-pixel camera array may be read out and digitized as a 256 x 256-pixel image by measuring the total accumulated charge in contiguous 2 x 2-pixel squares. Binning increases readout speed and (under some circumstances) signal-to-noise ratio at a cost of reduced resolution.

Bioluminescence: Light emission from certain living organisms caused by the enzyme-catalyzed oxidation of an organic substrate. In general, the enzyme is known as a luciferase, and the substrate is a luciferin. The catalyzed reaction between oxygen and luciferin leaves the oxidized molecule in an electronic excited state from which emission of a photon occurs.

Bit: A contraction of binary digit; the fundamental unit of digital computing. A bit is either 1 or 0, expressing the binary state of on or off, true or false, etc.

Blaze: A method of cutting the parallel grooves in a diffraction grating such that the spectrally dispersed light is concentrated in one diffraction order. In a blazed grating, the individual grooves are formed with flat smooth faces inclined to the surface by an angle known as the groove angle.

Birefringence: The optical property resulting from anisotropy of the refractive index in a material. In a birefringent medium, the effective refractive index depends on the direction of travel of a light wave and on its polarization. Many biological materials, crystals, liquid crystals, and synthetic polymers are birefringent (e.g., mitotic spindle, collagen fibrils, calcite, quartz, asbestos, cellophane, molded polystyrene). A thin flake of birefringent material placed between crossed polarizers will show brightness and possibly color variation as the flake is rotated relative to the polarizer axes. In crystals, birefringence gives rise to double refraction, in which a beam of incident light is separated into two beams of orthogonal polarization.

Bright field: Standard mode of transmitted-light microscopy in which the specimen is evenly back-illuminated through a condenser. If the specimen is pigmented or stained, or otherwise attenuates light, it appears in color or dark against a light background. In comparison, phase-contrast optics is used to view specimens where there is little or no attenuation of transmitted light.

Broadband: Indicating a capability to deal with a relatively wide spectral bandwidth.

Calcium flux: The net rate of calcium ion (Ca^{2+}) permeation of a membrane. Normally, this occurs through voltage-gated or receptor-type calcium channels.

Caged compound: A bioactive molecule or tracer that has been rendered inactive by chemical coupling to a light-sensitive residue. Excitation of the residue (usually by exposure to ultraviolet light) results in a photochemical reaction that regenerates the active form, in situ. Calcium ions, neurotransmitters, ATP, inositol phosphates, fluorescent dyes, and transcription factors have been caged and used in biological experiments.

Capacitance. *See* Membrane capacitance.

CARS: *See* Coherent anti-Stokes Raman scattering.

CCD: *See* Charge-coupled device.

Centration: The precision with which two rotationally symmetric optical elements are aligned on a common axis.

Charge-coupled device (CCD): A light-sensitive semiconductor detector array in the form of an integrated circuit that forms the basis for most modern video cameras and precision electronic imagers. In a CCD, each light-sensitive element is a capacitor, which stores a displaced electron for each incident photon that is absorbed within the confines of that element. The stored charge pattern in a CCD array is a pixelated linear representation of the image irradiance. Under control of a clock signal, the rows of the array can be read out in order through a charge-sensitive amplifier followed by an A/D converter.

Chromatic aberration: The lens aberration resulting from the normal increase in refractive index of all common lens materials toward the blue end of the spectrum. Change in best focus setting with color is known as longitudinal (axial) chromatic aberration. The change in image size with color is known as lateral (transverse) chromatic aberration, lateral color, or chromatic difference of magnification. Chromatic aberration gives rise to colored fringes or halos around points and edges in the image. *See also* Achromatic.

Coherence: The time-averaged degree of similarity in amplitude variation and relative phase between two locations in a wave field, or at a single location for a fixed time difference. In a highly coherent light source, such as a laser, all of the waves exit in phase. Temporal coherence, also known as longitudinal coherence, is a measure of the monochromaticity of the light source (i.e., the regularity of the wave train); white light has low time coherence, whereas laser sources usually have high time coherence. Spatial coherence, also known as transverse coherence, is a measure of the regularity in wave-front shape and is related to the effective size of the light source as seen from the illuminated field; single point sources necessarily produce spherical waves that have perfect spatial coherence (even if not monochromatic). A classic light source, such as a lamp, is composed of a large number of independent (incoherent) point sources spread out over a finite surface or volume. Therefore, waves exiting the lamp are highly irregular and variable. At a great distance from the lamp, however, the regularity becomes significant. Degree of coherence is usually measured by use of an interferometer. Highly coherent light will form interference fringes of high contrast. Most lasers produce light that is both temporally and spatially coherent to a high degree.

Coherent anti-Stokes Raman scattering (CARS): *See* Chapter 40 for a complete description.

Collimation: (1) The process of aligning the optical axes of optical systems to the reference mechanical axes or surfaces of an instrument. (2) The adjustment of two or more optical axes with respect to each other. (3) The process by which a divergent beam of radiation or particles is converted into a parallel beam. Collimation of an incoherent light source such as a lamp involves focusing an image of the lamp onto a small iris or pinhole placed at the focus of a lens. Spherical waves emanating from the other side of the pinhole are converted into plane waves (collimated rays) by the lens. Effectively, a plane-wave component of the lamp emission is selected by blocking out much of the full lamp output. The collimated light has increased transverse coherence.

Command voltage: The input signal to a voltage- or current-clamp amplifier system.

Condenser: A single positive lens or group of lenses used in a projection system to collect light from a source and cause it to evenly illuminate the object to be projected.

Confocal optics: A microscope optical system in which the condenser and objective lenses both focus onto one single point in the specimen, and the image of that point is focused onto a pinhole followed by a detector. An image is generated by scanning the beam over the object. The pinhole mask effectively blocks waves that originate at points other than the scan point (i.e., from out-of-focus or offset points in the object). Therefore, confocal optics improve both transverse and axial resolution. *See also* Descanning.

Conjugate plane: The idealized plane in which the image of a plane object is formed, or the idealized plane in which the object is located for a given image plane. Conjugate distances (object distance, image distance) are measured from each conjugate plane to the corresponding principal point of the lens system. For a simple thin lens, the conjugate distances are measured from the lens to the object and image points (s and s') and are related by $1/f = 1/s + 1/s'$, where f is the focal length.

Continuous-wave (cw) laser: A laser that emits radiation continuously rather than in short bursts, as in a pulsed laser.

Convolution: A linear process in which a variable input signal is represented as a sequence of scaled impulses. Each scaled impulse gives rise to a proportionally weighted and delayed impulse response. The system output is the summation or integral of the sequence of incrementally delayed impulse responses. A convolution integral or discrete convolution is the input–output relation for a linear system that is shift invariant. In this case, the impulse response, also known as the point-spread function (PSF), carries all of the information on the transformation caused by the system on the input signal. In fluorescence microscopy, the image is modeled as a convolution of the true object with the PSF of the microscope.

Dark-field: A microscope optical system in which the sample is illuminated by oblique light incident over a limited range of angles larger than the maximum acceptance angle of the objective. In this case only scattered light is seen, so that the features of the specimen appear bright against a dark background.

Deblur: Removal of the out-of-focus component in an image, or restoration of movement-blurred features.

Deconvolution: Computational procedure for estimating the input signal to a linear system when the output signal is the experimental data. The most common case in microscopy is computation of optical sections of a fluorescently labeled object from image data that contains out-of-focus features superimposed on in-focus features. Using the point-spread function (PSF) of the microscope as a known factor, deconvolution methods can significantly attenuate blurred features and sharpen in-focus features in each image plane of a serial-focus stack. In blind deconvolution, optical sections of both the object and the a priori unknown PSF are jointly estimated.

Depth of field: The range in the object field throughout which an optical system produces a sharp image. In photography, depth of field increases with f-number (f/#). In microscopy, depth of field decreases sharply with numerical aperture (NA). A shallow depth of field is essential for 3D imaging by serial-focus image acquisition.

Depth of focus: The range in image distance that corresponds to the range of object distance covered by the depth of field. In an image-forming instrument, the detector (film, video pickup tube, CCD array) must be located at the plane where the geometric image is formed. The depth of focus is the tolerance on that location.

Descanning: The process by which the time-dependent deflection of a light beam in a scanning optical system is compensated or nulled out in the back-reflected or returning light path. In many confocal scanning fluorescence microscopes, deflection of the laser beam is controlled by galvanometrically driven mirrors. Fluorescence captured by the objective travels the reverse path in the microsope, so that the mirrors exactly cancel the deflection. The emergent light can therefore be brought to a stationary focus on a detector.

Dichroism: (1) In reference to anisotropic materials, the selective absorption of light rays that are polarized in one particular plane relative to the crystalline axes or principal axes. (2) In reference to isotropic materials, the selective reflection and transmission of light as a function of wavelength, regardless of its polarization. The color of these materials varies with thickness. (3) An optical interference filter (dichroic reflector) that transmits one or more bands of wavelengths while reflecting the complementary bands.

Diffraction: The physical phenomena observed when any wave spreads out in space because of the apparent bending of that wave around small obstacles. Diffraction can also be visualized as the accumulation of the infinitesimal interference effects of a wave with itself.

Diffraction-limited: The property of an optical system whereby only the effects of diffraction determine the quality of the image it produces.

Digital: Denoting the use of binary notation; that is, the representation of data numerically by bits (1 or 0).

Digitize: To put in a digital representation. *See* Digital.

Drift: A gradual, uncontrolled change in the setting of an instrument or output of a circuit over time.

Dynamic range: The number of distinguishable values of a signal that may be measured by use of a particular detection device. In general, dynamic range is the number obtained by dividing the full output range of the device by the root-mean-square (rms) noise level. For example, an instrumentation amplifier with a full range of 10 V and an intrinsic noise level of 10 μV rms has a dynamic range of one million. Dynamic range is often expressed logarithmically, in bits or decibels.

Electron multiplier: A device containing a material that, when bombarded with a single electron, emits more than one electron. Many stages of such material are often combined in the presence of an electric field, which causes the emitted electrons to accelerate into subsequent stages, leading to an overall exponential multiplicative gain in the total number of electrons.

Emission: Conversion of the internal energy of an excited atom or molecule into a radiated photon. In general, the energy (color) of the radiation will be determined by the difference in energy between the initial and final quantum states. Emission can be spontaneous, as in fluorescence, or can be stimulated, as in a laser, by driving the excited atom or molecule at the resonant frequency for the transition. Absorption, the reciprocal process, is necessarily stimulated.

Emission spectrum: The wavelength composition of light emitted in the radiative de-excitation of atoms or molecules. For a fluorescent or phosphorescent dye, usually the graph showing the spectral composition of the emitted light versus wavelength. The absorption spectrum (and the closely related excitation spectrum) rise on the short-wavelength side of the emission spectrum and overlap it to some extent. Generally, the emission spectrum is insensitive to the exact choice of excitation wavelength, as long as that wavelength is not in the overlap region. *See also* Absorption, Excitation spectrum.

Excitation: (1) The process by which an atom or molecule acquires energy sufficient to raise it to a quantum state higher than its ground state. (2) More specifically with respect to lasers, the process by which the material in the laser cavity is energized by light or other means, so that atoms are converted to a semi-stable state, initiating the lasing process.

Excitation spectrum: The graph showing the brightness of fluorescence from a dye versus the excitation wavelength. Usually, the excitation spectrum is determined at the emission wavelength at which the dye is most fluorescent. The excitation spectrum clearly depends on the absorption spectrum but also includes the efficiency with which the absorbed energy is recovered as emission. This factor is the **quantum yield** of fluorescence. *See also* Absorption, Emission spectrum.

Extinction coefficient: The wavelength-dependent constant of proportionality between the absorbance and the molar concentration of a dye. The molar extinction coefficient (ε) is defined in Beer's Law, the relation between attenuation (I/I_0), concentration (c), absorbance (A), and path length (L): $A = \log I_0/I = \varepsilon cL$.

F-number (f/#): The focal length of a lens divided by the entrance pupil, or diameter, of the lens.

Ferroelectric: Materials that can be electrostatically polarized by an external electric field, and that retain the direction and degree of polarization. A polarized ferroelectric material is known as an electret (analogous to magnet), which has a positive pole, a negative pole, and establishes an electrostatic field in its vicinity.

Field of view: The maximum area that can be seen through a lens or an optical instrument.

Field stop: An iris located in an optical system such that its open diameter delimits the field of view. Generally, the field stop is located between the light source and the condenser, so that its image is formed in the specimen and therefore sets the diameter of the illuminated field. It is also possible to put a field stop between the objective and the detector, so that the iris directly masks the view. *See also* Aperture stop.

Flashlamp: A type of arc lamp that converts stored electrical energy into light by means of a sudden electrical discharge or spark. The discharge is generally between two electrodes in a gas or vapor contained in a glass or fused quartz tube. Depending on the needed spectral output, the lamp may be filled with air, xenon, mercury vapor, other gases, or gaseous mixtures.

Fluorescence: One of several processes in which there is emission of light by a molecule subsequent to the excitation of that molecule by absorption of light. Because of internal relaxation processes in which some energy is lost as heat, the emitted photon is of a lower energy (longer wavelength) than the initially absorbed photon. Fluorescence is characterized by conservation of the electronic spin state of the molecule (usually singlet/singlet), and therefore by strongly allowed transitions. The excited-state lifetime for fluorescence is generally in the nanosecond range. This is in contrast to phosphorescence, characterized by a spin change between excited state and ground state (usually triplet/singlet), weakly allowed transitions, and excited-state lifetimes ranging from microseconds to many minutes.

Focal length: For a simple thin lens, the focal length is the distance beyond the lens at which incoming collimated light is converged to a focused point. For diverging lenses the focal length is negative, with the virtual focus before the lens. In a lens system, the front focal length and the back focal length are measured from mathematically defined principal points on the axis of the lens group. In general, the front and back focal lengths will be in the ratio of the refractive indices of the media on either side of the lens; that is, for a lens group in air, the two focal lengths will be equal. For an immersion objective, the front focal length will exceed the back by a ratio equal to the immersion index.

Focal plane: (1) An idealized plane (including the focal point) normal to the axis of a lens or mirror. (2) It is most commonly used in microscopy to describe an idealized plane located at the position in an optical instrument at which the sharpest image is formed.

Frame: (1) To center an object in the field of a camera. (2) A single image from among the sequential images on motion-picture film. (3) In raster-scanned television, a complete 525-line image. In conventional video cameras, a frame is composed of two sequential interlaced 262.5-line fields. (4) The complete readout of a single image from a digital camera. *See also* Video.

Frame grabber: Electronic image acquisition device, usually in the form of a circuit card for a computer, that samples, digitizes, and stores a video camera frame in computer memory.

Full-well capacity: The number of electrons that each pixel element in a charge-coupled device (CCD) can hold without overflowing and causing image blooming. Generally, this will be in the range 10^4–10^5, depending on the size of the CCD elements.

Gain: Also known as amplification. (1) The numerical factor by which an amplifier or a concentrator (such as an antenna) increases signal strength. (2) In optics, a material that exhibits amplification rather than absorption of electromagnetic waves. This type of behavior provides the basis for laser operation. (3) In a light detector, the number of electrons or holes counted per detected photon; in a CCD, the gain is 1. In a photomultiplier, the gain may range from 10^2 to 10^6.

Gray scale: In image processing, the number of usable gray levels. In an 8-bit system, the gray scale runs from 0 to 255. The range is normally defined as running from black to white.

Group velocity dispersion: The time-dependent alteration of the amplitude and width of an electromagnetic wave pulse propagating through a material because of differences in the wave speed of the monochromatic components of the pulse. Perfectly monochromatic light is ideally a wave train of infinite duration; however, any finite pulse contains a band of monochromatic components (roughly, pulsewidth and bandwidth vary inversely) having specific relative phase. For femtosecond pulses, the bandwidth is large enough that dispersion in the refractive index of lenses, or even in the air path of the beam, may have significant effect on pulse shape and peak intensity.

Harmonic generation: In optics, the conversion of electromagnetic radiation at a given frequency into radiation at an integer multiple of the frequency. This can occur when an intense coherent light beam passes through a material having a nonlinear polarizability. The harmonic appears as a beam collinear with and coherent with the fundamental but can be separated by use of a dichroic filter, prism, or grating. The most common example of harmonic generation is frequency doubling; for example, the 1064-nm near-infrared output from a YAG laser can be frequency-doubled to 532-nm green light.

Holographic grating: A diffraction grating made by recording a holographic interference pattern in a photosensitive material rather than by the mechanical ruling of grooves. The medium is then processed so that the hologram is encoded as a refractive index pattern. Holographic gratings are phase gratings (i.e., block no light), and therefore have very high optical efficiency.

Image enhancement: A process by which an image is scaled or manipulated to increase the amount of information perceivable by the human eye. The simplest image enhancement operations are brightness and contrast adjustment.

Image intensifier: An electro-optic device that is essentially a vacuum-tube photomultiplier that preserves spatial information. When an intensifier is placed in the image plane of an optical instrument, each detected incoming photon causes ejection of an electron from the photocathode surface in the tube. Ejected electrons are accelerated in a high electric field within the tube and collide with a shielded phosphor screen at the output end. In the collision, a burst of photons is released, resulting in a bright secondary image that can be focused by conventional means onto a television or CCD camera. The field in the intensifier is shaped so that the electron trajectories preserve image coordinates. Modern intensifiers utilize microchannel plate (MCP) technology in which several stages of secondary electron multiplication precede the output phosphor screen. *See also* Photomultiplier tube.

Image plane: An idealized plane, perpendicular to the axis of a lens, in which an image is formed. A real image formed by a positive lens would be visible upon a screen located in this plane.

Incoherent: In optics, the term denoting the lack of a fixed phase relationship between two or more waves. If two incoherent waves are superimposed, interference effects are not stable, and no intereference pattern is observed. *See also* Coherence.

Iris diaphragm: Traditionally, the iris located in the back focal plane of the condenser. The open diameter of the iris diaphragm sets the illuminating numerical aperture (INA) of the condenser. *See also* Aperture.

Isosbestic wavelength: In spectroscopy, the occurrence of a wavelength where a family of spectral curves all intersect. Isosbestic points usually signify the proportional interconversion of two different molecular species, where the two extreme spectra in the family represent the two species. The most well-known example is the isosbestic point at 360 nm in the family of Fura-2 excitation spectra over the nanomolar–micromolar range in Ca^{2+} concentration. At this wavelength, the excitation of fluorescence is essentially independent of calcium concentration.

Köhler illumination: A two-stage illuminating system for a microscope, in which the light source is imaged in the back focal plane of the substage condenser by a collector lens, the substage condenser in turn uniformly illuminating the object. In the Köhler setup, an iris located in the back focal plane of the condenser controls the maximum angle at which light exits the condenser into the specimen. This iris is the aperture stop. A second iris, usually located closer to the collector lens, is imaged on the specimen by the condenser. This iris is the field stop.

Lag: A term applied to an electric charge image in a video camera tube that persists for a period of a few frames after the initial light exposure. Lag is particularly noticeable in high-gain video cameras operating at low light level.

Laser: An acronym for light amplification by stimulated emission of radiation. The variations in laser type have become numerous, but the basic device consists of an optical gain medium within an optical cavity composed of two or more reflectors or gratings. The gain medium may be a gas or vapor of excited atoms or ions, excited dye molecules in a solvent, excited atoms or molecules in a crystal or glassy solid, or excited charge carriers in a semiconductor. The excitation process serves to create an inverted population, in which more atoms are poised to emit a photon than are able to absorb; that is, emission can exceed absorption. The cavity serves to establish electromagnetic traveling or standing wave modes in the gain medium, thereby causing in-phase stimulated emission at specific wavelengths from the inverted population. The amplified light partially or completely escapes the cavity as an intense (usually monochromatic) coherent beam.

Liquid-crystal tunable filter (LCTF): An active optical filter composed of a variable liquid crystal retarder between polarizing filters. The retarder is composed of a liquid-crystalline medium sandwiched between two flat windows that support transparent electrode films. The birefringence of the liquid crystal can be adjusted by applying a voltage across the electrodes. By itself, the device can be used as a variable retarder. When inserted between polarizing filters, wavelength-dependent transmission results. To improve band-pass shape and extinction, several such elements are used in series in an LCTF, along with computer control of the individual stages. *See also* Retardation.

Magnification: The ratio of the size of the image of an object to the size of the object defines the transverse magnification. For a simple lens, transverse magnification varies with object distance. In a microscope, where the camera position is fixed relative to the optics and focus is set by moving the stage axially relative to the instrument, the transverse magnification is constant. Angular magnification is the ratio of the apparent angular size of the image observed through an optical device to that of the object viewed by the unaided eye. Axial magnification is defined in terms of focus setting as the ratio of the axial distance moved by a geometric image point (in image space) to the actual stage focus increment. Generally, axial magnification is proportional to the square of the transverse magnification.

Membrane capacitance: The charge-storage capability of a cell membrane due to its basic structure—a thin, highly insulating lipid bilayer separating two conductive ionic aqueous media, the cytoplasm and the extracellular fluid. When unbalanced charge (usually ions or, in some cases, electrons) is moved across a membrane, the excess ions of one polarity are attracted to ions of the opposite charge remaining on the other side of the membrane. This results in the unbalanced charge appearing in a pair of subnanometer aqueous layers, one associated with each face of the membrane. The energy of the separated charges is stored as an electric field within the bilayer, the constant of proportionality between charge and membrane voltage being the capacitance. For biological membranes, the capacitance is generally in the range 0.5–1.0 $\mu F/cm^2$.

Mode-locked laser: A laser that functions by modulating the energy content of each electromagnetic mode in its optical cavity to give rise selectively to energy bursts of high peak power and short duration. The pulse rate of a mode-locked laser is the round-trip time for a light pulse in the cavity.

Monochromator: Any device used for isolating portions of the spectrum of a light source. Monochromators generally contain a rotatable prism or diffraction grating, along with input and output masks (or slits) to limit the angular range of rays that enter or exit. The entering light is dispersed in angle by the prism or grating, and the exiting wavelength band is selected by rotating the dispersive element relative to the slits. For high spectral purity, two monochromators can be used in tandem in an additive-dispersion configuration.

Multiphoton absorption: The nonlinear optical process in which a molecular electronic excited state is produced by the simultaneous absorption of two or more photons. This differs from sequential absorption in that no intervening stationary quantum state exists as an intermediate. Therefore, extremely high irradiation intensities are required for significant multiphoton absorption rates to occur. Also, the absorption rate is proportional to the square of the intensity for two-photon excitation (the cube of the intensity for three-photon processes, etc.), in contrast to conventional one-photon excitation. In the case of fluorescent dyes, the summed energy of the absorbed photons approximately equals the energy of the single photon that would be required to cause the transition from ground state to the first excited singlet state. The most well-known example of multiphoton absorption is two-photon excitation fluorescence microscopy, in which pairs of near-infrared photons excite ultraviolet/visible-band dyes. This method combines the high intensity of a femtosecond-pulsed laser system with the sharp focusing of a microscope objective to achieve the excitation rates needed to produce photons in sufficient number for practical imaging.

NA: *See* Numerical aperture.

Neutral-density filter: Also known as a gray filter. A light filter that decreases the intensity of light without altering the relative spectral distribution of the energy.

Noise: The unwanted and unpredictable fluctuations that distort a received signal and hence tend to obscure the signal of interest. In microscopy, there is a fundamental statistical limit on the noise in a measurement of the number of photons in a given location. This noise is equal to the square root of the number of photons counted, which is confusingly referred to as shot noise, photon noise, counting noise, and even photon counting noise. In electronic equipment, noise is also generated by thermal processes. This noise is known as Johnson or thermal noise and can be reduced by lowering the temperature of the components performing the measurement. In addition, fluctuation in the illumination source such as the flicker or drift that can be produced by arc lamps will be reflected in the recorded signal as noise. Finally, read or readout noise in an electronic camera refers to the noise inherent in converting photons into electrons and the subsequent analog-to-digital conversion. Steps that can be taken to limit the different sources of noise include shielding, signal averaging or integration, cooling of measurement devices, tuned or phase-locked detection, and proper staging of amplification.

Numerical aperture (NA): A fundamental numerical quantity that is a measure of the range of angles over which a lens system can accept entering light rays. The transverse resolution of a microscope improves directly with NA, and axial resolution approximately with the square of NA. *See also* Aperture.

Objective: The optical element that receives light from the object and forms the first or primary image in telescopes and microscopes. In cameras, the image produced by the objective is the final image. In telescopes and microscopes, when used visually, the image formed by the objective is magnified by an eyepiece.

Observation volume: In a confocal or multiphoton microscope, the specimen volume from which emitted light is detected. For a high-NA diffraction-limited system, this volume can be as small as 0.044 μm^3 (0.044 fL).

Optical sectioning: The property of certain optical systems to attenuate or remove contributions to the image originating from out-of-focus parts of the specimen. In this case, the image contains only in-focus features and therefore represents a planar section of the object. Optical sectioning depends on the shallow depth of field in high-aperture systems. Confocal optical systems contain a spatial filter that blocks light waves that do not originate in the immediate neighborhood of the scan point. In multiphoton microscopes, nonlinear interaction of the focused scan beam with the fluorescent label brings about a similar spatial weighting. In conventional microscopes that form an image directly, quantitative optical sectioning is approached through computational deblurring.

Orthogonal: In comparing two vectors or functions, the property that the projection of one function on the other is 0. Mathematically, orthogonal objects have an inner product equal to 0. In the simplest case, two vectors are orthogonal in two or three dimensions when they are perpendicular. In the case of functions, sine and cosine are orthogonal, as are any two distinct harmonics of a sine or cosine wave. In quantum mechanics, the stationary state wave functions for an atomic or molecular system form a normalized orthogonal basis set.

Paraboloidal mirror: A concave mirror that has the form of a paraboloid of revolution. The paraboloidal mirror may have only a portion of a paraboloidal surface through which the axis does not pass, and is known as an off-axis paraboloidal mirror. All axial, parallel light rays are focused at the focal point of the paraboloid without spherical aberration, and conversely, all light rays emitted from an axial source at the focal point are reflected as a bundle of parallel rays without spherical aberration.

Paraxial optics: Ray diagrams and optical formulas derived by considering only rays that propagate close to the axis of an optical system and do not make a significant angle with respect to the axis. In this limit, the Law of Refraction can be linearized, greatly simplifying the derivation of conjugate distance formulas and other lens equations.

Phosphorescence: Emission of a photon of light by an atom or molecule when the transition is accompanied by a change in electron spin pairing. In biological imaging and spectroscopy, phosphorescence usually arises in the transition from an excited triplet state (unpaired spins) to a lower-energy singlet state (paired spins) of a molecule. Because singlet–triplet transitions are symmetry-forbidden in isolated systems of low atomic number, phosphorescence is a slow process relative to fluorescence or quenching. Molecular oxygen is a very effective quencher of triplet states, therefore phosphorescence spectroscopy or imaging is often most useful in systems where the oxygen partial pressure can be reduced. *See also* Fluorescence.

Photobleaching: The photophysical or photochemical conversion of a dye to a colorless form. In microscopy, photobleaching usually refers to the irreversible loss of fluorescence in dyes that are intensely irradiated. In some cases, photobleaching is reversible. Molecular oxygen is a common mediator of photochemical reactions leading to photobleaching, becoming activated upon quenching dye triplet states. Excited-state molecular oxygen can react not only with dye molecules in the specimen but also with other biochemicals and macromolecules in a cell. This type of activity can lead to phototoxicity.

Photodiode: (1) A two-electrode, light-sensitive junction formed in a semiconductor material in which the reverse current varies with illumination. Photodiodes are used for the detection of light and for the conversion of radiant energy to electrical power. (2) Originally, a vacuum tube diode in which the cathode was coated with a metal film such as sodium so as to make the cathode photo-emissive.

Photoelectric effect: The excitation of a bound electron in a material by absorption of a photon. In the classic photoelectric effect, one electron is ejected from a metal film into vacuum for each photon absorbed. The energy of the photon must exceed the potential energy needed to remove the electron from the metal, with the excess energy appearing as kinetic energy of the electron. The photoelectric effect is one of the primary pieces of experimental evidence for the quantum nature of the electromagnetic field.

Photomultiplier tube (PMT): A vacuum tube light detector consisting of a glass envelope with an internal photocathode followed by a series of dynodes (electrodes) and an anode. In operation, the dynodes and anode are held at sequentially more-positive voltages than the cathode. The cathode emits electrons via the photoelectric effect when exposed to light. The electrons are accelerated by the electrostatic field to collide with the first dynode where secondary electrons are ejected. The secondary group is accelerated to the next dynode, where there is another multiplicative gain in released charge. The dynode chain acts as an electron multiplier, providing a high charge gain at the anode output. The unit is essentially a photoelectric cell followed by a high-gain electron multiplier. PMTs can be used with an analog electrometer or with a discriminator and counter circuit.

Photon: The massless quantum particle of the electromagnetic field. The energy of a photon is given by Planck's formula, $E = h\nu$, where h is Planck's constant (6.626×10^{-34} J-sec) and ν is the frequency of the radiation in hertz. Photons propagate in free space at the speed of light ($c = 3.00 \times 10^8$ m/sec) and carry momentum, angular momentum, and a polarization. When an electromagnetic wave field gives up energy to an atomic or molecular system, the exchange occurs via the absorption of photons.

Pixel: Contraction of "picture element." A small element of a scene, often the smallest resolvable area, in which an average brightness value is determined and used to represent that portion of the scene. Pixels are usually arranged in a square array to form a complete image.

PMT: *See* Photomultiplier tube.

Point-spread function (PSF): In an optical system, the energy intensity distribution about the image of a point source. This function defines the resolution of the optical system as convolution of the PSF with a specimen mathematically defines the resulting image.

Polarized light: Light in which the electric field oscillates in a particular direction transverse to the direction of travel (linear polarization), or in which the field components rotate in a plane transverse to the direction of travel (elliptical or circular polarization). In an isotropic medium, light propagates as a transverse wave in which the electric field and the magnetic field.

PSF: *See* Point-spread function.

QE: *See* Quantum efficiency.

Quantum efficiency (QE): (1) With respect to a source of radiant flux, the ratio of the number of quanta of radiant energy (photons) emitted per second to the number of electrons flowing per second. (2) For a detector, the ratio of induced current to incident flux. Often measured in electrons per photon (dimensionless) or amperes per watt.

Quantum yield (QY): The fractional probability of a photophysical or photochemical event occurring subsequent to the absorption of a photon. In photosynthesis, the quantum yield is the probability that the energy from an absorbed photon gets coupled to the reaction center rather than lost as heat. For a fluorescent dye, the QY is the probability that a photon will be emitted for each photon absorbed. For different dyes and conditions, QY ranges from <0.1 to >0.95.

QY: *See* Quantum yield.

Raster: The pattern of lines traced by rectilinear scanning in display systems and scanning microscopes.

Ratiometric indicator: Fluorescent indicator dyes that undergo changes in the excitation or emission spectrum that can be utilized to approximately correct for unknown indicator concentration and the optical effect of irregularities in the shape of the specimen. The procedure, known as ratio imaging, involves normalization of one fluorescence image by a second image of the same object made with altered excitation or emission wavelength. In this way, the physiological response of the indicator can be estimated without inaccuracies due to concentration variations or cell shape–dependent brightness variations. The exact procedure used is dependent on the type of spectral changes that occur in the dye with changes in the physiologically relevant variable. The most well-known ratiometric indicator is the Ca^{2+}-sensitive dye Fura-2, which shows an altered excitation spectrum. In ratiometric Fura imaging, a single blue-green emission filter is used along with an alternate pair of ultraviolet excitation filters, usually 340 nm and 380 nm. The ratio image is computed digitally as the background-corrected 340-nm-excited image divided by the background-corrected 380-nm-excited image.

Refractive index: A fundamental unitless number that is the ratio of the speed of monochromatic light in a vacuum to the speed (phase velocity) in a particular material or medium: $n = c/v$. Ray paths are redirected at a refractive boundary according to the Law of Refraction; $n \sin \theta = n' \sin \theta'$, where n and n' are the refractive indices of the materials on either side of the boundary, and the angles are measured from the local normal to the boundary. In a material, the vacuum wavelength of the light is altered from λ to λ/n. In a more general sense, the refractive index is a complex number related to the phase of the optical response of a material to the light wave. The conventional index is the real part of the complex index, and the absorption (or stimulated emission) coefficient is the imaginary part.

Retardation: The phase change in a light wave, relative to a reference beam, due to propagation through a region of different refractive index. In phase-contrast microscopy, light propagating through an object of increased index will lag in phase relative to light that bypasses the structure. In a birefringent filter, two orthogonal linearly polarized light waves will propagate at different speeds, causing one to become retarded relative to the other. This causes the polarization to shift between linear, elliptical, and circular, depending on the thickness of the filter, the degree of birefringence, and the wavelength of the light. A retarder or wave plate is a birefringent flat optic, usually of mica, quartz, or calcite. The thickness of the plate is usually finished so that the retardance for one principal polarization relative to the other is either one-quarter or one-half wave. A properly oriented quarter-wave plate will convert linearly polarized light to circular, and vice versa. A half-wave plate will rotate linear polarization by 90° or will interconvert right-hand and left-hand circular polarizations. In a liquid-crystal retarder the birefringence can be controlled electronically. *See also* Liquid-crystal tunable filter.

Scattering (electromagnetic): A fundamental process that can be pictured as the result of a plane wave beam impinging on an atom, molecule, or small particle. The oscillating electromagnetic field polarizes the particle and causes it to act as a source of spherical waves. The scattered waves carry energy off in other directions and cause attenuation of the beam. When a beam of light propagates through a medium containing many such particles, the scatter makes the path of the beam visible. Light scattering is seen whenever there is refractive index heterogeneity within an otherwise transparent material, or on a surface. Milk appears white (in white light) because the numerous suspended lipoprotein particles backscatter essentially all the incident light. The sky appears blue because of the wavelength-dependent scatter of sunlight by air molecules. Refraction is the net result of scatter when the density of scatterers is both high and uniform on a scale comparable to the wavelength. A refracted wave can be pictured as the coherent superposition of the incident wave field and the scattered field.

Signal-to-noise ratio (SNR): The ratio of the power in a desired signal to the undesirable noise present in the absence of a signal.

SLM: *See* Spatial light modulator.

SNR: *See* Signal-to-noise ratio.

Spatial light modulator (SLM): *See* Chapter 47.

Superresolution: Any of a number of methods that produce an image with a resolution below the diffraction limit. Also referred to as subdiffraction microscopy. *See also* Chapters 31–35.

Video: The encoded picture signal output of a video camera, consisting of a sequential horizontal scan line analog signal interleaved with horizontal synchronization pulses and terminated by a vertical sync pulse sequence. In the U.S. system, a video picture consists of 525 horizontal scan lines, presented as 60 262.5-line interlaced fields/sec (30 frames/sec). In the European standard, the frame consists of 625 lines presented as 50 312.5-line interlaced fields/sec (25 frames/sec). The 1-V peak-to-peak (p-p) signal reserves 0.7 V for the monochrome image brightness signal and 0.3 V for the negative-going synchronization pulses. Video digitization is normally set at 8 bits, which can be increased by frame averaging.

Wavelength: Electromagnetic energy such as light is transmitted in the form of a periodic or quasiperiodic wave. In monochromatic light, the wave can be pictured as a sinusoidally varying electric field and magnetic field. The wavelength is the spatial dimension of one cycle of this wave field; wavelength is inversely proportional to frequency. The visible band of wavelengths runs from 390 nm (violet) to 750 nm (far-red).

Wave number: The frequency of a wave divided by its velocity of propagation; the reciprocal of the wavelength.

Appendix 6 Cautions

GENERAL CAUTIONS

Please note that the Cautions Appendix in this manual is not exhaustive. Readers should always consult individual manufacturers and other resources for current and specific product information. Chemicals and other materials discussed in text sections are not identified by the icon <!> used to indicate hazardous materials in the protocols. However, without special handling, they may be hazardous to the user. Please consult your local safety office or the manufacturer's safety guidelines for further information.

The following general cautions should always be observed.

- **Before beginning the procedure,** become completely familiar with the properties of substances to be used.
- **The absence of a warning** does not necessarily mean that the material is safe, because information may not always be complete or available.
- **If exposed** to toxic substances, contact your local safety office immediately for instructions.
- **Use proper disposal procedures** for all chemical, biological, and radioactive waste.
- **For specific guidelines on appropriate gloves to use,** consult your local safety office.
- **Handle concentrated acids and bases** with great care. Wear goggles and appropriate gloves. A face shield should be worn when handling large quantities.

  Do not mix strong acids with organic solvents because they may react. Sulfuric acid and nitric acid especially may react highly exothermically and cause fires and explosions.

  Do not mix strong bases with halogenated solvent because they may form reactive carbenes which can lead to explosions.
- **Handle and store pressurized gas containers** with caution because they may contain flammable, toxic, or corrosive gases; asphyxiants; or oxidizers. For proper procedures, consult the Material Safety Data Sheet that must be provided by your vendor.
- **Never pipette** solutions using mouth suction. This method is not sterile and can be dangerous. Always use a pipette aid or bulb.
- **Keep halogenated and nonhalogenated** solvents separately (e.g., mixing chloroform and acetone can cause unexpected reactions in the presence of bases). Halogenated solvents are organic solvents such as chloroform, dichloromethane, trichlorotrifluoroethane, and dichloroethane. Nonhalogenated solvents include pentane, heptane, ethanol, methanol, benzene, toluene, *N,N*-dimethylformamide (DMF), dimethyl sulfoxide (DMSO), and acetonitrile.
- **Laser radiation**, visible or invisible, can cause severe damage to the eyes and skin. Take proper precautions to prevent exposure to direct and reflected beams. Always follow the manufacturer's safety guidelines and consult your local safety office. See caution below for more detailed information.
- **Flash lamps**, because of their light intensity, can be harmful to the eyes. They also may explode on occasion. Wear appropriate eye protection and follow the manufacturer's guidelines.
- **Photographic fixatives, developers, and photoresists** also contain chemicals that can be harmful. Handle them with care and follow the manufacturer's directions.

- **Power supplies and electrophoresis equipment** pose serious fire hazard and electrical shock hazards if not used properly.
- **Microwave ovens and autoclaves** in the laboratory require certain precautions. Accidents have occurred involving their use (e.g., when melting agar or Bacto Agar stored in bottles or when sterilizing). If the screw top is not completely removed and there is inadequate space for the steam to vent, the bottles can explode and cause severe injury when the containers are removed from the microwave or autoclave. Always completely remove bottle caps before microwaving or autoclaving. An alternative method for routine agarose gels that do not require sterile agar is to weigh out the agar and place the solution in a flask.
- **Ultrasonicators** use high-frequency sound waves (16–100 kHz) for cell disruption and other purposes. This "ultrasound," conducted through air, does not pose a direct hazard to humans, but the associated high volumes of audible sound can cause a variety of effects, including headache, nausea, and tinnitus. Direct contact of the body with high-intensity ultrasound (not medical imaging equipment) should be avoided. Use appropriate ear protection and display signs on the door(s) of laboratories where the units are used.
- **Use extreme caution when handling cutting devices,** such as microtome blades, scalpels, razor blades, or needles. Microtome blades are extremely sharp! Use care when sectioning. If unfamiliar with their use, have an experienced user demonstrate proper procedures. For proper disposal, use the "sharps" disposal container in your lab. Discard used needles *unshielded*, with the syringe still attached. This prevents injuries and possible infections when manipulating used needles because many accidents occur while trying to replace the needle shield. Injuries may also be caused by broken pasteur pipettes, coverslips, or slides.
- **Procedures for the humane treatment of animals** must be observed at all times. Consult your local animal facility for guidelines. Animals, such as rats, are known to induce allergies that can increase in intensity with repeated exposure. Always wear a lab coat and gloves when handling these animals. If allergies to dander or saliva are known, wear a mask.

GENERAL PROPERTIES OF COMMON CHEMICALS

The hazardous materials list can be summarized in the following categories.

- Inorganic acids, such as hydrochloric, sulfuric, nitric, or phosphoric, are colorless liquids with stinging vapors. Avoid spills on skin or clothing. Spills should be diluted with large amounts of water. The concentrated forms of these acids can destroy paper, textiles, and skin and cause serious injury to the eyes.
- Inorganic bases, such as sodium hydroxide, are white solids that dissolve in water and under heat development. Concentrated solutions will slowly dissolve skin and even fingernails.
- Salts of heavy metals are usually colored, powdered solids that dissolve in water. Many of them are potent enzyme inhibitors and therefore toxic to humans and the environment (e.g., fish and algae).
- Most organic solvents are flammable volatile liquids. Avoid breathing the vapors, which can cause nausea or dizziness. Also avoid skin contact.
- Other organic compounds including organosulphur compounds, such as mercaptoethanol or organic amines, can have very unpleasant odors. Others are highly reactive and should be handled with appropriate care.
- If improperly handled, dyes and their solutions can stain not only your sample, but also your skin and clothing. Some are also mutagenic (e.g., ethidium bromide), carcinogenic, and toxic.
- Nearly all names ending with "ase" (e.g., catalase, β-glucuronidase, or zymolyase) refer to enzymes. There are also other enzymes with nonsystematic names such as pepsin. Many of them are provided by manufacturers in preparations containing buffering substances, etc. Be aware of the individual properties of materials contained in these substances.
- Toxic compounds are often used to manipulate cells. They can be dangerous and should be handled appropriately.
- Be aware that several of the compounds listed have not been thoroughly studied with respect to their toxicological properties. Handle each chemical with appropriate respect. Although the toxic effects of a compound can be quantified (e.g., LD_{50} values), this is not possible for carcinogens or mutagens where one single exposure can

have an effect. Also realize that dangers related to a given compound may also depend on its physical state (fine powder vs. large crystals/diethylether vs. glycerol/dry ice vs. carbon dioxide under pressure in a gas bomb). Anticipate under which circumstances during an experiment exposure is most likely to occur and how best to protect yourself and your environment.

HAZARDOUS MATERIALS

Note: In general, proprietary materials are not listed here. Kits and other commercial items as well as most anesthetics, sedatives, dyes, fixatives, embedding media, stains, herbicides, and fungicides are also not included. Anesthetics and antibiotics also require special care. Follow the manufacturer's safety guidelines that accompany these products.

Acetic acid (concentrated) must be handled with great care. It may be harmful by inhalation, ingestion, or skin absorption. Wear appropriate gloves and goggles. Use in a chemical fume hood.

Acetic acid (glacial) is highly corrosive and must be handled with great care. It may be a carcinogen. Liquid and mist cause severe burns to all body tissues. It may be harmful by inhalation, ingestion, or skin absorption. Wear appropriate gloves and goggles and use in a chemical fume hood. Keep away from heat, sparks, and open flame.

Acetone causes eye and skin irritation and is irritating to mucous membranes and upper respiratory tract. Do not breathe the vapors. It is also extremely flammable. Wear appropriate gloves and safety glasses. Keep away from heat, sparks, and open flame.

Alconox detergent is an irritant and may be harmful by inhalation, ingestion, or skin absorption. Wear appropriate gloves and safety glasses.

α-bungarotoxin is a potent neurotoxin and is harmful by inhalation, ingestion, or skin absorption. Wear appropriate gloves and safety glasses and always use in a chemical fume hood. Do not breathe the dust.

3-aminopropyltriethoxysilane (APTES), *see* **Silane**

Amino silane is an irritant and may cause severe corneal injury. It may also be harmful by inhalation, ingestion, and skin absorption. Wear appropriate gloves and safety goggles. Keep away from contact with water.

Anesthetics, *follow* manufacturer's safety guidelines

Antibiotics require special care. Follow the manufacturer's safety guidelines that accompany these products.

APTES, *see* **Silane**

Argon is a nonflammable high-pressure gas. It may be harmful by inhalation, ingestion, or skin absorption. Wear appropriate gloves and safety goggles. Use with sufficient ventilation and do not breathe the gas.

BAPTA, 1,2-*bis*(2-aminophenoxy)-ethane-*N,N,N′,N′*-tetraacetic acid, may be harmful by inhalation, ingestion, or skin absorption. It is irritating to the mucous membranes, upper respiratory tract, skin, and eyes. Wear appropriate gloves and safety glasses and use in a chemical fume hood.

Benzyl alcohol is an irritant and may be harmful by inhalation, ingestion, or skin absorption. Wear appropriate gloves and safety glasses. Keep away from heat, sparks, and open flame.

Benzyl benzoate is an irritant and may be harmful by inhalation, ingestion, or skin absorption. Avoid contact with the eyes. Wear appropriate gloves and safety glasses.

β-mercaptoethanol (2-mercaptoethanol), $HOCH_2CH_2SH$, may be fatal if inhaled or absorbed through the skin and is harmful if ingested. High concentrations are extremely destructive to the mucous membranes, upper respiratory tract, skin, and eyes. β-mercaptoethanol has a very foul odor. Wear appropriate gloves and safety glasses and always use in a chemical fume hood.

1,2-*bis*(2-aminophenoxy)-ethane-*N,N,N′,N′*-tetraacetic acid, *see* **BAPTA**

$CaCl_2$, *see* **Calcium chloride**

Calcium chloride, $CaCl_2$, is hygroscopic and may cause cardiac disturbances. It may be harmful by inhalation, ingestion, or skin absorption. Do not breathe the dust. Wear appropriate gloves and safety goggles.

Carbogen, *see* **Carbon dioxide**

Carbon dioxide, CO_2, in all forms may be fatal by inhalation, ingestion, or skin absorption. In high concentrations, it can paralyze the respiratory center and cause suffocation. Use only in well-ventilated areas. In the form of dry ice, contact with carbon dioxide can also cause frostbite. Do not place large quantities of dry ice in enclosed areas such as cold rooms. Wear appropriate gloves and safety goggles.

Carprofen is highly toxic if swallowed. It may be harmful by inhalation, ingestion, or skin absorption. Wear appropriate gloves and safety glasses. Do not breathe the dust.

CH_3CH_2OH, *see* **Ethanol**

CO_2, *see* **Carbon dioxide**

DABCO, *see* **1,4-diazabicyclo-[2,2,2]-octane**

DAPI, *see* **4,6-diamidine-2-phenylindole dihydrochloride**

Dexamethasone may be harmful by inhalation, ingestion, or skin absorption. Overexposure may cause reproductive disorder(s) and possible risks of harm to the unborn child. Wear appropriate gloves and safety glasses. Do not breathe the dust.

4,6-diamidine-2-phenylindole dihydrochloride (DAPI) is a possible carcinogen. It may be harmful by inhalation, ingestion, or skin absorption. It may also cause irritation. Avoid breathing the dust and vapors. Wear appropriate gloves and safety glasses and use in a chemical fume hood.

1,4-diazabicyclo-[2,2,2]-octane (DABCO) may be harmful by inhalation, ingestion, or skin absorption. Wear appropriate gloves and safety glasses and use in a chemical fume hood.

3,4-dihydroxybenzoic acid (PCA) may be harmful by inhalation, ingestion, or skin absorption. Wear appropriate gloves and safety glasses. Do not breathe the dust.

Dimethylformamide, *see* ***N,N*-dimethylformamide**

Dimethylsulfoxide (DMSO) may be harmful by inhalation or skin absorption. It easily penetrates the skin and anything dissolved or mixed with it will be absorbed. Wear appropriate gloves and safety glasses and use in a chemical fume hood. DMSO is also combustible. Store in a tightly closed container. Keep away from heat, sparks, and open flame.

DMF, *see* ***N,N*-dimethylformamide**

DMSO, *see* **Dimethylsulfoxide**

Dyes, *follow* manufacturer's safety guidelines

Epoxy and acrylic resins, *see* **Resins**

Ethanol (EtOH), CH_3CH_2OH, is highly flammable and may be harmful by inhalation, ingestion, or skin absorption. Wear appropriate gloves and safety glasses. Keep away from heat, sparks, and open flame.

Ether (diethyl ether, Et_2O, $[C_2H_5]_2O$) is extremely volatile and flammable. It is irritating to the eyes, mucous membranes, and skin. It is also a central nervous system depressant with anesthetic effects. It may be harmful by inhalation, ingestion, or skin absorption. Avoid breathing the vapors. Wear appropriate gloves and safety glasses. Always use in a chemical fume hood. Explosive peroxides can form during storage or on exposure to air or direct sunlight. Keep away from heat, sparks, and open flame.

EtOH, *see* **Ethanol**

Fluorescein phalloidin, *see* **Phalloidin**

Formaldehyde, HCHO, is highly toxic and volatile. It is also a possible carcinogen. It is readily absorbed through the skin and is irritating or destructive to the skin, eyes, mucous membranes, and upper respiratory tract. Avoid breathing the vapors. Wear appropriate gloves and safety glasses and always use in a chemical fume hood. Keep away from heat, sparks, and open flame.

Formalin is a solution of formaldehyde in water. *See* **Formaldehyde**

Formic acid, HCOOH, is highly toxic and extremely destructive to tissue of the mucous membranes, upper respiratory tract, eyes, and skin. It may be harmful by inhalation, ingestion, or skin absorption. Wear appropriate gloves and safety glasses (or face shield) and use in a chemical fume hood.

Glucose oxidase may be harmful by inhalation, ingestion, or skin absorption. Wear appropriate gloves and safety glasses.

Glutaraldehyde is toxic. It is readily absorbed through the skin and is irritating or destructive to the skin, eyes, mucous membranes, and upper respiratory tract. Wear appropriate gloves and safety glasses and always use in a chemical fume hood.

HCHO, *see* **Formaldehyde**

HCl, *see* **Hydrochloric acid**

H_3COH, *see* **Methanol**

HCOOH, *see* **Formic acid**

Heparin is an irritant and may act as an anticoagulant subcutaneously or intravenously. It may be harmful by inhalation, ingestion, or skin absorption. Wear appropriate gloves and safety glasses.

HNO_3, *see* **Nitric acid**

$HOCH_2CH_2SH$, *see* **β-mercaptoethanol**

Hydrochloric acid, HCl, is volatile and may be fatal if inhaled, ingested, or absorbed through the skin. It is extremely destructive to the mucous membranes, upper respiratory tract, eyes, and skin. Wear appropriate gloves and safety glasses and use with great care in a chemical fume hood. Wear goggles when handling large quantities.

Hydrogen peroxide, H_2O_2, is corrosive, toxic, and extremely damaging to the skin. It may be harmful by inhalation, ingestion, and skin absorption. Wear appropriate gloves and safety glasses and use only in a chemical fume hood.

6-hydroxy-2,5,7,8-tetramethylchromane-2-carboxylic acid (Trolox) is an irritant and may be harmful by inhalation, ingestion, or skin absorption. Wear appropriate gloves and safety glasses. Do not breathe the dust.

Isoflurane is an irritant and may be harmful by inhalation, ingestion, or skin absorption. Chronic exposure may be harmful. Wear appropriate gloves and safety glasses.

Kanamycin may be harmful by inhalation, ingestion, or skin absorption. Wear appropriate gloves and safety glasses. Use only in a well-ventilated area.

KCl, *see* **Potassium chloride**

KOH, *see* **Potassium hydroxide**

Magnesium chloride, $MgCl_2$, may be harmful by inhalation, ingestion, or skin absorption. Wear appropriate gloves and safety glasses and use in a chemical fume hood.

2-mercaptoethanol, *see* **β-mercaptoethanol**

MeOH, *see* **Methanol**

Methanol, MeOH or H_3COH, is toxic, can cause blindness, and is highly flammable. It may be harmful by inhalation, ingestion, or skin absorption. Adequate ventilation is necessary to limit exposure to vapors. Avoid inhaling these vapors. Wear appropriate gloves and safety goggles and use only in a chemical fume hood.

$MgCl_2$, *see* **Magnesium chloride**

Microtome blades are extremely sharp! Use care when sectioning. If unfamiliar with the use of a microtome, have someone demonstrate its use.

Mounting media may be harmful by inhalation, ingestion, or skin absorption. Follow manufacturer's instructions.

Na_2CO_3, *see* **Sodium carbonate**

NaN_3, *see* **Sodium azide**

NaOH, *see* **Sodium hydroxide**

Nitric acid, HNO_3, is volatile and must be handled with great care. It is toxic by inhalation, ingestion, and skin absorption. Wear appropriate gloves and safety goggles and use in a chemical fume hood. Do not breathe the vapors. Keep away from heat, sparks, and open flame.

Nitrogen (gaseous or liquid) may be harmful by inhalation, ingestion, or skin absorption. Wear appropriate gloves and safety glasses. Consult your local safety office for proper precautions.

***N,N*-dimethylformamide (DMF), $HCON(CH_3)_2$,** is a possible carcinogen and is irritating to the eyes, skin, and mucous membranes. It can exert its toxic effects through inhalation, ingestion, or skin absorption. Chronic inhalation can cause liver and kidney damage. Wear appropriate gloves and safety glasses and use in a chemical fume hood.

OCT is composed of polyvinyl alcohol, polyethylene glycol, and dimethyl benzyl ammonium chloride. Follow the manufacturer's guidelines for handling OCT.

Paraformaldehyde is highly toxic and may be fatal. It may be a carcinogen. It is readily absorbed through the skin and is extremely destructive to the skin, eyes, mucous membranes, and upper respiratory tract. Avoid breathing the dust or vapor. Wear appropriate gloves and safety glasses and use in a chemical fume hood. Keep away from heat, sparks, and open flame.

Para-phenylenediamine may be harmful by inhalation, ingestion, or skin absorption. Wear appropriate gloves and safety goggles.

PCA, *see* **3,4-dihydroxybenzoic acid**

Penicillin, *see* **Antibiotics**

Pentobarbital sodium is toxic and may be harmful by inhalation, ingestion, or skin absorption. It can induce respiratory depression and sedation and presents a risk to the unborn child. Do not breathe the dust. Wear appropriate gloves and safety glasses and use in a chemical fume hood.

Phalloidin is extremely toxic and may be fatal by inhalation, ingestion, or skin absorption. Great care should be taken when using these compounds. Wear appropriate gloves and safety glasses and use in a chemical fume hood. Do not breathe the dust. Do not use if skin is cut or scratched.

Potassium chloride, KCl, may be harmful by inhalation, ingestion, or skin absorption. Wear appropriate gloves and safety glasses.

Potassium hydroxide, KOH, and **KOH/methanol,** are highly toxic and may be fatal if swallowed. They may be harmful by inhalation, ingestion, or skin absorption. Solutions are corrosive and can cause severe burns. They should be handled with great care. Wear appropriate gloves and safety goggles.

***p*-phenylenediamine,** *see* **Para-phenylenediamine**

Resins are flammable and are suspected carcinogens. The unpolymerized components and dusts may cause toxic reactions, including contact allergies with long-term exposure. Avoid breathing the vapors and dusts. Wear appropriate gloves and safety goggles and always use in a chemical fume hood. Sensitivity to these chemicals may develop with repeated contact. Keep away from heat, sparks, and open flame.

Rhodamine phalloidin, *see* **Phalloidin**

SDS, *see* **Sodium dodecyl sulfate**

Silane is extremely flammable and corrosive. It may be harmful by inhalation, ingestion, or skin absorption. Keep away from heat, sparks, and open flame. The vapor is irritating to the eyes, skin, mucous membranes, and upper respiratory tract. Wear appropriate gloves and safety goggles and always use in a chemical fume hood.

Sodium azide, NaN_3, is highly poisonous. It blocks the cytochrome electron transport system. Solutions containing sodium azide should be clearly marked. It may be harmful by inhalation, ingestion, or skin absorption. Wear appropriate gloves and safety goggles and handle it with great care. Sodium azide is an oxidizing agent and should not be stored near flammable chemicals.

Sodium carbonate, Na_2CO_3, may be harmful by inhalation, ingestion, or skin absorption. Wear appropriate gloves and safety glasses and use in a chemical fume hood.

Sodium dodecyl sulfate (SDS) is toxic, an irritant, and poses a risk of severe damage to the eyes. It may be harmful by inhalation, ingestion, or skin absorption. Wear appropriate gloves and safety goggles. Do not breathe the dust.

Sodium hydroxide, NaOH, and **solutions containing NaOH,** are highly toxic and caustic and should be handled with great care. Wear appropriate gloves and a face mask. All other concentrated bases should be handled in a similar manner.

Streptomycin is toxic and a suspected carcinogen and mutagen. It may cause allergic reactions. It may be harmful by inhalation, ingestion, or skin absorption. Wear appropriate gloves and safety glasses.

Tris may be harmful by inhalation, ingestion, or skin absorption. Wear appropriate gloves and safety glasses.

Tris-Cl, *see* **Tris**

Tris-HCl, *see* **Tris**

Triton X-100 causes severe eye irritation and burns. It may be harmful by inhalation, ingestion, or skin absorption. Wear appropriate gloves and safety goggles. Do not breathe the vapor.

Trolox, *see* **6-hydroxy-2,5,7,8-tetramethylchromane-2-carboxylic acid**

Trypsin may cause an allergic respiratory reaction. It may be harmful by inhalation, ingestion, or skin absorption. Do not breathe the dust. Wear appropriate gloves and safety goggles. Use with adequate ventilation.

Xylene is flammable and may be narcotic at high concentrations. It may be harmful by inhalation, ingestion, or skin absorption. Wear appropriate gloves and safety glasses and use only in a chemical fume hood. Keep away from heat, sparks, and open flame.

Index

Page references followed by f denote figures; page references followed by t denote tables.

B

C

D

E

F

J

K

L

M

U